R. EDSON

Evolution of
Physical
Oceanography

The MIT Press
Cambridge, Massachusetts
and London, England

Evolution of Physical Oceanography

Scientific Surveys in Honor of Henry Stommel

Edited by
Bruce A. Warren and
Carl Wunsch

Publication of this volume was made possible in part by grants from the Office of Naval Research and the National Science Foundation.

Second printing, 1981

Copyright © 1981 by
The Massachusetts Institute of Technology

All rights reserved. No part of this book may be reproduced in any form or by any means, electronic or mechanical, including photocopying, recording, or by any information storage and retrieval system, without permission in writing from the publisher.

This book was set in VIP Trump Roman by DEKR Corporation and printed and bound by Halliday Lithograph in the United States of America.

Library of Congress Cataloging in Publication Data
Main entry under title:

Evolution of physical oceanography.
   Bibliography: p.
   Includes index.
   1. Oceanography—Addresses, essays, lectures. 2. Stommel, Henry M., 1920– I. Stommel, Henry M., 1920– II. Warren, Bruce Alfred, 1937– III. Wunsch, Carl.
GC150.7.E9   551.46   80-18452
ISBN 0-262-23104-2

Henry Stommel

Photograph by V. Cullen

Preface
xii

Henry Stommel
xiv

The Scientific Work of Henry Stommel
Arnold B. Arons — xiv

A Theoretical Model of Henry Stommel
George Veronis — xix

Notes Related to Stommel's Early Years in Woods Hole
Raymond B. Montgomery — xxiv

Henry Stommel
G.E.R. Deacon — xxv

Henry Stommel—On the Light Side
F.C. Fuglister — xxvi

Life and Work of Henry Stommel
xxviii

Introduction
2

Part One
General Ocean Circulation
5

I
Deep Circulation of the World Ocean
Bruce A. Warren
6

1.1 Introduction — 6

1.2 Historical Development of Ideas about the Deep Circulation — 7

1.3 A Dynamical Framework — 11

1.4 Sources of Deep Water — 15

1.5 Deep Western Boundary Currents in the World Ocean — 26

1.6 Why Is There a Deep Thermohaline Circulation At All? — 38

Notes — 40

2
The Water Masses of the World Ocean: Some Results of a Fine-Scale Census
L. V. Worthington
42

2.1 Introduction — 42

2.2 Methods of Describing the Oceans — 43

2.3 The World Water Masses As They Exist in the Second Half of This Century — 44

2.4 The Formation of Water Masses — 57

Appendix: Census of World-Ocean Water Masses with Division by Bivariate (°C × °/oo) Classes and Rank by Volume — 60

3
On the Mid-Depth Circulation of the World Ocean
Joseph L. Reid
70

3.1 Introduction — 70

3.2 The Circulation of the Upper Waters and Their Contribution to the Mid-Depths — 70

3.3 The Use of Geostrophy — 72

3.4 The Mid-Depth Circulation of the Atlantic Ocean from Core Analysis and Vertical Geostrophic Shear — 74

3.5 Studies of Total Transport and Layers — 79

3.6 Mid-Depth Studies Using Isopycnal Analysis — 81

3.7 Comparison of Relative Geostrophic Flow at Mid-Depth with Numerical Models of Transport — 85

3.8 Mid-Depth Patterns in the World Ocean — 91

3.9 Comparison of the Maps of Shear Field and Characteristics — 109

3.10 Conclusion — 110

4
The Gulf Stream System
N. P. Fofonoff
112

4.1 Introduction — 112

4.2 The Gulf Stream System — 113

4.3 The Florida Current — 113

4.4 The Gulf Stream — 123

4.5 The North Atlantic Current — 133

4.6 Summary and Conclusions — 137

5
Dynamics of Large-Scale Ocean Circulation
George Veronis
140

5.1 Introduction and Summary — 140

5.2 The Equations for Large-Scale Dynamics — 142

5.3 The Quasi-Geostrophic Equations and the β-Plane — 144

5.4 Ekman Layers — 147

5.5 Steady Linear Models of the Wind-Driven Circulation — 149

5.6 Preliminary Nonlinear Considerations — 153

5.7 Why Does the Gulf Steam Leave the Coast? — 157

5.8 Thermohaline Circulation — 158

5.9 Free Waves for a Constant-Depth Two-Layer Ocean on the β-Plane	164
5.10 Effect of Bottom Topography on Quasi-Geostrophic Waves	165
5.11 Baroclinic Instability	169
5.12 Effect of Nonlinearity and Turbulence	174
Notes	183

6
Equatorial Currents: Observations and Theory
Ants Leetmaa, Julian P. McCreary, Jr., and Dennis W. Moore
184

6.1 Introduction	184
6.2 Observations	185
6.3 Theories	188
6.4 Discussion	195

7
On Estuarine and Continental-Shelf Circulation in the Middle Atlantic Bight
Robert C. Beardsley and William C. Boicourt
198

7.1 Introduction	198
7.2 Estuarine Circulation in the Middle Atlantic Bight	199
7.3 Continental-Shelf Circulation	207
Appendix: Annual Air–Sea Interaction Cycles and Mean Runoff for the Middle Atlantic Bight	230
Notes	233

Part Two
Physical Processes in Oceanography
235

8
Small-Scale Mixing Processes
J. S. Turner
236

8.1 Introduction	236
8.2 Preliminary Discussion of Various Mechanisms	237
8.3 Vertical Mixing in the Upper Layers of the Ocean	240
8.4 Mixing in the Interior of the Ocean	245
8.5 Mixing near the Bottom of the Ocean	258

9
Internal Waves and Small-Scale Processes
Walter Munk
264

9.1 Introduction	264
9.2 Layered Ocean	268
9.3 Continuously Stratified Ocean	269

	9.4 Turning Depths and Turning Latitudes	271
	9.5 Shear	273
	9.6 Resonant Interactions	275
	9.7 Breaking	276
	9.8 Ocean Fine Structure and Microstructure	279
	9.9 An Inconclusive Discussion	283
	9.10 Conclusion	290
	Notes	290
10 Long Waves and Ocean Tides *Myrl C. Hendershott* 292	10.1 Introduction	292
	10.2 Astronomical Tide-Generating Forces	293
	10.3 Laplace's Tidal Equations (LTE) and the Long-Wave Equations	295
	10.4 Long Waves in the Ocean	297
	10.5 The Ocean Surface Tide	317
	10.6 Internal Tides	329
	10.7 Tidal Studies and the Rest of Oceanography	339
11 Low-Frequency Variability of the Sea *Carl Wunsch* 342	11.1 Introduction	342
	11.2 The Field of Variability of the Ocean	346
	11.3 Summary and Conclusions	373
12 Some Varieties of Biological Oceanography *J. H. Steele* 376	12.1 Introduction	376
	12.2 Space and Time Scales of Variation	377
	12.3 Ecological Variations	379
	12.4 Discussion	381
13 The Amplitude of Convection *Willem V. R. Malkus* 384	13.1 Introduction	384
	13.2 Basic Boussinesq Description	385
	13.3 Initial Motions	386

13.4 Quantitative Theories for High Rayleigh Number — 387

13.5 The Amplitude of Turbulent Convection from Stability Criteria — 389

Part Three
Techniques of Investigation
395

14 Ocean Instruments and Experiment Design
D. James Baker, Jr.
396

14.1 Observations and the Impact of New Instruments — 396

14.2 Instrument Development: Some Principles and History — 398

14.3 Examples of Modern Ocean Instruments — 402

14.4 Ocean Experiment Design — 429

15 Geochemical Tracers and Ocean Circulation
W. S. Broecker
434

15.1 Introduction — 434

15.2 Water-Transport Tracers — 435

15.3 Water-Mass Tracers — 448

15.4 Modeling Tracer Data — 448

15.5 Current Applications — 449

15.6 Ventilation of the Deep Sea — 450

15.7 Ventilation of the Main Oceanic Thermocline — 456

15.8 Formation of Deep Waters — 457

15.9 Vertical Mixing Rates — 459

16 The Origin and Development of Laboratory Models and Analogues of the Ocean Circulation
Alan J. Faller
462

16.1 A Brief Philosophy of Laboratory Experimentation — 462

16.2 Introduction — 463

16.3 The Experiments of W. S. von Arx — 465

16.4 The SAF Model — 466

16.5 Experiments with Rotating Covers — 468

16.6 A Variety of Interesting Experiments — 472

16.7 Concluding Remarks — 478

Part Four
Ocean and Atmosphere
481

17
Air–Sea Interaction
H. Charnock
482

17.1 Introduction — 482

17.2 The Surface Layer — 483

17.3 The Lower Boundary — 486

17.4 Waves — 490

17.5 The Atmospheric Boundary Layer — 495

18
Oceanic Analogues of Large-Scale Atmospheric Motions
Jule G. Charney and Glenn R. Flierl
504

18.1 Introduction — 504

18.2 The General Circulation of Oceans and Atmospheres Compared — 505

18.3 The Transient Motions — 506

18.4 The Geostrophic Formalism — 508

18.5 Linear Quasi-Geostrophic Dynamics of a Stratified Ocean — 520

18.6 Friction in Quasi-Geostrophic Systems — 525

18.7 Nonlinear Motions — 529

18.8 Summary Remarks — 544

Appendix: The Quasi-Geostrophic Equations — 546

Notes — 548

Acknowledgments and Permissions
550

Reference List
554

Index
612

Contributors
622

Preface

This book is a tribute to a prodigious man, Henry Stommel, on the occasion of his sixtieth birthday, September 27, 1980. In the course of his scientific career, his influence on the development of physical oceanography during nearly four decades has been immense. At the center of that influence has been the originality and penetration of his own research, which, in major part, has generated the modern concepts of ocean circulation. The fruits of his thinking are much more widespread than is apparent from his own work, however, for by his enthusiasm and intellectual vigor, he has stimulated many investigators to follow lines of inquiry that he foresaw as fruitful and revealing but that were too numerous to be pursued by himself. He is a theorist of extraordinary creativity, an astute observer willing to spend weeks at sea, and an ingenious laboratory experimentalist of the sealing-wax-and-string school, and he has been an inspiring, if sometimes bewildering, teacher to a generation of graduate students. Many have been drawn into his orbit simply by the fascination of his personality, for in company with these scientific activities, the variety of Stommel's avocations is also overwhelming. He pursues these interests with exuberance, whether it be as gentleman farmer, amateur painter, home printer, Oriental chef, marine antiquarian, musicboxicologist, or alarmingly casual experimenter with explosives. His pranks are a delight to most who witness them, his kindnesses have been felt by many, and his charm has won him widespread affection.

In considering a fit celebration of Henry Stommel's sixtieth birthday, we thought at first to solicit reports on research projects from his colleagues, students, and other friends, in the manner of the traditional *Festschrift*. We recognized almost immediately, however, that these contributors would be so numerous as to make the book impossibly large and forbidding. It seemed to us, moreover, that Stommel's preeminent position in oceanography warranted something more far-reaching and enduring than could likely be achieved by a collection of current research reports. We therefore suggested to certain of his colleagues that they prepare broad, comprehensive surveys of those aspects of oceanography with which Stommel has been concerned. Our aim was twofold: (1) to trace the development of these subjects since the time when Stommel entered the field; and (2) to lay out in specific terms the state of these subjects now—where we are and how we arrived, what is known and what seems probable, and what seem to be the key questions. The beginning of Stommel's career seemed a useful date from which to review the development of physical oceanography because it coincided roughly with the publication of *The Oceans* (1942), which had brought together much of what had gone before, and with World War II, which marked an important transitional period in the history of the subject. Furthermore, the scope of Stommel's research interests is so wide that surveys oriented around them would embrace nearly all of physical oceanography. For these reasons, we thought that a book so organized could serve at once as a very personal tribute to Henry Stommel and as a widely useful general account of physical oceanography in 1980.

The project turned out to be less easy to define and more difficult to carry out than we anticipated. Prescribing that the historical reviews cover the years since the early 1940s, for example, sometimes proved somewhat artificial and impractical, for a few topics warrant a longer perspective, while for a few others only work in the last one or two decades is really pertinent. We are deeply grateful to the contributors, who by intense efforts met this volume's demanding publication schedule, and to those individuals who provided useful criticism. We hope that these surveys will remain of both technical and historical interest for many years to come.

While scientific chapters constitute the bulk of this volume, they are preceded by a short section that attempts in various ways to capture something of Henry Stommel himself. Two friends and collaborators, Arnold Arons and George Veronis, have provided personal perspectives and assessments of Stommel's scientific contributions. Three other old friends, Raymond B. Montgomery, G. E. R. Deacon, and F. C. Fuglister, have written brief reflections that further illuminate facets of his personality and of the historical period in which he did his early work. The ways of doing oceanography have changed so much that the scientific world of that time might seem very strange to graduate students today. In addition we have included two informal photographs that hint at his approach to life, and reproductions of two emblems designed by him that exemplify one of the less heralded aspects of his creativity.

Owing to the plan of this volume, the number of authors is necessarily relatively small, and specific individuals were asked to contribute to the project as much for their expertise as for their personal associations with Stommel. We hope that his many friends who, for these reasons, were not invited to participate will appreciate the difficulty of preparing an anniversary volume that is suitable for so remarkable a man.

Our choice of title for this volume reflects our view that physical oceanography has not merely changed during the progress of Stommel's career, but has evolved in a certain general direction. Henry Stommel himself appealed for such an evolution, first in his pamphlet, "Why Do Our Ideas about the Ocean Circulation Have Such a Peculiarly Dream-Like Quality?" (1954), and again, with slight rewording, in his book *The Gulf Stream* (1958, 1965):

I should like to make it clear, finally, that I am not belittling the survey type of oceanography, nor even purely theoretical speculation. I am pleading that more attention be given to a difficult middle ground: the testing of hypotheses. I have not explored this middle ground very thoroughly, and the few examples given in this book may not even be the important ones; but perhaps they are illustrative of the point of view in which attention is directed not toward a purely descriptive art, nor toward analytical refinements of idealized oceans, but toward an understanding of the physical processes which control the hydrodynamics of oceanic circulation. Too much of the theory of oceanography has depended upon purely hypothetical physical processes. Many of the hypotheses suggested have a peculiar dream-like quality, and it behooves us to submit them to especial scrutiny and to test them by observation.

While the "testing of hypotheses" may not, strictly speaking, have become commonplace yet, nevertheless the convergence of theory and observation that is implied in the phrase is certainly growing closer and more evident. Indeed, it is an underlying theme of many of the chapters in this book, and to our eyes it seems to offer an overall definition of the evolution of physical oceanography during the latter half of the twentieth century.

The Office of Naval Research and the National Science Foundation have joined in this celebration by generously assuming partial costs for the publication of this volume. Their support will help to make the book accessible to a much larger group of scientists and students than otherwise would have been possible.

Bruce A. Warren
Carl Wunsch

Woods Hole, Massachusetts
August 1979

Henry Stommel

The Scientific Work of Henry Stommel
Arnold B. Arons

Tanquam ex ungue leonem.
Jean Bernoulli, 1697

Henry Stommel graduated from Yale University in 1942 with a major in astronomy and remained at Yale as an instructor in mathematics and astronomy through the wartime years of 1942–1944. In 1944 he took a job with Maurice Ewing, whose group was then located at the Woods Hole Oceanographic Institution. The job, however, initially did not involve full-time residence in Woods Hole. A Coast and Geodetic Survey chart of the Mississippi Delta can still draw Stommel into reminiscences about a period of instrument monitoring at an isolated station in the far reaches of the delta, finger pointing to the location of the remote outpost.

On taking up more continuous residence at Woods Hole, Stommel became one of a group of bachelors who formed a kind of informal fraternity residing in the old rectory of the Episcopal church on the east corner of Church Street and Woods Hole Road. All the inmates were members of the Institution staff, and to describe some of them as "characters" would be a form of gentle British understatement. The atmosphere of the rectory, however, was stimulating and congenial and constitutes another source of Stommel's anecdotes and reminiscences. Some of the more senior individuals of this group were quick to grasp the quality of their new young member's intellect. I recall two of them, at that early date, predicting his future leadership in research.

Toward the end of World War II, Stommel, becoming interested in the oceans, educated himself in basic oceanography. A by-product of this study was the charming little book *Science of the Seven Seas*, published in 1945. There followed a period of casting about for lines of research: a note on use of the T–S correlation in dynamic height anomaly computations, an exploration of the theory of convection cells, a sally into cloud physics. At one point in this interval he consulted Ray Montgomery about outstanding problems in oceanography, and Montgomery pointed to the dynamically unexplained phenomenon of the Gulf Stream.

In 1948 the *Transactions of the American Geophysical Union* carried a short paper entitled, "The Westward Intensification of Wind-Driven Ocean Currents." This was to become a classic, one of the most frequently cited papers in modern physical oceanography. Some years later I once heard Hans Panofsky say, "That paper? Oh, that's the paper which made Henry Stommel famous!" In an elegantly simple model—a plane, rectangular, homogeneous ocean driven by wind torque at the surface and braked by bottom friction—Stommel showed that the basic dynamical equations predicted a flow symmetrical about the central meridian if the Coriolis term is held constant over the plane but that a westward intensification (as in the Gulf Stream) emerges if the Coriolis term varies linearly with latitude (the so-called β-effect).

In this day, when the fundamental equations of geophysical hydrodynamics, including variation of the Coriolis parameter, are casually written down in text and lecture presentations for each new generation of students, it is probably difficult for many to grasp the research atmosphere of a time when the dynamical significance of variation of the Coriolis parameter had not yet been fully appreciated by either oceanographers or meteorologists. Though it all seems so obvious and compelling in retrospect, it is well to note that Bjerknes, Ekman, Defant, Sverdrup, and Rossby had not perceived the connection with westward intensification in a bounded basin; it was the youthful Henry Stommel who did. The 1948 paper coincided with the beginning of a new epoch of research in physical oceanography and constitutes a prototypical example of the Stommel style.

Reference to style calls for explanation of the Latin quotation at the head of this essay. In 1696 Leibniz and Jean Bernoulli, striving to demonstrate the power and significance of the new mathematical methods of "analysis" (the differential and integral calculus) as opposed to the ancient methods of "synthesis" (geometry), posed as a challenge to European mathematicians the now well-known brachistochrone problem.* They knew that this problem could be solved only by use of the new analytic methods, and they speculated that L'Hôpital, James Bernoulli, and Isaac Newton would be among the few likely to meet the challenge. When, early in 1697, Jean Bernoulli saw the correct and powerful solution published anonymously by Newton in *Philosophical Transactions of the Royal Society*, he is said to have remarked, "Tanquam ex ungue leonem"—literally translated, "As from the claw, the lion"; freely translated, "You can tell the lion by his claw."

This is a very fitting metaphor; in Henry Stommel's papers you can almost invariably tell the lion by his claw. He is diffident, almost apologetic, for what he regards as his "limited mathematical capacity" in dealing with the complexity of oceanographic problems, yet in this "limitation" perhaps lies much of his strength. With consummate artistry he constructs a model having just the right idealizations to make it tractable and just the right physical content to make it illuminating; then with the simplest mathematical methods he extracts the deep and significant physical insights that hitherto had not been attained.

At this point the most expert applied mathematicians take over, usually with Stommel's active encouragement and cooperation, and proceed to extend and refine the original picture. So it went with the westward intensification: Munk worked out a more sophisticated model with its multiplicity of "gyres" and with dissipation provided by a horizontal austausch coefficient rather than by bottom friction; Morgan and Charney, in continual personal contact with Stommel, examined the nonlinear aspects; Munk and Carrier worked out solutions for nonrectangular basins. But the deep physical insight opening up the entire field was in the 1948 paper. This paper, as well as the syntheses contained in the 1957 "Survey of Ocean Current Theory" and the 1958 book *The Gulf Stream*, continue to be deeply influential and widely cited.

Stommel has a way of looking at new papers published by others, frequently imposingly complex and difficult to penetrate, and, after a relatively short study, stripping away the complexity, revealing the essence of the paper in something of the form and style that he himself might have used had he formulated the problem *ab initio*. His unerring penetration of the essential physical content is steadily guided by his deep, reliable intuition for every aspect of fluid flow.

During a period in which we were deeply immersed in thinking about Rossby waves, long after Rossby's classic paper of 1939, I recall a moment at which Stommel emerged from the library where he had been reading Laplace. In a characteristically bubbly way he said, "You know, Laplace's tidal motions of the second class are simply Rossby waves; I hadn't realized this till now." Although we were aware of Haurwitz's 1940 paper in *Journal of Marine Research*, in which he had examined the formalism of Rossby waves in spherical coordinates and pointed out their identity with "motions of the second class" discussed by Margules, we were not fully sensitive to the latter comment. Stommel's insight quite independently penetrated the physical content of the arcane Laplace formulation.

His depth of intuition sustains another characteristic that I have frequently seen at play—an almost inarticulate, unswerving sense of when he is on the right track with some physical idea. When he has this sense, he will not be deflected, and he will not take his teeth out of a problem. Others will give up and fall by the wayside, but he persists until the initial hunch is brought to fruition.

In addition to his profound grasp of dynamics, Stommel has a broad descriptive knowledge of oceanic data and phenomena. If, however, he does not happen to have something you ask him about at his finger tips, he will vanish into the recesses of the library and

---

* To find the curve connecting two points, at different heights and not in the same vertical line, along which a body acted upon only by gravity will fall in the shortest time.

emerge a little while later with the crucial material in hand; he knows exactly where it is located.

The range of Stommel's more than 100 publications embraces not only almost all aspects of physical oceanography (both theoretical and observational) but also extends into cloud physics, limnology, and estuarine circulation. There are observational and theoretical papers on oceanic and limnological thermoclines, on time-series observations of thermal "unrest," on the formation and sinking of water cold enough to drop to the bottom, on oceanic Rossby waves, on monsoon effects in the Indian Ocean, on tidal mixing and density currents in estuaries, on the Kuroshio, on dynamical transients in the ocean—all containing some illuminating physical insight or significant observational data.

One characteristic of his work has been a steadfast concern for the most basic problems of interpretation of hydrographic data. He repeatedly returns, with increasing depth of insight, to matters such as the origin and significance of the T-S correlation, the problem of the depth of no motion, conservation of potential vorticity, formation of the main thermocline. Most recently his publications have dealt with the β-spiral—another effort to extract the absolute velocity field from hydrographic data.

As early as 1958, I heard Albert Defant, on a visit to Stockholm, characterize Stommel as the world's leading physical oceanographer, but during the 1950s Stommel, apparently diffident about not possessing the golden academic key, briefly toyed with the idea of going back to graduate school to earn a Ph.D. I was undoubtedly not the only one who was skeptical of this notion. When he mentioned it to me, I remarked that it was clearly far more appropriate for him to be directing the theses of others rather than pausing under someone else's direction.

Stommel's outside interests go through phases that range from the collection, repair, and playing of harmoniums and music boxes, to tracking down the literature on Harrison's invention of the chronometer and his winning of the Queen Anne prize, or to (a recent effort) tracking down the meteorological facts on the famous 1816 "year without a summer." At one time during the 1950s there was a printing-press phase, and the Institution suddenly blossomed with puckish printed notices, announcements, and invitations.

The printing-press phase, however, was characterized by another kind of output—a set of useful pamphlets, one, for example, detailing the hydrodynamical equations in finite-difference form. Another was titled "Why Do Our Ideas about the Ocean Circulation Have Such a Peculiarly Dream-Like Quality?" In the title alone you can tell the lion by his claw. Although never published other than by the Stommel underground press, this was a seminal paper. A penetrating, incisive critique of the status of ocean-current theory, it made the rounds of the active theoreticians and deeply influenced the direction of their thinking. I recall it being referred to repeatedly in seminars and colloquia of that period.

During the early 1950s, while engaging in a wide variety of studies ranging from estuarine dynamics, through the monitoring of a long series of temperature measurements on the bottom off Bermuda, to the use of submarine cables in measuring potential differences across oceanic currents, Stommel began to talk more and more explicitly about the need for time-series observations of pressure, temperature, and current in the deep ocean. He recognized the possibilities offered by moored-instrument strings and recoverable-instrument packages and envisaged the acquisition of synoptic data from such arrays. These ideas were, of course, "in the air" at that time, catalyzed by revolutionary progress in electronic instrumentation, but applications to oceanography were still almost nonexistent. It was with this need in mind that we brought David Frantz and then William S. Richardson to Woods Hole, and got them started on the buoy work that subsequently was massively extended and powerfully implemented by Nick Fofonoff and Ferris Webster.

During the mid-1950s, Stommel's imagination was strongly captivated by John Swallow's development of the neutrally buoyant float for the tracking of deep currents. Following Swallow's early observations off the Straits of Gibraltar, Stommel participated in organizing the *Aries* work off Bermuda in which Swallow and Crease provided some of the earliest indications of the now widely recognized deep oceanic fluctuations.

In 1956–1957 Stommel began thinking seriously about the abyssal circulation. Recognizing that geostrophic flow in the ocean basin, subject to planetary divergence (the β-effect), could not have an equatorward meridional component without an accompanying *downward* vertical velocity, and feeling compelled to reject a downward velocity in the abyssal layer, he reached the conclusion that little cold abyssal water from high latitudes would be able to enter the interior of an ocean basin directly. On this basis he predicted the existence of a deep equatorward boundary current along the western side of the ocean, transporting dense water from high to lower latitudes. The prediction was verified for the North Atlantic by the joint *Atlantis-Discovery II* expedition in the spring of 1957, an expedition in which the Swallow floats also played a key role.

Rather than being the more usual a posteriori rationalization of an observed phenomenon, this was one of the very first a priori predictions of a hitherto unobserved, major feature of oceanic circulation—a prediction derived in characteristic style from a deep understanding of geophysical fluid dynamics.

Arnold B. Arons

These insights motivated Alan Faller's beautiful rotating-tank experiments on stationary planetary flow patterns in bounded basins, exploring flows arising in the presence and absence of vertical velocities imposed on the basin by a distribution of sources and sinks. Faller compellingly demonstrated the refusal of the system to accommodate a meridional flow component in the interior geostrophic regime under conditions of zero vertical velocity. He demonstrated the intense southward western boundary current arising when a northern source imposed a vertically upward velocity (with its associated northward geostrophic flow in the interior) on the entire basin.

I vividly remember the morning on which Stommel stormed into the office and said, "There is more water being transported northward than is rising vertically." He was referring to the experiment with the northern source and the uniformly distributed upward velocity. In characteristic style, he had done what no one else had thought of doing. He had taken a careful look at the *continuity* of the flow in the bounded basin. The simplest kind of algebra reveals that, at any vertical zonal section, the horizontal poleward transport of the meridional geostrophic flow through the section is larger than the total vertical transport in the region north of the given section. The extra horizontal flow simply *had* to be recirculated in the southbound western boundary current. Faller quickly confirmed the hitherto unnoticed recirculation while Stommel and I showed that analogous conditions carried over to a spherical ocean and that recirculation was to be anticipated in real abyssal western boundary currents.

On another occasion, Stommel and I were engaged in one of our frequent discussions of how one might measure pressure variations at the bottom of the deep ocean over long time periods to an accuracy of a few centimeters of water. (Others finally did this more or less successfully; we did not.) In desperation we were considering the brute-force technique of making a 3-mile-long manometer, i.e., literally extending a tube from the ocean surface to the bottom and drawing abyssal water up into the tube. Since the salinity of the abyssal water is invariably lower than the average salinity of the column above it, the water in the tube, on coming to thermal equilibrium with its surroundings, would stand above the level of the surrounding ocean surface, and we could watch the level in the tube go up and down with variations of pressure at the bottom. We had this picture sketched on the chalkboard and were entirely focused on the manometric aspect when my own mind took a divergent turn. In some astonishment, I added a faucet to the upper level of our manometer and said, "Hank, if we open the faucet, it will run forever."

After we satisfied ourselves concerning the nature and temporal limitations of the physical phenomenon, Stommel ran down to Duncan Blanchard's laboratory and recruited this ready and skillful gadgeteer to our party. Blanchard quickly set up a large beaker with a layer of hot salty water floating on cold fresh, and we blissfully watched the little fountain that spurted for a long time out of the glass tube in which the cold fresh water had been drawn upward to start the sequence.

This was the genesis of our oft-cited short letter to *Deep-Sea Research* describing the "perpetual" salt fountain. We recognized that the key lay in blocking salt transfer while allowing thermal equilibrium; we recognized that if surface water were initially drawn downward in the tube, there would be a steady downward flow; but we did not perceive a deeper significance. We quickly convinced ourselves that this would not be a practicable way of inducing significant rates of upwelling of nutrient-rich water, and we dropped the subject.

Not long afterward, Melvin Stern, in his quite independent investigation of the stability problem, became aware of the dynamic significance of the huge difference between the molecular diffusion coefficients of heat and salt and thus discovered double-diffusive convection and "salt fingers." I believe that Stommel was probably a bit chagrined about having missed this himself, but he was strongly supportive of Stern's priority for the discovery and unstinting in his praise and enthusiasm.

In the third paper of our series on the abyssal oceanic circulation, Stommel and I were able to look at the distribution of oxygen and radiocarbon in the North Atlantic in a somewhat more sophisticated manner than that afforded by conventional "box models." We used the distribution of the chemical properties as an index to the dynamics. After completion of this paper, in the summer of 1966, Stommel assembled some of the leading ocean chemists who happened to be attending a National Academy of Sciences conference at Woods Hole and fired their interest and enthusiasm with a glimpse of the impact that modern, accurate, simultaneously made chemical measurements might have on all of physical, as well as chemical, oceanography. This seed, with subsequent watering by many other individuals, evolved into the Geochemical Ocean Sections Studies (GEOSECS), which was incorporated into the activities of the International Decade of Ocean Exploration (IDOE) and which is still bearing fruit throughout oceanographic science. In an even more direct fashion, Stommel catalyzed and helped sustain the Mid-Ocean Dynamics Experiment (MODE) and POLYMODE programs. During the 1960s, as recovery of buoyed instrumentation became increasingly reliable and as SOFAR float techniques looked promising, many individuals began to discuss large-scale observational efforts in the deep ocean. A principal objective was to describe in more complete detail the scale and periodicity of the eddy motions initially suggested by the results of the *Aries* expedition and imperfectly delineated by the buoy data acquired in subsequent years. Stommel was a participant, either in person or through radiated influence, in most of these discussions. When the MODE experiment was being organized in 1970, Stommel became cochairman (with Allan Robinson) of the steering committee.

"It is fair to say," Carl Wunsch, one of the principal participants, writes in 1978, "that it was Hank's presence and huge enthusiasm which allowed us to entrain a remarkable number of people.... The MODE experiment is notable in Hank's career in that he has never handed it off to others the way he has with most of the other major programs he was instrumental in launching. He is still officially the co-chairman of POLYMODE, eight years after it all began."

In essentially similar ways, Stommel's influence extended to work in the Kuroshio and to some of the Indian Ocean programs of the last decade. Throughout this continuing association with the genesis and fruition of large and *complicated* projects, Stommel has exhibited a pronounced talent for avoiding *complex* ones. (Complex projects, of course, are those for which costs are real and results imaginary.)

Most fine scientists almost automatically collect a cloud of individuals around them, forming something of a "school" of research. Stommel is no exception; the cloud collects wherever he is located. His prodigality of ideas is so vast that he cannot deal with all of them himself, and he hands them out right and left to other individuals. All who have worked in Stommel's vicinity are familiar with the explosive laughter that frequently reverberates in the course of conversations in his office. His own sense of humor (gentle and never at the expense of someone's feelings), his ebullience, his enjoyment of intellectual activity are infectious and pervade the atmosphere that surrounds him.

Discussions with him are invariably a chase and a challenge. He can be irascible with slow-wittedness. His quickness and penetration are such that he frequently leaves the verbal train of thought behind, but he makes things so interesting that individuals are attracted into an area of investigation largely because of the color and fascination he has infused.

When during the 1960s the wags that formed the American Miscellaneous Society used to make their annual Albatross Award, they would nominate a truly leading oceanographer and accord him a citation with irreverent and irrelevant content. In the citation for the 1966 Albatross Award, Stommel was twitted for his propensity toward "abandoning oceanography's most cherished chairs." This was an oblique reference to the sequence of changes of position that he had been making. In 1960, unhappy with some aspects of the administration at Woods Hole and attracted by a tempting academic offer, Stommel left the Oceanographic Institution for a professorship at Harvard. Unhappy at Harvard, he moved in 1963 to the Department of Meteorology at MIT; Cambridge and Lexington turned out to have little attraction as places of residence, and the Stommel family moved back to Sippewissett, eventually acquiring Sippewissett Farm, where one could ride a tractor, till a huge garden, and even raise sheep. During the interval, Stommel spent some of his time at the Oceanographic Institution, commuting to Cambridge to discharge his obligations at MIT. In 1978 he returned to a full-time position as Senior Scientist in Woods Hole, confirming the fundamental indestructability of his ties to the institution at which he began his scientific career.

No contributor to this volume is without some debt to Henry Stommel, whether it be in the way of an important scientific insight, a suggestion for fruitful activity, some kind of generous assistance, or some token of personal friendship. I am sure I speak for the contributors, as well as for the oceanographic constituency at large, in saluting him and his distinguished career with pleasure, respect, and enthusiasm on his sixtieth birthday.

Arnold B. Arons

A Theoretical Model of Henry Stommel
George Veronis

From a theoretician's viewpoint, the most significant contribution that Henry Stommel has made to oceanography is his development of simple physical models to demonstrate important processes in oceanic flows. His mastery of that approach, his innovative ideas, the generous sharing of his thoughts and his work, and his open-minded appreciation of new ideas by others are characteristics that have earned him wide recognition and many honors in oceanography. Yet he has always appreciated the awesome scope of science, and in that context he has retained a modest view of his own contributions.*

When he graduated from Yale near the beginning of World War II, Stommel planned to attend divinity school. What made him change his mind is not certain—perhaps he realized that there would be little opportunity to develop simple models in that area and even less likelihood that he could test his ideas against data. In any event, he dropped his sights from the spiritual to the celestial sphere when he entered the Yale Astronomy Department as a graduate student. He found little attraction in celestial mechanics, the main area of research in that department, and in 1944 at the suggestion of L. Spitzer he applied for a position at the Woods Hole Oceanographic Institution. Stommel was a conscientious objector to war, a position that was extremely difficult to maintain during World War II. Working at WHOI was considered an "acceptable" substitute for military service.

In 1945 he produced his first work, *Science of the Seven Seas*, a popular book on oceanography. The book was written in the evening during a 4-week period. Though the book was a modest financial success,† it was a mixed blessing. When he applied to Scripps for graduate school, he was turned down. He thinks this happened because H. U. Sverdrup was annoyed with him for writing a nonprofessional book on oceanography.

Intent on developing his scientific talents, Stommel bought Southwell's book on relaxation methods just after it appeared in 1946. He planned to apply the method to the solution of a system of elliptic equations that he had formulated for the tides. At about that time, R. B. Montgomery had mentioned to him the east-west asymmetry in the response of the ocean to a symmetric wind stress and Stommel proceeded to formulate that problem in terms of an elliptic partial differential equation. In solving it by relaxation, he noted that the β-term led to westward intensification. He realized the significance of that result and reformulated the problem in simpler form, one amenable to analytic solution. That was the first of his many simple and informative models of oceanic phenomena. It also influenced him strongly in realizing the significance of isolating the important physics in the simplest possible context.

His westward intensification paper was submitted to the American Geophysical Union in 1947 before Sverdrup published the relation between meridional transport and wind-stress curl. Today we look upon these papers as a logical sequence, with the Sverdrup transport relation providing a solution for the interior of the ocean and Stommel's frictional boundary layer closing the flow on the western side, but they were quite independent contributions. It was not until 1950 when G. F. Carrier, in collaboration with W. H. Munk, introduced boundary-layer methods to the circulation problem that the connection became clear and Stommel's model was fully appreciated. Initially the main reaction to it was that it was an interesting curiosity—shallow, homogeneous oceans with friction acting through a bottom drag had little to do with the "real" ocean.

During that period, Stommel was living in Woods Hole in the building known as the Rectory, a kind of rooming house shared by several bachelors.* His annual salary was $1,300. Out of that and the royalties on his book, he had saved $1,500. In early fall 1947, he obtained a leave of absence without salary from WHOI and used his savings to sail to England to spend the next half-year at Imperial College. In England he met Sheppard, Brunt, Francis, Deacon, Swallow, Charnock, Longuet-Higgins, and others who were active in meteorology and oceanography. He spent some time with Southwell learning about numerical relaxation procedures. He also journeyed to Scotland to meet and to collaborate with L. F. Richardson on an experiment to study turbulent eddy diffusion by measuring the separation with time of pairs of parsnip pieces thrown from a pier into a lake. He shared an interest with M. S. Longuet-Higgins in determining the electric field induced by ocean currents, but the two followed in-

---

* Hank has long been a voracious reader (sometimes three books in one evening) and the incredibly broad knowledge that he has acquired provided him with a realistic perspective on man. He told me once that he had learned to read fast at an early age because an optometrist had given him an incorrect prescription—the eyeglasses caused headaches if he read for any length of time—and he developed speed to get through books before a headache set in. He eventually prescribed enlarging spherical lenses for himself and got rid of the headaches, but his speed reading has remained.

† Two decades later Hank remarked to me that he had earned more money from that book (~$5,000) than from *The Gulf Stream*, a book that summarized his scientific efforts of 10 years.

* Hank clowned around a great deal in those days. I saw one of his acts during an evening at his house in the early 1950s. He played the organ and, with a tremulous voice, sang "Baby Hands," a very emotional funeral hymn that was sometimes sung at burial services for an infant.

dependent paths and did not collaborate on the topic until some years later.

A little over a year after returning to WHOI, Stommel bought a used 1946 Ford and drove to Scripps, where he spent the summer of 1949.* In the late 1940s and early 1950s, he continued to expand his interests in natural phenomena, and he collaborated with an increasingly diverse group of people† on topics including ecology, estuary circulation, hydraulic flows, and dynamic effects of rotation. As he developed his skills in constructing theoretical models, he familiarized himself with observations by working up charts of distributions of properties. The marriage of theory to data has characterized his work ever since.

In 1950, he began a collaboration (marriage) with Elizabeth Brown,‡ which culminated in two sons, a daughter, and most recently, a forthcoming book, *The Year without a Summer: 1816*, written with his wife. His enthusiastic concern for oceanography has never distracted him from an equally intense involvement with his family. He made most of his children's toys and the range was impressive. I remember an endless variety of tools made of wood, a 14-foot-high swing, and a one-passenger railway system around his property with the rails made of 2×4's and a locomotive powered by a lawn-mower engine.

The summer of 1953 was spent in Bermuda and Stommel returned there in 1954 to initiate the biweekly *Panulirus* hydrographic stations, which continue to the present time to provide an invaluable, nearly continuous record of "deep" ocean data.

As the first decade of his work in oceanography came to a close, Stommel began his synthesis of ideas and observations of ocean circulation and the Gulf Stream. When he visited the Institute for Advanced Study in Princeton in 1955 for a 3-month period, he had in hand a first-draft, typewritten manuscript of *The Gulf Stream*.* He had already started to put together his thoughts about the thermohaline circulation, a topic that claimed a good part of his time over the next decade. He was interested in the adjustment problem and during that visit, he and I collaborated on a paper that described the response of a two-layer, β-plane ocean to a variable wind stress.

Gnawing away at him at this time were the problems associated with an inertially generated Gulf Stream. A year earlier, N. P. Fofonoff had published his paper on steady frictionless flows showing the formation of an inertial western boundary current, and Stommel was trying to construct a forced model with the same dynamical control. Actually, he already had a solution for a model with constant potential vorticity, which was to appear in his book. In retrospect, it is clear that the essential physical processes were contained in these two simple models, but the ideas were new and the implications of the simpler models were not fully appreciated at the time. In any event, Stommel had discussed the problem with G. W. Morgan at Brown, and, while at Princeton, he outlined his thoughts to J. G. Charney. Shortly afterward, Morgan was invited to Princeton for a few days and a long discussion of the problem took place. Charney and Morgan became caught up in a fierce competition to outdo each other, and they eventually published separate papers containing essentially the same analysis of the problem. Stommel's input was lost sight of during that frenzied activity, and it reemerged only in the acknowledgements of the two papers.

A year later, he initiated the research into double-diffusive phenomena in a joint publication with A. B. Arons and D. Blanchard on the perpetual salt fountain, a phenomenon in which the salinity distribution in the thermocline (a destabilizing contribution to the density stratification) drives vertical motions. Sometime after that he tried without much success to demonstrate the principle using a pipe during a cruise near Bermuda. It was not until 1960 when M. E. Stern brought out his paper on salt fingers that it became evident that the process might be significant in the ocean without mechanical aids. What none of us knew at the time was that the original salt-finger experiment, with a proper explanation, had been reported a century before by D. Jevons (1857). In 1964, Stommel planted a second seed in the field of double-diffusive convection when he and J. S. Turner published a paper demonstrating the for-

---

* The speedometer of the Ford, with 59,000 miles on it, stopped functioning during that trip but Hank continued to drive it at ever decreasing speeds for another decade. He became known as the slowest and most careful driver in Woods Hole. It turned out, however, that his driving habits had been only temporarily conditioned by the reliability of that car, as we soon learned when he bought a more reliable one in 1960.

† As one can see from Hank's list of publications, his collaborators have ranged across the spectrum. He has more often provided the basic idea but he also acts as a catalyst for others' ideas. But whether the final publications are routine reports of data or new and exciting ideas, they all have been spiced with his ringing laughter, sometimes triggered by a humorous incident and sometimes by an unexpected development.

‡ At the beginning of his marriage Hank evidently decided to eat lunch at home with his wife whenever it was possible. In earlier years at about 11:50 A.M. each day an internal alarm must have been triggered and he would rush off for home irrespective of what he had been involved in moments before. Lately his lunchtime departures have been somewhat smoother but no less determined.

---

* J. G. Charney, who has been known to forget things now and then, borrowed the manuscript to read on the train during a 3-day trip. Hank aged 3 years during those 3 days, wondering whether the only copy of his manuscript would end up in some Lost and Found office of a disorganized railway system.

mation of layers when a fluid is stabilized by salinity and destabilized by temperature. Those brief forays into thermohaline convection provide an example of another of Stommel's characteristics. Having collaborated with an effective researcher in a particular subject, he withdraws from further active participation, assured that the development of the area is in competent hands.

The period 1955–1960 was an extremely prolific one, during which his own research embraced an ever expanding set of ideas. Noting that turbulent processes in the stably stratified upper ocean would cause a downward flux of heat, he assumed a deep upwelling of cold water to keep the temperature constant at a given level. The resultant circulation patterns of the abyssal waters had not been anticipated previously and were reported in short notes to *Nature* (1957a) and *Deep-Sea Research* (1958). He also drew attention to the fact that vertical velocities had the same dynamical effect on large-scale circulation whether they were induced by evaporation and precipitation, by thermohaline forcing, or by Ekman pumping. This idea was exploited in his survey on ocean current theory in *Deep-Sea Research* (1957b). In a publication with A. B. Arons and A. J. Faller in 1958, he made use of the same idea to suggest a laboratory analogue of ocean circulation. The first two papers with Arons on the abyssal circulation of the world oceans (1960a,b) provided a quantitative extension of some of his earlier work on the subject.

One consequence of Stommel's inquiries into the deep circulation was his theoretical prediction of a southward-flowing countercurrent under the Gulf Stream. This remarkable result was confirmed in 1957 by tracking neutrally buoyant floats off the coast south of Cape Hatteras (Swallow and Worthington, 1957). It is one of the few purely theoretical predictions of a significant oceanic phenomenon.

Along with these investigations, Stommel was working toward a theory of the thermohaline circulation. He and I published a linear study of the penetration of thermal anomalies from the surface into the deeper regions of a stratified ocean and derived an expression for the thermocline depth. Shortly after that, he and A. R. Robinson derived a similarity solution for a steady, nonlinear thermocline model with geostrophic dynamics and convective–diffusive thermal processes. P. Welander had been working on the same problem independently and produced a solution for an ideal-fluid model of the thermocline. Those two papers were published together in *Tellus* (1959) and formed the basis for a series of investigations into the effects of wind and thermal forcing on the interior of the ocean.

The thermocline model brought together many of the ideas that Stommel had suggested previously for slow, steady flows in the ocean, and to a certain extent, I think that it provides a test of the validity of those ideas. Although the model has inadequacies—e.g., it has never been satisfactorily closed on the west, and if it were, the interior solution would undoubtedly be modified—it contains satisfactory elements for a zero-order description of the oceanic interior. The overall advances leading up to the thermocline model are quite remarkable, especially when one remembers that when Stommel began his studies there were no theoretical models of the oceanic circulation. In any event, the end of the story is not yet in print. Welander has made several important advances in the development of the thermocline model since it was first introduced, and the β-spiral recently demonstrated by Stommel and Schott (1977) is a continuing part of the story.

An important development in the mid-1950s was the establishment of biweekly geophysical fluid dynamics (GFD) seminars between WHOI and MIT. When the numerical forecasting group left the Institute for Advanced Study in summer 1956, Charney and N. A. Phillips moved to MIT and I joined Stommel at WHOI. That fall, we set up the biweekly seminar, held on Fridays alternately at MIT and WHOI. Regular participants included Charney, Phillips, L. N. Howard, C. C. Lin, H.-L. Kuo, E. N. Lorenz, and J. T. Stuart (visiting from England) from MIT. WHOI participants included Stommel, W. Malkus, J. Malkus, A. J. Faller, M. E. Stern, F. Fuglister and W. S. von Arx. On occasion scientists from Harvard, the Geophysical Research Directorate, the Cambridge Air Force Research Center, and A. D. Little took part. One object of the time was to entice promising young scientists from other disciplines into GFD.* After R. Goody, A. Robinson, and Stommel joined the Harvard faculty, the seminar was held cyclically at MIT, Harvard, and WHOI and in later years occasionally at Brown, Yale, and the University of Rhode Island. The entire participating group was rather small during the early period, and nearly everyone attended nearly all the seminars. The after-seminar cocktail parties and dinners were important occasions for the exchange of ideas.†

In 1960, Stommel was elected to the National Academy of Sciences and he also accepted a position as

---

* For the first few years all notices were printed on Hank's printing press. We used a variety of types and different colored inks for the announcements.

† At one of the early seminars Hank talked on convection driven by a horizontally varying temperature field imposed at the top boundary. Starting from a purely diffusive basic state, he introduced the buoyancy term as a perturbation by expanding in powers of the thermal expansion coefficient. His argument was that the natural expansion parameter was the coefficient of expansion.

The Great Seal of the Summer Study Program in Geophysical Fluid Dynamics at the Woods Hole Oceanographic Institution; an emblem designed by Henry Stommel. The attention devoted to convection in the early years of the summer program inspired the figure of the dragon heated from below and cooled from above.

Professor of Oceanography at Harvard.* The move from a position of pure research to one involving teaching obligations and other academic duties was not a happy one for a man who had never felt comfortable giving a 1-hour seminar, let alone regularly scheduled courses. He never liked lecturing and it is one of the few activities in which he did not excel. Also, by that time Stommel was firmly committed to oceanography and he missed the daily discussions with observational oceanographers and the immediate access to oceanographic data that he had enjoyed at WHOI. In 1963, he left Harvard for MIT, where his position in the (graduate) Meteorology Department involved less emphasis on teaching. At about that time, he decided to maintain his residence near Woods Hole, and shortly afterward he bought a small farm there that he has operated ever since. For 15 years he commuted to Cambridge two or three times a week and spent the remainder of the work week at WHOI. He finally returned to a full-time position at WHOI in 1978.

On the whole, I think that he enjoyed his years at MIT. Certainly he accepted enormous responsibilities in the research effort there. At one point during his period at MIT in the mid-1960s, he was supporting 13 research students on his grants. The normal procedure was that at some time during a student's first year Stommel would suggest an idea by developing a simple model for a particular process or phenomenon, and he would then have the student develop the idea in his own fashion. He was a willing listener when a student had something to report but he never babied the students along.*

His interest in simple modeling of oceanic phenomena and in mapping of distributions of physical and chemical properties continued during the 1960s with the series of papers on the abyssal circulation of the world ocean, modeling of sinking and upwelling motions in the ocean, several articles on observations and analyses of motions in the Indian Ocean, and the work on double-diffusive phenomena. In addition to publications on a variety of other topics, he produced global charts relating currents to wind stresses, and together with different coworkers, he presented the results from several extended oceanographic cruises.

He was also becoming increasingly involved with oceanography on a grander scale, sparking much of the exploration into the physical oceanography of the Indian Ocean, and using his stature and influence to establish international observational programs. He played key roles in getting the Geochemical Oceans Sections Studies (GEOSECS) and the Mid-Ocean Dynamics Experiment (MODE) programs launched. His interest in making oceanographic data much more accessible was expressed in a publication with Pivar and Fredkin (1963) on the use of a computer to produce graphs of different oceanographic variables on command. At that time that approach was little more than a dream, but it has since been realized and is now operational on some computer systems.

In 1964 he was awarded an honorary Ph.D. by the University of Gothenberg, and a decade ago, Yale and the University of Chicago also recognized Stommel's achievements with honorary doctorates.†

In 1969, he participated in the multiship MEDOC campaign, which he had organized for the purpose of observing the formation of deep water in the western Mediterranean Sea. The success of that effort was reported by the MEDOC Group (1970). Actual instances of bottom water formation have rarely been observed,

---

* During the 1950s Hank had often talked about returning to graduate school for a Ph.D. He was very conscious of the fact that he did not have that degree and was wont to address even beginning graduate students as "Doctor." I think that the move to Harvard resolved the problem by placing him in the position of producing Ph.D.s. There were other reasons for his departure from WHOI but my own feeling is that the main reason was that the Harvard position provided academic recognition of his achievements.

---

* One student who listened to Hank go through one of these pilot studies emerged with a bewildered look and said to me, "I think that I understood what he did but what's the problem?"

† Alumni officials at Yale were surprised when Hank was proposed for an honorary degree because he had long since disappeared from their lists. Many years earlier, tired of being dunned for alumni contributions, he had written DECEASED on a pledge form and returned it to Yale. It worked, but he was reincarnated by the honorary degree.

George Veronis

and the MEDOC report contains a well-documented account, one that will be a useful starting point for theories. The treatise on the Kuroshio (1972), edited with K. Yoshida, occupied a good deal of his time in the late 1960s. He has continued to produce intriguing, theoretical models, as is evidenced by the last of the abyssal circulation papers with Arons (1972), showing how deep western boundary currents can be extensively broadened by topography. His lifelong interest in maps and charts is reflected in a guide to oceanographic atlases published with M. Fieux (1978). His faith in at least the statistical simplicity of ocean circulation appears in the study with A. Leetmaa and P. Niiler (1977) of the applicability of the Sverdrup relation to mid-Atlantic circulation and in the exploration of oceanographic data with F. Schott (1977) that led to the idea of the β-spiral. The latter study was produced during a sabbatical term at Kiel in 1976, and it has been extended in papers with Schott (1978) and D. Behringer (1980).

It is interesting to note that Stommel was a principal driving force in establishing the MODE program. He thought that a thorough study of oceanographic data in a restricted area would reveal strong mesoscale motions similar to cyclones and anitcyclones in the atmosphere, and this proved to be the case. Yet even though he remains associated with the successor POLYMODE program, he himself has not become involved in research on geostrophic eddies.

In looking back over Stommel's voluminous output, one is struck by his persistent search for simplicity in explaining observed phenomena. His models of the ocean vary depending on the problem. They may consist of infinitely deep or infinitely wide layers, one-dimensional channels or pipes and reservoirs.* He is much more likely to focus on some truth and an approximation to the truth rather than the whole truth and nothing but the truth. To my knowledge, he has never contrived ad hoc structure in his parameterizations in order to explain phenomena. His models are often constructed so that they can be analyzed with relatively modest mathematical tools, a characteristic that gives them a deceptively simple appearance. Not infrequently, his mathematical solutions have contained mistakes, but his superb intuition has invariably led him to the correct physical results.

With his move to WHOI in 1978, Stommel has returned to the research environment that is so dear to him. His concern for pertinent problems is evident in his increasing involvement with the climatic application of ocean–atmosphere interaction. But for most of us the best news is that once again he is immersed in the search for simple models to elucidate important oceanic phenomena.

---

* In the early 1950s W. Malkus, who was then at WHOI doing turbulence experiments with water, was frustrated by Hank's laminar, inviscid models of oceanic flow which nevertheless seemed to yield qualitatively correct results. For Stommel's benefit he had pasted the values of the viscosity coefficient ν on the fresh- and salt-water taps in his laboratory. He had marked the fresh water tap "$\nu \neq 0$" and the salt water tap "$\nu = 0$."

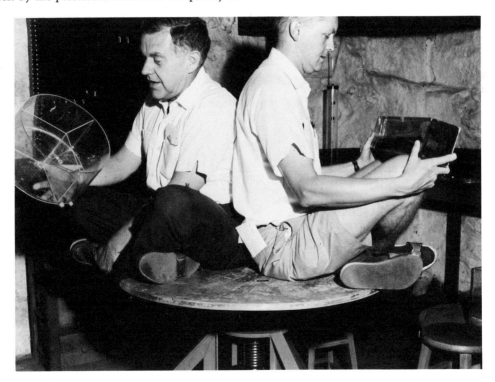

Henry Stommel and Louis N. Howard on a rotating table in the cellar of Walsh Cottage at WHOI, summer 1968.

Notes Related to Stommel's Early Years in Woods Hole
Raymond B. Montgomery

Henry Stommel's connection with Woods Hole Oceanographic Institution began when he came from Yale University in 1944. My association with him began in 1945. I want to mention some of the older scientists who were important to him and to me.

My own connection with WHOI had begun in 1931, and I should like to record the great benefit I derived from the leadership of H. B. Bigelow (1879–1967) and C. O'D. Iselin (1904–1971). As a student of C.-G. Rossby (1898–1957) at Massachusetts Institute of Technology, I had become familiar with the names of European, especially Scandinavian, oceanographers. I met H. U. Sverdrup (1888–1957) during his visits to Woods Hole. Rossby arranged that I work with A. Defant (1884–1974) in Berlin during the winter of 1938–1939 and with E. Palmén in Helsingfors during the summer of 1939. My return journey on the *Stavangerfjord* from Bergen to New York was made in company with B. Helland-Hansen (1877–1957) and H. Mosby. They were bound for the seventh general assembly, in Washington, of the International Union of Geodesy and Geophysics, Helland-Hansen being President of the Association d'Océanographie Physique. This assembly of IUGG was the first in America and was beclouded by the outbreak of World War II (1 September) just before the assembly (4–15 September). Attendance by Defant and other German oceanographers was canceled, and many more Europeans were absent. V. W. Ekman (1874–1954), during what must have been his only trip to America, lectured in Woods Hole and at the Washington assembly, and I was privileged to meet him in both places. He had been awarded the Agassiz Medal for Oceanography by the National Academy of Sciences in Washington 24 April 1928 but was not present in person.

In November 1945, following the close of the war, Iselin allowed me to return to Woods Hole, and that is when my thoroughly enjoyable acquaintance with Stommel began. Rossby again became a frequent visitor to Woods Hole, as he had been during his MIT years preceding the war. Attracted by Rossby, Stommel spent the spring of 1946 in Chicago.

Ekman had been strongly influenced by F. Nansen (1861–1930) and by V. Bjerknes (1862–1951), the two men who suggested to Ekman the topics for his early oceanographic studies. The progression of ideas is interesting. Ekman's work, especially his paper (1905) concerning the influence of the earth's rotation on wind-driven currents and his paper (1923) introducing the concept of the vertical component of vorticity, formed the background for Rossby's (1936) development and application of the vorticity equation. In turn, Stommel's influential paper (1948) on the westward intensification of wind-driven ocean currents resulted from his familiarity with the Rossby vorticity equation.

Stommel's paper was preceded by another very influential paper, that by Sverdrup (1947), and important features are common to both papers. While the casual reader might assume that Stommel had been helped by Sverdrup's slightly earlier work, I am convinced that the two papers were prepared quite independently. Sverdrup's was published in the November 1947 issue of the *Proceedings of the National Academy of Sciences*. Stommel had already presented his paper on 18 September 1947 on the unusual occasion of a Woods Hole meeting of the American Geophysical Union, and his paper was received for publication on 25 September. It appeared in the April 1948 issue of the AGU *Transactions*.

Henry Stommel in his typically cluttered office at WHOI, August 1979. Photograph by V. Cullen.

I am pleased to have played a part in the genesis of Stommel's paper. I think he and I were sharing an automobile trip between Woods Hole and Providence, probably in early 1947, when I recounted to him that Iselin had once pointed out to me an important problem: Why is the Gulf Stream narrow and swift and pressed against the western boundary of the North Atlantic Ocean? (This question, now so obvious, was novel then. And I do not pretend to reproduce the words that Iselin used; the term "boundary current" had not come into use.) If I remember correctly, Stommel answered the question qualitatively on a scrap of paper during a few minutes of discussion at a coffee stop during our short trip. My small part is generously acknowledged in his Gulf Stream book (1965; first edition 1958). The origin of the question is an example of the help Stommel and I and many others received from Iselin.

Stommel sought out stimulating scientists. Immediately following the September 1947 meeting already mentioned, he went to London to learn relaxation techniques from R. V. Southwell (1888–1970). Among the persons Stommel enjoyed meeting in London were D. Brunt (1886–1965), G. E. R. Deacon, and P. A. Sheppard (1902–1977). Stommel visited Cambridge to talk with G. I. Taylor (1886–1975). The highlight of Stommel's stay in Great Britain was the few days in January 1948 spent with L. F. Richardson (1881–1953) in Argyll, Scotland; the measurements they made together resulted in the joint paper in *Journal of Meteorology* that year. In 1949 Stommel spent several months at Scripps Institution of Oceanography. Besides meeting other oceanographers of his own generation, he became acquainted with C. Eckart (1902–1973) and E. C. Bullard (1907–1980). Stommel wrote me with special pleasure of time spent with Defant, whose delayed journey to America brought him to Scripps in October 1949 for several months and to Woods Hole in February 1950 for several days.

The eleventh general assembly of IUGG, the second in America, was held in Toronto on September 1957. Mosby was then President of the Association d'Océanographie Physique. As part of this assembly, Stommel arranged an outstanding Symposium on the General Circulation of the Ocean, with Particular Emphasis on the Deep-Water Movements. The symposium filled two entire days, 5–6 September. A feature memorable to many of us Americans was the opportunity to meet some of our Soviet counterparts for the first time. Despite the symposium's success, we keenly felt the absence of certain leaders. Stommel had invited Rossby and Sverdrup. But Rossby had died in Stockholm on 19 August and Sverdrup had died in Oslo on 21 August. Helland-Hansen, also unable to attend, died in Bergen on 7 September, while the assembly was in progress.

Henry Stommel
G. E. R. Deacon

Marine scientists all over the world hold Henry Stommel in great affection and esteem. His kindness and sincerity, supported by intellectual curiosity and clear thinking, are a universal source of inspiration.

He came to Woods Hole when oceanography was still in a primitive state. *The Oceans* by Sverdrup, Johnson, and Fleming had been published, war-time needs had promoted rapid expansion, and we had the bathythermograph, but most of the observations were made with methods nearly 50 years old. We had to titrate all our salinities, and our data, widely separated in space and time, had to be interpreted on the assumption that they represented steady-state conditions. It was not until 1950 that six ships working together between Cape Hatteras and Newfoundland gave us much idea of detail and variability, and not until 1957 that the joint British-American cruise in the *Aries,* sparked off by Stommel's ideas on deep-water movements, began to show that the deep circulation might consist of a wide spectrum of motions, some of them with velocities at least an order of magnitude faster than the mean velocities. It is only recently that we have been able to make the continuous, long-term recordings required for a realistic picture, and to apply theories that take reasonable account of both winds and density gradients.

Stommel can look back on this lively period with much satisfaction. The impressive list of his personal contributions is enhanced by so many joint publications that it reads like an author index to an "advances in oceanography."

His unpretentious approach and happy turn of phrase are magnetic. Who but he and L. F. Richardson would begin a serious study of eddy diffusion in the sea with the words "we have observed the relative motion of two floating pieces of parsnip." Other pleasing images are the salt fountain, varieties of oceanographic experience, submarine clouds, and the smallness of sinking regions. His surveys of progress, and his emphasis on the value of individual ideas, activity, and enthusiasm, and of directing expedition plans toward specific questions have been very timely. He was, I believe, the unintentional originator of the International Indian Ocean Expedition when he wrote round to all his friends to ask for information likely to be of help in a study of the time taken by the currents to respond to monsoonal reversal of the winds. But he was also, I am sure, editor of the five issues of *The Indian Ocean Bubble,* which "in gently pejorative tone" resisted too much regimentation and helped to formulate key problems.

I think some of us, rather overpowered by his achievements, have sometimes been revived at seeing him no less interested in simpler things: in the beginning of science, and in everyday life. He told us about William Leighton Jordan, the amateur oceanographer who received little help or sympathy from the professionals, about Whiston and Ditton, who suggested midnight guns fired from hulls moored along the shipping lanes as a means of determining longitude, and about Halley, the scientist and ship's captain, who, with good reason, had to court-martial all his officers.

It would be delightful to have an intimate record of his experiences, from the early postwar years in Woods Hole, when the Old Rectory was shared by a remarkable group of young scientists, into all the widening circle of his work and interests. He could write a fascinating autobiography.

The logo designed by Henry Stommel for INDEX, a program of the late 1970s focused on the monsoonal currents of the northwestern Indian Ocean. The name, *La Curieuse*, derived from a charter vessel that Stommel and colleagues used for a time out of the Seychelles, although that latter vessel was marginally more suited for oceanographic research than the one depicted on the logo.

Henry Stommel—On The Light Side
F. C. Fuglister

In the beginning, God said, "Let there be light." How long He had been stumbling around in the dark before He had this brilliant idea we are not told, but we can well imagine His delight when, at last, with the light on, He could see! The most impressive sight of all must have been His first glimpse of the world's ocean, that vast, awesome, mysterious deep. Later God created man to study this phenomenon and clear up the mysteries; and this leads us inevitably to Henry Melson Stommel, the man whose sixtieth anniversary is being celebrated with this volume.

To what extent Stommel has carried out God's plan is described elsewhere in this book; here I will just note that on some occasions he (Hank) may not have been entirely serious about his job and may even have slowed down the plan.

During the 1940s Stommel went to considerable effort to bring oceanographers together, using his own car to drive groups to Brown or Cambridge from Woods Hole. His car was old and he was a very careful driver, but the conversation was animated and the more spirited the discussion became, the slower the car moved. I remember, after one such trip to MIT, Carl Rossby, stepping out of the car, smiling, and stretching his arms wide, sighed, "It is hard to believe that only this morning we left Woods Hole."

It was on this same trip that we all had a thrilling experience. At one point, about halfway on our journey, we were amazed to find that we were behind a car that was traveling even slower than we were. The excitement ran high as Hank pulled over into the left lane and drew up abreast of the slow vehicle: he was going to pass him! Then we all saw who was driving the slow-moving car. It was George Veronis. He had a smile on his face as he waved to us, his thumb to his nose, and sped off for MIT. That is the closest we ever came to passing another car.

Fortunately, other cars and other drivers came along, and this regular movement of oceanographers back and forth, started by Stommel, continued and flourished; today, I understand that even going as far away as Yale is not considered unthinkable.

Over the years, Henry Stommel of course did more than drive cars (see the prior sketches by Arons and Veronis). He was the first President of SOSO, and he is the Special Committee for The William Leighton Jordan Esq. Award. Because of his retiring disposition, Hank has managed to keep these activities more or less secret.

In 1961, or thereabouts, L. V. Worthington (Val) showed me an item in an ONR newsletter about a new laboratory in Europe that was to be staffed by eight Ph.D.s and fifteen subprofessionals. "That is what we are," he said, "subprofessionals." and I had to agree with him. By the same token, we both realized that it made Stommel and others subprofessional, an astounding thought! That same evening I asked my young son to set up type for a letterhead for the Society of Subprofessional Oceanographers (SOSO). He asked me whether the society would have any administrators; I shuddered at the thought but I said, "Oh, make Hank president, me Vice-President, and Val Ambassador to the Court of St. James"; so he did and that was the birth of SOSO. Since the society never holds meetings or keeps any records, we don't know what that birth date is.

We know that in 1964, when Henry was off in California receiving the Sverdrup Medal of the American Meteorological Society, I, in a bold *coup d'état* took over the presidency of SOSO.

Stommel joined his membership in SOSO to his fascination with the nineteenth-century English amateur, William Leighton Jordan (M. Deacon, 1971, p. 376), in the following announcement, issued some years ago:

Society of Subprofessional Oceanographers
Special Committee
for

THE WILLIAM LEIGHTON JORDAN ESQ. AWARD

Announcement

The William Leighton Jordan Esq. Award is given annually to the Oceanographer who makes the most misleading contribution to his field. Ignorance and utter incompetence do not automatically qualify. The work cited must be distinguished not only by being in error, but it must be outstandingly bad: wrong both in principle and fact, and revealing the most mistaken intuition and the most faulty insight. It should be overambitious, and exhibit egregious error—willfully artful, well and plausibly presented, and totally misleading and false. It is not expected that every calendar year will be graced by so grand and profoundly negative an achievement deserving of this award commemorating our most illustrious and our deadest member. The author of many theoretical works on ocean currents, a fearless critic as in the pamphlet entitled: 'The Admiralty Falsification of the Challenger Record, exposed by William Leighton Jordan, Esq.', has few peers indeed. But the members of this Special Committee will remain alert to commemorate the truly deserving oceanographer with this newly established award.

Stommel and SOSO have never quite had the nerve—despite several temptations—actually to bestow this award on anyone.

Stommel puts out other announcements, as Arons has already recalled in this volume. Another example, advertising a seminar he organized on the newly rediscovered Equatorial Undercurrent, is reproduced here. These are all characteristic of his own appreciation for "the light side."

NOTICE
at the M.I.T. Faculty Club.
3:45 PM MAY 28, 1959 FRIDAY

A COMPETITION of
THEORIES.
of the
EQUATORIAL UNDERCURRENT
ALIAS "THE CROMWELL CURRENT"

FEATURING:
DR GEORGE VERONIS, PROF. J. G. CHARNEY & HENRY STOMMEL, ESQ. WITH THREE DIFFERENT MODELS EACH IN ONLY FIFTEEN MINUTES
and
PROFESSOR W. V. R. MALKUS PRESENTING THREE ANTIMODELS, EACH IN FIVE MINUTES
and
A SURPRISE APPEARANCE OF PROFESSOR R. S. ARTHUR WITH A DISCUSSION of SOME RELATED FACTS OBSERVED IN THE PACIFIC

BREATHLESS PERFORMANCE!

ADMISSION FREE

Life and Work of Henry Stommel

Born September 27, 1920

B.S., Yale University, 1942
M.A. (Hon.), Harvard University, 1961
Ph.D. (Hon.), Göteborgs Universitet, 1964
Ph.D. (Hon.), Yale University, 1970
Ph.D. (Hon.), University of Chicago, 1970

Instructor in Mathematics and Astronomy, 1942-1944, Yale University
Research Associate, 1944-1960, Physical Oceanographer (nonresident), 1960-1978, Woods Hole Oceanographic Institution
Professor of Oceanography, 1960-1963, Harvard University
Guest Lecturer, 1969-1970, Laboratoire d'Océanographie Physique du Muséum National d'Histoire Naturelle, Paris, France
Professor of Oceanography, 1963-1978, Massachusetts Institute of Technology
Senior Scientist, 1978-, Woods Hole Oceanographic Institution

Phi Beta Kappa
Sigma Xi
Fellow, American Academy of Arts and Sciences, 1959
Member, National Academy of Sciences, 1961
Sverdrup Medalist, American Meteorological Society, 1964
Albatross Award, American Miscellaneous Society, 1966
Fellow, American Geophysical Union, 1972
Henry Bryant Bigelow Medal, Woods Hole Oceanographic Institution, 1974
Elected to Soviet Academy of Sciences, 1976
Maurice Ewing Award, American Geophysical Union, 1977
Rosenstiel Award, American Association for the Advancement of Science, 1978
Alexander Agassiz Medal, National Academy of Sciences, 1979
Huntsman Award, Bedford Institute of Oceanography, 1980

Journal Publications

Stommel, Henry, 1947. A summary of the theory of convection cells. *Annals of the New York Academy of Sciences* 48: 715-726.

Stommel, Henry, 1947. Entrainment of air into a cumulus cloud. *Journal of Meteorology* 4: 91-94.

Stommel, Henry, 1947. Note on the use of the T-S correlation for dynamic height anomaly computations. *Journal of Marine Research* 6: 85-92.

Woodcock, A. H., and Henry Stommel, 1947. Temperatures observed near the surface of a fresh water pond at night. *Journal of Meteorology* 4: 102-103.

Stommel, Henry, 1948. The theory of the electric field induced in deep ocean currents. *Journal of Marine Research* 7: 386-392.

Stommel, Henry, 1948. The westward intensification of wind-driven ocean currents. *Transactions, American Geophysical Union* 29:202-206.

Richardson, L. F., and Henry Stommel, 1948. A note on eddy diffusion in the sea. *Journal of Meteorology* 5: 238-240.

Stommel, Henry, 1948. Theoretical physical oceanography. *Yale Scientific Magazine,* March 6, 14, 16.

Stommel, Henry, 1949. The trajectories of small bodies sinking slowly through convection cells. *Journal of Marine Research* 8: 24-29.

Riley, G. A., Henry Stommel, and Dean Bumpus, 1949. A quantitative ecology of the western North Atlantic. *Bulletin of the Bingham Oceanographic Collection* 12: 1-169.

Stommel, Henry, 1949. Horizontal diffusion due to oceanic turbulence. *Journal of Marine Research* 8: 199-225.

Bunker, Andrew, B. Haurwitz, Joanne Malkus, and Henry Stommel, 1949. The vertical distribution of temperature and humidity over the Caribbean Sea. *Papers in Physical Oceanography and Meteorology* 11: 1-82.

Stommel, Henry, 1950. Note on the deep circulation of the Atlantic Ocean. *Journal of Meteorology* 7: 245-246.

Stommel, Henry, 1950. An example of thermal convection. *Transactions, American Geophysical Union* 31: 553-554.

Stommel, Henry, 1950. The Gulf Stream: A brief history of the ideas concerning its cause. *The Scientific Monthly* 70: 242-253.

Stommel, Henry, 1951. An elementary explanation of why ocean currents are strongest in the west. *Bulletin of the American Meteorological Society* 32: 21-23.

Stommel, Henry, 1951. Entrainment of air into a cumulus cloud, II. *Journal of Meteorology* 8: 127-129.

Arons, A. B., and Henry Stommel, 1951. A mixing-length theory of tidal flushing. *Transactions, American Geophysical Union* 32: 419-421.

Klebba, Arthur A., and Henry Stommel, 1951. A simple demonstration of Coriolis force. *American Journal of Physics* 19: 247.

Stommel, Henry, and Alfred H. Woodcock, 1951. Diurnal heating of the surface of the Gulf of Mexico in the spring of 1942. *Transactions, American Geophysical Union* 32: 565-571.

Stommel, Henry, 1951. The determination of the lateral eddy diffusivity in the climatological mean Gulf Stream. *Tellus* 3: 43.

Stommel, Henry, 1951. Streaks on natural water surfaces. *Weather* 6: 72-74, plates 9+10.

Stommel, Henry, and Harlow G. Farmer, 1952. Abrupt change in width in two-layer open channel flow. *Journal of Marine Research* 11: 205-214.

Stommel, Henry, 1952. Small boat oceanography. *Rudder* September: 24-27, 28.

Stommel, Henry, 1952. Streaks of natural water surfaces. In *International Symposium on Atmospheric Turbulence in the Boundary Layer*, E. W. Hewson, ed., *Geophysical Research Papers* No. 19, Air Force Cambridge Research Center, Cambridge, Massachusetts, pp. 145-154.

Stommel, Henry, and Harlow G. Farmer, 1953. Control of salinity in an estuary by a transition. *Journal of Marine Research* 12: 12-20.

Stommel, Henry, W. S. von Arx, D. Parson, and W. S. Richardson, 1953. Rapid aerial survey of the Gulf Stream with camera and radiation thermometer. *Science* 117: 639-640.

Stommel, Henry, 1953. Computation of pollution in a vertically mixed estuary. *Sewage and Industrial Wastes* 25: 1065-1071.

Francis, J. R., and Henry Stommel, 1953. How much does a gale mix the ocean surface layers? *Quarterly Journal of the Royal Meteorological Society* 79: 534-536.

Stommel, Henry, 1953. Examples of the possible role of inertia and stratification in the dynamics of the Gulf Stream system. *Journal of Marine Research* 12: 184-195.

Stommel, Henry, 1953. The role of density currents in estuaries. In *Proceedings Minnesota International Hydraulics Convention*, International Association for Hydraulic Research, Minneapolis, Minnesota, pp. 305-312.

Stommel, Henry, 1954. Exploratory measurements of electrical potential differences between widely spaced points in the North Atlantic Ocean. *Archiv für Meteorologie, Geophysik und Bioklimatologie A* 7: 292-304.

Longuet-Higgins, M. S., M. E. Stern, and Henry Stommel, 1954. The electrical field induced by ocean currents and waves, with application to the method of towed electrodes. *Papers in Physical Oceanography and Meteorology* 13: 1-37.

Stommel, Henry, 1954. Circulation in the North Atlantic Ocean. *Nature* 173: 886-888.

Stommel, Henry, 1954. Serial observations of drift currents in the central North Atlantic Ocean. *Tellus* 6: 203-214.

Stommel, Henry, 1954. Direct measurement of subsurface currents. *Deep-Sea Research* 2: 284-285.

Stommel, Henry, 1954. An oceanographic observatory. *Research Reviews*, Office of Naval Research, January 11-13.

Stommel, Henry, 1955. Anatomy of the Atlantic Ocean. *Scientific American* 192: 30-35.

Stommel, Henry, 1955. Lateral eddy viscosity in the Gulf Stream. *Deep-Sea Research* 3: 88-90.

Deacon, G. E. R., H. U. Sverdrup, Henry Stommel, and C. W. Thornthwaite, 1955. Discussion on the relationships between meteorology and oceanography. *Journal of Marine Research* 14: 499-515.

Veronis, G., and Henry Stommel, 1956. The action of variable wind stresses on a stratified ocean. *Journal of Marine Research* 15: 43-75.

Stommel, Henry, Arnold B. Arons, and Duncan Blanchard, 1956. An oceanographical curiosity: The perpetual salt fountain. *Deep-Sea Research* 3: 152-153.

Stommel, Henry, 1956. On the determination of the depth of no meridional motion. *Deep-Sea Research* 3: 273-278.

Arons, A. B., and Henry Stommel, 1956. A β-plane analysis of free periods of the "second class" in meridional and zonal oceans. *Deep-Sea Research* 4: 23-31.

Stommel, Henry, 1956. Electrical data from cable may aid in hurricane prediction. *Western Union Technical Review* 10: 15-19.

Stommel, Henry, 1957. Florida Straits transports; 1952-1956. *Bulletin of Marine Science, Gulf and Caribbean* 7: 252-254.

Stommel, Henry, 1957. A survey of ocean current theory. *Deep-Sea Research* 4: 149-184.

Stommel, Henry, and George Veronis, 1957. Steady convective motion in a horizontal layer of fluid heated uniformly from above and cooled non-uniformly from below. *Tellus* 9: 401-417.

Stommel, Henry, 1957. The abyssal circulation of the ocean. *Nature* 180: 733-734.

Stommel, Henry, A. B. Arons, and A. J. Faller, 1958. Some examples of stationary planetary flow patterns in bounded basins. *Tellus* 10: 179-187.

Stommel, Henry, 1958. The abyssal circulation. *Deep-Sea Research* 5: 80-82.

Stommel, Henry, 1958. The circulation of the abyss *Scientific American* 199: 85-90.

Haurwitz, B., H. Stommel, and W. H. Munk, 1959. On the thermal unrest in the ocean. In *The Atmosphere and the Sea in Motion: The Rossby Memorial Volume*, Bert Bolin, ed., Rockefeller Institute Press, New York and Oxford University Press, London, pp. 74-94.

Schroeder, Elizabeth, Henry Stommel, David Menzel, and William Sutcliffe, Jr., 1959. Climatic stability of eighteen degree water at Bermuda. *Journal of Geophysical Research* 64: 363-366.

Stommel, Henry, 1959. Florida Straits transports, June 1956-July 1958. *Bulletin of Marine Science, Gulf and Caribbean* 9: 222-223.

Robinson, Allan, and Henry Stommel, 1959. The oceanic thermocline and the associated thermohaline circulation. *Tellus* 11: 295-308.

Robinson, Allan, and Henry Stommel, 1959. Amplification of transient response of the ocean to storms by the effect of bottom topography. *Deep-Sea Research* 5: 312-314.

Stommel, Henry, and A. B. Arons, 1960. On the abyssal circulation of the World Ocean—I. Stationary planetary flow patterns on a sphere. *Deep-Sea Research* 6: 140-154.

Stommel, Henry, and A. B. Arons, 1960. On the abyssal circulation of the World Ocean—II. An idealized model of the circulation pattern and amplitude in oceanic basins. *Deep-Sea Research* 6: 217-233.

Stommel, Henry, 1960. Wind drift near the equator. *Deep-Sea Research* 6: 298-302.

Stommel, Henry, 1960. An historical note. *Deep-Sea Research* 7: 222.

Bolin, Bert, and Henry Stommel, 1961. On the abyssal circulation of the World Ocean—IV. Origin and rate of circulation of deep ocean water as determined with the aid of tracers. *Deep-Sea Research* 8: 95-110.

Stommel, Henry, 1961. Thermohaline convection with two stable regimes of flow. *Tellus* 13: 224-230.

Stommel, Henry, 1961. Florida Straits transports: July 1958-March 1959. *Bulletin of Marine Science, Gulf and Caribbean* 11: 318.

Stewart, R. W., G. G. Carrier, A. R. Robinson, and H. Stommel, 1961. Heat flux from the ocean bed produced by dissipation of the tides. *Deep-Sea Research* 8: 275-278.

Stommel, Henry, and Jacqueline Webster, 1962. Some properties of thermocline equations in a subtropical gyre. *Journal of Marine Research* 20: 42-56.

Stommel, Henry, 1962. An analogy to the Antarctic Circumpolar Current. *Journal of Marine Research* 20: 92-96.

Stommel, Henry, 1962. Examples of mixing and self stimulated convection on the S,T diagram. (In Russian.) *Okeanologiya* 2: 205-209.

Stommel, Henry, 1962. On the cause of the temperature-salinity curve in the ocean. *Proceedings of the National Academy of Sciences, U.S.A.* 48: 764–766.

Stommel, Henry, 1962. On the smallness of sinking regions in the ocean. *Proceedings of the National Academy of Sciences, U.S.A.,* 48: 766–772.

Stommel, Henry, 1963. Varieties of oceanographic experience. *Science* 139: 572–576.

Pivar, Malcom, Ed Fredkin, and Henry Stommel, 1963. Computer-compiled oceanographic atlas: an experiment in man-machine interaction. *Proceedings of the National Academy of Sciences, U.S.A.,* 50: 396–398.

Stommel, Henry, 1964. Summary charts of the mean dynamic topography and current field at the surface of the ocean, and related functions of the mean windstress. In *Studies on Oceanography,* Kozo Yoshida, ed., University of Tokyo Press, Tokyo, pp. 53–58.

Turner, J. S., and Henry Stommel, 1964. A new case of convection in the presence of combined vertical salinity and temperature gradients. *Proceedings of the National Academy of Sciences, U.S.A.* 52: 49–64.

Stommel, Henry, and Warren S. Wooster, 1965. Reconnaissance of the Somali Current during the southwest monsoon. *Proceedings of the National Academy of Sciences, U.S.A.* 54: 8–13.

Stommel, Henry, 1965. Some thoughts about planning the Kuroshio Survey. In *Proceedings of Symposium on the Kuroshio, Tokyo, October 29, 1963,* Oceanographical Society of Japan and UNESCO, pp. 22–33.

Warren, Bruce, Henry Stommel, and J. C. Swallow, 1966. Water masses and patterns of flow in the Somali Basin during the southwest monsoon of 1964. *Deep-Sea Research* 13: 825–860.

Stommel, Henry M., 1966. The large-scale oceanic circulation. In *Advances in Earth Science,* P. M. Hurley, ed., MIT Press, pp. 175–184.

Stommel, Henry, and K. N. Fedorov, 1967. Small-scale structure in temperature and salinity near Timor and Mindanao. *Tellus* 19: 306–325.

Arons, A. B., and Henry Stommel, 1967. On the abyssal circulation of the World Ocean—III. An advection-lateral mixing model of the distribution of a tracer property in an ocean basin. *Deep-Sea Research* 14: 441–457.

Stommel, Henry, and Claes Rooth, 1968. On the interaction of gravitational and dynamic forcing in simple circulation models. *Deep-Sea Research* 15: 165–170.

Stommel, Henry, and Robert Frazel, 1968. Hidaka's onions (Tamanegi). *Records of Oceanographic Works in Japan* 9: 279–281.

Cooper, John, and Henry Stommel, 1968. Regularly spaced steps in the main thermocline near Bermuda. *Journal of Geophysical Research* 73: 5849–5854.

Reid, Joseph, Jr., Henry Stommel, E. Dixon Stroup, and Bruce A. Warren, 1968. Detection of a deep boundary current in the western South Pacific. *Nature* 217: 937.

Stommel, Henry, and Roberto Frassetto, 1968. The time of appearance of cold water off Somalia. *Proceedings of the National Academy of Sciences, U.S.A.* 60: 750–751.

Stommel, Henry, 1968. Kinematic waves in the Gulf Stream. *Proceedings of the National Academy of Sciences, U.S.A.* 60: 747–749.

Stommel, Henry, 1969. Horizontal variations in the mixed layer of the South Pacific Ocean. (In Russian, English abstract.) *Okeanologiya* 9: 97–102.

Stommel, Henry, Kim Saunders, William Simmons, and John Cooper, 1969. Observations of the diurnal thermocline. *Deep-Sea Research* 16 (Supplement): 269–284.

MEDOC Group (Lacombe, H., P. Tchernia, M. Ribet, J. Bonnot, R. Frassetto, J. C. Swallow, A. R. Miller, and H. Stommel), 1970. Observation of formation of deep water in the Mediterranean Sea, 1969. *Nature* 227: 1037–1040.

Anati, David, and Henry Stommel, 1970. The initial phase of deep water formation in the northwest Mediterranean during MEDOC '69 on the basis of observations made by Atlantis II, January 25, 1969 to February 12, 1969. *Cahiers Océanographiques* 22: 343–351, 24 charts.

Stommel, Henry, 1970. Future prospects for physical oceanography. *Science* 168: 1531–1537.

Stommel, Henry, and Kozo Yoshida, 1971. Some thoughts on the cold eddy south of Enshunada. *Journal of the Oceanographical Society of Japan* 27: 213–217.

Winterfeld, Thomas, and Henry Stommel, 1972. Distribution of stations and properties at standard depths in the Kuroshio area. In *Kuroshio: Its Physical Aspects,* Henry Stommel and Kozo Yoshida, eds. University of Tokyo Press, Tokyo, pp. 81–93.

Stommel, Henry, and Ants Leetmaa, 1972. Circulation on the continental shelf. *Proceedings of the National Academy of Sciences, U.S.A.* 69: 3380–3384.

Stommel, Henry, and A. B. Arons, 1972. On the abyssal circulation of the World Ocean—V. The influence of bottom slope on the broadening of inertial boundary currents. *Deep-Sea Research* 19: 707–718.

Stommel, Henry, 1972. Deep winter-time convection in the western Mediterranean Sea. In *Studies in Physical Oceanography, a Tribute to Georg Wüst on His*

80th birthday, Arnold L. Gordon, ed., Gordon and Breach, New York, Vol. 2, pp. 207-218.

Stommel, Henry, E. Dixon Stroup, Joseph L. Reid, and Bruce A. Warren, 1973. Transpacific hydrographic sections at Lats. 43°S and 28°S: the SCORPIO expedition—I. Preface. *Deep-Sea Research* 20: 1-7.

Stommel, Henry, Harry Bryden, and Paul Mangelsdorf, 1973. Does some of the Mediterranean outflow come from great depth? *Pure and Applied Geophysics* 105: 874-889.

Fieux, M., and Henry Stommel, 1975. Preliminary look at feasibility of using marine reports of sea-surface temperature for documenting climatic change in the western North Atlantic. *Journal of Marine Research* 33 (Supplement): 83-95.

Fieux, M., and Henry Stommel, 1977. Onset of the southwest monsoon over the Arabian Sea from marine reports of surface winds. *Monthly Weather Review* 105: 231-236.

Leetmaa, Ants, Pearn Niiler, and Henry Stommel, 1977. Does the Sverdrup relation account for the Mid-Atlantic circulation? *Journal of Marine Research* 35: 1-10.

Stommel, Henry, and Friedrich Schott, 1977. The beta-spiral and the determination of the absolute velocity field from hydrographic station data. *Deep-Sea Research* 24: 325-329.

Stommel, Henry, Pearn Niiler, and David Anati, 1978. Dynamic topography and recirculation of the North Atlantic. *Journal of Marine Research* 36: 449-468.

The MODE Group (H. Stommel and 69 others), 1978. The Mid-Ocean Dynamics Experiment. *Deep-Sea Research* 25: 859-910.

Schott, Friedrich, and Henry Stommel, 1978. Beta spirals and absolute velocities in different oceans. *Deep-Sea Research* 25: 961-1010.

Rooth, Claes, Henry Stommel, and George Veronis, 1978. On motions in steady, layered, geostrophic models. *Journal of the Oceanographical Society of Japan* 34: 265-267.

Stommel, Henry, 1979. Oceanic warming of western Europe. *Proceedings of the National Academy of Sciences, U.S.A.* 76: 2518-2521.

Stommel, Henry, and Elizabeth Stommel, 1979. The year without a summer. *Scientific American* 240:6, 176-185.

Stommel, Henry, 1979. Determination of water mass properties of water pumped down from the Ekman layer to the geostrophic flow below. *Proceedings of the National Academy of Sciences, U.S.A.* 76: 3051-3055.

Regier, Lloyd, and Henry Stommel, 1979. Float trajectories in simple kinematic flows. *Proceedings of the National Academy of Sciences, U.S.A.* 76:4760-4764.

Behringer, David, Lloyd Regier, and Henry Stommel, 1979. Thermal feedback on wind-stress as a contributing cause of the Gulf Stream. *Journal of Marine Research* 37:699-709.

Stommel, Henry M., and Gabriel T. Csanady, 1980. A relation between the T–S curve and global heat and atmospheric water transports. *Journal of Geophysical Research* 85:495-501.

Leetmaa, Ants, and Henry Stommel, 1980. Equatorial current observations in the western Indian Ocean in 1975 and 1976. *Journal of Physical Oceanography* 10: 258-269.

Behringer, David W., and Henry Stommel, 1980. The Beta Spiral in the North Atlantic subtropical gyre. *Deep-Sea Research* 27: 225-238.

Books

Stommel, Henry, 1945. *Science of the Seven Seas.* Cornell Maritime Press, New York, 208 pp.

Stommel, Henry (ed.), 1950. *Proceedings of the Colloquium on the Flushing of Estuaries, Cambridge, Massachusetts, September 7-8, 1950.* Woods Hole Oceanographic Institution, Ref. 50-37, Woods Hole, Massachusetts, 206 pp.

Stommel, Henry, 1958. *The Gulf Stream: A Physical and Dynamical Description.* University of California Press, Berkeley, and Cambridge University Press, London, 202 pp.

Stommel, Henry, 1965. *The Gulf Stream: A Physical and Dynamical Description*, 2nd ed. University of California Press, Berkeley, and Cambridge University Press, London, 248 pp.

Stommel, Henry, Bruce Warren, Mary Sears, and Mary Swallow (eds.), 1969. *Frederick C. Fuglister Sixtieth Anniversary Volume, Deep-Sea Research* 16 (Supplement), 470 pp.

Stommel, Henry, and Kozo Yoshida (eds.), 1972. *Kuroshio: Its Physical Aspects.* University of Tokyo Press, Tokyo, 517 pp.

Stommel, Henry and Elizabeth Stommel, 1980. *The Year without a Summer: 1816* (forthcoming).

Miscellaneous Publications

Stommel, Henry, 1950. Comments on the colloquium. In *Proceedings of the Colloquium on the Flushing of Estuaries, Cambridge, Massachusetts, September 7-8, 1950*, Henry Stommel, ed., Woods Hole Oceanographic

Institution, Ref. 50-37, Woods Hole, Massachusetts, pp. 194–199.

Stommel, Henry, 1954. Why do our ideas about the ocean circulation have such a peculiarly dream-like quality? Privately printed, 34 pp.

Stommel, Henry, 1955. On the present status of our physical knowledge of the deep ocean. Privately printed, 9 pp.

Stommel, Henry, 1956. Annual report to the National Oceanographic Council. *Nature* 177: 1025–1026.

Stommel, Henry, 1956. Talk at Washington conference on theoretical geophysics. *Journal of Geophysical Research* 61: 320–323.

Stommel, Henry, and Sloat F. Hodgson, 1956. Consecutive temperature measurements at 500 meters off Bermuda. *Technical report WHOI Ref. No. 56-43*, Woods Hole Oceanographic Institution, Woods Hole, Massachusetts, 12 pp.

Stommel, Henry, 1958. Review: *Meteor Report*, Band VI, Teil 2, Lief. 6, by G. Wüst. *Transactions of the American Geophysical Union* 39: 1171–1172.

Stommel, Henry, 1960. Impressions of the International Oceanographic Congress. *Oceanus* 6(3): 15–16.

Stommel, Henry, 1960. Review: *Atlantic Ocean Atlas* by F. C. Fuglister. *Oceanus* 7(6):16–17.

Howard, L. N., N. Phillips, and H. Stommel, 1961. Review: *Hydrodynamics of Oceans and Atmospheres* by C. Eckart. *Journal of Fluid Mechanics* 11: 317–319.

Stommel, Henry, 1962. Review: *The Tides and Kindred Phenomena in the Solar System*, by George Howard Darwin, [1962; 1st ed. 1898]. *Deep-Sea Research* 9: 153.

MacDonald, G. J. F., et al. (Henry Stommel on panel), 1966. Effective use of the sea. *Report of the Panel on Oceanography, President's Science Advisory Committee*. U.S. Government Printing Office, Washington, D.C.

Stommel, Henry, and Edward Goldberg, 1969. Oceanography: An international laboratory. *Science* 165: 751.

Stommel, Henry, 1969. Review: *Ocean Currents*, by G. Neumann. *Marine Geology* 8: 109–110.

Stommel, Henry, 1969. Frederick C. Fuglister. *Deep-Sea Research* 16 (Supplement): 1–3.

Stommel, Henry, 1974. Discussion finale. In *La Formation des eaux océaniques profondes en particulier en Mediterranée occidentale, Paris, 4–7 Octobre 1972*. Colloques International Centre National Recherche Scientifique, No. 215, pp. 271–273.

Munk, A. (sic) W., H. Stommel, A. S. Sarkisyan, and A. R. Robinson, 1975. Where do we go from here? In *Numerical Models of Ocean Circulation*, National Academy of Sciences, Washington, D.C., pp. 349–360.

Stommel, Henry, et al., 1975. Report of two 1973 workshops of the National Academy of Sciences, Ocean Affairs Board. In *The role of the ocean in climate.* National Academy of Sciences, Washington, D.C.

Stommel, Henry, and Michele Fieux, 1978. *Oceanographic Atlases: A Guide to Their Geographic Coverage and Contents.* Woods Hole Press, Woods Hole.

Addendum (works not included in the first printing)

Journal Publications

Stommel, Henry, and George Veronis, 1956. Comments on "Heat budget of a water column, autumn, North Atlantic Ocean". *Journal of Meteorology* 13: 222.

Schroeder, Elizabeth, and Henry Stommel, 1969. How representative is the series of *Panulirus* stations of monthly mean conditions off Bermuda? *Progress in Oceanography* 5: 31–40.

Stommel, Henry, Arthur Voorhis, and Douglas Webb, 1971. Submarine clouds in the deep ocean. *American Scientist* 59: 716–722.

Stommel, Henry, and John L. Bowen, 1971. How variable is the Antarctic Circumpolar Current? In *Research in the Antarctic*, L. O. Quam, ed., American Association for the Advancement of Science, Publication No. 93, Washington, D.C., pp. 645–650.

Miscellaneous Publications

Stommel, Henry, undated. Note on the vertical distribution of phytoplankton. Woods Hole Oceanographic Institution, Woods Hole, Mass., 19 pp.

Stommel, Henry, 1945. The theory of the time-lag thermocouple. Woods Hole Oceanographic Institution, Woods Hole, Mass., 7 pp.

Stommel, Henry, 1945. A bibliography on internal waves. Woods Hole Oceanographic Institution, Woods Hole, Mass.

Stommel, Henry, translator, 1945. Internal waves by Jonas Ekman Fjeldstad ["Interne Wellen", *Geofysiske Publikasjoner* 10 (1933), 6], Woods Hole Oceanographic Institution, Woods Hole, Mass., 51 pp.

Vine, Allyn C., G. A. Riley, Henry Stommel, and W. T. Edmondson, 1945. Preliminary report on acoustic location in shallow water. Woods Hole Oceanographic Institution, Woods Hole, Mass., 50 pp.

Stommel, Henry, 1949. The equations of physical oceanography in finite difference form. Woods Hole Oceanographic Institution, Woods Hole, Mass.

Stommel, Henry, 1949. Hydrography of the Western Atlantic; diffusion due to oceanic turbulence. Technical Report No. 11, *WHOI Ref. No. 49-1*, Woods Hole Oceanographic Institution, Woods Hole, Mass., 52 pp.

Stommel, Henry, 1951. Recent development in the study of tidal estuaries. *Technical Report WHOI Ref. No. 51-33*, Woods Hole Oceanographic Institution, Woods Hole, Mass., 16 pp.

Stommel, Henry, 1951. On the mutual adjustment of velocity and pressure fields in the ocean under the influence of a fluctuating wind stress. *Technical Report WHOI Ref. No. 51-57*, Woods Hole Oceanographic Institution, Woods Hole, Mass., 11 pp.

Stommel Henry, 1952. Bibliography on estuaries. Woods Hole Oceanographic Institution, Woods Hole, Mass., 32 pp.

Stommel, Henry, and Harlow G. Farmer, 1952. On the nature of estuarine circulation. Part I. Chaps. 1, 2. *Technical Report WHOI Ref. No. 52-51*, Woods Hole Oceanographic Institution, Woods Hole, Mass., pp. various.

Stommel, Henry, and Harlow G. Farmer, 1952. On the nature of estuarine circulation. Part III. *Technical Report WHOI Ref. No. 52-63*, Woods Hole Oceanographic Institution, Woods Hole, Mass., 53 pp.

Parson, Donald, Jr., Henry Stommel, and Sloat Hodgson, 1952. Airborne radiation thermometer. In Radiation Research Quarterly Progress Report, *Technical Report WHOI Ref. No. 52-77*, Woods Hole Oceanographic Institution, Woods Hole, Mass., pp. 6–10.

Stommel, Henry, and Harlow G. Farmer, 1952. On the nature of estuarine circulation. Part I. Chaps. 3, 4. *Technical Report WHOI Ref. No. 52-88*, Woods Hole Oceanographic Institution, Woods Hole, Mass., pp. various.

Francis, J. R. D., Henry Stommel, Harlow G. Farmer, and Donald Parson, Jr., 1953. Observation of turbulent mixing processes in a tidal estuary. *Technical Report WHOI Ref. No. 53-22*, Woods Hole Oceanographic Institution, Woods Hole, Mass., 20 pp.

Stommel, Henry, Robert G. Walden, Donald Parson, Jr., and Sloat F. Hodgson, 1954. A deep-sea radio telemetering oceanographic buoy. *Technical Report WHOI Ref. No. 54-61*, Woods Hole Oceanographic Institution, Woods Hole, Mass., 14 pp.

Wilson, K. G., A. B. Arons, and Henry Stommel, 1955. A simple electrical analog for the solution of the Ekman wind drift problem with the coefficient of eddy viscosity varying arbitrarily with depth. *Technical Report WHOI Ref. No. 55-52*, Woods Hole Oceanographic Institution, Woods Hole, Mass., 15 pp.

Evolution of
Physical
Oceanography

Introduction

The technical section of this book is divided into four general categories:

General Ocean Circulation (chapters 1–7);
Physical Processes in Oceanography (chapters 8–13);
Techniques of Investigation (chapters 14–16);
Ocean and Atmosphere (chapters 17–18).

Within each of these categories will be found several chapters, each of which discusses a broad aspect of physical oceanography or its relationship to an allied field. We asked the authors to respond to the question, "What is the state of the subject today and how did we arrive there?" Where appropriate, we sought emphasis on a description of the evolution of the subject over the past 40 years. Thus, most of these chapters differ from conventional review articles both in their historical sweep and in their depth of coverage. The book looks both backward and forward in time.

A substantial fraction of physical oceanography is covered in what follows, consistent with the diversity of Henry Stommel's interests. Indeed, we believe that this book could serve as a textbook in advanced courses on the subject. As with any writing that attempts to be up-to-date, some of the material here is speculative. Nonetheless, we believe that we have in this book a nearly comprehensive description of physical oceanography as it was understood toward the end of the twentieth century.

The first part of the text is devoted to what might be thought of as "large-scale" or perhaps, classical, oceanography. It deals, in several chapters, with the overall distribution and circulation of water properties—a problem that has its roots deep in the nineteenth century. How the different waters move, and why, is treated both in terms of observation and theory. The chapters range from poles to equator, treating the peculiar kinematics and dynamics of these extremes.

Much of dynamical oceanography has focused on the Gulf Stream System, and this book is no exception. The Atlantic is comparatively small, making this current accessible to European scientists from the earliest days, and the Gulf Stream immediately confronts oceanographers on the east coast of North America. Thus the subject has periodically obsessed many of the leading oceanographers of the past 100 years. Much of the flowering of circulation modeling, which began in the late 1940s and remains one of the great accomplishments of modern geophysical fluid dynamics, was motivated toward understanding the Gulf Stream. The section concludes with a chapter on estuaries and shelf circulations. The latter in particular was an area of great interest around the turn of the century and has only comparatively recently reacquired the attention of oceanographers, who for many years tended to focus upon the open sea. The reader will find that many of the most important unresolved questions pertain to the interaction of the shelves with the deep-sea, large-scale circulation.

The second part is devoted to chapters that examine some of the physical processes underlying the larger scale circulations and distributions treated in the first part. Oceanic mixing processes, of all kinds and on all scales, have been, and remain, among the most important and enigmatic parts of oceanography. Here there is often an intricate interaction between the time-dependent components of oceanic motion and the large-scale distribution of properties—including momentum and energy. With the development of modern electronic instrumentation in the past two decades, completely unsuspected physical processes have been discovered (for example, thermal and haline microstructure). The advent of computers and of instruments capable of measuring oceanic time series has permitted attempts at linking physical phenomena previously studied wholly independently (for example, internal waves and resultant mixing). Many of the quantitative relationships remain obscure.

Included in the second part is a discussion of problems of biological oceanography. This subject is clearly evolving in parallel with physical oceanography, displaying a need for better understanding of the physical and chemical context, and for an appreciation that the biological sampling problems are at least as difficult as those beginning to be solved by physical oceanographers.

The third part, techniques of investigation, contains a description of some of the major tools of physical oceanography. Unlike many sciences, but in common with the other earth and astronomical sciences, oceanography is a field in which experiments as commonly described to students cannot usually be done. One must make sense of observations in a completely uncontrolled environment. Thus it becomes important to examine the observational basis of this science and to understand how the evolution of the field has followed very closely the development of new instruments and techniques. This part also contains a description of those comparatively rare, but extremely important, laboratory experiments that have shed light on important oceanographic phenomena. We have also included a chapter on radioisotope tracers; this field is, of course, a branch of chemical oceanography, and it stands as a science on its own apart from its usefulness to physical oceanographers. But a discussion is included because exploitation of such tracers will probably be one of the most fruitful *future* techniques for understanding many of the phenomena described in this book—those ranging from the largest circulation scales down to the finest near-molecular mixing mechanisms.

The fourth part contains two chapters on oceans and atmospheres. The first is directed at the actual physical coupling between air and sea. This, one of the most difficult problems in oceanography, is an area of increasing activity and interest, with a growing concern for climatic interaction of the two fluids. The other chapter is directed at the large-scale dynamical analogues between atmospheric and oceanic motions. Oceanography and meteorology are sister sciences, with each contributing ideas and insights to the other. One day they will undoubtedly be treated as a unified field. But study of the atmosphere has greatly benefited from the abundance of observations required for practical weather forecasts and the comparative ease of measurement, and meteorology is the more advanced subject. Thus it seems fitting to end this volume with a chapter on the not-always-obvious analogues of atmospheric motions to be found in the sea.

We have tried to tie the individual chapters together in a variety of ways. The book includes a general index of subjects and names, and also a reference list that gives the page number for each citation in the text. We hope that these features will make possible a rapid entry into the book by anyone seeking a discussion of a particular piece of work. Special care was taken in compiling the reference list to correct many common miscitations, some of which extend back nearly 100 years. The reader will notice overlap between chapters and even some dispute among them. We regard this as inevitable and healthy in a field undergoing the ferment of active progress. A consequence of this activity is that we did not attempt to impose a common notation upon the book, but we did ask the authors to avoid idiosyncratic schemes.

In the context of the question to the authors posed above, we are impressed in reading these chapters with how far we have come. When Henry Stommel entered physical oceanography in the early 1940s, the three authors of *The Oceans*, H. U. Sverdrup, Martin W. Johnson, and Richard H. Fleming, could cover authoritatively, in one volume, the entirety of oceanography—physics, chemistry, biology, and geology. Today no single volume could cover one of these fields, and probably no three authors would have the temerity to attempt comprehensive descriptions of any. But we believe that the reader will find here a broad description of the present state and the historical evolution of physical oceanography.

Bruce A. Warren
Carl Wunsch

Part One

General Ocean Circulation

1
Deep Circulation of the World Ocean

Bruce A. Warren

1.1 Introduction

Historically, the deep circulation of the ocean has been viewed from the perspective of property fields, mainly the distributions of temperature, salinity, density, and dissolved-oxgyen concentration. The practical reason for not considering velocity measurements as well, of course, was a technical incapacity for making them until very recently. On the whole, this was probably not a bad thing: not merely because the property distributions are as interesting in themselves as the motion field, but also because the scalar fields are so much more stable than the velocity vectors—allowing spot measurements from different areas even years apart to be combined into coherent pictures that tell a good deal about general patterns of deep flow, albeit indirectly. The slight differences between corresponding hydrographic sections in the atlases by Fuglister (1960) and by Wüst and Defant (1936), when compared with the fluctuations much larger than the means of deep velocities observed by the MODE Group (1978), for example, demonstrate how much easier it is to obtain statistically significant information pertinent to the overall global deep circulation from water-property data than from current measurements.

On the other hand, the information gained from the property fields allows only a limited view of the deep motions, at very best some kind of long-term average. Although oceanographers have usually been mindful of variability in the deep flow, even if only to accomplish eddy mixing, it seems extremely unlikely that anyone imagined the highly energetic low-frequency mesoscale motions that current records have revealed. Instead, because of the stability of the property fields, it was stationarity rather than variability that was emphasized, however implicitly, in the circulation pictures derived from them. That stability also was surely the basis for the conceptual structure of water types and masses that has been so enormously useful in summarizing and comprehending the temperature-salinity structure of the ocean and in identifying features in the property fields that can be exploited as tracers for the flow. Without velocity information, though, such descriptions of oceans have sometimes degenerated into taxonomic sterility (naming something doesn't explain it), and perhaps sometimes there has been too elemental a character ascribed to water masses (as if they were truly building blocks rather than names for features), leading to pictures of the ocean more suggestive of rigid geological strata than of the real motion field that forms the distributions.

Plainly there cannot be a satisfying description of the deep ocean circulation that does not meld station data with current records. It does not seem to me, though, that such a description is yet possible. Far too few current records have been obtained to describe the deep, low-frequency motions in a global sense; it is only in the western North Atlantic that one can even contemplate making a basin-wide description. Moreover, we simply have not learned how to combine the stable station data with the fluctuating-velocity records to tell a story that is both consistent and informative. For example, Reid, Nowlin, and Patzert (1977) reported a record (Cato 2) from a current meter moored on the South American continental slope in the core of the North Atlantic Deep Water: for 2 weeks the daily-averaged velocity vectors were directed southwestward, parallel to the isobaths, as one would have expected in this particular deep western boundary current; but then the flow abruptly changed direction and went eastward for nearly 2 weeks. Nevertheless, with due regard for different density and accuracy of observations, the high-salinity core of the current looked very much as had been depicted by Fuglister (1960) and Wüst and Defant (1936). How are we to approach these two different sets of data, to reconcile the variability of the one to the steadiness of the other, and to learn something significant from their combination about that boundary current?

Finally, it is not at all clear what effect the low-frequency velocity fluctuations have on the long-term mean flow. There is enough theoretical reason (e.g., Rhines, 1977) to suspect that their role in its dynamics may be substantial, but measurements of deep Reynolds stresses are meager. In fact, values reported by Schmitz (1977) from the Sargasso Sea well south of the Gulf Stream actually favor a negligible contribution to the vorticity balance there, but those measurements are far too few to give a general characterization of the deep open ocean.

Consequently, although I recognize its incompleteness, the following account of the deep circulation of the world ocean is undertaken mainly from the traditional perspective of hydrographic station data, with reference to current measurements only where they seem helpful in estimating velocities and transports of the prominent currents. The emphasis is on mean thermohaline circulation. What has been learned about the low-frequency motions is described in detail by Wunsch in chapter 11 of this volume.

In section 1.2, I have attempted a historical review of what seem to me to be the important events and dates in the development of ideas about the deep circulation, from the first deep temperature measurements through Sverdrup's comprehensive synthesis in chapter XV of *The Oceans* (Sverdrup, Johnson, and Fleming, 1942). In section 1.3, I have discussed the dynamical ideas of Stommel and his colleagues that led to the overthrow of a substantial part of Sverdrup's picture, and its revision in contemporary thinking with dynamically consistent models of circulation. Section 1.4 is an account of the sinking processes that supply water to the deep ocean from the surface layer. Section 1.5 is a consideration of how well the kinds of deep-circulation patterns envisioned in dynamical theory stand up to observation; it is necessarily mainly a digest of the evidence for deep western boundary currents in the world ocean. Finally, in section 1.6, I have speculated about some fundamental aspects of the deep circulation that seem to me to be not very well understood at this time.

The focus throughout is more on the circulation of deep water than on the complementary problem of its properties, because the subject of deep-water characteristics has been treated in detail by Worthington in chapter 2 of this volume. There is, of course, some overlap with that chapter, as well as with Reid's general discussion of the mid-depth circulation (chapter 3). To the extent that our opinions are in conflict, we hope that readers will recognize subjects for further observation and thought.

1.2 Historical Development of Ideas about the Deep Circulation

In 1751 Henry Ellis, captain of the British slavetrader *Earl of Halifax*, wrote to the Reverend Stephen Hales to describe some deep temperature measurements that he had made at lat. 25°13'N, long. 25°12'W, with a "bucket sea-gage" devised and provided for him by Hales. This instrument, to be attached to a sounding line, was a "common houshold pail" covered at top and bottom by valves that would be forced open during descent and pushed shut by drawing the bucket back to the surface; it was furnished with a thermometer so that the temperature of the water sample thus trapped could be read when the bucket was returned to the ship. In his letter, which Hales transmitted to the Royal Society of London, Ellis (1751) reported:

Upon the passage, I made several trials with the bucket sea-gage, in latitude 25'-13" north; longitude 25'-12" west. I charged it and let it down to different depths, from 360 feet to 5346 feet; when I discovered, by a small thermometer of Fahrenheit's, made by Mr. Bird, which went down in it, that the cold increased regularly, in proportion to the depths, till it descended to 3900 feet: from whence the mercury in the thermometer came up at 53 degrees; and tho' I afterwards sunk it to the depth of 5346 feet, that is a mile and 66 feet, it came up no lower. The warmth of the water upon the surface, and that of the air, was at that time by the thermometer 84 degrees. I doubt not but that the water was a degree or two colder, when it enter'd the bucket, at the greatest depth, but in coming up had acquired some warmth.

Ellis's guess of one or two degrees warming during the ascent was based on changes observed in the temperature of the sample while on deck. Modern data, however, indicate that Ellis's values were some 10–

13°F too high. Perhaps he underestimated the conduction through the walls of the sampler when it was in the ocean, or perhaps the valves were not tight enough to prevent exchange of water while it was being raised (especially if raised unevenly).

Nevertheless, these are the earliest recorded subsurface temperature measurements in the open ocean—indeed, the first anywhere to a substantial depth—and they pointed to what must be the most fundamental and striking physical feature of the ocean: that deep water is all cold, and warm water is confined to a relatively thin layer near the surface in the tropics and subtropics.

Ellis himself did not seem to realize the far-reaching significance of his data, for he remarked later in his letter to Hales:

This experiment, which seem'd at first but mere food for curiosity, became in the interim very useful to us. By its means we supplied our cold bath, and cooled our wines or water at pleasure; which is vastly agreeable to us in this burning climate.

He does not seem to have made any further measurements, nor does his work appear to have stimulated immediate exploration of the deep water by others, for, according to Prestwich's (1875) tabulation, it was more than 60 years before any additional temperature measurements were made to the depth that Ellis reached (though others were made during this time at lesser depths).

What Ellis's observations meant, of course, was that deep water in the tropics must derive from polar regions, and that, accordingly, there must be a meridional circulation system in the ocean to carry deep water equatorward. Evidently (M. Deacon, 1971), the first recorded recognition of this profound implication was by Count Rumford in his essay, "The Propagation of Heat in Fluids," first published in 1797.[1] Rumford had made the experimental discovery of convection currents in liquids (the basic subject of this essay), and, in considering their possible role in nature, he reasoned (Rumford, 1800):

But if the water of the ocean, which, on being deprived of a great part of its Heat by cold winds, descends to the bottom of the sea, cannot be warmed *where it descends*, as its specific gravity is greater than that of water at the same depth in warmer latitudes, it will immediately begin to spread on the bottom of the sea, and to flow towards the equator, and this must necessarily produce a current at the surface in an opposite direction.

In advancing evidence for such a deep flow, he drew the correct inference from Ellis's measurements, in as clear and straightforward a fashion as one could want:

But a still more striking, and I might, I believe, say, an incontrovertible proof of the existence of currents of cold water at the bottom of the sea, setting from the poles towards the equator, is the very remarkable difference that has been found to subsist between the temperature of the sea at the surface and at great depth, at the tropic—though the temperature of the atmosphere *there* is so constant that the greatest changes produced in it by the seasons seldom amounts to more than five or six degrees; yet the difference between the Heat of the water at the surface of the sea, and that at the depth of 3600 [sic] feet, has been found to amount to no less than 31 degrees; the temperature above or at the surface being 84°, and at the given depth below no more than 53°.

It appears to me to be extremely difficult, if not quite impossible, to account for this degree of cold at the bottom of the sea in the torrid zone, on any other supposition than that of *cold currents from the poles*; and the utility of these currents in tempering the excessive heats of these climates is too evident to require any illustration.[2]

During the early nineteenth century, Humboldt (1814, 1831, 1845) popularized the notion of deep currents flowing from polar regions toward the equator. He mentioned no sources for the idea, but he did cite Ellis's measurements and later ones of better quality as supporting data.[3]

The physicist Lenz (1845) had interested himself in how the subsurface vertical temperature gradient varied with latitude. In the course of his investigation he discovered the shoaling of the thermocline at the equator in the Atlantic; and the much sparser data available from the Pacific suggested the same phenomenon to him there. He regarded that shoaling as evidence for upwelling of deep water to the sea surface, and, acquainted with Humboldt's work, he proposed a more specific scheme of closed meridional circulations than had been given hitherto. This scheme involved great convection cells symmetric about the equator, with water sinking in high latitudes, flowing equatorward at depth, and rising in the tropics to return poleward near the surface. He considered that this upwelling also contributed to the reduction in surface salinity that he had observed at the equator in the Atlantic. His understanding of the driving mechanism was not more profound than Rumford's, but he did recognize that wind stress and the earth's rotation would distort the motions substantially from the simple cellular form. He made no explicit reference to the Indian Ocean, but he discussed the symmetric cells in terms of general validity. Variants on Lenz's construction, some perhaps conceived independently, were to influence interpretations of the deep circulation into the first decade of the twentieth century (e.g., Schott, 1902; Brennecke, 1909).

Dr. Carpenter, for one (Carpenter, Jeffreys, and Thomson, 1869), invoked the general idea to account for differences in deep temperatures observed in the northern North Atlantic, and his resulting prolonged and contentious controversy with Croll attracted a great deal of attention to it (M. Deacon, 1971).[4] Carpenter had a more elastic conception of the convection cells than did Lenz, though, believing that their warm, poleward-directed limbs could touch bottom in places; nor was he rigid about the equatorial symmetry, considering that the northern cells would be weaker and less extended laterally than the southern ones, in conformity with the oceans being much more open to polar water in the south than in the north—whereby deep water from the Antarctic could be expected to penetrate into the northern hemisphere. He referred explicitly to recent subsurface temperature measurements in the Arabian Sea to argue that all deep water in the Indian Ocean derived from the Antarctic, and he implied a belief that most of the deep water in the Pacific would also be found to come from the south.

Nevertheless, Prestwich (1875) seems to have been the first to demonstrate from observations the nonexistence in the Pacific of relatively strong deep upwelling near the equator and the lack of a northern source for the deep water there. These facts showed that deep water in the North Pacific, as well as that in the South Pacific, was supplied from the Antarctic, and thus invalidated Lenz's picture of symmetric convection cells. Prestwich believed, however, that it was still a correct description of the meridional circulation in the Atlantic.

Following the *Challenger* expedition, Buchan (1895) produced world-ocean maps of the distribution of temperature at various levels, based mainly on the *Challenger* data. His map for 2200 fathoms illustrated the spreading of deep water from the Antarctic into the three oceans, including details of how its course is shaped by submarine ridges:

It is also to be noted that the lowest deep-sea temperatures are found in those parts of the ocean which lie in the Southern hemisphere, and that, on the whole, higher temperatures are encountered as we recede from the Antarctic region. It may also be pointed out that the lower deep-sea temperatures extend farther to the north from the Southern Ocean, just over those depths of the sea which appear to have, and probably do have, a direct communication with the south; that is, are not cut off by any intervening submarine ridge separating them from the cold waters of Antarctica.

... There can be no doubt that these very low deep-sea temperatures have their origin in the Southern or Antarctic Ocean, the icy cold waters of which are propagated northward, the rate of propagation being so slight as to be regarded rather as a slow creep than as a distinctly recognizable movement of the water.

Buchan (1895) also inferred a southward movement of deep water from the North Atlantic into the South Atlantic. As Merz and Wüst (1922) pointed out later, this perception, together with Buchanan's (1884) specific-gravity profile that showed the intermediate-water salinity minimum extending well north of the equator from high southern latitudes, would have been sufficient to disprove Lenz's equatorial upwelling from great depth to the sea surface—and thereby his symmetric convection cells for the Atlantic; but neither Buchan nor Buchanan attempted a comprehensive discussion of the meridional circulation.

Buchan formed his idea of southward flow after noticing in the *Challenger* data that the vertical temperature gradients at depths from 800 fathoms to 1500 fathoms and deeper (how much deeper not stated) were much smaller in the South Atlantic than in the North Atlantic. He linked this difference to the distribution of salinity (specific gravity):

The specific gravities at the bottom of the ocean afford a ready explanation for this remarkable distribution of temperature. Owing to the higher specific gravities of the North Atlantic, an extensive deep-sea current from the North to the South Atlantic, carrying a higher temperature with it, sets in at depths at which the influence of the surface currents is no longer felt, and becomes more pronounced as the depth below 1000 fathoms is increased. Hence the North Atlantic receives large accessions to its salinity from the surface currents, which the deep-sea currents again return to the South Atlantic.

This was probably the first recognition of deep southward flow across the equator, but unfortunately it is all that Buchan (1895) said on the subject. One cannot tell whether he also realized that the deep flow was layered, with North Atlantic water overriding the bottom water from the Antarctic, or whether he viewed "deep water" in the terms of his day as a single unit moving uniformly northward or southward, whereby the oppositely directed North Atlantic and Antarctic flows would simply collide and upwell at some middle southern latitude.

Whatever Buchan may have believed, the first unambiguous, well-documented statement of deep southward flow from the North Atlantic between the bottom and intermediate waters from the Antarctic was made by Brennecke (1911) on the basis of stations occupied by the *Deutschland* along the western rim of the South Atlantic:

Das Hauptergebnis unserer Reihenmessungen ist die Feststellung eines Tiefenstromes in etwa 1500 bis 3000 m Tiefe der vom Nordatlantischen Ozean nach Süden vordringt und durch hohe Temperatur und hohen Salzgehalt sich von der über- und unterlagernder Schicht abhebt. Dieser Tiefenstrom konnte bei allen Rehenmessungen von 5°S-Br. bis 40°S-Br. klar erkannt werden. Wenn bislang noch nicht erkannt worden ist, so liegt dies einerseits an der geringen Zahl der Messungen, die im diesen Schichten tatsächlich ausgeführt worden sind, anderseits an den frühen vielfach benutzten Maxima-Mimima-Thermometern, die eine Temperaturumkehr in der Tiefe nicht anzeigen. Soweit unserer Messungen Aufschluss geben, wird dieser Tiefenstrom bei seinem Vordringen nach Süden in grössere Tiefe gedrängt, bzw. in seiner Oberschicht mehr und mehr durch den schon von früheren Forschungen her bekannten, entgegengesetzt gerichteten, d.h. nordwärts vordringenden Tiefenstrom in 1000 m (ausgezeichnet durch das Minimum des Salzgehalts) gemischt.

The *Challenger* thermometers having been of the minimum-type, Buchan could not have detected the temperature inversion. Another factor that contributed to Brennecke's success was that, although he could not have known it, he actually made his stations in the swiftest part of the southward flow, where the inversion is much more markedly developed than to the east, and where, happily, the strongest evidence for the flow was available.

In the early 1920s, Merz amassed a file of all available deep observations of temperature and salinity, and began a program of systematic reexamination of these data to delineate the deep circulation of the whole world ocean. In the course of this study Merz and Wüst (1922) and Merz (1925) explicitly refuted the existence of Lenz's symmetric convection cells in the Atlantic, on the evidence already cited, and they proposed a comprehensive new picture of the meridional circulation there, whose essential feature was hemispheric exchange of water: ocean-wide northward flow across the equator in the upper kilometer, compensated by southward flow below to about 4000 m, with basin-wide northward and southward flows of bottom water in the western and eastern basins, respectively.[5] Their working materials were north-south sections of temperature and salinity, and they regarded the depression of isotherms and isohalines in the upper 2 km in the northern subtropics as indicating sinking from the surface in those latitudes, and thus formation of the North Atlantic deep water there. In making this misinterpretation of that feature, they evidently did not fully consider either the baroclinic character of the Gulf Stream gyre or the development of anomalously high temperatures and salinities at mid-depth from mixing of the Mediterranean outflow into the gyre.[6]

Following Merz's death in 1925 on the *Meteor* Expedition, Möller and Wüst undertook to complete his program for the Indian and Pacific Oceans, respectively. In her study of the Indian Ocean data set Möller (1929) joined with Schott (1926) to advance the Merz scheme of meridional circulation, emphasizing hemispheric exchange, for that ocean too—northward flow of intermediate and bottom water, southward flow of deep water in between—with the deep water sinking from the surface in the Arabian Sea, and receiving contributions from the Red Sea and Persian Gulf outflows. Thomsen (1933), however, found systematic error in some of these salinity values, and new data from the *Dana* showed much less extensive southward penetration of high-salinity deep water from the North Indian Ocean than Möller had described, implying a different character of deep-layer flow from that in the Atlantic: certainly weaker, and if directed across the equator at all, only in response to the northward movement of intermediate and bottom water. Implicit in this conclusion was rejection of the idea of deep water sinking from the surface of the Arabian Sea, although the influx of Red Sea water was recognized. From the distribution of bottom potential temperature, Wüst (1934) clarified the role of the Central Indian Ridge in dividing the bottom flow into at least two separate regimes (the Ninetyeast Ridge being then unknown), with bottom water entering from the south into both the western and the eastern basins of the Indian Ocean, as Schott (1902) had suggested in his report on the *Valdivia* observations.

For the Pacific, Wüst (1929) found that the general quality of data available was inferior to that for the other two oceans, and, with deep property differences being much smaller there anyway, he did not reach such definite conclusions as for the Atlantic. Nevertheless, his analysis suggested to him that the Pacific was different from the Atlantic, in that hemispheric exchange was insignificant, and that, in both the North and South Pacific, intermediate and bottom water spread equatorward and deep water poleward (i.e., that deep water sank from the surface layer near the equator, rather than in middle northern latitudes, as he thought it did in the Atlantic). In one respect, Wüst's (1929, 1938) work on the deep Pacific in this decade was actually retrogressive, because some temperature measurements made by the *Tuscarora* in 1874 that are incompatible with modern data led him to infer a northern source for bottom water in the Sea of Okhotsk, contrary to Prestwich's (1875) correct finding that sinking to great depth does not occur anywhere in the North Pacific.

From his study of the *Carnegie* observations, Sverdrup (1931) realized that there was not, in fact, any equatorial downwelling in the Pacific. He also recognized that Pacific bottom water did not derive from sinking directly to the south near Antarctica, as is the case for the Atlantic, but that all water below the low-salinity intermediate layers in the Pacific was, in effect, carried to it by the Circumpolar Current. He thought it likely that at least in the South Pacific there was some southward return flow of deep water, as in the other two oceans, but the very low oxygen concentration in the deep North Pacific indicated only small hemispheric exchange.

The idea of a deep-water connection among the Atlantic, Indian, and Pacific Oceans via the Circumpolar Current was reinforced by the systematic cruises of the *Discovery* in the Antarctic, which obtained evidence for the extension of high-salinity deep water all the way from the Atlantic to the Pacific (Deacon, 1937).

By the time Wüst began his comprehensive treatment of the *Meteor* results, Helland-Hansen and Nansen (1926) had shown clearly—and Defant (1931) had reaffirmed—that the high salinities at mid-depths in the subtropical North Atlantic were due to the Mediterranean outflow; and Wattenberg (1929) had traced the maximum in oxygen concentration below to high northern latitudes. Wüst (1935) therefore abandoned the broad mid-latitude source for the deep water that Merz and he (1922) had proposed, and that he (1928) had later modified somewhat. Instead, he distinguished three layers within the deep water, all spreading southward, and each with its own "formation" site: Upper North Atlantic Deep Water, characterized by the deep salinity maximum attributed to the Mediterranean outflow; Middle North Atlantic Deep Water, identified by a maximum in dissolved-oxygen concentration, and traced to the Labrador Sea; and Lower North Atlantic Deep Water, identical to Wüst's (1928) earlier "North Atlantic Bottom Water," defined by a second and deeper oxygen maximum, and thought to be formed somewhere in the waters south or southeast of Greenland by deep convection in winter.

The picture of deep circulation for the world ocean that Sverdrup (Sverdrup, Johnson, and Fleming, 1942, chapter XV) then constructed was broadly Merzian: ocean-wide or basin-wide flows in deep and bottom layers, the bottom water moving generally northward, the deep water southward—with the greatest hemispheric exchange in the Atlantic and the least in the Pacific. Sverdrup departed qualitatively from the Merzian schematic only in inferring that the principal movement of deep (and bottom) water in the North Pacific was a slow, nearly closed, clockwise circulation. By this time, though, the areal extent of regions where deep sinking was thought to occur had been much reduced from what Merz had supposed: sinking in the southern hemisphere limited mainly to the Antarctic continental slope in the Weddell Sea (Brennecke, 1915, 1921; Mosby, 1934), and sinking in the northern hemisphere confined to high latitudes in the western basin of the Atlantic, with probably smaller, subsurface influxes of relatively dense water from such marginal seas as the Mediterranean and Red Seas.

Even then, however, Wüst had already realized that the idea of deep and bottom flows in the Atlantic that were uniform in a basin-wide sense was oversimplified to the point of being misleading. His maps of tracer distributions from the *Meteor* data (Wüst, 1935) demonstrated more intense southward propagation of North Atlantic Deep Water in the western than in the eastern South Atlantic, and geostrophic velocity calculations (Wüst, 1938) showed relatively strong northward flow of Antarctic Bottom Water (at speeds of several centimeters per second) only close to the South American continental slope. In discussing property distributions, Wüst (1935) was careful to speak of the "spreading" ("*Ausbreitung*") of water types, stating that such spreading had a current-like character only on the western side of the Atlantic, and that elsewhere it was accomplished, in effect, by eddy fluxes. He regarded this distribution of currents as an extremely strange phenomenon, for which he was unable to give a satisfactory explanation; and, indeed, the real significance of these "deep western boundary currents" was not to be appreciated until dynamically consistent models of the deep circulation were developed.

It should be acknowledged, however, that, as penetrating as Wüst's insights were, it was not, in fact, he who first discovered the bottom-water boundary current, but Buchanan, who was presumably the chief contributor to chapter XXII of the *Challenger Report Narrative*. The *Challenger* stations were the first set that by virtue of their spacing could possibly have revealed such a feature, and Buchanan (Tizard, Mosely, Buchanan, and Murray, 1885a) did not fail to see it:

Another equally remarkable current is that which brings the cold water from the south polar regions up along the South American coast at the bottom as far as the Equator.... That this does take place is shown by the low bottom temperatures observed at every Station along the western side of the South Atlantic almost to the Equator, and that there is here a sensible current of cold water along the bottom is shown by the sudden change in the rate of decrease of temperature with increasing depth which is observed at depths of from 2000 to 2200 fathoms.

... The temperature of the water from 2400 fathoms to the bottom was uniform, the mean result of six observations being 32.43°. This water underruns the body of the Atlantic water, which at 1500 to 2000 fathoms has here a temperature of 37°, producing a temperature gradient of about 1.3° per hundred fathoms at the steepest. For the preservation of this gradient, a considerable supply of cold water is requisite, and it must be drawn from higher latitudes. But any motion of the water towards the Equator will be accompanied by a strong deflection to the westward (proportional to the change of the cosine of the latitude). A measure of this deflecting force is furnished by the rise of this cold water at the more inshore Station on the 28TH, where the maximum gradient is at about 1750 fathoms, while on the 29TH, at a distance of 120 miles,[7] it is at about 2100 fathoms.

In the last two sentences Buchanan seems to have been groping toward some notion of the thermal-wind relation, and to have correctly associated the lateral scale of the slope in the maximum-gradient surface with that of the northward flow beneath. He had bad luck with his discoveries, though. Not only was this one ignored [except for an allusion by Wüst (1933)], but, as will be recalled, his almost simultaneous discovery of the Atlantic Equatorial Undercurrent was completely forgotten after three decades.

1.3 A Dynamical Framework

Paralleling the intuitively appealing concept of ocean-wide or basin-wide meridional flows were two other ideas that seemed equally reasonable. One is that, in compensation for the known sinking of near-surface water in small regions, there is a general slow upward movement of deep water over most of the rest of the ocean, which balances the downward diffusion of heat from the upper layer and thus accounts for the existence of the main thermocline. The other is that the mean horizontal deep flow in the open ocean is strictly geostrophic, in the sense that streamlines parallel isobars, and the speed is inversely proportional to their spacing and to the sine of the latitude. As indicated in section 1.1, the latter may not seem so obvious now as it once did, but there is yet no proof to the contrary. Stommel (1965, manuscript of first edition completed in 1955) perceived the dynamical implication of these ideas—that the vertically integrated deep flow should be directed poleward—with the bizarre corollary that deep water in the open ocean of known polar origin should be moving on the whole *toward* its sources rather than away from them. Specifically, broad, deep equatorward flows in the North Atlantic, South Indian, and South Pacific Oceans, as envisaged in the Merz-Sverdrup picture, do not seem consistent with the peculiar dynamics of slow motions on a rotating sphere.

The physical argument is simple. The geostrophic vorticity balance is $\beta v = f w_z$, where f is the Coriolis parameter, β its meridional derivative, v the northward velocity component and w the vertical component, and z the vertical coordinate (positive upward). Integrating the equation vertically from the ocean bottom to the "top" of the deep water (however defined), and taking the floor of the ocean to be level in the large-scale mean, so that the vertical velocity at the bottom can be taken to be zero in a large-scale sense, then yields the result that, with upward velocities at the top of the deep water, the meridional component of flow is poleward, in both the northern and the southern hemispheres. This can also be seen in more rudimentary terms: because geostrophic flow cannot cross isobars, the transport per unit depth between any two streamlines of the horizontal flow increases downstream in flow toward lower latitudes, and decreases downstream in flow toward higher latitudes. *If* water is also moving upward out of the deep layer, then the associated horizontal flow must be the convergent one—directed poleward.

Clearly, however, there must be regions where this constraint breaks down, so that deep water of polar origin can reach the low latitudes where it is found. For the large-scale circulation generally, the regions where the rules of the open ocean are relaxed are the western boundaries of basins, where relatively narrow, intense currents exist to satisfy whatever boundary or continuity conditions are not fulfilled by the interior flow (see section 1.5.4). Stommel (1958) therefore hypothesized deep western boundary currents for all the oceans to carry deep water away from its sources, to supply it to the interiors of oceans from their western sides (rather than from their polar margins) and thus feed the poleward interior flows, and to correct ocean-wide continuity imbalances between the upward flux and the meridional flow.

The flow patterns thus constructed were so contrary to intuition that, to test their consistency, Stommel, Arons, and Faller (1958) developed a corresponding theory for circulation systems forced by a variety of prescribed distributions of sources and sinks in a rotating tank. The excellent agreement with the experiments [continued further by Faller (1960)] encouraged Stommel and Arons to rework the theory in spherical coordinates (1960a), and they used that theory to develop a model of the global deep circulation (1960b), including estimates of velocities and volume transports. The three oceans were represented as flat-bottom basins bounded by either parallels or meridians, and connected in the south by an analogue to the Antarctic Circumpolar Current. The flow was taken to be barotropic, driven by sources in the model counterparts of the Weddell Sea and northern North Atlantic (representing sinking from the surface layer), and by the compensatory upwelling, distributed uniformly over the rest of the world ocean. The resulting circulation pattern in a slightly modified version of the model (Kuo and Veronis, 1973) is sketched in figure 1.1. The analogue to the Circumpolar Current is indicated along the southern margin; its transport through the counterpart of the Drake Passage is not determined by the model dynamics, and was estimated roughly from observations. The layer thickness is 3 km, the strength of each source is about 17×10^6 m^3 s^{-1}, and the upwelling rate is 1.5×10^{-5} cm s^{-1}; the horizontal speeds in the interior are of order 10^{-2} cm s^{-1}, and the transports of the western boundary currents, of order 10×10^6 m^3 s^{-1}.

The interior flows are everywhere poleward and eastward, fed by intense currents along the western boundaries. In the Atlantic, the boundary current emanates from the northern source and flows to the southern limit of the South Atlantic, where it, the interior flow, and the Weddell Sea source increase the transport of the deep Circumpolar Current. That in turn is the source for northward-flowing western boundary currents in the Indian and Pacific Oceans, entering at their southwestern corners. At the northern margins of those two oceans the interior flow forced by the upwelling feeds into northern boundary currents, which flow westward into southward-directed western boundary currents.

This model was never intended as a realistic *description* of the deep-ocean circulation, and in at least two respects it would be a qualitatively bad one. By treating the ocean floor as level, the model gives single circulation systems in the northern and southern hemispheres of each ocean, whereas the mid-ocean ridges divide the oceans into separate basins with multiple circulation systems and multiple entrance points for deep water from the Antarctic. Also, because the model is barotropic, it cannot allow layered deep flow, with opposed boundary currents one above the other, as occur most spectacularly in the South Atlantic, where Antarctic Bottom Water flows northward beneath the southward current of North Atlantic Deep Water. (In fact, there could be no direct, boundary-current flow of Antarctic Bottom Water from the Weddell Sea into the South Atlantic anyway were it not that the Drake Passage sill is substantially shallower than the floor of the Atlantic to the east.)

Nevertheless, it was a dynamically consistent scheme, which offered some drastically different ideas about the deep circulation from those in the Merz-Sverdrup picture of basin-wide meridional flows, especially the new physical concept of deep western boundary currents. A search was immediately made for the predicted deep current off the east coast of the United States, by means of direct current measurements with neutrally buoyant floats; its discovery was reported by Swallow and Worthington (1957), and provided dramatic support for the essential truth of the new conceptual scheme. Considering what has been learned since then about the prevalence of strong, low-frequency motions in the deep water, one would probably question now how sound a "proof" of the current those measurements were, given their restricted geographical scope and brevity of duration (1 to 4 days); but they were consistent among themselves and with later observations (see section 1.5).

Actually, as pointed out in section 1.2, the relatively intense deep meridional flows adjacent to South America had been recognized long before, and Stommel (1950a) had already suggested that they might be dynamically akin to the Gulf Stream. Moreover, as the theory was being developed, Wüst (1955) constructed geostrophic velocity sections for two of the *Meteor* transects, one of which vividly illustrated these two currents, the one directed northward, the other southward. Thus the general idea of deep western boundary currents was not without observational support even before its confirmation by Swallow and Worthington (1957).

It was only for simplicity and clarity that the Stommel-Arons model was presented as a barotropic one. It can be brought somewhat closer to the real stratified ocean by regarding it instead as a theory for vertically integrated flow. Equatorward interior flow at some levels is therefore not excluded, so long as the vertical integral is poleward. Layered interior flows might be difficult to detect in tracer distributions, though, because mixing of properties can mask the flow patterns (see below, this section).

The theory is clear, however, in allowing no hemispheric exchange of water (of the sort that Merz emphasized) through the interior flow, because at all levels the meridional component of flow vanishes with the Coriolis parameter at the equator: hemispheric ex-

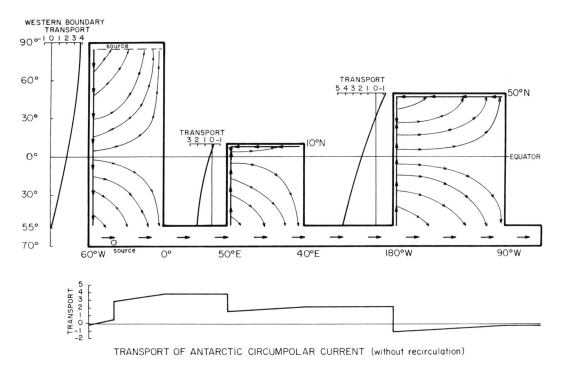

Figure 1.1 Deep circulation in a schematic world ocean driven by uniform upwelling with sources at the counterparts of the North Pole and Weddell Sea. See text. Transports measured in units of about 6×10^6 m^3 s^{-1}. (Kuo and Veronis, 1973.)

change by the mean flow is limited to the boundary currents. There can still be hemispheric exchange of *properties* all along the equator, of course, through lateral mixing.

A more fundamental novelty of the Stommel-Arons model is in its looser connection between specific horizontal flows and specific sinking phenomena than was implied in the earlier schemes. The net interior flow at any position is directly coupled to the local upward movement of the deep water, but the amplitude of that upwelling (as distinguished from its geographical variation) is related to the world-ocean integral of the rate of sinking into the deep water, rather than to the downward flux in any one sinking region. Nor is there any one-to-one relation between the transport of a deep western boundary current and the rate of sinking in a region from which it may emanate, even after subtracting losses to interior upwelling—as may be seen by considering the water budget for the portion of a deep ocean north of an arbitrary parallel of latitude. The boundary-current transport T out of that volume (positive southward) can be calculated as

$$T = \frac{f}{\beta} \int w\, dl - \int w\, d\sigma + S,$$

where S is the rate of sinking into that volume, and w is the vertical velocity at the top of the deep water; the first term on the right, expressing the net interior transport into the volume, includes an integral of w along the arbitrary latitude, and the second term is an integral over the top of the volume. Not only can T be less than S, as is obvious without the formula, but T can also be *greater* than S, signifying recirculation in an ocean or basin, because the boundary current must compensate not only the sinking but also any excess of interior transport into the volume over upward flux out of it. Extreme examples of this effect are illustrated in the northern Pacific and Indian areas of figure 1.1, where southward flowing boundary currents are required even though no sinking occurs to the north.

Despite the physical consistency of the deep-circulation model, the Merz-Sverdrup picture still looks intuitively more believable, in that poleward interior flows are not obvious in the property distributions. Stommel and Arons (1960b) suggested that lateral mixing of tracer properties might be so intense as to mask the patterns of such slow mean interior flow; to explore that effect, Kuo and Veronis (1973) made calculations of dissolved-oxygen distributions based on the circulation field of figure 1.1 and ranges of values for the mixing coefficient and oxygen-consumption rate. They prescribed the oxygen concentration at the two sinking points, and compared the resulting fields with the observed distribution at 4 km.

Two extreme cases are instructive. With no mixing, high oxygen concentrations are introduced to the oceans from their western boundaries, and the values diminish eastward on account of consumption, with the effect that the lowest values are found along the eastern boundaries; the western boundary currents are discernible in the field, as well as the eastward component of interior flow, but the meridional component of the latter is not particularly evident. With mixing but no advection, however, the oxygen concentration is simply diffused zonally in the circumpolar belt, and relatively high values are introduced to the Indian and Pacific Oceans all along their southern boundaries; the concentrations decrease northward and the lowest values in those oceans are obtained at their northern boundaries. (In the Atlantic, with northern and southern sources, the lowest values are found in the intertropical zone.)

The best match with observation (figure 1.2) was an intermediate case, representing approximately "equal" effects of advection and mixing. The mixing coefficient in this case was 6×10^6 cm^2 s^{-1}, and the oxygen consumption rate, 2×10^{-3} ml l^{-1} yr^{-1}—values roughly consistent with those from other studies. The patterns were found to depend on the speed of the Circumpolar Current, since that affected the "source" values for the Indian and Pacific Oceans; the optimal transport value for the Drake Passage was 35×10^6 m^3 s^{-1}. That figure from the model is difficult to compare precisely with observation because, given that the "top" of what ought to be considered deep water is much shallower in the Drake Passage than in the tropics, it is not clear what is the most appropriate depth interval to choose for comparison; density sections combined with year-long current measurements made during the ISOS program gave estimates of the net transport below 1000 m of about 47×10^6 m^3 s^{-1}, and below 2000 m of 12×10^6 m^3 s^{-1} (H. Bryden and R. D. Pillsbury, personal communication).

Certainly there is nothing in figure 1.2 to indicate the character of the underlying flow field, except perhaps some hint of the western boundary currents. The most obvious interpretation of the pattern—and one quite wrong—would be of slow, ocean-wide, transequatorial movement northward from the Antarctic and southward from the northern North Atlantic, in the manner of Merz and Sverdrup, though perhaps somewhat stronger in the west than in the east. Even though, as Kuo and Veronis (1973) pointed out, these calculations were based on a highly simplified flow scheme (especially as not constrained by the ridge system), the results do suggest that, with recognition of a moderate degree of lateral mixing in the ocean, observed tracer distributions can probably be rationalized in terms of more realistic flow fields constructed from the Stommel-Arons dynamics.

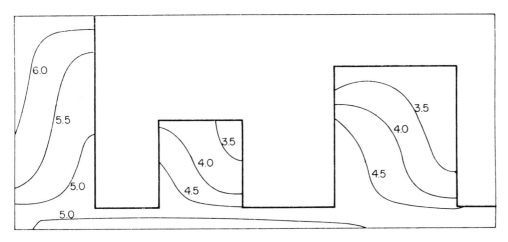

Figure 1.2 Model distribution of dissolved-oxygen concentration (ml l^{-1}) associated with the flow field in figure 1.1. See text for "best-fit" values of controlling parameters. (Kuo and Veronis, 1973.)

Although the Stommel-Arons (1960b) model itself is not a theory for the actual mean deep circulation, the model does provide a dynamical framework in which to think about aspects of the circulation. Evidence bearing on the verification of elements of the model as applied to the real ocean is discussed in section 1.5.

1.4 Sources of Deep Water

Even though the individual sites where water sinks to great depth do not seem so important for driving the overall deep circulation as they once did, it is desirable to know the rates at which sinking occurs, because the compensating upwelling from the rest of the deep water is thought to force its mean horizontal circulation. Moreover, the climatic conditions at the specific sinking regions determine the characteristics of the descending water, and thereby the properties and layering of deep water through the world ocean. The sinking phenomena are an integral part of the deep circulation as well, and the various processes by which near-surface water is brought to depth are of considerable interest in themselves.

As noted in section 1.2, in the southern hemisphere deep sinking is limited to the waters around Antarctica. In the northern hemisphere it occurs only in the northern North Atlantic, but there are also outflows from marginal seas: the Mediterranean Sea, the Red Sea, and the Persian Gulf. These outflows descend principally to mid-depths, but affect the water characteristics at deeper levels too.

1.4.1 Sinking around Antarctica

The only location in the Southern Ocean where sinking to the bottom through convective overturning has been identified is the Bransfield Strait (figure 1.3), which separates the northern tip of the Antarctic Peninsula from the South Shetland Islands (Clowes, 1934; Gordon and Nowlin, 1978). There a somewhat isolated trough of depth 1100-2800 m is filled with a mass of nearly homogeneous water very different from the surrounding Circumpolar Deep Water: much higher oxygen concentration, and notably lower temperature, salinity, and nutrient concentrations. Values of tritium concentration, increasing toward the bottom below 300 m in 1975, are further evidence for recent contact between the bottom water and the sea surface. Although no observations have been made there in the winter season, these circumstances strongly imply that local winter convection renews the water column through its full depth. So far as one can tell from the property distributions, however, none of this bottom water flows out of the trough into the rest of the ocean, probably because of the topographic barriers.

Reduced vertical stratification northeast of the Bransfield Strait, in a zone separating the Scotia and Weddell Seas, raises the possibility of some deep convection there too (G. E. R. Deacon and Moorey, 1975), but the evidence to date is less clear-cut than that from the Bransfield Strait.

The bottom water that does enter the rest of the ocean originates in several areas of the Antarctic continental shelf, where water is made sufficiently cold and saline that, in flowing down the continental slope and mixing with the surrounding deep water, it is dense enough to reach the floor of the ocean. In order of decreasing amount and extent of influence on the deep-water property distributions—and presumably, therefore, of rate of bottom-water production—these regions are the Weddell Sea, the Ross Sea, and the Adélie Coast (figure 1.3; A. L. Gordon, 1974); perhaps there is some production off Enderby Land too.

Because of its predominant role, the Weddell Sea has attracted the most attention, although the extreme harshness of working conditions there has discouraged

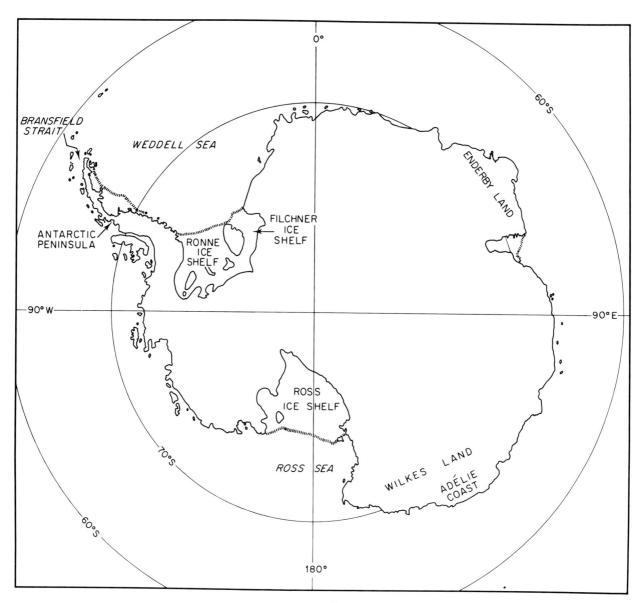

Figure 1.3 Index map identifying Antarctic place names mentioned in text.

any wintertime observations since the *Deutschland* was frozen into the pack ice in 1912 (Brennecke, 1921). Nevertheless, summer data suggest that flow down the continental slope takes place throughout the year. Sections of temperature, salinity, and dissolved-silica concentration, as constructed from observations made in 1968 on a station line running eastward from the Antarctic Peninsula (figure 1.5; see figure 1.4 for station positions) show a 200-m thick layer of relatively cold, fresh, low-silica bottom water on the slope, extending onto the floor of the Weddell Sea (Carmack, 1973). Although the station line did not reach the continental shelf (ice cover has prevented any shelf-water observations in that region), the evidence points plainly to a flow of shelf water down the slope, probably quite obliquely, entraining, and being diluted by, the surrounding water along the way. Similar sections near longs. 40°W and 29°W (Foster and Carmack, 1976a), as well as 10–20°W (Carmack and Foster, 1975), show diminishing evidence for downslope flow to the eastward, and the distribution of bottom potential temperature in the Weddell Sea (figure 1.6; Foster and Carmack, 1976a) demonstrates that the newly formed bottom water leaves the continental slope mainly at the northern tip of the Antarctic Peninsula, in lats. 63–65°S.

This Weddell Sea Bottom Water—defined by Carmack and Foster (1975) as of potential temperature $\theta < -0.7°C$—is widespread in the Weddell Sea, but since water of such extreme characteristics is not found to the north in the Argentine Basin (Reid, Nowlin, and Patzert, 1977), it must be mixed fairly rapidly into the water above. That overlying water in the Weddell Sea is itself also "Antarctic Bottom Water," but "older" water, whose properties have been modified by mixing elsewhere; it flows westward into the Weddell Sea close to the continent in the clockwise Weddell gyre (e.g., Brennecke, 1918; G. E. R. Deacon, 1976).

By combining station data with current records of duration between 3 and 4 weeks, Carmack and Foster (1975) estimated that the net flow of Weddell Sea Bottom Water out of the Weddell Sea is about 16×10^6 m^3 s^{-1}. At their stations, that water had a mean potential temperature of $-0.786°C$, brought about, presumably, by mixing overlying water of $\theta = -0.6$ to $-0.7°C$ into water that initially, at the edge of the continental shelf, was of $\theta = -1.2$ to $-1.4°C$. They estimated, therefore, that the rate of sinking from the shelf in the Weddell Sea is 2–5×10^6 m^3 s^{-1}.

In accounting for the formation of Weddell Sea Bottom Water, three distinct processes need to be treated: (1) how the density of the shelf water is increased to the point where it can sink through the relatively warm, saline water at mid-depths; (2) how the water moves off the shelf; and (3) how it descends with entrainment to the floor of the Weddell Sea. The first problem has been considered at length by Gill (1973). Although near-surface water in the open Weddell Sea has a temperature close to the freezing point, it is generally too fresh to be dense enough to sink through the water below. On the shelf, however, the salinity can be increased by salt release during ice formation, and there more saline water is, in fact, found, which is dense enough to sink to depth. By itself the annual cycle of freezing and melting at the sea surface cannot account for the greater salinity of the water on the shelf, because whatever salt were added to the water during winter freezing would be mixed back into the melt water during summer. Nor can it be supposed that the salt-enriched water formed during winter immediately leaves the shelf, because it is found during summer both on the shelf and flowing down the continental slope. Seabrooke, Hufford, and Elder (1971) therefore suggested that the salinity enhancement takes place year-round through freezing at the base of the Filchner and Ronne Ice Shelves (figure 1.3). Gill (1973) showed that this was unlikely, however: because the heat flux through the ice shelf is too small to allow enough salt release for a significant rate of bottom-water production; because the evidence is for melting rather than freezing under the ice shelf anyway; and because the temperature of the most saline shelf water is commonly well above the freezing point at the depth of the ice-shelf base. It seems instead that the higher

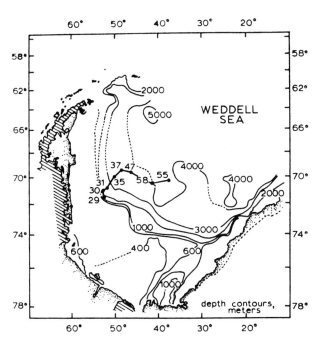

Figure 1.4 Positions of stations occupied by U.S.C.G.C. *Glacier* in the Weddell Sea, 26 February–13 March 1968, that are used for the construction of the sections in figure 1.5. Isobaths labeled in meters. (After Carmack, 1973.)

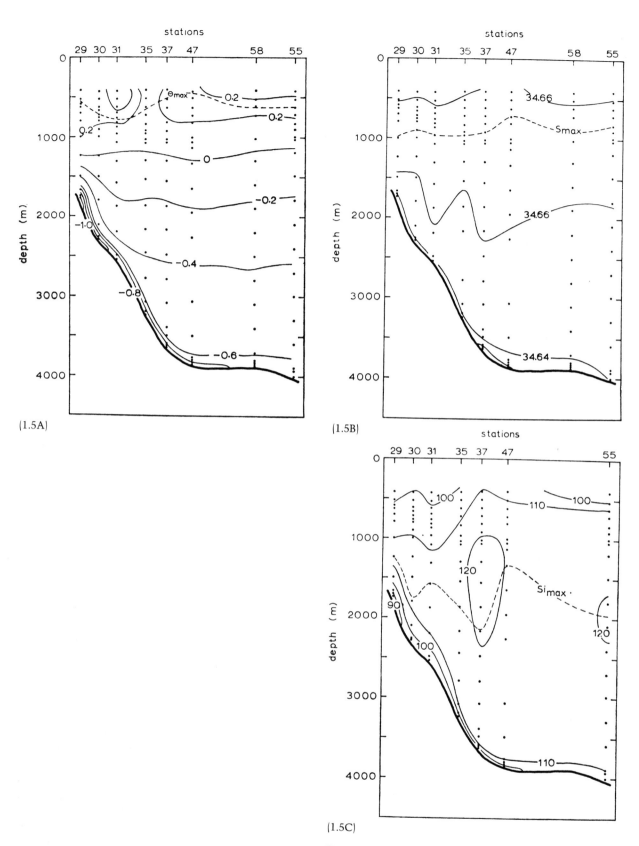

Figure 1.5 Sections of (A) potential temperature (°C), (B) salinity (‰), and (C) dissolved silica concentration ($\mu M l^{-1}$) along a line running eastward from the Antarctic Peninsula (left) in the southwestern Weddell Sea, illustrating the descent of dense shelf water along the Antarctic continental slope. (Carmack, 1973.) See figure 1.4 for positions of stations.

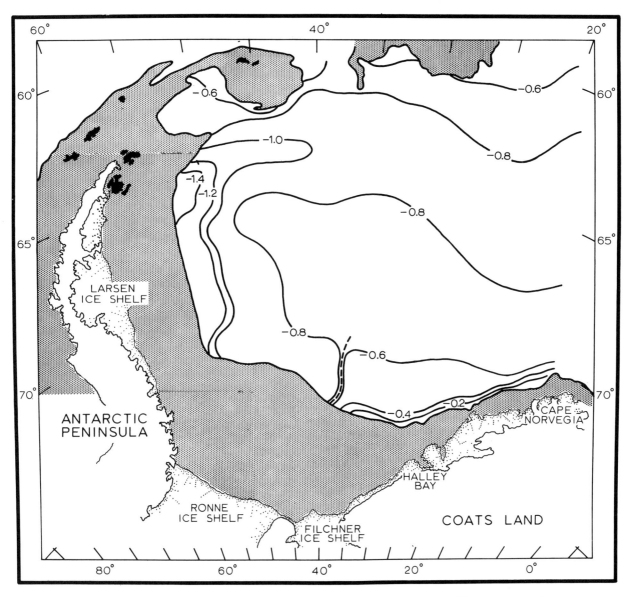

Figure 1.6 Distribution of bottom potential temperatures in the Weddell Sea, illustrating the eastward spreading of newly formed bottom water from the northern tip of the Antarctic Peninsula. (Foster and Carmack, 1976a.)

salinities are indeed due to the seasonal freezing-melting cycle at the sea surface: evidently there is a prevailing offshore movement of the pack ice that transports potential melt water locked up in ice out of the region, thus allowing a net brine production there. That movement also increases the wintertime brine release by opening up leads in the ice field where much more rapid freezing occurs. A plausibly estimated value for the net *effective* rate of ice formation on the continental shelf, through the complete seasonal cycle, is about 1 m yr^{-1} (Gill, 1973).[8]

Given denser water on the shelf than offshore, it is still not immediately obvious why it moves toward the continental slope, because even if the density difference were imparted impulsively, the water should only advance offshore a distance of order of the deformation radius (10–20 km), and the ensuing steady motion should be geostrophic flow *along* the isobaths, rather than across them. In zonally symmetric conditions without longshore pressure gradients, the only possible steady efflux of bottom water from the shelf would be in a frictional boundary layer, but the likely transport there is much too small to be realistic (Gill, 1973). Antarctica is far from zonally symmetric, of course, and the existence of large-scale indentations in the coastline, with much broadened continental shelves, as in the Weddell Sea, allows much more substantial transport off the shelf.

In particular, the salinity of the shelf water in the Weddell Sea increases markedly from east to west, by about 0.4‰, and the associated longshore density gradient implies a thermal-wind shear, with surface water moving onshore and the deeper shelf water moving offshore. Combining these density data with a value for the onshore Ekman flux associated with easterly winds, Gill (1973) estimated the offshore transport of dense water along the length of the shelf to be about 1×10^6 m^3 s^{-1}, of which half is compensation for the surface Ekman flux. The net offshore salt transport associated with this vertical circulation is only about half that required to remove the amount of salt added at the surface by net ice formation of 1 m yr^{-1}. It seems likely, though, that the longshore salinity gradient is itself maintained by a slow westward flow along the shelf—part of the Antarctic coastal current—whereby the salinity of the water is gradually raised to the westward by the widespread freezing there. There are no direct measurements of mean velocities on the shelf, but a longshore flow of 1×10^6 m^3 s^{-1}, carrying salt-enriched water off the shelf at its western end on the Antarctic Peninsula, would be adequate to remove the remaining salt produced during the annual ice-formation cycle (Gill, 1973). These estimates, derived entirely from conditioning on the shelf, thus give a figure for the total flux of dense water off the shelf of about 2×10^6 m^3 s^{-1}, which is in fairly good agreement with the value of 2–5 $\times 10^6$ m^3 s^{-1} obtained by Carmack and Foster (1975) from direct measurements of the outflow of Weddell Sea Bottom Water.

The offshore-moving shelf water encounters the waters of the open Weddell Sea in a frontal zone at the edge of the continental shelf. Detailed studies of this zone demonstrate a rather complicated mixing process, involving both the warm deep water of the Weddell Sea and the winter-cooled near-surface water, through which the characteristics of the shelf water are modified to those of the descending Weddell Sea Bottom Water (Foster and Carmack, 1976a).

As noted, there are subtleties as well in the dynamics of the descent on account of the rotational constraint, which tends to orient the flow westward along the isobaths rather than across them. In order to identify the conditions that bring about flow down the slope to the floor of the Weddell Sea, with the amount of entrainment observed, Killworth (1977) considered a series of simplified models of turbulent plumes in a stably stratified environment. Bottom friction was essential, of course, for any downslope flow at all, but the mere fact of shelf water being denser than its surroundings was not in itself sufficient for flow to reach the sea floor with the right entrainment, whether the plume was two-dimensional, three-dimensional, or intermittent. Because the thermal-expansion coefficient for sea water increases with pressure, however, very cold descending water can draw on its own *internal* energy to increase its kinetic energy; thus if a relatively dense falling parcel is sufficiently colder than its surroundings, the buoyancy force acting on it can *increase* during the descent, even though the parcel is falling in stably stratified water. With this effect taken into account, Killworth (1977) was able to construct model plumes (three-dimensional, steady as well as intermittent) that both descended far enough to reach the floor of the Weddell Sea and entrained roughly the observed amount of surrounding water along the way. These had a substantial component of flow along the isobaths too, consistent with the bulk of the bottom water entering onto the sea floor in the northwestern corner of the Weddell Sea.

The Weddell Sea Bottom Water is fresher than the overlying water, but in the southwestern Pacific—specifically, north of the Ross Sea (figure 1.3) to the mid-ocean ridge (about lat. 65°S), and close to Antarctica south of Australia—the salinity increases with depth near the bottom to values >34.72‰ (Gordon, 1971, 1975a; Gordon and Molinelli, 1975). This high-salinity bottom water must have a different source, and it has been traced to the Ross Sea (e.g., A. L. Gordon, 1974), where cold, saline shelf water has been observed to descend the western continental slope in a manner similar to that of the downslope flow in the Weddell Sea (Gordon, 1975a). Although the effect of the Ross Sea bottom water on the deep temperature distribution is barely noticeable, and its effect on the salinity distribution only became clear through the precision afforded by conductivity measurements, its production rate is not necessarily a great deal less than that of Weddell Sea Bottom Water, because the water with which the shelf water mixes to form bottom water is both more saline and much warmer than in the Weddell Sea. No transport measurements like those of Carmack and Foster (1975) have been made, however.

The oceanographic conditions on the Ross Sea continental shelf are roughly similar to those in the Weddell Sea, in that there is westward flow along the shelf, and the salinity of the shelf water increases to the west by 0.3–0.4‰ (Jacobs, Amos, and Bruchhausen, 1970; Jacobs, Gordon, and Ardai, 1979). This correspondence suggests that the mechanisms sketched by Gill (1973) are also operative in the Ross Sea in moving dense water off the shelf onto the continental slope. A vertical circulation (involving the Ekman transport and geostrophic shear) about 60% as large as that on the Weddell Sea shelf is indicated (Gill, 1973); that is, an offshore flux along the length of the Ross Sea shelf of roughly 0.6×10^6 m^3 s^{-1}. No estimate has been made of the rate of removal of shelf water by the longshore flow.

The water deeper than 300–500 m on the western part of the Ross Sea shelf is the most saline (34.8–35.0‰) and the most dense (σ_t = 28.0–28.1) water found in the Antarctic (Jacobs, Amos, and Bruchhausen, 1970; Gordon, 1971). One wonders to what extent it may be involved in the formation of bottom water in the Ross Sea, but no very satisfactory account seems to have been given for the existence of this very saline water. Because the high salinities of the layer are highly correlated with low temperatures, and because its thickness and breadth increase southward to the Ross Ice Shelf, it is tempting to think that the characteristics of the layer may be imparted through freezing at the base of the ice shelf. Near the front of the ice shelf, though, the evidence is of melting at the base (Jacobs, Amos, and Bruchhausen, 1970). Moreover, vertical variations of temperature and salinity observed beneath the ice shelf far to the south at 82°S, 169°W demonstrated melting there too at the time of observation, but because the lowest 6 m of ice was of marine origin rather than terrestrial, it is not certain what the prevailing condition is at that point (Gilmour, 1979; Jacobs, Gordon, and Ardai, 1979). Even if freezing is occurring, the maximum possible rate at the base of the ice shelf is very small (3.5 cm yr^{-1}; Jacobs, Gordon, and Ardai, 1979), and any outflow from beneath the ice shelf consistent with the required salinity enhancement would be much less than the estimated offshore transport of water on the continental shelf. The bulk of this very saline water is found in a large topographic depression on the shelf, whose sill depths near the edge of the shelf are not well known (Jacobs, Amos, and Bruchhausen, 1970); perhaps most of that dense layer is indeed renewed only slowly through freezing under the western Ross Ice Shelf, is trapped in the depression, and does not contribute significantly to the local bottom-water formation (Gordon, 1971).

On the other hand, the high salinity of this western shelf water might be due entirely to annual sea-ice freezing, as proposed by Gill (1973). Unusually dense water formed in that process could contribute substantially to bottom water, or it might form infrequently and merely accumulate in the topographic depression if the sill is actually shallow enough to allow much containment. The situation is unclear.

Several similar deep depressions—though of much smaller area—exist on the continental shelf off the Adélie Coast of Antarctica (figure 1.3). They are filled below their sill depths with water of temperature close to the freezing point, and of salinity 34.4–34.7‰, probably as a result of deep convection associated with winter sea-ice formation (Gordon and Tchernia, 1972). Evidently such water spills over the sills, because it is found as a layer a few tens of meters thick on the nearby continental slope, underlying the high-salinity bottom water from the Ross Sea. This low-salinity bottom water mixes into the Ross Sea water rapidly enough that it has not been detected far from the continental rise near its points of origin. No estimates of the rate of production of the Adélie Coast water have been made, though the rate must be relatively small, given the small volumes of such water that have been found away from the shelf. The flow down the continental slope appears intermittent, inasmuch as duplicated stations in February 1969 and December 1971 showed that the comparatively fresh layer was more widespread at the latter time (Gordon and Tchernia, 1972).

Evidence has also been cited by Jacobs and Georgi (1977) for some small production of bottom water off Enderby Land. They suggest that new source regions are revealed almost with each new exploration of the Antarctic continental slope.

Bottom water from these several sources—the Weddell Sea being the paramount one—mixes with the warmer, more saline water above to form the "Antarctic Bottom Water" of the world ocean. Since all the subtypes originate through freezing at the sea surface, it is the coldest, and thereby the densest, deep water in the open ocean, detectable into northern latitudes by low temperatures close to the bottom.

Not all deep water formed in the Antarctic is bottom water, however. Carmack and Killworth (1978) have described a layer of relatively cold, fresh water at depths of about 2 km off Wilkes Land (figure 1.3; long. 150–170°E), which they demonstrated through Killworth's (1977) plume theory to have originated most probably from shelf water that is not quite dense enough to sink to the bottom. Instead, it sinks to a depth where its density matches that of the surrounding water, in the manner of the Mediterranean and Red Sea outflows. This particular instance of sinking to mid-depths was discernible because the shelf water intruded into saline deep water from the Ross Sea, and hence could be identified by its relatively low salinity. Such phenomena seem possible elsewhere around Antarctica, but in regions where the surrounding deep water is fresher—more like that in the Weddell Sea— they would be far more difficult to detect in distributions of temperature and salinity, though perhaps less so in distributions of nutrients (e.g., silica). As Carmack and Killworth (1978) pointed out, if the salinity of shelf water really is increased almost everywhere around Antarctica by annual freezing, then some process must operate to remove the extra salt from the shelf; that process may be sinking of water to mid-depths, in the manner that was observed off Wilkes Land. If so, then the rate of deep sinking in the Antarctic might be much greater than that involved in the immediate production of bottom water; but there is no basis at present for estimating *how* much greater.

Furthermore, a set of closely spaced observations in the eastern Weddell Sea in February 1977 revealed a column of relatively cold, fresh water some 30 km in diameter, extending from near the surface through the warm, saline deep water to about 4000 m (Gordon, 1978). It suggests a relic of deep convective overturning, such as has been well documented in the northwestern Mediterranean Sea in regions of similarly small area (MEDOC Group, 1970). It may represent yet another mode of deep sinking in the Antarctic, if not to the bottom at least to mid-depths, that had been missed in the past, perhaps because of difficulty of access to critical areas or too coarse a spacing of stations. The specific conditions for overturning in the Weddell Sea have been carefully considered by Killworth (1979), but again there is unsufficient basis at present for estimating how significant a contribution such events may make.

1.4.2 Sinking in the Northern North Atlantic

Next to the Antarctic Bottom Water, the densest water in the open ocean is what Wüst (1935) called Lower North Atlantic Deep Water. It does not originate, however, as Nansen (1912) first proposed, and Wüst (1935) and Sverdrup (Sverdrup, Johnson, and Fleming, 1942) maintained, from wintertime convection to the bottom of the Irminger or Labrador Seas,[9] but from dense Norwegian Sea water that flows into the North Atlantic over three sills on the ridge connecting Greenland and the British Isles (figure 1.7). These overflows entrain resident North Atlantic water in the course of their descent, and join together to form the bottom water of the northern North Atlantic.

It is curious that the significance of these overflows for the large-scale circulation was not appreciated until the late 1950s. Dr. Carpenter (Carpenter, Jeffreys, and Thomson, 1869) had hypothesized long before that overflow east of the Faroe Islands was the source of bottom water in the northern North Atlantic, and Tizard (1883) actually observed some such flow over the Wyville Thomson Ridge, which is the southwestern boundary of the Faroe Bank Channel (figure 1.7). Knudsen (1899) discovered the western overflow, and also found overflow water on the southern side of the Iceland-Faroe Ridge, which he supposed had come south across that ridge; in hindsight, his station positions suggest that this water was really outflow from the Faroe Bank Channel, and had thus derived from the eastern, rather than the central, overflow. In any case, with better station coverage, Nielsen (1904) unambiguously established the existence of the central overflow, and Nansen (1912), with additional observations, discussed all three of them, presenting illustrative sections of water properties across the ridge. Yet somehow all four observers, with the possible exception of Nielsen, believed that the overflows could make only a minor contribution to the deep water of the North Atlantic—an opinion also expressed by Wüst (1935). Although Jacobsen (1916), Brennecke (1921), and Defant (1938) all briefly suggested somewhat greater significance for the overflows, they were generally ignored until Cooper (1952, 1955a,b) and Dietrich (1956, 1957a,b), more or less independently, recalled them to attention.[10] [Evidence for overflow was also noted by Vinogradova, Kislyakov, Litvin, and Ponomarenko (1959) in new data obtained at this time, but they, like the writers at the turn of the century, did not recognize its large-scale significance.]

The deepest passage across the ridge is the Faroe Bank Channel, which has a sill depth of about 800 m. Evidently some small amounts of Norweigan Sea water in this passage do spill intermittently over the Wyville Thomson Ridge (Ellett and Roberts, 1973), but most of this overflow continues northwestward through the channel to its sill, where the measured shear of geostrophic velocity in combination with analysis of water characteristics indicates an outflow of Norwegian Sea water of roughly 1×10^6 m^3 s^{-1} (Crease, 1965). After exiting from the Faroe Bank Channel, this current joins a second overflow, which passes over the ridge between the Faroe Islands and Iceland, where the sill depth is

300–400 m (figure 1.7).[11] By referencing geostrophic velocity calculations for sections southeast of Iceland according to the water characteristics, Steele, Barrett, and Worthington (1962) estimated the volume transport of this combined Iceland-Scotland overflow, after it has left the ridge, to be about 5×10^6 m^{3}s^{-1}. Analysis of that overflow water in terms of its component water types showed that about 3×10^6 m^{3}s^{-1} of the net flow is entrained Atlantic water, leaving 1×10^6 m^{3}s^{-1} for the transport of "pure" Norwegian Sea water across the Iceland-Faroe sill (Worthington, 1970). With different data, Hermann (1967) also estimated roughly 2×10^6 m^{3}s^{-1} for the net outflow of Norwegian Sea water between Iceland and Scotland.

This overflow continues southward against the eastern side of the Mid-Atlantic Ridge, and passes through the Gibbs Fracture Zone near lat. 53°N into the Labrador Basin (figure 1.7). Its westward volume transport there was calculated by Worthington and Volkmann (1965)—on the same basis as by Steele, Barrett, and Worthington (1962) farther north—to be still about 5×10^6 m^{3}s^{-1}, suggesting little direct flow into low latitudes of the eastern North Atlantic, although Lee and Ellett (1965) have detected overflow influence on water characteristics as far south as lat. 47°N in the eastern Atlantic. This current then flows northward in the eastern Labrador Basin, and, south of Greenland, joins the third overflow from the Norwegian Sea. That overflow comes through the Denmark Strait (Dietrich, 1957b; Mann, 1969), where the sill depth is about 600 m, and is easily identified as cold water banked against Greenland (figure 1.8).

Geostrophic velocity calculations, referenced by water characteristics as for the Iceland-Scotland overflow, gave a figure of about 10×10^6 m^{3}s^{-1} for the volume transport of the full overflow current south of Greenland (Swallow and Worthington, 1969), thereby indicating a contribution of 5×10^6 m^{3}s^{-1} from the

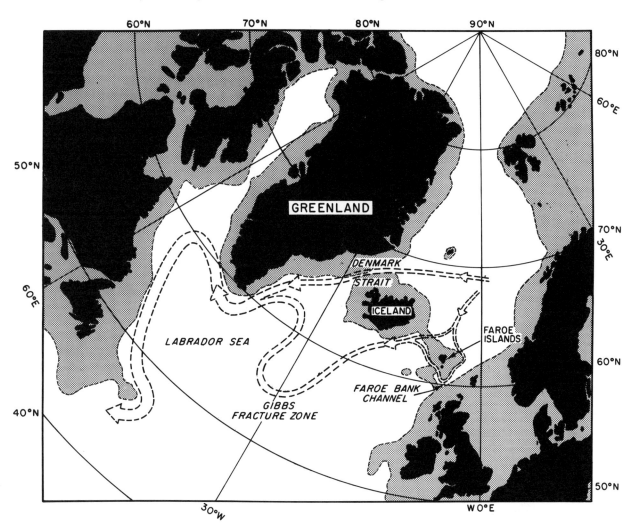

Figure 1.7 Index map identifying the overflow currents from the Norwegian Sea and local place names mentioned in the text. (After Worthington, 1970.)

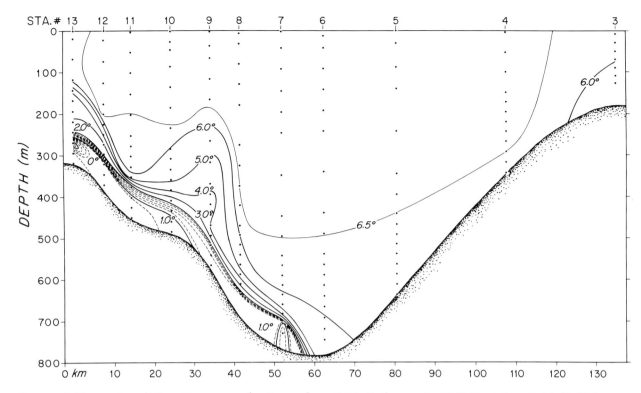

Figure 1.8 Temperature (°C) section across the Denmark Strait in lats. 65-66°N (left, west; see figure 1.7), illustrating the southward flow of cold water from the Norwegian Sea. C.S.S. *Hudson* cruise BI 0267, stations 3-13, 28-29 January 1967. (Worthington, 1969.)

Denmark Strait overflow. Worthington (1970) has estimated that about one-fifth of this is Atlantic water entrained by the overflow as it descends the Greenland continental slope, but the irregularity of the overflow is so great that the fraction may generally be much larger. For example, from a series of hydrographic sections reported by Mann (1969), Smith (1975) calculated an overflow transport of 1.3×10^6 m^{3}s^{-1} near the sill, which increased fourfold toward the southern tip of Greenland. Even though the overflows on either side of Iceland originate from the same Norwegian Sea water, the Denmark Strait overflow is distinctly colder and fresher than the Iceland-Scotland overflow (potential temperature 0.0-2.0°C and salinity 34.88-34.93‰, as contrasted with 1.8-3.0°C and 34.98-35.03‰; Worthington, 1976), mainly because the upper-kilometer Atlantic water that they entrain is warmer and more saline to the east.

Although the gross dimensions, course, and entrainment of the descending Denmark Strait overflow can be fairly well rationalized in terms of a steady turbulent plume model (Smith, 1975), it is, in fact, a highly variable flow: its water properties differ markedly from one time to another (e.g., Stefánsson, 1968; Lachenbruch and Marshall, 1968; Mann, 1969), and month-long near-bottom records of velocity and temperature (not taken, it was thought, from the core of the current) showed that overflow was occurring in bursts 1 or 2 days long, with speeds up to 140 cm s^{-1}, separated by intervals of 1 to several days (Worthington, 1969). Variability in water characteristics suggests that the Iceland-Scotland overflow is equally unsteady (Tait, 1967; Lee, 1967).

The process by which dense water is supplied to these overflows has not been investigated in as much detail as the formation of Antarctic Bottom Water. The Norwegian Sea appears to be an area of large mean-annual heat loss to the atmosphere, however, and Worthington (1970) worked out a heat budget for it in which he characterized it as a mediterranean basin where relatively light incoming water is made dense by strong local winter cooling, then sinks, and runs out beneath the inflow.[12] The overflow water is not *deep* Norwegian Sea water, which is colder and more dense still; tritium concentrations show that the overflow water comes from depths no greater than 1000 m, and probably from levels still shallower in the pycnocline (Peterson and Rooth, 1976). The outflow process thus seems to be separate from that of deep-water renewal; perhaps the overflow phenomenon is akin to the shallow buoyancy-driven flow over a sill modeled by O. M. Phillips (1966a) for the Red Sea. As with the saline water on the Antarctic continental shelf, the rotational constraint would tend to confine newly formed dense water to the Norwegian Sea, so that it would be expected to leak out throughout the year(s), rather than just when cooling was most intense.

The other location in the northern North Atlantic where deep sinking occurs is the Labrador Sea. North of roughly lat. 40°N there is a distinct salinity minimum at a potential temperature of about 3.5°C (Worthington and Metcalf, 1961), which grows more pronounced with approach to the Labrador Sea, where nearly homogeneous water (temperature ≈3.5°C, salinity ≈34.9‰) is found in winter from the sea surface to about 1500 m (e.g., Worthington and Wright, 1970). The salinity minimum is associated with an oxygen maximum, which was Wüst's (1935) diagnostic for his Middle North Atlantic Deep Water, and these characteristics must be imparted in the Labrador Sea, presumably through deep convection driven by intense winter cooling at the surface. Despite extensive wintertime surveys in the area, however, no well-documented deep-overturning events have yet been reported, but Lazier (1973a) has found a few examples in the records of Weather Ship *Bravo* (56°30'N, 51°00'W) of deep isopycnals rising to the surface layer for brief intervals of time, in association with virtually homogeneous water columns 1500 m deep. Most likely, the convection occurs like that in the northwestern Mediterranean Sea (MEDOC Group, 1970), in small patches and very intermittently (perhaps not even every winter), so that the convection could escape notice unless the observations were frequent and closely spaced.[13]

Because of these small space and time scales, it is difficult to make much more than a guess at the long-term mean rate of deep sinking in the Labrador Sea. From an estimate of the net annual heat flux across the sea surface, Wright (1972) suggested 3.5×10^6 m^3 s^{-1}.

Both the Labrador Sea water and the overflow water are more saline at their sources than the Antarctic Bottom Water, and their salinities are increased somewhat in middle latitudes of the North Atlantic through mixing with the outflow from the Mediterranean Sea (see below, this section). Although they are both made dense by cooling, their temperatures are not reduced to the freezing point before sinking, but remain high enough that, despite the salinity difference, they are less dense *in situ* than the Antarctic Bottom Water, and therefore override it.[14] This high-salinity water spreads throughout the South Atlantic, is carried eastward around Antarctica in the Circumpolar Current, and its elevated salinities can be followed into low latitudes of the South Indian and South Pacific Oceans (Reid and Lynn, 1971). The effect of North Atlantic Deep Water in maintaining high oxygen concentrations in the deep ocean is familiar, but its source waters are also exceptionally low in dissolved silica: <8 μM l^{-1} in the Norwegian Sea overflow (Stefánsson, 1968), roughly 10 μM l^{-1} in the Labrador Sea water (Mann, Coote, and Garner, 1973), and about 6 μM l^{-1} in the Mediterranean outflow (Schink, 1967). Consequently, the deep North Atlantic has the lowest silica concentrations of all the oceans (Metcalf, 1969), there is a pronounced deep silica minimum in the South Atlantic between the Antarctic Intermediate and Bottom Waters (Mann, Coote, and Garner, 1973), and there is even a weak silica minimum in the deep southwestern Pacific due to the influence of North Atlantic source waters (Warren, 1973).

1.4.3 Outflows from Marginal Seas

Excess evaporation makes water in the Mediterranean Sea, the Red Sea, and the Persian Gulf more saline and thereby more dense than that in the outlying ocean, and circulations are set up in the shallow connecting straits in which near-surface water flows in, and dense, saline water flows out beneath. The outflows do not fall to the floor of the open ocean, however, because in descending the continental slopes they entrain enough light thermocline water to become less dense than the deep water of the open ocean, and they therefore spread out at mid-depths. They are thus not important sources for the *deep* circulation, but through vertical mixing they do at least influence the properties of the deep water; hence they are noted briefly here.

The salinity maximum marking the core of the Mediterranean outflow [Wüst's (1935) Upper North Atlantic Deep Water] is found at a depth of about 1200 m in the southeastern North Atlantic (e.g., Fuglister, 1960), but relatively high salinities resulting from the outflow can be detected at least as deep as 3000 m (Worthington and Wright, 1970), and the effect of the outflow in raising the salinity of the Labrador Sea water and Norwegian Sea overflow water was remarked above. The rate of outflow through the Straits of Gibraltar has been estimated from the overall salt budget for the Mediterranean Sea and by current measurements of duration a few days to 2 weeks; both methods agree on a mean value of about 1×10^6 m^3 s^{-1} (Lacombe, 1971). Calculations by Smith (1975) show that by the time the outflow current completes its descent and departs from the continental slope, its volume transport has increased, by entrainment of Atlantic water, to roughly 10×10^6 m^3 s^{-1}. To what extent any of this transport contributes to the flow of the deep water below—as distinct from its properties—is not known.

The salinity maximum associated with the Red Sea outflow occurs at depths of 500–600 m in the northern Arabian Sea, but relatively high salinities, due presumably to mixing with the outflow water, are found there at least as deep as 2500 m (Wyrtki, 1971). A salt-budget calculation based on climatological data cited by Siedler (1968) gives a mean outflow through Bab-el-Mandeb of 0.4×10^6 m^3 s^{-1}, which is the same value that he derived from direct current measurements there of 2.5

days duration. Such close agreement was probably fortuitous, however, because his directly measured inflow value was 40% greater. The entrainment during the descent into the Gulf of Aden has not been estimated.

The Persian Gulf outflow (salinity maximum at depths of 300-400 m in the Gulf of Oman; Wyrtki, 1971) has a small, patchy effect on intermediate-depth property fields in the North Indian Ocean (Rochford, 1964), but any influence on the deeper distributions seems to be masked by that of the Red Sea outflow (Wyrtki, 1971). It is probably only slight, however, because the outflow through the Strait of Hormuz is very small: only $0.1 \times 10^6 \, m^3 \, s^{-1}$ according to the salt budget (Koske, 1972).

1.5 Deep Western Boundary Currents in the World Ocean

Many more hydrographic stations have been occupied since Sverdrup wrote chapter XV of *The Oceans*. They have filled in blank areas in the world-ocean coverage, added details to property distributions, and sharpened the definition of features, but our broad picture of the property fields has not changed qualitatively. What *has* changed greatly is the conception of their basis: that the underlying mean deep circulation, largely masked by lateral mixing of the water characteristics, is a system of western boundary currents and poleward interior flows, as in the dynamically consistent models of Stommel and Arons (1960a,b)—rather than the Merzian system of basin-wide, slow meridional flows from source regions.

Although this conception makes good sense, it is not so easy to tell from observation how correct it is. The key element in the circulation theory is the supposed general upward movement of deep water, but the hypothesized mean rate ($\sim 10^{-5} \, cm \, s^{-1}$) is so small as to be utterly beyond present measurement capability. Probably measurement of the deep interior flow, to see whether it is directed poleward in vertical average, *is* technically feasible, but certainly not economically and socially so, given the immensity of the current-meter program that would be required. Large-scale tracer fields are not of much help in this respect since mixing effects can make them ambiguous with regard to circulation patterns for the small mean speeds ($\sim 10^{-2} \, cm \, s^{-1}$) of the interior flow (section 1.3; Kuo and Veronis, 1973).

What *can* be observed, through current measurements and short lines of hydrographic stations, are the narrow, relatively swift, deep western boundary currents predicted by the circulation theory. To be sure, some other theory, with different forcing, might also require such boundary currents, but they are a necessary element of the Stommel-Arons dynamics, and demonstrations of their existence in much of the world ocean have contributed strong support—indeed, the primary observational support—for its essential ideas. This section will therefore assess the evidence for deep western boundary currents in the world ocean, with comment on a few related features of the circulation.

1.5.1 Atlantic Ocean

There are two deep boundary currents in the western Atlantic, the one flowing southward from the Norwegian and Labrador Seas, and the other flowing northward from the Antarctic [the latter not included in the barotropic model of Stommel and Arons (1960b) north of the counterpart of the Antarctic Peninsula].

After the Norwegian Sea overflow current passes the southern tip of Greenland, it takes a counterclockwise course around the Labrador Sea (figure 1.7) and flows southeastward along the North American continental slope by Labrador and Newfoundland (Swallow and Worthington, 1969). Figure 1.9, for example, is a temperature section running from southern Labrador northeast to the southern tip of Greenland. The cold overflow water may be seen pressed up against the lower Greenland slope, and the associated density gradient (mainly determined by temperature there) is consistent geostrophically with northwestward flow increasing in strength from mid-depths to the bottom. Adjacent to the North American continental slope, the southeastward-flowing deep boundary current is evident in similar fashion, 200-300 km wide, probably including Labrador Sea water above the overflow water. Three neutrally buoyant floats tracked nearby at depths of 1600-2400 m all registered southeastward flow and, in combination with station data, suggested extension of such flow up to about 1200 m. Though the measurements were of insufficient duration (12-31 hours) for very satisfactory results, nevertheless the indicated volume transport for the deep boundary current was roughly $10 \times 10^6 \, m^3 \, s^{-1}$ (Swallow and Worthington, 1969).

High-oxygen and low-silica values close to the continental slope in a similar depth interval show that this current continues around the Grand Banks of Newfoundland, at least in the long-term mean, but records from moored current meters in the area have not yet given clear, direct evidence for such flow (Clarke and Reiniger, 1973; Clarke, Hill, Reiniger, and Warren, 1980).

Westward and southward in the North Atlantic, pronounced tracer characteristics marking the deep boundary current are lost, and its cross-stream density gradient is not readily distinguishable from that of the nearby Gulf Stream; hence evidence for its existence comes mainly from velocity measurements. On the continental slope north of the Gulf Stream near long.

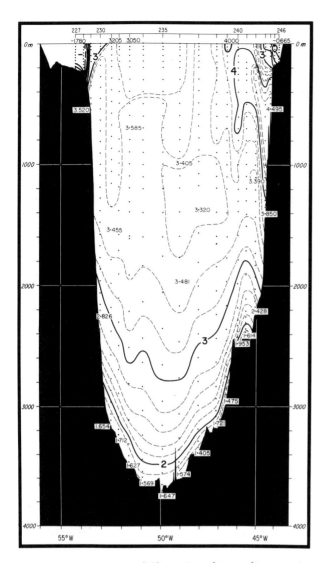

Figure 1.9 Temperature (°C) section along a line running northeastward across the Labrador Basin from southern Labrador (left) to the southern tip of Greenland (see figure 1.7), illustrating the deep western boundary current of the North Atlantic, flowing southwestward against the continental slope of Labrador. M.S. *Erika Dan* stations 227-246, 21-28 February 1962. (Worthington and Wright, 1970.)

70°W, current records of several months duration from five levels at a single site (Webster, 1969), deep float tracks of duration 1 to 2 days at several locations (Volkmann, 1962), and near-bottom current records of length 7-25 days, also from several locations (Zimmerman, 1971), all suggest a prevailing westward flow, surface to bottom, between the Stream and the continental shelf. What part of that should be considered the deep boundary current is unclear.

Clear evidence for the passage of the deep current under the Gulf Stream near Cape Hatteras (lat. 35°N) has been obtained from float tracks of 1 to 3 days duration (Barrett, 1965), from transport-float sections (Richardson and Knauss, 1971), and from current records of length 3 to 8 weeks (Richardson, 1977). These last were from six current meters moored 100 m above the bottom at depths of 1-4 km on a line crossing the mean axis of the Gulf Stream, and they showed a striking persistence of southwestward flow with speeds typically 10 cm s^{-1}. The mean velocities in combination with station data indicated a volume transport for the deep western boundary current of 24×10^6 m^3 s^{-1}.

The discovery of this current by means of neutrally buoyant floats farther south (lat. 33°N), offshore of the Gulf Stream, was mentioned in section 1.3; the width of the current appeared to be about 100 km there, and its volume transport was computed to be 7×10^6 m^3 s^{-1} (Swallow and Worthington, 1961). Using a similar zero-velocity surface to reference geostrophic calculations, Amos, Gordon, and Schneider (1971) estimated the transport of the current at lat. 30°N, near the Blake Bahama Outer Ridge, to be 22×10^6 m^3 s^{-1}.

Whether the differences in the cited transport estimates imply spatial variation is unclear. Probably they are related more to differences in methods and definitions of the current than to anything else.

Three SOFAR floats at depths of 1500-2000 m in the western Sargasso Sea were observed to drift westward in lats. 28-30°N, and then move rapidly southward (speeds roughly 10 cm s^{-1}) along the continental slope (Riser, Freeland, and Rossby, 1978). These tracks are consistent with the idea of a deep boundary current, but not with the specific Stommel-Arons (1960b) model, in which water flows *eastward* from the boundary current into the interior. Their significance for the basic ideas of the model is unclear (see Fofonoff's discussion in chapter 4 of this volume).

Some indications have been found for the continuation of the southward flow to about lat. 23°N, north of Hispaniola (Tucholke, Wright, and Hollister, 1973), but no clear evidence for a boundary current flow is available between there and the equator. [That there is some sort of deep southward movement in this latitude belt has long been recognized, of course, from large-scale property distributions (e.g., Wüst, 1935).]

Deep water from the North Atlantic is easily identified in the South Atlantic by its relatively high salinity at depths exceeding 1500 m, and relatively strong southward flow along the entire western boundary of the South Atlantic is evident from intensification close to South America of the salinity maximum (depth increasing southward from 1600 m to about 2500 m) and its associated temperature inversion (Wüst and Defant, 1936; Fuglister, 1960). The breadth of this zone of increased salinity maximum is 500–1000 km, which, however, is probably greater than the width of the boundary current because of lateral-mixing effects. Clear examples of the North Atlantic tracer characteristics are shown in figure 1.10: sections of potential temperature, salinity, and the concentrations of dissolved oxygen and silica on a line crossing the southern Brazilian Basin, roughly along lat. 30°S (Reid, Nowlin, and Patzert, 1977). The temperature inversion is seen at depths of 1600–1800 m within about 1000 km of South America; the strong salinity maximum and silica minimum are found at about 2000 m within 500 km of the continental slope, and the oxygen maximum lies deeper (2000–3000 m) and extends somewhat farther eastward with high values. The impression is of a concentration of southward flow near South America, from the "top" of the Antarctic Bottom Water at about 3500 m to roughly 1500 m.

Probably the most satisfactory estimates of volume transports for the deep meridional flows in the western South Atlantic are those by Wright (1970). He performed geostrophic calculations for the IGY sections (Fuglister, 1960) by assuming zero meridional velocity at the boundary between North Atlantic Deep Water and Antarctic Bottom Water, as defined by a break in the temperature-depth and salinity-depth curves near 2°C and 34.89‰, found at levels of 3400–4000 m [the same break that had been discovered by Buchanan (see section 1.2), and exploited also for velocity calculations by Wüst (1938)]. As an upper boundary for the North Atlantic Deep Water, he adopted the 1600-m level at 16°N and 8°S and the 1800-m level at 32°S (Wright, personal communication), and he derived thereby a figure of 9×10^6 m^{3}s^{-1} for the southward transport of North Atlantic Deep Water (including both boundary current and interior flow in the western trough).

Beneath the North Atlantic Deep Water, the cold, northward-flowing Antarctic Bottom Water is pressed up against the South American continental slope along the entire length of the South Atlantic, except close to the equator (Wüst and Defant, 1936; Fuglister, 1960). It is indicated in figure 1.10 by the relatively low temperature, salinity, and oxygen concentration and the high silica concentration (cf. figure 1.5) below 3500 m, the extreme values being found close against the western boundary. The isotherms (and isopycnals) slope downward to the east, consistent geostrophically with a northward-flowing current increasing in speed toward the bottom. The width of this current, as defined by the zone of sloping isotherms (Cato 6 stations 7–11), is about 500 km. Wright's (1970) calculations give a volume transport for the current of about 6×10^6 m^{3}s^{-1} in middle latitudes of the South Atlantic, diminishing to about 2×10^6 m^{3}s^{-1} near the equator.

The current continues northward across the equator, but in lats. 8–16°N it is found not beside the western boundary but against the Mid-Atlantic Ridge, which is the *eastern* boundary of the basin (Fuglister, 1960). Why this transposition occurs is not wholly clear, but perhaps it is related to the sharp northward increase in the depth of the western basin there, from about 4200 m at lat. 2°N to about 5500 m at lat. 16°N. The idea is familiar that meridional bottom slope has an effect on ocean circulation analogous to that of variation in the Coriolis parameter; poleward increase in depth tends to counteract the β-effect, and, if the rate of increase is large enough, boundary currents can be shifted from the western to the eastern sides of basins (e.g., Fandry and Leslie, 1972). The transposition should occur if, roughly speaking, the meridional bottom slope $S > H\beta/f$, where H is the thickness of the bottom current. At lat. 10°N, $f = 2.5 \times 10^{-5}$ s^{-1} and $\beta = 2.3 \times 10^{-13}$ cm^{-1} s^{-1}; for $H \approx 700$ m at lats. 8°N and 16°N (Fuglister, 1960), the critical value of the slope is 6.4×10^{-4}, which is less than the actual slope of about 8.4×10^{-4}.

Whatever the reason for the transposition, there is no indication of this boundary current on transatlantic sections at and north of lat. 24°N (Fuglister, 1960). Apparently, the water from it spreads out somehow over the floor of the basin (Worthington and Wright, 1970), and such a poleward interior flow would be qualitatively consistent with the level-bottom Stommel-Arons dynamics. Traces of Antarctic Bottom Water have been found in the lowest few hundred meters, for example, south of the Gulf Stream on the 50th meridian (e.g., Clarke et al., 1980), in the vicinity of the Blake-Bahama Outer Ridge near lat. 30°N (Amos, Gordon, and Schneider, 1971), and north of the Bahama Banks and Hispaniola (Tucholke, Wright, and Hollister, 1973). In the latter two instances these traces had fairly definitely been incorporated into the southward-flowing deep western boundary current, suggesting a general counterclockwise spreading and movement of Antarctic Bottom Water in the western North Atlantic.

The coldest water in the eastern trough of the Atlantic (north of the Walvis Ridge) is observed at the equator (potential temperature <1.7°C; Wüst and Defant, 1936; Fuglister, 1960). Minimum temperatures increase gradually both northward and southward, showing that the principal inflow of water below the general crest of the Mid-Atlantic Ridge (3500 m depth, say) occurs from the western Atlantic through the Romanche Fracture Zone, on the equator (e.g., Drygalski, 1904; Wüst, 1933; Metcalf, Heezen, and Stalcup, 1964). No estimate of the rate of this inflow has been made. Silica values on the surface where potential temperature is 2°C increase northward (Metcalf, 1969), as well as southward (Chan, Drummond, Edmond, and Grant, 1977), indicating further that the deep water "ages" poleward.

Many additional fracture zones cut across the Mid-Atlantic Ridge, but the existing evidence suggests that there are only three others through which noticeable deep-water transport may take place. As described in section 1.4, the Iceland-Scotland overflow passes westward into the Labrador Basin through the Gibbs Fracture Zone at lat. 53°N (Steele, Barrett, and Worthington, 1962). Indications of eastward flow through the Vema Fracture Zone near lat. 11°N were shown by Heezen, Gerard, and Tharp (1964), but the flux must be much less than that through the Romanche Fracture Zone, given its much smaller influence on property distributions in the eastern Atlantic. Perhaps there is also some slight westward flow through the Kane Fracture Zone near lat. 24°N (Purdy, Rabinowitz, and Velterop, 1979): high silica concentrations appear to extend westward across the Mid-Atlantic Ridge there (Metcalf, 1969), but the accuracy of the measurements is not entirely certain.

As noted in section 1.2, the Walvis Ridge blocks most direct deep connection between the Antarctic and the eastern trough of the Atlantic, but bottom temperatures nearly as low as on the equator at lat. 32°S in the southwestern corner of the Angola Basin (Fuglister, 1960) undoubtedly signify a small leakage of water from the south across the Walvis Ridge, probably through the Walvis Passage near long. 7°W (Connary and Ewing, 1974).

The dynamical fact that poleward movement from an equatorial source can be accomplished by an interior flow may explain to some extent why no deep boundary currents along the western margin of the eastern trough (Mid-Atlantic Ridge) have been observed in transatlantic hydrographic sections (Fuglister, 1960). On the other hand, as pointed out in section 1.3, the budgetary function of western boundary currents is not only to transport deep water into low latitudes, but also to correct imbalances between interior flows and upward fluxes; and it is not easy to see how continuity can be maintained in the eastern trough of the Atlantic without recirculation boundary currents. In the North Pacific and North Indian Oceans, where recirculation boundary currents are essential elements of the Stommel-Arons (1960b) model, convincing evidence for such currents has not been found either (see below, this section); but the observational material there is not really adequate to demonstrate whether or not they exist. In any case, there is no match yet between theory and observation in regard to recirculation currents.

1.5.2 Indian Ocean

The geometry of the Indian Ocean is more complicated than that of the other oceans, because several major ridges divide it into a multiplicity of separate basins (figure 1.11). The course of deep flow is correspondingly more tortuous and divided than in the other oceans, and further removed from the simple idealizations (Stommel and Arons, 1960b).

Since the Mozambique Basin (see figure 1.11 for geographical names) is closed off to the north at depths greater than 2500-3000 m, the main point where deep water enters the western Indian Ocean from the Circumpolar Current is the passage between the Crozet Plateau and the Kerguelan Ridge (Jacobs and Georgi, 1977). The pattern of potential temperature at 4000 m (Wyrtki, 1971) demonstrates equatorward flow on through the Crozet Basin, and this flow has been observed to pass through the Southwest Indian Ridge via the Atlantis II and Melville Fracture Zones (longs. 57°E and 61°E, respectively) into the Madagascar Basin beyond (Warren, 1978). On a broader scale, continuity of flow between the Crozet and Madagascar Basins is supported by the distributions of bottom potential temperature (Kolla, Sullivan, Streeter, and Langseth, 1976) and of clay-mineral fractions in the sediments of the two basins (Kolla, Henderson, and Biscaye, 1976).

This flow extends northward as a relatively narrow, intense, near-bottom current close against Madagascar, the effective western boundary for the deep South Indian Ocean (Warren, 1974). The distribution of temperature along lat. 12°S, from the northern tip of Madagascar to the Central Indian Ridge, is illustrated in figure 1.12. The coldest bottom water is seen pressed against the slope of Madagascar, and below 3600 m the isotherms slope downward to the east in a zone 400-500 km wide, indicative of the breadth of the boundary current. The sign of the associated horizontal density gradient is consistent with a northward geostrophic velocity increasing toward the bottom. Unlike the South Atlantic, water characteristics offer no clear demarcation here between southward flow above and northward flow below, but the volume transport of the current was estimated geostrophically, with reference to 3600 m, to be about 4×10^6 $m^3 s^{-1}$ at this section (Warren, 1974). A comparable estimate, 5×10^6 $m^3 s^{-1}$, was obtained at a similar section extending eastward from Madagascar along lat. 23°S. (No direct velocity measurements have ever been reported in deep boundary currents of the Indian Ocean.)

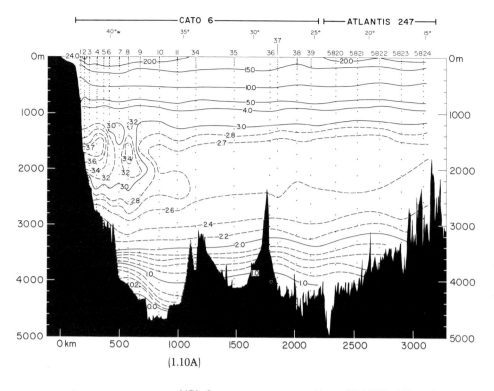

(1.10A)

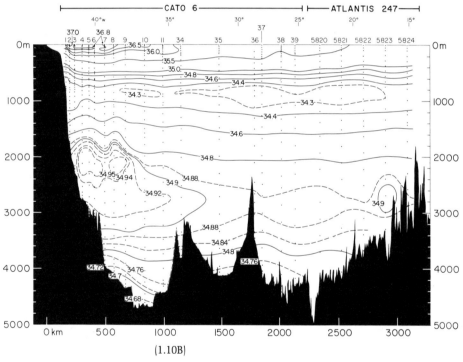

(1.10B)

30
Bruce A. Warren

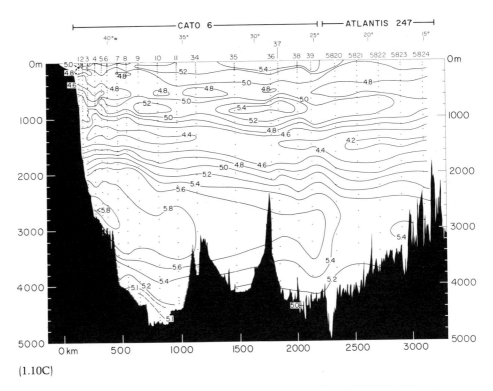

(1.10C)

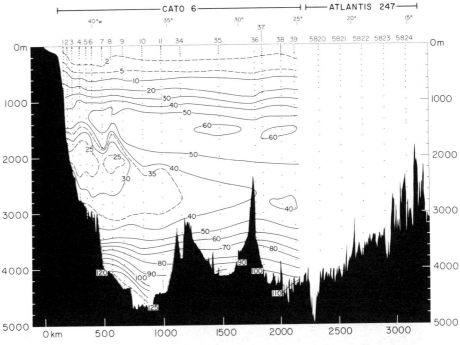

(1.10D)

Figure 1.10 Sections of (A) potential temperature (°C), (B) salinity (‰), and the concentrations of (C) dissolved oxygen (ml l^{-1}) and (D) silica (μM l^{-1}) along roughly lat. 30°S from South America (left) to the Mid-Atlantic Ridge, illustrating the two deep western boundary currents of the South Atlantic, namely, the northward-flowing Antarctic Bottom Water and the southward-flowing North Atlantic Deep Water above. Cato 6 (R.V. *Melville*) stations 1-11, 8-12 November 1972 and stations 34-39, 25-29 November 1972; R.V. *Atlantis* stations 5820-5824, 5-9 May 1959. (Reid, Nowlin, and Patzert, 1977.)

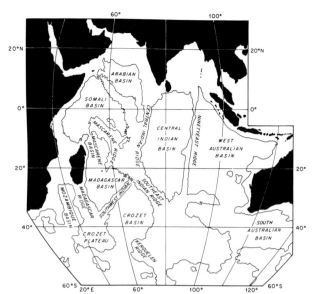

Figure 1.11 Index map identifying basins and ridges in the Indian Ocean, including an approximate representation of the 4-km isobath. (After Wyrtki, 1971.)

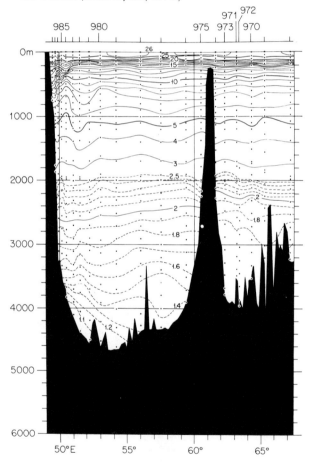

Figure 1.12 Temperature (°C) section along lat. 12°S between Madagascar (left) and the Central Indian Ridge, illustrating the deep western boundary current adjacent to Madagascar in the Mascarene Basin (see figure 1.11). R.V. *Chain* stations 968–988, 20–26 July 1970. (Warren, 1974.)

Evidently, some part of the current continues northward through the Amirante Passage (near 9°S, 52°E) into the Somali Basin (Johnson and Damuth, 1979), but the rate of inflow has not been estimated. During the southwest monsoon of 1964, deep salinity values near the continental slope off Somalia were found to be slightly lower at given potential temperatures than in the central Somali Basin, suggestive of a deep, northward-flowing boundary current (Warren, Stommel, and Swallow, 1966). The evidence was marginal, however; the observations have never been expanded or even repeated, and it is at least conceivable that the flow, if real, was related more to the seasonal Somali Current in the water above than to the global deep circulation.

In the Arabian Basin to the north, nothing is known about the deep circulation, except that the deep water must be renewed relatively slowly, given that its oxygen concentration is the lowest in the deep Indian Ocean (3.6 ml l^{-1} at 4000 m, as contrasted with 4.0–4.2 ml l^{-1} in the Somali Basin; Wyrtki, 1971).

A salinity maximum is found at about 2500 m in the southwestern Indian Ocean, and it is clearly due to North Atlantic Deep Water carried eastward by the Circumpolar Current. The maximum values decrease northward, but north of lat. 15°S, roughly, the salinity at these levels (and isopycnals) *increases* northward into the Arabian Sea (Wyrtki, 1971; see also figure 3.16B in this volume), plainly an effect of the salt source formed by the Red Sea outflow. Mixing thus seems to mask the large-scale field of motion at these levels, and it is uncertain what the sense and strength of the meridional flow are.

The western sequence of basins is separated from the Central Basin and West Australian Basin by the Central Indian Ridge and the Ninetyeast Ridge (figure 1.11). The deep Indian Ocean is open to the Antarctic, however, not only south of the Crozet Basin, but also just west of Australia; and the fact that 4000-m temperatures in the Central Basin and West Australian Basin are 0.2–0.3°C lower than in the Somali Basin and Arabian Basin (Wyrtki, 1971) indicates that the deep water in the former two basins is supplied through the eastern passage, as suggested long since by Schott (1902) and Wüst (1934). Consistent with the Stommel-Arons dynamics, this source water flows northward in the West Australian Basin as a deep western boundary current along the eastern flank of the Ninetyeast Ridge, observed at lat. 18° S (Warren, 1977). The field of specific volume anomaly (figure 1.13) shows a zone of sloping isopycnals at depths of 3000–4500 m, with the densest water against the Ninetyeast Ridge, and the breadth of the zone—indicating the width of the boundary current—being some 500–700 km. At depths of 3000–4000 m there is a slight silica maximum (concentrations >130 μM l^{-1}) in the basin, with values generally increasing toward the west; the maximum layer is separated from the Ninetyeast Ridge by a narrow zone about 30 km wide, however, where the values are lower and the maximum is absent. This distribution suggests: (a) northward flow of low-silica water from the Antarctic within the full boundary-current region deeper than 4 km, but only immediately adjacent to the Ninetyeast Ridge at lesser depths; and (b) southward flow of the high-silica water of the North Indian Ocean (Wyrtki, 1971) elsewhere in the boundary-current region. With reference to a zero-velocity surface constructed on that basis, the northward volume transport of the Ninetyeast Ridge current was estimated geostrophically to be about 4×10^6 m^{3}s^{-1} (Warren, 1977), like that of the deep Madagascar current. The course of the current north of lat. 18°S has not been observed.

In the distribution of dissolved-oxgyen concentration calculated by Kuo and Veronis (1973) for their model of the Indian Ocean, the lowest values occurred in the northeastern corner, because that was the region farthest removed, in the sense of combined advection-diffusion time scale, from the oxygen sources in their model: the circumpolar water in the south, and the single boundary current along the western margin of the ocean. The fact noted above that the lowest deep values are actually found in the Arabian Sea becomes intelligible in light of this second deep boundary current, as an oxygen source in the eastern Indian Ocean.

Only very tentative remarks can be made at this time about the deep water in the Central Basin. *Atlantis II* stations 2288-2306, occupied along lat. 18°S, show a temperature-minimum layer centered near 4000 m, most markedly developed toward the east (figure 1.14). This layer is also characterized by a salinity minimum and an oxygen maximum, both of which also decay westward from the Ninetyeast Ridge. Whether there are passages through the southern boundary of the Central Basin that permit exchange of deep water from high latitudes is uncertain, but it seems unlikely that this water with Southern Ocean characteristics derives directly from the south, because the dynamics requires *poleward* interior flow. The properties of the water below 3500 m are very much like those in the Ninetyeast Ridge current at 3500-4000 m on lat. 18°S, however, and it seems possible that the very deep water in the Central Basin is supplied across the Ninetyeast Ridge over known sills of depth 3500-4000 m at lats. 3°S and 10°S (Sclater and Fisher, 1974). Thus at lat. 18°S this water could be circulating in a southward interior flow. On a line of stations occupied by U.S.N.S. *Wilkes* in April 1979 along the western flank of the Ninetyeast Ridge between lat. 12°S and the equator, such an overflow was, in fact, observed near the 10°S sill, but no clear evidence for overflow was found at 3°S. Whether the 10°S

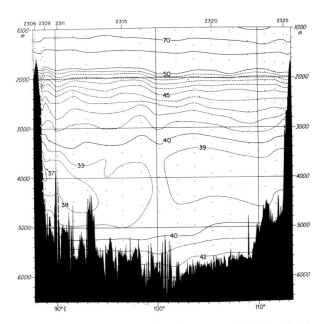

Figure 1.13 Section of specific volume anomaly (10^5 cm^3 g^{-1}) below 1000 m along lat. 18°S between the Ninetyeast Ridge (left) and Australia, illustrating the deep western boundary current of the West Australian Basin (see figure 1.11). R.V. *Atlantis II* stations 2306-2326, 7-17 August 1976. (Johnson and Warren, 1979.)

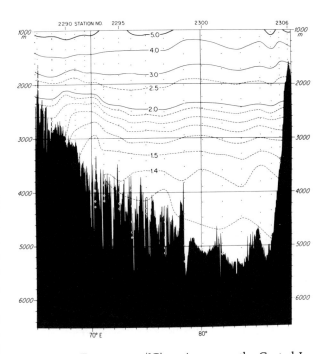

Figure 1.14 Temperature (°C) section across the Central Indian Basin along lat. 18°S from the crest of the Central Indian Ridge (left) to the Ninetyeast Ridge, illustrating the westward increase of temperature below 3500 m and the possible northward flow above the flank of the Central Indian Ridge at depths of 2000-3000 m. R.V. *Atlantis II* stations 2288-2306, 27 July-7 August 1976.

overflow is steady or intermittent is not certain, nor of course does this single set of observations establish that overflow never takes place across the 3°S sill.

At shallower levels, 2000–3000 m, isotherms sloping downward to the east above the flank of the Central Indian Ridge (figure 1.14; stations 2291–2296) suggest a third deep western boundary current in the Indian Ocean. The southern boundary of the Central Basin is deep enough to allow direct northward flow from the Antarctic at these levels, and the fact that, for given potential temperatures in this depth interval, oxygen values are higher by about 0.2 ml l^{-1}, and silica values lower by about 3 μM l^{-1}, at this group of stations than at those farther east hints of a more recent southern origin for the water in this current, consistent with northward flow there. The property differences are small, however, and observations are not yet numerous enough to define the deep flow pattern in the Central Basin with certainty.

1.5.3 Pacific Ocean

The Tasman Basin just east of Australia is the Pacific counterpart of the Mozambique Basin, in that it is closed off to the north at depths greater than 2850 m (Wyrtki, 1961a), and thus the effective western boundary for the deep South Pacific is not Australia but New Zealand and the Tonga-Kermadec Ridge. The deep current flowing northward along this boundary has been observed at lats. 28°S and 43°S on two transpacific hydrographic sections (Stommel, Stroup, Reid, and Warren, 1973). Portions of the temperature and salinity sections along lat. 28°S between the Tonga-Kermadec Ridge and the foot of the East Pacific Rise are illustrated in figure 1.15. The coldest water is again found toward the west, and below about 2500 m the isotherms slope downward to the east in a zone that, below 3500 m, is some 1000 km wide. At depths of 3000–4000 m a faint salinity maximum represents the last traces of deep water from the North Atlantic carried southward through the South Atlantic, eastward around Antarctica in the Circumpolar Current, and northward here into the Pacific. These traces are also registered as a slight minimum in the corresponding silica section (Warren, 1973). At both lats. 28°S and 43°S the mid-depth oxygen minimum is stronger and somewhat deeper (depths of 2000–2500 m) within 2000–3000 km of the effective western boundary than in the central South Pacific, indicating southward flow from the North Pacific at those levels, concentrated toward the western side of the basin (Reid, 1973a). It is perhaps analogous to the flow of North Atlantic Deep Water in the South Atlantic, though much weaker relative to the northward flow below.

The oxygen minimum is associated loosely with a silica maximum near the 2500-m level, but the silica-maximum layer is separated from the western bound-

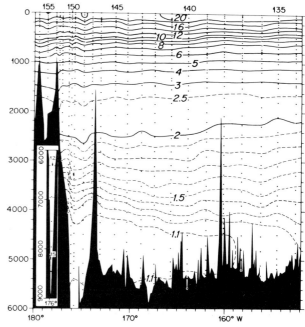

(1.15A)

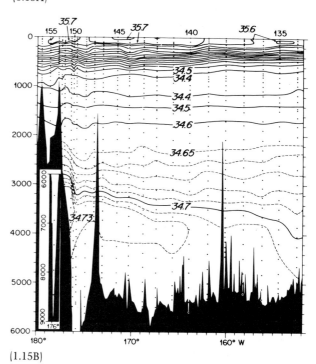

(1.15B)

Figure 1.15 Sections of (A) temperature (°C) and (B) salinity (‰) along lat. 28°S between the Tonga-Kermadec Ridge (left) and the base of the East Pacific Rise, illustrating the principal deep western boundary current of the South Pacific; the Tonga-Kermadec Trench is continued in the inset. Depth in meters. *Scorpio* stations 134–156, 3–18 July 1967, U.S.N.S. *Eltanin* cruise 29. Reproduced from *Scientific Exploration of the South Pacific*, ed. W. S. Wooster (Washington, D.C., National Academy of Sciences, 1970), pp. 45–46.

ary by a zone of low silica concentration about 100 km wide at lat. 28°S and some 1000 km wide at lat. 43°S (Stommel, Stroup, Reid, and Warren, 1973). The pattern of silica variation is strikingly like that at lat. 18°S in the West Australian Basin, but the features of the distribution in the South Pacific are much more markedly and clearly developed. Those features suggest northward flow close to the western boundary from about 2000 m to the bottom as well as in a much broader zone deeper than 3000–3500 m, with southward flow of low-oxygen, high-silica water from the North Pacific somewhat removed from the boundary at depths of 2000–3000 m. The southward flow seems to carry about 3×10^6 m^3 s^{-1} at lat. 28°S, and estimates for the volume transport of the northward current are roughly 20×10^6 m^3 s^{-1} at both latitudes (Warren, 1973, 1976). The latter is probably the largest of all the deep western boundary currents, consistent with its being the principal supplier of deep water to the largest ocean.

At lat. 22°S, nine neutrally buoyant floats were tracked for 4–6 days each in the boundary current (Warren and Voorhis, 1970). The averaged velocities (a few centimeters per second), in combination with thermal-wind calculations, suggested a surface of zero meridional velocity that descended eastward from 3100 m at the Tonga-Kermadec Ridge to 3600 m, and a volume transport for the underlying northward flow of 13×10^6 m^3 s^{-1}. On account of the brevity of the velocity measurements, it is not clear just how representative these figures are of the prevailing flow.

The only other velocity measurements in this current—lasting from a few hours to a few days—were made with current meters near the Samoan Passage (9°S, 169°W), whose sill (depth 4500–5000 m) separates the principal basins of the North and South Pacific (Reid and Lonsdale, 1974). The current measurements and density field indicated northward flow through the passage below roughly 3800 m and southward flow above.

The East Pacific Rise separates the main basin of the South Pacific from a sequence of smaller ones in the east. Water characteristics in the Southeast Pacific Basin at lat. 43°S (relatively low temperature and silica, high salinity and oxygen) demonstrate inflow of deep water to this sequence from the south, with a volume transport of perhaps 5×10^6 m^3 s^{-1} (Warren, 1973). The property fields do not give much impression of a *boundary* current, however, and while the characteristics farther north at lat. 28°S are still indicative of a relatively recent southern origin for deep water in the east, the flow must be very diffuse, and it certainly does not look "current-like." Probably the rugged topography in the region disperses the flow. From the temperature distribution Lonsdale (1976) has shown that the bottom water passes north from the Southeast Pacific Basin into the Chile Basin through faults in the Chile Rise near longs. 85–90°W (sill depth 3800–4000 m); further northward flow, into the Peru Basin, occurs via the Peru-Chile Trench and across a sill near long. 89°W. Bottom water also flows eastward across the East Pacific Rise, through several passages in lats. 4–10°S and at lats. 2°N and 8°N. The principal passage from the Peru Basin into the Panama Basin to the north is the northern extension of the Peru-Chile Trench; a 17-day record of bottom current in this passage gave an average velocity, northeastward, of 33 cm s^{-1}, implying a volume transport of inflowing bottom water to the basin (deeper than about 2500 m) of about 0.35×10^6 m^3 s^{-1} (Lonsdale, 1977). Such extraordinarily rapid inflow is thought to be forced, at least in part, by strong local geothermal heating in the Panama Basin (Detrick, Williams, Mudie, and Sclater, 1974).

Although, as noted, the Tasman Basin on the other side of the South Pacific is closed off by a sill of depth 2850 m, the temperature-salinity characteristics of water deeper than 3000 m in the Coral Sea Basin and Solomon Basin to the north demonstrate that this water is supplied from the Tasman Basin (Wyrtki, 1961a). By a heat-budget calculation, however, Wyrtki (1961a) estimated that this rate of inflow was roughly 0.04×10^6 m^3 s^{-1}, a trivial value in comparison with those for the deep northward flows farther east in the South Pacific.

From similar evidence Wyrtki (1961a) argued that deep water in the nearby New Hebrides Basin, New Caledonia Trough, and South Fiji Basin was renewed from the central Pacific to the north. Correspondence in water properties, however, suggests that the deep water of the South Fiji Basin might derive from the deep boundary current east of New Zealand and the Tonga-Kermadec Ridge, by spillage over a sill of depth 2500–3000 m just north of New Zealand (Warren, 1973).

In the northward course of the principal deep boundary current, the salinity maximum that is useful as a tracer characteristic is gradually diminished by mixing, and it disappears just north of the Samoan Passage (Reid and Lonsdale, 1974). South and east of Hawaii, though, Edmond, Chung, and Sclater (1971) identified a bottom layer a few hundred meters thick with temperature and silica concentration distinctly lower, and salinity and oxygen concentration distinctly higher, than in the water immediately above; this layer is undoubtedly sustained by a northeastward extension to the east of Hawaii of part of the flow through the Samoan Passage, that extension at levels of 4500 m and deeper having been demonstrated in broad terms by Knauss (1962a) and Mantyla (1975). The sharp vertical gradients defining the top of the layer might also mark a zero-velocity surface, but that is not self-evident: such gradients certainly do not do so in analogous features observed in the inflow to the Panama Basin (Lonsdale, 1977) and on the Blake-Bahama Outer Ridge in the western North Atlantic (Amos, Gordon, and Schneider, 1971). In any case, 4-day current records demonstrating eastward flow of bottom water at speeds of a few centimeters per second along the northern side of the Clipperton Fracture Zone—southeast of Hawaii—support the general course of water movement indicated by the large-scale property fields (Johnson, 1972). That movement extends eastward across the East Pacific Rise through the low-latitude faults noted above, and it can also be traced at least as far north as lat. 30°N (Mantyla, 1975). Such a poleward interior flow is consistent with the Stommel-Arons dynamics, and it is different from the southward flow in the eastern North Pacific hypothesized by Sverdrup (Sverdrup, Johnson, and Fleming, 1942, chapter XV; see also section 1.2).

Part of the flow through the Samoan Passage also extends northwestward into the North Pacific (e.g., Wooster and Volkmann, 1960; Knauss, 1962a; Mantyla, 1975), but there is no conclusive evidence for deep western boundary currents anywhere in that ocean. In low latitudes this lack of evidence is due at least in part to the island chains and undulating rises that make it difficult even to define a clear-cut western boundary; perhaps the very concept of interior flow plus boundary current is not a useful one there. In middle and high latitudes, Japan and the Kuriles form a distinct western boundary, but, because the deep North Pacific is so far removed from near-surface sources, the water is nearly homogeneous, and deep property maps near the western margin show mostly observational noise, rather than indications of prevailing currents (Moriyasu, 1972).

Nan'niti and Akamatsu (1966) have described deep velocity measurements made with neutrally buoyant floats tracked for 1-4 days from three sites along the Japan Trench near lats. 32, 38, and 40°N. Motions observed at 2-3 km were predominantly southward, with speeds of a few centimeters per second; floats at 1000 and 1500 m moved somewhat faster, and, near 32°N, more toward the west. Worthington and Kawai (1972) have reported additional float measurements, lasting about 1 day each, made just east of Honshu in lats. 34-35°N; two floats set to a depth of 1 km inshore of the Japan Trench moved westsouthwestward at speeds of order 10 cm s^{-1}, consistent with the observation above near 32°N, while three other floats, released offshore of the Japan Trench at depths of 1-3 km, moved northeastward. Each of the measurements cited, taken individually, is of much too short a duration for a statistically significant estimate of mean flow, but the ensemble hints of a deep, southward-flowing recirculation current close against Japan, as predicted by the Stommel-Arons model. Velocity measurements of very much greater duration are badly needed, however, to establish the existence of that current. Longer-term records have been obtained from current meters moored south of Honshu, but those results probably reflect the local deep flow in the Shikoku Basin, rather than the circulation in the open North Pacific, and the connection between the former and the latter has not been delineated (Taft, 1978).

In the central North Pacific, temperatures at 3500 m are slightly lower than those generally found at the same level in the South Pacific, suggesting widespread deep upwelling (Knauss, 1962a), and thus supporting that assumption of the Stommel-Arons model.

Along the northern rim of the deep North Pacific, near the Aleutian Trench, there is a band of water 100-200 km wide that is colder by about 0.05°C than the water to the south (Knauss, 1962a). It is unlikely that this feature is due to upwelling of deeper water because the anomaly persists all the way to the bottom (Reed, 1969). Reed (1970a) and Mantyla (1975) showed further that this band is continuous with cold water to the west, and inferred therefrom a small eastward flow of deep water along the northern margin. The inference seems unexceptionable, but it is puzzling in terms of the Stommel-Arons circulation model: as described in section 1.3, the model requires northern boundary currents, but they flow *westward* in consequence of being fed by the poleward interior flow; an eastward-flowing current diminishing to the east should be supplying an *equatorward* interior flow. It is conceivable that in this subpolar region the vertical velocity at mid-depths is downward rather than upward, so that the interior flow *would* be equatorward rather than poleward, but if so, then the local density-depth curves might be expected to have curvatures of opposite sign from those to the south, where the vertical velocity is thought to be upward. That the observed curvatures are in fact of the same negative sign suggests (but does not prove) that the upwelling is general. Reconciliation of this eastward current with a southward-flowing western boundary current (if the latter indeed exists) would also be problematical for any simple flow scheme. The cold band is a perplexing feature of the deep circulation.

In summary, when oceans have been reconnoitered with appropriate observational strategies, deep western boundary currents have been found where anticipated along the western margin of the North and South Atlantic, along the eastern sides of Madagascar and the Ninetyeast Ridge in the South Indian Ocean (perhaps also along the eastern flank of the Central Indian Ridge), and beside the New Zealand platform and Tonga-Kermadec Ridge in the South Pacific. Northward flow east of the East Pacific Rise is less like a boundary current than might have been imagined, perhaps because of the topographic complexity, and in the Atlantic the northward-flowing Antarctic Bottom Water current actually becomes an *eastern* boundary current in the southern North Atlantic, possibly on account of meridional bottom slope. The existence of deep western boundary currents in the North Pacific and North Indian Oceans has not been definitely established, probably for lack of suitable measurements. Finally, how eastward flow along the northern rim of the Pacific and the apparent *absence* of western boundary currents in the deep eastern Atlantic are to be reconciled with circulation dynamics is not obvious.

1.5.4 Structure of Deep Western Boundary Currents

When Stommel and Arons (1960b) constructed their model of a global deep circulation, they did not inquire into the character and dynamics of the deep western boundary currents, but simply hypothesized them as a closure device. It is clear from hydrographic sections that these currents differ among themselves, and differ considerably in form from the much swifter western boundary currents of the upper water like the Gulf Stream and Kuroshio. Most notably, perhaps, the deep currents are much wider, except in the North Atlantic. In the South Atlantic, Stommel and Arons (1972) recognized a striking parallelism between isopycnals in the Antarctic Bottom Water current and the bottom profile of the broad South American continental rise. They developed a two-layer model to show how such parallelism—and the associated greatly enhanced width of the current—could be consistent with uniform potential vorticity across the current. Undoubtedly a gentle bottom slope must act to broaden bottom currents, if only because geostrophic flow tends to follow isobaths, but in the vicinity of other deep currents that are at least as wide, particularly those in the South Pacific and the West Australian Basin of the Indian Ocean (figures 1.15 and 1.13), there is no broad continental rise, and the currents flow over ocean floor that is level (apart from small-scale features). Some other physics must be responsible for the large widths there.

In a different theory based on linear dynamics and the idea of lateral mixing of density (Warren, 1976), the velocity and water-property distributions are decomposed into interior and western-boundary fields, and the western-boundary density field is governed by a balance between lateral (zonal) diffusion of boundary-field density and vertical advection of the total density by the boundary current:

$$-w\rho_0 E = K\rho_{xx},$$

where ρ and w are the boundary fields of density and vertical velocity, E the static stability associated with the full density field, K the horizontal diffusion coefficient, and ρ_0 a mean density. Linear dynamics is appropriate to the deep currents because their speeds are small (in contrast to the Gulf Stream, say), but the extent to which *density* mixes horizontally in different parts of the ocean, as distinct from temperature and salinity individually, is uncertain. The theory is incomplete, moreover, in several respects; but, with application of upper and lower boundary conditions, it produces eigenfunctions that fairly successfully rationalize the observed structure of deep boundary currents.

Combining this density balance with the geostrophic vorticity equation and the thermal-wind relation gives a basic zonal scale for each eigenfunction [the "western scale" cataloged by Blumsack (1973), q.v. for parameter restrictions] of $l = (Kf^2)/(\beta g E_0 h^2)$, where h is the corresponding vertical scale and E_0 a typical value of E. In deep water E_0 is generally 10^{-9} cm^{-1}; for $K = 10^7$ cm^2 s^{-1}, and for an eigenfunction of vertical scale 1500 m, in midlatitudes ($f = 10^{-4}$ s^{-1}, $\beta = 2 \times 10^{-13}$ cm^{-1} s^{-1}, $g = 10^3$ cm s^{-2}) l would be 222 km, which is about right for the observed current widths, considered as two or three horizontal scales (e-folding distances).

The key property that l varies *inversely* with h (the eigenfunctions are tall and thin or short and broad) accounts for several features in the form of the currents. In the South Pacific, for example (Warren, 1976), a combination of two eigenfunctions is necessary for a good fit to the observed variation of density with depth at the western boundary on lat. 28°S (figure 1.15); the one generates the narrow zone of northward flow close to the boundary between 2000 m and the bottom that was inferred from the silica distribution, and the other requires the broad zone to the east with weak southward flow above 3500 m and northward flow below, as indicated by the oxygen, salinity, and silica distributions. A comparable construction (Johnson and Warren, 1979) accounts moderately well for the similar property distributions in the Ninetyeast Ridge current.

The scale relation also helps to explain why the deep boundary current of the North Atlantic (figure 1.9) is narrower than those in the southern hemisphere. The former is directed equatorward uniformly from 1000 m or shallower to the bottom, implying a large vertical scale, while the latter, being composed of equatorward flows near the bottom and poleward flows at mid-depths (some short distance from the western boundary, at least), have rather smaller vertical scales. For the North Atlantic current, with $f = 1.2 \times 10^{-4}$ s^{-1} (lat. 55°N) and $h = 3000$ m, l is 80 km; while in the South Atlantic (figure 1.10), with $f = 0.7 \times 10^{-4}$ s^{-1} (lat. 30°S) and $h = 1000$ m (quarter-wavelength of vertical variation), $l = 250$ km. It is not obvious, though, why poleward flow is part of the boundary-current systems in the southern hemisphere. It seems natural enough for the South Atlantic, where there is a northern source for deep water, but there are no northern sources in the Indian and Pacific Oceans. It must be a consequence somehow of the density stratification of deep water, in which case the existence of poleward flow along the boundary in the South Atlantic may not be linked so tightly to the northern source as intuition would suggest.

The boundary layer described by l is strictly geostrophic, in the sense that $\beta v = f w_z$. For example, by the density balance above, where isopycnals slope downward to the east near the bottom in the boundary region, the curvature of the density field demands a local downward vertical velocity, and thereby a negative value for w_z, which, through the geostrophic vorticity balance, necessitates the equatorward current. It is *this* departure from the Stommel–Arons open-ocean regime—downward vertical velocity—rather than a relaxation of geostrophy that allows this particular kind of equatorward boundary current.

In order to bring the diffusive density flux to zero at the western boundary, however, a nongeostrophic inner boundary layer, the so-called "hydrostatic layer," was fitted to the geostrophic layer (Warren, 1976). The scale of the inner layer $l_H = [(A g E_0 h^2)/(K f^2)]^{1/2}$, where A is the viscosity coefficient. Generally in the deep ocean $l_H \sim 10$ km and is less than l; when the scales are comparable, the two layers merge into a single layer of different form, not interpretable in so simple a fashion. [If $l_H \gg l$, it is a Munk boundary layer, of scale $(A/\beta)^{1/3}$, but for the moderate stratification and the vertical scales of the deep ocean, Munk layers do not apply.] The hydrostatic layer also completes the local vertical flow circuit by supplying and absorbing the vertical flow in the geostrophic layer. In addition, the inner layer sets the meridional velocity to zero at the western boundary, thereby providing a viscous shear-stress force to balance the meridional pressure gradient caused by the meridional variation in Coriolis parameter.

On the other hand, considering the thinness of the hydrostatic layer in relation to the large mixing coefficients on which it is based, the layer looks like an artificial construction, a concept that is probably internally inconsistent. Moreover, even apart from the generally questionable nature of the mixing-coefficient parameterization of diffusive flux, extrapolation of constant values of the coefficient into the coast is surely wrong. Processes somewhat different from those in the open ocean must be at work immediately adjacent to the boundary, and the hydrostatic layer is probably more a mathematical closure device than something one is likely to find at sea. Nonetheless, the layer is conceptually useful in that it illustrates specifically the incompleteness of the geostrophic layer, and shows what conditions must be fulfilled by the immediate boundary processes.

Boundary layers based on the same dynamics and density balance can also be constructed on eastern sides of oceans, but the eigenfunctions would have monotonic vertical variation, which is not compatible with realistic upper and lower boundary conditions. Furthermore, the condition that $w = 0$ at the floor of the ocean would require that the magnitude of the boundary fields increase upward to the sea surface, so that such layers could not be depth-intensified anyway. Hence the boundary layers that close deep circulations through these physical balances are necessarily *western* boundary currents.

The density-diffusive model requires that the fluctuating motions in the deep sea mix density horizontally, but that they do not, through divergence of Reynolds stress, drive the mean boundary currents directly. How true these assumptions are remains to be seen.

1.6 Why Is There a Deep Thermohaline Circulation At All?

Having discovered convection currents in the laboratory, Count Rumford argued that high-latitude cooling should force analogous currents, of global scale, in the ocean. With this motivation, he examined oceanographic data and disclosed the polar origin of deep water. In hindsight after nearly two centuries, however, Rumford's idea as to *why* there should be a deep meridional circulation seems too simple: too much an extrapolation of "everyday" experience, and too little informed both of how buoyancy flux is effected near the sea surface and of how the earth's rotation controls slow motions of such large scale.

To be specific, it is not really obvious that there should be a buoyancy flux at the sea surface to force sinking in high latitudes. The salinity of surface water is increased through evaporation and freezing, and its temperature is governed by short-wave solar radiation, outgoing long-wave radiation, sensible heat flux, and latent heat flux. The last three fluxes are determined as much by the sea-surface temperature as by external conditions, and it is easy to imagine situations in which the temperature of the water column adjusts to those conditions so that there is *no* net annual buoyancy flux across the surface, even in polar latitudes. To be sure, different temperatures in different latitudes would imply different densities, and therefore a field of meridional pressure gradient, but that need not drive a substantial meridional flow. As Ekman (1923) stressed, on the rotating earth the zeroth-order momentum balance would be geostrophic, and the pressure-gradient field would be associated wholly with *zonal* flow (except in the frictional boundary layers).

For an example, consider the earth to be entirely covered with an ocean of uniform depth and, to include only Rumford's essentials, disregard wind stress and salinity flux. Assume further that the external parameters controlling the components of heat flux depend only on latitude. Then an equilibrium field is possible in which the temperature is independent of depth and longitude, and is adjusted to the external conditions of each latitude to bring about zero net annual heat (and density) flux across the sea surface everywhere. In steady state the implied meridional density gradient is then balanced by a zonal thermal-wind shear, and the only meridional flow occurs in surface and bottom Ekman layers, as required by the conditions of no stress at the surface and no slip at the bottom. Demanding that the meridional fluxes in the two Ekman layers balance gives an eastward geostrophic flow (for density increasing poleward) that decreases linearly with depth from the surface to zero within the bottom Ekman layer. The surface speed u is typically $(gD\,\Delta\rho)/(\rho_0 fL)$, where $\Delta\rho$ is the meridional density difference, L the distance from equator to pole, and D the ocean depth; the Ekman-layer speeds are of order $u\alpha/D$, and the Ekman depth $\alpha \equiv (\nu/f)^{1/2}$, where ν is the vertical viscosity. If $\Delta\rho = 6 \times 10^{-3}$ g cm^{-3} (typical meridional difference in surface density), $D = 5$ km, $L = 9000$ km, and $f = 10^{-4}$ s^{-1}, then $u = 33$ cm s^{-1}, and the eastward volume transport, pole to pole, is about 1.5×10^{10} m^3 s^{-1}. For $\nu = 10^2$ cm^2 s^{-1}, the Ekman speeds are roughly 7×10^{-2} cm s^{-1}, and the meridional transports across the 45th parallels in the Ekman layers are merely 0.1×10^6 m^3 s^{-1}, poleward at the surface, equatorward at the bottom. This boundary-layer flow does tend to upset the imposed density field with which the geostrophic flow is associated, but the speeds are so small that the horizontal advection of density is easily balanced by vertical diffusion within the Ekman layer, without noticeably disturbing the interior density field.

This flow is *not* the response to meridional density forcing that Rumford envisaged, nor, of course, is it anything like the meridional circulation that actually occurs in the ocean. The circumstances of the model are far removed from reality in many respects, but it is informative to ask what are the *significant* differences that lead to such a different oceanic circulation system. Most fundamental, perhaps, is the existence of continents, which impose meridional barriers to zonal flow. Consequently, meridional pressure gradients cannot be balanced everywhere by Coriolis forces, and they must force meridional flow somewhere in the system. Meridional barriers are not *small* perturbations to the water-covered-globe model, however, and it is not clear what the different circulation pattern would be.

For example, merely introducing meridional barriers need not lead to *widespread* rising or sinking motions at a much greater rate than required by the upper and lower frictional boundary layers. It is conceivable, in fact, that a stationary density field, with associated zonal flows, could be achieved essentially through lateral and vertical diffusion, with no substantial vertical motion except in thin meridional-boundary layers [a variation without wind stress of a model developed by Rattray and Welander (1975)].

On the other hand, the horizontal circulation in the actual deep ocean is thought to be a consequence of localized sinking and general upwelling. The sinking that is known to take place, moreover, seems not to be merely a concomitant of the overall meridional density gradient, because most of it occurs from sheltered, semienclosed regions (Antarctic continental shelf, Norwegian Sea, low-latitude marginal seas) where near-surface water is driven in, contained long enough to become exceptionally dense, and then is forced back to the open ocean, sinking to depth because of its high density. Wind stress probably contributes to the forcing (e.g., on the Antarctic continental shelf), and certainly salinity enhancement through freezing and evaporation is the principal agent of densification in some cases. Nevertheless, the existence of such embayments where negative buoyancy flux can be sustained against the tendency for adjustment to a no-flux condition at the sea surface—a second geometric departure from the water-covered-globe model—appears to be important for "production" of deep water in the amounts observed.

In the Labrador Sea, sinking occurs through a different process, deep convective overturning, but that must be highly intermittent, associated with weather anomalies, because one would expect the effect of an annual heating and cooling cycle constant from one year to the next to be simply formation and destruction of a seasonal thermocline, with water properties just below adjusted to the winter conditions. Owing to the weakness of the stratification in the central Labrador Sea, however, severe winters or severe weather events within a single winter can, apparently, generate occasional convective overturn to great depth. Climatic unsteadiness thus seems to be another significant way in which the real world differs from the simple model sketched.

Given local, externally forced sinking in the deep ocean, there must be a compensating rising of deep water elsewhere. If this upwelling were confined to regions as small as those of the sinking, it should be discernible in property distributions; since it is not, one supposes that it is widespread over most of the rest of the ocean. This assumed large-scale character of deep upwelling is crucial to ideas about the horizontal circulation of deep water, but the physical basis for it has not been elaborated. Laboratory experiments (nonrotating) on convection forced by heating and cooling at the same level show an analogous asymmetry between the sizes of sinking and rising regions, the asymmetry being attributed to the relative efficiencies of advective and diffusive buoyancy flux (Rossby, 1965). The idea (in oceanic terms) is that although all density forcing occurs at the sea surface, density is added through vertical advection and withdrawn by vertical diffusion, and the total density flux across any level must be zero in the climatological mean. The advective flux varies essentially with the transport of the vertical circulation, while the total diffusive flux, depending as it does on the vertical density gradient, is proportional to the area across which it occurs. Consequently, to equalize the two fluxes, the area over which the upwelling limb of the vertical circulation feeds the upward diffusive flux of density may need to be very much greater than the area in which the downwelling limb occurs, and could, in fact, occupy most of the ocean. This is in the nature of an energy argument, however, and the dynamics of how the forcing generates the vertical motion has not been elucidated even for the laboratory experiments, let alone for the somewhat different oceanic problem.

These concluding remarks have been general and quite speculative, with little possibility of developing them to any satisfying conclusion. They point, rather, to several quite fundamental aspects of the deep ocean circulation that seem, even after 200 years of study, to be still only dimly understood. These basic questions, in contrast to those raised in sections 1.4 and 1.5, seem not likely to be answered by new observational programs, but by hard physical thought applied to data in hand.

Notes

1. I am indebted to Sanborn Brown, Rumford's editor and biographer, for explaining (personal communication) the circumstances in which this essay was first published: "In the late 18th century it was customary to publish the same paper in several journals, and, since the concept of science as an international effort was not yet accepted, it was also customary to publish articles in several different languages. The paper was first published as a separate pamphlet in London by his usual publisher, Cadell & Davies. Rumford sent the manuscript to Cadell & Davies on May 14, 1797 and it was published in July. On May 21 he sent a copy of the same paper to his physicist friend, Professor Auguste Pictet in Geneva, which at that time was part of France. Pictet was the editor of the *Bibliothèque Britannique*, and he translated the paper into French and published it later in that same year 1797. Count Rumford was at that time considered to be a German physicist since he was permanently settled in Munich and was a general in the Bavarian army. The paper was therefore translated into German, probably by one of his favorite mistresses, Countess Nogarola. It was published as a whole in the *Neues Journal der Physik* and in little pieces in the *Chemische Annalen für die Freunde der Naturlehre, Arzney, Gelahrtheit, Haushaltungskunst und Manufacturen*, also both in 1797. It was subsequently published a number of times both in English and in German. All of these publications have the same text."

2. Zöppritz and Krümmel (Krümmel, 1911) attributed the earliest concept of a polar origin for deep water to J. F. W. Otto (1800). In fact, Otto's discussion of the subject is a word-for-word transcription of selected passages from a German translation of the third chapter of Rumford's essay, published in 1799 in the *Annalen der Physik*—even to the extent of reproducing Rumford's erroneous figure of 3600 feet. Insasmuch as Otto did not even hint that he was quoting someone else, it is not surprising that Zöppritz and Krümmel failed to give Rumford the proper credit.

3. Although Humboldt (1831) asserted that he had proved in 1812 the existence of the deep equatorward flow, M. Deacon (1971) points out that there was enough stylistic similarity between Rumford's discussion and those of Humboldt and other later writers to justify the belief that Rumford was their ultimate source. Humboldt was certainly acquainted with Rumford's general idea of convective heat transfer [e.g., *Annalen der Physik* 24 (1803), 17], and it is difficult *not* to believe that he had also noticed Rumford's inference about deep currents, because three of his own papers were printed in the same number of the *Annalen der Physik* that contained the German translation of the pertinent section of Rumford's essay (see footnote 2).

4. Carpenter's source was partly Humboldt and partly Buff (1850), the latter not citing any specific antecedents for it. His discussion is reminiscent of Rumford's, however, in that he illustrated the convection phenomenon by a laboratory experiment: heating a glass vessel from below, which had been filled with water that had powder mixed into it to make motions visible. Rumford had stumbled onto convection currents by setting aside a large thermometer, strongly heated during the course of an experiment, on his window sill to cool; the "spirits of wine" in the thermometer had been contaminated with dust particles, which on being illuminated by the sunlight, revealed by their motion the ascending and descending currents set up in the thermometer as it cooled.

5. Thomson (1877) had long before recognized that bottom water in the South Atlantic was substantially warmer on the eastern side of the Mid-Atlantic Ridge than on the western, and, invoking his "doctrine of continuous barriers," he had hypothesized the existence of the then unknown Walvis Ridge to isolate the eastern basin from direct Antarctic influence.

6. That misinterpretation was originated by Schott (1902); earlier writers had put the sinking region farther north, where the surface temperature was closer to that of the deep water. A complementary misinterpretation must have been the root of Lenz's (1845) idea of strong equatorial upwelling of deep water.

7. *Challenger* stations 323 (28 February 1876, 35°19'S, 50°47'W) and 324 (29 February 1876, 36°09'S, 48°22'W).

8. An unresolved problem in Gill's (1973) analysis is how the ice is moved offshore, because the prevailing winds around the Antarctic coast are from the east, and thus tend to drive the pack ice together with the near-surface water onshore (Solomon, 1974). Yet the ice does seem to move away (Gill, 1973), and such movement appears essential for net annual brine production. Perhaps the coastal current helps to carry ice out of its formation regions.

9. Cooper (1952) pointed out that Nansen (1912) formed his hypothesis on the basis of observations that were not his own,

and were of uncertain quality; moreover, that his hypothesis was never actually vindicated, and no evidence can be found in modern data for convection to the bottom in that region—although observations made due south of Greenland suggested overturning to 2500 m in March 1935 (Dietrich, 1957a), and convection to mid-depths must certainly occur in the Labrador Sea. Cooper (1952) did leave open the possibility of unusually severe conditions when the data used by Nansen (1912) were collected, so that convection all the way to the bottom might conceivably have been occurring at that time.

10. My impression of the reason why the earlier writers underrated the importance of the overflows is that they found hardly any trace of overflow water in the open ocean just south of the Greenland-Scotland ridge. They do not seem to have suspected that such overflows should join together to form a narrow current along the northern and western boundaries of the North Atlantic (figure 1.7) rather than spread directly southward, and the early observations were much too sparse to reveal the existence of that current. Cooper (1955a), on the other hand, had access both to the general concept of western boundary currents and to more comprehensive observational coverage.

11. See the many reports of The Iceland-Faroe Ridge International (ICES) "Overflow" Expedition, May-June, 1960 (1967), *Rapports et Procès-Verbaux des Réunions. Conseil Permanent International pour l'Exploration de la Mer* 157, 274 pp.

12. Tizard (1883) had previously suggested such an analogy, but he abandoned it after concluding that the surface inflow was compensated mainly by outflow at the surface rather than at depth.

13. At the XVII General Assembly of the IUGG in Canberra, December 1979, Clarke and Gascard (1979) described a direct observation in March 1976 of convection penetrating to depths greater than 2000 m, with downward velocities as high as 9 cm s^{-1}, in an area of diameter 10 km near the western side of the Labrador Sea.

14. In much of the western Atlantic, the potential density referenced to the sea surface is actually *less* for the Antarctic Bottom Water than for the Lower North Atlantic Deep Water above it (e.g., Lynn and Reid, 1968). Because the thermal expansion coefficient for water increases with pressure, however, the potential density referenced to some appropriate deep level is indeed greater for the colder Antarctic Bottom Water than for the water above—a point first explained, apparently, by Ekman (1934).

2
The Water Masses of the World Ocean: Some Results of a Fine-Scale Census

L. V. Worthington

2.1 Introduction

In his original brief monograph, Helland-Hansen (1916) introduced the concept of a water mass as being defined by a temperature-salinity (T–S) curve. He found that over a large area of the eastern North Atlantic a "normal" T–S curve could be drawn. He showed that variations from this curve could be attributed to the intrusion of alien water masses that had originated elsewhere. The use of the T–S diagram has been almost universal in physical oceanography since Helland-Hansen introduced it. It is not only a powerful descriptive tool, but observers at sea routinely plot T–S diagrams and use them as a check on the tightness of their sampling bottles and the correct function of their thermometers.

The term "water mass" has been very loosely used by numerous authors. According to Sverdrup, Johnson, and Fleming (1942), a water mass is defined by a segment of a T–S curve, and a "water type" by a single value of temperature and salinity that usually falls on a T–S curve. Thus a T–S curve is made up of an infinite number of "water types." These definitions will be adhered to in this chapter as far as is possible. Oceanographers have used other methods to describe the ocean, both before and after Helland-Hansen's introduction of the T–S diagram, and these methods will be briefly discussed below, but I will deal primarily with the world water masses as defined by T–S diagrams.

The most important advance in water-mass analysis since Helland-Hansen came with the introduction by Montgomery (1958), Cochrane (1958), and Pollak (1958) of the *volumetric* T–S diagram, in which the volumes of all the world water masses were estimated. The volume of the world ocean, including adjacent seas, is 1369×10^6 km^3. Montgomery, Cochrane, and Pollak were able to divide the individual and world oceans into bivariate classes of temperature and salinity, each of which contained an assigned volume. For example, the most abundant class found by Montgomery (1958) in the world ocean was $T = 1.0$–$1.5°C$, $S = 34.7$–$34.8‰$; he calculated that this relatively small class contained 121×10^6 km^3, or 9% of the water in the ocean.

Wright and Worthington (1970) produced a volumetric census of the North Atlantic that was a direct descendant of Montgomery's (1958) work. This later census was motivated by the introduction, pioneered by Schleicher and Bradshaw (1956), of the very accurate salinometers based on the measurement of electrical conductivity. The precision of data obtained with these salinometers enabled Wright and Worthington (1970) to divide the North Atlantic into much smaller classes than Montgomery and his colleagues had used: Wright and Worthington's smallest class (below 2°C) was $0.1°C \times 0.01‰$. Fifty of these classes make up one of

Montgomery's classes (0.5°C × 0.1‰). This fine-scale census had clear advantages over the coarser-scale census that inspired it, and in consequence I undertook a census of the world-ocean water masses using the fine-scale classes that Wright and Worthington (1970) introduced. The greater part of this paper will be devoted to the presentation of the results of this census, with some discussion of the formation of these water masses.

2.2 Methods of Describing the Oceans

The simplest and the most universally used method of describing the oceans has been the preparation of vertical profiles of temperature, salinity, dissolved oxygen, or some other variables, constructed from oceanographic sections made across an ocean, or part of an ocean, from a ship or a number of ships. Ocean-wide temperature profiles have been drawn by oceanographers since Thomson's (1877) treatment of the *Challenger* sections, but the standard of excellence for this kind of presentation was set by Wüst and Defant (1936) in their atlas of the temperature, salinity, and density profiles from the Atlantic *Meteor* expedition of 1925–1927 and by Wattenberg (1939), who prepared the oxygen profiles. These vertical profiles were drawn in color, with detailed bottom topography provided. The atlas by Wüst and Defant (1936) provided the model for Fuglister's (1960) atlas of vertical profiles of temperature and salinity from the transatlantic sections made by various ships and observers during the International Geophysical Year. Later, Worthington and Wright (1970) drew similar profiles, for sections made by the *Erika Dan* in the northern North Atlantic in 1962. They also included dissolved-oxygen profiles modeled on those of Wattenberg, which Fuglister had been unable to do because of the poor quality of oxygen analyses made from Woods Hole ships during the International Geophysical Year.

Probably the finest example in this form is that of the vertical profiles by Stommel, Stroup, Reid, and Warren (1973) for the transpacific sections at 28°S and 43°S from *Eltanin* in 1967. These profiles are shown in six color plates; the variables are temperature, salinity, oxygen, phosphate, nitrate, and silicate. The station and sample-bottle spacing for these sections were carefully planned so as not to miss any important baroclinic gradient or variation in nutrient concentration.

Composite vertical profiles are often drawn from data provided by a number of ships from different years or even different decades. Such sections are, of course, less useful for dynamical studies, but sometimes provide an excellent description of the water. A fine example is that of Wüst's much cited north–south temperature, salinity, and oxygen profiles in the Atlantic (Wüst, 1935, plate XXIII). Reid (1965) used the same method to construct zonal and meridional profiles of temperature, salinity, oxygen, and phosphate across the Pacific.

Helland-Hansen and Nansen (1926) drew vertical profiles of temperature and salinity superimposed on each other in their work on the eastern North Atlantic. This method has been followed by others, notably Tait (1957). I find such profiles difficult to read, but that may be idiosyncratic. Helland-Hansen and Nansen (1926) also introduced vertical profiles of the anomaly of salinity—in this, they were followed by Iselin (1936). My own preference (Worthington, 1976, figures 18, 19, and 20) is to draw vertical salinity anomaly charts, preferably in color, with selected isotherms included. I feel that this method illustrates most clearly the vertical juxtaposition of different water masses.

As Wüst (1935) remarked, horizontal charts of variables are of limited use in water-mass analysis because the various layers rise and sink above and below a fixed level. Nevertheless, the best early example of such horizontal charts is their own atlas (Wüst and Defant, 1936). Such charts have other uses, however, and are of particular value in satisfying the queries of laymen about the ocean.

The use of the "core layer" ("*Kernschicht*") method to describe ocean waters is almost wholly due to Wüst (1935), his students, and, to a lesser extent, Defant (1936). In his classic description of the Atlantic, Wüst (1935) identified seven such core layers, characterized by maxima or minima of oxygen, salinity, or temperature. While the study of the spreading of these layers is of unquestionable value in descriptive oceanography, they have a drawback, pointed out by Montgomery (1938a), in that these layers are few in number, whereas the number of potential-density surfaces (which Montgomery prefers to core layers) is infinite. A further drawback, noted by Worthington (1976), is that core layers are often uncritically assumed to be the main paths of ocean circulation. He was struck by the paradox that most observers who had made hydrographic stations combined with direct current measurements in the northern North Atlantic (Steele, Barrett, and Worthington, 1962; Worthington and Volkmann, 1965; Swallow and Worthington, 1969) had been constrained by these measurements to place a level of zero motion in the salinity-minimum core layer of the Labrador Sea Water, which can be identified as Wüst's (1935) Middle North Atlantic Deep Water. The reasons for this seemed clear; the volume of Labrador Sea Water in the North Atlantic is about 6×10^6 km^3 (Wright and Worthington, 1970) and the formation rate was thought to be between 2 and 4×10^6 m^3 s^{-1}. This gives a mean residence time for Labrador Sea Water of between 50 and 100 years. Accordingly, a particle of Labrador Sea Water should move from its formation area to its geographical limit between Iceland and Scotland at a speed of only a fraction of a centimeter per second.

Montgomery (1938a) made the first really effective use of the examination of variable properties on surfaces of constant potential density in the ocean. He credits Shaw (1930) with the development of this method in the atmosphere. It was used by Montgomery (and others who followed) mainly in the study of ocean circulation; the *depth* of a potential-density surface can be used as a diagnostic tool for locating geostrophic currents since rapid changes in the depth of these surfaces occur where such currents are found. It has also been a useful qualitative tool for describing water masses, since waters of different origin on the same potential-density surface usually contain widely different concentrations of variables, such as salinity, dissolved oxygen, and nutrients. Montgomery's (1938a) study of the tropical North Atlantic illustrated (among other things) the strong contrast in most upper layers between the low-salinity waters of the South Atlantic and the high-salinity waters of the North Atlantic.

Taft (1963) used this method to describe the shallower circumpolar water masses (and their circulation), and Callahan (1972) the deeper circumpolar water masses. Reid (1965) prepared charts of salinity, oxygen, and phosphate on two potential-density surfaces in the upper Pacific. These charts provide the best qualitative description of the thermocline waters of the Pacific, and the accompanying charts of the depths and acceleration potential at these surfaces give an equally broad qualitative description of the circulation for the upper layers of that ocean.

Perhaps the finest use of this method was Tsuchiya's (1968) description of the upper Equatorial Pacific. He displayed (in color) the depth, acceleration potential, potential temperature, salinity, and dissolved-oxygen concentration on four potential-density surfaces. The effect of such features as the Pacific Equatorial Undercurrent on the water-mass distribution is seen very clearly (this current being characterized by high salinity and high oxygen). The spreading of the oxygen minimum on both sides of the equator is also shown. One could wish (in hindsight) that Tsuchiya had extended his work to greater depths in the light of the recent discovery (C. Eriksen, personal communication) of deep equatorial jet currents in the Pacific.

In his monumental atlas of the Indian Ocean, Wyrtki (1971) used *all* of the techniques listed (and more) to delineate the water masses of that ocean, including volumetric T-S diagrams of the kind introduced by Montgomery (1958), Cochrane (1958), and Pollak (1958). This atlas represents by far the most complete description we have of any ocean. It is marred by only one minor flaw; because of the international character of the atlas, Wyrtki (1971) included slightly inferior data from some ships that participated in the Indian Ocean expedition. This, however, does not present a problem to professional oceanographers, who can usually recognize and reject spurious data.

Readers who are interested in descriptive oceanography in general will find the atlas compendium of Stommel and Fieux (1978) an invaluable guide.

2.3 The World Water Masses As They Exist in the Second Half of This Century

The use of the conductivity properties of sea water to measure salinity was introduced in 1922 by Wenner (Wenner, Smith, and Soule, 1930; see chapter 14). The Wenner instrument was always calibrated by chemical titration using the Knudsen method, however, so that its full precision was never realized. In practical terms, the ocean-wide use of the conductivity method dates from 1954, when Schleicher and Bradshaw (1956) introduced their salinometer. The availability of this instrument was in large part responsible for the recommendation by Iselin in 1954 to reexamine the physical structure of the Atlantic Ocean using deep sections modeled on those of Wüst and Defant (1936). Subsequently, other oceanographic laboratories followed the lead of Schleicher and Bradshaw (1956), and a body of high-quality physical data began to appear on the world ocean; in 1973, this work on an ocean-wide census was undertaken at Woods Hole.

The coverage chart for this world-ocean water-mass census is shown in figure 2.1; it clearly requires some explanation. As in the North Atlantic census of Wright and Worthington (1970), the basic area unit used was 5° of latitude by 5° of longitude. For convenience this unit will be called a 5°-square.

If no high-quality, deep hydrographic station can be found in a 5°-square, that square is shown in black. There were certain standards that had to be met for a station to be considered of high quality. The most important of these was that salinity should have been recorded to three decimal places and that such recording should have been justified. A great deal of tiresome work went into deciding whether each station met this standard. All hydrographic stations in which salinity was recorded to three places were obtained from the National Oceanographic Data Center. Potential temperature-salinity diagrams were drawn for these deep stations, and if the scatter of points on any station in the deep water was more than about ±0.01‰, the station clearly did not qualify. Experience has taught us that *all* the deep water masses of the world have a very tight potential temperature-salinity correlation and that variations from a mean curve are due to genuine geographic differences in the water masses. (Hereafter, all temperatures referred to in the text will be *potential* temperatures unless *in situ* temperature is specified. The symbol θ, for potential temperature, is

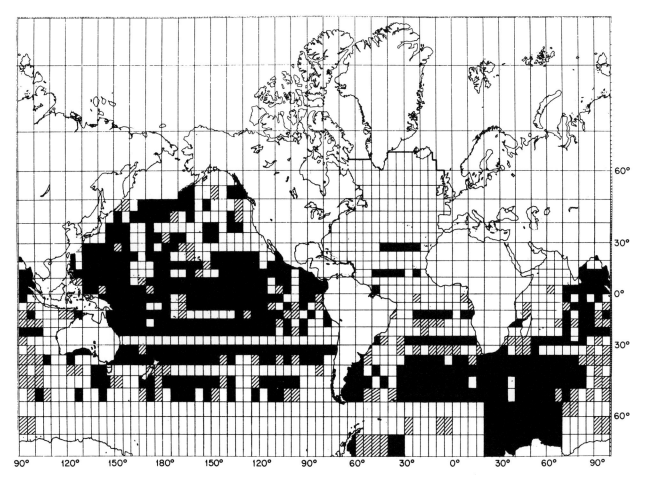

Figure 2.1 Coverage chart showing where high-quality deep stations are available (up to June 1977) in the world ocean. Unshaded 5°-squares contain at least one high-quality deep station. Crosshatched 5°-squares contain at least one high-quality station but in a shallow area of the square. Black 5°-squares contain no high-quality deep stations.

used in tables in case these should be reproduced elsewhere.) Occasionally, it appeared that a single ship, on a single cruise, had measured salinity incorrectly. This was determined by comparing the values obtained by this ship with those obtained by other ships where the data were plentiful. Usually, these systematic errors were of the order of 0.01‰. This was the most vexing determination in the census because the ship in question often traveled into virgin territory as well, but no empirical corrections were made to adjust the data from that ship to those from its fellow ships. Such data were simply omitted from the census.

Another, easier standard to be met was that the bottle spacing be sufficiently close that an unambiguous T-S curve could be drawn. Again, comparison with other stations in the same area was used to determine, for example, that no maxima or minima had been missed. The final standard involved the meaning of the word "deep." Generally, if a station extended to 90% of the water depth, it could be used in the census by extrapolation, but there are areas in the ocean where the temperature and salinity change rapidly near the bottom and extrapolation was not justified. Stations in such areas were also omitted from the census.

Accordingly, for a 5°-square to be unshaded in this coverage chart (figure 2.1), the square must contain at least one station of high quality that extended close enough to the bottom so that virtually no fictitious water could be assigned to it. Another qualification was that this station should be at, or close to, the greatest ocean depth in that 5°-square. If a 5°-square contained water somewhere that was, say, 5000 m deep, and the only station in that square did reach the bottom but only to some lesser depth (such as 3000 m) because it was taken over a seamount or on the continental slope, the 5°-square was not used in the census. Such 5°-squares are designated in figure 2.1 as crosshatched rather than black, to indicate that one or more high-quality stations exist in that square, but that they do not permit classification of all the water in that square.

It is possible that some fictitious water may exist in the census owing to systematic errors in thermometer calibration, but such volumes of water are thought to be small. Certainly no evidence for such errors was detected.

The census includes all high-quality data that were available from the National Oceanographic Data Center before June 1977. The North Atlantic census of Wright and Worthington (1970) was included unchanged. It can be seen (figure 2.1) that the North Atlantic is by far the best covered of the oceans, but as J. L. Reid (personal communication) has pointed out to me, nutrient data from the North Atlantic are sadly lacking.

The coverage chart (figure 2.1) is disappointing, coming as it does from the second half of a decade supposedly devoted to ocean exploration. In fact, by far the greater number of stations used in this census was made before this decade began. It is additionally disappointing that in spite of the large number of ocean-wide cruises made by ships of the Soviet Union before and during this decade, not a single station of theirs could be used in this census because the precision of their salinity data was not sufficient.

Certain features of this chart are worth mentioning. The North Atlantic was well covered, as we have seen, by purely physical data. Virtually all of the North Atlantic stations were made during the years 1954-1962. The excellent coverage of the Southern Ocean is almost entirely due to the work sponsored by the United States Antarctic Research Program aboard the U.S.N.S. *Eltanin*, under the leadership of Arnold Gordon. The lack of coverage in the southwest Indian Ocean and the southeast Atlantic Ocean is the result of a decision, made for financial reasons, to suspend temporarily the circumnavigation of Antarctica, combined with the cutoff date (June 1977) used for this census. The excellent coverage of the northwest Indian Ocean is due to the many participants in the International Indian Ocean Expedition whose interests lay chiefly in the glamorous monsoonal regions. Coverage in other parts of the Indian Ocean was poor except for the efforts of the Australians in the eastern part. In the South Pacific, the sections made by Stommel et al. (1973) at 28°S and 43°S are the most outstanding feature. In the North Pacific, one might deduce that there have been oceanographers working out of California and Japan, but deep coverage is generally poor.

None of these strictures applies to the upper layers of the oceans, where data are plentiful in nearly all oceans, and high precision of salinity analysis is, for the most part, not necessary to describe their broad features. However, only *precise* data have been used in this census at all depths. There are high-quality data in a number of the inland seas that connect to these oceans, but since not *all* these seas have been covered, it was decided to include none of them and deal only with the open ocean.

The volume of the oceans, excluding the adjacent seas, was found to be 1320×10^6 km^3. This figure was calculated from the following: the sounding (corrected for the speed of sound) at each deep station for the 1°-square in which that station fell; in 1°-squares where no station fell (the vast majority), the tables of Menard and Smith (1966). These tables give the mean elevation (above or below sea level) for each 1°-square from the equator to lat. 50°, for each area of 1° of latitude by 2° of longitude from lat. 50° to lat. 70°, and for each area of 1° of latitude by 5° of longitude above lat. 70°. These tables are, of course, useless near islands and land masses where a given 1°-square may contain a considerable volume of water while its *mean* elevation is above sea level.

A rough description of how this 1320×10^6 km^3 of water is distributed is shown in figure 2.2. This is a computer-simulated three-dimensional sketch of the world water masses. The apparent elevation in this diagram is proportional to the abundance of water that exists in the ocean in each class $0.1°C \times 0.01‰$. In the warmer parts of the ocean, where coarser classes were used, each class was prorated on the basis of this smallest class, as Wright and Worthington (1970) have described. For example, the classes in the temperature range 3-4°C have dimensions $0.2°C \times 0.02‰$. If such a class contained 4800×10^3 km^3, it would be prorated at 1200×10^3 km^3. In all important respects, the methods of Wright and Worthington (1970) have been followed in this work, except that a computer program has done most of the drudgery involved in separating the water at each station into the appropriate classes.

This diagram must be viewed with a certain amount of skepticism because the data on which it was based are so sparse. For example (by the standard shown in figure 2.1), only 28.7% of the North Pacific has been surveyed. For purposes of this diagram (figure 2.2), and the subsequent figures and tables, it has been assumed that the remaining 71.3% of the North Pacific is divided into the same T-S classes. The class that contains the largest volume of water, shown by the tallest peak in figure 2.2, is found in the North Pacific. This class is defined by $T = 1.1$-$1.2°C$, $S = 34.68$-$34.69‰$. According to this census, there are 26×10^6 km^3 of water in this class, but only 7.3×10^6 km^3 were actually observed in the white areas of the North Pacific (figure 2.1) in which reliable data were available.

Nevertheless, I feel that an image of the real ocean would not differ too greatly from that shown in figure 2.2. The most serious gap in the census is in the southeastern Atlantic and the southwestern Indian Ocean, where the transition between Atlantic and Indian Ocean water masses has not been properly observed. This diagram (figure 2.2) makes graphically clear what was already numerically clear from the work of Montgomery (1958), Cochrane (1958), and Pollak (1958), namely, that nearly all the water in the oceans is cold. The most abundant fine-scale class, occupying the bottom of the North Pacific, contains more water ($26 \times$

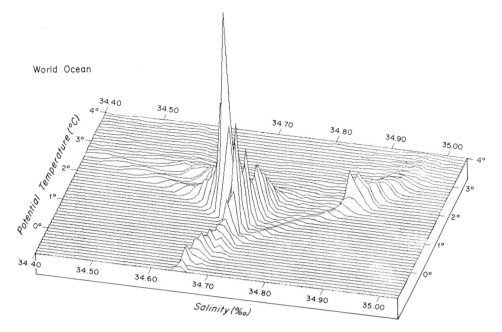

Figure 2.2 Simulated three-dimensional T-S diagram of the water masses of the world ocean. Apparent elevation is proportional to volume. Elevation of highest peak corresponds to 26.0×10^6 km³ per bivariate class $0.1°C \times 0.01‰$.

10^6 km³) than *all* the water in the world ocean warmer than 19°C (21.9×10^6 km³) and more than *all* the continental ice caps (23×10^6 km³). It is represented by the highest peak in figure 2.2. The coldest water (<0°C) is confined to the Southern Ocean (which has been defined as south of 55°S) except for a small volume in the South Atlantic that has penetrated north of 55°S. The Atlantic is the most eccentric of the oceans, standing aloof on the saline side of this diagram, connected to the Southern Ocean only by a narrow umbilicus that contains little water. The Indian and Pacific Oceans are much more similar to each other than to the Atlantic; in order to show this, the computer has drawn figures 2.3, 2.4, and 2.5 for the Pacific, Indian, and Atlantic Oceans, respectively. By referring to these figures, the reader should be able to identify, by ocean, the gross features that appear in the world-ocean diagram (figure 2.2).

Such well-known water masses as Antarctic Intermediate Water and Mediterranean Outflow Water are barely perceptible in these figures. These mid-depth water masses have little volume relative to the deep water, and moreover, their T-S characteristics are diverse, so that their volumes are spread over a wide area of figure 2.2 with little elevation.

For purposes of this census, the oceans have been subdivided: North Pacific, South Pacific, and Southern Ocean (Pacific); Indian and Southern Ocean (Indian); North Atlantic, South Atlantic and Southern Ocean (Atlantic). The Southern Ocean has also been totaled separately.

The volumes of all these subdivisions and the totals for each ocean and for the world ocean are listed in table 2.1, with the mean potential temperature and mean salinity of each. For reasons that will become clear when these results are compared with those of Montgomery, Cochrane, and Pollak, the accuracy of these means depends on the percentage of the ocean that has been surveyed. It has been assumed that all the water in the *unsurveyed* areas of each ocean listed in table 2.1 is divided into the same T-S classes as that in the surveyed areas and in the same proportion. This assumption has the effect of assigning artificially large volumes of water to classes that are found where coverage is good. In particular, the reader should be skeptical about my means for the Pacific, which has been less than one-third surveyed.

Mention should also be made of the two regions that I have designated as 100% surveyed. In the case of the Southern Ocean (Pacific) this designation is correct; the *Eltanin* stations have missed almost nothing. In the North Atlantic (figure 2.1), there are 11 5°-squares that have no reliable data, but in their census Wright and Worthington (1970) assigned these 5°-squares to neighboring stations in order to bring their total to 100% (a procedure not generally followed in this census). I have adopted the totals of Wright and Worthington (1970) uncritically.

In order to compare the new results with those of Montgomery (1958), Cochrane (1958), and Pollak (1958), it was necessary to remove the inland seas from the totals given by these authors and reaverage without them. The results of this comparison are shown in table 2.2, and the new means for the world ocean are remarkably close to the old ones, differing by only 0.03°C in temperature and 0.01‰ in salinity. Closer

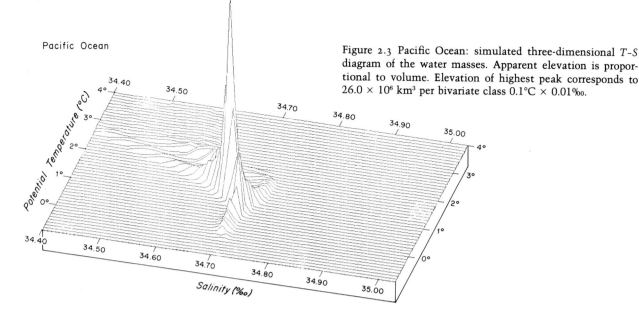

Figure 2.3 Pacific Ocean: simulated three-dimensional T-S diagram of the water masses. Apparent elevation is proportional to volume. Elevation of highest peak corresponds to 26.0×10^6 km^3 per bivariate class $0.1°C \times 0.01‰$.

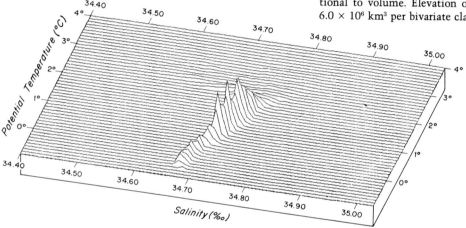

Figure 2.4 Indian Ocean: simulated three-dimensional T-S diagram of the water masses. Apparent elevation is proportional to volume. Elevation of highest peak corresponds to 6.0×10^6 km^3 per bivariate class $0.1°C \times 0.01‰$.

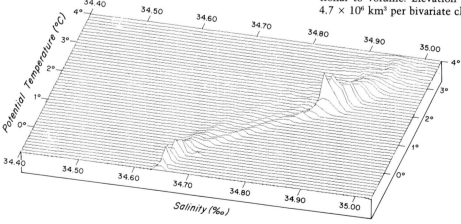

Figure 2.5 Atlantic Ocean: simulated three-dimensional T-S diagram of the water masses. Apparent elevation is proportional to volume. Elevation of highest peak corresponds to 4.7×10^6 km^3 per bivariate class $0.1°C \times 0.01‰$.

Table 2.1 Volumes (km³ × 10³) of the Oceans, Excluding Adjacent Seas

	Ocean volume	Volume surveyed	Percentage surveyed	Mean θ	Mean S (‰)
North Pacific	332,222	95,469	28.7	3.13	34.57
South Pacific	323,951	98,458	30.4	3.50	34.63
Southern Ocean (Pacific)	55,413	55,413	100.0	1.07	34.64
Pacific Ocean total	711,586	249,340	35.0	3.14	34.60
Indian	245,974	108,585	44.1	4.36	34.79
Southern Ocean (Indian)	36,730	17,163	46.7	0.72	34.66
Indian Ocean total	282,704	125,748	44.5	3.88	34.78
North Atlantic	137,222	137,222	100.0	5.08	35.09
South Atlantic	158,238	70,397	44.5	3.81	34.84
Southern Ocean (Atlantic)	30,738	24,980	81.3	0.04	34.64
Atlantic Ocean total	326,198	232,599	71.3	3.99	34.92
(Southern Ocean total)	122,881	97,556	79.4	0.71	34.65
World Ocean total	1,320,488	607,687	46.0	3.51	34.72

Table 2.2 Comparative Volumes and Means

	This census			Montgomery (1958), Cochrane (1958), and Pollak (1958), modified to exclude adjacent seas		
	Volume (km³ × 10⁶)	Mean θ (°C)	Mean S (‰)	Volume (km³ × 10⁶)	Mean θ (°C)	Mean S (‰)
North Pacific	332.2	3.13	34.57	346.1	3.39	34.59
South Pacific	379.4	3.14	34.63	376.5	3.36	34.64
Pacific	711.6	3.14	34.60	722.6	3.37	34.62
Indian	282.7	3.88	34.78	291.6	3.70	34.75
Atlantic	326.2	3.99	34.92	321.8	3.76	34.87
World Ocean	1320.5	3.51	34.72	1336.0	3.54	34.71

examination of the table will show that there is no cause for celebration of this similarity—in individual oceans, the new results differ by much larger amounts, particularly in temperature. The new-census mean temperature for the Pacific is lower than the old mean, and the new mean is higher for both the Atlantic and Indian Oceans.

Cochrane (table 2.2) has provided separate averages for the North and South Pacific, as I have, and it is instructive to examine the reasons for the discrepancy in the North Pacific (3.13°C, contrasted with 3.39°C) in some detail. The first step was to see whether Cochrane's (relatively) coarse intervals were responsible for the disagreement; it was reasoned that, in an ocean where the vertical temperature gradient generally decreases with depth, larger volumes of water would be found toward the colder side of each coarse temperature interval, resulting in artificially low averages. Accordingly, the *new-census* mean was recalculated, using Cochrane's coarse intervals. This resulted in an artificial increase in the *new* mean from 3.13°C to 3.15°C, a change in the expected direction but not nearly sufficient to account for the discrepancy. This step does suggest, however, that Montgomery, Cochrane, and Pollak may all have overestimated the mean temperatures of the oceans by a small amount due to the coarseness of their scale.

The real reason for the discrepancy can be seen in table 2.3, in which the volumes and mean temperatures found by Cochrane in four temperature ranges are compared with those of this census.

It can be seen that Cochrane's total volume is slightly larger than mine. This reflects the difference in Cochrane's depth estimates (which agree closely with those of Kossina (1921)) and those of Menard and Smith (1966). The table shows that there is little difference in the mean temperature for each of these layers, although the new census does show a higher mean in the 10–20°C layer. The main difference is that this census has counted (relatively) more cold water ($<4°C$) and much less warm water ($>20°C$). This is presumably due to the artificially small area of the North Pacific assigned to stations at *low* latitudes in this census (figure 2.1) relative to that of Cochrane (1958, his figure 1).

The conclusion to be drawn is that Cochrane's (1958) mean temperature for the North Pacific is closer to the truth than the mean obtained by this census, although it is probably very slightly too warm due to the coarse temperature interval used. By inference, the new means in other poorly surveyed oceans are also in doubt. This is regrettable since the whole object of this census was to describe, accurately, the temperature and salinity of the world ocean as they exist in the second half of this century in case this description should be useful to future oceanographers (if any) whose interests might lie in the climatic mean ocean and its changes. Nevertheless, I think I can show that this census has been worthwhile, especially in the deep ocean, by virtue of its finer scale.

Montgomery (1958) combined Cochrane's (Pacific Ocean) and Pollak's (Indian Ocean) totals in each bivariate class with his own Atlantic totals to provide a volumetric T-S diagram for the world ocean. In the deep ocean, he lists three bivariate classes (0.5°C × 0.1‰) that contain more than 100×10^6 km^3. These classes are listed in table 2.4 with the comparative values obtained by this census (given in italics below Montgomery's values).

The big difference here is that the new census has 43.3×10^6 km^3 more water in the class 1.0–1.5°C, 34.6–34.7‰, and 34.7×10^6 km^3 less in the class 1.0–1.5°C, 34.7–34.8‰. This, I believe, is almost entirely due to a disagreement between this census and Cochrane's about how much water there is in the North Pacific with a salinity greater than 34.7‰. Of Montgomery's 121.2×10^6 km^3 in the class 1.0–1.5°C, 34.7–34.8‰ (table 2.4), the North Pacific contributes 44.6×10^6 km^3, according to Cochrane, while my figures show only 4.6×10^6 km^3. In this case, I believe, my numbers are more correct than Cochrane's because his data are all from chemical titration. Nobody has claimed an accuracy better than ± 0.02‰ for salinity analyses made at sea, and experience has taught me that few of the old titrators could justifiably claim an accuracy of better than ± 0.03‰. It is easy to see how even a quantity as large as 45×10^6 km^3 in the North Pacific could be assigned to the more saline class since (as Cochrane himself has remarked) this water lies close to the borderline at 34.7‰.

My estimate of the distribution of water in the class 1.0–1.5°C, 34.6–34.7‰ is listed in table 2.5.

This table, I believe, justifies the fine scale used in this census because it shows that most of the water is concentrated in a few fine-scale classes on the salt-cold side of Montgomery's relatively coarse-scale class (0.5°C × 0.1‰). There are *small* amounts of water on

Table 2.3 Cochrane's North Pacific Volumes (km^3 × 10^6) Compared with Those of This Census in Four Layers, and the Ratio of These Volumes, with Mean Potential Temperature for Each Layer

	Total volume	<4°C	4–10°C	10–20°C	20–30°C
(A) Cochrane	346.1	274.2	48.9	15.6	7.4
(B) This census	332.2	271.1	41.8	15.0	4.3
(A)/(B)	1.04	1.01	1.17	1.04	1.72

	Mean T	Mean θ <4°	Mean θ 4–10°C	Mean θ 10–20°C	Mean θ 20–30°C
Cochrane	3.39	1.77	6.05	13.64	24.55
This census	3.15	1.76	6.08	13.93	24.48

Table 2.4 Montgomery's (1958) Volumes (km³ × 10⁶) in Three Principal Bivariate Classes of The World Ocean (Volumes from This Census in Italics)

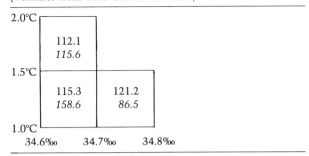

the fresh-cold side of this table, and two fine-scale classes contain no water whatsoever. The high concentrations of water that are numbered 1, 2, 4, 5, 6, and 9 fall on the T–S curve for the deep Pacific Ocean. These numbers represent their rank by volume in the world ocean. The class 1.1–1.2°C, 34.68–34.69‰, 26.0 × 10⁶ km³, is the most abundant class; it is represented, as we have seen, by the peak in figures 2.2 and 2.3. One hundred percent of this water (to the nearest percentage point) is found in the North Pacific, as is also 100% of the next most abundant class (1.2–1.3°C, 34.67–34.68‰).

The third-ranked class, which does not appear in table 2.4, is 0.7–0.8°C, 34.71–34.72‰. This is a circumpolar class, common to the South Pacific, 48%; the Southern Ocean (Pacific), 9%; the Indian Ocean, 35%; the Southern Ocean (Indian), 5%; the South Atlantic, 2%; and the Southern Ocean (Atlantic), 1%. This cosmopolitan water type is found in all the oceans except the North Pacific and the North Atlantic. The most abundant classes, those that, together, contain 50% of the world ocean (according to this census), all fall in the numbered squares (and rectangles) in figure 2.6. The numbers, 1–186, are their rank by abundance. The least of these, number 186, contains just under 1 × 10⁶ km³. All of them fall below 3°C. All these classes are also listed by rank in the appendix (23 pages at the end of this chapter), with their volumes and the percentages of these volumes that are found in each of the subdivisions of this census. Thus the reader can, for example, identify class number 80 in figure 2.6, turn to the appendix and find that it contains 2,449 × 10³ km³; he will see that it is a circumpolar class since it is found throughout the Southern Ocean (16% in the Pacific sector, 44% in the Indian Ocean sector, and 15% in the Atlantic sector), and that it is also found north of 55°S in the Indian Ocean (17%) and the South Atlantic (15%).

The classes that contain 50% of the volume of the world ocean (figure 2.6) are divided into two groups: a larger, Y-shaped group composed of the large-volume classes of the Southern Ocean, the Pacific and Indian Oceans (with some South Atlantic water present); and a smaller, lozenge-shaped group, somewhat warmer and about 0.2‰ more saline, that is exclusively Atlantic. A discussion of these two groups follows, but the reader is asked to bear in mind that the world-wide distribution of high-quality deep hydrographic data (figure 2.1) is far from satisfactory.

The deep North Pacific contains no water colder than 0.8°C, and so it is absent from the base of the Y-shape. It is the freshest of all the deep oceans, and we shall see that this relative freshness extends well up into the thermocline. It is also the most exclusive of the oceans—15 of its large-volume classes, which contain a total of 88.9 × 10⁶ km³, are shared with no other ocean. In many more, relatively fresh, classes, the contribution from other oceans is trivial. These classes occupy the fresher side of the left-hand arm of the Y-shape (figure 2.6). In the middle of this left-hand arm it shares a number of classes (e.g., 4, 5, 9, 15, and 40) with the South Pacific, but shares virtually no deep water with any other ocean.

The coldest class that contains South Pacific water is class 84; 0.2–0.3°C, 34.70–34.71‰. This class is cosmopolitan, being found in the Southern Ocean (Pacific and Indian), the Indian Ocean, and in the South Atlantic. It is lacking in the Southern Ocean (Atlantic). Considerable amounts of South Pacific water can be found in a number of circumpolar classes (3, 24, 32, 70, 89, and 94) that lie between 0.3 and 0.8°C and between

Table 2.5 Volumes (km³ × 10⁶) in the Class 1.0–1.5°C, 34.6–34.7‰ According to This Census[a]

Potential temperature (°C)	Salinity (‰)										
	34.60	34.61	34.62	34.63	34.64	34.65	34.66	34.67	34.68	34.69	34.70
1.5											
1.4			0.1	0.7	3.1	4.1	5.3	7.0	4.8	5.3	
1.3					0.5	3.0	6.1	9.8^9	5.5	3.8	
1.2						0.5	3.7	14.0^2	13.5^4	4.0	
1.1						0.1	0.5	6.0	26.0^1	12.9^5	
1.0								0.2	5.8	12.0^6	

a. Classes containing less than 0.05 × 10⁶ km³ have been omitted. Superscript numbers (1, 2, 4, 5, 6, and 9) indicate rank, by volume, in the world ocean.

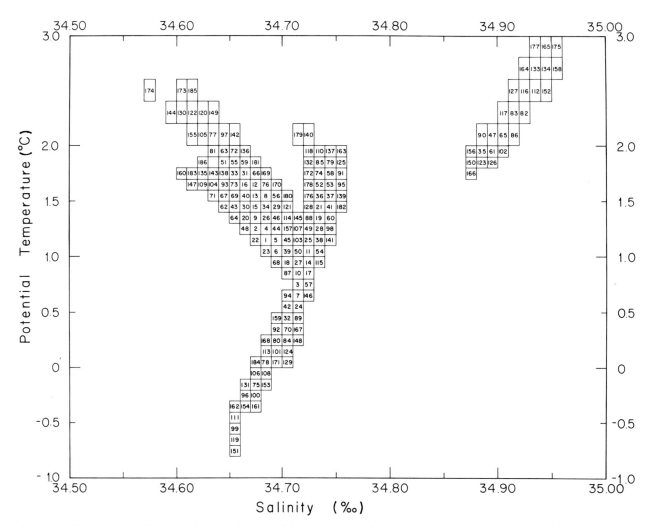

Figure 2.6 Catalog of the 186 most abundant fine-scale bivariate classes in the world ocean. These classes contain 50% of the world-ocean volume. The number in each bivariate class represents its ranking according to volume.

34.70 and 34.72‰. These circumpolar classes comprise the warmer portions of the base of the Y-shape (figure 2.6). The bulk of South Pacific water is found on the saline side of the left-hand arm of the Y-shape; the fresher side of this arm, as we have seen, is almost wholly North Pacific water. Large amounts of more saline water in the South Pacific are found on the fresher side of the right-hand arm of the Y-shape (e.g., classes 140, 118, 132, 172, 178, 76, and 128). These classes are all shared with the Indian Ocean. They represent, typically, the saline water found by Stommel et al. (1973) in the western South Pacific in the *Scorpio* sections at 28°S and 43°S. This water has entered the western South Pacific from the Indian Ocean. Traces of this more saline water are present in classes 137, 79, 58, 53, 37, 41, 60, 98, and 141. These classes lie between 1.0 and 2.0°C and between 34.74 and 34.75‰. (It is also of course found in the intermediate salinity range, 34.73-34.74‰; e.g., classes 110, 85, and 79.)

By far the greater part of the right-hand arm of the Y-shape is Indian Ocean water, although traces of South Atlantic water are found there, particularly on the saline side of the arm (e.g., classes 139, 182, 60, 98, 141, 54, and 115). Indian Ocean water (as can also be seen from figure 2.4) is represented by a nearly straight line from 2.0°C, 34.74‰ to −0.3-0.2°C, 34.66-34.68‰ (classes 154 and 161). The colder Indian Ocean classes in the base of the Y-shape are shared with the Southern Ocean classes from all the sectors.

The Atlantic Ocean *T-S* curve extends from about 3.0°C, 34.95‰ to class 166, 1.7-1.8°C, 34.87-34.88‰. All these classes (figure 2.6) are common to the North and South Atlantic; on the fresh side of the curve (e.g., classes 156, 90, 117, 127, 164, and 177), South Atlantic water predominates, and on the saline side, the North Atlantic water predominates. There are no large-volume classes that connect the Atlantic Ocean to the circumpolar oceans, but we shall see later that there is a thin group of relatively small-volume classes that does make this connection.

The coldest classes in the Southern Ocean (99, 119, and 151) belong exclusively to the Atlantic sector of that ocean, being found in the Weddell Sea. Class 111 contains a trace of South Atlantic water (north of 55°S). Class 162 (−0.4 to −0.3°C, 34.65 to 34.66‰) is exclusively in the Atlantic sector, but the two classes found on the saline side of this class (154 and 161) contain water from the Indian Ocean sector of the Southern Ocean and the Indian Ocean itself. Surprisingly, in the temperature range −0.3 to +0.3°C, the classes that are predominantly from the Atlantic sector of the Southern Ocean (classes 96, 131, 184, and 168) are found on the fresh side of the T-S curve; between 0.0 and 0.1°C (classes 184, 78, 171, and 129), what one considers the "normal" situation is exactly reversed. The Pacific sector dominates the saline side of the curve and the Atlantic sector the fresh side; 93% of class 129 is in the Pacific sector of the Southern Ocean, and 70% of class 184 is in the Atlantic sector.

There is a group of classes between 0.0 and 1.7°C that are truly circumpolar, being found in *all* sectors of the Southern Ocean. The warmer end of this group (classes 176 and 178) claim a niche on the *fresh* side of the right-hand arm of the Y-shape and contain water found in all oceans except the North Atlantic and the North Pacific. These circumpolar classes represent water carried by the Antarctic Circumpolar Current.

The water masses that embrace 50% of the world ocean (when ranked in order of volume) are diverse, owing to the widely different locations in which they have been formed and, presumably, to slow changes that have taken place in these water masses since their formation as the result of mixing with their neighbors. For example, all the relatively fresh North Pacific water (figure 2.6) must have been formed near the periphery of Antarctica, but the South Pacific water masses found in the warm half of the base of the Y-shape, which intervene between the North Pacific and the Southern Ocean (Pacific), are far more saline at the present time. There is thus no connection, at present, between the circumpolar water masses and the vast volumes of deep North Pacific water. The relatively low salinity of this North Pacific deep water is probably the result of slow mixing with the even fresher waters that lie above them. We have no conception of the time scale of such a mixing process, but I would hazard a guess that it is of the order of centuries or, conceivably, millennia.

By contrast with the world ocean (figure 2.6), the North Atlantic is far less diverse; it takes 186 fine-scale classes (ranked in order of volume) to contain 50% of the world ocean (figure 2.6), but Wright and Worthington (1970) show that 50% of the North Atlantic is contained in only 43 classes, 75% in 157 classes, and 90% in 295—in spite of the fact that they identify five separate sources of deep water for the North Atlantic.

The Sea of Japan is probably the least diverse of any considerable *inland* sea (this sea has not been included in this census, but its water masses have been tallied). The Sea of Japan contains $1{,}590 \times 10^3$ km³; 50% of it is found in only three fine-scale classes, 75% in 7 classes, and 90% in 28 classes. The sea surface within the Sea of Japan is its sole source of deep and bottom water, and this is undoubtedly the cause of its homogeneity.

The process of ranking the fine-scale classes of the world ocean for this census has been carried out until 75% of its volume was classified. To carry this process out to 90% (as Wright and Worthington (1970) have done for the North Atlantic) would not only be excessively time consuming but also unjustified by the distribution of reliable deep hydrographic data (figure 2.1). The classes that make up the third quarter of the volume of the world ocean are shown in figure 2.7 (included in the pocket at the back of the book). The first half of the ocean, the classes that are ranked by number in figure 2.6, are represented by the blacked-out areas in figure 2.7. One should always remember that the volume of water represented by these blacked-out classes is *double* that of the remainder of the water represented by the numbered classes in figure 2.7; these latter classes (187–693) are also listed in the appendix. Their volume diminishes with rank from class 187 (975×10^3 km³) to class 693 (192×10^3 km³). The *pro rata* volume of class 693 is 96×10^3 km³. As the volume of these classes diminishes, their diversity increases; 50% of the world ocean is in only 186 classes, while the next 25% requires 507 classes.

The T-S diagram for these 507 classes (figure 2.7) again resembles a primitively drawn, lopsided Y-shape. Roughly speaking, the Pacific Ocean dominates the fresher side of the Y-shape, and the Atlantic the more saline side. Indian Ocean water is found in both arms of the Y-shape, but its greater volume is found in the center, where the base and the arms of the Y-shape are joined.

In the temperature range 0.9–3.6°C, North Pacific water dominates the fresher edge of the Y-shape. From class 332 (0.9–1.0°C, 34.68–34.69‰) to class 627 (3.4–3.6°C, 34.24–34.26‰), all classes on this edge contain at least 50% North Pacific water. In the warmer part of the fresh edge, above 3.0°C, considerable volumes of (eastern) South Pacific and South Atlantic water are found. Virtually all the classes between 3.0 and 4.0°C and between 34.30 and 34.64‰ are shared among the North Pacific, the South Pacific, the Indian Ocean, and the South Atlantic; the exceptions are classes 673 and 565 (3.0–3.2°C, 34.30–34.34‰), which contain no Indian Ocean water. These classes represent an unnamed water mass of considerable volume that lies below the Antarctic Intermediate Water and above the Antarctic Bottom Water in the South Pacific and Indian Oceans, and between the Antarctic Intermediate Water and the

North Atlantic Deep Water in the South Atlantic. They are also found south of the southern limit of North Atlantic Deep Water. In the North Pacific they fall between the North Pacific Intermediate Water and the bottom water. This unnamed water mass presumably originates south of the Antarctic Polar Front. A typical class from the middle of this group is class 573—3.4-3.6°C, 34.44-34.46‰. It contains 594 × 10³ km³ of water: 31% in the North Pacific; 31% in the South Pacific; 17% in the Indian Ocean; and 21% in the South Atlantic.

There are only 10 classes above 4°C that are included in the third quarter of the world-ocean volume. These 10 classes are relatively coarse scale, each of them having the equivalent area on the T-S diagram of 25 of the finest-scale classes (those below 2°C). The combined volume is considerable—28,723 × 10³ km³; this is slightly more than the volume of the highest-ranking class (25,973 × 10³ km³), but the *pro rata* volume of these classes, that is, the volume contained in each 0.1°C × 0.01‰ bivariate fine-scale class is quite small. Class 586, the richest of these classes, contains 3,574 × 10³ km³, but its *pro rata* volume is only 143 × 10³ km³. Four of the saltier of these classes (586, 609, 631, and 677) are cosmopolitan, being found in all five of the major oceans, north of 55°S. These represent the Antarctic Intermediate Water, which is formed north of the South Polar Front and pushes across the equator into the North Pacific and the North Atlantic; these four classes are the only ones shared by these northern hemisphere oceans. Wright and Worthington (1970) show that none of this water penetrates north of 15°N in the North Atlantic. The South Pacific dominates the remaining five classes (688, 659, 669, 680, and 619).

The Indian Ocean occupies the center of the Y-shape (figure 2.7). The classes in the square representing the temperature range 2.0-3.0°C and the salinity range 34.70-34.80‰ are predominantly Indian Ocean classes, but in the warmer part of this square considerable volumes of South Atlantic water are present. For example, class 493 (2.6-2.8°C, 34.75-34.76‰; 392 × 10³ km³) contains 60% Indian Ocean water and 40% South Atlantic water. We have seen above that Indian Ocean water is found in nearly all the classes between 3 and 4°C and 34.30 and 34.64‰, in the left-hand arm of the Y-shape. Indian Ocean water is also found on the fresher side of the right-hand arm of the Y-shape (e.g., classes 665, 472, 676, 530, 553, 660, 516, and 657). These classes contain the most saline water found in the Indian Ocean between 3.2 and 4.0°C. This high salinity results from the contribution of saline overflow water from the Red Sea.

A closer examination of one of these classes (665) is instructive. This class contains 436 × 10³ km³; 37% of this volume is found in the Indian Ocean. Its *high* salinity relative to the rest of that ocean is the result of the contribution from the Red Sea outflow. The North Atlantic claims 39% of this class, and examination of the atlas of Wright and Worthington (1970) reveals that 100% of this North Atlantic water (172 × 10³ km³) is found in the Labrador Basin, where its *low* salinity is the result of the formation of Labrador Sea water, the freshest of the deep-water masses in that ocean. This illustrates how water masses of widely different origin that never connect geographically with each other can occupy the same class in the T-S diagram. The remaining 24% of this class is South Atlantic water that is also isolated from Red Sea water by water from the southern Indian Ocean that contains no perceptible fraction of Red Sea Outflow water.

The right-hand arm of the Y-shape (figure 2.7) is composed entirely of Atlantic water except for the few Indian Ocean classes that have just been discussed. As one would expect, the North Atlantic dominates the more saline side of this arm and the South Atlantic the fresher side. The coldest class in figure 2.7 that contains South Atlantic water is class 555 (0.0-0.1°C, 34.66-34.67‰). We have seen, however, that in the first 50% of the ocean (figure 2.6) traces of South Atlantic water colder than −0.4°C can be found.

The Atlantic T-S diagram extends from the coldest water in the Southern Ocean sector (classes 272 and 247) in a more-or-less straight line through the South Atlantic classes up to the large classes shown by the more saline blacked-out area (figure 2.7) that represents the North Atlantic Deep Water. All water on this curve is usually named the Antarctic Bottom Water. The warmest classes on this part of the T-S curve, those between 1.3 and 1.7°C, are common to the North and South Atlantic; the more saline of these warmer classes are found in the North Atlantic, and the fresher classes in the South Atlantic. Traces of Antarctic Bottom water as cold as 0.6°C (class 286: 0.6-0.7°C, 34.75-34.76‰) can be found in the North Atlantic, but these traces do not enter the deep North American Basin, according to Worthington and Wright (1970), but remain close to the equator. No classes in the Southern Ocean (Atlantic) can be traced to the North Atlantic. The "Antarctic Bottom Water" classes that are found in the North Atlantic are composed chiefly of the North Atlantic Deep Water with a very small amount of *true* Antarctic Bottom water mixed in. In other words, no "Antarctic Bottom Water" in the North Atlantic can be traced, unchanged, to its point of origin in the Weddell Sea. On the other hand, many North Atlantic Deep Water classes can be traced, unchanged, from their origin as overflow water from the Norwegian Sea all the way down into the South Atlantic. This argues either for a more rapid production rate for the North Atlantic Deep Water or (if production of both water masses is equal) for a higher mixing rate along the $T\text{-}S$ curve that represents the Antarctic Bottom Water.

In the Southern Ocean, the extended classification of figure 2.7 includes a number of classes not represented in figure 2.6. These less abundant classes confirm the curious fact that the "normal" distribution of salinity is reversed; the Atlantic sector of the Southern Ocean contains the freshest water (e.g., class 316: $-0.3\text{-}-0.2°C$, 34.65-34.66‰) and the Pacific sector the most saline (e.g., class 513: $-0.3\text{-}-0.2°C$, 34.71-34.72‰).

In general, expanding the description from 50% to 75% of the ocean results in the addition of few cold, deep classes, compared to the greater number of warmer classes. I believe that if we all made perfect observations of temperature and salinity, the number of deep classes would be even fewer than those listed in the appendix. The ocean is probably more finely stratified than we can observe at the present time.

One is tempted to play endless games with figures 2.6 and 2.7 and the appendix, and the reader is, of course, invited to do so, but, again, he should bear in mind that the coverage (figure 2.1) is really quite poor and the real rankings may be somewhat different from those listed in the appendix.

In order that the less abundant classes should not be wholly slighted, figures 2.8 and 2.9 have been constructed. These are contoured volumetric $T\text{-}S$ diagrams for the cold water ($<4°C$), and the thermocline and warm water ($>4°C$), respectively. The crosshatched area ($<4°C$) attached to the warm-water diagram (figure 2.9) encloses the cold water that is contoured in detail in figure 2.8. Each contour represents the volume in $km^3 \times 10^3$ that is found in each $0.1°C \times 0.01‰$ class, and the contours mapped are $2^n \times 10^5 \; km^3$, where $n = -2, -1, 0, 1, 2, \ldots, 8$; this last contour contains only the winning class from the North Pacific. Contours of 5 and $10 \times 10^3 \; km^3$ have been added in figure 2.9 because otherwise the thermocline and warm-water masses of the oceans would scarcely appear at all.

For example, the $5 \times 10^3\text{-}km^3$ contour in the upper right-hand corner of figure 2.9 contains two classes that comprise the 18° Water in the North Atlantic (Worthington, 1959). These classes are 17-18°C, 36.4-36.5‰, $827 \times 10^3 \; km^3$, and 18-19°C, 36.5-36.6‰, $556 \times 10^3 \; km^3$; there are 100 fine-scale classes in each of these relatively coarse-scale warm-water classes, so the *pro rata* volumes of these two classes are 8.3×10^3 and $5.6 \times 10^3 \; km^3$, respectively.

The analogy between these figures and a physical relief map is difficult to avoid because the dominant water masses in this presentation (and in figures 2.2-2.5) resemble mountain ranges. The highest of these ranges (below 34.8‰) is roughly Y-shaped. The left-hand branch of the Y-shape is composed predominantly of Pacific water (in which the highest peak is found),

and the right-hand branch of the Y-shape contains mostly South Pacific and Indian Ocean water. The far lower range centered on 2.0°C, 34.89‰ is the Atlantic. The highest peak in this range contains only $4659 \times 10^3 \; km^3$. The warmer extensions of these ranges fall off rapidly into what might be analogous to a coastal peneplain—however, no prominent monadnocks appear. An isolated elevation of $100 \times 10^3 \; km^3$ per class occurs near 6°C, 34.35‰. This represents part of the Subantarctic Mode Water described by McCartney (1977). The warm extension of this elevation ($>50 \times 10^3 \; km^3$ per class) toward the northeast also consists of Subantarctic Mode Water, mostly in the South Pacific and in the Indian Ocean.

The warmest water ($>10°C$) is divided into three low prongs. The freshest of these prongs consists principally of North Pacific water, but there is a considerable contribution from the eastern South Pacific. The central prong is cosmopolitan—it contains water formed in the subtropical convergences of the southern hemisphere. The saline side of the central prong consists principally of South Atlantic and Indian Ocean water, but a considerable volume was found in the Guiana and Guinea Basins of the North Atlantic by Wright and Worthington (1970). In the center of the central prong, South Atlantic, South Pacific, and Indian Ocean waters are found in roughly equal proportions. On the fresh side of the central prong, South Pacific water predominates, but there is a considerable volume of water from the subtropical North Pacific. The saline prong is composed of Western North Atlantic Water, but in temperature intervals below 13°C the Red Sea water in the Indian Ocean claims a significant proportion of the water ($>25\%$).

The contour of $5 \times 10^3 \; km^3$ per class encloses two warm islands isolated from the main prongs; one of these, an extension of the saline prong, comprises the 18° Water in the North Atlantic, as we have seen. The other, an extension of the central prong, is composed mainly of South Pacific water. This one is probably artificial because much of the surface mixed layer on the western half of the *Scorpio* section at 28°S was between 19 and 20°C and between 35.6 and 35.8‰, according to Stommel et al. (1973, plates 1 and 2). Undue weight was given to this section on account of the shortage of other high-precision hydrographic data in the South Pacific.

The thermocline and warm water-mass parts of this census, as they are illustrated here (figure 2.9), are merely a sketch, and no really reliable quantitative estimates of the volumes of these water masses in the world ocean exist. The reader is referred to Wyrtki's (1971, pp. 526-527) admirable volumetric diagrams for the Indian Ocean and to those of Wright and Worthington (1970) for the North Atlantic. For the other oceans, the warm-water diagrams (coarse scale) of Montgomery

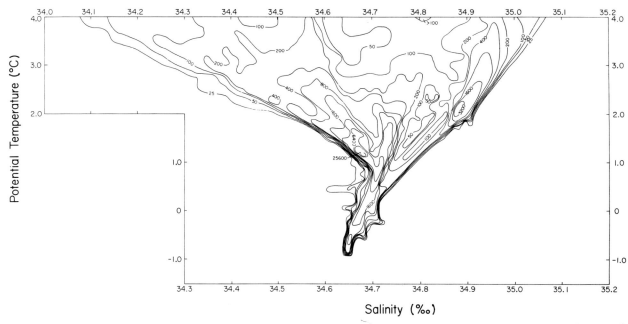

Figure 2.8 Contoured volumetric T-S diagram for the cold (<4°C) water in the world ocean. The values are volumes (km^3 × 10^3) that are found in each class of 0.1°C × 0.01‰.

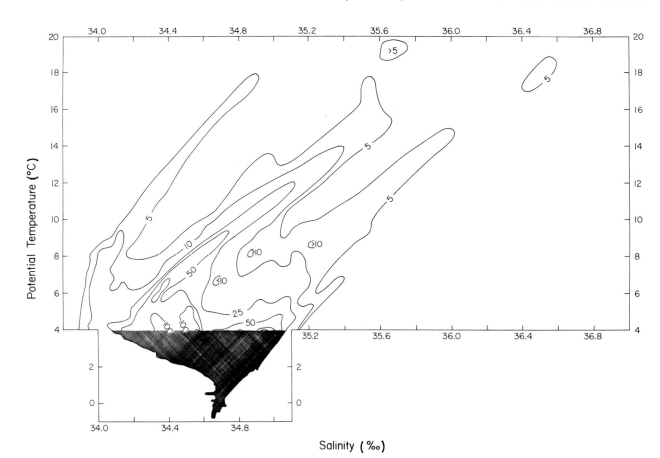

Figure 2.9 Contoured volumetric T-S diagram for the thermocline (>4°C) and warm (>16°C) waters of the world ocean. The values are the volumes (km^3 × 10^3) that are found in each class of 0.1°C × 0.01‰. Shaded area encloses the cold-water classes that appear in figure 2.8.

(1958), Cochrane (1958), and Pollak (1958) still stand, but I believe that their classes are too large. One can see that by merging the present classes into divisions as large as 0.5‰ of salinity, as Cochrane and Pollak have done, or 1.0‰, as Montgomery has done, the high-volume warm prongs (figure 2.9) would be artificially blurred and merged with one another.

One or two colleagues have asked whether I could not use a logarithmic volume scale in such presentations as figures 2.2-2.5 so that the warm water masses (if included) could be made to stand out more clearly, but one of the principal virtues of the volumetric T-S diagram is that it displays the relative abundances of the water masses as they actually exist. The *concentration* of water in the most abundant North Pacific class exceeds that in the warm-water prongs (shown in figure 2.9) by a ratio of about 25,973 to 10 or less. This is analogous to comparing the elevation of Mount Everest to that of Water Street, Woods Hole, near the original building of the Woods Hole Oceanographic Institution. In fact, if we were able to sample and measure salinity more perfectly, the apparent elevations shown in the deep water in figure 2.8 would probably be even higher.

The feelings I have about the census are compounded equally of fascination and frustration. The frustration is the result of the decrease in the rate of acquisition of new high-quality data. This decrease is due in part to the trends in modern physical oceanography in which the dramatic improvements in direct current measurements have understandably taken priority over routine measurements of water properties on a large scale. It is also clear that there is a long delay (as much as 5 years) between the time hydrographic data are obtained at sea and the time these data become available on tape from the National Oceanographic Data Center (in part because some investigators take a long time to turn their data in to the Data Center). I have been reluctant to obtain new data informally, from friendly colleagues, however, because I do not think that the Data Center should be bypassed at present; its function would be impaired if data were only exchanged between a cabal of skilled observers.

The fascination results from the precise but peculiar way in which the water masses of the oceans are arranged—particularly the deep water masses that make up the greater part of the oceans. Why, for instance, are the big, exclusive North Pacific classes fresher than existing circumpolar and South Pacific waters? Are they fossil water masses that were formed in some past millennium when the oceans were somewhat fresher, or are they still undergoing a change toward the fresher as the result of slow vertical mixing (across density surfaces) with the still fresher water that lies above them at the present time? I do not think that we can supply answers to such questions at present, and answers will not be available even in the future without painstaking observations. There are indications this style of observations may be coming back into vogue. The authorless Scripps data report of the INDOPAC expedition (Scripps Institution of Oceanography Reference: 78-21) is an excellent example. It should be worthwhile to reactivate this census (which was closed as of June 1977) when more high-quality data of this kind are available from NODC, and I shall probably propose to do so at some time in the future.

2.4 The Formation of Water Masses

There is only one hypothesis about water-mass formation that is universally agreed upon, that is, that the cold, dense water that fills the great ocean basins has been formed at high latitudes. The manner in which the thermocline-halocline is formed is under dispute, and there are almost as many notions of the *rate* at which all the various water masses are formed as there are investigators.

Given the extraordinary regularity of the T-S curves that are found in much of the oceans, it is natural to assume that these curves are the result of vertical mixing between two end water masses. Very simply stated, this assumption implies that the bottom water (as all agree) has been formed at high latitudes, that the surface water at middle and low latitudes has received its T-S characteristics from the atmosphere by the uneven processes of evaporation and heating, and that the remainder of the water column is a mixture of surface and bottom water. Wüst (1935) clearly recognized that this was an oversimplification, and his use of the "core-layer" method reflects his conviction that different water masses can be traced to a small number of more-or-less point sources at the sea surface over a wide range of latitude.

The notion that *all* the thermocline water masses can be traced to the sea surface is generally attributed to Iselin (1939a). He constructed a T-S diagram from winter observations at the surface of the western North Atlantic, and found that it corresponded closely to the T-S diagram obtained from a typical hydrographic station in that ocean. It is worth noting that Wüst (1935, p. 3) anticipated Iselin (in the South Atlantic) by 4 years. He wrote, "The vertical structure of the Subantarctic Intermediate Water, with its horizontal spreading at depths, is analogous to a vertical figure of the horizontal arrangement of temperature and salinity at the surface of the formation region." Wüst did not dwell on this subject further, and it is clear that he regarded core layers as more important as indices of ocean circulation.

In his 1939a paper, Iselin stressed "lateral mixing" as responsible for the T-S curve in the western North Atlantic. Sverdrup, in chapter XV of *The Oceans* (Sverdrup, Johnson, and Fleming, 1942) amplified Iselin's concept; he suggested that "subtropical convergences" were the dominant source of the waters in the thermocline-halocline. In these convergences, according to Sverdrup, surface water sinks, in late winter, over a wide range of latitude. He compared late-winter, sea-surface T–S points to the T–S curves obtained from subsurface hydrographic data and found a close correspondence in the south Indian Ocean, the eastern and western South Pacific, and the western North Pacific, just as Iselin (1939a) had done for the North Atlantic.

In this type of water-mass formation, very little change takes place in each individual water type—when 10°C water outcrops at the sea surface, it sinks again at 10°C or nearly so, not at 4°C. In vertical-mixing models, the water types are constantly changing. In the most violent of these, Stommel (1958) held that bottom water was produced at two sinks, the Weddell Sea in the south and the Irminger Sea in the north. He theorized that bottom water was produced constantly and moved upward through the thermocline. The thermocline in this theory was maintained by a downward diffusion of heat that balanced the upward advection of cold bottom water. Later, Stommel and Arons (1960b) produced a simple schematic model in which 20×10^6 m^3 s^{-1} was formed in each sink and flowed equatorward in the form of deep western boundary currents. Subsequently, this water upwelled through the thermocline (as outlined in Stommel's earlier paper) with an upward velocity of about 4 m yr^{-1}.

Cooper (1955a) refocused attention on the Norwegian Sea overflows as a source of the North Atlantic Deep Water, and this paper touched off a series of investigations into these overflows, using newly developed methods of direct current measurement such as the Swallow float (Swallow, 1955). I (1970) have summarized some of the earlier investigations (see also chapter 1, this volume). In essence, it was evident that if cold bottom water does in fact flow across the sills of the Norwegian Sea into the deep Atlantic Ocean, relatively warm surface water must be drawn into the Norwegian Sea to replace it. It was possible to write water and heat budgets for the Norwegian Sea; the cold outflows (9×10^6 m^3 s^{-1}) transported 63×10^{12} less calories per second than the warm inflows (also 9×10^6 m^3 s^{-1}). This excess heat must go to the atmosphere, and the calculated heat loss from the oceanographic data, 75 kcal cm^{-2} yr^{-1}, corresponded pretty well with Budyko's (1963) calculated heat loss from meteorological data—about 60 kcal cm^{-2} yr^{-1}.

Of the 9×10^6 m^3 s^{-1} of cold water flowing out of the Norwegian Sea, 3×10^6 m^3 s^{-1} is embodied in the shallow, fresh East Greenland Current. The remaining 6×10^6 m^3 s^{-1} is dense overflow water; the overflows entrain a further 4×10^6 m^3 s^{-1} of Atlantic water at their sills and a total of 10×10^6 m^3 s^{-1} flows southward along the western side of the North Atlantic as a narrow deep western boundary current. The fact that this current originates in the Norwegian Sea and not in the Irminger Sea does not affect the theory, but it is naturally of vital interest to the descriptive oceanographer.

Thus the formation of North Atlantic Deep Water follows a classic pattern—cold, dense water is formed at high latitudes, and warm surface water flows poleward to replace it. In this pattern, the process of water-mass formation changes the water characteristics radically. Much of the replacement water must come from the tropical South Atlantic since the North Atlantic Deep Water flows into the South Atlantic at a rate of 9×10^6 m^3 s^{-1} according to Sverdrup et al. (1942), or 7×10^6 m^3 s^{-1} according to my box model (Worthington, 1976, figure 11). As this tropical surface water flows north, it gradually transfers heat to the atmosphere and undergoes enormous changes in its T–S characteristics before it finally becomes North Atlantic Deep Water.

This pattern is clearly consistent with the Stommel (1958) theory since upwelling of the North Atlantic Deep Water must take place—provided that the volume of the North Atlantic Deep Water remains constant. One could conceive of a water mass that refuses to mix with its neighbors—in this case the Antarctic Intermediate Water, the Circumpolar Deep Water, and the Antarctic Bottom Water—but, instead, pushes them bodily aside, preserving the purity of its original T–S characteristics and increasing its own territory. In the case of the North Atlantic Deep Water, it seems unlikely that such behavior has taken place. The area of the Atlantic Ocean is 82.4×10^6 km^2. The North Atlantic Deep Water is found throughout two-thirds of this area. Its volume is 89×10^6 km^3—62×10^6 km^3 in the North Atlantic according to Wright and Worthington (1970) by their definition, and the remaining 27×10^6 km^3 in the South Atlantic according to this census (same definition). The mean thickness of the North Atlantic Deep Water is thus about 1500 m. The width of the South Atlantic at the southern limit of more or less pure North Atlantic Deep Water (35°S) is 6400 km. To advance 1° of latitude (110 km) southward, the North Atlantic Deep Water would have to increase in volume by 1.06×10^6 km^3. If 7×10^6 m^3 s^{-1} cross the equator, as I have suggested, this flux should result in an advance of 1° of latitude every 4.5 years if no upwelling were taking place. Since the Meteor expedition was made, over 50 years ago, the southern limit of the North Atlantic Deep Water would have had to advance 11° of latitude; it plainly has not done so. We should, however, be cautious even in crude calculations like these, since our numbers for the amount of water formed may not be even approximately correct. For example, most of the measurements made in the deep, dense overflows from the Norwegian Sea were made in the 1960s—a cold decade in the northern North Atlantic relative to the decades that preceded it (Colebrook, 1976). This could easily have resulted in an overestimation of the climatic mean flow of the dense overflows.

We should probably not regard the Iselin (1939a) theory of the thermocline-halocline formation as exclusive of the Stommel (1958) theory, or vice versa. The evidence that water sinks along density surfaces that have outcropped at the sea surface seems inescapable, yet there is equally strong evidence that upwelling of deep water through the thermocline must take place, at least in the Atlantic.

Recently (Worthington, 1977a), I have questioned the Stommel-Arons (1960b) model as it applies to the Pacific Ocean. The argument is a simple one; the deep water in the Pacific contains not less than 7 g of dissolved silicon per cubic meter, and the upper layers are very nearly silicon-free. If the deep water (formed in the Weddell Sea) upwells through the thermocline throughout the Pacific and flows southward in the upper layers, two processes must take place. First, the dissolved silicon must be removed from the deep water before it can return southward to the formation area, free of silicon. Second, an equal, or nearly equal, amount of silicon must be replaced before the surface water sinks below 2000 m in the Weddell Sea. The removal and replacement rate necessary to accommodate the model can be calculated to be about 30×10^{14} g of silicon per year. I believe that to ask the ocean to perform feats like this is unreasonable and that the most logical conclusion to be drawn is that the deep waters in the Pacific are not being renewed at the present time. One could guess that they might not be renewed until the time when the rate of accumulation of ice on the Antarctic continent begins to exceed that of ablation and runoff.

The *surface* waters around Antarctica are fairly rich in silicon, as are those of the northern North Pacific. Reid (1973b) has attributed this situation in the North Pacific to upwelling, and this also is the most reasonable explanation for the state of affairs around Antarctica. Nutrient-rich upwelled water, augmented by runoff from Antarctica, flows equatorward until the amount of sunlight is sufficient to cause photosynthesis; then the silicon (and other nutrients) is immediately used for plant growth—consistent with the rich biota in the zone around Antarctica and with the high rate of accumulation of siliceous sediments beneath this zone (Lisitzin, 1972).

By comparison, the North Atlantic surface waters (except in the coastal zones) are a desert because they consist of nutrient-stripped water from the middle and low latitudes, and Metcalf (1969) has shown that newly formed North Atlantic Deep Water is also silicon poor. The subject clearly invites further investigation, but, as a first approximation, one might suppose that in silicon-poor oceans such as the Atlantic, the Stommel (1958) theory holds true, but in silicon-rich oceans, such as the Pacific and Indian Oceans, the process of deep-water formation has been suspended and the Stommel mechanism awaits application for the next cold, climatic variation in the southern hemisphere. If such a cold variation should take place, and if the vast store of nutrients that has (I believe) been accumulating in the deep Pacific through the years should be brought up into the sunlight, the results, in terms of biological productivity, might be staggering.

If this hypothesis is even approximately correct, we are faced with the problem of how the thermocline-halocline is maintained in the Pacific and Indian Oceans. A clue to this has been provided by McCartney (1977). He has identified a family of water types in the South Atlantic, South Pacific, and Indian Oceans that he has termed Subantarctic Mode Water. These layers appear to originate by deep convective overturning immediately equatorward of the Antarctic Circumpolar Current. They can be identified at great distances from their formation regions by thermostads with T-S characteristics identical to those of the deep overturned waters, and usually by an oxygen maximum in these thermostads. These layers are strongly reminiscent of the 18°-Water thermostad formed south of the Gulf Stream (Worthington, 1959) and of the Subtropical Mode Water (Masuzawa, 1969) south of the Kuroshio and the Kuroshio Extension. They differ from 18° Water and the Subtropical Mode Water in that they are found at much greater distances (up to 3000 km) from their points of origin and that they occur over a wider range of temperature (14-4°C).

Although the matter is still under investigation (McCartney, 1980), it appears that most of the southern hemisphere thermocline waters (in terms of volume) originate in these thermostadal layers by means of convective overturn. Certainly, all the southern, high-volume classes between 4 and 14°C that appear at the base of the central prong in figure 2.9 can be traced to the deep mixed layers described by McCartney (1977).

Water-mass formation is a complicated process. There seems to have been no major hypothesis on the subject that can be entirely accepted or entirely rejected. The earth's geography and the atmospheric circulation encourage the production of a wild variety of water masses, and evidently these water masses are formed in a number of different ways. I have attempted to describe some of these. The study of water-mass formation is intimately tied to the study of the general ocean circulation. The most promising method for the future will be to combine these two studies by means of direct current measurements and the rigorous observation of the distribution of variables in the oceans and of the atmospheric processes that bring about these distributions.

Appendix: Census of World-Ocean Water Masses with Division by Bivariate (°C × ‰) Classes and Rank by Volume

Rank	T°	S, ‰	Volume	Total Volume	N. Pac.	S. Pac.	So. O. Pac.	Ind.	So. O. Ind.	N. Atl.	S. Atl.	So. O. Atl.
1	1.1- 1.2	34.68-34.69	25,973	25,973	100	-	-	-	-	-	-	-
2	1.2- 1.3	34.67-34.68	14,010	39,983	100	-	-	-	-	-	-	-
3	0.7- 0.8	34.71-34.72	13,571	53,554	-	48	9	35	5	-	2	1
4	1.2- 1.3	34.68-34.69	13,494	67,048	91	9	-	-	-	-	-	-
5	1.1- 1.2	34.69-34.70	12,905	79,953	82	18	-	-	-	-	-	-
6	1.0- 1.1	34.69-34.70	11,988	91,941	95	5	-	-	-	-	-	-
7	0.6- 0.7	34.71-34.72	11,930	103,871	-	55	12	24	6	-	2	1
8	1.5- 1.6	34.68-34.69	10,337	114,208	5	95	-	-	-	-	-	-
9	1.3- 1.4	34.67-34.68	9,826	124,034	99	1	-	-	-	-	-	-
10	0.8- 0.9	34.71-34.72	9,607	133,641	-	52	8	34	3	-	2	1
11	1.0- 1.1	34.72-34.73	9,252	142,893	-	24	9	63	2	-	1	1
12	1.6- 1.7	34.67-34.68	8,836	151,729	16	84	-	-	-	-	-	-
13	1.5- 1.6	34.67-34.68	7,881	159,610	22	78	-	-	-	-	-	-
14	0.9- 1.0	34.72-34.73	7,618	167,228	-	26	15	49	7	-	2	1
15	1.4- 1.5	34.67-34.68	7,027	174,255	49	51	-	-	-	-	-	-
16	1.6- 1.7	34.66-34.67	6,627	180,882	23	77	-	-	-	-	-	-
17	0.8- 0.9	34.72-34.73	6,487	187,369	-	28	17	40	11	-	3	1
18	0.9- 1.0	34.70-34.71	6,434	193,803	68	28	-	2	1	-	-	1
19	1.3- 1.4	34.73-34.74	6,387	200,190	-	10	9	78	2	-	-	1
20	1.3- 1.4	34.66-34.67	6,052	206,242	100	-	-	-	-	-	-	-
21	1.4- 1.5	34.73-34.74	6,030	212,272	-	10	7	81	2	-	-	-
22	1.1- 1.2	34.67-34.68	6,024	218,296	100	-	-	-	-	-	-	-
23	1.0- 1.1	34.68-34.69	5,844	224,140	100	-	-	-	-	-	-	-
24	0.5- 0.6	34.71-34.72	5,676	229,816	-	29	40	21	6	-	3	1
25	1.1- 1.2	34.72-34.73	5,497	235,313	-	23	12	61	1	-	1	2
26	1.3- 1.4	34.68-34.69	5,473	240,786	38	62	-	-	-	-	-	-
27	0.9- 1.0	34.71-34.72	5,382	246,168	-	60	9	27	1	-	2	1
28	1.2- 1.3	34.73-34.74	5,328	251,496	-	14	14	65	5	-	1	1
29	1.4- 1.5	34.69-34.70	5,322	256,818	-	100	-	-	-	-	-	-
30	1.4- 1.5	34.66-34.67	5,301	262,119	100	-	-	-	-	-	-	-
31	1.7- 1.8	34.66-34.67	5,065	267,184	19	80	-	-	1	-	-	-
32	0.4- 0.5	34.70-34.71	4,906	272,090	-	28	27	21	15	-	5	4
33	1.7- 1.8	34.65-34.66	4,844	276,934	34	65	-	-	1	-	-	-
34	1.4- 1.5	34.68-34.69	4,760	281,694	4	96	-	-	-	-	-	-
35	1.9- 2.0	34.88-34.89	4,659	286,353	-	-	-	-	-	19	81	-
36	1.5- 1.6	34.73-34.74	4,640	290,993	-	11	8	78	3	-	-	-
37	1.5- 1.6	34.74-34.75	4,550	295,543	-	11	10	72	7	-	-	-
38	1.1- 1.2	34.73-34.74	4,471	300,014	-	16	17	52	12	-	3	-
39	1.0- 1.1	34.70-34.71	4,384	304,398	58	39	1	-	-	-	-	2
40	1.5- 1.6	34.66-34.67	4,349	308,747	74	26	-	-	-	-	-	-
41	1.4- 1.5	34.74-34.75	4,337	313,084	-	12	11	65	11	-	1	-
42	0.5- 0.6	34.70-34.71	4,134	317,218	-	25	21	29	16	-	5	4
43	1.4- 1.5	34.65-34.66	4,065	321,283	100	-	-	-	-	-	-	-
44	1.2- 1.3	34.69-34.70	4,048	325,331	34	64	1	-	1	-	-	-
45	1.1- 1.2	34.70-34.71	3,843	329,174	53	44	1	-	1	-	-	1
46	1.3- 1.4	34.69-34.70	3,801	332,975	-	100	-	-	-	-	-	-
47	2.0- 2.2	34.89-34.90	7,579	340,554	-	-	-	-	-	21	79	-
48	1.2- 1.3	34.66-34.67	3,717	344,271	100	-	-	-	-	-	-	-
49	1.2- 1.3	34.72-34.73	3,655	347,926	-	20	14	62	1	-	1	2
50	1.0- 1.1	34.71-34.72	3,621	351,547	-	71	9	15	2	-	1	2
51	1.8- 1.9	34.64-34.65	3,566	355,113	43	56	1	-	-	-	-	-
52	1.6- 1.7	34.73-34.74	3,536	358,649	-	15	10	72	3	-	-	-
53	1.6- 1.7	34.74-34.75	3,504	362,153	-	12	10	72	6	-	-	-
54	1.0- 1.1	34.73-34.74	3,465	365,618	-	11	21	45	17	-	6	-
55	1.8- 1.9	34.65-34.66	3,404	369,022	14	85	1	-	-	-	-	-
56	1.5- 1.6	34.69-34.70	3,371	372,393	-	98	1	-	-	-	-	1
57	0.7- 0.8	34.72-34.73	3,330	375,723	-	9	22	52	11	-	6	-
58	1.7- 1.8	34.74-34.75	3,239	378,962	-	14	8	69	9	-	-	-
59	1.8- 1.9	34.66-34.67	3,211	382,173	13	85	1	-	1	-	-	-
60	1.3- 1.4	34.74-34.75	3,205	385,378	-	15	13	51	19	-	2	-
61	1.9- 2.0	34.89-34.90	3,141	388,519	-	-	-	-	-	85	15	-
62	1.4- 1.5	34.64-34.65	3,128	391,647	100	-	-	-	-	-	-	-
63	1.9- 2.0	34.64-34.65	3,064	394,711	24	74	1	-	1	-	-	-
64	1.3- 1.4	34.65-34.66	3,040	397,751	100	-	-	-	-	-	-	-
65	2.0- 2.2	34.90-34.91	5,958	403,709	-	-	-	-	-	75	25	-
66	1.7- 1.8	34.67-34.68	2,957	406,666	4	95	-	-	1	-	-	-
67	1.5- 1.6	34.64-34.65	2,896	409,562	100	-	-	-	1	-	-	-
68	0.9- 1.0	34.69-34.70	2,886	412,448	97	-	-	-	1	-	-	2

Rank	T°	S, ‰	Volume	Total Volume	N. Pac.	S. Pac.	So. O. Pac.	Ind.	So. O. Ind.	N. Atl.	S. Atl.	So. O. Atl.
69	1.5- 1.6	34.65-34.66	2,864	415,312	100	–	–	–	–	–	–	–
70	0.3- 0.4	34.70-34.71	2,829	418,141	–	21	37	14	17	–	8	3
71	1.5- 1.6	34.63-34.64	2,796	420,937	100	–	–	–	–	–	–	–
72	1.9- 2.0	34.65-34.66	2,708	423,645	15	83	1	–	1	–	–	–
73	1.6- 1.7	34.65-34.66	2,707	426,352	86	14	–	–	–	–	–	–
74	1.7- 1.8	34.73-34.74	2,698	429,050	–	14	12	67	7	–	–	–
75	-0.2--0.1	34.67-34.68	2,630	431,680	–	–	2	7	25	–	47	19
76	1.6- 1.7	34.68-34.69	2,609	434,289	–	98	1	–	–	–	–	1
77	2.0- 2.2	34.63-34.64	5,106	439,395	25	69	3	1	1	–	–	1
78	0.0- 0.1	34.68-34.69	2,485	441,880	–	–	4	10	49	–	21	16
79	1.8- 1.9	34.74-34.75	2,452	444,332	–	14	7	66	13	–	–	–
80	0.2- 0.3	34.69-34.70	2,449	446,781	–	–	16	17	44	–	15	8
81	1.9- 2.0	34.63-34.64	2,413	449,194	38	59	1	–	1	–	1	–
82	2.2- 2.4	34.92-34.93	4,788	453,982	–	–	–	–	–	89	11	–
83	2.2- 2.4	34.91-34.92	4,730	458,712	–	–	–	–	–	47	53	–
84	0.2- 0.3	34.70-34.71	2,361	461,073	–	23	50	6	2	–	19	–
85	1.8- 1.9	34.73-34.74	2,358	463,431	–	15	10	66	9	–	–	–
86	2.0- 2.2	34.91-34.92	4,671	468,102	–	–	–	–	–	95	5	–
87	0.8- 0.9	34.70-34.71	2,322	470,424	31	54	5	5	3	–	–	2
88	1.3- 1.4	34.72-34.73	2,319	472,743	–	16	20	59	–	–	2	3
89	0.4- 0.5	34.71-34.72	2,304	475,047	–	24	42	24	3	–	6	1
90	2.0- 2.2	34.88-34.89	4,597	479,644	–	–	–	–	–	2	98	–
91	1.7- 1.8	34.75-34.76	2,268	481,912	–	2	2	82	14	–	–	–
92	0.3- 0.4	34.69-34.70	2,195	484,107	–	–	17	24	32	–	12	15
93	1.6- 1.7	34.64-34.65	2,159	486,266	99	–	–	–	–	–	–	1
94	0.6- 0.7	34.70-34.71	2,157	488,423	–	23	30	14	20	–	7	6
95	1.6- 1.7	34.75-34.76	2,152	490,575	–	–	5	75	20	–	–	–
96	-0.3--0.2	34.66-34.67	2,145	492,720	–	–	–	1	–	–	35	64
97	2.0- 2.2	34.64-34.65	4,246	496,966	14	78	4	1	2	–	–	1
98	1.2- 1.3	34.74-34.75	2,116	499,082	–	8	18	43	24	–	7	–
99	-0.6--0.5	34.65-34.66	2,060	501,142	–	–	–	–	–	–	–	100
100	-0.3--0.2	34.67-34.68	2,037	503,179	–	–	1	5	45	–	38	11
101	0.1- 0.2	34.69-34.70	2,036	505,215	–	–	17	17	38	–	25	3
102	1.9- 2.0	34.90-34.91	2,014	507,229	–	–	–	–	–	97	3	–
103	1.1- 1.2	34.71-34.72	1,986	509,215	–	71	10	14	3	–	–	2
104	1.6- 1.7	34.63-34.64	1,974	511,189	99	–	1	–	–	–	–	–
105	2.0- 2.2	34.62-34.63	3,928	515,117	26	68	3	1	2	–	–	–
106	-0.1- 0.0	34.67-34.68	1,911	517,028	–	–	2	6	25	–	36	31
107	1.2- 1.3	34.71-34.72	1,909	518,937	–	85	7	7	–	–	–	1
108	-0.1- 0.0	34.68-34.69	1,900	520,837	–	–	4	11	58	–	20	7
109	1.6- 1.7	34.62-34.63	1,859	522,696	100	–	–	–	–	–	–	–
110	1.9- 2.0	34.73-34.74	1,857	524,553	–	21	9	60	10	–	–	–
111	-0.5--0.4	34.65-34.66	1,848	526,401	–	–	–	–	–	–	1	99
112	2.4- 2.6	34.93-34.94	3,691	530,092	–	–	–	–	–	84	16	–
113	0.1- 0.2	34.68-34.69	1,844	531,936	–	–	6	11	39	–	17	27
114	1.3- 1.4	34.70-34.71	1,836	533,772	–	97	1	–	–	–	–	2
115	0.9- 1.0	34.73-34.74	1,794	535,566	–	9	22	41	17	–	11	–
116	2.4- 2.6	34.92-34.93	3,547	539,113	–	–	–	–	–	33	67	–
117	2.2- 2.4	34.90-34.91	3,518	542,631	–	–	–	–	–	15	85	–
118	1.9- 2.0	34.72-34.73	1,744	544,375	–	16	11	67	6	–	–	–
119	-0.7--0.6	34.65-34.66	1,727	546,102	–	–	–	–	–	–	–	100
120	2.2- 2.4	34.62-34.63	3,440	549,542	27	67	3	2	1	–	–	–
121	1.4- 1.5	34.70-34.71	1,715	551,257	–	96	2	–	–	–	–	2
122	2.2- 2.4	34.61-34.62	3,396	554,653	22	72	4	1	1	–	–	–
123	1.8- 1.9	34.88-34.89	1,671	556,324	–	–	–	–	–	81	19	–
124	0.1- 0.2	34.70-34.71	1,644	557,968	–	–	79	5	–	–	16	–
125	1.8- 1.9	34.75-34.76	1,603	559,571	–	–	1	96	3	–	–	–
126	1.8- 1.9	34.89-34.90	1,564	561,135	–	–	–	–	–	98	2	–
127	2.4- 2.6	34.91-34.92	3,061	564,196	–	–	–	–	–	8	92	–
128	1.4- 1.5	34.72-34.73	1,515	565,711	–	16	21	54	–	–	3	6
129	0.0- 0.1	34.70-34.71	1,507	567,218	–	–	93	6	–	–	1	–
130	2.2- 2.4	34.60-34.61	2,998	570,216	24	69	4	1	1	–	–	1
131	-0.2--0.1	34.66-34.67	1,497	571,713	–	–	–	2	–	–	19	79
132	1.8- 1.9	34.72-34.73	1,489	573,202	–	23	18	50	9	–	–	–
133	2.6- 2.8	34.93-34.94	2,946	576,148	–	–	–	–	–	32	68	–
134	2.6- 2.8	34.94-34.95	2,939	579,087	–	–	–	–	–	82	18	–
135	1.7- 1.8	34.62-34.63	1,468	580,555	98	–	–	–	2	–	–	–
136	1.9- 2.0	34.66-34.67	1,459	582,014	9	83	3	–	2	–	1	2
137	1.9- 2.0	34.74-34.75	1,452	583,466	–	3	3	84	10	–	–	–
138	1.7- 1.8	34.64-34.65	1,448	584,914	90	8	1	–	1	–	–	–
139	1.5- 1.6	34.75-34.76	1,448	586,362	–	–	4	64	27	–	5	–

Rank	T°	S, ‰	Volume	Total Volume	N. Pac.	S. Pac.	So. Pac.	O. Ind.	So. O. Ind.	N. Atl.	S. Atl.	So. O. Atl.
140	2.0– 2.2	34.72–34.73	2,872	589,234	–	11	3	81	5	–	–	–
141	1.1– 1.2	34.74–34.75	1,404	590,638	–	8	16	45	20	–	11	–
142	2.0– 2.2	34.65–34.66	2,807	593,445	13	74	8	1	3	–	–	1
143	1.7– 1.8	34.63–34.64	1,393	594,838	98	–	–	–	1	–	–	1
144	2.2– 2.4	34.59–34.60	2,778	597,616	35	58	4	1	1	–	–	1
145	1.3– 1.4	34.71–34.72	1,385	599,001	–	89	6	4	–	–	–	1
146	0.6– 0.7	34.72–34.73	1,353	600,354	–	1	37	36	2	–	24	–
147	1.6– 1.7	34.61–34.62	1,347	601,701	100	–	–	–	–	–	–	–
148	0.2– 0.3	34.71–34.72	1,306	603,007	–	–	54	8	–	–	38	–
149	2.2– 2.4	34.63–34.64	2,591	605,598	19	73	3	4	1	–	–	–
150	1.8– 1.9	34.87–34.88	1,288	606,886	–	–	–	–	–	53	47	–
151	–0.8––0.7	34.65–34.66	1,284	608,170	–	–	–	–	–	–	–	100
152	2.4– 2.6	34.94–34.95	2,526	610,696	–	–	–	–	–	99	1	–
153	–0.2––0.1	34.68–34.69	1,259	611,955	–	–	2	4	84	–	8	2
154	–0.4––0.3	34.66–34.67	1,251	613,206	–	–	–	3	–	–	2	95
155	2.0– 2.2	34.61–34.62	2,493	615,699	45	47	4	1	3	–	–	–
156	1.9– 2.0	34.87–34.88	1,240	616,939	–	–	–	–	–	1	99	–
157	1.2– 1.3	34.70–34.71	1,236	618,175	–	90	2	–	5	–	–	3
158	2.6– 2.8	34.95–34.96	2,463	620,638	–	–	–	–	–	98	2	–
159	0.4– 0.5	34.69–34.70	1,226	621,864	–	–	17	13	31	–	10	29
160	1.7– 1.8	34.60–34.61	1,216	623,080	98	–	–	–	2	–	–	–
161	–0.4––0.3	34.67–34.68	1,205	624,285	–	–	–	6	85	–	–	9
162	–0.4––0.3	34.65–34.66	1,194	625,479	–	–	–	–	–	–	–	100
163	1.9– 2.0	34.75–34.76	1,177	626,656	–	–	–	100	–	–	–	–
164	2.6– 2.8	34.92–34.93	2,343	628,999	–	–	–	–	–	10	90	–
165	2.8– 3.0	34.94–34.95	2,339	631,338	–	–	–	–	–	56	44	–
166	1.7– 1.8	34.87–34.88	1,149	632,487	–	–	–	–	–	73	27	–
167	0.3– 0.4	34.71–34.72	1,134	633,621	–	2	57	9	–	–	32	–
168	0.2– 0.3	34.68–34.69	1,128	634,749	–	–	7	2	29	–	14	48
169	1.7– 1.8	34.68–34.69	1,120	635,869	–	94	2	–	3	–	–	1
170	1.6– 1.7	34.69–34.70	1,114	636,983	–	94	4	–	1	–	–	1
171	0.0– 0.1	34.69–34.70	1,109	638,092	–	–	35	16	32	–	17	–
172	1.7– 1.8	34.72–34.73	1,105	639,197	–	26	23	39	10	–	–	2
173	2.4– 2.6	34.60–34.61	2,202	641,399	33	55	–	10	–	–	2	–
174	2.4– 2.6	34.57–34.58	2,200	643,599	29	62	–	6	–	–	3	–
175	2.8– 3.0	34.95–34.96	2,194	645,793	–	–	–	–	–	87	13	–
176	1.5– 1.6	34.72–34.73	1,096	646,889	–	26	23	35	5	–	3	8
177	2.8– 3.0	34.93–34.94	2,190	649,079	–	–	–	–	–	11	89	–
178	1.6– 1.7	34.72–34.73	1,037	650,116	–	23	21	38	11	–	1	6
179	2.0– 2.2	34.71–34.72	2,062	652,178	–	19	7	66	8	–	–	–
180	1.5– 1.6	34.70–34.71	1,026	653,204	–	94	4	–	1	–	–	1
181	1.8– 1.9	34.67–34.68	1,019	654,223	–	93	4	–	2	–	1	–
182	1.4– 1.5	34.75–34.76	1,019	655,242	–	–	1	66	24	–	9	–
183	1.7– 1.8	34.61–34.62	1,018	656,260	97	–	–	–	2	–	–	1
184	0.0– 0.1	34.67–34.68	1,003	657,263	–	–	1	2	13	–	14	70
185	2.4– 2.6	34.61–34.62	1,996	659,259	35	48	–	14	–	–	3	–
186	1.8– 1.9	34.62–34.63	980	660,239	96	2	1	–	1	–	–	–
187	1.8– 1.9	34.63–34.64	975	661,214	88	9	2	–	1	–	–	–
188	2.2– 2.4	34.93–34.94	1,922	663,136	–	–	–	–	–	100	–	–
189	2.4– 2.6	34.58–34.59	1,892	665,028	37	51	–	8	–	–	4	–
190	1.8– 1.9	34.61–34.62	931	665,959	98	–	1	–	1	–	–	–
191	–0.1– 0.0	34.70–34.71	930	666,889	–	–	97	3	–	–	–	–
192	1.5– 1.6	34.62–34.63	925	667,814	99	–	1	–	–	–	–	–
193	–0.5––0.4	34.66–34.67	908	668,722	–	–	–	3	2	–	–	95
194	1.4– 1.5	34.71–34.72	889	669,611	–	84	9	4	–	–	–	3
195	1.9– 2.0	34.62–34.63	887	670,498	81	13	3	–	1	–	1	1
196	1.7– 1.8	34.59–34.60	877	671,375	97	–	1	–	2	–	–	–
197	2.2– 2.4	34.58–34.59	1,742	673,117	33	56	7	2	1	–	–	1
198	1.8– 1.9	34.90–34.91	869	673,986	–	–	–	–	–	100	–	–
199	1.8– 1.9	34.58–34.59	864	674,850	97	–	1	–	1	–	–	1
200	–0.3––0.2	34.68–34.69	857	675,707	–	–	1	–	99	–	–	–
201	0.7– 0.8	34.70–34.71	855	676,562	–	22	36	14	10	–	7	11
202	2.4– 2.6	34.59–34.60	1,710	678,272	27	59	–	11	–	–	3	–
203	2.8– 3.0	34.96–34.97	1,706	679,978	–	–	–	–	–	98	2	–
204	2.4– 2.6	34.56–34.57	1,679	681,657	17	72	–	7	–	–	4	–
205	1.9– 2.0	34.71–34.72	828	682,485	–	39	29	16	16	–	–	–
206	3.0– 3.2	34.94–34.96	3,289	685,774	–	–	–	–	–	60	40	–
207	1.8– 1.9	34.76–34.77	820	686,594	–	–	–	100	–	–	–	–
208	2.6– 2.8	34.59–34.60	1,638	688,232	35	50	–	9	–	–	6	–
209	0.1– 0.2	34.71–34.72	817	689,049	–	–	95	3	–	–	2	–
210	2.0– 2.2	34.60–34.61	1,611	690,660	70	18	5	2	4	–	–	1

Rank	T°	S, ‰	Volume	Total Volume	N. Pac.	S. Pac.	So. O. Pac.	Ind.	So. O. Ind.	N. Atl.	S. Atl.	So. O. Atl.
211	1.9- 2.0	34.76-34.77	801	691,461	-	-	-	100	-	-	-	-
212	1.8- 1.9	34.60-34.61	800	692,261	96	-	2	-	2	-	-	-
213	1.5- 1.6	34.85-34.86	798	693,059	-	-	-	-	-	87	13	-
214	2.2- 2.4	34.64-34.65	1,595	694,654	21	61	4	12	2	-	-	-
215	2.4- 2.6	34.62-34.63	1,582	696,236	21	54	-	21	-	-	4	-
216	1.8- 1.9	34.59-34.60	790	697,026	97	-	2	-	1	-	-	-
217	2.4- 2.6	34.90-34.91	1,578	698,604	-	-	-	-	-	-	100	-
218	1.9- 2.0	34.91-34.92	788	699,392	-	-	-	-	-	100	-	-
219	2.2- 2.4	34.89-34.90	1,563	700,955	-	-	-	-	-	-	100	-
220	2.0- 2.2	34.73-34.74	1,545	702,500	-	5	1	91	3	-	-	-
221	3.2- 3.4	34.96-34.98	3,062	705,562	-	-	-	-	-	71	29	-
222	2.2- 2.4	34.88-34.89	1,530	707,092	-	-	-	-	-	-	100	-
223	2.0- 2.2	34.74-34.75	1,477	708,569	-	-	-	97	-	-	3	-
224	2.4- 2.6	34.63-34.64	1,475	710,044	27	44	-	25	-	-	4	-
225	1.7- 1.8	34.88-34.89	732	710,776	-	-	-	-	-	98	2	-
226	1.9- 2.0	34.56-34.57	729	711,505	96	-	2	-	1	-	1	-
227	-0.1- 0.0	34.66-34.67	720	712,225	-	-	-	-	-	-	-	100
228	2.0- 2.2	34.70-34.71	1,433	713,658	-	37	16	36	11	-	-	-
229	0.3- 0.4	34.68-34.69	714	714,372	-	-	5	-	17	-	10	68
230	3.0- 3.2	34.96-34.98	2,745	717,117	-	-	-	-	-	89	11	-
231	2.4- 2.6	34.55-34.56	1,372	718,489	22	67	1	6	-	-	4	-
232	2.6- 2.8	34.60-34.61	1,367	719,856	45	37	-	10	-	-	8	-
233	1.9- 2.0	34.61-34.62	679	720,535	93	-	3	-	1	-	2	1
234	2.0- 2.2	34.92-34.93	1,332	721,867	-	-	-	-	-	98	2	-
235	1.6- 1.7	34.86-34.87	664	722,531	-	-	-	-	-	76	24	-
236	1.8- 1.9	34.57-34.58	663	723,194	98	-	1	-	1	-	-	-
237	2.6- 2.8	34.91-34.92	1,313	724,507	-	-	-	-	-	-	100	-
238	0.1- 0.2	34.67-34.68	656	725,163	-	-	-	1	8	-	15	76
239	1.4- 1.5	34.63-34.64	652	725,815	100	-	-	-	-	-	-	-
240	1.9- 2.0	34.60-34.61	649	726,464	95	-	2	-	1	-	1	1
241	1.9- 2.0	34.57-34.58	645	727,109	96	-	2	-	1	-	1	-
242	1.7- 1.8	34.71-34.72	642	727,751	-	33	34	-	16	-	4	13
243	3.4- 3.6	34.96-34.98	2,566	730,317	-	-	-	-	-	64	36	-
244	1.3- 1.4	34.75-34.76	639	730,956	-	-	2	64	11	-	23	-
245	-0.5--0.4	34.67-34.68	638	731,594	-	-	-	-	96	-	-	4
246	2.0- 2.2	34.75-34.76	1,275	732,869	-	-	-	99	-	-	1	-
247	-0.9--0.8	34.65-34.66	635	733,504	-	-	-	-	-	-	-	100
248	2.6- 2.8	34.56-34.57	1,247	734,751	53	30	-	11	-	-	6	-
249	1.9- 2.0	34.70-34.71	620	735,371	-	47	37	2	12	-	-	2
250	1.7- 1.8	34.69-34.70	620	735,991	-	82	6	-	9	-	2	1
251	-0.4--0.3	34.68-34.69	613	736,604	-	-	-	-	100	-	-	-
252	1.8- 1.9	34.71-34.72	613	737,217	-	40	48	-	9	-	-	3
253	2.6- 2.8	34.58-34.59	1,221	738,438	25	54	-	13	-	-	8	-
254	2.8- 3.0	34.92-34.93	1,208	739,646	-	-	-	-	-	1	99	-
255	0.3- 0.4	34.72-34.73	592	740,238	-	-	12	9	-	-	79	-
256	0.0- 0.1	34.71-34.72	576	740,814	-	-	100	-	-	-	-	-
257	2.6- 2.8	34.53-34.54	1,146	741,960	23	58	-	11	-	-	8	-
258	2.0- 2.2	34.68-34.69	1,144	743,104	-	56	24	6	13	-	-	1
259	2.0- 2.2	34.69-34.70	1,139	744,243	-	47	22	16	14	-	-	1
260	1.6- 1.7	34.71-34.72	567	744,810	-	40	35	-	6	-	7	12
261	2.6- 2.8	34.57-34.58	1,124	745,934	36	42	-	14	-	-	8	-
262	2.0- 2.2	34.54-34.55	1,112	747,046	92	-	2	2	4	-	-	-
263	2.6- 2.8	34.61-34.62	1,102	748,148	29	50	-	11	-	-	10	-
264	0.8- 0.9	34.73-34.74	550	748,698	-	10	14	53	8	-	15	-
265	2.8- 3.0	34.58-34.59	1,098	749,796	42	51	-	5	-	-	2	-
266	1.9- 2.0	34.69-34.70	548	750,344	-	42	41	-	10	-	1	6
267	2.0- 2.2	34.76-34.77	1,084	751,428	-	-	-	93	-	-	7	-
268	3.2- 3.4	34.94-34.96	2,148	753,576	-	-	-	-	-	60	40	-
269	1.9- 2.0	34.58-34.59	535	754,111	96	-	2	-	-	-	1	1
270	2.6- 2.8	34.54-34.55	1,059	755,170	18	61	-	12	-	-	9	-
271	2.6- 2.8	34.55-34.56	1,055	756,225	34	46	-	12	-	-	8	-
272	-0.9--0.8	34.64-34.65	527	756,752	-	-	-	-	-	-	-	100
273	3.0- 3.2	34.92-34.94	2,091	758,843	-	-	-	-	-	13	87	-
274	0.7- 0.8	34.76-34.77	522	759,365	-	-	-	-	-	1	99	-
275	1.6- 1.7	34.60-34.61	517	759,882	99	-	-	-	-	-	-	1
276	2.2- 2.4	34.70-34.71	1,032	760,914	-	2	-	98	-	-	-	-
277	1.2- 1.3	34.65-34.66	514	761,428	100	-	-	-	-	-	-	-
278	2.6- 2.8	34.52-34.53	1,026	762,454	17	62	-	12	-	-	9	-
279	2.0- 2.2	34.53-34.54	1,026	763,480	92	-	2	2	4	-	-	-
280	1.7- 1.8	34.86-34.87	503	763,983	-	-	-	-	-	25	75	-
281	1.1- 1.2	34.66-34.67	503	764,486	97	-	2	-	1	-	-	-

Rank	T°	S, ‰	Volume	Total Volume	N. Pac.	S. Pac.	So. O. Pac.	Ind.	So. O. Ind.	N. Atl.	S. Atl.	So. O. Atl.
282	1.9- 2.0	34.55-34.56	503	764,989	93	-	3	1	1	-	1	1
283	1.8- 1.9	34.68-34.69	501	765,490	-	85	8	-	5	-	2	-
284	1.9- 2.0	34.59-34.60	499	765,989	94	-	2	-	1	-	2	1
285	0.4- 0.5	34.72-34.73	496	766,485	-	-	21	25	-	-	54	-
286	0.6- 0.7	34.75-34.76	495	766,980	-	-	-	-	-	3	97	-
287	2.0- 2.2	34.67-34.68	989	767,969	-	53	29	4	13	-	-	1
288	2.6- 2.8	34.96-34.97	989	768,958	-	-	-	-	-	100	-	-
289	2.8- 3.0	34.97-34.98	986	769,944	-	-	-	-	-	100	-	-
290	2.2- 2.4	34.71-34.72	986	770,930	-	-	-	100	-	-	-	-
291	1.9- 2.0	34.67-34.68	491	771,421	-	66	15	-	8	-	2	9
292	0.5- 0.6	34.69-34.70	489	771,910	-	-	27	-	26	-	18	29
293	2.4- 2.6	34.54-34.55	967	772,877	26	63	1	6	-	-	4	-
294	3.2- 3.4	34.92-34.94	1,933	774,810	-	-	-	-	-	-	62	38	-
295	1.3- 1.4	34.64-34.65	482	775,292	100	-	-	-	-	-	-	-
296	2.0- 2.2	34.55-34.56	963	776,255	90	-	3	2	4	-	-	1
297	1.8- 1.9	34.70-34.71	478	776,733	-	35	36	-	8	-	3	18
298	1.5- 1.6	34.71-34.72	475	777,208	-	56	24	4	6	-	3	7
299	2.0- 2.2	34.77-34.78	938	778,146	-	-	-	91	-	-	9	-
300	2.8- 3.0	34.59-34.60	937	779,083	52	37	-	8	-	-	3	-
301	1.6- 1.7	34.70-34.71	468	779,551	-	78	13	-	7	-	-	2
302	2.2- 2.4	34.50-34.51	933	780,484	90	-	6	2	1	-	1	-
303	1.7- 1.8	34.76-34.77	465	780,949	-	-	-	93	-	-	7	-
304	1.6- 1.7	34.87-34.88	463	781,412	-	-	-	-	-	97	3	-
305	3.6- 3.8	34.96-34.98	1,845	783,257	-	-	-	-	-	62	38	-
306	2.2- 2.4	34.69-34.70	919	784,176	-	7	-	92	1	-	-	-
307	2.0- 2.2	34.66-34.67	918	785,094	1	54	26	4	12	-	-	3
308	-0.1- 0.0	34.69-34.70	454	785,548	-	-	40	26	34	-	-	-
309	2.6- 2.8	34.62-34.63	908	786,456	38	40	-	9	-	-	13	-
310	0.4- 0.5	34.68-34.69	453	786,909	-	-	3	-	17	-	11	69
311	2.0- 2.2	34.59-34.60	852	787,761	80	-	8	3	7	-	-	2
312	2.2- 2.4	34.72-34.73	849	788,610	-	-	-	89	-	-	11	-
313	2.8- 3.0	34.56-34.57	847	789,457	28	57	-	11	-	-	4	-
314	2.8- 3.0	34.55-34.56	828	790,285	44	37	-	15	-	-	4	-
315	0.4- 0.5	34.73-34.74	413	790,698	-	-	-	1	-	-	99	-
316	-0.3--0.2	34.65-34.66	410	791,108	-	-	-	-	-	-	-	100
317	2.6- 2.8	34.51-34.52	818	791,926	17	62	-	11	-	-	10	-
318	2.8- 3.0	34.60-34.61	818	792,744	37	51	-	9	-	-	3	-
319	2.2- 2.4	34.49-34.50	816	793,560	89	-	6	3	1	-	1	-
320	0.5- 0.6	34.74-34.75	405	793,965	-	-	-	-	-	-	100	-
321	2.8- 3.0	34.54-34.55	808	794,773	62	21	-	14	-	-	3	-
322	2.2- 2.4	34.57-34.58	802	795,575	45	32	14	4	3	-	1	1
323	2.2- 2.4	34.66-34.67	801	796,376	-	45	5	48	2	-	-	-
324	2.2- 2.4	34.65-34.66	799	797,175	1	52	8	36	3	-	-	-
325	-0.6--0.5	34.66-34.67	398	797,573	-	-	-	4	-	-	-	96
326	2.2- 2.4	34.51-34.52	796	798,369	88	-	7	3	1	-	1	-
327	2.8- 3.0	34.57-34.58	792	799,161	32	59	-	6	-	-	3	-
328	2.0- 2.2	34.56-34.57	790	799,951	86	-	5	3	5	-	-	1
329	2.4- 2.6	34.46-34.47	786	800,737	81	7	5	1	-	-	5	1
330	1.0- 1.1	34.74-34.75	393	801,130	-	-	13	50	13	-	24	-
331	2.6- 2.8	34.50-34.51	782	801,912	24	56	-	10	-	-	10	-
332	0.9- 1.0	34.68-34.69	390	802,302	88	-	1	-	8	-	-	3
333	3.0- 3.2	34.56-34.58	1,514	803,816	37	51	-	8	-	-	4	-
334	2.2- 2.4	34.76-34.77	756	804,572	-	-	-	97	-	-	3	-
335	0.5- 0.6	34.72-34.73	377	804,949	-	-	51	28	-	-	21	-
336	2.2- 2.4	34.67-34.68	754	805,703	-	29	1	69	1	-	-	-
337	1.7- 1.8	34.58-34.59	374	806,077	94	-	2	-	3	-	-	1
338	2.0- 2.2	34.52-34.53	747	806,824	89	-	3	3	4	-	-	1
339	1.9- 2.0	34.68-34.69	373	807,197	-	31	38	-	12	-	3	16
340	2.2- 2.4	34.68-34.69	739	807,936	-	10	-	89	1	-	-	-
341	3.0- 3.2	34.58-34.60	1,475	809,411	48	43	-	5	-	-	4	-
342	2.0- 2.2	34.58-34.59	734	810,145	80	-	7	4	7	-	-	2
343	1.4- 1.5	34.84-34.85	365	810,510	-	-	-	-	-	76	24	-
344	2.8- 3.0	34.51-34.52	715	811,225	28	56	-	13	-	-	3	-
345	-0.5--0.4	34.68-34.69	357	811,582	-	-	-	-	100	-	-	-
346	2.8- 3.0	34.61-34.62	706	812,288	43	45	-	9	-	-	3	-
347	0.5- 0.6	34.73-34.74	353	812,641	-	-	3	22	-	-	75	-
348	2.0- 2.2	34.57-34.58	705	813,346	82	-	6	4	6	-	-	2
349	1.8- 1.9	34.69-34.70	352	813,698	-	44	23	-	15	-	3	15
350	2.6- 2.8	34.42-34.43	701	814,399	76	11	3	2	-	-	7	1
351	2.4- 2.6	34.45-34.46	699	815,098	84	3	5	1	-	-	6	1
352	2.4- 2.6	34.53-34.54	697	815,795	28	57	2	6	-	-	6	1

Rank	T°	S, ‰	Volume	Total Volume	N. Pac.	S. Pac.	So. O. Pac.	Ind.	So. O. Ind.	N. Atl.	S. Atl.	So. O. Atl.
353	2.4- 2.6	34.88-34.89	697	816,492	-	-	-	-	-	-	100	-
354	-0.2--0.1	34.71-34.72	343	816,835	-	-	100	-	-	-	-	-
355	2.6- 2.8	34.43-34.44	686	817,521	70	16	3	2	-	-	8	1
356	2.6- 2.8	34.49-34.50	680	818,201	18	61	-	12	-	-	9	-
357	2.4- 2.6	34.47-34.48	678	818,879	77	10	5	1	-	-	6	1
358	3.2- 3.4	34.56-34.58	1,355	820,234	50	38	-	5	-	-	7	-
359	2.8- 3.0	34.50-34.51	677	820,911	16	67	-	13	-	-	4	-
360	2.4- 2.6	34.95-34.96	671	821,582	-	-	-	-	-	100	-	-
361	0.8- 0.9	34.77-34.78	333	821,915	-	-	-	-	-	2	98	-
362	2.2- 2.4	34.74-34.75	658	822,573	-	-	-	87	-	-	13	-
363	2.8- 3.0	34.53-34.54	653	823,226	46	33	-	17	-	-	4	-
364	-0.2--0.1	34.70-34.71	326	823,552	-	-	100	-	-	-	-	-
365	0.2- 0.3	34.67-34.68	326	823,878	-	-	-	-	2	-	15	83
366	2.8- 3.0	34.49-34.50	645	824,523	17	64	-	14	-	-	5	-
367	3.4- 3.6	34.92-34.94	1,286	825,809	-	-	-	-	-	55	45	-
368	0.6- 0.7	34.73-34.74	320	826,129	-	-	3	26	-	-	71	-
369	3.2- 3.4	34.58-34.60	1,280	827,409	42	42	-	9	-	-	7	-
370	-0.8--0.7	34.64-34.65	320	827,729	-	-	-	-	-	-	-	100
371	1.7- 1.8	34.70-34.71	319	828,048	-	40	19	-	24	-	6	11
372	2.2- 2.4	34.48-34.49	634	828,682	87	-	7	3	1	-	1	1
373	2.4- 2.6	34.76-34.77	633	829,315	-	-	-	89	-	-	11	-
374	2.6- 2.8	34.41-34.42	629	829,944	78	9	3	1	-	-	8	1
375	2.2- 2.4	34.52-34.53	615	830,559	81	-	11	4	2	-	1	1
376	2.2- 2.4	34.56-34.57	614	831,173	57	14	18	5	3	-	1	2
377	2.2- 2.4	34.78-34.79	612	831,785	-	-	-	95	-	-	5	-
378	1.4- 1.5	34.83-34.84	305	832,090	-	-	-	-	-	33	67	-
379	0.7- 0.8	34.73-34.74	302	832,392	-	-	5	14	-	-	81	-
380	3.8- 4.0	34.96-34.98	1,204	833,596	-	-	-	-	-	58	42	-
381	3.0- 3.2	34.98-35.00	1,192	834,788	-	-	-	-	-	99	1	-
382	2.8- 3.0	34.52-34.53	595	835,383	33	45	-	18	-	-	4	-
383	0.9- 1.0	34.78-34.79	297	835,680	-	-	-	-	-	2	98	-
384	-0.2--0.1	34.69-34.70	297	835,977	-	-	40	57	3	-	-	-
385	2.4- 2.6	34.52-34.53	591	836,568	30	51	5	4	-	-	8	2
386	2.8- 3.0	34.38-34.39	590	837,158	73	13	3	2	-	-	9	-
387	1.2- 1.3	34.75-34.76	295	837,453	-	-	-	67	4	-	29	-
388	1.6- 1.7	34.76-34.77	293	837,746	-	-	-	70	-	-	30	-
389	1.9- 2.0	34.77-34.78	289	838,035	-	-	-	90	-	-	10	-
390	2.6- 2.8	34.48-34.49	576	838,611	16	61	-	11	-	-	11	1
391	3.0- 3.2	34.54-34.56	1,147	839,758	33	56	-	7	-	-	4	-
392	2.8- 3.0	34.48-34.49	572	840,330	26	56	-	13	-	-	5	-
393	0.6- 0.7	34.74-34.75	284	840,614	-	-	-	2	-	-	98	-
394	2.2- 2.4	34.77-34.78	568	841,182	-	-	-	95	-	-	5	-
395	3.0- 3.2	34.46-34.48	1,135	842,317	17	63	-	14	-	-	6	-
396	0.7- 0.8	34.74-34.75	283	842,600	-	-	-	27	-	-	73	-
397	2.0- 2.2	34.78-34.79	560	843,160	-	-	-	75	-	-	25	-
398	2.4- 2.6	34.44-34.45	558	843,718	83	3	4	1	-	-	8	1
399	2.8- 3.0	34.91-34.92	556	844,274	-	-	-	-	-	-	100	-
400	1.5- 1.6	34.84-34.85	277	844,551	-	-	-	-	-	26	74	-
401	3.4- 3.6	34.94-34.96	1,108	845,659	-	-	-	-	-	53	47	-
402	1.3- 1.4	34.83-34.84	276	845,935	-	-	-	-	-	73	27	-
403	2.2- 2.4	34.75-34.76	552	846,487	-	-	-	93	-	-	7	-
404	2.8- 3.0	34.39-34.40	549	847,036	70	17	1	4	-	-	8	-
405	1.6- 1.7	34.85-34.86	271	847,307	-	-	-	-	-	20	80	-
406	2.4- 2.6	34.77-34.78	540	847,847	-	-	-	88	-	-	12	-
407	0.3- 0.4	34.67-34.68	269	848,116	-	-	-	-	3	-	-	97
408	2.4- 2.6	34.48-34.49	537	848,653	69	13	7	1	1	-	8	1
409	3.6- 3.8	34.98-35.00	1,068	849,721	-	-	-	-	-	96	4	-
410	2.2- 2.4	34.73-34.74	533	850,254	-	-	-	89	-	-	11	-
411	3.6- 3.8	34.56-34.58	1,054	851,308	37	44	-	11	-	-	8	-
412	3.2- 3.4	34.54-34.56	1,048	852,356	25	57	-	8	-	-	10	-
413	1.0- 1.1	34.79-34.80	261	852,617	-	-	-	-	-	1	99	-
414	2.2- 2.4	34.55-34.56	520	853,137	54	15	19	6	3	-	1	2
415	2.4- 2.6	34.89-34.90	517	853,654	-	-	-	-	-	-	100	-
416	3.2- 3.4	34.90-34.92	1,032	854,686	-	-	-	1	-	35	64	-
417	3.0- 3.2	34.52-34.54	1,031	855,717	58	30	-	8	-	-	4	-
418	3.4- 3.6	34.56-34.58	1,030	856,747	53	36	-	8	-	-	3	-
419	2.8- 3.0	34.45-34.46	514	857,261	16	66	-	14	-	-	4	-
420	2.8- 3.0	34.37-34.38	508	857,769	73	10	3	2	-	-	11	1
421	3.4- 3.6	34.54-34.56	1,015	858,784	36	54	-	6	-	-	4	-
422	2.6- 2.8	34.47-34.48	507	859,291	24	51	1	12	-	-	11	1
423	2.6- 2.8	34.76-34.77	505	859,796	-	-	-	69	-	-	31	-

Rank	T°	S, ‰	Volume	Total Volume	N. Pac.	S. Pac.	So. O. Pac.	Ind.	So. O. Ind.	N. Atl.	S. Atl.	So. O. Atl.
424	3.2- 3.4	34.44-34.46	1,010	860,806	19	57	-	16	-	-	8	-
425	1.3- 1.4	34.82-34.83	252	861,058	-	-	-	-	-	14	86	-
426	-0.1- 0.0	34.71-34.72	251	861,309	-	-	100	-	-	-	-	-
427	1.8- 1.9	34.56-34.57	248	861,557	95	-	2	-	3	-	-	-
428	3.4- 3.6	34.98-35.00	988	862,545	-	-	-	-	-	98	2	-
429	3.8- 4.0	34.56-34.58	986	863,531	39	42	-	6	-	-	13	-
430	1.5- 1.6	34.61-34.62	245	863,776	98	-	1	-	1	-	-	-
431	2.4- 2.6	34.87-34.88	490	864,266	-	-	-	-	-	-	100	-
432	2.8- 3.0	34.44-34.45	489	864,755	22	62	-	12	-	-	4	-
433	2.6- 2.8	34.40-34.41	485	865,240	77	6	4	1	-	-	11	1
434	3.0- 3.2	34.50-34.52	969	866,209	52	28	-	16	-	-	4	-
435	2.8- 3.0	34.46-34.47	482	866,691	18	63	-	15	-	-	4	-
436	2.2- 2.4	34.79-34.80	480	867,171	-	-	-	93	-	-	7	-
437	2.4- 2.6	34.79-34.80	479	867,650	-	-	-	74	-	-	26	-
438	2.8- 3.0	34.47-34.48	475	868,125	28	55	-	13	-	-	4	-
439	1.2- 1.3	34.81-34.82	237	868,362	-	-	-	-	-	6	94	-
440	1.8- 1.9	34.91-34.92	236	868,598	-	-	-	-	-	100	-	-
441	1.1- 1.2	34.80-34.81	235	868,833	-	-	-	-	-	1	99	-
442	2.2- 2.4	34.54-34.55	466	869,299	55	11	19	7	3	-	3	2
443	3.8- 4.0	34.94-34.96	928	870,227	-	-	-	2	-	50	48	-
444	1.7- 1.8	34.89-34.90	232	870,459	-	-	-	-	-	100	-	-
445	2.8- 3.0	34.43-34.44	463	870,922	19	65	-	12	-	-	4	-
446	0.7- 0.8	34.75-34.76	231	871,153	-	-	-	-	-	-	100	-
447	0.8- 0.9	34.76-34.77	230	871,383	-	-	-	-	-	1	99	-
448	3.4- 3.6	34.42-34.44	919	872,302	18	51	-	20	-	-	11	-
449	3.2- 3.4	34.52-34.54	911	873,213	32	48	-	7	-	-	13	-
450	3.2- 3.4	34.98-35.00	907	874,120	-	-	-	-	-	100	-	-
451	1.5- 1.6	34.76-34.77	223	874,343	-	-	-	48	-	-	52	-
452	2.2- 2.4	34.53-34.54	444	874,787	62	3	19	6	3	-	6	1
453	2.8- 3.0	34.36-34.37	441	875,228	74	5	4	2	-	-	14	1
454	2.4- 2.6	34.78-34.79	439	875,667	-	-	-	82	-	-	18	-
455	0.5- 0.6	34.68-34.69	219	875,886	-	-	2	-	2	-	-	96
456	2.4- 2.6	34.51-34.52	436	876,322	31	43	9	4	1	-	10	2
457	0.6- 0.7	34.69-34.70	218	876,540	-	-	32	-	3	-	11	54
458	3.4- 3.6	34.58-34.60	868	877,408	37	40	-	19	-	-	4	-
459	1.6- 1.7	34.77-34.78	217	877,625	-	-	-	47	-	-	53	-
460	1.9- 2.0	34.85-34.86	215	877,840	-	-	-	-	-	-	100	-
461	2.6- 2.8	34.85-34.86	430	878,270	-	-	-	1	-	-	99	-
462	2.6- 2.8	34.44-34.45	430	878,700	57	22	4	4	-	-	12	1
463	3.0- 3.2	34.34-34.36	856	879,556	72	15	1	5	-	-	7	-
464	2.4- 2.6	34.80-34.81	427	879,983	-	-	-	62	-	-	38	-
465	2.6- 2.8	34.80-34.81	426	880,409	-	-	-	61	-	-	39	-
466	3.0- 3.2	34.48-34.50	851	881,260	30	45	-	19	-	-	6	-
467	2.0- 2.2	34.51-34.52	425	881,685	82	-	5	5	7	-	-	1
468	3.2- 3.4	34.48-34.50	843	882,528	49	28	-	9	-	-	14	-
469	3.0- 3.2	34.60-34.62	835	883,363	38	38	-	18	-	-	6	-
470	2.4- 2.6	34.74-34.75	416	883,779	-	-	-	81	-	-	19	-
471	2.4- 2.6	34.49-34.50	415	884,194	56	20	10	2	1	-	10	1
472	3.2- 3.4	34.88-34.90	829	885,023	-	-	-	9	-	51	40	-
473	3.2- 3.4	34.38-34.40	829	885,852	15	75	-	8	-	-	2	-
474	1.0- 1.1	34.67-34.68	206	886,058	87	-	4	-	1	-	-	8
475	3.0- 3.2	34.40-34.42	820	886,878	17	72	-	10	-	-	1	-
476	0.9- 1.0	34.77-34.78	205	887,083	-	-	-	-	-	-	100	-
477	2.4- 2.6	34.64-34.65	410	887,493	3	12	-	70	-	-	15	-
478	2.6- 2.8	34.77-34.78	408	887,901	-	-	-	63	-	-	37	-
479	2.4- 2.6	34.68-34.69	407	888,308	-	-	-	83	-	-	17	-
480	2.0- 2.2	34.87-34.88	406	888,714	-	-	-	-	-	-	100	-
481	2.4- 2.6	34.72-34.73	403	889,117	-	-	-	65	-	-	35	-
482	2.0- 2.2	34.85-34.86	403	889,520	-	-	-	-	-	-	100	-
483	2.4- 2.6	34.75-34.76	403	889,923	-	-	-	80	-	-	20	-
484	2.8- 3.0	34.40-34.41	403	890,326	57	23	1	9	-	-	9	1
485	2.4- 2.6	34.71-34.72	401	890,727	-	-	-	79	-	-	21	-
486	2.4- 2.6	34.69-34.70	400	891,127	-	-	-	80	-	-	20	-
487	3.2- 3.4	34.50-34.52	797	891,924	47	28	-	9	-	-	16	-
488	2.6- 2.8	34.81-34.82	396	892,320	-	-	-	56	-	-	44	-
489	3.6- 3.8	34.40-34.42	792	893,112	8	55	-	21	-	-	16	-
490	1.9- 2.0	34.54-34.55	197	893,309	85	-	6	1	3	-	3	2
491	1.5- 1.6	34.86-34.87	197	893,506	-	-	-	-	-	97	3	-
492	3.8- 4.0	34.98-35.00	785	894,291	-	-	-	-	-	99	1	-
493	2.6- 2.8	34.75-34.76	392	894,683	-	-	-	60	-	-	40	-
494	3.6- 3.8	34.52-34.54	782	895,465	22	65	-	6	-	-	7	-

Rank	T°	S, ‰	Volume	Total Volume	N. Pac.	S. Pac.	So. O. Pac.	O. Ind.	So. O. Ind.	N. Atl.	S. Atl.	So. O. Atl.
495	-0.6--0.5	34.68-34.69	195	895,660	-	-	-	-	100	-	-	-
496	3.0- 3.2	34.42-34.44	779	896,439	29	58	-	12	-	-	1	-
497	2.6- 2.8	34.79-34.80	388	896,827	-	-	-	61	-	-	39	-
498	2.8- 3.0	34.98-34.99	388	897,215	-	-	-	-	-	100	-	-
499	0.8- 0.9	34.74-34.75	194	897,409	-	-	-	38	-	-	62	-
500	2.6- 2.8	34.78-34.79	387	897,796	-	-	-	64	-	-	36	-
501	2.6- 2.8	34.46-34.47	385	898,181	34	40	1	10	-	-	14	1
502	2.6- 2.8	34.71-34.72	384	898,565	-	-	-	57	-	-	43	-
503	3.0- 3.2	34.44-34.46	767	899,332	29	48	-	17	-	-	6	-
504	2.8- 3.0	34.42-34.43	382	899,714	39	41	-	13	-	-	7	-
505	2.4- 2.6	34.86-34.87	381	900,095	-	-	-	-	-	-	100	-
506	2.6- 2.8	34.45-34.46	381	900,476	43	32	2	8	-	-	14	1
507	3.6- 3.8	34.54-34.56	757	901,233	62	21	-	9	-	-	8	-
508	2.2- 2.4	34.80-34.81	378	901,611	-	-	-	56	-	-	44	-
509	2.2- 2.4	34.87-34.88	376	901,987	-	-	-	-	-	-	100	-
510	3.8- 4.0	34.54-34.56	749	902,736	50	22	-	10	-	-	18	-
511	1.8- 1.9	34.78-34.79	186	902,922	-	-	-	24	-	-	76	-
512	3.6- 3.8	34.94-34.96	744	903,666	-	-	-	-	-	56	44	-
513	-0.3--0.2	34.71-34.72	186	903,852	-	-	100	-	-	-	-	-
514	1.0- 1.1	34.78-34.79	186	904,038	-	-	-	-	-	-	100	-
515	1.7- 1.8	34.77-34.78	185	904,223	-	-	-	60	-	-	40	-
516	3.6- 3.8	34.90-34.92	735	904,958	-	-	-	18	-	35	47	-
517	3.4- 3.6	34.36-34.38	732	905,690	12	69	1	12	-	-	6	-
518	0.7- 0.8	34.69-34.70	183	905,873	-	-	45	-	7	-	-	48
519	2.2- 2.4	34.47-34.48	365	906,238	79	-	11	5	2	-	2	1
520	2.6- 2.8	34.74-34.75	361	906,599	-	-	-	60	-	-	40	-
521	2.4- 2.6	34.50-34.51	361	906,960	37	33	11	4	1	-	12	2
522	1.4- 1.5	34.76-34.77	179	907,139	-	-	-	38	-	-	62	-
523	1.1- 1.2	34.79-34.80	179	907,318	-	-	-	-	-	-	100	-
524	2.6- 2.8	34.66-34.67	358	907,676	-	-	-	64	-	-	36	-
525	2.4- 2.6	34.67-34.68	355	908,031	-	-	-	82	-	-	18	-
526	2.6- 2.8	34.73-34.74	354	908,385	-	-	-	61	-	-	39	-
527	3.8- 4.0	34.40-34.42	708	909,093	23	32	-	22	-	-	23	-
528	2.4- 2.6	34.65-34.66	353	909,446	-	1	-	81	-	-	18	-
529	3.2- 3.4	34.46-34.48	690	910,136	35	34	-	19	-	-	12	-
530	3.4- 3.6	34.88-34.90	689	910,825	-	-	-	21	-	46	33	-
531	1.8- 1.9	34.84-34.85	172	910,997	-	-	-	-	-	-	100	-
532	3.6- 3.8	34.42-34.44	688	911,685	32	29	-	20	-	-	19	-
533	2.6- 2.8	34.82-34.83	343	912,028	-	-	-	43	-	-	57	-
534	3.8- 4.0	34.38-34.40	683	912,711	15	59	-	13	-	-	13	-
535	2.2- 2.4	34.85-34.86	341	913,052	-	-	-	-	-	-	100	-
536	3.0- 3.2	34.36-34.38	681	913,733	56	28	1	10	-	-	5	-
537	2.4- 2.6	34.66-34.67	340	914,073	-	-	-	81	-	-	19	-
538	1.9- 2.0	34.84-34.85	170	914,243	-	-	-	-	-	-	100	-
539	3.2- 3.4	34.42-34.44	677	914,920	25	38	-	23	-	-	14	-
540	2.6- 2.8	34.70-34.71	337	915,257	-	-	-	55	-	-	45	-
541	1.2- 1.3	34.80-34.81	168	915,425	-	-	-	-	-	-	100	-
542	2.6- 2.8	34.86-34.87	335	915,760	-	-	-	-	-	-	100	-
543	3.2- 3.4	35.00-35.02	670	916,430	-	-	-	-	-	100	-	-
544	2.2- 2.4	34.81-34.82	334	916,764	-	-	-	14	-	-	86	-
545	3.4- 3.6	34.52-34.54	666	917,430	39	45	-	10	-	-	6	-
546	2.0- 2.2	34.80-34.81	332	917,762	-	-	-	49	-	-	51	-
547	2.8- 3.0	34.82-34.83	332	918,094	-	-	-	60	-	-	40	-
548	2.4- 2.6	34.70-34.71	331	918,425	-	-	-	77	-	-	23	-
549	3.4- 3.6	34.40-34.42	656	919,081	30	38	-	19	-	-	13	-
550	1.1- 1.2	34.75-34.76	164	919,245	-	-	-	55	-	-	45	-
551	3.8- 4.0	34.52-34.54	653	919,898	21	56	-	10	-	-	13	-
552	1.4- 1.5	34.82-34.83	163	920,061	-	-	-	-	-	-	100	-
553	3.4- 3.6	34.90-34.92	652	920,713	-	-	-	7	-	62	31	-
554	2.6- 2.8	34.90-34.91	319	921,032	-	-	-	-	-	-	100	-
555	0.0- 0.1	34.66-34.67	159	921,191	-	-	-	-	-	-	10	90
556	3.4- 3.6	34.50-34.52	631	921,822	38	46	-	8	-	-	8	-
557	1.3- 1.4	34.81-34.82	157	921,979	-	-	-	-	-	-	100	-
558	2.6- 2.8	34.67-34.68	312	922,291	-	-	-	58	-	-	42	-
559	3.6- 3.8	34.92-34.94	624	922,915	-	-	-	5	-	50	45	-
560	3.4- 3.6	34.46-34.48	623	923,538	47	29	-	9	-	-	15	-
561	3.4- 3.6	34.48-34.50	619	924,157	46	33	-	8	-	-	13	-
562	2.4- 2.6	34.73-34.74	309	924,466	-	-	-	66	-	-	34	-
563	2.8- 3.0	34.84-34.85	306	924,772	-	-	-	34	-	-	66	-
564	1.8- 1.9	34.86-34.87	153	924,925	-	-	-	-	-	-	100	-
565	3.0- 3.2	34.32-34.34	610	925,535	80	4	5	-	-	-	11	-

Rank	T°	S, ‰	Volume	Total Volume	N. Pac.	S. Pac.	So. O. Pac.	Ind.	So. O. Ind.	N. Atl.	S. Atl.	So. O. Atl.
566	3.2- 3.4	34.32-34.34	606	926,141	56	26	–	8	–	–	10	–
567	2.8- 3.0	34.77-34.78	302	926,443	–	–	–	74	–	–	26	–
568	2.4- 2.6	34.43-34.44	301	926,744	75	1	5	2	1	–	16	–
569	2.6- 2.8	34.39-34.40	300	927,044	69	2	8	4	–	–	17	–
570	3.4- 3.6	35.00-35.02	598	927,642	–	–	–	–	–	100	–	–
571	2.8- 3.0	34.79-34.80	298	927,940	–	–	–	66	–	–	34	–
572	2.0- 2.2	34.86-34.87	298	928,238	–	–	–	–	–	–	100	–
573	3.4- 3.6	34.44-34.46	594	928,832	31	31	–	17	–	–	21	–
574	1.8- 1.9	34.77-34.78	148	928,980	–	–	–	71	–	–	29	–
575	2.6- 2.8	34.63-34.64	296	929,276	27	1	–	32	–	–	40	–
576	1.0- 1.1	34.76-34.77	148	929,424	–	–	–	19	–	–	81	–
577	2.6- 2.8	34.89-34.90	295	929,719	–	–	–	–	–	–	100	–
578	1.7- 1.8	34.83-34.84	147	929,866	–	–	–	–	–	–	100	–
579	0.9- 1.0	34.75-34.76	147	930,013	–	–	–	33	–	–	67	–
580	2.8- 3.0	34.81-34.82	293	930,306	–	–	–	63	–	–	37	–
581	3.8- 4.0	34.50-34.52	583	930,889	36	49	–	6	–	–	9	–
582	1.5- 1.6	34.83-34.84	145	931,034	–	–	–	–	–	–	100	–
583	1.6- 1.7	34.59-34.60	144	931,178	94	–	1	–	4	–	–	1
584	2.6- 2.8	34.65-34.66	287	931,465	–	–	–	55	–	–	45	–
585	3.2- 3.4	34.40-34.42	573	932,038	25	56	1	9	–	–	9	–
586	4.0- 4.5	34.50-34.55	3,574	935,612	33	32	–	7	–	1	27	–
587	1.6- 1.7	34.84-34.85	141	935,753	–	–	–	–	–	–	100	–
588	2.8- 3.0	34.76-34.77	282	936,035	–	–	–	77	–	–	23	–
589	1.9- 2.0	34.78-34.79	141	936,176	–	–	–	46	–	–	54	–
590	2.8- 3.0	34.83-34.84	282	936,458	–	–	–	42	–	–	58	–
591	1.3- 1.4	34.76-34.77	140	936,598	–	–	–	27	–	–	73	–
592	1.5- 1.6	34.77-34.78	140	936,738	–	–	–	37	–	–	63	–
593	2.8- 3.0	34.78-34.79	278	937,016	–	–	–	66	–	–	34	–
594	2.2- 2.4	34.94-34.95	278	937,294	–	–	–	–	–	100	–	–
595	2.8- 3.0	34.80-34.81	278	937,572	–	–	–	57	–	–	43	–
596	3.2- 3.4	34.60-34.62	553	938,125	31	12	–	40	–	–	17	–
597	2.6- 2.8	34.68-34.69	276	938,401	–	–	–	51	–	–	49	–
598	3.2- 3.4	34.30-34.32	550	938,951	82	4	1	2	–	–	11	–
599	2.8- 3.0	34.41-34.42	275	939,226	42	29	1	15	–	–	12	1
600	1.7- 1.8	34.85-34.86	137	939,363	–	–	–	–	–	–	100	–
601	3.6- 3.8	34.46-34.48	548	939,911	42	43	–	7	–	–	8	–
602	3.8- 4.0	34.36-34.38	546	940,457	23	58	–	10	–	–	9	–
603	0.8- 0.9	34.75-34.76	136	940,593	–	–	–	17	–	–	83	–
604	2.6- 2.8	34.69-34.70	272	940,865	–	–	–	50	–	–	50	–
605	2.8- 3.0	34.35-34.36	272	941,137	70	4	5	2	–	–	17	2
606	2.6- 2.8	34.84-34.85	271	941,408	–	–	–	11	–	–	89	–
607	1.7- 1.8	34.78-34.79	135	941,543	–	–	–	21	–	–	79	–
608	2.4- 2.6	34.84-34.85	269	941,812	–	–	–	1	–	–	99	–
609	4.5- 5.0	34.50-34.55	3,356	945,168	33	31	–	2	–	7	27	–
610	1.4- 1.5	34.77-34.78	134	945,302	–	–	–	18	–	–	82	–
611	-0.7--0.6	34.64-34.65	134	945,436	–	–	–	–	–	–	–	100
612	2.4- 2.6	34.81-34.82	266	945,702	–	–	–	61	–	–	39	–
613	2.6- 2.8	34.97-34.98	266	945,968	–	–	–	–	–	100	–	–
614	1.6- 1.7	34.82-34.83	133	946,101	–	–	–	–	–	–	100	–
615	1.5- 1.6	34.81-34.82	133	946,234	–	–	–	–	–	–	100	–
616	3.6- 3.8	34.34-34.36	523	946,757	14	73	1	3	–	–	9	–
617	-0.3--0.2	34.69-34.70	129	946,886	–	–	10	87	3	–	–	–
618	2.6- 2.8	34.83-34.84	257	947,143	–	–	–	32	–	–	68	–
619	4.0- 4.5	34.35-34.40	3,210	950,353	18	52	–	16	–	–	14	–
620	3.6- 3.8	34.38-34.40	512	950,865	30	47	–	10	–	–	13	–
621	2.0- 2.2	34.84-34.85	256	951,121	–	–	–	–	–	–	100	–
622	0.8- 0.9	34.69-34.70	128	951,249	–	–	21	–	7	–	–	72
623	1.1- 1.2	34.77-34.78	128	951,377	–	–	–	13	–	–	87	–
624	3.6- 3.8	34.36-34.38	509	951,886	27	50	–	14	–	–	9	–
625	3.6- 3.8	34.58-34.60	509	952,395	50	21	–	14	–	–	15	–
626	3.6- 3.8	34.44-34.46	508	952,903	45	33	–	8	–	–	14	–
627	3.4- 3.6	34.24-34.26	508	953,411	70	1	5	–	–	–	24	–
628	2.8- 3.0	34.89-34.90	254	953,665	–	–	–	–	–	–	100	–
629	3.0- 3.2	34.38-34.40	507	954,172	27	51	–	18	–	–	4	–
630	3.2- 3.4	34.28-34.30	499	954,671	74	6	4	–	–	–	16	–
631	4.0- 4.5	34.55-34.60	3,116	957,787	40	31	–	4	–	3	22	–
632	3.6- 3.8	34.50-34.52	498	958,285	35	44	–	10	–	–	11	–
633	3.2- 3.4	34.34-34.36	498	958,783	33	46	–	13	–	–	8	–
634	2.2- 2.4	34.86-34.87	247	959,030	–	–	–	–	–	–	100	–
635	1.9- 2.0	34.79-34.80	123	959,153	–	–	–	25	–	–	75	–
636	3.0- 3.2	34.76-34.78	492	959,645	–	–	–	70	–	1	29	–

Rank	T°	S, ‰	Volume	Total Volume	N. Pac.	S. Pac.	So. O. Pac.	Ind.	So. O. Ind.	N. Atl.	S. Atl.	So. O. Atl.
637	1.4- 1.5	34.80-34.81	121	959,766	-	-	-	1	-	-	99	-
638	1.8- 1.9	34.80-34.81	121	959,887	-	-	-	48	-	-	52	-
639	2.6- 2.8	34.64-34.65	242	960,129	-	-	-	49	-	-	51	-
640	1.2- 1.3	34.78-34.79	121	960,250	-	-	-	9	-	-	91	-
641	1.4- 1.5	34.85-34.86	120	960,370	-	-	-	-	-	97	3	-
642	3.4- 3.6	34.26-34.28	478	960,848	75	7	1	2	-	-	15	-
643	0.4- 0.5	34.67-34.68	119	960,967	-	-	-	-	12	-	-	88
644	3.8- 4.0	34.34-34.36	476	961,443	33	57	-	6	-	-	4	-
645	1.4- 1.5	34.62-34.63	118	961,561	94	-	2	-	3	-	-	1
646	3.2- 3.4	34.36-34.38	468	962,029	22	61	-	13	-	-	4	-
647	3.0- 3.2	34.90-34.92	465	962,494	-	-	-	-	-	-	100	-
648	3.0- 3.2	34.78-34.80	464	962,958	-	-	-	58	-	12	30	-
649	3.0- 3.2	34.80-34.82	464	963,422	-	-	-	56	-	9	35	-
650	2.0- 2.2	34.83-34.84	232	963,654	-	-	-	19	-	-	81	-
651	1.3- 1.4	34.79-34.80	116	963,770	-	-	-	2	-	-	98	-
652	2.8- 3.0	34.75-34.76	231	964,001	-	-	-	76	-	-	24	-
653	3.0- 3.2	34.82-34.84	462	964,463	-	-	-	66	-	2	32	-
654	-0.7--0.6	34.66-34.67	114	964,577	-	-	-	11	-	-	-	89
655	2.8- 3.0	34.62-34.63	227	964,804	56	-	-	34	-	-	10	-
656	0.0- 0.1	34.65-34.66	113	964,917	-	-	-	-	4	-	-	96
657	3.8- 4.0	34.92-34.94	450	965,367	-	-	-	23	-	44	33	-
658	2.6- 2.8	34.72-34.73	225	965,592	-	-	-	43	-	-	57	-
659	5.5- 6.0	34.30-34.35	2,806	968,398	2	92	-	2	-	-	4	-
660	3.6- 3.8	34.88-34.90	444	968,842	-	-	-	20	-	25	55	-
661	3.8- 4.0	35.00-35.02	440	969,282	-	-	-	-	-	100	-	-
662	0.9- 1.0	34.74-34.75	110	969,392	-	-	-	65	-	-	35	-
663	3.8- 4.0	34.32-34.34	438	969,830	26	54	-	8	-	-	12	-
664	3.4- 3.6	34.38-34.40	437	970,267	26	45	1	7	-	-	21	-
665	3.2- 3.4	34.86-34.88	436	970,703	-	-	-	37	-	39	24	-
666	2.4- 2.6	34.85-34.86	217	970,920	-	-	-	-	-	-	100	-
667	2.8- 3.0	34.88-34.89	214	971,134	-	-	-	-	-	-	100	-
668	2.8- 3.0	34.74-34.75	214	971,348	-	-	-	76	-	-	24	-
669	4.5- 5.0	34.30-34.35	2,675	974,023	13	80	-	3	-	-	4	-
670	3.0- 3.2	34.62-34.64	426	974,449	23	-	-	61	-	-	16	-
671	3.0- 3.2	34.74-34.76	422	974,871	-	-	-	65	-	-	35	-
672	3.6- 3.8	34.48-34.50	422	975,293	39	41	-	9	-	-	11	-
673	3.0- 3.2	34.30-34.32	421	975,714	68	3	7	-	-	-	22	-
674	2.8- 3.0	34.34-34.35	208	975,922	62	4	8	2	-	-	22	2
675	3.2- 3.4	34.26-34.28	416	976,338	58	-	7	-	-	-	35	-
676	3.4- 3.6	34.86-34.88	414	976,752	-	-	-	32	-	19	49	-
677	5.0- 5.5	34.50-34.55	2,586	979,338	30	30	-	3	-	7	30	-
678	2.8- 3.0	34.87-34.88	203	979,541	-	-	-	-	-	-	100	-
679	3.4- 3.6	34.30-34.32	405	979,946	33	51	-	6	-	-	10	-
680	4.0- 4.5	34.30-34.35	2,513	982,459	25	67	-	4	-	-	4	-
681	1.0- 1.1	34.75-34.76	101	982,560	-	-	-	72	-	-	28	-
682	3.6- 3.8	35.02-35.04	402	982,962	-	-	-	-	-	100	-	-
683	3.8- 4.0	34.80-34.82	402	983,364	-	-	-	32	-	1	67	-
684	-0.3--0.2	34.70-34.71	100	983,464	-	-	100	-	-	-	-	-
685	2.8- 3.0	34.64-34.65	198	983,662	-	-	-	89	-	-	11	-
686	3.8- 4.0	34.42-34.44	395	984,057	38	37	-	18	-	-	7	-
687	3.4- 3.6	34.34-34.36	394	984,451	27	63	-	8	-	-	2	-
688	6.5- 7.0	34.35-34.40	2,456	986,907	3	94	-	1	-	-	2	-
689	3.4- 3.6	35.02-35.04	392	987,299	-	-	-	-	-	100	-	-
690	3.8- 4.0	34.44-34.46	392	987,691	36	49	-	7	-	-	8	-
691	4.0- 4.5	34.45-34.50	2,431	990,122	25	38	-	11	-	-	26	-
692	2.0- 2.2	34.50-34.51	194	990,316	62	-	11	10	15	-	-	2
693	2.8- 3.0	34.65-34.66	192	990,508	-	-	-	86	-	-	14	-

3
On the Mid-Depth Circulation of the World Ocean

Joseph L. Reid

3.1 Introduction

There is a large part of the ocean circulation for which we have very little information and very vague concepts. This is the great domain of the mid-depth ocean. We have considerable information about the flow at and quite near the sea surface, and some inferences about the abyssal flow derived mostly from the traditional patterns of characteristics at the bottom. Recently, some attention has been focused on the deep western boundary currents, where the flow is strong enough to be detected both in the density field and in some cases by direct measurement. But for the greater part of the volume of the ocean—beneath the upper kilometer and away from the western boundary currents and above the abyssal waters—we have little information on, or understanding of, the circulation. Most treatments of the deep water as well as the abyssal water have dealt in terms of the western boundary flow, and a general meridional flow is all that has emerged from most of the studies. Wüst (1935), for example, assumed a principally thermohaline meridional flow to obtain from the abyssal layer up through his Subantarctic Intermediate Water, at depths above 1 km, with no recognizable pattern of gyral flow analagous to the surface circulation.

It seems worthwhile to consider what information there is for this great volume of water. This study will begin with a general discussion of the earlier ideas on this problem. It will review briefly the recent work (of the last 10 years or so), which has begun to make substantial contributions, and will display and discuss some world-wide mid-depth patterns of characteristics and of geostrophic vertical shear.

There is no simple distinction between the upper waters, the deep and abyssal waters, and what I shall call the mid-depth waters. A working definition will be that the mid-depth waters are those that are found between about 1 and 3 km in middle and low latitudes and their source waters, which are shallower in high latitudes. Warren's study (this volume, chapter 1) of the deep circulation includes some of these waters, of course, and I have tried to avoid duplication. Some duplication remains, however, in part for immediate clarity and in part for different emphasis.

3.2 The Circulation of the Upper Waters and Their Contribution to the Mid-Depths

Our first information about general ocean circulation came from the experience of mariners crossing the great oceans. They found the best routes for eastward travel to be in the zone of the west winds and for westward travel in the trades, and noted early the western boundary currents. As the information accumulated, these findings led, by the middle of the nineteenth century, to the general concept of subtropical anticyclonic gyres, subarctic gyres, and various zonal flows near the equator.

The variability of this general pattern was learned early and is most clearly presented in the sailing directions, coast pilots, and atlases prepared by the various hydrographic offices. For example, the typical atlas of surface currents of the northwestern Pacific Ocean (U.S. Navy Hydrographic Office, 1944) provides information by averages in 1° × 1° squares, but for 5° × 5° areas provides summations by octants in direction, with average speed and fractions of time for each octant. While this can give no information on the frequency of the variations (each measurement represented a mean of 12 to 24 hours or longer), the presence of variation is clearly shown everywhere, and the general findings of Fuglister (1954), Dantzler (1976) and Wyrtki, Magaard, and Hager (1976) are to some degree anticipated.

But in spite of the variability and the smoothing effects in taking its mean, certain major features of the gross field stand out. On this particular atlas the strongest of these are the Kuroshio and the North Equatorial Current. The West Wind Drift around 40°N is also clear, though weaker. But in the area between the Kuroshio–West Wind Drift and the North Equatorial Current, the return flow from the Kuroshio toward the southwest described by Sverdrup, Johnson, and Fleming (1942) is only marginally discernible. In a later compilation of the average drift (Stidd, 1974), it is somewhat clearer.

This surface circulation had been generally accepted as wind driven, but the depth to which it extended, or to which any wind-driven current extended, was not known. It is not clear what was generally believed, or why, but the impression left from reading the various papers on this subject is that it was very shallow over most of the ocean.

Information about the subsurface circulation arose from a different source. Measurements of water characteristics began in the eighteenth century. Prestwich (1875) reviewed them and the various interpretations that had been made. The measurements were mostly of temperature with some of salinity. Very few had reached abyssal depths, though there were enough to identify the Antarctic and Greenland Seas as sources of abyssal water. He concluded that all of the water, from top to bottom, is in a state of movement, and that high-latitude cold waters flow equatorward at abyssal depths from both north and south in the Atlantic, but only from the south in the Pacific and Indian Oceans, and that these sources account for the low subsurface temperatures of the central oceans.

He did not, however, consider only such a simple convection model, but worked out some more detailed parts of the system as well. His most interesting interpretations are of the details of the shallower subsurface flows. He noted that zones of maximum surface temperature and salinity in the Atlantic and Pacific are not exactly at the equator but in two zones roughly parallel to it, north and south; that the waters between 10°N and 10°S in the upper 200 m are colder than those to the north and south; and that this must result from a rising of the deeper, colder waters in that zone, where they are moved poleward as they are warmed.

He noted the excessive salinity of the Mediterranean and the very high temperature at great depth. He explained the high temperature compared to that in the Atlantic by the presence of the sill at the Straits, which excluded the colder waters of high latitudes, and winter overturn within the Mediterranean that gave the bottom waters the same temperature as the surface minimum value. He noted that the salt balance had been explained by surface inflow and subsurface outflow and noted that water with characteristics similar to those within the Mediterranean had been found at mid-depth outside the Straits.

Most important, he concluded that warmer waters are conducted into higher latitudes not by shallow surface currents alone, but by substantially thicker subsurface flows, which provide a thick, warmer subsurface layer in the polar regions. He found two channels of flow from the Arctic Ocean to the Atlantic, via Baffin Bay and the East Greenland Current, and noted that in the eastern Norwegian Sea thick layers of warmer water were found, having entered from the Atlantic. He states:

There is every reason to believe that the open seas of the north polar regions are due, as suggested by Maury and others, to the influence of warm southern waters, though this is not, as supposed by those authors, owing to the action of the Gulf Stream, but by the surging-up of these deeper warm strata; and in the same way the open sea found by Cook, Weddell, Ross, and others, after passing the first barrier of ice in the south polar seas, may be due to a similar cause. [Prestwich (1875, p. 635).]

He concludes:

Some of the great surface currents, which originate or acquire additional force in the equatorial and polar seas, are intimately connected with the surging-up of polar waters in the great oceans and of tropical waters in Arctic and Antarctic seas, although the ultimate course of these currents may be influenced and determined by the action of the prevailing winds and by the movement of rotation of the earth. [Prestwich (1875, p. 638).]

Further information on the mid-depth circulation was provided by Buchanan (1877), who noted the great intermediate-depth salinity minima of the North and South Pacific and of the South and Equatorial Atlantic.

He noted the generally higher salinity of the North Atlantic and ascribed it in part to the exchange with the Mediterranean.

Nansen (1902) had confirmed the presence of subsurface warmer waters from the Atlantic over a much larger area of the Arctic than that known to Prestwich, and had proposed (1906) that convection takes place to the bottom only in the Greenland Sea; but the only recognized outflow was of water of low salinity and low density through the Denmark Strait, and this outflow did not contribute directly to mid-depth circulation. Instead, the colder abyssal waters of the northern North Atlantic were attributed (Nansen, 1912; Wüst, 1935) to overturn in the Irminger and Labrador Seas. Brennecke (1921) and Defant (1938) had suggested that this overturn and formation of deep water might possibly contain a mixture of water that had overflowed from the Norwegian-Greenland Sea through the Denmark Strait or east of Iceland, but Wüst (1943) in his discussion of the subarctic bottom flow of this water apparently did not accept their conjecture, which was finally argued convincingly by Cooper (1955a).

For the Antarctic component, Brennecke's (1921), Mosby's (1934), and Deacon's (1937) work had shown the presence of a subsurface warm layer nearly everywhere throughout the Antarctic Ocean, as surmised by Prestwich, and had identified the southwestern Weddell Sea as the area where this layer was penetrated, leading to the formation of the coldest abyssal layer from the Antarctic. Deacon's (1937) study of the Southern Ocean, in particular, gave the first description of the subsurface temperature and salinity maximum, the "warm deep water," or circumpolar water extending throughout the Antarctic region. His interpretation that the meridional exchange with lower latitudes takes place in alternating directions in various strata in all oceans, which was developed further by Sverdrup et al. (1942, figure 164), was a substantial step beyond Prestwich's model. Taking as a starting point Prestwich's (1875) argument for a thick subsurface poleward flow instead of a surface flow alone, the extra layer of low salinity described very roughly by Buchanan (1877) and by Brennecke (1921) is clearly identified and accounted for, and a deep poleward flow above the equatorward abyssal flow is clearly seen and described (Sverdrup et al., 1942, figure 164).

Wüst (1933) had shown that the abyssal layer from the south extends well north of the equator in the Atlantic, with the northern component much smaller in lateral extent. He found the major product of the North Atlantic to be a thick deep-water layer of high salinity and oxygen, extending southward to the Antarctic Circumpolar Current.

3.3 The Use of Geostrophy

Sandström and Helland-Hansen (1903) provided methods for calculation of vertical shear from the density field by use of the geostrophic approximation. Helland-Hansen and Nansen (1909) used this method to calculate the shear in the upper 200 m of the Norwegian Sea in the area between Norway and Iceland and, comparing it with the information about sources of water of various characteristics, used it in their interpretation of the circulation of the Norwegian Sea.

From the data they collected they saw at once that the variability already found at the sea surface occurred also at greater depths. They noted, and discussed at length, variations in temperature, salinity, and density in the upper strata:

At any rate down to 600 m, and probably much deeper ... such irregularities, great or small, are seen in most vertical sections where the stations are sufficiently numerous and not too far apart. The equilines ... form bends or undulations like waves, sometimes great, sometimes small. When, in 1901, Helland-Hansen first found a great wave of this kind in the sections across the Norwegian Atlantic Current, he thought that it indicated some kind of permanent division of the current.... But by continued research with more stations, even several "waves" were sometimes found in the same sections, and it soon became evident that they could not indicate any such division as he had at first thought, but must have some hitherto unknown causes. [Helland-Hansen and Nansen (1909, p. 87).]

They considered the possibility of these waves in intermediate depths as pulsations in the currents, periodic variations, temporary disturbances, or cyclonic and anticyclonic vortices, and remarked upon the necessity of numerous and closely spaced observations if the density field were to be described in detail. Their map of the geopotential anomaly of the sea surface relative to the 200-db (decibar) surface (steric height 0–200 db) is reproduced as figure 3.1, illustrating one of the irregularities they encountered. Concluding that the method was good enough, in spite of the noted variability, to give useful results, they continued to use it. Later, Nansen (1913) concluded that the waters between about 200 and 1500 m off the coast of Europe and Ireland did not originate in the Gulf Stream but derived from the waters to the south and indicated a substantial mixture of highly saline water from the Mediterranean, and Helland-Hansen and Nansen (1926) mapped salinity, temperature, density σ_t, and steric height of various depths down to 2500 m, illustrating the pattern of characteristics and the geostrophic shear. Their plates 48 and 68, of temperature and salinity at 1000 m and of steric height 1000–2000 db, are reproduced as figures 3.2 and 3.3.

Ekman (1923) mapped the steric height relative to various pressures down to 1000 db and showed the

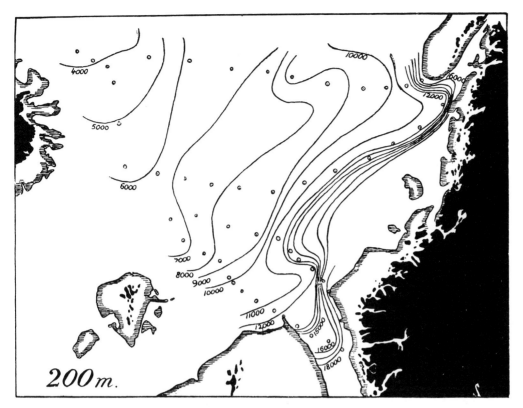

Figure 3.1 Geopotential anomaly (steric height) 0-200 db (10^{-5} dynamic m). (Helland-Hansen and Nansen, 1909.)

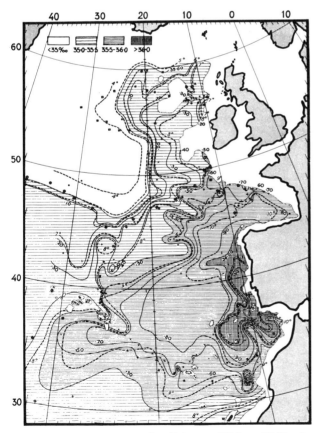

Figure 3.2 Temperature and salinity at 1000 m. (Helland-Hansen and Nansen, 1926.)

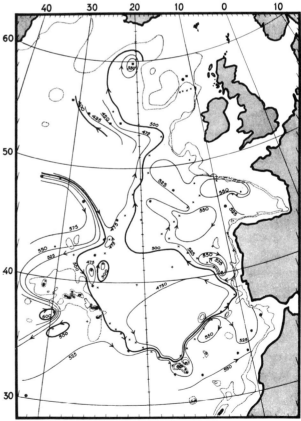

Figure 3.3 Steric height 1000-2000 db (dynamic mm). (Helland-Hansen and Nansen, 1926.)

relation of the density field to the deep southern extension of the Grand Banks (figure 3.4). Geostrophic shear was used extensively in the work of the International Ice Patrol (figure 3.5, from Smith, Soule, and Mosby, 1937) from the early 1920s and appeared to give valuable indications of surface currents when compared with the drift of icebergs. Parr (1938) reviewed various doubts about the validity of the "dynamic method" to obtain such trajectories and pointed out especially that the "method" eliminated any trajectory of a water parcel other than in the horizontal.

Jacobsen (1929) mapped the geostrophic shear between various pressure surfaces and the 1000-db surface for the central North Atlantic and the Caribbean, using data from the *Dana* expedition of 1920-1922 and adding in the northwest the results of Ekman's (1923) study of the surface flow. He found, with his limited data, quite a fair picture of the general surface circulation as supposed at present (figure 3.6A): a Gulf Stream-North Atlantic Current with a return flow of the Gulf Stream along 70°W turning toward an eastward flow north of 20°N. This, with sparse and less accurate data, is remarkably similar to the more recent work of Leetmaa, Niiler, and Stommel (1977). He showed also a map of steric height 500-1000 db (figure 3.6B). It is not detailed enough to be useful except for one interesting feature: the east-west axis of the anticylconic gyre at the sea surface appears to lie at about 25°N in midocean, but at greater depths it lies considerably farther north, past 30°N.

Koenuma (1939) examined the flow of the southwestern part of the North Pacific Ocean and found clear evidence in the density field (geostrophic shear) of a return flow just southeast of the Kuroshio, in spite of the strong eddy field.

3.4 The Mid-Depth Circulation of the Atlantic Ocean from Core Analysis and Vertical Geostrophic Shear

Apparently, the *Meteor* expedition to the South and Equatorial Atlantic was planned to study deeper circulation, which was assumed to be mostly meridional. The east-west lines were not suited for studying the gyral patterns of middle latitudes or the zonal flows in the equatorial area. Indeed, the separation of the ocean, as discussed in the *Meteor* reports, into a troposphere and stratosphere, or warm- and cold-water bodies, implies that an entirely different pattern of flow was expected below the troposphere.

Wüst's (1935) major work on the subsurface waters of the Atlantic dealt mostly with this very saline, oxygen-rich North Atlantic Deep Water (figure 3.7) and with the overlying salinity minimum, which he called Subantarctic Intermediate Water. In this subsurface

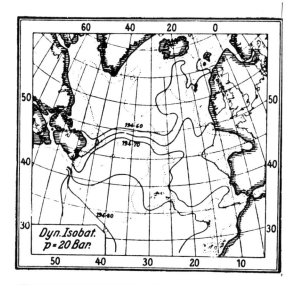

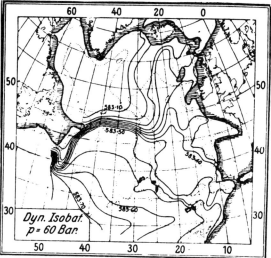

Figure 3.4 Steric height at 0-200, 0-600, and 0-1000 db (dynamic m). (Ekman, 1923.)

domain he examined maxima and minima of salinity and oxygen (the core method) under the assumption that beneath a shallow wind-driven layer the circulation was almost entirely meridional, with major flow along the western boundary except for the confluence with the Antarctic Circumpolar Current. His choice of the core-layer method limited his examination of a layer to those areas where it contained an extremum, and this was particularly limiting in the case of his Upper North Atlantic Deep Water, from the Mediterranean: where it joined other saline layers, its "core" was no longer recognizable.

His assumption of predominantly meridional, thermohaline-driven circulation was also limiting. He made no use of the density field in calculations of geostrophic shear, although the density field as mapped in the *Meteor* atlas (Wüst and Defant, 1936) clearly indicated substantial zonal patterns as well, and indeed indicated that a circulation pattern very like that recognized at the surface extended at least into the depth range of the Intermediate and Deep Water. His single consideration of the density field in any context except that of thermohaline meridional flow lies in his reference to the *Meteor* atlas map of density σ_t at 200 m in the Atlantic. He proposed that it is useful to infer the sense of flow along isopleths of σ_t at 200 m, but not below, and inferred a countercurrent on the southeastern side of the Gulf Stream that could be followed westward from the Azores to Bermuda and from there southward toward the Antilles. It is curious that he chose to limit this flow to the upper 200 m, as the density field retains that pattern to much greater depths, clearly to at least 1000 m in the *Meteor* atlas maps, and the salinity maps might have led him to propose this flow as the reason for the westward extension of the Mediterranean outflow water. The map of intermediate flow for the North Atlantic prepared by Sverdrup et al. (1942, figure 188; reproduced here as figure 3.8) makes this point quite clear. Presumably, it was the emphasis on meridional flow and his reluctance to make use of geostrophy at greater depths that caused him to consider only the upper 200 m.

The earliest portrayal of the density distribution over a large area was that of the *Meteor* atlas (Wüst and Defant, 1936). Vertical sections of σ_t as well as of temperature and salinity were prepared for all of the *Meteor* stations and, by selection of other data in the North Atlantic Ocean, maps of these quantities were prepared at selected depths from 200 m to the bottom. Wattenberg (1939) prepared corresponding maps and sections from the *Meteor* data for dissolved oxygen but did not attempt to treat the North Atlantic. It was not until 1957 that Wattenberg's phosphate atlas was published.

It was Defant (1941a,b) who first used the density field from the *Meteor* expedition and other data avail-

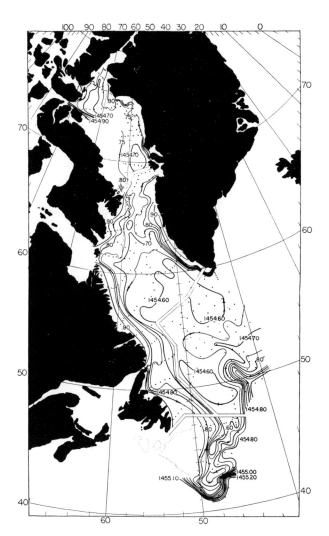

Figure 3.5 Steric height 0–1500 db (dynamic m). (Smith et al., 1937.)

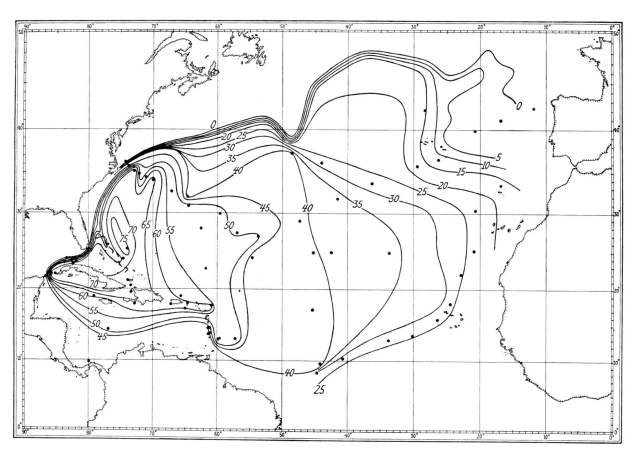
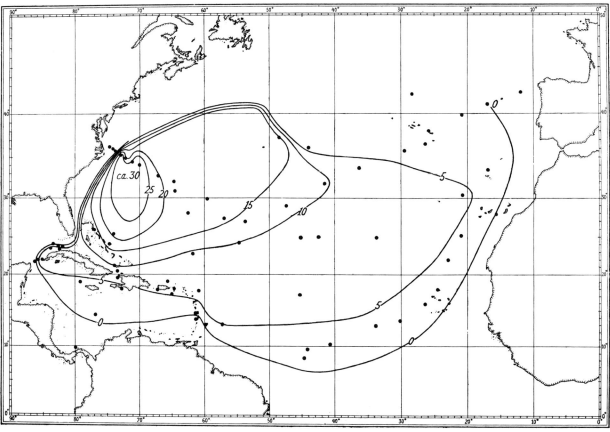

Figure 3.6 Steric height (dynamic cm): (A) 0-1000 db; (B) 500-1000 db. (Jacobsen, 1929.)

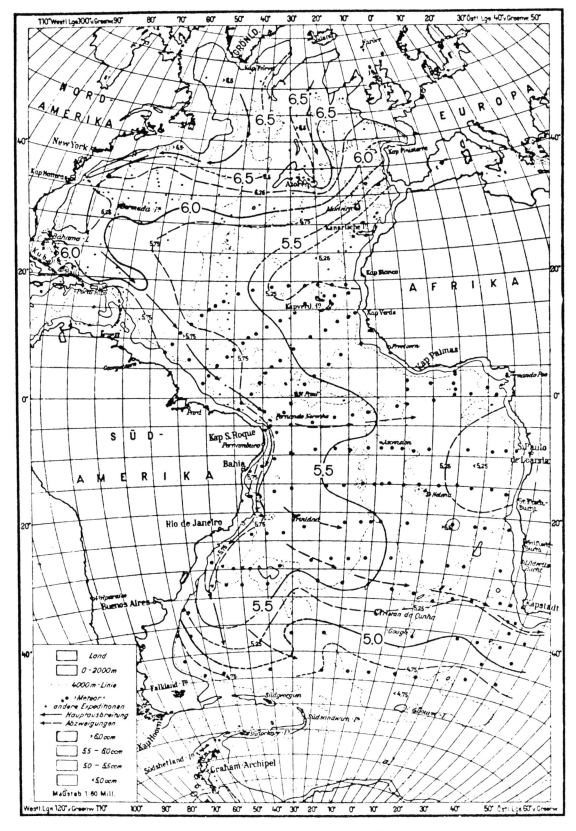

Figure 3.7 Oxygen (ml l^{-1}) of the core layer (intermediate oxygen maximum) of the Middle North Atlantic Deep Water. (Wüst, 1935.)

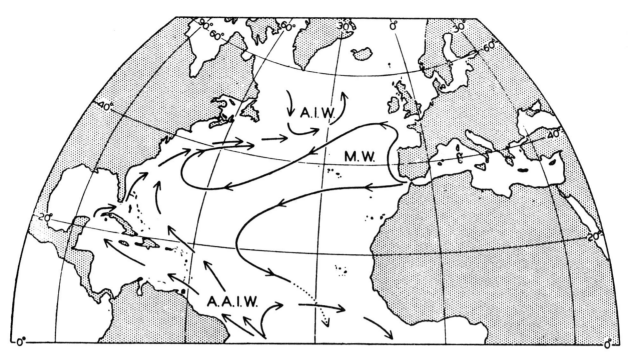

Figure 3.8 Approximate directions of flow of the intermediate water masses of the North Atlantic. A.I.W., Arctic Intermediate Water; M.W., Mediterranean Water; A.A.I.W., Antarctic Intermediate Water. (Sverdrup et al., 1942.)

able in the North Atlantic, with the geostrophic approximation, to investigate the circulation of the Atlantic Ocean at depths down to 2000 db. He prepared (1941a) maps of the steric height at the sea surface relative to various standard pressures down to 3000 db but prepared none of the steric height between two subsurface isobars. His surface maps, especially those referred to the deeper isobaric surfaces, show about the same pattern as seen in more recent data (Leetmaa et al., 1977), except for some rather severe limitations imposed by the quality and number of the data available in the North Atlantic Ocean. In particular, the curious high cell that appears on all his maps at 19–30°N along about 30–41°W stems from the *Carnegie* stations 18, 19, and 20, which he must have received before final processing, unaware of the errors. [Wüst (1935, footnote, p. 230) had noted that the *Carnegie* salinity values in the North Atlantic were too low by 0.03 to 0.04‰ on the average and had apparently either adjusted or deleted them in his work on the *Meteor* atlas, but Defant used them for his maps of steric height without adjustment. When the data were published, the error was noted (Fleming et al., 1945, p. 1).] The effect of a few errors of this sort and the general sparsity of data, and the predominantly zonal track lines in the South Atlantic, apparently tended to detract from the significance of his results. His maps of shear between the surface and 1000 or more db show clearly, however, the Gulf Stream return flow as a much sharper feature than in Iselin's (1936) map. The maps identify the major surface anticyclonic and cyclonic gyres and the North Equatorial Countercurrent, but the zonal track lines obscure the details of the equatorial circulation.

Later (1941b), in order to study subsurface flow patterns, he proposed and used a method for deriving a reference surface for the geostrophic shear. The method of derivation has not been accepted as valid by most physical oceanographers, and the resulting "absolute topography" maps at various pressures have been discounted. Indeed, one might argue that his presentations set back for some years the whole concept of geostrophy as a useful means of examining large-scale circulation through the density structure. This is unfortunate, because his maps, however referenced, are still of geostrophic shear, and provide the first evidence of the horizontal and vertical extent of the return flow just southeast of the Gulf Stream.

In his choice of a reference surface, it seems that he had hoped to satisfy continuity. His reference surface lies shallower near the equator (about 400 to 700 m) and sinks monotonically to 2500 m at 50°S. In the north, the reference extends downward to 1900 m at 50°N, but with a minor shoaling to 500 m at 12°N in mid-ocean and an abrupt rise from 1900 m to 1000 m along the western boundary in the zone from 20 to 45°N. This general shape might appear to take some account of the latitudinal variation of the Coriolis acceleration in calculations of geostrophic transport and to permit the presence of a southward flow beneath the Gulf Stream. In that area, the geostrophic shear retains the same sign to great depth, as shown by his maps of the relative flow, and any calculation of flow relative to a shallower depth would provide such a counterflow. His work with Wüst on the *Meteor* atlas provided maps at 1500 m and 2000 m (and at all depths down to 3500 m) with colder, less saline waters against the coast of North America than could be accounted for by horizontal flow from the south. On these maps, these low values connect directly to the Labrador Sea, and this may have suggested to Defant the presence of a Gulf Stream undercurrent, which would require a reference surface somewhere near 1500–2000 m. [This is remarkably consonant with the direct measurements of Swallow and Worthington (1961).] Furthermore, Wüst's (1935) maps of the spreading of the upper and middle deep water of the North Atlantic (figure 3.7), which were based upon the oxygen core as well as upon salinity and temperature, indicated such a deep return flow along the coast of the American continents.

Defant's map of "absolute" flow [figure 3.9, reproduced from Defant (1941b, Beilage XXIX)] does, indeed, show such a flow extending from the Labrador Sea southward along the western boundary to about 35 to 40°S, where it turns eastward with the West Wind Drift. This is quite like Stommel's (1965) results. This is not meant to imply that these investigators had solved conclusively, with these methods, the problem of the deep flow beneath the Gulf Stream. Defant's colder water on the west could be interpreted as another aspect of geostrophic shear, with isotherms rising to the west, rather than his interpretation of southward flow, and Wüst's core-layer method, while useful, may also yield different interpretations. The problem of the Gulf Stream Undercurrent, or deep western boundary current, has not yet been solved in a generally satisfactory manner. Richardson (1977) finds a strong southwestward flow extending past Cape Hatteras, including a coastal component as shallow as 1200 m. Worthington (1976) allows only a small southwestward component, and at the greater depths and lower temperatures. In any case, the schemes presented by Wüst and by Defant seem to have got some parts of it right by present standards.

It is noteworthy, however, that Sverdrup et al. (1942) make little reference to these studies. While recognizing the North Atlantic as the source of the deep saline waters extending into the other oceans, the undercurrent along the western boundary proposed by Wüst and by Defant was rejected, and no reference was made to a Gulf Stream return flow on its southeastern side. It is perhaps because of this that the western boundary undercurrent was not addressed again until reintroduced by Stommel (1957b), nor the Gulf Stream recirculation until the work of Worthington (1976).

3.5 Studies of Total Transport and Layers

Sverdrup et al. (1942) give little information about subsurface circulation. Aside from the meridional flow of the abyssal waters, generally northward along the western boundary, and the overlying southward flow of the deep waters that they described for all of the oceans that exchange with the Antarctic, and the poleward subsurface currents along the eastern boundary of the North and South Pacific, they provide only one other pattern of flow that is different from the surface pattern. That is the schematic pattern for the intermediate-depth circulation of the North Atlantic [figure 3.8, from Sverdrup et al. (1942, figure 188)]. This is notably different from the transport pattern they produced for the North Atlantic. It shows a westward return flow across the Atlantic just south of the Gulf Stream-North Atlantic Current, much like the pattern Wüst (1935) had proposed from the map of σ_t at 200 m in the *Meteor* atlas. It seems likely that they inferred this flow from the pattern of salinity created by the Mediterranean outflow, but this is not mentioned explicitly.

Their major flow patterns for both the North Atlantic and North Pacific are presented as calculations of transport and do not provide details of the vertical structure. Later maps (Fleming et al., 1945) of the geostrophic shear between various pressure surfaces were prepared from the *Carnegie* data in the Pacific. Though the data set was very sparse for such a presentation, the patterns clearly vary with depth. The poleward shift of the anticyclonic gyres and some return flow south of the Kuroshio are indicated.

The next major contributions on circulation, after publication of *The Oceans*, were from theory and dealt with transport rather than vertical structure. Concepts of the Sverdrup transport (Sverdrup, 1947) and westward intensification (Stommel, 1948) were followed by a study by Reid (1948) of equatorial circulation and by Munk's (1950) treatment of the large-scale wind-driven circulation. Munk's results matched the recognized circumstances of the surface flow remarkably well, accounting for the subpolar cyclonic and subtropical anticyclonic gyres and the system of zonal flows near the equator. They set in train a series of studies by several investigators [some of which were republished together by Robinson (1963)] that dealt with wind-driven circulation and western boundary currents in particular. The idealized oceans were mostly homogeneous, though some two-layer models were discussed as well, and the results were usually given as transports, without consideration of depth variations in the flow patterns other than a downward decrease of velocity.

A massive investigation of the Atlantic Ocean between about 50°N and 50°S, employing both conservative characteristics and oxygen and nutrients, relative

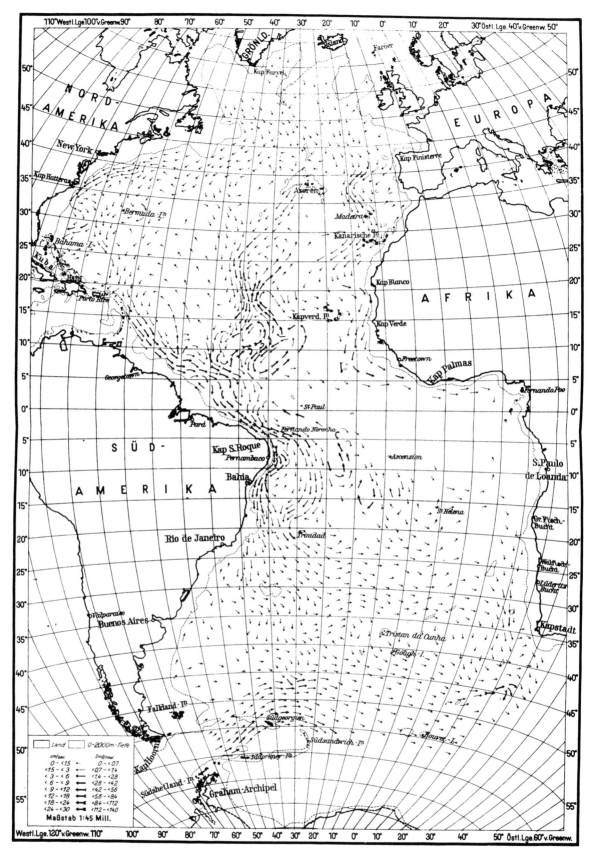

Figure 3.9 Flow field at 2000 m. (Defant, 1941b.)

geostrophic flow, and isopycnal analysis, was carried out by Riley (1951). Although his purpose was to study the nonconservative concentrations and to derive rates of oxygen utilization at various depths, he cast his study in the framework of the general circulation and provided flow patterns along various σ_t-surfaces (from 26.5 through 27.7), the deepest extending to about 2100 m in the central South Atlantic. These flow patterns, while not carried as far as they might have been had they been the principal purpose of investigation, are based upon salinity, oxygen and nutrient data (gridded by 10° intervals of latitude and longitude), as well as the density field, and show some remarkable features.

The scale of the grid he used eliminated some features, of course, and he missed the Gulf Stream return flow except in his deepest layers, and for the upper circulation in the subtropical zone he found only the large anticyclonic gyre that Sverdrup et al. (1942) had mapped. He did show that a branch of the subarctic cyclonic gyre extends southward from the Labrador Sea, forming a substantial Gulf Stream undercurrent as far south as Cape Hatteras, carrying waters of high-oxygen and low-nutrient concentration from the Labrador Sea along the western boundary. He found a poleward subsurface flow along the eastern boundary of the North Atlantic carrying northward waters of high salinity from the Mediterranean outflow and of lower oxygen from the eastern tropical zone. He found that some part of the southward-flowing North Atlantic Deep Water turns eastward near the equator, carrying waters of higher oxygen content eastward between the two eastern tropical zones of low oxygen. Much of this, of course, was quite similar to the earlier results of Wüst and Defant. He did not attempt to carry their studies far forward, but to array them better for his particular study. His principal interests were in estimating the utilization of oxygen and the regeneration of nutrients, and the depth ranges and rates at which these processes occur. He found, using his estimated circulation patterns, that the total oxygen consumption and phosphate regeneration below the σ_t-surface 26.5 (average depth about 200 m) represent the utilization of about one-tenth of the surface production of organic matter by phytoplankton. This is consonant with later findings of Menzel and Ryther (1968), using measurements of dissolved organic carbon, that nearly all regeneration of nutrients takes place above about 500 m, and that in the deeper waters oxygen and nutrients are much more nearly conservative characteristics than in the upper levels. They are not entirely conservative, of course, even at great depth, as Fiadeiro and Craig (1978) have emphasized.

It is, perhaps, unfortunate that Riley's (1951) work came later than Munk's (1950) study. Otherwise, it might have stimulated a more thorough investigation, even with those limited data, of the variation of flow patterns with depth that might have been carried out concurrently with the studies of total transport. Both approaches merited further investigation, but it appears that the impact of Munk's very exciting paper, using one approach, had already engaged the attention of many investigators, and Riley's approach was not so quickly followed, even as more adequate data and methods became available. It has not been ignored or forgotten, of course; one very intensive continuation of his study, attempting to derive rates of deeper circulation, is the GEOSECS program.

It is also worthwhile to note that the differences in approach were not only conceptual but also practical. The Sverdrup transport concept allowed investigators to perform complex studies upon an idealized (homogeneous, steady, two-layer, flat-bottomed, etc.) ocean under an idealized or realistic wind field and to achieve important results in terms of total transport. Riley's sort of approach required the assimilation and manipulation of large quantities of data (though still perhaps too few and of uncertain quality) in order to perform the calculations, and to achieve quite different sorts of results (subsurface flow patterns, for example, instead of total transport).

Wyrtki (1961b), in a study of the thermohaline circulation and its relation to the general circulation, emphasized more clearly the density stratification of the ocean and the necessity that models should include not just surface and abyssal flow, but at least two additional layers, the intermediate and deep waters. These two layers have circulations quite different from the other layers and from each other, and their flow patterns obviously cannot be derived from the various assumptions of purely wind-driven homogeneous or two-layer oceans. And even the four layers discussed by Wyrtki (1961b) are simplifications, as he recognized.

3.6 Mid-Depth Studies Using Isopycnal Analysis

The concept that buoyancy forces in a stratified fluid may influence flow and mixing to conserve density more than other characerictics has been a topic of interest for a long time. Examination of characteristics along surfaces defined by various density-related parameters began in the 1930s, both for the atmosphere and the oceans. Various quantities ($\sigma_t, \sigma_\theta, \delta_T, \delta_\theta$, and $\sigma_1, \sigma_2, \sigma_3, \ldots$, referring the density to 1000, 2000, 3000 db), ... have been employed, and the method has been called "isentropic," "isosteric," "isanosteric," "isopycnic," and "isopycnal." (Hereafter, I shall refer to all the investigations as isopycnal and to mixing along any of the surfaces defined by these parameters as lateral mixing.) None of these quantities is entirely satisfactory because surfaces so defined can represent mixing or spreading surfaces only in various approximations. The problem, of course, is that while such spreading may take place predominantly along such definable surfaces it need not, and indeed cannot, exactly preserve any chosen density parameter. Density is also altered by mixing processes, as examination of the characteristics along such isopycnals makes obvious. This assumption of maximum mixing and flow along such isopycnals remains an assumption, but it has been accepted as one of the useful concepts in studying the ocean.

Isopycnal analysis can lead to some understanding of the mid-depth flow, but until recently most such studies were of the upper waters. The first major studies using the methods of isopycnal analysis were those of Montgomery (1938a) and Parr (1938). Both of these dealt with the upper levels of the ocean, where a simple density parameter, such as σ_t, could be used. At greater depths the choice becomes more difficult. It is interesting to note that in Wüst's (1933) study of the deep Atlantic he showed a vertical section of potential density σ_θ in the deep water. Was he led to do this by some consideration of isopycnal analysis? In any case, he found an inversion—a maximum in σ_θ well above the bottom—and supposed that this would imply a hydrostatic instability. He concluded that either the equation of state or the salinity-chlorinity ratio was not correct and never again dealt with potential density. Instead, he used (Wüst, 1935) the core-layer method in his analysis of the Atlantic stratosphere.

Would the course of investigation of the deep waters of the world ocean have taken a different turn if Wüst had found a way around the inversion in potential density? Ekman immediately provided the proper resolution of the σ_θ inversion in his review (1934) of Wüst's paper. He showed that the vertical gradient of σ_θ is not identical with stability and is not even a useful approximation (in some cases different in sign) below a depth of a few hundred meters. He first proposed the use of different sorts of potential density referred to pressures of 1000, 2000, 3000, . . . db, designated $\sigma_0, \sigma_1, \sigma_2, \ldots$. This is the concept and notation that Kawai (1966) and Reid and Lynn (1971) have used. Veronis (1972) has provided a thorough exposition of the problems in dealing with density.

Montgomery (1938a) derived a method for calculating the geostrophic shear between a surface of constant specific volume anomaly and a deeper isobaric surface (or a deeper surface of constant specific volume anomaly). This has been used in the upper ocean by various investigators (Reid, 1965; Tsuchiya, 1968; Buscaglia, 1971). At greater depths, however, the specific volume anomaly term (which is in itself not a significant physical quantity, but an offset from an arbitrary standard that varies with pressure) does not correspond to any of the various quantities used in isopycnal interpretations and cannot be used below the upper few hundred meters.

Montgomery's (1938a) study of the upper layers (mostly above 500 m) of the Atlantic Ocean between the equator and 30°N was the first attempt to discuss the concept of isopycnal analysis and its practical application, and to implement it over a substantial area of the ocean. It was based upon almost the same data set that Defant (1936) had used to derive circulation from the salinity maximum and the vertical density gradients (used to determine "discontinuity levels"). Both of these studies were hampered by the lack of north-south lines of stations in a predominantly zonal circulation system (still a limitation in much of the ocean), but the isopycnal method seemed to be the more fruitful. Later work, with a more complete data set (Cochrane, 1963, 1969) appears to support Montgomery's interpretations. It is interesting to compare Montgomery's maps with those recently produced in Merle's (1978) atlas, which uses the larger data base now available.

Clowes (1950), in a study of the waters surrounding southern Africa, used maps on σ_t-surfaces reaching as deep as 1200 m to identify saline waters flowing southward from the Indian Ocean with the Agulhas Current and then westward into the Atlantic. At the deeper surfaces he used, such as σ_t of 26.5 to 27.25, the contrast between the high-salinity west-Indian source and the lower-salinity eastern Atlantic (Intermediate) waters is particularly marked: both the inflow to the Atlantic and the mingling eddy patterns where the Agulhas Current meets the West Wind Drift are well delineated.

Taft (1963) used isopycnal δ_θ-surfaces to discuss the distribution of salinity and oxygen south of the equator in all the oceans. The greatest depth reached by his isopycnals was about 1500 m. Although the data set he used was sparse, it was well chosen and representative, and he was able to show, among other features, the southward extension from the Arabian Sea of high-salinity, low-oxygen water through the Mozambique Channel, with a part entering the Atlantic. In particular, the maps of the depth of the isopycnals are the first to show for all three southern oceans that the great lens of low-density water corresponding to the subtropical anticyclonic gyre at the surface shifts poleward at greater depths. This had been apparent for both the North and South Atlantic from the *Meteor* atlas and for the North Pacific from the NORPAC atlas (NORPAC Committee, 1960), but had not been mapped for the South Pacific and Indian Oceans. He remarked also upon the differences in salinity (and thus temperature) between the various oceans on the isopycnals chosen. The influence of low-salinity water from the Pacific, entering the Indian Ocean north of

Australia, is clearly evident in the upper 300 m, and some effect may be detected as deep as 1200 m in the eastern area. Wyrtki's (1971) atlas, with additional data, confirms this effect to at least 1000 m in the eastern area, and Sharma (1972) has carried out a more detailed study of this low salinity in the upper waters (at a σ_t of 26.02).

Kirwan (1963), in a study of the circulation of the Antarctic Intermediate Water of the South Atlantic, calculated relative geostrophic flow along isopycnal σ_t-surfaces relative to 2000 db. He showed not only that the subtropical anticyclonic gyre shifts poleward at greater depth, but also that an eastward flow appears north of it, at about 10°S, in the entire depth range he considered (about 300 to 1200 m). While this is consonant with the results of Defant (1941b), Defant did not explicitly accept such a flow pattern as real. Riley (1951) has maps of flow on various σ_t-surfaces, some of which indicate such a flow, but his others do not, and he does not remark upon it. Not surprisingly, the three investigators, using the same data base, though with different methods, derived somewhat similar results in that area. But only Kirwan was concerned with details of flow in that area and explicitly pointed out the eastward flow beneath the surface. Later, Reid (1964a) noted it at the surface across one meridian, and Mazeika (1968) mapped the surface flow over a larger area. Lemasson and Rébert (1973) found it both in the geostrophic shear and by direct measurement. The most recent and comprehensive maps are those in the atlas prepared by Merle (1978).

A similar feature was noted in the Pacific Ocean (Reid, 1961a) when the data were being arrayed to perform a larger-scale isopycnal analysis of the intermediate-depth low-salinity water. Investigations using isopycnal surfaces have nearly always led to consideration of geostrophic shear as well, as the depth patterns of the isopycnals seem to be defined largely by the geostrophic balance to the flow.

Such isopycnal studies have also led to a consideration of vertical diffusion as well, as nearly all large-scale examinations show some obvious evidence of its effect. A particular case is that of the salinity minimum that lies within the subtropical anticyclonic gyre of the North Pacific. Kuksa (1962) had examined the salinity minimum of the North Pacific and used the geostrophic shear between 400 and 1000 db in studying the circulation. Later (Kuksa, 1963), he examined the salinity along an isopycnal ($\sigma_t = 26.75$) and concluded that the origin of the low-salinity water coincides with the shallow subsurface temperature minimum found in the northwestern Pacific in summer (that is, from the winter mixed layer, of which the temperature minimum in summer is a remnant). Reid (1965) has concluded, however, that while the isopycnal range near the core of the minimum does not outcrop in the North Pacific, yet the source of the minimum is at the sea surface there. Vertical diffusion from the low-salinity waters in the mixed layer north of 45°N, through the pycnocline and into the higher-density underlying water, is the only process that can account for the salinity pattern to the south (Reid, 1965).

There have been more isopycnal studies of the Pacific Ocean than of the other oceans, though most of them have dealt with the waters above a depth of 1000 m. A series of papers began with the work of Austin (1960) and Bennett (1963) in the eastern tropical area and continued with Reid's (1965) study of the Intermediate Waters, Cannon's (1966) study of the Tropical Waters, Barkley's (1968) atlas on a set of surfaces from σ_t-values 23.00 to 27.70, and Tsuchiya's (1968) analysis of the upper waters of the intertropical zone. Tsuchiya's study was particularly important in that it investigated the longitudinal extent and depth range of the Equatorial Undercurrent and revealed the sources of its waters. Employing isopycnal distributions and geostrophy, he showed that the major source of the waters of the Equatorial Undercurrent, which are highly saline and oxygen rich, is the South Pacific, though the Undercurrent and North Equatorial Countercurrent in the western area are not separated by a westward flow. It is in the west that the South Pacific waters appear to cross the equator, as Sverdrup et al. (1942) had suggested, though this pattern is stronger at greater depths. He examined also the North Equatorial and South Equatorial Countercurrents, pointing out that Yoshida (1961) had given some explanation of the latter in terms of the curl of the wind stress. Tsuchiya described the flow as lying near 5°S at depths of 200 to 300 m and shifting toward 10°S at shallower depths. The feature had been examined, and a few direct measurements made, by Burkov and Ovchinnikov (1960) and Koshlyakov and Neiman (1965). Tsuchiya (1975) has presented additional information from the more recent EASTROPAC expedition (Love, 1972).

In a study of the Subtropical Mode Water of the Pacific, analogous to the 18° water of the North Atlantic described by Worthington (1959), the characteristics were examined on isopycnal surfaces from the equator to 45°N by Masuzawa (1969), who made some use of the deeper density field in discussing the circulation of this slowly moving central-ocean water mass.

An attempt to approximate an isopycnal surface over a wide pressure range and to map the characteristics along such a surface over the world ocean was made by Reid and Lynn (1971). The spreading of a layer of water formed in the North Atlantic Ocean from a mixture of the Denmark Strait overflow water with the ambient, more saline waters south of Greenland was traced along such an approximated surface by changing the reference pressure wherever the depth of the isopycnal changed by some chosen amount. In this case the reference pressures were 0, 2000, and 4000 db of hydrostatic pressure. The warm, saline water found at the chosen isopycnal in the North Atlantic could be traced southward in the Atlantic, along the Circumpolar Current into the Indian and Pacific Oceans, and back into the southwestern Atlantic. The effect of vertical mixing in altering the density as well as other characteristics is particularly obvious in the deep North Pacific, where salinity and temperature decrease northward along a deep isopycnal ($\sigma_4 = 45.92$) in the absence of any new lateral source of cold water.

Callahan (1972) used a pair of isopycnals (δ_θ values of 50 and 30 cl ton^{-1}, corresponding to σ_θ-values of 27.60 and 27.81) lying somewhat shallower than that chosen by Reid and Lynn (1971) for examining the ocean south of about 40°S. His upper surface corresponds closely to the vertical oxygen minimum of the upper Circumpolar Deep Water, the lower surface approximately to the vertical salinity maximum associated with the Warm Deep Water of the Antarctic that originates from the North Atlantic. He accounted for the low oxygen of the upper circumpolar waters partly by lateral exchange with the very low-oxygen waters of the western Indian Ocean and southeastern Pacific Ocean and found evidence of a western boundary current flowing poleward at depths of 1600-1800 m east of New Zealand.

The isopycnal method has also been used in studies of the Atlantic Ocean. Buscaglia (1971) examined the Intermediate Water in the western South Atlantic using isopycnal distributions and relative geostrophic flow and concluded, as had Martineau (1953), that the low-salinity water from the Antarctic does not extend northward all along South America, as Wüst (1935) had supposed. He found, instead, that it extends along the coast only to about 40°S with the Malvinas (Falkland) Current, and then turns eastward and flows around the anticyclonic gyre, whose axis at the depth of the Intermediate Water lies at about 35°S. The concept of thermohaline flow as a western boundary current does not apply at the depth range of the Intermediate Water in either the South Atlantic or the Pacific. Instead, the circulation seems more like that of the wind-driven anticyclonic gyres recognized in the upper layer, though they appear to be shifted poleward in this depth range.

Similarly, Ortega (1972) used both isopycnal distributions and relative geostrophic flow in his study of the Caribbean. He was able to show that the Intermediate Waters enter only through the southern passages, where the isopycnals lie shallower, in balance with a generally westward transport, and that the warmer and more saline, less dense waters are the dominant part of the incoming water farther north, where the isopycnals lie deeper.

Lazier (1973a) examined the renewal of Labrador Sea Water and showed that renewal takes place in the central part of the sea in winter, and that the cooled, less saline waters spread outward along isopycnals that rise to meet the sea surface near the axis of the cyclonic circulation that obtains in the northern North Atlantic.

Pingree (1972) first studied mixing processes in the deep ocean and found evidence of isopycnal mixing in the small-scale structures. He defined a potential density over a small depth range in terms of one intermediate pressure and called such isopycnals neutral surfaces. Later, Pingree (1973) examined larger patterns and showed that cold, low-salinity waters from the Labrador Sea may extend along such neutral surfaces into the Bay of Biscay. Pingree and Morrison (1973) examined the northward extension of Mediterranean outflow water along isopycnals variously defined and found that this water has maxima in buoyancy frequency both above and below it. [This sort of structure was examined later in the South Atlantic by Reid, Nowlin, and Patzert (1977).]

Ivers (1975) used a variant of Pingree's neutral-surface concept in examining the deep circulation of the northern North Atlantic. His neutral surfaces are normal at every point to the gradient of potential density, such potential density being referred to the pressure at the point in question. He examined five such neutral surfaces, all of which outcrop in the north. The shallowest outcrops in the Labrador Sea and extends to about 900 m near 38°N. The deepest outcrops only in the Norwegian-Greenland Sea and lies as deep as 3000 m along 40°N. From this examination he concluded that there is a cyclonic gyre to the north of the Gulf Stream-North Atlantic Current, and that this gyre, though distorted by the Rockall Bank, Reykjanes Ridge, and Greenland, extends throughout the northern North Atlantic down to 3000 m at least. He found that most of the features of the salinity distribution can be explained by processes of lateral flow and mixing: the notable exceptions are the vertical mixing induced by cooling at the outcrops and by the rapid flow over the sills and through the narrow channels of the Greenland-Scotland Ridge. He attempted an interpretation of the geostrophic shear, not assuming a level or even continuous layer of zero flow, but by entering various constraints upon the velocity wherever measurements of speed were available or where qualitative reasoning, based upon the characteristics along the neutral surfaces, indicated a sense of flow. The result is a qualitative, but internally consistent, circulation field. In addition to deriving flow patterns within the four embayments of the North Atlantic (Labrador Sea, the troughs east and west of the Reykjanes Ridge, and the Rockall Channel), he shows a strong return flow south of the Gulf Stream and a subsurface poleward eastern boundary current.

More recently, Clarke, Hill, Reiniger, and Warren, (1980) have examined the area just south and east of the Grand Banks of Newfoundland, using isopycnal distributions, geostrophic shear, and current measurements. Their purpose was to determine whether there is a branching of the Gulf Stream near the Grand Banks, with a part flowing into the Newfoundland Basin, or whether a separate anticyclonic gyre exists within the Newfoundland Basin, as Worthington (1976) had proposed. They concluded, on the basis of the distribution of characteristics and the relative geostrophic flow, that the branching does occur.

3.7 Comparison of Relative Geostrophic Flow at Mid-Depth with Numerical Models of Transport

There is an interesting correspondence between some of the newer patterns of mid-depth circulation based upon geostrophic flow relative to some deep isobaric surface and the total transports calculated in some of the recent numerical models. This seems especially significant in that the newer pattern seems to have been arrived at quite independently through the two different methods of investigation.

3.7.1 The Density Field

The density field in the North Atlantic Ocean can be represented in part by a map of the depth of a σ_t-surface (figure 3.10). On this map the major trough has the shape of the letter C, with the two arms extending eastward from the western boundary. An analogous pattern exists in the South Atlantic but was not apparent in the zonal data array of the *Meteor* expedition, from which this map was made.

The density pattern in the North Atlantic has about the same shape as that in figure 3.10 over a substantial depth range, as can be seen in the *Meteor* atlas, and is reflected in the appropriate maps of steric height (Leetmaa et al., 1977; Reid, 1978; Stommel, Niiler and Anati, 1978). The trough in figure 3.10 appears as a continuous ridge on the maps of steric height. This C-shaped pattern thus suggests the Gulf Stream and its westward return flow, which turns southward near 70°W, eastward along about 30°N, and finally westward along about 25°N.

The poleward arm of the C is much more clearly defined in these data than the equatorward limb. It was suggested in Jacobsen's (1929) maps (figure 3.6A) of steric height and was very clear in Defant's (1941a) maps of relative flow, but apparently, it was not accepted by most investigators as a real feature until the recent work of Worthington (1976).

A similar pattern appeared marginally in the Pacific from the *Carnegie* maps (Fleming et al., 1945), though the data were sparse. It has already been pointed out that Taft (1963) showed for all three southern oceans that the great lens of low-density water corresponding to the subtropical anticyclonic gyre shifts poleward at greater depths, and that Kirwan (1963) mapped it in detail in the South Atlantic and noted an eastward flow appearing on the equatorward side. All subsequent treatments of the data (Muromtsev, 1958; Reid, 1965; Barkley, 1968; Reed, 1970b; Burkov, Bulatov, and Neiman, 1973; Reid and Arthur, 1975) have shown this shift.

It is illustrated clearly in the world maps of Burkov et al. (1973) of geopotential anomaly at the sea surface and 500 db relative to 1500 db (reproduced here as figures 3.11A and 3.11B). They note that the maximum values of steric height along the axes of the subtropical anticyclonic gyres are farther poleward in the deeper field in all oceans.

The equatorward limb of the C-shape is not such a strong feature and has been much harder to detect. Of course, it must appear to some extent in surface maps that show the return flows of the Gulf Stream (Defant, 1941a; Reid et al., 1977), Kuroshio (Koenuma, 1939; Wyrtki, 1975a; Reid and Arthur, 1975), East Australian Current (Wyrtki, 1975a), Brazil Current (Reid et al., 1977) and Agulhas Current (Wyrtki, 1971). Its eastward part was discussed by Yoshida and Kidokoro (1967) and Hasunuma and Yoshida (1978) for the Pacific.

At depths below the sea surface, the split of the anticyclonic high cell into two parts, connected in the west into a C-shape, can be seen in the South Pacific Ocean on the map of steric height at 800 db relative to 1500 db, reproduced as figure 3.12 from the publication of the Academy of Sciences of the U.S.S.R. (Kort, 1968). At the sea surface the axes of the subtropical gyres in the North and South Pacific lie near 20°N and 20°S in mid-ocean (figure 3.11A). At the 500-db surface they lie near 30°N and 35°S. At the 800-db surface the northern axis is near 35°N and the southern axis near 40–45°S. A new high lies along about 30°S, connected to the other feature along about 180°: This is the C-shape, which is defined by the present data bank better in the South Pacific than in any other ocean.

The C-shaped deep pattern has also been shown in the North Atlantic by Leetmaa et al. (1977). They take into account not only the westward Gulf Stream recirculation but its southward component across 24°N; from their map of steric height, this would appear to be southeastward, as part of the equatorward arm of the C.

Reid (1978) used a somewhat more detailed interpretation of the relative geostrophic flow field in an attempt to account for the distribution of the Mediterranean outflow water in the North Atlantic, and Reid

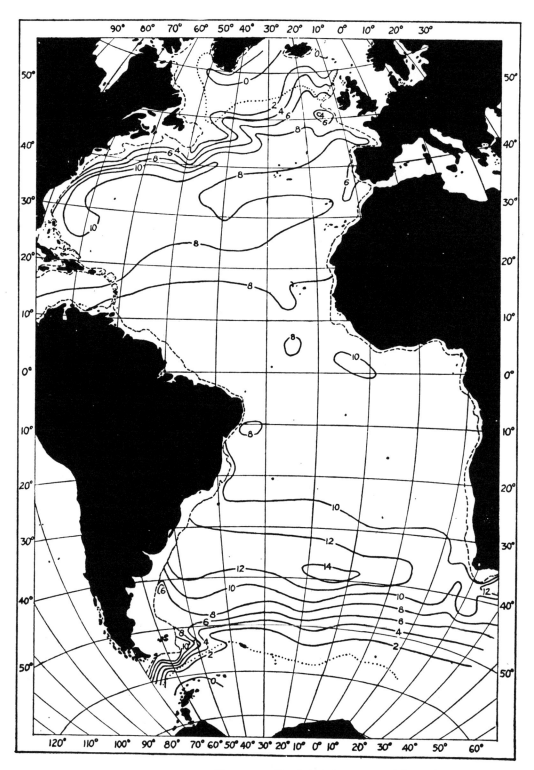

Figure 3.10 Depth in hektometers of the surface where $\sigma_t = 27.4$. (Montgomery and Pollak, 1942.)

(3.11A)

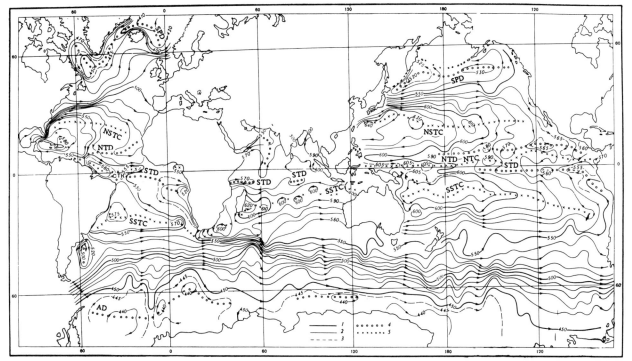

(3.11B)

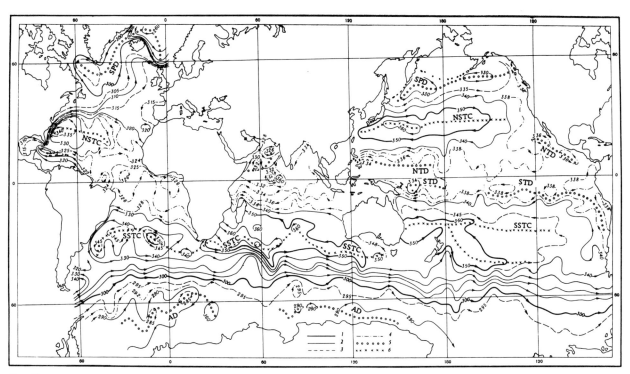

Figure 3.11 Steric height (dynamic cm) (A) at the sea surface, and (B) at 500 db, with respect to the 1500-db surface. (Burkov et al., 1973.)

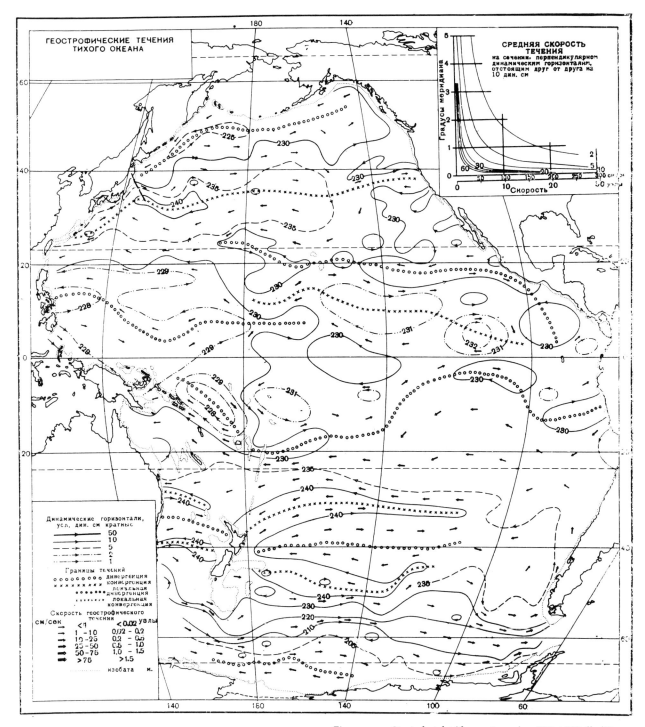

Figure 3.12 Steric height (dynamic cm) at 800–1500 db. (Kort, 1968.)

and Mantyla (1978) presented a map of steric height in examining the mid-depth oxygen pattern in the North Pacific: in both cases the C-shape is clear, and is consonant with the distribution of those characteristics.

3.7.2 Numerical Models
Most of the flow patterns illustrated by the numerical models are of total transport or of surface flow. Models of flow fields at mid-depth (1000–2500 m) are presented but rarely. This may be because there has been so little information with which such results could be compared. Total transports need not correspond precisely to the mid-depth flow, but are probably weighted toward the near-surface flow, which is usually much stronger. It is still useful to compare them with the steric maps, as there seems to be some correspondence between the patterns.

The first of the numerical models to give some indication of this C-shaped pattern was that of Bryan (1963), followed by such studies as those of Veronis (1966b), Blandford (1971), Gill and Bryan (1971), Holland and Lin (1975a), Semtner and Mintz (1977), and Robinson, Harrison, Mintz, and Semtner (1977). The first and last of these are reproduced as figures 3.13 and 3.14.

In the first of these (figure 3.13), only the beginning of the deformation is seen, but it very quickly expands (Veronis, 1966b) and the pattern of figure 3.14 is typical of most of the subsequent studies. I shall not try to describe the bases of such models, but shall borrow from Robinson et al. (1977, p. 191), who stated that their streamfunction (figure 3.14) is "qualitatively similar to nonlinear single gyre experiments by Veronis (1966b) and Holland and Lin (1975a).... The interior circulation is qualitatively similar to the traditional linear Sverdrup interior." However, the poleward shift does not require a nonlinear model; for example, it appears in Kuo's (1978) model of a nonhomogeneous ocean, both with and without consideration of bottom topography.

Only Gill and Bryan (1971), who modeled the South Pacific Ocean, noted that the poleward shift at increasing depth bore some resemblance to the actual fields of density and relative geostrophic flow in the South Pacific and cited some supporting evidence. Their model showed an eastward flow equatorward of the shifted anticyclone and connected to it in the west, but did not complete the C-shape with a westward flow farther equatorward.

From an examination of figures 3.13 and 3.14 it is unclear what to call these features. Is what has been referred to as a poleward shift really a shift, or is it that only the higher-latitude part of the C-shaped pattern was recognized at first and a phrase prematurely applied? Do we really have a poleward shift of the anticyclonic gyre with a new but weaker gyre added on the equatorward side and connected in the west, or is it merely that one large gyre, thought at first to be elliptical, now is seen to have a huge and asymmetric dent in its eastward side?

Two studies of the pattern in relation to the distribution of characteristics bear upon this. In one (Reid, 1978), the position of the highly saline waters of the Mediterranean outflow at a depth of 1000 m is interpreted as extending westward across the North Atlantic between the two arms of the C-shaped pattern. Lower values of salinity are seen to the north, west, and south of the Mediterranean outflow. From the salinity pattern alone, one might imagine that there are two separate anticyclonic gyres, that in the north containing Labrador Sea water and that in the south made up of low-salinity water from the South Atlantic. There is no immediate reason from the salinity pattern to suppose that the two anticyclonc gyres are connected in the west, though the pattern of relative geostrophic flow suggests that they are and that the C-shaped steric high surrounds the Mediterranean outflow.

In a similar treatment of the North Pacific (Reid and Mantyla, 1978), however, the oxygen pattern does require a connection in the west between the two anticyclonic arms, and the C-shaped pattern must be complete. The higher oxygen values in the far north must derive from a lower latitude and extend continuously through both arms of the C-shaped pattern.

3.7.3 The Diagnostic Models
Various investigators have used compilations of data to describe the density field and used these as inputs to their models. Arraying a proper set from the available data bank has not been an easy task. Substantial editing of the materials would be required to eliminate biases and random errors from a data base that extends over a period of more than 50 years, and has been collected with some differences in techniques and instrumentation, with most expeditions based on an ad hoc plan to address a particular problem rather than to contribute to a coherent ocean-wide program. Accuracy is most limiting at the greater depths, where both horizontal and vertical gradients of salinity are small, and errors can often exceed the local time-and-space variation.

As a result, most of the diagnostic models have used simplified or smoothed fields and have not yet made extensive use of the density field at great depths. If they do use unedited materials at great depths, a few measurement errors may lead them seriously astray. Work is under way on the preparation of such an edited data set by, for example, Levitus and Oort (1977).

Kozlov (1971), using density at only two levels in the Pacific, found only the poleward shift in the North Pacific, but found the full C-shape in the South Pacific. Holland and Hirschman (1972) used a partly diagnostic

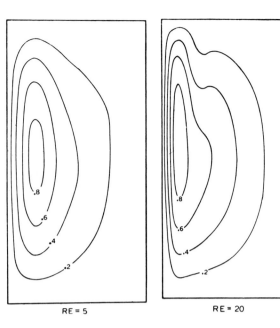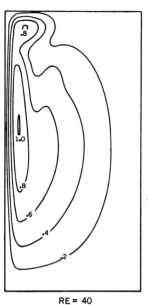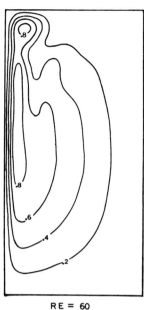

Figure 3.13 Patterns of transport stream functions for various values of Reynolds number: 5, 20, 40, 60. (Bryan, 1963.)

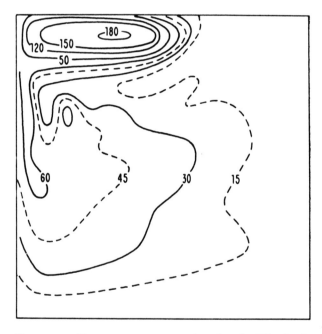

Figure 3.14 Mean transport stream function, in 10^6 m^3 s^{-1}. (Robinson et al., 1977.)

model for the North Atlantic. Their data base for the upper 1000 m matched fairly well the patterns of Defant (1941a), Reid et al. (1977) and Stommel et al. (1978) and showed the C-shape in the density field. The derived surface flow of the diagnostic model showed it somewhat better, but their mass transport showed it less clearly. Their deep flow field (their figure 11) appears to have too many separate gyres to allow for interpretations of the C-shape, but clearly shows the poleward shift. (The feature in their figures 11 and 12 along 15-25°N at about 40°W, referred to as a complex set of currents west of the Mid-Atlantic Ridge, is remarkably like the *Carnegie* effect referred to earlier that obscured Defant's maps, but not Wüst's.)

Cox (1975) used a data set averaged or interpolated to a 1° × 1° grid for the world ocean. His flow field indicates a weak poleward shift, but only in the southern oceans; in only the South Pacific is there a suggestion of the C-shape.

Sarkisyan and Keonjiyan (1975) produced various diagnostic models of total transport and free-surface elevation (surface flow) in the North Atlantic, both with and without bottom relief, and emphasized that consideration of the effects of baroclinicity and bottom relief is essential. All of their results give the poleward shift, or Gulf Stream recirculation, with an eastward flow south of the recirculation, but only one of their maps (surface topography) suggests the complete C-shape.

3.7.4 Other Approaches
It has been supposed that over much of the ocean the speed of flow varies with depth, usually decreasing. The poleward shift with depth and the other limb of the C-shape indicate that direction also changes with

depth. Stommel and Schott (1977) also find a change in direction with depth, which they call the β-spiral, when applying their method of calculation of absolute velocity from the density field. The results of their calculations from data surrounding 28°N, 36°W give a flow at 200 m directed about south by west. This is consonant with the typical large elliptical anticyclonic flow at the surface and with the various maps of relative geostrophic flow (Defant, 1941a; Reid et al., 1977; Leetmaa et al., 1977; Reid, 1978). Their calculated flow spirals clockwise with increasing depth and is toward about east by north at 1000 m. This is quite consonant with Reid's (1978) map of flow at 1000 m (geostrophic flow at 1000 db relative to 2000 db): it corresponds to the interior side of the southern arm of the C-shaped gyre.

In a later study, Schott and Stommel (1978) used this method to calculate the absolute geostrophic flow from the density field at several other positions. At 20°N, 54°W they find a weak southwestward flow at 100 m depth, where Leetmaa et al. (1977) and Reid et al. (1977) find a weak southeastward flow. This is very near the axis of the equatorward arm of the C and may not be significantly different. Near 1000 m depth they find a stronger flow toward east by north, matching fairly well Reid's (1978) relative-flow map. In the Pacific, at 22°N, 151°W, they calculate a weak south-southwestward flow at 100 m, consonant with the various maps of surface flow relative to 1000 or 500 db (Reid, 1961a; Wyrtki, 1975a). At 1000 m their calculated flow is directed east by north, consonant with the 1000-3500-db map of the North Pacific prepared by Reid and Mantyla (1978).

A similar calculation made in the North Pacific from data along 35°N and 155°W has been made by Coats (1979). He points out that this is farther north than the 28°N Atlantic position used by Stommel and Schott (1977) and in a different part of the circulation system. He finds the near-surface flow directed east-southeast, turning clockwise downward, and toward west-southwest at 1000 m; he notes that these directions are in agreement with the maps of flow at the surface relative to 1000 db (Reid, 1961a) and at 1000 db relative to 3500 db (Reid and Mantyla, 1978). These agreements may signify only that in these areas the velocity decreases monotonically with depth through much of the depth range considered. If the unknown absolute values at the reference surfaces for the relative-flow patterns, while not zero, are small, then the relative flows are not offset very much from the absolute. In such particular cases, the two methods would agree closely, as they use the same geostrophic shear field.

This could, perhaps, partly explain the discrepancy in the South Atlantic between the calculations of Stommel and Schott (1977) and the relative maps. They all (between 20 and 30°S, 10 and 20°W) indicate northwestward flow at the surface, consonant with the 0-2000-db map (Reid et al., 1977), but southeastward flow near 1000 m, quite different from the westward relative flow at 1000-2000 db (Reid et al., 1977). This implies that if there is no error in either the data bank or in its use, there must be a strong eastward flow at 2000 m depth. Unfortunately, there are no north-south data sets in that area that allow this to be examined properly through the density field, though both Wüst and Defant (figures 3.7 and 3.8) found some evidence for such an eastward flow.

Wunsch (1978a) has used selected data to examine the general circulation of the North Atlantic west of 50°W. He has used linear, geostrophic, and mass-conserving models and inverse methods. As in the diagnostic and the β-spiral models, the geostrophic shear is preserved, with some smoothing. In the areas where the shear is strongest, it remains monotonic with depth, and his particular constraints include very low velocity at the bottom. It is not surprising, then, that his total transport pattern (figure 3.15A) shows some resemblance to the maps of relative flow. His map (figure 3.15B) of the transport in the depth range of 17 to 12°C (about 400 to 700 m southeast of the Gulf Stream) shows the return flow westward but not the eastward flow near 25°N seen on the relative maps at 100-1500 db (Stommel et al., 1978) or at 0-1000, 0-2000, or 1000-2000 db (Reid et al., 1977; Reid, 1978). He does show this feature in his map (figure 3.15D) in the range of 4 to 7°C (about 1000 to 1600 m), though it is nearer 20°N. His deepest map (figure 3.15E), at temperatures less than 4°C (below 1600 m), indicates no Gulf Stream south of 35°N and no flow west of 70°W, but a southward turn of the return current east of 70°W extending past 10°N. This is remarkably like Defant's (1941b) map of "absolute" topography of the 2000-db surface (figure 3.8), except north of 35°N in the west. There Defant had inserted a zero surface sloping southeastward from 1000 m to 1900 m across the Gulf Stream, and at greater depths this provided a southwestward flow from the Labrador Sea as far south as Cape Hatteras.

None of Wunsch's maps indicates the full C-shaped pattern, including a westward flow along the southern arm, except possibly in the range from 4 to 7°C, which indicates such flow at 10°N. This is farther south, however, than the relative flow maps of Stommel et al. (1978) and Reid (1978).

3.8 Mid-Depth Patterns in the World Ocean

In the hope of bringing some coherence to the preceding descriptions of the various expositions, I have pre-

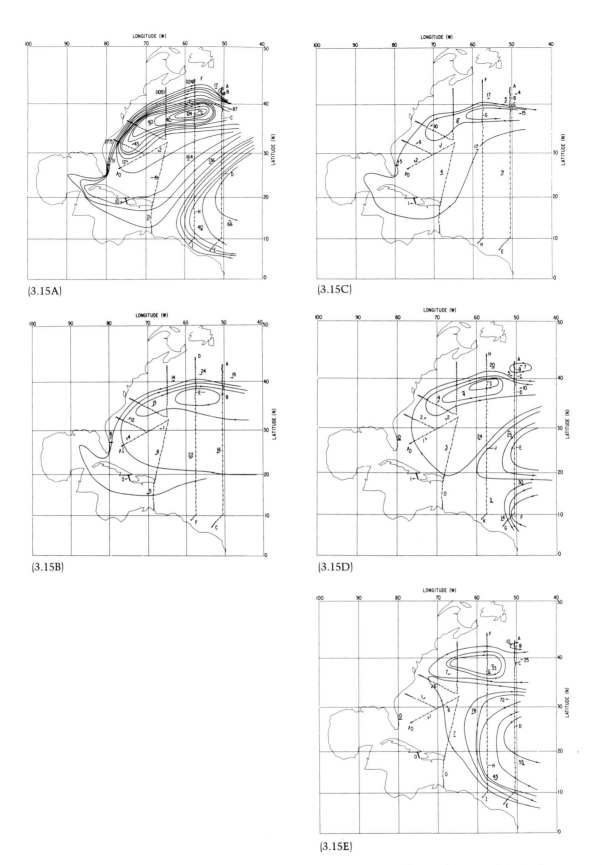

Figure 3.15 Circulation diagrams: (A) Total transport; (B) transport, 12-17°C level; (C) transport, 7-12°C level; (D) transport, 4-7°C level; (E) transport, <4°C level. (Wunsch, 1978a.)

pared a set of maps (figures 3.16A-3.16E, 3.17). They are meant to show the patterns of characteristics in the world ocean that are produced by the various sources and the processes of advection, diffusion, consumption, and regeneration.

The data set used is not necessarily the best selection possible from the present data bank, but represents the present stage in such a selection. Station positions are indicated by dots (though along some tracks the stations were so close together that I omitted alternate dots to avoid creating a solid line on the figure). The various fields are not equally well represented because the measurements vary in both quantity and quality. The selection is best in the Pacific because I am more familiar with the data there. Also, oxygen and nutrient data are more numerous there and perhaps of better quality. In the Atlantic, both oxygen and nutrients appear to show some biases among different expeditions. I have taken some liberties in adjusting the values for these maps: perhaps new data sets, or data I have not yet considered, will make substantial changes. The maps of the Indian Ocean show a denser station coverage in mid-ocean than the others. This may be because I have begun the selection there only recently and have not yet had time to identify those of highest quality.

For substantial areas of the ocean, these maps must be recognized as preliminary, and some alterations of the patterns will be made, perhaps from further consideration of these data alone, and certainly as other data are made available. Indeed, the mapping may be premature, but we must start somewhere.

3.8.1 Distributions on an Isopycnal Surface

The maps prepared consist of various characteristics along a chosen density parameter and of steric height 2000-3500 db. The surface represented here was chosen, almost at random, from the range of density that lies near 2000 m in low latitudes and extends throughout most of the ocean, intersecting the sea surface only in very high latitudes. It was not meant to represent any particular source, but merely the general patterns and processes that obtain throughout the world ocean. The density parameter (henceforward called isopycnal) is defined by a value of 37.0 in σ_2 at depths below 1500 m. Where it rises to the north in the Atlantic, it is defined in the manner of Reid and Lynn (1971) by a value of 32.47 in σ_1 between 1500 and 500 m and by a value of 27.845 in σ_0 where it lies above 500 m. Toward the Antarctic the values are 32.44 in σ_1 and 27.76 in σ_0: these differ from those in the North Atlantic because of the lower temperature and salinity. The surface does not extend as shallow as 1500 m in the North Indian or North Pacific Oceans, and the original parameter of 37.0 in σ_2 applies there.

Although the isopycnal was not chosen to correspond to any particular layer, its position does correspond to, or lie near, some layers that have been named or discussed. In the northwestern Atlantic, it lies within the lower part of the Labrador Sea Water. Farther south, it lies near the layers that Wüst (1935) had called the "Upper and Middle North Atlantic Deep Water." It lies beneath the salinity minimum of the Subantarctic Intermediate Water and in the Antarctic it is included within the Circumpolar Water. In the Indian and Pacific Oceans, it lies well beneath the Intermediate Water but above the Common Water described by Montgomery (1958).

Depth The depth of this isopycnal (figure 3.16A) is greatest in the Pacific and least in the Atlantic. Values near the equator are above 1900 m in the Atlantic, 2250 m in the Indian, and 2600 m in the Pacific: the Pacific is the least dense of the oceans. In the Atlantic, the isopycnal lies deepest in the northwest and near 30 to 40°S, consonant with a poleward shift of the deep anticyclonic circulation. In the North Atlantic, it extends across the Faroe-Iceland Ridge and through the Faroe Channel and the Denmark Strait, into the Norwegian-Greenland Sea, where it outcrops. In the south, it lies at depths less than 100 m in the central Weddell Sea gyre and probably outcrops there in winter. It lies shallow all around Antarctica and probably outcrops at various places on the shelf.

In the Indian Ocean it lies deepest between 30 and 40°S in a long trough that extends eastward south of Australia and into the Tasman Sea. In the Pacific, which has a better set of north-south expedition tracks than the other oceans, the shape can be defined better, and it shows a series of zonal troughs and ridges reflecting the vertical shear patterns that will be seen on the map of steric height.

Salinity The highest salinity values (figure 3.16B) are seen in the central North Atlantic, directly beneath the vertical salinity maximum originating from the Mediterranean outflow. Another high is seen just south of Iceland, stemming from the flow from the Norwegian Sea over the Iceland-Scotland Ridge (Worthington and Wright, 1970; Ivers, 1975). Low values are found within the Labrador Sea, where the overturn and vertical diffusion provide a low salinity whose lateral extension separates vertically the Mediterranean and Norwegian sources, as seen in the atlases of Fuglister (1960) and Worthington and Wright (1970). The Denmark Strait outflow does not provide an extreme in salinity at this density. In the North Atlantic, the depth of this isopycnal lies near the depth of Wüst's Middle North Atlantic Deep Water, which he defined by an intermediate oxygen maximum, and the salinity pattern is very similar. In the South Atlantic, the isopyc-

Figure 3.16A (Facing pages) Depth (hm) of the isopycnal on which the characteristics are shown on figures 3.16B–3.16E. In the shaded areas, all the water is less dense than the isopycnal chosen. The isopycnal outcrops around the hatched area in the Norwegian–Greenland Sea, and in the following figures, the sea-surface values are contoured there but the area is not outlined.

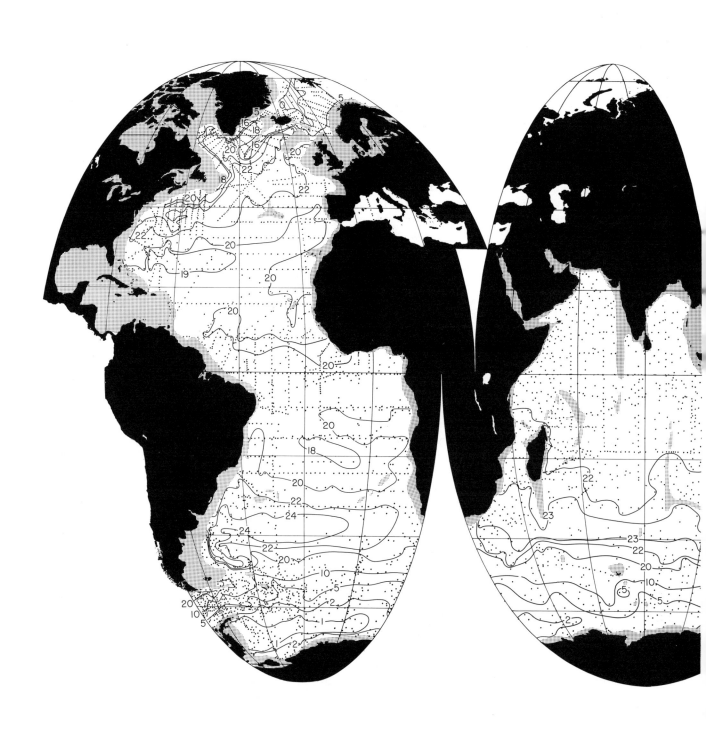

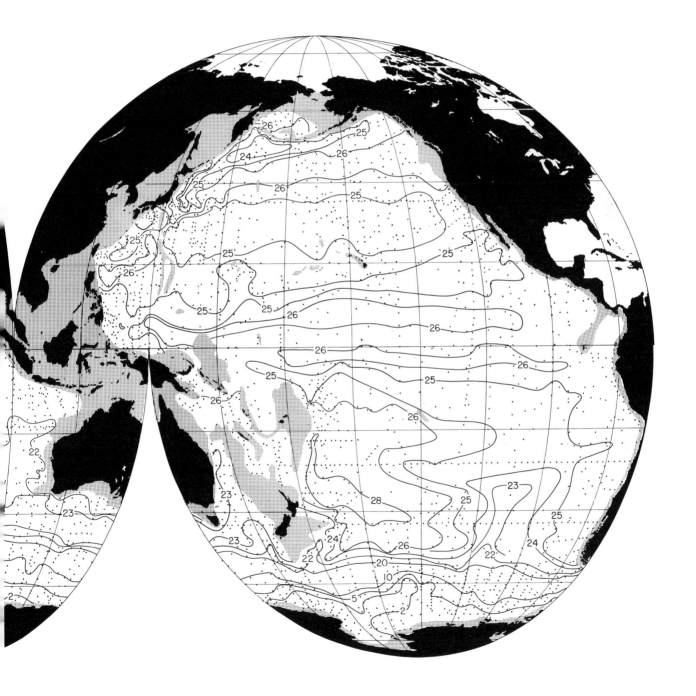

On the Mid-Depth Circulation of the World Ocean

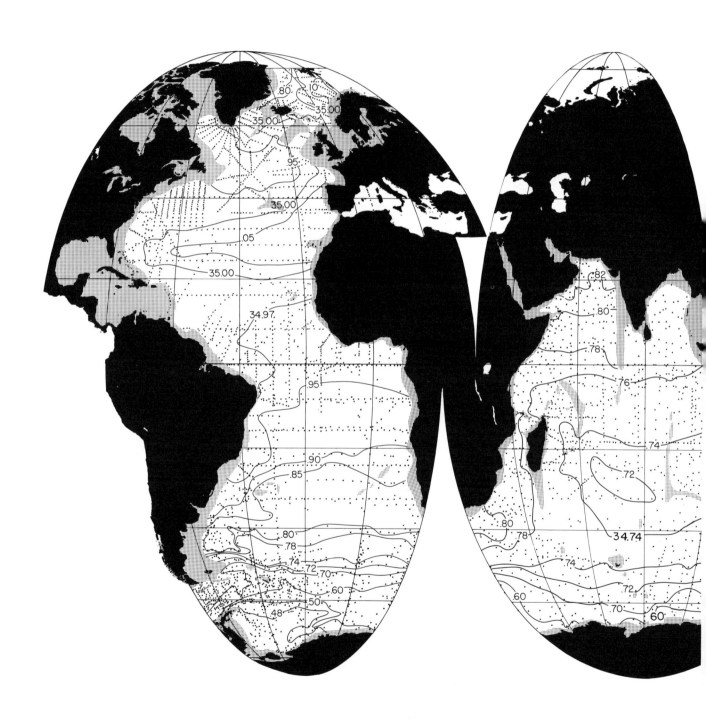

Figure 3.16B (Facing pages) Salinity (‰) on the isopycnal.

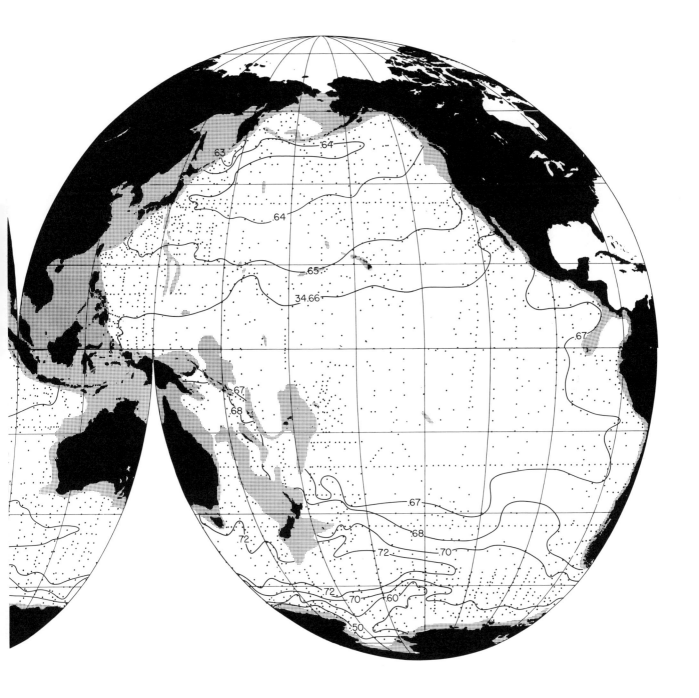

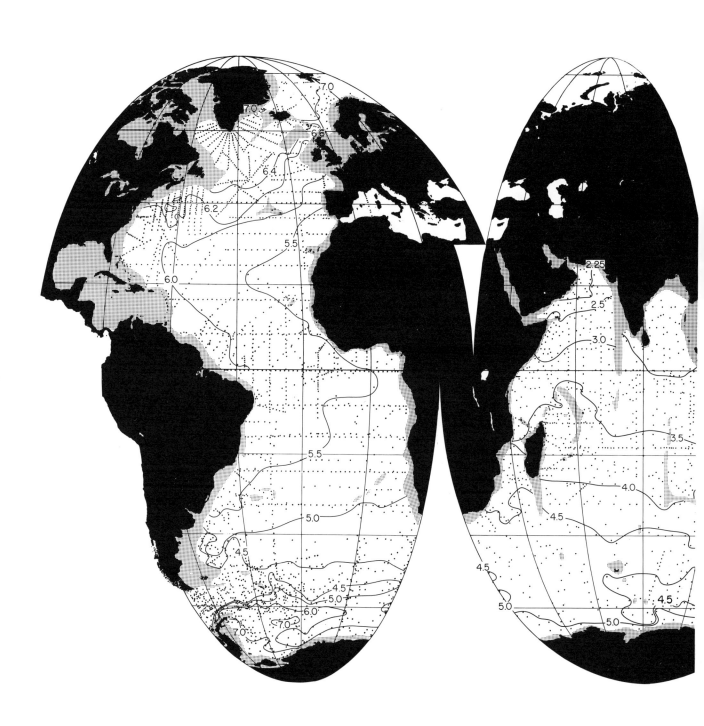

Figure 3.16C (Facing pages) Dissolved oxygen (ml l^{-1}) on the isopycnal.

Joseph L. Reid

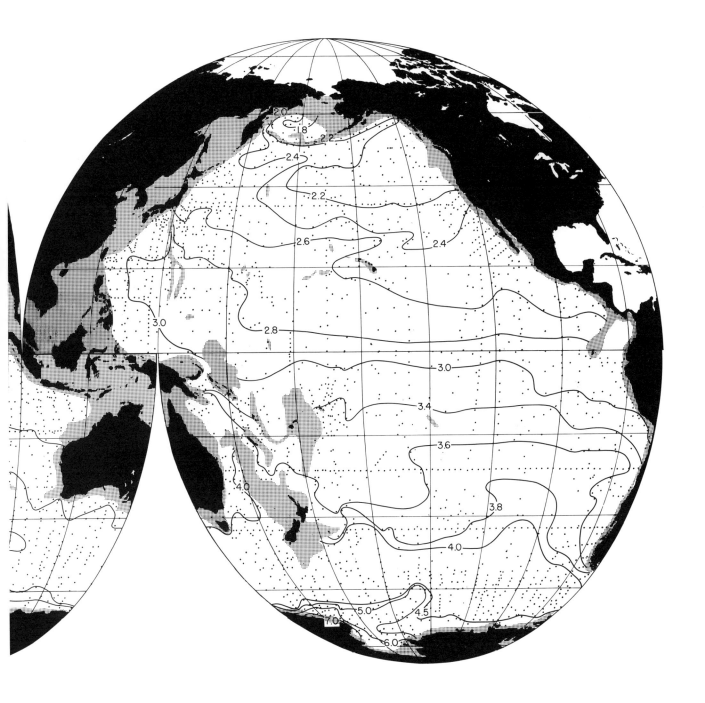

On the Mid-Depth Circulation of the World Ocean

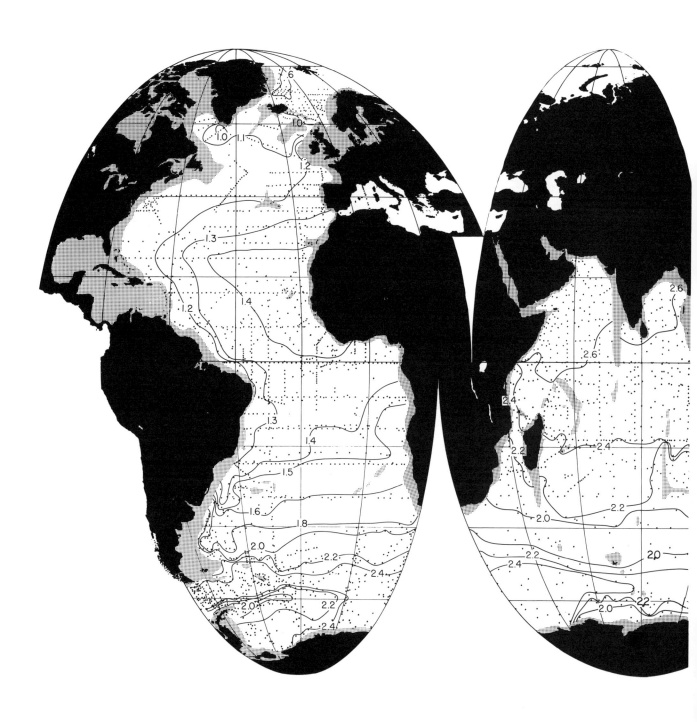

Figure 3.16D (Facing pages) Phosphate ($\mu M l^{-1}$) on the isopycnal.

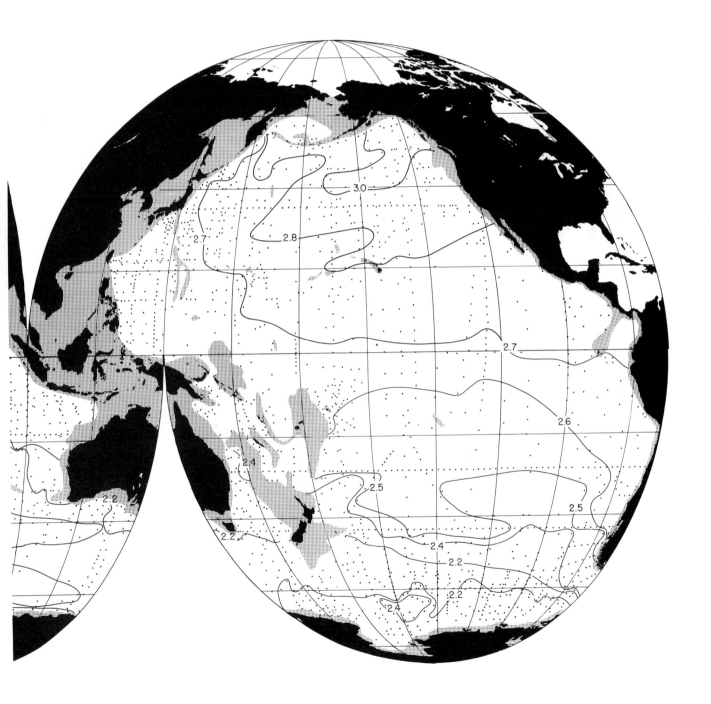

On the Mid-Depth Circulation of the World Ocean

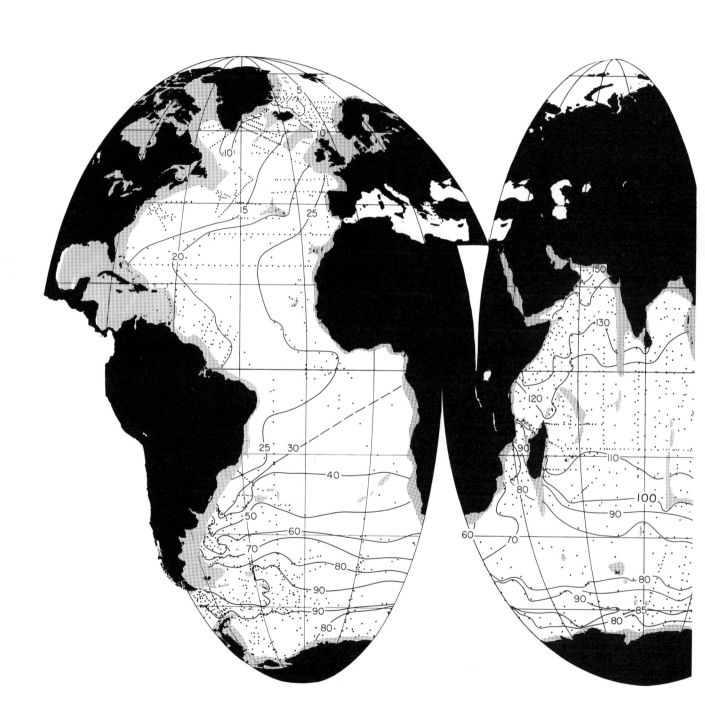

Figure 3.16E (Facing pages) Silica ($\mu M l^{-1}$) on the isopycnal.

Joseph L. Reid

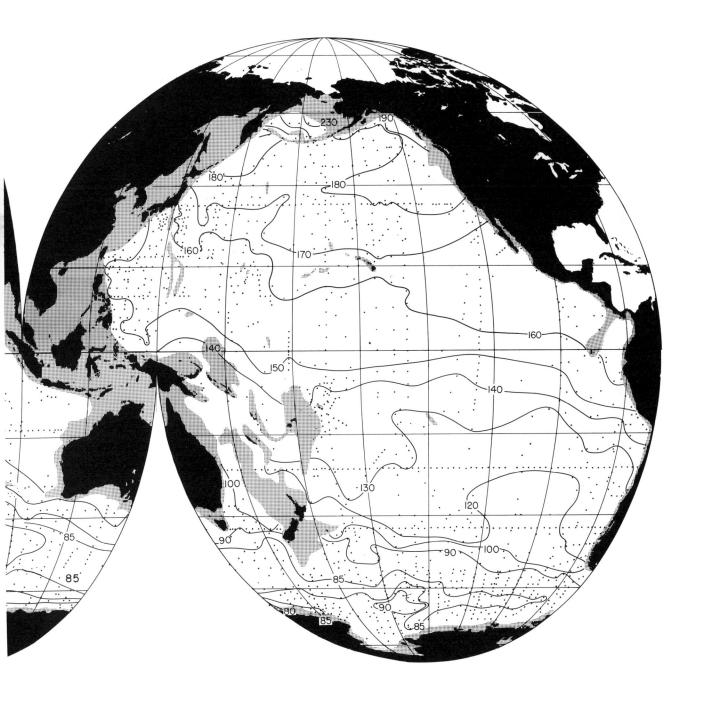

Figure 3.17 (Facing pages) Steric height at 2000-3500 db [dynamic m ($10\,\mathrm{m^2\,s^{-2}}$ or $10\,\mathrm{J\,kg^{-1}}$)]. Shaded area is less than 3500 m depth.

Joseph L. Reid

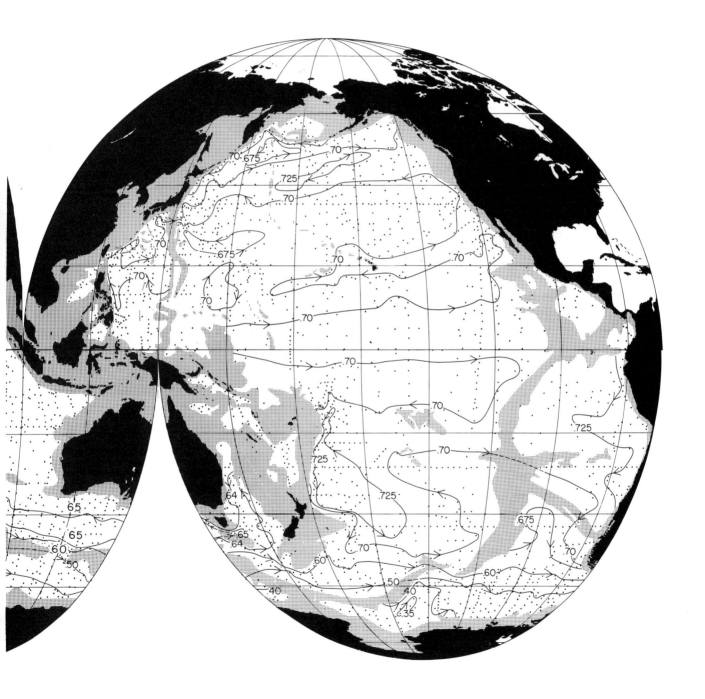

On the Mid-Depth Circulation of the World Ocean

nal lies below but near the depth of his Upper North Atlantic Deep Water, defined by an intermediate salinity maximum, and the salinity pattern here is similar. The three maps of salinity [Wüst's (1935) plates XIII and XIX and figure 3.16B herein] show roughly similar patterns because three of the principal sources of salinity extrema—the Mediterranean outflow, the Labrador Sea, and the Antarctic—extend through a thick depth range. The core method, however, which follows extrema, did not show the effect of the Iceland-Scotland overflow seen in figure 3.16B. The vertical extremum in salinity associated with this overflow lies deeper and is not connected with that of the Mediterranean outflow. The vertical oxygen maximum rises to the surface in the north, and Wüst ascribed the high values east of the Reykjanes Ridge to overturn and vertical mixing south of Iceland. This figure (3.16B) and that of oxygen (3.16C) suggest an outflow from the Norwegian Sea rather than overturn and mixing alone.

The saline mid-depth waters of the North Atlantic extend as a subsurface salinity maximum through most of the world ocean (Wüst, 1951; Stommel and Arons, 1960b; Reid and Lynn, 1971). This maximum lies beneath the less saline Intermediate Water and above the less saline but denser Antarctic waters wherever they obtain; in the North Pacific it provides the bottom layer. In the North Indian Ocean, which lacks the Intermediate Water salinity minimum, the North Atlantic waters lie beneath the more saline waters derived from the Red Sea and the Persian Gulf, which provide a maximum in salinity at lower density and shallower depth.

The southward extension of the North Atlantic high salinity is seen along the western boundary, turning eastward near 40-50°S and extending as a tongue of high salinity between the less saline Antarctic waters to the south and the less saline waters, influenced by the overlying Antarctic Intermediate Water, to the north. The tongue extends through the Indian and Pacific Oceans into the Drake Passage. As it emerges from the Pacific, it flows between the more saline North Atlantic Water and the less saline Weddell Sea Water and is no longer a lateral maximum. Some part of it turns back with the Weddell Sea gyre, into the southern Weddell Sea.

In the Indian Ocean, the area of high salinity is seen clearly in the Arabian Sea, resulting from vertical diffusion from the very saline outflows from the Red Sea and Persian Gulf, that affect this deep layer in the same way as does the Mediterranean outflow in the Atlantic. The vertical salinity maxima of these sources, reflecting the depth of their direct input, are much shallower, well above 1000 m; yet there is a clear contribution of high salinity to much greater depths.

The source of the low salinity to the north of the eastward-extending maximum near 40 to 60°S is also demonstrated by the isolated lateral salinity minimum in the central South Indian Ocean: there is no source possible for this minimum other than vertical diffusion from the overlying low-salinity Intermediate Water. This process creates the low salinity of the central Indian Ocean that separates the Atlantic high salinity at the tip of Africa from the Arabian Sea high.

In the South Pacific, there is a northward excursion of higher salinity across 40°S in the east. There is also a suggestion of some water of higher salinity flowing westward south of 60°S as part of the Ross Sea gyre.

Over most of the Pacific, the salinity on this isopycnal decreases to the north. As in the Indian Ocean, there is no other source for this freshening than vertical exchange with the overlying low-salinity Intermediate Water. The very lowest values are found near the Kuril Islands.

Oxygen The highest values of dissolved oxygen concentration (figure 3.16C) are found at the outcrop in the Norwegian-Greenland Sea. Values only a little lower are found in the Weddell Sea gyre and probably along the Antarctic coast in winter. The major source of oxygen to this isopycnal in the Atlantic is in the north, where vertical diffusion in the Labrador Sea and overflow from the waters north of the Greenland-Scotland Ridge provide concentrations of more than 6.4 ml l^{-1}. High values from this area extend southward along the western boundary, much as Wüst (1935) showed for the core of the Middle North Atlantic Deep Water, which he defined by an intermediate-depth oxygen maximum, at depths slightly greater than this isopycnal. As the vertical maximum ends near 50 to 55°S, he was unable to follow the water further. Along this isopycnal, a tongue of high oxygen extends eastward across the Indian Ocean, as Callahan (1972) showed for a slightly deeper isopycnal. Farther east, its value is intermediate between the high values around Antarctica and the lower values of the Pacific to the north, and its final features in the south are the excursion into the Southeast Pacific Basin and around the Ross Sea gyre.

During the passage around Antarctica, the concentration of oxygen has been reduced so much, in spite of the Antarctic source, that in its reentry into the Atlantic through the Drake Passage, it appears as a lateral minimum, between the waters from the North Atlantic and the fresher waters of the Weddell Sea gyre. Its decrease can be the result of respiration and decay at this density or by vertical diffusion to the overlying strong oxygen minima of the Indian and Pacific Oceans, which are generally both deeper and of lower concentration than those in the Atlantic. As Callahan (1972) has shown, the lower values of oxygen within the Indian and Pacific Oceans contribute substantially to the decrease of oxygen along the Circumpolar Current by lateral exchange. It is noteworthy that the high salinity introduced by vertical diffusion in the Arabian Sea is not accompanied by an increased oxygen there.

Phosphate and Silica The other nonconservative concentrations (figures 3.16D, 3.16E) show slightly more detail than the oxygen. They are lowest in the Labrador and Norwegian-Greenland Seas and highest in the Arabian Sea and northeastern Pacific. Phosphate differs by a factor of about 2.5 between the northwestern Atlantic and northeastern Pacific, silica by a factor of about 12. These ranges are large enough to provide patterns more detailed than, as well as different from, those of salinity.

Following the pattern of salinity and oxygen, a simple picture emerges of low nutrient values in the Labrador Sea extending southward along the western boundary to the West Wind Drift, or Circumpolar Current. A tongue of low nutrients extends eastward from the South Atlantic all across the Indian and Pacific Oceans and through the Drake Passage. Concentrations within the lateral minimum increase to the east, and as the waters enter the Atlantic, they appear as a high in nutrients between the nutrient-poor waters of the western South Atlantic and the shallower, somewhat depleted waters of the Weddell Sea gyre. That the highest values in the Indian Ocean are found in the Arabian Sea may be a consequence of the ridges; the narrow gaps may limit the lateral exchange.

A somewhat clearer pattern is seen in the North Pacific. In both phosphate and silica, the highest open-ocean values (excluding the almost enclosed Bering Sea) are in the northeast, with slightly lower values to the north. The resulting tongue of low nutrient extending northeastward from the western boundary suggests an advective feature. In phosphate, there are two of these tongues (possibly two in silica also but the data are too few to be certain), consonant with the shallower flow pattern proposed by Reid (1978).

Sources of the Characteristics The maps make fairly clear the sources of the high and low values of the characteristics that appear in various areas. The lateral sources are the Norwegian-Greenland Sea, which provides shallower waters of this density that extend downward into the open Atlantic, and the long zone around Antarctica, where this isopycnal lies shallow or outcrops. Both lateral sources provide high oxygen and low nutrients (though the lowest-nutrient waters are from the north), but the northern sources are warm and saline, the southern sources colder and fresher.

A combination of convection and vertical diffusion in the Labrador Sea extends high-oxygen and low-nutrient concentrations down to this layer. Over most of the ocean, where this layer lies at mid-depth, the characteristics appear to be modified only by vertical diffusion and by consumption of oxygen and regeneration of nutrients. In the northeastern Atlantic and northwestern Indian Oceans, the salinity is made high by vertical diffusion from the overlying layers of outflow from the Mediterranean Sea and the Red Sea and Persian Gulf. In the southeastern Indian Ocean and the North Pacific, the low-salinity values must be the consequence of vertical diffusion from the overlying Intermediate Water; in neither case is there a lateral source. Over most of the Indian and Pacific Oceans, this isopycnal lies beneath a thick oxygen minimum and a thick maximum in phosphate, and such vertical diffusion as takes place does not raise the oxygen concentration or lower that of the nutrients on this isopycnal.

3.8.2 The Steric Height at 2000 db Relative to 3500 db

It seems worthwhile to examine the density field to inquire whether the geostrophic vertical shear is consistent in any simple way with the mid-depth patterns that have been illustrated herein.

The upper surface of 2000 db was selected to correspond roughly to the depth of the chosen isopycnal over most of the ocean. The 3500-db surface was selected not because of any special assumptions of minimal flow near that pressure, but because the shear field is weak over much of the area in this depth range and it seemed useful to consider a thick layer. A substantial part of the western boundary flow takes place in areas shallower than 3500 m and cannot be represented on such a map, but we already have some information about the boundary flow there, at least in some cases.

This map (figure 3.17) has been made from a selection of stations rather than from a set of averages. It is, of course, not synoptic, even to season, and though most of the data are from the later period when salinometers have been available, it has been necessary to include some earlier, less accurate measurements as well. As in the set used in figure 3.16, to which it corresponds closely, it does not represent the best possible selection, but the present stage of progress toward it.

The map has been made not only to illustrate such features as it can of the deep density field but also to indicate what is lacking in the bank of good-quality data. One striking feature is that although the contours extend mostly east-west, indicating a zonal flow that requires north-south sections of data for proper resolution, most of the major track lines (IGY, Scorpio, INDOPAC, etc.) extend east-west, except in the Antarctic zone. This has certainly compounded the difficulty in interpreting the density field and limited my own confidence in many of the features I have drawn. We must, however, start somewhere. Though the earlier interest in abyssal meridional flow has dictated the east-west station arrays, there seems to be enough zonal pattern in the density field to merit investigation. A few more north-south lines would certainly help.

The Range of Steric Height The range of values of steric height is very small. A difference of about 30-35 dynamic cm (hereafter cm) is seen across the Antarctic Circumpolar Current, but north of 40°S the range is about 7 cm in the Atlantic Ocean, 6 cm in the Indian, and 7 cm in the Pacific. The western boundary currents and the Antarctic Current have the strongest gradients, and variability will be strongest there, but in much of the ocean the data indicate that in a 10° × 10° area, for example, the combination of time and space variability must be very small—little more than 3 cm, and much less than Wyrtki (1975a) has shown in the upper 500 m alone.

The surface of the Pacific Ocean stands on the average about 40 cm higher than the Atlantic with respect to the 1000-db surface, and the North Atlantic and North Pacific stand, respectively, about 14 and 17 cm higher than the South Atlantic and South Pacific; referred to 4000 db, the Pacific stands about 68 cm higher than the Atlantic (Reid, 1961b; Lisitzin, 1974). The upper-level differences are reflected in figure 3.11A (from Burkov et al., 1973), relative to 1500 db, and the Indian Ocean seems to be intermediate in steric height.

These differences are not only reduced in the 2000-3500-db layer, but in some cases reversed. The 2000-db surface is about 5 cm higher relative to 3500 db in the North Pacific than in the North Atlantic, but it stands highest in the South Pacific (excluding the areas poleward of 50°). The North and South Atlantic and the Indian Ocean values do not appear to differ very much from each other.

The Flow Interpreting the pattern of steric height as representing relative geostrophic shear, the most striking feature is the predominance of zonal relative flow, except in the eastern Indian Ocean, where no clear pattern can be seen in the present data set.

This pattern, of course, does not necessarily represent the sense and magnitude of the absolute flow, but only the geostrophic shear between two isobaric surfaces. Where a deep or abyssal boundary current is moving faster than the flow at 2000 m, the relative flow may be in the wrong sense. Fairly rapid flow may occur at 2000 m in areas where the water depth is less than 3500 m: flow in these areas will of course not appear on the map. The deep current from the northern North Atlantic that flows cyclonically around the Labrador Sea and southwestward along the continental slope of North America does not appear on this map. This is because most of the 2000-m flow of that current takes place in water depths less than 3500 m (Worthington, 1970; Swallow and Worthington, 1961; Luyten, 1977; Richardson, 1977). The map begins, instead, with the deep Gulf Stream, which it portrays clearly, and the broader return current (Defant, 1941b; Worthington, 1976). This is quite consistent with Richardson's (1977) interpretation of the direction of the flow at 2000 m over deep water and is remarkably consistent with the float trajectories near that depth reported by Riser, Freeland, and Rossby (1978) near 20°N, 74°W, which extended westward and southward, paralleling the 0.65-m contour in that area.

It is also true, as Defant (1941a) pointed out, that along much of the western boundary of the North Atlantic the vertical shear field is such that the calculated flow will be northward relative to the deeper water. If a surface shallower than 3500 db had been chosen, values could be calculated in shallower water but would not have shown a southward flow relative to the underlying water (Swallow and Worthington, 1969; Ivers, 1975).

Along the western boundary of the South Pacific (just east of the Tonga-Kermadec Ridge, near long. 180°), a deep northward flow of Antarctic water has been detected (Warren and Voorhis, 1970) with a southward flow above it as part of the subtropical anticyclonic gyre (Reid and Lonsdale, 1974). At about 22°S, Warren and Voorhis (1970) propose a surface of zero meridional velocity that varied from 3100 to 4300 m, and in this region the sense of flow given by the shear map appears to be correct and the relative speed might not be biased very much. Farther south, Warren (this volume, chapter 1) finds the northward flow extending up to about 2000 m near 43°S. The shear map, referred to the stronger flow at 3500 db, extends the southward flow too far south before turning eastward. The characteristics along the isopycnal, which lies near 2700 m at 43°S, suggest a northward extension of the circumpolar waters at this depth to about 30°S before turning southeastward and around the anticyclonic gyre.

A case of steep continental slope and broad western boundary currents is seen in the South Atlantic near 40°S (Reid et al., 1977), where the strong shear between the equatorward abyssal flow and the poleward mid-depth flow is reflected in figure 3.17. Warren's studies of the Indian Ocean were directed toward the deep western boundary currents, but he also provides information about the choice of references and the overlying flow in most of the areas he considered. With measurements along 12°S and 23°S in the western Indian Ocean, Warren (1974) found a deep (3000-3500 m) northward western boundary current flowing close against Madagascar. He found no boundary current near 2000 m but proposed the weak northward flow between 2000 and 3000 m that appears in figure 3.17.

Later, Warren (1977), with the section along 18°S in the eastern Indian Ocean, showed evidence for a deep northward flow just east of the Ninetyeast Ridge, and indicated that the deeper flow is somewhat stronger than that at 2000 m. The shear field is weak and the other data are confusing, so it has not been possible to provide contours there. Across the Southwest Indian Ridge (near 30°S, 60°E), he tentatively set 3500 m as a zero reference level and calculated northward flow below it (Warren, 1978). This reference surface also provides northward flow at 2000 m.

3.9 Comparison of the Maps of Shear Field and Characteristics

Over those parts of the ocean where direct measurements or other sorts of information are available, the shear map (figure 3.17) is at least not in severe disagreement as to the direction of flow. Over the rest of the area, particularly the central ocean, we must simply recognize that this map can at best represent only the vertical geostrophic shear, and inquire how it relates to the patterns of figure 3.16.

3.9.1 Flow across the Equator

The transequatorial flows will not be seen in the geostrophic shear field, but the characteristics suggest a southward extension across the equator in the western Atlantic and a northward extension in the western Pacific. A corresponding case might be made for northward extension in the eastern Atlantic and southward extension in the eastern Pacific, but this is not so clear. In the Indian Ocean, the characteristics suggest a southward flow across the equator in the west.

3.9.2 Atlantic Ocean

In the North Atlantic, the shear field shows westward flow between 30 and 50°N in the eastern area, and eastward flow near 15-25°N. This is consonant with the pattern of high salinity extending westward across the North Atlantic. The lower-salinity waters from the north may simply flow southwestward to the western boundary near 25°N, where they turn southeastward, surrounding the area of high salinity. These low-salinity waters may also move southwestward along the continental slope from Greenland to Cape Hatteras, in waters less than 3500 m deep and hence not represented on the shear map. The Gulf Stream at this depth appears here as simply part of the anticyclonic flow, with its return circulation perhaps carrying southward a major part of the lower-salinity northern waters.

This scheme is also consistent with the distribution of oxygen and nutrients. Their patterns differ from that of salinity only in that the source of the high salinity is clearly the Mediterranean outflow, but the low oxygen and high nutrients have sources at a lower latitude, in the equatorial and tropical eastern Atlantic as well as from the Mediterranean, and their westward extensions cover a broader zone. They are all consistent with a westward flow north of 25°N in mid-ocean, as the shear map suggests, and of course with the southwestward recirculation of the Gulf Stream at this depth.

In the South Atlantic, the shear map suggests some segments of a western boundary current, and all of the characteristics extend extrema southward along the boundary. Just north of the Falkland Plateau (50°S, 50°W), where the abyssal flow is strongly northwestward, the relative flow (figure 3.17) may appear much stronger than the actual flow. At the confluence of the western boundary current on this map with the Circumpolar Current, they both turn back northwestward and then eastward at about 40°S. All of the characteristics suggest such a bight, with Antarctic waters turning sharply northwestward near 40°W.

The zonal flow patterns between the equator and 40°S in the eastern Atlantic in figure 3.17 are very poorly defined by the available data, which are all from east-west tracks. Likewise, the characteristics are ill-defined there, and no useful comparison can be made.

3.9.3 Indian Ocean

In the Indian Ocean, the shear field as interpreted here is too weak north of 30°S for a useful comparison. Some suggestion of a northward western boundary current is seen in the anticyclonic feature at the tip of Africa, and perhaps of a northward flow between Madagascar and the Mascarene Ridge. The oxygen and nutrients partly support this, but their dominant feature is a southward extension along the western boundary from the Arabian Sea toward Madagascar. The major extension, however, of the Atlantic characteristics seen entering at the tip of Africa is clearly eastward, and this is consonant with the shear field.

3.9.4 Pacific Ocean

It is in the Pacific that data are most nearly satisfactory for this study. They allow a much better resolution of the patterns than in any other area than the Circumpolar Current. The shear field in the South Pacific shows an anticyclonic gyre near 35-40°S, an eastward flow leaving this gyre along 25°S in the east, a westward flow along 15°S, and an eastward flow along the equator: this last feature is weak and must remain uncertain. This is the complete C-shape, roughly circumscribed by the 0.70-dynamic-m contour. There is also a southward flow along the eastern boundary.

All of these features are matched in the North Pacific—an anticyclonic gyre near 40-50°N, an eastward flow near 20-30°N, a westward flow from 20°N in the east to 15°N in the west, and the weak eastward flow near the equator. The 0.70-dynamic-m contour that circumscribes the C-shape in the South Pacific circumscribes only the two arms of the C; it is broken apart in the west, perhaps by the bottom topography. Whether the Philippine Sea is a major part of this pattern or is largely cut off at this depth, as are the Caribbean Sea and Gulf of Mexico from the open Atlantic, is uncertain.

There is a suggestion of a northward extension of characteristics near the western boundary of the North Pacific. A tongue of high salinity extends from Japan northeastward to mid-ocean, and this is even clearer in some of the nonconservative characteristics. The extension is consonant with the eastward flow of the subarctic cyclonic gyre and the adjacent anticyclonic gyre. The nonconservative characteristics are also consonant with a return flow southwestward across 40°N in mid-ocean. Oxygen and phosphate also indicate an eastward flow at about 20–25°N in mid-ocean, corresponding to the shear field. For silica, the pattern is less clear there.

The correspondence between the shear field and the characteristics is clearest in the South Pacific. From the Circumpolar Current, tongues of high salinity and oxygen, and low-nutrient concentration, extend northward around the eastern part of the anticyclonic gyre of the shear field. From the equatorial zone, low-oxygen and high-nutrient concentrations extend southward around the western part of the gyre. The shear field suggests that the eastward flow from mid-ocean at about 20–30°S carries the water from both these sources eastward toward South America. From there, they are carried southward to the Circumpolar Current and through the Drake Passage. Over this area the contours of the nonconservative characteristics and of steric height are nearly parallel.

It appears to be along the route defined by the shear field in the South Pacific that the extreme characteristics created in the North Pacific—the thick, mid-depth oxygen-minimum and nutrient-maximum layers described by Reid (1973a)—leave the Pacific and enter the circumpolar system.

3.9.5 Antarctic Ocean

In the Antarctic Circumpolar Current, all of the fields are fairly well defined and in simple consonance. The shear is eastward, with excursions only into the South Atlantic around the Falkland Plateau (50°S, 40–50°W), in the Weddell and Ross sea gyres, and the Southeast Pacific (Bellingshausen) Basin. It carries the saline, high-oxygen, low-nutrient water from the western Atlantic all around Antarctica and back into the Atlantic. Salinity is still a lateral maximum as it emerges from the Drake Passage, but the nonconservative characteristics have changed from one extreme to the other during their long subsurface passage.

3.9.6 Westward Flow in the Deep Anticyclonic Gyres

There is one other noteworthy feature of the poleward shift of these gyres. They may shift, at some depths, south of Africa and Australia. The westward return flow of the West Wind Drift currents can be seen clearly in all oceans, and it seems to be returning from all across the ocean. Indeed, part of the return flow (if that is a correct term) in the South Indian Ocean may originate from the Tasman Sea, and some part of the Agulhas Current system may extend across the Atlantic. South of Australia, the characteristics rather suggest that such flow may occur at depths near this isopycnal. South of Africa, only the oxygen and phosphate might suggest such a flow. It would be interesting to explore this possibility at greater depths.

3.10 Conclusion

Mid-depth circulation has received much less attention than the upper wind-driven layer (of uncertain or at least disputed thickness) and the abyssal, presumed thermohaline flow. Some evidence for the direction of these flows was available from ship's drift at the surface and such quantities as potential temperature patterns at the bottom, but little evidence of the details of any mid-depth patterns was available to the early investigators other than the long meridional extensions of salinity and oxygen extrema. Theoretical, numerical, and descriptive studies have dealt mostly with surface or abyssal flow, or with total transport.

That there might be flow patterns at mid-depth that are significantly different from these has been considered possible, at least since the work of Prestwich (1875), but the means of working on these were few. The major attempts to date have been made through the descriptive studies.

I have cited numerous studies here that have involved mid-depth circulation through the examination of the distribution of conservative and nonconservative characteristics along surfaces of constant depth or along extrema in the characteristics, or along some sort of density surface, or by consideration of relative geostrophic flow, or through some combination of these. (The list is of course incomplete, partly because of space, but also because I am not as familiar as I could wish to be with the work of many of the investigators, especially those in Japan and the Soviet Union: I am sure that I have left out some important studies.)

I have made some remarks also about the trends—the abrupt change in emphasis from flow patterns at depth to models of total transport that occurred in the early 1950s, and the more recent trend back toward models of nonhomogeneous oceans with irregular bottoms. This has brought the density field back into the foreground; with all its limitations, it still seems to be one of the strong ocean signals and one that can no longer be neglected.

I cannot claim to have established the correct mid-depth flow pattern from the materials cited or from the new maps presented here. I have, however, made a preliminary case for a general pattern of mid-depth circulation. It involves substantial zonal flow in mid-latitudes, and is perhaps more analogous to the overlying wind-driven gyres than to the recognized thermohaline deep and abyssal flow. The poleward shift in the density field cannot be questioned: it is a real feature of all the oceans and is seen clearly on every density field prepared from data. The equatorward arm of the C-shape appears in some of these maps, and in some models, both prognostic and diagnostic.

The flow field considered was calculated, quite arbitrarily, to the 3500-db surface. It is strong in the higher latitudes, but within about 20° of the equator, the field is weak and the shear uncertain. This is especially so in the South Atlantic, where data are lacking, and in the Indian Ocean, where the present selection of data gives no resolution. Taking the sense of flow from the shear map does not make for any obvious difficulties, except for the narrow zone near New Zealand that has been discussed. This does not mean that it is therefore correct everywhere else, but it shows, I believe, that such considerations of the shear field may serve as a useful way to start work on the problem. Over the greater part of the ocean, where no direct information is available, the only tests we can apply are to compare it with the distribution of the characteristics and with the various models.

The shear field, treated as flow, seems to be supported fairly well by the patterns of characteristics. This was to be expected in the Circumpolar Current, which is deep, broad, and in high latitudes, where the shear signal is well defined. It might not have been expected that the patterns of characteristics and the shear field would match (at least qualitatively) to the degree that may be seen in the middle latitudes of the North Atlantic and the Pacific. In the South Pacific, in particular, the fields are defined clearly, and the coherence between the patterns of shear and characteristics is remarkably good.

On the deeper isopycnal chosen to represent the Lower North Atlantic Deep Water (Reid and Lynn, 1971) only the depth, salinity, and potential temperature were illustrated. It lay in greater depths (3-4 km in middle and low latitudes) than the isopycnal mapped herein, and, with only the conservative characteristics illustrated, little suggestion of zonal flow was suggested except in the Circumpolar Current. Western boundary currents at these depths were suggested in the southern Indian and Pacific Oceans, and a mid-ocean southwestward flow across 40°N in the Atlantic; this latter seems to correspond to the deep return flow of the Gulf Stream.

Including the nonconservative characteristics on the isopycnal presented herein has added considerable detail to the pattern and allowed for a broader interpretation. Some support is given to Prestwich's (1875) concept of poleward extensions of warmer waters well beneath the sea surface, even down to the depths of the isopycnal illustrated here. In the North Atlantic, there is a northward extension of highly saline water along the eastern boundary (on the isopycnal, of course, this is also warmer water). The deep shear field shown here does not show this flow, but the earlier work of Helland-Hansen and Nansen (1926) and Reid (1978) at depths near 1000 m does show it. A more recent study (Reid, 1979) discusses the extension of some of the deeper warmer waters through the Faroe-Shetland Channel into the Norwegian Sea, where they contribute to the warm layer of the eastern and northern Norwegian Sea and the Arctic Ocean.

The warm water (the high-salinity tongue) extending along the western boundary of the South Atlantic into the Circumpolar Current is well known, of course. And along this isopycnal, an extension of the warmer waters into the North Pacific, at both the eastern and western boundaries, is seen, and is roughly consistent with the shear field. In neither case, however, is the flow only meridional: if the oxygen and nutrient patterns and the shear field have been interpreted correctly, this water may reach the higher latitudes, at least in part, by a gyre-to-gyre transport, involving substantial zonal flow.

If diagnostic models are to proceed usefully, then a better density field must be provided for much of the ocean. Substantial improvements over the selections shown here can certainly be made, given enough time, but a set of north-south station lines must also be obtained if the density field is to be made more useful.

4
The Gulf Stream System

N. P. Fofonoff

4.1 Introduction

In the two decades since the first publication of Stommel's (1965) monograph on the Gulf Stream, our knowledge of the Gulf Stream System has been expanded dramatically through the development and application of new, powerful measuring techniques. Multiple ship surveys of the type organized by Fuglister (1963) provided the first systematic descriptions of the spatial structure between Cape Hatteras and the Grand Banks that included the surrounding slope waters to the north and the Sargasso Sea waters to the south of the Gulf Stream. Several major theoretical and interpretative studies grew from the base of data and descriptions provided by this study. During the same period, instrumented buoys, both moored and drifting, were beginning to reveal some of the complexities of the subsurface and deep fields of temperature and currents. Among the new techniques implemented in the 1960s was infrared-radiation imaging to map the thermal patterns of the ocean surface from satellites orbiting the earth (Legeckis, 1978). The two-dimensional surface thermal maps that have been obtained have added rich detail to our knowledge of the strongly varying thermal structure associated with the Gulf Stream throughout its path. Yet, despite these advances in our ability to measure, our understanding of the dynamic mechanisms by which the Gulf Stream forms, develops in intensity, decays, and finally merges into the large-scale circulation of the North Atlantic have not evolved as satisfactorily. Even the mechanism controlling the position of the Gulf Stream after leaving the continental shelf at Cape Hatteras has not yet been firmly established. Is the Gulf Stream controlled by bottom topography, by the distribution of mean wind stress, or by a mechanism yet to be determined? The dynamics by which meanders of the Gulf Stream amplify and develop into large rings and eddies and the subsequent movement and evolution of these entities are not well understood. In spite of the impressive progress of the past decades, much remains to be done to resolve and understand the particular mechanisms that determine the character and behavior of the Gulf Stream along its entire path from the Gulf of Mexico into the central North Atlantic.

In preparing material for this review, I concluded that my initial plans for a comprehensive discussion of the literature since 1958 were unrealistic. As over 200 references plus numerous technical reports and articles were identified, it became obvious that only a few aspects of the Gulf Stream System could be covered in a single short review. Given the necessity for choice, it is clear that the selection must reflect my preferences, interests, and perhaps, biases. I hope my effort to trace particular lines of research in the literature will prove of interest to readers and will serve as a guide to a part of the rapidly growing body of literature that represents our collective knowledge of the Gulf Stream System. That other, equally important, aspects of research are omitted is unfortunate but inevitable.

4.2 The Gulf Stream System

The subdivisions of the *Gulf Stream System* proposed by Iselin (1936), reproduced in figure 4.1, although not entirely accepted in practice, serve as a convenient framework for grouping the research literature. Starting from the Gulf of Mexico, the *Florida Current* was labeled by Iselin as the portion of the Gulf Stream System flowing through the Florida Straits northward past Cape Hatteras to the point where the flow leaves the continental slope. Objections had been raised by Nielson (1925) and Wüst (1924) to using the word "Gulf" in reference to the Florida Current, as they considered that the water flowed directly across from the Yucatan Channel into the Florida Straits rather than from the Gulf of Mexico. This distinction seems less justified now because the flow through the Yucatan Channel has been observed to loop well into the Gulf of Mexico on occasion (Leipper, 1970; Behringer, Molinari, and Festa, 1977), although the Florida Current does not originate there. After leaving the Florida Straits, the Florida Current presses close to the continental slope and in the upper layers forms a relatively continuous system. The flow is augmented on the seaward side by inflow of water of essentially the same characteristics as the Florida Current. Iselin included both sources under the same label. Oceanographers frequently refer to the Florida Current between the Florida Straits and Cape Hatteras as the Gulf Stream. However, because measurements and theoretical studies have tended to relate this portion of the Gulf Stream System more closely to the current in the Straits rather than to the currents downstream from Cape Hatteras, Iselin's nomenclature is more convenient in the present review.

North of Cape Hatteras, the current begins to flow seaward off the slope into deeper water. Freed of the constraints of the shelf, the Gulf Stream develops meanders of increasing amplitude downstream. Bowing to popular usage, Iselin retained the name *Gulf Stream* for the section between Cape Hatteras and the Grand Banks. The name *North Atlantic Current* had already been widely accepted for easterly flows at mid-latitudes beyond the Grand Banks. Even though an extension of the Gulf Stream, the North Atlantic Current, according to Iselin, becomes separated into branches and eddies to form a distinctly different regime of flow. Its eastern limit is not clearly defined, though Iselin assumed that the branches extended into the eastern North Atlantic. A composite view of the western portion of the Gulf Stream System (figure 4.2) has been assembled by Maul, deWitt, Yanaway, and Baig (1978) from infrared satellite images and surface tracking by ships and aircraft. The sharp thermal contrasts between the warm currents and the neighboring waters are detectable from space and reveal the variability of the Gulf Stream System throughout its length. The complexities introduced by the near-surface spatial and temporal variability of the Gulf Stream System are only beginning to be described. Comparatively little is known of the variable subsurface and deep structure, particularly downstream of Cape Hatteras.

4.3 The Florida Current

Iselin (1936) defined that Florida Current as all the northward-moving waters with velocities exceeding 10 cm s^{-1} starting along a line south of Tortugas and extending to the point past Cape Hatteras where the current ceases to follow the continental shelf. The three chief characteristics of the Florida Current noted by Iselin are that it greatly increases in volume as it flows north, that it flows most swiftly along the continental slope, and that over most of its length it is relatively shallow, transporting water no colder than 6.5°C until passing the northern limit of the Blake Plateau. The surface thermal structure of the Florida Current between Miami and Cape Hatteras can be seen in the infrared image reproduced in figure 4.3.

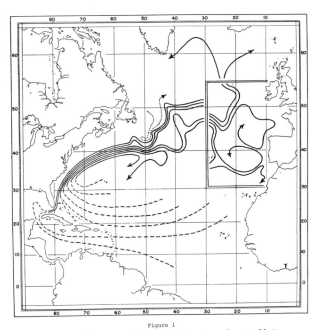

Figure 4.1 A schematic diagram showing the Gulf Stream System as described by Iselin (1936). Each streamline represents a transport interval of about 12×10^6 m^{3}s^{-1}.

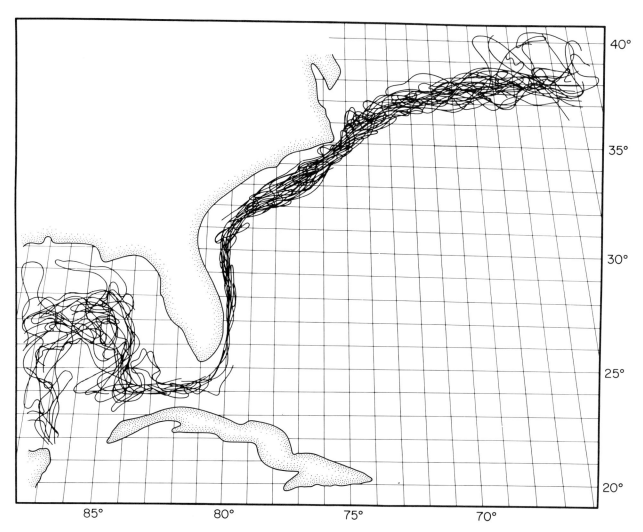

Figure 4.2 A composite of thermal fronts of the Gulf Stream System showing variability for a 9-month period between February and November 1976. [Courtesy of G. Maul from Geostationary Operational Environmental Satellite (GOES) infrared observations.]

4.3.1 Sea-Level Slope from Tide Gauges

Montgomery (1938b) assumed that the intensification of the Florida Current as it flowed into the Straits of Florida is produced by a hydraulic, or pressure, head between the Straits and the Gulf of Mexico. Although the differences in sea level corresponding to the pressure head could not be measured directly, tide-gauge measurements could show variations in the slope and hence, in the Florida Current itself. The recent development of precision altimetry from satellites capable of resolving the shape of the sea surface has renewed interest in the possibility of monitoring major ocean currents remotely. A brief review is given of the use of sea-level records to infer variations of the Florida Current (see also chapter 11).

Iselin (1940a) in his report on the variations of transport of the Gulf Stream noted that sea-level measurement by tide gauges "provides a continuous and inexpensive record of the variations in the cross-current density gradient, if it is assumed that the average surface velocity varies with the total transport of the current." The relation between sea-level and ocean currents had been used to infer variations of ocean currents (and vice versa) much earlier by Sandström (1903). Montgomery (1938b) first applied the method to the Florida Current using data from tide-gauge stations at Key West and Miami, Florida, and at Charlestown, South Carolina, from the eastern coast of the United States and from St. Georges harbor in Bermuda on the seaward side of the Florida Current. Variations in relative differences (absolute differences in heights of tide gauges were not determined) were examined as indicators of the strength of the mean surface current. Sea level, although reflecting tidal variation primarily and to a lesser extent local atmospheric pressure and winds, contains significant contributions also from the cross-stream slope necessary to balance Coriolis forces acting at the surface and the downstream slopes associated

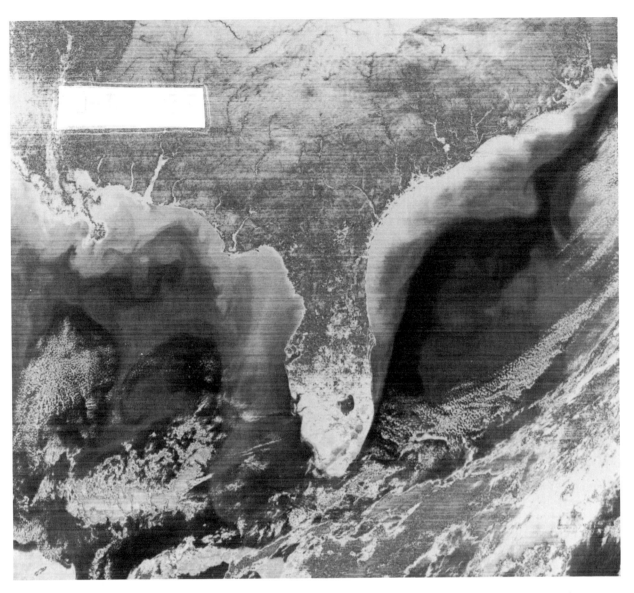

Figure 4.3 The Florida Current between Miami and Cape Hatteras as seen in the infrared on February 26, 1975, from NOAA-4 satellite. A large eastward deflection occurs south of 32°N, possibly as a result of a topography feature. (Courtesy of R. Legeckis, NOAA-NESS.)

with accelerations or decelerations between stations. Montgomery (1938b) concluded from a 47-month study of mean differences of Bermuda minus Charleston that a seasonal cycle was present with the maximum difference and, hence, maximum surface current occurring in July and a minimum in October. The downstream difference of Key West minus Miami, based on 67 monthly values, showed a maximum hydraulic head in July with minima in November and February. As the gauges were not connected by geodetic leveling, the total hydraulic head was not known. Montgomery noted that the difference of 19 cm measured by leveling across the northern part of the Florida Peninsula would be adequate to accelerate the current off Miami to 193 cm s^{-1}, corresponding to maximum speeds observed. When leveling data were obtained, however, the drop in mean sea level from Key West to Miami was found to be only 4.9 cm, too small to account for the observed increase of speed between the two stations. Stommel (1953a) estimated that a difference of 20 cm is required to produce the observed acceleration to satisfy simple geostrophy and Bernoulli's equation. The lack of confirmation of the driving head by direct leveling forced Montgomery (1941) to conclude that the downstream differences between Key West and Miami could not be regarded as an indicator of the strength of the Florida Current. The cross-stream differences, however, clearly indicated a seasonal variation.

Schmitz (1969) reexamined Stommel's estimate using data from free-fall instruments obtained in the Florida Straits off Miami. He noted that the measured relative vorticity was considerably smaller than the value used by Stommel ($0.1f$ rather than $0.4f$, where f is the Coriolis parameter). Furthermore, the change of the Coriolis parameter between the Key West–Havana and the Miami–Bimini sections is approximately $0.1f$, offsetting the change in layer thickness necessary to conserve potential vorticity. Based on vorticity estimates, it apparently is not necessary to have a drop in head much larger than that found by land leveling. However, the observed maximum surface speeds in the Straits would indicate a considerably larger drop of 20 cm or more. It seems likely that the horizontal pressure gradient does not vanish with depth, so that the two-layer assumption of both Stommel (1953a) and Schmitz (1969) of a resting lower layer appears to be overly restrictive. Furthermore, the geodetic leveling may contain errors, and the actual drop in sea level could be larger than reported.

The disagreement between land-leveling and sea-level differences expected from the distribution of currents and density was examined by Sturges (1968). Using historical surface-current and wind observations, he calculated a surface topography for the western Atlantic near the Gulf Stream that represented a best fit to the slopes estimated from the data. He concluded that the northward rise in sea level found by precise geodetic leveling along the east coast of the United States was inconsistent with his results. Sea level within the Gulf Stream must drop northward to maintain the northward flow. In a later paper, Sturges (1974) estimated the north-south slope from hydrographic-station data to be 0.8 cm deg^{-1} (centimeters per degree of latitude) upward to the north seaward of the Florida Current. From the estimated downstream increase in transport and the increase in magnitude of the Coriolis parameter, the cross-stream difference in level would require the inshore edge of the Gulf Stream to slope down 2.8 cm deg^{-1} relative to the seaward edge or a net downward slope of 2.0 cm deg^{-1} in the direction opposite to the land-survey results. Sturges concluded that the precise leveling surveys must contain systematic errors of undetermined nature that gave rise to the slight bias in meridional geodetic leveling.

4.3.2 Variability of the Florida Current

Speculation about the variability of the Florida Current was inspired not only by evidence from tide gauges but also from measurements of electrical potential using a telegraph cable from Key West to Havana, Cuba (Wertheim, 1954). The electrical potential induced by flow of sea water through the earth's magnetic field, shunted partially by the conducting sea floor, provides a signal that is correlated with the transport. The variations of nontidal flow appear to be exaggerated in the electrical potential (Schmitz and Richardson, 1968) because of shifts of the Florida Current relative to the bottom topography. The cause of these shifts was not determined. Maul et al. (1978) have speculated that meanders of the Loop Current in the Gulf of Mexico may affect the Florida Current. Sanford and Schmitz (1971) concluded that induced electrical potential was more closely correlated with the transport at the Miami section. The estimated error was found to be about 10% of the mean compared to a factor of two changes for the Key West section.

Supporting evidence for the seasonal variation of the Florida Current found by Montgomery (1938b) in the tide-gauge data came from other sources. Fuglister (1951) used monthly mean current speed and direction charts from an atlas published by the U.S. Navy Hydrographic Office (1946) to estimate seasonal variability in 10 regions following the Gulf Stream System from Trade Wind Latitudes to beyond the Grand Banks. He found that the maximum currents occurred in summer (July) in southern portions and in winter in northern portions, while the minimum tended to occur in fall (September to November) in all regions. The seasonal variability first seen on tide gauges was also confirmed by direct measurements. Niiler and Richardson

(1973), using a 7-year study of volume-transport measurements at a cross-stream section at Miami, found that the annual variation accounted for 45% of the total variability, with transports ranging from an early winter low of 25.4×10^6 m^{3}s^{-1} to a summer high of 33.6×10^6 m^{3}s^{-1}. The average transport was 29.5 m^{3}s^{-1}. They also reported a strong short-period (2 weeks or less) modulation of the velocity and density structure that was as strong as the seasonal change. These rapid fluctuations have been studied intensively since (Brooks, 1975; Wunsch and Wimbush, 1977; Schott and Düing, 1976).

Wunsch, Hansen, and Zetler (1969) extended the analysis of sea-level record and sea-level differences over the remainder of the frequency range contained in the tide-gauge data. They examined simultaneous records from Key West and Havana (1953–1956), Key West and Miami (1957–1962), and Miami and Cat Cay (1938–1939). The power levels below tidal frequencies were low at all sites. About half the power was in seasonal variations with no detectable peaks between seasonal and tidal frequencies. Coherences between stations were small and, where detectable, showed zero phase across the Florida Current, consistent possibly with a Bernoulli effect or large-scale atmospheric forcing. Lack of downstream coherence was attributed to Doppler shifting because of different mean speeds between stations. An inverted-barometer response to atmospheric-pressure forcing could be detected at lower frequencies (periods 10–128 days), and a direct response at higher frequencies.

4.3.3 Eddy–Mean Flow Interaction

A downstream pressure gradient is called for in inertial models of westward intensification. In the simplest model of this type, the frictionless, homogeneous circulation on a β-plane described by Fofonoff (1954), the pressure and free surface drop along the western boundary as the flow intensifies. The lowest pressure is found at the boundary where the highest speeds are attained. Along each streamline, pressure is related to speed by the Bernoulli equation. In the model, the highest speeds and lowest pressures (and hence sea levels) are found at the boundary. In the real ocean, the sea surface within the Florida Current has to be matched to a coastal boundary region, implying that the pressure gradient is continued into the coastal region. This, in turn, implies an active dynamic regime inshore of the Florida Current. Several studies have described the fluctuations within and adjacent to the Florida Current in some detail.

Von Arx, Bumpus, and Richardson (1955) observed a succession of short, overlapping segments that they described as "shingles" extending from the Florida Straits past Cape Hatteras. These shingles were first noted during an attempt to follow the Florida Current with an airborne infrared radiometer. The inshore edge was found not to be continuous in its thermal structure but made up of a series of fronts. They speculated that the cause might be tidal modulation of the Florida Current emerging from the Florida Straits, but admitted that no sound basis had been found for a relation between tides and the short-term fluctuations observed. These structures in the thermal field could be interpreted as instabilities of the Florida Current and evidence for exchange of energy between the mean flow and a time-dependent field.

Webster (1961a, 1961b, 1965) analyzed these and other surface-velocity measurements made during repeated crossings of the Florida Current at sections off Miami and Jacksonville, Florida, and off Onslow Bay and Cape Hatteras, North Carolina, to estimate Reynolds stresses associated with the nontidal velocity fluctuations present in the flow. One of the objectives of the study was to evaluate the magnitude of eddy–mean flow interactions within the Florida Current. The surface currents were estimated using a towed GEK (Von Arx, 1950), which responds to the current component perpendicular to the ship's track. In some cases, currents were inferred from the ship's set during crossings. The repeated crossings enabled Webster to compute means and fluctuations of the cross stream $(\bar{u}, u')$, the long-stream $(\bar{v}, v')$ components of surface flow, and the momentum-flux component $\rho \overline{u'v'}$ in several zones across the Florida Current. The Reynolds-stress component τ_{xy} corresponding to this eddy momentum flux is $-\rho \overline{u'v'}$. At nearly all sections, the velocity correlations were positively correlated $(\rho \overline{u'v'} > 0)$, implying that momentum was being transported offshore and that the Florida Current was exerting a negative (southward) stress on the coastal region. As the northward flow $\bar{v}$ increases offshore, the momentum has to be transported into regions of increasing mean flow against the mean-velocity gradient. Thus, the slower-moving coastal waters appear to exert an accelerating stress on the swiftly moving Florida Current offshore, a result that is opposite to the intuitive expectation that the Florida Current might tend to be retarded by the coastal boundary and lose momentum to it. Webster calculated also the rate of work W done by the Reynolds stresses on the mean flow from the term

$$W = \bar{v}\frac{\partial \tau_{xy}}{\partial x} = \frac{\partial \bar{v}\tau_{xy}}{\partial x} - \tau_{xy}\frac{\partial \bar{v}}{\partial x}.$$

Integration from a straight coast $x = 0$ to the axis of the current $x = L$ (Webster integrated across the entire current) yields the total work per unit time within the coastal strip inshore of the current axis:

$$\int_0^L W\,dx = \bar{v}\tau_{xy}|_{x=L} - \int_0^L \tau_{xy}\frac{\partial \bar{v}}{\partial x}\,dx.$$

Assuming $\bar{v}$ to be zero at the coast, the total work done in the coastal strip is equal to the work done on the seaward boundary $(\bar{v}\tau_{xy})$ plus the eddy work on the mean flow within the strip. For a steady state to exist, the two terms must balance; otherwise, the mean flow in the strip would have to gain or lose energy at a rate equal to the difference between the two terms, assuming that other terms, such as work against pressure gradients, neglected above, remain small. Webster examined the eddy-mean flow term at each section and concluded that the net energy transfer was from eddy to mean flow for all sections. As a consequence of the eddy-mean flow interaction, the inshore strip is doing work on the Florida Current and therefore must contain an energy source to supply the offshore flux of momentum and energy. Schmitz and Niiler (1969) reexamined Webster's estimates and analyzed additional measurements made by free-fall instruments that confirmed the earlier conclusions about significant eddy-to-mean energy flux within the coastal region of cyclonic shear. They found, in addition, a region of negative velocity correlation in shallow depths close to shore, indicating a region of retarding stress and flow of momentum to the shore. This feature was not observed by Webster in Onslow Bay presumably because his sections did not approach close enough to the coast. Lee (1975) and Lee and Mayer (1977) describe recent measurements in this dissipative near-shore strip in the Florida Straits. Schmitz and Niiler (1969) found that the total energy flux integrated across the entire width of the current was not significantly different from zero within each section. They concluded that although a region of intense energy transfer from eddy to mean flow existed, it was offset by a wider region of mean-flow-to-eddy transfer over the rest of the current, resulting in a redistribution of energy that required no external energy source.

Brooks and Niiler (1977) carried out a comprehensive study of historical and new transport-profile data for a section across the Florida Current in the vicinity of Miami. Their estimates showed that statistically significant conversions of kinetic and potential energy between fluctuations and mean flow occurred in either direction in parts of the section, but the net conversion rates were too small to be dynamically important. Based on these rates, the decay time for the total perturbation energy was about 50 days, much longer than the residence time for the Florida Current in the Florida Straits. They concluded that pressure gradients must be present to balance the energy flow. The coupling between mean flow and fluctuations may, in fact, be rather weak compared to the major energy conversion between mean potential energy and mean kinetic energy, with the fluctuations playing a minor or negligible role. Such a model is also suggested by the distribution of surface velocity and kinetic energies of the mean and eddy flow tabulated for the Florida Current by Hager (1977) from ship-drift reports collected by the U.S. Hydrographic Office for the period 1900-1972. While these data are not of the same quality as direct measurements, they reveal clearly the spatial extent of the Florida Current and its region of intensification as it flows into the Florida Straits. The peak currents and kinetic energies appear to be underestimated by the dead reckoning used to compute ship drift because of the spatial averaging involved. Hager found that the eddy kinetic energy was comparable to the mean-flow kinetic energy in the Loop Current and in the Gulf Stream past Cape Hatteras. However, within the Florida Straits, the eddy kinetic energy $(1-2 \times 10^3 \text{ cm}^2 \text{ s}^{-2})$ was much smaller than the mean-flow kinetic energy $(>10^4 \text{ cm}^2 \text{ s}^{-2})$ and showed little similarity in its spatial distribution. These results suggest that the fluctuations are not essential to the intensification of the mean flow in the Florida Straits.

4.3.4 The Downstream Pressure Gradient

The downstream pressure gradient is important to the energetics of the Florida Current because it provides the simplest mechanism for converting potential to kinetic energy within the Florida Current (Webster, 1961a). The development of satellite radar altimetry with precision and resolution capable of detecting differences in surface elevation of less than a meter (Vonbun, Marsh, and Lerch, 1978) has created the opportunity to use sea-level slopes to infer behavior of current fluctuations in considerably greater detail than possible with surface observations only. The interpretation of surface slopes and internal pressure gradients related to these slopes will become increasingly important as altimetry measurements are accumulated (see figure 4.11).

The Florida Current increases in transport as it flows northward along the continental slope to Cape Hatteras (Iselin, 1936; Richardson, Schmitz, and Niiler, 1969; Knauss, 1969). Its momentum, energy content, and flux increase, implying the presence of strong energy sources within the Florida Current and perhaps the surrounding regions. As the increasing momentum and energy within the Florida Current is most likely produced by a downstream pressure gradient acting to accelerate the flow, the most probable source of energy for the inshore region is a continuation of this downstream gradient into the coastal region.

Godfrey (1973) has given a clear physical interpretation of the effects of a downstream (northward) pressure gradient based on an examination of a six-layer numerical model reported by Bryan and Cox (1968a,b). The longshore pressure gradient was well developed in the upper layers and weakened with depth along the western wall. The drop was equivalent to about 1 m at the 100-m level and had reversed sign at 1600 m. Because a balancing geostrophic flow would have to be outward from the coast, a complete geostrophic balance is impossible. The gradient must be balanced partly by an outflow, causing upwelling along the boundary, and partly by downstream acceleration. The upwelling along the coastal boundary implies shoreward motion at depth. Godfrey used the model to interpret eddy formation in the East Australian Current, but it had been developed originally by Bryan and Cox with application to the Gulf Stream in mind.

Blanton (1971, 1975) presented evidence for a vigorous movement of shelf water into the Florida Current and intrusion of Gulf Stream Water from the Florida Current along the bottom onto the North Carolina Shelf off Onslow Bay in summer. A section taken on July 22, 1968, showed Gulf Stream Water covering the entire shelf, with shelf water forming an isolated lens in the upper layer at mid-shelf. A month earlier, Gulf Stream Water had shown only a slight intrusion at the shelf break (40-m depth). Many other factors may be present. The driving mechanism, whether dominated by pressure gradients originating in the Florida Current as described by Godfrey (1973) or by local winds, has not been clearly established. The occurrence of strong upwelling and exchange with the coastal region is apparent, however, and may be evidence of a current-induced pressure gradient on the shelf.

The mechanism by which the pressure gradient can supply momentum to the eddies and not to the mean flow remains obscure. Because the meanders described by Webster (1961b) move downstream in the direction opposite to the propagation of topographic Rossby waves, the mechanism of wave-momentum transport suggested, for example, by Thompson (1971, 1978) does not appear to be appropriate.

Pedlosky (1977) studied the radiation conditions for a linear two-layer ocean model to propagate waves away from a forcing region consisting of a sinusoidal moving zonal boundary. Eastward-moving meanders can radiate into the ocean interior only if their phase speed is less than the local interior velocity. If the local interior velocity is westward, the eastward-moving meanders cannot radiate in either baroclinic or barotropic Rossby-wave modes. For nondivergent flows over a sloping north–south boundary, such as the continental shelf, these results seem to imply that topographic waves in the shallow coastal region cannot be coupled to northward-moving meanders. Other mechanisms may be possible. Webster (1961a) noted that "each of the meanders resembles a sort of skewed wave motion and consists of an intense offshore running current (time 1 to 4 days) followed by a broad diffuse flow onshore time 4–7 days then followed by another intense offshore current." The intense offshore jets shown in figure 4.4 may be similar to inertial jets formed along western boundaries. The sloping shelf provides a strong topographic β-effect in the coastal strip. Currents that flow offshore down a pressure gradient and across depth contours on the shelf so as to conserve potential vorticity would intensify into narrow jets with strong cyclonic relative vorticity that may be incorporated into the cyclonic inshore region of the Florida Current. Such jets could carry momentum and energy offshore. If the instantaneous downstream (northward) pressure gradient were concentrated across a narrow jet, the transfer of momentum into the Florida Current would be readily accomplished. The existence of intensifying jets detached from solid boundaries has not been established, however, so that this line of reasoning must be considered speculative. Most theoretical studies applicable to the Florida Current assume that the basic flow is nondivergent with zero downstream pressure gradient. It is possible that the neglect of the pressure gradient excludes relevant mechanisms of meander formation and may exaggerate the role of eddy-mean flow interactions in numerical models.

4.3.5 Stability and Atmospheric Forcing
Mechanisms for the conversion of kinetic and potential energy associated with the mean flow to perturbations have been examined in several studies of the stability of the Florida Current.

Orlanski (1969) developed a two-layer model for two cases of bottom topography in the lower layer resembling the continental slope under the Gulf Stream and

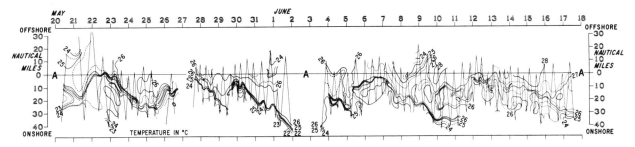

Figure 4.4 Space-time variation of temperature off Onslow Bay, showing movement of temperature fronts with 4–7-day time scale. Note that the offshore motion was discontinuous or more rapid than the onshore motion of the front. (Webster, 1961b.)

the continental rise in the open ocean further downstream. The model has a constant Coriolis parameter with no downstream variation of the basic current, pressure fields, or topography. Orlanski found that a necessary condition for instability to occur is that the gradient of potential vorticity of the basic flow be of opposite sign in the two layers. As only cross-stream variation occurs, the stability depends critically on the slope of the interface between the two layers relative to the bottom slope. The change of thickness of the bottom layer across the current can determine the sign of the potential-vorticity gradient and hence the stability. The most unstable modes found by Orlanski are given in table 4.1. Orlanski and Cox (1973) reexamined the stability of the western boundary current in a three-dimensional numerical model. The model had better resolution in the vertical (15 levels) but was periodic along the coast, thus excluding a downstream pressure gradient. Nonlinear terms and a β-effect were included in the model. Instabilities developed as predicted by linear theory but with a growth rate about double that of the simpler two-layer model. The growth rate decreased by an order of magnitude as finite amplitude was attained.

Niiler and Mysak (1971) analyzed a barotropic, constant-f model in which the velocity distribution and bottom topography of the continental shelf were approximated by segments of constant potential vorticity and depth. Unstable barotropic waves were possible in the model because the potential vorticity was chosen to contain maxima in its distribution across the current. The arguments for these extrema are that the cyclonic shear in the inshore region raises the relative vorticity sufficiently to overcome the opposing effect of increasing depth of the shelf and slope. Thus if the slope is small enough, a maximum occurs in potential vorticity. A region of anticyclonic shear on the seaward side of the Florida Current over a slowly varying depth yields a minimum in the cross-stream distribution.

These extrema in the potential-vorticity distributions imply the existence of unstable barotropic modes. With no basic current, the solutions are shelf and topographic waves already discussed by Robinson (1964) and Rhines (1969a) (see chapters 10 and 11). With a basic northward current, the southward-traveling waves can be reversed and made unstable. The most unstable barotropic mode on the Blake Plateau has a period of about 10 days and a wavelength of 140 km and can reach finite amplitude in a few wavelengths downstream. Because these barotropic waves require bottom topography to induce regions of instability, their growth is not sustained in deep water. Here the unstable waves were found to have a period of 21 days and a wavelength of 195 km. The authors suggest that the unstable shelf modes can be triggered by narrow fast-moving frontal systems. These short-period waves increase in amplitude as they move northward to deep water, where they are no longer unstable because of the change in potential-vorticity structure of the basic deep flow but may persist as a smaller-scale structure on the longer and slower meanders that develop downstream.

Brooks (1978) has also pointed out the importance of wind stress and its curl as a forcing mechanism for shelf waves. He concludes that strong coupling can occur for periods that are less than or greater than the zero group-velocity period of barotropic shelf waves for the continental shelf off Cape Fear (i.e., 2.5–3.5 and >10 days, respectively). The model was used to interpret correlations between atmospheric-pressure variations and winds and sea-level variations from tide gauges at Beaufort and Wilmington, North Carolina. Recently, Brooks and Bane (1978) reported that deflections of the Florida Current are induced by a small topographic feature in the continental slope off Charlestown, South Carolina. Satellite observations of thermal patterns (figure 4.3, for example) show considerable difference in amplitude upstream and down-

Table 4.1 Characteristics of Perturbations Found for the Florida Current and Gulf Stream

Author	Wavelength (km)	Period (days)	Growth rate (days^{-1})	Phase speed (cm/s)	Type
Orlanski (1969)					
Shelf waves	220	10	1/5		Baroclinic
Deep ocean	365	37.4	1/7.23		Baroclinic
Orlanski and Cox (1973)	246		1/12.1		
Niiler and Mysak (1971)					
Shelf	140	10	1/13	14	Barotropic
Ocean	195	21	1/13	9	Barotropic
Brooks (1975)	190	12	0	South	
Schott and Düing (1976)	170	10–13	0	South (17 cm s^{-1})	From current measurements

stream of the "Charlestown Bump" located near 32°N. Stumpf and Rao (1975) suggested possible topographic influences in studying a sequence of infrared images of meanders off Cape Roman and Cape Fear. They point out that a well-coordinated field experiment would be necessary to distinguish wind forcing from topographic influences or instability of the Florida Current.

Schott and Düing (1976) found southward-traveling waves in the Florida Straits based on velocity measurements from three moored buoys located close to the same isobath at 335 m near the "approximate location of the axis of Gulf Stream" according to nautical charts. Records were obtained for a duration of 65 days from a depth of about 300 m. The most likely wave parameters were fitted by least-square methods to 36 independent auto- and cross-spectra. A significant fit was found in the 10–13-day spectral band for a wavelength of 170 km traveling south at 17 cm s^{-1}. These are identified as stable continental-shelf waves probably generated by passage of atmospheric cold fronts. The parameters obtained agree well with a model by Brooks (1975) that included realistic topography and horizontal current shear to yield a southward-propagating wave of 12-day period and 190-km wavelength at maximum response to forcing by cold fronts. The characteristics of these wave models are summarized in table 4.1.

The observed coherence with meteorological events noted by Wunsch and Wimbush (1977) and Düing, Mooers, and Lee (1977) may be a consequence of the weak coupling between mean flow and the fluctuations. The perturbations apparently can receive a significant fraction of their energy from atmospheric forcing rather than from the mean flow and consequently show measurable correlation with wind events.

Several mechanisms for generation of meanders in the Florida Current have been identified: barotropic and baroclinic instability in the presence of topography; bottom features forcing deflections and downstream lee waves; and excitation of propagating waves by atmospheric forcing. Nonlinear mechanisms are yet to be explored, as are the effects of the downstream inhomogeneity of the Florida Current.

Richardson et al. (1969) found that the transport of the Florida Current increased relatively slowly (17%) through the Florida Straits from Miami to Jacksonville with a slight increase in surface speeds and a shift to westward of the current axis. A larger increase in transport (67%) was found from Jacksonville to Cape Fear with a slight decrease of maximum surface velocity and a broadening of the current. The effects on the instability modes of the downstream increase in transport are not known. Other intermittent perturbations to the Florida Current are the passage of rings and eddies seaward of the Florida Current. These apparently can be swept into the current (Richardson, Strong and Knauss, 1973). The consequences of such events are not known. The deep western boundary current predicted by Stommel (1957b) and the first observed by Swallow and Worthington (1957, 1961) has been found in recent studies to contain strongly time-dependent components (Riser, Freeland, and Rossby, 1978). The effects on the Florida Current are not adequately known at present but may be profound.

4.3.6 The Deep Western Boundary Current

If an upwelling velocity of broad horizontal scale is assumed in the deep interior flow of an ocean, Stommel (1957b) showed that the conservation of mass and potential vorticity cannot be satisfied by geostrophic flow alone. A deep western boundary current is necessary to allow both constraints on the deep flow to be met. In the North Atlantic, Stommel concluded that a southward flow should be present along the continental slope. This prediction together with the development of the neutrally buoyant float for measuring current by Swallow (1955) led Swallow and Worthington (1957, 1961) to measure the deep flow off Cape Romain, South Carolina, near the northern end of the Blake Plateau, where the flow was expected to lie seaward of the strong surface current over the Blake Plateau. Southward flows of 9–18 cm s^{-1} measured over a period of a month led them to conclude that the deep western boundary current is a persistent feature of the circulation along the continental slope. The transport of the undercurrent was estimated to be 6.7×10^6 m^3 s^{-1}.

Subsequent measurements by a number of investigators (Volkmann, 1962; Barrett, 1965; Worthington and Kawai, 1972; Richardson and Knauss, 1971; Amos, Gordon, and Schneider, 1971; Richardson, 1977) reported transports ranging from 2 to 50×10^6 m^3 s^{-1} with an average of 16×10^6 m^3 s^{-1}. The flow is persistent, though apparently quite variable. Westward and southward deep currents along the continental rise north of the Gulf Stream have been reported by Webster (1969), Zimmerman (1971) and Luyten (1977). Recent measurements using SOFAR floats (Riser, Freeland, and Rossby, 1978) show flow south of the Blake-Bahama Outer Ridge along the Blake Escarpment. The southward flow may be simply a consequence of a deep westward flow onto the sharply rising topography. Steady slow geostrophic flows are constrained to follow contours of f/h, where h is depth, which are concentrated along the slope. Near Cape Hatteras, where the Gulf Stream crosses over the deep current (Richardson, 1977), the combined effect of vortex stretching within the northward-moving current and the deep flow crossing the bottom slope would be felt. Holland (1973) has examined the enhancement of transport in the western boundary current in a numerical model including baroclinicity and topography. The vortex stretching in the stratified upper layers must counteract changes of f only to conserve potential vorticity, whereas the deep water can be subjected to a much larger stretching by crossing depth contours. Thus a relatively weak deep flow crossing depth contours can have a vertical velocity equal and opposite to a large meridional baroclinic flow. This type of flow was described briefly by Fofonoff (1962a) under a general class of thermohaline transports.

Slow steady barotropic flow must be along contours of f/h or must cross contours of f/h at a rate that balances the combined divergence of baroclinic and Ekman flow. For deep flow onto the continental slope, the intensification of the current on the slope can be estimated from the potential-vorticity equation along each streamline:

$$\frac{f_0}{h_0} = \frac{f_0 + \beta y}{h_0 - s_x x - s_y y},$$

where f_0, h_0 are open-ocean values of Coriolis parameter and depth, and s_x, s_y the (constant) bottom slopes. For a slope width Δx, the deep streamlines are displaced equatorward by an amount Δy, where

$$\Delta y = \frac{\beta_x}{\beta_y} \Delta x = \frac{f_0 s_x / h_0}{(\beta + f_0 s_y / h_0)},$$

and β_x, β_y are the horizontal gradients of f/h_0. From the sketch in figure 4.5, it is seen that the narrowing or intensification over the slope is

$$U_s = U_0 \frac{h_0}{h} \frac{\sqrt{\Delta x^2 + \Delta y^2}}{\Delta x} = U_0 \frac{h_0}{h} \sqrt{1 + \left(\frac{f_0 s_x}{\beta h_0}\right)^2}$$

$$\sim \frac{U_0 f_0 s_x}{\beta h_0} \quad \text{(for } s_y = 0\text{)}.$$

For $h_0 = 2500$ m, $f_0 = 10^{-4}\,\text{s}^{-1}$, $\beta = 2 \times 10^{-13}\,\text{cm}\,\text{s}^{-1}$, $s = 1/100$ (continental rise),

$$U_s = 20 U_0.$$

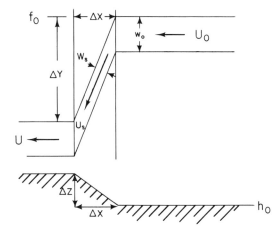

Figure 4.5 Displacement Δy of a current flowing over a bottom slope of width Δx on a β-plane. The velocity U_s on the slope is magnified by the ratio of widths w_0/w_s. Relative vorticity is assumed to be small.

For the continental slope (e.g., $s = 1/15$, $h = 1500$ m),

$$U_s = 200 U_0.$$

The intensification even over the gentle continental rise is sufficient to magnify flows U_0 that are below a measurable level in the interior to observable velocities on the rise and slope. Thus, it is very difficult to determine by direct measurement whether the flow over the continental rise is being forced by an upslope or downslope component.

The main thermocline deepens northward (Iselin, 1936) on the seaward side of the Florida Current, intensifying the apparent β-effect below the thermocline. The deep flow must move southward to conserve potential vorticity. Within the Gulf Stream itself, the thermocline rises sharply downstream. The rise is equivalent to $s_y < 0$ in the lower layer. Furthermore, the thermocline slopes sharply downward in the x-direction because of the shear across the thermocline. If the thermocline slopes are denoted by T_x, T_y, the lower-layer-displacement equation becomes

$$\Delta y = \frac{f_0 (T_x - s_x)/h_0}{\beta - f_0 (T_y - s_y)/h_0} \Delta x.$$

The displacement Δy is no longer necessarily southward along the western boundary. Northward deep flows are permitted by the vorticity equation if

$$T_x - s_x < 0 \quad \text{or} \quad T_y - s_y > \frac{\beta h_0}{f_0}.$$

These flows would likely be unstable because the potential-vorticity gradient would then be of opposite sign in the two layers.

According to simple potential-vorticity conservation, westward deep flow on reaching the continental rise should turn southward and continue to have a southward component as long as the main thermocline slopes downward to the north. The decreasing thickness of the deep-water layer has to be compensated by decreasing the Coriolis parameter. In the region of the accelerating western boundary current, the thermocline slope is reversed and the constraint on the deep flow is altered. The current may then turn northward if the thermocline slope is sufficiently large. The circulation diagram given by Worthington (1976) for the deep water (potential temperature $\theta < 4°C$) reproduced in figure 4.6 has southward flow along the continental slope with northward flow further to the east opposite to the deep flow expected based on the elementary potential-vorticity arguments given here. The present interpretation of the deep circulation in western North Atlantic and its interaction with the deep boundary current is not consistent with potential-vorticity conservation and needs further development.

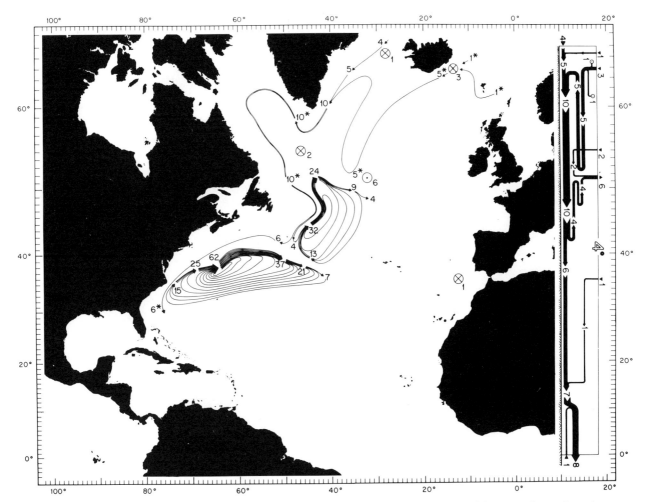

Figure 4.6 Circulation diagram for the deep ($\theta < 4°C$) circulation in the North Atlantic. Worthington estimates a flow of 62×10^6 m³ s⁻¹ in the recirculation gyre with 6×10^6 m³ s⁻¹ flowing southward inshore of the recirculation. (Worthington, 1976.)

4.4 The Gulf Stream

The portion of the Gulf Stream System from Cape Hatteras to the Southeast Newfoundland Rise is the *Gulf Stream* according to the nomenclature introduced by Iselin (1936). The Gulf Stream is separated from the continental shelf to the north by a band of westward-flowing Continental Slope Water (McLellan, 1957; Webster, 1969) and bounded to the south by the westward recirculation gyre described by Worthington (1976) and Schmitz (1978). Thus, the Gulf Stream is an eastward-flowing current flanked on either side by broader regions of westward flow. As the Gulf Stream is assumed not to be locally driven, enough energy and momentum must be carried by the flow into the region to maintain the eastward motion and the eddy and circulation fields in the surrounding areas. Fuglister (1963) noted that the Gulf Stream leaving Cape Hatteras flows approximately along a great circle while the continental slope turns northward. There is no pronounced curvature of the Gulf Stream on entering deep water, as might be expected from the increasing depth. The lack of a mean curvature at a point of rapid deepening of the ocean bottom is interpreted as an indication that the Florida Current is injected into the upper layer above and into the main thermocline and may only contact the bottom intermittently. Richardson (1977) found that the Gulf Stream did not extend to the bottom off Cape Hatteras for direct current measurements from six moorings over periods ranging from 5 to 55 days. Other measurements (Barrett, 1965; Richardson and Knauss, 1971) show northeast flow near the bottom under the axis of the Gulf Stream. After leaving Cape Hatteras, the Gulf Stream gradually develops meanders, clearly visible in the infrared image in figure 4.7. The meanders become progressively larger downstream (Hansen, 1970), but are especially marked after crossing the New England Seamounts (Fuglister, 1963; Warren, 1963). The most intense horizontal thermal and density structures are found between Cape Hatteras and the Seamounts. Strong horizontal density gradients are found throughout the entire depth. These

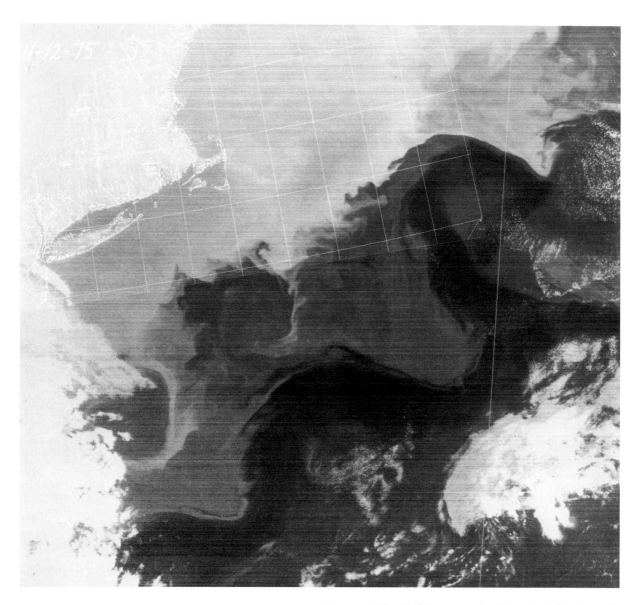

Figure 4.7 The Gulf Stream south of Cape Cod, showing well-developed meanders with several eddies formed in the slope water to the north (Courtesy of R. Legeckis, NOAA-NESS, from NOAA-4 satellite November 12, 1975.)

gradients begin to weaken in the deep water after crossing the Seamount chain (Fuglister, 1963). The meanders become sufficiently large to form detached cold eddies (rings) to the south as shown in figure 4.8 (Fuglister, 1972; Parker, 1971) and warm eddies to the north (Saunders, 1971) of the Gulf Stream at irregular intervals. The meander amplitudes probably do not continue to grow eastward because of the confining effects of the Grand Banks and Southeast Newfoundland Rise. These topographic features appear to lock the Gulf Stream into quasi-stationary spatial patterns similar to those described by Worthington (1962) and Mann (1967) that are more constrained than the meanders farther to the west.

4.4.1 Gulf Stream Separation Mechanisms

The mechanism of separation of the Gulf Stream from the continental slope at Cape Hatteras remains ambiguous. The early theories of mean ocean circulation developed by Stommel (1948) and Munk (1950) required an intensifying current along the western boundary only to the latitude of maximum wind-stress curl. Poleward of the maximum, the current weakened and returned into the ocean interior as a broad slow flow specified by the meridional scale of the wind-stress curl field. It was soon recognized that the lack of even qualitative agreement with the Gulf Stream could be attributed to the neglect of nonlinear terms in the western boundary. Munk, Groves, and Carrier (1950) showed by a perturbation analysis that the nonlinear terms acted to shift the point of maximum velocity downstream past the maximum in the stress curl.

The inertial models that were developed subsequently indicated that an intensifying current with westward flow from the interior (Charney, 1955b; Morgan, 1956) could be extended well past the latitude of maximum curl of the wind stress by inertial recirculation. In two-layer inertial models, the northward extent is limited by surfacing of the inshore isopycnal and ending of the potential to kinetic-energy conversion in the boundary current [Veronis (1973a) and chapter 5]. By increasing the size of the recirculation gyre, the boundary current can be extended to the latitude with zero wind-stress curl that separates the major ocean gyres (Munk, 1950). Leetmaa and Bunker (1978) recomputed the curl of wind stress from recent data and found that the zero stress-curl contour lies near Cape Hatteras on the average. Thus the separation may be a consequence of the larger-scale mean wind field rather than the local dynamics at Cape Hatteras. Moreover, Stommel, Niiler, and Anati (1978) point out that all of the transport in excess of about 30–36 × 10^6 m^3 s^{-1} required by the wind-stress-curl distribution can be attributed to recirculation without violating conservation of mass and heat. The possibility that the

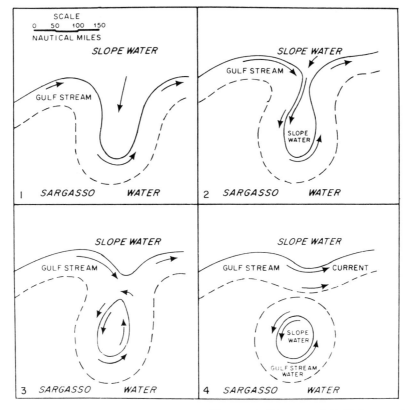

Figure 4.8 Diagram of Gulf Stream ring generation from meander formation to separation. (Parker, 1971.)

mean Gulf Stream path may be determined by the mean-wind pattern deserves further study. It is not obvious why the Gulf Stream should separate from the continental slope to allow formation of the slope-water region extending from Cape Hatteras to the Grand Banks, rather than flow along the slope to the Banks and beyond (for further discussion, see chapter 5).

4.4.2 Gulf Stream Trajectory Models

Different mechanisms for determining the shape of the Gulf Stream have been proposed in terms of path or trajectory models. These will be examined briefly.

The systematic measurements in a series of 11 meridional hydrographic sections through the Slope Water and Gulf Stream carried out in 1960 under Fuglister's guidance (Fuglister, 1963) provided the basis for a number of theoretical studies in the following years, including the development of the trajectory or path theories. The observations during the 3-month duration of GULF STREAM '60 showed rather small changes of the large meander found in the Stream path. Fuglister stated that no evidence was found in the data for shifts in the meanders by more than the Gulf Stream width. Moreover, neutrally buoyant floats tracked at 2000 m depth to provide reference velocities for the computation of geostrophic currents yielded currents extending to the ocean bottom flowing in the same direction as the surface Gulf Stream. These characteristics of the Gulf Stream prompted Warren (1963) to develop a model based on bulk or integrated properties of the Gulf Stream. By assuming that the Gulf Stream could be treated as a narrow current or jet and integrating between streamlines over a section across the current, the vorticity equation was converted to a form relating path curvature to vortex stretching by the changing depth along the path of the current and by changes in the Coriolis parameter resulting from a change of latitude. Given initial conditions of position and direction, as well as the bulk properties of volume transport, momentum transport, and volume transport per unit depth, the subsequent path is determined by the topography and change of latitude encountered enroute. The simple model applied to five observed paths exhibited remarkable agreement in shape in the region of longitude between 65 and 73°W. The path computation could not be continued eastward because of the obvious breakdown of the model in describing meanders over the New England Seamounts. Warren noted, as did Fuglister (1963), that the New England Seamount Arc underlies the region where large meanders develop.

The model possessed several attractive features. The separation from the continental shelf at Cape Hatteras occurred as a natural consequence of the topography and was not related to a wind-stress mechanism as suggested by earlier theories. Moreover, the meanders could develop as a consequence of the initial angle of injection relative to the topography and would not necessarily indicate an instability of the current. Subsequent elaboration of the theory by Niiler and Robinson (1967) brought to light several shortcomings of the approach. The narrow-jet trajectory theory assumed a steady state, whereas later observations revealed the Gulf Stream to exhibit strong time dependence in its meanders. Neither the simple model studied by Warren nor the more elaborate models developed later could be fitted simultaneously to the mean-path and meander data (Robinson, 1971). Robinson concluded that "vortex-line stretching will undoubtedly play some role in the vorticity balance" in a properly posed nonlinear time-dependent theory of meanders. Hansen (1970) obtained a series of measurements of Gulf Stream paths to describe the occurrence and progressive development of meanders in an effort to discriminate between the inertial-jet theories and dynamic-wave models with possible unstable modes that can extract energy from the basic flow. The paths were mapped over a period of a year by towing a temperature sensor along the 15°C isotherm at 200 m depth supplemented as necessary by bathythermograph observations. Hansen concluded that although no model then available could account for all of the major features of the Gulf Stream, the most likely models would have to include topographic influences that are clearly seen in some, but not all, observed paths as well as energy conversion processes such as baroclinic instability necessary to account for meander development at least, where topography is too weak to influence the path. Path models alone were not adequate to account for the meanders.

Time-dependent extensions of the path model have been given by Luyten and Robinson (1974) and Robinson, Luyten, and Flierl (1975). A consistent dynamic quasi-geostrophic model was developed in which the velocity field is resolved into a jet velocity, a velocity of the jet axis, and a transient adjustment velocity assumed small relative to the geostrophic velocities. The model was applied to Gulf Stream data collected during 1969 near 70°W (Robinson, Luyten, and Fuglister, 1974). Using parameters appropriate to the Gulf Stream for 70°W, Robinson et al. (1975) found that an inlet period of 31 days had a spatial wavelength of 560 km and a downstream growth (e-folding) scale of 200 km, in agreement with the observed large-scale meanders of the Gulf Stream. In the local vorticity balance, advection and transient terms dominated the topographic and β-effects. The model contains mechanisms analogous to ring or eddy formation. Because the path displacements are not constrained to be small, the path equations can, at least in the case of the purely baroclinic limit and no β-effect, yield solutions in which the meanders grow spatially and close upon themselves to form isolated eddies. For the thin jet models to be applicable, however, the transient flows must be small.

4.4.3 Deep Currents of the Gulf Stream

The vertical coherence of the thermal and density field within the Gulf Stream and the slow evolution of meanders noted by Fuglister (1963) made plausible the assumption that the Gulf Stream extended to, and interacted with, the ocean bottom. A vertically coherent Gulf Stream, however, was not substantiated by subsequent direct current measurements. These showed a vigorous velocity field in the deep water (Schmitz, Robinson, and Fuglister, 1970) that was not a simple downward extension of the near-surface flow.

Luyten (1977) designed a current-meter array consisting of 15 moorings deployed in two meridional lines about 50 km apart at 70°W to measure the deep flow on the continental rise and beneath the Gulf Stream. Current meters were placed 1000 m above the ocean bottom on each mooring with a second instrument 200 m above bottom on some moorings as shown in figure 4.9. Data for 240 days were obtained on 30 of the 32 current meters. The measurements revealed a remarkable and unexpected feature of the deep flow in that the downstream coherence scale was very small (less than 50 km) while the meridional or cross-stream scale was found to be several times as large (≈ 150 km).

The currents contained strong meridional bursts of speeds reaching 40 cm s^{-1} and lasting for about a month. The highest intensity of flow occurred nearest the Gulf Stream, with four to six bursts in individual records. Hansen (1970) calculated (figure 4.10) an average eastward phase speed of 8 cm s^{-1} and a mean wavelength of 320 km for meanders of the 15°C isotherm at 200 m depth. This corresponds to a period of 46 days, agreeing approximately with the interval between events (four to six events in 240 days) in Luyten's array shown in figure 4.9. Because the measurements were not simultaneous, a correlation between the motion in the deep water and the upper levels has not been firmly established. The agreement of time scales is suggestive of a strong coupling between them.

The mean flow for depths shallower than 4000 m on the continental rise is westward with speeds 2-5 cm s^{-1} and apparently dominated by the bottom slope. Deeper than 4000 m, the mean flow vectors tend to have a small eastward component with an erratic or perhaps rapid spatial variation of the meridional component, reflecting the burst structure of the variable flow. The Gulf Stream, if it exists near the bottom above 70°W as suggested by Fuglister (1963) and Robinson, et al. (1974), is nearly completely masked by the strong deep meridional eddy field. The lack of coherence downstream is puzzling. It may be a consequence of having only two points of measurement along the Gulf Stream direction. The deep fluctuations may be locked in some manner by local topography, resulting in an inhomogeneous spatial pattern of coherence. Luyten found the interaction between eddies and mean flow to be small north of the Gulf Stream on the continental rise. The horizontal eddy-stress divergence vector was nearly perpendicular to the mean flow. Under the Gulf Stream itself, the mean flow appeared to be gaining energy from the eddies (figure 4.11).

4.4.4 Dynamical Gulf Stream Models

The instability and subsequent evolution of a thin zonal jet in the upper layer of a two-layer model described by Rhines (1977) has several points of similarity with the observations taken by Luyten (1977) and Schmitz (1978) of the velocity field near the Gulf Stream. Rhines investigated numerically the evolution of an eastward jet imbedded in a broad westward flow in the upper layer of a two-layer ocean model. The initial field was perturbed by broadband noise. The sequence of development is shown in figure 4.12. In less than 20 days, organized meanders become visible in the upper layer and elliptical eddies with predominantly north-south motion in the lower layer resembling the meridional motions observed by Luyten (1977). In the early stages, the pattern moves downstream with the motion in the deep layer leading the upper layer as required by baroclinic instability. After the meander exceeds unit steepness (about 42 days), the nonlinear eddy-eddy interactions become evident, causing distortion and stretching of vorticity contours by horizontal velocity shear. In the lower layer, eddies of like sign begin to coalesce, producing a north-south displacement and creating an abyssal flow in the same direction as the upper layer jet.

As the eddy field develops, the jet is no longer recognizable in the velocity or streamfunction field, but is still visible in the topography of the interfacial surface. At later stages ($t = 62$ days), the eddies in the upper and lower layer begin to lock together and become more barotropic in structure. Rhines suggests, as illustrated in figure 4.12, that the time evolution of the model resembles the downstream development of the Gulf Stream with the injection of the jet into the upper layers of the ocean corresponding to initial conditions in the model. Downstream migration of the eddies in the lower layer, however, was not observed by Luyten (1977). Instead, the phase propagation appeared to be southward, probably dominated by the bottom slope. The downstream motion of the surface meander has been documented by several authors, notably Hansen (1970), as seen in figure 4.10. Coupling of the surface meanders and eddies to the lower layer drives a mean flow below the main thermocline and the development of barotropic eddies. Schmitz (1977, 1978) has found an eastward deep flow and increased barotropic signature

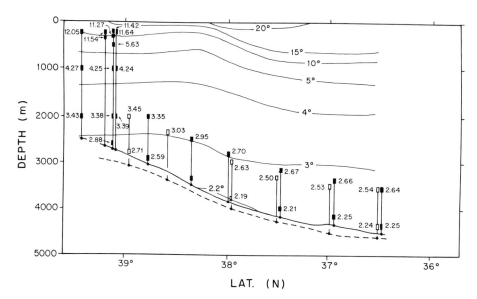

Figure 4.9A Distribution of current meters and temperature-pressure recorders on the Luyten (1977) Continental Rise array. The solid lines refer to 70°W, dashed to 69°30'W. (Luyten, 1977.)

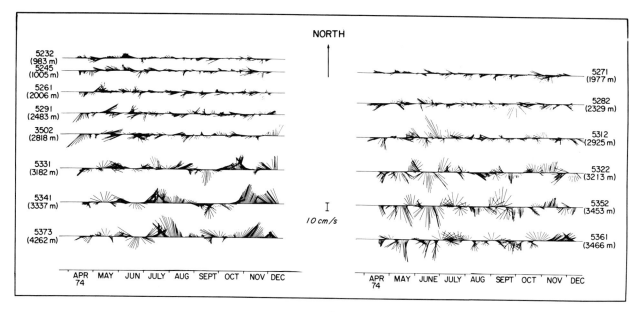

Figure 4.9B Time sequence of 1-day averaged current. The numbers identify mooring and instrument. Instrument depths in meters are shown in parentheses. (Luyten, 1977.)

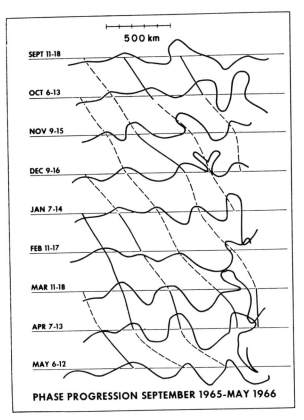
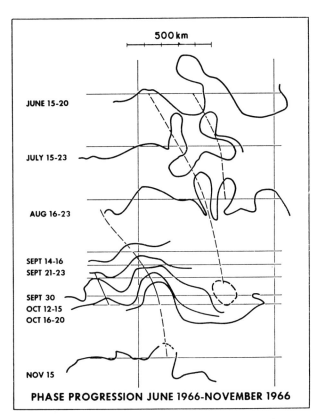

Figure 4.10 Inferred progression and evolution of meanders relative to the mean Gulf Stream path. The diagonal lines show phase propagation (solid where supported by other evidence). (Hansen, 1970.)

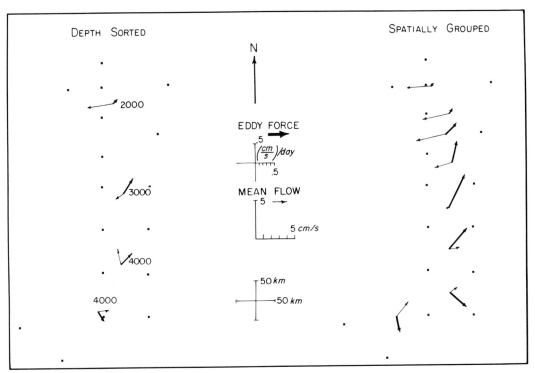

Figure 4.11 Vector acceleration of mean flow by eddy Reynolds stress gradients. The "eddy forces" tend to oppose the mean flow on the continental rise and to accelerate it under the Gulf Stream at the southern portion of the array. The eddy forces are estimated by grouping the data in depth intervals and from neighboring values in the array at a common depth for the spatial grouping. (Luyten, 1977.)

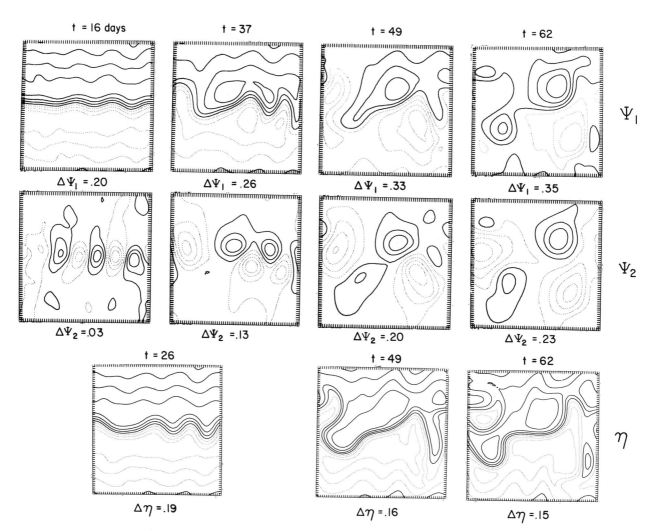

Figure 4.12 Evolution in time of a two-layer spatially periodic model. The figure shows upper-layer streamfunction ψ_1, lower-layer streamfunction ψ_2, and interface height η. The interface height retains the strong slope across the initial jet long after the jet has been obscured by barotropic flows in the velocity field. Contour intervals are indicated for relative comparison. Dimensions are 1250 km on each side with a speed of 50 cm s^{-1} averaged across the jet. (Rhines, 1977.)

to the velocity field (figure 4.13) in a long-term moored-array experiment along 55°W designed to examine the mesoscale eddy field in the vicinity of the Gulf Stream. [The array is part of a long-range program developed under the cooperative experiments MODE carried out in 1973 and in POLYMODE (1974-1979). The reader is referred to chapter 11 for a discussion of mesoscale eddies in the ocean.]

Although it is evolving and not stationary in time, Rhines's model is attractive and suggestive of processes that may be acting in the actual Gulf Stream. However, the model lacks an explicit recirculation mechanism and has no bottom topography. Both may affect the behavior significantly and may need to be incorporated as essential elements in a more complete Gulf Stream model. The simpler models must be explored and understood before the combined effects of several mechanisms can be interpreted in the more complex general circulation models.

The mass flux of the Gulf Stream has been estimated to be in the range of $100-150 \times 10^6 \, \text{m}^3 \, \text{s}^{-1}$ south of Cape Cod (Fuglister, 1963; Warren and Volkmann, 1968; Knauss, 1969). Stommel et al. (1978) concluded that most of the transport in the Gulf Stream is recirculated in the western North Atlantic. The net transport northeast, assumed to be wind driven in the ocean interior, is only $38 \times 10^6 \, \text{m}^3 \, \text{s}^{-1}$. Thus the recirculating portion may be as much as $60-110 \times 10^6 \, \text{m}^3 \, \text{s}^{-1}$. The high recirculation rate implies that a relatively small fraction of the energy is lost by the Gulf Stream in flowing between Cape Hatteras and the Grand Banks. Most of the kinetic energy is converted back to potential energy to form the recirculation to the south. Just as the conversion from potential to kinetic energy requires an accelerating pressure gradient along the western boundary, the conversion from kinetic to potential energy requires an opposing or decelerating pressure gradient. The details of the conversion process must be complicated because of the large-amplitude meandering and eddy formation that takes place between Cape Hatteras and the Grand Banks. Because it can be assumed that the recirculation is relatively stationary and does not change its size and intensity rapidly, its energy content is essentially constant. Energy is converted from potential to kinetic in flowing through the Gulf Stream and is converted back to potential energy on entering the westward gyre. As the recirculation transport is larger (perhaps twice) than the wind-driven transport, the kinetic-energy flux in the Gulf Stream probably contains only a minor contribution, depending on the velocity profile, attributable to direct forcing, and only this amount of energy needs to be dissipated to maintain a steady state. The magnitudes involved can be estimated very roughly from available data. The average wind-stress magnitude over the subtropical North Atlantic is about $0.5 \, \text{dyn cm}^{-1}$ (Leetmaa and Bunker, 1978). Assuming that the curl of the wind stress drives a transport of $40 \times 10^6 \, \text{m}^3 \, \text{s}^{-1}$ into the western boundary current through a section of average depth 500 m deep and 2000 km in length, the average inflow current would be $4 \, \text{cm s}^{-1}$. Assuming that $4 \, \text{cm s}^{-1}$ is also typical of interior geostrophic velocities and that wind stress-eddy forcing is small, the work done by wind stress would be $|\tau| \cdot |\mathbf{v}| \approx 2 \, \text{erg cm}^{-2} \, \text{s}^{-1}$. Over the entire basin (area $\approx 1 \times 10^{17} \, \text{cm}^2$), the rate of energy input by the wind is $2 \times 10^{17} \, \text{ergs s}^{-1}$; it is converted entirely to potential energy. The interior flux of kinetic energy by the mean flow is negligible. No net work is done by Ekman velocities in the frictional layer because the work by surface stresses must be dissipated within the Ekman layer for a steady state to exist. Only the surface geostrophic currents need be

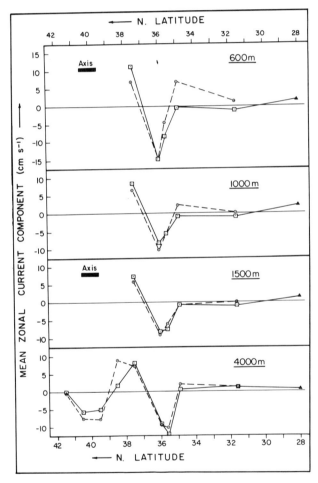

Figure 4.13 Meridional distribution of time-averaged zonal current at four depths along 55°W from the first POLYMODE array shown by circles ○ and from the combined first and second setting of the array shown by squares □. The approximate mean axis of the Gulf Stream is indicated. The westward mean flow shows little depth dependence. The mean flow is eastward under the Gulf Stream at 4000 m depth. (Schmitz, 1978.)

considered in estimating net wind work (Stern, 1975a).

If τ is wind stress and

$$\mathbf{v}_g = \frac{1}{\rho f}(\mathbf{k} \times \nabla_H p)$$

the surface geostrophic current, the net rate of wind work on the ocean surface is

$$\mathbf{v}_g \cdot \boldsymbol{\tau} = \frac{1}{\rho f}(\mathbf{k} \times \nabla_H p) \cdot \boldsymbol{\tau}$$

$$= -\frac{1}{\rho f}(\mathbf{k} \times \boldsymbol{\tau}) \cdot \nabla_H p$$

$$= \frac{1}{\rho}\mathbf{V}_e \cdot \nabla_H p$$

where $\mathbf{V}_e = -1/f\,(\mathbf{k} \times \boldsymbol{\tau})$ is the Ekman mass transport, $\nabla_H p$ the horizontal pressure gradient, and ρ density of the surface layer. The net work by wind stress can be interpreted as the rate at which mass is transported up the pressure gradient in the Ekman layer. Because pressure gradients are produced hydrostatically, the "uphill" Ekman flow represents an increase of potential energy at a rate equal to the wind work. Thus, except for the portion dissipated in the Ekman layer, the wind work is converted entirely to potential energy in the ocean interior. If no interior dissipation is present, all of the input wind energy must be removed through the western boundary current and dissipated within the recirculation region. Assuming a width of the boundary current of 80 km and a depth of 500 m, the energy and mass fluxes require a mean speed of 100 cm s^{-1} in the western boundary current. These are joined near Cape Hatteras by recirculation fluxes of mass and energy, resulting in a deepening and intensification of the flow, but leaving the scale width of the Gulf Stream unchanged. Suppose the recirculating transport is 80×10^6 m^3 s^{-1}, making a total transport of 120×10^6 m^3 s^{-1}. The total kinetic energy flux would be three times the interior value or 6×10^{17} ergs s^{-1} if the section deepened with no change of mean velocity. Only 2×10^{17} ergs s^{-1} must be dissipated and 40×10^6 m^3 s^{-1} transport returned to the interior circulation to maintain a steady mean state. The kinetic-energy fluxes into the western boundary layer and out of the dissipation region are assumed to be negligible compared to conversion rates of potential energy to kinetic energy. The kinetic-energy flux is proportional to the cube of speed and is, therefore, very sensitive to the velocity profile in the intensified Gulf Stream. For a given mean flux, the kinetic-energy flux is least if the flow is spread throughout the entire water column. A more detailed calculation is required to improve the estimate. An essential point of the argument is that the conversion of kinetic energy to and from potential energy is an exchange between mean fields. These conversions are not usually examined in numerical models because their basin average is zero.

An approximate estimate of the rate of energy dissipation in the Gulf Stream can be obtained from the initial decay rate of Gulf Stream rings, as these are formed from segments of the Gulf Stream itself (Fuglister, 1972). Cheney and Richardson (1976) found decay rates in Gulf Stream rings of $1.7 \pm 0.2 \times 10^{21}$ ergs day^{-1}, or about 37 ergs cm^{-2} s^{-1} averaged over the ring area, from the observed decrease of available potential energy of the ring. If the same dissipation rate is assumed in the Gulf Stream, it would require over 7000 km of path or about 80 days to dissipate the kinetic energy transported in the boundary region from the ocean interior. The distance from Cape Hatteras to the Grand Banks is about 2300 km—too short to get rid of the kinetic energy by internal dissipation and radiation. Meandering will increase the Gulf Stream path significantly, perhaps by a factor of two or more. Some of the energy is removed into the rings and eddies.

The flux of mass and energy in the Gulf Stream is sufficient to form a cyclonic ring of 100-km radius per week or about 50 rings per year. Many fewer are believed to form. Fuglister (1972) and Lai and Richardson (1977) estimated that as many as 10 to 16 rings of either anticyclonic or cyclonic type may form annually north and south of the Gulf Stream. If so, the formation of rings, although a spectacular manifestation of the Gulf Stream decay mechanism, is not the major mechanism for dissipating the kinetic energy. Some, possibly a considerable fraction, of the kinetic energy is transferred to the lower layers to replace the energy loss in the recirculation gyre through instability, radiation, and dissipation from the gyre, as indicated in some numerical models or by other dispersive processes within the gyre. The bulk of the kinetic-energy flux, however, appears to be converted back into potential energy.

4.4.5 Numerical Gulf Stream Models

The complete energetics of the Gulf Stream are far from obvious. The development of numerical models with sufficient spatial resolution to permit significant eddy-mean flow and eddy-eddy interactions to take place has revealed energy conversion and dissipation modes to help interpret Gulf Stream behavior.

The first series of ocean-scale general circulation models to exhibit baroclinic instability in the western boundary current and recirculation gyre were described by Holland and Lin (1975a,b). They found that if lateral friction was taken sufficiently small or wind forcing sufficiently strong, a steady-state flow was not attained in the numerical integration. The flow remained time dependent but statistically stationary, in that means and variances approached constant values with time.

Thus, the time-varying eddy flow appeared to be an essential part of the momentum-transfer mechanism in the model. Their model was chosen with a "single-gyre" wind-stress distribution so that the western boundary current was constrained by both western and northern boundaries and did not exhibit strong instabilities. The next step was taken by Semtner and Mintz (1977), who developed a five-level primitive equation model with shelf topography and surface heat exchange to simulate the Gulf Stream and mesoscale eddies in the western North Atlantic. Their model contained two novel features not included in the Holland and Lin model: a biharmonic friction to prevent a "violet catastrophe" resulting from a transfer of mean square vorticity (enstrophy) to high wavenumbers (enstrophy cascade), and a bottom frictional Ekman layer to allow dissipation at the ocean bottom. These mechanisms had been introduced and explored earlier by Bretherton and Karweit (1975) and Owens and Bretherton (1978) in the study of open-ocean mesoscale eddy models. The primitive equation model showed that the dominant instability occurred within the simulated Gulf Stream over the continental rise. Over the flat abyssal plain, energy was transferred from the eddies to the mean flow.

Reduction of the effective lateral friction using the biharmonic dissipation allowed the eastward jet (Gulf Stream) to develop intense meanders, some of which formed ringlike eddies that separated from the jet and drifted westward in the recirculation gyres. Strong deep gyres developed in the vicinity of the meandering eastward jet as a consequence of downward flux of momentum associated with the meanders. The energetics of the primitive equation model were studied in detail by Robinson et al. (1977) to evaluate the types and rates of energy transfers in several regions of the basin. The primitive equation models are expensive to run and Semtner and Holland (1978) concluded after comparison that most of the behavior of the western boundary current and the free eastward jet used to simulate the Gulf Stream is contained in the simpler quasi-geostrophic two-layer model. Holland (1978) carried out a series of experiments using this latter model to explore the effects of horizontal diffusion and bottom friction on energy flow to the eddy field. For low lateral diffusion, most of the energy input by wind stress was transferred via upper-layer eddies to eddies in the lower layer, to be dissipated by bottom friction. For high lateral viscosity, the energy is dissipated by diffusion in the mean flow, with much smaller fractions being transferred to eddies in the upper and lower layers. An ocean basin with a "double-gyre" wind forcing produced a free mid-ocean jet with behavior recognizably closer to that of the Gulf Stream. Momentum and energy were transferred to the lower layer through the coupling of eddies across the interface, to be dissipated by friction at the bottom. A strong recirculating gyre is developed in the lower layer as seen in the streamfunction and interface topography and total transports shown in figure 4.14. The results are sufficiently promising to attempt a comparison (figure 4.15) with current measurements along 55°W by Schmitz (1977). Good agreement was obtained with the mean zonal currents in deep water. However, the north-south variations of model eddy kinetic energy $\overline{u'^2 + v'^2}$ and Reynolds-stress term $\overline{u'v'}$ were considerably broader than measured by Schmitz. Holland attributed the broader distribution of eddy variables to the absence of bottom topography in the model. A similar comparison for the Semtner-Mintz model was given by Robinson et al. (1977). Further comparisons with measurements are necessary to define the limits of applicability and to locate the parameter ranges for best fit of the numerical models to the Gulf Stream. The initial results are encouraging.

4.5 The North Atlantic Current

The Gulf Stream undergoes a radical change on reaching the Southeast Newfoundland Rise, a ridge running toward the Mid-Atlantic Ridge from the Tail of the Grand Banks. According to Iselin (1936), the Gulf Stream divides, with the major branch flowing north across the Southeast Newfoundland Rise parallel and opposite to the cold Labrador Current running south along the eastern face of the Grand Banks. Another part of the Gulf Stream continues eastward to divide further into northward and southward branches before crossing the Mid-Atlantic Ridge. The southern branch recirculates into the Sargasso Sea. The northern branch turns eastward after reaching a latitude of about 50°N to become the North Atlantic Current. Details of the separation into branches were not available to Iselin. Even today little is known of the time-varying features of this critical-breakdown region of the Gulf Stream. The sharp thermal contrast between the Labrador Current and the warm waters originating from the Gulf Stream is seen vividly in the infrared images reproduced in figures 4.16 and 4.17. The ribbon of cold (white) Labrador Water marks the eastern wall of the Grand Banks. A narrow strip of cold water (figure 4.17) is entrained along the edge of the northward-turning current, heightening the contrast.

Worthington (1962, 1976) proposed a fundamentally different interpretation of the hydrographic data, in which the Southeast Newfoundland Rise acts to separate the circulation into two independent anticyclonic gyres (figure 4.6). A trough of low pressure near to, and parallel with, the Southeast Newfoundland Rise, assumed to be continuous, marks the boundaries between the gyres. Worthington reached this conclusion

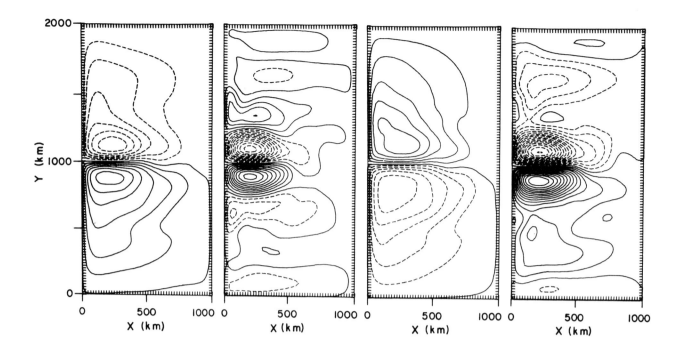

Figure 4.14 The mean fields for Holland experiment 3: (A) upper-layer streamfunction (contour interval CI = 5000 m²s⁻¹); (B) lower-layer streamfunction (CI = 1000 m²s⁻¹); (C) interface height (CI = 20 m); (D) total transport (CI = 50 × 10⁶ m³ s⁻¹). (Holland, 1978.)

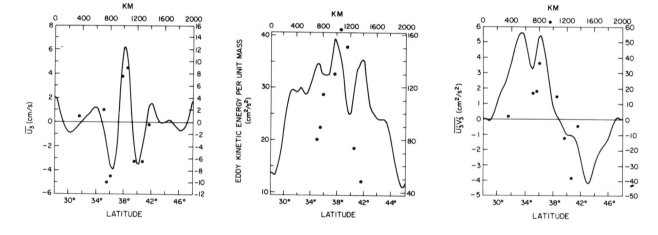

Figure 4.15 Comparison of meridional distributions of (A) zonally averaged mean flow $\bar{u}_3$, (B) eddy kinetic energy $\overline{u_3'^2 + v_3'^2}$, and (C) Reynolds stress $\overline{u_3'v_3'}$ in the lower layer (subscript 3) for experiment 3 compared with observed distribution (Schmitz, 1977) along 55°W. (Holland, 1978.)

Figure 4.16 The Gulf Stream and North Atlantic Current as seen from an infrared image taken November 2, 1977. The figure shows the colder shelf and slope water responding to a complex meandering of the Gulf Stream. The cold ribbon of Labrador Water is seen flowing south along the eastern edge of the Grand Banks. (Courtesy of R. Legeckis.)

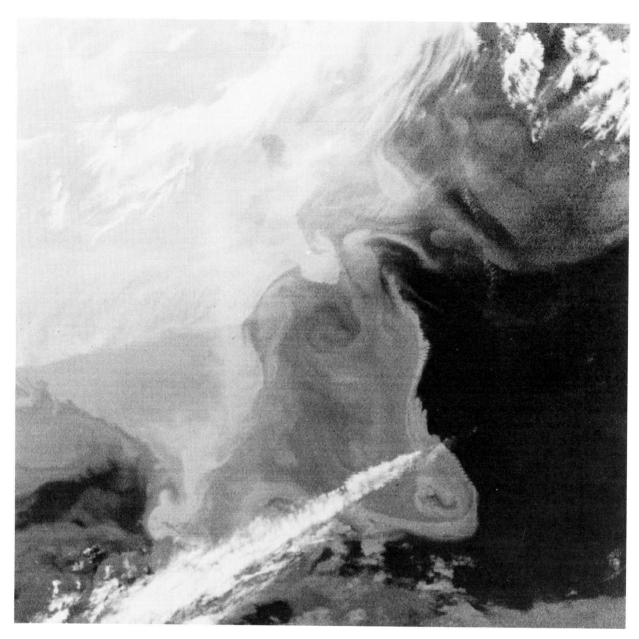

Figure 4.17 A detailed infrared image of the circulation in the vicinity of the Grand Banks showing the southward flowing cold Labrador Water. (Courtesy of P. La Violette, NORDA.)

from the observation that the northern gyre transported water about $1\,ml\,l^{-1}$ richer in dissolved oxygen than the water of the same temperature and salinity type carried by the Gulf Stream. The source of water of the same density with higher oxygen content is available to the west but it is fresher and colder and would have to be entrained in large amounts to produce the observed concentration of oxygen in the North Atlantic Current. Moreover, a large flow into the Northern Labrador Basin from the Gulf Stream would require a compensating return flow of equal magnitude passing east and south of the Gulf Stream. Worthington argued that the close proximity of the saline Mediterranean Water anomaly is evidence against such a strong return flow from the Northern gyre. Mann (1967), using more recent data obtained on cruises of C.S.S. *Baffin* during April–May 1963 and June–July 1964, disagreed with Worthington's interpretation and proposed a splitting of the current analogous to Iselin's (1936) scheme. Analysis of the hydrographic stations indicated that a branch transporting about $20 \times 10^6\,m^3\,s^{-1}$ joined by about $15 \times 10^6\,m^3\,s^{-1}$ originating in the slope water formed the branch of the North Atlantic Current to the north. A southward-flowing branch of about $30 \times 10^6\,m^3\,s^{-1}$ carried the remainder of the Gulf Stream back into the Sargasso. A weak anticyclonic eddy centered over the Labrador Basin appeared to be a persistent feature of the circulation. Mann suggested that mixing between waters in the eddy and the currents to the west could supply the additional dissolved oxygen observed in the northern basin. A similar interpretation has been constructed using data from subsequent cruises to the area in 1972 (Clarke, Hill, Reiniger, and Warren, 1980). These authors estimate that the oxygen gradient of $2\,ml\,l^{-1}$ per 100 km across the northward current is sufficient to enrich the waters north of the Southeast Newfoundland Rise by horizontal diffusion along isopycnals given an eddy diffusion coefficient of $10^7\,cm^2\,s^{-1}$ acting along the 700-km length of the northward current. They point out also that the separate gyres postulated by Worthington would require significant departures from geostrophic flow. The dynamic topography relative to 2000 db shown in figure 4.18 is consistent with the branching hypothesis described by Mann (1967).

Although the weight of evidence available today seems to favor the general interpretation given by Mann, the definitive answer is not yet in. Evidence from the hydrographic cruises and infrared surface thermal structure indicate strong time dependence in the region. The flow across the Southeast Newfoundland Rise may be intermittent, so that the dynamic topography may show a varying degree of coupling across the Southeast Newfoundland Rise. The interpretation of mean flow across the Southeast Newfoundland Rise can be modified considerably by time dependence. It seems less likely that the interpretation of water-type characteristics will be altered significantly as these are inherently conservative and less affected by time variations than the spatial distributions.

4.6 Summary and Conclusions

The growing literature describing the Gulf Stream System and its dynamics is impressive in its diversity and detail. The author recognizes that the treatment of many of the topics included in this review is superficial and may not reflect accurately all of the accomplishments and directions of current research. The number of papers and their detail and complexity necessitated rather brutal simplification to reduce the length of the review and the time needed for its preparation. Many topics could not be discussed at all. The general circulation of the North Atlantic leading into the Gulf Stream System is bypassed. The reader is referred to Stommel's (1965) monograph on the Gulf Stream, in which much of the classical oceanographic material is summarized, and to Worthington (1976) for his detailed examination of the water masses and their sources and sinks in the North Atlantic (see also chapter 2). The Loop Current in the Gulf of Mexico has been excluded from the Gulf Stream System. Yet the dynamics of the Loop Current may have significant downstream effects. The exclusion seems arbitrary.

The deep western boundary current remains a mystery. How does it coexist with the Gulf Stream and the recirculation gyre? The topic of warm- and cold-core rings has been omitted. Both theory and observations of rings are being pursued vigorously at present and a substantial body of literature has accumulated (Lai and Richardson, 1977; Flierl, 1977, 1979a). Their formation by the Gulf Stream is of obvious importance for removing mass, momentum, and energy from the Gulf Stream and for exchanging Continental Slope Water and Sargasso Water across the Gulf Stream, even though their overall contribution to the kinetic-energy flow from the Gulf Stream may prove to be relatively small compared with the energy loss by other processes.

The transport of heat, salt, and other quantities by the Gulf Stream is omitted. The literature on transport of heat by the Gulf Stream is surprisingly meagre. Heat exchange for the North Atlantic has been estimated by Bunker and Worthington (1976). Newton (1961) concluded that rings represented the principal mechanism for transporting heat between the Sargasso and the Continental Slope Water. Vonder Haar and Oort (1973) estimate that 47% of poleward transport of heat in the northern hemisphere at latitudes 30–35°N is carried by ocean currents such as the Gulf Stream. It is expected

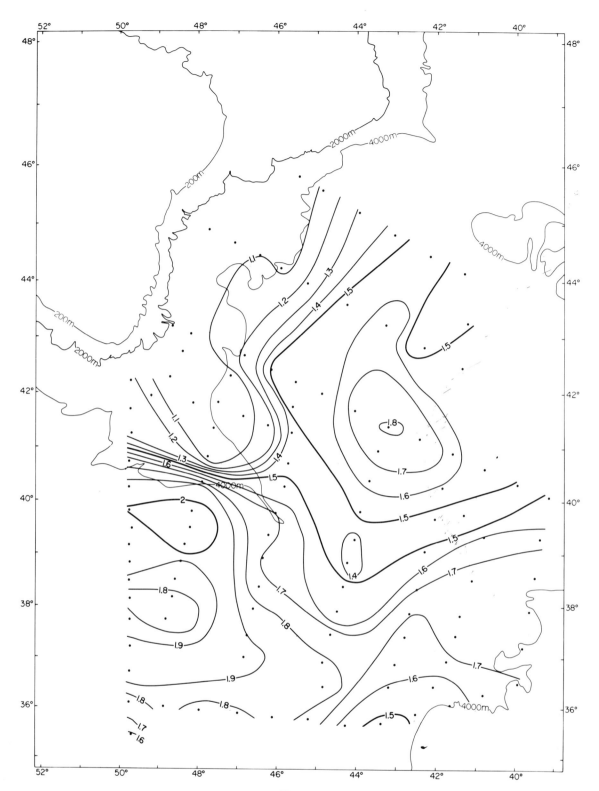

Figure 4.18 Dynamic topography from the North Atlantic Current showing the separation of the Gulf Stream into a northern branch flowing around a weak, anticyclonic eddy in the Labrador Basin. (Clarke et al., 1980.)

that interest in heat transport by the Gulf Stream will grow because of the need to develop more comprehensive and realistic models of world climate.

The Continental Slope Water to the north of the Gulf Stream is not adequately explained. Is the separation of the Gulf Stream from the continental slope at Cape Hatteras a consequence of a local dynamic process related to local topography, as suggested by many of the inertial models (Greenspan, 1963; Pedlosky, 1965a; Veronis, 1973a)? Is the separation a consequence of the large-scale wind-stress pattern? Leetmaa and Bunker (1978) show that the mean curl of the wind stress reverses sign near Cape Hatteras and is zero over a path that is surprisingly like the mean Gulf Stream across the Western North Atlantic. Is the path simply determined by the mean wind field? It is conceivable that the presence of the Gulf Stream with its strong lateral thermal contrast may significantly affect the wind-stress gradients in its vicinity. The position of the line of zero curl of the wind stress may be a consequence, as well as a cause, of the observed Gulf Stream location. Another possibility is an upstream influence of the Grand Banks jutting southward into the path of the Gulf Stream and possibly forcing it away from the continental slope as far back as Cape Hatteras.

The Florida Current emerges as the part of the Gulf Stream System that is best documented, analyzed, and understood. The Gulf Stream itself is likely to be more complex, but is as yet poorly measured by comparison. The systematic program of moored current and temperature measurements developed by Schmitz (1976, 1977, 1978) are slowly building a foundation of time-series data that will enable the next interpretive steps to be taken. Because of the complexities of the Gulf Stream System, understanding of the behavior in terms of dynamics will rely heavily on numerical modeling and analysis. It is essential that the observational and numerical studies of the Gulf Stream System proceed cooperatively.

5
Dynamics of Large-Scale Ocean Circulation

George Veronis

5.1 Introduction and Summary

The past 30 years have witnessed a rapid evolution of circulation theory. Much of the progress can be attributed to the intuition and physical balance that have emerged from the use of simple models that isolate important processes. Major contributions along these lines were made by Stommel, Welander, and others. An excellent presentation of the ideas together with a number of significant advances appears in Stern's (1975a) book. More recently numerical simulations have provided a different attack on the problem. Processes that are difficult to study with analytical models become accessible through the latter approach. Early, climatological-type studies by Bryan have now been supplemented by numerical models oriented toward the isolation of the effects of individual mechanisms. The papers of Rhines and Holland cited below have been especially instructive.

The development of the theory for the dynamics of large-scale oceanic flows is very recent. One has only to look at the chapter on dynamics in Sverdrup, Johnson, and Fleming (1942) to realize how primitive the theory was in the mid-1940s. Sverdrup's (1947) important demonstration of the generation of planetary vorticity by wind stress was the first step in obtaining explicit information about oceanic flow from a simple external observable. Until that time the dynamic method (i.e., geostrophic-hydrostatic balance) was used to obtain flow information, but this hardly constitutes a theory since one internal property must be used to determine another.

Ekman's (1905) theory for what we now call the Ekman layer was a significant early contribution, but its application to large-scale theory was not understood until Charney and Eliassen (1949) showed the coupling to large-scale flows via the spin-up mechanism. Actually, the generation of large-scale flow by Ekman suction in the laboratory was observed and described by Pettersson (1931), who repeated some of Ekman's (1906) early experiments with a stratified fluid to determine the inhibition of vertical momentum transport by stratification. Pettersson found the large-scale circulation to be an annoying interference, however, in his primary objective, determining vertical transfer of momentum by turbulence, and he discarded the approach as unpromising.

Shortly after Sverdrup's paper Stommel (1948) produced the first significant, closed-basin circulation model showing that westward intensification of oceanic flow is due to the variation of the Coriolis parameter with latitude. Hidaka (1949) proposed a closed set of equations for the circulation including the effects of lateral (eddy) dissipation of momentum. Munk (1950) continued the development by obtaining a solution that resembled Stommel's except for details in the boundary layers near the eastern and western sides of the basin. He applied his solution to an idealized ocean basin with observed wind stresses and related a number of observed oceanic gyres to the driving wind patterns. The first nonlinear correction to these linearized models (Munk, Groves and Carrier, 1950) showed that inertia shifts positive vortices to the south and negative vortices to the north. Nonlinear effects thus introduce the observed north-south asymmetry into a circulation pattern that is predicted by steady linear theory to be symmetric about mid-latitude when the wind driving is symmetric.

Fofonoff (1954) approached the problem from the opposite extreme, treating a completely inertial, nondriven model. His solution exhibits the pure effect of inertia for steady westward flows. The circulation pattern is symmetric in the east-west direction and closes with the center of a cyclonic (anticyclonic) vortex at the south (north) edge of the basin. When linear, frictional effects perturb the nonlinear pattern (Niiler, 1966), the center of the vortex shifts westward. Niiler's model had been proposed independently by Veronis (1966b) after a numerical study of nonlinear effects in a barotropic ocean, and Niiler's solution had been suggested heuristically by Stommel (1965).

The theoretical models leading to these results for wind-driven circulation are discussed below in sections 5.5 and 5.6. More general considerations in section 5.2, based on conservation integrals for the nondissipative equations (Welander, 1971a), prepare the way for the ordered system of quasi-geostrophic equations that are presented in section 5.3. The latter are derived for a fluid with arbitrary stable stratification and for a two-layer approximation to the stratification.[1] A large portion of the remainder of the paper reports results obtained with the simpler two-layer system.[2]

Section 5.7 concludes the discussion of simple models of steady, wind-driven circulation with a suggested simple explanation of why the Gulf Stream and other western boundary currents leave the coast and flow out to sea (Parsons, 1969; Veronis, 1973a). Separation of the Gulf Stream from the coast occurs within an anticyclonic gyre at a latitude where the Ekman drift due to an eastward wind stress in the interior must be returned geostrophically in the western boundary layer. If the mean thermocline depth is sufficiently small, i.e., if the amount of upper-layer water is sufficiently limited, the thermocline surfaces on the onshore side of the Gulf Stream and separation occurs. The surfacing of the thermocline is enhanced by the poleward transport by the Gulf Stream of upper-layer water that eventually reaches polar latitudes and sinks.

A review of models of thermohaline circulation is given in section 5.8. The open models introduced by Welander (1959) and Robinson and Stommel (1959) and the subsequent developments by them as well as other authors are described. The section concludes with a description of a closed, two-layer model in which the heating and cooling processes are parameterized by an assumed upwelling of lower-layer water across the thermocline (Veronis, 1978). The closure of the model leads to an evaluation of the magnitude of upwelling of 1.5×10^{-7} m s^{-1}, in agreement with values obtained from chemical tracers and the estimated age of deep water.

The normal modes for a two-layer system are derived in section 5.9 and the free-wave solutions are obtained for an ocean of constant depth. The derivation is a generalization of the treatment by Veronis and Stommel (1956) but the method is basically the same. The results include barotropic and baroclinic modes of inertiogravity and quasi-geostrophic Rossby waves. Brief mention is made of observations of these waves and the roles they play in developed flows.

Topography introduces a new class of long-period wave motions. Quasi-geostrophic analysis leads to the three types of waves described by Rhines (1970, 1977) as topographic-barotropic Rossby waves, fast baroclinic (bottom-trapped) waves, and slow baroclinic (surface-trapped) waves. The properties of slow baroclinic waves are independent of topography, yet the creation of these waves may be facilitated by steep topography that inhibits deep motions. For purposes of comparison the analysis is carried out with stratification approximated by two layers and by a vertically uniform density gradient.

Baroclinic instability in a two-layer system is described in section 5.11. The model (Phillips, 1951; Bretherton, 1966a) has convenient symmetries (equal layer depths and equal and opposite mean flows in the two layers) that simplify the analysis and show the nature of the instability more clearly. The stabilizing effect of β is evident after the simpler model has been analyzed. After a discussion of the energetics and of the relative phase of the upper- and lower-layer motions required for instability, the study of linear processes ends with a brief review of the stability study made by Gill, Green, and Simmons (1974) for a variety of mean oceanic conditions.

The last section extends the discussion to include the effects of turbulence and strong nonlinear interactions. Batchelor's (1953a) argument that two-dimensional turbulence leads to a red cascade in wavenumber space is followed by a description of several of Rhines's (1977) numerical experiments exhibiting the red cascade for barotropic quasi-geostrophic flow and the inhibition of the red cascade by lateral boundaries and topography. An initially turbulent flow in a two-layer fluid will evolve toward a barotropic state followed by the red cascade when nonlinear interactions or baroclinic instability generate motions on the scale of the internal radius of deformation. The latter scale is the window leading to barotropic behavior. Rough topography can inhibit the tendency toward barotropy by scattering the energy of the flow away from the deformation scale.

The generation of deep motions in wind-driven flows by upper-layer eddies that evolve from barotropic and baroclinic instabilities leads to a mean flow that is very different from the one predicted by the linear theories of the earlier sections. The closed-basin circulation obtained in a two-layer quasi-geostrophic numerical experiment by Holland (1978) and analyzed by Holland and Rhines (1980) shows how many of the processes described earlier come together to generate the mean flow. Simple balances for some of the results are suggested. A significant result of this experiment (and others mentioned) is the enhancement of the mean transport by the circulation resulting from the eddy interactions. A similar enhancement is made possible when topography and baroclinic effects are present (Holland, 1973). A brief discussion of several other numerical studies concludes the review.

Most of the emphasis in this paper is on linear processes and on the remaining features of the dynamics that can be used as building blocks to synthesize the involved, interactive flows observed in the ocean. Only a selected few of the many numerical studies that have emerged in the past few years are discussed, and even for those only some of the generalizable results are mentioned. Some important topics, such as the use of diagnostic models (Sarkisyan, 1977) and the generation of mean circulation by fluctuating winds (Pedlosky, 1964a; Veronis, 1970; Rhines, 1977), are omitted only because time limits forced me to draw the line somewhere. Most of the references are to the literature in the English language because that is the literature with which I am most familiar.

5.2 The Equations for Large-Scale Dynamics

The complete equations for conservation of momentum, heat, and salt are never used for studies of large-scale oceanic dynamics because they are much too complicated, not only for analytical studies but even for numerical analyses. Justification for use of an appropriate set of simplified equations requires a much more extensive argument than is feasible here so we shall confine ourselves to a short discussion with references to publications that discuss the different issues. It is appropriate, however, to mention a general result for a fluid with a simple equation of state.

If dissipative processes are ignored, the conservation of momentum for a fluid in a rotating system can be written as

$$\frac{\partial \mathbf{v}}{\partial t} + \mathbf{v}\cdot\nabla\mathbf{v} + 2\mathbf{\Omega}\times\mathbf{v} = -\frac{1}{\rho}\nabla p - \nabla\Phi, \quad (5.1)$$

or equivalently as

$$\frac{\partial \mathbf{v}}{\partial t} + (2\mathbf{\Omega} + \nabla\times\mathbf{v})\times\mathbf{v}$$
$$= -\frac{1}{\rho}\nabla p - \nabla\left(\frac{1}{2}\mathbf{v}\cdot\mathbf{v}\right) - \nabla\Phi \quad (5.2)$$

where $\mathbf{v}$ is the three-dimensional velocity vector, $\mathbf{\Omega}$ is the angular rotation vector of the system, ρ the density, p the pressure, and $\nabla\Phi$ the total gravity term (Newtonian plus rotational acceleration).

Conservation of mass is described by

$$\frac{d\rho}{dt} + \rho\nabla\cdot\mathbf{v} = 0, \quad \frac{d}{dt} \equiv \frac{\partial}{\partial t} + \mathbf{v}\cdot\nabla. \quad (5.3)$$

Furthermore, if a state variable $s(p,\rho)$ is conserved along a trajectory, it satisfies the equation

$$\frac{ds}{dt} = 0. \quad (5.4)$$

These equations can be combined to yield the conservation of potential vorticity (Ertel, 1942):

$$\frac{d}{dt}\left[\frac{(2\mathbf{\Omega} + \nabla\times\mathbf{v})\cdot\nabla s}{\rho}\right] = 0. \quad (5.5)$$

This general result for a dissipation-free fluid does not apply precisely to sea water where the density is a function not only of temperature and pressure but also of the dissolved salts. The effect of salinity on density is very important in the distribution of water properties. However, for most dynamic studies the effect of the extra state variable is not significant and (5.5) is valid.

Circulation of waters in the world ocean involves trajectories from the surface to the deep sea and from one ocean basin to another. The relative densities of two parcels of water formed at the surface in different locations can be inverted when the parcels sink to great depths. Thus, surface water in the Greenland Sea is denser than surface water in the Weddell Sea; yet when these water masses sink and flow to the same geographic location, the latter (Antarctic Bottom Water) is denser and lies below the former (North Atlantic Deep Water). This inversion is due in large part to the different amounts of thermal expansion of waters of different temperatures and salinities.[3]

Neither compressibility nor individual effects of temperature and salinity on the density are included in the treatment that follows. Use of potential density (not only in the equations but in boundary conditions as well) together with the Boussinesq approximation (Spiegel and Veronis, 1960) makes it possible to treat the dynamic effects of buoyancy forces in a dynamically consistent fashion. Comparison of observed motions (especially long- and short-period waves) with those deduced when potential density is used yields good qualitative, and often quantitative, agreement. But it is clear that some phenomena, such as the relative layering of water masses and small-scale mixing related to double-diffusive processes, cannot be analyzed without the use of a more extended thermodynamic analysis. Therefore, although the present discussion allows a treatment of inertially controlled flows, it does not admit the interesting array of phenomena associated with tracer distributions, except in the crudest sense. By implication, motions related to the largest time and space scales are not accessible either.

In those cases where a homogeneous fluid model is invoked the effects of stratification are implicitly present since the basic equations would be different for a truly homogeneous fluid (where the direction of the rotation axis could be more important than the local vertical). The fluid is sometimes assumed to be homogeneous only because the feature that is being emphasized is independent of stratification or because the simplified analytical treatment is a helpful preliminary for the more complicated stratified system.

The effects of rotation and Newtonian gravitation lead to an equilibrium shape for the earth that is nearly a planetary ellipsoid. For earth parameters the ellipticity is small (1/298) and an expansion in the ellipticity yields a spherical system with a mean (rather than variable) radius to lowest order (Veronis, 1973b). An additional simplification is to neglect the horizontal component of the earth's rotation. This assumption is not entirely separate from the use of a mean radius (N. A. Phillips, 1966a). It is normally valid for the types of motion treated here, though the effect of the neglected term is discussed for certain physical situations by Needler and LeBlond (1973) and by Stern (1975a). Grimshaw (1975) has reexamined the β-plane approximation and gives a procedure in which the horizontal rotation is retained.

With all these simplifications the foregoing equations simplify to

$$\frac{d\mathbf{v}}{dt} + \mathbf{f} \times \mathbf{v} = -\frac{1}{\rho_m}\nabla p - g\frac{\rho}{\rho_m}\hat{\mathbf{k}}, \quad (5.6)$$

$$\frac{d\rho}{dt} = 0, \quad (5.7)$$

$$\nabla \cdot \mathbf{v} = 0, \quad (5.8)$$

$$\frac{d}{dt}[(\mathbf{f} + \nabla \times \mathbf{v})\cdot\nabla\rho] \equiv \frac{dq}{dt} = 0, \quad (5.9)$$

where $\mathbf{f} = 2\Omega\sin\phi\hat{\mathbf{k}}$ is twice the locally vertical (direction $\hat{\mathbf{k}}$) component of the earth's rotation, ϕ is the latitude, g is gravity, ρ_m is a mean (constant) density, and ρ is the deviation of density from the mean. The hydrostatic pressure associated with the mean density has been subtracted from the system. Equations (5.7) and (5.8) describe the incompressible nature of this Boussinesq fluid. The quantity s in (5.4) can then be replaced by ρ, and the potential vorticity q in (5.9) is simplified accordingly [note the change of dimensions of potential vorticity as defined in (5.5) and (5.9)].

For steady or statistically steady flows we can multiply (5.6) by $\mathbf{v}$ to obtain a kinetic energy equation which can be written as

$$\mathbf{v}\cdot\nabla\left(\frac{\rho_m}{2}\mathbf{v}\cdot\mathbf{v} + p + g\rho\right) \equiv \mathbf{v}\cdot\nabla B = 0, \quad (5.10)$$

where B is the Bernoulli function. In this case, since q, ρ, and B are each conserved along flow paths, any one of them can be expressed in terms of the other two and we obtain

$$\rho = \rho(B,q), \quad B = B(\rho,q), \quad q = q(B,\rho). \quad (5.11)$$

Even though the distributions of the surfaces cannot be determined without knowledge of the flow field, the relationship between ρ, B and q is conceptually useful.

The quantities B, q, and ρ are specified by their values in certain source regions where dissipation, mixing, and other physical processes are important. (Obvious source regions are Ekman layers, areas of convective overturning, and boundary layers near coasts.) Having acquired values of B, q, and ρ at the sources, fluid particles will retain these values along their flow paths. If particles from different sources and with different values of B, q, and ρ converge to the same geographical location, regions of discontinuity will develop, and mixing, dissipation or some other non-ideal fluid process will be required. The locations of these discontinuous regions can be determined only from a solution to the general problem, and, in general, we may anticipate new sources of B, q, and ρ to develop there. Hence, the system becomes a strongly implicit one and the closure of the problem is very complicated.

Even though a solution to the general problem may be impossible, these general considerations are important. We should be prepared for the likelihood that the solution at a particular location will not be simply determined by values at solid boundaries that are easily specified. The ocean is more likely a collection of dynamically self-contained pools (some subsurface) that interact along open-ocean boundaries where they join. Perhaps only the most persistent of these are statistically steady features. It is possible that locally the flow is relatively laminar. In that case the solution would be accessible once the source regions were identified and the values of B, q, and ρ in these regions could be specified.

5.3 The Quasi-Geostrophic Equations and the β-Plane

Even with the simplifications made in the previous section the equations are more general than required for a study of large-scale dynamics. We shall therefore simplify them further by invoking geostrophic and hydrostatic balances at lowest order and by restricting attention to spatial scales of interest. In so doing we shall derive an appropriate β-plane approximation for the study of oceanic waves and mesoscale motions. A similar procedure is followed by N. A. Phillips (1963).[4]

5.3.1 Continuous Stratification

The spherical components of (5.6) take the form

$$\frac{du}{dt} - \frac{uv \tan \phi}{a} + \frac{uw}{a} - 2\Omega \sin \phi \, v$$
$$= -\frac{1}{a \cos \phi} \frac{dP}{d\lambda}, \tag{5.12}$$

$$\frac{dv}{dt} + \frac{u^2 \tan \phi}{a} + \frac{vw}{a} + 2\Omega \sin \phi \, u = -\frac{1}{a} \frac{\partial P}{\partial \phi}, \tag{5.13}$$

$$\frac{dw}{dt} - \frac{u^2 + v^2}{a} = -\frac{\partial P}{\partial z} - g \frac{\rho}{\rho_m}, \tag{5.14}$$

$$\frac{1}{a \cos \phi}\left[\frac{\partial u}{\partial \lambda} + \frac{\partial}{\partial \phi}(v \cos \phi)\right] + \frac{\partial w}{\partial z} + \frac{2w}{a} = 0, \tag{5.15}$$

$$\frac{d\rho}{dt} = 0, \tag{5.16}$$

$$\frac{d}{dt} \equiv \frac{\partial}{\partial t} + \frac{u}{a \cos \phi} \frac{\partial}{\partial \lambda} + \frac{v}{a} \frac{\partial}{\partial \phi} + w \frac{\partial}{\partial z}, \tag{5.17}$$

where (λ, ϕ, z) are longitude, latitude, and upward and have respective velocities (u, v, w); P is p/ρ_m, a is the mean radius of the earth, and ρ is the total density minus ρ_m.

Center attention on a latitude ϕ_0, write $\phi = \phi_0 + \phi'$, and consider flows with north–south scale L substantially smaller than a. Then with $a\phi' = y$, we can expand the trigonometric functions in y, keeping only terms of $O(L/a)$, to obtain

$$\sin \phi \approx \sin \phi_0 (1 + \cot \phi_0 \, y/a),$$
$$\cos \phi \approx \cos \phi_0 (1 - \tan \phi_0 \, y/a), \tag{5.18}$$
$$f_0 = 2\Omega \sin \phi_0,$$

$$\frac{\partial}{\partial x} \equiv \frac{1}{a \cos \phi_0} \frac{\partial}{\partial \lambda}, \quad \frac{\partial}{\partial y} \equiv \frac{1}{a} \frac{\partial}{\partial \phi}. \tag{5.19}$$

To first order in y/a the equations become

$$\frac{du}{dt} + \frac{y}{a} \tan \phi_0 u \frac{\partial u}{\partial x} + \frac{uw}{a}$$
$$- \frac{uv}{a} \tan \phi_0 \left(1 + \frac{2}{\sin 2\phi_0} \frac{y}{a}\right) - f_0 v \left(1 + \frac{y}{a} \cot \phi_0\right)$$
$$= -\frac{\partial P}{\partial x}\left(1 + \frac{y}{a} \tan \phi_0\right), \tag{5.20}$$

$$\frac{dv}{dt} + \frac{y}{a} \tan \phi_0 u \frac{\partial v}{\partial x} + \frac{vw}{a}$$
$$+ \frac{u^2}{a} \tan \phi_0 \left(1 + \frac{2}{\sin 2\phi_0} \frac{y}{a}\right) + f_0 u \left(1 + \frac{y}{a} \cot \phi_0\right)$$
$$= -\frac{\partial P}{\partial y}, \tag{5.21}$$

$$\frac{dw}{dt} + \frac{y}{a} \tan \phi_0 u \frac{\partial w}{\partial x} - \frac{u^2 + v^2}{a} = -\frac{\partial P}{\partial z} - \frac{g\rho}{\rho_m}, \tag{5.22}$$

$$\frac{\partial u}{\partial x} + \frac{\partial v}{\partial y} + \frac{\partial w}{\partial z} - \frac{\partial}{\partial y}\left(\frac{y}{a} v \tan \phi_0\right)$$
$$- \frac{y}{a} \tan \phi_0 \frac{\partial w}{\partial z} + \frac{2w}{a} = 0, \tag{5.23}$$

$$\frac{d\rho}{dt} + \frac{y}{a} \tan \phi_0 u \frac{\partial \rho}{\partial x} = 0, \tag{5.24}$$

$$\frac{d}{dt} \equiv \frac{\partial}{\partial t} + u \frac{\partial}{\partial x} + v \frac{\partial}{\partial y} + w \frac{\partial}{\partial z}. \tag{5.25}$$

Flows with a primary geostrophic balance will satisfy

$$f_0 v \sim \frac{\partial P}{\partial x}, \quad -f_0 u \sim \frac{\partial P}{\partial y}. \tag{5.26}$$

Hydrostatic balance yields

$$\frac{\partial P}{\partial z} \sim -g \frac{\rho}{\rho_m}. \tag{5.27}$$

Variations over the depth H of the ocean are described by

$$\frac{\partial}{\partial z} \sim \frac{1}{H}, \tag{5.28}$$

so the "pressure" scale derived from (5.27) is

$$P \sim \frac{gH \Delta \rho}{\rho_m}. \tag{5.29}$$

Geostrophic balance then suggests the velocity scale

$$V \sim \frac{gH}{f_0 L} \frac{\Delta \rho}{\rho_m}. \tag{5.30}$$

If these scales are used as orders of magnitudes for the respective variables and if we also take

$$\frac{\partial}{\partial x} \sim \frac{1}{L}, \quad \frac{\partial}{\partial y} \sim \frac{1}{L}, \quad \frac{\partial}{\partial t} \sim f_0,$$

$$\frac{y}{a} \sim \frac{L}{a}, \quad \delta = \frac{H}{L}, \quad w \sim V\delta,$$

we note the following.

Relative to the lowest order (in y/a) Coriolis terms, the nonlinear terms in d/dt in (5.19) and (5.20) are $O(Ro)$ where $Ro = V/f_0L$. The remaining nonlinear terms are

$$O\left(\frac{L}{a}Ro\right) \quad \text{or} \quad O\left(\delta\frac{L}{a}Ro\right).$$

In the vertical equation of motion the acceleration terms are

$$O(\delta^2), \quad O(\delta^2 Ro), \quad \text{or} \quad O\left(\delta\frac{L}{a}Ro\right)$$

when compared to the terms on the right. Observations of the flows of interest support the inequalities

$$Ro \ll 1, \quad L/a \ll 1, \quad \delta \ll 1. \tag{5.31}$$

Rather than expand the equations in powers of the small parameters we shall simply make use of (5.31) and drop all terms which involve products of Ro, δ and L/a. Also, rather than give a relative ordering of these three parameters we keep all terms up to first order in Ro, δ, and L/a, a procedure that yields the following general system of equations

$$\frac{du}{dt} - f_0 v\left(1 + \frac{y}{a}\cot\phi_0\right)$$
$$= -\frac{\partial P}{\partial x}\left(1 + \frac{y}{a}\tan\phi_0\right), \tag{5.32}$$

$$\frac{dv}{dt} + f_0 u\left(1 + \frac{y}{a}\cot\phi_0\right) = -\frac{\partial P}{\partial y}, \tag{5.33}$$

$$\frac{\partial P}{\partial z} = -g\rho/\rho_m, \tag{5.34}$$

$$\frac{\partial u}{\partial x} + \frac{\partial v}{\partial y} + \frac{\partial w}{\partial z} - \frac{\partial}{\partial y}\left(\frac{y}{a}v\tan\phi_0\right) - \frac{y}{a}\tan\phi_0\frac{\partial w}{\partial z}$$
$$= 0, \tag{5.35}$$

$$\frac{d\rho}{dt} + \frac{y}{a}\tan\phi_0 u\frac{\partial\rho}{\partial x} = 0, \tag{5.36}$$

keeping in mind that the nonlinear terms in (5.32) and (5.33) are $O(Ro)$ compared to the lowest-order Coriolis terms.

Now write

$$\mathbf{v} = \mathbf{v}_0 + \mathbf{v}_1, \quad P = P_0 + P_1, \quad \rho = \rho_0 + \rho_1, \tag{5.37}$$

where $(\mathbf{v}_1, P_1, \rho_1)$ are $O(Ro)$ or $O(L/a)$. We shall also assume that time variations appear at first order, i.e., $\partial/\partial t = O(Ro)$ or $O(L/a)$. Then at lowest order we obtain the expected geostrophic hydrostatic system:

$$f_0 v_0 = \frac{\partial P_0}{\partial x}, \tag{5.38}$$

$$f_0 u_0 = -\frac{\partial P_0}{\partial y}, \tag{5.39}$$

$$\frac{\partial P_0}{\partial z} = -g\rho_0/\rho_m, \tag{5.40}$$

$$\frac{\partial u_0}{\partial x} + \frac{\partial v_0}{\partial y} = 0, \quad \frac{\partial w_0}{\partial z} = 0, \tag{5.41}$$

If w_0 vanishes at any level or if it is required to satisfy inconsistent (with $\partial w_0/\partial z = 0$) boundary conditions at top and bottom, it will vanish everywhere. One or the other of these two conditions is satisfied for all of the flows that we shall consider, so we obtain the result

$$w_0 = 0. \tag{5.42}$$

This means that the scaling $w \sim V\delta$ suggested by the geometry is inappropriate and that a factor L/a or Ro should be included on the right-hand side. In other words, quasi-geostrophic flows are quasi-horizontal and the convective derivative in (5.32) reduces to

$$\frac{d}{dt} = \frac{\partial}{\partial t} + u_0\frac{\partial}{\partial x} + v_0\frac{\partial}{\partial y}. \tag{5.43}$$

The restriction to flows with less than global scales precludes a treatment leading to the basic stratification. Since vertical density changes $\Delta\rho$ are generally much larger than the horizontal changes, say $\Delta\rho'$, generated by the motion field, we must account for the difference in (5.36). In particular, we write $\rho = \bar{\rho}(z) + \rho'(x, y, z, t)$ so that

$$\frac{\partial\rho'}{\partial t} + u\frac{\partial\rho'}{\partial x} + v\frac{\partial\rho'}{\partial y} + w\frac{\partial\rho'}{\partial z}$$
$$+ w\frac{\partial\bar{\rho}}{\partial z} + \frac{y}{a}\tan\phi_0 u\frac{\partial\rho'}{\partial x} = 0. \tag{5.44}$$

The considerations leading to (5.43) apply here as well for the terms involving ρ'. Accordingly, at lowest order we can drop the terms $w\,\partial\rho'/\partial z$ and $(y/a)\tan\phi_0 u(\partial\rho'/\partial x)$ to end up with

$$\frac{d\rho'}{dt} + w\frac{\partial\bar{\rho}}{\partial z} = 0, \tag{5.45}$$

where $\Delta\rho'$ is assumed to be $O(Ro)$ or $O(L/a)$ relative to $\Delta\rho$. Since w is correspondingly smaller than u or v, the two terms balance. In terms of our ordering, therefore, we can write

$$\frac{d\rho_0}{dt} + w_1\frac{\partial\bar{\rho}}{\partial z} = 0, \tag{5.46}$$

where we have used the fact that the density used in the hydrostatic equation is really ρ' (since the balance $\partial\bar{P}/\partial z = -g\bar{\rho}$ is valid when there is no motion and hence can be subtracted from the system).

At next order we have

$$\frac{du_0}{dt} - f_0 v_1 - \frac{y}{a} f_0 \cot \phi_0 v_0$$
$$= -\frac{\partial P_1}{\partial x} - \frac{y}{a} \tan \phi_0 \frac{\partial P_0}{\partial x}, \qquad (5.47)$$

$$\frac{dv_0}{dt} + f_0 u_1 + \frac{y}{a} f_0 \cot \phi_0 u_0 = -\frac{\partial P_1}{\partial y}, \qquad (5.48)$$

$$\frac{\partial u_1}{\partial x} + \frac{\partial v_1}{\partial y} + \frac{\partial w_1}{\partial z} - \frac{\partial}{\partial y}\left(\frac{y}{a} \tan \phi_0 v_0\right) = 0. \qquad (5.49)$$

These equations, in addition to providing the balances for first-order quantities, serve the important function of closing the zero-order system when first-order terms are eliminated. Thus, cross differentiating (5.47) and (5.48) and making use of (5.38) to (5.41) and (5.49), we obtain

$$\frac{d\zeta_0}{dt} + \beta v_0 = f_0 \frac{\partial w_1}{\partial z}, \qquad \beta = \frac{f_0 \cot \phi_0}{a},$$
$$\zeta_0 = \frac{\partial v_0}{\partial x} - \frac{\partial u_0}{\partial y}. \qquad (5.50)$$

From (5.46) we observe

$$\frac{\partial w_1}{\partial z} = -\frac{\partial}{\partial z}\left[\frac{1}{\partial \bar{\rho}/\partial z}\frac{d\rho_0}{dt}\right] = -\frac{d}{dt}\frac{\partial}{\partial z}\left[\rho_0 \Big/ \frac{\partial \bar{\rho}}{\partial z}\right]. \qquad (5.51)$$

Also,

$$\beta v_0 = \frac{df}{dt}, \qquad (5.52)$$

where $f = f_0(1 + \cot \phi_0 (y/a))$. Then using (5.38) to (5.41) to express u_0, v_0, ρ_0 in terms of p_0 we obtain the lowest-order closure

$$\frac{d}{dt}\left[\nabla^2 P_0 + f f_0 + \frac{\partial}{\partial z}\left(\frac{f_0^2 \partial P_0/\partial z}{N^2}\right)\right] = 0, \qquad (5.53)$$

where $N^2 = -g(\partial \bar{\rho}/\partial z)/\rho_m$ is the square of the buoyancy frequency. Equation (5.53) describes the conservation of quasi-geostrophic potential vorticity. It is sometimes written in terms of the stream functions $\psi = P_0/f_0$,

$$\frac{d}{dt}\left[\nabla^2 \psi + f + \frac{\partial}{\partial z}\left(\frac{f_0^2 \partial \psi/\partial z}{N^2}\right)\right] = 0. \qquad (5.54)$$

The derivation given here has been carried out in dimensional form. It is as rigorous, though not as formal, as derivations with nondimensional variables (e.g., Pedlosky, 1964a) and has the advantage of including the intermediate equations in dimensional form. Obviously, the equations are valid only for those motions (smaller than global scale, low frequency, etc.) that satisfy the assumptions.

5.3.2 Equations More Commonly Encountered

Instead of the set (5.47)–(5.49) one more often encounters the equations with rectangular cartesian coordinates, no subscripts, and with $f = f_0 + \beta y$, i.e.,

$$\frac{du}{dt} - fv = -\frac{\partial P}{\partial x}, \qquad (5.55)$$

$$\frac{\partial v}{\partial t} + fu = -\frac{\partial P}{\partial y}, \qquad (5.56)$$

$$\frac{\partial u}{\partial x} + \frac{\partial v}{\partial y} + \frac{\partial w}{\partial z} = 0. \qquad (5.57)$$

This system is often used even when the flow is not quasi-geostrophic. For quasi-geostrophic flows, particularly if one makes use principally of the vorticity equation and the fact that w is really a higher order quantity, one can avoid serious errors.

For flows at low latitudes (small ϕ_0) the neglected terms ($\sim \tan \phi_0$) are small and (5.55) to (5.57) may be adequate. But errors notwithstanding, a large part of the literature deals with this more approximate system, and we shall have to refer to it frequently.

5.3.3 Layered Stratification

Continuous density stratification is frequently approximated by a series of discrete layers each of uniform density. The derivation parallels the one just given but it is easier to make use of what we have done and to note the following.

Number the layers sequentially downward from the top so that the ith layer has thickness $h^{(i)}$, density $\rho^{(i)}$, and mean thickness (for linearized cases) $H^{(i)}$. Furthermore, write $h^{(i)} = \eta^{(i)} + H^{(i)} - \eta^{(i+1)}$ so that $\eta^{(i)}$ and $\eta^{(i+1)}$ are the deviations of the top and bottom surfaces of the layer from the mean. Integrate the hydrostatic relation downward from the top surface to layer i to derive the horizontal pressure gradient in terms of gradients of thicknesses

$$\nabla p^{(i)} = g \sum_{n=1}^{i-1} \rho^{(n)} \nabla h^{(n)} + g \rho^{(i)} \nabla \eta^{(i)}. \qquad (5.58)$$

Conservation of mass for each homogeneous layer is $\nabla_3 \cdot \mathbf{v}^{(i)} = 0$, where ∇_3 is the three-dimensional operator. The horizontal velocities are independent of z because the flow is hydrostatic. Therefore, integrating over the depth of the layer yields

$$h^{(i)} \nabla \cdot \mathbf{v}^{(i)} + \frac{dh^{(i)}}{dt} = 0, \qquad (5.59)$$

where we have used the free surface conditions

$$w^{(i)}(x, y, \eta^{(i)}, t) = \frac{d\eta^{(i)}}{dt},$$
$$w^{(i)}(x, y, \eta^{(i+1)}, t) = \frac{d\eta^{(i+1)}}{dt}. \qquad (5.60)$$

Also, since $H^{(i)}$ is constant

$$\frac{d}{dt}(\eta^{(i)} - \eta^{(i+1)}) = \frac{dh^{(i)}}{dt}.$$

We can thus integrate (5.50) over the depth of each layer to obtain the conservation of potential vorticity for the layered system

$$\frac{d}{dt}\left(\frac{\zeta_0 + f}{h_0^{(i)}}\right) = 0. \tag{5.61}$$

The velocity in the convective derivative is $(u_0^{(i)}, v_0^{(i)})$.

In subsequent treatments of the two-layer, β-plane, inviscid, momentum equations, we shall use the approximate form (5.55) and (5.56) together with the vertically integrated form of (5.57). The equations are

$$\frac{du_1}{dt} - fv_1 = -g\frac{\partial \eta_1}{\partial x}, \tag{5.62}$$

$$\frac{dv_1}{dt} + fu_1 = -g\frac{\partial \eta_1}{\partial y}, \tag{5.63}$$

$$\frac{dh_1}{dt} + h_1\left(\frac{\partial u_1}{\partial x} + \frac{\partial v_1}{\partial y}\right) = 0, \tag{5.64}$$

$$\frac{du_2}{dt} - fv_2 = -g\left[(1-\epsilon)\frac{\partial \eta_1}{\partial x} + \epsilon\frac{\partial \eta_2}{\partial x}\right], \tag{5.65}$$

$$\frac{dv_2}{dt} + fu_2 = -g\left[(1-\epsilon)\frac{\partial \eta_1}{\partial y} + \epsilon\frac{\partial \eta_2}{\partial y}\right], \tag{5.66}$$

$$\frac{dh_2}{dt} + h_2\left(\frac{\partial u_2}{\partial x} + \frac{\partial v_2}{\partial y}\right) = 0, \tag{5.67}$$

where, $\epsilon \rho_2 = \rho_2 - \rho_1$, $h_2 = \eta_2 + H_2 - \eta_3$, and η_3 is the height of the bottom above an equilibrium level. The subscripts in (5.62) to (5.67) identify the layer rather than the order of L/a or Ro.

For linear steady flows the above system is sometimes used with spherical coordinates.

5.4 Ekman Layers

The equations derived above do not contain friction explicitly. However, when the variables are written in terms of a mean (ensemble, time average, etc.) plus a fluctuation and the equations are averaged, Reynolds stresses emerge and these are often parameterized in frictional form through the use of Austausch or eddy coefficients. Though this procedure is often questionable, it may not be a bad approximation near the top surface where wind stresses impart momentum to the ocean and near the bottom where frictional retardation brakes the flow. This was the view taken by Ekman (1905), who introduced the model for what is now called the Ekman layer.[5]

5.4.1 Pure Ekman Layers

Ekman first applied the theory to the wind-driven layer near the surface of the ocean. It is preferable to introduce the subject by investigating how a horizontally uniform geostrophic flow given by $f\hat{k} \times \mathbf{v}_g = -\nabla P$ in a fluid occupying the region $z > 0$ is brought to rest by frictional processes acting near the bottom.[6] Assuming that horizontal variations of the stress are small compared to vertical variations (easily verified a posteriori), we can write

$$-fv = -fv_g + \nu u_{zz}, \tag{5.68}$$

$$fu = fu_g + \nu v_{zz}, \tag{5.69}$$

where subscript z corresponds to $\partial/\partial z$ and the pressure gradient is written in terms of the geostrophic velocity. The velocity vanishes at the (flat) bottom

$$\mathbf{v} = 0 \quad \text{at} \quad z = 0. \tag{5.70}$$

The method of solution is well-known (Lamb, 1932, p. 593). Combining u and v as $u + iv$, $i = \sqrt{-1}$, the equations (5.68) and (5.69) take the form

$$(u + iv)_{zz} = \frac{if}{\nu}(u + iv) - \frac{if}{\nu}(u_g + v_g) \tag{5.71}$$

and the solution satisfying (5.70) with $\mathbf{v} \to 0$ as $z \to \infty$ is

$$u + iv = (u_g + iv_g)(1 - e^{-\sqrt{i\delta z}}), \tag{5.72}$$

where $\delta = \sqrt{f/\nu}$. Accordingly, the flow vanishes at $z = 0$, tends to $\mathbf{v}_g$ for large z and is predominantly to the left of $\mathbf{v}_g$ in between.

The vertically integrated transport of the exponentially decaying part of (5.71) is $(-1 + i)(u_g + iv_g)(\nu/2f)^{1/2}$, which suggests $h_e = (\nu/2f)^{1/2}$ as the scale of the Ekman layer. If we integrate the geostrophic part over the depth, h_e, we obtain the transport $(u_g + iv_g)h_e$. Hence, the *net* transport is $i(u_g + iv_g)h_e$, which is to the left of the geostrophic current, i.e., down the pressure gradient required to support $\mathbf{v}_g$, as we would expect. In vector form the net transport is $(-v_g, u_g)h_e$.

Next consider Ekman's problem, with fluid occupying the region $z < 0$ and with the flow driven by the spatially uniform wind stress (divided by the density) given by (τ^x, τ^y) acting at $z = 0$. With $\mathbf{v}_g = 0$, the solution is

$$u + iv = \frac{e^{\delta z}}{\sqrt{f\nu}}\{\tau^x \sin(\delta z + \pi/4) - \tau^y \sin(\delta z - \pi/4) + i[\tau^x \sin(\delta z - \pi/4) + \tau^y \sin(\delta z + \pi/4)]\}. \tag{5.73}$$

In the hodograph (u, v)-plane the solution has the form of a spiral (called the Ekman spiral). Just as rotation generates a velocity component to the right (for $f > 0$) of the (pressure) force for geostrophically balanced flow, a flow to the right of the tangential-stress force is generated in the Ekman spiral solution. In contrast to geostrophic flow, however, the present system is dissipative, and a velocity component parallel to the force is also present. At the surface the magnitudes of the components are equal so the flow is directed 45° to the right of the wind stress. The velocity component parallel to τ decreases with depth but near the surface the normal component does not (it cannot since τ has the same direction as $\partial\tau/\partial z$). But below that the stress veers to the right as does the velocity vector.

Though Ekman's solution provided a satisfactory explanation of Nansen's observation of surface velocity, the spiral is not normally observed in the field. Ekman failed to observe it in spite of repeated attempts. Hunkins (1966) reported measuring a well-defined Ekman spiral (ironically, in the Arctic Ocean, where Nansen's first observations were made). The spiral structure depends on the form of the stress term, and since the stresses near the surface are turbulent (due to thermal convection, surface waves, and other small-scale processes) and therefore not necessarily of Navier–Stokes form, it is not surprising that the observed current structure differs from the theoretical one. Also, the mixed layer at the surface sits on a stably stratified fluid and the depth h of the former often does not exceed h_e when a turbulent eddy viscosity is used. Gonella (1971) showed that when a stress-free condition is applied at the base of the mixed-layer the solution is a function of h_e/h. For shallow ($h \ll h_e$) mixed layers there is essentially no spiral. Csanady (1972) reported that field measurements in the mixed layer in Lake Huron support Gonella's findings. He also reformulated the problem in terms of external parameters of the system instead of using an eddy viscosity.

In contrast to the detailed velocity structure, the vertically integrated transport of the wind-driven Ekman layer is independent of the form of vertical variation of the stress. If the stress terms in (5.68) and (5.69) are written as $\partial\tau/\partial z$ and if we integrate the equations vertically, the transports are given by $(\tau^y, -\tau^x)/f$. Thus, the total transport is to the right of the wind stress irrespective of the form of τ and subject only to these conditions: $\tau = (\tau^x, \tau^y)$ at the surface and $\tau = 0$ at the bottom. In vector form, with $V_e = \int_{-h}^{0} \mathbf{v}\, dz$ (where h is a depth—finite or infinite—at which τ vanishes), the result (called the Ekman drift) is

$$\mathbf{V}_e = (\boldsymbol{\tau} \times \hat{\mathbf{k}})/f. \tag{5.74}$$

where τ is now the wind stress vector and $\hat{\mathbf{k}}$ is the vertical unit vector.

5.4.2 Effect of Ekman Layers on Interior Flows

Although the pure Ekman layer theory given above requires horizontally uniform conditions, the theory is valid with horizontal variations as long as the horizontal scale is substantially larger than h_e. The neglected horizontal variations of the stress terms are smaller than $\partial\tau/\partial z$ by the ratio of the squares of vertical to horizontal scales. Furthermore, for the mixed layer near the surface the vertical pressure gradient in the vertical equation of motion vanishes (as long as we consider scales larger than the small-scale turbulence which generates the mixed layer). Hence, the horizontal pressure gradients associated with Ekman layer processes are negligible at lowest order, and the original equations, and therefore the results given by (5.73), are still applicable.

Accordingly, suppose that τ in (5.74) varies horizontally. When the continuity equation $\boldsymbol{\nabla}\cdot\mathbf{v} = 0$ is integrated in the vertical over the depth of the Ekman layer and the boundary condition ($w = 0$ at the top) is applied, we find (Charney, 1955a)

$$w_e = \boldsymbol{\nabla}\cdot\mathbf{V}_e, \tag{5.75}$$

where w_e is the vertical velocity at the base of the Ekman layer. With (5.74) this becomes

$$w_e = \frac{\partial}{\partial x}\left(\frac{\tau^y}{f}\right) - \frac{\partial}{\partial y}\left(\frac{\tau^x}{f}\right) = \hat{\mathbf{k}}\cdot\boldsymbol{\nabla}\times\left(\frac{\boldsymbol{\tau}}{f}\right). \tag{5.76}$$

Thus, horizontal variations in τ generate vertical motions which penetrate into the fluid below. Since the Ekman layer is thin relative to the depth of the ocean, this forced vertical velocity (called *Ekman pumping*) can be applied as a boundary condition (approximately at the surface) for the underlying inviscid fluid.

The same analysis can be applied to the bottom (subscript b) Ekman layer, where the vertically integrated transport was found to be $\mathbf{V}_b = (-v_g, u_g)h_b$. If the bottom is flat, so that $w = 0$ there, the vertically integrated continuity equation yields

$$w_b = -\boldsymbol{\nabla}\cdot\mathbf{V}_b,$$

where w_b is the vertical velocity induced at the top of the Ekman layer. Substituting for $\mathbf{V}_b$ we obtain

$$w_b = \frac{\partial}{\partial x}(v_g h_b) - \frac{\partial}{\partial y}(u_g h_b)$$

or taking h_b constant,

$$w_b = h_b \hat{\mathbf{k}}\cdot\boldsymbol{\nabla}\times\mathbf{v}_g. \tag{5.77}$$

This value for w serves as a boundary condition (approximately at the bottom of the ocean) for the overlying inviscid fluid.

5.4.3 Additional Considerations

Only the simplest results of Ekman layer theory have been given here. A number of important extensions are discussed by Stern (1975a, chapters 7 and 8). Horizontal momentum is imparted to the ocean by the wind stress acting at the surface; yet the momentum vanishes at the base of the Ekman layer. Stern answers the question where the momentum goes by analyzing the angular momentum balance about the axis of rotation for a cylindrical system. The analysis is carried out in an inertial frame of reference where the torque of the wind stress is balanced by the absolute angular momentum of the fluid. The latter is proportional to the absolute vorticity of the undisturbed (no wind stress) vortex flow (in our case, solid-body rotation). The correspondence between the cylindrical problem and the rectilinear system [with (5.74) as the result] is that for large radii the angular momentum argument is equivalent to saying that the rate of momentum imparted by the wind stress is balanced by the divergence of the radial flux of absolute azimuthal momentum.

Though the Ekman layer depth h_e is clearly defined for laminar boundary layers, the value for turbulent boundary layers is not. Caldwell, van Atta, and Holland (1972) formed the boundary layer scale $\tau^{1/2}/f$ from the (only) external parameters τ and f. Assuming that the molecular scale $(\nu/f)^{1/2}$ is not likely to affect the turbulent scale, they suggest that $\tau^{1/2}/f$ is the turbulent Ekman boundary layer thickness. Stern (1975a, §8.1) carried out a crude stability analysis to conclude that a layer thicker than $h_e \sim \tau^{1/2}/f$ will radiate energy to the deep water. He surmised that nonlinear modifications will show that the turbulent energy is thereby reduced as the thickness shrinks to $\tau^{1/2}/f$, where the system will stabilize. For typical values of τ, the value of h_e (so defined) is $O(100 \text{ m})$ at mid-latitudes. These considerations are based on the assumption of a homogeneous fluid. For a stratified fluid like the ocean the stratification may be decisive in determining the boundary layer thickness as Csanady's (1972) report of observed velocities in Lake Huron indicates.

As we saw from the simple analysis presented above, the effect of the top Ekman layer on the underlying water is determined completely by the wind stresses, whereas in the bottom Ekman layer the condition is expressed in terms of the velocity of the overlying water. More generally there will be a nonlinear coupling between the Ekman layer and the interior which can alter the results significantly. Fettis (1955) carried out the analysis for a laboratory model of a nonlinear Ekman layer to show that the results can be approximated by (5.74) but with the absolute vertical vorticity replacing f. Stern (1966; 1975a, §8.3) and Niiler (1969) have investigated the effect of coupling of Ekman layer flow with geostrophic vorticity (eddies) and have shown that the latter can have a dominant influence since coupling with the interior can occur even for a uniform wind stress.

5.5 Steady Linear Models of the Wind-Driven Circulation

For steady, linear flow of moderate scale we have $Ro \ll L/a$ so the term $d\zeta_0/dt$ in (5.50) can be neglected. The resulting equation is

$$\beta v_0 = f_0 \frac{\partial w_1}{\partial z}. \tag{5.78}$$

When integrated vertically from $z = -h$ to $z = 0$ this yields

$$\beta V = f_0 w_1 \Big|_{-h}^{0}, \quad V = \int_{-h}^{0} v_0 \, dz$$

or

$$\beta V = \hat{\mathbf{k}} \cdot \nabla \times \boldsymbol{\tau} - f_0 w_1(x, y, -h), \tag{5.79}$$

where the variation of f in $\hat{\mathbf{k}} \cdot \nabla \times (\boldsymbol{\tau}/f)$ is (consistently) neglected at lowest order.

5.5.1 Sverdrup Transport
If the stratification is strong enough so that distortion of the density surfaces is negligible at some depth above the bottom, the last term in (5.79) vanishes and we obtain the Sverdrup transport

$$\beta V = \hat{\mathbf{k}} \cdot \nabla \times \boldsymbol{\tau}. \tag{5.80}$$

Thus, the vertically integrated north-south transport is determined by the curl of the wind stress. Sverdrup (1947) introduced this relation to estimate transports in the eastern equatorial Pacific (see chapter 6). Physically, the interpretation of (5.80) is straightforward. With βV written as $h \, df/dt$ we see that a column of fluid moves to a new latitude (new value of planetary vorticity f) with a speed that compensates for the rate at which the wind stress imparts vorticity to the ocean.

The continuity equation (5.41) can be integrated in the vertical and in x to yield

$$U = -\int \frac{\partial V}{\partial y} \, dx + F(y)$$

or

$$U = -\int \frac{1}{\beta} \frac{\partial}{\partial y} (\hat{\mathbf{k}} \cdot \nabla \times \boldsymbol{\tau}) \, dx + F(y), \tag{5.81}$$

where $F(y)$ is arbitrary. The most common procedure for theoretical analyses is to assume that the foregoing is valid eastward to a meridional boundary $x = L$, where U must vanish. Then

$$U = \int_{x}^{L} \frac{1}{\beta} \frac{\partial}{\partial y} (\hat{\mathbf{k}} \cdot \nabla \times \boldsymbol{\tau}) \, dx, \tag{5.82}$$

and the transport is determined in the entire region in which the assumptions are valid. In general, the theory does not determine the flow in a basin bounded on the west as well since it is not possible to satisfy the zero normal flow condition there.

5.5.2 Stommel's Frictional Model
If the fluid motion penetrates to the (flat) bottom, the last term in (5.79) is given by (5.75) with $\mathbf{v}_g = \mathbf{v}_0|_{z=-h}$ and (5.79) becomes

$$\beta V = \hat{\mathbf{k}} \cdot \nabla \times \boldsymbol{\tau} - f_0 h_b \left(\frac{\partial v_0}{\partial x} - \frac{\partial u_0}{\partial y} \right)_{z=-h}. \tag{5.83}$$

Thus, one must supplement this equation with additional ones that determine the vertical structure of the velocity field. However, if the fluid is assumed to be homogeneous so that v_0 is independent of z and if one then writes

$(U, V) = (u_0, v_0)h,$

the system closes with

$$\beta V = \hat{\mathbf{k}} \cdot \nabla \times \boldsymbol{\tau} - \frac{f_0 h_b}{h}\left(\frac{\partial V}{\partial x} - \frac{\partial U}{\partial y}\right). \quad (5.84)$$

Introducing the transport stream function $\mathbf{V} = \mathbf{k} \times \nabla \psi$, yields

$$\beta \psi_x = \hat{\mathbf{k}} \cdot \nabla \times \boldsymbol{\tau} - K \nabla^2 \psi, \quad (5.85)$$

where $K = f_0 h_b/h$. Stommel (1948) obtained (5.85) by assuming a bottom drag law for the friction term. The derivation using Ekman layer theory makes the assumptions more evident.

The solution with $\tau^x = -T\cos(\pi y/M)$, $\tau^y = 0$ and with $\psi = 0$ at $x = 0, L$ and $y = 0, M$ is

$$\psi = \frac{MT}{\pi K}\left\{1 - \frac{(1 - e^{D_2 L})e^{D_1 x} - (1 - e^{D_1 L})e^{D_2 x}}{e^{D_1 L} - e^{D_2 L}}\right\}$$

$$\times \sin\frac{\pi y}{M}, \quad (5.86)$$

where

$$D_1 = -\frac{\beta}{2K} + \sqrt{\left(\frac{\beta}{2K}\right)^2 + \left(\frac{\pi}{M}\right)^2},$$

$$D_2 = -\frac{\beta}{2K} - \sqrt{\left(\frac{\beta}{2K}\right)^2 + \left(\frac{\pi}{M}\right)^2}.$$

Values of ψ versus x are shown in figure 5.1 for the case with $L = 6000$ km, $M = 3000$ km, $\beta = 2 \times 10^{-11}$ m^{-1} s^{-1}, and $K = 2 \times 10^{-6}$ s^{-1}. Stommel's model was the first to exhibit the westward intensification of the oceanic response to a symmetric wind-stress curl.

With $K/\beta L \ll 1$, (5.85) is a boundary-layer problem, where the highest derivative term (the bottom frictional effect) is important only in a narrow region near the western boundary where the flow is northward. In the remainder of the basin the Sverdrup balance (5.80) is approximately valid (but see below), the flow is slow and southward, and friction is unimportant. The negative vorticity injected into the ocean by the wind is eventually dissipated in the western boundary layer, where the induced northward flow deposits columns of fluid at their original latitudes with the original planetary vorticity restored. Detailed balances and a fairly comprehensive discussion are given by Veronis (1966a).

The westward intensification is normally explained in terms of the vorticity balance, but a qualitative discussion in terms of momentum balance is also possible. Thus, we note that the Ekman wind drift in the northern half-basin is southward whereas that of the southern half-basin is northward. Water piles up at mid-latitude, raising the free surface level and creating a high pressure ridge at mid-latitude (H in figure 5.2). The induced eastward geostrophic flow in the northern half-basin requires a low pressure along the northern boundary. In the southern half-basin a westward flow of the same magnitude requires less of a north–south pressure difference (because the Coriolis parameter is smaller) so the low pressure (HL in figure 5.2) at the south is higher than the low pressure (LL in figure 5.2) at the north. The solid boundaries at the east and west will divert the flow. A narrow frictional boundary layer at the east would require flow from the low low pressure at the north to the high low pressure at the south, i.e., flow up the (gross) pressure gradient. On the western side, on the other hand, a narrow frictional boundary layer supports flow from high to low pressure. Hence, if a thin frictional boundary layer exists, it must be on the western side. This "explanation" ignores a lot of important details, but the reasoning is consistent with the roles that rotation and friction play in balancing the pressure gradient.

If a system without meridional boundaries (a zonal channel) were subjected to a zonal stress, a zonal flow would be generated (apart from the Ekman drift). Hence, the Sverdrup transport of Stommel's model must depend on the presence of meridional boundaries. Yet it seems likely that if the meridional boundaries are far enough apart, the system should resemble a zonal channel more than an enclosed ocean except in relatively narrow regions near the east and west where meridional flow takes place. Welander (1976) showed that that is the case. With the zonal wind stress given above one can substitute $\psi = \Phi(x)\sin(\pi y/M)$ to derive

$$K\Phi'' - K\frac{\pi^2}{M^2}\Phi + \beta\Phi' = -\frac{\pi T}{M}. \quad (5.87)$$

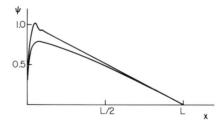

Figure 5.1 The transport streamfunction, normalized with respect to the Sverdrup transport and divided by $\sin \pi y/M$, is shown for Munk's solution with lateral diffusion (top curve) and Stommel's solution with bottom friction. The nominal boundary layer thickness is $L/60$. Stommel's solution shows the decreased transport because of the effect of friction in a basin with $\pi L/M \gg 1$. Munk's solution oscillates near the western boundary, giving rise to a weak countercurrent to the east of the main northward flow.

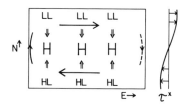

Figure 5.2 A cosine wind stress τ^x causes an Ekman drift (double arrows) toward mid-latitude where the free surface is elevated and a high pressure region (H) is created. A geostrophically balanced current flows eastward in the north half-basin and westward in the south. Because of the larger Coriolis parameter a lower low pressure (LL) is required along the north boundary than along the south (HL) to support the same transport geostrophically. If the zonal transport is deflected southward in a frictional boundary layer near the eastern side (dashed curve), the flow must go against the gross pressure difference (from LL to HL). If the flow is in a western boundary layer (solid curve), the gross pressure difference drives the flow against frictional retardation. The latter is a consistent picture.

As we have seen, the second-derivative term is important only in the western boundary layer where the scale of variation is $K/\beta = 100$ km. North–south diffusion (the undifferentiated Φ term) is unimportant when the geometry is square. But when the zonal separation is large ($\pi L/M \gg 1$), the balance is between wind-stress curl and north–south diffusion, $\Phi \approx MT/\pi K$, and the flow is zonal. The Sverdrup transport relation holds in an eastern boundary layer with the east–west scale $\beta M^2 \pi^2 / K$. Bye and Veronis (1979) pointed out that the northward transport in the western boundary layer is much smaller than the transport calculated by the Sverdrup balance if the aspect ratio $\pi L/M$ is large, as is the case for nearly all wind-driven oceanic gyres. Of course, these results are contained in the complete solution of the simple model discussed here. But when relatively modest refinements are introduced (e.g., spherical geometry), a complete solution is no longer possible and boundary layer methods must be used. It is then necessary to recognize the correct approximate balance in the different regions of the basin.

5.5.3 Topography and Lateral Friction

The principal result of the foregoing analysis, viz., the westward intensification of an oceanic gyre, is verified both by observations and by much more sophisticated analyses. Hence, it is a feature that appears to be insensitive to the drastic simplifications that were made. But it is a simple matter to change the result by relaxing one of the simplifications and then restoring the result with a second, seemingly unrelated, assumption. In other words, the simple model is not as crude as it appears to be.

For example, introduce realistic topography (Holland, 1967; Welander, 1968). Then on vertical integration, we see from (5.60) that, in addition to the effects of wind stress and bottom friction, the vertical divergence term in (5.78) will also contribute the term $f\,dh/dt$ to the right-hand side of (5.84). If the latter is combined with the β term, the result is

$$h^2 \frac{d}{dt}\left(\frac{f}{h}\right) = \hat{\mathbf{k}} \cdot \nabla \times \boldsymbol{\tau} - K\left(\frac{\partial V}{\partial x} - \frac{\partial U}{\partial y}\right). \qquad (5.88)$$

Hence, the driving and dissipative forces on the right will cause a fluid column to respond by moving to points determined by the value of f/h rather than f as before. Since the contours of f/h are sometimes strongly inclined to latitude circles (Gill and Parker, 1970), the transport pattern is very different from (in fact, less realistic than) Stommel's. Thus, the effect of topography is exaggerated in a homogeneous model.

Stratification can reduce the topographic effect. In fact, if the density surfaces adjust so that the pressure gradient in (5.55) vanishes at and below a given level, there will be no driving force to support a flow. If topography does not project above this level of density compensation, it has no effect on the flow. In an intermediate situation, the density distribution can compensate for part of the pressure gradient so that at the level where it interacts with the bottom the velocity is considerably weaker than the surface velocity. A treatment of the latter case would necessarily incorporate convective processes in some form.

When complete compensation takes place in a steady model, the topographic influence is eliminated, but our derivation of bottom friction is no longer valid because it is no longer possible to parameterize the frictional processes at the bottom in terms of the mean velocity. The essential results of the model can be preserved, however, by parameterizing frictional effects in terms of an assumed lateral eddy diffusion. The last term in (5.83) is then replaced by a lateral frictional term so that the vorticity equation, in terms of the transport stream function becomes

$$\beta \frac{\partial \psi}{\partial x} = \hat{\mathbf{k}} \cdot \nabla \times \boldsymbol{\tau} + A\,\nabla^4 \psi, \qquad (5.89)$$

where A is the magnitude of eddy viscosity based on the intensity of eddy processes at scales smaller than those being analyzed. Hidaka (1949) introduced this equation together with the vertically integrated continuity equation

$$\frac{\partial U}{\partial x} + \frac{\partial V}{\partial y} = 0. \qquad (5.90)$$

A convenient set of boundary conditions where the wind stress curl is proportion to $\sin(\pi y/M)$ is

$$U = 0 = V \quad \text{at} \quad x = 0, L,$$

$$V = 0 = \frac{\partial U}{\partial y} \quad \text{at} \quad y = 0, M. \qquad (5.91)$$

The solution is easily obtained (Munk, 1950) and is included in figure 5.1. It contains a Sverdrup transport in the interior; a narrow eastern boundary layer in which V decreases to zero at the eastern wall; and a western boundary layer with no tangential velocity at the western wall, a northward flow near the wall and a weak, narrow countercurrent just east of the northward flow. Because frictional processes are now associated with higher derivatives, the effect of friction in the interior is considerably weaker than in Stommel's model, and the Sverdrup balance is valid throughout the interior. Accordingly, in this case the aspect ratio of the basin has little influence on the magnitude of the transport. Because the zonal velocity increases linearly with distance from the eastern boundary, for broad ocean basins the flow has a strongly zonal appearance. In Stommel's model the north–south flow essentially vanishes in the western portions of the basin and the flow is truly zonal there.

Although these formal models are steady, the application is to flows that are transient but statistically steady. Transient motions can have a strong barotropic component even when the statistically steady flow is largely baroclinic. With that in mind we may still use a bottom frictional drag for the stratified steady model, though the connection to the mean flow will then be not through the coupling to a steady Ekman layer but through a time averaging of interacting transient motions. Rooth (1972) has made such an estimate for K and obtains a value considerably smaller than the one normally used.

5.5.4 Laboratory Models

Though these steady, linear models can provide only the crudest approximation to real oceanic flows, they have served an important function in the development of oceanic theory. Stommel (1957b) put together the important components (Ekman suction and β-effect) to construct a comprehensive picture of ocean current theory as determined by these simple processes. The ideas were tested in a laboratory model of ocean circulation (Stommel, Arons, and Faller, 1958) in which the β-effect was simulated by the paraboloidal depth of a homogeneous layer of water in a pie-shaped basin rotating about the apex (see chapter 16). The equivalence of β and variable depth is suggested by the linearized form of potential vorticity,

$$(\zeta + f)/h \simeq (\zeta + f)/H_0 - \frac{f_0 \eta}{H_0^2},$$

where η is the deviation of the free surface from its mean value H_0, so that a change in $f_0\eta/H_0^2$ is equivalent to a change in f, i.e., to β. When water is being added at the apex, the free surface in the interior rises not by a direct vertical motion but by a radially uniform inward movement of columns of fluid (figure 5.3). The circulation generated in this way simulates the Sverdrup transport, the inward radial direction corresponding to north (increasing f or decreasing depth).

In the experiment, boundary layers near the "western" boundary and the rim and apex are required to complete the circulation pattern (figure 5.3). The azimuthal flow and the rising free surface needed to feed the interior radial flow are generated in the rim boundary layer. Near the apex the flow is diverted to the western boundary layer to join the fluid being injected. It is interesting to note that the radially inward flow that causes the free surface to rise is *toward* the source of fluid. Thus, the transport in the western boundary layer is twice that of the source. Half of the former goes to raise the free surface; the other half serves as the vehicle for the indirect circulation. (Also see Figures 16.1 and 16.2 and the accompanying discussion.)

Additional experiments and a rigorous analysis using rotating-fluid theory to treat the various boundary layers were subsequently provided by Kuo and Veronis (1971), who showed that for different parametric ranges the experiment could be used to simulate Stommel's model with a bottom frictional boundary layer or the Hidaka-Munk model with a lateral frictional boundary layer. Veronis and Yang (1972) provided a perturbation treatment of the nonlinear effects and verified the results with a series of experiments. Pedlosky and Greenspan (1967) proposed an alternative laboratory model with the depth variation provided by an inclined boundary at the top and/or bottom of a rotating cylinder. The flow was driven by the differential rotation of the top plate. For this model Beardsley (1969, 1972) carried out a comprehensive set of experiments and

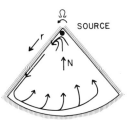

Figure 5.3A A weak source of fluid at the apex of a rotating pie-shaped basin will cause flow toward the rim in a "western" boundary layer. Fluid flows from the rim boundary layer radially inward toward the apex as shown.

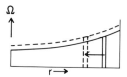

Figure 5.3B A vertical cross section through the apex. The basin is filled in the interior by the inward movement of columns of fluid as shown.

extended the theory analytically and numerically to include inertial effects.

The foregoing experiments and theories are more appropriate areas of application than the real ocean is for the ideas introduced by Sverdrup, Stommel, and Munk. At the time that they were introduced, however, these ideas were remarkable advances into unknown territory. They have provided a framework for further development and some of them persist as important elements in more extensive theories.

5.6 Preliminary Nonlinear Considerations

The first perturbation analysis of nonlinear effects in a wind-driven gyre was by Munk, Groves, and Carrier (1950), but it is easier to see the qualitative changes by looking at Stommel's model (Veronis, 1966a). From the linear problem we saw that the vorticity and its zonal variation are largest in the western boundary layer, so we expect the largest nonlinearities there. The wind stress is not important in that region, and we start with the vorticity equation, including inertial terms but not the wind-stress curl:

$$\frac{d}{dt}(\zeta + f) = \mathbf{v}\cdot\nabla\zeta + \beta v = -\epsilon\zeta. \quad (5.92)$$

In the southern half of the basin the flow is westward $(u < 0)$ into the boundary layer where it is diverted northward. Thus, a fluid particle is carried from the interior, where ζ vanishes, into the boundary layer, where ζ is large and negative, so $d\zeta/dt < 0$. Northward flow implies $df/dt > 0$. Hence, the convective term balances part of the β-effect and $-\epsilon\zeta$ must consequently decrease in size. Since the vorticity is essentially $\partial v/\partial x$, it will decrease if v decreases or if the horizontal scale increases. But from $v = \partial\psi/\partial x$ we see that a decrease in v also corresponds to an increase in the horizontal scale. Therefore, we conclude that inertial effects weaken the flow by broadening the scale. This effect will also decrease the dissipation in the inflow region. The same considerations apply to the case with lateral friction.

In the northern half of the basin where the flow emerges $(u > 0)$ from the western boundary layer, fluid is carried from a region of negative vorticity to the interior where ζ vanishes, so $d\zeta/dt > 0$. Since the flow is northward in the boundary layer, df/dt is also positive. Therefore, the amplitude of the vorticity must be larger since the dissipation $-\epsilon\zeta$ must be larger than in the linear case. Hence, the horizontal scale of variation must decrease.

The net effect of inertial processes is thus to broaden the boundary layer thickness and to reduce the dissipation in the region of inflow, and to sharpen the boundary layer thickness and increase the dissipation in the region of outflow.

For more nonlinear flows the dissipation takes place largely in the northern half of the boundary layer. Furthermore, the excess inertia of the particles causes them to overshoot their original (interior) latitudes so there must be an additional region where inertial processes and friction restore the particles (southward) to their starting points. The effect is to spread the region of inertial and frictional control first to the north and eventually eastward from the northwest corner of the basin. A discussion of the successively stronger effects of nonlinear processes and a division of the basin into regions where different physical balances obtain is given by Veronis (1966b).

This argument strongly suggests that it may be possible to analyze the region of formation of western boundary currents in terms of a frictionless inertial model. Stommel (1954) proposed such an analysis which he subsequently included in his book (Stommel, 1965).

Fofonoff (1954) focused his attention on nonlinear processes by treating the steady circulation in a frictionless, homogeneous ocean. The starting point is the conservation of potential vorticity in a basin of constant depth, viz.,

$$\frac{d}{dt}(\zeta + f) = 0, \quad (5.93)$$

together with the two-dimensional continuity equation. These equations are satisfied by $u = -\partial\psi/\partial y =$ constant or

$$\psi = -uy, \quad (5.94)$$

but boundary conditions are not, so it is necessary to add boundary layers at the eastern and western sides of the basin.

A first integral of (5.93) is

$$\nabla^2\psi + f = F(\psi) \quad (5.95)$$

and in the interior where the relative vorticity vanishes

$$F(\psi) = f = f_0 + \beta y. \quad (5.96)$$

But (5.94) yields $y = -\psi/u$ there, so that $F(\psi) = f_0 - \beta\psi/u$ and (5.95) becomes

$$\nabla^2\psi + \frac{\beta\psi}{u} = -\beta y. \quad (5.97)$$

This equation is satisfied nearly everywhere by $\psi = \phi(x)y$ so that

$$\phi'' + \frac{\beta\phi}{u} = -\beta. \quad (5.98)$$

The north-south flow near the meridional boundaries is thus geostrophic. A boundary layer solution with $\phi = 0$ at $x = 0, L$ is possible for $u < 0$ if $\epsilon = (-\beta/u)^{1/2} \gg L$. It is

$$\phi = \frac{u}{\sinh \epsilon L} [\sinh \epsilon x - \sinh \epsilon L + \sinh \epsilon (L - x)]$$
$$\times [y - M e^{-(M-y)\epsilon}]. \tag{5.99}$$

This yields a uniform, westward flow in the interior, and boundary layers of thickness ϵ^{-1} with northward flow at the west, southward flow at the east, and a jet across the northern edge (figure 5.4).

It is possible to have the eastward jet at any latitude by adding an appropriate constant to ψ in (5.94). With $u > 0$ the system does not have a boundary layer solution but oscillates across the basin (Fofonoff, 1962a).

Although Fofonoff's solution appears to be very artificial, it is one of the survivors of the earlier theories. The strongly nonlinear version of Stommel's model leads to a solution that looks remarkably like Fofonoff's (Veronis, 1966b; Niiler, 1966). The recirculation region just south of the Gulf Stream after the latter has separated from the coast has the appearance of a local inertial circulation. Thus, it is likely that some version of the latter will be part of any successful model of large-scale ocean circulation.

Shortly after Fofonoff's analysis and following Stommel's (1954) suggestions, Charney (1955b) and Morgan (1956) produced models of the Gulf Stream as an inertial boundary layer. By using observed or simulated conditions at the inflow edge of the Gulf Stream to fix the form of $F(\psi)$, and working with a two-layer model with potential vorticity $(f + \zeta)/h$ and geostrophic balance for the northward flow, they were able to calculate the streamfunction pattern and the thermocline depth distribution in the formation region of the Gulf Stream. Charney showed that in a two-layer ocean inertial forces can cause the thermocline to rise to the surface at a latitude corresponding to Cape Hatteras. His solution could not extend beyond that point.

Morgan began his analysis by dividing the ocean into an interior with a Sverdrup balance, a formation region for the western boundary current, which he analyzed using the same model that Charney did, and a northern region. He speculated that friction and inertial and transient processes would interact in the north, but he did not attempt to analyze that region. He was one of the first to point out that pressure torques at the bottom and sides of the ocean can help to balance the torque exerted by the wind stress about a mid-ocean axis.

In contrast to the demonstration following (5.92) that inertial effects are consistent with the formation of a western boundary layer by the interior flow, a similar argument for the formation of an eastern boundary layer is not possible. For example, consider an anticyclonic gyre when an eastward interior flow generates an eastern boundary layer with southward flow. The vorticity in the boundary layer is negative, so $-\epsilon \zeta$ is positive. For southward flow df/dt is negative and therefore $d\zeta/dt$ must be positive. But that is not possible since ζ must change from a nearly zero value in the interior to a large negative value in the boundary layer. An analysis of the various possibilities for both cyclonic and anticyclonic gyres shows that it is generally not possible to form eastern boundary layers from eastward interior flows (Veronis, 1963). The actual existence of eastern boundary layers means that the necessary physical processes (in my opinion horizontal advection of density must be included) are missing from these simple models.

In an important model of a steady wind-driven gyre in a homogeneous ocean of constant depth, Derek Moore (1963) produced a complete circulation pattern with contributions from frictional and inertial processes in both inflow and outflow regions of the western boundary layer. Moore combined boundary-layer arguments from classical fluid mechanics with most of the features given above. Using a Navier-Stokes form for friction, he proved that frictional and inertial processes cannot be combined consistently to produce a boundary layer confined to the eastern side. In the vorticity equation of his model inertia is included as an east-west convection of the vorticity with a zonal velocity, $U(y) = U_0 \cos(\pi y/M)$, consistent with the form of the wind stress. In the southern half-basin (figure 5.5) the incoming (westward) flow forms an inertially controlled western boundary current. In the northern half-basin the emerging flow oscillates eastward and has the appearance of standing, damped Rossby waves imbedded in an eastward current. The center of the gyre is north of mid-latitude, consistent with the effects of inertia mentioned earlier. His results depend on the magnitude of a Reynolds number defined by $Re = U_0^{3/2}/\nu\beta^{1/2}$, which can be looked upon as the ratio of the inertial boundary layer scale $(U_0/\beta)^{1/2}$ to the viscous scale ν/U_0. The result is shown for $Re = 5$. As Re is decreased, the flow tends toward the Munk pat-

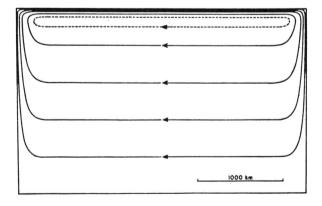

Figure 5.4 Fofonoff's (1954) inertial flow pattern for steady westward flows in the interior. An inertial boundary layer at the west diverts the flow northward and an eastward jet is formed. The latter feeds into an inertial boundary layer on the east that supplies the steady westward flow of the interior.

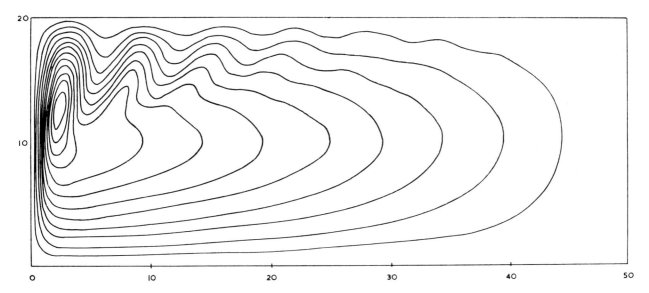

Figure 5.5 Contours of the streamfunction in a homogeneous ocean driven by a wind stress of the form $-\cos \pi y/M$ as derived by Moore (1963). An Oseen approximation for the nonlinear terms with a mean current $U(y) \propto \cos \pi y/M$ was used. The wavy contours in the north half-basin are standing Rossby waves imbedded in the mean velocity field.

tern. With larger Re the oscillations extend farther to the east and eventually fill the northern half of the basin. In the latter case there is a rapid transition across mid-latitude in the interior and the oscillatory flow becomes unstable. Qualitatively this homogeneous model contains a remarkably realistic array of features of oceanic flow, though the observed recirculation in the northwest corner is missing.

We turn to a discussion of stronger nonlinear effects in Stommel's model. As Ro is increased (Veronis, 1966b), the western boundary layer in the southern half-basin broadens and dissipative effects are more confined to the north. Inertial effects also intensify in the north so that a particle overshoots the northernmost latitude that it had in the interior. Hence, a new boundary layer region must be generated (offshore of the original one) where friction and inertia force the particle southward to its original latitude. In this latter region the relative vorticity is actually positive because the return flow to the south is stronger close to the boundary layer than it is farther to the east. The overshoot can be seen in figure 5.6A.

With even stronger driving the overshoot is larger and eventually the particle is driven close to the northern boundary and then eastward before it starts its southward return to its original latitude (figure 5.6B). Thus, the frictional-inertial region is broadened. In an extreme case (figure 5.6C) fluid particles move eastward in a jet at the north and reach the eastern boundary before turning south. In the latter case, there is essentially no Sverdrup interior, and the flow pattern resembles Fofonoff's free inertial flow with a mild east-west asymmetry as the only evidence that the flow is wind driven. An interesting fact here is that the northward transport in the western boundary layer does not increase beyond the Sverdrup transport until the eastward moving inertial jet reaches the eastern boundary. In the calculations cited, that happens when the inertial scale $(U_0/\beta)^{1/2}$ $(\equiv Ro^{1/2}L)$ exceeds the viscous scale K/β by a factor of 2 or so. Here, U_0 is a measure of the Sverdrup velocity. Qualitatively, at least, the observed recirculation to the south and east of the Gulf Stream after it has separated from the coast is simulated by this model. The separation from the coast is not. An analytic model of the highly nonlinear case was suggested by Veronis (1966b) and independently carried out by Niiler (1966). The resulting pattern is consistent with the one shown in figure 5.6C. Stommel (1965) guessed a similar pattern.

Bryan (1963) carried out an extensive set of numerical calculations in a rectangular basin for the nonlinear Hidaka-Munk model with $\hat{\mathbf{k}} \cdot \nabla \times \tau \sim \sin \pi y/M$, zero velocity boundary conditions at east and west, and zero-shear conditions at north and south. He presented his results in terms of a Reynolds number Re essentially the same as Moore's, and the Rossby number, Ro. The results differ greatly from those with bottom friction because for $Re > 60$ a barotropic (Rayleigh-type) instability can occur near the western boundary where the tangential velocity must vanish. Figure 5.7 illustrates his results for three values of Re, with $Ro = 1.28 \times 10^{-3}$ for figures 5.7A and 5.7B and $Ro = 3.2 \times 10^{-4}$ for figure 5.7C. The first two cases, with $Re = 20$ and $Re = 60$, show the development of the flow with increasing nonlinearity. Only a mild, steady, oscillatory pattern is present with $Re = 20$, whereas with $Re = 60$ the oscillations are more intense and a closed eddy (recirculation) is present near the northwest corner. For $Re = 100$ the flow is transient with a barotropic

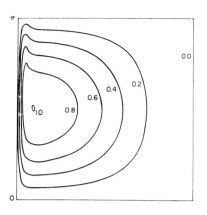 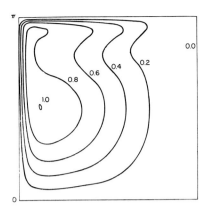 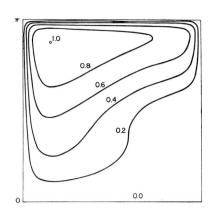

Figure 5.6 Three streamfunction patterns by Veronis (1966b) for an ocean basin with varying degrees of intensity of wind stress. (A) shows the perturbation effect of nonlinearity with fluid particles in the western boundary layer overshooting their equilibrium latitudes. (B) shows a much stronger inertial effect. In (C) inertia dominates the system, creating an eastward jet along the north reminiscent of Fofonoff's solution.

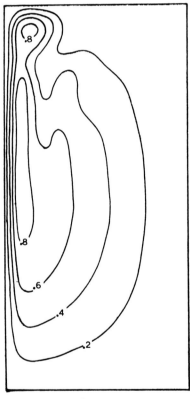 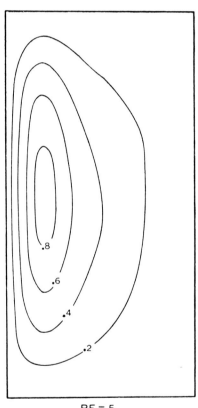 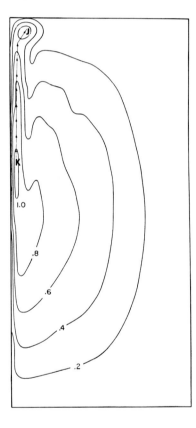

RE = 60   RE = 5

Figure 5.7 Bryan's (1963) streamfunction contours for a homogeneous ocean with lateral friction. The circulation in (A) is nearly linear; that of (B) is near the limit of forcing that still leads to a steady circulation. With even more intense driving a barotropic instability occurs as in (C), where a time-average field is shown after the system approaches a statistically steady state. See also Figure 3.13 and discussion there. *Re* is 5 for (A), 60 for (B), and 100 for (C).

instability induced in the northern half of the intense northward jet. Figure 5.7C shows the time-averaged flow after the transients have settled down. In this case also there is an offshore region in the north with positive vorticity where particles return southward to their starting latitudes. It is not possible to obtain an intense recirculation with this model because of the barotropic instability.

Bryan also calculated the flow for a basin with a western boundary directed north, then due east, and then north again. The break was north of mid-latitude. The object was to see whether the break in the boundary would force the western boundary current out to sea. The flow pattern was modified mildly, but the stream turned the corner and hugged the coast.

5.7 Why Does the Gulf Stream Leave the Coast?

The Gulf Stream flows along the coast from Florida to Cape Hatteras, where it parts from the coast and flows slightly north of eastward out to sea (see chapter 4). The Kuroshio and all other western boundary currents also separate. The phenomenon is explained here by a very simple argument. Although processes more complicated than the ones discussed below are also present, I believe that the argument given here contains the essential features even though the local dynamical details are not included.

Consider a two-layer system with the lower layer at rest. Then from equations (5.65) and (5.66) it follows that

$$\nabla \eta_2 = -\frac{\rho_1}{\Delta \rho}\nabla \eta_1, \qquad \nabla \eta_1 = \frac{\Delta \rho}{\rho_2}\nabla h_1. \qquad (5.100)$$

If the motion is geostrophic ($Ro \ll 1$) except for the vertical stress term near the surface, equation (5.62) upon vertical integration over the depth h_1 of the top layer becomes, with $g' = g\Delta\rho/\rho_2$,

$$-fV_1 = -\frac{g'h_1}{a\cos\phi}\frac{\partial h_1}{\partial \lambda} + \tau, \qquad (5.101)$$

where spherical coordinates have been retained so there is no geometrical distortion. Here the stress at the interface is assumed negligible and τ corresponds to the zonal wind stress. Multiply (5.101) by $a\cos\phi$ and apply the operator $\int_\lambda^{\lambda_e}(\ )d\lambda$, where λ_e is the meridian of the eastern boundary, to obtain

$$h_1^2 = h_{1e}^2 - \frac{2f}{g'}T_1 - \frac{2f}{g'}T_E, \qquad (5.102)$$

where subscript e denotes a value at λ_e, $T_1 = \int_\lambda^{\lambda_e} a\cos\phi\, V_1 d\lambda$ is the meridional transport, and $T_E = \int_\lambda^{\lambda_e} a\cos\phi\, \tau\, d\lambda/f$ is the Ekman drift.

In all of the calculations reported in the previous section, the downstream velocity in the western boundary layer is geostrophic to a very good approximation. Hence, (5.102) is valid not only for interior flow but for the entire basin from west to east. Therefore, if we evaluate (5.102) at the western edge λ_w, T_1 represents the total meridional transport. If the ocean basin is enclosed to the north of the latitude in question, T_1 must vanish in the steady state and (5.102) becomes

$$h_{1w}^2 = h_{1e}^2 - \frac{2f}{g'}T_E. \qquad (5.103)$$

Now, for $\tau > 0$ the Ekman drift, T_E is toward the south (positive as defined above) and the depth of the upper layer at the western boundary h_{1w} will be less than h_{1e}. For sufficiently large T_E, h_{1w} will vanish, i.e., the thermocline (interface) rises to the surface. With observed values for $\Delta\rho/\rho_2$, τ, and h_{1e} for the North Atlantic, h_{1w} vanishes at about the latitude of Cape Hatteras.

North of that latitude τ is even larger and (5.103) cannot be satisfied because T_E is too large. However, the solution can be extended northward by setting h_{1w} equal to zero at a new longitude ($>\lambda_w$) which is chosen to reduce T_E so that the terms on the right of (5.103) balance. This new longitude marks the westernmost edge of the warm-water mass and is the longitude of the Gulf Stream. But $\lambda > \lambda_w$ means that the Gulf Stream must separate from the coast and extend out to sea. This argument alone does not suffice for higher latitudes where τ eventually becomes negative. We shall return to that issue presently.

Before doing so, however, we discuss the simple physical balances given above. The meridional flow in the interior is a combination of geostrophically balanced motion and Ekman drift. If the flow were completely geostrophic, vanishing T_1 would require equal values of h_1 at the eastern and western edges. But the Ekman wind drift, which does not involve a pressure gradient, accounts for part of the southward transport when $\tau > 0$. Therefore, since the total transport vanishes, there is a net northward geostrophic transport, of magnitude T_E, which requires $h_{1w}^2 < h_{1e}^2$. Thus, the Ekman drift causes the thermocline to rise to the surface. Separation of the Gulf Stream from the coast simply moves the western edge of the warm-water mass (upper layer) eastward so that the smaller Ekman drift acting on that water mass of more limited east–west width can just balance the geostrophic flow determined by h_{1e}^2 (since h_{1w}^2 vanishes).

It is also interesting to note that the Coriolis parameter does not appear in (5.103). In fact, the result is exactly the one obtained for a nonrotating lake where the wind blows the warm water to the leeward edge and causes the thermocline to rise on the windward side. The principal difference between the two phenomena is that the induced pressure gradient drives a vertical circulation in the lake, whereas it is geostrophically balanced in the rotating ocean, thereby generating a horizontal cell. But the leeward piling up of water is the same in the two cases.

Returning to the problem at high latitudes, we note first that the analysis given above must be supplemented by the remaining dynamic balances. The reader is referred to Veronis (1973a) for the details for the wind-driven model. The qualitative discussion given here is simpler and clearer than in the original paper.

The first problem is that the Sverdrup transport for the interior vanishes with $\hat{\mathbf{k}} \cdot \nabla \times \tau$, and without adding to the simple argument there is no way of supplying warm water to the north of the latitude ($\approx 40°$N in the North Atlantic) where the curl vanishes. Second, even supposing that warm water has somehow been supplied to the north, the Sverdrup transport there is northward ($\hat{\mathbf{k}} \cdot \nabla \times \tau > 0$), so the southward return of the flow by a western boundary current would require that the thermocline be deeper on the western side of the boundary layer. That is not possible with the boundary current in mid-ocean.

Both of these issues can be resolved by considering what happens even farther to the north where warm water flows northward and impinges on the northern boundary. In the real ocean and in a model including thermal driving (Veronis, 1978), this water will sink and give rise to a deep circulation and an overturning cell. In a wind-driven model the water travels counterclockwise as an isolated warm boundary current and rejoins the stream at the point of separation. In the analysis given above, this recirculating current represents an excess transport in the separated boundary current. Because its transport does not depend on local winds, it can transport water past the latitude of vanishing wind-stress curl and supply warm water to the interior at high latitudes. When it is included in the analysis, a revised longitude for the separated boundary current is obtained. The calculation, which can be made consistent and quanitative for both the wind-driven model and the one including thermal driving, is contained in the two papers cited above. The path of the separated Gulf Stream is reproduced in figure 5.8. It is especially interesting to note that the vestigial current in the northeastern corner of the basin corresponds to the Norwegian Current (the Alaskan Current in the Pacific) and that its transport is important for the separation of the Gulf Stream and also for the determination of the longitude of the current after it has separated.

The analysis leading to the separation of the Gulf Stream from the coast is contained in a quasi-geostrophic model by Parsons (1969). It was derived independently by Veronis (1973a) as part of a study of the circulation of the World Ocean. The extension poleward of the latitude where the wind-stress curl vanishes is contained in the latter paper. Kamenkovich and Reznik (1972) included a (bottom friction) analy-

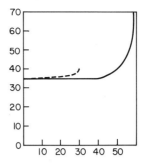

Figure 5.8 The path (solid curve) of the Gulf Stream after it has separated from the coast [from a reduced gravity model by Veronis (1973a)]. The zonal wind stress that drives the system is taken from observations and has zero curl at 40°N. The Norwegian Current is the narrow jet in the northeast. The dashed curve is the prediction for an isolated anticyclonic wind gyre (Parsons, 1969). The latter solution cannot be extended north of the latitude of zero wind-stress curl. Axes are latitude and longitude.

sis of the deep circulation induced by the separated current.

All of the above make use of a steady, linear, quasi-geostrophic model, and it is certain that the details (e.g., the longitude of the separated current) will be altered when a more complete dynamic model is used. The key elements of the argument, however, are the geostrophic balance of downstream velocity in the western boundary current, the Ekman wind drift, and a limited amount of upper-layer water. As long as a different dynamic model does not drastically change those three features (they are pretty rugged and can withstand a lot of battering) the more complicated dynamics can be incorporated to change the details of the results, leaving the main argument unchanged.

By the same token, the present analysis suggests that an explanation of the separation of western boundary currents from the coast must necessarily include the surfacing of the thermocline (with a possible mixed layer at the surface). Western boundary currents can be forced out to sea between wind-driven gyres of opposite sign, but that occurs at low latitudes as well where the phenomenon is qualitatively different because the thermocline does not surface.

In addition, the argument given here depends on properties of global scale. A more precise dynamic treatment based on local properties can lead to a better understanding of the detailed mechanistic balances of the separated current, but the cause of separation seems to be based on global properties.

5.8 Thermohaline Circulation

The physical processes that are involved in the formation of the thermocline have been studied as a separate part of the general circulation. The models incorporate geostrophic dynamics and steady convection of density, the latter often including vertical diffusion. Though the analyses sometimes make use of the β-plane, the scales are really global and spherical coordinates are more appropriate. The real difficulty is the nonlinearity in the convection of density, and as it turns out, the limited successes of the analyses have been achieved as often in the spherical system as in the β-plane. None of the nonlinear investigations treats a closed basin, though a single eastern boundary is sometimes included. A closed two-layer basin is tractable (see section 5.8.2).

5.8.1 Continuous Models for an Open Basin

The starting point for these studies is the simplified set of equations in spherical coordinates

$$fv = \frac{1}{a \cos\phi} \frac{\partial P}{\partial \lambda}, \tag{5.104}$$

$$fu = -\frac{1}{a} \frac{\partial P}{\partial \phi}, \tag{5.105}$$

$$\frac{\partial P}{\partial z} = -g \frac{\rho}{\rho_m}, \tag{5.106}$$

$$\frac{\partial u}{\partial \lambda} + \frac{\partial}{\partial \phi}(v \cos\phi) + a \cos\phi \frac{\partial w}{\partial z} = 0, \tag{5.107}$$

$$\frac{u}{a \cos\phi} \frac{\partial \rho}{\partial \lambda} + \frac{v}{a} \frac{\partial \rho}{\partial \phi} + w \frac{\partial \rho}{\partial z} = K \frac{\partial^2 \rho}{\partial z^2}, \tag{5.108}$$

where the last term in (5.108) contains the only dissipative process, vertical diffusion of density. These equations cannot be used to analyze the balances for a closed basin because there is not enough flexibility to satisfy even the condition of no normal flow through the boundaries. Essentially all past efforts have been restricted to this open system.

In principle, there is enough flexibility to satisfy four boundary conditions in the vertical (three if diffusion is omitted). These must be chosen to be consistent with the form of the solution that is obtained; therefore, much of the flexibility is lost. Still, it is possible to obtain interesting, if limited, information about the thermal structure.

Results based on linearized models by Lineikin (1955) and Stommel and Veronis (1957) were superseded by the nonlinear models of Welander (1959), who treated the ideal fluid system ($K = 0$), and Robinson and Stommel (1959), who obtained a similarity solution with K included. Stommel and Webster (1962) made use of the latter model to determine the dependence of the vertical structure of w and T on the value of K and on surface boundary values of w and T. Exact solutions were obtained by Fofonoff (1962a), Blandford (1965), Kozlov (1966), Needler (1967, 1972) and Welander (1959, 1971a). More recently the problem has been reformulated by Hodnett (1978) with density instead of vertical distance as an independent coordinate. A review of the earlier papers is given by Veronis (1969).

The variables u, v, and ρ are given in terms of P by (5.104) to (5.106). These can be substituted in (5.108) to give w in terms of P and the continuity equation then yields Needler's pressure equation

$$\kappa \sin\phi \cos\phi (P_{zz}P_{zzzz} - P_{zzz}^2)$$
$$= P_{zzz} \frac{\partial(P_z, P)}{\partial(\lambda, \phi)} + P_{zz} \frac{\partial(P, P_{zz})}{\partial(\lambda, \phi)} + \cot\phi P_\lambda P_{zz}^2, \tag{5.109}$$

where $\kappa = 2\Omega K a^2$.

Welander (1959, 1971b) defined the variable

$$M = \int_0^z P \, dz + a^2 f \sin\phi \int_0^\lambda w(\lambda, \phi, 0) \, d\lambda \tag{5.110}$$

(so that $P = M_z$), to obtain the simpler equation

$$\kappa \sin\phi \cos\phi M_{zzzz} + \frac{\partial(M_{zz}, M_z)}{\partial(\lambda, \phi)}$$
$$- \cot\phi M_\lambda M_{zzz} = 0. \tag{5.111}$$

By integrating (5.104) from λ to 0 and setting $P(0, \phi, z) = 0$, it is easy to see that M/f is the geostrophic, wind-driven, meridional transport between λ and 0 and below level z. $P(0, \phi, z) \neq 0$ means that the reference pressure is not passive (there must be density anomalies at $\lambda = 0$) and it gives rise to an added transport. In interpreting the system, however, it is best to think in terms of $P(0, \phi, z) = 0$.

Needler (1967) derived a solution that had been obtained previously by Blandford (1965) under more restrictive conditions and by Welander (1959), who ignored K. In his analysis Needler proposed the following form with three arbitrary functions of λ and ϕ:

$$P(\lambda, \phi, z) = A(\lambda, \phi) + B(\lambda, \phi) e^{zC(\lambda, \phi)}. \tag{5.112}$$

This is a solution to (5.109) provided that A, B, and C are independent of λ or that C is given by

$$C(\lambda, \phi) = c/\sin\phi, \tag{5.113}$$

where c is a constant. Only the latter case seems to have received much attention, even though the case with $A_\lambda = B_\lambda = C_\lambda = 0$ could be a zero-order solution to which necessary corrections could be made (away from the coasts the oceans exhibit a quasi-zonal distribution of properties).

With (5.113) the remaining variables are given by

$$\frac{\rho}{\rho_m} = -\frac{cB}{g \sin\phi} e^{zc/\sin\phi}, \tag{5.114}$$

$$u = -\frac{1}{fa} \left\{ A_\phi + \left[B_\phi - \frac{zc \cos\phi}{\sin^2\phi} B \right] e^{zc/\sin\phi} \right\}, \tag{5.115}$$

$$v = \frac{1}{fa \cos\phi} \left[A_\lambda + B_\lambda e^{zc/\sin\phi} \right], \tag{5.116}$$

$$w = \frac{1}{fa^2}\left[\frac{B_\phi}{c}e^{zc/\sin\phi} + \frac{\tan\phi}{cB}\frac{\partial(A,B)}{\partial(\lambda,\phi)} + \frac{A_\phi}{c} + \frac{zA_\phi}{\sin\phi}\right]$$
$$+ \frac{Kc}{\sin\phi}. \quad (5.117)$$

The general functions A and B and the constant c are available to satisfy boundary conditions.

Before proceeding further, it is worth noting that the diffusivity K appears only in the last term of w. In fact, with $w_d \equiv Kc/\sin\phi$ it is evident that

$$w_d \frac{\partial \rho}{\partial z} = K \frac{\partial^2 \rho}{\partial z^2}. \quad (5.118)$$

If we write the vertical velocity as $w = w_a + w_d$ (the subscripts a and d correspond to *advective* and *diffusive*, respectively) we see that w_d absorbs the diffusive effect and w_a satisfies the ideal fluid system with $K = 0$. Hence, with this solution diffusion plays a minor role in the dynamic balances, and the essential balances coincide with those in Welander's (1959) solution.

It would appear at first sight that the general functions A and B and the constant c are available to satisfy boundary conditions. We note, however, that only B multiplies the exponential and that the properties described by A penetrate undiminished to the bottom. Since we expect neither the surface density nor the Ekman pumping to generate effects that penetrate undiminished to the bottom, we can discard A for the time being and concentrate on $B(\lambda, \phi)$. Evaluating (5.114) at $z = 0$ yields

$$\frac{\rho(\lambda,\phi,0)}{\rho_m} = -\frac{cB}{g\sin\phi}, \quad (5.119)$$

so that the product cB is determined by the surface density distribution. The constant c can be evaluated by matching the e^{-1} decay depth with the middle of the thermocline at one latitude. This is not really a boundary condition, but it is forced on us if we want the solution to generate a realistic vertical density profile. With $c^{-1} = 1500$ m, a surface temperature (a linear measure of density by the Boussinesq approximation) proportional to $\cos(\phi + 10°)$ and a reference temperature of 2.45°C at 5000 m depth at $\phi = 10°$, Needler constructed the isotherm pattern in a vertical (ϕ versus z) section shown in figure 5.9. The choice of c corresponds to a thermocline depth of 750 m at $\phi = 30°$.

It is not possible to satisfy any more conditions. Hence, the Ekman pumping velocity is determined by the surface density. One could specify w_E instead and then the surface density would be determined. We shall return to this point at the end of this section.

The pattern shown in figure 5.9 reproduces the observed density minimum at mid-latitude for levels near the thermocline. The same feature appears in the so-

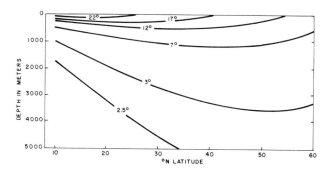

Figure 5.9 Isotherms, or isopycnals, from Needler's (1967) thermocline model with vertical dependence $\propto \exp(cz/\sin\phi)$. The observed maximum vertical penetration of warm water at mid-latitude is a feature of almost all of the thermocline models.

lutions of Welander (1959) [from the M equation (5.111)] and of Robinson and Stommel (1959). The pattern lacks the abrupt variations with latitude of the observed system, though that could be remedied by choosing the surface conditions more realistically, a procedure that is equivalent to introducing higher-order dynamic effects through the boundary conditions. But we should remember that this is an open system and that introducing more realistic surface conditions to obtain a more pleasing pattern requires that fluid leave the region governed by the simple, assumed balances and reenter it after substantial changes in the properties have taken place. This would amount to rationalizing the data without really learning anything about the processes that bring about the desired change.

The vertical velocity w_a decays exponentially with depth so this advective solution is a surface boundary layer solution. The analysis described is valid as long as the boundary layer thickness (determined by c) is small compared to the depth of the ocean. Below the boundary layer the vertical velocity is given by w_d and is induced by vertical diffusion. Once c is chosen w_d is determined by K. However, the value (and even the form!) of K is as unknown and as unmeasurable as w_d. In fact, the balance given by (5.118) is often used to obtain an estimate of the scale depth K/w_d, and when it is used in conjunction with measured vertical profiles of tracers with a known decay rate, individual estimates of w_d and K can be made. But the whole procedure makes use of purely vertical balances and is itself questionable. In the present context we can only conclude that the model is not sufficiently constrained by K to enable us to determine its effects. Typical values used for K and w_d in deep water (Munk, 1966) are $K = 10^{-4}$ m^2 s^{-1} and $w_d = 10^{-7}$ m s^{-1}.

The finite depth of the ocean requires that the vertical velocity w_d at the base of the thermocline region match the vertical velocity at the top of the layer below. Within the framework of the present approach that means that a barotropic mode must be included to satisfy the boundary condition of zero normal velocity at the bottom. Needler (1967, 1972) gives a thorough discussion of the issue. In his second paper he seeks the conditions under which $P_1(\lambda, \phi, z) = P(\lambda, \phi, z) + D(\lambda, \phi)$ is a solution where P itself is a solution, i.e., what are the restrictions on P for an arbitrary barotropic mode $D(\lambda, \phi)$ to be added as part of the solution? It is easy to see from (5.109) that, since D_z vanishes and P must satisfy the equation, one is left with

$$D_\phi \frac{\partial(P_{zz} P_z)}{\partial(\lambda, z)} - D_\lambda \left[\frac{\partial(P_{zz} P_z)}{\partial(\phi, z)} + \cot\phi P_{zz}^2 \right] = 0. \quad (5.120)$$

Furthermore, with P independent of D, and D an arbitrary function, each of the P expressions must vanish. A straightfoward argument then shows that P must be of the form

$$P = \sin\phi \, \Phi(\eta) + E(\lambda, \phi), \quad (5.121)$$

$$\eta = \frac{cz}{\sin\phi} + F(\lambda, \phi), \quad (5.122)$$

where E, F, and Φ are arbitrary functions of their arguments. Hence $E(\lambda, \phi)$ can be absorbed into $D(\lambda, \phi)$. With $K \neq 0$ the solution reduces essentially to the exponential one given earlier. For $K = 0$ it can be shown that the conditions given by (5.120) are equivalent to the statement that the density and potential vorticity ($f\rho_z$ in this case) are functions of each other. N. A. Phillips (1963) had already shown that Welander's (1959) (hence, Needler's) exponential solution satisfied $f\rho_z = 2\Omega c\rho$, a special case of the above. We shall return to this point shortly when we discuss Welander's more general solutions for an ideal fluid thermocline.

Needler (1972) satisfies the bottom boundary condition of zero normal flow by using the arbitrary barotropic mode introduced above with $K = 0$. In addition, he shows that the consistency conditions required in order to add an arbitrary barotropic mode make it possible to satisfy only two of the three independent conditions: $w(\lambda, \phi, 0)$, $T(\lambda, \phi, 0)$, and zero normal flow at the bottom. Once two of them are satisfied, the third is determined.

Needler's two papers are highly recommended reading. He discusses both the possibilities and the inadequacies of this approach to the thermocline circulation and he gives a sound analysis of some very difficult problems.

Welander has spearheaded perhaps the most significant advances in the theory of the thermohaline circulation. His first paper on the problem contained the exponential solution given above with an arbitrary function available to satisfy a general surface boundary condition. The next paper (Robinson and Welander, 1963) merged his approach with that of Robinson and Stommel (1959). In his third paper for steady ideal flows (Welander, 1971a) he first derived, and then applied, the general relationship (5.11) between potential vorticity q, density ρ, and the Bernoulli function B to the geostrophic, hydrostatic system of equations (5.104) to (5.108). The latter yields the simplified forms $q = f\rho_z$ and $B = p + g\rho z$ so that equation (5.11) reduces to

$$\sin\phi \, \rho_z = F(\rho, p + g\rho z), \quad (5.123)$$

where F is an arbitrary function.

Linearization of F yields

$$\sin\phi \, \rho_z = a\rho + b(p + g\rho z) + c, \quad (5.124)$$

where a, b, and c are arbitrary constants. Upon differentiation with respect to z and use of the continuity equation, we obtain

$$\sin\phi \, \rho_{zz} = (a + bgz)\rho_z, \quad (5.125)$$

and two integrations yield

$$\rho(\lambda, \phi, z) = \rho(\lambda, \phi, 0) - C(\lambda, \phi) \int_z^0 e^{bg(\zeta + z_0^2)/(2\sin\phi)} d\zeta, \quad (5.126)$$

where $z_0 = a/bg$ and C is an arbitrary function of λ and ϕ. It is evident that b must be negative; otherwise the integral grows indefinitely with z. Furthermore, z is negative, so $a > 0$ implies a monotonic profile. With $a < 0$ an inflection point occurs at $z = -a b/g$. Thus, the constants a and b can be chosen to give an inflection point at a desired depth and a desired thickness to the thermocline. The latter varies inversely as $\sin^{1/2}\phi$. Welander fitted the constants to match the observed density profile along 160°W in the South Pacific (Reid, 1965) shown in figure 5.10A. His theoretical solution (figure 5.10B) captures the general structure of the observed profile, though it is smoother, as one would expect. In the construction Welander used the observed surface density for $\rho(\lambda, \phi, 0)$, and $C(\lambda, \phi)$ was chosen to give a deep constant density. He gives no other details for the construction.

This solution is a remarkable step forward. It takes advantage of only the simplest of the possibilities that the general conservation integrals contain and it justifies Welander's faith in the use of ideal-fluid theory to obtain realistic results. Welander also presented more general solutions to the system, but the latter are quite formal and no detailed results from them have been reported. Making use of this first integral to the general system is very promising and it is surprising that this path has not been pursued more actively.

In a subsequent paper on this topic Welander (1971b) explored the possible balances in the M equation (5.111) by means of a scale analysis. His conclusions can be summarized without detailed analysis by making use of the results already found. In regions of Ekman suction ($w > 0$) diffusive processes adjust the density to surface values, a simple possible balance

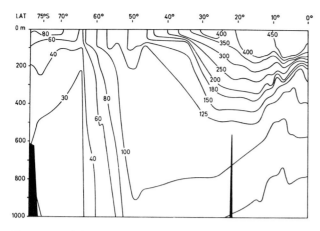
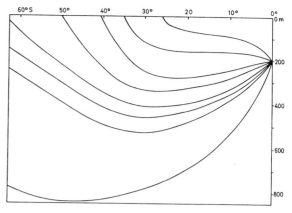

Figure 5.10 (A) Contours of thermosteric anomaly (units of 10^{-5} cm³ g⁻¹) in the upper kilometer of the South Pacific along 160°W from Reid (1965). (B) Isopycnal contours from Welander's (1971a) ideal-fluid thermocline model.

being given by (5.118) with a scale depth $H \sim K/w$. With K fixed this diffusive depth decreases with increasing w. Ekman pumping ($w < 0$), on the other hand, forces lighter water into the oceanic interior so that the surface value of ρ extends to some depth. With ρ constant, (5.118) is satisfied trivially and advection must be important so that the order-of-magnitude balance is $V/a \sim W/H_a$, where we use the global scale a in the horizontal and H_a is the vertical (advective) scale. W and V are velocity scales. Geostrophic balance yields $fV/H_a \sim g\rho/a\rho_m$ and eliminating V then yields $H_a \sim (fWa^2\rho_m/g\rho)^{1/2}$, which increases with W. Therefore, more intense surface forcing gives rise to a deep advective layer where $w < 0$ and an advective layer under a thin diffusive layer where $w > 0$. The geostrophic transport is carried by the advective layer, the diffusive process serving simply to adjust the density to surface values. Welander gives a more detailed analysis of the possibilities to show that an advective layer must be present. He also points out that a deep diffusive layer with a balance like (5.118) but with w_d different from W_E is also likely.

5.8.2 Layered Models

Stommel (1957b), noting that upwelling suggested by (5.118) would produce a vertical divergence from the (level) bottom to the base of the thermocline, assumed that the deep ocean is homogeneous and used the planetary divergence relation $v = a \tan\phi \, \partial w/\partial z$ to determine the meridional velocity. With a uniform upwelling at the base of the thermocline v is poleward everywhere in the interior. Then with $u = 0$ at all eastern boundaries (taken along meridians) he calculated the zonal velocities by integrating the continuity equation with respect to longitude to obtain a trajectory pattern for the interior of the world ocean.

Interior upwelling of deep water requires that sources of deep water be present somewhere. Stommel chose sources of equal strength in the North Atlantic and in the Weddell Sea (South Atlantic). He assumed that these source waters flowed along western boundary layers and then eastward to supply the upwelling flow in the interior. The transports in the western boundary layers were obtained by requiring mass conservation for a basin bounded by two meridians, a northern boundary and the latitude in question. The pattern of flow that results from these considerations is shown in figure 5.11.

Veronis (1978) has combined this reasoning with an analysis similar to that of section 5.7 to construct a two-layer model of the thermohaline circulation in the world ocean with wind stress acting on the surface. The intensity of the upwelling and the locations and intensities of the sources of deep water can be deduced from the model. The reasoning is as follows.

On the basis of an expected balance like that of (5.118) in deep water, assume a vertical flux of water through the interface from the lower to the upper layer. The amplitude of upwelling is taken to be horizontally uniform but of unknown magnitude. The height h_2 of the interface above the level bottom is determined by the wind stress acting on the surface and the upwelling through the interface. The two-layer, steady, linear system of equations (5.62) to (5.67) on a sphere can be manipulated to yield a first-order partial differential equation for h_2 with coefficients depending on h_2. This quasi-linear equation can be integrated along characteristics from (assumed) known values on the eastern boundary to give h_2 throughout the interior.

Here, too, the assumed upwelling will require sources of deep water that will flow along the western boundaries to supply the oceanic interior. The downstream flow in the western boundary layers is assumed to be geostrophic. Mass conservation of water in both layers is required in the region bounded by boundaries at the sides and along the north and by the latitude in question. As in section 5.7 this will lead to an expression for the depth of the thermocline (or the height h_2)

at the western boundary. This expression will depend not only on the value of h_2 at λ_e, however, but will be a function also of the unknown amplitude of upwelling and the unknown sources of deep water. If the latter quantities were known, it would be possible to determine h_2 and, in particular, the latitude at which the thermocline rises to the surface. North of this latitude the western boundary current will flow eastward and poleward across the open ocean. Since it represents the boundary between upper- and lower-layer water, it will also determine the area covered by upper-layer water, and therefore, the total amount of upwelling (w times the area) that occurs north of any latitude. So we have an implicit problem with h_2, w, and the strengths and locations of the sources interrelated.

As stated earlier, obtaining an estimate for the upwelling is not straightforward, depending as it does on complex, turbulent, convective processes. Therefore, the problem is inverted. Instead of assuming values for w and for the strengths and locations of the sources to determine the surfacing latitude, the latter, a simple observable, is taken from observation and the former quantities are determined by the model. It turns out that to evaluate w *and* the sources requires more than one piece of information. For example, in the Pacific the surfacing latitudes of the Kuroshio (35°N) and the East Australian Current (31°S) are specified and these yield an upwelling velocity w of magnitude 1.5×10^{-7} m s^{-1} and a distribution of sources of deep water at the northern boundary (along the Alaskan–Aleutian current system), at the latitude of separation of the Kuroshio, and along the Australian coast from 31 to 35°S. The circulation patterns and the details of the calculations are given in the paper cited.

Some major features (e.g., deepest penetration of light water at mid-latitude) are consistent with those obtained by the continuous thermocline models discussed earlier. However, the present model also allows one to close the circulation with boundary layers, and in particular, to determine the open-ocean path of the separated boundary current. For the continuous model that possibility would enable one to adjust surface boundary conditions as part of the analysis in order to obtain a more realistic vertical density distribution with latitude.

Most noteworthy of the results obtained with this two-layer model is the deduced magnitude of the assumed upwelling. Most estimates for w are made from observed tracer distributions by assuming that vertical advective and diffusive processes balance locally. They yield values between 10^{-7} and 2×10^{-7} m s^{-1} (see chapter 15). The present value lies midway in the range cited and is based on global circulation processes with no reference to the vertical diffusive process.

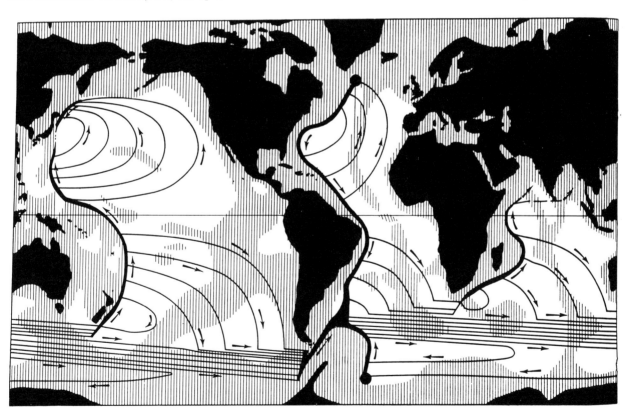

Figure 5.11 The abyssal circulation obtained by Stommel (1958) and generated by equal sources in the North Atlantic and in the Weddell Sea with uniform upwelling elsewhere.

5.9 Free Waves for a Constant-Depth Two-Layer Ocean on the β-Plane

The linear equations for a two-layer ocean on the β-plane with constant depth are (5.62) to (5.67) (with $Ro \ll 1$):

$$u_{1t} - fv_1 = -g\eta_{1x}, \tag{5.127}$$

$$v_{1t} + fu_1 = -g\eta_{1y}, \tag{5.128}$$

$$\frac{1}{H_1}(\eta_1 - \eta_2)_t + u_{1x} + v_{1y} = 0, \tag{5.129}$$

$$u_{2t} - fv_2 = -g[(1-\epsilon)\eta_1 + \epsilon\eta_2]_x, \tag{5.130}$$

$$v_{2t} + fu_2 = -g[(1-\epsilon)\eta_1 + \epsilon\eta_2]_y, \tag{5.131}$$

$$\frac{1}{H_2}\eta_{2t} + u_{2x} + v_{2y} = 0, \tag{5.132}$$

where $\epsilon = \Delta\rho/\rho_2$ and H_1, H_2 are the constant mean depths of the two layers. Elimination of all but one variable leads to a sixth-order equation. However, it is possible to simplify the mathematics to a third-order system by introducing normal modes (Veronis and Stommel, 1956). We do so by multiplying (5.130) to (5.132) by a constant α and adding to (5.127)-(5.129), respectively, to derive

$$(u_1 + \alpha u_2)_t - f(v_1 + \alpha v_2)$$
$$= -g\{[1 + \alpha(1-\epsilon)]\eta_1 + \alpha\epsilon\eta_2\}_x, \tag{5.133}$$

$$(v_1 + \alpha v_2)_t + f(u_1 + \alpha u_2)$$
$$= -g\{[1 + \alpha(1-\epsilon)]\eta_1 + \alpha\epsilon\eta_2\}_y, \tag{5.134}$$

$$\left[\frac{\eta_1}{H_1} + \left(\frac{\alpha}{H_2} - \frac{1}{H_1}\right)\eta_2\right]_t$$
$$+ (u_1 + \alpha u_2)_x + (v_1 + \alpha v_2)_y = 0. \tag{5.135}$$

The velocities appear in the same combination $\mathbf{v}_1 + \alpha\mathbf{v}_2$, everywhere, but the surface deviations appear in two different forms. If the latter are the same except for a multiplicative constant, the three equations will involve only three variables (plus parameters). So let us define

$$\mathbf{V}_1 = \mathbf{v}_1 + \alpha\mathbf{v}_2, \tag{5.136}$$

$$\Phi = [1 + \alpha(1-\epsilon)]\eta_1 + \alpha\epsilon\eta_2, \tag{5.137}$$

$$\frac{\Phi}{h} = \frac{\eta_1}{H_1} + \left(\frac{\alpha}{H_2} - \frac{1}{H_1}\right)\eta_2, \tag{5.138}$$

where h is a constant to be determined.
Equating Φ in (5.137) and (5.138) yields

$$[1 + \alpha(1-\epsilon)\eta_1] + \alpha\epsilon\eta_2$$
$$= h\left[\frac{\eta_1}{H_1} + \left(\frac{\alpha}{H_2} - \frac{1}{H_1}\right)\eta_2\right]. \tag{5.139}$$

Since this must be valid for arbitrary η_1 and η_2, the coefficients of η_1 and η_2 must be the same, i.e.,

$$1 + \alpha(1-\epsilon) = \frac{h}{H_1}, \qquad \alpha\epsilon = \frac{h\alpha}{H_2} - \frac{h}{H_1}. \tag{5.140}$$

Eliminating h yields

$$\frac{H_1}{H_2}(1-\epsilon)\alpha^2 + \left(\frac{H_1}{H_2} - 1\right)\alpha - 1 = 0. \tag{5.141}$$

For small ϵ the two values for α are

$$\alpha_1 \simeq \frac{H_2}{H_1}, \qquad \alpha_2 \simeq -1. \tag{5.142}$$

Corresponding values of h and the variables are

$$h_1 = H_1 + H_2, \qquad h_2 = \epsilon H_1 H_2/(H_1 + H_2), \tag{5.143}$$

$$\mathbf{V}_1 = \mathbf{v}_1 + \frac{H_2}{H_1}\mathbf{v}_2, \qquad \mathbf{V}_2 = \mathbf{v}_1 - \mathbf{v}_2, \tag{5.144}$$

$$\Phi_1 = \frac{H_1 + H_2}{H_1}\eta_1 + \epsilon\frac{H_2}{H_1}\eta_2,$$

$$\Phi_2 = \frac{\epsilon H_2}{H_1 + H_2}\eta_1 - \epsilon\eta_2. \tag{5.145}$$

Then the six equations can be reduced to the following two independent sets ($i = 1,2$)

$$U_{it} - fV_i = -g\Phi_{ix}, \tag{5.146}$$

$$V_{it} + fU_i = -g\Phi_{iy}, \tag{5.147}$$

$$\Phi_{it} + h_i(U_{ix} + V_{iy}) = 0, \tag{5.148}$$

which describe the linear, time-dependent motions on a β-plane for a barotropic ocean of depth $H = H_1 + H_2$ when $i = 1$ and for a baroclinic ocean of depth $\epsilon H_1 H_2/H$ when $i = 2$. When Φ_2 is replaced by Φ_2/ϵ and g by $g' = \epsilon g$, the internal (baroclinic) case is then called the reduced-gravity system with depth $H_1 H_2/H$. If U_i and Φ_i are eliminated from (5.146) to (5.148) the resulting equation in V_i is

$$(\partial_{ttt}^2 + f^2 \partial_t - gh_i \partial_{xxt}^3$$
$$+ gh_i \beta \partial_x - gh_i \partial_{yyt}^3)V_i = 0. \tag{5.149}$$

If f is now replaced by its reference value, f_0 (this is lowest order in L/a), all coefficients are constant. By substituting $V_i \sim e^{i(-\omega t + kx + ly)}$, we obtain the frequency equation for free waves,

$$\omega_i^3 - (f_0^2 + gh_i K^2)\omega_i - gh_i\beta k = 0,$$
$$K^2 = k^2 + l^2. \tag{5.150}$$

This yields the (approximate) dispersion relations[7]

$$\omega_{i1} = f_0\sqrt{1 + \lambda_i^2 K^2}, \qquad \omega_{i2} = -\omega_{i1},$$
$$\omega_{i3} = -\frac{\beta k \lambda_i^2}{1 + \lambda_i^2 K^2}. \tag{5.151}$$

A plot of ω versus K is shown in figure 5.12 for $k = l$ at mid-latitude ($f_0 = 10^{-4}$ s^{-1}, $\beta = 2 \times 10^{-11}$ m^{-1} s^{-1}) and for the λ values given below.

The first two of these waves for each mode reduce to long gravity waves for $\lambda_i K \gg 1$ (small wavelength) and to inertial waves for $\lambda_i K \ll 1$ (large wavelength). The dividing scale is λ_i, which is about 2000 km for the barotropic mode and about 36 km for the baroclinic. The third wave for each case is a westward-traveling Rossby wave that arises because of conservation of potential vorticity $(\zeta + f)/h$ on the β-plane. As a fluid column changes latitude or moves into a region with different depth, a relative vorticity is generated to keep $(\zeta + f)/h$ constant. The distortion of the free surface or of the density interface has an effect for scales larger than λ_i and the waves become nondispersive $(\omega_{i3} \sim \beta k \lambda_i^2)$. For the baroclinic mode the period of this wave at mid-latitude is of the order of years for scales of the size of an ocean basin, making the linear baroclinic response of the ocean very slow (Veronis and Stommel, 1956; see chapters 10 and 11).

Lighthill (1967, 1969) applied wave theory in the limit of vanishing frequency to study the development of forced, steady flows. In his analysis of the responses of the equatorial Indian Ocean he used the fact that λ_i becomes very large near the equator to conclude that the baroclinic response there is much faster—of the order of weeks—than it is at mid-latitudes.

A comprehensive account of barotropic Rossby waves is given by Longuet-Higgins (1964, 1966). Here we shall make use only of a simple property that relates directly to large-scale circulation. The dispersion relation (5.151) in the (ω_{i3}, k)-plane in figure 5.12 shows that the phase velocity ω_{i3}/k is westward and that a given value of ω_{i3} corresponds to two wavelengths, a short wave with small velocity and a long wave with a fast velocity. The zonal group velocity $\partial \omega_{i3}/\partial k$, which transports energy zonally, is westward for the fast waves and eastward for the slow ones. Thus, at a meridional boundary, where the energy flux must vanish, the energy of an incoming wave will be reflected quickly at an eastern boundary and will accumulate at a western boundary.

Pedlosky (1965b) offered this as an alternative explanation of westward intensification, and N. A. Phillips (1966a) used these reflected properties to account for the more frequent observation of intense eddy motions near the western versus the eastern regions of the North Atlantic. Ibbetson and Phillips (1967) report a laboratory confirmation of this east-west distribution of energetic eddy motions. Observed long barotropic Rossby waves in the ocean are suggested in the bottom pressure records in the MODE area by Brown et al. (1975). Freeland, Rhines, and Rossby (1975) show longitude versus time plots of the streamfunction inferred from objective maps of currents at 1500 m depth in the

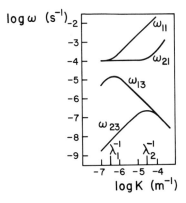

Figure 5.12 Frequency-wavenumber diagram for waves in a two-layer, constant-depth ocean in the β-plane. The upper two curves are inertiogravity waves, the two lower are Rossby waves. Transitions in the dispersion curves occur at the deformation scales shown on abscissa.

MODE area (figure 10.6). There is a definite tendency for constant phase lines to move westward, with phase speeds ranging from 0.02 to 0.12 m s^{-1} (average 0.05 m s^{-1}). With $l = k$ this suggests wavelengths clustering around 400 km. A time versus latitude plot shows no definite north-south propagation.

The slow phase velocities deduced for baroclinic Rossby waves make the linear theory less reliable because particle motions equal to and exceeding the wave speeds occur in all parts of the ocean. Rhines (1977) identifies thermocline eddies (intense baroclinic modes, principally confined to the waters above the thermocline) with these baroclinic Rossby waves and offers evidence of their existence in observed records from open ocean regions. These noisy, nearly stationary modes make the determination of the slow mean flow in the open ocean a difficult task.

5.10 Effect of Bottom Topography on Quasi-Geostrophic Waves

The results of the previous section were extended by Rhines (1970) to include simple bottom topography. Though the general normal-mode procedure does not work in this case, Rhines modified it for quasi-geostrophic wave motions when topography varies linearly in y. The multiplicative constant α is a function of wavelength in that case and the method is difficult to interpret when topography varies with x as well. Because the equations in terms of the surface height lead to a quadratic dispersion relation in a straightforward manner even in the latter case, we shall not use normal modes.

5.10.1 Two-Layer Model
The linearized, potential vorticity equation for each layer becomes

$$\zeta_t + \beta v - \frac{fh_t}{H} + \frac{f\mathbf{v}\cdot\nabla h}{H} = 0. \tag{5.152}$$

For quasi-geostrophic motions we can substitute the geostrophic velocity in $\zeta = v_x - u_y$ and in $\mathbf{v}$. Furthermore, we have $h_{2t} = \eta_{2t}$, $h_{1t} = (\eta_1 - \eta_2)_t$, $\mathbf{v}_1\cdot\nabla h_1 \simeq 0$ (for linear flows), $\mathbf{v}_2\cdot\nabla h_2 \simeq -\mathbf{v}_2\cdot\nabla \eta_3$ where bottom topography η_3 is defined by a linear function of x and y and $\eta_3 \ll H_2$ so that the constant depth H_2 can be used in the coefficients. Then we obtain

$$\frac{f^2}{gH_1}(\eta_1 - \eta_2)_t - \nabla^2 \eta_{1t} - \beta \eta_{1x} = 0, \tag{5.153}$$

and

$$\frac{f^2}{gH_2}\eta_{2t} - \nabla^2 p_{2t} - \beta p_{2x} + \frac{f}{H_2}\frac{\partial(\eta_3, p_2)}{\partial(x,y)} = 0, \tag{5.154}$$

where $p_2 = (1 - \epsilon)\eta_1 + \epsilon\eta_2$. The substitution $e^{i(-\omega t + kx + ly)}$ for η_1 and η_2 then leads to the quadratic frequency equation

$$(1 + \lambda^2 K^2 + \lambda_1^2\lambda_2^2 K^4)\omega^2 + \{(\lambda^2 + 2\lambda_1^2\lambda_2^2 K^2)k\beta$$
$$+ (1 + \lambda_1^2 K^2)\lambda_2^2 kb\}\omega + \lambda_1^2\lambda_2^2 k^2\beta(b + \beta) = 0, \tag{5.155}$$

where $\lambda^2 = gH/f^2$, $\lambda_2^2 = gH_2/f^2$, $\lambda_1^2 = \epsilon gH_1/f^2$, $b = f(k\eta_{3y} - l\eta_{3x})/H_2 k$. The three λ's correspond, respectively, to the radii of deformation for a barotropic fluid with the total depth, a barotropic fluid of depth H_2, and a reduced gravity fluid with depth H_1. The quantity b is a "topographic β-effect" that simply reenforces β when the depth shallows northward and $\eta_{3x} = 0$. More generally, it combines with β [through $(d/dt)(f/h)$] to determine a new, pseudonorth direction.

For the ocean λ is large (2000 km), λ_2 is nearly as large ($\sim$1600 km) and λ_1 is small ($\sim$40 km). Hence, for large wavelengths (small K) (5.155) yields the barotropic solution

$$\omega_1 = -\frac{k\beta + (kbH_2/H)}{K^2 + (1/\lambda^2)}, \tag{5.156}$$

where kbH_2/H is independent of H_2. This dispersion relation reduces to the one for ordinary barotropic Rossby waves when b vanishes. With $b \neq 0$, it represents barotropic Rossby waves with both direction and frequency modified by topography. Where the topography is strong, it yields topographic Rossby waves with the direction of propagation to the left of upslope. For negative b (depth decreasing southward) it is possible for the two restoring mechanisms to cancel each other almost completely.

The second solution for large wavelengths is the baroclinic mode

$$\omega_2 = -\frac{k\beta\lambda_1^2\lambda_2^2}{\lambda^2}\frac{\beta + (bH/H_2)}{\beta + b}$$
$$= -\frac{k\beta g'H_1 H_2}{f^2 H}\frac{\beta + (bH/H_2)}{\beta + b}. \tag{5.157}$$

When topography is weak ($b \ll \beta$) this reduces to the nondispersive wave

$$\omega_2 = -\frac{k\beta g' H_1 H_2}{f^2 H} \tag{5.158}$$

[limit of (5.151) for small K]. In this case, both layers are in motion. With strong topography ($b \gg \beta$) the bottom depth drops out of the dispersion relation and (5.157) becomes

$$\omega_2 = \frac{k\beta g'H_1}{f^2}. \tag{5.159}$$

The point here is that the bottom slope is so large that only a small excursion by a column of fluid in the bottom layer is required to bring about vortex stretching, so that most of the motion is confined to the upper layer where the only restoring force is (the relatively weak) β. Thus, strong topography acts to decouple the layers and to increase the frequency, hence the phase speed, in the upper layer by a factor of H/H_2 over that with weak topography. Rhines (1977) emphasized this detuning effect of topography and showed that it applies for flows of much larger amplitudes.

For small wavelengths, $\lambda_1^2 K^2 \gg 1$, (5.155) reduces to

$$\omega^2 + \frac{k}{K^2}(2\beta + b)\omega + \frac{k^2}{K^2}\beta(\beta + b) = 0, \tag{5.160}$$

which yields a nondivergent baroclinic Rossby wave confined to the upper layer with (low) frequency $\omega = -k\beta/K^2$. The second solution has the frequency

$$\omega = -\frac{k}{K^2}(\beta + b) \tag{5.161}$$

and is a barotropic mode that feels both β and the topography. In the limit of strong topography it reduces to $\omega = -kb/K^2$, and the motion is confined to the lower layer alone. This is a new type of motion, bottom trapped by topography, and does not occur in the flat-bottom case. Rhines calls it a fast baroclinic mode since it appears as an evanescent mode of relatively high frequency in the continuously stratified case, which is presented below.

5.10.2 Uniform Stratification

We noted earlier that effects from upper and lower boundaries are transmitted throughout the respective layers in the two-layer model. The vertical structure of the modes in the real ocean is represented somewhat more realistically in a model with uniform stratification, which represents the opposite extreme in modeling the stratification; the density gradient is smeared uniformly over the depth instead of being squeezed into a layer of infinitesimal thickness.

Dropping the subscript zero we can write the linearized potential vorticity equation (5.53) as

$$\nabla^2 p_t + \frac{1}{S^2} p_{zzt} + \beta p_x = 0, \quad (5.162)$$

where $S = N/f_0$.

At the upper boundary we assume

$$w = 0 \quad \text{or} \quad p_z = 0 \quad \text{at} \quad z = 0. \quad (5.163)$$

This is a "rigid-lid" condition that makes the barotropic radius of deformation infinite. The lower boundary is taken with a uniform slope α in the y direction only, and we write $w = \alpha v$ there, or in terms of p,

$$p_{zt} = -HbS^2 p_x \quad \text{at} \quad z = -H, \quad (5.164)$$

where $b = f_0 \alpha / H$ is the topographic β-effect.

Substitution of $p \sim e^{i(-\omega t + kx + ly)}$ in (5.162) leads to

$$\omega p_{zz} - S^2(\omega K^2 + k\beta)p = 0, \quad (5.165)$$

where $K^2 = k^2 + l^2$, and (5.164) becomes

$$\omega p_z = HbS^2 kp \quad \text{at} \quad z = -H. \quad (5.166)$$

Solutions with sinusoidal and with exponential vertical variation are both possible and are discussed below.

(1) $p \sim \cos mz/H$: This is the form for pure Rossby waves ($\beta \neq 0$, $b = 0$) but is not a solution for pure topographic waves ($\beta = 0$, $b \neq 0$). It satisfies (5.163) and then (5.165) and (5.166) yield

$$\omega = -\frac{k\beta}{K^2 + (m^2/H^2 S^2)} = -\frac{k\beta \lambda_1^2}{\lambda_1^2 K^2 + m^2} \quad (5.167)$$

and

$$m \tan m = \lambda_1^2 kb/\omega \quad (5.168)$$

where[8]

$$S^2 H^2 = -\frac{g \, \partial \bar{\rho}/\partial z \, H^2}{\rho_m f_0^2} = \frac{g \Delta \rho H}{\rho_m f_0^2} = \lambda_1^2. \quad (5.169)$$

and $\Delta \rho$ is the density difference from bottom to top.

Eliminating ω from (5.167) and (5.168) leads to

$$-\gamma m \tan m = m^2 + \lambda_1^2 K^2, \quad (5.170)$$

where $\gamma = \beta/b$ measures the relative roles of β and topography as restoring mechanisms when fluid moves to different latitudes or depths.

If the depth is constant ($\gamma = \infty$), the solution to (5.170) is $m = n\pi$ ($n = 0,1,2,...$). This yields pure Rossby waves with the lowest mode ($m = 0$) describing the barotropic wave with infinite deformation radius and $\omega = -k\beta/K^2$. Higher modes correspond to baroclinic Rossby waves, which have low frequencies and are nondispersive for $\lambda_1 K \ll 1$ (weak stratification or wavelengths much larger than λ_1). As in the two-layer case, with $\lambda_1 K \ll 1$, the β-effect can be taken up by vertical divergence ($\beta v \simeq f w_z$) and only a small change in the relative vorticity is required; hence ω is small. In the opposite extreme, with $\lambda_1 K \gg 1$, vertical motions are inhibited and the frequency is approximately the barotropic value $-k\beta/K^2$ for the lower range of m.

For finite positive slopes ($\gamma > 0$), when topography enhances β, the lowest mode solutions to (5.170) are modified forms of the first baroclinic Rossby wave, which in its pure form has $m = \pi$ and a node at mid-depth for the horizontal velocity. When a sloping bottom is present, fluid flowing north or south will have to move vertically against the constraint of stratification. If the slope is small (large γ), the induced vertical motion is small and the solution is a modified baroclinic Rossby wave. If the slope is large, the required vertical motion may be larger than the stratification will allow.

How does the system enable the fluid to negotiate a large slope? It does so by moving the node in the horizontal velocity from mid-depth to the bottom ($m \to \pi/2$). Thus with vanishing v at the bottom no vertical velocity is required. The mode is then simply the upper half of the first baroclinic mode for a constant-depth ocean with twice the depth H and a corresponding higher frequency. For moderate slopes the node is between mid-depth and the bottom, so that the required vertical velocity at the bottom is reduced to a value that can be sustained. The values of m and ω together with sketches of the vertical structure of the horizontal velocity are shown in figure 5.13 for the lowest-order mode with $l = k$ and for a range of values of γ. For $\gamma > 0$, ω is normalized with respect to the frequency of the first baroclinic mode (to which it tends as $\gamma \to \infty$). Strong stratification (large λ) or small wavelength (large K) have the same effect as a large slope.

It should be noted that when topography reenforces β there is no vertically oscillatory solution that yields a barotropic Rossby wave in the limit $\gamma \to \infty$. The reason is that the slope has its strongest effect on water near the bottom. These cosine solutions have a maximum amplitude at the surface and can do no better than extend that maximum to the bottom (if $m = 0$), which gives no enhancement. On the other hand, in the baroclinic mode the flow at the bottom is reversed (negative maximum) and the bottom slope can help by making the flow there less negative, i.e., somewhat more barotropic, though as we have seen, only to the point where the node is at the bottom. For $\gamma > 0$ the barotropic limit is described by evanescent modes [see (2) below].

When the depth decreases to the south ($b < 0$, hence $\gamma < 0$), the solution for large $|\gamma|$ is a barotropic Rossby

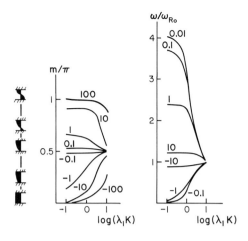

Figure 5.13A Values of the vertical wavenumber m for the gravest quasi-geostrophic mode in a uniformly stratified ocean as functions of horizontal wavenumber. Each curve is marked with value of γ. Topographic effects dominate at small γ; β dominates at large γ. The vertical structure of the modes is sketched at the left for the different values of m.

Figure 5.13B Ratio of frequency to pure barotropic Rossby wave frequency for $\gamma < 0$ and to baroclinic Rossby wave frequency for $\gamma > 0$.

wave modified by topography. In this case the sloping bottom works against β and tends to reduce the frequency. The effect is achieved with a partial cosine wave of one sign throughout the depth. The results for m and ω are shown in figure 5.13. For $\gamma < 0$ figure 5.13B shows ω *normalized with respect to the barotropic Rossby wave frequency* (to which it tends as $\gamma \to -\infty$). A larger slope gives more vertical structure until the node reaches the bottom for small $|\gamma|$ and the form is again a half-baroclinic mode.

It is interesting to note that the frequency is always negative, i.e., the phase velocity is westward even when the bottom slope is large and works against β. These oscillatory modes are not possible without β so the most that the bottom slope can do is to change the magnitude but not the sign of the frequency. When it helps β, it makes the response less baroclinic. When it opposes β, it makes the response less barotropic.

(2) $p \sim \cosh \mu z/H$: This is the form for pure topographic waves ($\beta = 0$, $b \neq 0$) but is not a solution for pure Rossby waves ($\beta \neq 0$, $b = 0$). It satisfies (5.163), and then (5.165) and (5.166) yield

$$\omega = -\frac{k\beta}{K^2 - (\mu^2/S^2H^2)} = -\frac{k\beta\lambda_1^2}{\lambda_1^2 K^2 - \mu^2} \quad (5.171)$$

and

$$-\mu \tanh \mu = \frac{\lambda_1^2 bk}{\omega}. \quad (5.172)$$

Eliminating ω gives

$$\gamma\mu \tanh \mu = \lambda_1^2 K^2 - \mu^2. \quad (5.173)$$

For pure topographic waves ($\gamma = 0$) we have $\mu = \lambda_1 K$ and (5.172) becomes

$$\omega = -\frac{\lambda_1^2 bk}{\lambda_1 K \tanh \lambda_1 K}. \quad (5.174)$$

Strong stratification or short wavelength gives $\tanh \lambda_1 K \to 1$ and

$$\omega = -\frac{\lambda_1 bk}{K} = -\frac{\alpha N k}{K}. \quad (5.175)$$

The frequency has a maximum value for waves traveling parallel to bottom contours since the associated particle velocity is transverse, i.e., up or down the slope, and has maximum vertical displacement; hence it is subjected to maximum restoring force. The wave amplitude decays exponentially upward from the bottom.

For weak stratification or long waves ($\lambda_1 K \ll 1$), $\tanh \lambda_1 K \simeq \lambda_1 K$ and

$$\omega = -\frac{bk}{K^2}. \quad (5.176)$$

These are topographic, barotropic Rossby waves with b replacing β.

All of these evanescent waves, with or without β, have phase velocity always to the left of upslope, since they are basically topographic waves and can be only modified by β. When β and b work together, the phase velocity is westward. When they are opposed, the phase velocity is eastward.

With $\gamma > 0$ we have already seen that the oscillatory solution has no barotropic mode. That function is taken over by this evanescent form, which reenforces β and tends always to increase the frequency above the value for barotropic Rossby waves. For $\gamma \gg 1$ the change is small, but as the slope increases, the frequency does too. As $\lambda_1 K$ increases, the wave is confined closer to the bottom (figure 5.14A) and the frequency rises even more (figure 5.14B). This is Rhines's fast baroclinic mode referred to earlier. The frequencies in figures 5.14B and 5.14C are normalized with respect to the barotropic Rossby wave frequency.

When the slope is negative and small ($-\gamma \gg 1$), the mode is confined to a shallow layer near the bottom where the effect of the slope dominates (figure 5.14A). The frequency is small and tends to zero as $-\gamma$ increases. When the slope is large ($-\gamma \ll 1$), β is a perturbation and the frequency ratio exceeds unity (figure 5.14C). Strong stratification or short wavelength ($\lambda_1 K \gg 1$) also serves to confine the mode to the bottom and increase the frequency.

Rhines (1977) summarized the observational evidence of the existence of these bottom-trapped waves in current-meter records taken at site D (39°10'N, 70°W) by Luyten, Schmitz, and Thompson of the Woods

Hole Oceanographic Institution. Rhines (1971a) and Thompson and Luyten (1976) used spectral analysis to show that the kinetic energy increases toward the bottom and that the horizontal velocities are negatively correlated, as they should be for these waves. The evidence is particularly striking in the high-pass filtered records by Luyten that Rhines showed.

McWilliams and Flierl (1976) have described a number of mesoscale features observed during the MODE-0 and MODE-1 programs in terms of the linear, quasi-geostrophic waves that can exist given the mean properties in the observational regions. Although nonlinear effects seem to have been present, modified forms of the linear wave features can be identified (see chapter 11).

5.11 Baroclinic Instability

The wave solutions in the foregoing two sections are valid when the phase velocity is large compared to the ambient fluid velocity. The condition is not really satisfied either by baroclinic waves or by short barotropic waves. Furthermore, the steady wind-driven and thermohaline velocities derived earlier are equilibrium solutions but they are not necessarily stable. Here, we shall extend our study by considering quasi-geostrophic perturbations on a mean flow and exploring the question of stability. (Also see the discussion in chapter 18.)

In pursuing this approach one should allow for both horizontal and vertical variations in the basic velocity field. Although both types of variations can lead to instabilities, it is known from earlier studies, particularly for atmospheric motions (Kuo, 1949), that the observed fluctuating motions generally have a structure characteristic of barotropically stable modes. In other words, fluctuations do not seem to draw their energy from the horizontal variation in the basic velocity field. For that reason we shall explore a mean velocity that has a variation only in the vertical direction and refer the reader to the published literature for studies involving horizontal variations (Kuo, 1949; Lipps, 1963).

5.11.1 Linear Theory

The problem that we consider is baroclinic instability of the basic velocity field. The topic has an extensive literature in meteorology, beginning with the work of Charney (1947) and of Eady (1949) and extended, inter alios, by Kuo (1952), Phillips (1951, 1954), Green (1960), Charney and Stern (1962), Pedlosky (1964a), and Bretherton (1966a,b). As is the case with the development of a topic, the earliest papers explore the basic concepts and the later ones help to illuminate the issues raised. I have found Bretherton's two articles to be especially enlightening, though each of the ones mentioned, and some others too, discuss important aspects of the problem. The mathematical development given below follows Bretherton (1966a).

Since our intent is to discuss the dynamical balances, we do so with the simplest model that contains the essential elements. Additional effects are mentioned later. So consider the very special case of a uniformly rotating channel (f = constant) with two layers of equal thickness, $H_1 = H_2$, and a total depth $H = H_1 + H_2 = 2H_1$. The upper layer has a uniform basic current U, and the lower layer an equal but opposite current $-U$. Lateral walls at $y = 0, M$ are sufficiently close so that the depth of each layer in the coefficients can be considered constant. Geostrophic balance of the basic state yields

$$fU_1 = fU = -gH_y,$$
$$fU_2 = -fU = -g[(1 - \epsilon)H_y + \epsilon H_{2y}], \quad (5.177)$$

and this basic state is perturbed by infinitesimal quasi-geostrophic motions.

We first linearize the potential vorticity $(\zeta + f)/h$ of each layer by writing

$$\frac{\zeta + f}{h} = \frac{\zeta + f}{H(1 + \eta/H)} \approx \frac{\zeta + f}{H}\left(1 - \frac{\eta}{H}\right) = \frac{f}{H} + \frac{\zeta}{H} - \frac{f\eta}{H^2},$$

where f/H is the mean and $\zeta/H - (f\eta/H^2)$ the perturbation potential vorticity, and η the fluctuating disturbance of the layer depth.

The linearized equation for conservation of potential vorticity becomes

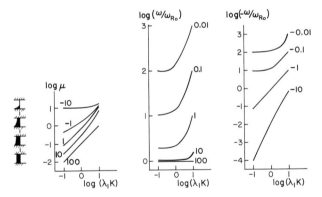

Figure 5.14A Vertical scale μ for the evanescent mode $\cosh \mu z/H$ as a function of horizontal wavenumber. Each curve is marked with value of γ. The vertical structure of the modes is sketched at the left for the different values of μ.

Figure 5.14B Ratio of frequency with $\gamma > 0$ to pure barotropic Rossby wave frequency. This mode tends to the limit of pure barotropic Rossby wave for vanishing topography. Decreasing γ corresponds to increasing bottom slope and serves to increase frequency.

Figure 5.14C Ratio of $-\omega$ to frequency of pure barotropic Rossby wave frequency. Frequency decreases as bottom slope decreases because mode is not possible without topography.

$$(\partial_t + \bar{u}\,\partial_x)\left(\zeta - \frac{f\eta}{H}\right) - \frac{vf}{H}H_y = 0, \quad (5.178)$$

where we have multiplied through by H which is taken as a constant in the coefficients and where $\bar{u}$ is the mean velocity of the layer.

We use the following definitions and relations for the two layers:

Upper Lower
─────────────

$H_1 \qquad\qquad H_2(=H_1)$

$\eta = \eta_1 - \eta_2 \qquad \eta = \eta_2$

$\bar{u}_1 = U \qquad\qquad \bar{u}_2 = -U \qquad (5.179)$

$Q_{1y} \equiv -\dfrac{f}{H_1}H_{1y} = \dfrac{2U}{\lambda_1^2} \qquad Q_{2y} \equiv -\dfrac{f}{H_2}H_{2y}$

$\qquad\qquad\qquad\qquad\qquad = -\dfrac{(2-\epsilon)}{\lambda_1^2}U \simeq -\dfrac{2U}{\lambda_1^2}$

$q_1 \equiv \zeta_1 - f(\eta_1 - \eta_2)/H_1 \qquad q_2 \equiv \zeta_2 - f\eta_2/H_2$

Here,

$$\lambda_1^2 = \frac{g\epsilon H_1}{f^2}$$

is the reduced gravity radius of deformation based on H_1. For convenience we have defined Q and q as H_1 times the potential vorticities. Using (5.179) in (5.178) we obtain

$$\begin{aligned}(\partial_t + U\,\partial_x)q_1 + v_1 Q_{1y} &= 0,\\ (\partial_t - U\,\partial_x)q_2 - v_2 Q_{2y} &= 0,\end{aligned} \quad (5.180)$$

where we have neglected the $O(\epsilon)$ term in Q_2 to write $Q_{2y} = -Q_{1y}$.

The q's can be related to the η's, or more conveniently to "pressure" potentials ϕ, though the geostrophic relations

$$\begin{aligned}v_1 &= \phi_{1x}, \quad u_1 = -\phi_{1y}, \quad v_2 = \phi_{2x},\\ u_2 &= -\phi_{2y},\end{aligned} \quad (5.181)$$

where

$$\phi_1 \equiv g\eta_1/f, \qquad \phi_2 \equiv g[(1-\epsilon)\eta_1 + \epsilon\eta_2]/f. \quad (5.182)$$

Then

$$\zeta_1 = v_{1x} - u_{1y} = \nabla^2\phi_1, \qquad \zeta_2 = v_{2x} - u_{2y} = \nabla^2\phi_2,$$

$$\frac{f}{H_1}(\eta_1 - \eta_2) = \frac{1}{\lambda_1^2}(\phi_1 - \phi_2),$$

$$\frac{f\eta_2}{H_2} = \frac{1}{\lambda_1^2}(\phi_2 - \phi_1 - \epsilon\phi_1) \approx \frac{1}{\lambda_1^2}(\phi_2 - \phi_1), \quad (5.183)$$

$$q_1 = \nabla^2\phi_1 - \frac{1}{\lambda_1^2}(\phi_1 - \phi_2), \qquad q_2 = \nabla^2\phi_2 + \frac{1}{\lambda_1^2}(\phi_1 - \phi_2),$$

where we have neglected the $O(\epsilon)$ term in the expression for $f\eta_2/H_2$. The neglect of the $O(\epsilon)$ terms is equivalent to assuming a rigid lid at the top since it corresponds to neglecting the barotropic radius of deformation gH/f^2.

Equations (5.180) take the form

$$\begin{aligned}(\partial_t + U\,\partial_x)q_1 + \phi_{1x}Q_{1y} &= 0,\\ (\partial_t - U\,\partial_x)q_2 - \phi_{2x}Q_{1y} &= 0,\end{aligned} \quad (5.184)$$

with ϕ and q given by (5.182) and (5.183). We now add and subtract the two equations in (5.184) to obtain the symmetric set

$$(q_1 + q_2)_t + U(q_1 - q_2)_x + Q_{1y}(\phi_1 - \phi_2)_x = 0, \quad (5.185a)$$

$$(q_1 - q_2)_t + U(q_1 + q_2)_x + Q_{1y}(\phi_1 + \phi_2)_x = 0. \quad (5.185b)$$

From (5.183)

$$\begin{aligned}q_1 + q_2 &= \nabla^2(\phi_1 + \phi_2),\\ q_1 - q_2 &= \nabla^2(\phi_1 - \phi_2) - \frac{2}{\lambda_1^2}(\phi_1 - \phi_2),\end{aligned} \quad (5.186)$$

and we have a system in which only sums and differences of the variables appear. The sum is related to the vertical mean, the difference to the baroclinic contribution. Thus, the potential vorticity for the entire depth is just the sum of relative vorticities because the height adjustments are equal and opposite. The differential potential vorticity involves twice the effect of the interface. The symmetry of the set is made possible by the equal mean depths, and equal and opposite mean velocities and mean potential vorticities.

Since all coefficients are constant, the system has solutions of the form

$$e^{-i\omega t + ikx}\sin ly, \qquad l = \frac{m\pi}{M}, \qquad m = 1,2,\ldots, \quad (5.187)$$

which satisfies the lateral boundary condition of zero normal flow at $y = 0, M$. Then (5.186) can be used to express the amplitudes of $q_1 \pm q_2$ in terms of $\phi_1 \pm \phi_2$ as

$$\begin{aligned}q_1 + q_2 &= -K^2(\phi_1 + \phi_2), \qquad K^2 = k^2 + l^2,\\ q_1 - q_2 &= -(K^2 + 2/\lambda_1^2)(\phi_1 - \phi_2).\end{aligned} \quad (5.188)$$

For the moment we keep the x and t derivatives on the ϕ's but use (5.188) to write

$$K^2(\phi_1 + \phi_2)_t$$
$$+ [U(K^2 + 2/\lambda_1^2) - Q_{1y}](\phi_1 - \phi_2)_x = 0, \quad (5.189a)$$

$$(K^2 + 2/\lambda_1^2)(\phi_1 - \phi_2)_t$$
$$+ (UK^2 - Q_{1y})(\phi_1 + \phi_2)_x = 0. \quad (5.189b)$$

Then substituting for Q_{1y} yields

$$K^2(\phi_1 + \phi_2)_t + K^2 U(\phi_1 - \phi_2)_x = 0, \quad (5.190a)$$

$$(K^2 + 2/\lambda_1^2)(\phi_1 - \phi_2)_t + UK^2(\phi_1 + \phi_2)_x$$
$$-\frac{2U}{\lambda_1^2}(\phi_1 + \phi_2)_x = 0. \tag{5.190b}$$

Use of (5.187) in (5.189) and in (5.190) leads to the frequency equation

$$\omega^2 = \frac{k^2}{K^2(K^2 + 2/\lambda_1^2)}[U(K^2 + 2/\lambda_1^2) - Q_{1y}][UK^2 - Q_{1y}]$$
$$= kU^2 \frac{K^2 - (2/\lambda_1^2)}{K^2 + (2/\lambda_1^2)}. \tag{5.191}$$

When ω^2 is negative, the perturbations grow exponentially and instability occurs. In terms of potential vorticity, instability requires

$$U(K^2 + 2/\lambda_1^2) > Q_{1y} > UK^2. \tag{5.192}$$

The right-hand side in (5.191) shows that long waves, with $K^2 < 2/\lambda_1^2$, are unstable. For shorter wavelengths the system is oscillatory with no growth. For $l = k$ the maximum growth rate is $U(3 - 2\sqrt{2})^{1/2}/\lambda_1$ and occurs for $\lambda_1^2 K^2 = 2(\sqrt{2} - 1)$.

We can get some insight into the instability mechanism by noting the following. The local change of the average potential vorticity $q_1 + q_2$ is balanced by the *sum* of the mean advections of q and by the *sum* of the perturbation advections of Q in the two layers. Because both the mean advection velocity and mean potential vorticity have opposite signs in the two layers, these two balancing quantities involve the *differences rather than the sums* of the perturbation quantities, as we can see in (5.185a). But we would not expect internal adjustments of the interface to affect the average potential vorticity since the effect in one layer is cancelled by the effect in the other. This expectation is borne out when we evaluate the different terms to get (5.190a), where only the mean advection of the relative vorticity is left to balance $(\phi_1 + \phi_2)_t$. Thus, if at some instant of time we had $q_1 - q_2 \sim \sin kx$ and $q_1 + q_2 \sim 0$, the quantity $q_1 + q_2$ would be $\sim -\cos kx$ at the next instant.

Now consider the local change in the *differential* potential vorticity $q_1 - q_2$. Again, because of the equal but opposite U and Q_y in the two layers, $(q_1 - q_2)_t$ is balanced by terms involving the *sums* of the perturbation quantities, as we see from (5.185b). The mean convective part $U(q_1 + q_2)_x$ gives rise to a change in $q_1 - q_2$ like $-\sin kx$ after the initial instant and acts as a restoring force, tending to cancel the initial $q_1 - q_2$ ($\sim \sin kx$). The perturbation advection of Q has the opposite sign as we see from (5.190b) and gives rise to a change in $q_1 - q_2$ like $\sin kx$, i.e., it reenforces the initial distribution. The latter, destabilizing, effect will dominate for large wavelengths, i.e., for $K^2 < 1/\lambda^2$. We expect this to be the case because *internal* adjustments

of the interface should cause significant changes in the *differential* potential vorticity.

A second way of looking at the problem is to observe that the perturbation advection of mean potential vorticity vQ_y has the form of mean advection of perturbation potential vorticity Uq_x when we substitute the expressions for v and Q. Because vQ_y involves advection of the layer thickness, only the part due to the interface adjustment appears in Uq_x. When the terms are combined as mean advections in (5.190), we see that the first involves advection in the positive direction by U but the second can have either sign depending on the sign of $K^2 - (2/\lambda_1^2)$. For small K the phase is appropriate for reenforcement and instability occurs.

We can understand the energetics of the instability by writing the disturbances in the more convenient form

$$\phi_1 + \phi_2 = Ae^{\sigma t}\sin kx \sin ly,$$
$$\phi_1 - \phi_2 = Be^{\sigma t}\cos kx \sin ly, \tag{5.193}$$

where the solution of the stability problem gives the growth rate σ as

$$\sigma = \pm kU \frac{(2/\lambda_1^2 - K^2)^{1/2}}{(2/\lambda_1^2 + K^2)^{1/2}}. \tag{5.194}$$

Then (5.190a) yields

$$\sigma A = UkB$$

or

$$B = \pm \left(\frac{2/\lambda_1^2 - K^2}{2/\lambda_1^2 + K^2}\right)^{1/2} A. \tag{5.195}$$

For very long waves ($K^2 \ll 2/\lambda_1^2$) we have $A \simeq B$ for the growing mode and $A \simeq -B$ for the decaying mode. For $k = l$ the corresponding results are $B = \pm(\sqrt{2}-1)^{1/2}A$.

Now consider the case with $U > 0$ so that the upper layer is thicker toward $y = 0$ (south). The quantity η_2 is a measure of the perturbation thickness of the lower layer so $-\eta_2$ is a measure of the excess heat. Then $-v_1\eta_2$ and $-v_2\eta_2$ correspond to northward transports of heat in the respective layers. Using an overbar for an average over x and y, we have $-\overline{(v_1 + v_2)\eta_2}$ as the northward heat transport in the two layers. Eqs. (5.181) and (5.183) can be used to write

$$-\overline{(v_1 + v_2)\eta_2} = \frac{H_2}{f\lambda_1^2}\overline{(\phi_1 + \phi_2)_x(\phi_1 - \phi_2)}$$
$$= \frac{kH_2}{f\lambda_1^2} AB, \tag{5.196}$$

so we have northward heat transport for the unstable mode $A \simeq B$ and southward heat transport for the stable mode $A \simeq -B$.

The vertical velocity at the interface is $w = (\partial_t + U\partial_x)\eta_2 + v_1 H_{2y}$ and the upward heat transport is given by $-\overline{w\eta_2}$. When the foregoing relations are used to evaluate the latter quantity, we find that the unstable mode transports heat upward and the stable mode transports heat downward. For the unstable mode the northward and upward heat fluxes combine to exchange a particle of fluid initially at the south just above the interface with a particle at the north that was initially just below the interface. Thus, the net effect is to level the interface and the instability grows by drawing on the potential energy of the sloping interface. It can also be shown that the stable mode tends to increase the slope of the interface, and there is no energy available for such a change. Furthermore, the oscillatory, neutral modes that occur for $K^2 > 2/\lambda_1^2$ have zero northward heat flux.

From (5.193) the values of ϕ_1 and ϕ_2 are

$$\phi_2 = \tfrac{1}{2}(A \sin kx + B \cos kx)e^{\sigma t} \sin ly,$$
$$\phi_1 = \tfrac{1}{2}(A \sin kx - B \cos kx)e^{\sigma t} \sin ly, \quad (5.197)$$

Hence, with $K^2 \ll 2/\lambda_1^2$ the unstable mode $A \simeq B$ yields

$$\phi_1 = \frac{Ae^{\sigma t}}{\sqrt{2}} \sin\left(kx + \frac{\pi}{4}\right) \sin ly,$$
$$\phi_2 = \frac{Be^{\sigma t}}{\sqrt{2}} \sin\left(kx - \frac{\pi}{4}\right) \sin ly \quad (5.198)$$

and we note that the upper-layer disturbance is displaced to the west of (lags) the lower-layer disturbance by $\pi/2$. For $k = l$ the phase lag of the upper layer is about 66°. Observed baroclinic instabilities in the atmosphere are characterized by such a phase shift. The phases are reversed for the stable mode.

At this point it is a simple matter to take into account the effect of variable f by noting

$$\frac{\partial}{\partial y}\left(\frac{f}{H}\right) = \frac{\beta}{H} - \frac{fH_y}{H^2}, \quad (5.199)$$

so that the term βv must be added to each layer in (5.180). The result is to add $\beta(\phi_1 + \phi_2)_x$ to (5.185a) and $\beta(\phi_1 - \phi_2)_x$ to (5.185b) and the system (5.190) becomes

$$(\phi_1 + \phi_2)_t + \frac{\beta}{K^2}(\phi_1 + \phi_2)_x + U(\phi_1 - \phi_2)_x = 0,$$

$$\left(K^2 + \frac{2}{\lambda_1^2}\right)(\phi_1 - \phi_2)_t - \beta(\phi_1 - \phi_2)_x \quad (5.200)$$

$$+ U\left(K^2 - \frac{2}{\lambda_1^2}\right)(\phi_1 + \phi_2)_x = 0.$$

The frequency equation is

$$\omega^2 + \frac{2k\beta(K^2 + 1/\lambda_1^2)}{K^2(K^2 + 2/\lambda_1^2)}\omega + \frac{k^2\beta^2}{K^2(K^2 + 2/\lambda_1^2)}$$

$$- \frac{k^2 U^2}{K^2 + 2/\lambda_1^2}(K^2 - 2/\lambda_1^2) = 0 \quad (5.201)$$

and the condition for instability is changed to

$$\lambda_1^8 K^8 - 4\lambda_1^4 K^4 + \frac{\beta^2 \lambda_1^4}{U^2} < 0 \quad (5.202)$$

or

$$2 - 2\sqrt{1 - \frac{\beta^2 \lambda_1^4}{4U^2}} < \lambda_1^4 K^4 < 2 + 2\sqrt{1 - \frac{\beta^2 \lambda_1^4}{4U^2}}. \quad (5.203)$$

Thus, the wavelength is bounded at the long end as well and the unstable range is cut down. A necessary condition on U is

$$2|U| > \beta \lambda_1^2. \quad (5.204)$$

With $\lambda_1 = 36$ km this gives a velocity difference between the layers of 2.5 cm s^{-1}.

These results were first derived by Phillips (1951). The condition on $|U|$ is a particular example of the more general necessary condition for instability (Charney and Stern, 1962; Pedlosky, 1964a) that the mean potential vorticity gradient (including β) must change sign in the region. If β dominates, Q_y is positive everywhere and the flow is stable.

In two-layer models we note particularly that eastward flow in the upper layer requires $-H_{1y} > 0$ so that the stabilizing β-effect is reenforced in (5.199). Hence, instability can occur only if $-fH_{2y}/H_2$ (which is negative) is large enough to offset β. If the lower layer is very deep, the system is stable. On the other hand, westward flow in the upper layer is destabilizing ($-H_{1y} < 0$) and a shallow upper layer is more conducive to instability since fH_{1y}/H_1 increases with decreasing H_1. Two-layer oceanic models normally have a shallow upper layer so instability sets in first in regions where the surface velocity is to the west. Since geostrophic westward flow is associated with a thermocline that deepens to the north, the instability serves to transport heat to the south when the thermocline is flattened. Thus, the primary instability tends to resist the effect of the thermal driving.

A bottom slope also affects the stability. We can make use of the general condition in the present case to observe that if the bottom slope is in the same direction as, and exceeds, the slope of the interface it stabilizes the flow because Q_{2y} has the same sign as Q_{1y}. More generally, when β is included, a stable flow can be destabilized (and vice versa) by changing the sign of U and/or the slope.

These results can be generalized to flows with arbitrary, stable stratification (Charney and Stern, 1962; Pedlosky, 1964a). Necessary conditions for instability are that (1) Q_y change sign somewhere between the surface and the bottom, (2) $Q_y U_z < 0$ at $z = 0$, or (3) $Q_y(U_z + N^2 H_y/f) > 0$ at $z = -H$.

Bretherton (1966a) compared the two-layer model with the continuously stratified (constant-N) model of

Eady (1949) to explain the similarity in the results of two systems that have obvious differences. In the two-layer case the vertical difference in Q_y is necessary for instability whereas in the Eady model Q is constant. The latter system becomes unstable because the sloping isopycnals intersect the bottom and introduce destabilizing boundary effects. When the boundary conditions are incorporated into the potential vorticity of the fluid as infinitesimally thin sheets adjacent to the boundaries, their effects can be interpreted as internal differences in Q_y and the similarity between the two models is evident. The two models are opposite extremes in the simulation of the variable stratification of the ocean. In the two-layer model the stratification is squeezed into a discontinuous layer and boundary effects are distributed uniformly throughout each layer. The constant-N model smears out vertical differences in the stratification, thereby exaggerating the stratification at the boundaries, but boundary effects appear properly as boundary conditions.

The question of critical layer instability, where the phase speed of the disturbance is equal but opposite to the convective velocity, is addressed by Bretherton (1966b), who showed that the existence of a critical layer in a fluid makes it highly unlikely that the flow will be stable. If Q_y vanishes or if certain other conditions are satisfied at or near the critical layer, stability is possible, but generally instability will occur. Bretherton used his analysis to show that the addition of β to Eady's problem (Green, 1960) yields $Q_y = \beta/H$, i.e., Q_y does not vanish, and since a critical layer exists, the flow is unstable. This destabilizing effect of β is in direct contrast to the result that we found for the two-layer case. For a two-layer system, however, there is no critical layer (it is buried in the discontinuity of the interface), and the corresponding mode is stable. The instability that does occur with two layers is due to the change of sign of Q_y between the two layers. We have already pointed out that the equivalent effect is due to boundary effects in the Eady problem without β.

Stommel (1965) suggested that the enormous amount of potential energy present in the stratification of the ocean and associated with the mean circulation may be a possible source of energy for instabilities. The obvious mechanism to tap that energy is baroclinic instability. Gill, Green, and Simmons (1974) have explored the stability of several combinations of vertical profiles of density and of vertical shear of horizontal velocity. Using exponential approximations to mean observed vertical profiles, they concluded that westward flows[9] (isopycnals sloping up toward the equator) are the most likely unstable ones, and they analyzed one such case with a velocity maximum at the surface and two with the maximum at 100 m depth. For one of the latter the slope of isopycnals reverses (up toward the pole) at 100 m depth. The most unstable modes have wavenumbers close to the reciprocal of the internal radius of deformation (wavelength 190 km) and e-folding time of 80 days with velocities significant in the upper kilometer or so. When U has a subsurface maximum where the isopycnals reverse slope, the growth rate decreases, suggesting that seasonal changes that can affect the isopycnal slope can also affect the stability. The maximum growth rate occurs for profiles with monotonic U_z. Secondary instabilities with minimum amplitude at 1000 m depth, smaller wavelength, and smaller growth rate also occur. Since the amplitudes of these secondary instabilities are large at depth, they are strongly affected by topography. The conversion of available potential energy to eddy energy for all of the primary instabilities is confined to the upper half-kilometer.

Because linear instability theory yields no information about absolute amplitudes, Gill, Green, and Simmons (1974) assumed that the eddies draw energy from the mean field at the rate at which it is supplied to the mean field by the wind. That sets the amplitudes for the disturbances. By assuming further that the primary and secondary instabilities draw equal amounts of available potential energy, they obtain maximum eddy velocities of 0.14 m s^{-1} and wavelengths of 200 km in the upper kilometer and corresponding values of 0.05 m s^{-1} and 300-500 km in deep water. The latter values are close to the observed but the amplitudes of the upper velocities are on the small side.

The analysis by Gill, Green, and Simmons (1974) supplies qualitative, and even some quantitative, support for the pertinence of linear baroclinic-instability theory to observed eddy motions in the ocean. Linear growth rates have the right magnitude and the scales and distributions of the disturbances are also approximately correct. However, more detailed features and quantitative results require additional considerations.[10] Nonlinear processes must be important, particularly in altering the structure of the mean field, which is assumed known in the linear theory. Friction, too, must have at least a quantitative effect over the lifetime of these flows.

5.11.2 Finite-Amplitude Effects

Phillips (1954) calculated the finite-amplitude effects of the baroclinically unstable modes to determine the lowest-order corrections to the assumed mean properties. He derived the heat fluxes given above, the induced meridional circulation, and the changes in the zonal momentum wrought by nonlinear corrections. The altered fields are in qualitative agreement with observed structures in the atmosphere.

This use of the most rapidly growing eigenfunctions of linear theory to determine nonlinear corrections to the system has been applied and extended by a number of other investigators. In part, the purpose is to enable one to take into account the effects of these smaller-scale processes in larger-scale models by a consistent parameterization. Nongeostrophic processes can also be included to give an idea of the effects on the evolution of the flow.

In an excellent review article on finite-amplitude baroclinic instability Hart (1979a) gives an account of the different approaches to the treatment of the nonlinear system and the reader is referred to Hart's paper for a more detailed discussion and an extensive bibliography. The different treatments summarized by Hart include simple physical effects of the first nonlinear interaction (e.g., Phillips, 1954), weakly nonlinear interactions between the perturbation and the mean field (e.g., Stuart, 1960), a truncated set of normal modes with amplitudes that must be determined by the dynamical system (e.g., Lorenz, 1963a), and a mean-field, single-wave approach (e.g., Herring, 1963). Pedlosky (1975) gives a discussion appropriate to oceanic eddies.

These studies of finite-amplitude effects are important in providing an understanding of the interplay of different parts of the system and especially for determining which processes are pertinent to the behavior of the observed system. For example, observed asymmetries in the structure of the evolved eddies can be traced to nongeostrophic effects [see Hart (1979a) for references], and inclusion of those effects may be necessary to recover the asymmetries, even in strongly nonlinear models. But other features, such as the occlusion observed in fully developed cyclones or eddies and the greater intensity of eddy amplitudes as compared to mean-flow velocities, do not emerge from these local, finite-amplitude treatments, and recourse to more nonlinear (numerical) models is suggested if the results are to be applicable to oceanic flows.

One such treatment, a numerical study of baroclinic instability by Orlanski and Cox (1973), using Bryan's (1969) numerical model of the general circulation, focused on the stability of an intense, confined, stream along a coast. The application is to the Florida Current (Miami to Cape Hatteras). The effects of a bounding coast, bottom topography, nonlinearity, friction, and diffusion are all included in this study, which shows that an evolved eddy field is established 10 days after the current, is initially uniform in the downstream direction, and is randomly perturbed in density and velocity. The disturbance of maximum growth has a scale of about 20 km, which seems to be the deformation radius (not reported but estimated from the published temperature pattern). The growth rate decreases by about a factor of 10 from the initial value when nonlinear effects become important. This seems to be a characteristic effect of nonlinearity; similar results are reported for meteorological studies (Gall, 1976). The initially rapid growth of the disturbances is much faster than one obtains for a current in an unbounded ocean, so there is a strong suggestion that bottom topography and the nearby coast are important factors in determining the growth rate. An energy diagram shows that the flux of energy is from mean to eddy potential energy, then to eddy and finally mean kinetic energy. These transfers are all consistent with baroclinic instability and its evolution.

5.12 Effect of Nonlinearity and Turbulence

Linear theory provides at least a qualitative explanation of some observed dynamic features in the ocean, including westward intensification of boundary currents, the several types of long-period wave motions, and the existence of baroclinically unstable modes. It is also clear, however, that nonlinear processes provide more than merely quantitative corrections. For example, the observed Gulf Stream transport is several times the value predicted by linear theory, and the recirculating gyre associated with the increased transport (see chapter 4) not only involves flows that are baroclinically unstable, but the unstable modes seem to be fully developed as well. The effects of smaller-scale processes cannot generally be parameterized in terms of eddy coefficients, so an analysis of the evolution of the flow requires analysis of strong, interacting features and the use of concepts and relationships developed for turbulent flows.

Because of the difficulties inherent in nonlinear studies, many of the explorations have used numerical models, some of which have required extensive calculations. Rhines (1977) has made a significant effort to construct a comprehensive account of the emerging dynamic picture. Only an outline of that remarkable synthesis can be given here.

Most of the numerical experiments discussed in this section are oriented toward a study of processes. Except for the calculations by Bryan (1969) they were not intended to be simulations of ocean circulation. Therefore, the models are often very idealized (two-layer, rectangular basins with cyclic boundary conditions, etc.). At times the results can only be suggestive of what happens in nature, but they should be of great help in indicating what should be included in more realistic, predictive models.

The linear theories discussed in the earlier sections must be supplemented by a few additional observations. One is that baroclinically unstable motions evolve into fully developed, closed eddies. A description of the latter is not accessible with theories based on extensions from linear treatments and one must resort to numerical experiments to obtain the evolved fields. A second observation is that mature eddies are often (but not always) in an occluded state in which the motion has penetrated vertically to form a relatively barotropic gyre. In horizontal wavenumber space the window for vertical penetration is at or near the internal radius of deformation. This tendency toward barotropy in a Boussinesq fluid means that certain features of the system can be described as if the flow were two-dimensional, a significant simplification for both analytic and numerical studies.

5.12.1 Nonlinear Effects with Constant Depth

Batchelor (1953a) points out that for nonlinear, two-dimensional flow in an inviscid fluid the time rates of change of integrated energy and enstrophy (squared relative vorticity) vanish:

$$\frac{\partial}{\partial t}\int_0^\infty E\,dk = 0, \qquad \frac{\partial}{\partial t}\int_0^\infty k^2 E\,dk = 0, \qquad (5.205)$$

where E is the kinetic energy in wavenumber space. (For a fluid with small viscosity the same relations hold approximately for times short compared to the transfer time to high wavenumbers where viscous dissipation is significant.) If one assumes that an initially narrow spectrum spreads in time about its mean wavenumber,

$$\frac{\partial}{\partial t}\int_0^\infty (k-k_1)^2 E\,dk > 0, \qquad (5.206)$$

where $k_1 = \int kE\,dk/\int E\,dk$, then use of (5.205) yields

$$\frac{\partial k_1}{\partial t} < 0. \qquad (5.207)$$

Thus, the mean wave number k_1 decreases. Furthermore, since $\int k^2 E\,dk$ is conserved, a transfer of energy from, say, k_0 to $2k_0$ must be balanced by a transfer of four times that energy to $k_0/2$ or an equivalent change. In order to achieve this red cascade of energy, fluid elements with vorticity of like sign must coalesce into larger eddies. Hence, a small number of strong, isolated vortices will emerge and they will continue to tend to coalesce. Nonlinearity will generate small scales as well, and since enstrophy involves a weighting by k^2, enstrophy will peak at smaller scales (see chapter 18).

The red cascade relies on nonlinear processes to provide the transfer to different scales. Furthermore, the turbulent field must be spatially homogeneous and isotropic. A local concentration of turbulent eddies will spread spatially as well as in wavenumber and the conditions for the red cascade can be violated. For example, an initially isolated cluster of eddies will spread to the point where nonlinear transfers are so weak that the system no longer acts as turbulence and the evolution to larger scales stops.

The relations (5.205) apply to an unbounded fluid on the β-plane as well. Yet in the vorticity balance the planetary vorticity term βv becomes comparable to the nonlinear term $\mathbf{v}\cdot\nabla\zeta$ when the Rossby number is small enough. If U is the rms turbulent velocity, and if motions cluster around wave number k_1, the pertinent measure of the Rossby number is $2Uk_1^2/\beta$. As the red cascade proceeds, k_1 becomes smaller and $2Uk_1^2/\beta$ drops toward unity. Energy and vorticity can then be radiated away by Rossby waves, decreasing the intensity of the motion still further. Thus, the red cascade is halted as turbulence gives rise to waves. It is interesting to note that the larger scales generated by the turbulence are the ones that radiate away most quickly. Hence, the end state is toward a pattern of zonal flows with scale $k_\beta^{-1} \sim (\beta/2U)^{-1/2}$ (Rossby number ~1).

Rhines (1975) ran a numerical calculation for a basin with periodic boundary conditions in which an initially turbulent, barotropic fluid with energy in a narrow band of wavenumbers evolves toward the red cascade with and without β. His results are illustrated in figure 5.15, where contours of ψ as a function of longitude x are shown as time progresses. The relatively small initial scales expand in time in both cases. With β, propagation to the west occurs shortly after the start of the experiment and the phase speed increases until k approaches k_β toward the end of the run.

A schematic illustration of the results is shown in figure 5.16B. The initially turbulent field with energy in a narrow band of wavenumbers starts at point a with frequency kU and evolves toward lower wavenumbers, hence lower frequencies. When it reaches the frequency corresponding to barotropic Rossby waves, evolution of the turbulent field is halted and gives way to an evolving Rossby wave field with scale $(\beta/2U)^{-1/2}$.

Another obstacle to the red cascade is afforded by a western meridional boundary, where, as we have seen, reflection of Rossby waves favors the generation and accumulation of smaller scales. These modes tend to line up meridionally. Nonlinear flows also concentrate energy and enstrophy near the western boundary.

The qualitative effect of stratification can be introduced via a quasi-geostrophic two-layer model on the β-plane. Rhines carried out a series of initial-value calculations varying both the intensity and the scale of the initial turbulence. In these experiments the internal radius of deformation $\lambda_i = (g'H_1H_2/Hf^2)^{1/2}$ is a critical parameter. When the initial, turbulent scale L is small compared to λ_i, the two layers are effectively decoupled, as they are for linear waves. If the intensity of the turbulent flow is not large, the red cascade will develop in each layer until the frequency and scale of the barotropic Rossby wave for that layer is reached. This case is shown in figure 5.16A by the arrow from point b to a point on the wave dispersion curve where $\lambda_i k > 1$.

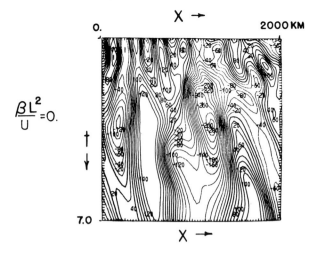
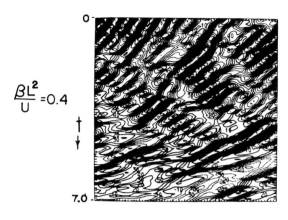

Figure 5.15A Contour diagrams for ψ as function of longitude and time for Rhines's numerical barotropic experiment with initially turbulent field without β. Cascade to large scales occurs as time increases.

Figure 5.15B Same experiment with β. Westward propagation increases with time as cascade to larger scales occurs. Cascade is halted as $k \to \sqrt{\beta/2U}$.

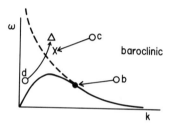

Figure 5.16A Frequency–wavenumber diagram for baroclinic modes. Solid curve shows linear baroclinic Rossby wave, dashed curve linear barotropic Rossby wave. For turbulent motions frequency is rms velocity U times k. Initial field of small-scale motions evolves by nonlinear interactions toward larger scales. If turbulent intensity is small (from point b), frequency arrives at linear wave values before deformation scale is reached and each layer experiences "barotropic" westward propagation. If intensity is larger (from point c), evolution to large scales proceeds until deformation scale is reached (point $\times$); then layers couple and barotropic development in figure 5.16B takes place. If initial turbulence has large scale (from point d), smaller scales are generated by nonlinear transfer and by baroclinic instability and system evolves to point Δ, where barotropic mode is generated and behavior shifts to figure 5.16B from point Δ.

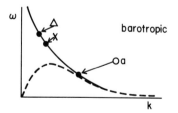

Figure 5.16B From point a nonlinear interactions of initially small-scale turbulent field produce red cascade until frequency reaches linear barotropic value (solid curve), when westward propagation disperses the energy to reduce nonlinear interactions. Same development from points $\times$ and Δ after barotropic mode is generated from initially baroclinic turbulent field.

If the intensity of the turbulence (in either layer) is increased to point c in figure 5.16A, the red cascade will take place again, but now the scale of the internal radius of deformation is reached (at point $\times$) before the wave steepness $U/\beta L^2$ has dropped to unity. As k decreases toward λ_i^{-1} the two layers will begin to interact with each other. The intensity of the flow allows a full evolution of the eddying motion to the occluded state described earlier and a barotropic response sets in. The latter is shown by the point $\times$ in figure 5.16B. From this point the system behaves like the barotropic system described earlier and the barotropic red cascade proceeds until the dispersion curve for the barotropic Rossby wave is reached. These results can be extended to the three-dimensional system.

If the initial, turbulent eddies have scales larger than λ_i, the flow can be treated locally as a large-scale basic flow field that becomes baroclinically unstable to disturbances with scale λ_i that derive their energy from the large-scale potential-energy field. Hence, the system will quickly develop *smaller* scales of motion. In this case the release of potential energy is responsible for the "blue" cascade. Once instability sets in at $k \simeq \lambda_i^{-1}$, the layers will couple, a barotropic response will be generated, and the barotropic red cascade will occur. The generation of smaller scales is shown in figure 5.16A from point d to Δ, where the layers couple; the barotropic red cascade from that point is shown in figure 5.16B. In Rhines's numerical experiment for this case the upper layer is shallow and the instability sets in where the large-scale eddy has a westward component, consistent with our observation that in a shallow upper layer westward flow is more unstable than eastward flow. At a later stage the eastward flow also becomes unstable. This particular case leads more quickly to the banded zonal flow described earlier because the baroclinic instability enhances the process by destabilizing the initially large-scale flow (particularly the large north–south scale) and by generating eddies of size λ_i, followed by a quick vertical coupling and then occlusion. If the red cascade leads to a zonal flow at an early stage, some of the potential energy in the initial, large-scale field remains to support a baroclinic zonal flow, i.e., one with a vertical shear. This is possible for $\Delta U/\lambda_i^2 \beta < 1$ where ΔU is the velocity difference between the two layers.

Additional initial-value, run-down experiments were carried out by Rhines to study special features. One is the stability of a meridional baroclinic flow (essentially a westward-traveling Rossby wave in the upper layer). Eddies with scales slightly exceeding λ_i form and grow to larger amplitude, interacting laterally as they lock together vertically. The barotropic red cascade then sets in and a barotropic Rossby wave field again radiates energy away, leaving a zonal flow with scale k_β^{-1} as in the purely barotropic case discussed earlier.[11] Essentially all of the initial potential energy is released in the evolution of this flow.

Another experiment explores the instability of an eastward jet (an open-ocean "Gulf Stream") with $\beta = 0$. The flow is initially confined to the upper layer with a transverse scale of about $2\lambda_i$. A superposed weak disturbance develops rapidly via baroclinic instability and eddies are generated in both layers. The evolution of the flow involves an interaction between the (initially intense) mean flow and the growing waves as well as wave–wave interactions within each of the layers. The reader is referred to Rhines's paper for the details of the developing system, but the final result is mentioned here because of its pertinence to both observations and theories of the Gulf Stream. The flow that emerges has a strong barotropic component with a much increased "Gulf Stream transport," but there is little evidence in the flow field of an identifiable Gulf Stream. The density field (expressed as the interface height), however, preserves much more of the character of the observed Gulf Stream, with a meandering structure, detached rings, and a crowding of isopycnals along the axis. Thus, the dynamic picture involves two dissimilar modes, one barotropic, the other baroclinic, both of which are essential for a satisfactory description of the pertinent physical processes.

5.12.2 Effects of Topography in Two-Layer Flow

The conservation of potential vorticity requires a balance between $\partial \zeta/\partial t$, $\mathbf{v} \cdot \nabla \zeta$, df/dt, and $-f\,dh/h\,dt$. Nondimensional measures of these terms are a frequency ω/f, the Rossby number $Ro = Uk/f$, the β-effect β/fk, and the topographic effect $\delta = \Delta H\, k_T/Hk$. Here k and k_T are the horizontal scales of the flow and of the topography respectively, ΔH the amplitude of the topography, and H the depth.

For transient flows the local change ω/f will be balanced by a combination of the other parameters. If β/fk is larger than Ro or δ (weak flow over small topography), the system will respond with Rossby waves. In this case, if $k_T \gg k$ (small-scale topography), the Rossby waves will be weakly scattered by topography (Thompson, 1975). For large-scale topography $(k_T \ll k)$ the response will be Rossby waves modified by topography as described in sections 5.9 and 5.10. For intermediate scales $(k_T \sim k)$ the oscillations are irregular and will reflect the geometrical complexity of the topography (Rhines and Bretherton, 1974).

Nonlinear flows $(Ro > \delta, \beta/fk)$ will exhibit the spectral broadening (red cascade) described above. Large topography $(\delta > Ro, \beta/fk)$ of small horizontal scale $(k_T \gg k)$ leads to generation of small scales through topographic scattering and refraction. If the topographic scale is large $(k_T \ll k)$, a new "westward" direction is defined to the left of upslope.

With stratification the interplay of the three effects can be intricate. For example, if the flow is dominated by nonlinearity near the surface, spectral broadening will occur, and if the deep flow is weak, scattering will lead to small scales and the pseudowestward direction can generate fast baroclinic waves.

A series of initial-value numerical experiments by Rhines (1977) made use of random topography of rms amplitude $\delta = 0.053$ generated with a spectrum of scalar wavenumber $k^{-1.5}$ with scales $k \leq 8$ over a periodic domain of width 2000 km. The internal deformation radius λ_i was 40 km, so the topography cutoff scale was at the deformation radius.

A first experiment started off with turbulent eddies of scale $k \sim \lambda_i^{-1}$ and $Ro > \beta/fk$. The initial flow was confined to the upper layer. The deformation scale of the flow leads to coupling of the two layers and deep motion is generated. However, as soon as the deep water is set into motion, it interacts with the irregular topography which determines the direction of flow. Thus, topography detunes the layers and barotropic development is impeded.

When the initial flow includes deep motion as well so that $Ro > \delta$ there, the inertia of the flow eventually overcomes the relatively weak effect of the topography allowing coupling of the two layers to proceed. When the flow has evolved to expanded scales, the barotropic mode that emerges responds to the larger scales of the topography and the flow follows f/h contours. During the time that vertical adjustment is taking place, a more complicated, quasi-local relation among vertical structure, energy level (Ro) and topography (δ) may exist.

In both of these cases the initial flow has a small scale so that it contains only a limited amount of potential energy. If the initial eddies have a larger scale, there is more potential energy available to the disturbances and faster growth via baroclinic instability is possible. When there is no topography, deformation scale disturbances will grow, the two layers will couple and occlude, and the barotropic red cascade will then take place. With rough topography, deformation-scale eddies are still generated and the growth is enhanced by the topography. This leads to an even larger release of the potential energy of the initial, larger-scale flow, though the locations of these enhanced deformation-scale eddies are determined by the topography so the two layers are detuned. To offset this detuning, continuous spectral broadening by topography takes place in deep water and the larger scales so generated can couple with the larger scales of the surface eddies. Hence, there is a tendency toward barotropy once again. If the flow is sufficiently nonlinear, topography is not so important and the development is more like that of a flat-bottom ocean.

What emerges from these considerations is that the response of the ocean depends in a complicated manner on the intensity and scale of the energy input and on the topographic structure in the region of response. However, the fact that the ratio of barotropic to baroclinic energy is a monotonic function of the ratio Ro/δ, as suggested by the calculations, seems to be a zero-order description of the observations. Rhines provides a much more detailed discussion of these issues as well as other considerations, such as the propagation of energy from a source area to distant regions with no input and the trapping of certain modes by topography. Several other investigators, including Bretherton and Karweit (1975) and Bretherton and Haidvogel (1976), have made important contributions to the study of the effects of different physical features on mid-ocean eddies. Holloway (1978), Salmon, Holloway, and Hendershott (1976), and Herring (1977) have applied closure modeling to the problem to enrich the story (see chapters 11 and 18).

5.12.3 Closed Basin Circulation

It is evident from the results reported above that the character of the circulation can undergo qualitative local changes because of nonlinear interactions and effects of bottom topography. When land boundaries are added, the circulation can be very different from that obtained by linear theory or even by nonlinear steady theory. Significant differences emerge if nonlinear, transient effects can develop fully. The latter are not easily treated analytically and most of the useful results have emerged from numerical experiments.

Some of the numerical studies (e.g., Robinson, Harrison, Mintz, and Semtner, 1977) use the primitive equations rather than the quasi-geostrophic set that we have been discussing. Although there are important nongeostrophic effects, many of the important features emerge from a study of the quasi-geostrophic system.

For a closed, forced system it is necessary to add dissipation. This can be done by introducing the terms $\partial \tau^x/\partial z + A \nabla^2 u$ and $\partial \tau^y/\partial z + A \nabla^2 v$ to the right-hand sides of (5.32) and (5.33), respectively, and $K_v \partial^2 \rho_0/\partial z^2 + K_H \nabla^2 \rho_0$ to the right-hand side of (5.44). The eddy coefficients A, K_v, and K_H are taken to be constant.[12] The vertical stress terms $\partial \tau^x/\partial z$ and $\partial \tau^y/\partial z$ contribute only near top and bottom boundaries, where they are normally evaluated to take into account the effect of wind stress and bottom drag via Ekman-layer processes.

Equation (5.54) is changed to

$$\frac{dq}{dt} = \frac{\partial}{\partial z}(\hat{\mathbf{k}} \cdot \nabla \times \boldsymbol{\tau}) + A\nabla^4\psi + K_H \nabla^2 \frac{\partial}{\partial z}\left(\frac{f_0^2}{N^2}\psi_z\right)$$
$$+ K_v \frac{\partial}{\partial z}\left(\frac{f_0^2}{N^2}\psi_{zzz}\right), \qquad (5.208)$$

where $q = \nabla^2\psi + f + \partial/\partial z(f_0^2 \psi_z/N^2)$ is again the potential vorticity. If we write the variables in terms of an ensemble mean (overbar) and a perturbation (primed), we can write the equation for the mean potential vorticity as

$$\frac{\partial \bar{q}}{\partial t} + \bar{\mathbf{v}} \cdot \nabla \bar{q} + \overline{\mathbf{v}' \cdot \nabla q'}$$
$$= \frac{\partial}{\partial z}(\hat{\mathbf{k}} \cdot \nabla \times \boldsymbol{\tau}) + A\nabla^4\bar{\psi} + K_H \nabla^2 \frac{\partial}{\partial z}\left(\frac{f_0^2}{N^2}\bar{\psi}_z\right)$$
$$+ K_v \frac{\partial}{\partial z}\left(\frac{f_0^2}{N^2}\bar{\psi}_{zzz}\right). \qquad (5.209)$$

For a two-layer model the equivalent system before averaging is

$$\frac{d}{dt}\left[\nabla^2\psi_1 + f - f_0 \ln\left(\frac{h_1}{H_1}\right)\right] = \frac{\hat{\mathbf{k}} \cdot \nabla \times \boldsymbol{\tau}_0}{h_1} + A\nabla^4\psi_1,$$
$$\frac{d}{dt}\left[\nabla^2\psi_2 + f - f_0 \ln\left(\frac{h_2}{H_2}\right)\right] = -K\frac{\nabla^2\psi_2}{h_2} + A\nabla^4\psi_2. \qquad (5.210)$$

Here, we have integrated vertically over each layer. The effect of topography is in the term η_3 (in h_2). Bottom friction is written explicitly as $-K\zeta_2 = -K\nabla^2\psi_2$ and $\hat{\mathbf{k}} \cdot \nabla \times \boldsymbol{\tau}_0$ is the wind-stress curl. The convective derivative for each layer involves the horizontal velocity components for that layer.

In (5.210) the thicknesses h_1 and h_2 of the two layers include the variations of the free surface, the interface, and the topography. In general, the h_i should be allowed to vanish, if need be, but no calculations have been made with variable h_i because following a material surface is very difficult to do numerically. The usual procedure is to set the h_i at their mean values in the coefficients. That means that surfacing of the thermocline is not permitted, so the separation mechanism of section 5.7 is not included. This is a serious omission for the complete problem. The numerical calculations have been oriented toward determining the effects of eddies and topography on the vertical transfer of vorticity for the generation of abyssal circulation. The results are applicable so long as h_1 does not vanish.

Assuming a rigid lid at the top and a mean depth for the h_i in the coefficients, we can write (5.210) as

$$\frac{dq_1}{dt} = \frac{\hat{\mathbf{k}} \cdot \nabla \times \boldsymbol{\tau}_0}{H_1} + A \nabla^4 \psi_1,$$

$$\frac{dq_2}{dt} = -K \frac{\nabla^2 \psi_2}{H_2} + A \nabla^4 \psi_2, \quad (5.211)$$

where $q_1 = \nabla^2 \psi_1 + f + f_0 \eta_2/H_1$, $q_2 = \nabla^2 \psi_2 + f = -f_0(\eta_2 - \eta_3)/H_2$, where η_3 describes the topography. Again writing mean and fluctuation variables, we obtain for the mean equations

$$\frac{\partial \bar{q}_1}{\partial t} + \bar{\mathbf{v}}_1 \cdot \nabla \bar{q}_1 + \overline{\mathbf{v}_1' \cdot \nabla q_1'} = \frac{\hat{\mathbf{k}} \cdot \nabla \times \boldsymbol{\tau}_0}{H_1} + A \nabla^4 \bar{\psi}_1,$$

$$\frac{\partial \bar{q}_2}{\partial t} + \bar{\mathbf{v}}_2 \cdot \nabla \bar{q}_2 + \overline{\mathbf{v}_2' \cdot \nabla q_2'} = -\frac{K \nabla^2 \bar{\psi}_2}{H_2} + A \nabla^4 \bar{\psi}_2, \quad (5.212)$$

where

$$\bar{q}_1 = \nabla^2 \bar{\psi}_1 + f + \frac{f_0}{H_1} \bar{\eta}_2, \quad q_1' = \nabla^2 \psi_1' + f_0 \eta_2'/H_1,$$

$$\bar{q}_2 = \nabla^2 \bar{\psi}_2 + f - \frac{f_0(\bar{\eta}_2 - \eta_3)}{H_2}, \quad q_2' = \nabla^2 \psi_2' - f_0 \eta_2'/H_2. \quad (5.213)$$

The mean properties in each layer depend on the correlations between fluctuating velocity and potential vorticity in addition to the remaining mean variables. We mention results from two numerical experiments.

The first is a numerical calculation by Holland (1973) of the continuous system [(5.209) plus the remaining equations of the system] for a rectangular basin 45° wide and extending from the equator (where symmetry is assumed) to about 65°N. A three-gyre wind system drives the circulation in a basin with topography shown in figure 5.17. Transport streamline patterns for a baroclinic ocean without topography, for a barotropic ocean with topography, and for a baroclinic ocean with topography are illustrated in figure 5.17. The first of the figures shows a structure not terribly different from the one obtained from linear theory with oceanic gyres determined by the wind pattern. The second exhibits the strong effect of topography on homogeneous water columns. The third contains response due to the combined effects of eddies and topography. The pattern is very different from the other two but the result of greatest significance is the enhanced transport generated by the eddies in conjunction with topography. The transport is increased by more than a factor of two over the case without topography.[13]

A more recent numerical experiment, analyzed by Holland and Rhines (1980), is based on the two-layer system (5.212) with a sinusoidal, cyclonic, wind stress in the north half-basin of amplitude 1 dyn cm^{-2} and an anticyclonic wind stress in the south. The basin is small, extending 1000 km eastward and 2000 km northward. The depth of the upper layer is 1 km, that of the lower 4 km, the bottom is flat, and the density difference is $\Delta \rho = 0.002 \rho_2$. Instead of the Navier–Stokes form for lateral friction, a biharmonic form is used so that ∇^4 is replaced by $-\nabla^6$ in (5.212). At the sides normal flow and tangential stress vanish.

The experiment was run until it approached a statistically steady state, and time averages of the fields were then determined. For a statistically steady state equations (5.212) and (5.213) can be rewritten as

$$\bar{\mathbf{v}}_1 \cdot \nabla \bar{q}_1 + \overline{\mathbf{v}_1' \cdot \nabla \nabla^2 \psi_1'} + \frac{f_0}{H_1} \overline{\mathbf{v}_1' \cdot \nabla \eta_2'} + A \nabla^6 \bar{\psi}_1 - \frac{\hat{\mathbf{k}} \cdot \nabla \times \boldsymbol{\tau}_0}{H_1}$$
$$= 0. \quad (5.214)$$

$$\bar{\mathbf{v}}_2 \cdot \nabla \bar{q}_2 + \overline{\mathbf{v}_2' \cdot \nabla \nabla^2 \psi_2'} - \frac{f_0}{H_2} \overline{\mathbf{v}_2' \cdot \nabla \eta_2'} + A \nabla^6 \bar{\psi}_2 + \frac{K \nabla^2 \bar{\psi}_2}{H_2}$$
$$= 0,$$

Statistically steady streamlines for the two layers are shown in figure 5.18. The upper layer consists of two gyres, one anticyclonic, one cyclonic, with their centers displaced toward mid-latitude as is characteristic of circulations with strong inertial effects. The smaller gyres C and D contain the nonlinear recirculation. The upper-layer pattern is not so different from the one for a homogeneous ocean with strong inertial effects (figure 5.6). The lower layer, on the other hand, exhibits entirely different behavior. Small intense gyres exist under, and have flows parallel to, the small-surface gyres C and D. Counterrotating gyres lie to the north and south of the center two.

Now suppose that equations (5.214) are integrated over the depth of each layer and over an area bounded by the streamlines of the respective gyres. Then the mean convective derivatives in (5.214) will vanish [each of them can be rewritten in divergent form $\nabla \cdot (\mathbf{v} q)$ and vanishes on horizontal integration because the boundary is a streamline]. The remaining terms will then balance and we have

$$H_1 \overline{\mathbf{v}_1' \cdot \nabla \nabla^2 \psi_1'} + f_0 \overline{\mathbf{v}_1' \cdot \nabla \eta_2'} + \overline{H_1 A \nabla^6 \bar{\psi}_1} - \hat{\mathbf{k}} \cdot \nabla \times \boldsymbol{\tau}_0 = 0$$
$$U_1 \quad + \quad U_2 \quad + \quad U_3 \quad + \quad U_4 \quad = 0,$$
$$(5.215)$$
$$H_2 \overline{\mathbf{v}_2' \cdot \nabla \nabla^2 \psi_2'} - \overline{f_0 \mathbf{v}_2' \cdot \nabla \eta_2'} + \overline{H_2 A \nabla^6 \bar{\psi}_2} + \overline{K \nabla^2 \bar{\psi}_2} = 0$$
$$L_1 \quad + \quad L_2 \quad + \quad L_3 \quad + \quad L_4 \quad = 0,$$

where the subscripted capital letters (U for upper layer, L for lower) identify the different terms. For the upper layer from left to right these vorticity balances are due

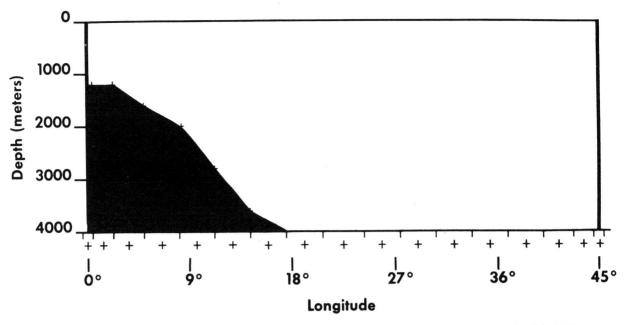

Figure 5.17A Topography used by Holland (1973) in ocean circulation calculation.

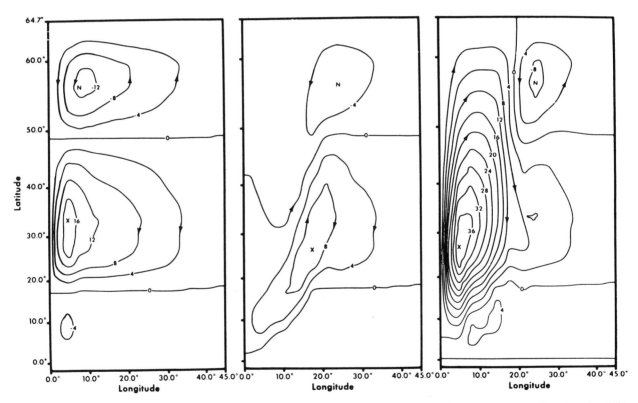

Figure 5.17B Horizontal transport streamfunction for (left) baroclinic ocean with no topography, (middle) barotropic ocean with topography, (right) baroclinic ocean with topography.

to Reynolds stress, interfacial stress, lateral friction, and wind stress, respectively. For the lower layer the first three are the same and the last is due to bottom friction.

The statistically steady streamlines fall between grid points in the numerical experiment, so the different quantities can be evaluated only approximately. Furthermore, the mean vorticities showed time dependence even for the long times of the averages. Hence, the numerical results can provide only an approximation to (5.215).

Holland and Rhines evaluated the integrals for each gyre. Their convention uses a positive sign for the integral if it serves to drive the vortex and a negative value if it opposes the vorticity. The results are shown in table 5.1. We try here to interpret the results in terms of the processes described in earlier sections, though such balances tell only part of the story.

In the upper layer the large gyres, A and B, are driven primarily by the wind. Resistance is effected via Reynolds stresses though there is a small retardation by the interfacial stresses. The assumed biharmonic lateral "friction" helps to drive the flow, but the magnitude of this driving is smaller than the errors due to the approximate evaluation of the integrals, so even the sign is not reliable.

The smaller gyres, C and D, are driven more by Reynolds stresses than by the winds (which is not surprising since these gyres are inertially controlled and have their centers in the wrong place for directly wind-driven gyres). The largest single contribution for these smaller gyres comes from the retarding effect of interfacial stress. The interacting fluctuations of velocity and interface height (temperature) in gyre C depress the interface, thus tending to stretch vortex lines and weaken the anticyclonic vortex. In the cyclonic gyre D the interface is raised by $v'_1 \cdot \nabla \eta'_2$ and the vortex lines are compressed, thus generating anticyclonic vorticity. Hence, in both small gyres the mean of the fluctuating interactions serves to weaken the prevailing vorticity. The approximation errors for these smaller gyres are smaller but still annoying.

The difference between the integral values for the small and large gyres in the upper layer gives the corresponding values for the region between the small and large gyre in each basin—or it would if there were no error. It is especially important to note that U_2 for the small gyres is larger than the value for the large gyres. In other words, in the regions between the streamlines in each half-basin the interfacial stress term *drives* the gyre (against the retarding effect of Reynolds stresses). This serves to raise the thermocline in the south (between C and A) and lower it in the north (between D and B).

The lower layer, which would have no average flow in a real steady state or for linear transient motions, is

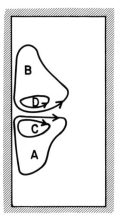

Figure 5.18A Circulation gyres generated in top layer of ocean model with cyclonic wind stress in north, equal-strength anticyclonic wind stress in south. Gyres C and D contain recirculating water.

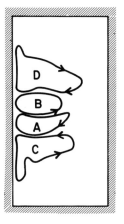

Figure 5.18B Lower-layer gyres near mid-latitude have same sign as those of upper layer. Other gyres opposite.

Table 5.1 Values of U_i and L_i in Equation (5.215): Units m³ s⁻²

Upper layer	U_1	U_2	U_3	U_4	Imbalance[a]
Gyre A	−61.2	−17.6	5.1	53.9	−19.8
Gyre B	−49.2	−10.2	5.7	57.9	5.2
Gyre C	11.0	−23.8	−0.3	7.9	−5.2
Gyre D	11.6	−21.2	0.4	3.3	−5.9

Lower layer	L_1	L_2	L_3	L_4	Imbalance[a]
Gyre A	−19.9	44.6	−0.7	−24.1	−0.1
Gyre B	−18.8	42.8	−0.6	−22.3	1.1
Gyre C	−11.4	44.5	−7.3	−25.0	0.8
Gyre D	−9.2	40.5	−6.5	−22.0	2.8

a. Imbalances are due to residual time dependence and approximation errors for evaluating integrals.

driven by the interfacial stresses (table 5.1). Thus, the depression of the interface under upper-level gyre C causes vortex shrinking in the lower layer and generates the anticyclonic gyre there. Similarly, the elevation of the interface under upper-level gyre D causes vortex stretching in the lower layer and generates the cyclonic gyre. We noted above that the interfacial stresses change sign in the region between the small and large gyre of each half-basin, so the same mechanism leads to vortices of opposite sign, i.e., in the lower layer D is opposed to B and C is opposed to A. These results are all consistent with the simple idea of vortex generation by vertical divergence—at least, in the mean. Local balances could well be different, especially if there are large horizontal variations in the different quantities, as there undoubtedly are.

The vertically integrated transport of the northward boundary current near mid-latitude exceeds the Sverdrup transport by a factor of three. Thus the effect of eddies and the corotating deep gyre contribute greatly to the transport. The transport of the counterrotating deep gyres can overwhelm the value due to the surface gyres above them.

An analysis of the results of the numerical experiment (Holland, 1978) indicates that there is no region where the Sverdrup balance is locally valid for the vertically integrated flow. However, the nature of eddy effects on the mean flow is such that spatial averaging tends to reduce their importance. Hence, Sverdrup balance may apply approximately to a spatially averaged region even when it does not apply at a point. Furthermore, the east–west scale of the basin is small (1000 km), and the recirculating gyres extend across almost the entire basin. In a wider basin the recirculating gyre is probably not much larger and a region of local Sverdrup balance can exist.

An interesting observation about this experiment is that if one calculates the northward transport from hydrographic data (the interface height) across the boundary current, assuming zero deep flow, one obtains a transport about 50% larger than the Sverdrup transport and about half of the actual transport in the two layers. Thus, the deeper thermocline gives a correction in the proper sense.

The Eulerian mean circulation in these models is driven by a combination of wind-stress curl and the divergence of the eddy flux of potential vorticity. In some circumstances this eddy effect can be rewritten in terms of an eddy diffusion coefficient based on Lagrangian dispersion of fluid particles. Taylor (1915) first proposed such an approach for nondissipative systems. Bretherton (1966b) made use of the argument and pursued the issue in subsequent papers. Rhines (1977) (see also Rhines and Holland, 1979) has been developing the theory for more general use. A parameterization of the effect of oceanic eddy processes is extremely important, but it is too early to try to summarize the theory.

Holland's (1978) paper describing the numerical solutions gives more details. An important point is that the free jet (Gulf Stream) is barotropically unstable in this case and the baroclinic instability of the westward return flows is weaker. The calculation is intended to study processes, however, rather than the observed phenomenon, so the relative strengths of the instabilities of the model are not necessarily a prediction of what happens in the ocean.

A linear stability analysis of the mean and instantaneous flows in the experiment was carried out by Haidvogel and Holland (1978). Many of the unstable features that emerge from the full experiment could be correctly accounted for by linear theory using the flows of the experiment. In some cases, however, use of the instantaneous velocity fields led to better results than did the use of time-averaged or mean flows. Since the time scale of the former was sometimes shorter than the time scale of the deduced unstable motions, the significance of the results is not clear. One point made by Haidvogel and Holland is that the Reynolds stresses deduced from linear theory have the wrong sign. Hence, finite-amplitude effects must change the interactions in the developed flow. We have already remarked on the importance of large amplitudes in the occlusion phenomenon.

Semtner and Holland (1978) compare the results of the quasi-geostrophic model with the results obtained from the more complete (and much more expensive) runs using a five-level primitive-equation model. The two-layer quasi-geostrophic model can simulate many of the features of the latter, particularly if the parameters, such as the depth ratio, are adjusted for optimum fit. Gill et al. (1974) and Flierl (1978) have also made such comparative studies but for more restricted physical models.

Large-scale numerical models to determine climatological oceanic behavior have been carried out by the NOAA group at Princeton [see Bryan, Manabe, and Pacanowski (1975) for earlier references]. These models are sometimes run together with meteorological models, allowing an interaction of ocean and atmosphere, and have the greatest potential scope of any of the numerical experiments. Because of the grand scope there is little opportunity to vary the parameters. As the simpler, process-oriented models map out the more realistic domain of parameter space, larger models based on more optimal parameters may achieve a predictive level.

Notes

1. Approximating the stratification by two layers is an idealization. Even a cursory examination of the data (Veronis, 1972) shows that it is not possible to simulate the real stratification with two layers each of constant density separated by an interface that corresponds to the thermocline. The idealization is useful, however, in developing an intuition about the effects of stratification, and the results obtained are suggestive as guides when one is working with more realistic models.

2. Flierl (1978) has analyzed a variety of flows using a two-layer fluid, two vertical normal modes and many vertical normal modes. In each case he has calibrated the parameters in the two-layer and two-mode systems to obtain the best fit with the results derived from the more complete system containing many normal modes. He has shown that the best-fit parameters for the two-layer system vary widely depending on the phenomenon under investigation. For example, a study of topographic effects in a two-layer model requires an optimal choice of mean depths for the upper and lower layers quite different from the optimal choice required for a study of nonlinear effects. The optimal parameters for a model using two normal modes are much less sensitive to the process being studied.

3. Fofonoff (1962b) and Kamenkovich (1973) give detailed discussions of the general thermodynamic properties of sea water. Lynn and Reid (1968) and Veronis (1972) discuss the vertical stability characteristics of water columns. Small-scale mixing in waters stratified by heat and salt is discussed by Turner (1973a) and by Stern (1975a). The latter raises the question of the possibly crucial role played by the salinity balance in global circulation.

4. Phillips first transforms the system to Mercator coordinates and then makes the β-plane approximation. That leads to a rectangular coordinate system with eastward distance $x = (a \cos \phi_0)\lambda$ and northward distance $y = (a \cos \phi_0)\phi$. Therefore, in polar latitudes, where linear distance between meridians decreases, consistency requires that a correspondingly smaller range of latitude be chosen for an equivalent measure of distance. The two expansions lead to the same vorticity equation obtained at first order, but his first-order momentum equations are symmetric whereas our set is not.

5. It is well known that Ekman developed the analysis to explain Nansen's observation that ice floes in the Arctic drifted at an angle of 45° to the right of the wind. Nansen, in fact, had a good physical argument to describe the process that Ekman subsequently quantified.

6. The present analysis is for an idealized case. In nature other processes, such as lateral penetration from distant topographic irregularities (Armi, 1978), may affect the vertical structure near the bottom.

7. $\lambda_i = (gh_i/f_0^2)^{1/2}$ is the radius of deformation for a rotating system with a free surface defined by Rossby (1938) as the distance over which a disturbance will be transmitted by longitudinal pressure forces before the (transverse) effect of rotation takes over. It emerges naturally from (5.146) to (5.148) for longitudinal (Kelvin) waves near a wall, say along the y direction, with no transverse (U) flow. Then (5.147) and (5.148) yield the dispersion relation $\omega^2 = gh_i l^2$ for long gravity waves and (5.146) gives a transverse decay scale of λ_i.

8. The parameter $B = NH/f_0 L$, where L is the horizontal dimension, is frequently used with continuously stratified fluids (N. A. Phillips, 1963) and may be preferred because it contains only external parameters. The use of $\lambda_1^2 K^2$ here means that we measure the wave scale K relative to the internal radius of deformation λ_1.

9. Baroclinic instability of a westward current implies equatorward heat transport by the finite-amplitude field. Thus, the required poleward heat transport in the ocean must be supplied by the mean circulation.

10. Gall (1976) shows that the vertical structure of finite-amplitude motions that appear to evolve from baroclinic instability differs significantly from the structure of the unstable infinitesimal motions.

11. Gareth Williams (1978, 1979) has carried out extensive numerical experiments to explain Jupiter's bands and has obtained strong zonal jets from both barotropic and baroclinic eddy fields.

12. The assumption of constant eddy coefficients is less serious here than in the linear theories discussed earlier if the numerical calculations use a grid that is fine enough to resolve the eddies that arise from baroclinically unstable modes. The effect of the eddies can be taken into account explicitly and the eddy coefficients then include processes on even smaller (and hopefully dynamically less important) scales.

13. As can be seen in the first of the figures, the driving in this case is not very strong. Perhaps stronger motions in the deeps could override some of the topographic effect, as Rhines (1977) showed in the calculations reported above. This experiment is one of the earlier eddy-resolving calculations and the dependence on the parameters was only marginally explored.

6
Equatorial Currents: Observations and Theory

Ants Leetmaa
Julian P. McCreary, Jr.
Dennis W. Moore

6.1 Introduction

Historically, our knowledge of the circulation patterns in the tropics was derived from compilations of ship-drift data and so was restricted to a description of the surface currents. Although information of this type is crude, a picture of the spatial and temporal structure of the surface flow field was deduced over the years, and it has been little improved upon in the modern era of instrumentation. By contrast, almost all of the information about subsurface equatorial flows has been acquired recently. It is remarkable that one of the major ocean currents, the Pacific Equatorial Undercurrent, was not discovered until 1952.

For many reasons, progress in understanding equatorial circulations has been slow. The equatorial regions are vast and remote. The swift currents and high vertical shears put special demands on instrumentation. The geostrophic approximation, which is so useful at mid-latitudes, breaks down close to the equator and cannot be relied on to give accurate information about the currents. Finally, the flows seem much more time dependent than at mid-latitudes. Hence, on the basis of individual cruises, haphazardly taken in time and space, it is difficult to develop a consistent picture of the circulation patterns.

The variability of equatorial circulations has only recently been appreciated. This is partly because a great deal of the information about equatorial circulations comes from the central Pacific, where historically the mean appears to dominate the transient circulation; recent NORPAX (North Pacific Experiment) observations (Wyrtki, McLain, and Patzert, 1977; Patzert, Barnett, Sessions, and Kilonsky, 1978), however, suggest that the variability can be quite large even there. In other regions such as the western or eastern Pacific or the Indian Ocean, the fluctuating components are as large as or larger than the means.

The goal of this chapter is to give a short overview of the outstanding features of the equatorial ocean circulation patterns, the dominant spatial and temporal structures of the Pacific equatorial wind field (as an example of the kinds of driving mechanisms that need to be considered), and a summary of some of the theoretical ideas that have been developed to explain the ocean circulation and its relation to the wind field. No attempt is made to be comprehensive because in recent years there have been numerous excellent reviews of equatorial phenomena and theories for them. These include articles by Knauss (1963), Tsuchiya (1970), Rotschi (1970), Philander (1973), Gill (1975a), and Moore and Philander (1977). A collection of papers discussing various topics of equatorial oceanography is contained in the proceedings of the FINE (1978) workshop, held at Scripps Institution of Oceanography during the summer of 1977. A comprehensive discussion of analytic techniques for studying forced baroclinic ocean motions in the equatorial regions is presented in a three-part paper by Cane and Sarachik (1976, 1977, 1979). In discussing the theories, we shall stress the important physical ideas of each model (avoiding whenever possible the use of mathematics), put them in historical perspective, and relate them to the observations. The objective here is to identify the observations for which we have physical theories, and thereby indicate where further work is needed.

The importance of knowing the detailed time and spatial structure of the wind field is emphasized throughout this chapter. The reason is that in the tropics, the characteristic response times for baroclinic oceanic processes are much shorter than they are at mid-latitude and are much closer to the time scales characterizing the wind variations. Therefore the baroclinic response to atmospheric forcing is expected to be much stronger than at mid-latitude. The implication is that in order to arrive at a satisfactory explanation of the oceanic features, the temporal and spatial structure of the atmospheric forcing must be known accurately.

6.2 Observations

6.2.1 The Ocean

The surface currents are characterized by zonal bands in which the flow is alternately eastward or westward (Knauss, 1963). The eastward flows are referred to as *countercurrents* because they flow counter to the direction of the easterly trade winds. The westward flows are referred to as *North and South Equatorial Currents.* In the Atlantic and the Pacific, the North Equatorial Countercurrent (NECC) is approximately located between 5° and 10°N with westward flow to the north of this region in the North Equatorial Current (NEC) and westward flow to the south of it in the South Equatorial Current (SEC). There is also evidence for a South Equatorial Countercurrent (SECC) in both oceans between 5 and 10°S (Reid, 1964b; Tsuchiya, 1970; Merle, 1977); these flows are not as well developed, however, as the NECC. Both the intensity and location of the various currents vary seasonally (Knauss, 1963; Merle, 1977). The SEC and NECC are strongest during July and August. In the northern winter and spring the SEC generally vanishes and the NECC is weak; in the eastern Pacific there is some evidence that during this time the NECC is discontinuous at some longitudes or is entirely absent (Tsuchiya, 1974). In the northern summer the Pacific NECC assumes its northernmost position, whereas in the northern winter the current lies closest to the equator. The data base is insufficient to show an analogous migration of the Atlantic NECC.

The structure of the surface currents in the Indian Ocean differs markedly from those in the other two oceans (*African Pilot,* 1967). In the Indian Ocean the SEC usually lies totally south of 4°S. The predominantly eastward flow in the Indian Ocean is almost totally confined between the equator and the SEC. North of the equator the flow direction varies seasonally. During the northeast monsoon it is to the west.

The different circulation pattern in the Indian Ocean is no doubt related to the nature of the wind field. In the Atlantic and the Pacific, the southeast and the northeast trades are well developed over most of the ocean throughout the year. The mean stress is generally greater than the annual or semiannual components. In the Indian Ocean south of about 10°S, the southeast trades are reasonably steady. North of 10°S, the mean winds are weak and the stress field is dominated by the strong, regular forcing of the southwest and northeast monsoons.

The earliest measurements that indicated that the currents at depth might behave differently from those at the surface were made in 1886 by Buchanan (1888; see discussion by Montgomery and Stroup, 1962). Using drogues, he found that the subsurface flow on the equator in the Atlantic at 14°W was toward the southeast, while the surface flow had a slight westward set. These unusual measurements were thought not to be representative until after the discovery of the Equatorial Undercurrent in the Pacific in 1952 (Cromwell, Montgomery, and Stroup, 1954). The Equatorial Undercurrent has been found to be a subsurface eastward flow that is about 100-200 m thick and 200-300 km wide. It is centered approximately on the equator. Its core lies just beneath the base of the mixed layer in the top of the equatorial thermocline. Undercurrents are found in all three oceans. The one in the Indian Ocean, however, appears to be present primarily during the northeast monsoon, and most observations of it have been made during the spring. A thorough discussion of early evidence for the Equatorial Undercurrent in all three oceans is given in Montgomery (1962); see also Philander (1973).

Evidence for additional subsurface countercurrents in the Pacific is discussed in detail by Tsuchiya (1975). He finds narrow subsurface countercurrents in the Pacific Ocean symmetrically located about the equator. The jets occur at a depth of 200-300 m, and are associated with the poleward limits of the thermostad (a subsurface, relatively homogeneous layer of equatorial 14°C water). In the eastern Pacific, they are located roughly at 5°N and 5°S and are distinct from the surface countercurrents as well as from the Equatorial Undercurrent itself. Further to the west, the jets move closer to the equator (Taft and Kovala, 1979) and have been observed to merge with the Equatorial Undercurrent (Hisard, Merle, and Voituriez, 1970). The jets appear in virtually all central and eastern Pacific Ocean hydrographic sections, and thus appear to be permanent features of the equatorial current system there. There is also evidence that similar currents exist in the Atlantic Ocean as well (Tsuchiya, 1975; Cochrane, Kelly, and Olling, 1979).

Although something is known about the temporal fluctuations of the surface currents from ship-drift observations, little systematic information exists about the time and space variations at any frequency. The recent survey article by Wunsch (1978b) discusses the observational evidence for equatorial motions with periods from a few days to several months. Fluctuations have been observed with periods of 4 to 5.5 days (Wunsch and Gill, 1976; Weisberg, Miller, Horigan, and Knauss, 1980; Weisberg, Horigan, and Colin, 1979) Harvey and Patzert (1976) present evidence that suggests waves with a period of about 25 days. Features with a similar time scale and wavelength of about 1000 km are observed in the satellite infrared images of the equatorial front by Legeckis (1977b). In the Atlantic, Düing et al. (1975) found evidence for a meandering of the Equatorial Undercurrent with a "period" of 2-3 weeks and a "wavelength" of 3200 km during GATE (GARP Atlantic Tropical Experiment).

Neumann, Beatty, and Escowitz (1975) and Katz et al. (1977) have shown that the east-west slope of the thermocline in the Atlantic varies on a seasonal time scale. This variation appears to be in phase with changes in the zonal component of the wind stress. When the trades are strongest the tilt of the thermocline is the greatest, and when the stress is at a minimum so is the slope. Meyers (1979), in a similar analysis for the Pacific, finds strong annual and interannual variations in the depth of the 14° isotherm. This isotherm was chosen because it is representative of movements of the thermocline as a whole. The east-west slope was a minimum in May and June and a maximum during October and November. The easterly wind stress is weakest in March-May; there is some phase change, however, in the annual component with longitude. Hence there appears to be a time lag between the minimum east-west slope and the minimum wind. The exact relation between the winds and the variations in the slope was difficult to deduce because the variations about the mean slope are small and because smaller spatial scales than the basin width were present in the slope of the 14° isotherm. These smaller scales appear to propagate westward. In the Indian Ocean the mean east-west stress is small. In the transition between the monsoons, however, westerly winds appear along most of the equator. Associated with these is an eastward oceanic jet and simultaneously a rise in the sea level off Sumatra and a rise of the thermocline off East Africa (Wyrtki, 1973a). The oceanic response appears to occur coincidently with the winds.

6.2.2 The Wind Field

To date, most theoretical treatments of equatorial flows consider the applied wind stress to be some simple functions of time and latitude, whereas even the most general meteorological description of the trades indicates that there is considerable spatial and temporal structure (Riehl, 1954). In recent years, there have been several thorough analyses of the stress field over the ocean as derived from historical merchant-vessel reports. In the Pacific, this analysis was done for the whole ocean by Wyrtki and Meyers (1975) and for the eastern part by Hastenrath and Lamb (1977). The latter authors and Bunker (1976) also analyzed the winds over the tropical Atlantic. From these studies, it is possible to look in detail at the long-period temporal fluctuations and the large-scale spatial structure of the stress field. For the sake of brevity, we shall limit our discussion to the Pacific data. Some similar features are evident in the Atlantic, while the Indian Ocean wind field is dominated by the monsoon circulation.

Values of the zonal stress within 4° of the Pacific equator are shown in figure 6.1. These curves are derived from the mean monthly values of Wyrtki and

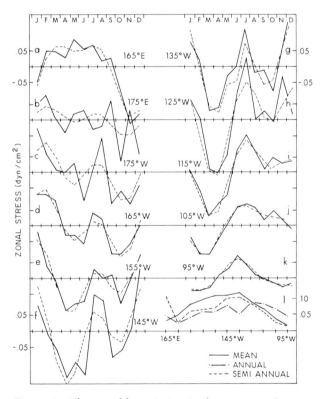

Figure 6.1 The monthly variation in the mean zonal stress between 4°N and 4°S. In figures 6.1a-6.1k, the solid curves display the average stress for blocks of 10° of longitude centered at the indicated values. The dashed curves show the proportion that is fitted by the annual and semiannual components. Figure 6.1l shows the amplitude of the mean, annual, and semiannual components as a function of longitude. The amplitude of the mean component has been reduced by a factor of five, i.e., at 145°W it is about 0.5 dyn cm^{-2}.

Meyers (1975), which are tabulated for areas of 2° of latitude by 10° of longitude. The monthly values of the equatorial zonal stress with the yearly mean removed are shown in figures 6.1a-6.1k. As can be seen, the annual and semiannual components account for almost all of the annual variation. The amplitudes and phases of these components are tabulated in table 6.1 The phase of the annual component decreases rapidly, although irregularly, to the east. The phase of the semiannual component also decreases to the east, but more slowly. The amplitudes of the mean, annual, and semiannual components as a function of longitude are shown in figure 6.1l. The mean stress over most of the Pacific is about a factor of five larger than the annual and semiannual components. The amplitude of each component varies strongly in the zonal direction. Thus we expect that models driven by zonally uniform stress distributions may not be adequate for describing the oceanic response to the wind.

Meyers (1979) presents evidence that the longitudinal structure of the wind field must be properly accounted for. He finds significant energy at the semiannual period in the eastern Pacific Ocean even though there is little energy in the wind stress there at that period. A study of the EASTROPAC data supports his findings. Figure 6.2 shows the average rate of change of the depth of the 20° isotherm (m/60 days) between 1°N and 1°S at four different longitudes. As can be seen, there is a pronounced semiannual variation, and the changes occur almost simultaneously at each longitude. The large amplitude of the changes (~50 m/60 days) is surprising considering the small amplitude of the semiannual component of the stress at these longitudes (0.02-0.03 dyn cm^{-2}). We suggest that these fluctuations are caused by baroclinic waves that have propagated eastward from a region where the semiannual component of the stress is much larger. Meyers reached the same conclusion.

So far we have considered only winds in the vicinity of the equator and have ignored the transient changes in the stress, and the curl of the stress, that are related to seasonal movements of the Intertropical Convergence Zone (ITCZ). To investigate these effects, the monthly values of the stress averaged between 100 and 120°W were examined. To emphasize the meridional structure, the weak monthly mean stress at the equator (~0.1 dyn cm^{-2}) was subtracted from the monthly mean value of the stress at each latitude. Resulting monthly values are shown in figure 6.3. The annual movement of the ITCZ has little effect close to the equator, with the largest amplitude of the variation in the stress occurring at about 9°N. There the amplitude is about 0.5 dyn cm^{-2}. The variation south of the equator is considerably less. But it is evident that the annual variation in wind-stress curl can be very large and must be taken into account in the models.

Table 6.1 Amplitude and Phase of Annual and Semiannual Components of Zonal Stress[a]

	Annual		Semiannual	
	Amplitude (dyn cm^{-2})	Phase	Amplitude (dyn cm^{-2})	Phase
165°E	0.11	308°	0.05	77°
175°E	0.03	263°	0.03	46°
175°W	0.04	239°	0.06	39°
165°W	0.06	224°	0.07	34°
155°W	0.06	153°	0.07	13°
145°W	0.08	122°	0.10	356°
135°W	0.05	113°	0.11	331°
125°W	0.09	72°	0.10	337°
115°W	0.08	55°	0.06	333°
105°W	0.07	34°	0.03	287°
95°W	0.04	342°	0.02	291°

a. Amplitude $= a^2 + b^2$, phase $= \tan^{-1}(b/a)$, and

$$\tau = \tau_0 + a_1 \cos\frac{2\pi}{12}t + b_1 \sin\frac{2\pi}{12}t + a_2 \cos\frac{2\pi}{6}t + b_2 \sin\frac{2\pi}{6}t.$$

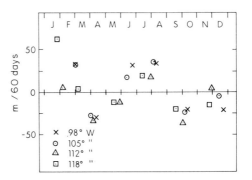

Figure 6.2 The rate at which the depth of the 20°C isotherm was observed to change during the EASTROPAC expedition in 1967 and early 1968.

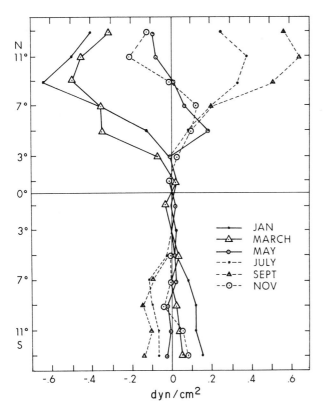

Figure 6.3 Profiles of zonal wind stress from 100 to 120°W, illustrating the annual cycle. To emphasize the meridional structure, the weak monthly mean value of the equatorial stress (the average of figures 6.1i and 6.1j) was subtracted from the monthly mean value at each latitude.

6.3 Theories

6.3.1 Integrated Theories

The earliest theoretical attempts sought an explanation for the North Equatorial Countercurrent. This current defies intuition since it flows opposite to the prevailing winds. Montgomery and Palmén (1940) suggested that the easterly wind stress in the equatorial zone is balanced by the vertically integrated zonal pressure gradient. They presented supporting observational evidence in the Atlantic. One interesting result was their demonstration that the baroclinic pressure gradients generally were confined to the top few hundred meters of the water column. As an explanation for the North Equatorial Countercurrent, they hypothesized that in the doldrums, i.e., in the vicinity of the ITCZ, where the magnitude of the zonal stress is greatly reduced, the pressure gradient would maintain the value that it had on either side of this region and hence no longer be balanced by the wind stress. As a result, an eastward flow would develop, which they suggested would be retarded by lateral friction. Charts of dynamic topography (Tsuchiya, 1968) indicate that the zonal pressure gradient in this region is, in fact, less than it is farther to the north and south. Furthermore, since 1° or 2° off the equator Coriolis terms cannot be neglected and the depth-integrated value of the meridional velocity usually does not vanish, the suggestion of Montgomery and Palmen (1940) does not represent a proper explanation for the North Equatorial Countercurrent.

Sverdrup (1947) suggested that the North Equatorial Countercurrent was not only a consequence of the zonal wind stress but also was related to the curl of the wind stress and continuity requirements. He assumed in his model that the currents were steady and vanished at a deep level. The equations were integrated from this depth to the sea surface; hence the solution gave no information about the vertical structure of the motion field. The apparent success of Sverdrup's theory in the eastern Pacific is often cited as observational endorsement of its widespread use in large-scale ocean-circulation theory.

In light of the earlier discussion about the large variability in the position and amplitude of the NECC, and the considerable monthly variation in the winds in this region, it is surprising that this steady theory should be applicable there. In fact, Sverdrup (1947) and Reid (1948), instead of using the mean wind stress in their computations, used the October-November values. They also used October-November oceanographic data to compute vertically integrated pressure. Implicit in their theory, then, was the assumption that the oceanic response to the fluctuating winds was sufficiently rapid that the ocean was almost always in equilibrium with the instantaneous winds (i.e., quasi-steady).

Therefore, it is of some interest to redo the computations of Sverdrup and Reid using the values for the monthly mean stress field from Wyrtki and Meyers (1975) and to compare the results with oceanographic data from different times of the year to see whether Sverdrup balance really is quasi-steady. The monthly values of the zonal Sverdrup transport are shown in figure 6.4. The October and November curves look quite similar to those of Sverdrup. The overall spatial and temporal evolution of the currents throughout the year resembles the picture that has been derived for the surface flow field from ship-drift observations. It is possible to check to see whether the historical integrated geostrophic transports follow a similar pattern of change.

During 1967 and the first part of 1968, hydrographic sections at four different longitudes between 100 and 120°W were occupied across the equator at seven different times. The geostrophic velocity computations relative to 500 db for these sections are given by Love (1972). Using these figures, the geostrophic transport for 2° bands of latitude from 2 to 12°N was computed by a planimetric integration. The minimum contour interval for the velocities was 5 cm s^{-1}; this interval introduces an uncertainty of about 3×10^9 kg s^{-1} into

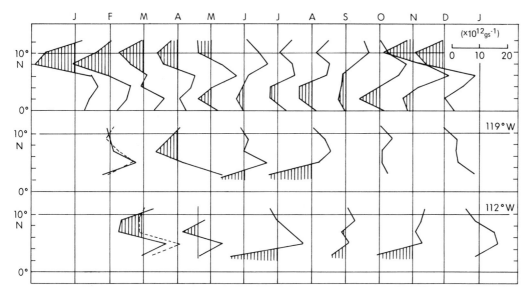

Figure 6.4 Upper panel depicts monthly Sverdrup transport. Lower two panels show geostrophic transports observed during the EASTROPAC expedition.

each transport computation (2.5 cm s^{-1} × 500 m × 2° × 1 g cm^{-3}). The results of these integrations for the sections at 119°W and 112°W are shown in figure 6.4. The tendency at both sections is for the transport in the south equatorial current to be strongest during the summer. The NECC lies close to the equator early in the year and moves northward during the summer and southward again during late fall. The transport patterns at both sections during the first part of 1967 and 1968 are quite similar. This agreement could be coincidence or indicate that these fluctuations perhaps have a regular annual cycle. The observed geostrophic transport patterns are visually similar to the theoretical pattern; there clearly are quantitative differences, however.

The important conclusion from this study is that it cannot be said whether the Sverdrup relation provides an accurate description of the currents in the eastern tropical Pacific. Sverdrup and Reid were probably fortunate in that their results seemed to agree so well with observations. The real problem with testing Sverdrup theory, in light of what we now know about barotropic and baroclinic adjustment, is associated with the assumed level of no motion. For an ocean with a free-slip flat bottom, Sverdrup balance should hold for the integrated flows and pressures, as long as the integral goes all the way to the bottom and barotropic adjustment has had time to occur (approximately a few days). If the integration extends only over the surface layers (upper 500 or 1000 m, say), however, then the integrated quantities depend on both the barotropic and baroclinic components, and a quasi-steady theory will apply only if the dominant baroclinic modes have also reached equilibrium. In general for the annual cycle, there is no reason this should be true since baroclinic adjustment even near the equator takes at least a few months. Clearly what is needed is an understanding of baroclinic adjustment processes.

6.3.2 Baroclinic Theories

The discovery of the Equatorial Undercurrent in the Pacific in 1952 (Cromwell, Montgomery, and Stroup, 1954) triggered a great deal of equatorial modeling activity in the latter part of that decade. Almost all of these early models include baroclinic effects in a crude way by assuming the surface layer is essentially decoupled from the deep ocean below (Yoshida, 1959; Stommel, 1960; Charney, 1960). Another assumption common to these models is that all fields except the pressure are assumed to be independent of x, the east-west coordinate; the zonal pressure gradient $\partial p/\partial x$ is taken to be a constant related to the zonal wind stress that is driving the motion.

The simplest and most elegant of these models is that of Stommel (1960). This model is the first successful extension of classical Ekman theory to the equator. Stommel's model equations balance the Coriolis force with vertical diffusion of momentum (the Ekman balance) while retaining horizontal pressure gradients. It is the zonal pressure gradient which allows him to avoid the equatorial singularity of previous Ekman theory, and which provides a source of eastward momentum to drive the undercurrent. The wind forcing is put into the ocean as a surface stress, and a zero-stress condition is imposed at the bottom of the layer. The vertical structure of the zonal velocity at the equator is parabolic, with surface flow in the direction of the wind and subsurface flow (undercurrent) in the opposite direction. In addition, the model develops a meridional circulation very similar to that hypothesized by Cromwell to account for the distribution of tracers in the equatorial Pacific (Cromwell, 1953). Figure 6.5, by Stommel (1960), illustrates the circulation pattern developed by his model. There is surface divergence of fluid from the equator, subsurface convergence, and equatorial upwelling. A major drawback of this theory is that the response is too closely related to the choice of eddy viscosity ν. Let U and L represent the velocity and depth scales of the flow and τ measure the equatorial zonal wind stress. Let H be the depth of the layer and β be the meridional derivative of the Coriolis parameter. Then in the Stommel model

$$U = \frac{H\tau}{\nu} \quad \text{and} \quad L = \frac{\nu}{\beta H^2}. \tag{6.1}$$

It is not possible to select a value for ν which simultaneously sets a realistic magnitude as well as width scale for the flow. In particular, if ν is adjusted so that the current has a reasonable speed, then it is far too narrow.

Charney's (1960) nonlinear model differs fundamentally from Stommel's in that he requires a no-slip bottom boundary condition. Because of this stringent condition, the linear model cannot produce any flow counter to wind, and so it is clearly the presence of nonlinearities that allows the model to generate an undercurrent. Charney computed the zonal and vertical velocity only on the equator. Later, Charney and Spiegel (1971) extended the model off the equator to determine the flow field in a meridional section and also obtained the same general pattern of meridional circulation as that proposed by Cromwell. They explained the existence of an undercurrent in both models in the following qualitative manner [first suggested by Fofonoff and Montgomery (1955)]. Fluid converging toward the equator conserves absolute vorticity. As a result, planetary vorticity is converted to relative vorticity, and this conversion provides an additional source of eastward momentum to drive the undercurrent. These studies show that nonlinearities are likely to play an important role in equatorial dynamics. On the other hand, the nonlinear solutions are even more sensitive to the choice of ν than Stommel's linear one. For example, when ν decreases from 17 to 14 cm^{2}s^{-1}, the speed of the Equatorial Undercurrent more than doubles, and the e-folding width of the jet narrows significantly.

Other workers subsequently considered similar layer models and explored the importance of terms in the equations of motion that were ignored in the earlier studies (Philander, 1973; Cane, 1979b). Despite their successes, the layer models have obvious limitations. The assumption of an x-independent flow field is questionable, the decoupling of the lower layer is highly artificial, and the solutions are extremely dependent on the choice of mixing coefficients. Finally, the observations themselves suggest the most serious drawback. Although an undercurrent has been observed in the mixed layer (see for example Hisard, Merle, and Voituriez, 1970), the strongest eastward flow is almost always found in the thermocline below the surface mixed layer. Therefore, an undercurrent model that treats the effects of stratification in a more realistic way is required. It took another decade for such a model to be developed, and the development proceeded from an entirely different direction, namely, from attempts to understand time-dependent processes in the equatorial region.

Initial interest in equatorial waves was largely theoretical, and grew out of attempts to extend the theory of midlatitude gravity and Rossby waves to the equator. Longuet-Higgins (1964, 1965, 1968a) undertook a detailed investigation of planetary waves on a rotating sphere (cf. chapters 10 and 11). At about the same time, Blandford (1966) and Matsuno (1966) both used equatorial β-plane models (i.e., the Coriolis parameter is approximated by $f = \beta y$) to describe free baroclinic waves. [Groves (1955) had earlier derived the same β-plane equatorially trapped solutions as Blandford and Matsuno, but he was considering barotropic tides.] All of these studies are linear and inviscid, and consider perturbations on a background state consisting of a stably stratified ocean at rest. Normal modes for the vertical structure of the fields are derived just as Fjeldstad (1933) originally did for internal waves. The eigenvalues for the vertical separation problem are usually presented as equivalent depths h' for the resulting normal modes. In general, for a finite depth ocean, there is one barotropic mode for which the equivalent depth is nearly equal to the actual depth of the ocean, and the corresponding eigenfunction describing the depth dependence of the horizontal velocities and pressure perturbation is nearly depth independent. There is also a denumerably infinite set of baroclinic equivalent

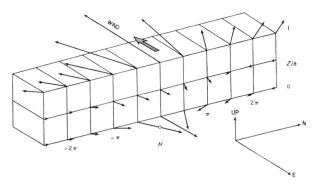

Figure 6.5 Schematic three-dimensional diagram of the flow field in the neighborhood of the equator, not showing the vertical component of velocity, which can in principle be determined from continuity. (After Stommel, 1960.)

depths. The vertical structures of the baroclinic eigenfunctions depend on the details of the background density distribution. For each equivalent depth h', there is a corresponding phase speed c defined by $c^2 = gh'$.

With this separation, the horizontal and time dependence of each vertical mode of the system can now be investigated independently of the z-dependence. If c is the phase speed for the mode in question, then free oscillations at frequency ω can exist in a latitudinal band, $|y| \leq y_T$. In the equatorial β-plane model, the turning latitude is given by

$$y_T = \left(\frac{\omega^2}{\beta^2} + \frac{c^2}{4\omega^2}\right)^{1/2} \qquad (6.2)$$

and is strongly dependent on ω. For the first baroclinic mode, c is typically about 250 cm s^{-1}. Then at the annual cycle, $y_T = 6250$ km; at the 5-day period $y_T = 640$ km, a much smaller trapping scale. [For a complete discussion of free equatorial trapped waves see Moore and Philander (1977) and chapter 10.]

Moore (1968) investigated the effects of oceanic boundaries on free waves at frequencies corresponding to strong equatorial trapping (near-minimum y_T for a given value of c). He showed that a Kelvin wave at frequency ω hitting an eastern boundary will generally reflect some of its energy back in the form of equatorially trapped Rossby waves, and the rest of the energy propagates poleward as coastally trapped Kelvin waves. These results were later extended to the case of an incoming Kelvin wave of step-function form by Anderson and Rowlands (1976a).

The first observational evidence—in Pacific sea-level variations—for the existence of strongly equatorially trapped baroclinic waves in the ocean was presented by Wunsch and Gill (1976). Spectral peaks corresponding to periods in the range 2-7 days were explained in terms of equatorially trapped inertial-gravity waves with vanishing east-west group velocity. The variation in the strength of the spectral peaks from islands at different latitudes was consistent with the expected latitudinal structure of the theoretical wave modes (see figure 10.11). Additional observational evidence for equatorially trapped waves was provided by Harvey and Patzert (1976) from deep mooring data in the eastern equatorial Pacific. Weisberg, Miller, Horigan, and Knauss (1980) and Weisberg, Horigan, and Colin (1979) did the same using Atlantic GATE data and later Gulf of Guinea observations.

An interesting property of the low-frequency large-scale baroclinic waves (nearly nondispersive Rossby waves) is that their phase speed increases markedly the more trapped they are to the equator. The swiftest wave, the equatorial Kelvin wave, is also the most strongly trapped. This property suggests that the forced baroclinic response of the equatorial ocean need not remain locally confined to the forcing region, but can radiate swiftly away. That is to say, baroclinic effects may be remotely as well as locally forced.

In 1969, Lighthill published a seminal paper entitled "Dynamic Response of the Indian Ocean to the Onset of the Southwest Monsoon." He was motivated by the idea that the Somali Current might be remotely forced by winds in the interior of the Indian Ocean. He coupled the wind stress to the model ocean in a novel way, not as a surface condition, but rather as a body force uniformly distributed throughout an upper mixed layer. This approach avoids the details of the frictional processes by which the stress actually enters the ocean and allows the use of the inviscid normal modes to study the oceanic response to variable wind forcing. The model ocean is forced by an impulsive wind-stress distribution concentrated away from the coast in the center of the Indian Ocean. In addition to the locally forced response, long-wavelength nondispersive Rossby waves are generated at the edge of the forcing region and propagate westward to the coast. They are reflected there as short-wavelength dispersive Rossby waves that remain trapped to the coast. He interpreted the coastal response as the onset of the Somali Current.

One of the most prominent features of the equatorial oceans is the presence of a strong near-surface pycnocline. Observations show that changes in the pycnocline depth can be large and occur very rapidly. It is sensible to model this motion as simply as possible, with either a one-and-a-half-layer model (a single upper moving layer overlying a slightly denser inert layer) or with a two-layer model; in both cases the layer interface plays the role of the pycnocline. Wind stress enters the ocean as a body force in the surface layer, and so such models are analogous to the Lighthill model if it is limited to a single baroclinic mode. Not surprisingly, a large number of these models, both linear and nonlinear, have been used in the past decade to investigate a variety of equatorial adjustment problems.

O'Brien and Hurlburt (1974) used a two-layer model to study the interior response of the equatorial Indian Ocean to change in zonal wind stress. The model produces an eastward equatorial jet with many of the observed features documented by Wyrtki (1973a). The meridional structure of the accelerating jet produced in the O'Brien and Hurlburt model is basically the same as the structure of the surface flow given in Yoshida's (1959) undercurrent paper and has been called the Yoshida jet. See Moore and Philander (1977) for a detailed description of the spin-up of the Yoshida jet from rest for a single inviscid baroclinic mode.

The idea that the Somali Current might be remotely forced stimulated a number of additional studies. Anderson and Rowlands (1976b) used a one-and-a-half-layer analytical model to investigate the relative importance of local and remote forcing. They concluded that local forcing is initially dominant but that remotely forced effects will ultimately become important. Hurlburt and Thompson (1976) used a two-layer model forced by a basin-wide northward stress τ^y to show that nonlinear advective effects rapidly become important in the developing boundary flow. Cox (1976) addressed the same problem using a more sophisticated 12-level numerical model. His findings corroborate the conclusions of the simpler models.

Hickey (1975) and Wyrtki (1975b) suggested that El Niño occurs in conjunction with a relaxation of the southeast trades in the central and western Equatorial Pacific. Hurlburt, Kindle, and O'Brien (1976) applied substantially the same model as O'Brien and Hurlburt (1974) to simulate the onset of El Niño. They spun-up the model with a uniform westward wind for 50 days, and then turned off the wind. Major features of the response included an eastward-traveling Kelvin wave originating at the western boundary, westward-traveling nondispersive Rossby waves originating at the eastern boundary, and poleward-traveling Kelvin waves along the eastern boundary. The thickening of the warm layer near the eastern boundary in response to the relaxation of the winds demonstrated the plausibility of Wyrtki's hypothesis. Although the model of Hurlburt, Kindle, and O'Brien was nonlinear, the response was completely dominated by linear dynamics. The only discernible nonlinear effect was the steepening front of the poleward-propagating coastal Kelvin wave.

McCreary (1976) used a one-and-a-half-layer linear model in a similar study of El Niño. He followed the initial response of the eastern ocean to various zonally uniform distributions of wind stress. Important conclusions are the following. Changes in the meridional wind field cannot cause an El Niño event. Changes in the zonal wind field outside the equatorial band (roughly ±5° of latitude) are not important in generating El Niño. Finally, model results suggest that the meridional profile of the wind field can significantly affect the resulting flow field. For example, the southward motion of the ITCZ can force southward cross-equatorial transport of water in the eastern Pacific. It is interesting that in this model the deepening of the pycnocline in the eastern Pacific is initiated by the weakening of the local zonal winds, and is stopped by the arrival of an equatorially trapped Kelvin wave from the western ocean.

McCreary (1977, 1978), realizing that the changes in the Pacific wind field during El Niño are not zonally uniform, but rather are confined in the central and western ocean, extended the model of the 1976 study so that a wind-stress patch of limited longitudinal extent could be imposed. Figure 6.6 shows the adjustment of the model thermocline topography after a weakening of the westward trade winds by an amount $+0.5$ dyn cm^{-2}. The dotted lines in the upper left panel of the figure delineate the region of weakened trade winds; the wind-stress change is a maximum in the central portion and weakens markedly in the surrounding areas. After 35 days, a downwelling signal rapidly radiates from the region of the wind patch as a packet of equatorially trapped Kelvin waves and already the equatorial thermocline (and also sea level) begins to tilt to generate a zonal pressure gradient that balances the wind there. In the region of strongest wind curl, the thermocline begins to move closer to the surface, and this upwelling signal propagates westward as a packet of Rossby waves. After 69 days, the packet of Kelvin waves has already reflected from, and spread along, the eastern boundary. As time passes, this downwelling signal propagates back into the ocean interior as a packet of Rossby waves, and more and more upper-layer water piles up in the eastern ocean. As shown in the lower two panels, the source of the water is a broad area of the western and central ocean where the thermocline is uplifted; in some regions the thermocline rises over 65 m. Note that in this case it is the arrival of the Kelvin wave that causes the initial deepening of the pycnocline in the eastern ocean. The figure also further demonstrates the importance of knowing the horizontal structure of the wind field; the rapid tilting of the equatorial thermocline, and the form of the extraequatorial upwelling are a direct result of the particular structure chosen here.

Moore et al. (1978) suggested that the equatorial upwelling in the eastern Atlantic and coastal upwelling in the Gulf of Guinea might be a remote response to an increase in the strength of the westward wind stress in the western Atlantic. The variation of $\partial p/\partial x$ in the western Atlantic reported by Katz et al. (1977) was interpreted as the response to such an increase. O'Brien, Adamec, and Moore (1978) and Adamec and O'Brien (1978) used the same basic adjustment model as had been applied earlier to the Indian and Pacific Oceans to study Atlantic adjustment. The ocean geometry is rectangular with a rectangular notch cut out of the northeast corner of the basin to simulate West Africa. The model calculations showed that an increase in strength of the westward wind stress in the western Atlantic generates an equatorial-upwelling response that propagates eastward, leaving behind a thermocline tilt to balance the stress. When it hits the coast, the upwelling signal propagates away from the equator along the boundary and travels westward along the coast of the Gulf of Guinea.

The important role that equatorially trapped waves play in all these models has stimulated several studies

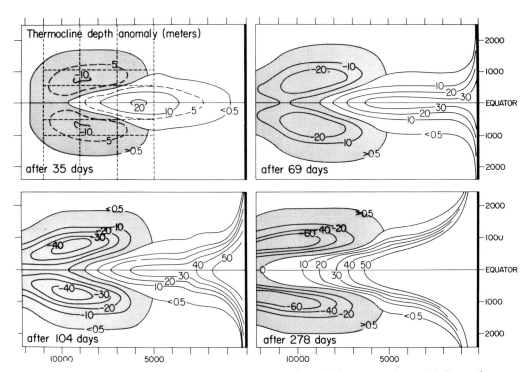

Figure 6.6 Time development of the thermocline depth anomaly to a longitudinally confined zonal wind stress. The dotted lines in the upper left panel indicate the region of the wind. The solid lines in each panel indicate the presence of an eastern ocean boundary. Horizontal distances are in kilometers. (After McCreary, 1977.)

of the interaction of the waves with the mean zonal currents. Hallock (1977) used a two-and-a-half-layer model forced by meridional surface winds to investigate the Equatorial Undercurrent meanders observed during GATE. A meridional stress τ^y with a zonal wavelength of 2400 km is turned on gradually in a few days and then held steady. Part of the response is an equatorially trapped Yanai wave of about 16-day period. The characteristics of this Yanai wave were somewhat, but not drastically, modified by including mean surface currents in the upper layer and a mean undercurrent in the second layer. Thus it was suggested that the GATE observations of Equatorial Undercurrent meanders might be explained in terms of passive advection of the Equatorial Undercurrent by a wind-generated Yanai wave. McPhaden and Knox (1979) and Philander (1979) used one-and-a-half-layer models to study the modifications of other free waves in the presence of various meridional profiles of zonal mean currents. McPhaden and Knox concentrate on the equatorial Kelvin and inertial-gravity waves (including the Yanai wave), and conclude that the zonal velocity associated with a given mode can be strongly modified by the background currents. Philander also considers equatorial Rossby waves and suggests that these waves may be influenced strongly by the presence of the Equatorial Undercurrent.

Philander (1976) used a two-and-a-half-layer model to show that background zonal currents also allow unstable waves and predicted a possible instability of the surface currents in the tropical Pacific. The most unstable waves which propagate westward have a wavelength of about 1000 km and a period of approximately 25 days. The satellite observations of Legeckis (1977b) seem consistent with the instability scales Philander predicted. A later numerical study by Cox (1980) confirmed a suggestion of Philander's that oscillations generated by such surface-current instabilities could propagate zonally for large distances while propagating vertically into the deep ocean. The results indicate that the Harvey and Patzert (1976) observations in the eastern Pacific near the sea floor could, in fact, be such remote effects of surface-current instabilities generated farther to the west.

It is now apparent that simple layer models have been remarkably successful in reproducing many of the obvious features of time-dependent equatorial ocean circulation. These models are intrinsically limited, however, in their ability to describe the vertical structure of the ocean, and recently theoreticians have begun to study more sophisticated models that allow for high vertical resolution. There are several reasons for this interest. In 1975, Luyten and Swallow (1976) discovered high-vertical-wavenumber equatorially trapped multiple jets in the Indian Ocean, and in 1978 similar structures were observed in the Pacific (C. Eriksen, personal communication). In addition to this observational impetus, general theoretical questions arise in trying to understand in greater detail just how the wind stress enters the ocean. For example, how valid is the assumption that the wind enters as a body force distributed over a surface mixed layer? Finally, there is the perennial interest in understanding the dynamics of the Equatorial Undercurrent.

In order to model the multiple jets, Wunsch (1977) considered an inviscid continuously stratified model, which he took to be horizontally unbounded and infinitely deep. (At the end of his study he also reported effects introduced by a western ocean boundary.) Since the model was inviscid, it was not possible to introduce the wind stress as a surface boundary condition. Instead, Wunsch assumed that a thin wind-driven surface boundary layer pumped the deeper ocean with a surface distribution of vertical velocity. For convenience, he chose for the form of the distribution a westward-propagating sinusoidal wave. To specify the solution, he required it to match this vertical velocity distribution at the ocean surface and also to exhibit upward phase propagation everywhere in the upper ocean (a radiation condition). Although the north-south scale of the imposed surface w-field was broad, the significant response was strongly equatorially trapped and showed an oscillatory structure in the vertical, just as in the observations.

In an effort to understand better how wind stress enters the equatorial ocean and to explore further the vertical structures that might be possible in stratified viscid equatorial oceans, Moore (unpublished) has studied the simplest linear model that includes the effects of continuous stratification and vertical mixing on an equatorial β-plane. The buoyancy frequency N is taken as constant, constant eddy coefficients for vertical mixing of heat κ and momentum ν are used, and, for convenience, unit Prandtl number is assumed ($\nu = \kappa$). The ocean is laterally unbounded and infinitely deep. The simplest problem to treat is the analog of the Yoshida jet, in which the system is driven by a uniform zonal wind stress τ started impulsively at $t = 0$. The problem as posed has a natural or canonical scaling. This means that for suitable choices of velocity scale U, time scale T, horizontal length scale L, and depth scale H, the resulting nondimensional equations contain no parameters. The scales are

$$L = (\nu N^2/\beta^3)^{1/2},$$
$$T = (\nu \beta^2 N^2)^{-1/5},$$
$$H = (\nu^2/\beta N)^{1/5}, \qquad (6.3)$$
$$U = \tau(\beta N \nu^3)^{-1/5}.$$

Note that equations (6.3) are much less dependent on ν than are equations (6.1), essentially because here the depth scale is no longer an externally set constant! This weak dependency on ν is a characteristic advantage that all stratified models have over their homogeneous counterparts. For $N = 2 \times 10^{-3}$ s^{-1}, $\nu = 20$ cm^2 s, and $\tau = 1$ dyn cm^{-2}, the typical values are $L = 60$ km, $T = 8$ days, $H = 40$ m, and $U = 2$ m s^{-1}. Figures 6.7A-6.7C show (y, z)-profiles of the zonal velocity u at times $t = 0.25$, 1.0, and 5.0 in Moore's model (the equator is at the right). The ranges are $0 \le y \le 5$ positive to the left and $0 \le z < 10$ positive down. Right on the equator in this x-independent model, the zonal velocity is simply governed by downward diffusion of momentum: $u_t = \nu u_{zz}$. The contour plot at time $t = 0.25$ shows that initially this balance is nearly true everywhere. At later times, the Coriolis effects have a distinct influence on the solution. The equatorial jet is well developed by $t = 1$. The solution serves to illustrate typical time and space scales of equatorial processes, as well as to point out their dependence on model parameters. Moreover, the model clearly suggests that the diffusion of zonal momentum into the deep ocean may play an important role in equatorial dynamics. However, in this x-independent and bottomless model the jet continues to intensify and deepen indefinitely. The only way to reach a steady state is to require a non-slip bottom on the ocean or to allow zonal pressure gradients, either by adding zonal structure to the wind field or by introducing boundary effects.

In the past 2 years, several fully three-dimensional and continuously stratified ocean models have been used to study wind-driven equatorial ocean circulation. The numerical models of Semtner and Holland (1980) and Philander and Pacanowski (1980a,b) are nonlinear. The analytical model of McCreary (1980) is linear; it can be regarded as an extension of the Lighthill model that allows the diffusion of heat and momentum into the deeper ocean, and also as an extension of the Stommel model into a stratified ocean. All three models are remarkably successful in reproducing many of the observed features of the Equatorial Undercurrent. For example, figure 6.8 [taken from McCreary (1980)] shows the flow field in the center of the ocean basin (10,000 km wide), where the zonal wind stress reaches a maximum (-0.5 dyn cm^{-2}). The model Equatorial Undercurrent is situated in the main thermocline beneath a surface mixed layer (75 m). The width and depth scales, as well as the speed, of the zonal flow are realistic. In addition, the pattern of meridional circulation resembles quite well the classic pattern suggested by Cromwell (1953); note that there is a strong convergence of fluid slightly above the core of the Equatorial Undercurrent, strong surface upwelling, and weaker downwelling below the core. The presence of

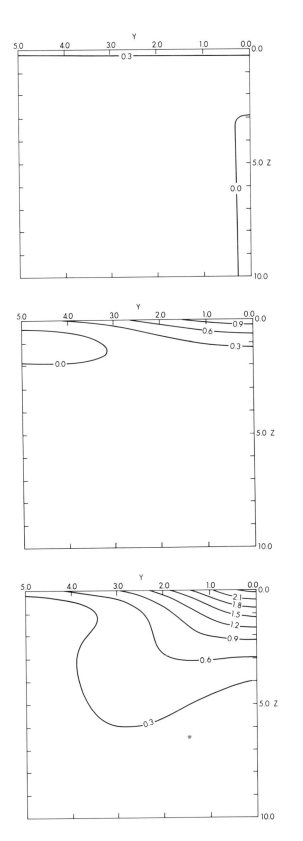

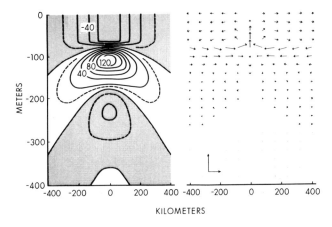

Figure 6.8 Vertical sections of zonal velocity (left panel) and meridional circulation (right panel) at the center of the ocean basin. The contour interval is 20 cm s^{-1}; dotted contours are ±10 cm s^{-1}. Calibration arrows in the lower left corner of the right panel have amplitudes 0.005 cm s^{-1} and 10 cm s^{-1}, respectively. (After McCreary, 1980.)

zonal baroclinic pressure gradients and the diffusion of momentum into the deep ocean are crucial for the development of this undercurrent. In fact, at the equator the balance of terms is simply $p_x = (\nu u_z)_z$. Just as in the x-independent model discussed in the preceding paragraph, stratification weakens considerably the dependence of the solution on ν. It is this property that allows a realistic undercurrent to be generated for a wide range of choices of ν. The success of the linear model suggests that nonlinearities are not essential for maintaining the Equatorial Undercurrent. They apparently act only to modify the linear dynamics, not to destroy it.

6.4 Discussion

Observations of the equatorial oceans during the past 30 years have shown the existence of surprisingly strong subsurface currents and greatly increased our knowledge of the mean structure and time variability of the surface flows. Zonal currents, both surface and subsurface, tend to occur as narrow jets several hundred kilometers wide, several hundred meters deep, and thousands of kilometers long. The observations also suggest that many tropical ocean events are remotely forced. A prime example is the phenomenon of El Niño, where wind changes in the central and western Pacific are associated with strong changes in the thermal structure of the eastern Pacific.

Perhaps because the signal-to-noise ratio is so large, theroeticians have had remarkable success in providing explanations for many equatorial phenomena. All the models suggest that stratification is an essential ingredient in equatorial dynamics. Width, depth, and time scales of model response are all intimately related to

Figure 6.7 (y,z)-contour plots of zonal velocity u in x-independent stratified spin-up on an equatorial β-plane, including the effects of vertical mixing. (A) t = 0.25. Diffusion effects are dominant. (B) t = 1.0. Equatorial jet starts to develop. (C) t = 5.0. Jet continues to deepen.

the background buoyancy frequency distribution [e.g., see equation (6.3)]. The presence of stratification also allows the existence of swiftly propagating equatorially trapped baroclinic waves. These waves have been used to account for remotely forced events as well as the meandering of equatorial currents.

Additional experimental and theoretical work needs to be done. It is now known that the tropical ocean responds readily and coherently to large-scale low-frequency changes in the wind field. Moreover, the nature of the response depends critically on the spatial distribution of the wind field. It is important, then, to monitor the detailed structure of the wind field on the seasonal, annual, and interannual time scales. For the first time, such observations may be practical due to rapidly improving satellite technology. Long time series of the variability of the tropical ocean flow field must also be obtained. Equatorial observations made in 1979 as part of the Global Weather Experiment should add substantial new knowledge about the circulation in all three equatorial oceans. However, the duration of this experiment is too short; it will provide at most a single realization of the annual cycle.

A major advance in our understanding of equatorial dynamics occurred when simple layer models began to be studied. These models were successful because their solutions could be easily and convincingly compared with observations; the layer interface is interpretable as the motion of the pycnocline. The recent use of equatorial models with high vertical resolution suggests that other advances may soon be forthcoming. Such models can produce a much more detailed picture of the flow field, and so can be quantitatively compared to a greater degree with the observations. Obvious problems to which these models should be applied include the existence of the thermostad and the associated subsurface countercurrents and the presence of multiple jet structures at the equator. In addition, even more sophisticated models that include realistic surface mixed layers need to be developed.

7
On Estuarine and Continental-Shelf Circulation in the Middle Atlantic Bight

Robert C. Beardsley
William C. Boicourt

7.1 Introduction

We shall attempt in this chapter to trace the development of ideas about circulation over the continental shelf in the Middle Atlantic Bight and in the major estuaries which drain into the Middle Atlantic Bight. The term *Middle Atlantic Bight* refers to the curved section of the continental shelf off the eastern United States stretching between Cape Hatteras to the south and Cape Cod and Nantucket Shoals to the northeast. The New York Bight is a subsection of the Middle Atlantic Bight and refers to the shelf region stretching between the New Jersey and Long Island coasts. A schematic version of Uchupi's (1965) topographic map is shown in figure 7.1, indicating both the general shape of this shelf region plus the names and locations of the major estuaries and key positions discussed in the text.

We have decided to focus this review of estuarine and shelf cirulation on the Middle Atlantic Bight and its estuaries for several reasons. The major Middle Atlantic Bight estuaries have been extensively examined and have provided several important case studies in the development of new ideas about circulation and turbulent-mixing processes in moderately stratified coastal-plain-type estuaries. These estuaries and the adjacent continental shelf border on one of the world's largest urban complexes; a better description and understanding of the circulation and dominant mixing processes occurring in this particular region is clearly needed for a more effective management of the regional estuarine and shelf resources in the face of man's many conflicting uses of the water bodies. Fostered in part by increased environmental concerns, more adequate research funding, and the availability of new instrumentation and observational techniques, many new circulation and related physical studies have been undertaken in the last two decades, and a synthesis of both old and new material into a review of the regional estuarine and shelf circulation seems particularly appropriate at this time. While a few scientists have made important contributions in both fields and have thus helped to carry new ideas and techniques from one field into the other, basic research on problems concerning estuarine and continental-shelf circulation have evolved more or less independently in time, so that we will present here separate reviews of the historical and the modern ideas about estuarine and shelf circulation in the Middle Atlantic Bight. One important research objective in the 1980s will be to develop a better kinematic and dynamic description of the physical coupling between estuarine and shelf waters. Our present meager knowledge about the different physical processes that connect the shelf and estuary together prevent a more unified discussion.

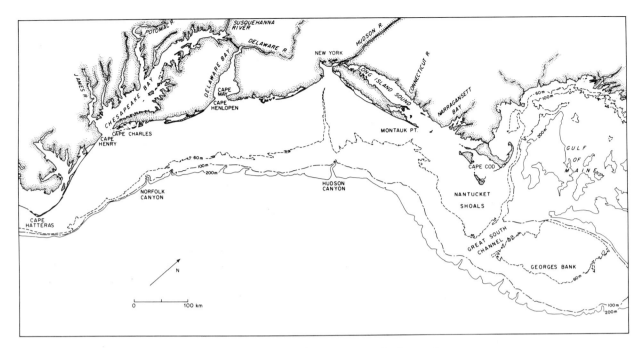

Figure 7.1 A topographic map for the Middle Atlantic Bight and a western section of the Gulf of Maine. The 60-, 100-, and 200-m isobaths are shown.

7.2 Estuarine Circulation in the Middle Atlantic Bight

We shall trace here the evolution of estuarine-circulation ideas. We will use the Middle Atlantic Bight estuaries as a focus because physical studies of these water bodies have played a major role not only in the development of the early concepts of the physics of estuaries but also in the recent refinement and reformulation of these ideas. Narragansett Bay, the Hudson River, Delaware Bay, and especially the Chesapeake Bay system—all have provided case studies from which significant advances have been made in our understanding of "coastal bodies of water having free connection with the open sea and within which sea water is measurably diluted by fresh water drainage" [Pritchard (1967a)].

7.2.1 The Development of Estuarine-Circulation Concepts

The idea that the introduction of fresh water into the sea can produce an oppositely directed, two-layer circulation has existed for a long time. Some early oceanographers understood that the lighter, fresher water spreads away from the source along the surface and the heavier, more saline water moves toward the source underneath. The kinematic details and the dynamics of this process have, however, remained elusive until recent years. On a scale as large as the Mediterranean Sea, early oceanographers correctly held that the circulations were driven by density differences. Two-layer flows observed at the Kattegat in the Baltic (Ekman, 1876), the Strait of Gibraltar in the Mediterranean Sea (Douglas, 1930), and the Strait of Bab el-Mandeb at the mouth of the Red Sea (Buchan, 1897) all reflect the balance between evaporation, runoff, and precipitation in the enclosed seas. On a smaller estuarine scale, the formulation of clear circulation ideas has been hindered by the difficulty of unraveling the different circulation components in a variety of estuarine geometries. In addition to oscillatory tidal currents and currents driven by density differences, there has been the possibility of "reaction currents" as suggested by F. L. Ekman (1876). These were countercurrents and undercurrents associated with the entrainment of ambient water by the discharge of a river into the sea. Although the idea of reaction currents seems to have been supported in Helland-Hansen and Nansen's (1909) discussion of the rivers entering the Norwegian Sea and in Buchanan's (1913) description of the flow off the mouth of the Congo River, F. L. Ekman's son, V. W. Ekman, showed that significant reaction currents were unlikely at the mouths of rivers (Ekman, 1899). The terms "reaction current," "induction current" (Cornish, 1898), "undercurrent" (Dawson, 1897), and "compensatory bottom current" (Johnstone, 1923) were often employed without detailed discussion of the physics. In some cases, these terms were used simply to refer to low-salinity water flowing seaward over more saline water flowing landward. In other cases, however, the usage harked back to F. L. Ekman's sense of a countercurrent associated with the river outflow jet.

Reaction currents were invoked to explain the two-layer circulation phenomena found in the early studies of the Middle Atlantic Bight estuaries. Although Harris (1907) correctly described the distribution of pressure surfaces in an estuarine situation, he interpreted Mitchell's (1889) observations on the Hudson River as illustrative of F. L. Ekman's countercurrent. Mitchell, in contrast, had interpreted the observed "underrun" of salt water below the outflowing fresher water as a response to a decrease in fresh-water discharge in the Hudson River. He calculated that "the surface of the river water would have to stand 2-1/2 feet above the ocean to prevent the salt water from running in along the bottom; and the sea-water would creep into the basin as soon as the head fell below this." If there were not mixing between the salt water and the fresh water, Mitchell's analysis would explain the movement of the density interface between the upper and lower layer and the position of the "neutral plane" where "inflow and outflow balance."

Although we now know there is significant mixing between the upper and lower layers in the Hudson, Mitchell's interpretation of his remarkably good current and density observations provided a reasonable explanation for the seasonal variation in salt intrusion. R. P. Cowles (1930) in his monograph on the Bureau of Fisheries' studies of the Chesapeake Bay waters noted that such behavior was not adequately explained by reaction currents. He wondered "why the undercurrent (as deduced by salt intrusion) moving in an ingoing direction is so marked during the winter months, when the discharge from the rivers is not ordinarily at its height?" That the surface layers were moving seaward was evident from the U.S. Coast and Geodetic Survey data collected by Haight, Finnegan, and Anderson (1930). They report 320 days of current pole measurements at the Chesapeake Bay mouth (Tail of the Horseshoe Lightship), where the mean current flowed out of the bay at 13 cm s^{-1}. At Thomas Point in the upper Bay, a 27-day record showed an 8-cm s^{-1} mean flow seaward. Cowles mentioned many possible mechanisms for moving water and salt in Chesapeake Bay, and remarked on the difficulty in analyzing the resultant complexity. He did succeed in documenting the distribution and seasonal progression of temperature and salinity. Of particular note was the lateral gradient in salinity whereby the eastern side of the Bay was markedly saltier than the western side. Cowles ascribed this feature to the "fact that the deep-water channel which contains the most saline bottom water lies on (the eastern) side throughout most of its extent and to the fact that a large volume of fresh water from the rivers of the western shore presses the more saline water toward the eastern shore." He did not mention the possibility that the rotation of the earth played a role. Wells, Bailey, and Henderson (1929), who titrated the salinity samples for Cowles's surveys, did suggest that although the lateral gradient in salinity "had been ascribed to the fact that the principal rivers enter the Bay on its west side, the rotation of the earth may also be a factor."

Marmer (1925) thought that the greater nontidal surface flow (and greater ebb duration) along the western shore of the Hudson River was due to "the effect of the deflecting force of the earth's rotation." He noted that, at mid-depth, the flood-current velocities were greater on the eastern side of the river, and that the ebb velocities were greater on the western side. Marmer in his Hudson River study, and Zeskind and LeLacheur (1926) in their study of Delaware Bay, pointed out the decrease with depth in ebb-current duration in the estuary, and ascribed this decrease to the river discharge. Although they noted that the duration of flood may be greater than ebb near the bottom of the estuary, neither Marmer nor Zeskind and LeLacheur conveyed the sense that there is an internal, nontidal circulation present and that there is net up-estuary motion in the lower layer.

Haight's (1938) review of current measurements in Narragansett Bay contains little discussion of the nontidal flows. Although the U.S. Coast and Geodetic Survey's interest was to define and predict the tidal currents, Haight may have disregarded reporting the mean flows because the collected measurements were taken by a variety of methods under a variety of conditions, and because Narragansett Bay displayed such "irregularity of currents." Hicks (1959) later noted that Pillsbury's 1889 measurements [summarized by Haight (1938)] did show nontidal flow into the estuary in the lower layer.

Early current-measurement techniques did not allow easy determination of flow direction at depth. Mitchell's vertical profiles and Pillsbury's 5.5-day time series appear as notable achievements. Mitchell (1859) developed an apparatus to measure the "countercurrent" at depth in the Hudson River by modifying a device used to measure subsurface currents in European canals. Two floats (copper globes) were connected by a wire, one weighted to sink to a depth and act as a drogue. In order to reduce the errors resulting from the drag of the surface float, Mitchell added a third float attached to the surface float and weighted to have the same cross-sectional area. The attachment line of this additional float was connected to a reel to allow the two surface floats to separate freely. Mitchell argued that the original pair of floats would move at the mean of the surface and subsurface velocities and that the free float would move with the surface velocity. The Price current meter used by the U.S. Coast and Geodetic Survey had no provision for measuring direction. The common practice was to assume the current direction at depth corresponded to that indicated by the drift pole at the surface, a practice that could create substantial errors in nontidal-flow determinations in the estuary. Marmer's (1925) observations were successful because he employed a "bifilar direction indicator" in conjunction with the Price meter. This direction sensor, developed by Otto Petterson (Witting, 1930), consisted of a set of three vanes that were positioned at various depths and that transmitted their alignment to the surface by wires (Zeskind, 1926). Petterson also developed an internally recording current meter that could record speed and direction at 30-minute intervals for 2 weeks. The Petterson meter became available to the U.S. Coast and Geodetic Survey in 1925. The majority of current measurements reported in the cited survey reports on the Middle Atlantic Bight estuaries were made by current pole (often from anchored light ships) and Price current meters. Although Haight, Finnegan, and Anderson (1930) reported measurements made in the early 1920s by the U.S. Fisheries Commission employing Ekman current meters, their use does not seem to have been widespread.

The Coast and Geodetic Survey's collected current measurements in Long Island Sound were reported by LeLacheur and Sammons (1932). Again, they were primarily interested in tidal currents and they did not report the mean currents obtained from the long time series at the lightship. Prytherch (1929) released 500 drift bottles with drogues in his study of oyster-larvae transport and setting. With the 300 returns and a few Ekman and Price current-meter measurements, he deduced a net outflow from the Long Island Sound on the surface.

Important estuarine research was also being conducted elsewhere during the early decades of the twentieth century. Europeans and Canadians were active in the coastal regions where river runoff affects the regional general circulation. Palmén (1930) employed Bjerknes' (1898) solenoid method in a study of the wind-driven circulation in the Gulf of Finland. He demonstrated an oppositely directed two-layer flow and determined a wind-stress coefficient.

Jacobsen (1930) provided further details of the two-layer flow near the Kattegat through an analysis of current time-series measurements made from two lightships. Jacobsen also examined current and density measurements made in Randersfjord on the east coast of Jutland. From these data, which showed clearly the estuarine outflow and inflow (of the order of 10 cm s^{-1}), he calculated the surface slope along the axis of the fjord and determined coefficients of viscosity and mixing. He also considered the problem wherein a concentration of plankton was placed at the level of no net motion and allowed to disperse, illustrating the interaction of vertical diffusion and horizontal advection in such a two-layer system.

Following a suggestion by A. G. Huntsman that Watson's (1936) observations from Passamaquoddy Bay in the Gulf of Maine could be interpreted as a three-layer flow driven by tidal mixing, Hachey (1934) conducted a series of tank experiments in which he produced both two-layer and three-layer flows (figure 7.2). He concluded that

the mixing of stratified water sets up dynamic gradients causing the following differential movements:

(a) where, through the addition of fresh water at the mixing point, the mixed water is of a density which is less than that of the waters otherwise available for mixing, the mixed water is carried away from the mixing area in the upper layers, while a compensating current carries water to the mixing area in the lower layers; and

(b) where the mixed water is of a density which is intermediate between the densities of the surface and bottom waters available for mixing, the mixed water is carried away from the mixing area at some intermediate level, and surface and bottom waters are carried to the mixing area to compensate for the waters entering into the mixing.

A steady wind blowing towards the area of mixing is responsible for considerable modification of the above systems of currents. Such a wind seems to offer some resistance to the system outlined in (a), but considerably enhances a system of currents outlined in (b).

While the current measurements (in Digdeguash Harbor off Passamaquoddy Bay) offered by Hachey as an example of the three-layer flow may not be convincing because of their short duration and the uncertainties in density structure and in the strength of the wind-driven component, his conclusions from the tank ex-

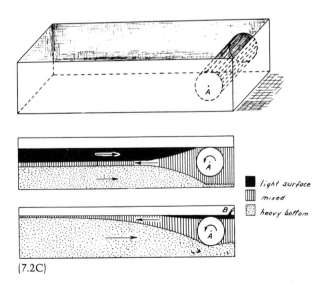

Figure 7.2 Diagram of Hachey's (1934) experimental approach (A), and resultant three-layer (B) and two-layer (C) circulation patterns. Mixing was provided by rotor A and fresh water was introduced by pipe B.

periments were correct and later substantiated by observations made in Baltimore Harbor. Hachey's work is especially significant because it was one of the first explanations of a density-driven circulation in stratified water generated by wind and tidal mixing.

In the years immediately following World War II, there was a marked increase in interest in the circulation of estuarine waters, caused in part by military needs and a heightened sense that the resources of the estuary and coastal waters were threatened by the nearby activities of man and should be protected.[1] Many of the papers from this period address the *flushing* characteristics of the estuary rather than the circulation per se. In addition to the Office of Naval Research's interest in basic research in the oceans, the Navy recognized a need for shallow-water studies to aid in mine warfare, amphibious warfare, and submarine-detection problems (Solberg, 1950). The Navy was also concerned with the possible environmental threat from nuclear submarine activity in bays, harbors, and estuaries. An indication of the scientific interest and talent dedicated to estuarine-circulation studies in the late 1940s is given in the proceedings (Stommel, 1950b) of the Colloquium on the Flushing of Estuaries held at the Massachusetts Institute of Technology in September 1950 and sponsored by the Office of Naval Research. The papers and discussion show not only that oceanographers were beginning to model the mixing processes in the estuary, but also that they were beginning to understand the possibility of an internal estuarine circulation.

The task at hand during the late 1940s was to determine the flushing mechanisms for estuaries. After Tully's (1949) extensive work on Alberni Inlet in British Columbia, much of the observational study was carried out on Middle Atlantic Bight estuaries. Ketchum (1950, 1951) sought to improve the tidal prism model whereby the sea water brought into the estuary on flood tide is assumed to mix completely with the water in the estuary. In addition, the water flushed out of the estuary on the following ebb is assumed lost to the system and does not reenter on the subsequent flood. Estuaries, however, do not mix completely on each tide. Ketchum therefore proposed to divide the estuary into successive volume segments the lengths of which were determined by tidal excursions. Within each segment complete mixing is assumed at high tide. Ketchum applied this concept to Tully's observations on Alberni Inlet and to his own study of Raritan Bay, New Jersey, and of Great Pond in Falmouth, Massachusetts, and achieved good agreement with the observed salinity distribution. Ketchum's success prompted Arons and Stommel (1951) to translate his segmented model into a continuous-mixing-length model. They produced a family of curves that were also successful in describing the salinity distribution in Alberni Inlet and Raritan Bay. The constant of proportionality relating eddy diffusivity to the tidal excursion and the tidal current amplitude differed, however, by an order of magnitude between the two estuaries. Pritchard (1965b) pointed out that these two treatments were applicable only to vertically homogeneous estuaries in which tidal mixing was sufficiently intense to eliminate vertical stratification. Stommel (1953b) later applied both Ketchum's model and that of Arons and Stommel to the Severn estuary, which has small vertical stratification. He showed that neither hypothesis worked for the Severn and mentioned that "it does not appear likely that any good purpose can be served at present by making *a priori* suppositions about the turbulent mixing process."

The obvious differences in the topography and salinity distributions in various estuaries led Stommel (1950b) to call for an estuarine-classification system employing differences in morphology and mixing processes as criteria. Stommel (1951) began the process with a classification scheme based primarily upon the "predominant physical causes of movement and mixing of water in the estuary," identifying river flow and tidal and wind mixing as the important processes. Pritchard (1952a, 1955, 1967b) and Cameron and Pritchard (1963) developed and refined Stommel's initial scheme. Hansen and Rattray (1966) later advanced the classification scheme by suggesting a two-parameter system that includes the stratification and the ratio of the net nontidal velocity at the surface to the river flow divided by the cross-sectional area of the estuary. An attempt at further refinement of these classification schemes has not been fruitful because of the difficulty in quantifying the parameter-selection process for a particular estuary. Many estuaries exhibit a variety of estuarine types.

In his discussion of estuarine classification, Pritchard (1952a, 1967a, 1967b) proposed the definition of an estuary quoted earlier as "a semi-enclosed coastal body of water which has a free connection with the open sea and within which sea water is measurably diluted with fresh water derived from land drainage." While this definition excludes inverse estuaries such as Laguna Madre, Texas, and San Diego Harbor, which are driven by evaporation, it is the most useful yet proffered because it sets the scale and the important elements controlling the characteristic estuarine circulation—lateral boundaries, the transmission of tidal energy and salt between the open sea and the estuary, and the introduction of sufficient fresh water to provide density gradients driving the currents. The Baltic Sea, for instance, would not be considered an estuary under this definition because its large scale renders the lateral boundaries less important to the kinematics and dynamics of water movement than they are in a true estuary.

One of the notable aspects of the development of estuarine-circulation concepts in the active decade following World War II was the extensive (and successful) use of laboratory models in deciphering mixing and transport processes. The first problems addressed with these models involved the simplest of estuarine types—the highly stratified or salt-wedge estuary. As physical oceanographers began to exchange ideas in meetings such as the 1950 colloquium at MIT, they became aware that the U.S. Army Corps of Engineers had been working with flumes and physical models for over 10 years at the U.S. Waterways Experiment Station in Vicksburg, Mississippi (Simmons, 1950). Among the earliest salt-intrusion studies were the Army Corps of Engineers' investigations of water-supply problems in the lower Mississippi River. The Army Corps of Engineers recognized a need for analytic help and in 1945 requested the aid of hydrodynamicists at the National Bureau of Standards "to investigate and establish the basic laws of similitude for models involving a study of density currents and the mixing of salt water and fresh water."[2] Keulegan provided this help and addressed many problems concerning the laboratory modeling of salt-wedge circulation. Keulegan (1949) produced salt wedges in flumes in which there was almost no mixing between the upper and lower layer. When he increased the flow of the upper layer, however, breaking internal waves formed on the fluid interface. Keulegan noted that, in this entrainment process, the waves only broke upward, carrying fluid from the lower layer to the upper layer.

Stommel and Farmer (1952) also examined the salt-wedge estuary with the aid of a laboratory flume. They showed that an abrupt widening in a channel can produce a stationary internal wave on the interface if the internal Froude number equals a critical value. This internal wave acts as a control on the outflow of the upper layer by restricting the thickness of the upper layer. Stommel and Farmer (1953) later noticed that if they added mixing to their flume, there was a point beyond which increased mixing has no effect on the outflow of the upper layer. Dyer (1973) explains that this "overmixing" mechanism is a result of the downward erosion of the density interface reaching the level where it restricts the compensating inflow in the lower layer. Model analyses were also conducted for wider, well-mixed estuaries such as Delaware Bay. Pritchard (1954a) studied flushing in the Army Corps of Engineers' Delaware model at Vicksburg, Mississippi, by employing dye as a tracer. Pritchard found that the eddy diffusivity was spatially scale dependent approximately in the proportion suggested by Stommel (1949).

The Chesapeake Bay Institute began a study of the moderately stratified James River in the summer of 1950 to examine the influence of the salinity and currents on the oyster seed-bed region in the middle reaches of the James River. The recent development of techniques for rapid sampling of currents and salinity from an anchored vessel allowed for the first time the collection of continuous detailed measurements for periods of 3 days or more. Current velocity was measured with a biplane drag (Pritchard and Burt, 1951), a modification of a method used by Jacobsen (1909) and apparently by Nansen (Witting, 1930), and salinity profiles were obtained with *in situ* conductivity and temperature sensors (Schiemer and Pritchard, 1957). These new observational tools were used to collect a data set sufficiently extensive in both time and space that meaningful temporal and spatial averages could be computed. This averaging procedure further minimized the (apparently low) variability due to local wind-driven currents and variations in the river flow that occurred during the sampling periods. Pritchard (1952b, 1954b) used this data set to evaluate the terms in the averaged salt-balance equation and to conclude that the horizontal advective flux and the vertical diffusive flux of salt were the most important in maintaining the balance. Pritchard (1956) also examined the momentum balance in the James River, determining the unknown terms in the equation of motion from the observations of the mean distribution of temperature, salinity, and current velocity. He evaluated the important Reynolds-stress terms and described the topography of the pressure surfaces, which sloped down toward the sea in the upper layer and down toward the head of the estuary in the lower layer. He found, as did Cameron (1951), that the cross-estuary pressure gradient and the Coriolis force were in approximate balance.

Rattray and Hansen (1962) used the James River observations and Pritchard's analysis to develop a theoretical steady-state circulation model for a moderately stratified estuary. Employing similarity transformations, whereby functional forms were assumed for the dependence of the stream function and salinity defect on the longitudinal position in the estuary, Rattray and Hansen reduced the two-dimensional partial differential equations governing the stream function and salinity defect to a pair of simultaneous ordinary differential equations. Under the conditions specified by Pritchard for the James River (in which the field accelerations and the vertical advective and horizontal diffusive fluxes were unimportant), and given the surface salinity distribution, Rattray and Hansen produced vertical profiles of salinity and velocity that matched those observed in the James River. Hansen and Rattray (1965) later relaxed some of the restrictive assumptions to retain the river-forced component of circulation.

Their work also found good agreement with the James River observations.

While this matching between these two theoretical treatments and Pritchard's analysis of the James River data has enhanced the attention paid to these studies, the value of these analytic models lies less in the agreement per se with observations than in the insight they provide into fundamental estuarine processes. These solutions to Pritchard's (1956) dynamic equations were the first to show clearly the interdependence of salinity and velocity fields in the estuary. While Agnew (1961) had considered two separate aspects of this interdependence in the free-convection part of the problem, Hansen and Rattray (1965) solved the coupled equations, including both the free-convection and forced-convection modes. Hansen and Rattray (1965) not only delineated the effects of fresh-water discharge and wind stress on the gravitational circulation in the James River, but also considered the interrelationships in an estuary such as the Mersey, which has a well-developed gravitational circulation despite the fact that tidal mixing nearly eliminates the vertical salinity gradient.

The concept and description of the internal circulation in a moderately stratified estuary evolved primarily from these studies of the James River data set. This circulation differed from the salt-wedge circulation, not only because the lower layer moved strongly toward the head of a moderately stratified estuary, but also because the transport in the individual layers was much greater than in the salt wedge. Near the mouth of a moderately stratified estuary, the upper-layer net (nontidal) transport can be an order of magnitude greater than the river flow entering the estuary. In describing the driving mechanism for this internal circulation, Pritchard (1967b) stated:

It has been attributed to the increased potential energy of the system which follows from increased exchange between the fresh-water and saltwater layers. More accurately, tidal mixing produces horizontal density gradients of increased strength, which in turn produce horizontal pressure gradients of sufficient magnitude and extent to maintain the relatively higher velocities even in the face of increased eddy friction. Tidal mixing is responsible for both the increase in potential energy and the distribution of potential energy within the estuary.

7.2.2 Recent Developments in the Study of Estuarine-Circulation Processes

While the contributions of Pritchard, Rattray, and Hansen represent significant advances in our understanding of estuarine-circulation processes, fundamental questions remain as to the nature of the transport of salt and momentum, the role of the wind in the transport processes, and the effects of topography in producing both order and disorder. In spite of an improved ability to attain spatial coverage and resolution with modern instrumentation, our ability to describe and model estuarine physics is still limited by inadequate parameterizations of friction and turbulent mixing. Longer current-meter records are showing that the current variability due to wind forcing is more complex than previously thought. As more detailed information on the circulation becomes available, there is a growing conviction on the part of estuarine investigators that the variations in the lateral direction are significant in the dynamics, and that bottom topography can generate both secondary flows and residual circulations.

Three observational methods have been employed to separate and examine mixing processes in an estuary: (a) the evaluation of terms in the temporally and spatially averaged salt-balance equation; (b) the direct measurement of turbulent fluctuations in velocity and salinity; and (c) the observation of dispersion by an introduced tracer. The first method is the analytic technique employed by Pritchard (1952b, 1954b) on the James River data set. His conclusion that the horizontal advective flux of salt and the vertical diffusive flux are the dominant terms is based on an analysis that assumes lateral homogeneity. For estuaries such as Delaware Bay, which can exhibit vertical homogeneity but have lateral gradients in salinity and velocity, Pritchard (1955) suggests that, by analogy with the James River, the dominant salt-flux terms are probably the lateral-diffusive and the longitudinal-advective terms. These analyses involved tidally and spatially averaged values of salinity and velocity, but the averaging process is not explicitly developed in the salt-balance equation. Pritchard (1958) begins the rigorous averaging of the three-dimensional salt-balance equation, expressing salt and velocity variables as sums of time-mean values and deviation terms. Bowden (1963) and Cameron and Pritchard (1963) further decompose the variables into a time mean, a turbulent fluctuation, and a single oscillatory term varying sinusoidally over the tidal cycle. Bowden employs this decomposition to examine the effect of vertical shear on the longitudinal transport of salt in a laterally homogeneous estuary. He finds that, for the Mersey River, there are occasions when the advective flux of salt out of the estuary (driven by the river discharge) is approximately balanced by the transport associated with the vertical shear in velocity and vertical variations in salinity. On other occasions, Bowden finds that the up-estuary transport is shared between the "shear effect" and the transport arising from the correlation between the harmonically varying terms of the depth-mean velocity and salinity. There are also times when this tidal-correlation term dominates and times when the upstream and downstream salt transports do not balance.

Okubo (1964) has carefully examined the averaging process for an estuary with lateral as well as vertical variations, specifying the assumptions under which his salt-balance equation is appropriate. He uses the salt equation averaged over the cross section in a successful analysis of measurements made at the Delaware estuary model at the U.S. Waterways Experiment Station. Hansen (1965) also considers variations over the cross section of the estuary. He decomposes the cross-sectional mean variables into a mean, a harmonic tidal variation, and a turbulent fluctuation. For the Columbia estuary, which has a large river flow, a large tidal range, and a weak gravitational circulation, Hansen finds that the advection of salt driven by the river discharge is balanced primarily by fluxes associated with the correlation of velocity and salinity fluctuations of the tidal period and with the shear effect. Fischer (1972) argues that for the Mersey, the salt flux associated with the lateral shear is not only larger than that associated with the vertical shear, but that it is dominant. He proposes decomposing the deviations from the cross-sectional mean into variations in the vertical and lateral directions. Fischer's conclusions for the Mersey stand in contrast to the analysis by Bowden and Gilligan (1971) of Mersey observations made in the reach where density currents are significant. There may be agreement for the reaches seaward of this region. While Fischer's point that the lateral shear can make a significant contribution to the longitudinal transport of salt is well taken, his estimates of the terms in the salt-flux equation are based primarily on parameterizations of the dispersion coefficients and not on direct computations using the salinity and velocity observations in the manner of Pritchard, Bowden, Okubo, or Hansen. Dyer (1973) states:

So far it is not possible to define precisely which are the dominant factors since different investigators have used slightly different methods of analysis; they split up their components in a variety of ways with certain implicit assumptions. Consequently, the results of differences in tidal response and topography between estuaries are not clear.

The approach of averaging the salt-balance equation does seem to offer a promising means of attaining an explicit separation of the flux components. Dyer (1973) suggests combining Hansen's (1965) and Fischer's (1972) schemes, and applies (Dyer, 1977) the full set of terms to a salt-wedge, a partially mixed, and a well-mixed estuary. He finds that for Southampton Water, a partially mixed estuary, the salt flux associated with the vertical shear and the lateral shear are of the same order. It is clear that great care is required to avoid dependence on the observational scheme and method of data handling. In light of the recently observed wind-driven variability in estuaries and the importance of topographic effects, proper evaluation of this approach will require long record lengths, good spatial coverage and resolution, and shrewdness in averaging procedures. Rattray (1977) calls for both better methods of integrating the governing equations in conjunction with field programs and more extensive and elaborate field observations.

While averaging and evaluating the various terms in the salt-balance equation provides insight into the spatially integrated mixing processes in the estuary, the direct measurement of turbulent fluctuations provides a unique look at mixing on a small scale. This complementary method is particularly suited for the determination of the source(s) of mixing, about which little is now known. The various roles of wind stress, shear at the pycnocline, bottom stress, and surface and internal waves in providing the turbulent mixing of salt have not yet been evaluated. Bowden (1977) reviews turbulence measurements made in estuaries [including a noteworthy early attempt by Francis, Stommel, Farmer, and Parson (1953)] and the subsequent attempts to parameterize the observed fluctuations for the construction of models. The measurement of turbulent velocity fluctuations has required many innovative techniques. Bowden and Fairbairn (1952) mounted two Dodson-propeller current meters on a rigid stand on the bottom of the Mersey estuary. This device employed a spring-loaded propeller, which enabled a rapid response to the turbulence. Bowden and Howe (1963) were the first to employ an electromagnetic flow sensor [developed earlier by Bowden and Fairbairn (1956)] to measure turbulent fluctuations in an estuary. Many of the subsequent turbulent-velocity measurements were made in estuaries tributary to the Chesapeake Bay. Cannon (1971) used a biaxial current meter (Cannon and Pritchard, 1971) to measure intermediate-scale turbulence from a tower erected in the Patuxent River. Seitz (1973) also measured turbulence in the Patuxent River, using an acoustic Doppler-shift current meter (Wiseman, Crosby, and Pritchard, 1972). He showed the approach to isotropy of the three Cartesian velocity components at high wavenumber, and also showed the spectral distribution of horizontal shearing stress, which reaches a peak at intermediate wavenumbers. Pronounced intermittency in the Reynolds stresses in the Choptank River was reported by C. M. Gordon (1974). J. D. Smith (1978) has developed a profiling system which is particularly suited for measuring turbulent fluctuations of temperature, conductivity, and velocity in an estuary. With this instrumentation, Gardner and Smith (1978) investigated mixing events in the Duwamish salt-wedge estuary in Washington that are apparently triggered by a hydraulic jump that occurs at a sharp change in river depth.

Tracking the dispersion of an introduced dye tracer provides a third method for examining mixing processes in an estuary. Pritchard and Carpenter (1960) developed a technique for detecting a concentration of 0.04 parts per billion of Rhodamine dye. The three-layer circulation of Baltimore Harbor was discovered through the use of this dye-tracer technique. An application of particular interest is the examination of the shear effect by Wilson and Okubo (1978) in the York River, off Chesapeake Bay. They employed Okubo's (1967, 1969) theoretical methods to analyze the dispersion of a dye release in the lower layer of the York estuary. They provide a model for separating the longitudinal dispersion in a stratified estuary due to horizontal turbulence and to the interaction of vertical shear with vertical mixing. They also include the modifications to the shear effect caused by the nontidal upward advection.

Although early investigators showed a keen awareness of the effects of strong wind forcing on estuarine circulation, and they often invoked wind effects to explain discrepancies that arise in interpretations that ignore wind driving, the significance of wind-driven circulations in estuaries has only been recently discovered through the analysis of long current observations. Pickard and Rodgers (1959) show a wind-induced shift in the mean velocity profile in Knight Inlet, British Columbia. Hansen and Rattray (1965) suggest that even a small wind stress could have a marked influence on the gravitational circulation. Weisburg and Sturges (1976) were among the first, however, to examine the wind transport in a partially mixed estuary using long-term current-meter data. They show, using month-long current measurements from Narragansett Bay, that wind transients can easily dominate the longitudinal flux of water in an estuary and that a proper separation of the gravitational circulation is difficult with short records (Weisberg, 1976a). Weisberg (1976b) has developed a stochastic model for the wind-driven longitudinal flow at one position in the Providence River in Narragansett Bay. Farmer and Osborn (1976) describe the wind circulation in Alberni Inlet. Up-estuary winds can reverse the current in the upper low-salinity layer and cause a deepening of this layer near the head. Farmer (1976) presents a simple model for the freshwater thickness in the upper reaches of the inlet, and produces an accurate simulation of the wind-driven behavior.

A year-long series of current measurements made in the Potomac River estuary led Elliott (1978) to the discovery that the estuarine circulation was not only affected by local wind forcing, but also by sea level in the Chesapeake Bay proper. This nonlocal forcing is examined by Wang and Elliott (1978), who find that the nonlocal forcing extends to the continental shelf. The dominant sea-level fluctuations in the Chesapeake Bay have a period of 20 days and are the result of up-estuary propagation of coastal sea-level fluctuations. Local winds operate on a shorter time scale, driving seiche oscillations in the bay at a period of 2.5 days. Wang (1979a,b) has examined this wind driving further, and finds that the predominant current fluctuations in the lower Chesapeake Bay are barotropic.

The increasing evidence for wind control on time scales of 10 days or less leads to speculation on the role of wind mixing versus tidal mixing in providing the energy source for the gravitational circulation. While we do not have sufficient evidence at present to decide this question, investigators are beginning to reveal both the mode and details of the topographic effects on the tidal mixing process. The oscillatory movement of the tides acting on shoreline irregularities and complex bottom topography is known to increase the longitudinal dispersion in estuaries (Pritchard, 1953; Holley, Harleman, and Fischer, 1970; Okubo, 1973). Sugimoto (1975) and Zimmerman (1978) describe the production of residual vortices by propagation of the tidal wave over a complicated topography. Ianello (1977) and Zimmerman (1979) remind us that, for transport processes, careful consideration of the Stokes drift must be given. The inherent errors and logistical difficulty of Lagrangian current measurements as yet leave us with Eulerian measurements as the only means of spatial and temporal coverage. Rattray's (1977) call for elaborate and extensive measurement programs should be repeated if we are to consider the measurement of the Stokes drift by Eulerian means.

The generation of secondary flows by bends in rivers is well known to fluid dynamicists and geologists. Secondary flows in estuaries that have stratification and tidal oscillation, however, are less well understood, partly because observational evidence is scanty. Dyer (1977) outlines the expected cross-estuary flow pattern for various degrees of stratification and shows, as does Stewart (1957), that the field-acceleration terms cannot be neglected in the lateral dynamic equation when there is curvature in the estuary.

Episodic tidal-mixing events may occur not only on a time scale of the semidiurnal tide, but also on a scale of the fortnightly variation in tidal range. Haas (1977) reports observations from the lower York River and Rappahannock River on the Chesapeake Bay and suggests that, in these rivers, the increase in tidal-mixing energy from neaps to springs provides sufficient increase in tidal mixing to eliminate the vertical stratification. Cannon and Ebbesmeyer (1978) and Cannon and Laird (1978) describe fortnightly salinity intrusions in the fjordlike Puget Sound estuary in Washington. These events are associated with the large spring tides over the entrance sill.

Garvine (1977) shows that lateral fronts in estuaries may be important to both vertical and lateral mixing.

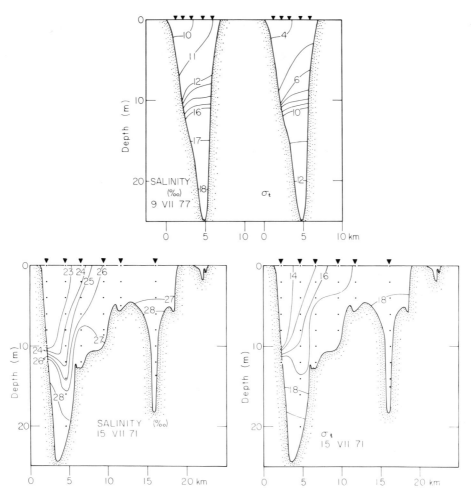

Figure 7.3 Salinity and density (σ_t) distributions at two cross sections of the Chesapeake Bay. Top section is located near Annapolis, Maryland, approximately 220 km up the estuary from the mouth, and bottom section is located at the mouth of the Bay between the Virginia Capes.

These fronts may be tidally time dependent or occur when the pycnocline breaks the surface, as in lower Chesapeake Bay. Figure 7.3 illustrates salinity and density distributions at two positions in Chesapeake Bay; one section (top) is near Annapolis, Maryland, approximately 220 km up the estuary from the mouth, and the other section (bottom) is between the Virginia capes at the mouth of the Bay. The cross-estuary tilt of the pycnocline is evident in the Annapolis section. If this cross-estuary tilt is approximately in geostrophic balance, the increase in transport in the gravitational circulation toward the mouth of the bay requires a corresponding increase in tilt. In the lower Chesapeake Bay, the tilt increases to the point where the pycnocline breaks the surface, often appearing as a series of strong lateral fronts. This observed increase in tilt is probably the combined result of the increase in the geostrophically balanced gravitational flow, the addition of fresh water by rivers on the western side of the bay, and by the widening of the bay in the lower reaches.

The salinity and density sections shown in figure 7.3 serve to illustrate the inherent three-dimensionality of the flow near the mouths of estuaries. This three-dimensionality and complexity near the mouth often makes it difficult to formulate realistic boundary conditions for numerical circulation models of the estuary. While the estuary does not often strongly affect the circulation on the adjacent continental shelf, the estuary often dominates the flow in the mouth and in the nearshore regions.

7.3 Continental-Shelf Circulation

We shall discuss in this section some ideas and observations about the general circulation over the continental shelf in the Middle Atlantic Bight. Bumpus (1973) and Beardsley, Boicourt, and Hansen (1976) have presented recent reviews on the circulation within the Middle Atlantic Bight. Bumpus (1973) describes some of the historical ideas about the Middle Atlantic Bight circulation and gives a summary interpretation of the large amount of surface-drift bottle and sea-bed-drifter data acquired during the 1960s over the eastern United States continental shelf. In the 1970s, moored arrays of self-contained current meters and other *in situ* instrumentation have been deployed in the Middle Atlantic Bight, and Beardsley, Boicourt, and Hansen (1976) present some of the preliminary results from these new field programs. In the 4 years since that review, longer current-meter records have been obtained and other descriptive and theoretical advances have occurred, making it seem both worthwhile and appropriate for us to attempt here to update the preliminary physical picture presented in 1976.

We shall begin with a brief physiographic description of the Middle Atlantic Bight and then present a review of the early observational work and ideas about water structure and the general circulation in the Middle Atlantic Bight. This review is presented both for completeness and to give the reader a sense of the origin and evolution of key ideas and observational methods used to study the shelf circulation. We shall next describe the nature and structure of atmospheric forcing over the Middle Atlantic Bight because the early moored-array work demonstrated that much of the subtidal current variability observed in the Middle Atlantic Bight is directly wind driven. We shall next describe what is known about the temporal and spatial structure of the wind-driven subtidal transient circulation on both the synoptic (2-to-10-day) time scale and the longer monthly time scale. The observed mean current field and ideas about how it is driven and maintained will be discussed at the end.

7.3.1 Physiographic Setting
Uchupi's (1965) bathymetric map shows that the shelf topography within the Middle Atlantic Bight is relatively simple and smooth in comparison to the more complex topography within the Gulf of Maine and Scotian Shelf region. The depth within the Middle Atlantic Bight generally increases in a monotonic fashion from shore out to the shelf break. The depth of the shelf break decreases from about 150 m south of Georges Bank to about 50 m off Cape Hatteras. The width of the shelf from shore to shelf break is generally about 100 km except near Cape Hatteras, where the shelf becomes very narrow (about 50 km), and near New York, where the New Jersey and Long Island coasts form a corner region making the shelf there about 150 km wide. Both the mean depth and cross-sectional area of the shelf decrease roughly by a factor of two from the New England shelf to off Cape Henry. The continental slope is indented by many submarine canyons, but only a few penetrate up onto the outer shelf. Several drowned river channels partially cross the shelf, the most notable being the Hudson River Channel off New York (see figure 7.1).

Milliman, Pilkey, and Ross (1972) have mapped the superficial sediments over the eastern United States continental margin and find the Middle Atlantic Bight to be covered mostly with medium-sized sand. Finer-grained sediments are found in a large region southwest of Nantucket, near the major estuaries, and generally seaward of the shelf break. A wide spectrum of small-scale morphological features exists over much of the shelf, ranging from wave-formed ripples 10 to 15 cm long and 1 to 10 cm high up to large-scale ridges 2 to 4 km long and up to 10 m high. Intermediate-scale features such as sand waves of varying size are frequently superimposed on the larger-scale features. The topography in the transition region between estuary and inner shelf is complex and most estuaries within the Middle Atlantic Bight have at least one relatively deep channel connecting the estuary and shelf. These larger-scale topographic features can influence currents through both topographic steering and generation of horizontal eddies, while the smaller-scale features can exert a significant form drag on the flow. More detailed descriptions of the Middle Atlantic Bight bottom topography, superficial-sediment distribution, and ideas about the formation of these features are given by Emery and Uchupi (1972), Swift, Duane, and McKinney (1973), Swift et al. (1976), Freeland, Swift, Stubblefield, and Cok (1976), and Freeland and Swift (1979).

7.3.2 Early Development of Ideas about the Shelf Circulation
It was considered accepted knowledge before 1915 that a rather sharp transition zone existed near the shelf break between the generally cooler and fresher "coastal" water found over the shelf in the Middle Atlantic Bight and the generally warmer and more saline "Gulf Stream" water found offshore.[3] The textbooks and ocean atlases of this early period [e.g., Findlay (1853), Maury (1855), and the current chart of the U.S. Navy published by Soley (1911)] showed the coastal water to be generally moving slowly toward the southwest along the shelf from Nova Scotia to Cape Hatteras. The low temperature and salinity of the coastal water suggested a northern origin, and Verrill (1873), among others, emphasized that the coastal currents supported a boreal littoral fauna rather than the warm-water fauna characteristic of the Gulf Stream. The cold coastal water had been mapped as far north as Newfoundland and most oceanographers like Libbey (1891, 1895) and Sumner, Osburn, and Cole (1913) believed that the Labrador Current flowed along the coast from the Grand Banks past Nova Scotia and the Gulf of Maine into the Middle Atlantic Bight and perhaps even as far south as Florida. This belief was modified when Schott (1897) and Dawson (1913) showed,

using direct-current as well as temperature and salinity measurements, that the outflow of the Gulf of St. Lawrence via the Cabot Straits is the primary source of coastal water on the Scotian Shelf. The British Admiralty (1903) charts show this coastal water flowing toward the southwest into the Gulf of Maine at Cape Sable, where the current either turned northward toward the Bay of Fundy, or became too diffuse to determine from the mariner reports.

In 1912, H. B. Bigelow began a remarkable series of cruises that provided the first comprehensive description of the hydrography, circulation, and biology of the Middle Atlantic Bight and Gulf of Maine region. Bigelow had first gone to sea as a college undergraduate with Alexander Agassiz in 1902 (Schlee, 1973), and after finishing his doctorate at Harvard in 1906, he joined Agassiz as a research assistant at the Museum of Comparative Zoology, where he spent much of his time describing and classifying jellyfish collected on Agassiz's expeditions. In 1908, an ailing Agassiz directed Bigelow to conduct a short cruise of his own across the continental shelf to collect animals from the Gulf Stream, which Bigelow did aboard the Bureau of Fisheries' 90-foot schooner *Grampus*. Agassiz died in the summer of 1910, and Bigelow spent the next year working on jellyfish at the Museum and reading about the research being conducted in the eastern North Atlantic by Scandinavian scientists under the guidance of J. Hjort, the Director of the Norwegian Board of Sea Fisheries. Then Sir John Murray visited Harvard in 1911 and convinced Bigelow to leave the laboratory for a time and launch his own expedition, which he eagerly did the following summer aboard the *Grampus* (Schlee, 1973).

It seems clear that Bigelow, in developing his own field program, was strongly influenced by both Hjort's systematic approach to oceanographic research (see Schlee, 1973) and several key technological advances made by the Scandinavians in the period 1900–1910 (see chapter 14). Knudsen (1901) had prepared tables for conveniently calculating salinity and density at atmospheric pressure (σ_t) from values of temperature and chlorinity, Ekman had developed a mechanically recording propeller-type current meter [see von Arx (1962) for a description] that could be used from an anchored ship, and Nansen had perfected a practicable reversing water sampler with an attached thermometer. Equipped with these new tools, plus a variety of improved biological and geological sampling gear, and sponsored by the Bureau of Fisheries and the Museum of Comparative Zoology, Bigelow and his coworkers set sail on the *Grampus* in July 1912 to study the hydrography, currents, and biology of the Gulf of Maine. Bigelow conducted a similar research cruise in the next summer.

In 1915, Bigelow published his first tentative chart (shown here in figure 7.4) of the summer surface circulation for the Gulf of Maine and the Middle Atlantic Bight region, as inferred from the July 1913 cruise. Bigelow stated that "the combined evidence of the various records of ocean currents, our own included, points to the conclusion that the dominant drift over the continental shelf south of New York is to the southwest; and this is certainly the prevalent opinion of practical navigators and hydrographers" [p. 232]. His chart suggested the importance of runoff from the major estuaries within the Middle Atlantic Bight in both the surface salinity and current patterns. Bigelow also speculated that the cold bottom water found in the Middle Atlantic Bight was formed locally in the previous winter and was essentially static and not advected into the Middle Atlantic Bight from the east. He (1922) found further support for this idea in the August 1916 data. While the idea of a mean near-surface drift toward the southwest in the Middle Atlantic Bight has been confirmed by more modern measurements, the concept of the cold bottom water as static was clearly refuted when direct-current measurements began in the 1970s.

Bigelow was primarily interested in the Gulf of Maine during this period, however, and after a brief interruption due to World War I, he resumed his field work and began to focus more on the circulation there. He began to release surface drift bottles along strategic sections within the Gulf of Maine and also experimented with E. Smith with the Scandinavian method for geostrophic-current computation.[4] In 1927 Bigelow's monograph on the physical oceanography of the Gulf of Maine was published by the Bureau of Fisheries. Using hydrographic data and geostrophic computations as well as current information inferred from the movement of fish eggs and larvae and drift bottles, Bigelow developed a rather accurate conceptual model of the general circulation of the Gulf of Maine on a seasonal time scale, which has become the foundation for all subsequent work in this region. He described the springtime formation of a counterclockwise circulation around the basin (called the Gulf of Maine gyre) and a clockwise circulation around Georges Bank (the Georges Bank gyre). He recognized that slope water characterized by $S \geq 35\textperthousand$ penetrated through the Northeast Channel into the deeper basins of the Gulf of Maine and that this water mixed with very fresh shelf water from the Scotian Shelf and farther north to form the intermediate salinity water found in the Gulf. Bigelow's schematic near-surface circulation diagram (1927, p. 973) showed that at least during the summer (when drift-bottle returns were highest), some shelf water flowed westward past Nantucket Shoals into the Middle Atlantic Bight.

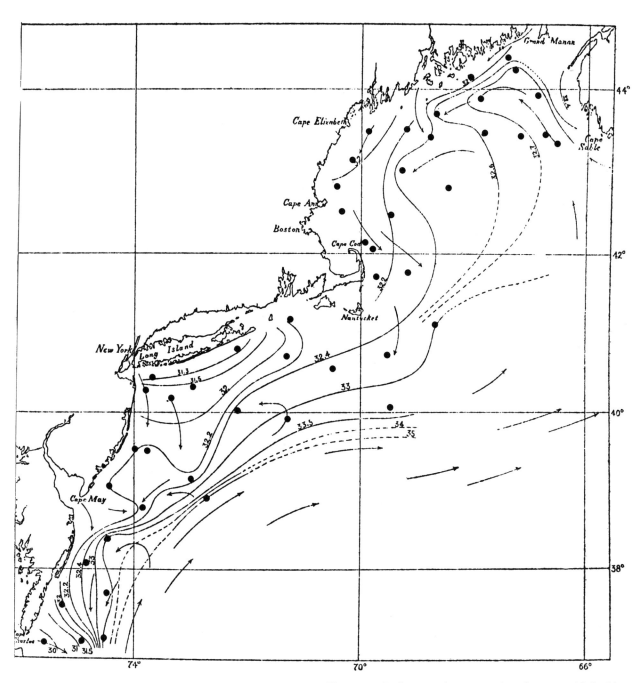

Figure 7.4 Surface circulation map for July 1913 published by Bigelow (1915). Surface salinities are shown and dots have been added to show hydrographic station locations.

Summarizing their past work and incorporating some new measurements made aboard the *Atlantis*,[5] Bigelow (1933) and Bigelow and Sears (1935) produced the first complete description of the seasonal temperature and salinity fields within the Middle Atlantic Bight. They found that vernal warming and fresh-water runoff built a strong stratification during the late spring and summer months, which was subsequently destroyed in the fall and early winter by surface cooling and winter storms. Bigelow and Sears recognized that shelf water represented a mixture of continental runoff and the more saline slope water and documented the basic structure of the transition zone between these two water masses. The transition from shelf to slope water often occurred as a sharp outward-sloping front located near the shelf break during winter, while the front was less distinct in summer because of the development of a seasonal thermocline in the adjacent slope water. Large temperature and salinity gradients still persisted in the offshore direction below the seasonal thermocline on account of a band of cold, low-salinity shelf water that was located near the bottom on the outer shelf and was described by Bigelow (1915, 1922, 1933) as a remnant from the previous winter cooling. Bigelow incorrectly visualized an essentially static pool of cold bottom water extending from south of Long Island to Cape Henry that was entirely surrounded by warmer water and persisted without replenishment through the summer.

In the late 1930s, Bigelow's personal research returned to fish and he did not write further about coastal circulation per se. Bigelow and C. Iselin did encourage a 3-year interdisciplinary field study of the Georges Bank region and its high biological productivity (Schlee, 1978), and, although it was stopped early in 1941 by World War II, this study did produce a number of biological and ecological papers, including one on ecosystem modeling (Riley, Stommel, and Bumpus, 1949). Based on his own work on slope water and the Bigelow-Sears picture of the Middle Atlantic Bight hydrography, Iselin (1939b) stated without discussion that "the coastal waters, because of their relative freshness, are at most times of the year less dense than the corresponding layer offshore and consequently a current is maintained which for some reason not clearly understood, tends to have its greatest strength just outside the 100-fathom curve." The idea that the geostrophic balance represented a driving mechanism was apparently a common misconception. Iselin clearly believed that the density distribution over the shelf and slope was the principal driving mechanism of the shelf circulation. The maximum horizontal density gradients occurred in the frontal zone near the shelf break, so with an assumed level-of-no-motion near the bottom, as suggested by Bigelow (1915, 1922, 1933), the surface geostrophic current would be a maximum, and directed toward the southwest along the shelf break. Iselin (1939b, 1940b) did correctly point out that the generally observed increase of salinity with depth over the shelf in the Middle Atlantic Bight implies an offshore motion near the surface and an onshore flow at depth.

The first dynamic model for the Middle Atlantic Bight circulation was published in *The Oceans* by Sverdrup, Johnson, and Fleming (1942). The circulation scheme shown in figure 7.5 is taken from *The Oceans* and indicates a drift of coastal (meaning shelf) and slope water along the continental margin towards the southwest in the correct sense. According to Sverdrup, Johnson, and Fleming (1942, pp. 677-680), precise geodetic leveling experiments conducted in the early 1930s indicated that mean coastal sea level rose between Cape Hatteras and Cape Cod by some 10 cm. The north-south gradient of mean atmospheric pressure was known to be small enough that oceanographers believed that these measurements indicated a *real* northward rise in the absolute sea-surface topography. Since the Gulf Stream presumably did not run uphill, and Dietrich (1937) had "showed" that the northward surface slope was *not* caused by a northward decrease in mean density along the slope, Sverdrup inferred that the sea surface had the profile labeled 2 in figure 7.5, which would imply a southwestward geostrophic current over the shelf with a maximum near the shelf break as argued by Iselin (1936, 1939b). Sverdrup stated that "a current to the south must also flow over the

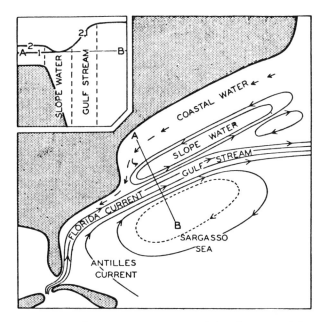

Figure 7.5 Schematic representation of the character of the Gulf Stream, taking results of precise leveling into account. Inset: Profiles of the sea surface along the line A-B. Profile 1 derived from oceanographic data only; Profile 2, from these data and the results of precise leveling. [Circulation scheme given by Sverdrup, Johnson, and Fleming (1942).]

shallow portion of the shelf where it flows downhill and where the balance of forces is maintained by the effect of friction" [Sverdrup, Johnson, and Fleming (1942, p. 678)]. Sverdrup speculated that this surface topography pattern was caused by large-scale wind forcing but his reasoning was vague. Even though the accuracy of the geodetic leveling has been disputed by Sturges (1968) and others, recent circulation models also invoke a mean alongshore pressure gradient. This point will be discussed again below (and see chapter 4).

Haight (1942) published the first long-term *surface*-current observations made in the Middle Atlantic Bight and Gulf of Maine region. The U.S. Coast and Geodetic Survey had a 30-year-long cooperative program with the Lighthouse Service and the Coast Guard to measure surface currents using the current drift-pole technique at lightships and other stations on the shelf. Haight presented quite accurate charts for the tidal currents and summary charts for the nontidal or mean and wind-driven currents. This tidal-current information and other direct measurements made in the estuaries and harbors form the basis for the current roses found on today's navigation charts.

World War II stopped active research on the Middle Atlantic Bight and Gulf of Maine, and although a number of useful instruments like the bathythermograph (BT) and Loran A were developed and perfected, and many BT profiles were taken over the shelf, most oceanographers were busy with defense-related research, and active work on the Middle Atlantic Bight did not resume until the late 1940s. Spilhaus and Miller (1948) modified the BT to obtain discrete water samples while ascending, and Spilhaus, Ehrlich, and Miller (1950) and Miller (1950) then used this new instrument to examine the shelf-slope water front south of New England. Miller (1950) found evidence for significant mixing across the deeper σ_t-surfaces near the shelf break, which he attributed to internal wave breaking in the frontal zone. Ford, Longard, and Banks (1952) found narrow filaments of relatively cold and fresh water along the shoreward edge of the Gulf Stream north of Cape Hatteras, and Ford and Miller (1952) correctly surmised that this water was, in fact, shelf water from the Middle Atlantic Bight entrained along the edge of the Gulf Stream near Cape Hatteras.

In 1950, the National Lead Company began to dump acid-iron waste from barges in the New York Bight and a number of oceanographers were asked by the National Research Council to examine the environmental effects and estimate the flushing time for the New York Bight. The results of this work were reported by Redfield and Walford (1951) and Ketchum, Redfield, and Ayers (1951). A concerted effect was also made in the early 1950s to understand and model the circulation and mixing within estuaries. This effort was in part stimulated by concern over the environmental impact of waste disposal within rivers and estuaries, and it produced a number of key ideas [e.g., Ketchum's (1950) tidal prism method to compute flushing times, and Stommel's (1953b) method for estimating the longitudinal diffusion coefficient in a well-mixed river or estuary from the observed salinity field]. Ketchum and Keen (1955) segmented the Middle Atlantic Bight from Cape Hatteras to Cape Cod and computed the flushing times for each segment, assuming only cross-shelf mixing and advection. They concluded that a considerable amount of cross-shelf transport of salt and river water must occur in both winter and summer to account for the observed mean salinity field. Ketchum and Corwin (1964) later examined the water structure south of Long Island over the period 1956–1959 and incorrectly concluded that the cool bottom water was formed only by local winter cooling, and then was warmed up by mixing with either warmer surface water or warmer slope water.

Both Bigelow and Iselin had long been aware of eddy-like features in the near-surface water structure over the shelf and slope region. In his interpretation of drift-bottle data obtained in the Middle Atlantic Bight in the spring of 1951, Miller (1952) suggested that a series of distinct current branches or eddies was superimposed on the general southwest alongshore drift. This work, plus the growing evidence of current variability in the Gulf Stream obtained by Fuglister and Worthington (1951) in Operation Cabot, led Iselin (1955) to urge that new observational methods be developed to study the Middle Atlantic Bight circulation. Iselin suggested a several-year program of continuous measurements of meteorological and oceanographic variables using new instruments deployed in moored arrays. Although others besides Iselin had also considered the potential of long-term moored-array measurement programs, the instrumentation and mooring technology required for such a program were simply not available yet. In 1954, D. Bumpus, C. Day, and J. Chase did start a cooperative program (Bumpus, 1955) with the U.S. Coast Guard to collect daily temperature and salinity measurements as well as meteorological observations at lightships and light stations in the Middle Atlantic Bight and Gulf of Maine. This collection program ran through the 1960s and provided the data used by Chase (1959), Howe (1962), Chase (1969), and Bumpus (1969) to examine the influence of runoff and wind on nearshore salinities and surface currents.

Bumpus, Miller, and others had frequently released drift bottles during hydrographic and other cruises in the Middle Atlantic Bight, and Bumpus and Lauzier (1965) summarized the results of both American and Canadian drift-bottle work conducted from 1948 to 1962. Frustrated by the fact that a standard drift bottle provides only a "birth-and-death" notice and little hard information about the drift in between, Bumpus (1956) and Bumpus et al. (1957) experimented with radio-tracked surface drifters (the "talking drift bottle") in the mid-1950s, but this effort was dropped by Bumpus as too expensive and inefficient (Bumpus, personal communication). Howe (1962) did use radio-tracked buoys with parachute drogues to make some short-term current measurements over the middle and outer shelf. In 1960, Bumpus launched a massive 10-year surface drift-bottle program (some 150,000 bottles released) over the eastern United States shelf, and in 1961 he began an equally massive 10-year bottom-drifter program (75,000 drifters released), using the newly developed Woodhead seabed drifter described by Lee, Bumpus, and Lauzier (1965). The essential idea behind this work was to seed the shelf water with a network of drifters at least monthly over a 10-year period in order to determine the annual cycle of surface and bottom drift from the inferred trajectories of the field of drifters. At about the same time, Lauzier began a separate, more modest long-term drifter program over the eastern Canadian shelf, and he and Bumpus decided to share their data collected in the Gulf of Maine region. The results of this work were presented in stages by Bumpus (1965, 1969), and in his final summary report (Bumpus, 1973).

Bumpus found that throughout the year there was a strong nearshore movement of bottom drifters into or at least toward the mouths of the major estuaries within the Middle Atlantic Bight. He also found that bottom drifters over the mid-shelf region in the Middle Atlantic Bight primarily moved southwestward with a mean speed of a few centimeters per second, with no significant seasonal variation in either pattern or inferred speed. Because very few bottom drifters deployed at depths greater than about 60 to 80 m over the outer shelf were recovered, Bumpus concluded that a line of divergence existed in the bottom flow. The recovery rate in the surface-drifter program was more meager because offshore winds in the fall through early spring dramatically reduced the percentage of drifters returned except from very near shore. The observed summer surface drift was southwest all along the coast except during prolonged periods of strong northward winds and low runoff, when the nearshore surface flow was reversed and became northeastward.

This drifter work summarized by Bumpus (1973) provides the first general picture of the mean and seasonal surface and bottom circulation in the Middle Atlantic Bight. Bumpus (1973) concluded that a mean longshore flow of order 5 cm s^{-1} occurs between Cape Cod and Cape Hatteras, and that Nantucket Shoals and Cape Hatteras appear to be oceanographic barriers that limit in some sense the alongshore flow. Near Cape Hatteras the alongshore flow turns seaward and becomes entrained in the Gulf Stream, a process discussed first by Ford and Miller (1952) and more recently by Fisher (1972) and Kupferman and Garfield (1977).

7.3.3 Recent Developments

The drifter work undertaken by Bumpus and others in the 1960s provided the first quantitative description of the mean and seasonal surface and bottom circulation in the Middle Atlantic Bight. In the 1970s, moored arrays of self-contained current meters and other *in situ* instrumentation have been deployed as part of several new field programs, and these new *direct*-current measurements are providing the first detailed description of the regional circulation in the Middle Atlantic Bight. It is important to note that much of the new *in situ* instrumentation used in the 1970s evolved from development efforts begun in the 1950s and 1960s, and it was not until the late 1960s and early 1970s that U.S. oceanographers began to use instrumented moored arrays as a routine and reliable "tool" to measure current, temperature, and other physical variables in both the deep ocean and the shallower continental shelf and lakes.[6] The development of solid-state electronics, digital computers, and time-series analysis methods in the late 1950s and 1960s especially helped make the current-meter development efforts successful. This instrumentation revolution in the 1950s and 1960s has had a tremendous impact on the field of physical oceanography in the 1970s, for it has allowed in many cases a first direct look at a wide spectrum of oceanic motions and phenomena [see Gould (1976) and chapter 14].

The availability of these new observational tools plus an increased public concern over environmental issues helped motivate much of the new field work begun in the Middle Atlantic Bight in the 1970s. National concern over the environmental impact of marine waste disposal in the Middle Atlantic Bight became paramount in the late 1960s and early 1970s. The public media characterized the New York Bight—the coastal ocean between Long Island and New Jersey—as a "dead sea" caused by decades of sewage discharge and marine dumping of dredge spoils, building rubble, sewage sludge, industrial wastes, and other materials (Gross, Swanson, and Stanford, 1976). The NOAA Marine EcoSystems Analysis (MESA) Program was formed in 1972 to help focus both government and nongovernment research on regional problems caused by man's use of marine and estuarine resources, and, in 1973, the MESA New York Bight Project was started to develop a comprehensive research program to understand the New York Bight as a productive marine ecosystem. Other environmental concerns also helped initiate new field programs within the Middle Atlantic Bight. The New Jersey Public Service and Electric and Gas Company and the Long Island Electric Power Company both began to consider building floating nearshore nuclear-power plants, and New Jersey Public Service sponsored a 5-year field study of the physical oceanography off Little Egg Inlet, New Jersey, a potential power-plant site, while the Brookhaven National Laboratory began in 1973 a long-term program to study the nearshore currents, pollutant dispersion, and primary productivity off the southern coast of Long Island. The Bureau of Land Management of the Department of Interior also sponsored new field programs to identify and study the dominant sediment-transport processes in the Baltimore Canyon and Georges Bank regions in order to assess better the possible physical hazards and environmental impact of potential offshore petroleum development there.

These specific field programs plus others undertaken with National Science Foundation, Office of Naval Research, and other federal support have provided most of our new knowledge about the mean circulation and low-frequency current variability in the Middle Atlantic Bight. Using moored instrumentation, Boicourt (1973) and Beardsley and Butman (1974) showed that strong winter storms could produce alongshore currents of order 20 to 50 cm s^{-1} in the mid-shelf region, and Boicourt and Hacker (1976) demonstrated that the low-frequency alongshore-current fluctuations were spatially coherent over a 200-km separation along the 38-m isobath in the southern Middle Atlantic Bight. Some preliminary results from several of these field programs were then presented by Beardsley, Boicourt, and Hansen (1976), who demonstrated that much of the subtidal current variability over the shallower portion of the Middle Atlantic Bight was directly wind driven in the synoptic (2- to 10-day) band. Their map of the directly observed mean subsurface currents demonstrated that at least during the 1-to-3-month duration of the initial moored-array experiments, the subsurface currents did flow alongshore toward the southwest, and that the average currents generally increased in magnitude offshore and decreased with closeness to the bottom. At most sites, the mean current veered toward shore with increasing depth. The summer current measurements made off New York and Cape Henry showed that the alongshore currents in the near-bottom cold pool equaled or exceeded the alongshore currents in the surrounding warmer water, clearly indicating that the summer cold pool is not the static feature originally envisaged by Bigelow and others.

Beardsley et al. (1976) also used the mean-current observations to estimate crudely the alongshore volume transport of shelf water through three transects across the Middle Atlantic Bight to the 100-m isobath. They found that the three estimates were surprisingly uniform considering the different time periods of the measurements and the different instrumentation used, and they suggested a value of 250×10^3 m^3 s^{-1} for the average alongshore transport of shelf water within the 100-m isobath through the Middle Atlantic Bight. This mean value, when divided into the volume of the Middle Atlantic Bight, implied a mean residence time of the order of 0.75 years. This value, based on alongshore advection, is somewhat less than the value of 1.3 years based solely on cross-shelf mixing given by Ketchum and Keen (1955). Beardsley et al. (1976) also speculated that most of the shelf water observed flowing westward south of New England must, by continuity, flow into the Middle Atlantic Bight around Nantucket Shoals from the southern flank of Georges Bank and the Gulf of Maine region.

Longer current records have now been obtained in the Middle Atlantic Bight, and, in the rest of this section, we shall reexamine the preliminary circulation picture described in 1976. Since much of the subtidal current variability over the shelf is wind driven, we shall first describe the nature and structure of the atmospheric forcing found over the Middle Atlantic Bight in some detail, and then describe what is known about the wind-driven shelf response on three time scales: the synoptic 2-to-10-day time scale, the monthly-mean time scale, and the long-term mean time scale.

Atmospheric Forcing The temporal and spatial structure of the surface-wind-stress and pressure fields found over the Middle Atlantic Bight will be described next. Atmospheric motions are commonly classified into micro-, meso-, or synoptic-scale motions, depending on their characteristic time and space scales. While the exact limits of these scales are somewhat ill defined (Fiedler and Panofsky, 1970; Orlanski, 1975), synoptic-scale motions are generally characterized by periods in excess of 2 days and horizontal-length scales in excess of 500 km. The synoptic-scale weather disturbances are generally caused by baroclinic instability, while the principal mesoscale phenomena in the Middle Atlantic Bight—the summer seabreeze and cellular convection—are primarily associated with mechanical and hydrostatic instabilities (Mooers, Fernandez-Partagas, and Price, 1976; Mayer, Hansen, and Ortman, 1979). Because the early direct oceanic wind measurements by Millard (1971) and more recent work demonstrate that the synoptic-scale disturbances cause most of the observed surface-wind variance over the shelf and open ocean, we shall focus this discussion on the synoptic-scale surface weather over the Middle Atlantic Bight.

Synoptic-scale surface forcing: Mooers, Fernandez-Partagas, and Price (1976) have examined in great detail a variety of atmospheric-data sets to construct the first comprehensive picture of synoptic-scale atmospheric forcing over the Middle Atlantic Bight. The Atlantic coastal region between North Carolina and New England is well known for intense cyclogenesis. The combination of cool continental air over the eastern United States and warm moist maritime air offshore causes the synoptic-scale low-pressure disturbances or cyclones both to intensify with time and to tend to propagate toward the northeast along the coast. The temperature contrast between continental and maritime air masses is greatest in winter, so that winter cyclones are usually the most intense and frequent (averaging 5 per month), and cause most of the synoptic-scale variability. The summer cyclones are generally much weaker and less frequent (averaging 2.5 per month), though strong storms can occasionally occur in summer (see Mather, Adams, and Yoshioka, 1964). Hurricane-strength winds occur in the Middle Atlantic Bight on average once in 6 years.

Mooers et al. (1976) have constructed a model winter cyclone based on a case study of 34 synoptic disturbances occurring in the Middle Atlantic Bight during the winters of 1972–1974. While quite simplified, their model cyclone exhibits a number of features that appear to characterize many winter cyclones. The model cyclone forms near Cape Hatteras and propagates toward the east and north as shown in figure 7.6. The central pressure deepens, and both the spatial scale of the cyclone and the strenth of the surface wind stress increase with time. The surface wind stress and the surface heat flux into the atmosphere are a maximum in the western sector of the storm, where cold continental air flows out across the shelf and pronounced mesoscale cellular convection occurs (Burt and Agee, 1977). Pronounced warm-front-type precipitation occurs in the northern sector and causes the net precipitation to exceed evaporation along the storm track. The model cyclone develops a surface wind-stress pattern which is coherent in both time and space over the extent of the Middle Atlantic Bight, and the maximum wind stresses are larger than the winter mean stress in the Middle Atlantic Bight by a factor of 2 to 4. While real storms are more complex and follow a variety of different paths both north and south of the Middle Atlantic Bight, the model cyclone provides a useful conceptual framework of the spatial and temporal evolution of the surface wind stress and wind-stress curl fields from an event point of view.

Mooers et al. (1976) also examined surface pressure and wind data collected at coastal stations along the Middle Atlantic Bight and the Gulf of Maine region, and found, using coherence and correlation analysis, that (a) the surface atmospheric pressure (p_a) field was spatially coherent over the entire Middle Atlantic Bight for time scales greater than 2 days, (b) the p_a and east and north wind-stress components τ_E and τ_N have (zero-time lag) correlation scales of about 1600, 600, and 800 km, respectively, and (c) the p_a-, τ_E-, and τ_N-fields generally propagate toward the northeast with speeds of 15, 5, and 10 m s^{-1}. They argue that the stress field is not "frozen" into the cyclone as defined by the surface pressure field, and that the difference in phase speeds between p_a and the stress components is associated with the tendency for winter storms to intensify in time, as illustrated in the model cyclone.

Two NOAA environmental buoys designated EB34 and EB41 were deployed in the central Middle Atlantic Bight in late 1975. EB34 (40.1°N, 73.0°W) was located 50 km southeast of New York City and EB41 (38.7°N, 73.6°W) was located 100 km off the New Jersey coast (see figure 7.7). Both buoys were equipped with vortex-shedding anemometers at a height of 5 m, and routinely transmitted 8.5-minute averages of wind speed and direction to shore. The initial wind data collected with these buoys have been studied by Mooers et al. (1976), Williams and Godshell (1977), Overland and Gemmill (1977), Noble and Butman (1979), and Mayer et al. (1979); Wang (1979c) has examined wind variability at the Chesapeake Lighttower. These studies collectively demonstrate three key features of the synoptic-scale surface pressure and wind-stress fields: (a) the winter-buoy data are highly coherent with coastal data, a result which is consistent with the correlation-scale information obtained by Mooers et al. (1976), and implies that their results apply to the surface pressure and wind-stress fields over the shelf; (b) the synoptic-scale surface wind stress increases in magnitude by a factor of 2 to 3 across the shelf; and (c) a significant cyclonic verring of the surface stress field by as much as 30° may occur across the shelf.

We show in the upper half of figure 7.8 wind-stress spectra obtained at Atlantic City, EB41, and at Ocean Weathership C (52.5°N, 35.5°W) located east of the Grand Banks in the western North Atlantic. The wind

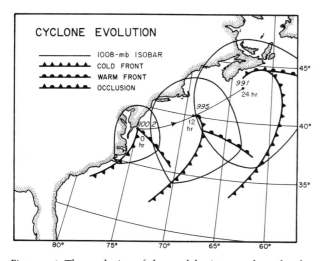

Figure 7.6 The evolution of the model winter cyclone developed by Mooers, Fernandez-Partagas, and Price (1976). The 1008-mb contour outlining the cyclone is shown at 0, 12, and 24 hours after the storm center has left the coast. The deepening of the central pressure is also shown along the storm track.

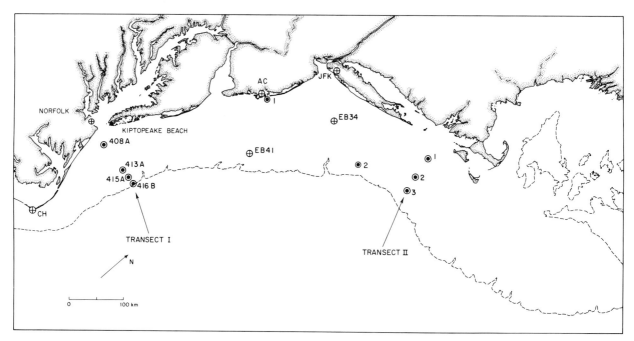

Figure 7.7 Map of the Middle Atlantic Bight showing the 200-m depth contour (dashed) and the wind (⊕), sea-level (▲), and current-meter (◉) stations discussed in the text.

stress has been computed using the quadratic drag law, the observed or adjusted 10-m-high wind vector, and a constant drag coefficient of 1.5×10^{-3}. The Atlantic City and EB41 spectra are computed by Noble and Butman (1979) from 6-month-long time series obtained from December 1975 through May 1976, while the Weathership C spectrum reported by Willebrand (1978) is based on a 25-year-long time series. See table 7.1 for more details. We have included the Weathership C spectrum since the weathership was located some 2500 km northeast of the Middle Atlantic Bight along the principal storm track followed by winter cyclones that develop along the Atlantic coastal region, and it represents the closest station (known to us) for which a well-resolved wind-stress spectrum has been published. The spectra have been smoothed within the estimated statistical uncertainty to simplify the graphical presentation. (See the additional spectra in chapter 11.)

The three spectra demonstrate that most of the wind-stress variability is caused by rather broadband synoptic-scale atmospheric transients with characteristic periods between 2 and 10 days. The spectra are red (with increasing power densities at decreasing frequencies), and show an approximate −2 power-frequency decay at frequencies above about 0.5 cpd. The two winter shelf spectra do not exhibit a diurnal peak since the sea breeze is a summer phenomenon (Mayer et al., 1979). While the Atlantic City and EB41 time series are too short to determine clearly the very low-frequency dependence, the Weathership C spectrum follows a well-defined −0.4 power-frequency dependence between 0.2 cpd and the annual frequency 0.003 cpd.

The two Middle Atlantic Bight spectra demonstrate the marked increase in wind-stress magnitude from nearshore toward the shelf break. The EB41 power density is about a factor of 8 larger than the Atlantic City density over most of the frequency bands resolved. The Weathership C power density is even a factor of 5 larger than the EB41 power density. Willebrand (1978) has examined surface pressure and wind-stress data obtained at both Weathership C and Ocean Weathership D (44°N, 41°W), located roughly about half-way between the Middle Atlantic Bight and Weathership C. He finds that synoptic-scale disturbances in the 2-to-10-day band do propagate toward the northeast in the western North Atlantic but that at periods greater than about 10 days, atmospheric pressure fluctuations seem to have no preference for east-west phase propagation. Between the two weatherships, τ_N was coherent at all periods greater than 1 day, while τ_E was incoherent at periods greater than 10 days. These results suggest that the larger synoptic-scale transients continue to intensify as they move northeastward toward the Grand Banks region.

We have focused on the traveling winter cyclone as the predominant synoptic-scale disturbance that influences the Middle Atlantic Bight, and have described the frequency structure and correlation spatial-scale information presently known in some detail. The surface pressure and wind-stress-correlation space scales are sufficiently large in comparison to the cross-shelf and alongshelf dimensions of the Middle Atlantic Bight

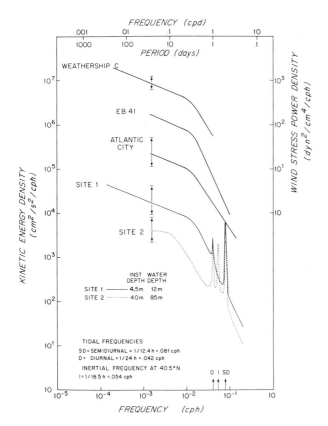

Figure 7.8 Wind-stress and current spectra. The three upper curves represent the wind-stress spectra for Ocean Weathership C, environmental buoy EB41, and Atlantic City, New Jersey. The two lower curves are the kinetic-energy spectra obtained at a nearshore site (site 1) off New Jersey and a deeper site (site 2) located near the shelf break south of New England (see table 7.1). The vertical brackets indicate the 95% confidence limits.

that the synoptic-scale surface forcing should be spatially coherent over much of the shelf.[7] Frankignoul and Müller (1979) have recently reviewed the meager information available on the wavenumber structure of the surface wind-stress field over the ocean, and suggest several forms for the wavenumber spectra for the synoptic-scale surface pressure, wind-stress, and wind-stress-curl fields. Their model wind-stress spectrum is essentially white for wavenumbers $k \ll k_b = 2\pi/5000$ km^{-1} (which is a wavenumber magnitude characteristic of mid-latitude baroclinic instability), and decays at smaller scales like k^{-2} for $k > k_b$. The smaller-scale surface wind-stress fluctuations appear to be horizontally isotropic at higher wavenumbers $k > 2\pi/200$ km^{-1}. If we assume that the shape of the frequency spectrum obtained at Weathership C and the model wavenumber spectrum suggested by Frankignoul and Müller (1979) are both applicable to the Middle Atlantic Bight region, then we find that about 50% of the total wind-stress variance in the 1-day-to-1-year band is caused by synoptic-scale transients concentrated in the 2-to-10-day band, and some 70% of the wind-stress variance in the $2\pi/200$-to-$2\pi/10,000$-km^{-1} wavenumber band is caused by atmospheric motions with scales larger than 1600 km, which is twice the alongshelf length of the Middle Atlantic Bight.

Seasonal and mean surface forcing: Saunders (1977) has computed the seasonal surface wind-stress pattern with a 1°-square resolution over the eastern continental shelf of North America using 32 years of ship wind reports. His annual mean and three-monthly mean wind-stress maps show that, except in summer, the mean and seasonal wind stresses in the Middle Atlantic Bight are generally to the east and southeast and exhibit a significant increase in magnitude and some cyclonic veering with increasing distance from the coast. The relatively weak summer wind stress is directed toward the northeast and exhibits some anticyclonic veering relative to the coast. The standard deviation of the stress in nearshore and offshore regions is 1.5 and 2.5 dyn cm^{-2}, respectively, in winter, and 0.5 and 1.5 dyn cm^{-2} in summer. The offshore increase in the stress is most obvious in winter and spring seasons, and, since these periods dominate the mean stress, they cause the pronounced offshore increase observed in the mean stress pattern. The mean and seasonal wind-stress pattern over the adjacent western North Atlantic is described by Leetmaa and Bunker (1978) and will be discussed below. The annual air–sea interaction cycles and continental runoff are described in the appendix for completeness.

The Synoptic-Scale Shelf Circulation We shall now focus on the response of the Middle Atlantic Bight to synoptic-scale (2-to-10-day) atmospheric forcing. We shall first examine what is known about the temporal and spatial structure of the transient wind-driven shelf circulation, and then describe a conceptual model suggested by the existing current and sea-level data.

Temporal structure: We show in the bottom half of figure 7.8 kinetic-energy spectra computed from current records 6 months or longer obtained at two representative sites in the Middle Atlantic Bight. The two sites are labeled 1 and 2 and the locations and other pertinent information for each site are given in table 7.1. Site 1 is located 4.5 km off Little Egg Inlet, New Jersey, and site 2 is located on the outer New England shelf about 50 km east of the Hudson Canyon (see figure 7.7). The kinetic-energy spectra for sites 1 and 2 are from EG&G (1978) and Ou (1979) respectively, and have been smoothed within the estimated uncertainties to simplify the graphical presentation. It is important to remember that a significant spectral gap exists between energetic high-frequency motions characterized by periods from several seconds to minutes (associated with surface and internal gravity waves, and related wave and turbulent phenomena) and energetic lower-frequency motions characterized by periods

Table 7.1 Location and Other Pertinent Information for the Three Wind-Stress and Two Current-Kinetic-Energy Spectra Shown in Figure 7.8

Site	Location	Time span	Inst. height-depth (m)	Water depth (m)	Data source
Wind stress					
Atlantic City	39°27'N, 74°34'W	Dec. 75–May 76	6	—	Noble & Butman (1979)
EB41	38°42'N, 73°36'W	Dec. 75–May 76	5	—	Noble & Butman (1979)
Weathership C	52°30'N, 35°30'W	Jan. 45–Dec. 71	10	—	Willebrand (1978)
Current					
1	39°28'N, 74°15'W	Apr. 73–Dec. 76	4.5	12	EG&G (1978)
2	39°59'N, 71°54'W	Feb. 76–Aug. 76	38	83	Ou (1979)

greater than several hours. We do not show here the higher-frequency end of the kinetic-energy spectra, but simply note that significant kinetic energy can exist at high frequencies, primarily associated with gravity-wave phenomena. While the importance of surface waves on beach erosion and nearshore sediment transport is generally appreciated (Lavelle et al., 1976), the recent field observations by Lavelle, Young, Swift, and Clarke (1978) and Butman, Noble, and Folger (1979) demonstrate sediment resuspension by storm-generated surface waves in depths out to 85 m in the Middle Atlantic Bight, and Grant and Madsen (1979) describe how the combined motion of waves and a lower-frequency current over a rough bottom can lead to an increased bottom drag on the current. While the existence of the spectral gap allows the high-frequency current signal to be removed by vector averaging, the 5- to-15-second oscillatory currents and the mooring motion associated with surface gravity waves are the primary source of contamination in the measurement of lower-frequency currents with existing mechanical current meters and standard mooring techniques.

The two kinetic-energy spectra shown in figure 7.8 illustrate several fundamental features of the transient-current variability observed in the Middle Atlantic Bight. Both spectra are inherently red, with the kinetic-energy density generally increasing with decreasing frequency below 0.5 cpd. The marked similarity in shape between the three wind-stress spectra and the kinetic-energy spectrum at site 1 (especially the -0.4 frequency dependence below the break near 0.3 cpd) suggests that the subtidal current fluctuations over the shallow inner shelf are directly wind driven over a very wide frequency band, from 0.01 to 1.0 cpd. Approximately 50% of the subtidal current variance observed at site 1 occurs in the synoptic-scale (2-to-10-day) band. At site 2 near the shelf break, the subtidal kinetic-energy density is smaller because of the increased water depth and measurement level, and the spectral shape is different in that the transition or break in the subtidal portion of the spectrum occurs at a lower frequency, near 0.1 cpd. This indicates that the subtidal current variability observed over the outer shelf is caused by both direct wind forcing and by the transmission or leakage of lower-frequency motion onto the outer shelf from the deeper ocean. The propagation of topographic Rossby waves up the continental rise and slope, the meandering of the Gulf Stream, and the passage of anticyclonic warm-core eddies near the shelf break can conceivably generate strong low-frequency currents over the outer shelf. These three different mechanisms and some supporting observations are described by P. C. Smith (1978), Ou (1979), Halliwell (1978), and Scarlet and Flagg (1979).

Flagg (1977), EG&G (1978), Bennett and Magnell (1979), Mayer et al. (1979), Butman et al. (1979), and Chuang, Wang, and Boicourt (1979) have examined the coherence between the local wind stress and the alongshelf and cross-shelf current components and all find that alongshelf currents at all observed levels are significantly coherent with the local alongshelf wind stress over the synoptic-scale band. The reported coherence squared generally varies from about 0.5 to 0.8, with a poorly defined time lag of 4 to 10 hours between the alongshelf wind-stress and current components. To a lesser extent the cross-shelf current component is also significantly coherent with the local alongshelf wind-stress component. Neither current component is significantly coherent with the local cross-shelf wind stress except perhaps very near the surface (within a few meters) and near the coast in the New York Bight (EG&G, 1978; Csanady, 1980), where cross-shelf winds can set up trapped pressure fields that drive alongshore currents.

To illustrate the strong coherence between alongshore wind and currents, we show in figure 7.9 the coherence and phase between the alongshelf wind or wind stress and the alongshelf current observed at the two shelf sites just discussed. The coherences shown are significant and relatively high over the synoptic (2-to-10-day) band at both sites, and the phases indicate that the alongshelf currents lag the alongshelf wind or wind stress by about 4 hours at site 1 and about 10 hours at site 2. While the coherence at site 1 exhibits some seasonal change [with a tendency for higher coherence in the synoptic-scale band during less stratified winter periods (EG&G, 1978)], the mean coherence squared is relatively constant over a wide frequency band between 0.03 and 0.5 cpd, indicating the alongshore currents just off New Jersey are directly wind driven at very low frequencies down to the monthly time scale. The current record at site 2 is too short to resolve accurately the lower-frequency coherence cutoff.

Spatial structure: We have pictured next in figures 7.10 and 7.11 composite vector diagrams to illustrate the vertical and cross-shelf structure of the synoptic-scale current fluctuations in the Middle Atlantic Bight. In figure 7.10 the adjusted sea level at Kiptopeake Beach near Cape Charles, the Norfolk wind stress, and the subtidal currents observed on a cross-shelf transect off Cape Henry are shown for the summer period 24 July to 22 August 1974. In figure 7.11 the local wind stress (computed from surface pressure charts) and the subtidal currents observed on a cross-shelf transect south of New England are shown for the winter period 1 March to 31 March 1974. The locations of the two transects are shown in figure 7.7. The mooring number and local water depth are given on the right of each figure and the depth of the current measurements is given on the left. The local alongshelf direction is vertical in both figures and the current time series have been low-pass filtered to remove the tidal and higher-frequency components above about 0.7 cpd. The means have not been removed from each time series.

These two vector diagrams illustrate the following common features of the synoptic-scale current fluctuations. There is a clear coherence between synoptic-scale wind-stress events and current fluctuations, with both strong up- and downshelf currents being driven by up- and downshelf wind stresses.[8] The alongshelf current fluctuations are themselves visually coherent and roughly in phase in both the vertical and cross-shelf directions. The amplitudes of the alongshelf current fluctuations do not vary significantly in the cross-shelf plane except within the bottom boundary layer, and the empirical orthogonal-function computations of Flagg (1977) and Chuang et al. (1979) indicate that a single barotropic vertical mode can account for approximately 90% or more of the subtidal alongshelf-current variance in both winter and summer. The observed cross-shelf currents have a more complicated vertical structure, with strong upshelf wind stress driving offshelf flow near the surface and onshelf flow near the bottom, and strong downshelf wind stress driving onshelf flow near the surface and offshelf flow near the bottom. This tendency for the profile of the cross-shelf-current fluctuation to reverse with depth has been noted by Boicourt and Hacker (1976), Scott and Csanady (1976), Flagg (1977), EG&G (1978), Mayer et al. (1979), and Chuang et al. (1979). Flagg (1977) and Chuang et al. (1979) find that the lowest baroclinic empirical orthogonal mode and the barotropic mode must both be used to account for more than 90% of the cross-shelf-current variance. Mayer et al. (1979) and Chuang et al. (1979) find that the depth of the cross-shelf-velocity node varies with stratification over

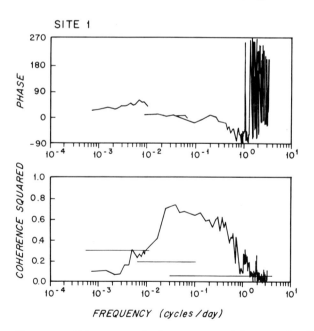
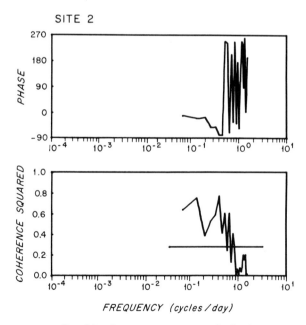

Figure 7.9 Coherence and phase computed between local alongshelf wind and current components at site 1 and between local alongshelf wind-stress and current components at site 2. Locations of the two shelf sites are given in Table 7.1 and the corresponding kinetic energy spectra at both sites are shown in figure 7.8. The horizontal lines indicate the 95% confidence limit for zero true coherence, and a negative phase indicates that the current lags the wind or wind-stress.

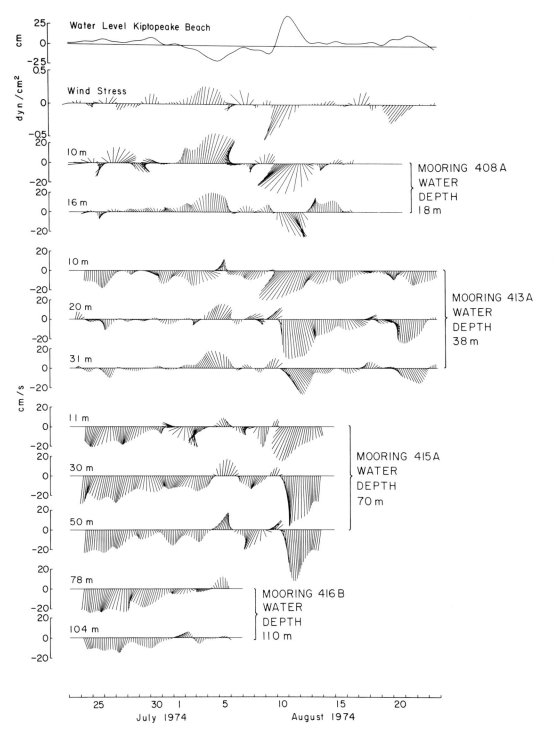

Figure 7.10 Summer vector time series of adjusted sea level at Kiptopeake Beach, Norfolk wind stress, and subtidal currents measured along the cross-shelf transect I located off Cape Henry. Current-measurement depths shown to the left and mooring numbers and local-water depths shown to the right. The transect mooring locations are shown in figure 7.7. North is upward, approximately parallel with the local along-shelf direction.

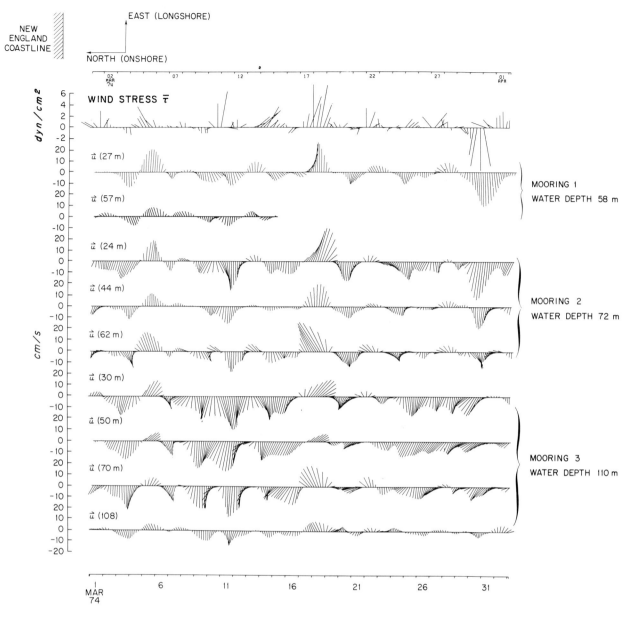

Figure 7.11 Winter vector time series of local wind-stress and subtidal currents measured along the cross-shelf transect II located approximately 100 km west of Nantucket Shoals. Current-measurement depths shown to left and mooring numbers and local-water depths shown to right. The transect and the mooring locations are shown in figure 7.7. East is oriented upward, approximately parallel with the local alongshelf direction.

the mid-shelf. The node is relatively deep during winter, and is significantly shallower in the summer when the seasonal pycnocline is well established.

Much less is directly known about the alongshelf structure of the synoptic-scale current fluctuations within the Middle Atlantic Bight. Boicourt and Hacker (1976), Butman et al. (1979), and Ou (personal communication) find that along an isobath over the middle and outer shelf, the alongshelf current component is highly coherent over alongshore separations up to 235 km, the largest separation examined, while the cross-shelf current component is incoherent over a 70-km separation, the smallest separation examined.

In summary, the synoptic-scale alongshelf-current fluctuations are generally coherent with the local alongshelf wind stress, and appear to have a relatively simple spatial structure throughout the year. The alongshelf-current fluctuations are essentially barotropic and spatially coherent in the cross-shelf plane. The alongshelf currents are also coherent over alongshelf separations up to 235 km, although the structure of the synoptic-scale atmospheric forcing suggests that significant coherence should be found over much larger separations. The generally weaker cross-shelf current component, although spatially coherent in the cross-shelf plane, appears to have a much smaller alongshelf coherence, of order 50 km or less.

A conceptual model: These current observations can be interpreted within the conceptual framework provided by continental shelf-wave theory. It is now recognized from a number of theoretical and experimental studies [see reviews by Mysak (1980) and Allen (1980) and chapters 10 and 11] that continental margins can act as effective waveguides for the alongshelf propagation of subinertial current fluctuations. The sloping topography of the continental margin and the density stratification over the shelf and slope are two basic conditions that lead to coastally trapped wave motions. The offshore increase in depth can support barotropic vorticity waves known as continental shelf waves in a homogeneous fluid, while internal Kelvin waves can propagate along a vertical boundary in a stratified fluid. Since the internal Rossby radius of deformation over the inner shelf in the Middle Atlantic Bight is of order 10 km or less, and is thus considerably smaller than the width of the shelf, the internal Kelvin wave activity, if present, should be trapped in a relatively thin coastal boundary layer. The theoretical and numerical calculations made by A. J. Clarke (1977) and Wang and Mooers (1977) imply that even with realistic stratification, the alongshelf currents for both forced and free continental shelf waves in the Middle Atlantic Bight should be essentially barotropic over the shelf, and coherent with the local coastal sea level.[9] The cross-shelf momentum balance is essentially geostrophic in both free and forced shelf waves, so that in theory the subsurface pressure fluctuations should have a simple monotonic cross-shelf structure, with a maximum amplitude at the coast and a vanishing amplitude off the shelf. Beardsley et al. (1977) have examined the spatial structure of the synoptic-scale subsurface pressure fluctuations observed over the northern half of the Middle Atlantic Bight, and they found that (a) the subsurface pressure fluctuations observed over the shelf were coherent and in phase with coastal sea level, and (b) the subsurface pressure fluctuations had a monotonic cross-shelf structure, with very small amplitudes observed near the shelf break. This last result has been further substantiated with bottom pressure measurements made by Brown (personal communication) over the outer New England shelf and upper slope. Flagg (1977) found significant coherence (coherence squared $\simeq 0.7$–0.8) between local subsurface pressure and alongshelf-current fluctuations at a mid-shelf site on the New England shelf, and Chuang et al. (1979) also found high coherence (coherence squared $\simeq 0.7$–0.8) between synoptic-scale alongshelf currents and coastal sea level off Cape Henry.

These direct-current and pressure observations are consistent with the conceptual model of coastally trapped continental shelf waves. In view of the clear relation observed between the essentially barotropic alongshelf-current and coastal-sea-level fluctuations, the coastal-sea-level studies of Wang (1979c) and Noble and Butman (1979) can be used to infer the alongshelf structure of the synoptic-scale transient shelf circulation over larger spatial scales than have been examined yet through direct-current measurements. Wang (1979c) and Noble and Butman (1979) have investigated the relation between local wind stress and sea level in the Middle Atlantic Bight; they find that north of Cape May coastal sea level and the alongshelf wind stress are highly coherent over the synoptic-scale band, and coastal sea level lags the local alongshelf wind stress by 8 to 12 hours, indicating that the alongshelf-current and coastal-sea-level fluctuations are in phase to within a few hours. Both coastal wind stress and coastal sea level move slowly upshelf toward the northeast, and the very slow spatial decay in coastal-sea-level coherence suggests that the synoptic-scale alongshelf-current fluctuations are spatially coherent over the entire northern section of the Middle Atlantic Bight.

The observed coastal-sea-level response south of Cape May is more complex. Wang (1979c) finds that local alongshore-wind-stress and coastal-sea-level coherence is only high at frequencies above 0.3 cpd, while the lower-frequency coastal-sea-level fluctuations appear to propagate downshelf toward the south (like free shelf waves) with a phase speed of order 600 km day^{-1}. Downshelf propagation of the band-averaged (0.04 to 0.4 cpd) coastal-sea-level fluctuations is also reported by Noble and Butman (1979).

These results suggest the following physical interpretation. Free continental shelf waves have a downshelf or southward phase velocity in the Middle Atlantic Bight, while forced shelf waves driven by propagating atmospheric disturbances tend to move in either alongshelf direction, in phase with the forcing. The nature of the forced shelf-wave response depends critically on several factors: the shelf topography and coastline geometry, the spatial and temporal structure and propagation characteristics of the wind-stress pattern, and the effective frictional-adjustment time scale of the shelf. In the limit of no friction, both free and forced shelf waves are generated by wind-stress patterns moving along the shelf, and the alongshelf current generally lags the local wind stress by 90° (Gill and Schumann, 1974). In the limit of steady forcing and/or rapid frictional adjustment only the forced wave exists [called an *arrested topographic wave* in this limit by Csanady (1978)], and the phase lag between the alongshelf current and wind-stress components is significantly reduced (Gill and Schumann, 1974; Hsueh and Peng, 1978; Brink and Allen, 1978). The frictional time scale for the shelf north of Cape May appears to be sufficiently short that a quasi-steady-state response to synoptic-scale atmospheric forcing is observed. The analytic and numerical-model calculations of Csanady (1974) and Beardsley and Haidvogel (1980) suggest that the frictional-adjustment time scale for much of the Middle Atlantic Bight is roughly $T_f = 10$ hours. Thus the ratio of T_f to the characteristic time scale T_a of the forcing ($T_a = L/c$ where L and c are the alongshelf length scale and phase speed of the forcing) is small enough (≤ 0.2) for most storms that the wind-driven response has much of the character of a heavily damped or arrested shelf wave, slowly moving along the shelf in phase with the forcing. The shelf south of Cape May appears to exhibit both this quasi-steady-state response and a southward-moving shelf-wave response at lower frequencies due to upshelf generation. Why this occurs over the southern region of the Middle Atlantic Bight is not yet clear, although it must be related to the three-dimensional character of the shelf topography north of Cape Hatteras, as well as to the spatial structure of the atmospheric forcing. The scattering of kinetic energy from longer to shorter shelf waves by topographic perturbations like canyons and capes may explain the small alongshelf coherence scale for the cross-shelf current field (Wang, 1980).

We have focused in this section on the large-scale response of the Middle Atlantic Bight to synoptic-scale atmospheric forcing and have attempted to interpret recent current and pressure observations in light of continental shelf-wave theory (with its implicit three-dimensionality). It seems clear that more observational and theoretical work is needed before the synoptic-scale response is thoroughly understood. It does seem feasible, however, that the alongshelf current fluctuations within the Middle Atlantic Bight may be predictable from the atmospheric forcing and coastal sea level once the relative roles of free and forced shelf waves and their generation and dissipation mechanisms in this particular shelf domain are clarified. Coastally trapped pressure fields and the transient circulations associated with them also occur on shorter space and time scales. These fields are generally created by either sharp spatial or temporal variations in the wind-stress pattern or apparent spatial changes in the coastal wind-stress pattern due to changes in the coastline orientation. The almost 90° change in the coastline orientation at New York causes a smaller-scale trapped pressure and current field to exist within the New York Bight (Wang, 1979c; Csanady, 1980).

The Monthly-Mean Circulation We focus in this section on the monthly-mean current field in the Middle Atlantic Bight. We have chosen to use one month as a convenient averaging period for the following reason. In the previous section, we noted that the local alongshore wind-stress and current components over much of the shelf are coherent and nearly in phase over the synoptic-scale frequency band from 0.1 to 0.5 cpd. At lower frequencies, the alongshore wind-stress and current components generally become incoherent although off New Jersey the nearshore wind stress and currents remain significantly coherent at lower frequencies down to at least 0.03 cpd (a period of 1 month). From this we should expect to see some coherence between the monthly-mean wind stress and currents over the middle and inner shelf and we should be able, perhaps, to attribute some of the observed monthly-mean current variability to the monthly-mean wind-stress fluctuations.

Mayer et al. (1979) have discussed the monthly-mean current variability observed at the MESA long-term mooring site in the New York Bight. We show here in figure 7.12 a composite vector diagram incorporating both the data given by Mayer et al. (1979) and other current data obtained during the same time span at a long-term site located off Cape Henry by Boicourt and Chuang (personal communication). The additional wind-stress data shown for the two environmental buoys and Cape Hatteras have been supplied by Halliwell (personal communication) and Noble and Butman (personal communication). The wind-stress and current vectors at the different measurement levels are plotted with true north directed upward, and the

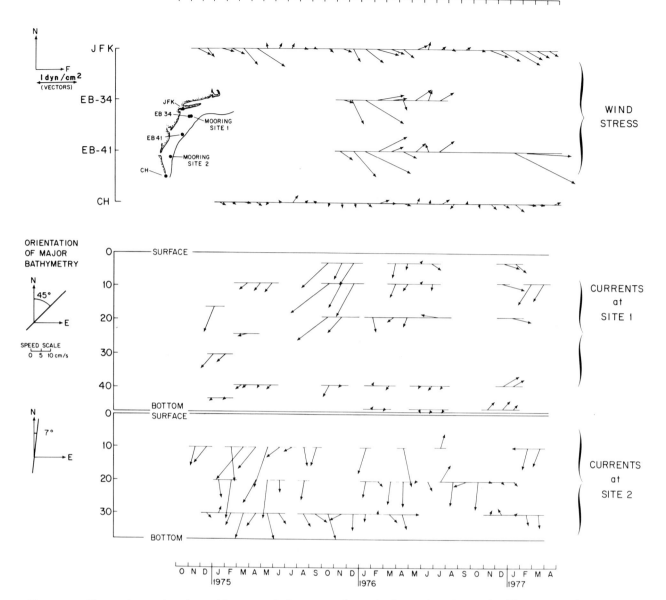

Figure 7.12 Vector time series of monthly-mean wind-stress and subsurface currents observed within the Middle Atlantic Bight during the period October 1974 to April 1977. The measurement locations are shown in the inset map. The monthly-mean wind-stress at two coastal and two mid-shelf sites are shown in the top four time series. The currents shown at site 1 are taken from Mayer, Hansen, and Ortman (1979). Boicourt and Chuang (personal communication) supplied the current data shown at site 2. The depth scale at left is shown in meters.

orientation of the regional topography at each mooring site is indicated on the left of the figure. The locations of the four meteorological stations and the two current measurement sites are also shown in figure 7.12.

This composite figure illustrates several key features of the lower-frequency current variability of the Middle Atlantic Bight. The monthly-mean wind-stress data show both a definite offshore increase in strength and a clear tendency for cyclonic veering in most months. The wind-stress vectors are generally aligned to within 20° although the wind-stress vectors at EB41 and JFK differ in orientation by 40° in February 1976. The monthly-mean currents are generally directed downshelf at all measurement levels except during a few periods when the alongshore flow is reversed for 1 to 3 months. While the magnitude of the possible errors in these current measurements is not clear [see Mayer et al. (1979)], the alongshore current components at both sites exhibit a vertical shear consistent with an offshore increase in the mean density field, so that the upshelf flow during the reversals appears to be strongest near the bottom. While strong upshelf flows occur on shorter time scales, the submonthly current fluctuations are generally comparable to or less than the long-term mean currents at both sites. Since a similar picture is presented for the nearshore monthly-mean currents off New Jersey and Long Island by EG&G (1978) and Scott (personal communication), we conclude that the monthly-mean subsurface currents over most of the Middle Atlantic Bight are directed downshelf except for relatively infrequent reversals of one to several months duration. The vertical and horizontal structure of these current reversals is clearly complex although the preliminary data shown in figure 7.12 suggest that the monthly-mean current fluctuations over the middle and outer shelf may have a large alongshelf coherence scale.

We shall now focus on the monthly-mean wind stress and currents observed during two specific periods—the spring and summer of 1976 and the winter of 1976-1977. Winds for the period May through July 1976 were more persistent and stronger than normal, and the monthly mean wind-stress vectors were directed toward the north and northeast (Diaz, 1980). This wind stress drove a definite upshelf and onshore surface flow over the nearshore region from New York to Cape Cod (Frey, 1978), which, coupled with a higher-than-average river discharge in May and June, led directly to the severe pollution of the western Long Island beaches (Swanson, Stanford, and O'Conner, 1978). This wind stress pattern also apparently caused the reduced downshelf flow observed at both mid-shelf sites (figure 7.12), and the weak upshelf near surface current in June at site 1, and the strong reversal in July at site 2. The dissolved-oxygen concentration in the near-bottom water over the shallower half of the New Jersey shelf reached very low values (less than 2 ml l^{-1}) during June through August, resulting in an extensive mortality of shellfish valued at $60 million. This major anoxic event and the environmental conditions that may have caused the severely depleted oxygen levels off New Jersey and not elsewhere in the Middle Atlantic Bight are examined in the comprehensive report edited by Swanson and Sinderman (1980). We note here that Armstrong (1980), Walsh, Falkowski, and Hopkins (1980), and others attribute the 1976 anoxic event to: (a) the early development of the seasonal pycnocline [this is clearly shown in the temperature time series shown for site 1 by Mayer et al. (1979)], which reduced the initial dissolved-oxygen concentration in the deeper water and inhibited later oxygen replenishment; and (b) an excessive local oxygen (respiration) demand created by an unusual abundance of the dinoflagellate *Ceratium tripos* advected onto the New Jersey shelf by the weak upshelf and onshore deep flow. The monthly-mean current measurements for this period presented here and by Mayer, Hansen, and Minton (1980), Han, Hansen, and Cantillo (1980), EG&G (1978), Butman et al. (1979), and Ou (1979) suggest that the persistent northward wind stress caused an upshelf flow over the shallower New Jersey shelf with perhaps an offshelf and enhanced downshelf flow of shelf water over the outer New Jersey shelf, producing in essence a persistent mesoscale clockwise gyre during June. Han et al. (1980) have used a diagnostic numerical model together with observed density and current data to predict the quasi-steady current field over the New York Bight during the 1976 anoxic event. The computed deep transport fields indicate both a net convergence of deep water over the inner New Jersey shelf and strong cross-shelf flow occurring in the Hudson shelf valley.

A more dramatic case for atmospheric forcing of shelf circulation on the monthly time scale occurred during the winter period, November 1976 through January 1977. Winds during this 3-month period were both more persistent and stronger than normal, and the monthly-mean wind-stress vectors were generally directed toward the east-southeast (Wagner, 1977). This wind-stress pattern produced strong upshelf currents near the bottom at both sites (thus providing an effective bottom stress in *opposition* to the upshelf wind-stress component), and strong upshelf and offshore flow in the upper 20-30 m at site 1. The monthly-mean downshelf subsurface flow on the southern side of Georges Bank was less than average in November and December 1976, and the mean alongshelf current was essentially zero in January 1977 (Folger, Butman, Knebel, and Sylvester, 1978). The downshelf mean flow was observed over the shelf in February. These limited observations strongly imply that a sufficiently strong and persistent adverse wind-stress pattern can effectively stop and reverse the normally downshelf monthly-mean flow over much of the Middle Atlantic Bight. On December 15, 1976, the tanker *Argo Merchant* ran aground on Nantucket Shoals, and later broke apart on December 22 during a major storm, causing one of the largest oil spills off the east coast of the United States. It was indeed fortunate that almost all of the oil remained on the surface and was blown off the continental shelf by the unusually strong and persistent winds occurring during late December and early January (Grose and Mattson, 1977; Mattson, 1978).

In summary, the sparse current observations discussed here suggest that the monthly-mean currents over most of the Middle Atlantic Bight are primarily downshelf except during infrequent periods of strong and persistent adverse wind forcing. The spatial structure of the monthly-mean current field is quite complex, especially during periods of upshelf flow.

The Annual Mean Circulation We have plotted in figure 7.13 the long-term mean currents observed at four shelf sites in recent moored-array programs. Only records of duration 1 year or longer have been used, and information about the individual current measurements is listed by site number in table 7.2. The mean currents are plotted as vectors with the magnitude equal to the average speed. The rectangle shown around the head of each current vector represents the statistical uncertainty in the computed mean. Each side of the rectangle corresponds to the standard error ϵ for the current component parallel to that side, where ϵ has been computed using the formula $\epsilon = \sigma_{LF}/\sqrt{T/\tau_c}$,

where σ_{LF} is the low-frequency (subtidal) standard deviation of that current component, T the length of the record in days, and τ_c the correlation time scale in days. The quantity T/τ_c is an estimate of the number of independent observations in the current time series, based on the assumption that the time series is stationary. We have used here $\tau_c = 5$ days for both the alongshore and cross-shore components. This is a conservative estimate since Mayer et al. (1979) and Flagg (1977) found somewhat shorter correlation time scales. However, the use of a more accurately determined correlation time scale would not substantially change the size of the standard errors shown in figure 7.13. Two wind-stress vectors taken from Saunders (1977) have also been plotted in figure 7.13 to illustrate the orientation of the mean wind stress with respect to the shelf break over the southern and northern sections of the Middle Atlantic Bight.

These measurements of the annual mean current field on the continental shelf clearly demonstrate that the mean subsurface alongshelf flow is directed toward the southwest through the Middle Atlantic Bight. While there are not enough long-term measurements to define the spatial structure of the mean circulation in detail, this picture when combined with other current information and hydrographic evidence substantiates the previous concept that over much of the shelf (except perhaps very near the major estuaries), the long-term subsurface currents are downshelf towards the southwest. The mean vectors at sites 1, 2, and 3 show a definite tendency for onshore veering of the mean velocity vector with increasing depth that is consistent

Table 7.2 Tabulation of Long-Term Mean Currents, Standard Errors, and Subtidal Standard Deviations Obtained at Four Sites in the Middle Atlantic Bight and Georges Bank Region[a]

Site number	Location	Orientation of topography (θ_T)	Coordinate orientation (θ_c)	Measurement time span	Record length (days)	Water depth (m)	Nominal instrument depth (m)	E' (cm s^{-1}) Mean	St. error	St. dev.	N' (cm s^{-1}) Mean	St. error	St. dev.
1	36°52'N, 75°03'W	7°	0°	Nov. 74– April 77	469	38	10	2.1	±1.3	13.0	−5.8	±2.5	24.7
					573		20	0.3	±0.9	9.2	−4.4	±2.1	22.3
					668		30	0.5	±0.6	6.9	−2.6	±1.5	17.5
2	39°28'N, 74°15'W	36°	36°	May 72– Dec. 76	1089	12	4.5	1.6	±0.2	2.7	−3.9	±0.8	12.2
					1082		10	−1.1	±0.2	3.0	−2.1	±0.6	8.2
3	40°07'N, 72°54'W	45°	45°	June 74– March 77	482	47	10	1.3	±0.7	7.0	−5.2	±0.9	9.0
					575		20	0.5	±0.7	7.0	−3.9	±1.1	12.0
					528		41	0.2	±0.2	2.0	−0.9	±0.5	5.0
4	40°51'N, 67°24'W	58°	58°	May 75– Dec. 78	742	85	45	−0.8	±0.3	3.9	−8.8	±0.6	7.6
					692		75	0.4	±0.3	3.4	−3.6	±0.5	5.5

a. Each velocity has been decomposed into an offshore component E' and an orthogonal alongshore component N' where the orientation of the alongshore direction relative to north is indicated by θ_c. The orientation of the regional topography relative to north is denoted by θ_T. Except at site 1, θ_c and θ_T are equal. [Boicourt and Chuang (personal communication), EG&G (1978), Mayer et al. (1979), and Butman and Noble (1978, 1979) provided the data for sites 1–4, respectively.]

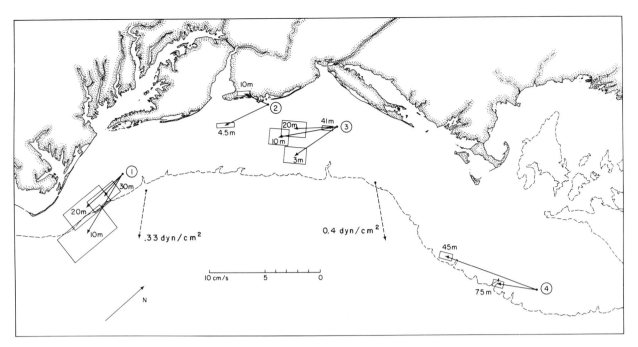

Figure 7.13 Map of long-term mean currents computed from 1-year or longer current time series with moored current meters in the Middle Atlantic Bight and Georges Bank region. The individual sites are circled and numbered according to table 7.2; the measurement depth in meters is shown next to the mean-current vector. The standard error for each mean-current computation is indicated by the rectangle around the head of the current vector. Two representative mean wind-stress vectors taken from Saunders (1977) are also shown to illustrate the relative orientation of the mean wind-stress near the shelf break.

with the mean surface- and bottom-drifter results reported by Bumpus (1973). This mean-current veering is most pronounced at site 2 off the New Jersey coast where nearshore upwelling must occur on average.

Beardsley, Boicourt, and Hansen (1976) speculated that the mean alongshore transport of shelf water through the Middle Atlantic Bight might be approximately constant throughout the year, so that the mean velocities should increase over the southern section of the Middle Atlantic Bight where the cross-sectional area of the shelf is smaller. This mechanism may account for the somewhat larger mean currents found at site 1 off Cape Henry. The observed low-frequency current variances are also largest at site 1, where the typical standard deviation of the alongshelf current component is about 20 cm s^{-1}, or a factor of two larger than the typical values found at the other sites. Although more measurements are needed to determine accurately the spatial distribution of the subtidal current variance over the Middle Atlantic Bight, the few values given here do suggest that the wind-driven-current variability is significantly larger over the shallower southern section of the Middle Atlantic Bight.

In summary, these long-term direct measurements demonstrate that the alongshelf flow, if averaged over a year or longer, is directed toward the southwest at all levels throughout the water column except perhaps near the surface, and even there the shorter-term measurements at 3 m reported by Mayer et al. (1979) suggest that the long-term mean surface flow is also directed towards the southwest with an offshore component. The long-term current measurements made on the southern flank of Georges Bank help substantiate the notion from hydrographic evidence that the southern side of Georges Bank is the immediate upstream source region for much of the shelf water found in the Middle Atlantic Bight. We shall now discuss the dynamic ideas that have been put forth to explain the observed mean circulation over the Middle Atlantic Bight.

A dynamical discussion: As mentioned above in section 7.3.3, Sverdrup, Johnson, and Fleming (1942) suggested at a very early date that the southwestward flow over the shelf was driven against friction by an alongshore pressure gradient, but this dynamic explanation was ignored by Bumpus (1973) and others as too vague and speculative, especially after the accuracy of the original coastal geodetic leveling results was questioned by Sturges (1968) and Montgomery (1969).

Some 30 years later, Stommel and Leetmaa (1972) constructed the first theoretical model for the mean winter circulation on a flat two-dimensional continental shelf driven by a steady, uniform wind stress and a distributed fresh-water source located at the coast. The model incorporated linear Ekman dynamics with constant vertical friction and diffusion coefficients, while the cross-shelf salt balance was maintained by an advective-diffusive Taylor shear-dispersion mechanism.

Stommel and Leetmaa (1972) applied their model to the Middle Atlantic Bight and concluded that an alongshore pressure gradient, or surface slope of order 10^{-7}, must exist to drive the mean alongshore flow toward the southwest against the mean wind stress, which (in their model computations) had an alongshore component toward the northeast.

Using a similar dynamic model for the steady flow over a sloping two-dimensional shelf, Csanady (1976) also concluded that an alongshore surface slope must exist to account for the observed circulation within the Middle Atlantic Bight. The essential physical argument for the alongshore pressure gradient can be found in the alongshore momentum balance. The net Coriolis force on the onshore flow is negligible since the depth-averaged onshore flow is very small, and the divergence of the onshore flux of alongshore momentum (or onshore Reynolds stress) is also negligible. Since the mean windstress and the bottom stress have alongshore components directed toward the northeast, the boundary stresses must then be balanced by an alongshore pressure-gradient force directed toward the southwest in the direction of the mean alongshore flow. Csanady (1976) noted that if the alongshore surface slope remained approximately constant across the shelf, then the total pressure-gradient force associated with this surface slope, the alongshore bottom stress required to partially balance this total pressure-gradient force, and the alongshore velocity should all increase as the local water depth increases offshore. The early mean-current data summarized by Beardsley et al. (1976) clearly show such an offshore increase in the alongshore currents. Csanady (1976) also pointed out that the alongshore pressure gradient is required to account for the line of bottom-drift divergence reported by Bumpus (1973) to occur near the 60-m isobath. Scott and Csanady (1976) then offered some supporting indirect evidence for the existence of the alongshore pressure gradient off Long Island, although their estimated value of the gradient has been questioned by Wang (1979c) and Beardsley and Winant (1979).

Csanady (1978) then presented a simple linear model for the frictionally damped steady depth-averaged flow driven over a sloping shelf by a uniform alongshore pressure gradient externally imposed at the shelf break. He found that the surface elevation over the shelf is governed by a parabolic differential equation, and that downstream from the initial (spatial) transient behavior, the alongshore surface slope or pressure gradient remained constant across the shelf. Since in this model the total alongshelf pressure gradient is balanced solely by bottom stress, the alongshore velocity increased with increasing depth. Csanady (1978) concluded that the open ocean was not "inert," but instead imposed the mean pressure gradient along the outer edge of the Middle Atlantic Bight that was required to drive the observed mean southwestward flow on the shelf.

This suggestion has been strengthened by Beardsley and Winant (1979), who found supporting evidence in the numerical simulations by Semtner and Mintz (1977) of the circulation in the western North Atlantic. Semtner and Mintz (1977) constructed a multilayered numerical model, using a rectangular basin oriented with the long side of the basin adjacent to the east coast of the United States. The model basin had a flat bottom except along the western boundary where a simple continental shelf and slope were located. The model ocean is driven by steady zonal surface temperature and wind-stress patterns, and Leetmaa and Bunker (1978) indicate that the model wind-stress curl is reasonably realistic over the northern region of the model. The model topography and wind-stress distribution are shown in figure 7.14. Semtner and Mintz (1977) first conducted an initial highly damped spin-up experiment and then two shorter experiments with reduced dissipation in which the model Gulf Stream became unstable and mesoscale eddies developed. The mean volume transport, surface pressure, and temperature fields obtained in the three experiments are quite similar in the northern shelf and slope region (the model equivalent to the Middle Atlantic Bight) and so only the mass transport and surface elevation patterns found in the spin-up experiment are shown in figure 7.14. In all three experiments, a mean alongshore pressure gradient corresponding to a surface slope of order 10^{-7} was imposed over the northern shelf and upper slope by the large wind-driven cyclonic gyre formed north of the Gulf Stream. Because continental runoff was not included in the model, Beardsley and Winant (1979) concluded that these model results demonstrated that the shelf circulation in the Middle Atlantic Bight could be considered a boundary-layer component of the offshore large-scale ocean circulation.

Beardsley and Winant (1979) also discussed (without reaching any conclusions) the possibility that a major fresh water source like the St. Lawrence system could produce a coastally trapped pressure-gradient field that might extend downstream through the Middle Atlantic Bight. This other possible driving mechanism has been recently ruled out by Csanady (1979), who concluded from the steric set-up distribution over the continental margin from Cape Hatteras to the Grand Banks that the effects of the St. Lawrence outflow are primarily confined to the Scotian shelf and the northern section of the Gulf of Maine.

The mean southwestward flow of shelf water through the Middle Atlantic Bight thus appears to be driven primarily by an alongshore pressure gradient imposed at the shelf break by the deep-water cyclonic gyre found between the continental shelf and the Gulf Stream. The magnitude of this mean alongshore pres-

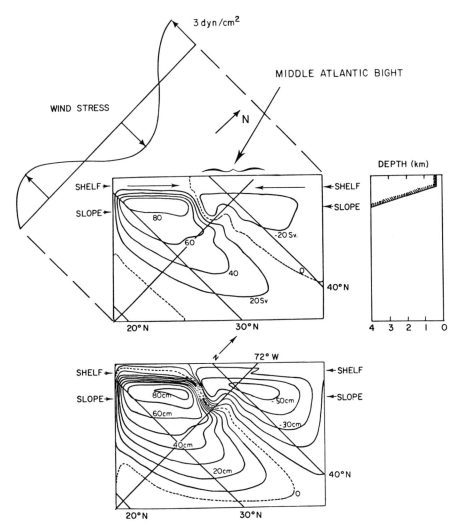

Figure 7.14 Steady mass transport (top) and surface elevation (bottom) fields obtained by Semtner and Mintz (1977) in their initial spin-up experiment. The contour intervals for top and bottom are 20×10^6 m³ s⁻¹ and 10 cm. The zonal wind-stress pattern and the model topography is shown to the upper left and right of top. The model equivalent of the Middle Atlantic Bight is indicated at top.

sure gradient can be expected to vary along the shelf break due to changes in both the strength of the regional wind stress over the shelf and in the relative orientation of the wind stress with respect to the regional shelf topography. The calculations of Bush and Kupferman (1980) suggest that the relative magnitude of the alongshore pressure gradient may be quite small off the southern section of the Middle Atlantic Bight where the regional mean wind stress is directed primarily offshore with a weak alongshore component toward the southwest. While local wind stress and runoff definitely influence the mean nearshore circulation, the model results of Semtner and Mintz (1977) indicate that the mean circulation over much of the Middle Atlantic Bight may be viewed as a boundary-layer component of the large-scale general circulation of the western North Atlantic. Continental runoff injects fresh water into the boundary layer, which is then mixed with an onshore flux of upper slope water to produce shelf water of intermediate salinity, while entrainment by the Gulf Stream provides the principal downstream sink of this shelf water.

What does the dynamic model presented here imply about the seasonality of currents in the Middle Atlantic Bight? Both Saunders (1977) and Leetmaa and Bunker (1978) indicate that the wind-stress and wind-stress-curl distributions vary significantly over the western North Atlantic on a seasonal basis, yet the phase of these seasonal fluctuations is not stable enough from year to year to produce a very significant annual peak in the surface wind-stress spectra at Weathership C reported by Willebrand (1978). The volume transport of shelf water through the Middle Atlantic Bight corresponds to just a few percent of the mean volume-transport estimates for the slope water gyre given by Worthington (1976) and Leetmaa and Bunker (1978). The numerical ocean-model computations made with realistic time-dependent winds by Willebrand, Philander, and Pacanowski (1980) suggest that the volume transport of the slope water gyre should *not* vary significantly over the year, and, in fact, Thompson (1977) found no evidence of an annual signal in the long-term deep-current measurements made near site D on the New England continental rise. These two results imply that the magnitude of the alongshore pressure gradient imposed at the shelf break should remain nearly constant over the year. Since the long-term current measurements made just off the New Jersey coast by EG&G (1978) also exhibit no significant annual cycle, we suggest that the very low-frequency currents over most of the Middle Atlantic Bight will reflect broadband forcing and not exhibit a significant annual variation.

7.3.4 Summary and Some Remaining Problems

We have attempted here to describe the shelf response to wind forcing on three time scales. Synoptic-scale atmospheric disturbances and in particular winter cyclones can drive strong transient current fields that tend to move along the shelf in phase with the forcing. The cross-shelf momentum balance is approximately geostrophic and both the current and subsurface pressure fluctuations are generally coherent over much of the Middle Atlantic Bight, reflecting the relatively small size of this shelf region with respect to the atmospheric forcing. The synoptic-scale shelf response appears to be consistent with continental shelf-wave theory. The direct effect of wind forcing is also evident in the monthly-mean shelf circulation, although the observed currents have a complex spatial structure and other processes like runoff and offshelf forcing contribute to the variability on this time scale. The mean flow over the Middle Atlantic Bight is primarily driven not by local runoff and wind stress but by the large-scale wind stress and heat-flux patterns over the western North Atlantic. Thus the observed currents can be decomposed into a mean component driven by a steady offshelf forcing and a fluctuating component driven by the regional wind stress field acting over the shelf.

This review has focused on the wind-driven-circulation components in the Middle Atlantic Bight. The actual influence of density stratification on the different components of the general circulation is still not clear, and only a crude estimate of the flushing rate of the shelf is available. The processes controlling the local position and movement of the shelf-slope-water front are poorly known. As noted by Fofonoff (chapter 4, this volume), satellite infrared mapping of the sea surface temperature has greatly added to our perception of the spatial variability and complexity of the near-surface current and thermal fields. The satellite infrared photograph shown here in figure 7.15 illustrates thermal structures or fronts on a wide range of scales throughout the Middle Atlantic Bight. The most pronounced thermal front is the shelf-slope-water front, and while it crudely follows the shelf break, some of the colder shelf water does extend far offshelf into the slope water or along the northern side of the Gulf Stream. Several anticyclonic Gulf Stream eddies are also shown in the slope water north of the Gulf Stream. It remains to be determined whether Gulf Stream eddies and other very low-frequency phenomena in the slope water have much real influence on the flow of shelf water through the Middle Atlantic Bight.

Appendix: Annual Air–Sea Interaction Cycles and Mean Runoff for the Middle Atlantic Bight

The mean and average monthly heat-flux cycles over the southern section of the Middle Atlantic Bight have been described by Bunker (1976). We show here his results, plus additional data kindly supplied by Bunker (personal communication) for the northern section, to give a complete picture of the seasonal heat flux cycles for the entire Middle Atlantic Bight. The monthly air and sea surface-temperature and heat-flux cycles for both sections shown in figure 7.16 and the annual mean values are listed in table 7.3. We note that the sea surface-temperature cycle lags the net heat-flux cycle by about 90°, and that the approximately 17°C increase in sea surface-temperature from March to August is consistent with a uniform mixing of the net internal-energy gain by the shelf water during that period to a mean depth of 30 m.

Average precipitation and evaporation data for the Middle Atlantic Bight are also shown in figure 7.16 and in table 7.3. The evaporation rates are computed from Bunker's heat-flux data. Although few precipitation data are available over the Middle Atlantic Bight, precipitation along the coast of the Middle Atlantic Bight is roughly uniform and exhibits relatively little seasonality [see Geraghty et al. (1973) and Lettau, Brower, and Quayle (1976)]. We thus have used data from New York City as an estimate for the precipitation cycle over the entire Middle Atlantic Bight. Mean steamflow of fresh water entering the Middle Atlantic Bight along the coast via the major estuaries is also given in table 7.4. Most of the fresh water is contributed by a few major sources, especially the Chesapeake Bay. The mean precipitation crudely balances evaporation over the Middle Atlantic Bight and the net input of fresh water via local precipitation minus evaporation into the Middle Atlantic Bight is minor in comparison to runoff at the coast. Advection of fresh water (meaning here zero salinity) from the Gulf of Maine and Georges Bank region accounts for most of the total fresh water flux into the Middle Atlantic Bight.

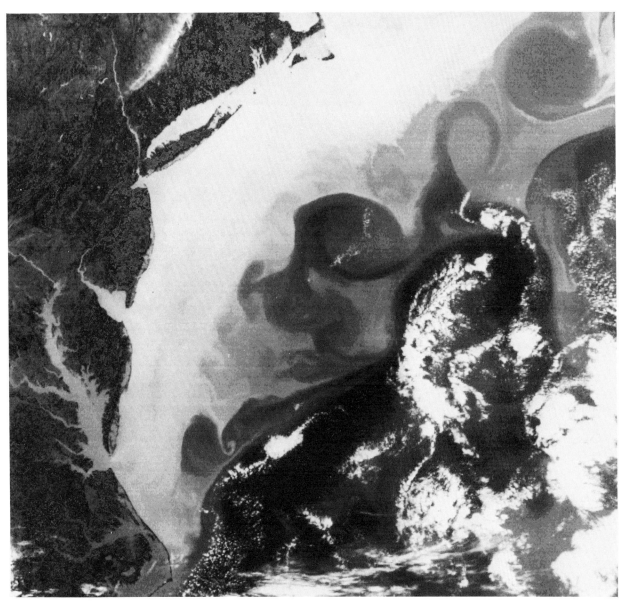

Figure 7.15 The thermal infrared image of the Middle Atlantic Bight was obtained on 1430 GMT, May 1, 1977, by the NOAA-5 polar orbiting environmental satellite. The sea surface temperature is indicated by a grey-scale code with darker tones corresponding to warmer water. The purely white areas indicate clouds. This image is shown courtesy of R. Legeckis, National Environmenttal Satellite Service.

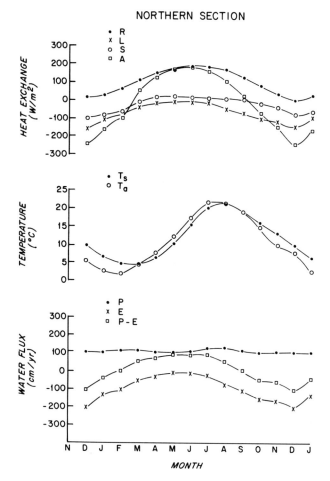

Figure 7.16 Seasonal heat flux, surface temperature, and water-flux cycles for the southern and northern sections of the Middle Atlantic Bight.

Table 7.3 Mean Surface Heat and Water Fluxes in the Middle Atlantic Bight[a]

	Units	Southern[b] section	Northern[b] section
Air temperature	°C	15.0	11.6
Sea surface temperature	°C	15.5	12.3
Dew point temperature	°C	11.2	8.2
Net radiation flux into ocean R	W m^{-2}	126.5	109.1
Heat loss by evaporation L	W m^{-2}	88.7	72.3
Sensible heat loss S	W m^{-2}	23.9	27.3
Net heat flux into ocean $A = R - L - S$	W m^{-2}	13.9	+9.5
Precipitation P[c]	cm yr^{-1}	111.2	111.2
Evaporation E	cm yr^{-1}	113.9	93.3
Net $P - E$ flux into ocean	cm yr^{-1}	−2.7	+17.9
Surface area of section[b]	km^2	5.9 × 10^4	4.8 × 10^4

a. All data, unless otherwise noted, are given by Bunker (1976) or Bunker (personal communication). Useful conversion factors: E (cm yr^{-1}) = 1.29 × L (W m^{-2}); 1 kcal cm^{-2}/30 days = 16.15 W m^{-2}.
b. The southern section is that part of the Middle Atlantic Bight lying west of 72°W and between 36° and 40°N. The northern section is the remaining part of the Middle Atlantic Bight lying north of 40°N and west of 70°W. Both sections are shown in Bunker (1976, figure 3).
c. The mean monthly precipitation values given here and in figure 7.16 are for New York City, 1939–1978.

Table 7.4 Mean Stream Flow into the Middle Atlantic Bight between Cape Cod and Cape Hatteras, 1931–1960[a]

Segment	Stream flow (m^3 s^{-1})	Stream flow–drainage area (cm yr^{-1})	Principal source	Stream flow (m^3 s^{-1}) due to principal source	Percentage of total stream flow due to principal source
Cape Cod, MA, to CT-NY state line	888	60.3	Connecticut River	627	71
CT-NY state line to Cape May, NJ	873	54.8	Lower New York Bay (including Hudson, Hackensack, Passaic, and Raritan Rivers)	748	86
Cape May, NJ, to Cape Charles, VA	622	54.8	Delaware Bay	587	94
Cape Charles, VA, to Cape Hatteras, VA	2651	39.0	Chesapeake Bay	2154	81

a. All data from Bue (1970).

Notes

1. An example of the increased concern for resource and environmental management of estuaries is the formation in 1948 of the Chesapeake Bay Institute at The Johns Hopkins University. R. Revelle was instrumental in bringing together the interests of the Office of Naval Research, the State of Maryland, and the Commonwealth of Virginia into a consortium that provided the necessary support for the founding of the Institute. The primary focus of the new laboratory was the physical oceanography of estuaries and D. W. Pritchard was chosen as its first director.

2. Letter from the Office of Chief of Engineers, U.S. Army, to the Director, the National Bureau of Standards, December 19, 1945, Army file number CE-SPEWE.

3. Early observers in the Middle Atlantic Bight generally recognized only one transition or frontal zone between relatively fresh coastal water with salinities S less than 34 or 35‰ and Gulf Stream water with S greater than 35‰. The term *slope water* was first used by Bjerkan (1919) to describe water between 33.0 and 35.0‰ over the Scotian Shelf. Huntsman (1924), Bigelow (1927), and Bigelow and Sears (1935) next used the same term to describe the more saline water with $S \geq 35.0$‰ found over the slope off the Scotian Shelf, Gulf of Maine, and the Middle Atlantic Bight regions, respectively. Iselin (1936) then firmly established the use of the term slope water for water with salinities between 35.0 and 36.0‰ found between the Gulf Stream (then characterized by surface $S \geq 36.0$‰) and the continental shelf from Cape Hatteras east to longitude 65°W.

   Both Bigelow in his various reports and Bigelow and Sears (1935) referred to water fresher than 35‰ found in the Gulf of Maine and the Middle Atlantic Bight as both *coastal* and *shelf water* although they clearly preferred the former term, which was then in common use. Iselin (1939b) later called water fresher than 34‰ coastal water, although some, like Ford, Longard, and Banks (1952) and Ford and Miller (1952), chose shelf water as a more accurate and descriptive term for water fresher than 35‰. Both terms have been used rather interchangeably in the literature since then, although the term shelf water is now more popular. Wright and Parker (1976) have constructed a seasonal volumetric temperature-salinity census for the Middle Atlantic Bight and found a clear distinction between the shelf- and slope-water masses in both winter and summer seasons. They advocate the general term shelf water for all water on or near the Middle Atlantic Bight shelf fresher than 35‰. We shall use this term here.

4. The dynamic or geostrophic method was developed in northern Europe by Sandström and Helland-Hansen (1903), who derived the formula for computing relative geostrophic currents from the observed density field on the basis of V. Bjerknes's (1898) circulation theorem, and by Helland-Hansen and Nansen (1909), who applied the method to hydrographic data taken in the Norwegian Sea under Hjort's direction (Schlee, 1973). In 1911 a report entitled *Dynamic Meteorology and Hydrography*, written by Bjerknes and others, was printed in the United States to help introduce the geostrophic method to the American scientific community. It appears, however, that Bigelow first learned of this method from Sandström's chapter in the report of the Canadian Fisheries Expedition of 1914-1915, published in 1919. Hjort had been invited by the Biological Board of Canada to lead this expedition, and he brought Sandström along to work up the dynamic computations. Sandström's chapter described the dynamic method and presented two spectacular plates showing the geostrophic currents off eastern Canada in perspective view. Shortly after World War I, the U.S. Coast Guard sent E. Smith of the International Ice Patrol to study oceanography with Bigelow. With Bigelow's encouragement, Smith began to study the geostrophic method and, after a year spent with Helland-Hansen in Norway, Smith (1926) wrote a Coast Guard manual on practical dynamic computation that was immediately adopted and put to use by the Ice Patrol to predict the drift of icebergs. During this period, Smith worked closely with Bigelow and helped him prepare the geostrophic-current maps for the Gulf of Maine presented in Bigelow's (1927) monograph. The estimated geostrophic surface flow was consistent enough with his own drift-bottle data that Bigelow became an early and influential exponent of the dynamic method in the United States (Schlee, 1973).

5. In 1930 the Woods Hole Oceanographic Institution (WHOI) was established through the joint efforts of F. R. Lillie of the Marine Biological Laboratory in Woods Hole and Bigelow, and Bigelow was appointed director, a position he held for the next decade. Bigelow's research had been focused on coastal problems and he helped select Woods Hole as the site for the new institution since it offered easy access to two very different and contrasting coastal regimes (the Gulf of Maine and the Middle Atlantic Bight) as well as the open sea (Bigelow, 1929). Tired of getting seasick on the *Grampus* and other small coastal vessels, and believing that a sailing vessel would be both more stable and less expensive to operate than a motor vessel, Bigelow had the 142-foot *Atlantis*, the large steel-hulled ketch, built as the first research vessel for the new institution (Schlee, 1978).

6. Bumpus himself experimented with moored instrumentation in the early 1960s, but abandoned this approach when he found that the early Richardson current meters were too fragile to withstand mooring motion (Bumpus, personal communication).

7. Mooers et al. (1976) noted that the winter weather systems are sufficiently well organized that objective hindcast methods may be used to predict the synoptic-scale surface pressure and wind-stress fields over the Middle Atlantic Bight. They demonstrated this idea by using the environmental buoy and coastal data to hindcast successfully (using a Cressman phase-lagged method) the surface pressure and vector wind stress at the two buoys. Overland and Gemmill (1977) made a less successful prediction of surface winds at the two buoys using just coastal data. This work suggests that the synoptic-scale surface-pressure and wind-stress fields could be accurately predicted over the entire Middle Atlantic Bight using just coastal data once sufficient buoy data are obtained in different regions to calibrate the prediction scheme.

8. Alongshelf currents directed toward the northeast are referred to here as *upshelf*, while alongshelf currents directed towards the southwest are referred to as *downshelf*.

9. The term *coastal sea level* is used in this chapter to mean the adjusted coastal sea level, i.e., the sum of observed coastal sea level and local atmospheric pressure.

Part Two

Physical Processes in Oceanography

8
Small-Scale Mixing Processes

J. S. Turner

8.1 Introduction

Forty years ago, the detailed physical mechanisms responsible for the mixing of heat, salt, and other properties in the ocean had hardly been considered. Using profiles obtained from water-bottle measurements, and their variations in time and space, it was deduced that mixing must be taking place at rates much greater than could be accounted for by molecular diffusion. It was taken for granted that the ocean (because of its large scale) must be everywhere turbulent, and this was supported by the observation that the major constituents are reasonably well mixed. It seemed a natural step to define eddy viscosities and eddy conductivities, or mixing coefficients, to relate the deduced fluxes of momentum or heat (or salt) to the mean smoothed gradients of corresponding properties. Extensive tables of these mixing coefficients, K_M for momentum, K_H for heat, and K_S for salinity, and their variation with position and other parameters, were published about that time [see, e.g., Sverdrup, Johnson, and Fleming (1942, p. 482)]. Much mathematical modeling of oceanic flows on various scales was (and still is) based on simple assumptions about the eddy viscosity, which is often taken to have a constant value, chosen to give the best agreement with the observations. This approach to the theory is well summarized in Proudman (1953), and more recent extensions of the method are described in the conference proceedings edited by Nihoul (1975).

Though the preoccupation with finding numerical values of these parameters was not in retrospect always helpful, certain features of those results contained the seeds of many later developments in this subject. The lateral and vertical mixing coefficients evaluated in this way differ by many orders of magnitude, and it was recognized that the much smaller rates of vertical mixing must in some way be due to the smaller scale of the vertical motions. Qualitatively, it was also known that the vertical eddy coefficients tended to be smaller when the density gradients were larger. The analysis of Taylor (1931) had shown that in very stable conditions K_S was smaller than K_M, which he interpreted to mean that the vertical transport of salt requires an intimate mixing between water parcels at different levels, whereas momentum can be transported by wave motion and is less affected by a strong vertical density gradient.

In contrast to these direct considerations of vertical mixing, Iselin (1939a) introduced the far-reaching idea that, because of the vertical stability, virtually all the large-scale mixing in the ocean might be accounted for in terms of lateral mixing (along isopycnals, rather than horizontally). In particular, he pointed to the striking similarity of the T-S relations for a vertical section and a surface section in the North Atlantic, each of which crossed the same isopycnals.

A strong constraint on achieving a fuller understanding of the small-scale mixing processes implicit in early measurements, was the lack of suitable instruments to resolve the scales that are directly involved. Most of the data came from water-bottle samples and widely spaced current meters, and it was tacitly assumed that the smooth profiles drawn through the discrete points actually represented the state of the ocean. Even when continuous temperature profiles became available in the upper layers of the ocean through the development of the bathythermograph, there was a tendency to attribute abrupt changes in slope to malfunctions in the instrument. The parameterization in terms of eddy coefficients implied that turbulence is distributed uniformly through depth, and is maintained by external processes acting on a smaller scale than the flows of interest; but in the absence of techniques to observe the fluctuations, and how they are maintained, little progress could be made.

Many such instruments are already in existence (see chapter 14); their use has rapidly transformed our view of the ocean, and in particular the understanding of the nature of the mixing processes. Temperature, salinity, and velocity fluctuations can be measured down to centimeter scales, and these records show that the distribution of properties is far from smooth. Rapid changes of vertical gradients are common, amounting in many cases to "steps" in the profiles. At some times the temperature and salinity variations are nearly independent, while at others they are closely correlated in a manner that has a profound effect on the vertical fluxes of the two properties (see Section 8.4.2). Viewed on a small scale, the ocean is *not* everywhere turbulent: on the contrary, turbulence in the deep ocean occurs only intermittently and in patches (which are often thin, and elongated horizontally), while the level of fluctuations through most of the volume is very low for most of the time. This is now more clearly recognized to be a consequence of the stable density gradient, which can limit the vertical extent of mixing motions and thus keep the relevant Reynolds numbers very small.

The newly acquired ability to study various mixing processes in the ocean has produced a corresponding increase in activity by theoretical and laboratory modelers in this field. The stimulation has been in both directions: theoreticians have been made aware of striking new observations requiring explanation, and they have developed more and more sophisticated theories and experiments that in turn suggest new observations to test them. Some of the work has required subtle statistical analysis of fluctuating signals, while many of the most exciting developments have been based on identifying individual mixing events (in the laboratory or the ocean), followed by a recognition of their more general significance.

Perhaps the most important factor of all has been the change in attitude to observational oceanography which took place in the early 1960s. Henry Stommel in particular advocated an approach more akin to the formulation and testing of hypotheses in other experimental sciences. Experiments designed to test specific physical ideas in a limited geographical area are now commonplace; but it is easy to forget how recently such uses of ship time have replaced the earlier "expedition" approach, in which the aim was to explore as large an area as possible in a given time (chapter 14).

This chapter will concentrate on the scales of mixing in the ocean, ranging from the smallest that have been studied to those with vertical dimensions of some tens of meters. Vertical-mixing processes will be emphasized, though the effects of quasi-horizontal intrusions near boundaries and across frontal surfaces will also be considered. After a preliminary section introducing ideas that are basic to the whole subject, various mixing phenomena will be identified and discussed in turn, starting with the sea surface and continuing into the interior and finally to the bottom. The grouping of topics within each depth range will be on the basis of the physical processes on which they depend. We shall not attempt to follow the historical order of development or to discuss observations in detail. Where there is a recent good review of a topic available, the reader will be referred to it. The major aims have been to describe the interrelation between theory, observation, and laboratory experiments that has led to the present state of understanding of each process, and to identify the areas still most in need of further work.

8.2 Preliminary Discussion of Various Mechanisms

8.2.1 Classification of Mixing Processes

It is important to keep clearly in mind the various sources of energy that can produce the turbulent motions responsible for mixing in the ocean. The first useful contrast one can make is between mechanically generated turbulence, i.e., that originating in the kinetic energy of motion, by the breakdown of a shear flow for example, and convective turbulence, produced by a distribution of density that is in some sense top heavy. The latter may occur in situations that seem obviously unstable, as when the surface of the sea is cooled [section 8.3.2.(d)], or more subtly, in the interior of the ocean when only one component (salt or heat) is unstably distributed (section 8.4.2) while the net density distribution is "hydrostatically stable."

A second informative classification depends on whether the energy comes from an "external" or "internal" source. In the first case, energy put in at a boundary is used directly to produce mixing in a region extending some distance away from the source. An example is the mixing through the upper layers of the ocean, and across the seasonal thermocline, caused by the momentum and heat transfers from the wind blowing over the surface. By "internal mixing" is implied a process in which the turbulent energy is both generated and used in the same volume of fluid, which is in the interior well away from boundaries. The mechanisms whereby the energy is ultimately supplied to the interior region must then also be considered carefully. Reviews of mixing processes based on the above classifications or a combination of them have been given by Turner (1973a,b) and Sherman, Imberger and Corcos (1978).

8.2.2 Turbulent Shear Flows

The maintenance of turbulent energy in a shear flow will be introduced briefly by summarizing the results for a "constant stress layer" in a homogeneous fluid flowing over a fixed horizontal boundary. [For a fuller treatment of this subject see chapter 17 and Turner (1973a, chapter 5).]

The boundary stress τ_0 is transmitted to the interior fluid by the so-called Reynolds stresses $\rho\overline{u'w'}$, which arise because of the correlation between the horizontal and vertical components of turbulent velocity u' and w'. The velocity gradient responsible for maintaining the stress can be related to the "friction velocity" u_* defined by $\tau_0 = \rho u_*^2$ $(= -\rho\overline{u'w'}$ in a constant stress layer) using dimensional arguments:

$$\frac{du}{dz} = \frac{u_*}{kz}, \qquad (8.1)$$

where z is the distance from the boundary and k a universal constant (the von Karman constant: $k = 0.41$ approximately). Integrating (8.1) leads to the well-known logarithmic velocity profile, which for an aerodynamically rough boundary becomes

$$u = \frac{u_*}{k} \ln \frac{z}{z_0}, \qquad (8.2)$$

where z_0, the roughness length, is related to the geometry of the boundary.

Using (8.1), one can also find the rate of production of mechanical energy per unit mass, ϵ say (which is equal to the rate of dissipation in a locally steady state):

$$\epsilon = u_*^2 \frac{du}{dz} = \frac{u_*^3}{kz}. \qquad (8.3)$$

It is also possible to define an eddy viscosity

$$K_M \equiv \frac{\tau_0}{\rho} \bigg/ \frac{du}{dz}, \qquad (8.4)$$

which is equal to ku_*z for the logarithmic profile. The relation between the flux and the gradient of a passive tracer can in some circumstances be predicted by assuming that the turbulent diffusivity (say K_H for heat) is equal to K_M: this procedure makes use of "Reynolds analogy." Notice, however, that the assumption of a *constant* value of K_M or K_H, by analogy with laminar flows, is already called into question by the above analysis. The logarithmic profile implies that these coefficients are proportional to the distance z from the boundary. In practice they can turn out to be more complicated functions of position, if observations are interpreted in these terms.

8.2.3 Buoyancy Effects and Buoyancy Parameters

As already outlined in section 8.1, vertical mixing in the ocean is dominated by the influence of the (usually stable) density gradients that limit vertical motions. The dynamic effect of the density gradient is contained in the parameter

$$N = \left(-\frac{g}{\rho_0}\frac{\partial \rho}{\partial z}\right)^{1/2}, \qquad (8.5)$$

the Brunt-Väisälä or (more descriptively) the *buoyancy frequency*, which is the frequency with which a displaced element of fluid will oscillate. The corresponding periods $2\pi/N$ are typically a few minutes in the thermocline, and up to many hours in the weakly stratified deep ocean.

In a shear flow, the kinetic energy associated with the vertical gradient of horizontal velocity du/dz has a destabilizing effect, and the dimensionless ratio

$$Ri = N^2 \bigg/ \left(\frac{du}{dz}\right)^2 = -g \frac{d\rho}{dz} \bigg/ \rho_0 \left(\frac{du}{dz}\right)^2, \qquad (8.6)$$

called the *gradient Richardson number*, gives a measure of the relative importance of the stabilizing buoyancy and destabilizing shear. Of more direct physical significance is the *flux Richardson number* Rf, defined as the ratio of the rate of removal of energy by buoyancy forces to its production by shear. It can be expressed as

$$Rf = \frac{g\overline{\rho'w'}}{\overline{\rho}u_*^2(du/dz)} = \frac{K_H}{K_M} Ri \quad = \frac{-B}{\epsilon} \qquad (8.7)$$

where ρ' is the density fluctuation and $B = -g\overline{\rho'w'}/\overline{\rho}$ is the buoyancy flux. Note that K_H may be much smaller than K_M in a stratified flow, so that while there is a strict upper limit of unity for Rf in steady conditions with stable stratification, turbulence can persist when $Ri > 1$.

Another "overall" Richardson number expressing the same balance of forces, but involving finite differences rather than gradients, can be written in terms of the overall scales of velocity u and length d imposed by the boundaries:

$$Ri_0 = g\frac{\Delta\rho}{\rho} d \bigg/ u^2. \qquad (8.8)$$

One must always be careful to define precisely what is meant when this term is used.

Two parameters commonly used to compare the relative importance of mechanical and buoyancy terms are expressed in the form of lengths. When the vertical fluxes of momentum $\tau_0 = \rho u_*^2$ and buoyancy B are given, then the Monin-Obukhov length (Monin and Obukhov, 1954)

$$L = \frac{-u_*^3}{kB} \qquad (8.9)$$

(where k is the von Karman constant defined in (8.1)) is a suitable scaling parameter; it is negative in unstable conditions and positive in stable conditions. This is a measure of the scale at which buoyancy forces become important; as B becomes more negative (larger in the stabilizing sense) buoyancy affects the motions on smaller and smaller scales. If, on the other hand, the rate of energy dissipation ϵ and the buoyancy frequency are known, then dimensional arguments show that the length scale

$$L_0 = \epsilon^{1/2} N^{-3/2}, \qquad (8.10)$$

first used by Ozmidov (1965), is the scale of motion above which buoyancy forces are dominant.

The molecular diffusivity κ (of heat, say) and kinematic viscosity ν do not appear in the parameters introduced above (though the processes of dissipation of energy and of buoyancy fluctuations ultimately depend on molecular effects at the smallest scales [section 8.4.1(d)]. When convectively unstable conditions are considered, however, the relevant balance of forces is between the driving effect of buoyancy and the stabilizing influence of the two diffusive processes that act to retard the motion. The parameter expressing this balance, the Rayleigh number

$$Ra = g\frac{\Delta\rho}{\rho} d^3 \bigg/ \kappa\nu \qquad (8.11)$$

does therefore involve κ and ν explicitly. Here $\Delta\rho/\rho = \alpha\Delta T$ is the fractional (destabilizing) density difference between the top and bottom of a layer of fluid of depth d (often due to a temperature difference ΔT, where α is the coefficient of expansion). The Reynolds number $Re = ud/\nu$ and the Prandtl number $Pr = \nu/\kappa$ can also be relevant parameters in both the stable and convectively unstable cases. In particular, it is clear that the Reynolds number based on internal length scales such as (8.9) and (8.10) is much smaller than that defined using the whole depth, which would only be appropriate if the ocean were homogeneous.

8.2.4 Turbulent Mixing and Diffusion in the Horizontal

We mention now several ideas which will not be followed up in detail in this review but which have had an important effect on shaping current theories of mixing in the ocean.

Eckart (1948) pointed to the distinction which should be made between stirring and mixing processes. The former always increase the gradients of any patch of marker moving with the fluid, as it is sheared out by the larger eddies of the motion. True mixing is only accomplished when molecular processes, or much smaller-scale turbulent motions, act to decrease these gradients and spread the marker through the whole of the larger region into which it has been stirred.

The total range of eddy scales, up to that characteristic of the current size of the patch, also play a part in the "neighbor separation" theory of diffusion due to Richardson (1926), which was originally tested in the atmosphere, and applied to the ocean by Richardson and Stommel (1948). Their results imply that the separation of pairs of particles l or equivalently the dispersion $\Sigma = \overline{y^2}$, of a group of particles about its center of gravity, satisfies a relation of the form

$$\frac{\partial \Sigma}{\partial t} \propto \Sigma^{2/3} \propto l^{4/3}. \qquad (8.12)$$

The rate of separation and hence the effective diffusivity increases with increasing scale because a larger range of eddy sizes can act on the particles. As Stommel (1949) pointed out, there is no way that (8.12) can be reconciled with an ordinary gradient (Fickian) diffusion theory using a constant eddy diffusivity. He showed, however, that it *is* explicable using Kolmogorov's theory of the inertial subrange of turbulence. Ozmidov (1965) has demonstrated that (8.12) can be written down directly using a dimensional argument, assuming that the rate of change of Σ (which has the same dimensions as an eddy diffusivity) depends only on the value of Σ and the rate of energy dissipation ϵ:

$$\frac{\partial \Sigma}{\partial t} \propto \epsilon^{1/3} \Sigma^{2/3}. \qquad (8.12a)$$

Bowden (1962) summarized the evidence available up to that time to show that this $\Sigma^{2/3}$ (or $l^{4/3}$) relation described the observations well over a wide range of scales, from 10 to 10^8 cm. The dependence on ϵ was not then investigated explicitly, but later experiments have shown that this factor (and hence the rate of diffusion) can vary greatly with the depth below the surface.

The apparent "longitudinal diffusion" produced by a combination of vertical (or lateral) shear and transverse turbulent mixing is another important concept. This originated in a paper by Taylor (1954), who showed that the downstream extension of a cloud of marked particles due to shear, followed by cross-stream mixing, leads to much larger values of the longitudinal dispersion coefficient D than can be produced by turbulence alone. For homogeneous fluid in a two-dimensional channel it is given by

$$D = 5.9du_*, \qquad (8.13)$$

where d is the depth and u_* the friction velocity.

Fischer (1973, 1976) has written two excellent reviews of the application of these (and other) results to the interpretation of diffusion processes in open channels and estuaries. The second in particular contains much that is of direct interest to oceanographers (cf. chapter 7).

8.3 Vertical Mixing in the Upper Layers of the Ocean

The major inputs of energy to the ocean come through the air–sea interface, and it is in the near-surface layers that the concept of an "external" mixing process is most clearly applicable. Because of the overall static stability, the direct effects of the energy input is to produce a more homogeneous surface layer, with below it a region of increased density gradient. Thus one must consider in turn the nature of the energy sources at the surface (whether the mixing is due to the mechanical effect of the wind stress, or to the transfer of heat), the influence of these on the motion in the surface layer, and finally the mechanism of mixing across the thermocline below. The variation of these processes with time, which leads to characteristic daily or seasonal changes of the thermocline, will also be considered. Most of the models are one-dimensional in depth, implying that the mixing is uniform in the horizontal, but some individual localized mixing processes will be discussed (see chapter 9 for additional discussion).

8.3.1 Parameterization of Stratified Shear Flows

The earliest theories of flow driven by the stress of the wind acting on the sea surface were based on an extension of the eddy-viscosity assumption. The vertical transports of heat and momentum can be related to the mean gradients of these properties using eddy coefficients that have an assumed dependence on the stability. Munk and Anderson (1948), for example, took K_H and K_M to be particular (different) functions of the gradient Richardson number (8.6), assumed that the (stabilizing) heat flux was constant through the depth, and solved the resultant closed momentum equation numerically to derive the velocity and temperature profiles. More recently, this method has been taken up again, but using more elaborate second-order closure schemes to produce a set of differential equations for the mean fields and the turbulent fluxes in the vertical (see, e.g., Mellor and Durbin, 1975; Launder, 1976; Lumley, 1978).

In this writer's opinion, this method has not yet proved its value for stratified flows in the ocean. The method is relatively complex, and the necessary extrapolations from known flows are so hard to test that the integral methods described in the following sections are at present preferred because they give more direct physical insight. It will, however, be worth keeping in touch with current work in this rapidly developing field.

8.3.2 Mixed-Layer Models

Thorough reviews of the various one-dimensional models of the upper ocean that have appeared (mainly in the past 10 years) have been given by Niiler (1977), and Niiler and Kraus (1977) [and the whole volume edited by Kraus (1977) is very useful]. The treatment here will be more selective: certain features of the models will be isolated, and their relative importance assessed. Laboratory experiments that have played an important role in the development of these ideas will also be discussed (though these have not always proved to be as directly relevant as was originally thought).

The starting point of all these models, which is based on many observations in the ocean and the laboratory, is that the mean temperature and salinity (and hence the density) are nearly uniform within the surface layer. In the case where mixing is driven by the action of the wind stress at the surface, the horizonal velocity in this layer is also assumed to be constant with depth, implying a rapid vertical exchange of horizontal momentum throughout the layer. Another assumption used in most theories is that there is an effectively discontinuous change of these variables across the sea surface and across the lower boundary of the mixed layer (see figure 8.1).

It is implied in what follows that the surface mixing is limited by buoyancy effects, not by rotation (i.e., the "Ekman layer depth" v_*/f is not a relevant scale), but rotation still enters into the calculation of the velocity step across the interface (see section 8.3.3). The Ekman layer depth will be considered explicitly in the context of bottom mixing (section 8.5.1).

Integration of the conservation equations for the heat (or buoyancy) content, and the horizontal momentum equation (in rotating coordinates where appropriate) across such a well-mixed layer, gives expressions for the temperature and bulk horizontal velocity in terms of the exchanges of heat and momentum across the sea surface and the interface below. The surface fluxes are in principle calculable from the external boundary conditions, but the interfacial fluxes are not known a priori. They depend on the rate of deepening, i.e., the rate at which fluid is mixed into the turbulent layer from the stationary region below. A clear understanding of the mechanism of *entrainment* across a density interface, with various kinds of mechanical or convec-

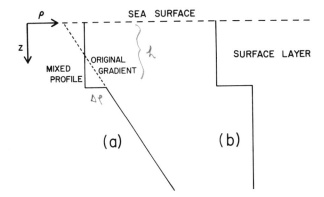

Figure 8.1 Sketch of the two types of density profiles discussed in the text: (a) a well-mixed surface layer with a gradient below it, (b) a two-layer system, with homogeneous water below.

tive energy inputs near the surface, is therefore a crucial part of the problem.

(a) Mixing Driven by a Surface Stress Purely mechanical mixing processes will be discussed first, starting with the case where a constant stress $\tau_0 = \rho_w v_*^2$ is applied at the surface (for example, by the wind), without any transfer of heat. When the water is stratified, a well-mixed layer is produced (see figure 8.1), which at time t has depth h and a density step below it, $\Delta\rho$ say. A dimensional argument suggests that the entrainment velocity $u_e = dh/dt$ can be expressed in the form

$$E_* = \frac{u_e}{v_*} = E_*(Ri_*), \qquad (8.14)$$

where

$$Ri_* = \frac{g\Delta\rho h}{\rho_w v_*^2} = \frac{C^2}{v_*^2} \qquad (8.15)$$

is an overall Richardson number based on the friction velocity v_* in the water. For the moment dh/dt is taken to be positive, but the general case where h can decrease will be considered in section 8.3.2(d).

Two related laboratory experiments have been carried out to model this process and test these relations. Kato and Phillips (1969) started with a linear salinity gradient (buoyancy frequency N) and applied a stress τ_0 at the surface (figure 8.1a). The depth h of the surface layer is related to $\Delta\rho$ and N by

$$C^2 = g\frac{\Delta\rho}{\rho}h = \frac{1}{2}N^2h^2. \qquad (8.16a)$$

Kantha, Phillips, and Azad (1977) used the same tank (annular in form, to eliminate end effects), but filled it with two layers of different density (figure 8.1b) rather than a gradient. In this case, conservation of buoyancy gives

$$C^2 = g\frac{\Delta\rho}{\rho}h = \text{a constant}. \qquad (8.16b)$$

Thus C^2 and Ri_* are clearly increasing as the layer deepens in the first case (8.16a) whereas they are constant when (8.16b) holds.

Equation (8.14) does not, however, collapse the data of these two experiments onto a single curve: at a given Ri_* and h, E_* was a factor of two larger in the two-layer experiment than with a linear gradient. Price (1979) has proposed that the rate of entrainment should be scaled instead with the mean velocity V of the layer (or more generally, the velocity difference across the interface):

$$\frac{u_e}{V} = E(Ri_V), \qquad (8.17)$$

where Ri_V is also defined using V. This is a more appropriate way to describe the process, since other physical effects may intervene between the surface and the interface and V need not be proportional to v_*. (In the experiments described above, side wall friction can dominate, though the method used to correct for this will not be discussed explicitly here.)

The form of Ri_V deduced from the experiments agrees with that previously obtained by Ellison and Turner (1959), which is shown in figure 8.2. [This will also be discussed in another context in section 8.5.2(a).] For the present purpose, we note only that E falls off very rapidly between $0.4 < Ri_V < 1$, so that $Ri_V \approx 0.6$ is a good approximation for Ri_V over the whole range of experimental results. The *assumption* that Ri_V is constant (which was based on an argument about the stability of the layer as a whole) was made by Pollard, Rhines, and Thompson (1973). This was used as the basis of the closure of their mixed-layer model which will be referred to in section 8.3.3.

The formulation in terms of V implies that conservation of momentum is the most important constraint

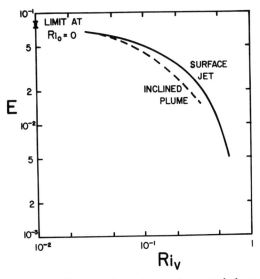

Figure 8.2 The rate of entrainment into a turbulent stratified flow as a function of overall Richardson number, for two types of experiments described by Ellison and Turner (1959).

on the entrainment in these laboratory experiments (and in the analogous oceanic case). The momentum equation

$$\frac{d(hV)}{dt} = v_*^2 = \text{constant}, \qquad (8.18)$$

together with the conservation of buoyancy relations (8.16), gives

$$E_* = \frac{u_e}{v_*} = nRi_V^{1/2}Ri_*^{-1/2} \qquad (8.19)$$

where

$$n = \begin{cases} \frac{1}{2} & \text{if } C^2 = \frac{1}{2}N^2h^2 \text{ (linear stratification)} \\ 1 & \text{if } C^2 = \text{constant (homogeneous lower layer).} \end{cases}$$

The predicted entrainment rate is a factor of two smaller in the linearly stratified case, as is observed. This difference arises because, in order to maintain Ri_V constant as C^2 increases, the whole layer, as well as the entrained fluid, must be accelerated to a velocity $V = (C^2/Ri_V)^{1/2}$. In the two-layer case, the stress is only required to accelerate entrained fluid to velocity V, which is constant.

Nothing has been said yet about the detailed mechanism of mixing across the interface, in the laboratory or in the ocean. Thorpe (1978a) has shown from measurements in a lake under conditions of surface heating that Kelvin–Helmholtz instability dominates the structure soon after the onset of a wind [section 8.4.1(c)]. Dillon and Caldwell (1978) have demonstrated the importance of a few "catastrophic events," relative to the much slower continuous entrainment processes. We tentatively suggest that these events could be the breakdown of large-scale waves on the interface, a mechanism that has been proposed for the benthic boundary layer (see section 8.5.1).

(b) The Influence of Surface Waves The experiments just described, and the models based on them, imply that the whole effect of the wind stress on the surface is equivalent to that produced by a moving plane, solid boundary. There is only one relevant length scale, the depth of the well-mixed layer. In the ocean, of course, there is a free surface on which waves are generated as well as a current, and this can introduce entirely new physical effects. The presence of waves can modify the heat, momentum, and energy transfer processes [as reviewed by Phillips (1977c)] and individual breaking waves can inject turbulent energy at a much smaller scale. Longuet-Higgins and Turner (1974) and Toba, Tokuda, Okuda, and Kawai (1975) have taken the first steps toward extending wave theories into this turbulent regime.

The interaction between a wind-driven current and surface waves can produce a system of "Langmuir cells," aligned parallel to the wind. These circulations, extending through the depth of the mixed layer, have long been recognized to have a significant effect on the mixing, and there are many theories purporting to explain the phenomenon [see Pollard (1977) for a review]. It now appears most likely that the generation depends on an instability mechanism, in which there is a positive feedback between the wind-driven current and the cross-wind variation in the Stokes drift associated with an intersecting pattern of two crossed wave trains. Physically, this implies that the vorticity of the shear flow is twisted by the presence of the Stokes drift into the vorticity of the Langmuir circulations. A heuristic model of the process was given by Garrett (1976), but the most complete and satisfactory theory is that of Craik (1977), who has clarified the differences between earlier related models. Faller (1978) has carried out preliminary laboratory experiments that support the main conclusions of this analysis (see also chapter 16).

The effect of density gradients on these circulations has not yet been investigated, so it is not clear whether a mixed layer can be set up by this mechanism under stable conditions. At high wind speeds, however, it seems likely that these organized motions, with horizontal separation determined by the wave field, will have a major influence on both the formation and rate of deepening of the mixed layer. This is an important direction in which the detailed modeling of thermocline mixing processes should certainly be extended.

(c) Input of Turbulent Energy on Smaller Scales The case where kinetic energy is produced at the surface, with turbulence scales much less than that of the mixed layer, has been modeled in the laboratory using an oscillating grid of solid bars. An early application to the ocean was made by Cromwell (1960). The more recent experiments by Thompson and Turner (1975), Hopfinger and Toly (1976), and McDougall (1979), have shown that the turbulent velocity decays rapidly with distance from the grid (like z^{-1}), while its length scale l is proportional to z. The experiments also suggest that the entrainment across an interface below is most appropriately scaled in terms of the velocity and length scales u_1 and l_1 of the turbulence near the interface, rather than using overall parameters such as the velocity of the stirrer and the layer depth. When this is done, the laboratory data are well-described by the relation

$$\frac{u_e}{u_1} = f(Ri_0, Pe), \qquad (8.20)$$

where

$$Ri_0 = \frac{g\,\Delta\rho}{\rho}\frac{l_1}{u_1^2},$$

$Pe = u_1 l_1/\kappa$ is a Peclét number.

The general form of the curves is similar to those shown in figure 8.2, but there is a distinct difference between the results of experiments using heat and salt as the stratifying agent (reflecting the different molecular diffusivity κ in Pe). They tend to the same form with a small slope at low Ri_0, where neither buoyancy nor diffusion is important, but diverge at larger Ri_0 to become approximately $u_e/u_1 \propto Ri_0^{-1}$ (heat) and $u_e/u_1 \propto Ri_0^{-3/2}$ (salt). The first form is attractive since it seems to correspond to the prediction of a simple energy argument (see Turner 1973a, chapter 9), but it now appears that one must accept the complications of the general form (8.20), including the fact that molecular processes can affect the structure of the interface and hence the entrainment at low values of Pe (Crapper and Linden, 1974).

The laboratory experiments of Linden (1975) have also shown that the rate of entrainment across an interface, due to stirring with a grid, can be substantially reduced when there is a density gradient below, rather than a second homogeneous layer. This is due to the generation, by the interfacial oscillations, of internal gravity waves that can carry energy away from the interface. The process can have a substantial effect on the mixing in the thermocline below the sharp "interface" itself. It is also a source of wave energy for the deep ocean, though the existence of a mean shear flow in the upper layer probably has an important influence on wave generation as well as on wave breaking (Thorpe, 1978b). It seems likely too that the process of mixing itself is affected in a significant way by the presence of a mean shear. Certainly organized motions in the form of Kelvin-Helmholtz billows occur only with a shear [see section 8.4.1(c)].

(d) The Effect of a Surface Heat Flux Only mechanical energy inputs have so far been considered; the effect of a buoyancy flux will now be added. The discussion will be entirely in terms of heat fluxes, but clearly the increase in salinity due to evaporation should also be taken into account (see, e.g., Niiler and Kraus, 1977).

When there is a net (equivalent) heat input to the sea surface, a stabilizing density gradient is produced that has an inhibiting effect on mixing. Note, however, that penetrating radiation, with simultaneous cooling by evaporation and long-wave radiation right at the surface, produces a localized convective contribution to the turbulence (see Foster, 1971). When there is a net cooling, at night or in the winter, convective motions can extend through the depth of the mixed layer and contribute to the entrainment across the thermocline below. In the latter case, detailed studies of the heat transfer and the motions very near the free surface have been carried out by Foster (1965), McAlister and McLeish (1969) and Katsaros et al. (1977) for example, and a comparison between fresh and salt water has recently been made by Katsaros (1978).

When there is a constant stabilizing buoyancy flux $B = g\overline{\rho'w'}/\bar\rho$ from above (i.e., a constant rate of heating, assumed to be right at the surface), and simultaneously a fixed rate of supply of kinetic energy, there can be a balance between the energy input and the work required to mix the light fluid down. Assuming, as did Kitaigorodskii (1960), that the friction velocity u_* is the parameter determining the rate of working, the depth of the surface layer can become steady at

$$h = au_*^3/B \tag{8.21}$$

while it continues to warm. This argument is closely related to that leading to the Monin-Obukhov length (8.9), and is also a statement of the conservation of energy.

During periods of increasing heating, the equilibrium depth achieved will be continually decreasing, so that the bottom of the mixed layer will rise and leave previously warmed layers behind. The minimum depth coincides with the time of maximum heating (assuming a constant rate of mechanical stirring). As the rate of heating decreases, the interface will descend slowly, while the temperature of the upper layer increases. Finally, when the surface is being cooled, there will be a more rapid cooling and also a deepening of the upper layer. Whether this mixing is "penetrative" or "nonpenetrative," i.e., the extent to which convection contributes to entrainment across the thermocline, is the subject of a continuing debate that is summarized in the following section. Molecular effects can also affect the rate of entrainment at low Pe, by altering the shape of the density profile on which the convective turbulence acts.

8.3.3 Energy Arguments Describing the Behavior of the Thermocline

Virtually all the models of the upper mixed layer in current use are based on energy arguments that balance the inputs of kinetic energy against changes in potential energy plus dissipation. They vary in the emphasis they put on different terms in the conservation equations, but some of the conflict between alternative models is resolved by recognizing that different processes may dominate at various stages of the mixing.

Niiler (1975) and de Szoeke and Rhines (1976), for instance, have shown that four distinct dynamic stages can be identified in the case where a wind stress begins to blow over the surface of a linearly stratified ocean (assuming there is no heating or energy dissipation). Using standard notation, the turbulent kinetic energy equation can then be summarized as

$$\frac{1}{2}\frac{\partial h}{\partial t}\left(\underset{A}{\alpha u_*^2} + \underset{B}{\frac{N^2 h^2}{2}} - \underset{C}{|\delta V|^2}\right) = \underset{D}{mu_*^3}. \tag{8.22}$$

The terms represent

- A the storage rate of turbulent energy in the mixed layer;
- B the rate of increase of potential energy due to entrainment from below;
- C the rate of production of turbulent mechanical energy by the stress associated with the entrainment across a velocity difference δV;
- D the rate of production of turbulent mechanical energy by surface processes.

Initially, there is a balance between A and D and the depth of the layer grows rapidly to a meter or so. After a few minutes, a balance between B and D is attained, and the mixed layer under typical conditions can grow to about 10 meters within an hour. In the meantime, the mean flow has been accelerating, and the velocity difference across the base of the layer increasing. Following Pollard et al. (1973), it can be deduced from the momentum equations in rotating coordinates that

$$(\delta V)^2 = \frac{2u_*^4}{h^2 f^2}[1 - \cos(ft)], \qquad (8.23)$$

where f is the Coriolis parameter. The turbulent energy produced at the base of the mixed layer by this shear can dominate on a time scale of half a pendulum day, and the balance is between B and C. The depth after this time is of order $u_*/(Nf)^{1/2}$, determined both by the stratification and rotation. [A modification of this argument by Phillips (1977b) leads to the alternative form $u_*/f^{2/3}N^{1/3}$, but observationally it could be difficult to choose between them.] In order to close this model, one still needs to add an independent criterion to relate the entrainment to the shear across the interface. As discussed in section 8.3.2(a), this can be taken as $Ri_V \approx 0.6$ since the mixing rate falls off sharply at this value of the overall Richardson number.

Finally, the intensity of the inertial currents decreases and a slow erosion continues, again as a balance between B and D. It is this final state that corresponds to the model of the seasonal thermocline introduced by Kraus and Turner (1967), and modified by Turner (1973b). Without considering any horizontal motion, they used the one-dimensional heat and mechanical energy equations, assuming that all the kinetic energy is generated at the surface, and that a constant fraction of this is used for entrainment across the interface below. This model can also deal with the case where the surface is being heated. Gill and Turner (1976) have carried the discussion a stage further, and shown that the cooling period is much better described by assuming that the convectively generated energy is nonpenetrative, i.e., that it contributes very little to the entrainment, only to the cooling of the layer.

As will be apparent from the earlier discussion of individual mixing mechanisms, it is by no means easy to identify which sources of energy are important, much less to quantify their effects. The most arbitrary feature of mixed-layer models at present is the parameterization of dissipation. It is usually assumed (or implied) that the energy available for entrainment is some fixed fraction of that produced by each type of source, and laboratory and field data are used to evaluate the constants of proportionality. This is the view adopted in the review by Sherman et al. (1978), for example, who added a term representing radiation of wave energy from the base of the layer. They compared the coefficients chosen by various authors, as well as suggesting their own best fit to laboratory experimental data. Another such comparison with laboratory and field data has been made by Niiler and Kraus (1977).

Processes such as the decay of turbulent energy with depth, and the removal by waves propagating through a density gradient below, are not adequately described by such constant coefficients. It is not clear that the shear through the layer is small enough to ignore as a source of extra energy, nor that two-dimensional effects (due to small-scale intrusions in the thermocline) are unimportant. In short, the models seem to have run ahead of the physical understanding on which they should be based. In the following section, some of the theoretical predictions will be compared with observations, consisting usually only of the mixed-layer depth and temperature. There are some good measurements of fluctuating temperature and salinity (Gregg 1976a) but to make further progress, and to distinguish between alternative models, much more detailed measurements of turbulent velocity and structure through the surface mixed layer and the thermocline below will be required. Thorpe (1977) and his coworkers have made the best set of measurements to date, using a Scottish Loch as a large-scale natural laboratory. (The JASIN 1978 experiment should contibute greatly to our understanding of these processes, but none of those results were available at the time of writing.)

8.3.4 Comparison of Models and Observations

In the absence of detailed measurements, the best hope of testing theories lies in choosing simple experimental situations in which as few as possible of the competing processes are active, or where they can be clearly distinguished. It is pointless, for example, to apply a model based on surface inputs alone if the turbulent-energy generation is in fact dominated by shear at the interface.

Price, Mooers, and van Leer (1978) have reported detailed measurements of temperature and horizontal-velocity profiles for two cases of mixed-layer deepening due to storms. They estimated the surface stress from wind observations, and compared model calculations based directly on u_*^2 with those based on the velocity difference (8.23). The predicted responses are quite different in the two cases, and the latter agreed well with the observations. The deepening rate accelerated during the initial rise in wind stress, but decreased abruptly as δV was reduced during the second half of the inertial period, even though u_*^2 continued to increase. They thus found no evidence of deepening driven by wind stress alone on this time scale, although turbulence generated near the surface must still have contributed to keeping the surface layer stirred.

A particularly clear-cut series of observations on convective deepening was reported by Farmer (1975). He has also given an excellent account of the related laboratory and atmospheric observations and models in the convective situation. The convection in the case considered by Farmer was driven by the density increase produced by surface heating of water, which was below the temperature of maximum density in an ice-covered lake. Thus there were no horizontal motions, and no contribution from a wind stress at the surface. From successive temperature profiles he deduced the rate of deepening, and showed that this was on average 17% greater than that corresponding to "nonpenetrative" mixing into a linear density gradient. Thus a small, but not negligible, fraction of the convective energy was used for entrainment. [The numerical values of the energy ratio derived in this and earlier studies will not be discussed here; but note that the relevance of the usual definition has been called into question by Manins and Turner (1978).]

In certain well-documented cases, models developed from that of Kraus and Turner (1967) (using a parameterization in terms of the surface wind stress and the surface buoyancy flux) have given a good prediction of the time-dependent behavior of deep surface mixed layers. Denman and Miyake (1973), for example, were able to simulate the behavior of the upper mixed layer at ocean weather station P over a 2-week period. They used observed values of the wind speed and radiation, and a fixed ratio between the surface energy input and that needed for mixing at the interface.

On the seasonal time scale, Gill and Turner (1976) have systematically compared various models with observations at a North Atlantic weathership. They concluded that the Kraus-Turner calculation, modified to remove or reduce the penetrative convective mixing during the cooling cycle, gives the best agreement with the observed surface temperature T_S of all the models so far proposed. In particular, it correctly reproduces the phase relations between the dates of maximum heating, maximum surface temperature, and minimum depth, and it predicts a realistic hysteresis loop in a plot of T_S versus total heat content H (i.e., it properly incorporates the asymmetry between heating and cooling periods). This behavior is illustrated in figure 8.3. The model also overcomes a previous difficulty and allows the potential energy to decrease during the cool-

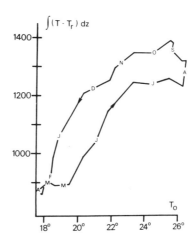

Figure 8.3 The heat content in the surface layer as a function of surface temperature T_0 at ocean weather station Echo. (After Gill and Turner, 1976.) The reference temperature T_r is the mean of the temperature at 250 m and 275 m depth, and the months are marked along the curve.

ing period, instead of increasing continuously as implied by the earlier models.

The mixed-layer depth and the structure of the thermocline are not, however, well predicted by these models; this fact points again to the factors that have been neglected. Niiler (1977) has shown that improved agreement is obtained by empirically allowing the energy available for mixing to decrease as the layer depth increases [though a similar behavior is implied by the use of (8.23); see Thompson (1976) for a comparison of the two types of model]. Direct measurements of the decay of turbulent energy with depth in the mixed layer will clearly be important. In many parts of the ocean it may also be necessary to consider upwelling.

Perhaps the most important deficiency is the neglect of any mixing below the surface layer. There is now strong evidence that the density interface is never really sharp, but has below it a gradient region that is indirectly mixed by the surface stirring. At greater depths too, the density profile is observed to change more rapidly than can be accounted for by advection, so that mixing driven by internal waves, alone or in combination with a shear flow, must become significant. These internal processes are the subject of the following section.

8.4 Mixing in the Interior of the Ocean

The overall properties of the main thermocline apparently can be described rather well in terms of a balance between upwelling w and turbulent diffusion K in the vertical. Munk (1966), for example, after reviewing earlier work, summarized data from the Pacific that show that the T and S distributions can be fitted by exponentials that are solutions of diffusion equations, for example

$$K\frac{d^2T}{dz^2} - w\frac{dT}{dz} = 0, \quad (8.24)$$

with the scaleheight $K/w \approx 1$ km. By using distributions of a decaying tracer ^{14}C, he also evaluated a scale time K/w^2, and the resulting upwelling velocity $w \approx 1.2$ cm day^{-1} and eddy diffusivity $K \approx 1.3$ cm^2 s^{-1} have been judged "reasonable" by modelers of the large-scale ocean circulation (chapter 15). Munk found the upwelling velocity consistent with the quantity of bottom water produced in the Antarctic, but he was not able to deduce K using any well-documented physical model. The most likely candidate seemed to be the mixing produced by breakdown of internal waves, but other possibilities are double-diffusive processes, and quasi-horizontal advection following vertical mixing in limited regions (such as near boundaries or across fronts).

Some progress has been made in each of these areas in the past 10 years, and they will be reviewed in turn. First, however, we shall discuss a set of interrelated ideas about the energetics of the process that are vital to the understanding of all types of mixing in a stratified fluid.

8.4.1 Mechanical Mixing Processes

(a) Energy Constraints on Mixing The overall Richardson number Ri_0 [defined by equation (8.8)] based on the velocity and density differences over the whole depth of the ocean, is typically very large, implying that the associated flow is dynamically very stable. But a second important fact is that the profiles of density (and other properties) are now known to be very nonuniform, with nearly homogeneous layers separated by interfaces where the gradients are much larger. Is it possible that a discontinuous structure of this kind (figure 8.4) is less stable, allowing turbulence to exist when it could not do so in the smooth average conditions?

Part of the answer was given by Stewart (1969), whose argument was developed by Turner (1973a, chapter 10). It can be shown that nonuniform profiles (different for velocity and density) can be chosen such that *any* value of the gradient Richardson number is attained everywhere in the interior, whatever the value of Ri_0—essentially because

$$(\Delta U/\Delta z)^2 \le \overline{(\partial u/\partial z)^2}. \quad (8.25)$$

Thus a redistribution of properties can always reduce the gradient Ri to a value at which turbulence can be maintained.

But a crucial question remains: how is this redistribution actually produced? Consider the energy changes associated with a transition from linear gradients of velocity and density $u = \alpha z$, $\rho = -\beta z + \rho_0$ say, to a well-mixed layer of depth H (see figure 8.4). The change

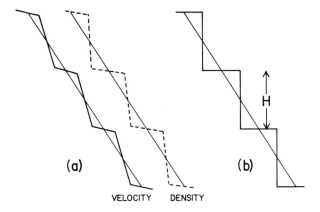

Figure 8.4 Discontinuous profiles produced by mixing, from initially linear density and velocity distributions. In (a) the final profiles, different for density and velocity, correspond to a constant gradient Richardson number everywhere, and (b) is the simpler model of homogeneous layers and thin interfaces used to derive (8.26).

in kinetic energy is $(1/24)\alpha^2 H^3$ and in potential energy $(1/12)g\beta H^3$; the two are equal when

$$Ri_0 = g\beta/\alpha^2 = 1/2. \quad (8.26)$$

This argument implies that for all $Ri_0 \gg 1$ there is *not* enough kinetic energy in the local mean motion to produce the observed, nonuniform profiles, even when dissipation is neglected entirely. In the absence of sources of convective energy due to double-diffusive processes (see section 8.4.2), the general conclusion is inescapable: extra energy must be propagated into the region from the boundaries in the form of inertial or internal gravity waves if mixing is to be sustained.

The role of internal waves and their relation to the nonuniform density structure may be approached in another way, using the argument set out by Turner (1973a, p. 137), and extensions of it. Consider a deep region of stable fluid, having linear profiles of both density and velocity through it. Suppose there is a constant stress (momentum flux) $\tau_0 = \rho u_*^2$ and buoyancy flux B through this region, sustained by small-scale turbulent motions. Mixing occurs only with fluid immediately above and below any level, so that only the internal lengthscale L defined by (8.9) will be relevant, not the overall depth or the distance from the boundaries. It follows on dimensional grounds that

$$\frac{du}{dz} = k_1 \frac{-B}{u_*^2} = k_1 \frac{u_*}{L}, \quad (8.27)$$

$$N^2 = -\frac{g}{\rho}\frac{d\rho}{dz} = k_2^2 \frac{B^2}{u_*^4} = k_2^2 \frac{u_*^2}{L^2} \quad (8.28)$$

where k_1 and k_2 are constants (which have not been determined experimentally). This is thus an equilibrium, self-regulated state, in which there is a unique relation between the gradients and the fluxes. The flux

Richardson number (8.7) has a fixed value $Rf = k_1^{-1}$, and so does the gradient Richardson number

$$Ri = k_2^2/k_1^2 = Ri_e, \qquad (8.29)$$

which has been called the equilibrium Richardson number.

Only in rather special cases can this equilibrium state be maintained—one good example is the edge of a turbulent gravity current, which is treated in section 8.5.2. When density and velocity differences are imposed over a given depth, the only equilibrium state is $Ri_0 = Ri_e$. If $Ri_0 < Ri_e$, the shear will dominate, and mixing will soon be influenced directly by the boundaries. If $Ri_0 > Ri_e$, as it is in the case of most interest here, then the stratification will dominate, though we have already seen how a nonuniform stratification allows the local Ri to be much smaller than Ri_0, so that turbulence can persist.

In this nonuniform state, however, (8.27) shows that the transport of momentum by turbulent processes is much less efficient in the interfaces where du/dz is larger, and it is impossible to have constant purely turbulent fluxes of both buoyancy and momentum through the whole depth. But the existence of interfacial waves provides a complementary mechanism to transport momentum across the steep gradient regions without a corresponding increase in the buoyancy flux.

There have been other suggestions about the mechanism of formation of layers from a linear gradient that can be related to the above ideas. Posmentier (1977), extending an idea formulated by Phillips (1972), suggested that if the vertical turbulent flux of buoyancy decreases as the vertical density gradient increases, any perturbation causing an increase in the gradient will be amplified. This occurs because the local decrease in flux leads to an accumulation of mass, which increases the density gradient further. This behavior is in contrast to the more familiar case, described by an eddy diffusivity, where an increase in gradient increases the flux, thus tending to smooth out any irregularity.

Linden (1979) has recently reviewed a wide range of laboratory experiments on "mechanical" mixing across a density interface, including those that use a shear flow, or stirring with oscillating grids [cf. sections 8.3.2(a) and 8.3.2(c)], and has suggested how they can be unified in terms of an energy argument. Briefly, he has shown that as the overall Richardson number Ri_0 increases from zero, the flux Richardson number Rf at first increases, reaches a maximum, and then falls as Ri_0 becomes even larger (see figure 8.5). This form can most readily be understood in terms of the grid-stirring results already described in section 8.3.2(c). The rate of increase in potential energy is $\frac{1}{2}g\Delta\rho u_e D^2$, where u_e is the entrainment velocity, and the rate of supply of kinetic energy is $\frac{1}{2}\rho u^3 D$. Thus by definition

$$Rf = \frac{u_e}{u}\frac{g\,\Delta\rho\, D}{\rho u^2} = \frac{u_e}{u} Ri_0. \qquad (8.30)$$

Using the power-law fit to the experiments $u_e/u \propto Ri_0^{-n}$, we find

$$Rf \propto Ri_0^{1-n}. \qquad (8.31)$$

In fact, the experimental results with salinity differences, described in section 8.3.2(c) imply that Rf is an increasing function of Ri_0 at low Ri_0 and a decreasing function at high Ri_0 (when $n = \frac{3}{2}$). The point where $n = 1$ corresponds to the simple overall energy argument, with a constant fraction of the energy supply being used for mixing.

Relating this now to the earlier argument, the maximum on figure 8.5 (which is schematic, but has the same form for the grid-stirred and shear-driven experiments) corresponds to the "equilibrium" conditions, where the gradients and fluxes are in balance [equations (8.27) and (8.28)]. If there is a self-balancing mechanism operating in which the rate of energy supply is itself regulated by the mixing it produces [cf. section 8.5.2(a)], then this is the state attained. If there is an excess of mechanical energy and a weak gradient (to the left of the maximum), mixing acts throughout the depth to reduce the gradient and spread out the interface. When the density gradient is the dominant factor (to the right of the maximum), turbulence is suppressed in an interface but can remain unaffected elsewhere, so that it acts to sharpen incipient interfaces. The relation to Phillips's and Posmentier's stability argument becomes clear once we note that, for a fixed rate of kinetic energy supply, Rf is proportional to the buoyancy flux and Ri_0 to the density gradient.

(b) Instability of Waves in a Smoothly Stratified Fluid Next, we consider the mechanisms of instability in a stratified fluid that can lead to local mixing, and thus produce or accentuate nonuniformities of the gradient. All of these involve waves propagating in from the boundaries, with or without a large-scale background shear set up by horizontal pressure gradients. When interfaces are already present, these will be the

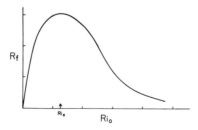

Figure 8.5 Schematic relation between the flux Richardson number Rf and the overall Richardson number Ri_0 for experiments on mixing across a density interface. (After Linden, 1979.) The maximum of the curve corresponds to the "equilibrium" condition.

first regions to become unstable [see section 8.4.1(c)], but it is logical first to describe how such a structure can be set up.

It is only recently that precisely what is meant by the "breaking" of internal waves has been properly investigated (see chapter 9). Some mechanisms are clearly related to localized sources of wave energy at a nearby boundary. Lee waves can be generated by the flow over bottom topography, and when the amplitude becomes large, overturning and the production of "rotors" is possible. When the mean horizontal velocity u varies in the vertical, the "critical-layer" mechanism also leads to the growth of the waves and the local absorption of energy near the level where u equals the horizontal phase velocity of the waves (Bretherton 1966c). Small-scale "jets" attributable to this mechanism have been reported in the ocean.

Nearly always, however, the source of wave energy at a point in the interior of the ocean is not clearly identifiable, and the motion is the result of the superposition of many waves. Energy can then be concentrated in limited regions through two types of interaction. Strong interactions between an arbitrary pair of waves of large amplitude can feed energy rapidly into small-scale forced waves that overturn locally. Resonant interactions are more selective, and require two waves to be such that the sum or difference of their wavenumbers is related to the sum or difference of their frequencies by the same dispersion relation as the individual waves. These are discussed in detail by Phillips (1977a).

Various laboratory experiments have played an important part in illuminating these processes; these (and many other experiments relevant to the subject of this chapter) have been reviewed by Maxworthy and Browand (1975), and by Sherman et al. (1978). McEwan (1971) generated a single low-mode standing wave, and showed that for sufficiently large amplitudes, the original waveform became modulated with two higher modes that formed a resonant triplet with the forced wave. These grew by extracting energy from the original mode until the superposition of the several motions produced visible local disturbances of the smooth gradient, and eventually turbulent patches that were attributed to a shear-breakdown in regions of enhanced density gradient. Orlanski (1972) carried out a similar experiment, but concluded that local overturning was responsible for the production of turbulence. McEwan (1973) used two traveling internal waves of different frequency, interacting in a limited volume of an experimental tank, to examine the local conditions just before breakdown, but he was unable to say definitely whether the primary mechanism for the production of turbulence was shear breakdown or overturning.

During the experiments reported in 1971, McEwan found that patches of turbulence could also be formed under conditions such that no resonant interaction was predicted [see also Turner (1973a, plate 24)]. More recently, this case has been studied in detail by McEwan and Robinson (1975), who explained it in terms of a "parametric" instability, which is, in fact, another resonant mechanism that had not previously been considered. This one is less selective, and gives rise to waves within a large range of much shorter wavelengths than the forcing wave, as follows. The original long wave produces a modulation of the effective component of gravity acting on shorter waves propagating through the same volume of fluid. When the forcing frequency is nearly twice the frequency of the growing disturbance, energy is fed into this disturbance through a mechanism analogous to that which causes the sideways oscillations of a pendulum to grow when the support is oscillated vertically. The major predictions of the theory, which include an estimate of the amplitude of the forcing wave required for the disturbances to overcome internal viscous dissipation and grow, were accurately verified in a most elegant laboratory experiment.

The application of this mechanism to the ocean has not yet been thoroughly tested, though McEwan and Robinson have extended Garrett and Munk's (1972a) ideas (based on their universal internal wave spectrum) to compute a mean-square slope of the isopycnals, which they deduce is large enough to excite the parametric instability. Much more work on this process is indicated; it certainly seems capable in principle of transferring energy directly from a broad range of large-scale internal waves to much smaller scales and thus creating patches of mixing in an otherwise smoothly stratified ocean.

(c) Mixing Due to Interfacial Shears Once sharp transition regions exist, across which both density and velocity vary markedly, it is easier to understand how local instabilities arise. The now extensive literature in this field has been well reviewed by Maxworthy and Browand (1975), and it will be treated only briefly here.

When the velocity and density profiles are similar, and the shear is gradually increased, a parallel stratified flow becomes unstable when the minimum-gradient Richardson number falls below 1/4. The fastest-growing instability takes the form of regular Kelvin–Helmholtz (K–H) "billows," with a wavelength that can be predicted knowing the profiles, and that is about six times the interface thickness. Experiments by Thorpe (1971) (see figure 8.6) and Scotti and Corcos (1972) confirmed the linear-stability theory for this case in great detail. On the other hand, when the density profile is much thinner than that for velocity, some wavelengths are unstable at larger values of Ri and interfacial second-mode waves of another type have been observed at Richardson numbers up to 0.7.

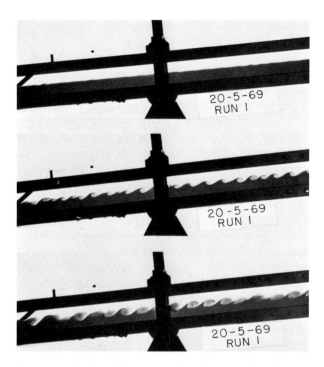

Figure 8.6 The breakdown of an interface in a shear flow to produce an array of Kelvin-Helmholtz billows. (Thorpe, 1971.)

The growth beyond the stage of initial instability has also been studied experimentally [see Thorpe (1973a) for a good review]. When the shear is increased, and then kept constant, the array of billows becomes unstable to a subharmonic disturbance, which leads to a two-dimensional rolling-up and merging of alternate vortices, a process that continues until limited by an energy constraint (as discussed below). Small-scale turbulence is produced by the concentration of vorticity into discrete lumps along the interface, and by gravitational instability within the overturned regions. The system of vortices stops growing, and then collapses, with much horizontal interleaving of mixed regions and a rapid dampening of the turbulence. This leaves behind a smoothly varying, nearly linear mean gradient of density, with thin higher gradient regions superimposed on it. Woods and Wiley (1972) suggested, however, on the basis of measurements in the ocean, that this overturning process should produce a well-mixed layer bounded by sharp interfaces. The implied "splitting" of interfaces to form new regions of high gradient does not seem to be borne out by the subsequent detailed laboratory experiments.

Thorpe (1978a) has reported observations of the mixing across the interface bounding a near-surface layer in a lake under stable conditions. Detailed measurements of temperature profiles as a function of time at one station contain all the features, including overturning and small scale mixing, described for the laboratory experiments. Thorpe concluded that the K-H instability was the dominant mechanism for mixing in this observation period, and it is likely to be equally important in the ocean under comparable conditions.

The maximum thickening of the interface, due to mixing following the K-H instability, is limited by energy considerations closely related to those set out in section 8.4.1(a). If an initial discontinuity is transformed by this process into linear gradients of velocity and density over the same interfacial depth δ, then equating the changes in kinetic and potential energies gives

$$\delta_{max} = 2\rho_0 u^2/g\,\Delta\rho. \tag{8.32}$$

The process is not perfectly efficient, however, and energy dissipation leads to much smaller limiting values. The numerical factor varies with the initial Richardson number of the interface, but Sherman et al. (1978) suggest using $\delta = 0.3\rho_0 u^2/g\,\Delta\rho$ as a typical value.

There are two important implications of this result. First, the instability is self-limiting. Unless the shear is increased, no further instability can occur, because the Richardson number in the thickened state is above that needed for instability. Second, the amount of vertical mixing that K-H instabilities alone can account for is small. Some other mechanism is needed to produce the turbulence in the well-mixed layers, which is essential both to transport heat and salt across them and produce the thinning of the interfaces required before further shear instabilities will be possible.

The shear needed to reduce Ri and so lead to instability at an interface can often be produced by internal waves. When a long internal wave propagates through the ocean, vorticity is concentrated at density interfaces. The sharper the interface, the more unstable it will be (i.e., the smaller the wave amplitude at which billows will form at the crests and troughs). Thorpe (1978b) has recently studied the interaction between finite-amplitude waves and an interfacial shear flow in the laboratory, and has shown that the slope at which breaking occurs can be significantly reduced. Direct visual observations of billows in the ocean generated in this way were made by Woods (1968a), using skin-diving techniques and dye tracers. Those observations had a great influence on subsequent work, by concentrating attention on the need to understand individual mixing *events* and processes in some detail, rather than always thinking in statistical terms. They also clearly demonstrated the relevance of simple experiments in the ocean and in the laboratory.

Recent, more sophisticated work has confirmed the importance of fine structure as a means for producing mixing in an internal wave field. Eriksen (1978) has described measurements made with an array of moored instruments, which he interprets in terms of large-scale waves "breaking." (This paper also contains a good summary of the relevant wave theory, and references to related work.) He has shown that the appearance of local temperature inversions (overturning) is associated with high shears, and that these are dominated by the fine-structure contribution. Moreover, there is a cutoff in the measured values of Ri at $Ri = 1/4$, indicating that regions with lower values of Ri are continuously becoming unstable, and implying some kind of saturation of the wave spectrum (see figure 9.28). Breaking is equally likely at any internal-wave frequency. These deductions were made using differences over 7 m, and it seems probable that the actual mixing events were unresolved at a smaller scale.

(d) Microstructure in Turbulent Patches The breakdown of internal waves by the mechanisms described above leaves behind a turbulent patch of fluid that tends to be thin, but very elongated in the horizontal. Such "blini" or pancakes of turbulence are distributed very intermittently in space and time, and are surrounded by fluid in which the level of fluctuations is very low. Measurements using towed instruments have shown that sometimes the turbulence is "active," i.e., there are both velocity and temperature-salinity fluctuations, but there can also be "fossil turbulence," or T-S microstructure remaining after the velocity fluctuations have decayed. This specialized field can only be mentioned briefly here, though it is important enough to deserve a full-scale review [see Phillips (1977a, chapter 6)]. It has developed somewhat independently, along lines established from the statistical measurements of turbulence properties in laboratory wind tunnels and in the atmosphere, and groups in the U.S.S.R. have been particularly active [see Monin, Kamenkovich, and Kort (1974, chapter 3); Grant, Stewart, and Moilliet (1962); Gargett (1976)]. Recently, other groups have become involved, and more measurements will be summarized in section 8.4.3. In this section we just refer to two results relating to the smallest scales of motion, where the turbulence is isotropic and decaying.

For active three-dimensional turbulence to persist, it is found that the Ozmidov length scale (8.10) must be larger than about 60 times the Kolmogoroff dissipation scale $(\nu^3/\epsilon)^{1/4}$; for typical conditions this implies $L_0 \approx 1$ m. When this is so, the form of the velocity, temperature and salinity-fluctuation spectra can be predicted from the local similarity theory (Batchelor, 1959), using a scaling that does not depend on the buoyancy frequency. When an actively turbulent patch is damped by stratification, however, the form of the fossil (T or S) turbulence is clearly affected by N, and a different scaling is appropriate. The cutoff length scale in the latter case is larger, and in principle the two can be distinguished. The important point made here, and reinforced below, is that fluctuation measurements can only be properly interpreted with a full knowledge of various other parameters, relating to large as well as small scales.

Osborn and Cox (1972) introduced a method (which is now widely used) for estimating the vertical flux of heat from measurements of temperature fluctuations T' in the dissipation range. They suggested that there is a balance between the production of small-scale variance by turbulent velocities acting in a mean vertical-temperature gradient $\partial \overline{T}/\partial z$ and the destruction of variance by molecular processes acting on sharpened microscale gradients. An effective vertical eddy diffusivity K_z can be defined by

$$K_z = \kappa \overline{(\partial T'/\partial z)^2}(\partial \overline{T}/\partial z)^{-2} = \kappa C, \qquad (8.33)$$

where the overbar denotes an average taken over the whole record, and C has been called the Cox number. There is an uncertainty of a factor between 1 and 3 because of the unknown degree of isotropy, but there are also some more fundamental constraints on the use of this idea [e.g., see Gargett (1978)]. Particularly when there are horizontal intrusions, with associated T and S anomalies that can produce correlations between microscale temperature and salinity fluctuations (see section 8.4.3), it is not appropriate to think in terms of a gradient diffusion process based on temperature alone. Stern (1975a, chapter 11) has derived more general thermodynamic relations involving both T and S variances, but these have not yet been properly tested by detailed measurements.

8.4.2 Convective Mixing

(a) Double-Diffusive Instabilities The most dramatic change in the whole field of oceanic mixing has come about through the recognition that molecular processes can have significant effects, even on scales of motion larger than those over which molecular diffusion can act directly. It is not sufficient just to know the net density distribution: the separate contributions of S and T are also important, and when these have opposing effects on the density, the transports of the two properties are quite different, and certainly cannot be described in terms of a single eddy diffusivity. Forty years ago this was unsuspected; twenty years ago a first consequence of unequal transports was recognized but regarded as "an oceanographical curiosity" (Stommel, Arons, and Blanchard, 1956), while in recent years the examples and literature documenting coupled molecular effects has multiplied rapidly.

When temperature and salinity both increase or both decrease with depth, one of the properties is "unstably" distributed, in the hydrostatic sense. The basic fact about double-diffusive convection is that the difference in molecular diffusivities allows potential energy to be released from the component that is heavy at the top, even though the mean density distribution is hydrostatically stable. Stommel (1962a) was one of the first to recognize that this convective source of energy implies that the potential energy is decreased, and the density difference between two vertically separated regions is increased following mixing—just the opposite to the changes occurring during mechanical mixing (see figure 8.7).

The interaction between theory and laboratory experiments on the one hand, and ocean observations on the other, has played a particularly important role in this field. The early work has been reviewed by Turner (1973a, chapter 8; 1974) and Stern (1975a, chapter 11), and more recent developments by Sherman et al. (1978). It is nevertheless worth repeating a description of two basic experiments that illustrate the different mechanisms of instability in the cases where the temperature and salinity distributions, respectively, provide the potential energy to drive the motion.

When a linear stable salinity gradient is heated from below (Turner and Stommel, 1964; Turner, 1968) the bottom boundary layer breaks down to form a convecting layer of depth d that grows in time as $d \propto t^{1/2}$. The experiments show that there is no discontinuity of density at the top of this layer, i.e., the temperature and salinity steps are compensating. When the thermal boundary layer ahead of the convecting region reaches a critical Rayleigh number Ra_c, it too becomes unstable. The first layer stops growing when d reaches

$$d_c = (\nu Ra_c/4K_T)^{1/4}B^{3/4}N_s^{-2}, \qquad (8.34)$$

and a second convecting layer is formed. Here d_c is the critical depth, $B = -g\alpha F_H/\rho C$ the imposed buoyancy flux corresponding to a heat flux F_H (α being the coefficient of expansion and C the specific heat), and N_s the initial buoyancy frequency of the salinity distribution. Huppert and Linden (1979) recently have extended this work to describe the formation of multiple layers as heating is continued. Linden (1976) used the analog system of salt–sugar solutions to study the case where there is a destabilizing salt (T) gradient partially compensating the stabilizing sugar (S) gradient in the interior. He found that as the density gradients become nearly equal, the properties of the layers depend mostly on the internal properties of the gradient region, with the boundary flux just acting as a trigger. [This analog has been much used for laboratory work, since it eliminates unwanted heat losses, and more experiments using this device will be discussed later. Salt is here the analog of heat, or temperature T, since it has a higher diffusivity than sugar (S).]

This first group of experiments illustrates well an important general consequence of opposing distributions of S and T: smooth gradients of properties are often unstable, and can break up to form a series of convecting layers, separated by sharper interfaces. In the case described, where the hotter, saltier water is

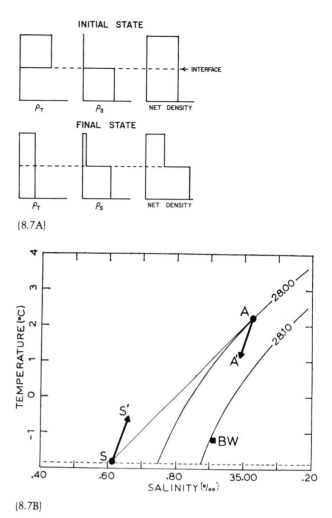

Figure 8.7 The changes in the separate concentrations and in the net densities produced by double-diffusion in a two-layer system: (A) schematic diagram of the initial and final properties, with a flux ratio $\beta F_S/\alpha F_T = 0.2$; (B) the water properties in the Greenland Sea on a θ–S correlation diagram. (After Carmack and Aagaard, 1973.) A = Atlantic water, S = Polar water, BW = Bottom water. The double-diffusive flux alters A in direction A' and S in direction S'.

below, the convection is driven by the larger vertical flux of heat relative to salt through the "diffusive" interfaces, which are in turn kept sharpened by the convection in the layers. Many examples of such interfaces are now known in the ocean. They are distinguished from layers formed in other ways [by internal wave breaking, for instance; see section 8.4.1(b)] by the regularity of the steps and the systematic increase of both S and T with depth. For example, Neshyba, Neal, and Denner (1971) have observed such layers under a drifting ice island in the Arctic; they have been found in various lakes that are hotter and saltier near the bottom (Hoare, 1968; Newman, 1976; see figure 8.8), and they occur in various Deeps in the Red Sea (Degens and Ross, 1969).

Now consider the second type of double-diffusive process, that for which the potential energy comes from the salinity distribution (or, more generally, from the component having the *lower* molecular diffusivity). When a small amount of hot, salty water is poured on top of cooler, fresh water, long narrow convection cells or "salt fingers" rapidly form. The alternating upward and downward motions are maintained by the more rapid horizontal diffusion of heat relative to salt, which leaves behind salinity anomalies to drive the motion. The form of motion and the scale was predicted by Stern (1960a) using linear stability theory, and a description of the finite amplitude state has since been given by Linden (1973) and Stern (1975a).

At first sight, there is a very great difference between the finger structure and a series of horizontal convecting layers seen in the "diffusive" case, but Stern and Turner (1969) showed that layers can form in the finger case too. [See also Linden (1978) for a recent experiment of this kind.] A sufficiently large flux of S acting on a smooth gradient of T can cause a deep field of salt fingers to break down into a series of convecting layers, with fingers confined to the interfaces. The mechanism appears to be a "collective instability" (Stern 1969), feeding potential energy from the salt fingers into a large-scale nearly horizontal wave motion that grows in amplitude and leads to overturning. When viewed on the scale of the convecting layers, there is a close correspondence between the two cases; an unstable buoyancy flux across a statically stable interface drives convection in layers, and only the mechanism of interfacial transport differs. Many examples of layering in the ocean due to the fingering process are now known, and they often occur under warm, salty intrusions of one water mass into another. The first observations were made by Tait and Howe (1968, 1971) under the Mediterranean outflow, and a summary of other measurements is given by Fedorov (1976). The direct detection of salt fingers in the interfaces between convecting layers using an optical method (Williams, 1974a, 1975) and conductivity probes (Magnell, 1976) has now given strong support to these ideas.

Another kind of instability that is potentially important is the merging of double-diffusive layers once they have formed. Turner and Chen (1974) and Linden (1976) have shown that this can occur either by the migration of an interface, so that one layer grows at the expense of its neighbor, or by a breakdown of an interface without migration. The possibility of merging implies that one cannot always interpret observed layer scales in terms of the initial mechanism of formation—subsequent events may have changed that scale. Recent theoretical and laboratory work on double-diffusive instabilities, including finite-amplitude effects, has been summarized by Sherman et al. (1978). Much of this has continued to concentrate on one-dimensional effects, though it is difficult to find situations in the ocean where one can be sure that the *formation* of layers and interfaces has been the result of one-dimensional processes. Nevertheless, as is discussed in section 8.4.2(c), the fluxes through such interfaces can probably be adequately described in these terms. The strongest layering is associated with large horizontal gradients of temperature and salinity, and the work that takes this fact explicitly into account will now be presented.

(b) Two- and Three-Dimensional Effects It became clear in early laboratory experiments that layers are readily produced in a smooth salinity gradient if it is heated from the side. Thorpe, Hutt, and Soulsby (1969) and Chen, Briggs, and Wirtz (1971) showed that a series of layers forms simultaneously at all levels by the following mechanism. A thermal boundary layer grows by conduction at the heated wall, and begins to rise. Salt is lifted to a level where the net density is close to that in the interior, and fluid flows away from the wall. The layer thickness is close to

$$l = \frac{\alpha \Delta T}{\beta \, dS/dz}, \qquad (8.35)$$

the height to which a fluid element with temperature difference ΔT would rise in the initial salinity gradient. More recent work has shown that similar layers are formed when the salinity as well as the temperature of the vertical boundary does not match that in the interior, for example, when a block of ice is inserted into a salinity gradient and allowed to melt. Huppert and Turner (1978) have demonstrated that when there is a salinity gradient in the environment, the melt water also spreads out into layers in the interior rather than rising to the surface. This will clearly influence the way icebergs affect the water structure in the Antarctic Ocean, and it also needs to be taken into account when assessing the feasibility of using towed icebergs as a source of fresh water.

Various two-dimensional processes were explored by Turner and Chen (1974) using a tank stratified with opposing vertical gradients of sugar and salt. When an inclined boundary is inserted into a stable "diffusive" system [i.e., one having a maximum salt (T) concentration at the top, and a maximum of sugar (S) at the bottom], a series of layers forms by a closely related mechanism to that of side-wall heating. Both depend on there being a mismatch between conditions at the boundary and in the interior, and again a series of extending "noses" forms, and extends out into the interior. With countergradients in the "finger" sense, disturbances can propagate much more rapidly across the tank in the form of a wave motion, which leads to nearly simultaneous overturning and convection by a mechanism reminiscent of Stern's (1969) collective instability.

The intrusion of one fluid into a gradient of another has been treated explicitly by Turner (1978). The basic intrusion process with which other phenomena can be compared is the two-dimensional flow of a uniform fluid at its own density level into a linear gradient (buoyancy frequency N) of the same property S, for example, salt solution into a salinity gradient. In that case, detailed studies (Maxworthy and Browand, 1975; Manins, 1976; Imberger, Thompson, and Fandry, 1976) show that the intruding fluid remains confined to a thin layer by the density gradient (see figure 8.9). For large Reynolds numbers, there is a balance between inertia and buoyancy forces, and the velocity U of the nose is constant, at

$$U \propto Q^{1/2} N^{1/2}, \tag{8.36}$$

where Q is the volume flux per unit width. At later stages, viscosity dominates and

$$U \propto N^{1/3} Q^{2/3} \nu^{-1/6} t^{-1/6} \tag{8.37}$$

(where ν is the kinematic viscosity), so the velocity decreases in time. These results can be used to describe the flow into the interior of fluid mixed by processes occurring near solid boundaries (section 8.5.4). The related unsteady process, the collapse of a fixed volume of homogeneous fluid into a density gradient, has been studied by Wu (1969a) and Kao (1976); this should model the subsequent spreading of an interior mixed region produced, for example, by breaking waves [section 8.4.1(b)]. Note that relatively sharp density gradients are maintained above and below the intruding fluid (because of the way it distorts the environmental density distribution), so this is another mechanism for producing and extending layered structures. Little research has been done on the corresponding three-dimensional flows, though these are worth further study. More attention should also be paid to the possible effects of rotation in limiting the amount of spreading (cf. Saunders, 1973).

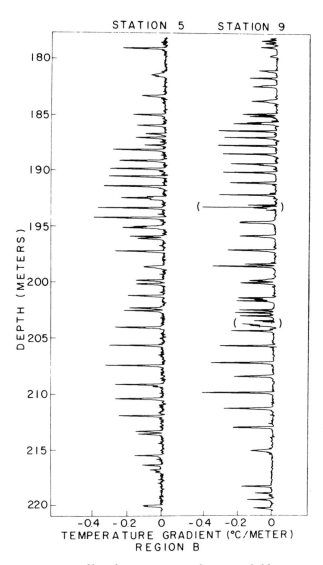

Figure 8.8 Profiles of temperature gradient recorded by Newman (1976) in Lake Kivu, showing a series of homogeneous layers separated by interfaces in which the gradient is much larger.

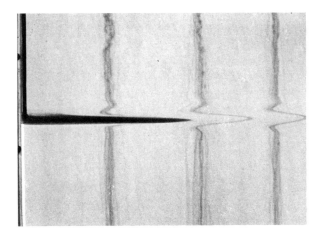

Figure 8.9 A two-dimensional intrusion of dyed salt solution into a salinity gradient at its own density level. (Turner, 1978.) Note the distortion of initially vertical dye streaks, even ahead of the injected fluid.

When the source fluid has different T-S properties from its surroundings, but still the density appropriate to its depth, the behavior is very different. (The laboratory experiments were carried out using a source of sugar in a salinity gradient, but they will be described in terms of the oceanic analog of warm, salty water released into temperature-stratified fresher water.) As shown in figure 8.10, there is a strong vertical convection near the source; this is limited by the stratification, and "noses" begin to spread out at several levels above and below the source. Further layers appear, and the volume of fluid affected by mixing with the input is many times the original volume. Each individual nose as it spreads is warmer and saltier than its surroundings, so "diffusive" interfaces form above and fingers below, and there a local decrease of T or an inversion through each layer. Note too the slight upward tilt of each layer as it extends. This implies that the net buoyancy flux [see section 8.4.2(c)] through the finger interface is greater than that through the diffusive interface, so that the layer becomes lighter and moves across isopycnals. These conclusions have been supported by experiments using a source of salt in a gradient of sugar in which the sense of the interfaces, and the tilt, are just the inverse of those just described. Strong systematic shears are also associated with the layers, and the sense of these motions has been explained in terms of the horizontal density anomalies set up by the net buoyancy flux.

Another geometry of direct relevance to the ocean is a discontinuity of T-S properties in the horizontal over a narrow frontal surface. In the present context, we consider only "fronts" across which the net density difference is small, and neglect rotational effects. [The larger-scale (baroclinic) instabilities that could lead to enhanced horizontal mixing in other circumstances will not be discussed here.] To model this case, Ruddick and Turner (1979) have set up identical vertical density distributions on two sides of a barrier, using sugar (S) in one-half of a tank and salt (T) in the other. When the barrier is withdrawn, a series of regular, interleaving layers develops (figure 8.11) whose depth and speed of advance are both proportional to the horizontal property differences, and therefore increase with depth. The scale is of the form (8.35), where $\alpha \Delta T$ is now the horizontal anomaly across the front (though a rather different energy argument has been used to derive this result).

A general conclusion to be drawn from all the experiments just described is that the formation and propagation of interleaving double-diffusive layers is a *self-driven* process, sustained by *local* density anomalies due to the quasi-vertical transports across the interfaces. Thus, however layers have formed, whether through strictly one-dimensional processes or by interleaving, it is important to understand the mechanism and magnitude of the fluxes of S and T through them.

(c) Double-Diffusive Fluxes through Interfaces
Quantitative laboratory measurements have been made of the S and T fluxes across an interface between a hot, salty layer below a cold, fresh layer. They can be interpreted using an extension of well-known results for pure thermal convection at high Rayleigh number Ra. Explicitly, Turner (1965), Crapper (1975), and Marmorino and Caldwell (1976) have shown that the heat flux αF_H (in density units) is well described by $Nu \propto Ra^{1/3}$, where Nu is the Nusselt number. This may be expressed in the form

$$\alpha F_H = A_1(\alpha \Delta T)^{4/3}, \qquad (8.38)$$

where A_1 has the dimensions of a velocity. For a specified pair of diffusing substances, A_1 is a function only of the ratio R_ρ of contributions of S and T to the density difference

$$R_\rho = \beta \Delta S / \alpha \Delta T, \qquad (8.39)$$

where β is the factor relating salinity to density. When $R_\rho < 2$, $A_1 > A \approx 0.1(g\kappa^2/\nu)^{1/3}$, the corresponding constant for solid boundaries, and for $R_\rho > 2$, A_1 falls progressively below A as R_ρ increases and more energy is used to transport salt across the interface. As discussed further by Turner (1973a), the empirical form

$$A_1/A = 3.8(\beta \Delta S/\alpha \Delta T)^{-2} \qquad (8.40)$$

(Huppert, 1971) gives a good fit to the observations.

The salt flux also depends systematically on R_ρ and has the same dependence on ΔT as does the heat flux. Thus the ratio of salt to heat fluxes should be a function of R_ρ alone:

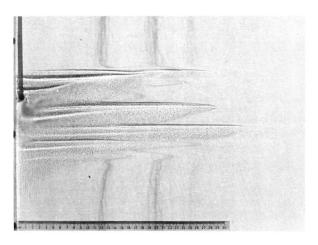

Figure 8.10 The flow produced by releasing sugar solution into a salinity gradient at its own density level. (Turner, 1978.) The gradient and the flow rate are exactly the same as for figure 8.9, but because of the double-diffusive effects, there is now strong convection and mixing near the source, followed by intrusion at several levels.

Figure 8.11 A system of interleaving layers produced by removing a barrier separating sugar solution (left) and salt solution (right), which have identical linear vertical density distributions. (Ruddick and Turner, 1979.)

$$R_F = \beta F_S/\alpha F_H = f_*(\beta \Delta S/\alpha \Delta T). \quad (8.41)$$

The first two papers cited above suggest that R_F falls rapidly from unity at $R_\rho = 1$ to 0.15 at $R_\rho = 2$, and then stays constant at $R_F = 0.15 \pm 0.02$ for $2 < R_\rho < 7$. (It must always be less than 1 for energetic reasons, and this implies that the density difference between the two layers will always be increasing in time.) The more recent paper of Marmorino and Caldwell (1976) suggests that the flux ratio can be as high as 0.4 with much smaller heat fluxes, and also gives different values of the normalized heat flux, for reasons that are as yet unresolved. The discrepancy merits further study, since Huppert and Turner (1972) applied the earlier laboratory values to explain the temperature structure of a salt-stratified Antarctic lake, with an accuracy that seemed to make an error of a factor of two unlikely.

Linden and Shirtcliffe (1978) have extended the "thermal-burst" model of Howard (1964a) to calculate fluxes and flux ratios in the two-component case. Transports through the center of the interface are by pure molecular diffusion, while the outer edge becomes intermittently unstable when the Rayleigh number based on its thickness reaches a critical value. [This process had previously been discussed by Veronis (1968a) in a more qualitative way.] The constancy of R_F over a certain range can be predicted by assuming that boundary layers of T and S grow by diffusion to thicknesses proportional to $\kappa_T^{1/2}$ and $\kappa_S^{1/2}$, and then break away together, down to the level where $\alpha \Delta T = \beta \Delta S$. The fluxes will then be in the ratio $\tau^{1/2} = (\kappa_S/\kappa_T)^{1/2}$, a result in reasonable agreement with experiments using both salt–heat and sugar–salt systems. The agreement with the individual flux measurements is much less impressive.

For finger interfaces, the condition for fingers to form in the first place has been examined by Huppert and Manins (1973). They showed that when a hot, salty layer is placed on a cold, fresh layer (or the equivalent in the analogous system), fingers can form in the interface, as it thickens by diffusion, provided

$$\beta \Delta S/\alpha \Delta T > \tau^{3/2}. \quad (8.42)$$

Since $\tau \approx 10^{-2}$ for heat–salt fingers, only very small destabilizing salinity differences are needed for them to form, and this suggests that salt fingers will be ubiquitous phenomena in the ocean.

Fluxes have also been measured across finger interfaces, and relations like (8.38) and (8.41) are again found to hold. Both the salt flux and the flux ratio are systematic functions of the density ratio, now more conveniently defined in the inverse sense as $R_\rho^* = \alpha \Delta T/\beta \Delta S$. In particular, Turner (1967) found $\alpha F_H/\beta F_S = 0.56$ for heat–salt fingers over the range $2 < R_\rho^* < 10$. This result seems to be confirmed by recent work due to Schmitt (1979) and by the author (unpublished), though a much lower value of the flux ratio obtained by Linden (1973) remains unexplained. No experiments have convincingly achieved values of R_ρ^* very close to 1, where an increase in flux ratio might be expected, but this range could be of great importance for the ocean.

The experiments of Linden (1974) are also of interest. He applied a shear across a finger interface and showed that a steady shear has little effect on the fluxes, though it changes the (nearly square) fingers into two-dimensional sheets aligned down shear. Thus fluxes of S and T will be expected to persist in spite of the shears set up by interleaving motions across a front (figure 8.11). Unsteady shears, on the other hand (i.e., mechanical stirring on both sides of the interface) can rapidly disrupt the interface and decrease the salt flux.

Stern (1975a, 1976) has extended his collective-instability model (in two different ways) to describe the breakdown of fingers at the edge of an interface. He supposes that instability sets in when the salt flux becomes too large for the existing temperature gradient, a condition that can be expressed in terms of a Reynolds number of the finger motions, and that is achieved first near the edges. This is consistent with a steady state through the interface in which the flux is

$$\beta F_S \sim C(g\kappa_T)^{1/3}(\beta\Delta S)^{4/3}, \qquad (8.43)$$

where C is a function of R_ρ^* only when $\tau = \kappa_S/\kappa_T$ is small. When $R_\rho^* < 2$, various laboratory experiments agree in giving $C = 0.1$, and this form is in better accord with experiments than previous expressions, which involved κ_S explicitly. Griffiths (1979) has recently proposed a model based on the intermittent growth and instability (in the Rayleigh-number sense) of the edge of an interface, the model previously applied only to diffusive interfaces. He has been able to predict a flux ratio of about 0.6, and also various properties of the fingers in the interface, including the relation between the width h and length L of the fingers $(h \propto L^{1/4}$ approximately) that was observed by Shirtcliffe and Turner (1970) for sugar–salt fingers.

(d) Multiple Transports through Diffusive Interfaces It is clear from all the results described in the previous section that the "transport coefficients," defined as the vertical fluxes divided by the corresponding mean gradients, are inevitably different for heat and salt when the transports are due to double-diffusive processes acting across interfaces. For a diffusive interface, for example,

$$\frac{K_S}{K_H} = \frac{F_S}{\Delta S}\frac{\Delta T}{F_H} = R_\rho^{-1}\tau^{1/2}. \qquad (8.44)$$

The effective "eddy diffusivity" of the driving (unstably distributed) component will always be larger than the driven (see figure 8.7): $K_H > K_S$ in the diffusive case, and $K_S > K_H$ when there are salt fingers. Both T and S are transported down their respective gradients, at different rates, but the inappropriateness of the eddy diffusivity approach becomes obvious when one considers the net density flux. This is transported against the gradient, since the potential energy is decreasing and the density difference tending to increase. Moreover, the largest individual transports occur when the density gradient is weakest, i.e., when $R_\rho \to 1$.

When there are several different stabilizing salts in a hot, salty layer below a cold, fresh layer, the relative transports of each across the diffusive interface can, however, be usefully described in terms of the ratio of transport coefficients K_i. Turner, Shirtcliffe, and Brewer (1970) suggested that the individual K_i can depend on the molecular diffusivities, and Griffiths (1979) has recently examined this more carefully, both theoretically and experimentally. He has predicted, using a further extension of Linden and Shirtcliffe's (1978) "intermittent instability" model, that K_1/K_2 should be proportional to $\tau^{1/2} = (\kappa_1/\kappa_2)^{1/2}$ at low total solute-heat density ratios R_ρ^T, and to τ at higher R_ρ^T, where a steady diffusive core dominates. The results are insensitive to the relative contributions of each component to the total density difference.

Experiments at interfacial density ratios between 2 and 4 are consistent with $K_1/K_2 = \tau$, but Griffiths finds an even greater separation of components at higher R_ρ^T, for reasons that are so far unexplained. The results of one of his experiments are shown in figure 8.12. He has also shown that the separation of different salts during transport through a finger interface is relatively unimportant.

These ideas have been tested on a geophysical scale using available data for Lake Kivu, a salt-stratified lake that is heated geothermally at the bottom. As illustrated in figure 8.8, this contains many well-mixed layers separated by diffusive interfaces, and Newman (1976) showed that the upward salt flux, calculated using a heat flux derived from laboratory results, is in satisfactory agreement with the salinity in the river flowing out of the top of the lake. Griffiths has used the geochemical data of Degens et al. (1973) to estimate the fluxes and gradients, and hence K_i, for several ions (potassium, sodium, and magnesium) separately. In the range $R_\rho^T \approx 2.0$, the effective transport coefficients decrease in order of decreasing molecular diffusivity, and the ratios are consistent with the laboratory measurements.

The above results have far-reaching, but as yet hardly explored, implications for the ocean. It is tacitly assumed that a tracer may be used to mark a water mass, and that its changing concentration is a measure of the

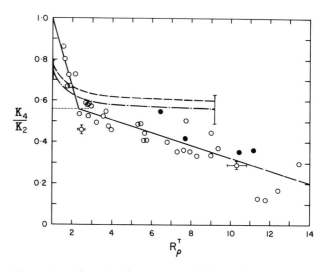

Figure 8.12 The ratio of transport coefficients for magnesium (K_4) and potassium (K_2) measured by Griffiths (1979) across a diffusive interface, with heating below. The ratio of molecular coefficients $K_4/K_2 \approx 0.60$, and the upper curves are Griffiths's theoretical predictions.

"mixing rate" for the water mass as a whole. But if diffusive interfaces are important, the transport of a tracer having a different molecular diffusivity is not necessarily a good indicator of the transport of a major component, much less of heat. In the absence of definite knowledge of the mixing mechanisms operating between the source and the sampling point, a single "eddy diffusivity" must be used with great caution.

(e) Cabbeling and Related Instabilities Another kind of convective instability that can lead to internal mixing depends on the nonlinear-density behavior of sea water. Particularly at low temperatures, the mixing of two parcels of water with the same density but different T-S properties produces a mixture with a greater density than that of the constituents. This will sink, generating additional mixing, and the whole process is called cabbeling (various other spellings appear in the literature). Even when S and T are not quite compensating, so that the density decreases upward, a finite amplitude vertical displacement, followed by mixing, can lead to the effect described.

This possibility was first recognized at the turn of the century. Fofonoff (1956) showed that the formation of Antarctic bottom water is probably influenced by this process, and Foster (1972) has given a good account of the history, as well as a stability analysis for the case of superimposed water masses. He has applied his results to the Weddell Sea, in which the surface is generally colder and fresher than the underlying deep water. When the salinity at the surface increases due to sea-ice formation in the winter, mixtures of surface and deep water may become denser than the deep water, and thus sink through it and contribute to bottom-water formation.

Foster and Carmak (1976b) have since applied related ideas to the explanation of layers at mid-depth in the center of the Weddell Sea. They note that where the T and S gradients are weak and nearly compensating, deep well-mixed layers are formed, and they attribute this to cabbeling. But the sense of the two opposing gradients is just that required for double-diffusive instabilities to produce layers, separated by "diffusive" interfaces. At shallower depths, the gradients are larger and the layers thinner; no cabbeling instability appears to be possible, and layer formation due solely to double-diffusive effects is postulated. A closer study of the conditions separating these regimes would be instructive.

Gill (1973) has shown that when parcels of water are given finite vertical displacements, instabilities can arise due to the different compressibility of sea water at different temperatures and pressures. The compressibility of cold water is generally greater than that of warm, so in the situation discussed above, a cold parcel displaced downward could in principle become heavier than its new surroundings. The displacements required, however, are rather large, and though the effect may be significant for bottom-water formation (see Killworth, 1977), no evidence has been found that it can influence the formation of layers in the interior, or mixing on a smaller scale.

8.4.3 Observations of Fine Structure and Microstructure

There are now many observations in the deeper ocean in which the influence of the processes described in sections 8.4.1 and 8.4.2 can be identified. Most of these have been made using vertical profiles from lowered or freely falling instruments, with a few significant contributions from towed sensors [see section 8.4.1(d)]. Temperature and salinity fluctuations are the most commonly measured quantities, though small-scale velocity shear measurements are just becoming available (Simpson, 1975; Osborn, 1978).

A useful summary of the observations up to about 1974 has been given by Fedorov (1976) (with extra references in the English translation to mid-1977). He emphasizes the fine structure, or nonuniformities, of vertical gradient associated with a "layer-and-interface" structure, which needs to be known before microstructure measurements in the water column can be understood properly. The strongest layering is found near boundaries between water masses of different origin, and it is most prominent when there is a large horizontal contrast in T-S but a small net density difference. Interleaving motions, with associated temperature inversions, readily develop in these circumstances, and the double-diffusive processes described in section 8.4.2(b) become especially relevant.

Only two recent examples will be cited here: profiles across the Antarctic polar front (Gordon, Georgi, and Taylor, 1977) reveal inversions that decrease in strength with increasing distance away from the front. Joyce, Zenk, and Toole (1978) have made a more detailed analysis of observations in this area, and have concluded that double-diffusive processes are significant. Coastal fronts between colder, fresh water on a continental shelf and warmer, salty water offshore also exhibit strong interleaving (Voorhis, Webb and Millard 1976). More observations and laboratory experiments related to such intrusions have been reviewed and compared by Turner (1978). It must be emphasized that double-diffusive processes can be important even in regions where the mean S increases and T decreases with depth, and both distributions are stabilizing [e.g., off the coast of California (Gregg, 1975)]. Horizontal interleaving organizes the gradients so that double-diffusive convection can act: it is a *self-driven* process, sustained by local density anomalies set up by the quasi-vertical fluxes. In this way double-diffusion serves both to produce the layering and to dissipate the energy within it. Observations also show (Howe and Tait, 1972; Gargett, 1976) that the density gradient above a warm intrusion is typically much larger than that below, in accord with the laboratory observation of sharp diffusive interfaces above and more diffuse finger interfaces below such an intrusion.

More detailed measurements of microstructure in relation to the fine structure also show the importance of intrusions, though the interpretation of the detailed mechanism involved is sometimes ambiguous. We have already discussed [section 8.4.1(c)] the observations of Eriksen (1978), who related wave-breaking events to the fine structure, so there are certainly some occasions on which shear-generated turbulence is important. Gregg (1975) concluded from T and S microstructure profiles measured in the Pacific that the regions of most intense activity are the upper and lower boundaries of intrusions produced by interleaving, and suggested alternative explanations in terms of shear-generated turbulence and double-diffusive phenomena. The undersides of temperature-inversion layers were found to have the highest level of activity, which we can now attribute to salt fingers. Williams (1976) also found, using thermal sensors mounted on a mid-water float, that the regions of most intense mixing are closely associated with intrusive features, and he was able to distinguish occasions when one or other mechanism was dominant.

Gargett (1976) has shown that higher levels of small-scale temperature fluctuations are invariably found in areas where the vertical profiles of T and/or S have fine-structure inversions. The highest percentage of the sampled water volume was found to be turbulent when the local T and S gradients are in the finger sense. So we come back to the point made by Gargett (1978), and mentioned in section 8.4.1(d); double-diffusive processes, associated with intrusions, are very often important in producing fine structure and microstructure. Thus deductions made on the basis of temperature fluctuations alone, which, moreover, imply that the transport is entirely vertical, are not likely to be valid.

We turn now to examples of the large-scale effects of vertical double-diffusive transports across the boundaries of intrusions. Lambert and Sturges (1977) have shown that the decrease in salinity in and below the core of a warm saline intrusion can be explained in terms of the downward flux of salt in fingers. They observed a series of stable layers separated by finger interfaces, through which the flux (calculated using laboratory results) was sufficient to account for the observed rate of decrease of S with distance. Voorhis et al. (1976) used neutrally buoyant floats to record the change in T and S in the same water mass over a period of days. They found evidence of rapid vertical fluxes of heat and salt between layers, at different rates consistent with $\alpha F_H / \beta F_S \approx 0.5$, approximately the laboratory value for salt fingers.

Schmitt and Evans (1978) have shown that salt fingers grow rapidly enough to survive even in an active internal wave field. They have calculated salt fluxes for measured profiles of S and T, using laboratory data and assuming that fingers are intermittently active on the high gradient regions. The calculated flux of salt is comparable to the surface input of salt due to evaporation, i.e., they deduce that fingers can account for all the vertical flux in the ocean. Carmack and Aagaard (1973) have given an example of the large-scale importance of vertical transports in the "diffusive" sense. From the changes in S and T in the deep water of the Greenland Sea, they suggest that bottom water is not formed at the sea surface, but by a subsurface modification across an interface between a colder, fresher surface layer of Polar water and a warm salty lower layer of Atlantic water (see figure 8.7B). Their deduced ratio of $K_S/K_T \approx 0.3$, showing that heat is definitely transported faster than salt, supports this view.

Thus, far from being an amusing curiosity, double-diffusive convection is playing a significant, and in some regions dominant, role in the vertical mixing of heat and salt in the world's oceans. Its overall importance relative to other processes such as wave breaking and boundary mixing (reviewed in the following section) has not yet been assessed adequately.

8.5 Mixing near the Bottom of the Ocean

Compared to that in the atmospheric boundary layer, or even the surface layers of the ocean, work on the ocean-bottom layer has been very sparse. The early research, summarized by Bowden (1962), concentrated on shallow seas, but the measurements of heat fluxes through the deep ocean bottom made it desirable to know more about the flows in those regions (Wimbush and Munk, 1970). More sophisticated instruments have now been developed to allow more detailed measurements, but as the proceedings of a recent conference on the subject show (Nihoul, 1977), there is as yet no clear consensus in this field.

8.5.1 Mixing Induced by Mean Currents

Most measurements of the depth of the benthic boundary layer have been referred to the "Ekman depth"

$$h_e = 0.4 u_* / f, \tag{8.45}$$

which is the scale appropriate to an unstratified, turbulent flow in a rotating system. A logarithmic layer, described by (8.2), is contained within the lowest part of this, where the stress can be regarded as constant and rotation is unimportant. The Ekman theory also predicts a veering of the current with height above the boundary.

But the data suggest (in various ways) that this may not be a directly relevant scale, because of the stable stratification of the water column. For example, figure 8.13 shows profiles of θ and S measured by Armi and Millard (1976) on an abyssal plain. The well-mixed layer, bounded by a sharper interface, strongly suggests that this structure has been formed by stirring up the bottom part of the gradient region above (cf. figure 8.1a). This stirring is therefore an "external" mixing process, driven by turbulent energy put in at the boundary. The buoyancy flux associated with the heat flux through the bottom has a negligible effect, except when the speed of the current is very low. The layer depth h in this case is about six times h_e, and the mean depth on different days was correlated with the mean current velocity U. Armi and Millard showed that

$$F = U \bigg/ \left(g \frac{\Delta\rho}{\rho} h\right)^{1/2},$$

a Froude number, was approximately constant at ~ 1.7. This led to the hypothesis that the layer depth is controlled by the instability of large-scale waves traveling along the interface. No direct evidence of such a wave-breaking process has been reported, though the temporal variation of layer depth in these and other measurements [see, e.g., Greenewalt and Gordon (1978)] indicate that waves of large amplitude are often present. Note that this Froude-number criterion is closely related to the "constant-overall-Richardson-number" hypothesis used in surface mixed-layer models [sections 8.3.2(a) and 8.3.3].

A different picture has been developed by Weatherly and van Leer (1977), on the basis of their measurements of temperature and current profiles on a continental shelf. Their boundary-layer thicknesses (defined from current profiles) were substantially smaller than (8.45), and they attributed this to the effect of stable stratification. They did, however, observe large changes of current direction in a sense consistent with Ekman veering, particularly when the stratification was large, and they have described their results in terms of a stably stratified turbulent Ekman layer. Relatively few of their profiles had well-mixed layers at the bottom, and even those suggested an advective origin, rather than local turbulent mixing. When the bottom was sloping, and the flow was along the isobaths, the observations contain a systematic increase or decrease of temperature in time, which is consistent with downwelling or upwelling along the slope produced by the Ekman transport.

It seems likely that the difference between these observations and those of Armi and Millard (1976) lies in the much weaker stratification in the deep ocean,

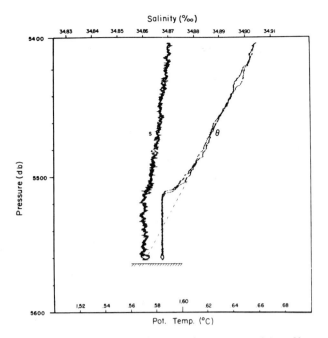

Figure 8.13 Salinity (S) and potential temperature (θ) profiles measured by Armi and Millard (1976) in the middle of the Hatteras Abyssal Plain. The dashed line indicates that the structure could have formed by mixing up a stratified region above the bottom.

and the consequently larger values of F there. But before this can be regarded as certain, more measurements at other sites will be needed. Theoreticians should also look more carefully at the properties of mixed layers shallower and deeper than the Ekman depth, and systematically compare the bottom-layer results with surface mixed layers.

In shallow seas, mixing driven by turbulence produced at the bottom can extend to the surface. This tendency is counteracted by the input of heat at the surface, which produces a stabilizing temperature gradient. Qualitatively, one can see that the larger the current and the shallower the depth, the more likely is the water to be uniformly mixed for a given surface flux [cf. section 8.3.2(d)].

Simpson and Hunter (1974) showed, in fact, that there are marked frontal structures in the Irish Sea, separating well-mixed from stratified regions. The location of these fronts is determined by the parameter h/u^3, where h is the water depth and u the amplitude of the tidal stream. The choice of this form can be justified using an energy argument, related to that used to obtain (8.21). The relevant dimensionless parameter must include the buoyancy flux B, and is Bh/u^3. Thus the only extra assumption is that B varies little over the region of interest. Simpson and his coworkers have now extended this model to include the effects of wind stress as well as tidal currents, and find that this has a significant, though less important, influence.

8.5.2 Buoyancy-Driven Bottom Flows

(a) Turbulent Gravity Currents The flow of heavier water down a slope under lighter layers is important in many oceanic contexts. The density differences may be due to T and S differences, or to suspended sediment (as in turbidity currents). The velocity of such flows is strongly influenced by the mixing between the current and the water above, and mixing can determine the final destination of the flowing layer. For example, the water flowing out through the Strait of Gibraltar is denser than water at any depth in the Atlantic, but mixing with lighter water near the surface eventually makes its density equal to that of its surroundings, so that it flows out into the interior at middepth (see figure 8.14).

Turner (1973a, chapter 6) has shown how a nonrotating turbulent gravity current can be treated as a special case of a two-dimensional plume, rising vertically through its environment. [This more general problem is also relevant to the disposal of waste water in the ocean, which will not be treated here; see Koh and Brooks (1975) for a review.] The "entrainment assumption," that the rate of inflow u_e is proportional to the local mean velocity u, must be modified to take account of the stabilizing effect of buoyancy normal to the plume edge. Explicitly, it is found that

$$\frac{u_e}{u} = E(Ri_0), \qquad (8.46)$$

i.e., the entrainment ratio is a function of an overall Richardson number

$$Ri_0 = \frac{g(\Delta\rho/\rho)h\cos\theta}{u^2} = \frac{A\cos\theta}{u^3}, \qquad (8.47)$$

where h is the thickness of the layer, θ the slope, and $A = g(\Delta\rho/\rho)hu$ the buoyancy flux per unit width [cf.

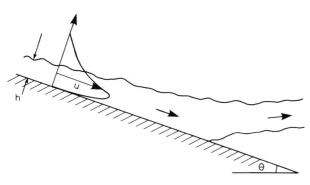

Figure 8.14 Sketch of a steady gravity current on a slope; the edge shown represents the level of most rapid variation of density. The outer part of the velocity profile sketched is linear, and so is the density profile at this level. (Ellison and Turner, 1959.) In stratified surroundings, the plume leaves the slope at a depth given by equation (8.48).

(8.8)]. E is a strong function of Ri_0 (see figure 8.2), and it becomes very small at low slopes.

When θ is small, the stress across the interface is therefore negligible, and the velocity of the layer is determined by friction at the solid boundary. At high slopes, on the other hand, the stress due to entrainment dominates, and in the steady state u = constant $\propto A^{1/3}$ from (8.47), and the rate of increase of depth with distance $dh/dx = E$. In this state, the turbulence is both generated and used for mixing at the interface, and this flow is thus a good example of the equilibrium "internal"-mixing process referred to in sections 8.2.1 and 8.4.1(a). The profiles of both velocity and density through the outer edge of the interface are observed to be linear, in agreement with an argument equivalent to that which led to (8.27) and (8.28) (figure 8.14).

The results for plumes rising or falling through a stratified environment can also be adapted to describe gravity currents along a slope, just by using the appropriate (smaller) value of E. For example, with a constant slope θ and density gradient (specified by N), the depth at which a two-dimensional current will reach the density of its surroundings and move out into the interior will be

$$Z_{\max} \propto E^{-1/3}A^{1/3}N^{-1}, \qquad (8.48)$$

where E is an (empirical) function of θ.

The effects of rotation can be added to these plume theories, and Smith (1975) described a three-dimensional rotating model that fits the observations of outflows from the Norwegian and Mediterranean Seas very well. When entrainment is dominant, rotation makes the flow tend to move along bottom contours, whereas strong bottom friction allows a larger excursion downslope. Killworth (1977) has discussed and extended rotating two- and three-dimensional models, with the flow on the Weddell Sea continental slope in mind. In order to explain both the depth of penetration and the dilution, he also needed to include the change of buoyancy flux resulting from the increase of thermal expansion coefficient with depth.

(b) Buoyancy Layers In a stably stratified fluid, motions along a slope can in principle be set up by diffusion near the solid boundary, which results in the surfaces of constant concentration (of S say) being bent so as to become normal to the slope. This distortion of the density field means that fluid against the boundary will be lighter than that in the interior, and there will be an upslope flow in a thin layer where changes due to advection are balanced by diffusion. Phillips (1970) and Wunsch (1970) showed that with these boundary conditions, the thickness l and upslope velocity w are constant and given by

$$l \sim (\nu \kappa_s)^{1/4} N^{-1/2}, \quad w \sim (\nu \kappa_s)^{1/4} N^{1/2}. \tag{8.49}$$

Under laboratory conditions these are very small, but Wunsch (1970) proposed that (8.49) could be extended to oceanic slopes by using "eddy" values for ν and κ rather than molecular coefficients. With $\nu, \kappa \sim 10^4 \text{ cm}^2 \text{ s}^{-1}$, $l \sim 20$ m, $w \sim 5$ cm s^{-1}, and more intense mixing will drive a stronger upslope current. There are several difficulties with this interpretation. It is implied that the larger mixing coefficients must be driven by some external mixing process, which is most likely to be associated with currents against the slope. This being so, it seems more appropriate to regard these "mechanical" processes as the cause, not the effect, of the near-slope motions and to investigate directly their effects on mixing. Second, the presence of two stratifying components, in the interior, with compensating effects on the density, changes the behavior markedly. As discussed in section 8.4.2(b) [see also Turner (1974)], counterflows along the slope are then produced, with much larger velocities than in the single-component case. These cannot remain steady, however, and the net result is the formation of a series of layers, extending out into the interior (cf. figure 8.10). When conditions near the slope are quiet, this mechanism could produce enhanced mixing and fine structure, but again it is likely to be overwhelmed by the mixing produced by currents.

8.5.3 Mixing Due to Internal Waves

Internal waves impinging on a sloping boundary can provide enough energy to cause significant mixing. The conditions under which this occurs in a continuously stratified fluid have been convincingly illustrated in the laboratory experiments of Cacchione and Wunsch (1974).

When waves of lowest mode propagate into a wedge-shaped region bounded by a solid sloping boundary and a free surface (or interface), three types of behavior are possible. These depend on the relative magnitudes of the slope β of the boundary and the wave-characteristic slope $\alpha = \sin^{-1}(\omega/N)$, which is the direction of the group velocity, and of the particle motions [see figure 8.15 and Wunsch (1969)]. If $\beta > \alpha$, then energy can be reflected back into the interior. If $\beta < \alpha$, which occurs only at sufficiently high frequencies ω for a given β and stratification N, the horizontal component of the group velocity after reflexion is still directed toward the slope. Energy thus cannot escape backward, and is fed into the corner region. For example, when the deeper layers are stratified, and there is a well-mixed layer above, the amplitude will build up in the thermocline and strong local mixing can occur there. When $\beta = \alpha$, the particle motions become parallel to the slope, and this strong shearing motion becomes unstable to form a series of periodic vortices. Overturning

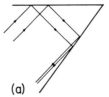 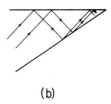

Figure 8.15 Propagation of waves into a wedge of angle β. The angle α that rays make with the horizontal stays constant, so that when $\beta > \alpha$ [case (a)] energy can be reflected, and when $\beta < \alpha$ [case (b)] energy is trapped in the corner. The critical case $\beta = \alpha$ produces strong shearing motions against the slope.

associated with these produces mixed fluid that propagates into the interior as regularly spaced layers all along the slope.

Though there is a suggestion in these and other experiments that the layer spacing is related to the amplitude of the excursion along the slope, they do not provide a definite length scale that can be used for predictions in the ocean. Nor do there yet seem to be any oceanic measurements that are detailed enough to distinguish the structure resulting from this mechanism from other possibilities.

Waves formed on density interfaces can also produce mixing when they approach a sloping boundary. For example, in a fjord that has a well-mixed surface layer and a strong pycnocline at sill depth, Stigebrandt (1976) showed that interfacial waves generated at the sill can propagate toward the landward end, where they break on the sloping shore. Using field data and a laboratory experiment, he described the vertical mixing in the lower layer in terms of this wave-breaking process, followed by the flow of mixed fluid into the interior. Similar observations have been reported by Perkin and Lewis (1978), who concluded that this mechanism probably dominates the transport between the surface and bottom layers of fjords for most of the year.

8.5.4 The Effect of Bottom Mixing on the Interior

There is no doubt that mixing near the bottom is much stronger than it is in the interior of the ocean. Hogg, Katz, and Sanford (1978), continuing a series of measurements near Bermuda initiated by Wunsch (1972a), recently have documented a close relation between the distribution of temperature fine structure and strong currents associated with large eddies near the island. They are cautious about identifying the precise mechanism of interaction (from among those described above, and others not discussed here), but the generation of the structure at the island slope and its decay with distance away from Bermuda is very clear.

Armi (1978) has used the contrast between vertical temperature profiles near topographic features and in the interior of an ocean basin to support one of the mechanisms for vertical mixing discussed by Munk

(1966): that the largest cross-isopycnal mixing occurs in boundary-mixed layers, and that these are then advected into the interior and so influence the structure there as well. The single well-mixed bottom layer discussed in section 8.5.1 and shown in figure 8.13 is characteristic of a smooth bottom on an abyssal plain, but over rougher topography a number of steps is often observed, suggesting bottom mixing at several depths, followed by spreading out along isopycnals that intersect the slope. The horizontal variability of such layering indicates that the process is patchy and intermittent.

The layer structure decays and the profiles become smoother as the water moves out into the interior. The various mechanisms that could play a part at this stage have been discussed in section 8.4. Some layers of water with distinctive T-S properties are identifiable, however, over large distances. Armi (1978) has shown that Norwegian Sea water can be followed as a 20-m-thick layer for over 3000 km into the North Atlantic, and cites this as evidence both for large-scale advection and slow vertical mixing. Carmack and Killworth (1978) have identified a layer with anomalously low T and S characteristics that interleaves along a surface of constant potential density with Antarctic bottom water near the Ross Sea. They also suggest that the sinking of water in the form of plumes along the continental margin, followed by an outflow at mid-depth, is possible nearly everywhere round Antarctica, although water masses that are so clearly distinguishable from their surroundings are rather rare.

In summary, the available evidence supports the view that the bottom of the ocean, particularly the sloping bottom around coasts or topographic features, plays an essential role in the internal-mixing process. Near the topography, the dominant mixing mechanisms are probably mechanical, driven by large-scale currents, though gravity currents can sometimes be important. The main way in which the resulting mixed layers are carried into the interior of the ocean must be by large-scale advection, associated with processes that are nearly independent of the layers themselves. The extra spreading and interleaving due to local horizontal density anomalies [described by equations (8.36) or (8.37)] occur on a longer time scale, though these processes will also affect the final profiles in the interior. Direct vertical mixing driven by internal waves is probably active too, and bottom topography enters here in another way as a mechanism for generating the internal wave field.

The only other regions where the deduced mixing rates are comparable with those at solid boundaries are boundaries between different water masses. The evidence presented in section 8.4.3, for example, shows that frontal surfaces with large horizontal T and S anomalies but a small net density difference are particularly active. The primary process envisaged in that case is double-diffusive transport in the vertical, producing local density anomalies that drive quasi-horizontal interleaving. Double-diffusive convection can also be significant when water masses with very different T-S properties lie one on top of the other.

We conclude on a cautionary note. Though there have been rapid advances in the observation and understanding of many individual physical processes, particularly in the past 10 years, these have not yet been put together to give a satisfactory, unified picture of mixing in the ocean. We must now seek ways to distinguish between the effects of the diverse vertical and horizontal processes that have been reviewed, and to assess their relative importance in controlling the vertical distributions of temperature and salinity in the ocean as a whole.

9
Internal Waves and Small-Scale Processes

Walter Munk

9.1 Introduction

Gravity waves in the ocean's interior are as common as waves at the sea surface—perhaps even more so, for no one has ever reported an interior calm.

Typical scales for the internal waves are kilometers and hours. Amplitudes are remarkably large, of the order of 10 meters, and for that reason internal waves are not difficult to observe; in fact they are hard *not* to observe in any kind of systematic measurements conducted over the appropriate space-time scales. They show up also where they are not wanted: as short-period fluctuations in the vertical structure of temperature and salinity in intermittent hydrocasts.

I believe that Nansen (1902) was the first to report such fluctuations;[1] they were subsequently observed on major expeditions of the early nineteen hundreds: the *Michael Sars* expedition in 1910, the *Meteor* expeditions in 1927 and 1938, and the *Snellius* expedition in 1929–1930. [A comprehensive account is given in chapter 16 of Defant (1961a)]. In all of these observations the internal waves constitute an undersampled small-scale noise that is then "aliased" into the larger space-time scales that are the principal concern of classical oceanography.

From the very beginning, the fluctuations in the hydrocast profiles were properly attributed to internal waves. The earliest theory had preceded the observations by half a century. Stokes (1847) treated internal waves at the interface between a light fluid overlaying a heavy fluid, a somewhat minor extension of the theory of surface waves. The important extension to the case of a vertical mode structure in *continuously* stratified fluids goes back to Rayleigh (1883). But the discreteness in the vertical sampling by hydrocasts led to an interpretation in terms of just the few gravest modes, with the number of such modes increasing with the number of sample depths (giving j equations in j unknowns). And the discreteness in sampling time led to an interpretation in terms of just a few discrete frequencies, with emphasis on tidal frequencies.

The development of the bathythermograph in 1940 made it possible to repeat soundings at close intervals. Ufford (1947) employed three vessels from which bathythermograph lowerings were made at 2-minute intervals! In 1954, Stommel commenced three years of temperature observations offshore from Castle Harbor, Bermuda, initially at half-hour intervals, later at 5-minute intervals.[2] Starting in 1959, time series of isotherm depths were obtained at the Navy Electronics Laboratory (NEL) oceanographic tower off Mission Beach, California, using isotherm followers (Lafond, 1961) installed in a 200-m triangle (Cox, 1962).

By this time oceanographers had become familiar with the concepts of continuous spectra (long before routinely applied in the fields of optics and acoustics), and the spectral representation of surface waves had proven very useful. It became clear that internal waves, too, occupy a frequency continuum, over some six octaves extending from inertial to buoyant frequencies. [The high-frequency cutoff had been made explicit by Groen (1948).] With regard to the vertical modes, there is sufficient energy in the higher modes that for many purposes the discrete modal structure can be replaced by an equivalent three-dimensional continuum.

We have already referred to the measurements by Ufford and by Lafond at horizontally separated points. Simultaneous current measurements at vertically separated points go back to 1930 (Ekman and Helland-Hansen, 1931). In all these papers there is an expression of dismay concerning the lack of resemblance between measurements at such small spatial separations of oscillations with such long periods. I believe (from discussions with Ekman in 1949) that this lack of coherence was the reason why Ekman postponed for 23 years (until one year before his death) the publication of "Results of a Cruise on Board the 'Armauer Hansen' in 1930 under the Leadership of Björn Helland-Hansen" (Ekman, 1953). But the decorrelation distance is just the reciprocal of the bandwidth; waves separated in wavenumber by more than Δk interfere destructively at separations exceeding $(\Delta k)^{-1}$. The small observed coherences are simply an indication of a large bandwidth.

The search for an analytic spectral model to describe the internal current and temperature fluctuations goes back over many years, prompted by the remarkable success of Phillips's (1958) saturation spectrum for surface waves. I shall mention only the work of Murphy and Lord (1965), who mounted temperature sensors in an unmanned submarine at great depth. They found some evidence for a spectrum depending on scalar wavenumber as $k^{-5/3}$, which they interpreted as the inertial subrange of homogeneous, isotropic turbulence. But the inertial subrange is probably not applicable (except perhaps at very small scales), and the fluctuations are certainly not homogeneous and not isotropic.

Briscoe (1975a) has written a very readable account of developments in the early 1970s. The interpretation of multipoint coherences in terms of bandwidth was the key for a model spectrum proposed by Garrett and Munk (1972b). The synthesis was purely empirical, apart from being guided by dimensional considerations and by not violating gross requirements for the finiteness of certain fundamental physical properties. Subsequently, the model served as a convenient "strawman" for a wide variety of moored, towed and "dropped" experiments, and had to be promptly modified [Garrett and Munk (1975), which became known as GM75 in the spirit of planned obsolescence]. There have been further modifications [see a review paper by Garrett and Munk (1979)]; the most recent version is summarized at the end of this chapter.

The best modern accounts on internal waves are by O. M. Phillips (1966b), Phillips (1977a), and Turner (1973a). Present views of the time and space scales of internal waves are based largely on densely sampled moored, towed, and dropped measurements. The pioneering work with moorings was done at site D in the western North Atlantic (Fofonoff, 1969; Webster, 1968). Horizontal tows of suspended thermistor chains (Lafond, 1963; Charnock, 1965) were followed by towed and self-propelled isotherm-following "fishes" (Katz, 1973; McKean and Ewart, 1974). Techniques for dropped measurements were developed along a number of lines: rapidly repeated soundings from the stable platform FLIP by Pinkel (1975), vertical profiling of currents from free-fall instruments by Sanford (1975) and Sanford, Drever, and Dunlap (1978), and vertical profiling of temperature from a self-contained yo-yoing capsule by Cairns and Williams (1976). The three-dimensional IWEX (internal wave experiment) array is the most ambitious to date (Briscoe, 1975b). These experiments have served to determine selected parameters of model spectra; none of them so far, not even IWEX, has been sufficiently complete for a straightforward and unambiguous transform into the multidimensional $(\omega, \mathbf{k})$-spectrum. The FLIP measurements come closest, giving an objective spectrum in the two dimensions ω, k_z, with fragmentary information on k_x, k_y. Otherwise only one-dimensional spectra can be evaluated from any single experiment, and one is back to model testing. Yet in spite of these observational shortcomings, there is now evidence for some degree of universality of internal wave spectra, suggesting that these spectra may be shaped by a saturation process (the interior equivalent of whitecaps), rather than by external generation processes.

Internal waves have surface manifestations consisting of alternate bands of roughened and smooth water (Ewing, 1950; Hughes, 1978), and these appear to be visible from satellites (figure 9.1). High-frequency sonar beams are a powerful tool for measuring internal wave related processes in the upper oceans (figures 9.2, 9.3). The probing of the deep ocean interior by acoustics is ultimately limited by scintillations due to internal waves (Flatté et al., 1979; Munk and Wunsch, 1979) just as the "diffraction-limited" telescope has its dimensions set by the small-scale variability in the upper atmosphere.

It will be seen that internal waves are a lively subject. The key is to find the connections between internal waves and other ocean processes. The discovery of ever finer scales, down to the scale of molecular processes,

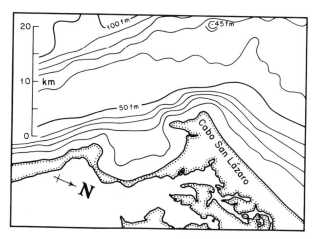

Figure 9.1 SEASAT synthetic aperture radar image off Cabo San Lázaro, Baja California (24°48′N, 112°18′W) taken on 7 July 1978. Scale of image nearly matches that of bathymetric area. The pattern in the right top area is most likely formed by internal waves coming into the 50 fathom line. (I am indebted to R. Bernstein for this figure.)

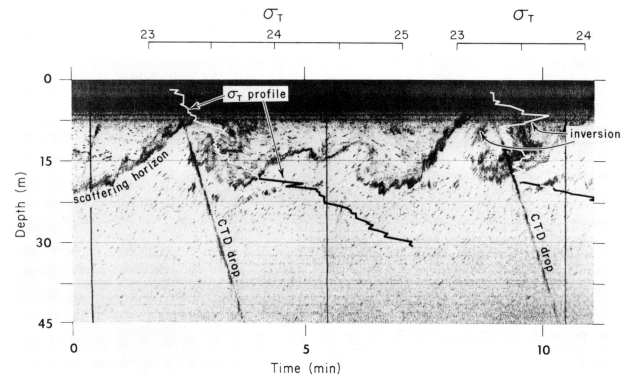

Figure 9.2 The water column is insonified with a narrow downward sonar beam of 200 kHz (wavelength 0.75 cm). The dark band is presumably a back-scattering layer convoluted by shear instabilities. In a number of places the instabilities have created density inversions. This is confirmed by the two σ_t-profiles. The acoustic reflection from the sinking CTD along the steeply slanting lines shows the depth–time history of the σ_t-profiles. The profiling sound source was suspended from a drifting ship. The horizontal distance between overturning events was estimated to be 60–70 m. (I am indebted to Marshall Orr of Woods Hole Oceanographic Institution for this figure; see Haury, Briscoe, and Orr, 1979.)

has been a continuing surprise to the oceanographic community for 40 years. Classical hydrographic casts employed reversing (Nansen) bottles typically at 100-m intervals in the upper oceans beneath the thermocline, and half-kilometer intervals at abyssal depths. Only the gross features can be so resolved. Modern sounding instruments (BT, STD, CTD) demonstrated a temperature and salinity[3] fine structure down to meter scales. An early clue to microstructure was the steppy traces on the smoked slides of bathythermographs. These steps were usually attributed to "stylus stiction," and the instruments suitably repaired.

Free-fall apparatus sinking slowly (~ 0.1 m s^{-1}) and employing small, rapid-response (~ 0.01 s) transducers, subsequently resolved the structure down to centimeter scales and beyond. The evolving terminology

gross structure: larger than 100 m vertical
fine structure: 1 m to 100 m vertical
microstructure: less than 1 m vertical

is then largely based on what could be resolved in a given epoch (see chapter 14). The fine-structure measurements of temperature and salinity owe much of their success to the evolution of the CTD (Brown, 1974). The pioneering microscale measurements were done by Woods (1968a) and by Cox and his collaborators (Gregg and Cox, 1972; Osborn and Cox, 1972). Measurements of velocity fine structure down to a few meters have been accomplished by Sanford (1975) and Sanford, Drever, and Dunlap (1978). Osborn (1974, 1980) has resolved the velocity microstructure between 40 and 4 cm. Evidently velocity and temperature structure have now been adequately resolved right down to the scales for which molecular processes become dominant. At these scales the dissipation of energy and mean-square temperature gradients is directly proportional to the *molecular* coefficients of viscosity and thermal diffusivity. The dissipation scale for salinity is even smaller (the haline diffusivity is much smaller than the thermal diffusivity) and has not been adequately resolved. The time is drawing near when we shall record the entire fine structure and microstructure scales of temperature, salinity and currents [and hence of the buoyancy frequency $N(z)$ and of Richardson number $Ri(z)$] from a single free-fall apparatus.

Perhaps the discovery of very fine scales could have been anticipated. There is an overall ocean balance between the generation and dissipation of mean-square gradients. Eckart (1948) refers to the balancing processes as *stirring* and *mixing*. Garrett (1979) has put it succinctly: "Fluctuations in ocean temperature produced by surface heating and cooling, and in salinity due to evaporation, precipitation, run-off and freezing, are stirred into the ocean by permanent current systems and large scale eddies." Mixing ultimately occurs through dissipation by "molecular action on small-scale irregularities produced by a variety of processes." The microstructure (where the mean-square gradients largely reside) are then a vital component of ocean dynamics. This leaves open the question whether mixing is important throughout the ocean, or whether it is concentrated at ocean boundaries and internal fronts, or in intense currents (an extensive discussion may be found in chapter 8).

What are the connections between internal waves and small-scale ocean structure? Is internal wave breaking associated with ocean microstructure? Is there an associated flux of heat and salt, and hence buoyancy? Does the presence of internal waves in a shear flow lead to an enhanced momentum flux, which can be parameterized in the form of an eddy viscosity? What are the processes of internal-wave generation and decay? I feel that we are close to having these puzzles fall into place (recognizing that oceanographic "breakthroughs" are apt to take a decade), and I am uncomfortable with attempting a survey at this time.

Forty years ago, internal waves played the role of an attractive nuisance: attractive for their analytical ele-

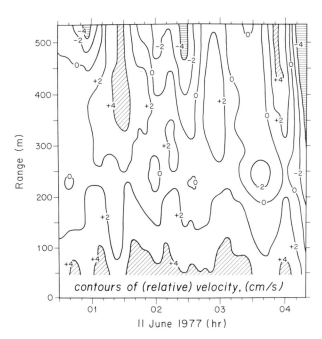

Figure 9.3 Measurements of Doppler vs. range were made at 2-minute intervals with a quasi-horizontal 88-kHz sound beam mounted on FLIP at a depth of 87 m. Bands of alternating positive and negative Doppler (in velocity contours) are the result of back scatter from particles drifting toward and away from the sound source (the mean drift has been removed). The velocities are almost certainly associated with internal wave-orbital motion. The range-rate of positive or negative bands gives the appropriate projection of phase velocity. The measurements are somewhat equivalent to successive horizontal tows at 3000 knots! (I am indebted to Robert Pinkel of Scripps Institution of Oceanography for this figure.)

gance and their accessibility to a variety of experimental methods, a nuisance for their interference with what was then considered the principal task of physical oceanography, namely, charting the "mean" density field. Twenty years from now I expect that internal waves will be recognized as being intimately involved with the vertical fluxes of heat, salt, and momentum, and so to provide a vital link in the understanding of the mean fields of mass and motion in the oceans.

9.1.1 Preview of This Chapter

We start with the traditional case of a two-layer ocean, followed by a discussion of continuous stratification: constant buoyancy frequency N, N decreasing with depth, a maximum N (thermocline), a double maximum. Conditions are greatly altered in the presence of quite moderate current shears. Short (compliant) internal waves have phase velocities that are generally slower than the orbital currents associated with the long (intrinsic) internal waves, and thus are subject to critical layer processes. There is further nonlinear coupling by various resonant interactions.

Ocean fine structure is usually the result of internal-wave straining, but in some regions the fine structure is dominated by intrusive processes. Microstructure is concentrated in patches and may be the residue of internal wave breaking. Little is known about the breaking of internal waves. Evidently, there are two limiting forms of instability leading to breaking: advective instability and shear instability.

The chapter ends with an attempt to estimate the probability of wave breaking, and of the gross vertical mixing and energy dissipation associated with these highly intermittent events. An important fact is that the Richardson number associated with the internal wave field is of order 1. Similarly the wave field is within a small numerical factor of advective instability. Doubling the mean internal wave energy can lead to a large increase in the occurrence of breaking events; halving the wave energy could reduce the probability of breaking to very low levels. This would have the effect of maintaining the energy level of internal waves within narrow limits, as observed. But the analysis is based on some questionable assumptions, and the principal message is that we do not understand the problem.

9.2 Layered Ocean

We start with the conventional discussion of internal waves at the boundary between two fluids of different density. The configuration has perhaps some application to the problem of long internal waves in the thermocline, and of short internal waves in a stepwise fine structure.

Following Phillips (1977a), this can be treated as a limiting case of a density transition from ρ_u above $z = -h$ to ρ_l beneath $z = -h$, with a transition thickness δh (figure 9.4). The vertical displacement $\zeta(z)$ has a peak at the transition, and the horizontal velocity $u(z)$ changes sign, forming a discontinuity (vortex sheet) in the limit $\delta h \to 0$. For the second mode (not shown), $\zeta(z)$ changes sign within the transition layer and $u(z)$ changes sign twice; this becomes unphysical in the limit $\delta h \to 0$. For higher modes the discontinuities are even more pathological, and so a two-layer ocean is associated with only the gravest internal mode.

For the subsequent discussion it is helpful to give a sketch of how the dependent variables are usually derived and related. The unknowns are u,v,w,p (after eliminating the density perturbation), where p is the departure from hydrostatic pressure. The four unknowns are determined by the equations of motion and continuity (assuming incompressibility). The linearized x,y equations of motion are written in the traditional f-plane; for the vertical equation it is now standard [since the work of Eckart (1960)] to display the density stratification in terms of the buoyancy (or Brunt-Väisälä) frequency

$$N(z) = \left\{ -\frac{g}{\rho}\left[\frac{d\rho}{dz} - \left(\frac{d\rho}{dz}\right)_{\text{adiabatic}}\right] \right\}^{1/2}, \quad (9.1)$$

thus giving

$$\frac{\partial w}{\partial t} = \frac{1}{\rho_0}\frac{\partial p}{\partial z} - N^2 \zeta = 0.$$

The last term will be recognized as the buoyancy force $-g\,\delta\rho/\rho_0$ of a particle displaced upwards by an amount $\zeta = \int w\,dt$.

For propagating waves of the form $\zeta(z)\exp i(kx - \omega t)$ the equations can be combined (Phillips, 1977a, §5.2 and §5.7) into

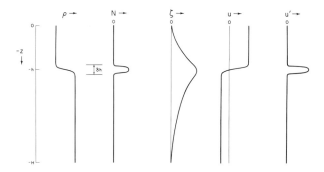

Figure 9.4 A sharp density transition from ρ_u to ρ_l takes place between the depths $-z = h - \frac{1}{2}\delta h$ and $-z = h + \frac{1}{2}\delta h$. This is associated with a delta-like peak in buoyancy frequency $N(z)$. Amplitudes of vertical displacement $\zeta(z)$, horizontal velocity $u(z)$, and shear $u'(z) = du/dz$ are sketched for the gravest internal wave mode.

$$\frac{d^2\zeta}{dz^2} + k^2 \frac{N^2(z) - \omega^2}{\omega^2 - f^2} \zeta = 0. \qquad (9.2)$$

The linearized boundary conditions are $\zeta = 0$ at the surface and bottom.

A simple case is that of $f = 0$ and $N = 0$ outside the transition layer. We have then

$$\zeta_u = A \sinh kz,$$

$$\zeta_l = B \sinh k(z + H),$$

above and below the transition layer, respectively. The constants A and B are determined by patching the vertical displacement at the transition layer:

$$\zeta_u = \zeta_l = a \quad \text{at} \quad z = -h.$$

The dispersion relation is found by integrating Eq. (9.2) across the transition layer:

$$\zeta'_u - \zeta'_l = -k^2 \zeta \int_{\delta h} dz \frac{N^2(z) - \omega^2}{\omega^2}$$

$$\approx -\frac{k^2 \zeta}{\omega^2} \left(g \frac{\delta \rho}{\rho} - \omega^2 \delta h \right) \quad \text{at} \quad z = -h,$$

where $\zeta' \equiv d\zeta/dz$. In the limit of small $k\,\delta h$, that is, for waves long compared to the transition thickness, the foregoing equations lead to the dispersion relation

$$\omega^2 = \frac{g(\delta\rho/\rho)k}{\coth kh + \coth k(H - h)}.$$

For a lower layer that is deep as compared to a wavelength, the denominator becomes $\coth k(h - H) + 1$. If the upper layer is also deep, it becomes $1 + 1$, and

$$\omega^2 = \frac{1}{2} gk \frac{\delta\rho}{\rho} = \frac{1}{2} gk \frac{\rho_l - \rho_u}{\frac{1}{2}(\rho_l + \rho_u)}.$$

As $\rho_u \to 0$, $\omega^2 \to gk$, which is the familiar expression for surface waves in deep water.

The case of principal interest here is that of an isolated density transition $\delta\rho \ll \rho$ and $k\,\delta h \ll 1$. Then $\omega^2 = \frac{1}{2} gk\,\delta\rho/\rho$. The vertical displacement is a maximum at the transition and dies off with distance δz from the transition as $a \exp(-k|\delta z|)$.

A question of interest is the variation of Richardson number across the transition layer. We know from the work of Miles and Howard [see Miles (1963)] that for a transition $\rho(z)$ and a *steady* $u(z)$ of the kind shown in figure 9.4, the flow becomes unstable to disturbances of length scale δh if $Ri < \frac{1}{4}$. I find it convenient to refer to the *root-reciprocal* Richardson number

$$Ri^{-1/2} = |u'/N|,$$

so that large values imply large instabilities (as for Reynolds numbers); the critical value is $|u'/N| = 2$. One would think offhand that $|u'/N|$ is a minimum at the transition where N reaches a maximum, but just the opposite is true. To prove this, we use the condition of incompressibility, $iku(z) - i\omega\zeta' = 0$, and equation (9.2) to obtain

$$u'(z) = \frac{\omega}{k} \zeta'' = -\frac{N^2(z) - \omega^2}{\omega} ka, \qquad (9.3)$$

and so $u' \sim N^2$ for small ω/N; accordingly u'/N varies as N. Thus the layers of largest gravitational stability (largest N) are also the layers of largest shear instability (largest $|u'/N|$).

9.3 Continuously Stratified Ocean

The simplest case is that of constant N. The solution to (9.2) is

$$\zeta(z) = a \sin mz, \qquad m^2 = k^2 \frac{N^2 - \omega^2}{\omega^2 - f^2} \qquad (9.4)$$

with m so chosen that ζ vanishes at $z = -H$. Solving for ω^2,

$$\omega_j^2 = \frac{k^2 N^2 + m_j^2 f^2}{m_j^2 + k^2}, \qquad m_j H = j\pi, \qquad j = 1,2,\ldots. \qquad (9.5)$$

This dispersion relation is plotted in figure 9.5. The vertical displacements for the first and third mode are shown in figure 9.6. Very high modes (and the ocean is full of them) in the deep interior are many wavelengths removed from the boundaries, and we can expect the waves to be insensitive to the precise configuration of top and bottom. The discrete dispersion $\omega_j(k)$ is then replaced by an equivalent continuous dispersion $\omega(k, m)$.

The standard expressions for the particle velocities u, w and the group velocities $\mathbf{c}_g$ with components $\partial\omega/\partial k$, $\partial\omega/\partial m$ as functions of the propagation vector

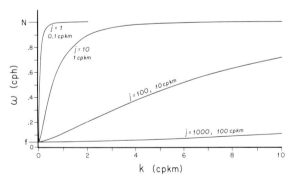

Figure 9.5 The dispersion $\omega_j(k)$ [equation (9.5)], for modes $j = 1, 10, 100, 1000$, corresponding to vertical wavenumbers $m = 0.1, 1, 10, 100$ cpkm in an ocean of depth 5 km. The inertial frequency is taken at $f = 0.0417$ cph (1 cpd), and the buoyancy frequency at $N = 1$ cph.

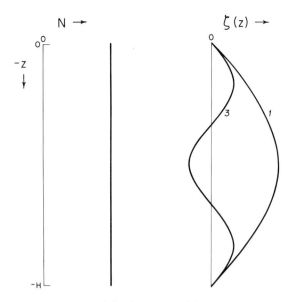

Figure 9.6 Vertical displacements $\zeta(z)$ in a constant-N ocean, for modes $j = 1$ and $j = 3$.

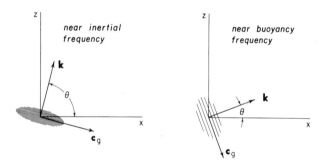

Figure 9.7 The wavenumber vector $\mathbf{k} = (k, m)$ and group velocity $\mathbf{c}_g$ near the inertial frequency ($\omega = f + \epsilon$) and near the buoyancy frequency ($\omega = N - \epsilon$), respectively. A packet of wave energy is projected on the (x, z)-plane. Crests and troughs in the wave packet are in a plane normal to $\mathbf{k}$, and travel with phase velocity $\mathbf{c}$ in the direction $\mathbf{k}$. The wave packet travels with group velocity $\mathbf{c}_g$ at right angles to $\mathbf{k}$, thus sliding sideways along the crests and troughs. The particle velocity $\mathbf{u}$ (not shown) is in the planes at right angles to $\mathbf{k}$.

$\mathbf{k} = (k, m)$ are easy to derive, but hard to visualize. Consider a wave packet (figure 9.7) with crests and troughs along planes normal to the paper and inclined with respect to the (x, z)-axis as shown. The phase velocity is in the direction $\mathbf{k}$ normal to the crests, but the group velocity $\mathbf{c}_g$ is parallel to the crests, and the wave packet slides sideways. $\mathbf{k}$ is inclined to the horizontal by

$$\tan \theta = \frac{m}{k} = \left(\frac{N^2 - \omega^2}{\omega^2 - f^2} \right)^{1/2} \qquad (9.6a)$$

and so the angle is steep for inertial waves ($\omega = f + \epsilon$) and flat for buoyancy waves ($\omega = N - \epsilon$). The energy packet is propagated horizontally for inertial waves, and vertically for buoyancy waves, but the group velocity goes to zero at both limits.

The flow $\mathbf{u} = (u, w)$ takes place in the plane of the crest and troughs. For inertial waves, particles move in horizontal circles. The orbits become increasingly elliptical with increasing frequency, and for buoyancy waves the particle orbits are linear along the z-axis, in the direction of $\mathbf{c}_g$. The wavenumber $\mathbf{k}$ is always normal to both $\mathbf{c}_g$ and $\mathbf{u}$. [The nonlinear field accelerations $(\mathbf{u} \cdot \nabla)\mathbf{u}$ vanish for an isolated elementary wave train, leading to the curiosity that the linear solution is an *exact* solution.] Readers who find it difficult to visualize (or believe) these geometric relations should refer to the beautiful laboratory demonstrations of Mowbray and Rarity (1967).

It is not surprising, then, that internal waves will do unexpected things when reflected from sloping boundaries. The important property is that the inclination θ relative to the x-axis depends only on frequency [equation (9.6a)]. Since frequency is conserved upon reflection, incident and reflected θ must be symmetric with respect to a level surface rather than with respect to the reflecting surface. At the same time the flow $\mathbf{u}$ for the combined incident and reflected wave must be parallel to the reflecting boundary. For a given ω, there is a special angle for which the orbital flow is parallel to the boundary. This requires that the boundary be inclined at a slope

$$\tan \beta(z) = \tan(90° - \theta) = \left[\frac{\omega^2 - f^2}{N^2(z) - \omega^2} \right]^{1/2}. \qquad (9.6b)$$

It can be shown that for slopes steeper than β, the energy of "shoreward" traveling internal waves is reflected "seaward"; for slopes of less than β, the energy is forward reflected. Repeated reflections in a wedge-shaped region such as the ocean on the continental slope can lead to an accumulation of energy at ever smaller scales (Wunsch, 1969). For a given slope, we can expect an amplification of the internal waves at the frequency ω determined by (9.6b). Wunsch (1972b) has suggested that a peak in the spectrum of temperature fluctuations measured southeast of Bermuda could be so explained. Pertinent values are $N = 2.6$ cph, $f = 0.045$ cph, and $\beta \approx 13°$. Equation (9.6b) gives $\omega = 0.59$ cph, in agreement with the observed spectral peak at 0.5 cph.

9.4 Turning Depths and Turning Latitudes

Figure 9.8 shows the situation for an ocean with variable $N(z)$. For frequencies that are less than N throughout the water column, the displacements are similar to those for constant N (figure 9.6) except that the positions of the maxima and zeros are displaced somewhat upward, and that the relative amplitudes are somewhat larger at depth. The important modification occurs for frequencies that exceed $N(z)$ somewhere within the water column. At the depths z_T where $\omega = N(z_T)$, called turning depths, we have the situation shown to the right in figure 9.8. Equation (9.2) is locally of the form $\zeta'' + z\zeta = 0$ where z is now a rescaled vertical coordinate relative to z_T. The solution (called an Airy function) has an inflection point at the turning depth (here $z = 0$), is oscillatory above the turning depth, and is exponentially damped beneath. The amplitudes are somewhat larger just above the turning depth than at greater distance, but nothing very dramatic happens.

The refraction of a propagating wave packet is illustrated in figure 9.9. As the packet moves into depths of diminishing $N(z)$ the crests and troughs turn steeper, and the direction of energy propagation becomes more nearly vertical. The waves are totally reflected at the turning depth z_T where $\omega = N(z_T)$. Modal solutions $\zeta_j(z) \times \exp i(kx - \omega_j t)$ with $\zeta_j(z)$ as illustrated in figure 9.8 can be regarded as formed by superposition of propagating waves with equal upward and downward energy transport. The wave energy remains trapped between the surface and the turning depth.

The common situation for the deep ocean is the main thermocline associated with a maximum in $N(z)$. Internal waves with frequencies less than this maximum are in a waveguide contained between upper and lower turning depths. For relatively high (but still trapped) frequencies the sea surface and bottom boundaries play a negligible role, and the wave solutions can be written in a simple form (Eriksen, 1978). The bottom boundary condition (9.5) for a constant-N ocean, e.g., $m_j H = j\pi$, $j = 1, 2, \ldots$, is replaced in the WKB approximation by

$$m_j b = j\pi \left(\frac{N^2 - \omega^2}{N_0^2 - \omega^2} \right)^{1/2} \approx j\pi N/N_0, \tag{9.7}$$

where b is a representative thermocline (or stratification) scale. Equation (9.7) assures an exponential attenuation outside the waveguide. For the case of a double peak in $N(z)$ with maxima N_1 and N_2, the internal wave energy is concentrated first at one thermocline, then

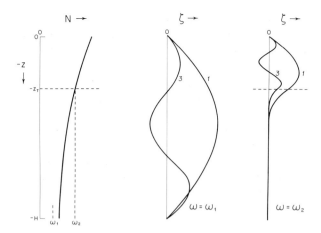

Figure 9.8 Vertical displacements $\zeta(z)$ in a variable-N ocean, for modes $j = 1$ and $j = 3$. ω_1 is taken to be less than $N(z)$ at all depths. ω_2 is less than $N(z)$ in the upper oceans above $z = -z_p$ only.

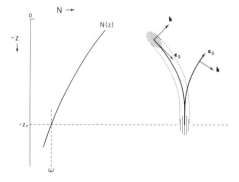

Figure 9.9 Propagation of a wave packet in a variable-$N(z)$ ocean without shear (U = constant). The turning depth z_T occurs when $\omega = N(z_T)$.

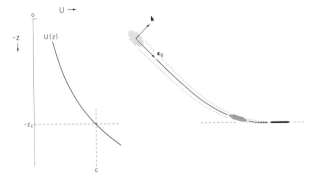

Figure 9.10 Propagation of a wave packet in a constant-N ocean with shear. The critical depth z_C occurs where $U = c(z_C)$.

the other, migrating up and down with a frequency $|N_1 - N_2|$ (Eckart, 1961). This is similar to the behavior of two loosely coupled oscillators. The quantum-mechanical analogy is that of two potential minima and the penetration of the potential barrier between them.

There is a close analogy between the constant- and variable-N ocean, and the constant- and variable-f ocean (the f-plane and β-plane approximations). For a fixed ω, the condition $\omega = f = 2\Omega \sin\phi_T$ determines the turning latitude ϕ_T. Eastward-propagating internal gravity waves have solutions of the form $\eta(y)\zeta(z) \exp i(kx - \omega t)$. The equation governing the local north-south variation is (Munk and Phillips, 1968)

$$\eta'' + y\eta = 0, \qquad \eta'' = d^2\eta/dy^2,$$

where y is the poleward distance (properly scaled) from the turning latitude. This is in close analogy with the up-down variation near the turning depth, which is governed by

$$\zeta'' + z\zeta = 0, \qquad \zeta'' = d^2\zeta/dz^2.$$

Thus $\eta(y)$ varies from an oscillatory to an exponentially damped behavior as one goes poleward across the turning latitude. Poleward-traveling wave packets are reflected at the turning latitude.

From an inspection of figure 9.7, it is seen that the roles of horizontal and vertical displacements are interchanged in the $N(z)$ and $f(y)$ turning points. In the $N(z)$ case the motion is purely vertical; in the $f(y)$ case the motion is purely horizontal (with circular polarization).

It has already been noted that nothing dramatic is observed in the spectrum of vertical displacement (or potential energy) near $\omega = N$—only a moderate enhancement, which can be reconciled to the behavior of the Airy function (Desaubies, 1975; Cairns and Williams, 1976). Similarly we might expect only a moderate enhancement in the spectrum of horizontal motion (or kinetic energy) near $\omega = f$. In fact, the spectrum is observed to peak sharply. If the horizontal motion is written as a sum of rotary components (Gonella, 1972), it is found that the peak is associated with negative rotation (clockwise in the northern hemisphere).

I have made a parallel derivation of the spectra at the two turning points (figure 9.11), assuming horizontally isotropic wave propagation within the entire equatorial waveguide. It turns out that the buoyancy peak is in fact much smaller than the inertial peak at moderate latitudes. But at very low latitude the inertial peak vanishes. This is in accord with the equatorial observations by Eriksen (1980). Fu (1980) gives an interesting discussion of the relative contributions to the spectral peak at the local inertial frequency $\omega = f_{\text{local}}$ from two processes: (1) *local* generation of resonant inertial

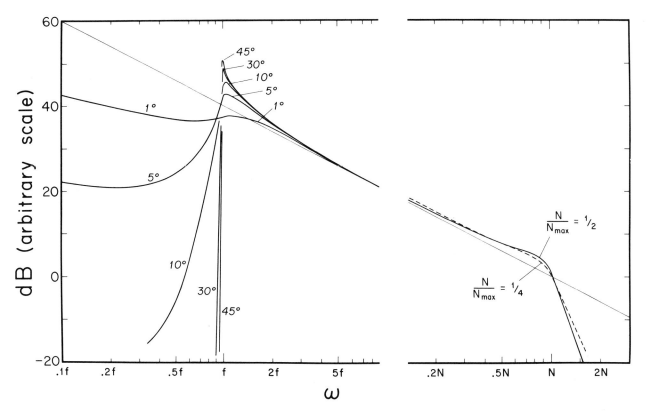

Figure 9.11 Enhancement of the kinetic-energy spectrum (left) and of the potential-energy spectrum (right) at the inertial and buoyancy frequencies, respectively. The inertial spectrum is drawn for latitudes 1°, 5°, 10°, 30°, 45°. The buoyancy spectrum is drawn for two depths, corresponding to $N = \frac{1}{2}, \frac{1}{4}$ times the maximum buoyancy frequency.

waves $\omega = f_{\text{local}}$; and (2) *remote* generation of waves of the same frequency $\omega = f_{\text{local}}$ at lower latitudes (where $f < f_{\text{local}}$). Figure 9.11 is drawn for case 2 under the assumption that the equatorial waveguide is filled with horizontally isotropic, freely propagating radiation. Take the curve marked 30°, say. Then for $\omega > f$ a station at lat. 30° is within the equatorial waveguide; for $\omega < f$ the spectrum is the result of evanescent extensions from a waveguide bounded by lower latitudes. Over rough topography and in regions of strong surface forcing, the case can be made for local generation of the inertial peak. It would seem that the buoyancy peak at mid-depth must always be associated with remote generation.

9.5 Shear

Internal waves are greatly modified by an underlying shear flow.[4] A variable $U(z)$ can have a more traumatic effect on internal waves than a variable $N(z)$. For ready comparison with figure 9.9 showing the effect of a variable $N(z)$ on a traveling wave packet, we have sketched in figure 9.10 the situation for a wave packet traveling in the direction of an increasing $U(z)$. As the wave packet approaches the "critical depth" z_C where the phase velocity (in a fixed frame of reference) equals the mean flow, $c = U(z_C)$, the vertical wavenumber increases without limit (as will be demonstrated).

For the present purpose we might as well avoid additional complexities by setting $f = 0$. The theoretical starting point is the replacement of ∂_t by $\partial_t + U\partial_x + w\,\partial_z$ in the linearized equations of motion. The result is the Taylor–Goldstein equation [Phillips (1977a, p. 248)]:

$$\frac{d^2\zeta}{dz^2} + \left(\frac{N^2}{(U-c)^2} - \frac{U''}{U-c} - k^2\right)\zeta = 0, \quad (9.8)$$

$$U'' = \frac{d^2U}{dz^2},$$

where c is the phase velocity in a fixed reference frame. (This reduces to

$$\frac{d^2\zeta}{dz^2} + k^2\frac{N^2 - \omega^2}{\omega^2}\zeta = 0 \quad (9.9)$$

for $U = 0$.) The singularity at the critical depth where $U = c$ is in contrast with the smooth turning-point transition at $N = \omega$; this is the analytic manifestation of the relative severity of the effect of a variable $U(z)$ versus that of a variable $N(z)$.

Thorpe (1978c) has computed the wave function $\zeta(z)$ for (1) the case of constant N and U' and (2) the case where N and U' are confined to a narrow transition layer. The results are shown in figures 9.12 and 9.13. The profiles are noticeably distorted relative to the case

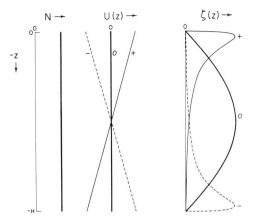

Figure 9.12 First mode vertical displacements $\zeta(z)$ in a Couette flow (constant U' and constant N), for $U'/N = 0, \pm 1$. Waves move from left to right, and U is positive in the direction of wave propagation. (Thorpe, 1978c.)

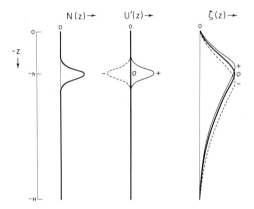

Figure 9.13 Similar to figure 9.12, but with U' and N confined to a narrow transition layer.

of zero shear, with the largest amplitudes displaced toward the level at which the mean speed (in the direction of wave propagation) is the greatest. Finite-amplitude waves have been examined for the case 2. Where there is a forward[5] flow in the upper level (including the limiting case of zero flow), the waves have narrow crests and flat troughs, like surface waves; with backward flow in the upper layer, the waves have flat crests and narrow troughs. Wave breaking is discussed later.

9.5.1 Critical Layer Processes[6]

The pioneering work is by Bretherton (1966c), and by Booker and Bretherton (1967). Critical layers have been associated with the occurrence of clear-air turbulence; their possible role with regard to internal waves in the oceans has not been given adequate attention.

Following Phillips (1977a), let

$$\omega_F = kU + \omega, \quad \frac{\omega}{N} = \frac{k}{(k^2 + m^2)^{1/2}} = \cos\theta \quad (9.10)$$

designate the frequency in a *fixed* reference frame. $U(z)$ is the *mean current* relative to this fixed frame, and $\omega_F - kU = \omega$ is the *intrinsic* frequency [as in (9.5)], as it would be measured from a reference frame drifting with the mean current $U(z)$.

Bretherton (1966c) has given the WKB solutions for waves in an ocean of constant N and slowly varying U. [It is important to note the simplification to (9.8) when $U'' = 0$ at the critical layer.] Near the critical layer depth z_C, the magnitudes of w, u, and of the vertical displacement ζ vary as

$$w \sim |z - z_C|^{1/2}, \quad u \sim |z - z_C|^{-1/2}, \quad \zeta \sim |z - z_C|^{-1/2}.$$

The quantities ω_F and k are constant in this problem, but m and ω are not. The vertical wavenumber increases, whereas the intrinsic frequency decreases as a wave packet approaches its critical layer:

$$m \sim |z - z_C|^{-1}, \quad \omega \sim |z - z_C|.$$

A sketch of the trajectory is given in figure 9.10. Waves are refracted by the shear and develop large vertical displacements ζ (even though $w \to 0$), large horizontal velocities u, and very large induced vertical shears u'. This has implications for the dissipation and breaking of internal waves.

For $Ri > \tfrac{1}{4}$, Booker and Bretherton (1967) derived an energy transmission coefficient

$$\rho = \exp(-2\pi\sqrt{Ri - \tfrac{1}{4}}). \quad (9.11)$$

In the usual case, $U' \ll 2N$ so that $Ri \gg \tfrac{1}{4}$ and ρ is small. This is interpreted as wave energy and momentum being absorbed by the mean flow at z_C. As $Ri \to \infty$, $\rho \to 0$, consistent with the WKB prediction of Bretherton (1966c) that a wave packet approaches but never reaches the critical layer.

The small coefficient of transmission for Richardson numbers commonly found in the ocean implies that the critical layer inhibits the vertical transfer of wave energy. This effect has been verified in the laboratory experiments of Bretherton, Hazel, Thorpe, and Wood (1967). When rotation is introduced, the energy and momentum delivered to the mean flow may alternatively be transferred from high-frequency to low-frequency waves (if the time scales are appropriate). Thus it is possible that some sort of pumping mechanism may exist for getting energy into, for example, the high-mode, quasi-inertial internal waves. This mechanism can be compared with McComas and Bretherton's (1977) parametric instability, a weakly nonlinear interaction (section 9.6).

The work of Bretherton and of Booker and Bretherton has prompted a great number of critical-layer studies. One of the most interesting extensions was done by Jones (1968). Whereas Booker and Bretherton found the critical layer to be an absorber, not a reflector, when $Ri > \tfrac{1}{4}$, Jones found that reflection from the critical layer is possible when $Ri < \tfrac{1}{4}$; in fact, the reflected wave amplitude can exceed that of the incident wave. Jones called these waves "overreflected," their energy being enhanced at the expense of the mean flow. This is illustrated in figure 9.14, based on a solution for a hyperbolic-tangent profile intended to display the results of linear theory. Transmission and reflection ratios at $z = \pm\infty$ were derived using definitions of wave energy density appropriate to moving media. "Overtransmission" as well as overreflection occurs at very small Richardson numbers, with the internal waves gaining energy from the mean flow on both counts.

We shall now consider the condition for critical-layer absorption. Let ω_1 designate the intrinsic frequency of a wave packet at some depth z_1 with a mean flow U_1 in the direction of wave propagation. According to (9.10),

$$\omega_F = kU_1 + \omega_1.$$

Let U increase to some value U_2 at z_2. Then since ω_F and k are conserved along the trajectory of the wave packet,

$$\omega_F = kU_2 + \omega_2.$$

For the special case that z_2 is to be a critical depth, we have $\omega_F = kU_2$, hence $\omega_2 = 0$, and so

$$\omega_1/k = U_2 - U_1.$$

The vertical wavenumber of internal waves is given by the dispersion relation

$$m = k\left(\frac{N^2 - \omega^2}{\omega^2 - f^2}\right)^{1/2}$$

$$\approx kN/\omega \quad \text{for} \quad f \ll \omega \ll N. \quad (9.12)$$

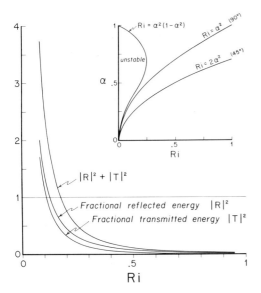

Figure 9.14 Fractional internal wave energy reflected and transmitted through a mean shear flow $U = U_0 \tanh(z/d)$ at constant N, as a function of the minimum Richardson number $Ri = N^2 d^2/U_0^2$. Internal wave energy is lost to the mean flow for $|R|^2 + |T|^2 < 1$, or $Ri > 0.18$; internal wave energy is gained from the mean flow for $Ri < 0.18$. The plot is drawn for $Ri = 2\alpha^2$, where $\alpha = kd$ is the dimensionless horizontal wavenumber. This corresponds to a wave packet traveling at an inclination of 45° at $z = \pm\infty$. ($Ri = \alpha^2$ corresponds to the limiting case of vertical group velocity to $\pm\infty$.) (I am indebted to D. Broutman for this figure.)

For critical absorption within the interval Δz over which the mean flow varies by ΔU, we replace ω/k by ΔU, and obtain the critical vertical wavenumber

$$m_C = N/\Delta U. \tag{9.13}$$

R. Weller (personal communication) has analyzed a month of current measurements off California for the expected *difference* $\Delta U = |U_2 - U_1|$ in a velocity component (either of the two components) at two levels separated by $\Delta z = |z_2 - z_1|$. The observations are, of course, widely scattered, but the following values give representative magnitudes:

Δz in m	0	10	25	50	100	
ΔU in cm s^{-1}	0	4	7	10	15	(upper 100 m)
	0	7	10	12	15	(100–300 m depth)

For $\Delta U = 10$ cm s^{-1} and $N = 0.01$ s^{-1} (6 cph), (9.13) gives $m_C = 10^{-3}$ cm^{-1} (16 cpkm). Internal waves with vertical wavelengths of less than 60 m are subject to critical-layer interactions.

A large fraction of the measured velocity difference ΔU can be ascribed to the flow field $u(z)$ of the internal waves themselves, and deduced from the model spectra. The expected velocity difference increases to $\sqrt{2}$ times the rms value as the separation increases to the vertical coherence scale, which is of order 100 m. Here most of the contribution comes from low frequencies and low wavenumbers. I am tempted to interpret ΔU for critical layer processes as rms u from the internal waves themselves. The internal wave spectrum is then divided into two parts: (1) the *intrinsic* part $m < m_C$, which contains most of the energy, and (2) the *compliant* part $m > m_C$, which is greatly modified by interaction with the intrinsic flow field. The phase speed for critical reflection is

$$c_C = \mathrm{rms}\, u, \tag{9.14}$$

and the critical wavenumber is

$$m_C = N/\mathrm{rms}\, u. \tag{9.15}$$

There is the separate question whether the internal waves at the critical layer will be underreflected, just reflected, or overreflected, and this depends on the ambient Richardson number. In the underreflected case there is a flux of energy from the compliant to the intrinsic waves. In the overreflected case the flow is the other way. For an equilibrium configuration, one may want to look for a transmission coefficient ρ near unity, and the exponential behavior of $\rho(Ri)$ will then set narrow bounds to the ambient spectrum. But this gets us into deep speculation, and had better be left to the end of this chapter.

9.6 Resonant Interactions

Up to this point the only interactions considered are those associated with critical layers. In the literature the focus has been on the resonant interaction of wave triads, using linearized perturbation theory. There are two ways in which critical layer interactions differ from resonant interactions: (1) compliant waves of *any* wavenumber and *any* frequency are modified, as long as c equals u somewhere in the water column; and (2) the modification is apt to be large (the ratio u/c being a very measure of nonlinearity). For the wave triads, the interaction is (1) limited to *specific* wavenumbers and frequencies, and (2) assumed to be small in the perturbation treatment.[7] To borrow some words of O. M. Phillips (1966b), the contrast is between the "strong, promiscuous interactions" in the critical layer and the "weak, selective interactions" of the triads.

The conditions for resonance are

$$\mathbf{k}_1 \pm \mathbf{k}_2 = \mathbf{k}_3, \qquad \omega_1 \pm \omega_2 = \omega_3,$$

where $\mathbf{k}_i = (k_i, l_i, m_i)$, and all frequencies satisfy the dispersion relation $\omega_i(\mathbf{k}_i)$. Resonant interactions are well demonstrated in laboratory experiments. For a transition layer (as in figure 9.4), Davis and Acrivos (1967) have found that a first-order propagating mode, which alternately raises and lowers the transition layer, was unstable to resonant interactions, leading to a rapid growth of a second-order mode, which alternately thickens and thins the transition layer like a propagating link sausage. Martin, Simmons, and Wunsch (1972) have demonstrated a variety of resonant triads for a constant-N stratification.

Among the infinity of possible resonant interactions, McComas and Bretherton (1977) have been able to identify three distinct classes that dominate the computed energy transfer under typical ocean conditions. Figure 9.15 shows the interacting propagation vectors in (k, m)-space. The associated frequencies ω are uniquely determined by the tilt of the vectors, in accordance with (9.4). Inertial frequencies (between f and $2f$, say) correspond to very steep vectors, buoyancy frequencies (between $\frac{1}{2}N$ and N) to flat vectors, as shown.

Elastic scattering tends to equalize upward and downward energy fluxes for all but inertial frequencies. Suppose that $\mathbf{k}_3$ is associated with waves generated near the sea surface propagating energy downward (at right angles to $\mathbf{k}_3$, as in figure 9.7). These are scattered into $\mathbf{k}_1$, with the property $m_1 = -m_3$, until the upward energy flux associated with $\mathbf{k}_1$ balances the downward flux by $\mathbf{k}_3$. The interaction involves a near-intertial wave $\mathbf{k}_2$ with the property $m_2 \approx 2m_3$. (The reader will be reminded of Bragg scattering from waves having half the wavelength of the incident and back-scattered radiation.) Similarly, for bottom-generated $\mathbf{k}_1$ waves with upward energy fluxes, elastic scattering will transfer energy into $\mathbf{k}_3$ waves.

Induced diffusion tends to fill in any sharp cutoffs at high wavenumber. The interaction is between two neighboring wave vectors of high wavenumber and frequency, $\mathbf{k}_1$ and $\mathbf{k}_3$, and a low-frequency low-wavenumber vector $\mathbf{k}_2$. Suppose the $\mathbf{k}_2$ waves are highly energetic, and that the wave spectrum drops sharply for wavenumbers just exceeding $|\mathbf{k}_3|$, such as $|\mathbf{k}_1|$. This interaction leads to a diffusion of action (energy/ω) into the low region beyond $|\mathbf{k}_3|$, thus causing $\mathbf{k}_1$ to grow at the expense of $\mathbf{k}_2$.

Parametric subharmonic instability transfers energy from low wavenumbers $\mathbf{k}_2$ to high wavenumbers $\mathbf{k}_1$ of half the frequency, $\omega_1 = \frac{1}{2}\omega_2$, ultimately pushing energy into the inertial band at high vertical wavenumber. The interaction involves two waves $\mathbf{k}_1$ and $\mathbf{k}_3$ of nearly opposite wavenumbers and nearly equal frequencies. The periodic tilting of the isopycnals by $\mathbf{k}_2$ varies the buoyancy frequency at twice the frequency of $\mathbf{k}_1$ and $\mathbf{k}_3$. (The reader will be reminded of the response of a pendulum whose support is vertically oscillated at twice the natural frequency.)

The relaxation (or interaction) time is the ratio of the energy density at a particular wavenumber to the net energy flux to (or from) this wavenumber. The result depends, therefore, on the assumed spectrum. For representative ocean conditions, McComas (in preparation) finds the relaxation time for elastic scattering to be extremely short, of the order of a period, and so up- and downgoing energy flux should be in balance. This result does not apply to inertial frequencies, consistent with observations by Leaman and Sanford (1975) of a downward flux at these frequencies. The relaxation time for induced diffusion is typically a fraction of a period! (This is beyond the assumption of the perturbation treatment.) Any spectral bump is quickly wiped out. The conclusion is that the resonant interactions impose strong restraints on the possible shapes of stable spectra.

In a challenging paper, Cox and Johnson (1979) have drawn a distinction between radiative and diffusive transports of internal wave energy. In the examples cited so far, energy in wave packets is radiated at group velocity in the direction of the group velocity. But suppose that wave-wave interactions randomize the direction of the group velocity. Then eventually the wave energy is spread by diffusion rather than radiation. The relevant diffusivity is $\kappa = \frac{1}{3}\langle c_g^2 \rangle \tau$, where τ is the relaxation time of the nonlinear interactions. Cox and Johnson have estimated energy diffusivities and momentum diffusivities (viscosities); they find that beyond 100 km from a source, diffusive spreading is apt to dominate over radiative spreading. There is an interesting analogy to crystals, where it is known that energy associated with thermal agitation is spread by diffusion rather than by radiation. The explanation lies in the anharmonic restoring forces between molecules, which bring about wave-wave scattering at room temperatures with relaxation times in the nanoseconds.

9.7 Breaking

This is the most important and least understood aspect of our survey. Longuet-Higgins has mounted a broadly based fundamental attack on the dynamics of breaking surface waves, starting with Longuet-Higgins and Fox (1977), and this will yield some insight into the internal-wave problems. At the present time we depend on laboratory experiments with the interpretation of the results sometimes aided by theoretical considerations.

Figure 9.16 is a cartoon of the various stages in an experiment performed by Thorpe (1978b). A density transition layer is established in a long rectangular tube. An internal wave maker generates waves of the first vertical mode. Before the waves have reached the far end of the tube, the tube is tilted through a small angle to induce a slowly accelerating shear flow. The underlying profiles of density, shear, and vertical displacement correspond roughly to the situation in figure 9.13.

For relatively steep waves in a weak positive[8] shear, the waves have sharpened crests. At the position of the crest, the density profile has been translated upward and steepened (B_1). There is significant wave energy loss in this development (Thorpe, 1978c, figure 10).

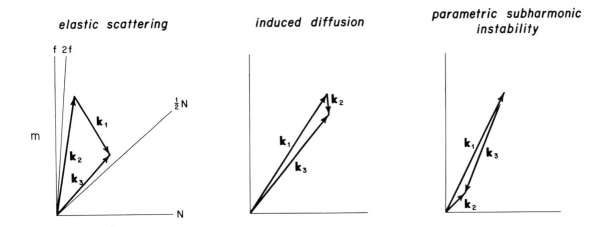

Figure 9.15 Resonant triads for three limiting classes of interaction, according to McComas and Bretherton (1977). The propagation vectors are drawn in (k, m)-space. Radial lines designate the tilt of the **k** vectors for $\omega = f, 2f, \frac{1}{2}N, n$. taking $N = 24f$.

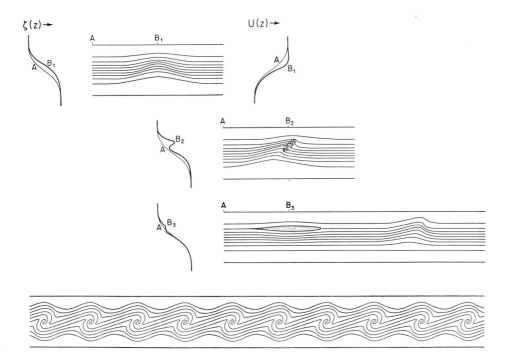

Figure 9.16 Cartoon for various stages of Thorpe's experiment. The early stages lead to the development of *advective* instability (upper three sketches), and the final stage to *shear* instability (bottom). Waves are traveling from left to right; the mean flow is forward (in the direction of wave propagation) above the density transition layer and backward below the transition layer. The density profiles along the indicated vertical sections are shown to the left; a velocity profile is shown to the top right (thin lines give the undisturbed profiles).

With increased positive shear, or with increased time, the particles at the crest accelerate, the isopycnal wave front becomes momentarily vertical, and a jet of fluid moves forward of the crest (B_2). The resulting density inversion gives rise to a Rayleigh-Taylor instability, forming a turbulent patch (shaded) whose turbulent energy is irretrievably lost to the organized wave motion. The turbulent patch becomes fairly well mixed, and introduces a steplike feature into the density profile (B_3). The patch spreads horizontally under the influence of the ambient stratification, forming *blini*, or pancakes. The detailed dynamics are complicated (Barenblatt and Monin, 1979); it is possible that in the oceans the spreading of the patches is eventually retarded by geostrophic confinement.

In Thorpe's laboratory experiment, the later stages of horizontal spreading are interrupted by the sudden formation of billows that grow rapidly, extracting energy from the mean shear flow (bottom of figure). Their wave length is quite short, only several times the thickness of the transition layer.

Hence, Thorpe (1978b, 1979) distinguishes between two types of instability leading to internal wave breaking. In the case of *advective instability*, breaking grows out of existing large-amplitude internal waves: more precisely, waves associated with steep isopycnal slopes. Eventually the particles in the crest are advected forward of the crest, leading to a local density inversion with the potential for a Rayleigh-Taylor instability. Advective instability can take place in the absence of ambient shear, though it is advanced by shear. The second type is induced *shear instability* (Kelvin-Helmholtz instability in the limit of an abrupt density transition), and can take place even in the absence of any (finite) wave disturbance, but is catalyzed by an existing wave background.

The two types appear as end points on a stability curve in slope-shear space, constructed by Thorpe (1978b, 1979) from theory and experiment (figure 9.17). Under the conditions described by the author, internal waves on a transition layer are unstable if their slope exceeds 0.34 in the absence of ambient shear, and if the shear exceeds $2N$ in the absence of slope. Away from the end points, there is advective instability modified by shear, and shear (K-H) instability modified by advection. The stability curve for the transition profile is not symmetric, implying that (under the prescribed geometry) negative shear delays instability.

The essential feature of advective instability is that the particle speed at the crest eventually exceeds the wave speed. The stability curve in figure 9.16 has been constructed from $u_{crest} = c$ (carrying the theory to third order in wave slope). This is in fair agreement with experiment. From a similar point of view, Orlanski and Bryan (1969) had previously derived the required critical amplitude for advective instability in the oceans, and have checked their analysis with numerical experiments. They conclude that more than enough internal wave energy exists for this type of instability to occur. They also conclude that conditions favor advective instability over shear instability, by the following very simple argument. From (9.12),

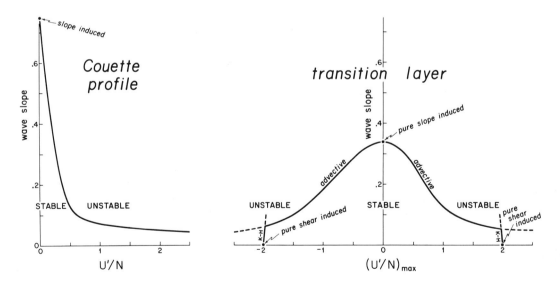

Figure 9.17 Stability diagram for internal waves in a shear flow for a Couette profile (as in figure 9.12) and for a transition layer (as in figure 9.13). Slope is defined as $\pi \cdot$ wave height/wavelength. The ordinate is U'/N (or $\pm Ri^{-1/2}$), with U in the direction of wave propagation. For the transition profile, U' and N^2 are proportional to $\text{sech}^2(z - h)$, and hence $U'/N \sim \text{sech}(z - h)$ has its maximum value at the transition $z = h$. The curves are drawn for the specific dimensions described by Thorpe (1978a, 1979).

$$\frac{mu}{N} = \frac{u}{\omega/k}\left(\frac{1-\omega^2/N^2}{1-f^2/\omega^2}\right)^{1/2}$$

$$\approx \frac{u}{\omega/k} \quad \text{for} \quad f \ll \omega \ll N. \quad (9.16)$$

But $mu/N = Ri^{-1/2} = 2$ for shear instability, and $u/(\omega/k) = 1$ for advective instability. The linearized[9] treatment then says that waves have to be twice as high to be shear unstable than to be advectively unstable. The trouble with this argument is that it is limited to the self-shear of an elementary wave train, and does not take into account the imposed ambient shear (possibly due to other components of the internal wave spectrum). Thorpe's stability plot (figure 9.17) shows that the instabilities can go either way, depending on wave slope and the ambient shear.

McEwan (1973) has generated breaking internal waves in the laboratory by crossing two internal wave beams from separate sources. He finds that the breaking is associated with localized, abruptly appearing intensification in density gradient and shear. These "traumata" persist and spread, and become the locus of incipient turbulence. True turbulent disorder was always preceded by the sudden and widespread occurrence of the traumata.

In some further laboratory experiments with breaking internal waves, McEwan (personal communication) has estimated separately the work done in generating internal waves (allowing for wall friction), and the fraction of this work going into mixing, e.g., going into the increase in the potential energy of the mean stratification. The remaining energy is dissipated into heat. McEwan finds that something less than ¼ of the input energy goes into mixing, in support of an estimate by Thorpe (1973b). (In the ocean, the mixing of salt and heat may proceed at different rates because of the disparity in the diffusivities.) Thompson (1980) argues that this ratio is, in fact, the critical Richardson number.

All of this points toward a strong connection between breaking internal waves and the microstructure of density and velocity. Evidently, breaking internal waves can modify a density profile, reducing gradients in turbulent patches and sharpening them elsewhere. This can lead to a steppy fine structure. But we have shown that internal wave shear is concentrated at the steps, thus producing conditions for shear instability, and renewed breaking. This is like the chicken and the egg: which comes first?

9.8 Ocean Fine Structure and Microstructure

Measurements by Gregg (1975) off Cabo San Lucas and in the North Pacific gyre (figure 9.18 and table 9.1) speak for great geographic variability in the mixing processes. (This is apart from the *local* patchiness in microstructure even in regions of strong mixing.) Three water masses intermingle off Cabo San Lucas: the saline outflow from the Gulf of California, the relatively fresh waters being brought in from the northwest by the California Current, and Equatorial Water of intermediate salinity from the eastern tropical Pacific. MR6 remains in Equatorial Water. MR7 is from a shallower drop taken the next day within a few kilometers of MR6. Here we see the intermingling of the three water masses, each jostling for a level appropriate to its density.

Temperature inversions (negative dT/dz) are generally balanced by positive salinity gradients, so that the density increases with depth, and N^2 is positive. The temperature inversions have typical vertical scales of 5 m, with a step structure (e.g., just beneath feature D) attributed to the *diffusive* regime of double diffusion. The underside of temperature inversions (just above E) is often characterized by strong salinity inversions (positive dS/dz), and by prominent microstructure attributed to the *fingering* regime of double diffusion. Double-diffusive processes can be very important locally; they are discussed by J. S. Turner in Chapter 8.

Occasional density inversions (such as at 13 m depth in MR7) are accompanied by intense microstructure. These inversions are very local, and they disappear in a plot of 3-m averages. We are tempted to attribute the density inversions and associated intermittent microstructure to internal wave breaking.

MR7 is a good example of *intrusive* fine structure. Stommel and Fedorov (1967) gave the first discussion of such features based on their measurements near Timor and Mindanao. At the bottom of a well-mixed layer they found a pronounced temperature inversion (balanced by high salinity) that could be traced for 200 km! Evidently the warm saline water was formed 1 or 2 months earlier over the Australian continental shelf at a distance of 500 km, sliding down along an isopycnal surface. The thickness of the inversion layer varied from 20 to 40 m. Beneath the inversion layer, a number of warm, saline lamina of typically 5-m thickness could be traced over 5 km. All these features are associated with horizontal pressure gradients that must be geostrophically balanced. The authors made some calculations of the rate of lamina spreading associated with frictional dissipation in Ekman spirals above and beneath the lamina boundaries. Once the lamina are thinner than 1 m, they are swiftly conducted away. I refer the reader to Stommel and Fedorov's stimulating discussion.

Table 9.1 summarizes some statistical parameters. For comparison we have included MSR4 from the mid-gyre of the central North Pacific (Gregg, Cox, and Hacker, 1973). The three stations MR7, MR6, and MSR4 characterize strongly intrusive, weakly intrusive

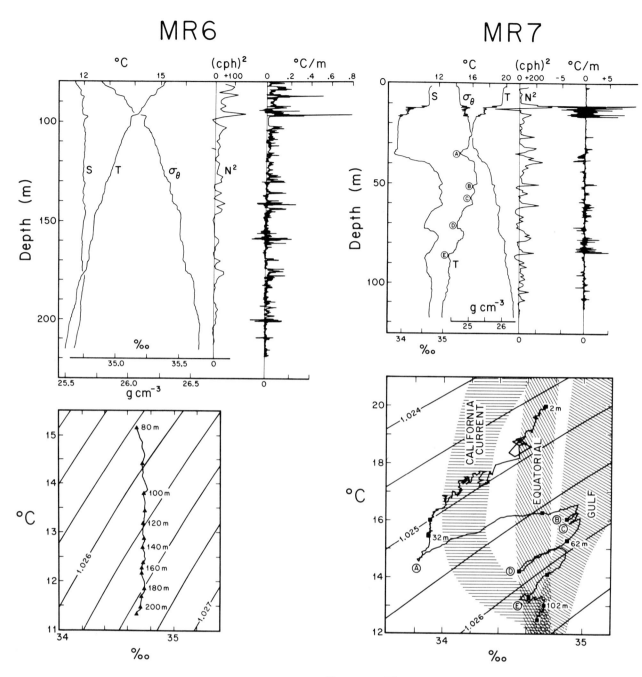

Figure 9.18 The water structure at two stations 60 km southwest of Cabo San Lucas (the southern tip of Baja California), and the associated T-S diagrams (from Gregg, 1975). The measurements have been processed to give the fine structure of S, T, σ_θ, N and the microstructure of dT/dz. Note differences in scale. The *cuspy* T-S diagram for MR7 is an indication of intrusive fine structure.

Table 9.1 Variances and Spectra of Vertical Gradients in the Ocean Fine Structure and Microstructure at Two Stations off Cabo San Lucas (MR7 and MR6) and in the Mid-Gyre of the Central North Pacific (MSR4)

	MR7			MR6			MSR4[a]		
	Strongly intrusive						Nonintrusive		
$\langle (\partial_z T - \partial_z \overline{T})^2 \rangle$ in $\left(\frac{°C}{m}\right)^2$	26			9×10^{-3}			1×10^{-3}		
$\langle \partial_z T \rangle^2$ in $\left(\frac{°C}{m}\right)^2$	4×10^{-3}			8×10^{-4}			6×10^{-4}		
Thermal diffusivity in $cm^2 s^{-1}$	8^b			0.015			0.002		
Cycles per meter	0.1	1	10	0.1	1	10	0.1	1	10
Spectrum of									
$\partial_z T$ in $\left(\frac{°C}{m}\right)^2$/cpm	2×10^{-1}	1×10^{-2}	1×10^{-2}	2×10^{-3}	2×10^{-4}	7×10^{-5}	7×10^{-4}	1×10^{-4}	3×10^{-5}
$\partial_z S$ in $\left(\frac{‰}{m}\right)^2$/cpm	2×10^{-2}	1×10^{-3}	8×10^{-4}	5×10^{-5}	1×10^{-5}	?	4×10^{-6}	8×10^{-6}	2×10^{-5}
$\partial_z \rho$ in $\left(\frac{g\,cm^{-3}}{m}\right)^2$/cpm	2×10^{-9}	7×10^{-10}	7×10^{-10}	2×10^{-10}	2×10^{-11}	?	2×10^{-11}	1×10^{-11}	9×10^{-12}
Spectrum of									
$a \, \partial_z T$ in $\left(\frac{g\,cm^{-3}}{m}\right)^2$/cpm	2×10^{-8}	1×10^{-9}	1×10^{-9}	2×10^{-10}	2×10^{-11}	?	3×10^{-11}	4×10^{-12}	1×10^{-12}
$b \, \partial_z S$ in $\left(\frac{g\,cm^{-3}}{m}\right)^2$/cpm	1×10^{-8}	0.6×10^{-9}	0.5×10^{-9}	0.3×10^{-10}	0.6×10^{-11}	?	0.3×10^{-11}	5×10^{-12}	13×10^{-12}
$\partial_z \rho$ in $\left(\frac{g\,cm^{-3}}{m}\right)^2$/cpm	0.2×10^{-8}	0.7×10^{-9}	0.7×10^{-9}	2×10^{-10}	2×10^{-11}	?	2×10^{-11}	10×10^{-12}	9×10^{-12}

a. MSR4 is not necessarily representative for the mid-gyre; subsequent cruises have given larger mean-square gradients.
b. The vertical heat flux for MR7 can probably not be modeled by an eddy coefficient (Gregg, 1975).

and nonintrusive situations, respectively. The conclusions are: (1) The ratio of the mean-square gradient to the mean gradient squared (the "Cox number") for temperature is highly variable, from 5000 at MR7 to 2 in the mid-gyre. Under certain assumptions (Osborn and Cox, 1972), the eddy diffusivity is the molecular diffusivity times this ratio, giving values all the way from 8 to 0.002 $cm^2 s^{-1}$ (but see the footnote to table 9.1). The canonical value of 1 $cm^2 s^{-1}$ [for which I am partly responsible (Munk, 1966)] is of no use locally. (2) Spectral levels in vertical gradients diminish with increasing vertical wavenumber up to 1 cpm, and then level off. (3) The relative contributions to the density-gradient spectrum has been estimated from $\partial_z \rho = -a \, \partial_z T + b \, \partial_z S$, with $a = 1.7 \times 10^{-4} \, g\,cm^{-3}(°C)^{-1}$, $b = 8 \times 10^{-4} \, g\,cm^{-3}(‰)^{-1}$. For MR7 at 0.1 cpm, the measured density gradient is much smaller than that inferred from either temperature alone or salinity alone. This is consistent with the near cancellation between temperature and salinity for intrusive features. (4) At higher wavenumbers for MR7, and at all wavenumbers of MR6 and MSR4, the density gradient spectrum is of the same order as that inferred from temperature or salinity alone, thus implying the dominance of internal waves.

Probability densities of the temperature gradients are highly non-Gaussian with an enormous flatness factor (138 for MR7, 55 for MR6) attesting to the patchiness (Gregg, 1975). The construction of meaningful ensemble averages in a highly intermittent environment (space and time) is an important task for the future.

To return now to internal waves, we can distinguish between two quite different effects on vertical profiles: (1) an (irreversible) microstructure and fine structure associated with intermittent internal wave breaking, and (2) a (reversible) fine structure due to the vertical straining of an otherwise smooth profile by internal waves of short vertical wavelength. The reversible contribution to fine structure by internal waves was first noticed by Lazier (1973b) and Garrett (1973). How are we to distinguish it from diffusive (and other irreversible) fine structure?

Let δT and δS designate departures in (potential) temperature and salinity from some long-time or long-distance averages $T_0(z)$, $S_0(z)$ at the same depth. Then

$$\delta \rho = -a \, \delta T + b \, \delta S$$

is the associated density departure, with $a(T, S, p)$ and $b(T, S, p)$ designating the (positive) coefficients of thermal expansion and haline contraction. Take first the case of an intrusion only (figure 9.19). If it is totally compensated,

$$\delta\rho = 0,$$

and if it is not totally compensated, it soon will be (in a time of order N^{-1}). It follows that any vertical displacement of the isopycnals is not intrusive but due to a vertical displacement of the water, which we associate with internal waves (see next section). The vertical displacement ζ can be found from conservation of potential density:

$$\rho(z + \zeta) = \rho_0(z).$$

For the case of internal waves only, conservation of potential temperature and salinity give

$$T(z + \zeta) = T_0(z), \qquad S(z + \zeta) = S_0(z),$$

and the ζ-values from the three preceding equations should be the same:

$$\zeta_\rho = \zeta_T = \zeta_S. \tag{9.17}$$

Then in general, ζ_ρ gives the vertical displacement by internal waves, and $\zeta_T - \zeta_\rho$ and $\zeta_S - \zeta_\rho$ are measures of intrusive activity.

Figure 9.20 shows the situation in (T, S)-space. In the combined case, a projection parallel to the isopycnals can separate the two effects. For constant a and b, it is convenient to introduce a family of lines that are orthogonal to the lines of equal potential density (Veronis, 1972). They are here designated by π, for "spiciness" (hot and salty[10]), and they give a measure of the strength of the intrusion. The construction in (ρ, π)-space has some convenient properties. If the x- and y-axes are scaled in equivalent density units, bS and aT, then

$$\delta\rho = -aT + bS, \qquad \delta\pi = aT + bS. \tag{9.18}$$

Figure 9.21 shows plots of the inferred vertical displacements in an area 200 miles southwest of San Diego (Johnson, Cox, and Gallagher, 1978). ζ_T, ζ_S, and ζ_ρ should all be alike for the case of a fine structure due to internal waves only [equation (9.17)], and this turns out to be the case down to a depth of 225 m. There is a broad intrusion between 225 and 260 m, and a narrow intrusion at 275 m. From a spectral analysis it was found that internal waves dominated the fine structure for all vertical scales that could be resolved, that is, down to 5 m.

The displacement spectrum in vertical wavenumber m steepens from approximately m^{-2} for $m < m_u$ to m^{-3} for $m > m_u$, with m_u near 0.6m^{-1} (~ 0.1 cpm). This kink appears to be a common feature in temperature spectra (Gregg, 1977; Hayes, 1978), and is most clearly portrayed in the temperature-gradient spectra (figure 9.22). A similar steepening is found in the spectrum of currents and current shear, but at a somewhat lower vertical wavenumber (Hogg, Katz, and Sanford, 1978).

A free-fall instrument called the "camel" for measuring the velocity microstructure has been developed by Osborn (1974; see chapter 14). Figure 9.23 presents measurements in the Atlantic Equatorial Undercurrent during the GATE experiment (Crawford and Osborn, 1980). The most intense microstructure of temperature and current was found above the velocity core. The microstructure in the core was weak and intermittent. Moderately intensive microstructure was found below the core, near the base of the thermocline. This is shown in great detail in figure 9.24. As an example, between 81 and 82 m there is an active temperature microstructure with positive and negative $\partial_z T$, accompanied by an active velocity microstructure. Similar evidence is found in horizontal tows, as for example in the upper part of figure 9.25 (Gibson, Schedvin, and Washburn, personal communication; see Gibson, 1980). The important conclusion is that velocity microstructure and small-scale temperature inversions must be closely linked, for one is not found without the other.

Occasionally one encounters patches of temperature microstructure without velocity microstructure. The inference is that these patches are the remains of a mixing event for which the velocity microstructure has decayed (fossil turbulence). Examples are found in the vertical profiles (figure 9.24 between 69 and 70 m depth), and in the horizontal tows (figure 9.25, bottom). But for the vertical profiles the temperature microstructure is here limited to only positive $\partial_z T$; the authors suggest that this might be a peculiarity of the core (velocity and salt) of the undercurrent.

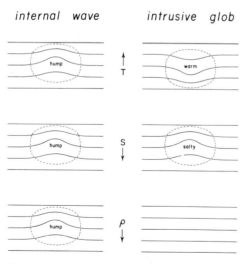

Figure 9.19 Contours of potential temperature, salinity and potential density in a vertical section (x, z) for an internal wave hump and a compensated warm and salty intrusive glob.

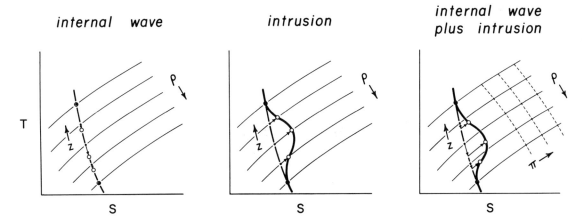

Figure 9.20 T–S relations for an internal wave hump and a compensated intrusive glob. The dots (●) correspond to the undisturbed positions of the five contours in figure 9.19. The open circles (○) give the positions through the center of the disturbance. The "isospiceness" lines (constant π) are orthogonal to the isopycnals (constant ρ).

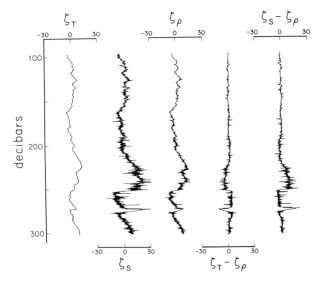

Figure 9.21 Displacement profiles in meters as inferred from the temperature, salinity, and density profiles. These should be alike for internal wave produced fine structure. The mean T and S gradients were of opposite sign (as in figure 9.19), hence the opposite signs of $\zeta_T - \zeta_\rho$ and $\zeta_S - \zeta_\rho$. (Johnson, Cox, and Gallagher, 1978.)

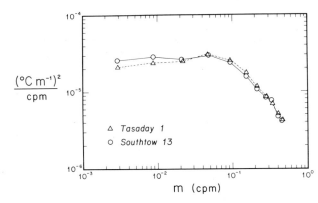

Figure 9.22 Temperature gradient spectra for two stations in the North Pacific. (Gregg, 1977.)

Crawford and Osborn have calculated the dissipation rates. Typical values just beneath the core of the undercurrent range from $\epsilon = 10^{-4}$ to 10^{-3} cm^2 s^{-3} (10^{-5} to 10^{-4} W m^{-3}). (Measurements away from the equator fall within the same limits.) But Belyaev, Lubimtzev, and Ozmidov (1975) obtain dissipation rates in the area of the undercurrent from horizontal tows that are higher by two orders of magnitude.[11]

The question of the relative magnitudes of vertical and horizontal scales has been examined by Hacker (1973) from a comparison of wing tip and nose temperatures of a rotating free-fall instrument. The horizontal separation is 1.7 m. The two records are coherent for vertical wavelengths down to 1 m. At smaller wavelengths the analysis is made difficult by the random tilts (5° rms) associated with internal waves. By selectively analyzing depth ranges of small tilts, Elliott and Oakey (1975) found coherence over a horizontal spacing of 0.5-m down to 10-cm vertical wavelengths. The conclusion is that anisotropy extends beyond the fine structure into the microstructure, perhaps as far as the dissipation scale (~1 cm).

The picture that emerges is one of a fine structure that is usually dominated by internal wave straining and is fairly uniform, in contrast to a microstructure that is extremely patchy and variable even in the mean. Patches of temperature microstructure without velocity microstructure ("fossil turbulence") evidently mark the demise of internal waves that had previously broken.

9.9 An Inconclusive Discussion

Is there a connection between internal wave activity, dissipation, and buoyancy flux? What is the explanation for the seeming steadiness of the internal wave field? Having gone this far, I cannot refrain from continuing with some speculation. The reader is encouraged to go no further (if he has gotten this far).

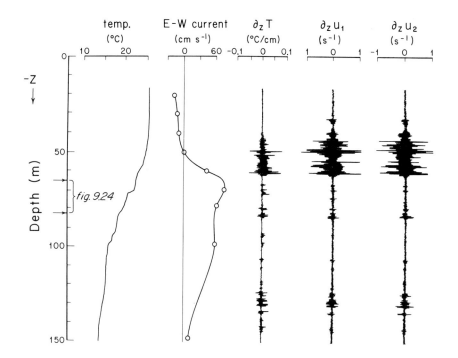

Figure 9.23 Temperature and velocity microstructure in the Atlantic Equatorial Undercurrent at 0°18′S, 28°01′W. (Crawford and Osborn, 1980.) The large scale-current profile was measured by J. Bruce. The region between 65 and 82 m is shown on an enlarged scale in figure 9.24.

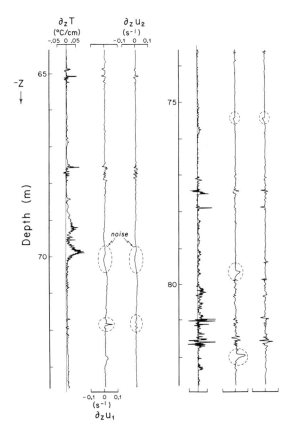

Figure 9.24 An enlarged section of the microstructure profile shown in figure 9.23. The encircled features have been attributed to various sources of instrumental noise.

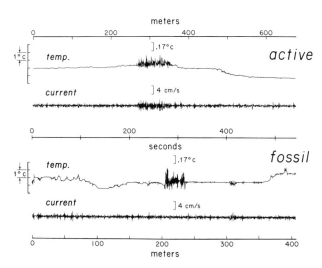

Figure 9.25 Active and fossil turbulence from towed body measurements during the mixed-layer experiment (MILE) in September 1977 near ocean station PAPA. The body was towed in the seasonal thermocline at a depth of 33 m; note the horizontal temperature change by about 0.3°C across the patches. The 1°C scale (left) refers to frequencies $f < 1$ H_z; the 0.17°C scale is for 1–12 H_z.
(I am indebted to C. Gibson, J. Schedvin, and L. Washburn for permission to show these measurements. See also Gibson, 1980.)

We shall need some quantitative guidance for internal wave intensities. For that purpose I shall use the model spectrum of Garrett and Munk (1972b, 1975, 1979), with slight modifications. The model fails near the ocean boundaries[12] (Pinkel, 1975). The spectrum was developed on the basis of rank empiricism, with no trace of underlying theory. But it has since gained some respectability by the theoretical findings of Watson and collaborators that the shape of the GM spectrum is stable to nonlinear interactions, except for the lowest modes and near-inertial frequencies [Meiss, Pomphrey, and Watson (1979); Pomphrey, Meiss, and Watson (1980); see also McComas (1977)].

9.9.1 Model Spectrum GM79

The internal wave energy is assumed to be equally distributed in all horizontal directions, so that only a single horizontal wavenumber, $k = (k_1^2 + k_2^2)^{1/2}$, is used. Upward and downward energy flux are taken as equal. The spectra of vertical displacement, horizontal velocity, and energy per unit mass are[13]

$$F_\zeta(\omega,j) = b^2 N_0 N^{-1}(\omega^2 - f^2)\omega^{-2} E(\omega,j), \quad (9.19)$$

$$F_u(\omega,j) = F_{u_1} + F_{u_2} = b^2 N_0 N(\omega^2 + f^2)\omega^{-2} E(\omega,j), \quad (9.20)$$

$$F_e(\omega,j) = \tfrac{1}{2}(F_u + N^2 F_\zeta) = b^2 N_0 N E(\omega,j), \quad (9.21)$$

where j is the vertical mode number, $b \approx 1.3$ km the e-folding scale of $N(z)$, with $N_0 \approx 5.2 \times 10^{-3}$ s^{-1} (3 cph) the surface-extrapolated buoyancy frequency and $f = 7.3 \times 10^{-5}$ s^{-1} the Coriolis frequency at lat. 30°. We can ignore F_w compared to F_u. At high frequencies, $\omega \gg f$, kinetic and potential energy densities are equal: $\tfrac{1}{2}F_u = \tfrac{1}{2}N^2 F_\zeta$. $E(\omega,j)$ is a dimensionless energy density that is factored as follows:

$$E(\omega,j) = B(\omega)\cdot H(j) \cdot E,$$

$$B(\omega) = 2\pi^{-1} f \omega^{-1}(\omega^2 - f^2)^{-1/2}, \quad \int_f^{N(z)} B(\omega)d\omega = 1,$$

$$H(j) = \frac{(j^2 + j_*^2)^{-1}}{\sum_1^\infty (j^2 + j_*^2)^{-1}}, \quad \sum_{j=1}^\infty H(j) = 1.$$

The factor $(\omega^2 - f^2)^{-1/2}$ in the expression for $B(\omega)$ is a crude attempt to allow for the peak at the inertial turning frequency (see figure 9.11); $j_* = 3$ is a mode scale number, and E is the internal wave "energy parameter." We set

$$E = 6.3 \times 10^{-5} \quad \text{(dimensionless)}. \quad (9.22)$$

There is a surprising universality[14] to the value of E (mostly within a factor of two).

The transfer into (ω,k)- or (ω,m)-space is accomplished by setting $F(\omega,j)\delta j = F(\omega,k)dk = F(\omega,m)dm$, with

$$m = k\left(\frac{N^2 - \omega^2}{\omega^2 - f^2}\right)^{1/2} = \pi b^{-1}\left(\frac{N^2 - \omega^2}{N_0^2 - \omega^2}\right)^{1/2} j \quad (9.23a)$$

for a slowly varying $N(z)$, in accord with the WKB approximation [equations (9.4) and (9.7)]. For most purposes we can ignore the situation near the buoyancy turning frequency,[15] so that

$$m \approx kN(\omega^2 - f^2)^{-1/2} \approx \pi b^{-1}(N/N_0)j. \quad (9.23b)$$

For the sake of simplicity, the energy spectrum has been factored into $B(\omega)\cdot H(j)$. But there is evidence from Pinkel (1975) and from the IWEX measurements (Müller, Olbers, and Willebrand, 1978) that there is relatively more energy in the low modes at high frequency, and this could account for the astounding vertical coherences found by Pinkel in the upper 400 m at high frequencies.

I have no doubt that further discrepancies will be found; still, I believe that the model can now give useful quantitative estimates. For example, according to "Fofonoff's rule" (he disclaims ownership), the mean-square current within a 1-cph band centered at 1 cph is 1 cm^2 s^{-2}; this compares to F_u(1 cph) = 0.8 cm^2 s^{-2} (cph)$^{-1}$ from (9.20). This agreement is not an accident, of course, the GM model having been based, in part, on the site D measurements (Fofonoff, 1969).

The mean-square quantitites are likewise in accord with the usual experience. From (9.19), (9.20), and (9.21),

$$\begin{aligned}
\langle \zeta^2 \rangle &= \int d\omega \sum F_\zeta(\omega,j) \\
&= \tfrac{1}{2}b^2 E N_0 N^{-1} = 53(N/N_0)^{-1} \text{ m}^2, \\
\langle u^2 \rangle &= \langle u_1^2 \rangle + \langle u_2^2 \rangle = \int d\omega \sum F_u(\omega,j) \\
&= \tfrac{3}{2}b^2 E N_0 N = 44(N/N_0) \text{ cm}^2 \text{ s}^{-2}, \\
\hat{E}(z) &= \int d\omega \sum F_e(\omega,j) \\
&= b^2 E N_0 N = 30(N/N_0) \text{ cm}^2 \text{ s}^{-2},
\end{aligned} \quad (9.24)$$

giving 7 m for the rms vertical displacement and 7 cm s^{-1} for the rms current in the upper oceans beneath the mixed layer. The energy can be written alternatively $\hat{E}(z) = \tfrac{1}{2}[\langle u^2 \rangle + N^2 \langle \zeta^2 \rangle]$ so that the *total* kinetic energy is three times the total potential energy in the GM model. $\rho \hat{E}(z)$ is the energy per unit volume; the energy per unit area is

$$\begin{aligned}
\rho \hat{E} &= \int \rho \hat{E}(z)\, dz = \rho b^2 E N_0 \int N\, dz = \rho b^2 E N_0 \int b\, dN \\
&\approx \rho b^3 N_0^2 E \\
&= 3.8 \times 10^6 \text{ erg cm}^{-2} = 3800 \text{ J m}^{-2},
\end{aligned}$$

using $b^{-1} = N^{-1} \cdot dN/dz$ as definition for the e-folding scale b.

9.9.2 Universality

It has turned out, quite unexpectedly, that the intensities are remarkably uniform in space and time.

Wunsch (1976) has made a deliberate attempt to find systematic deviations for a variety of deep water locations in the North Atlantic, with the purpose of identifying sources and sinks of internal wave energy. Using the frequency band $\frac{1}{8}$ to $\frac{1}{4}$ cph as a standard, the only clear deviations he could find were associated with topographic features, particularly Muir Seamount, and even these were inconspicuous at short distance. In a further study (Wunsch and Webb, 1979) some evidence is presented for deviations on the equator and in regions of high mean shear.

Figure 9.26 shows a continuing spectral display over an 18-day period, and this is found consistent with a stationary Gaussian process (Cairns and Williams, 1976). [The mean distribution is in accord with the equation (9.19) for $F_\zeta(\omega) = \Sigma F_\zeta(\omega, j)$, setting $\Sigma H(j) = 1$.] These observations were taken during 21 days of mild to moderate winds. Davis (personal communication) has recorded currents in the seasonal thermocline over a 19-day interval with two periods of heavy winds (figure 9.27). The first event is followed in about 2 days by an increase in mean-square currents, the second event in somewhat less time. Energy enhancement is by a factor of three or less. Johnson, Cox, and Gallagher (1978) found a temporarily elevated spectral level on a windy day. Following these events, the intensities rapidly relax to their normal state.

9.9.3 Generation

The observed growth times are consistent with a theory for the generation of internal waves by resonant interaction with surface waves (Brekhovskikh, Goncharov Kurtepov, and Naugol'nykh, 1972; Watson, West, and Cohen, 1976).

But there are other means of generating internal waves. Garrett (1979) has reviewed a variety of contenders, all of which fall (surprisingly) into the right order of magnitude. For reference, he takes 7×10^{-3} W m^{-2} for the internal wave dissipation (corresponding to a relaxation time of one week). Globally, this amounts to 2 TW (terawatts: tera = 10^{12}). The total loss of energy of the earth–moon system is known from the moon's orbit to be 4 TW, mostly by tidal dissipation in the oceans. It is not impossible that surface tides pump significant amounts of energy into internal wave motions via internal tides (cf. chapter 10). Other contenders are surface forcing by traveling fluctuations of wind stress and buoyancy flux, currents over bottom topography, and extraction from the mean current shear. There is no problem with supplying internal waves with 2 TW of power; the problem is rather to eliminate some of the potential donors.

9.9.4 Instability

We can now derive some numerical estimates for a variety of instability parameters. The spectrum $F_{u'/N}$ of

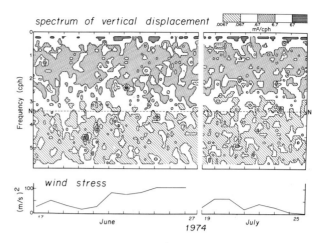

Figure 9.26 Time-frequency display of $F_\zeta(\omega)$ from MISERY 1 and MISERY 3. (Cairns and Williams, 1976.) $\zeta(t)$ is the depth of the 6.60° isotherm (at a mean depth of 350 m) in a location 800 km offshore of San Diego, California, measured with a yo-yoing midwater capsule. The squared wind (bottom) shows light winds at the start and end of the experiments.

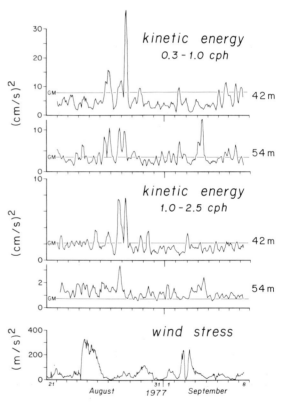

Figure 9.27 Kinetic energy $\langle u^2 \rangle = \langle u_1^2 \rangle + \langle u_2^2 \rangle$ during the mixed-layer experiment (MILE) on station PAPA (50°N, 145°W) at 42 and 54 m depths. The upper two plots refer to a frequency band of 0.3 to 1.0 cph, the next two plots to a band 1.0 to 2.5 cph. The squared wind (bottom) shows two episodes of large wind stress. The GM model levels are indicated [using $N = 0.023, 0.0096$ s^{-1} (13, 5.5 cph) at 42, 54 m], though the model is not really applicable to such shallow depths and sharp N-gradients. (I am indebted to R. Davis for permission to use his measurements.)

the reciprocal Richardson number is defined by

$$Ri^{-1} = \langle (u')^2 \rangle / N^2 = \int d\omega \sum_j F_{u'/N}(\omega, j),$$

where $\langle (u')^2 \rangle = \langle (\partial_z u_1)^2 \rangle + \langle (\partial_z u_2)^2 \rangle = m^2 \langle u^2 \rangle$. But $m = \pi b^{-1}(N/N_0)j$ (9.23b) and so

$$F_{u'/N} = 2\pi E(N/N_0)f(\omega^2 + f^2)\omega^{-3}(\omega^2 - f^2)^{-1/2}j^2 H(j). \quad (9.25)$$

The principal contribution comes from the inertial frequencies. Performing the ω-integration,

$$Ri^{-1} = \tfrac{3}{2}\pi^2 E(N/N_0)\sum j^2 H(j). \quad (9.26)$$

We now perform the mode summations (subscript u for upper)

$$\sum_{j=1}^{j_u} j^2(j^2 + j_*^2)^{-1} \approx j_u,$$

$$\sum_{j=1}^{j_u} (j^2 + j_*^2)^{-1} \approx \tfrac{1}{2}j_*^{-2}(\pi j_* - 1) \equiv J^{-1} = 0.47$$

for $j_u \gg j_* = 3$. The spectrum $F_{u'/N}(j)$ is white (except for the lowest few mode numbers) and Ri^{-1} depends on the choice of the upper cutoff j_u. In terms of the limiting vertical wavenumber $m_u = \pi b^{-1}(N/N_0)j_u$ we have, finally,

$$Ri^{-1} = \tfrac{3}{2}\pi J E m_u b. \quad (9.27)$$

If we identify m_u with the kink (figure 9.22) at $0.6 m^{-1}$ (0.1 cpm), then $Ri^{-1} = 0.52$. If the spectrum is extended beyond the break, with a slope m^{-1} to some new upper limit $m_{uu} = 10 m_u$ (say), then $Ri^{-1} = 0.52(1 + \ln m_{uu}/m_u)$ = 1.72. Hogg, Katz, and Sanford (1978) find Ri^{-1} near 0.5 in the open ocean near Bermuda, including m up to 0.2 cpm. An interesting scatter plot has been produced by Eriksen (1978) and is shown in figure 9.28. The conclusion is that the internal-wave shear field is associated with Richardson numbers of order 1:

$$Ri^{-1} = \text{order}(1). \quad (9.28)$$

We can proceed in a similar manner with regard to advective instability. The simplest generalization of the earlier discussion (section 9.7) is to derive the spectral decomposition of $\langle u^2/c^2 \rangle$, $c = \omega/k$. (Would $\langle u^2 \rangle / \langle c^2 \rangle$ be better?) The equations of continuity and of dispersion (away from the turning frequencies) can be written

$$k\xi + m\zeta = 0, \quad c = \omega/k = N/m,$$

so that

$$\frac{u}{c} = \frac{mu}{mc} = \frac{u'}{N}.$$

Similarly, for horizontal and vertical strain,

$$\partial \xi/\partial x = -\partial \zeta/\partial z = ik\xi = (k/\omega)u = u/c.$$

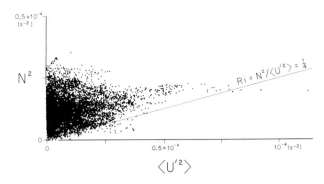

Figure 9.28 A scatter plot of squared shear (over 6.3 m vertical separation) versus N^2 (over 7.1 m) of estimates made every 40 s for 78 hours. (Eriksen, 1978.) Eriksen finds that Ri rarely falls below the critical value $\tfrac{1}{4}$, and that $\phi = \tan^{-1} Ri$ is uniformly distributed for ϕ greater than $\tan^{-1}(\tfrac{1}{4})$.

Thus shear, advection, and longitudinal strains all have similar conditions for instability, and we can write

$$\langle \phi^2 \rangle = CEm_u b \quad (9.29)$$

for any of these, without having to go into gruesome details. [Integrations yield $C = 2.5, 5.8, 3.3, 3.3$ for $u'/2N$, u/c, $\partial_x \xi$, $\partial_z \zeta$, respectively; this is a spectrally weighted version of the Orlanski and Bryan argument [equation (9.16)] that advective instability is the most likely to occur.]

9.9.5 Compliant Wave Cutoff

The instability condition $\langle \phi^2 \rangle = \text{order}(1)$ is an argument for a universal value of the product Em_u. To account for a universal E we need some additional condition.

I propose that the upper cutoff m_u is related to the transition at $c_c \approx N/m_c = \text{rms}\, u$ from the intrinsic to the compliant parts of the internal wave spectrum:

$$m_u = C' m_c = C' N / \text{rms}\, u, \quad (9.30)$$

where C' is a constant of order (1). It stands to reason that the strongly interacting high wavenumbers have a different spectral form from the intrinsic waves. If we identify m_u with the kink (figure 9.22) at $0.6\,\text{m}^{-1}$ (0.1 cpm), then for $N = 0.01\,\text{s}^{-1}$ and $\text{rms}\, u = 10\,\text{cm s}^{-1}$ this gives $C' = 6$ (somewhat large for comfort).

There is an equivalent way of postulating the upper cutoff. The ϕ-spectrum is white up to some limit which is the reciprocal of the vertical extent Δ of the smallest ϕ-features. One might suppose this vertical extent to be some given fraction of the rms *amplitude* of the internal waves. (White caps occupy some fraction of the surface wave crests; the distance between crests is not a critical factor.) Using the foregoing numbers, we can write

$$\Delta = m_u^{-1} = 1.37(C')^{-1} \text{rms}\, \zeta$$
$$= 0.23\, \text{rms}\, \zeta = 1.7\,\text{m}. \quad (9.31)$$

From the expressions (9.24) for $\langle \zeta^2 \rangle$ and $\langle u^2 \rangle$, either condition (9.30) or condition (9.31) leads to an energy parameter

$$E = [\langle \phi^2 \rangle / CC']^2 N_0/N. \quad (9.32)$$

For $C = 5$, $C' = 6$, $E = 10^{-3} \langle \phi^2 \rangle^2 N_0/N$. Then with $\langle \phi^2 \rangle$ a moderate fraction of its critical value 1, we can recover the numerical value $E = 6 \times 10^{-5}$. This is not to say that E has been calculated from first principles; it is only to say that acceptable values for the various coefficients lead to a small numerical value of the dimensionless energy parameter, as observed.

9.9.6 Dissipation

For small numerical values of the instability parameter $\langle \phi^2 \rangle$ we are in a regime of *sparse* instabilities in space and time (such as incipient whitecaps in light winds). When $\langle \phi^2 \rangle$ is near 1 the probability for instabilities is high. For an a priori estimate of $\langle \phi^2 \rangle$ we require (1) a model to relate $\langle \phi^2 \rangle$ to the internal wave energy dissipation, and (2) an estimate of the rate of dissipation (or generation for a given steady state). This is essentially the procedure followed by Longuet-Higgins (1969a) in his stimulating attempt to interpret the Phillips saturation constant for surface waves.

Perhaps the simplest scheme is to relate the dissipation to the probability for $\phi > 1$.[16] The variance of ζ associated with $\phi > 1$ is $\langle \zeta^2 \rangle \cdot p (\phi > 1)$ for uncorrelated ζ and ϕ (as when $\phi = \partial_z \zeta$). Potential energy is proportional to $\langle \zeta^2 \rangle$; accordingly the rate of fractional energy dissipation can be written

$$\frac{1}{E} \frac{dE}{dt} = -\sigma p(\phi > 1), \quad (9.33)$$

where σ^{-1} is the characteristic interval during which the energy associated with $\phi > 1$ is lost to the organized wave field and renewed by generation processes. For a rough estimate (Garrett and Munk, 1972a),

$$\sigma^2 = \pi^{-2} \int_f^N \omega^2 F_\phi(\omega) d\omega \bigg/ \int_f^N F_\phi(\omega) d\omega$$

$$\approx \pi^{-2} f N \quad (9.34)$$

for any of the ϕ-spectra [such as (9.25)]. A Gaussian ϕ-distribution $p(\phi) = \pi^{-1/2} \beta \exp(-\beta \phi^2)$ leads to

$$p (\phi > 1) \approx \pi^{-1/2} \beta^{-1/2} \exp(-\beta), \quad \beta \equiv \frac{1}{2\langle \phi^2 \rangle}, \quad (9.35)$$

provided β is large. In the upper ocean $\sigma = 20$ per day, and $\beta = 2.1, 3.7$, for relaxation times of 1 day, 1 week, respectively. The foregoing numerical values are not important; what is significant is that a tenfold increase in the rate of dissipation (and generation) is accompanied by only a threefold increase in β^2 (and hence in wave energy). Thus the energy level stays within rather narrow limits even though generation and dissipation processes may vary widely, particularly for large β. We propose for a "universality hypothesis" that the energy level responds only logarithmically to variable forcing. We shall examine this situation in more detail.

9.9.7 The Energy Balance

The differential equation of wave energy can be written

$$dE/dt = G(t) - D(t),$$

where $G(t)$ and $D(t)$ are the rates of energy generation and dissipation. We use the notation $\tilde{E}, \tilde{G}, \tilde{D}$ to represent the "normal" state of internal wave statistics. From (9.32) and (9.35)

$$E \sim \beta^{-2}(N_0/N), \quad D \sim \beta^{-1/2} e^{-\beta},$$

with $\tilde{D} = \tilde{G}$. We define the relaxation time

$$\tilde{t} = \tilde{E}/\tilde{G};$$

accordingly $\tilde{t}^{-1}$ is the initial rate of decay for a wave field in equilibrium with $\tilde{G}$ if the generation is suddenly turned off. The differential energy equation can now be written

$$\frac{d\mathcal{E}}{d\tau} + \mathcal{E}^{1/4} \exp[\tilde{\beta}(1 - \mathcal{E}^{-1/2})] = g(\tau),$$

$$\tilde{\beta} = \tilde{E}^2 (N/N_0)^{-1/2}, \quad (9.36)$$

where

$$\mathcal{E}(\tau) = E/\tilde{E}, \quad g(\tau) = G/\tilde{G}, \quad \tau = t/\tilde{t},$$

with

$$\langle \mathcal{E} \rangle = 1, \quad \langle g \rangle = 1.$$

For large $\mathcal{E}$, the dissipation is $\mathcal{E}^{1/4} \exp \tilde{\beta}$ and thus large; for small $\mathcal{E}$, it is $\mathcal{E}^{1/4} \exp -(\tilde{\beta} \mathcal{E}^{-1/2})$ and thus very small.

The problem is to derive properties of the energy statistics for given generation statistics. (This is related to the fluctuation-dissipation theorem in the study of Brownian motion.) Two special solutions are easily found. For an equilibrium situation, (9.36) with $d\mathcal{E}/d\tau = 0$ gives the values $\mathcal{E}(g; \tilde{\beta})$ in table 9.2. Departures from the normal state in E are much smaller than those in g, particularly at large β and for small g's.

To obtain some feeling for the nonlinear response time, let $g(\tau)$ go abruptly from 1 to g at time 0, and set $\mathcal{E}(\tau) = 1 + \epsilon(\tau)$, with $\epsilon \ll 1$ (but not $\epsilon \tilde{\beta} \ll 1$). Equation (9.36) becomes

$$dz/d\tau' + z^2 = gz, \quad z = \exp(\tfrac{1}{2} \tilde{\beta} \epsilon), \quad \tau' = \tfrac{1}{2} \tilde{\beta} \tau,$$

with the solution

$$\tau = \frac{2}{\tilde{\beta} g} \ln \frac{(g-1)z}{g-z}.$$

For the case $g \to 0$, $\tau' \to z^{-1} - 1$, and

$$d\epsilon/d\tau = 2\tilde{\beta}^{-1} d \ln z/d\tau \to -z, \quad (9.37)$$

Table 9.2 Energy Equilibrium $\mathscr{E}(g;\beta)$ from (9.36) for $d\mathscr{E}/d\tau = 0$ (top), Dimensionless Response Time $\tau_{1/2}$ (9.38) for an Abrupt Change in Generation from 1 to g (center), Relaxation Time $\tau_{1/2}$ (9.39) for a Change from g to 1 (bottom)[a]

		g (dB)				
		-10	-5	0	5	10
	Energy levels in dB					
	2	-6.0	-3.4	0	4.9	13.0
$\tilde{\beta}$	5	-3.0	-1.7	0	2.0	4.6
	10	-1.7	-0.9	0	1.0	2.1
	Response time (dimensionless)					
	2	2.75	1.41	0.69	0.32	0.14
$\tilde{\beta}$	5	1.10	0.56	0.28	0.13	0.06
	10	0.55	0.28	0.14	0.06	0.03
	Relaxation time (dimensionless)					
	2	1.43	1.02	0.69	0.45	0.27
$\tilde{\beta}$	5	0.57	0.41	0.28	0.18	0.11
	10	0.29	0.20	0.14	0.09	0.05

a. Generation g and energy $\mathscr{E}$ are in decibels relative to normal levels.

in accord with $d\epsilon/d\tau = -1$ for the initial rate of decay. As τ goes from 0 to ∞, z goes from 1 to g, and ϵ from 0 to $2\tilde{\beta}^{-1} \ln g$. Half energy response is for $\epsilon = \tilde{\beta}^{-1} \ln g$, $z = g^{1/2}$, and

$$\tau_{1/2} = 2\tilde{\beta}^{-1} g^{-1} \ln(1 + g^{1/2}). \qquad (9.38)$$

For energy *relaxation*, going abruptly from g to 1 at time 0, the solution is

$$\tau = 2\tilde{\beta}^{-1} \ln \frac{(g-1)z}{g(z-1)}, \qquad \tau_{1/2} = 2\tilde{\beta}^{-1} \ln(1 + g^{-1/2}). \qquad (9.39)$$

Half-times $\tau_{1/2}$ are given in table 9.2. For a linear system these would all be the same. Here the times are shorter for internal wave storms ($g > 1$) and longer for calms ($g < 1$), and this variation is more pronounced in response to a change in generation from a normal to a perturbed level than for relaxation back to normal generation. (The assumption of weak nonlinearity for computing τ is violated in the columns for $g = \pm 10$ dB.)

The imperceptible decay of internal wave intensities during relatively low winds (figure 9.26) is consistent with half the normal energy, and the rapid decay following a blow (figure 9.27) requires perhaps three times normal energy. If we can count on a storm or some other generation event to "top up" the internal wave energy once every hundred days, then we may expect the wave energies to remain generally within a factor of two.

I have paid no attention to depth dependence. The dissipation can be written

$$\hat{D} = \sigma(\pi\beta)^{-1/2} \exp(-\beta)\hat{E},$$

where $\rho\hat{D}(z)$, $\rho\hat{E}(z)$ are the dissipation and energy per unit volume, respectively. The dependence on depth is through

$$\sigma \sim n, \qquad \beta \sim n^{-1/2}, \qquad \hat{E} \sim n,$$

with $n(z) = N(z)/N_0$; hence

$$\hat{D} = \hat{D}_0 \beta^{-9/2} \exp(\beta_0 - \beta), \qquad \hat{E} = \hat{E}_0 \beta^{-2}.$$

Writing $dz = b\, dn/n = -2b\, d\beta/\beta$,

$$\int \hat{D}\, dz = 2b\hat{D}_0 \int \beta^{-11/2} \exp(\beta_0 - \beta)\, d\beta,$$

$$\int \hat{E}\, dz = 2b\hat{E}_0 \int \beta^{-3}\, d\beta.$$

Integrating from the surface ($\beta = \beta_0$) to the bottom ($\beta \approx \infty$), the integrals for large β_0 (as previously assumed) give

$$\int \hat{D}\, dz = 2b\hat{D}_0 \beta_0^{-11/2}, \qquad \int \hat{E}\, dz = b\hat{E}_0 \beta_0^{-2}$$

with an "integral relaxation time"

$$\tfrac{1}{2}\beta_0^{7/2} \hat{E}_0/\hat{D}_0 = \tfrac{1}{2}\pi^{1/2}\beta_0^4 e^{\beta_0}\sigma_0^{-1}. \qquad (9.40)$$

Suppose the generation takes place in the upper few hundred meters, so that the dissipation in the interior ocean is compensated by downward radiation of internal wave energy. From a rotary decomposition of current profiles, Leaman (1976) estimates a downward flux of order $10^{-4}\,\text{W m}^{-2}$. We compare this with the integrated dissipation beneath a scale depth $z = -b = -1.3$ km (where $n = e^{-1}$ and $\beta = \beta_0 \sqrt{e}$):

$$\rho \int_{\beta_0\sqrt{e}}^{\infty} \hat{D}(z)\, dz = 2\pi^{-1/2} e^{-11/4} \beta_0^{-6} e^{-\beta_0} \sigma_0 N_0^2 \rho b^3 \tilde{E},$$

where $\tilde{E}$ is the "normal" dimensionless energy parameter. The result is $\beta_0 = 2.1$. The corresponding integral relaxation time [equation (9.40)] is 7 days. The normal surface relaxation time is

$$\tilde{t}_0 = \hat{E}_0/\hat{D}_0 = (\pi\beta_0)^{1/2} e^{\beta_0} \sigma_0^{-1} = 1.1 \text{ days}.$$

The dimensionless times in table 9.2 can be interpreted as shallow response times in days.

The strong dependence of dissipation on depth is an inherent feature of the proposed phenomenology.

9.9.8 Mixing

The balance between production and dissipation of turbulent energy can be represented by

$$\mu = \epsilon = \epsilon_p + \epsilon_k. \qquad (9.41)$$

$\rho\epsilon_p$ is the rate of production of potential energy, e.g., the *buoyancy flux* $g\langle w'\rho'\rangle$, with the primes designating the fluctuating components; $\rho\epsilon_k$ is the dissipation of kinetic energy into heat. The fraction of work going into potential energy is the flux Richardson number Rf:

$$Rf = \epsilon_p/\mu = \epsilon_p/\epsilon. \tag{9.42}$$

$Rf < 1$ in order for $\epsilon_k > 0$. We have previously discussed the evidence that something less than $\frac{1}{4}$ of the kinetic energy dissipated appears as potential energy. Taking a typical value $\epsilon_p/\epsilon_k = \frac{1}{5}$ (Thorpe, 1973b, p. 749) gives $Rf = \frac{1}{6}$. The eddy flux of density can be written in terms of the eddy diffusivity A: $\langle w'\rho' \rangle = A \cdot d\rho/dz = AN^2(g/\rho)^{-1} = \rho\epsilon_p/g$, and so

$$A = \frac{\epsilon_p}{N^2} = \frac{\epsilon}{N^2} Rf = \frac{\epsilon_k}{N^2} \frac{Rf}{1 - Rf}. \tag{9.43}$$

Our procedure is to estimate ϵ_p from internal wave breaking, and to compute A and ϵ from (9.43), using $Rf = \frac{1}{6}$. There are, of course, other sources of turbulence; Osborn (1980) stresses the work done by the ambient turbulence in a mean shear: $\mu = \langle u'w' \rangle \, du/dz$. His procedure is to estimate ϵ from the measured mean-square shear, and to compute A from (9.43).

The random superposition of internal waves leads to the intermittent occurrence of "traumata"[17] associated with $\phi > 1$. The traumata are the locus of incipient turbulence, and quickly spread to some thickness Δ within which the average ϕ is reduced from 1 to about 0.9 (Thorpe, 1973b). Subsequently the patch continues to grow to some maximum thickness Δ_{max} at the time the surrounding ϕ is largest, always keeping ϕ within the patch to 0.9.

The change of potential energy per unit surface area associated with perfect mixing over a depth Δ in a density gradient $d\rho/dz$ is $(1/12)g(d\rho/dz)\Delta^3 = (1/12)\rho N^2 \Delta^3$. (Imperfect mixing just reduces the factor 1/12.) The average change of potential energy per unit time per unit volume is then

$$\rho\epsilon_p = \frac{1}{12} \rho N^2 \Delta^3 \nu,$$

where ν is the number of traumata per unit (t,z)-space. $\rho\epsilon_p$ equals the buoyancy flux $\rho N^2 A$ by definition of the eddy diffusivity A; hence

$$A = \frac{1}{12} \Delta^3 \nu.$$

I have previously identified Δ with m_u^{-1} [equation (9.31)]. For ν we write

$$\nu = m'\sigma p(\phi > 1), \quad m' = 0.2 m_u, \quad \sigma \approx \pi^{-1}(fN)^{1/2},$$

where m' is the rms spacial frequency derived from an equation analogous to (9.34). Putting all this together,

$$A \approx 10^{-2} \Delta^2/T, \quad T \equiv \sigma^{-1}\beta^{1/2}e^\beta, \tag{9.44}$$

where T is the expected time interval between events over a distance $1/m'$. This is of similar form as the result of Stommel and Fedorov (1967), as is inevitable for what is, after all, a mixing-length theory.

The principal conclusion is a strong dependence of A on depth, and on any departures from normal generation. The numerical value for the normal state is $\bar{A} = 10^{-2} \, \text{cm}^2 \, \text{s}^{-1}$, much lower than the global $1 \, \text{cm}^2 \, \text{s}^{-1}$.

9.9.9 Saturation Spectra

There is an essential distinction between the usual formulation of turbulence and the saturation processes as here envisioned. We consider the regime of *sparse* instabilities in space and time (such as incipient whitecaps in light winds). Then the ϕ-field consists of scattered and uncorrelated spikes, and the ϕ-spectrum is accordingly white up to some limit that is the reciprocal of the vertical extent of the spikes. The dissipation is localized in physical space, and therefore broadly distributed in wavenumber space. In the usual turbulent situation, the dissipation is confined to a narrow (dissipation) region in wavenumber space, and spread in physical space.

A white spectrum in any of the ϕ-spectra, whether shear, advection or strain, implies an m^{-2} energy spectrum. The energy spectrum steepens (perhaps to m^{-3}) in the transition from the intrinsic to the compliant waves. Presumably the m^{-3} energy spectrum extends to the Ozmidov (or Richardson or Monin–Obukov) scale $m_0 = (N^3/\epsilon)^{1/2} \approx 4 \, \text{m}^{-1}$ (about 0.6 cpm), which is conveniently close to the definition of the microstructure boundary, then flattens out to the Kolmogorov dissipation scale $m_k = (\epsilon/\nu^3)^{1/4} \approx 3 \, \text{cm}^{-1}$ (0.5 cpcm), and finally cuts off exponentially (Gregg, Cox, and Hacker, 1973, figure 11). But such a description of "in-the-mean" scales may not be appropriate to a patchy environment, and is anyway beyond the scope of this survey.

9.10 Conclusion

I shall end as I started: the connection between internal waves and small scale processes—that is where the key is. I feel that we are close to having these pieces fall into place, and I am uncomfortable with having attempted a survey at this time.

Notes

1. They were found at Loch Ness at about the same time (Watson, 1904; Wedderburn, 1907).

2. The temperature measurements were made using a submarine cable from a recording Wheatstone bridge on the shore at Bermuda to two resistance thermometers offshore: both lay on the bottom, one at a depth of 50 m and the other at 500 m (Haurwitz, Stommel, and Munk, 1959).

3. Salinity is not directly measured, but has to be inferred from conductivity or sound speed, which are primarily responsive to temperature. This large temperature "correction" has been a source of some difficulty.

4. The notation $U' = \partial_z U$ refers to the ambient shear, and $u' = \partial_z u$ to the shear induced by the orbital wave motion. The distinction is not always so clear.

5. In general, *forward* flow refers to positive $\mathbf{c} \cdot (\mathbf{g} \times \text{curl}\,\mathbf{U})$, where $\mathbf{g}$ and $\mathbf{U}$ are the vectors of gravity and ambient velocity, respectively.

6. I am indebted to D. Broutman for very considerable improvements of this section, and for the preparation of figure 9.14.

7. The linearized calculations indicated that they are in fact not small.

8. Backward breaking occurs for negative shear; this can be visualized by turning the figure upside down.

9. Frankignoul's (1972) treatment of *finite* amplitude waves [his equations (24) and (28)] lead to precisely the same result.

10. Garrett points out that a lot of laboratory experiments have been sweet-and-sour rather than spicy.

11. It has been suggested that vibration and temperature contamination contributes to the high values from the towed devices; it has also been suggested that the dropped devices have inadequate dynamic range to measure ϵ in the highly active patches where most of the dissipation takes place.

12. A normal-mode formulation is applicable near the boundaries (Watson, Siegmann, and Jacobson, 1977).

13. Frequencies are in rad s^{-1}, wavenumbers in rad m^{-1}. For comparison with computed spectra we sometimes include (in parenthesis) the values in cycles per hour (cph) and cycles per meter (cpm).

14. We note that $F_e(\omega \gg f) \propto E(\omega, j) \propto \omega^{-2} fE$. There is some evidence that the spectral energy density is independent of latitude (Wunsch and Webb, 1979; Eriksen, 1980), and we should probably replace fE by $N_0 E'$, with $E' = (f_{30°}/N_0)E = 8.8 \times 10^{-7}$ the appropriate f-scaled energy parameter.

15. Desaubies (1973, 1975) explains the observed N-peak in the spectrum and the vertical coherence of vertical displacement.

16. This is related to the "intermittency index" evaluated by Thorpe (1977) from temperature inversions in Loch Ness.

17. "A disorderly state resulting from stress." This descriptive terminology is due to McEwan (1973).

10
Long Waves and Ocean Tides

Myrl C. Hendershott

10.1 Introduction

The main purpose of this chapter is to summarize what was generally known to oceanographers about long waves and ocean tides around 1940, and then to indicate how the subject has developed since then, with particular emphasis upon those aspects that have had significance for oceanography beyond their importance in understanding tides themselves. I have begun with a description of astronomical and radiational tide-generating potentials (section 10.2), but say no more than is necessary to make this chapter self-contained. Cartwright (1977) summarizes and documents recent developments, and I have followed his discussion closely.

The fundamental dynamic equations governing tides and long waves, Laplace's tidal equations (LTE), remained unchanged and unchallenged from Laplace's formulation of them in 1776 up to the early twentieth century. By 1940 they had been extended to allow for density stratification (in the absence of bottom relief) and criticized for their exclusion of half of the Coriolis forces. Without bottom relief this exclusion has recently been shown to be a good approximation; the demonstration unexpectedly requires the strong stratification of the ocean. Bottom relief appears able to make long waves in stratified oceans very different from their flat-bottom counterparts (section 10.4); a definitive discussion has not yet been provided. Finally, LTE have had to be extended to allow for the gravitational self-attraction of the oceans and for effects due to the tidal yielding of the solid earth. I review these matters in section 10.3.

Laplace's study of the free oscillations of a global ocean governed by LTE was the first study of oceanic long waves. Subsequent nineteenth- and twentieth-century explorations of the many free waves allowed by these equations, extended to include stratification, have evolved into an indispensible part of geophysical fluid dynamics. By 1940, most of the flat-bottom solutions now known had, at least in principle, been constructed. But Rossby's rediscovery and physical interpretation, in 1939, of Hough's oscillations of the second class began the modern period of studying solutions of the long-wave equations by inspired or systematic approximation and of seeking to relate the results to nontidal as well as tidal motions. Since then, flat-bottom barotropic and baroclinic solutions of LTE have been obtained in mid-latitude and in equatorial approximation, and Laplace's original global problem has been completely solved. The effects of bottom relief on barotropic motion are well understood. Significant progress has been made in understanding the effects of bottom relief on baroclinic motions. I have attempted to review all those developments in a self-contained manner in section 10.4. In order to treat this vast subject coherently, I have had to impose my own view of its development upon the discussion. I have cited observations when they appear to illustrate some property of the less familiar solutions, but the central theme is a description of the properties of theoretically possible waves of long period (greater than the buoyancy period) and, consequently, of length greater than the ocean's mean depth.

Although the study of ocean surface tides was *the* original study of oceanic response to time-dependent forcing, tidal studies have largely proceeded in isolation from modern developments in oceanography on account of the strength of the tide-generating forces, their well-defined discrete frequencies, and the proximity of these to the angular frequency of the earth's rotation. A proper historical discussion of the subject, although of great intellectual interest, is beyond the scope of this chapter. To my mind the elements of such a discussion, probably reasonably complete through the first decade of this century, are given in Darwin's 1911 *Encyclopedia Britannica* article "Tides." Thereafter, with a few notable exceptions, real progress had to await modern computational techniques both for solving LTE and for making more complete use of tide gauge observations. Cartwright (1977) has recently reviewed the entire subject, and therefore I have given a discussion in section 10.5 that, although self-contained, emphasizes primarily changes of motivation and viewpoint in tidal studies rather than recapitulates Cartwright's or other recent reviews.

This discussion of tides as long waves continuously forced by lunar and solar gravitation logically could be followed by a discussion of tsunamis impulsively forced by submarine earthquakes. But lack of both space and time has forced omission of this topic.

Internal tides were first reported at the beginning of this century. By 1940 a theoretical framework for their discussion had been supplied by the extension of LTE to include stratification, and their generation was (probably properly) ascribed to scattering of barotropic tidal energy from bottom relief. The important developments since then are recognition of the intermittent narrow-band nature of internal tides (as opposed to the near-line spectrum of surface tides) plus the beginnings of a statistically reliable characterization of the internal tidal spectrum and its variation in space and time. The subject has recently been reviewed by Wunsch (1975). Motivation for studying internal tides has shifted from the need for an adequate description of them through exploration of their role in global tidal dissipation (now believed to be under 10%) to speculation about their importance as energy sources for oceanic mixing. In section 10.6 I have summarized modern observational studies and their implications for tidal mixing of the oceans.

Many features of the presentday view of ocean circulation have some precedent in tidal and long-wave studies, although often unacknowledged and apparently not always recognized. The question of which parts of the study of tides have in fact influenced the subsequent development of studies of ocean circulation is a question for the history of science. In some cases, developments in the study of ocean circulation subsequently have been applied to ocean tides. In section 10.7 I have pointed out some of the connections of which I am aware.

10.2 Astronomical Tide-Generating Forces

Although correlations between ocean tides and the position and phase of the moon have been recognized and utilized since ancient times, the astronomical tide-generating force (ATGF) was first explained by Newton in the *Principia* in 1687. Viewed in an accelerated coordinate frame that moves with the center of the earth but that does not rotate with respect to the fixed stars, the lunar (solar) ATGF at any point on the earth's surface is the difference between lunar (solar) gravitational attraction at that point and at the earth's center. The daily rotation of the earth about its axis carries a terrestrial observer successively through the longitude of the sublunar or subsolar point [at which the lunar (solar) ATGF is toward the moon (sun)] and then half a day later through the longitude of the antipodal point [at which the ATGF is away from the moon (sun)]. In Newton's words, "It appears that the waters of the sea ought twice to rise and twice to fall every day, as well lunar or solar" [Newton, 1687, proposition 24, theorem 19].

The ATGF is thus predominantly semidiurnal with respect to both the solar and the lunar day. But it is not entirely so. Because the tide-generating bodies are not always in the earth's equatorial plane, the terrestrial observer [who does not change latitude while being carried through the longitude of the sublunar (solar) point or its antipode] sees a difference in amplitude between the successive semidiurnal maxima of the ATGF at his location. This difference or "daily inequality" means that the ATGF must be thought of as having diurnal as well as semidiurnal time variation.

Longer-period variations are associated with periodicities in the orbital motion of earth and moon. The astronomical variables displaying these long-period variations appear nonlinearly in the ATGF. The long-period orbital variations thus interact nonlinearly both with themselves and with the short-period diurnal and semidiurnal variation of the ATGF to make the local ATGF a sum of three narrow-band processes centered about 0, 1, and 2 cycles per day (cpd), each process being a sum of motions harmonic at multiples 0,1,2 of the frequencies corresponding to a lunar or a solar day plus sums of multiples of the frequencies of long-period orbital variations.

A complete derivation of the ATGF is beyond the scope of this discussion. Cartwright (1977) reviews the subject and supplies references documenting its modern development. For a discussion that concentrates upon ocean dynamics (but not necessarily for a practical tide prediction), the most convenient representation of the ATGF is as a harmonic decomposition of the tide-generating potential whose spatial gradient is the ATGF. Because only the horizontal components of the ATGF are of dynamic importance, it is convenient to represent the tide-generating potential by its horizontal and time variation U over some near-sea-level equipotential (the geoid) of the gravitational potential due to the earth's shape, internal mass distribution, and rotation. To derive the dynamically significant part of the ATGF it suffices to assume this surface spherical. U/g (where g is the local gravitational constant, unchanging over the geoid) has the units of sea-surface elevation and is called the *equilibrium tide* ζ. Its principle term is (Cartwright, 1977)

$$\zeta(\phi,\theta,t) = U(\phi,\theta,t)/g$$

$$= \sum_{m=0,1,2} (A_2^m \cos m\phi + B_2^m \sin m\phi) P_2^m, \quad (10.1)$$

in which ϕ,θ are longitude and latitude, the $P_2^m(\bar\theta)$ are associated Legendre functions

$$P_2^0 = \tfrac{1}{3}(3\cos^2\bar\theta - 1),$$
$$P_2^1 = 3\sin\bar\theta\cos\bar\theta, \quad (10.2)$$
$$P_2^2 = 3\sin^2\bar\theta$$

of colatitude $\bar\theta = (\pi/2) - \theta$, and A_2^m, B_2^m are functions of time having the form

$$A_n^m(t) = \sum_i M_i \, {\cos \atop \sin} \left[\sum_{j=1}^{6} N_j^{(i)} S_j(t)\right]. \quad (10.3)$$

The M_i are amplitudes obtained from Fourier analysis of the astronomically derived time series $U(\phi,\theta,t)/g$; the $N_j^{(i)}$ are sets of small integers (effectively the Doodson numbers); and the $S_j(t)$ are secular arguments that increase almost linearly in time with the associated periodicity of a lunar day, a sidereal month, a tropical year, 8.847 yr (period of lunar perigee), 18.61 yr (period of lunar node), 2.1×10^4 yr (period of perihelion), respectively.

The frequencies of the arguments $\Sigma N_j^{(i)} S_j(t)$ fall into the three "species"—long period, diurnal, and semidiurnal—which are centered, respectively, about 0, 1, and 2 cpd ($N_1 = 0,1,2$). Each species is split into "groups" separated by about 1 cycle per month, groups are split into "constituents" separated by one cyle per year, etc. Table 10.1 lists selected constituents. In the following discussion they are referred to by their Darwin symbol (see table 10.1).

An important development in modern tidal theory has been the recognition that the ATGF is not the only important tide-generating force. Relative to the amplitudes and phases of corresponding constituents of the equilibrium tide, solar semidiurnal, diurnal, and annual ocean tides usually have amplitudes and phases quite different from the amplitudes and phases of other nearby constituents. Munk and Cartwright (1966) attributed these anomalies in a general way to solar heating and included them in a generalized equilibrium tide by defining an ad hoc radiational potential (Cartwright, 1977)

$$U_R(\phi,\theta,t) = \begin{cases} S(\xi/\bar\xi)\cos\alpha, & 0 < \alpha < \pi/2: \\ 0, & \pi/2 < \alpha < \pi: \end{cases} \quad (10.4)$$

which is zero at night, which varies as the cosine of the sun's zenith angle α during the day, and which is proportional to the solar constant S and the sun's parallax ξ (mean $\bar\xi$). Cartwright (1977) suggests that for the oceanic S_2 (principal solar) tide, whose anomalous portion is about 17% of the gravitational tide (Zetler, 1971), the dominant nongravitational driving is by the atmospheric S_2 tide. Without entering further into the discussion, I want to point out that the global form of

Table 10.1 Characteristics of Selected Constituents of the Equilibrium Tide

Darwin symbol	N_1, N_2, N_3, N_4				Period (solar days or hours)	Amplitude M (m)	Spatial variation
S_{sa}	0	0	2	0	182.621 d	0.02936	
M_m	0	1	0	-1	27.55 d	0.03227	
M_f	0	2	0	0	13.661 d	0.0630	$\tfrac{1}{3}(3\cos^2\bar\theta - 1)$
O_1	1	-1	0	0	25.82 h	0.06752	
P_1	1	1	-2	0	24.07 h	0.03142	$3\sin\bar\theta\cos\bar\theta \times \sin(\omega_1 t + \phi)$
K_1	1	1	0	0	23.93 h	0.09497	
N_2	2	-1	0	1	12.66 h	0.01558	
M_2	2	0	0	0	12.42 h	0.08136	$3\sin^2\bar\theta \times \cos(\omega_2 t + 2\phi)$
S_2	2	2	-2	0	12.00 h	0.03785	
K_2	2	2	0	0	11.97 h	0.01030	

the atmospheric S_2 tide is well known (Chapman and Lindzen, 1970), so that the same numerical programs that have been used to solve for global gravitationally driven ocean tides could easily be extended to allow for atmospheric-pressure driving of the oceanic S_2 tide.

Ocean gravitational self-attraction and tidal solid-earth deformation are quantitatively even more important in formulating the total tide-generating force than are thermal and atmospheric effects. They are discussed in the following section, since they bring about a change in the form of the dynamic equations governing ocean tides.

10.3 Laplace's Tidal Equations (LTE) and the Long-Wave Equations

Laplace (1775, 1776; Lamb, 1932, §213-221) cast the dynamic theory of tides essentially in its modern form. His tidal equations (LTE) are usually formally obtained from the continuum equations of momentum and mass conservation (written in rotating coordinates for a fluid shell surrounding a nearly spherical planet and having a gravitationally stabilized free surface) by assuming (Miles, 1974a)

(1) a perfect homogeneous fluid,
(2) small disturbances relative to a state of uniform rotation,
(3) a spherical earth,
(4) a geocentric gravitational field uniform horizontally and in time,
(5) a rigid ocean bottom,
(6) a shallow ocean in which both the Coriolis acceleration associated with the horizontal component of the earth's rotation and the vertical component of the particle acceleration are neglected.

The resulting equations are

$$\frac{\partial u}{\partial t} - 2\Omega \sin\theta\, v = -\frac{\partial}{\partial \phi}(\zeta - \Gamma/g)/a\cos\theta, \quad (10.5a)$$

$$\frac{\partial v}{\partial t} + 2\Omega \sin\theta\, u = -\frac{\partial}{\partial \theta}(\zeta - \Gamma/g)/a, \quad (10.5b)$$

$$\frac{\partial \zeta}{\partial t} + \frac{1}{a\cos\theta}\left[\frac{\partial}{\partial \phi}(uD) + \frac{\partial}{\partial \theta}(vD\cos\theta)\right] = 0. \quad (10.5c)$$

In these, (ϕ, θ) are longitude and latitude with corresponding velocity components (u, v), ζ the ocean surface elevation, Γ the tide-generating potential, $D(\phi, \theta)$ the variable depth of the ocean, a the earth's spherical radius, g the constant gravitational attraction at the earth's surface, and Ω the earth's angular rate of rotation.

Two modern developments deserve discussion. They are a quanitative formulation and study of the mathematical limit process implicit in assumptions (1) through (6), and the realization that assumptions (4) and (5) are quantitatively inadequate for a dynamic discussion of ocean tides.

It has evidently been recognized since the work of Bjerknes, Bjerknes, Solberg, and Bergeron (1933) that assumption (6) (especially the neglect of Coriolis forces due to the horizontal component of the earth's rotation) amounts to more than a minor perturbation of the spectrum of free oscillations that may occur in a thin homogeneous ocean. Thus Stern (1963) and Israeli (1972) found axisymmetric equatorially trapped normal modes of a rotating spherical shell of homogeneous fluid that are extinguished by the hydrostatic approximation. Indeed, Stewartson and Rickard (1969) point out that the limiting case of a vanishingly thin homogeneous ocean is a nonuniform limit: the solutions obtained by solving the equations and then taking the limit may be very different from those obtained by first taking the limit and then solving the resulting approximate (LTE) equations. Quite remarkably, it is the reinstatement of realistically large stratification, i.e., the relaxation of assumption (1), that saves LTE as an approximate set of equations whose solutions are uniformly valid approximations to some of the solutions of the full equations when the ocean is very thin.

The parodoxical importance of stratification for the validity of the ostensibly unstratified LTE appears to have been recognized by Proudman (1948) and by Bretherton (1964). Phillips (1968) pointed out its importance at the conclusion of a correspondence with Veronis (1968b) concerning the effects of the "traditional" approximation (Eckart, 1960; N. A. Phillips, 1966b), i.e., the omission of the Coriolis terms $2\Omega\cos\theta\, w$ and $-2\Omega\cos\theta\, u$, in (10.6)-(10.8) below. But it was first explicitly incorporated into the limit process producing LTE by Miles (1974a) who addressed all of assumptions (1) through (6) by defining appropriate small parameters and examining the properties of expansions in them. He found that the simplest set of uniformly valid equations for what I regard as long waves in this review are

$$\frac{\partial u}{\partial t} - 2\Omega\sin\theta\, v + 2\Omega\cos\theta\, w = -\frac{\partial p}{\partial \phi}\frac{1}{\overline{\rho}_0 a\cos\theta}, \quad (10.6a)$$

$$\frac{\partial v}{\partial t} + 2\Omega\sin\theta\, u = -\frac{\partial p}{\partial \theta}\frac{1}{\overline{\rho}_0 a}, \quad (10.6b)$$

$$N^2 w - 2\Omega\cos\theta\,\frac{\partial u}{\partial t} = -\frac{\partial^2 p}{\partial z\, \partial t}\frac{1}{\overline{\rho}_0}, \quad (10.6c)$$

$$\frac{\partial u}{\partial \phi} + \frac{\partial(v\cos\theta)}{\partial \theta} + a\cos\theta\,\frac{\partial w}{\partial z} = 0, \quad (10.6d)$$

with boundary conditions

$$w = 0 \quad \text{at} \quad z = -D_*$$

(for uniform depth D_*), (10.7)

$$p = -\bar{\rho}_0 g \zeta_0 \quad \text{at} \quad z = 0. \tag{10.8}$$

In these, z is the upward local vertical with associated velocity component w, ρ the deviation of the pressure from a resting hydrostatic state characterized by the stable density distribution $\rho_0(z)$, $\bar{\rho}_0$ a constant characterizing the mean density of the fluid (the Boussinesq approximation has been made in its introduction); the buoyancy frequency

$$N(z) = \left(-\frac{g}{\rho_0} \frac{\partial \rho_0}{\partial z} - \frac{g^2}{c^2} \right)^{1/2} \tag{10.9}$$

is the only term in which allowance for compressibility (c is the local sound speed) is important in the ocean.

Miles (1974a) further found that when $N^2(z) \gg 4\Omega^2$, free solutions of LTE for a uniform-depth (D_*) ocean covering the globe also solve (10.6)–(10.8) with an error which is of order $(\sigma/2\Omega)(4\Omega^2 a/g) \ll 1$, where σ is the frequency of oscillation of the free solutions. LTE surface elevations ζ are consequently in error by order $(D_*/a)(D_* N^2/g)^{-3/4}$, while LTE velocities u,v may be in error by $(D_*/a)(D_* N^2/g)^{-1/4}$. Miles (1974a) obtained this result by taking the (necessarily barotropic) solutions of LTE as the first term of an expansion of the solutions of (10.6)–(10.8) in the parameter $(\sigma/2\Omega)(4\Omega^2 a/g)$, which turns out to characterize the relative importance of terms neglected and terms retained in making assumption (6). The next term in the expansion consists of internal wave modes (see section 10.4.3). Because their free surface displacements are very small relative to internal displacements, the overall free surface displacement remains as in LTE although in the interior of the ocean, internal wave displacements and currents may well dominate the motion.

The analysis is inconclusive at frequencies or depths at which the terms assumed to be correction perturbations are resonant. Finding expressions for all the free oscillations allowed by Miles's simplest uniformly valid system (10.6)–(10.8) involves as yet unresolved mathematical difficulties associated with the fact that these equations are hyperbolic over part of the spatial domain when the motion is harmonic in time (Miles, 1974a).

Application of this analysis to ocean tides is further circumscribed by its necessary restriction to a global ocean of constant depth. I speculate that if oceanic internal modes are sufficiently inefficient as energy transporters that they cannot greatly alter the energetics of the barotropic solution unless their amplitudes are resonantly increased beyond observed levels (section 10.6), and if they are sufficiently dissipative that they effectively never are resonant, then an extension of this analysis to realistic basins and relief would probably confirm LTE as adequate governors of the surface elevation. The ideas, necessary for such an extension, that is, how variable relief and stratification influence barotropic and baroclinic modes, are beginning to be developed (see section 10.4.7).

Miles (1974a) discusses assumptions (3)–(5) with explicit omission of ocean gravitational self-attraction and solid-earth deformation. Self-attraction was included in Hough's (1897, 1898; Lamb, 1932, §222–223) global solutions of LTE. Thomson (1863) evidently first pointed out the necessity of allowing for solid-earth deformation. Both are quantitatively important. The latter manifests itself in a geocentric solid-earth tide δ plus various perturbations of the total tide-generating potential Γ. Horizontal pressure gradients in LTE (10.5) are associated with gradients of the geocentric ocean tide ζ, but it is the observed ocean tide

$$\zeta_0 = \zeta - \delta \tag{10.10}$$

that must appear in the continuity equation of (10.5c).

All these effects are most easily discussed (although not optimally computed) when the astronomical potential U, the observed ocean tide ζ_0, the solid earth tide δ, and the total tide-generating potential Γ are all decomposed into spherical harmonic components U_n, ζ_{0n}, δ_n, and Γ_n. The Love numbers k_n, h_n, k'_n, h'_n, which carry with them information about the radial structure of the solid earth (Munk and McDonald, 1960), and the parameter $\alpha_n = (3/2n + 1)(\bar{\rho}_{\text{ocean}}/\bar{\rho}_{\text{earth}})$ then appear naturally in the development. The total tide-generating potential Γ_n contains an astronomical contribution U_n (primarily of order $n = 2$), an augmentation $k_n U_n$ of this due to solid earth yielding to $-\nabla U_n$, an ocean self-attraction contribution $g\alpha_n \zeta_{0n}$, and a contribution $k'_n g \alpha_n \zeta_{0n}$ due to solid-earth deformation by ocean self-attraction and tidal column weight. Thus

$$\Gamma_n = (1 + k_n) U_n + (1 + k'_n) g \alpha_n \zeta_{0n}. \tag{10.11}$$

There is simultaneously a geocentric solid-earth tide δ made up of the direct yielding $h_n U_n g$ of the solid earth to $-\nabla U_n$ plus the deformation $h'_n g \alpha_n \zeta_{0n}$ of the solid earth by ocean attraction and tidal column weight. Thus

$$\delta_n = h_n U_n/g + h'_n g \alpha_n \zeta_{0n}. \tag{10.12}$$

For computation, Farrell (1972a) has constructed a Green's function such that

$$\sum_n (1 + k'_n - h'_n) \alpha_n \zeta_{0n}$$
$$= \iint_{\text{ocean}} d\theta' d\phi' \cos \theta' \, G(\phi', \theta' | \phi, \theta) \zeta_0(\phi', \theta'). \tag{10.13}$$

With $U_n = U_2$ and with (10.13) abbreviated as $\iint G\zeta_0$, LTE with assumptions (4) and (5) appropriately relaxed become

$$\frac{\partial u}{\partial t} - 2\Omega \sin\theta\, v = -\frac{g}{a\cos\theta}\frac{\partial}{\partial \phi}[\zeta_0 - (1 + k_2 - h_2)U_2/g$$

$$-\iint G\zeta_0], \qquad (10.14a)$$

$$\frac{\partial v}{\partial t} + 2\Omega \sin\theta\, u = -\frac{g}{a}\frac{\partial}{\partial \theta}[\zeta_0 - (1 + k_2 - h_2)U_2/g$$

$$-\iint G\zeta_0], \qquad (10.14b)$$

$$\frac{\partial \zeta_0}{\partial t} + \frac{1}{a\cos\theta}\left[\frac{\partial (uD)}{\partial \phi} + \frac{\partial (vD\cos\theta)}{\partial \theta}\right] = 0. \qquad (10.14c)$$

The factor $1 + k_2 - h_2 \simeq 0.69$ is clearly necessary for a quantitatively correct solution. The term $\iint G\zeta_0$ was first evaluated by Farrell (1972b). I wrote down (10.14) and attempted to estimate the effect of $\iint G\zeta_0$ on a global numerical solution of (10.14) for the M_2 tide by an iterative procedure which, however, failed to converge (Hendershott, 1972). Subsequent computations by Gordeev, Kagan, and Polyakov (1977) and by Accad and Pekeris (1978) provide improved estimates of the effects (see section 10.5.3).

With appropriate allowance for various dissipative processes (including all mechanisms that put energy into internal tides), I regard (10.14) as an adequate approximation for studying the ocean surface tide. Oceanic long waves should really be discussed using (10.6) but allowing for depth variations by putting

$$w = \mathbf{u}\cdot\nabla D \quad \text{at} \quad z = -D(\phi,\theta) \qquad (10.15)$$

in place of (10.7). Miles (1974a) derives an orthogonality relationship that could be specialized to (10.6)–(10.8) in the case of constant depth, but even then the nonseparability of the eigenfunctions into functions of (ϕ,θ) times functions of z has prevented systematic study of the problem. Variable relief compounds the difficulty. Most studies either deal with surface waves over bottom relief, and thus start with LTE (10.5), or else with surface and internal waves over a flat bottom. In the latter case, (10.6)–(10.8) are solved but with the Coriolis terms $2\Omega\cos\theta\, w$ and $-2\Omega\cos\theta\, u$ arbitrarily neglected (the "traditional" approximation). Miles' (1974a) results appear to justify this procedure for the former case, but Munk and Phillips (1968) show that the neglected terms are proportional to (mode number)$^{1/3}$ for internal modes so that the traditional approximation may be untenable for high-mode internal waves. The following discussion (section 10.4) of oceanic long waves relies heavily upon the traditional approximation, but it is important to note that its domain of validity has not yet been entirely delineated.

10.4 Long Waves in the Ocean

10.4.1 Introduction

The first theoretical study of oceanic long waves is due to Laplace (1775, 1776; Lamb, 1932, §213–221), who solved LTE for an appropriately shallow ocean covering a rotating rigid spherical earth by expanding the solution in powers of $\sin\theta$. For a global ocean of constant depth D_*, Hough (1897, 1898; Lamb, 1932, §222–223) obtained solutions converging rapidly for small

$$\Lambda = 4\Omega^2 a^2/gD_* \qquad (10.16)$$

(sometimes called Lamb's parameter) by expanding the solution in spherical harmonics $P_n^l(\sin\theta)\exp(il\phi)$. He found the natural oscillations to be divided into first- and second-class modes whose frequencies σ are given, respectively, by

$$\sigma = \pm[n(n+1)gD_*/a^2]^{1/2},$$

$$\sigma = -2\Omega l/[n(n+1)] \qquad (10.17)$$

as $\Lambda \to 0$. But $\Lambda \simeq 20$ for $D_* = 4000$ m, and is very much larger for internal waves (see section 10.4.3). A correspondingly complete solution of Laplace's problem, valid for large as well as small Λ, was given only recently (Flattery, 1967; Longuet-Higgins, 1968a). Physical understanding of the solutions has historically been developed by studying simplified models of LTE.

10.4.2 Long Waves in Uniformly Rotating Flat-Bottomed Oceans

Lord Kelvin (Thomson, 1879; Lamb, 1932, §207) introduced the idealization of uniform rotation, at Ω, of a sheet of fluid about the vertical (z-axis). LTE become

$$\frac{\partial u}{\partial t} - f_0 v = -g\frac{\partial \zeta}{\partial x}, \qquad (10.18a)$$

$$\frac{\partial v}{\partial t} + f_0 u = -g\frac{\partial \zeta}{\partial y}, \qquad (10.18b)$$

$$\frac{\partial \zeta}{\partial t} + \frac{\partial (uD)}{\partial x} + \frac{\partial (vD)}{\partial y} = 0. \qquad (10.18c)$$

f_0, here equal to 2Ω, is the Coriolis parameter. Lord Kelvin's plane model (10.18) is often called the f-plane.

The solid earth is nearly a spheroid of equilibrium under the combined influence of gravity g and centrifugal acceleration $\Omega^2 a$; the earth's equatorial radius is about 20 km greater than its polar radius. Without water motion, the sea surface would have a congruent spheroidal shape. Taking the depth constant in LTE models this similarity; the remaining error incurred by working in spherical rather than spheroidal coordinates is (Miles, 1974a) of order $\Omega^2 a/g = 10^{-3}$. It is correspondingly appropriate to take the depth constant in Lord Kelvin's plane model (10.18) in order to obtain planar solutions locally modeling those of LTE with constant depth. But the laboratory configuration corresponding to constant depth in (10.18c) is a container with a paraboloidal bottom $\frac{1}{2}\Omega^2(x^2 + y^2)/g$ rather than a flat bottom [see Miles (1964) for a more detailed analysis].

Without rotation ($f_0 = 0$) and with constant depth ($D = D_*$) Lord Kelvin's plane model reduces to the linearized shallow-water equations (LSWE). For an unbounded fluid sheet, they have plane gravity-wave solutions

$$\zeta = a \exp(-i\sigma t + ilx + iky), \quad (10.19a)$$

$$u = (gl/\sigma)\zeta, \quad (10.19b)$$

$$v = (gk/\sigma)\zeta, \quad (10.19c)$$

$$w = -i\sigma(z/D_* + 1)\zeta, \quad (10.19d)$$

$$\sigma^2 = gD_*(l^2 + k^2), \quad (10.19e)$$

which are dispersionless [all travel at $(gD_*)^{1/2}$] and are longitudinal [(u, v) parallel to (l, k)]. Horizontal particle accelerations are exactly balanced by horizontal pressure-gradient forces while vertical accelerations are negligible. Such waves reflect specularly at a straight coast with no phase shift; thus

$$\zeta = a \exp(-i\sigma t + ilx + iky)$$
$$+ a \exp(-i\sigma t - ilx + iky) \quad (10.20)$$

satisfies $u = 0$ at the coast $x = 0$, the angle $\tan^{-1}(k/l)$ of incidence equals the angle of reflection, and the (complex) reflected amplitude equals the incident amplitude a. The normal modes of a closed basin with perimeter P are the eigensolutions $Z_n(x,y)\exp(-i\sigma_n t)$, σ_n of

$$\nabla_2^2 Z_n + (\sigma_n^2/gD_*)Z_n = 0 \quad (10.21)$$

with

$$\partial Z_n/\partial(\text{normal}) = 0 \quad \text{on} \quad P. \quad (10.22)$$

If P is a constant surface in one of the coordinate systems in which ∇_2^2 separates, then the normal modes are mathematically separable functions of the two horizontal space coordinates and so are readily discussed in terms of appropriate special functions. Even in general basin shapes, the existence and completeness of the normal modes are assured (Morse and Feshbach, 1953).

With rotation, plane wave solutions of (10.18) with constant depth D_* are

$$\zeta = a \exp(-i\sigma t + ilx + iky), \quad (10.23a)$$

$$u = g[(l\sigma + ikf_0)/(\sigma^2 - f_0^2)]\zeta, \quad (10.23b)$$

$$v = g[(k\sigma - ilf_0)/(\sigma^2 - f_0^2)]\zeta, \quad (10.23c)$$

$$w = -i\sigma(z/D_* + 1)\zeta, \quad (10.23d)$$

$$\sigma^2 = gD_*(l^2 + k^2) + f_0^2. \quad (10.23e)$$

These are often called Sverdrup waves (apparently after Sverdrup, 1926). Rotation has made them dispersive and they propagate only when $\sigma^2 > f_0^2$. The group velocity $\mathbf{c}_g = (\partial\sigma/\partial l, \partial\sigma/\partial k)$ is parallel to the wavenumber and rises from zero at $\sigma = f_0$ toward $(gD_*)^{1/2}$ as $\sigma^2 \gg f_0^2$. Dynamically these waves are LSW waves perturbed by rotation. Particle paths are ellipses with ratio f/σ of minor to major axis and with major axis oriented along (l, k). Particles traverse these paths in the clockwise direction (viewed from above) when $f > 0$ (northern hemisphere). Sverdrup waves are reflected specularly at a straight coast but with a phase shift; the sum

$$\zeta = a \exp(i\sigma t + ilx + iky)$$
$$+ a[(l\sigma + ikf_0)/(l\sigma - ikf_0)]$$
$$\times \exp(-i\sigma t - ilx + iky) \quad (10.24)$$

of two Sverdrup waves satisfies $u = 0$ at the coast $x = 0$, the angle $\tan^{-1}(k/l)$ of incidence equals the angle of reflection, and the reflected amplitude differs from the incident amplitude by the multiplicative constant $(l\sigma + ikf_0)/(l\sigma - ikf_0)$, which is complex but of modulus unity. The sum (10.24) is often called a *Poincaré wave*. The normal modes of a closed basin with perimeter P are the eigensolutions $Z_n(x,y)\exp(-i\sigma_n t)$, σ_n of

$$\nabla_2^2 Z_n + [(\sigma_n^2 - f_0^2)/gD_*]Z_n = 0 \quad (10.25)$$

with

$$-i\sigma_n \partial Z_n/\partial(\text{normal})$$
$$+ f_0 \partial Z_n/\partial(\text{tangent}) = 0 \quad \text{at} \quad P. \quad (10.26)$$

On account of the boundary condition (10.26), they are not usually separable functions of the two horizontal space coordinates. The circular basin (Lamb, 1932, §209-210) is an exception. Rao (1966) discusses the rectangular basin, but the results are not easily summarized. A salient feature, the existence of free oscillations with $\sigma^2 < f_0^2$, is rationalized below.

With rotation, Lord Kelvin (Thomson, 1879; Lamb, 1932, §208) showed that a coast not only reflects Sverdrup waves for which $\sigma^2 > f_0^2$, but makes possible a new kind of coastally trapped motion for which $\sigma^2 \gtrless f_0^2$. This Kelvin wave has the form

$$\zeta = a \exp[-i\sigma t + iky + k(f_0/\sigma)x], \quad (10.27a)$$

$$u = 0, \quad (10.27b)$$

$$v = (gk/\sigma)\zeta, \quad (10.27c)$$

$$w = -i\sigma(z/D_* + 1)\zeta, \quad (10.27d)$$

$$\sigma^2 = gD_* k^2 \quad (10.27e)$$

along the straight coast $x = 0$. The velocity normal to the coast vanishes everywhere in the fluid and not only at the coast. The wave is dispersionless and propagates parallel to the shore with speed $(gD_*)^{1/2}$ for $\sigma^2 \gtrless f_0^2$ just like a longitudinal gravity wave but with an offshore profile $\exp[k(f_0/\sigma)x]$ that decays or grows exponentially seaward depending upon whether the wave propagates with the coast to its right or to its left (in the northern hemisphere, $f_0 > 0$). For vanishing rotation, the offshore decay or growth scale becomes infinite and the Kelvin wave reduces to an ordinary gravity wave propagating parallel to the coast. The Kelvin wave is dynamically exactly a LSW gravity wave in the longshore direction and is exactly geostrophic in the cross-shore direction.

A pair of Kelvin waves propagating in opposite directions along the two coasts of an infinite canal (at, say, $x = 0$ and $x = W$) gives rise to a pattern of sealevel variation in which the nodal lines that would occur without rotation shrink to amphidromic points, at which the surface neither rises nor falls and about which crests and troughs rotate counterclockwise (in the northern hemisphere) as time progresses. For equal-amplitude oppositely propagating Kelvin waves, the amphidromic points fall on the central axis of the canal and are separated by a half-wavelength (πk^{-1}). When the amplitudes are unequal, the line of amphidromes moves away from the coast along which the highest-amplitude Kelvin wave propagates. For a sufficiently great difference in amplitudes, the amphidromes may occur beyond one of the coasts, i.e., outside of the canal.

Such a pair of Kelvin waves cannot by themselves satisfy the condition of zero normal fluid velocity in a closed canal (say at $y = 0$). Taylor (1921) showed how this condition could be satisfied by adjoining to the pair of Kelvin waves an infinite sum of channel Poincaré modes

$$\zeta = [\cos(m\pi x/W) - (f_0/\sigma)(kW/m\pi)$$
$$\times \sin(m\pi x/W)] \exp[-i\sigma t \pm iky], \quad (10.28)$$
$$\sigma^2 = (m^2\pi^2/W^2 + k^2)gD_* + f_0^2, \quad m = 1, 2, \ldots,$$

each of which separately has vanishing normal fluid velocity at the channel walls $(x = 0, W)$. These are just the waveguide modes of the canal. Mode m decays exponentially away from the closure (is evanescent) if

$$\sigma^2 < f_0^2 + (m^2\pi^2/W^2)gD_*. \quad (10.29)$$

If σ is so low, W so small, or D so great that all modes $m = 1, 2, \ldots$ are evanescent, then the Kelvin wave incident on the closure has to be perfectly reflected with at most a shift of phase. When (10.29) is violated for the one or more lowest modes, then some of the energy of the incident Kelvin wave is scattered into traveling Poincaré modes.

If (10.29) is satisfied for all m and if the decay scale $(gD_*)^{1/2}/f_0$ of the Kelvin waves is a good deal smaller than the channel width W, then the Poincaré modes sum to an appreciable contribution only near the corners of the closure. The Kelvin wave then proceeds up the channel effectively hugging one coast, turns the corners of the closure with a phase shift [evaluated by Buchwald (1968) for a single corner], and returns back along the channel hugging the opposite coast. Now it becomes apparent that, with allowance for corner phase shifts, closed rotating basins have a class of free oscillation whose natural frequencies are effectively determined by fitting an integral number of Kelvin waves along the basin perimeter. Such free oscillations may have $\sigma^2 \lesssim f_0^2$. They are readily identified in Lamb's (1932, §209–210) normal modes of a uniform-depth circular basin.

10.4.3 The Effect of Density Stratification on Long Waves

All of the foregoing solutions are barotropic surface waves. Stokes (1847; Lamb, 1932, §231) pointed out that surface waves are dynamically very much like waves at the interfaces between fluid layers of differing densities. Allowance for continuous vertical variation of density was made by Rayleigh (1883). Lord Kelvin's plane model (10.18) must be extended to read

$$\frac{\partial u}{\partial t} - f_0 v = -\frac{1}{\bar{\rho}_0}\frac{\partial p}{\partial x}, \quad (10.30a)$$

$$\frac{\partial v}{\partial t} + f_0 u = -\frac{1}{\bar{\rho}_0}\frac{\partial p}{\partial y}, \quad (10.30b)$$

$$N^2 w = -\frac{1}{\bar{\rho}_0}\frac{\partial^2 p}{\partial z\, \partial t}, \quad (10.30c)$$

$$\frac{\partial u}{\partial x} + \frac{\partial v}{\partial y} + \frac{\partial w}{\partial z} = 0, \quad (10.30d)$$

with

$$w = 0 \quad \text{at} \quad z = -D_* \quad (10.31)$$

and

$$p = \bar{\rho}_0 g \zeta \quad \text{at} \quad z = 0. \quad (10.32)$$

Notation is as in (10.6).

For the case of constant depth D_*, the principal result is that the dependent variables have the separable form

$$(u, v, w, p) = \{U(x,y,t), V(x,y,t), W(x,y,t), Z(x,y,t)\}$$
$$\times \{F_u(z), F_u(z), F_w(z), F_p(z)\}^T \quad (10.33)$$

where

$$W = \partial Z/\partial t,$$
$$F_u = F_p/\bar{\rho}_0 g = D_n\, \partial F_w/\partial z, \quad (10.34)$$

and

$$\frac{\partial U}{\partial t} - f_0 V = -g \frac{\partial Z}{\partial x}, \quad (10.35a)$$

$$\frac{\partial V}{\partial t} + f_0 U = -g \frac{\partial Z}{\partial y}, \quad (10.35b)$$

$$\frac{\partial Z}{\partial t} + D_n \left(\frac{\partial U}{\partial x} + \frac{\partial V}{\partial y} \right) = 0, \quad (10.35c)$$

with

$$\frac{\partial^2 F_w}{\partial z^2} + \frac{N^2(z)}{g D_n} F_w = 0, \quad (10.36)$$

$$F_w = 0 \quad \text{at} \quad z = -D_*, \quad (10.37)$$

$$F_w - D_n \partial F_w / \partial z = 0 \quad \text{at} \quad z = 0. \quad (10.38)$$

According to (10.35), the horizontal variations of all quantities are exactly as in the homogeneous flat bottom case except that now the apparent depth D_n is obtained by solving the eigenvalue problem (10.36) for the vertical structure. Typically (10.36) yields a barotropic mode $F_{w0} \simeq z + D_*$, $D_0 \simeq D_*$ plus an infinite sequence of baroclinic modes F_{wn} characterized by n zero crossings (excluding the one at $z = -D_*$) and by very small equivalent depths D_n. Baroclinic WKB approximate solutions (10.36)–(10.38) are

$$F_{wn}(z) = N^{-1/2}(z) \sin \left[(1/g D_n)^{1/2} \int_{-D_*}^{z} N(z') dz' \right],$$

$$D_n = \left[\int_{-D_*}^{0} N(z') dz' \right]^2 / (g n^2 \pi^2). \quad (10.39)$$

These are exact for constant buoyancy frequency N_0. For $D_* = 4000$ m and $N_0 = 10\Omega$, $D_n = (0.1/n^2)$ m. When $N(z) = N_0$, it is easy to show that

[w at the free surface/interior maximum value

of w] $= N_0^2 D_* / n \pi g \ll 1. \quad (10.40)$

Free surface variation is thus qualitatively and quantitatively unimportant for baroclinic modes. They are therefore usually called internal modes.

In a flat-bottomed ocean, stratification is thus seen to make possible an infinite sequence of internal replicas of the barotropic LSW gravity waves, Sverdrup waves, Poincaré waves, Kelvin waves, and basin normal modes discussed above. All these except the Kelvin waves have $\sigma^2 > f_0^2$. They must also have $\sigma^2 < N^2(z)$ over part of the water column, although the present treatment does not make this obvious because $\sigma^2 \ll N^2$ is always assumed. The horizontal variation of these replicas is governed by the equations describing the barotropic mode, except that the depth D_n is a small fraction of the actual (constant) depth D_*. Without rotation, the speed of barotropic long gravity waves is $(gD_*)^{1/2} \simeq 200$ m s^{-1} in the deep sea. Long internal gravity waves move at the much slower $(gD_n)^{1/2} \simeq$ (1/n) m s^{-1}. For comparable frequencies, the internal waves thus have much shorter wavelength than the surface wave.

An important point is that the separation of variables (10.33) works in spherical coordinates as well as in Cartesian coordinates, provided only that the depth is constant. The horizontal variation of flow variables is then governed by LTE with appropriate equivalent depth D_n given by (10.36)–(10.38).

For plane waves of frequency σ, relaxation of $N^2 \gg \sigma^2$ leads to the replacement of (10.36) by

$$\frac{\partial^2 F_w}{\partial z^2} + \frac{[N^2(z) - \sigma^2]}{g D_n} F_w = 0. \quad (10.41)$$

High-frequency waves, for which $[N^2(z) - \sigma^2]$ changes sign over the water column, are discussed in chapter 9.

The simplicity of these flat-bottom results is deceptive, because they are very difficult to generalize to include bottom relief. The reason for this is most easily seen by eliminating (u, v) from (10.30) for harmonic motion [$\exp(-i\sigma t)$] to obtain a single equation in w:

$$\frac{\partial^2 w}{\partial z^2} - \left(\frac{N_0^2}{\sigma^2 - f_0^2} \right) \left(\frac{\partial^2 w}{\partial x^2} + \frac{\partial^2 w}{\partial y^2} \right) = 0. \quad (10.42)$$

This equation is hyperbolic in space for internal waves (for which $f_0^2 < \sigma^2 < N_0^2$). Its characteristic surfaces are

$$z = \pm (x^2 + y^2)^{1/2} (\sigma^2 - f_0^2)^{1/2} / N_0. \quad (10.43)$$

Solutions of (10.42) may be discontinuous across characteristic surfaces, and they depend very strongly upon the relative slope of characteristics and bounding surfaces. Without rotation, internal waves of frequency σ are solutions of the hyperbolic equation

$$\frac{\partial^2 p}{\partial z^2} - \frac{(N_0^2 - \sigma^2)}{\sigma^2} \left(\frac{\partial^2 p}{\partial x^2} + \frac{\partial^2 p}{\partial y^2} \right) = 0 \quad (10.44)$$

with the simple condition

$$\frac{\partial p}{\partial n} = 0 \quad \text{at solid boundaries.} \quad (10.45)$$

For closed boundaries (i.e., a container filled with stratified fluid) this is an ill-posed problem in the sense that tiny perturbations of the boundaries may greatly alter the structure of the solutions. Horizontal boundaries, although analytically tractable, are a very special case.

10.4.4 Rossby and Planetary Waves

In an influential study whose emphasis upon physical processes marks the beginning of the modern period, Rossby and collaborators (1939) rediscovered Hough's second-class oscillations and suggested that they might be of great importance in atmospheric dynamics (see also chapters 11 and 18).

Perhaps of even greater influence than this discovery was Rossby's creation of a new plane model. It amounts to LTE written in the Cartesian coordinates

$$x = (a \cos \theta_0)(\phi - \phi_0), \qquad y = a(\theta - \theta_0) \tag{10.46}$$

tangent to the sphere at (ϕ_0, θ_0). Rossby's creative simplification was to ignore the variation of all metric coefficients $(\cos \theta = \cos \theta_0)$ and to retain the latitude variation of

$$f = 2\Omega \sin \theta \simeq 2\Omega \sin \theta_0 + y(2\Omega/a) \cos \theta_0$$
$$= f_0 + \beta y \tag{10.47}$$

only when f is explicitly differentiated with respect to y. Rossby's notation

$$\beta \equiv \partial f/\partial y \tag{10.48}$$

has since become almost universal. Such Boussinesq-like approximations to the spherical equations are usually called β-plane equations.

Rossby's original approximation, which I shall call Rossby's β-plane, further suppressed horizontal divergence ($\zeta = 0$). It yields rational approximations (Miles, 1974b) for second-class solutions of LTE whose horizontal scale is much smaller than the earth's radius. A quite different approximation yielding rational approximations for both first- and second-class solutions of LTE when they are equatorially trapped is the equatorial β-plane (see section 10.4.5). Both Rossby's β-plane equations and the equatorial β-plane equations differ from those obtained by the often encountered procedure of making Rossby's simplification but retaining divergence. This results in what I shall call simply the β-plane equations, in conformity with common usage. It is a rational approximation to LTE only at low ($\sigma^2 \ll f_0^2$) frequencies and is otherwise best regarded as a model of LTE.

Without divergence, the homogeneous LTE (10.5) may be cross-differentiated to yield a vorticity equation

$$\frac{\partial \nabla_{2s}^2 \psi}{\partial t} + 2\Omega \frac{\partial \psi}{\partial \phi} = 0$$
$$\left[\nabla_{2s}^2 \equiv \frac{1}{a^2 \cos^2 \theta} \frac{\partial^2}{\partial \phi^2} + \frac{1}{a^2 \cos \theta} \frac{\partial}{\partial \theta} \left(\cos \theta \frac{\partial}{\partial \theta} \right) \right] \tag{10.49}$$

here written in terms of a streamfunction ψ defined by

$$v \cos \theta = \partial \psi / \partial \phi, \qquad u = -\partial \psi / \partial \theta. \tag{10.50}$$

Rossby's β-plane approximation to this, obtained by locally approximating the spherical Laplacian ∇_{2s}^2 as the plane Laplacian and going to the locally tangent coordinates (10.46), is

$$\frac{\partial \nabla_2^2 \psi}{\partial t} + \beta \frac{\partial \psi}{\partial x} = 0, \tag{10.51}$$

where

$$v = \partial \psi / \partial x, \qquad u = -\partial \psi / \partial y. \tag{10.52}$$

The spherical vorticity equation (10.49) has as everywhere bounded solutions

$$\psi = S_n(\phi', \theta'), \qquad S_n(\phi, \theta) \equiv P_n^l(\sin \theta) \exp(il\phi), \tag{10.53}$$

where (ϕ', θ') are spherical coordinates relative to a pole P' displaced an arbitrary angle from the earth's pole P of rotation and rotating about P with angular velocity

$$c = -2\Omega/[n(n + 1)] \tag{10.54}$$

(Longuet-Higgins, 1964). When $P' = P$, (10.54) is exactly the second part of (10.17); these *are* Hough's second-class oscillations. Rossby's β-plane equivalents are

$$\psi = Z_n(x - ct, y), \qquad \nabla_2^2 Z_n + \lambda_n^2 Z_n = 0, \tag{10.55}$$

where $c = -\beta/\lambda_n^2$.

Plane Rossby waves are the particular case

$$\psi = a \exp(-i\sigma t + ilx + iky), \tag{10.56}$$

$$\sigma = -\beta l/(l^2 + k^2). \tag{10.57}$$

They are transverse [(u, v) perpendicular to (l, k)] and dispersive, with the propagation of phases always having a westward component ($\sigma/l < 0$). Their frequencies are typically low relative to f_0. Since σ depends both on wavelength and wave direction, these waves do not reflect specularly at a straight coast. Longuet-Higgins (1964) gives an elegant geometrical interpretation (figure 10.1) of their dispersion relation (10.57) and shows that it is the group velocity $(\partial\sigma/\partial l, \partial\sigma/\partial k)$ that reflects specularly in this case (figure 10.2).

Rossby-wave normal modes of a closed basin of perimeter P have the form (Longuet-Higgins, 1964)

$$\psi = Z_n(x,y) \exp[-i\sigma_n t - i(\beta/2\sigma_n)x], \tag{10.58}$$

$$\sigma_n = \beta/2\lambda_n \tag{10.59}$$

where

$$\nabla_2^2 Z_n + \lambda_n^2 Z_n = 0, \qquad Z_n = 0 \quad \text{on} \quad P. \tag{10.60}$$

These Rossby-wave normal modes are in remarkable contrast with the gravity-wave normal modes of the same basin without rotation:

$$\zeta_n = Z_n(x,y) \exp(-i\sigma_n t), \tag{10.61}$$

$$\sigma_n = \lambda_n (gD_*)^{1/2} \qquad [\partial Z_n/\partial(\text{normal}) = 0 \text{ on } P]. \tag{10.62}$$

The gravity-wave modes have a lowest-frequency ($n = 1$) grave mode and the spatial scales of higher-frequency modes are smaller. The Rossby-wave modes have a highest frequency ($n = 1$) mode and the spatial scales of lower frequency modes are smaller.

The physical mechanism that makes Rossby waves possible is most easily seen for nearly zonal waves ($\partial/\partial y \ll \partial/\partial x$). Then Rossby's vorticity equation (10.51) becomes

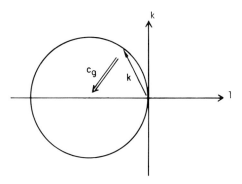

Figure 10.1 The locus of wavenumbers $k = (k,l)$ allowed by the Rossby dispersion relation (10.57)

$$(l + \beta/2\sigma)^2 + k^2 = (\beta/2\sigma)^2$$

is a circle of radius $\beta/2\sigma$ centered at $(-\beta/2\sigma, 0)$. The group velocity vector $\mathbf{c}_g = (\partial\sigma/\partial l, \partial\sigma/\partial k)$ points from the tip of the wavenumber vector toward the center of the circle and has magnitude $|\mathbf{c}_g| = \beta/(l^2 + k^2)$.

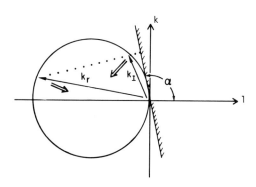

Figure 10.2 A Rossby wave with wavenumber $\mathbf{k}_i$ is incident on a straight coast inclined at an angle α to the east-west direction. The wavenumber $\mathbf{k}_r$ of the reflected wave is fixed by the necessity that $\mathbf{k}_i$ and $\mathbf{k}_r$ have equal projection along the coast. The group velocity reflects specularly in the coast.

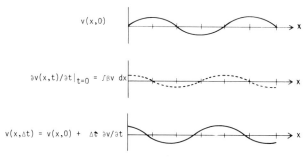

Figure 10.3 The flow $v(x, t)$ evolving from the initial flow $v(x, 0) \sim \sin(lx)$ as fluid columns migrate north-south (and so exchange planetary and relative vorticity) is a westward displacement of the initial flow. Notice that although parcels take on clockwise-counterclockwise relative vorticity as they are moved north-south, the westward displacement is not the result of advection of vorticity of one sign by the flow associated with the other as is the case in a vortex street.

$$\frac{\partial^2 v}{\partial x \, \partial t} + \beta v = 0. \tag{10.63}$$

North-south motions v result in changes in the local vorticity $\partial v/\partial x$. When the initial north-south motion is periodic in x, then examination (figure 10.3) of (10.63) shows that the additional north-south motion generated by the vorticity resulting from the initial pattern of north-south motion combines with that pattern to shift it westward, in accordance with (10.57).

Rossby's vorticity equation (10.51) corresponds to the plane equations

$$\frac{\partial u}{\partial t} - fv = -\frac{1}{\rho_0} \frac{\partial p}{\partial x}, \tag{10.64a}$$

$$\frac{\partial v}{\partial t} + fu = -\frac{1}{\rho_0} \frac{\partial p}{\partial y}, \tag{10.64b}$$

$$\frac{\partial u}{\partial x} + \frac{\partial v}{\partial y} = 0, \tag{10.64c}$$

$$f = f_0, \qquad \partial f/\partial y = \beta, \tag{10.64d}$$

so that Rossby's solutions are almost geostrophic $(\sigma \ll f_0)$ and perfectly nondivergent. The absence of divergence and vertical velocity is an extreme of the tendency, in quasigeostrophic flow, for the vertical velocity to be order Rossby number $(\ll 1)$ smaller than a scale analysis of the continuity equation would indicate (Burger, 1958). This tendency is absent at planetary length scales, and Rossby's β-plane (10.64) correspondingly requires modification.

Remarkably, Rossby and collaborators (1939) prefaced their analysis with a resumé of a different physical mechanism due to J. Bjerknes (1937), a mechanism that also results in westward-propagating waves but for a different reason, and that supplies the modification of Rossby's β-plane required at planetary scales. The plane equations corresponding to Rossby's summary of Bjerknes' arguments are

$$-fv = -g\frac{\partial \zeta}{\partial x}, \tag{10.65a}$$

$$fu = -g\frac{\partial \zeta}{\partial y}, \tag{10.65b}$$

$$\frac{\partial \zeta}{\partial t} + D_*\left(\frac{\partial u}{\partial x} + \frac{\partial v}{\partial y}\right) = 0, \tag{10.65c}$$

$$f = f_0, \qquad \partial f/\partial y = \beta. \tag{10.65d}$$

By physical arguments (figure 10.4), Bjerknes deduced that an initial pressure perturbation would always propagate westward. The corresponding analysis of (10.65) is to form an elevation equation

$$\frac{\partial \zeta}{\partial t} - (gD_*\beta/f_0^2)\frac{\partial \zeta}{\partial x} = 0 \tag{10.66}$$

and to note that it has the dispersionless solutions

302
Myrl C. Hendershott

$$\zeta = F[x + (gD_*\beta/f_0^2)t, y],$$

where $F(x, y)$ is the initial pressure perturbation. In particular,

$$\zeta = a \exp(-i\sigma t + ilx + iky), \quad (10.67)$$

$$\sigma = -(gD_*\beta/f_0^2)l. \quad (10.68)$$

All solutions travel westward at $(gD_*\beta/f_0^2)^{1/2}$. These motions, according to (10.65), are perfectly geostrophic but divergent.

More complete analysis (Longuet-Higgins, 1964) shows that the two dispersion relations (10.57) of Rossby and (10.68) of Bjerknes are limiting cases of the β-plane dispersion relation

$$\sigma = -\beta l/(l^2 + k^2 + f_0^2/gD_*) \quad (10.69)$$

for second-class waves displayed in figure 10.5. It would be appropriate to call the two kinds of second-class waves Rossby and Bjerknes waves, respectively, but in practice both are commonly called Rossby waves. I shall distinguish them as short, nondivergent and long, divergent Rossby waves.

When divergence is allowed, the (constant) depth D_* enters the dispersion relation (10.69) in the length scale

$$a_R = (gD_*/f_0^2)^{1/2}, \quad (10.70)$$

usually called the Rossby radius. There is not one Rossby radius, but rather there are many, since the constant-depth barotropic second-class waves so far discussed have an infinite sequence of baroclinic counterparts with $D_* = D_n$, $n = 1, \ldots$, given by (10.36)-(10.39). Waves longer than the Rossby radius are long, divergent Rossby waves; those shorter than the Rossby radius are short, nondivergent Rossby waves.

The barotropic Rossby radius $a_{R0} = (gD_0/f_0^2)^{1/2}$ has $D_0 \simeq D_*$ and is thus the order of the earth's radius. Barotropic Rossby waves are consequently relatively high-frequency (typically a few cycles per month) waves and they are able to traverse major ocean basins in days to weeks. Baroclinic Rossby radii $a_{Rn} = (gD_n/f_0^2)^{1/2}$ are the order of 10^2 km or less in mid-latitudes. Baroclinic mid-latitude Rossby waves are consequently relatively low-frequency waves and would take years to traverse major mid-latitude basins. In the tropics, f_0 becomes small and baroclinic Rossby waves speed up to the point where they could traverse major basins in less than a season. But a different discussion is really necessary for the tropics (see chapter 6).

Rossby advanced his arguments to rationalize the motion of mid-latitude atmospheric pressure patterns. In both atmosphere and ocean, the slowness and relatively small scale of most second-class waves must make their occurrence in "pure" form very rare. Oceanic measurements from the MODE experiment

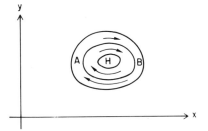

Figure 10.4 If the flow is totally geostrophic but the Coriolis parameter increases with latitude, then the flow at A converges because the geostrophic transport between a pair of isobars south of H is greater than that between the same pair north of H. By (10.65c), pressure thus rises at A. Similarly, the flow at B diverges and pressure there drops. The initial pattern of isobars is then shifted westward.

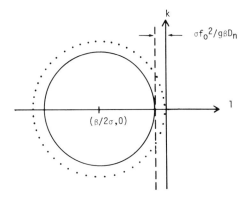

Figure 10.5 The locus of wavenumbers (l, k) allowed by the β-plane dispersion relation (10.69) for second-class waves

$$(l + \beta/2\sigma)^2 + k^2 = (\beta/2\sigma)^2 - (f_0^2/gD_n)$$

is a circle(———) whose radius is $[(\beta/2\sigma)^2 - (f_0^2/gD_n)]^{1/2}$ centered at $(-\beta/2\sigma, 0)$. Dotted circle $(\cdots)$ is the Rossby dispersion relation (10.57) for short waves. Dashed line (---) is the Bjerknes dispersion relation (10.68) appropriate for long waves. The scale a_R dividing short and long waves is

$$a_R = [2(\beta/2\sigma)(\sigma f_0^2/\beta gD_n)]^{-1/2}.$$

do show, however, characteristics of both baroclinic (figure 10.6) and long barotropic (figure 10.7) Rossby waves.

The oscillations having the two dispersion relations (10.23e) with $D_* = D_n$ for first-class waves and (10.69) for second-class waves are mid-latitude plane-wave approximations of solutions of LTE. Figure 10.8 plots the two dispersion relations together. A noteworthy feature is the frequency interval between f_0 and $(\beta/2f_0)(gD_n)^{1/2}$ within which no plane waves propagate. Taken at face value, this gap suggests that velocity spectra should show a valley between these two frequencies with a steep high-frequency [f_0] wall and a rather more gentle low-frequency [$(\beta/2f_0)(gD_n)^{1/2}$, $n = 0, 1, 2, \ldots$] wall. Such a gap is indeed commonly observed; but the dynamics of the low frequencies are almost surely more complex than those of the linear β-plane. The latitude dependence implicit in the definition of f_0 and β is consistent with equatorial trapping of low-frequency first-class waves and high-frequency second-class waves. This is more easily seen in approximations, such as the following, which better acknowledge the earth's sphericity.

10.4.5 The Equatorial β-Plane

For constant depth D_*, the homogeneous LTE (10.5) may be equatorially approximated by expanding all variable coefficients in θ and then neglecting $\theta^2, \theta^3, \ldots$. The resulting equatorial β-plane equations are

$$\frac{\partial u}{\partial t} - \beta y v = -g \frac{\partial \zeta}{\partial x}, \quad (10.71a)$$

$$\frac{\partial v}{\partial t} + \beta y u = -g \frac{\partial \zeta}{\partial y}, \quad (10.71b)$$

$$\frac{\partial \zeta}{\partial t} + D_* \left(\frac{\partial u}{\partial x} + \frac{\partial v}{\partial y} \right) = 0, \quad (10.71c)$$

where $x = a\phi$, $y = a\theta$, and $\beta = 2\Omega/a$. They govern both barotropic and baroclinic motions provided that D_* is interpreted as the appropriate equivalent depth D_n defined by (10.36)–(10.38). Moore and Philander (1977) and Philander (1978) give modern reviews.

Solutions of these equations can be good approximations to solutions of LTE only when they decay very rapidly away from the equator. But the qualitative nature of their solutions, bounded as $y \to \pm\infty$, closely resembles solutions of LTE bounded at the poles, even

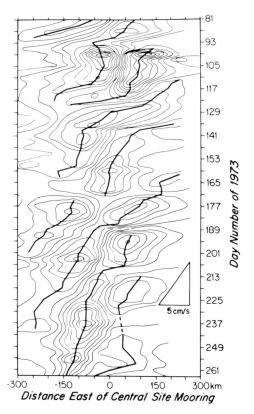

Figure 10.6A Time–longitude plot of streamfunction inferred from objective maps of 1500-m currents along 28°N (centered at 69°49′W) by Freeland, Rhines and Rossby (1975). There is evidence of westward propagation of phases. Currents at this depth are not dominated by "thermocline eddies" (section 10.4.7) but are representative of the deep ocean.

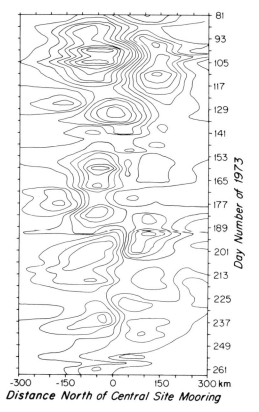

Figure 10.6B As figure 10.6A but in time–latitude plot along 69°40′W. There is no evidence for a preferred direction of latitudinal phase propagation. (Rhines, 1977.)

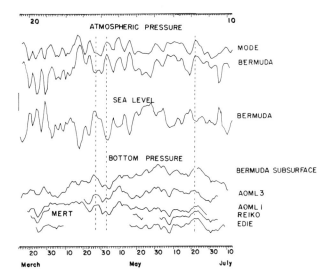

Figure 10.7 Time series of bottom pressure in MODE (Brown et al., 1975). The cluster of named gauges centered at 28°N, 69°40'W show remarkable coherence despite 0 (180-km) separation, and all are coherent with the (atmospheric pressure corrected) sea level at Bermuda (650 km distant, labeled Bermuda bottom). (Brown et al., 1975.)

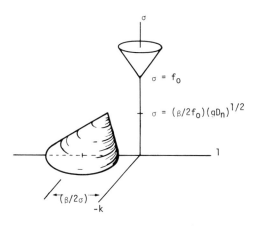

Figure 10.8 The f-plane dispersion relation

$$\sigma^2 = f_0^2 + gD_n(l^2 + k^2)$$

for first-class waves allows no waves with $\sigma^2 < f_0^2$. The β-plane dispersion relation

$$\sigma = -\beta l/[l^2 + k^2 + f_0^2/gD_n]$$

for second-class waves allows no waves with $\sigma > (\beta/2f_0)(gD_n)^{1/2}$.

when the equatorial approximation is transgressed. Historically these approximate solutions provided a great deal of insight into the latitudinal variation of solutions of LTE.

Most of the solutions are obtainable from the single equation that results when u, ζ are eliminated from (10.71). With

$$v = V(y)\exp(-i\sigma t + ilx) \tag{10.72}$$

that equation is

$$\frac{\partial^2 V}{\partial y^2} + \left[\left(\frac{\sigma^2}{gD_*} - l^2 - \frac{l\beta}{\sigma}\right) - \frac{\beta^2}{gD_*}y^2\right]V = 0. \tag{10.73}$$

It also occurs in the quantum-mechanical treatment of the harmonic oscillator. Solutions are bounded as $y \to \pm\infty$ only if

$$\left(\frac{\sigma^2}{gD_*} - l^2 - \frac{l\beta}{\sigma}\right) = (2m+1)\frac{\beta}{(gD_*)^{1/2}}, \tag{10.74}$$

$$m = 0, 1, 2, \ldots,$$

and they are then

$$V(y) = H_m[y\beta^{1/2}/(gD_*)^{1/4}]\exp[-y^2\beta/2(gD)^{1/2}], \tag{10.75}$$

wherein the H_m are Hermite polynomials ($H_0(z) = 1$, $H_1(z) = z, \ldots$).

The remaining solution may be taken to be $v = 0$ with $m = -1$ in (10.74). It is obtained by solving (10.71) with $v = 0$. The solution bounded as $y \to \pm\infty$ is

$$\zeta = \exp[-i\sigma t + ilx - (\beta l/\sigma)y^2/2] \tag{10.76}$$

with

$$l = \sigma/(gD_*)^{1/2} \tag{10.77}$$

[(10.77) is (10.74) with $m = -1$].

The very important dispersion relation (10.74) with $m = -1, 0, 1, \ldots$ thus governs all the equatorially trapped solutions of (10.71). Introducing the dimensionless variables $\omega, \lambda, \tau, \eta$ defined by

$$\sigma = \omega(2\Omega\Lambda^{-1/4}), \quad l = \lambda(a^{-1}\Lambda^{1/4}), \tag{10.78}$$

$$(x, y) = (\xi, \eta)(a\Lambda^{-1/4}), \quad t = \tau[(2\Omega)^{-1}\Lambda^{1/4}]$$

$(\Lambda = 4\Omega^2 a^2/gD_*)$ allows us to rewrite (10.73) and its solutions (10.72), (10.75), (10.76) as

$$\frac{\partial^2 V}{\partial \eta^2} + [(\omega^2 - \lambda^2 - \lambda/\omega) - \eta^2]V = 0, \tag{10.79}$$

$$v = H_m(\eta)\exp(-i\omega\tau + i\lambda\xi - \eta^2/2), \tag{10.80}$$

$$m = 0, 1, \ldots,$$

$$\zeta = \exp[-i\omega\tau + i\lambda\xi - (\lambda/\omega)\eta^2/2], \quad m = -1, \tag{10.81}$$

while the dispersion relation (10.74) becomes

$$\omega^2 - \lambda^2 - \lambda/\omega = 2m + 1, \quad m = -1, 0, 1, \ldots . \quad (10.82)$$

These forms allow easy visualization of the solutions and permit a concise graphical presentation of the dispersion relation (figure 10.9).

The dispersion relation is cubic in σ (or ω) for given values of l (or λ) and m. For $m > 1$ the three roots correspond precisely to two oppositely traveling waves of the first class plus a single westward-traveling wave of the second class. The case $m = 0$ (Yanai, or Rossby-gravity, wave) is of first class when traveling eastward but of second class when traveling westward. The case $m = -1$ is an equatorially trapped Kelvin wave, dynamically identical to the coastally trapped Kelvin wave (10.27) in a uniformly rotating ocean.

The most useful aspect of these exact solutions is their provision of a readily understandable dispersion relation [(10.74) or (10.82)]. The latitudinal variation of flow variables is more readily discussed in terms of WKB solutions of (10.73). One can easily see the salient feature of the solutions, a transition from oscillatory to exponentially decaying latitudinal variation as the turning latitudes y_T of (10.73) (at which the coefficient [ ] of that equation vanishes), are crossed poleward. For waves of the first class, the term $l\beta/\sigma$ is small relative to the other terms in the dispersion relation and in the coefficient [ ]. The corresponding turning latitudes $y_T^{(1)}$ are therefore approximately given by

$$[y_T^{(1)}]^2 = (\sigma/\beta)^2[1 - l^2(gD_*)/\sigma^2] < (\sigma/\beta)^2. \quad (10.83)$$

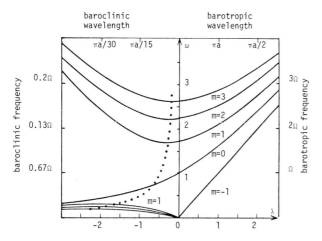

Figure 10.9 The equatorial β-plane dispersion relation (10.82) $\omega^2 - \lambda^2 - \lambda/\omega = 2m + 1$.

Dimensional wavelengths and frequencies are obtained from the scaling (10.78) and are given for the barotropic mode ($D_0 = 4000$ m, $\Lambda = 20$) and for the first baroclinic mode ($D_0 = 0.1$ m, $\Lambda = 10^6$). For all curves but $m = 0$, intersections with dotted curve are zeros of group velocity.

For waves of the second class, the term σ^2/gD_* is small relative to the other terms in the dispersion relation and in the coefficient [ ]. The corresponding turning latitudes $y_T^{(2)}$ are therefore approximately given by

$$[y_T^{(2)}]^2 = (gD_*/\beta^2)(-l^2 - l\beta/\sigma) < gD_*/4\sigma^2. \quad (10.84)$$

Increasingly low-frequency waves of the first class and increasingly high-frequency waves of the second class are thus trapped increasingly close to the equator.

Only first-class waves having frequency greater than the inertial frequency βy penetrate poleward of latitude y [by (10.83)]. Only second-class waves having frequency below the cutoff frequency $(gD_*/4y^2)^{1/2}$ penetrate poleward of latitude y [by (10.84)]. This frequency-dependent latitudinal trapping corresponds to the mid-latitude frequency gap between first- and second-class waves discussed in the previous section and illustrated in figure 10.8. The correspondence correctly suggests that trapping and associated behavior characterize slowly varying (in the WKB sense) packets of waves propagating over the sphere as well as the globally standing patterns corresponding to the Hermite solutions (10.75). Waves thus need *not* be globally coherent to exhibit trapping and the features associated with it.

Near the trapping latitudes, (10.73) becomes

$$\frac{\partial^2 V}{\partial y^2} + (-2\beta^2 y_T/gD_*)(y - y_T)V = 0. \quad (10.85)$$

The change of variable $\eta = (2\beta^2 y_T/gD_*)^{1/3}(y - y_T)$ reduces this to Airy's equation

$$\frac{\partial^2 V}{\partial \eta^2} - \eta V = 0, \quad (10.86)$$

whose solution $Ai(\eta)$ bounded as $\eta \to \infty$ is plotted in figure 10.10. This solution has two important features: (1) gentle amplification (like $\eta^{-1/4}$) of the solution as the turning latitude ($\eta = 0$) is approached from the equator; and (2) transition from oscillatory to exponentially decaying behavior in a region $\Delta\eta$ of order roughly unit width surrounding the turning latitude. Consequently the interval Δy over which the solution of (10.85) changes from oscillatory to exponential behavior is $\Delta y = [2\beta^2 y_T/gD_*]^{-1/3}\Delta\eta$ or, since $\beta = 2\Omega/a$ and $\Delta\eta \approx 1$,

$$\Delta y = a(\Lambda\sigma/2\Omega)^{-1/3} \quad (10.87)$$

$$= a(2^{-1/2}\Lambda^{-1/2}\sigma/2\Omega)^{1/3} \quad (10.88)$$

for maximally penetrating first- and second-class waves. For diurnal ($\sigma = \Omega$) first-class barotropic ($\Lambda = 20$) waves, $\Delta y \approx 0.5a$; for diurnal first-class baroclinic ($\Lambda \approx 10^6$) waves, $\Delta y \approx 0.013a$. For 10-day ($\sigma = 0.1\Omega$) second-class barotropic waves, $\Delta y \approx 0.25a$. For 1-month

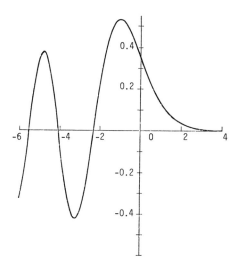

Figure 10.10 The Airy function Ai.

($\sigma \simeq .03\Omega$) second-class baroclinic waves, $\Delta y = 0.03a$. We thus obtain the important result that barotropic modes are not noticeably trapped (Δy is a fair fraction of the earth's radius and the Airy solution is only qualitatively correct anyway) but baroclinic modes are abruptly trapped (Δy is a few percentage points of the earth's radius).

The abrupt trapping of baroclinic waves at their inertial latitudes means that the Airy functions may describe quite accurately the latitude variation of near-inertial motions. Munk and Phillips (1968) and Munk (chapter 9) discuss the structure.

The clearest observations of equatorial trapping are by Wunsch and Gill (1976), from whose paper figure 10.11 is taken. Longer-period fluctuations at and near the equator have been observed, but their relation to the trapped solutions is not yet clear.

When an equatorially trapped westward-propagating wave meets a north-south western boundary (at, say, $x = 0$) it is reflected as a superposition of finite numbers of eastward-propagating waves including the Kelvin ($m = -1$) and Yanai ($m = 0$) waves (Moore and Philander, 1977). But when an equatorially trapped eastward-propagating wave meets a north-south eastern boundary, some of the incident energy is scattered into a poleward-propagating coastal Kelvin wave (10.29) and thus escapes the equatorial region (Moore, 1968). In latitudinally bounded basins, the requirement that solutions decay exponentially away from the equator is replaced by the vanishing of normal velocity at the boundaries. Modes closely confined to the equator will not be greatly altered by such boundaries; modes that have appreciable extraequatorial amplitude will behave like the β-plane solutions of sections 10.4.1–10.4.4 near the boundaries. A theory of free oscillations in idealized basins on the equatorial β-plane could be constructed on the basis of such observations, but powerful techniques for dealing with the spherical problem now exist (section 10.4.8).

10.4.6 Barotropic Waves over Bottom Relief

Stokes (1846; Lamb, 1932, §260) had shown that shoaling relief results in the trapping of an edge wave whose amplitude decays exponentially away from the coast, but the motion was not thought to be important.

Eckart (1951) solved the shallow-water equations [(10.18) with $f_0 = 0$] with the relief $D = ax$. Solutions of the form

$$\zeta = h(x)\exp(-i\sigma t + iky) \qquad (10.89)$$

are governed by

$$x\frac{\partial^2 h}{\partial x^2} + \frac{\partial h}{\partial x} + [\sigma^2/(ag) - xk^2]h = 0. \qquad (10.90)$$

Solutions of this are bounded as $x \to \infty$ only if

$$\sigma^2 = k(2n+1)ag, \quad n = 0, 1, \ldots, \qquad (10.91)$$

and they are then

$$h(x) = L_n(2kx)\exp(-kx), \qquad (10.92)$$

where the L_n are Laguerre polynomials [$L_0(z) = 1$, $L_1(z) = z - 1, \ldots$]. The $n = 0$ mode corresponds to Stokes's (1846) edge wave.

Eckart's solutions are LSW gravity waves refractively trapped near the coast by the offshore increase in shallow water wave speed $(gax)^{1/2}$. The Laguerre solutions (10.92) are correspondingly trigonometric shoreward of the turning points x_T at which the coefficient [ ] of (10.90) vanishes, and decay exponentially seaward.

Eckart's use of the LSW equations is not entirely self-consistent, since $D = ax$ increases without limit. Ursell (1952) removed the shallow-water approximation by completely solving

$$\frac{\partial^2 \phi}{\partial z^2} + \frac{\partial^2 \phi}{\partial x^2} - k^2\phi = 0 \qquad (10.93)$$

subject to

$$\frac{\partial \phi}{\partial z} = (\sigma^2/g)\phi \quad \text{at} \quad z = 0 \qquad (10.94)$$

and

$$\frac{\partial \phi}{\partial \eta} = 0 \quad \text{at} \quad z = -ax \qquad (10.95)$$

plus boundedness of the velocity field ($\partial\phi/\partial x, ik\phi, \partial\phi/\partial z$) as $x \to \infty$. He found (1) a finite number of coastally trapped modes with dispersion relation

$$\sigma^2 = kg\sin[(2n+1)\tan^{-1}a], \qquad (10.96)$$
$$n = 0, 1, \ldots < [\pi/(4\tan^{-1}a)-1/2]$$

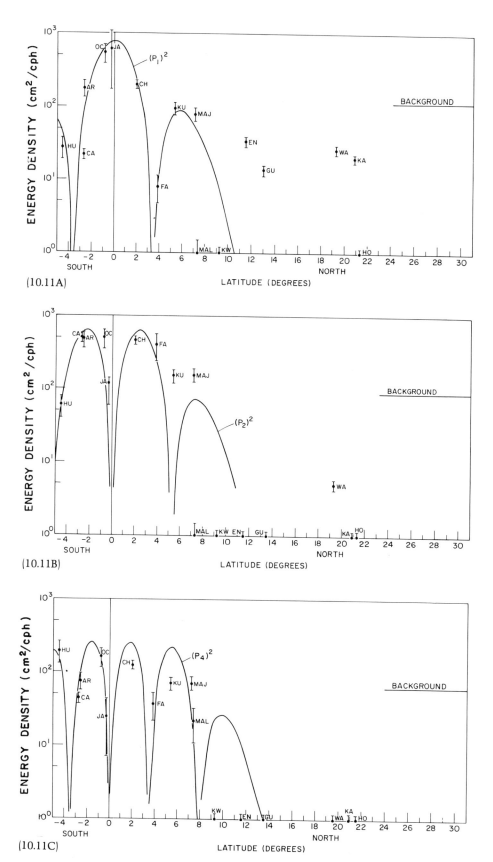

Figure 10.11 Energy at periods of 5.6d, $m = 1$ (A); 4.0d, $m = 2$ (B); 3.0d, $m = 4$ (C); in tropical Pacific sea-level records as a function of latitude. A constant (labeled BACKGROUND) representing the background continuum has been subtracted from each value. Error bars are one standard deviation of χ^2. The solid curves are the theoretical latitudinal structure from the equatorial β-plane. (Wunsch and Gill, 1976.)

corresponding, for low n, to Eckart's results, plus (2) a continuum of solutions corresponding to the coastal reflection of deep-water waves incident from $x = \infty$ and correspondingly not coastally trapped. Far from the coast, the continuum solutions have the form $\phi = \cos(lx + \text{phase}) \exp[-i\sigma t + iky + (l^2 + k^2)^{1/2}z]$ and their dispersion relation must require $\sigma^2 \geq gk$. They are filtered out by the shallow-water approximation. Figure 10.12 compares Eckart's (1951) and Ursell's (1952) dispersion relations.

With rotation f_0 restored to (10.18), (10.90) becomes (Reid, 1958)

$$x\frac{\partial^2 h}{\partial x^2} + \frac{\partial h}{\partial x} + \left[\left(\frac{\sigma^2 - f_0^2}{ag} - \frac{f_0 k}{\sigma}\right) - xk^2\right]h = 0. \quad (10.97)$$

Solutions still have the form (10.89), (10.92) but now the dispersion relation is

$$\sigma^2 - f_0^2 - f_0 kag/\sigma = k(2n + 1)ag, \quad (10.98)$$

which is cubic in σ, whereas with $f_0 = 0$ it was quadratic. Rotation has evidently introduced a new class of motion.

That this should be so is clear from the β-plane vorticity equation [obtained by cross-differentiating (10.30a,b) and with $\partial f_0/\partial y = \beta$]:

$$\frac{\partial}{\partial t}\left(\frac{\partial v}{\partial x} - \frac{\partial u}{\partial y}\right) - \frac{f_0}{D}\frac{\partial \zeta}{\partial t} - u\frac{f_0}{D}\frac{\partial D}{\partial x} - v\left(-\beta + \frac{f_0}{D}\frac{\partial D}{\partial y}\right)$$

$$= 0. \quad (10.99)$$

We have already seen (section 10.4.4) that the term βv gives rise to short and long Rossby waves (with the vortex-stretching term $f_0/D \, \partial \zeta/\partial t$ important only for the latter) when the depth is constant. But in (10.99), the topographic vortex-stretching term $\mathbf{u} f_0/D \cdot \nabla D$ plays a role entirely equivalent to that of $\beta v = \mathbf{u} \cdot \nabla f_0$. Hence, we expect it to give rise to second-class waves, both short and long, even if $\beta = 0$. Such waves are called *topographic Rossby waves*. Over the linear beach $D = -ax$ they are all refractively trapped near the coast.

The nondimensionalization

$$s = \sigma/f, \quad K = kag/f^2 \quad (10.100)$$

casts the dispersion relation into the form

$$s^3 - s[1 + (2n + 1)K] - K = 0, \quad (10.101)$$

which is remarkably similar to (10.82) and plotted in figure 10.13.

The linear beach $D = ax$ is most unreal in that there is no deep sea of finite depth in which LSW plane waves can propagate. When the relief is modified to

$$D(x) = \begin{cases} ax, & 0 < x < D_0 a^{-1} \quad \text{(shelf)} \\ D_0, & D_0 a^{-1} < x < \infty \quad \text{(sea)}, \end{cases} \quad (10.102)$$

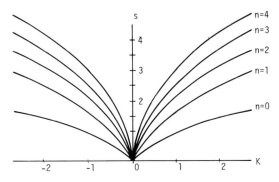

Figure 10.12A Eckart's (1951) dispersion relation (10.91)

$$s^2 = K(2n + 1)$$

for the shallow-water waves over a semi-infinite uniformly sloping, nonrotating beach. For convenience in plotting, $s = \sigma/f$ and $K = gak/f^2$ even though problem is not rotating.

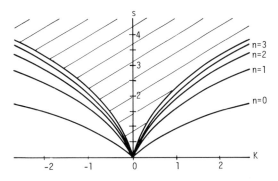

Figure 10.12B Ursell's (1952) dispersion relation (10.96)

$$s^2 = Ka^{-1}\sin[(2n + 1)\tan^{-1}a]$$

for edge waves, and the continuum

$$s^2 > Ka^{-1}$$

of deep-water reflected waves. For convenience in plotting, s and K are defined as above. Plot is for $a = 0.2$.

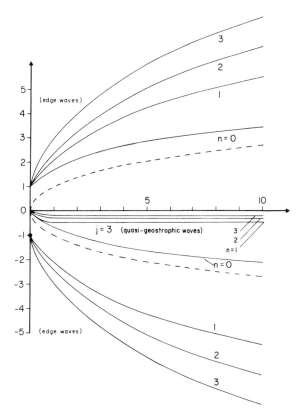

Figure 10.13 The dispersion relation (10.91) for edge and quasi-geostrophic shallow-water waves over a semi-infinite uniformly sloping beach. The dashed curves are the Stokes solution without rotation [(10.91) with $n = 0$]. Axes are as in figure 10.12. (LeBlond and Mysak, 1977.)

the most important alteration of the dispersion relation is an "opening up" of the long-wavelength part of the dispersion relation to include a continuum analogous to that of Ursell (1952) but now consisting of LSW first-class waves incident from the deep sea and reflected back into it by the coast and shelf. These waves are not coastally trapped. They are often called *leaky modes* because they can radiate energy that is initially on the shelf out into the deep sea. There are no second-class counterparts because the deep sea with constant depth and $\beta = 0$ cannot support second-class waves.

The dispersion relation corresponding to (10.102) is plotted in figure 10.14. All of Eckart's modes are modified so that $\sigma^2 > f_0^2$ save one ($n = 0$, traveling with the coast to its right), which persists as $\sigma \to 0$ and is a Kelvin-like mode. The others cease to be refractively trapped at superinertial ($>f_0$) individual cutoff frequencies bordering the continuum of leaky modes. At subinertial frequencies there is an infinite family of refractively trapped topographic Rossby waves, all traveling with the coast to their right (like the Kelvin mode) and all tending toward the constant frequency $s = -1/(2n + 1)$ at small wavelengths. This dispersion relation is qualitatively correct for most other shelf shapes. It differs from its equatorial β-plane counterpart (10.82) only in the absence of a mixed Rossby-gravity (Yanai) mode and in the tendency of short second-class modes to approach constant frequencies.

Topographic vortex stretching plus refraction of both first- and second-class waves are effective over any relief. Thus islands with beaches, submerged plateaus, and seamounts can in principal trap both first- and second-class barotropic waves (although these topographic features may have to be unrealistically large for their circumference to span one or more wavelengths of a trapped first-class wave). A submarine escarpment can trap second-class waves (then called *double Kelvin waves*; Longuet-Higgins, 1968b). Examples of such solutions are summarized by Longuet-Higgins (1969b) and by Rhines (1969b).

First-class waves trapped over the Southern California continental shelf have been clearly observed by Munk, Snodgrass, and Gilbert (1964), who computed the dispersion relation for the actual shelf profile and found (figure 10.15) sea-level variation to be closely confined to the dispersion curves thus predicted for periods of order of an hour or less. Both first- and second-class coastally trapped waves may be variously significant in coastal tides [Munk, Snodgrass, and Wimbush (1970) and section 10.5.2]. At longer periods, a number of observers claim to have detected coastally trapped second-class modes (Leblond and Mysak, 1977). A typical set of observations is shown in figure 10.16, after R. L. Smith (1978).

10.4.7 Long Waves over Relief with Rotation and Stratification

The two mechanisms of refraction and vortex stretching that govern the behavior of long waves propagating in homogeneous rotating fluid over bottom relief are sufficiently well understood that qualitatively correct dispersion relations may be found intuitively for quite complex relief even though their quantitative construction might be very involved. Stratification complicates the picture greatly. In this section, emphasis is upon problems with stratification that may be solved with sufficient completeness that they augment our intuition.

By appealing to the quasi-geostrophic approximation, Rhines (1975; 1977) has given a far-reaching treatment of the interplay between beta, weak bottom slope, and stratification for second-class waves. If equations (10.30a) and (10.30b) are cross-differentiated to eliminate pressure, and continuity (10.30d) is then invoked, the result is

$$\frac{\partial}{\partial t}\left(\frac{\partial v}{\partial x} - \frac{\partial u}{\partial y}\right) + \beta v - f_0 \frac{\partial w}{\partial z} = 0$$

in the β-plane approximation (section 10.4.4) $f = f_0$, $\beta = \partial f/\partial y$. Now this equation is recast as an approxi-

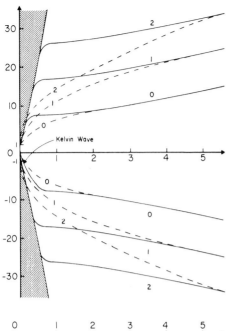

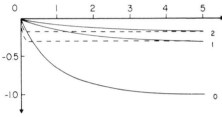

Figure 10.14 The dispersion relation (solid lines) for edge and quasi-geostrophic shallow-water waves over a uniformly sloping beach (slope $a = 2 \times 10^{-3}$) terminating in a flat ocean floor at a depth of $D = 5000$ m. The dotted lines are for the semi-infinite uniformly sloping beach. The shaded region is the continuum of leaky modes. Axes are as in figure 10.12. (Leblond and Mysak, 1977.)

mate equation in p by using (10.30c) plus the geostrophic approximation to obtain

$$\frac{\partial}{\partial t}\left(\frac{\partial^2 p}{\partial x^2} + \frac{\partial^2 p}{\partial y^2} + \frac{f_0^2}{N_0^2}\frac{\partial^2 p}{\partial z^2}\right) + \beta\frac{\partial p}{\partial x} = 0. \qquad (10.103)$$

This result emerges from a more systematic treatment (Pedlosky, 1964a) as the linearized quasi-geostrophic approximation. It is here specialized to the case of constant buoyancy frequency N_0. The free surface may be idealized as rigid without loss of generality; the corresponding condition on p is

$$\frac{\partial p}{\partial z} = 0 \quad \text{at} \quad z = 0. \qquad (10.104)$$

The inviscid bottom boundary condition $w = av$ at the north–south sloping bottom $z = -D_0 + ay$ becomes, in quasi-geostrophic approximation,

$$-\frac{\partial^2 p}{\partial t\,\partial z} + a\frac{N_0^2}{f_0}\frac{\partial p}{\partial x} = 0 \quad \text{at} \quad z = -D_0, \qquad (10.105)$$

and is linearized about $z = -D_0$ for sufficiently small a.

With no bottom slope, solutions of (10.103)–(10.105) are

$$P = \cos(\lambda z)\exp(-i\sigma t + ilx + iky), \qquad (10.106)$$

$$\sigma = -\beta l/(l^2 + k^2 + \lambda^2 f_0^2/N_0^2), \qquad (10.107)$$

with λ given by (10.105) as a solution of

$$\sin(\lambda D_0) = 0 \qquad (10.108)$$

i.e.,

$$\lambda = n\pi/D_0, \qquad n = 0, 1, 2, \ldots. \qquad (10.109)$$

These correspond to the barotropic ($n = 0$) and baroclinic ($n = 1, 2, \ldots$) Rossby waves of section 10.4.4.

With no beta but with bottom slope, solutions of (10.103)–(10.105) are

$$P = \cosh(\lambda z)\exp(-i\sigma t + ilx + iky), \qquad (10.110)$$

$$\lambda = (N_0/f_0)(l^2 + k^2)^{1/2}, \qquad (10.111)$$

$$\sigma = \frac{aN_0^2\coth(\lambda D_0)}{\lambda f_0}. \qquad (10.112)$$

If $\lambda D_0 \ll 1$, p is virtually depth independent and the dispersion relation (10.112) becomes

$$\sigma = D_0^{-1}af_0 l/(l^2 + k^2). \qquad (10.113)$$

This is a barotropic topographic Rossby wave with vortex stretching over the relief playing the role of beta. If $\lambda D_0 \gg 1$, p decays rapidly away from the bottom and the dispersion relation (10.112) becomes

$$\sigma = N_0 l/(l^2 + k^2)^{1/2}. \qquad (10.114)$$

Such *bottom-trapped* motions are of theoretical importance because they allow mid-latitude quasi-geostrophic vertical shear and density perturbations at periods much shorter than the very long ones predicted by the flat-bottom baroclinic solutions (10.106)–(10.109). Rhines (1970) generalizes this bottom-trapped solution to relief of finite slope and points out that it reduces to the usual baroclinic Kelvin wave at a vertical boundary. Figure 10.17 shows what appear to be motions of this type.

With both beta and bottom slope, solutions of (10.103) and (10.104) are

$$P = {\cos \atop \cosh}(\lambda z)\exp(-i\sigma t + ilx + iky) \qquad (10.115_b^a)$$

$$\sigma = -\beta l/(l^2 + k^2 \pm \lambda^2 f_0^2/N_0^2), \qquad (10.116_b^a)$$

with λ given by (10.105) as

$$\lambda\tan(\lambda D_0) = -\frac{aN_0^2}{f_0}\frac{l}{\sigma} \qquad (10.117a)$$

311
Long Waves and Ocean Tides

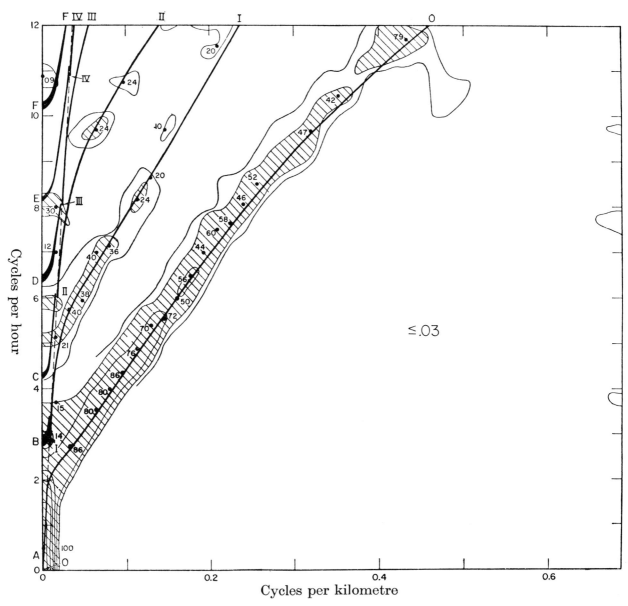

Figure 10.15 Comparison of theoretical and observed dispersion for the California continental shelf. Heavy lines 0–IV correspond to theoretical dispersion relations for the first five trapped first-class modes. The dashed line bounds the continuum of leaky modes. The observed normalized two-dimensional cospectrum of bottom pressure is contoured for values of 0.03, 0.05, 0.10, 0.25, 0.50, 0.75, and 0.90 with the area above 0.05 shaded. (Munk, Snodgrass, and Gilbert, 1964.)

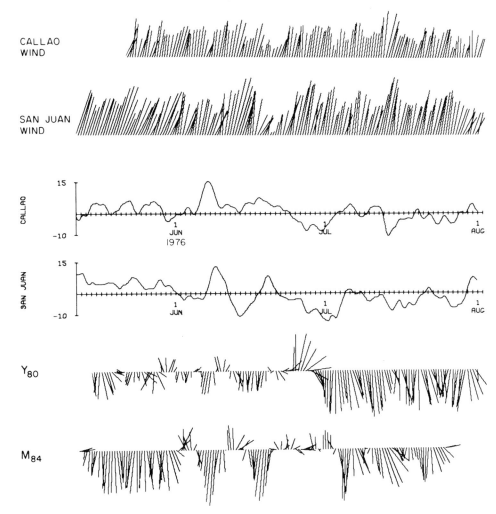

Figure 10.16 Low-passed wind vectors and sea-level records from Callao (12°04′ S) and San Juan (15°20′ S), Peru, and current vectors from Y (80 m below surface off Callao) and from M (84 m below surface off San Juan). Sea level and currents show propagation of events along the coast; wind records do not. (R. L. Smith, 1978.)

for case (a) and

$$\lambda \tanh(\lambda D_0) = \frac{aN_0^2}{f_0} \frac{1}{\sigma} \quad (10.117b)$$

for case (b). Equation (10.117b) has one root λ corresponding to a bottom-trapped wave for large a and to a barotropic β-wave for vanishing a. For vanishing a, (10.117a) reproduces the familiar flat-bottom barotropic and low-frequency baroclinic modes $\lambda = n\pi/D_0$, $n = 0, 1, \ldots$. When a is large the baroclinic roots are shifted toward $\lambda \simeq (\pi/2 + n\pi)/D_0$, so that the pressure (10.115a) and hence the horizontal velocity have a node at the bottom. There is thus a tendency for relief to result in the concentration of low-frequency baroclinic energy away from the bottom. With more realistic stratification this concentration is increasingly in the upper ocean. Rhines (1977) therefore calls such motions "thermocline eddies" and suggests that they are relevant to the interpretation of the observations of figure 10.18. It is straightforward to allow for an arbitrary direction of the bottom slope, but the results are not easy to summarize. Rhines (1970) gives a complete discussion.

A powerful treatment of second-class motion in a rotating stratified fluid over the linear beach $D = ax$ has been provided by Ou (1979), Ou (1980), and Ou and Beardsley (1980). They have generously permitted me to make use of their results in this discussion. Neglecting free surface displacement (so that $w = 0$ at $z = 0$) and eliminating u,v,w from (10.30) in favor of p yields

$$\frac{\partial^2 p}{\partial z^2} + \frac{N^2}{f_0^2 - \sigma^2}\left(\frac{\partial^2 p}{\partial x^2} + \frac{\partial^2 p}{\partial y^2}\right) = 0, \quad (10.118)$$

$$\frac{\partial p}{\partial z} = 0 \quad \text{at} \quad z = 0, \quad (10.119)$$

$$a\left(i\sigma\frac{\partial p}{\partial x} - f_0\frac{\partial p}{\partial y}\right) + i\sigma\frac{f_0^2 - \sigma^2}{N_0^2}\frac{\partial p}{\partial z} = 0$$

at $z = -ax$ $\quad (10.120)$

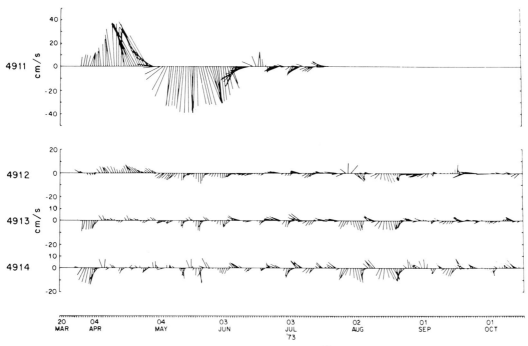

Figure 10.17A Currents at 39°10'N, 70°W (site D) at 205, 1019, 2030, and 2550 m. The total depth is 2650 m. A "thermocline" eddy initially dominates the upper flow.

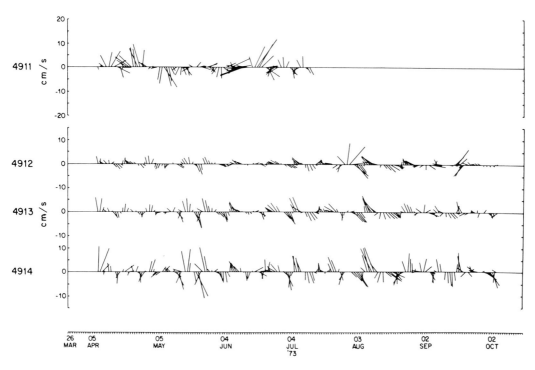

Figure 10.17B A high-passed version of figure 10.17A. The lower layers are now dominated by fast-bottom intensified oscillations. (Rhines, 1977.)

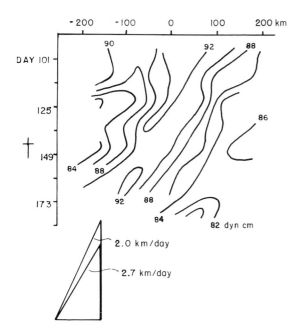

Figure 10.18 Time-longitude plot of 501–1500-db dynamic height along 28°N from MODE showing westward propagation of "thermocline eddies" but at a rate significantly slower than that observed in 1500-m currents (figure 10.6). (Rhines, 1977.)

as the boundary-value problem governing periodic $[\exp(-i\sigma t)]$ second-class (and low-frequency first-class internal) waves over the linear beach. Ou (1979) saw that the affine transformation

$$z = \hat{z}(f_0^2 - \sigma^2)^{1/2}/N_0 \qquad (10.121)$$

followed by a rotation of coordinates from $(x, y, \hat{z})$ to (x', y, z') such that the beach $z = -ax$ now becomes $z' = 0$ leads to

$$\frac{\partial^2 p}{\partial z'^2} + \frac{\partial^2 p}{\partial x'^2} - k^2 p = 0, \qquad (10.122)$$

$$\frac{\partial p}{\partial n} = \frac{k f_0}{\sigma} \frac{a'}{(1 + a'^2)^{1/2}} p \quad \text{at} \quad z' = 0, \qquad (10.123)$$

$$\frac{\partial p}{\partial n} = 0 \quad \text{at} \quad z' = a'x' \qquad (10.124)$$

for motions periodic in $y\,[\exp(iky)]$, and furthermore that this is *exactly* Ursell's (1952) problem just turned upside down; (10.122), (10.123), and (10.124) correspond to (10.93), (10.94), and (10.97). The transformed beach slope a' is given by

$$a' = aN(f_0^2 - \sigma^2)^{-1/2}. \qquad (10.125)$$

For second-class waves

$$\sigma^2 < f_0^2, \qquad (10.126)$$

a' is real and all of Ursell's (1952) results are immediately available. There are thus coastally trapped waves whose dispersion relation is

$$\sigma = f_0 a'/\{(1 + a'^2)^{1/2} \sin[(2n + 1)\tan^{-1} a']\} \qquad (10.127)$$

as well as a continuum of bottom-trapped waves that have the form

$$P = \cos(lx' + \text{phase}) \exp[-i\sigma t + iky - (l^2 + k^2)^{1/2} z']$$

as $x' \to \infty$ and that have their dispersion relation included in

$$\sigma = -N \sin(\tan^{-1} a)\left[\frac{k}{(l^2 + k^2)^{1/2}}\right]. \qquad (10.128)$$

The latter are just the bottom-trapped waves (10.110)–(10.112) of Rhines (1970).

Equation (10.128) is their dispersion relation in a half-plane bounded by the sloping bottom (Rhines, 1970). The frequency may be either sub- or super-inertial. The coastally trapped waves have $\sigma > N\sin(\tan^{-1}a)$ [by analogy with the fact that for Ursell's (1952) edge waves $\sigma^2 < gk$] and there are a finite number of them:

$$n = 1, 2, \ldots < [\pi/(4\tan^{-1}a') - 1/2]. \qquad (10.129)$$

Note that $n = 0$ would imply $\sigma = f_0$ but this does not solve the full equations (10.30) and associated boundary conditions.

In the limit of decreasing slope a', (10.129) allows ever more modes, and the dispersion relation (10.127) for coastally trapped waves simplifies to

$$\sigma = -f_0(2n + 1)^{-1}. \qquad (10.130)$$

This is the low-frequency, short-wavelength second-class limit of the barotropic dispersion relation (10.98). We thus identify Ou's coastally trapped modes as the stratified analog of the already familiar refractively trapped second-class topographic Rossby waves.

Over the linear beach, then, stratification limits the number of second-class refractively trapped topographic waves and opens up a new continuum of bottom-trapped waves. Figure 10.19 compares barotropic and baroclinic dispersion relations when the ocean surface is rigid. Further results are given by Ou (1979).

Suppose now that the linear beach terminates in a flat bottom of depth D_0, as in (10.102). Ou's (1979) transformation allows us to deal efficiently with the stratified problem. Figure 10.20 summarizes the boundary-value problem and its alteration by Ou's transformation into an equivalent problem in deep-water waves (figure 10.20D). In this latter problem, the deep-water continuum that existed for the linear beach must now be quantized into an infinite family of modes by repeated reflection between the shoaling and the

315
Long Waves and Ocean Tides

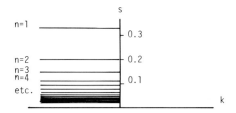

Figure 10.19A The dispersion relation (10.130), viz., $s = -1/(2n + 1)$ for topographic Rossby waves refractively trapped over the linear beach $D = -ax$ beneath rotating homogeneous fluid. There are an infinite number $n = 1, 2,...$ of trapped modes.

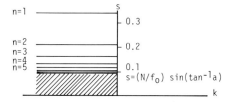

Figure 10.19B The dispersion relation (10.127)

$$s = a'/\{(1 + a'^2)^{1/2} \sin[(2n + 1)\tan^{-1} a']\},$$

where $a' = aN_0/[f_0(1-s^2)^{1/2}]$, for topographic Rossby waves refractively trapped over the linear beach $D = -ax$ beneath uniformly stratified (buoyancy frequency N_0) rotating fluid. There are a finite number $n = 1, 2,... < [\pi/(4 \tan^{-1} a') - \frac{1}{2}]$ of trapped modes all with frequencies $s > N_0 f_0^{-1} \sin[\tan^{-1} a]$. At lower frequencies a continuum of bottom-trapped modes reflected from $x = \infty$ exists. Sketch is for $N_0 = 10 f_0$, $a = 10^{-2}$.

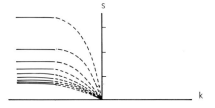

Figure 10.19C Sketch of dispersion relation for topographic Rossby waves in rotating stratified fluid over a linear beach that terminates in a uniform-depth ocean. The continuum of bottom-trapped modes that existed over the semi-infinite beach is quantized. All dispersion curves pass through $s = 0$ as $k \to 0$.

overhanging coasts. Low-mode edge waves have small amplitude at the overhanging coast and are not much affected by it. But higher-mode edge waves have appreciable amplitude at the overhanging coast and they blend smoothly into the infinite family of modes made up of waves repeatedly reflected between beach and the overhanging coast. All these results have direct analogs in the original stratified problem (figure 10.20A). There are a number of second-class topographic waves that are refractively trapped near the coast and that have decayed to very small amplitudes at the seaward termination of the beach. The continuum of bottom-trapped waves present over the unending linear beach is replaced by an infinite family of bottom-trapped waves reflected repeatedly between the coast and the seaward termination of the beach. In the special case $a = \infty$ of a perpendicular coast, their dispersion relations are easily seen to be

$$\sigma = ND_0 k/n\pi \qquad (10.131)$$

and they then correspond to ordinary internal Kelvin waves (in this case there are no refractively trapped modes). Other shelf geometries invite similar treatment. Figures 10.20E and 10.20F show the equivalent deep-water wave problem for a step shelf. Deep-water waves on surface 1 of figure 10.20F that are short enough that their particle displacements at the level of surface 2 are negligible correspond in the stratified problem (figure 10.20E) to internal Kelvin waves trapped against the coast. Deep-water waves on surface 2 of figure 10.20F correspond to baroclinic counterparts of the double Kelvin wave that, in homogeneous fluid, may be trapped along a discontinuity in depth (Longuet-Higgins, 1968b; see also section 10.4.6).

In general, if the equivalent deep-water wave problem has the waveguide-like dispersion relations $\sigma^2/g = F_n(H, \gamma, k)$, $n = 1, 2, \ldots$, with H, γ as defined in figure 10.20D, then the corresponding stratified shelf problem must have the dispersion relation

$$\frac{kf}{\sigma} \times \frac{a'}{(1 + a'^2)^{1/2}} = F_n[((D_0 a'/a)^2 + D_0 a^{-2})^{1/2},$$

$$\tan^{-1} a', k] \qquad (10.132)$$

so that for all the dispersion curves $\sigma \to 0$ as $k \to 0$. The dispersion relation must thus qualitatively look like figure 10.19C. Wang and Mooers (1976) have generalized the problem numerically to more complex shelf profiles with nonuniform stratification $N(z)$.

Ou's (1979) transformation is most useful for second-class motions (10.125) because for them the transformed coordinates are real. For first-class motions with $\sigma^2 > f_0^2$ it still produces Laplace's equation but now z' is imaginary. Wunsch (1969) has nonetheless been able to use it to discuss first-class internal waves obliquely incident on the linear beach without rota-

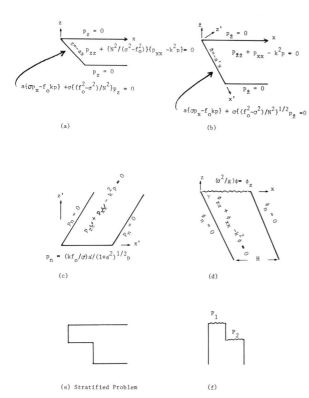

Figure 10.20 The deep-water surface-wave analog (d,f) of two shelf problems involving topographic Rossby waves in uniformly stratified rotating fluid: (a) stratified problem; (b) result of affine transformation; (c) result of rotation; (d) equivalent deep-water problem (velocity potential ϕ); (e) stratified problem; (f) equivalent deep-water problem (atmospheric pressure P_1 must be maintained lower than P_2 for physical realizability).

tion. This is the stratified analog of Eckart's (1951) nonrotating LSW study of waves over a sloping beach (section 10.4.6).

For beach slopes much smaller than the slope (σ/N) of (low-frequency) internal wave characteristics (10.43), Wunsch thus found that internal waves are refracted just like surface gravity waves by the shoaling relief and that refractively trapped edge modes occur. From the dispersion relation

$$\left(\frac{n\pi}{D_0}\right)^2 = \frac{N^2 - \sigma^2}{\sigma^2}(l^2 + k^2)$$

for plane internal waves of the form

$$w = \sin\left(\frac{n\pi z}{D_0}\right)\exp(-i\sigma t + ilx + iky)$$

over a uniform bottom D_0, l must ultimately become imaginary if D_0 is allowed to grow parametrically offshore while n and k are held fixed. One would therefore expect a WKB treatment of internal waves over gently shoaling relief to result in refraction and refractive trapping *provided* that the mode number n does not change, i.e., provided that the relief does not scatter energy from one mode into others. Constancy of n is indeed a feature of Wunsch's solutions but it cannot be expected to hold for more abrupt relief, especially if the relief slope exceeds the characteristic slope. If the relief couples modes efficiently, then scattering into higher modes allows l to remain real even in deep water far from shore so that energy is not refractively trapped near the coast. In principle, scattering into internal modes thus even destroys the perfect trapping of long surface gravity waves predicted by LSW theory over a step shelf, but in practice appreciable trapping is often observed. The efficiency of mode coupling depends both on the relief and on the vertical profile $N(z)$ of the buoyancy frequency, so that a general result for internal waves is difficult to formulate.

10.4.8 Free Oscillations of Ocean Basins

Finding the free oscillations allowed by LTE in rotating ocean basins is difficult even in the f-plane (section 10.4.2). Platzman (1975, 1978) has developed powerful numerical techniques for finding the natural frequencies and associated flow fields of free oscillations allowed by LTE in basins of realistic shape and bottom relief. The general classification of free oscillations into first- and second-class modes characteristic of the idealized cases discussed in sections 10.4.2 and 10.4.5 (effectively for a global basin) persists in Platzman's (1975) calculations. For a basin composed of Atlantic and Indian Oceans, there are 14 free oscillations with periods between 10 and 25 hours. Some of these are very close to the diurnal and semidiurnal tidal periods, and all of them, being within a few percentage points of equipartition of kinetic and potential energies, are first-class modes. There are also free oscillations of much longer period, for which potential energy is only about 10% or even less of kinetic energy; they are second-class modes.

I know of no extratidal peaks in open-ocean sea-level records that correspond to these free oscillations. There is some evidence in tidal admittances for the excitation of free modes but the resonances are evidently not very sharp (see section 10.5.1). Munk, Bryan, and Zetler (private communication) have searched without success for the intertidal coherence of sea level across the Atlantic that the broad spatial scale of these modes implies. The modes are evidently very highly damped.

10.5 The Ocean Surface Tide

10.5.1 Why Ocean Tides Are of Scientific Interest

The physical motivation for studying and augmenting the global ensemble of ocean-tide records has expanded enormously since Laplace's time. In this section I have tried to sketch the motivating ideas without getting involved in the details of theoretical models; some of these receive attention in subsequent sections.

Certain of the ancients knew a great deal about tides [see, e.g., Darwin's (1911a) summary of classical references], but the first extant reduction of observations made explicitly for predictive purposes may be the table of "flod at london brigge" due to Wallingford who died as Abbot of St. Alban's in 1213 (Sager, 1955). Making practical tide predictions was probably *the* preoccupation of observers for the next 500 years.

In 1683, Flamsteed (Sager, 1955) produced a table of high waters for London Bridge as well as, in the following year, corrections making it applicable to other English ports. Darwin (1911a) quotes Whewell's description, written in 1837, of how successors to Flamsteed's tables were produced:

The course . . . would have been to ascertain by an analysis of long series of observations, the effects of changes in the time of transit, parallax, and the declination of the moon and thus to obtain the laws of phenomena.
. . . Though this was not the course followed by mathematical theorists, it was really pursued by those who practically calculated tide tables. . . . Liverpool London, and other places had their tables, constructed by undivulged methods . . . handed down from father to son.
. . . The Liverpool tide tables . . . were deduced by a clergyman named Holden, from observations made at that port . . . for above twenty years, day and night. Holden's tables, founded on four years of these observations, were remarkably accurate.
At length men of science began to perceive that such calculations were part of their business. . . . Mr. Lubbock . . . , finding that regular tide observations had been made at the London docks from 1795, . . . took nineteen years of these . . . and caused them to be analyzed. . . . In a very few years the tables thus produced by an open and scientific process were more exact than those which resulted from any of the secrets.

Quite aside from its proprietary aspects, Darwin (1911b) explicitly notes the synthetic nature of this process; it at least conceptually represents "the oscillation of the sea by a single mathematical expression" provided by Bernoulli in 1738 for an inertialess ocean (the equilibrium tide), by Laplace for a global ocean obeying Newton's laws of motion, and assumed to exist for actual oceans even if too complex to represent in simple form.

Kelvin, in about 1870 (Darwin, 1911b) introduced the harmonic method, which Darwin (1911b) calls "analytic" because synthesis of the entire tide into one dynamically derived form is abandoned and instead the tide at any given place is regarded as a sum of harmonic oscillations whose frequencies are determined from astronomy (section 10.2) but whose amplitudes and phases must be determined from analysis of *in situ* sea-level observations. Prediction is then carried out by recombining the harmonic oscillations at future times.

Kelvin's suggested procedure was made feasible by the introduction of recording tide gauges in which the motion of a float in a well, insulated from short-period waves but otherwise freely connected with the sea, drives a pencil up and down a paper wrapped on a drum rotated by clockwork, thus producing a continuous plot of sea level versus time. [Darwin (1911a) describes contemporary instruments.] Harmonic analysis of this record *at relatively few astronomically determined frequencies* was feasible by judicious sampling and manual calculation. The recombination of harmonics at future times was then carried out mechanically by means of a series of pulleys, movable at frequencies corresponding to the astronomical ones, that drove a pencil over a paper wrapped on a drum rotated by clockwork, thus ultimately providing a plot of predicted sea level versus time. The design of such a machine was due to Kelvin, and elaborations were in regular use until the mid-1960s (Zetler, 1978).

Even before Kelvin's introduction of the harmonic method, Lubbock and Whewell (Darwin, 1911b) had begun to combine observations at different ports into cotidal maps showing the geographical variation of sea level associated with tides. Thus Airy in 1845 gave a chart (modified by Berghaus in 1891) of locations of high water at different times of day in the North Sea (figure 10.21). Concerning this, Darwin (1911b) remarks, "It will be noticed that between Yarmouth and Holland the cotidal lines cross one another. Such an intersection of lines is in general impossible; it is indeed only possible if there is a region in which the water neither rises nor falls. . . . A set of observations by Captain Hewitt, R.N. made in 1840 appears to prove the existence of a region of this kind." This is probably the first recorded observation of an amphidromic point.

But hourly maps of high-water locations change throughout the month. Kelvin's harmonic analysis decomposed the tide into harmonic components for which a single cotidal map, with cotidal lines drawn at fractions of the period of the component, can represent the entire spatial variation of that component forever. Time series at thousands of ports may thus be reduced to a handful of global maps that are ideal summaries of observations for comparison with solutions of LTE forced by the different harmonic components of the ATGF. Kelvin's abandonment of the "synthetic" viewpoint thus in effect provided the means for its reinstatement.

Of the handful of such maps constructed empirically for global tides, Dietrich's (1944) are perhaps the most widely quoted. Villain (1952) gives an extensive discussion of the observations leading to his global M_2 cotidal map (figure 10.22). Much modern tidal research has consisted of attempts to apply the principles of

Figure 10.21 Airy's chart of cotidal lines in British seas. (Darwin, 1911b.)

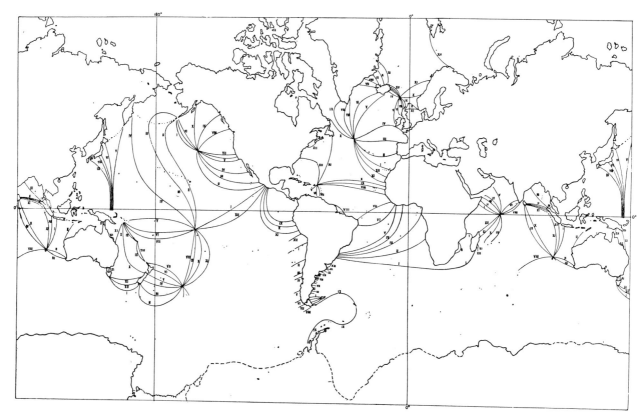

Figure 10.22 Cotidal lines for M_2 (in lunar hours relative to moon's transit over Greenwich). (Villain, 1952.)

dynamics to reproduce and hence to "explain" the global distribution of tides as suggested by such empirical maps. But the degree of success achieved to date, as well as insight into the variation of the dynamics of tides over the globe, has required thinking about how the response of the ocean would change if tidal frequencies could be varied. Since they cannot be, this implies comparing global tidal maps at different tidal periods. The origins of this viewpoint are found in the work of Munk and Cartwright (1966), who were enabled by the advent of modern computers to analyze 19 years of hourly tide readings at Honolulu and Newlyn "without astronomical prejudice as to what frequencies are present and what are not, thus allowing for background noise."

Their work has been influential in two very general ways quite apart from the improvement in tide prediction that it introduced. First, it provided a clear distinction between sea-level fluctuations due to TGF and those of similar period due to nontidal agents, a distinction crucial in establishing the significance of any geophysical interpretation of all but the strongest constituents of ocean or solid-earth tides. Second, it introduced the idea of oceanic admittance, the (possibly complex) ratio between ocean response and forcing, as a continuous function of frequency that can be estimated from tidal observations and that summarizes the dynamic response of the ocean to time-variable forcing in a manner easily related to the properties of free solutions of LTE by an expansion in eigenfunctions.

If the ocean had many sharp resonances within the frequency bands spanning the three species, the tidal admittance would have amplitude peaks and rapid phase shifts. Typical deep-sea admittances tend to be smooth across a species but are far from constant. Admittance curves for the Coral Sea (Webb, 1974) and at Bermuda (Wunsch, 1972c) are shown in figures 10.23A and 10.23B, respectively. The Coral Sea admittance is unusual in its very sharp sudden variation between M_2 and S_2, apparently showing the existence of a sharp local resonance. The amplitude of the Bermuda admittance rises smoothly, by 400% toward lower frequencies over the semidiurnal band; Wunsch's result is consistent with Platzman's (1975) prediction of an Atlantic resonance of roughly 14-hour period, but one appreciably broadened by dissipation.

Smoothness of the admittance across tidal bands was anticipated by Munk and Cartwright (1966) in their "credo of smoothness": "We do not believe, nor will we tolerate, the existence of very sharp resonance peaks." In part, this credo had its origin in the prevailing beliefs, since then largely confirmed, that ocean

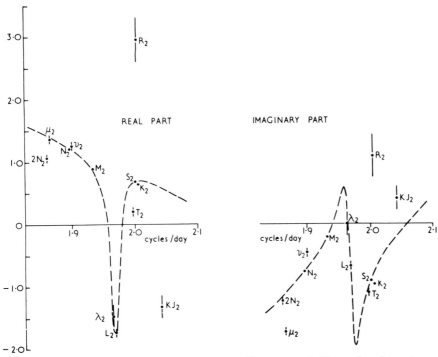

Figure 10.23A The real and imaginary parts of the response function at Cairns (16°55'S, 145°47'E), showing a resonance in the Coral Sea. (Webb, 1974.)

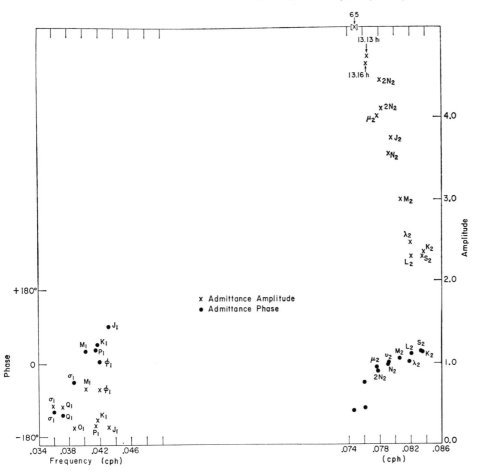

Figure 10.23B Amplitude and phase of the admittance at Bermuda. (Wunsch, 1972c.)

321
Long Waves and Ocean Tides

tides must be of rather low Q. Evaluation of Q requires knowledge of the total energy E stored in the tide as well as the rate $\dot{E}$ at which it is dissipated; then

$$Q = \frac{2\pi E}{\dot{E}T}.$$

Estimation of the stored energy E was difficult before modern numerical solutions of LTE because of the open-ocean detail required. Earliest estimates assumed the tide to be in equilibrium; allowance for (likely) equipartition between potential and kinetic energy and for the area of the oceans led to an estimate of 5.6×10^{16} J for M_2 (Garrett and Munk, 1971). I interpolated coastal M_2 harmonic constants over the globe by solving LTE with these as boundary values and thus obtained (Hendershott, 1972) an estimate of 7.29×10^{17} J. But my kinetic energy was over twice my potential energy and I now believe this to have been a numerical artifact, especially since Platzman's (1975) near-tidal normal modes are within a few percentage points of equipartition. My estimate should thus be revised to 5.14×10^{17} J. Parke and Hendershott (1980) improved the interpolation by taking island data into account and found 2.68×10^{17} J (assuming equipartition).

The estimation of $\dot{E}$ historically has been of importance in cosmology. Halley in 1695 first discovered that the apparent position of the moon is *not* that predicted by (frictionless) Newtonian mechanics. The discrepancy is real; Munk (1968) outlines ultimately unsuccessful attempts to resolve it by appealing to perturbation of the moon's orbit by changes in the earth's orbit around the sun. Much of the discrepancy is now believed to be due to tidal friction. As Immanuel Kant noted in 1754, tidal friction must slow the earth's daily rate of rotation; this alone gives rise to an apparent perturbation of the moon's mean longitude. By the conservation of angular momentum, the moon's angular velocity about the earth-moon center of mass is also altered and (for the present prograde rotation of the moon about the earth) the moon recedes from the earth (by about 6 cm yr^{-1}; Cartwright, 1977). Müller (1976) reviews astronomical data both ancient (eclipse observations) and modern, and analyzes them simultaneously to estimate $\dot{n}_t$ (the tidal acceleration of the moon's longitude n), $\dot{\Omega}/\Omega$ (the observed apparent acceleration of the earth's rotation frequency Ω), $\dot{\Omega}/\Omega_{TNT}$ (the nontidal part of $\dot{\Omega}/\Omega$), and $\dot{G}/G$ (the possible rate of change of the gravitational constant G). He finds

$\dot{n}_t = -27.2 \pm 1.7''$ cy^{-2},

$\dot{\Omega}/\Omega = -22.6 \pm 1.1 \times 10^{-11}$ yr^{-1},

the latter corresponding to a lengthening of day of 2.0×10^{-3} s cy^{-1}. If he assumes $\dot{G}/G = 0$, then $\dot{\Omega}/\Omega_{TNT}$ becomes $9.2 \pm 2.5 \times 10^{-11}$ yr^{-1}, a sizable portion of $\dot{\Omega}/\Omega$ that demands geophysical explanation. Various cosmological theories have $\dot{G}/G$ of order 5×10^{-11} yr^{-1}; $\dot{\Omega}/\Omega_{TNT}$ then becomes zero with an uncertainty of order 5×10^{-11} yr^{-1}. "It appears that either we really have a (non zero) cosmological constant $\dot{G}/G$ consistent with the Hubble constant, or we have a significant $\dot{\Omega}/\Omega_{TNT}$" [Müller (1976)].

Lambeck (1975) gives expressions for the tidally induced rates of change of the semimajor axis a of the moon's orbit, of its eccentricity e, and of its orbital inclination i in terms of a spherical harmonic decomposition of the ocean tide ζ_0. For semidiurnal tides only the second harmonic is important. Once these rates of change have been estimated, then $\dot{n}_t$ and $(\dot{\Omega}/\Omega)_t$ (i.e., the tidal part of $\dot{\Omega}/\Omega$) may be estimated from, respectively, Kepler's law (Cartwright, 1977, equation 8.3) and from the conservation of angular momentum (Lambeck, 1977, equations 2). $\dot{n}_t$ and $(\dot{\Omega}/\Omega)_t$ imply a rate $\dot{E}_t$ of tidal energy dissipation in the earth-moon system (Lambeck, 1977, equations 3). Using global calculations of ζ_0 for the M_2 ocean tide by Bogdanov and Magarik (1967), by Pekeris and Accad (1969) and by myself (Hendershott, 1972), Lambeck (1977) thus estimates for M_2

$\dot{n}_t = -27.8 \pm 3''$ cy^{-} (his table 7),

$(\dot{\Omega}/\Omega)_t = -25.8 \times 10^{-11}$ yr^{-1} (his table 8),

$\dot{E}_t = 3.35 \times 10^{19}$ erg s^{-1} (his equation 3b).

Since his work, new M_2 calculations by Accad and Pekeris (1978) and by Parke and Hendershott (1980) have appeared. These calculations include ocean self-attraction and loading (section 10.5.3) and are not unrealistically resonant. Accad and Pekeris (1978) directly evaluate the flow of M_2 energy out of the numerical ocean and obtain 2.44-2.79×10^{19} erg s^{-1}. Parke and Hendershott (1980) evaluate the rate $\langle \dot{W} \rangle$ at which the M_2 tide generating forces (potential Γ) and ocean floor (solid earth M_2 tide δ) do work on the ocean averaged ($\langle \ \rangle$) over a tidal period

$$\langle \dot{W} \rangle = \iint_{\text{ocean}} \left(\rho \left\langle \Gamma \frac{\partial \zeta_0}{\partial t} \right\rangle + \rho g \left\langle \zeta_0 \frac{\partial \delta}{\partial t} \right\rangle \right) dA$$

(Hendershott, 1972) and obtain 2.22×10^{19} erg s^{-1}. All this work is lost in tidal friction. If these results are taken as an improved estimate $\dot{E}'_t$ of E_t for the M_2 tide

$\dot{E}'_t = 2.2$-2.8×10^{19} erg s^{-1},

then Lambeck's (1975) procedure would yield

$\dot{n}'_t = -(18.3$-$23.2)''$ cy^{-2},

$(\dot{\Omega}/\Omega)'_t = -(16.9$-$21.6) \times 10^{-11}$ yr^{-1}

for M_2. If we retain unaltered Lambeck's (1977) estimate of the contribution $\Delta \dot{n}$ and $\Delta(\dot{\Omega}/\Omega)$ of all remaining tides to $\dot{n}_t$ and $(\dot{\Omega}/\Omega)_t$,

$\Delta \dot{n} = -3.1'' \text{ cy}^{-2}$,

$\Delta(\dot{\Omega}/\Omega) = -6.9 \times 10^{-11} \text{ yr}^{-1}$,

then we obtain the revised estimates for all tides:

$\dot{n}_t = (21.4-26.3)'' \text{ cy}^{-2}$,

$(\dot{\Omega}/\Omega) = -(23.8-28.5) \times 10^{-11} \text{ yr}^{-1}$.

These are to be compared with Müller's (1976) estimates from astronomical data:

$\dot{n}_t = -27.2'' \text{ cy}^{-2}$

and

$(\dot{\Omega}/\Omega)_t = -(13.4-22.6) \times 10^{-11} \text{ yr}^{-1}$

for

$\dot{G}/G = -(0-6.9) \times 10^{-11} \text{ yr}^{-1}$.

The comparison is worst if $\dot{G}/G$ is taken zero and becomes rather good if $\dot{G}/G$ is allowed to differ from zero.

There is thus some interest in estimating $\dot{E}$ for ocean tides but, as indicated above, results differ significantly depending on details of the estimation procedure. The estimates referred to above (except for that of Accad and Pekeris, 1978) essentially use global cotidal maps to find the part of the ocean tide in phase with the tide generating forces. The resulting rate of working $\dot{W}$ is then attributed to friction without having to localize it anywhere. Indeed, the long waves making up the tide transmit energy over the globe so readily that we may expect no correlation between where the moon and sun work hardest on the sea and where the energy thus put into the sea is dissipated.

It may be that little of that dissipation occurs in the open ocean. Taylor (1920) estimated tidal friction in the Irish Sea and showed that most of the energy thus lost comes from the adjacent deep ocean with little direct input due to local working by moon and sun. His methods were extended to the world's coasts and marginal seas by Jeffreys (1921), Heiskanen (1921), and Miller (1966). Miller finds $\dot{E} = 0.7-2.5 \times 10^{19} \text{ erg s}^{-1}$, two-thirds of which occurs in the Bering Sea, the Sea of Okhotsk, the seas north of Australia, the seas surrounding the British Isles, the Patagonian shelf, and Hudson Bay. This is below all but the most recent estimates of $\langle \dot{W} \rangle$. It should be, by perhaps 10%, because of open-ocean internal tidal dissipation not consistently or completely taken into account (section 10.6). It is now difficult to say whether or not the difference indicates an important omission of some dissipative mechanism.

Additional information about tidal dissipation is contained in the width of conjectured or observed peaks in the admittance amplitude and in shifts in phase of the admittance from one constituent to another. Thus the width of the amplitude-response curve at Bermuda (figure 10.23B; Wunsch, 1972c) suggests a local Q exceeding about 5. Garrett and Munk (1971) surveyed the difference in admittance phase between M_2 and S_2 (the age of the tide) and concluded that worldwide semidiurnal tides had a Q of order 25. Webb (1974) argued that such age-derived estimates of Q primarily reflect localized resonances. It is thus difficult to compare such results with the global Q, for M_2, with a Q of 17 emerging from the most recent cotidal chart of Parke and Hendershott (1980).

Astronomical and oceanographic interest in the amount and geographical distribution of tidal friction constitutes one of the principle modern motivations for studying ocean tides. The other principle motivation is the need, by solid-earth tidalists (Farrell, 1979) and satellite geodesists (Marsh, Martin, McCarthy, and Chovitz, 1980) for a very accurate map of the global distribution of ocean tides. Significant improvement of the most recent numerical maps is going to require extensive new observations.

The technology of deep-ocean pressure sensors suitable for gathering pelagic tide records was pioneered by Eyriés (1968), F. E. Snodgrass (1968) and Filloux (1969). The latest compilation of such results (Cartwright, Zetler, and Hamon, 1979) summarizes harmonic constants for 108 sites irregularly distributed around the world. Cartwright (1977) reviews the history and considerable accomplishments of pelagic tide recording but concludes that economic and political difficulties as well as rapidly evolving research priorities make it an unlikely method for detailed global tide mapping.

Several alternative methods are beginning to be studied. Given a sufficient number of measurements of the solid-earth tide, it is possible to construct estimates of the ocean tide that (in part) generated the solid-earth tide. But high precision earth-tide measurements are needed, and ocean tides in the vicinity of coastal earth-tide stations must be accurately known in order to perceive global ocean tide contributions (Farrell, 1979). Kuo and Jachens (1977) document attempts along these lines.

Satellites may be employed to study ocean tides in two ways. First, the periodic tidal deformation of earth and ocean results in significant perturbation in the orbits of close satellites (Cazenave, Daillet, and Lambeck, 1977). The lowest-order spherical-harmonic components of the tide are the most accessible by this method. It therefore complements the second possibility, direct measurement of satellite-to-sea-surface altitude. The greatest obstacle to extraction of ocean tides from such altimetry is not the error in the altitude measurement but rather the error in our knowledge of where the satellite is relative to the center of the earth. This "tracking" or "orbit" error is greatest at a spatial scale corresponding to the earth's circumference but decreases rapidly at smaller spatial scales. It probably makes the large-scale features of ocean tides inaccessible from the GEOS-3 altimetry-data set. But smaller-scale tidal systems appear to be directly observable from the later SEASAT-1 altimetry. Parke has permitted me to reproduce (figure 10.24) his recovery of tides along the Patagonian shelf from SEASAT-1 altimetry (Parke, 1980) as an example.

Determining the combination of ocean-tide gauge data (coastal, island, and pelagic), of earth-tide data, of satellite-orbit perturbations, and of satellite altimetry optimal for mapping ocean tides and localizing their dissipation is now perhaps the outstanding theoretical problem in ocean tides.

10.5.2 Partial Models of Ocean Tides

Introduction In his George Darwin lecture "The Tides of the Atlantic Ocean," Proudman (1944) stated, "I shall mainly be concerned with the discovery of the distribution of tides over the open Atlantic Ocean, by the application of the principles of dynamics."

This was, of course, Laplace's goal for global tides. From Laplace's time until now, many researchers have pursued this goal with dogged persistence by solving LTE with astronomical forcing for oceans having shape and relief sufficiently idealized that existing methods of solution could produce an evaluable answer. With hindsight, the properties of these solutions may be appreciated by regarding the solution as eigenfunction expansions in which the various eigenfunctions $Z_n(\phi,\theta) \exp[-i\sigma_n t]$ or free oscillations allowed by LTE have the properties summarized in section 10.4. The frequency σ_n of oscillation is the most natural eigenparameter, but the eigenfunction expansion $\zeta(\phi,\theta,t) = \sum_n a_n Z_n \exp[-\sigma_T t]$ for a tide forced at frequency σ_T is *not* of the usual form in which (in the absence of dissipation) $a_n \sim (\sigma_T^2 - \sigma_n^2)^{-1}$. If however, for a given frequency σ_T of forcing, the inverse Δ^{-1} of the mean depth

$$\Delta \equiv (4\pi)^{-1} \iint D(\phi,\theta) \cos\theta \, d\theta \, d\phi$$

$[D(\phi,\theta) = 0$ on land$]$ is regarded as the eigenparameter with "resonant" depths Δ_n, then $a_n \sim (\Delta^{-1} - \Delta_n^{-1})^{-1}$. Nothing restricts Δ_n to positive values. Indeed, negative-depth modes having $\Delta_n < 0$ often exist and may be important in the eignefunction expansion of forced solutions. This evidently was pointed out first by Lindzen (1967) for atmospheric tides.

Direct numerical solution of LTE in realistically shaped basins may be viewed as summation of this eigenfunction expansion, and has gone some distance toward attaining Proudman's stated goal. But Proudman's George Darwin lecture marked an important break with the sequence of dynamic studies that have since culminated in modern numerical solutions.

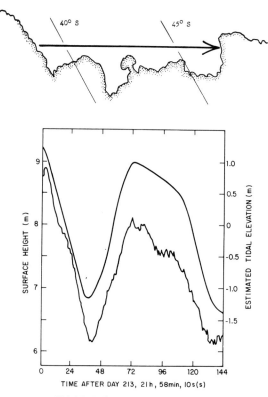

Figure 10.24 SEASAT altimeter record (wiggly line) and a reconstruction (smooth lines) from coastal harmonic constants of nearshore Patagonian shelf tides at the subsatellite point for the SEASAT pass whose path is shown in the upper panel. (Parke, 1980.)

Rather than solving LTE for Atlantic tides, Proudman computed free and forced M_2 solutions of LTE for a portion of the Atlantic and fitted their sum to observations. Subsequent studies carried out in the same spirit but for more simple continental-shelf and marginal-sea geometries have provided dynamically understandable rationalizations for the distribution of tides in these regions and have led to a reappraisal of both observations and of global solutions of LTE. Discussion of these matters occupies the remainder of this section.

Tides in the Gulf of California Godin (1965) and Hendershott and Speranza (1971) noted that (10.29) is satisfied for all the Poincaré channel modes $n = 1,2,\ldots$ in many of the world's long and narrow marginal seas. In these, then, all Poincaré modes are evanescent so that the tide away from the ends of the basin must be mainly a sum of two oppositely traveling Kelvin waves, usually of unequal amplitude. Friction in the basin (or a net rate of working on the tide-generating body by tides in the basin) will make the outgoing Kelvin wave of lower amplitude than the incoming one and will shift amphidromic points (at which the two Kelvin waves interfere destructively) toward the "outgoing" coast.

Figure 10.25 illustrates application of these ideas to the M_2 tide in the Gulf of California. The westward displacement of the amphidromes points to substantial dissipation in the upper reaches of the Gulf. But this Kelvin wave fit does not well represent the tide there. On the basis of his extensive network of tide-gauge observations, Filloux (1973a) was able to estimate the tidal prism and mass transport for six sections along the length of the Gulf and could thus directly evaluate stored energy and energy flux along the Gulf, and energy flux from the moon into dissipation. About 10% of the energy entering the mouth from the Pacific $(4.7 \times 10^{16} \mathrm{~erg~s^{-1}})$ is lost as the Gulf M_2 tide works on the moon; the remainder is dissipated frictionally (over 80% northward of the islands in figure 10.25).

Elementary considerations suggest that the Gulf of California has a resonance fairly close to the semidiurnal tidal frequency. Filloux (1973a) estimates a Q of about 13 for the thus nearly resonant M_2 tide. Stock (1976) constructed a finite-difference model of Gulf tides using a very fine (10-km) mesh. His solutions effectively sum both of the Kelvin waves and all the evanescent Poincaré modes as well as allowing for their distortion by the irregular shape of the basin. He included dissipative effects and specified the elevation across the mouth of the Gulf in accordance with observations. His model is resonant at about 1.8 cpd with a Q sufficiently high that different discretizations of the problem, all a priori equally reasonable, can give very different Gulf tides. He found it necessary to force his model to have a realistic resonant frequency—fixed by arbitrarily varying the mean depth—before it would produce realistic cotidal maps (figure 10.26). Once this had been done, he found small but nevertheless significant sensitivity of the solution to the localization of dissipation; the solution agreeing best with Filloux's data was that in which most of the dissipation took place around the islands in the upper portion of the Gulf.

The Boundary-Value Problem for Marginal Sea Tides The Gulf of California is one of many marginal seas that connect with the global ocean across a relatively small mouth. Dynamic models of tides in such regions have generally been constructed by solving LTE in the region subject to the condition that the elevation

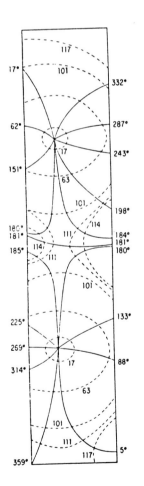 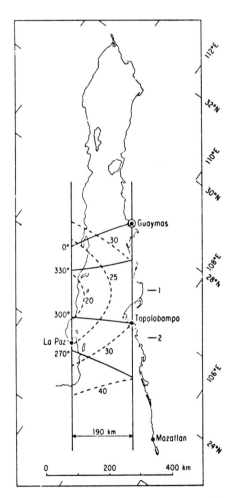 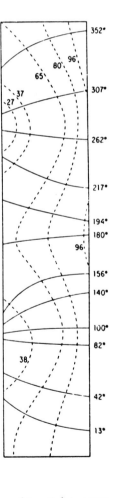

Figure 10.25 Left and right panels are co-oscillating tides in a rectangular gulf with little (left panel) and much (right panel) absorption at upper boundary. Center panel is a Kelvin wave fit to M_2 as observed in the Gulf of California.

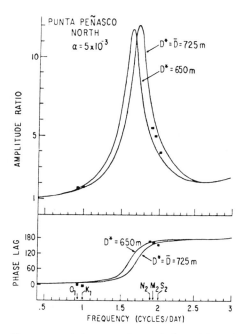

Figure 10.26 A comparison of the tidal response at Punta Penasco (solid squares) with the tide at the mouth of the Gulf of California for two numerical models with different mean depths.

across the open mouth is equal to that actually observed. This is disadvantageous for two reasons. First of all, it eliminates damping of the marginal sea tide by radiation into the deep sea; second, it results in solutions that cannot predict the effects of changes in basin geometry (i.e., installation of causeways, etc.) on the tides because the tide across the open mouth is not allowed to respond to them.

Garrett (1974) pointed out that in many cases these difficulties may be resolved partially by allowing the marginal sea to radiate into an idealized deep sea. For a given constituent, suppose that, with forcing included and all other boundary conditions (i.e., no mass flux through coasts) satisfied, the mass flux $\alpha \delta(S - S')$ normal to the mouth (across which distance is measured by S) would result in the tide $\zeta_G(S) + \alpha K_G(S,S')$ across the mouth when the marginal-sea problem is solved and would result in $\zeta_D(S) + \alpha K_D(S,S')$ when the deep-sea problem is solved. The tides $\zeta_G(S)$ and $\zeta_D(S)$ are thus those that would result just inside and just outside across the mouth if it were closed by an imaginary impermeable barrier. In the real world, the mass flux $U(S)$ across the mouth is fixed by the necessity that its incorporation into either the marginal-sea or the deep-sea problem give the same tide $\zeta_M(S)$ across the mouth:

$$\zeta_M(S) = \zeta_G(S) + \int U(S')K_G(S,S')dS'$$
$$= \zeta_D(S) + \int U(S')K_D(S,S')dS. \quad (10.133)$$

The latter half of this relation is an integral equation to be solved for $U(S)$. Once $U(S)$ has been found, the problem for the marginal-sea tide is well posed. If the deep sea is, for example, idealized as an infinite half-plane ocean, then $K_D(S,S')$ can be constructed by imposing the radiation condition far from the mouth. The boundary condition across the mouth for marginal sea tides, the specified mass flux $U(S)$, will thus incorporate radiative damping into the solution for marginal-sea tides. Garrett (1974) has discussed limiting cases of (10.133). Garrett and Greenberg (1977) have used the method to discuss possible perturbations of tides by construction of a tidal power station in the Bay of Fundy.

$U(S)$ as given by (10.133) is also the correct marginal-sea boundary condition for models of deep-ocean tides. Its application could allow optimal coupling of finely resolved marginal-sea models to more coarsely resolved global ones, but the methodology requires further development.

Continental Shelf Tides When the tide progresses parallel to a fairly long, straight continental shelf, then the free waves of section 10.4.6 are natural ones in terms of which to expect an economical representation of the tide. Munk, Snodgrass, and Wimbush (1970) analyzed California coastal tides in this way. In addition to the free waves capable of propagating energy at tidal frequencies, they introduced a forced wave to take local working by TGF into account. For the M_2 tide, the Kelvin wave, the single representative member of the Poincaré continuum, and the forced wave have coastal amplitudes of 54, 16, and 4 cm, respectively. The coastal tide is dominated by the northward-propagating Kelvin mode, but further at sea the modes unexpectedly combine to yield an amphidrome (figure 10.27) whose existence was subsequently confirmed by Irish, Munk, and Snodgrass (1971). For the K_1 California tide, the corresponding amplitudes are 21, 24, and 9 cm; the Kelvin wave is not nearly as important. Platzman (1979) has shown how this local representation is related to the properties of eigensolutions of LTE for the world ocean.

The California coast is too low in latitude for second-class shelf modes (section 10.4.6) to propagate energy at tidal frequencies. At higher latitudes, however, Cartwright (1969) has found evidence of their excitation; strong diurnal tidal currents without correspondingly great diurnal surface tides. At very low latitudes, low-mode edge waves could be resonant at tidal frequencies. Stock (private communication) has applied these ideas to the west coast of South America and to the Patagonian shelf. Geometrical difficulties prevent quantitative results in the latter case but the qualitative prediction that the coastally dominant Kelvin

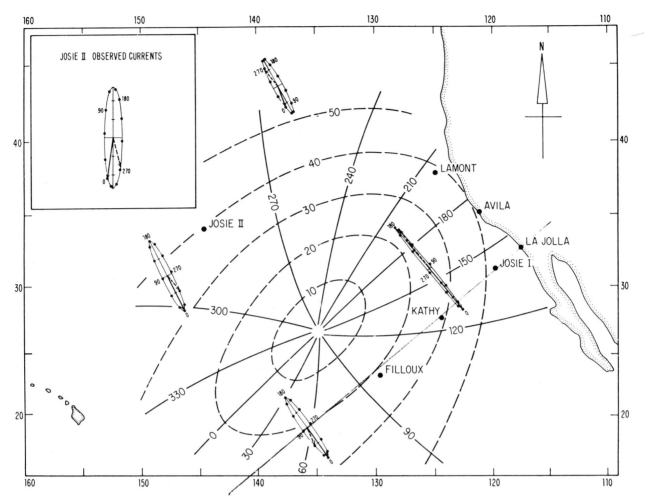

Figure 10.27 M$_2$ cotidal chart from Munk, Snodgrass, and Wimbush (1970) (amplitudes in cm, phases relative to moon's transit over Greenwich). Ellipses show computed currents at ellipse center (ticks on ellipse axis correspond to 1 cm s^{-1}). Modal fit was to coastal stations plus Josie I, Kathy, and Filloux. Subsequent observations at Josie II confirmed phase shift across predicted amphidrome (Irish, Munk, and Snodgrass, 1971.)

mode decays by e^{-1} across the broad and shallow Patagonian shelf and that the low speed of long-wave propagation over the shallow shelf so compresses the length scale of the tides that a complex system of several amphidromes fits over the shelf are nonetheless important.

On all the shelves so far mentioned, the tide advances parallel to the shelf so that decomposition into modes traveling parallel to the coast is natural. But not all shelf tides are of this nature. Redfield (1958) has summarized observations of United States east coast continental shelf tides (figure 10.28). There the salient features are a very close correspondence between local shelf width and the coastal amplitude and phase of the tide. Tides are nearly coincident over the entire length and width (Beardsley et al., 1977) of the shelf, in marked contrast with the California case.

Island Modification of Tides Island tide records have been prized in working out the distribution of open-ocean tides not only because of their open-ocean location but also because they have been supposed more representative of adjacent open-ocean tides than are coastal records.

Nevertheless, they are not entirely so. Tsunami travel-time charts suggest that tides in island lagoons may be delayed by as much as 20 minutes; harmonic constants for open-ocean tide charts correspondingly may need revision (Parke and Hendershott, 1980). Pelagic records (section 10.5.1) do not, of course, present this problem.

Diffractive effects near island chains may result in appreciable local modification of the tides. Larsen (1977) has studied the diffraction of an open-ocean plane wave of tidal frequency by an elliptical island (intended to model the Hawaiian Island plateau). A typical cotidal chart is shown in figure 10.29. Diffraction alters the time of high water by as much as an hour.

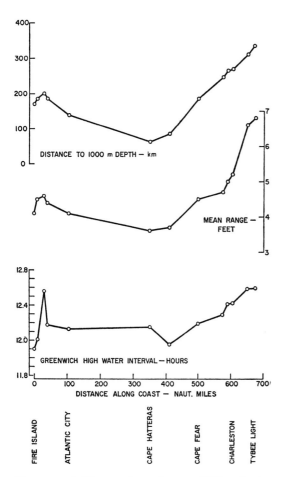

Figure 10.28 Distance from shore to 1000 m depth contour, mean coastal tidal range, and Greenwich high-water interval for selected outlying stations along the eastern coast of the United States. (Redfield, 1958.)

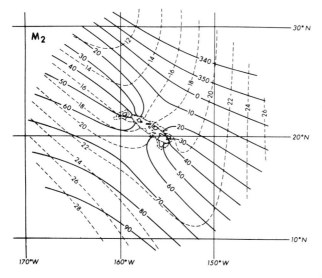

Figure 10.29 Theoretical cotidal chart for an M_2 plane wave in a uniformly rotating ocean of 5000 m depth incident from the northeast on an elliptical island modeling the Hawaiian Chain. (Larsen, 1977.)

10.5.3 Global Tidal Models

The shape of the world's oceans is so complicated that realistic solutions of LTE must be numerical. Pioneering studies were made by Hansen (1949) and by Rossiter (1958). The first global solution was presented by Pekeris and Dishon at the 1961 IUGG Assembly in Helsinki. I have reviewed subsequent developments elsewhere (Hendershott and Munk, 1970; Hendershott, 1973, 1977) and so will not attempt a comprehensive discussion.

Generally, numerical tidalists have solved (often by time-stepping) the forced LTE (10.5) with adjoined dissipative terms, and taking the numerical coasts as impermeable, or else they have solved the elliptic elevation equation [obtained by eliminating the velocities from LTE (10.5) without dissipative terms] for individual constituents (most often M_2) with elevation at the numerical coast somehow specified from actual coastal observations. Combinations of these approaches have also been employed.

The first procedure yields solutions that may be thought of as a weighted sum of the dissipative analogs of Platzman's (1975) normal modes (section 10.4.8). Neither mass nor energy flows across the numerical coast. If the dissipation is modeled accurately (a matter of real concern since the smallest feasible global mesh spacing of about 1° cannot adequately resolve many marginal-sea and shelf tides), then such models should have fairly realistic admittances.

The second procedure attempts to circumvent this difficulty by allowing most or all dissipation to occur beyond the numerical coasts in regions that thus do not have to be resolved. It yields solutions that may be thought of as a tide reproducing the prescribed coastal tide plus a superposition of eigensolutions [of LTE (10.5) or of the elevation equation] that have vanishing elevation at the numerical coast. These eigensolutions have no simple oceanic counterparts since their coastal boundary condition does not require vanishing coastal normal velocity. The full solution satisfies the forced LTE and reproduces the prescribed coastal tide but also generally does not have vanishing normal velocity at the numerical coasts. Consequently there may be at any instant a net flow of water through the numerical coastline, and the flux of energy (averaged over a tidal period) through the numerical coastline need not be zero.

This flux of energy through the numerical coast is a realistic feature since the numerical coast is not intended to model the actual coast but, instead, crudely models the seaward edges of the world's marginal seas and shelves. The same is true of the mass flux, although, in using the solution to estimate ocean-tide perturbations of gravity, etc., the water that thus flows through the numerical coast must somehow be taken into account (Farrell, 1972b). Perhaps the greatest drawback of the second procedure is the possibly resonant forcing of the unphysical zero-coastal-elevation eigensolutions. This can cause the model to have a very unrealistic admittance even though it is in principle capable of correctly reproducing all constituents. In practice, it often causes the model to be unrealistically sensitive to the way in which discretization of the equations or of the basin has been carried out. Thus Parke and Hendershott (1980) encountered resonances in solving for semidiurnal constituents by the second procedure and were forced to appeal to island observations in the manner described below in order to obtain realistic results. They encountered no similar resonances when solving for the diurnal K_1 constituent, perhaps because the artificial coastal condition filters out the Kelvin-like modes that could be resonant at subinertial frequencies in the f-plane (section 10.4.2). All these remarks also apply to marginal-sea-tide models (section 10.5.2): when the elevation at the connection to the open ocean is specified *ab initio*.

These two procedures and variants of them have resulted in global solutions (most for M_2) that show good qualitative agreement (Hendershott, 1973, 1977). The most recent published global models are by Zahel (1970), Parke and Hendershott (1980), and Accad and Pekeris (1978). I know of new calculations by Zahel, by Estes, and by Schwiderski (Parke, 1979) as well, but have not been able to examine them in detail. When all have been published, a careful comparison of these models with one another, with island and pelagic tidal data, with gravity data, and with tidal perturbations of satellite orbits ought to be carried out.

All the most recent solutions include effects of ocean loading and self-attraction (section 10.3). Many of them have been published since Cartwright (1977) and I (Hendershott, 1977) reviewed the tidal problem. The varying methods of solution may be summarized by abbreviating LTE (10.5) or the elevation equation as in Hendershott (1977):

$$\mathscr{L}[\zeta_0] = \mathscr{L}'[\iint G\zeta_0] + \mathscr{L}'[(1 + k_2 - h_2)U_2/g]. \quad (10.134)$$

Here U_2 is the tide-generating potential (a second-order spherical harmonic) for a given constituent, (k_2, h_2) are Love numbers (section 10.3), $\mathscr{L}$ and $\mathscr{L}'$ are operators elliptic in space with $\mathscr{L}$ representing LTE (10.5) or the elevation equation, and $\iint G\zeta_0$ abbreviates the global convolution expressing effects of loading and self-attraction as in (10.14).

I attempted to solve (10.134) for M_2 using the second procedure iteratively,

$$\mathscr{L}[\zeta_0^{(i+1)}] = \mathscr{L}'[\iint G\zeta_0^{(i)}] + \mathscr{L}'[(1 + k_2 - h_2)U_2/g], \quad (10.135)$$

(Hendershott, 1972) but the iteration did not look as though it would converge. Gordeev, Kagan, and Polyakov (1977) found that inclusion of dissipation could result in convergence. Parke (1978) used the iterates $\zeta_0^{(i)}$ as a basis set for a least-squares solution $\hat{\zeta}_0$ of (10.134) of the form

$$\hat{\zeta}_0 = \Sigma A_i \zeta_0^{(i)} \quad (10.136)$$

in which the A_i are found by solving

$$\frac{\partial}{\partial A_i} \{E \equiv \iint_{\text{ocean}} |\mathscr{L}(\hat{\zeta}_0) - \mathscr{L}'[\iint G\hat{\zeta}_0]$$

$$- \mathscr{L}'[(1 + k_2 - h_2)U_2/g]|^2 \} = 0. \quad (10.137)$$

He obtained solutions that evidently were quite accurate [E as defined in (10.137) was small], but their realism was marred by the unphysical resonances of the second procedure. Parke and Hendershott (1980) therefore effectively adjusted the locations of these resonances to yield realistic global results by getting the A_i from a least-squares fit of (10.136) to island and pelagic observations.

Accad and Pekeris (1978) noticed that $\iint G\zeta_0^{(i)}$ was very similar to $\zeta_0^{(i+1)}$. They therefore put

$$\iint G\zeta_0^{(i)} = K\zeta_0^{(i)} + \iint \Delta\zeta_0^{(i)}, \quad (10.138)$$

where K is a constant evaluated empirically at each iteration by

$$K = \iint [\zeta_0^{(i)*} \iint G\zeta_0^{(i)}] / \iint [\zeta_0^{(i)*}\zeta_0^{(i)}] \quad (10.139)$$

and then iterated not (10.135) but

$$\mathscr{L}[\zeta_0^{(i+1)}] - K\mathscr{L}'[\zeta_0^{(i+1)}]$$

$$= \mathscr{L}'[\iint \Delta\zeta_0^{(i)}] + \mathscr{L}'[(1+k_2 - h_2)U_2/g]. \quad (10.140)$$

This greatly accelerated the slow convergence of (10.135), presumably already established by dissipation in their calculations.

Figure 10.30 shows two M_2 global cotidal maps of Accad and Pekeris (1978), which differ only in the inclusion of the convolution terms $\iint G\zeta_0$. These terms do not result in an order of magnitude alteration of the computed tide but their effects are large enough that they must be included in any dynamically consistent model aiming at more than order-of-magnitude correctness. These solutions and others like them are obtained solely from a knowledge of the tidal potential and are, in that sense, as close as modern investigators have come to attaining Laplace's original goal.

10.6 Internal Tides

10.6.1 Introduction

Internal tides have long been recognized as internal waves somehow excited at or near tidal periods. Their potential as a source of error in hydrographic casts seems to have been recognized since their earliest re-

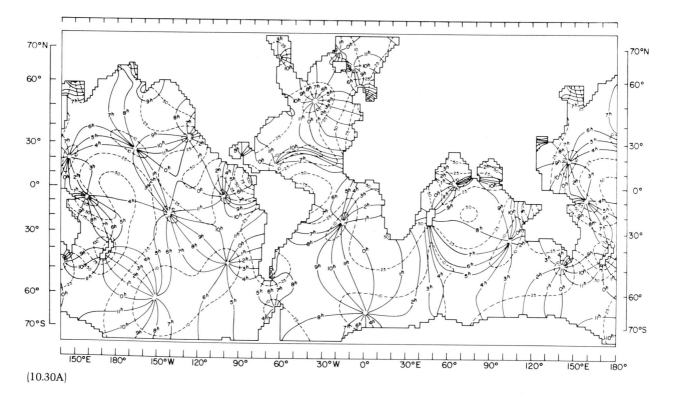

(10.30A)

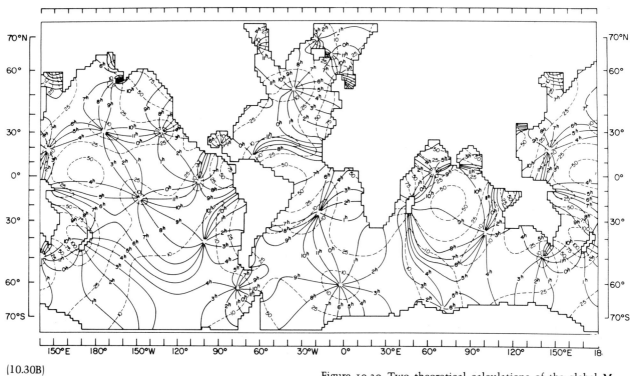

(10.30B)

Figure 10.30 Two theoretical calculations of the global M_2 tide obtained solely from a knowledge of the astronomical tide-generating forces (A) and differing only in the inclusion (B) of the effects of loading and self-attraction. (Accad and Pekeris, 1978.)

ported observation by Nansen (1902). Because constant-depth internal-wave modes are almost orthogonal to the ATGF (they would be completely so if the sea surface were rigid and the ATGF exactly depth independent) it has always been difficult to see why internal tides exist at all. The work of Zeilon (1911, 1912) appears to be the precursor of the now generally accepted explanation—energy is scattered from surface to internal tides by bottom roughness—but there has been a history of controversy. The lack of correlation between internal tides at points separated vertically by O (100 m) or horizontally by O (100 km) puzzled early observers. Subsequent observations showed semidiurnal and diurnal internal tides to be narrow-band processes each with a finite band width $\Delta\sigma$ of order several cycles per month. This property manifests itself both in a decay of spatial coherence of internal tides over a length associated with the spread of spatial wavenumbers corresponding to $\Delta\sigma$ and in temporal intermittency over times $\Delta\sigma^{-1}$, as well as in a corresponding lack of coherence with either the surface tide or the ATGF. Typical observations are shown in figure 10.31.

10.6.2 Generation Mechanisms

Zeilon (1934) carried out laboratory experiments showing that a step in bottom relief could excite internal waves in a two-layer fluid when a surface tidal wave passed overhead. Two-layer models are attractive analytically because each layer is governed by a well-posed boundary-value problem; such experiments have been studied theoretically by Rattray (1960) and many others.

Haurwitz (1950) and Defant (1950) noticed that in the f-plane both the horizontal wavelength and the phase speed of plane internal waves grow very large as $\sigma \to f_0$ [(10.23e) with $D_* = D_n$]. Resonance with the ATGF might thus be possible near the inertial latitudes corresponding to tidal frequencies. But the equatorial β-plane solutions (section 10.4.5) (even though only qualitatively applicable at tidal inertial latitudes) show that this apparent possibility of resonance is an artifact of the f-plane, which provides WKB solutions of LTE, and so cannot be applied at the inertial latitudes.

Miles (1974a) has shown that the Coriolis terms customarily neglected in the traditional approximation scatter barotropic energy into baroclinic modes (section 10.3). Observations of internal tides (section 10.6.3) appear to favor bottom relief as the primary scatterer, but this may be because steep bottom relief is spatially localized whereas the "extra" Coriolis terms are smoothly distributed over the globe. Further theoretical work is needed to suggest more informative observations.

For a continuously stratified ocean, Cox and Sandstrom (1962) calculated the rate of energy flow from surface to internal tides due to single scattering from small-amplitude, uniformly distributed, open-ocean-bottom roughness $\epsilon D_1(x, y)$ [where $\epsilon \ll 1$, D_1 is $O(1)$]. Their calculation is most succinctly summarized by specializing to one-dimensional relief $D_1(x)$ and constant buoyancy frequency N_0. If the incident surface tidal-velocity field is idealized as $U \exp(-i\sigma_T t)$, with no space dependence, then the singly scattered internal-tide field $u^{(1)}$ is obtained by solving (10.45)

$$\frac{\partial^2 w^{(1)}}{\partial z^2} - \left(\frac{N_0^2}{\sigma_T^2 - f_0^2}\right)\frac{\partial^2 w^{(1)}}{\partial x^2} = 0 \quad (10.141)$$

and

$$\frac{\partial u^{(1)}}{\partial x} + \frac{\partial w^{(1)}}{\partial z} = 0 \quad (10.142)$$

subject to

$$w^{(1)} = 0 \quad \text{at} \quad z = 0, \quad (10.143)$$

$$w^{(1)} = \epsilon U \, \partial D_1/\partial x \quad \text{at} \quad z = -(D_* + \epsilon D_1) \quad (10.144)$$

plus a radiation condition as $|x| \to \infty$.

Equation (10.143) idealizes the free surface as rigid (adequate for internal waves); (10.144) is the $O(\epsilon)$ expansion about the mean relief $z = -D_*$ of the condition (10.15) of zero normal flow at the actual relief:

$$w = u\frac{\partial}{\partial x}(D_* + \epsilon D_1) \quad \text{at} \quad z = -D_* - \epsilon D_1. \quad (10.145)$$

The solution of (10.141)–(10.144) for $w^{(1)}$ is

$$w_1(x,z) = \int_{-\infty}^{\infty} \overline{w}_1(l,z) \exp(ilx) \, dl, \quad (10.146)$$

where

$$\overline{w}_1(l,z) = \overline{(\epsilon U \, \partial D_1/\partial x)} \frac{\sin[lzN_0/(\sigma_T^2 - f_0^2)^{1/2}]}{\sin[-lD_*N_0/(\sigma_T^2 - f_0^2)^{1/2}]} \quad (10.147)$$

with $\overline{(\epsilon U \, \partial D_1/\partial x)}$ defined as the Fourier transform of $(\epsilon U \, \partial D_1/\partial x)$. The integrand of (10.14.6) has simple poles at $[-lD_*N_0/(\sigma_T^2 - f_0^2)^{1/2}] = n\pi$, i.e., at horizontal wavenumbers l satisfying the internal wave dispersion relation $\sigma_T^2 - f_0^2 = gD_n l^2$, $D_n = N_0^2 D_*^2/gn^2\pi^2$. Equatorward of the tidal inertial latitude, $\sigma_T^2 > f_0^2$, so that each pole is real and corresponds to an internal wave traveling away from the scattering relief. Poleward of the tidal inertial latitude, $\sigma_T^2 < f_0^2$, so that each pole is imaginary and the corresponding internal mode decays exponentially away from the scattering roughness without carrying energy away. The sum of all evanescent modes also decays in the vertical away from the scattering relief. Wunsch (1975) reports the existence of observations showing this evanescent behavior for diurnal internal tides.

When $\sigma_T^2 > f_0^2$, (10.141) is hyperbolic in (x, t) with characteristic slope $(\sigma_T^2 - f_0^2)^{1/2}/N_0$. Baines (1971) solved

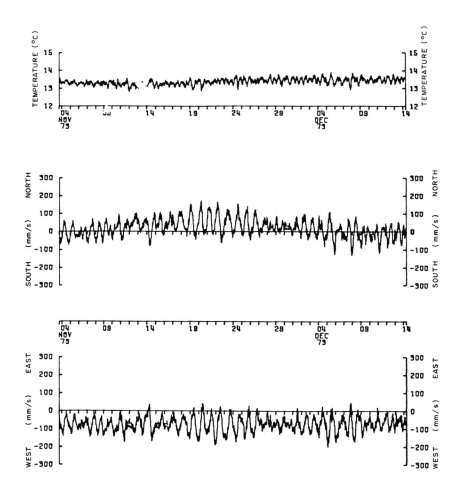

Figure 10.31A Time series of temperature and velocity at the IWEX mooring (Hatteras Abyssal Plain) at 640 m depth (Briscoe, 1975b).

(10.141) and (10.145) exactly, by the method of characteristics, thus eliminating the restriction to weakly sloping relief. The analytical novelty of his work was the imposition of the radiation condition on the characteristic form

$$F(x - Rz) + G(x + Rz), \quad R = N_0/(\sigma_T^2 - f_0^2)^{1/2}$$

of the solutions of (10.141) by, for example, choosing

$$F(x) = \int_0^\infty e^{ilx} \overline{F(l)} dl$$

so that $F(x \pm Rz) \exp(-i\sigma_T t)$ contains only outgoing plane waves. Laboratory work (Sandstrom, 1969) and analysis (Wunsch, 1969) showed that when bottom and tidal characteristic slopes coincide, the near-bottom motion is strongly intensified. Wunsch and Hendry (1972) show evidence for such intensification over the continental slope south of Cape Cod (figure 10.32).

The general possibility that diurnal tides enhance diurnal inertial motion by some mechanism has been suggested by Ekman (1931), Reid (1962) and Knauss (1962b). I (Hendershott, 1973) estimated the amplitude of motion if the mechanism is scattering of surface tides into internal tides by open-ocean bottom roughness, but obtained a result sufficiently small that it would not stand out noticeably against the high level of inertial motion found at all mid-latitudes (Munk and Phillips, 1968).

Thus far, the discussion is in terms of linearly scattered linear waves. Bell (1975) considers the formation of internal lee waves on periodically varying barotropic tidal currents. This process could generate a complex spectrum of internal waves even with a monochromatic surface tide. What actually occurs when laboratory or ocean stratified flow passes over relief is complicated. In Massachusetts Bay, Halpern (1971) has observed that tidal flow over a ridge generates a thermal front that propagates away as a highly nonlinear internal wavetrain (Lee and Beardsley, 1974) or as an internal bore. Such bores are commonly observed along the Southern California coast (Winant, 1979).

Maxworthy (1979) emphasizes the importance of the collapse of the stirred region that initially develops over relief in the subsequent generation of laboratory internal waves. The relative importance of all these processes near the sea floor is unknown. If separation

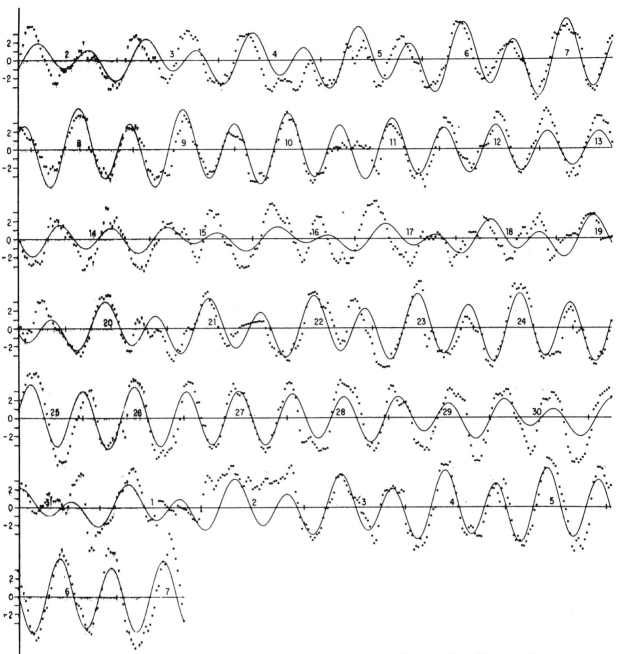

Figure 10.31B Observed (dotted) and predicted barotropic (solid) longshore bottom velocity at Josie I (figure 10.27) off the southern California coast. (Munk, Snodgrass, and Wimbush, 1970.)

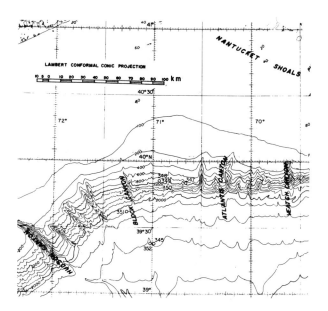

Figure 10.32A Current meter mooring positions and shelf topography. (Wunsch and Hendry, 1972.)

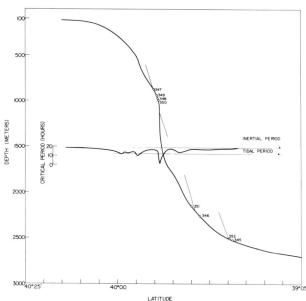

Figure 10.32B Profile of topography through the array along the dashed line of figure 10.32B with mooring positions indicated. Several internal wave characteristics for the M_2 tide are shown, and the critical period (at which internal-wave characteristics are locally tangent to the relief) is plotted across the profile. (Wunsch and Hendry, 1972.)

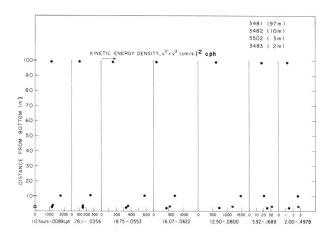

Figure 10.32C Vertical profiles of kinetic energy density (observed values are solid dots) for various periods over the slope where the tidal characteristic is locally tangent to the relief (moorings 347–350). (Wunsch and Hendry, 1972.)

over abyssal relief does occur in tidal currents, it could contribute to abyssal mixing by helping to form the near-bottom laminae observed by Armi and Millard (1976). Wunsch (1970) made a somewhat similar suggestion based on laboratory studies.

Besides this potential complexity of generation, the medium through which the internal tide moves is strongly inhomogeneous in space and time. The overall result is the complicated and irregularly fluctuating internal tide observed. Still, away from generation regions, some features of the linear theory shine through.

10.6.3 Observations

In linear theory, breaks in the slope of the relief and extended regions where that slope coincides with a tidal characteristic slope make themselves felt in the body of the ocean as narrow-beam disturbances concentrated along the characteristics (Rattray et al., 1969). The beams are typically narrow (figure 10.33) and their (characteristic) slope in the presence of mean currents varies both with local stratification and shear. This suggests that, especially near generation regions, the internal tide will have a complex spatial structure and that its amplitude at a given point may vary markedly as nearby stratification and mean flow change. Thus Hayes and Halpern (1976) document very large variability of semidiurnal internal tidal currents during a coastal upwelling event; they account for much of it by appealing to the deformation of characteristics as vertical and horizontal density gradients change during the upwelling. Regal and Wunsch (1973) find internal tidal currents at site D, over the continental slope south of Cape Cod, to be concentrated near the surface and there highly (and uncharacteristically) coherent

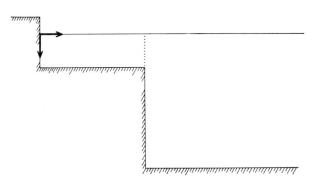
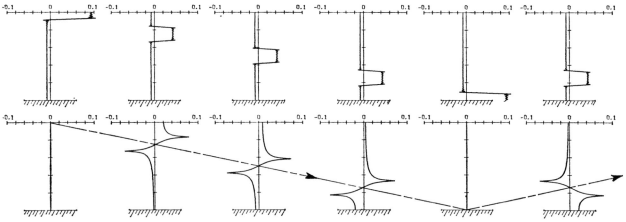

Figure 10.33 Depth distribution of horizontal internal tidal currents away from a step shelf (top) at two times (center and bottom) separated by a quarter-wave period. Ratio of deep sea to shelf depth is 12.5, characteristic slope is 1/715; current profiles begin at shelf edge and are separated horizontally by 1.8 of the deep-sea (first-mode) internal-tide wavelength. (Rattray et al., 1969.)

with the surface tide. The result is consistent with generation where tidal characteristics graze the slope perhaps 60 km to the north followed by propagation along characteristics that leave the region of tangency and bounce once off the ocean floor before passing through the near surface part of the water column at site D (figure 10.34). Observations at site L, some 500 km to the south, show no evidence of propagation along beams.

Beams are a coherent sum of many high internal modes. We intuitively expect that high modes are more rapidly degraded by whatever processes ultimately lead to dissipation than are low modes, and that they are more sensitive to medium motion and fluctuation than are low modes because they propagate so slowly. We thus do not expect beamlike features in the deep sea, and they are not observed. Instead, we expect a few low modes to dominate in a combination of arrivals from distant steep relief. These will have made their way through significant oceanic density fluctuations and through fluctuations of mean flows often at an appreciable fraction of internal-wave-phase speeds. The line spectrum characteristic of the ATGF and the surface tide will thus be so broadly smeared into semidiurnal and diurnal peaks that individual constituents or even the spring-neap cycle are at best very difficult (Hecht and Hughes, 1971) to perceive.

The most complete description of open-ocean internal tides is due to Hendry (1977), who used the western central Atlantic Mid-Ocean Dynamics Experiment (MODE) data. Figure 10.35 summarizes the results. M_2 tends to dominate semidiurnal temperature variance over the water column. Adjacent N_2 and S_2 variances are nearly equal, and the vertical variation of N_2, S_2 variance generally follows M_2 (with qualifications near the bottom). M_2 likewise dominates horizontal semidiurnal current variance over the water column. At subthermocline depths M_2 variance approaches estimates of the barotropic M_2 tidal current variance while S_2 and N_2 variances exceed their barotropic counterparts by an order of magnitude. All this suggests that much of what appears in the N_2 and S_2 bands has really been smeared out of M_2. The vertical distribution of variance is broader than the two WKB profiles $(\partial\bar{\theta}/\partial z)^2 N(z)^{-1}$ ($\bar{\theta}$ is mean potential temperature) and $N(z)$, for temperature and horizontal current variance, respectively. This indicates that the lowest vertical modes dominate. M_2 temperature variance has a coherence of about 0.7 with the ATGF in the upper thermocline while N_2 and S_2 are far less coherent with the

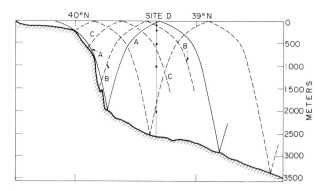

Figure 10.34A Profile of relief along 70°W (see figure 10.32 for local isobaths) together with selected semidiurnal characteristics passing near site D. (Regal and Wunsch, 1973.)

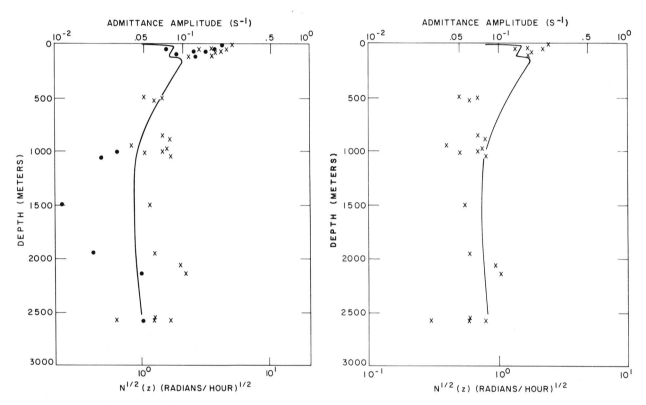

Figure 10.34B Admittance amplitude x for semidiurnal tidal currents together with buoyancy frequency $N(z)$ at site D. Near-surface admittances are strongly intensified; currents there are highly coherent with the surface tide. (Regal and Wunsch, 1973.)

Figure 10.34C A similar display at site L, 599 km south of site D, shows no comparable surface intensification. (Regal and Wunsch, 1973.)

ATGF but yet not totally incoherent. The M_2 first internal mode dominates and propagates to the southeast; this plus the (significantly not random) phase lag between M_2 and S_2 in the thermocline (the age of the internal tide) point to the 700-km distant Blake escarpment as a generating region. Other discussions of open-ocean internal tides are consistent with the foregoing picture although necessarily based upon less extensive observations.

Wunsch (1975) reviews observations allowing estimation of the energy density (in units of ergs per squared centimeters) of internal tides and suggests that it is from 10 to 50% of the corresponding energy density of the barotropic tide, albeit with wide and unsystematic geographic variation.

10.6.4 Internal Tides and the Tidal Energy Budget

It thus appears that barotropic tides somehow give up energy to internal tides. Return scattering is probably unimportant. It is important to know the rate at which this energy transfer occurs because (section 10.5) the energy budget for global tides may not yet be closed. Wunsch (1975) reviews estimates arising from the various scattering theories outlined above (section 10.6.2); typical values are 0.5×10^{19} erg s^{-1} from deep-sea roughness [using the theory of Cox and Sandstrom as rediscussed by Munk (1966)], 6×10^7 erg cm s^{-1} from continental shelves [using the theory of Baines (1974) and also from independent measurements by Wunsch and Hendry (1972)]. The latter value extrapolates to 5.6×10^{15} erg s^{-1} over the globe.

A bound on this estimate independent of scattering theories was pointed out by Wunsch (1975). Internal-tide energy densities E_I are order 0.1 to 0.5 times surface-tide energy densities E_S. Group velocities c_{gI} of internal waves are order $(D_n D_0)^{1/2} \simeq (N^2 D_0/gn^2\pi^2)^{1/2}$ times group velocities c_{gS} of long-surface gravity waves. If open-ocean tidal energy is radiated toward shallow seas (or any other dissipation region) at rates $c_{gS} E_S$ and $c_{gI} E_I$, then internal tides can never account for more than $O(10\%)$ of the total energy lost from surface tides.

Wunsch's (1975) discussion of the caveats to this result has not been substantially altered by subsequent developments. Nonlinear interactions that drain internal energy from the tidal bands to other frequencies and scales certainly do occur but their rates are not yet accurately estimable. Such rates as have been calculated [Garrett and Munk (1972a) calculated the energy loss due to internal wave breaking; McComas and Bretherton (1977) the time scale for the low-frequency part of the internal wave spectrum to evolve by resonant interactions]; they are small, but the problem is not closed.

10.6.5 Internal Tides and Ocean Stirring

Even if internal tides turn out to be a minor component of the global tidal-energy budget, they could be an im-

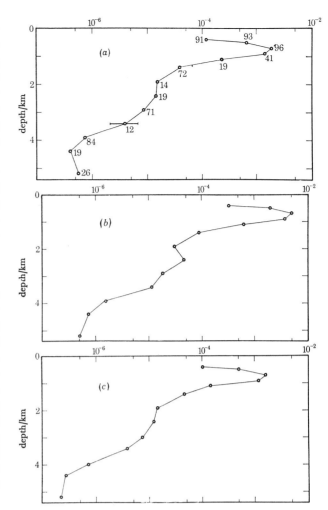

Figure 10.35 A Vertical profile of squared temperature fluctuations in the (a) S_2 band (°C), averaged at depth levels over the entire array; the number of 15-day-piece lengths at each level is indicated. Vertical profile of average squared temperature fluctuations in the (b) M_2 band. Vertical profile of average squared temperature fluctuations in the (c) N_2 band. (Hendry, 1977.)

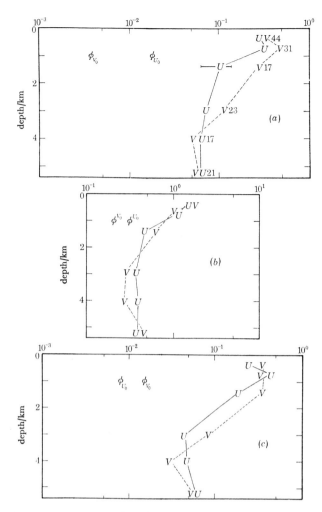

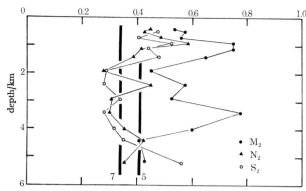

Figure 10.35C Vertical profile of average coherence amplitude of temperature fluctuations and the equilibrium tide for three semidiurnal frequency bands. The averages are taken over the whole array at depth levels, and include individual cases with both five and seven degrees of freedom. The expected values of coherence amplitude for zero true coherence are shown for each case, and while the central M_2 band shows a definite determinism, the adjacent frequency bands are much more dominated by randomly phased temperature fluctuations. (Hendry, 1977.)

Figure 10.35B Vertical profile of squared horizontal current (cm s^{-1}) for U (east) and V (north) in the (a) S_2 band, averaged at depth levels over the entire array; estimates of squared amplitude for the barotropic current components U and V are given, showing that the currents are dominated by internal waves at all depths. Similar estimates for the (b) M_2 band currents. Here the deep currents are greatly influenced by the barotropic mode. Similar estimates are given for the (c) N_2 band; internal waves appear to dominate at all depths. (Hendry, 1977.)

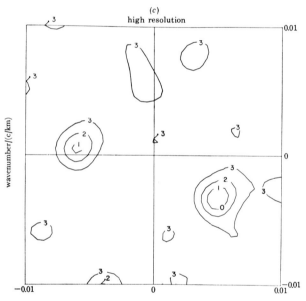

Figure 10.35D Conventional wavenumber spectrum of first-mode M_2 temperature fluctuations from MODE. The peak in the southeast quadrant has wavenumber 1/163 cpkm and represents a wave propagating from northwest to southeast. A secondary peak in the northwest quadrant is interpreted as an alias of the main peak. (Hendry, 1977.)

portant source of energy for ocean stirring and mixing. About 10% of the total tidal dissipation would be en-energetically adequate (Munk, 1966). The hypothesis is that internal tides somehow (by bottom turbulence, by nonlinear cascade) cause or enhance observed fine structure and microstructure events in which mixing is believed to be occurring. It thus is pertinent to examine observations for any correlation between tidal phenomena and smaller-scale events. Such a correlation could be temporal [with the intensity of small structure modulated at semidiurnal, diurnal, fortnightly (i.e., the spring-spring interval) or even longer tidal periods] or spatial (with small structure near generators different from that far away).

Although definitive studies have yet to be made, preliminary indications are that little such correlation exists. Cairns and Williams (1976) contour the spectrum of vertical displacement in a frequency-time plane for 17 days over the frequency band 0.2-6.0 cpd (figure 10.36) but see no modulation at tidal or fortnightly periods of any part of the spectrum. Wunsch (1976) finds no correlation between the overall spectral level of internal waves (specifically, the spectral intensity at 5-hour period of a model fitted to observed spectra) and the intensity of the internal tidal peak for observations from the western North Atlantic (figure 10.37).

The demonstration that tidal contributions to ocean mixing are significant will thus involve subtle measurements. Perhaps something may be learned by comparing internal waves, fine structure, and microstructure in the open ocean with their counterparts in the relatively tideless Mediterranean Sea or in the Great Lakes (see chapters 8 and 9).

10.7 Tidal Studies and the Rest of Oceanography

Although tidal studies were the first dynamic investigation of oceanic response to forcing, insight into wind- and thermohaline-driven ocean circulation developed largely independently of them. In the case of semidiurnal and diurnal tides, the reason is primarily dynamic. But the dynamics of long-period tides are likely to be much more like those of the wind-driven circulation (both steady and transient) than like those of semidiurnal and diurnal tides. It is possible that, if the long-period components of the ATGF had been large enough to make the long-period tides stand out recognizably above the low-frequency noise continuum, then the early tidalists might have recognized, in the low-frequency tides, features such as westward intensification also evident in the general circulation. They might then have been forced into the recognition that a linear superposition of (possibly damped) first-class waves could not account for long-period tides, as it seems able to do for semidiurnal and diurnal tides. As things are, however, long-period tides are so near to the noise level that their observation did not provide a global picture clear enough to force tidalists out of the semidiurnal-diurnal framework.

It was, in fact, insight into the problem of time-dependent wind-driven ocean circulation that led Wunsch (1967) to provide the modern view of long-period tides: a superposition of damped second-class waves (section 10.4.4) whose horizontal length scales are only $O(10^3$ km$)$ and whose amplitudes and phases are likely to undergo substantial fluctuations in time on account of the overall time variability of the ocean currents through which they propagate. Laplace had supposed that a small amount of dissipation would bring the long-period tides into equilibrium, i.e., the geocentric sea surface ζ would be an equipotential of the total tide-generating potential Γ (Lamb, 1932, §217). The most recent elaboration of this view is by Agnew and Farrell (1978), who solved the integral equation

$$\zeta = \zeta_0 + \delta = \Gamma/g$$

[with δ (the solid-earth tide) and Γ given by (10.11) and (10.12) as functions both of the observed tide ζ_0 and of the long-period astronomical potential U_2] for equilibrium global-ocean fortnightly and monthly tides ζ_0 subject to the conservation of mass. Wunsch's (1967) analysis and dynamic model of the fortnightly tides suggest that they are not in equilbrium with either the astronomical potential U_2 or with the full potential Γ of Agnew and Farrell (1978). Their Pacific averaged admittance has magnitude 0.69 ± 0.02 relative to Γ with significant island-to-island variation. The Pacific averaged monthly tide admittance has magnitude 0.90 ± 0.05 relative to Γ, but island-to-island fluctuations vanish only by pushing individual island admittances to the very end of their error bands. Whether the kinds of dissipation and of nonlinear interaction between low-frequency motions that occur in the real ocean favor an equilibrium tidal response at sufficiently low frequencies is not yet known either observationally or theoretically. The 14-month pole tide is known to be significantly non-equilibrium in the shallow seas of Northern Europe (Miller and Wunsch, 1973) but nearly invisible elsewhere.

Observations of long-period tides have thus been too noisy to exert real influence on tidal studies and hence on dynamic oceanography. But the theoretical ideas emerging from study of the low-frequency solutions of LTE have been of great importance for dynamic oceanography (even though they are likely to be inadequate to model the full dynamics of the general circulation).

Steady (as opposed to wavelike) solutions of LTE obtained by active tidalists (Hough, 1897; Goldsbrough, 1933) had a profound effect on dynamic oceanography through the review by Stommel (1957b) of

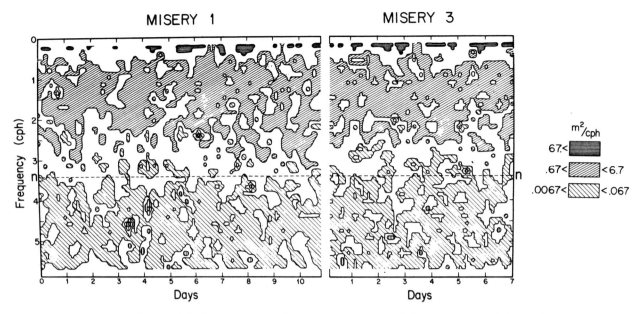

Figure 10.36 Contours of vertical-displacement spectral energy vs. elapsed time for the 6.60°C isotherm off the coast of southern California. Spatial estimates with 2 df are made for successive 5.8-h data segments and the results are contoured. (Cairns and Williams, 1976.)

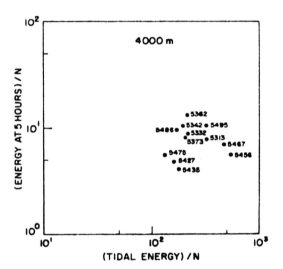

Figure 10.37 Five-hour internal wave energy at various locations vs. semidiurnal tidal energy. (Wunsch, 1976.)

ocean-current theory and through the work by Stommel and Arons (1960a,b) on abyssal circulation. Goldsbrough (1933) studied nonperiodic solutions of LTE driven by global patterns of evaporation and precipitation. The solutions are steady, provided that the precipitation–evaporation distribution vanishes when integrated along each parallel of latitude between basin boundaries. Stommel (1957b) pointed out that Ekman suction and blowing due to wind-stress convergence and divergence could effectively replace the evaporation-precipitation distribution, while the introduction of ageostrophic western boundary currents allowed the solutions to remain steady even when the integral constraint on the evaporation-distribution function was violated. The resulting flows display the main dynamic features of the theory of wind-driven circulation due to Sverdrup (1947), Stommel (1948), and Munk (1950). When the evaporation–precipitation function is viewed as modeling the high-latitude sinking of deep water and its mid-latitude subthermocline upwelling, the abyssal circulation theories of Stommel and Arons (1960a,b) result.

The seminal work on low-frequency second-class motions was the study (section 10.4.4) by Rossby and collaborators (1939), ironically inspired by meteorological rather than tidal studies. It led, through the studies by Veronis and Stommel (1956) and Lighthill (1969) of time-dependent motion generated by a fluctuating wind to the very different views of mid-latitudes and tropical transient circulation that prevail today (although, especially in mid-latitudes, linear dynamics are now generally acknowledged to be inadequate for a full description; see chapter 5). Pedlosky (1965b) showed how the steady western boundary currents of Stommel (1948), Munk (1950), and Fofonoff (1954) could be viewed as Rossby waves reflected from the western boundary and either damped by friction or swept back toward the boundary by the interior flow that feeds the boundary current; Gates's (1968) numerical examples showed clearly the development of a frictional western boundary current as a group of short Rossby waves with seaward edge propagating away from the western boundary at the appropriate group velocity.

Modern interest in estimating the role of direct transient wind forcing in generating mesoscale oceanic variability (see chapter 11) calls for an up-to-date version of N. A. Phillips's (1966b) study of mid-latitude wind-generated Rossby waves using more realistic wind fields and taking into account new insight into the combined effects of bottom relief and stratification (section 10.4.7). Such a calculation would closely resemble a proper (linear) dynamic theory of long-period tides. But similar caveats apply to uncritically accepting either as representing an actual flow in the ocean.

11
Low-Frequency Variability of the Sea

Carl Wunsch

11.1 Introduction

The purposeful study of the time-dependent motion of the sea having periods longer than about 1 day is comparatively recent. In the classic *Handbuch* of the early 1940s, Sverdrup, Johnson, and Fleming (1942), one searches in vain for more than the most peripheral reference to temporal changes on the large scale (one of the few examples is their figure 110 showing the California Current at two different times). Until very recently, the ocean was treated as though it had an unchanging climate with no large-scale temporal variability. The reason for this is compelling and plain: until the electronics revolution of the past 30 years, the major oceanographic observational tool was the Nansen bottle; using slow, uncomfortable ships, it took essentially 100 years to develop a picture of the gross characteristics of the mean ocean. The more recent period, 1947 (Sverdrup, 1947) through about 1970 (Stommel, 1965; Veronis, 1973b; and see chapter 5), was one of the intensive development of the theory of large-scale, steady models of the ocean circulation. The methods were initially analytic, later numerical. Most of these models were essentially low-Reynolds-number, steady, sluggish, sticky, climatic oceans. In them, the role (if any) of small-scale, time-dependent processes is simply parameterized by a positive eddy coefficient (*Austauch*) implying a down-the-mean-gradient flow of energy, momentum, heat, etc. The westward-intensification theories (Stommel, 1948; Munk, 1950) imply that any strong influence of such eddy coefficients would be confined to the western boundaries and could be ignored in the interior ocean, except possibly in the immediate vicinity of the eastward-moving free-jet "Gulf Stream" (see Morgan, 1956). The resulting models bear a remarkable resemblance to many of the gross features of the large-scale mean ocean circulation (see chapter 5).

The culmination of these analytic and numerical models of the large-scale circulation coincided with a number of developments that ultimately undermined the momentary confidence that the models represented the correct dynamics of the ocean circulation. These developments were of two kinds—instrumental and intellectual.

By 1970 instruments had been developed that made it possible to obtain *time-series* measurements in the open sea for periods far longer than a ship could possibly remain in one location. These instruments included moored current meters, drifting neutrally buoyant floats, pressure gauges, and many others (Gould, 1976; and see chapter 14). An additional "instrument" was the computer, which made it possible both to handle the large data sets generated by time-series instruments and to explore new ideas by nonanalytic means. This computer impact has been felt, of course, in most branches of science.

The intellectual developments that shifted the focus from the mean circulation to the time-dependent part were also of various kinds. The analytic models seemingly had reached a plateau at which their increasingly intricate features [e.g., essentially laminar boundary layers of higher and higher order as in Moore and Niiler (1975)] seemed untestable and intuitively implausible outside the laboratory. Physical oceanography is also to some extent a mirror of meteorology; by 1970 most oceanographers were at least vaguely familiar with the picture of the atmosphere that had emerged over the previous decades. In that fluid system, the view of the role of eddies had shifted from a passive means of dissipating the mean flows (through purely down-gradient fluxes of momentum, energy, etc.) to a much more interesting and subtle dynamic linkage in which the mean flows (the climate) were in at least some parts of the system *driven* by the eddy fluxes (Jeffreys, 1926; Starr, 1968; Lorenz, 1967). Because many of the meteorological results would apply to any turbulent fluid, there was reason to believe that the ocean could also exhibit such intimate dynamic linkages. But we should note that even now much work is still directed at studying the mean circulation by essentially classical (though improved) means, as if the variability were not dynamically important (e.g., Schott and Stommel, 1978; Wunsch, 1978a; Reid, 1978). The extent to which such pictures of the mean circulation of the large-scale tracers will survive complete understanding of variability dynamics is not now clear.

In this chapter we shall review what is known about the variability of the ocean. The expression "low-frequency variability," which is part of the title of this chapter, is a vague one used in a variety of ways by oceanographers, and encompassing a wide range of things. Here we mean by it anything with a time scale longer than a day out to the age of the earth, although we cannot really study by instrumental means phenomena with time scales longer than about 100 years. In spatial scale, it means phenomena ranging from some tens of kilometers to the largest possible global ocean oscillations. We shall, in common with recent practice, also refer to the "eddy" field in the ocean. This word is often prefixed by "mesoscale" and is used loosely to denote the subclass of variability encompassing motions occurring on scale of hundreds of kilometers with time scales of months and longer. It is a convenient shorthand and is meant to imply neither any particular dynamics nor only flows with closed streamlines. (The equivalent Soviet term is "synoptic scale").

There is little doubt that oceanographers were quite aware, from the very beginning, of time variability in the ocean. Maury (1855, p. 358) remarked that in drawing his charts he had disregarded "numerous eddies and local currents which are found at sea." He also notes in particular (p. 188) the highly variable equatorial currents of the Pacific Ocean. Even earlier, Rennel (1832) had quoted another observer (C. Blagden), as referring to North Atlantic currents as "casual" (Swallow, 1976).

Most of the astute observers who worked at sea since Maury were very conscious of the difficulties of drawing conclusions about the mean circulation in the presence of a highly time-dependent field. Figure 11.1, taken from Helland-Hansen and Nansen (1909), clearly depicts what one suspects to be a time-dependent eddy field. Sverdrup et al. (1942) make the statement that determining the mean is difficult in the presence of the time variations, and that the closer the station pairs are together, the greater is the requirement of simultaneity in hydrographic measurements. This is, of course, a statement about the frequency-wavenumber character of the baroclinic variability.

It is possible to give many instances of references to ocean variability and eddies throughout the history of observational oceanography. But it is also fair to say

Figure 11.1 Chart of Norwegian Sea surface currents as constructed by Helland-Hansen and Nansen (1909); reproduced by Sverdrup et al. (1942). One presumes the small-scale currents are in fact time-dependent features.

that little attention was paid to the phenomenon per se; it was a nuisance—a noise-contaminating determination of the time mean flow. There are some major exceptions, including Pillsbury's (1891) heroic efforts in the Florida Current, the 400-page work by Helland-Hansen and Nansen (1920), and somewhat later, Iselin's (1940a) attempts at monitoring the western North Atlantic. The question of the physical significance of a weak mean flow in the presence of strong variability has rarely been addressed even now.

11.1.1 Early Theory

The first theoretical attempts to study the purely time-dependent oceanic motions at low frequency seem to be outgrowths of the papers by Rossby and collaborators (1939) and by Haurwitz (1940a). These two studies, while directed primarily at the atmosphere, nonetheless addressed themselves to the large-scale time-dependent wave motions of a rotating hydrostatic fluid—a characterization applying equally well to the ocean. The Rossby paper in particular introduced the β-plane approximation. These early efforts, and the large number that followed, examined the wave motions known much earlier. Indeed Laplace (1775) had discussed motions that we now would call Rossby or planetary waves (or in Hough's terminology, "tidal motions of the second class"). The study of these motions has a long and distinguished history (e.g., Darwin, 1886; Rayleigh, 1903; Poincaré, 1910), culminating in Hough's (1897, 1898) remarkable study of the solutions of the Laplace tidal equations on a sphere. [Lamb (1932), in his chapter on tides gives a good summary of this work. He also thoroughly discusses (§206 and §212) what we call "topographic Rossby waves" in which topographic gradients play a role analogous to the variation of the Coriolis parameter with latitude on a sphere.]

But it was Rossby's β-plane that demonstrated the physics in the simplest form and permitted an escape from the geometrical complexities of spherical coordinates. Arons and Stommel (1956), Veronis and Stommel (1956), Rattray (1964), Rattray and Charnell (1966), and others made explicit attempts to understand the possible role of Rossby waves in the ocean. Longuet-Higgins in a series of papers (1964, 1965) justified the β-plane approximation and carried out a modern exhaustive search of the solutions on a sphere for a complete range of parameters far beyond what Hough could do in his time (Longuet-Higgins, 1968a, Longuet-Higgins and Pond, 1970). Most of this work was done in the absence of any direct observational base in the ocean. (For further discussion of these waves, see chapters 10 and 18).

Observations, which will be discussed at length below, suggest that linear wave models are inadequate to describe much of the actual time-dependent motion in the ocean. Nonetheless, as in the atmosphere (Holton, 1975), many of the features of the observations are qualitatively similar to those deducible from the linear theories. That is, the physics is modified by the nonlinearity, but many of the linear features persist into the nonlinear range. The precise extent to which this is true is a function of the periods and spatial scales of the motions and is not really understood. As a generalization, it may be safe to assert that the largest oceanic scales of fluctuation are most likely to be dominated by linear dynamics. A linear description becomes increasingly doubtful for smaller scales, and barotropic motions ought to be more nearly linear than baroclinic ones (Rhines, 1979).

The postulate of a time-dependent field in the interior ocean immediately calls into question (Stommel, 1965, p. 221) one of the fundamental deductions of the steady-ocean models—that a Sverdrup balance applies in the interior ocean.

Consider, for example, Stommel's (1948) model of a homogeneous flat-bottom ocean on a β-plane. Let ψ_0 be the time-mean transport streamfunction and let ψ_1 be the time-dependent part. Then the time average vorticity balance may be written

$$J(\nabla^2\psi_0, \psi_0) + \langle J(\nabla^2\psi_1, \psi_1)\rangle$$
$$+ \beta\frac{\partial \psi_0}{\partial x} + R\nabla^2\psi_0 = -\hat{\mathbf{k}}\cdot\nabla\times\boldsymbol{\tau}, \quad (11.1)$$

where the bracket denotes a temporal average, $\hat{\mathbf{k}}\cdot\nabla\times\boldsymbol{\tau}$ is the vertical component of the mean wind-stress curl, J denotes the Jacobian operator, and R is the coefficient of bottom friction. Let us assume that the mean field varies over scales of 10^4 km, and that the time-dependent eddy field varies over 10^2 km. Let both the mean flows and the time-dependent part have magnitude $10\,\text{cm}\,\text{s}^{-1}$. Scaling, we obtain roughly, in nondimensional form,

$$10^{-4}\nabla^2\psi_0 + \frac{\partial\psi_0}{\partial x} + 10^{-3}J(\nabla^2\psi_0,\psi_0)$$
$$+ 10^1\langle J(\nabla^2\psi_1,\psi_1)\rangle = -\hat{\mathbf{k}}\cdot\nabla\times\boldsymbol{\tau}. \quad (11.2)$$

Away from the western wall, the first term is negligible; hence if we ignore both nonlinear terms, an interior balance is

$$\frac{\partial\psi_0}{\partial x} = -\hat{\mathbf{k}}\cdot\nabla\times\boldsymbol{\tau}, \quad (11.3)$$

which is the conventional Sverdrup balance. But the nonlinear term

$$10^1\langle J(\nabla^2\psi_1,\psi_1)\rangle \quad (11.4)$$

will be of the same order as the Sverdrup terms if the

correlation in the bracket is no greater than 0.1, and hence the Sverdrup balance would be upset. The ease with which one could destroy the simple relation (11.3) has motivated many of the recent studies of mesoscale eddies. It is fair to state, however, that we are still not in a position to compute terms like (11.4) (or comparable terms in more sophisticated models) with sufficient accuracy to assess the adequacy of the Sverdrup relation. There is some evidence (Leetmaa, Niiler and Stommel, 1977) that (11.3) is qualitatively correct away from the eddy-rich area near the Gulf Stream itself, but no quantitative assessment has yet been possible. One would anticipate, based upon the known wide variability in eddy energy levels (see discussion below) that there is a wide geographical variability in (11.4).

Constructs such as equation (11.2), which suggest that eddies may be important in the open sea, are, in effect, a reopening of the question that appeared to have been answered by Stommel (1948) and Munk (1950), where the first-order effects of eddies—in the guise of eddy viscosities—were confined to the western boundaries of the ocean. Webster (1961a, 1965), following Starr's lead, showed that at least in some regions the sign of the eddy flux of momentum might be opposite to that assumed in the viscous models, and now we are at the stage of being unsure even whether the regional confinement to the west, which had seemed so clear in 1950, is valid.

Equation (11.2) and more realistic formulations imply that an energetic eddy field could upset the lowest-order open-ocean mean-vorticity balance. But eddies also can carry mass, momentum, heat, and other variables. It is not difficult to show the potential importance and confusion that can arise from the presence of a strong time-dependent flow field possessing small scales. A simple example was presented by Longuet-Higgins (1969c), who considered a weakly nonlinear wave. In the weak-interaction limit, one can write the time-mean particle velocities as

$\langle \mathbf{U}_E \rangle = \langle \mathbf{U}_L \rangle + \langle \mathbf{U}_S \rangle,$

where $\mathbf{U}_E$ is the Eulerian velocity, $\mathbf{U}_L$ the Lagrangian velocity, and $\mathbf{U}_S$ the Stokes velocity, which derives from the wave Reynolds stresses. The Eulerian velocity is the value that would be measured by a current meter at a fixed point; the Lagrangian velocity would be measured by tracking a dyed particle. In the absence of imposed exterior flows, Longuet-Higgins (1969c) demonstrated that the Eulerian and Lagrangian flows need not have the same magnitude or direction, and indeed can yield values differing by 180°. In the highly nonlinear limit it can be extremely difficult to find any simple relation between the Eulerian and Lagrangian flow fields. Such possibilities call into question the entire notion that there is some unique "general circulation" of the ocean. Presumably one must carefully define the quantity whose overall circulation is desired, be it heat, mass, momentum, energy, passive tracer, etc., and seek the dynamic balance that will govern its flux. Doing this represents one of the most important problems facing oceanographers, who only recently have come to grips with the existence of oceanic time variability. One anticipates that over the next decade the problem will be solved, but it is impossible at the present time to perceive the details of the actual dynamic and kinematic balances in the ocean.

In the past, there have been some attempts to study special situations in which time variability was modeled in simple ways in order to seek an understanding of its potential role in the mean circulation. In particular, Pedlosky (1965c), Veronis (1966c), and Munk and Moore (1968) all examined the possible role of weakly nonlinear Rossby waves in generating large-scale mean flows. All of these models used an Eulerian frame; as noted above, obtaining an Eulerian means does not necessarily imply the presence of an actual net water movement. Often one can demonstrate the actual impossibility of a Lagrangian mean (e.g., Moore, 1970) and the question often hangs on subtle questions of dissipation (Eliassen and Palm, 1960; Charney and Drazin, 1961) and the existence of critical levels (Andrews and McIntyre, 1978a).

11.1.2 More Recent Theory

Many of the second-order effects of linear Rossby-wave motions have been studied. Longuet-Higgins and Gill (1967), Lorenz (1972), Kim (1978), and Jones (1979) have shown that the waves themselves are unstable. The scattering of the waves by random currents was examined by Keller and Veronis (1969), and scattering from topography was studied by Hall (1976) among others. McKee (1972) examined the diffraction limit of the waves. Interaction of Rossby waves with mean shears has been studied primarily in a meteorological context by Charney and Drazin (1961), Holton (1975, chapter 4), and finite-amplitude waves in steady flows were analyzed by Pedlosky (1970).

Finite-amplitude "soliton" solutions have been constructed by Flierl (1979b) and Redekopp (1977) as an effort to model the extreme form of ocean variability represented by Gulf Stream rings. These models have usually been based upon some form of Korteweg-de Vries equation and share many of the same properties as other known solitary waves (Whitham, 1974, chapters 16, 17; also see chapter 18, this volume).

Partly in response to the observational data base, which suggests that linear dynamics cannot be wholly adequate, there is a recent and growing literature of the opposite extreme, that is, based on the assumption of a completely turbulent motion. In the weak-interaction theories of Rossby waves (e.g., Longuet-Higgins and Gill, 1967) the energy transformation of a wave occurs on a time scale long compared to a wave period. In the turbulence models, this is no longer true; the interactions are rapid and strong and one is forced into a statistical framework. An important step was taken by Charney (1971a), who showed that "geostrophic turbulence" in a stratified fluid would behave much like strictly two-dimensional turbulence known in other contexts (Fjørtoft, 1953; Kraichnan, 1967). This follows from the requirement that the fluid satisfy two independent quadratic constraints—conservation of energy and conservation of potential vorticity (enstrophy).

These ideas have been much elaborated since then (Rhines, 1979; Holloway and Hendershott, 1977; Salmon, Holloway, and Hendershott, 1976; Salmon, 1978). The problems are subtle and difficult and the dominant mechanisms not yet completely sorted out. Numerical experiments have been done to study several different physical processes that can govern the evolution of given initial conditions. The pure barotropic nonlinear cascade process moves energy toward larger spatial scales. Another nonlinear process transfers baroclinic energy toward the Rossby radius of deformation and thence to the barotropic mode. Topographic influences can scatter energy toward either larger or smaller scales, depending upon the wavenumber spectrum of the initial conditions and of the bottom topography. The relative importance of each of these processes depends on the energy level and spatial structure of the initial conditions. The balances are often subtle and no single process seems to dominate throughout the plausible range of initial conditions.

The advent of observations of mesoscale motions in the early 1970s also stimulated attempts to make numerical oceanic general-circulation models that could resolve an interior eddy field rather than parameterizing it simply as a constant, positive eddy viscosity. By restricting the calculations to limited ocean areas, it was possible to have many forms of time variability appear explicitly in the models (Holland and Lin, 1975a,b; Robinson, Harrison, Mintz, and Semtner, 1977). As computers have become larger and faster, the models have become ever more complex and realistic, including all of the physical mechanisms of the simpler-process models. However, it is fair to say that even the most sophisticated models now extant cannot fully model all of the observed physical complexity of the ocean (Schmitz annd Owens, 1979). But this is not to denigrate the models. They are beginning to show qualitatively many of the observed features of the oceans and doubtless rapidly will become much more realistic. Nonetheless, the ocean is very complex; numerical models that are intended to be realistic need also to be complex. As more and more physics is added to the computer codes the results become increasingly difficult to understand. Indeed it may be that understanding a fully realistic numerical model of the ocean requires nearly as much time, effort, and ingenuity as understanding the real ocean.

We shall not dwell further here on the modern theories because there have been a number of recent reviews of the state of the art. Rhines (1977, 1979) has discussed many of the basic ideas; the numerical models have been examined and compared by Robinson, Harrison, and Haidvogel (1979) and Harrison (1979b). As we proceed to discuss the observations, we shall introduce additional theoretical hypotheses as needed.

11.2 The Field of Variability of the Ocean

Perhaps the most sweeping generalization that can be made about the known time variability in the ocean is this: the ocean is filled with time-varying features, with all space and time scales, whose energy levels vary by orders of magnitude over the ocean basins. To state it slightly differently, the field of variability is locally representable by a continuous frequency-wavenumber spectrum; but the underlying process is not spatially stationary in the statistical sense and this vitiates much of the utility of the spectral description. The use of the word "continuous" for the spectrum is deliberate. Through much of the history of oceanography, as in many fields, there has been a search for simple line processes, "cycles," which simply do not exist.

We know from the past decade of observation that simple universal parameterizations of the variability are not valid. The upper ocean is different (at least superficially) from the lower ocean and the gyre centers are different from the gyre boundaries. Eastern walls differ in their variability from western walls. It is probably also true that the dynamics, as well as the kinematics, of these regions differ. At this time, it is difficult to give more than a fragmentary picture of open-ocean low-frequency variability because our observational tools are still not adequate for the job of measuring a global ocean.

11.2.1 The Meteorological Forcing Function

The ocean is driven by the heating of the atmosphere and by direct momentum transfer from the winds. The details of these processes are not completely clear (e.g., Phillips, 1977a; Kraus, 1977), involving as they do small-scale turbulent transfer processes within the marine atmospheric and oceanic boundary layers. Nonetheless it is probable that the larger scales of these forcing functions are able to communicate themselves from the atmosphere to the ocean. That is, the details of the transfer of momentum to the ocean from the

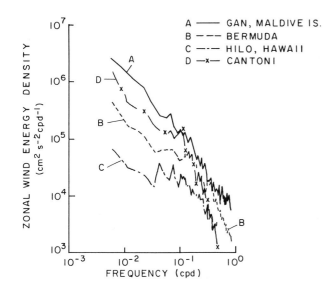

Figure 11.2A Zonal wind spectra from a few representative open-ocean islands.

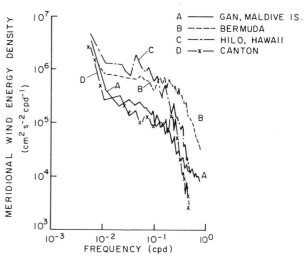

Figure 11.2B Meridional wind spectra from open-ocean islands.

winds involves such small-scale processes as ordinary ripples, but we anticipate that if the wind varies over a 1000-km scale, then it is this variability scale that is relevant to understanding the oceanic response to winds, and it is thus meaningful to seek a description of the forcing function in frequency-wavenumber space.

It is comparatively easy (Wunsch and Gill, 1976; Philander, 1978) to show that direct atmospheric-pressure forcing is a very inefficient process compared to wind-stress forcing. Direct thermal forcing is likely also (Frankignoul and Müller, 1979) to be comparatively weak except on the very largest time scales that determine the *mean* thermohaline general circulation.

The description of how the atmosphere forces the ocean is of considerable interest in studying the field of variability, but it may not be a decisive factor. The reason is that theoretically one can drive oceanic variability indirectly through instabilities of the strong "mean" current systems (Lipps, 1963), and of the interior ocean (Gill, Green, and Simmons, 1974; Robinson and McWilliams, 1974). Nonetheless in some regions at least—the continental shelves (see chapters 7 and 10), in the open sea far from intense currents (Brown et al., 1975) and in regions like the Florida Current (Düing, Mooers, and Lee, 1977; Wunsch and Wimbush, 1977)—there does seem to be some direct response to atmospheric forcing, although it seems to be at the short-period end of the spectrum, namely, periods shorter than about 10 days.

Some representative wind spectra are displayed in figure 11.2. It should be noted that stress is usually

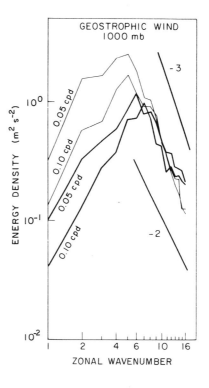

Figure 11.2C Estimated wavenumber spectrum of meridional geostrophic wind at 30°N (heavy line) and 50°N (light line) for two frequencies at 1000 mb. Notice steep slope at high wavenumbers. (Frankignoul and Müller, 1979.)

computed from the two winds components, (w_x, w_y) by the formula

$$(\tau_x, \tau_y) = C(w_x^2 + w_y^2)^{1/2}(w_x, w_y), \qquad (11.5)$$

where C is a parameter depending upon the drag coefficient and which in general (Bunker, 1976) depends upon the air-sea temperature difference and possibly, upon the wind speed itself. But the spectrum of stress will strongly resemble that of the wind because the expression (11.5) preserves the zero crossings of the wind components. The spectrum is distorted relative to that of the wind by the amplitude modulation factor $(w_x^2 + w_y^2)^{1/2}$.

The spectra tend to become white (or less red) at periods longer than a few days, reflecting the unpredictability of weather (by linear methods at least), and then redden again at the longer (and here unresolved) periods (but see figure 11.9A). The spectra show a great variety of geographical effects, viz., latitude changes, proximity to continental influences, topography (the Hilo spectrum in particular seems greatly affected by the presence of high mountains, being highly anisotropic at low frequency), and sea-land contrasts. In the mid-latitude spectra, most of the energy is found in the 4-10-day band characteristic of the weather systems. At periods longer than those displayed, the wind spectra tend to become white (see Willebrand, 1978). Frankignoul and Müller (1978) have computed estimates of the 1000-mb zonal wavenumber spectra for a variety of latitudes, displayed here in figure 11.2C. One sees a distinct concentration in the low wavenumber bands. Extrapolation to the sea surface is not straightforward, however.

The response of the ocean to forcing by fluctuating wind fields has been considered by Phillips (1966), Frankignoul and Müller (1979), Leetmaa (1978), and Harrison (1979a) in the period range of days to months. Although the final word has not been spoken, it appears that over most of the ocean direct wind forcing is unlikely to compete with internal instability processes. The weak seafloor pressure fluctuations measured by Brown et al. (1975) are spatially coherent on the large scale only at periods of 10 days and shorter, where current meter records show very little energy. These may in fact be wind-forced barotropic modes but they are energetically unimportant. In what follows the reader may want to compare the shape of the wind spectra displayed in figure 11.2 with those of the other variables discussed later.

Frequency spectra of atmospheric-pressure fluctuations (not shown) also tend to show a whitening at low frequency although Madden and Julian (1972) and Luther (1980) have found some large-scale organized motions at long periods, circa 50 days.

Sea-level measurements of atmospheric variability are sparse and inadequate for making definitive statements. Frankignoul and Müller (1979) attempted to construct a model spectrum by synthesizing the available data. They assumed spatial homogeneity and isotropy in the wind field. The two assumptions are ultimately untenable, but their model is probably the best that can be constructed at the present time.

11.2.2 Interannual Fluctuations in the Ocean

These are changes with periods longer than 1 year. Our major emphasis will be on those motions deduced in the modern era of instrumentation, excluding periods accessible only through essentially geological methods. Thus with one exception (see below) we will not treat what is best called paleo-oceanography, which deserves a full treatment by itself.

Before attempting to describe what is known about the very long-period changes in the ocean, there are two points to be made. Consider first figure 11.3. Figure 11.3A is a section made by the vessel *Challenger* from New York to Bermuda to St. Thomas, Virgin Islands, in 1873. The figure at bottom is a section from the Grand Banks to Bermuda to the Mona Passage obtained in 1954 and 1958 (Fuglister, 1960). The qualitative resemblance is very close; clearly the fundamental assumption of large-scale physical oceanography—that at least some aspects of the overall circulation, in particular the large-scale baroclinic structure, have remained stable for 100 years—is correct. That is consistent with the statement of Sverdrup et al. (1942) noted above about combining nonsimultaneous stations if they are sufficiently widely spaced, and is a (usually) unstated assumption in discussions of the mean fields to this day. It is well justified, for example, by comparing the high-quality *Meteor* sections in the South Atlantic made in 1925-1927 and reported by Wüst and Defant (1936) with those made in 1954-1958 and reported by Fuglister (1960). One is hard pressed to detect any significant differences on the large scale. (Of course, from these data one is able to say nothing regarding barotropic changes.)

One result of the paleo-oceanographic studies is useful here too. Figure 11.4 displays the reconstruction from paleontological data of sea-surface temperature 18,000 years ago, and the annual mean today (from Gates, 1976). Eighteen thousand years BP was the height of last major glaciation (CLIMAP Project Members, 1976). What is so striking is how *little* the ocean surface changed away from the immediate proximity of the edge of the ice sheet. Indeed the change appears to be less than the present seasonal range (Fuglister, 1947). That changes in the ocean under the impact of such a large disturbance as a glaciation are so minor suggests that seeking changes in the ocean owing to present

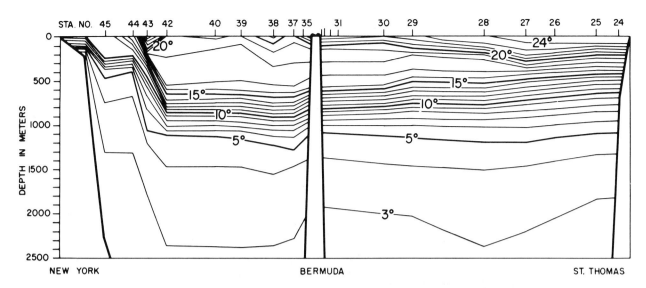

Figure 11.3A *Challenger* section (1873) from New York to Bermuda to Virgin Islands. (Tizard et al., 1885b, p. 135.)

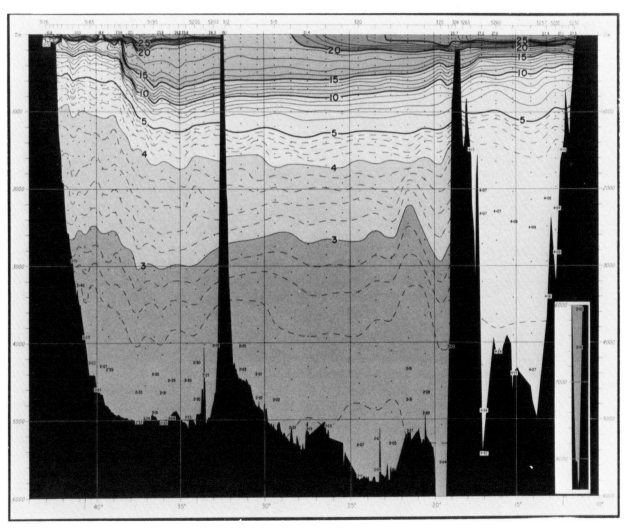

Figure 11.3B Section made in the fall, 1954 (from Nova Scotia to Bermuda and Mona Passage to South America), and in winter, 1958 (from Bermuda to Mona Passage). (Fuglister, 1960.)

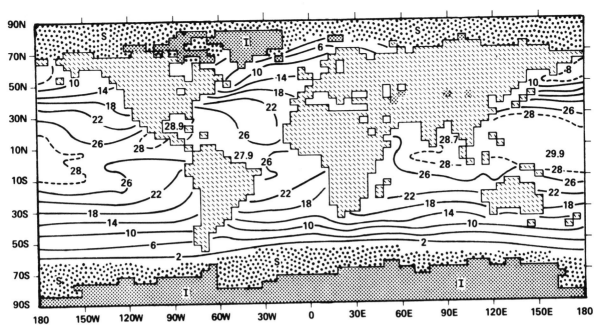

Figure 11.4A Present annual mean sea-surface temperature in North Atlantic. (Gates, 1976.)

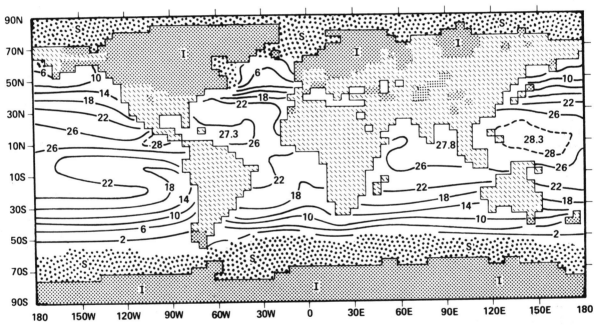

Figure 11.4B Estimated sea-surface temperature 18,000 years ago at height of glaciation as determined by CLIMAP study of fossil assemblages in deep-sea cores.

minor climatic fluctuations may be fruitless. The changes 18,000 years ago could have been much greater at depth than at the surface; whether this is indeed true is now the subject of study.

The success of oceanographers in obtaining at least qualitative descriptions of the large-scale circulation by combining observations made many years, or even decades, apart in time is, as noted, a reflection of some underlying truth about the frequency-wavenumber content of the baroclinic variability in the ocean. The most intense fluctuations seem to occur on a spatial scale small compared to the large-scale mean gyres. On the other hand, if one is to seek the role of the ocean in climatic changes, it is likely to be reflected in very large-scale low-frequency oceanic fluctuations, on the intuitive assumption that perturbations to the atmosphere owing to small-scale fluctuations of the ocean will tend to be "integrated out" by the atmosphere. In a nonlinear system like the atmosphere, the validity of this assumption is by no means obvious, but it is a reasonable initial hypothesis. One can then attempt to search for very large-scale changes in ocean circulation on time scales of years and space scales of thousands of kilometers.

For a number of reasons, most of the work on this hypothesis has been conducted in the Pacific Ocean. First, it has long been hypothesized (Bjerknes, 1969, Namias, 1972) that the U.S. continental weather may be sensibly modified by large-scale thermal anomalies over the Pacific Ocean. Bjerknes provided some specific hypotheses now generally called "teleconnections." Beginning in the early 1960s investigators at the Scripps Institution of Oceanography began a series of investigations to attempt to define, and ultimately to understand, both the apparent anomalies themselves and the role they might play in U.S. climate and weather. It seems clear that the anomalies are real; but it also seems fair to state at many of the links between the anomalies and weather are the result of wishful thinking rather than evidence (e.g., Davis, 1976, 1978a).

Sea Level Few extended time series for studying very long-period motions are available; the only real data consists of sea-level measurements. The idea of using tide-gauge records for studying fluctuations of geostrophically balanced currents evidently dates back to Sandström (1903), although Montgomery (1938b) seems to have been the first to actually attempt it.

The spectra presented here in figure 11.5 (see also figure 11.9), in Wyrtki (1979), in Munk and Cartwright (1966) and in other places, of the longest available records (circa 100 years—the longest record may be the one from Brest analyzed by Cartwright (1972), which runs from 1856 to date) are red, although decreasingly so beyond periods of 1 year (see Figure 11.9).

The limits of this increasing power with decreasing frequency are unknown. In some regions, one may be seeing slight fluctuations in the geodetic levels of the tidal gauges; in other regions this seems implausible. Taken at face value, the red spectra suggest that there indeed may be barotropic, large-scale fluctuations of the oceanic gyres. One infers that they are barotropic because of the decrease in temperature variance with lengthening period appearing in the temperature spectra (see section 11.2.4), and large scale because of the decreasing variance in the moored-current spectra at long periods. But the evidence is ambiguous.

A number of attempts to understand the physics of long-period fluctuations in sea-level records have been made (Wunsch, 1972c, Groves and Hannan, 1968; Groves and Zetler, 1964; Shaw and Donn, 1964; Schroeder and Stommel, 1969). The major difficulty is that not only are long records few, but the number of long simultaneous records, which are vital for understanding spatial correlations and possible propagation, are even rarer.

Sea-level fluctuations at a point are a complex summation of many different physical phenomena. Much of the work cited above was dedicated to attempting to unravel the role of local weather variables in sea-level fluctuations. In the open sea, using island data, Groves and Hannan (1968), Wunsch (1972c), and others found an inverted barometer response at periods longer than about a day. At periods of months and longer, the effects of wind tend to dominate. The procedure for deducing the relative role of the two wind components and pressure is not straightforward because the weather variables are themselves coherent; a multiple regression procedure described in the cited papers is required. With *localized* weather effects removed one can then ask whether fluctuations from location to location are coherent and can be ascribable to any particular known physics. A major problem has been (e.g., Groves and Hannan, 1968) the absence of measurable coherences between the few simultaneous island records available. The entire procedure is made very difficult because the prior removal of the fluctuations coherent with local weather may remove global phenomena that are forced by the meteorology.

The clearest picture of large-scale fluctuations of sea level stems from the work of Wyrtki (1974, 1975a) in the Pacific. Previous work in the Atlantic by Schroeder and Stommel (1969) and Wunsch (1972c) had suggested that on the time scale of months that sea level was to a large extent responding to fluctuations in dynamic height relative to reference levels at about 1500 db (decibars). Wyrtki (1979) has shown in the tropical and equatorial Pacific Ocean that he could find large-scale coherent patterns of dynamic height variation that also

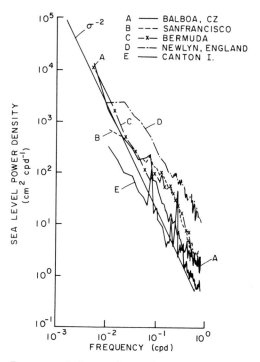
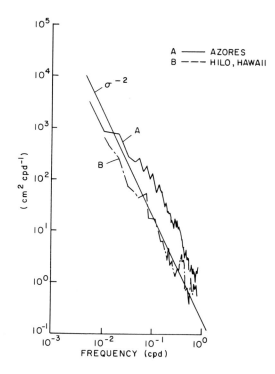

Figure 11.5A Spectra from a few representative sea-level records (log-log form).

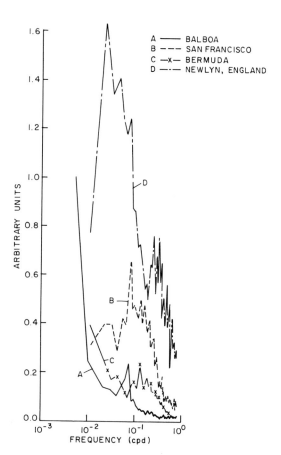
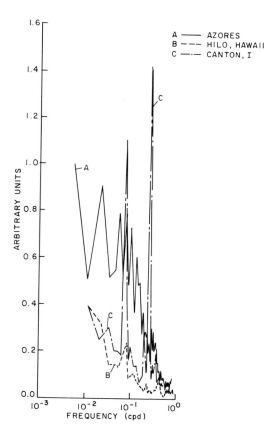

Figure 11.5B Spectra from sea-level records in an energy- (variance-) conserving form; notice that units are arbitrary—only relative spectral shapes are comparable. Canton I spectrum contains sharp peaks at 4 and 5 days, described by Wunsch and Gill (1976). The fortnightly tide is also apparent in the spectrum.

shows up convincingly in the Pacific tide gauges. He has been able to relate his observations to fluctuations of the large-scale tropical current systems of the Pacific. Much of this work has been directed toward understanding of the relationship between the ocean fluctuations and the catastrophic economic and climatic effects of the El Niño phenomenon (see chapter 6).

There does seem to be a link (or at least a correlation) between trans-Pacific sea-level fluctuations, the occurrence of warm water on the coast of Peru, and gross changes in the wind field over the Pacific. These latter changes are supposed to be part of the so-called southern oscillation and the Walker cell circulation (Bjerknes, 1969) in the atmosphere. Thus there is some indication of an actual coupling of large-scale oceanic and atmospheric fluctuations, but the global extent of the phenomenon and the causal links are very obscure. The data base is inadequate to be truly definitive, but the apparent patterns are plausible and are a highly promising line for future research into gyre-wide fluctuations, at least in the near-equatorial oceans.

For the purely oceanic phenomenon one needs ultimately to understand the extent to which one is seeing dynamic topography variations relative to a fixed reference level (a difficult idea to rationalize) or fluctuations possibly representable as vertical normal modes. In this latter case, the changes in sea-surface topography imply equivalent deep-water fluctuations, which are, however, at this time totally unknown.

Presumably similar fluctuations occur in the Atlantic and Indian Oceans. But the absence of many islands in these oceans has largely precluded the determination of sea-level fluctuations there. The Azores and Bermuda records were examined by Wunsch (1972c) and the Iceland record by Donn, Patullo, and Shaw (1964). Most of the low-frequency variability of the Indian Ocean, especially in the western portions, is obscured by the very large monsoonal signals.

Thermal Record The most conspicuous low-frequency phenomenon in the ocean is the sea-surface temperature anomalies of the Pacific that have been studied intensively the past two decades. Barnett (1978) has reviewed this work. A primary motive for the interest was the Bjerknes (1969) teleconnection hypothesis. There is little doubt that extensive changes in sea-surface temperature do exist and can persist for months and years. Figure 11.6, taken from Barnett (1978), is a multiple-year record of temperature (fluctuations about the long-term mean) at Talara, Peru, and Christmas Island. These "anomalous" temperatures occupy major areas of the Pacific Ocean. Figure 11.7 displays the first three empirical normal modes of Pacific sea-surface temperature (from Barnett and Davis, 1975). These modes describe slightly under half the total variance of the Pacific sea-surface temperature fluctuations; their immense scale is apparent.

Much of the interest in the anomalies has been in the possibility that in changing the lower boundary condition of the atmosphere, fluctuations in the atmosphere might be induced on the long oceanic time scale rather than on the intrinsically short atmospheric time scale. Studies of the question have been hampered

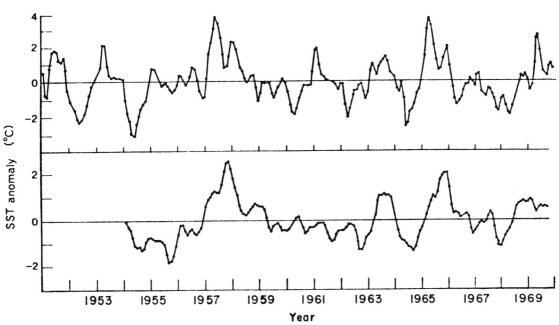

Figure 11.6 Sea-surface temperature anomalies at Talara, Peru (upper) and at Christmas Island (lower). (Barnett, 1978.)

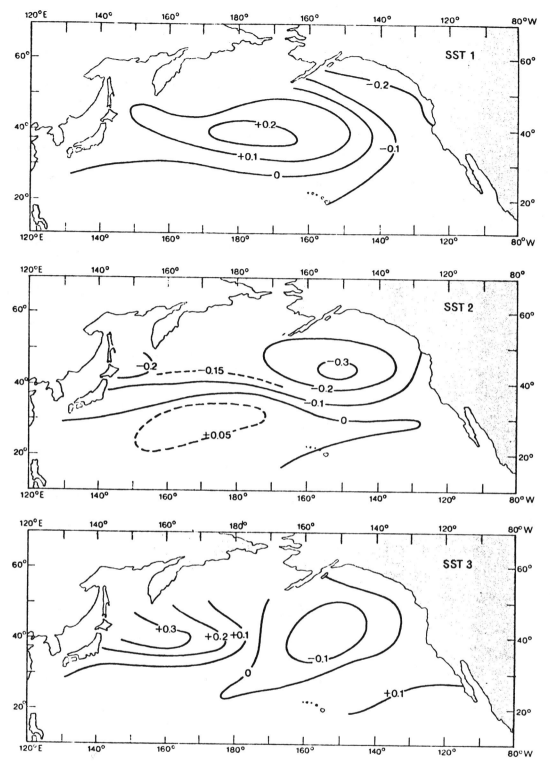

Figure 11.7 Three lowest empirical normal modes of sea-surface temperature anomaly field in North Pacific Ocean. Notice very large scales involved. (Barnett and Davis, 1975.)

by the inherent noisiness of the atmosphere. But Davis (1976) showed that changes in the ocean tended to lag those in the atmosphere, thus implying that the anomalies were being driven by the atmosphere rather than the reverse. In a later paper, Davis (1978a) obtained some atmospheric predictability from sea-surface temperature anomalies stratified by season. But the same predictability was found using atmospheric sea-surface pressure anomalies. The cause-and-effect relationships thus remain unknown. Rowntree (1972) and Kutzbach, Chervin, and Houghton (1977) have studied the reaction of the atmosphere to imposed sea-surface temperature anomalies. With unrealistically large anomalous values, an atmospheric reaction can be detected, but its significance is still not understood.

Frankignoul and Hasselmann (1977) have shown that purely random forcing of the ocean by the atmosphere can plausibly generate anomalies. The degree to which the surface features represent anomalies of heat content, that is, represent subsurface features as well, also is not clear. White and Walker (1974) have displayed time–depth diagrams for temperature anomaly at three positions in the Pacific Ocean (figure 11.8). Gill (1975b) shows similar data and perhaps the simplest conclusion to be drawn is that the physics governing sea-surface temperature anomalies is a combination of interaction with the atmosphere and with deeper ocean dynamics in a form that varies in space and time. It is difficult to relate the gyre scale fluctuations described by Wyrtki, to the anomalies, except in the case of El Niño.

Observations in the Atlantic similar to those made in the Pacific have been described by Rodewald (1972). On a much longer time scale, there are variations in the extent of the pack ice in the vicinity of Iceland that seem relatable to atmospheric changes. Oceanic changes that may accompany the atmospheric fluctuations beyond those locally involved with the sea ice are unknown.

The significance and meaning of apparent large-scale gyre-wide fluctuations is obscure. Attempts at demonstrating massive fluctuations in western boundary current flows that might be of real climatic significance have generally not been convincing. The wild variability in estimates of the volume transport of the Gulf Stream north of Cape Hatteras (see Worthington, 1976) has generally been due to varying methods of estimating reference levels, and there are no convincing demonstrations of nonseasonal volume-transport fluctuations. Similar remarks apply to the Antarctic Circumpolar Current, where 100% changes in estimates of the volume transport essentially disappear when appropriate reference levels are used (Nowlin, Whitworth, and Pillsbury, 1977).

Much of the problem of documenting real low-frequency large-scale fluctuations in the major ocean gyres may be looked upon as a classical problem of aliasing. Momentary fluctuations in the mass field, which owe their existence to short-period time variability, may inappropriately be interpreted as representing much lower-frequency phenomena (Worthington, 1977b). The large-scale baroclinic structure of the ocean is remarkably stable and one cannot store large amounts of water anywhere in the system without raising havoc with sea level. Real low-frequency oceanic changes on large scales appear to be subtle and difficult to detect in the presence of the much stronger short-period fluctuations discussed below in section 11.2.4.

11.2.3 Annual Variability

The entire atmosphere-ocean system is subject to a large thermal forcing with an annual period as the sun migrates meridionially throughout the year. In the atmosphere the seasonal change in character is of course enormous at mid- and high-latitudes. The ocean would respond both to the direct thermal forcing of the sun and to the indirect thermal forcing due to the change in air–sea temperature difference. One anticipates, from calculations like those of Gill and Niiler (1973) and Frankignoul and Müller (1979), that except in the mixed layer and on the equator the response to both would be quite small. Even in the region of intense wintertime air–sea heat exchange that occurs in the northwest regions of the northern hemispheric subtropical gyres and that leads to the formation of 18°C water in the Atlantic and 16°C water in the Pacific, the oceanic response seems localized and nearly passive (Warren, 1972; Worthington, 1959). More generally Gill and Niiler (1973) show that the heat budget is essentially locally controlled with lateral advection comparatively unimportant. Only the upper 100–200 m of ocean is involved.

It is the oceanic response to the annual cycle in the wind field that is likely to dominate the annual period. The annual line in the wind field at the ocean surface is superimposed upon a background continuum, and is surprisingly weak (figure 11.9) even in the monsoon regions of the Indian Ocean. The annual line at mid-latitudes, e.g., at the location of Bermuda, seems to carry no more than about 1% of the total wind-field variance in periods between 10 years and 1 day.

Patullo, Munk, Revelle, and Strong (1955) analyzed the available tide-gauge records for the annual variability in sea level and provide an interesting global map. Most of the inverted barometer-corrected sea-level signal between 40°N and 40°S is in "steric" height, i.e., the essentially passive change resulting from the warming or cooling of the upper water col-

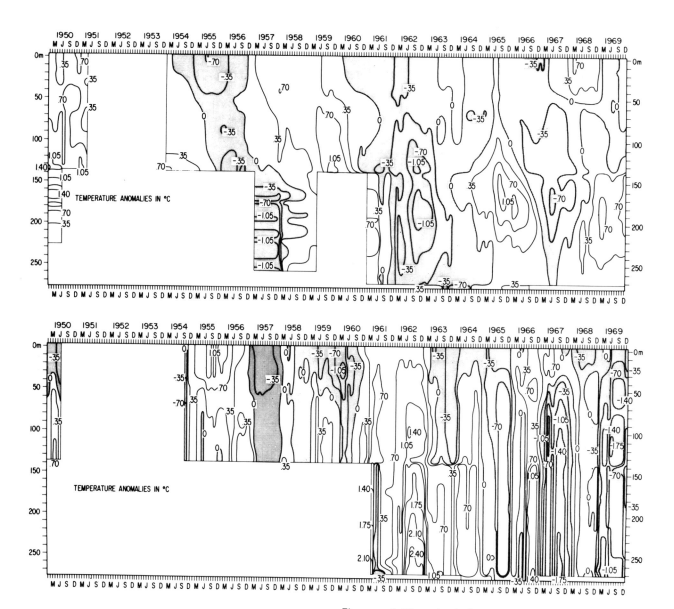

Figure 11.8 Time-depth diagram of subsurface anomalies of temperature at ocean station N (upper at 30°N, 140°W) and at station V (lower at 34°N, 164°E). (White and Walker, 1974.)

umn. The International Geophysical Year (IGY) data were subsequently examined by Donn, Patullo, and Shaw (1964), resulting in much the same conclusions. The effects of the direct steric changes do not appear to be visible below about 300 m in subtropical latitudes (Wunsch, 1972c).

The change in elevation of the sea surface owing to these direct heating effects should give rise to a weak annual surface circulation; this circulation would be weak because of the very large scale over which the steric effects are induced.

In their analysis of the annual variability, Gill and Niiler (1973) concluded that over most of the ocean the major baroclinic effects of the wind would be to induce an Ekman suction at the annual period with a consequent response of the main thermocline. This signal seems to be lost in the background movement of the thermocline in response to other forces. By and large, deep current meter or temperature records of a few years duration do not display any recognizable annual variability above the continuum [although White (1977) claims he sees annual-period baroclinic Rossby waves generated at the eastern boundary of the Pacific].

The western boundary currents do show some form of annual cycle; in the Florida Current the cycle manifests itself over the entire water column from top to bottom and has been convincingly documented by Niiler and Richardson (1973) as an essentially barotropic change. Farther north, Fuglister (1947, 1951) documented a cycle in the surface temperatures and flow by using ship-drift reports. The extent to which this annual cycle penetrates into the deep water north of the Florida Straits is unknown. Much earlier, Iselin (1940a) had attempted to study the annual cycle of the Gulf Stream between Woods Hole and Bermuda. Over a 3-year period, he found transport fluctuations between 76 and 93×10^6 m$^{3-1}$, with the system strongest in summer. In view of the problems of aliasing referred to above, the necessity for a fixed reference level, and the short record duration, it is difficult to assess the significance of this result. Wyrtki (1974) has shown an annual cycle in the surface circulation of the equatorial Pacific in response to the annual trade-wind changes. Elsewhere (Wyrtki, 1975a) he finds little annual signal except in the immediate vicinity of the Kuroshio.

There is a fundamental problem in using the annual cycle to understand the dynamics of the ocean. Because virtually all physical variables (wind, temperature, sea level, etc.) show an annual line, it is difficult to sort out cause and effect. Narrow-band processes tend to contain much less useful information than random broad-band ones.

11.2.4 The Mesoscale
The ocean appears to be in a turbulent state with a very high Reynolds number, and as that fact has come

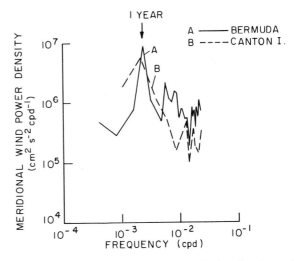

Figure 11.9A Spectra of wind records (displayed in figure 11.2 but at much higher resolution), showing the annual peak.

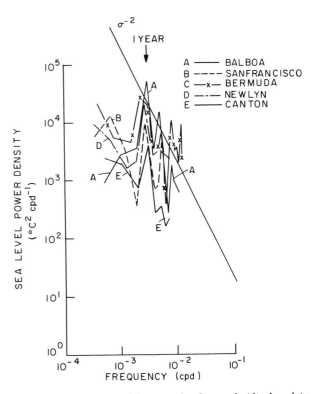

Figure 11.9B Spectra of long sea-level records (displayed in figure 11.5 but at much higher resolution), showing resolved annual peak.

to be more widely appreciated attention has shifted from the very large-scale mean flows and large-scale variability discussed above, to the nature of the turbulence itself. Much of this shift in focus has come about because of technical innovations. These made possible multimonth *in situ* time series independent of ship endurance limitations. The presence of disturbances of spatial scales of O(100 km) is obvious in conventional hydrographic sections (figure 3b from Fuglister, 1960) but studies of these undulations in the isopycnals were impossible before about 1970. In discussing the mesoscale, we include all fluctuations with the appropriate space and time scales, including continental shelf waves even though these latter motions are essentially nonturbulent in character.

Shelf Waves These waves were discovered by Hamon (1962), who found that sea-level fluctuations on the east coast of Australia were not in isostatic balance with atmospheric pressure (i.e., the inverted barometer effect failed to apply). Robinson (1964) pointed out that Hamon's results could be rationalized in terms of topographic Rossby waves excited by atmospheric pressure forces [the basic physics of the phenomenon is discussed by Lamb (1932, §212)]. Subsequently Adams and Buchwald (1969) demonstrated that wind stress was a much more efficient driving mechanism and obtained reasonable agreement between theory and observation.

Since the original observations of Hamon and the later theoretical work much attention has been paid to these waves for a number of reasons. Unlike most open-ocean variability, linear theory seems to apply reasonably accurately, the shallow water of the continental shelves is comparatively easily accessible, conventional harbor tide gauges can often be used, and the waves presumably play a major role in the time-dependent response of the low-frequency shelf circulation to external forces.

We shall not dwell on the subject here because a number of comprehensive reviews have appeared recently (e.g., LeBlond and Mysak, 1977; Munk, Snodgrass, and Wimbush, 1970; Mysak, 1980; and see chapter 10). The detailed connection between open-ocean variability and shelf waves is unknown. It is possible, however, that, in some situations, the waves generate open-ocean effects rather than merely being the response to local winds or forcing from the open sea. An example of this may be in the Gulf Stream System (Düing, Mooers, and Lee, 1977; Wunsch and Wimbush, 1977), where waves of several days period are observed in the Florida Current region. A number of authors have suggested that these waves may trigger meanders and other motions of the Gulf Stream System in the region north and east of Cape Hatteras. No actual cause and effect has ever been demonstrated; the topic is discussed in chapter 4.

Gill and Clarke (1974) made plausible the hypothesis that shelf waves (they specifically emphasized the baroclinic Kelvin wave) play a fundamental role in the special subclass of mesoscale variability called coastal upwelling. If this is correct (and it permits remote forcing to give rise to strong localized effects) there can be a dramatic influence of shelf waves on local climatology. In some cases (e.g., the El Niño phenomenon), coastal upwelling may be part of an ocean-wide, and perhaps even global, system involving both the ocean and atmosphere. The extent to which such events linking shelf waves (which may be thought of as a local focusing and amplification of comparatively weak forcing) are part of a global system is the subject of much work at the present time. That such links exist seems clear; their extent is obscure. Upwelling deserves a special discussion of its own; the reader is referred to O'Brien et al. (1977).

In our more general context, shelf waves provide a simple (i.e., nearly linear) example of mesoscale variability that in some cases exhibits the combined effect of rotation, topography, and baroclinicity. The literature contains a large number of special cases of topography for which analytical solutions are available. The basic physical phenomena are Kelvin waves trapped against the coast, double Kelvin waves at the shelf edge (Longuet-Higgins, 1968c), barotropic topographic Rossby waves (Rhines, 1969a), and baroclinic topographic Rossby waves (Rhines, 1970). In the presence of stratification finite bottom slopes couple the different wave types together and one is generally driven to numerical solutions (Wang and Mooers, 1976), but the underlying physics may be understood from the simpler physical situations. Ou (1979) has significantly advanced the analytic treatment of the problem. Hendershott in chapter 10 discusses these topics further.

Very similar effects occur at mid-ocean islands (Longuet-Higgins, 1969b; Wunsch, 1972b; Hogg, 1980); both barotropic and baroclinic trapped waves can occur and particular bottom-slope configurations can modify qualitatively the dispersion relations. The barotropic case with varying topography had been treated by Lamb (1932, §212), for the *inside* of a cylinder, rather than for the outside, as is appropriate to the oceanic case (the mathematical differences are slight).

Mesoscale Eddies Much of what we now call the *mesoscale variability*, or the *eddy field*, is plainly what early investigators like Maury had in mind when they described their difficulties with time-dependent motion. In a modern context, these motions were first quantitatively measured in 1959 (Crease, 1962; Swallow, 1971) in a noteworthy experiment. Using the newly developed neutrally buoyant floats (Swallow, 1955) they detected motions west of Bermuda with velocities of up to 40 cm s^{-1} and apparent time scales of roughly 40 days. Their measurements were analyzed in an important paper by N. A. Phillips (1966b), who suggested that they could be treated as a sum of linear Rossby waves driven by the wind. Phillips's model could not reproduce the high velocities (but see Harrison, 1979a). Later, Rhines (1971b) suggested that they were intensified by bottom trapping on the Bermuda Rise, but this too seems inadequate.

By 1970, the combination of the knowledge of the existence (at least in the area studied by Swallow and Crease) of very intense time-dependent motions, of the theoretical work on turbulent interaction, and of the meteorological picture alluded to, led to a series of experiments aimed at elucidating the nature of the mesoscale in the ocean [Koshlyakov and Monin (1978) discuss the history of parallel Soviet efforts, which seemingly had a very different intellectual content]. It seems fair to call the 1970s the "decade of the mesoscale" in physical oceanography both for the intensive efforts that went into it and for the striking advances in knowledge that occurred. The initial tentative observational programs (e.g., Gould, Schmitz, and Wunsch, 1974) had first to convince the investigators of the reality of the phenomenon; it was only then that serious efforts could be mounted to understand mesoscale physics. Although 100-day time-scale fluctuations are much more humanly accessible than the interannual frequencies described above, it is still important to note that to obtain 20 degrees of freedom at a resolved 100-day period for a spectrum of velocity (or anything else) from a point measurement requires 1000 days of data. Thus studies of the mesoscale are necessarily in their infancy.

It is probably a mistake to overemphasize the mesoscale as an isolated phenomenon. There is reason to believe that the very large-scale phenomena described previously and the mesoscale are to a great extent simply manifestations of the extremes of a continuum. Ocean variability occupies a broad space-time spectrum; for purely experimental reasons it has been convenient to study the 100-km scale separately from the 5000-km scale. But in the long term, it will be necessary to understand their linkages.

On the other hand, there does appear to be a "bulge" or plateauing of low-frequency spectral variability in the very roughly defined period range of 50-150 days. It is apparent even in the sea-level spectra, and as nearly as one can tell this plateau is a global phenomenon. We shall follow Richman, Wunsch, and Hogg (1977) in calling this the *eddy-containing* band. At the low-frequency end this band fuzzily merges into the annual and interannual variability. We shall refer to frequencies above the eddy-containing band as the *isotropic band* because of the energy isotropy there. For the moment this distinction into separate bands is kinematically descriptive; there may also be a dynamic distinction.

Much of the work on mesoscale variability was focussed in the Mid-Ocean Dynamics Experiment (MODE-1) which ran from about 1971 to 1973. Many of the results have been summarized elsewhere (MODE Group, 1978; Mode-1 Atlas Group, 1977; Richman, Wunsch, and Hogg, 1977). The effort was continued in POLYMODE—as an amalgam of the early Soviet efforts Polygons (Brekhovskikh, et al., 1971) and the MODE experiment. By the time of POLYMODE, however, the mesoscale field was widely recognized as a universal, important phenomenon and was being studied as part of many aspects of oceanography, including the biological ones.

The detailed results of these experiments are too extensive to be described here, and, indeed, their full implications are not understood. But it appears that an eddylike field of variability is nearly universal in the ocean (MODE Group, 1978; Schmitz, 1976, 1978; Bryden, 1979; Bernstein and White, 1974, 1977; Hunkins, 1974) and is normally much more energetic than the local mean flow (see figure 11.10).

The time variability of the North Atlantic ocean on the large scale is represented by figure 11.11, taken from Dantzler (1977) [see Bernstein and White (1977) for a similar discussion of the North Pacific and Lutjeharms and Baker (1979) for the Southern Ocean]. It is a compilation of the temperature variance in the North Atlantic from expendable bathythermograph (XBT) measurements. The variance of temperature, which may be thought of as a crude measure of potential energy of the fluctuation field, shows a large-scale midocean trough, with steep gradients toward the boundaries, and especially toward the Gulf Stream System. This figure, while extremely suggestive, also illustrates several of the caveats that must be applied to discussions of the field of variability.

The variability field is commonly referred to as the *eddy problem*; the implication is that eddies have an identifiable scale, nearly unit aspect ratios and possibly even closed streamlines. But figure 11.11 is simply a representation of the time variability of the North Atlantic irrespective of its source. For example, it is well known that the Gulf Stream moves and meanders (Fuglister, 1963; and see chapter 4). Such variability will appear in figure 11.11 and presumably represents much of the energy appearing in the northwest region. The large-scale, low-frequency variability described in the previous sections would also be included but probably is only a small fraction of the variance. It is more than a semantic distinction to wish to distinguish the variability of migrating jets from that of features with

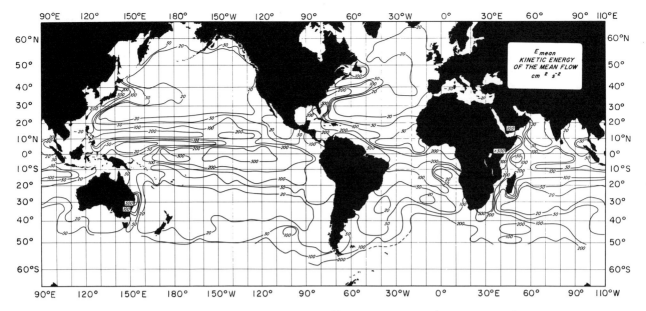

Figure 11.10A Five-degree-average chart, based upon ship-drift reports of mean surface kinetic energy. (Wyrtki, Magaard, and Huyer, 1976.)

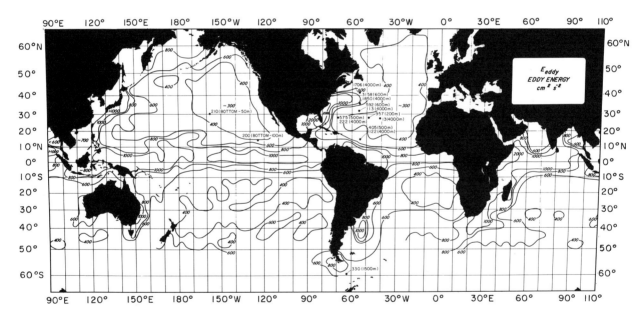

Figure 11.10B Five-degree-average chart, based upon ship-drift reports of total fluctuation kinetic energy (Wyrtki, Magaard, and Hager, 1976). Superimposed upon the chart are kinetic energy densities ($cm^2 s^{-2} cpd^{-1}$) from moored current meters, at a 50-day period at nominal depths of 500 m and 4000 m. Actual depths are shown.

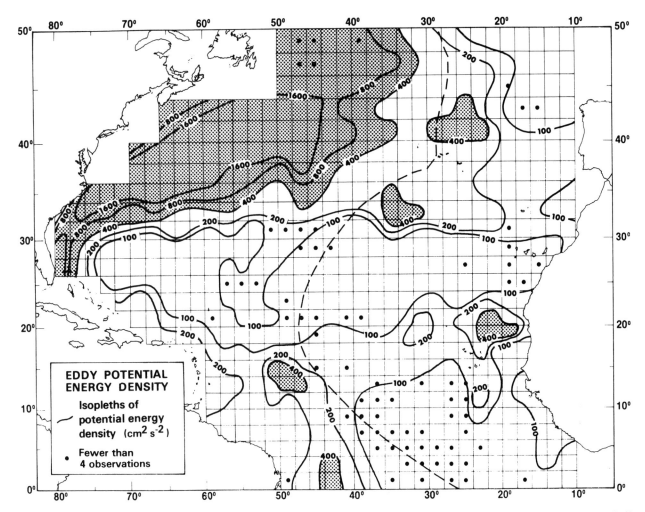

Figure 11.11 North Atlantic eddy potential energy as compiled from data by Dantzler (1976). Notice strong intensification in Gulf Stream region and general increase toward all boundaries.

relative isotropy of scale. Another difficulty is that XBTs measure only the baroclinic contribution to the eddy field. If there is a significant velocity field that might be described as barotropic, it will be missed by this form of data. Finally, XBTs reach only to some hundreds of meters depth. There is every reason to believe that the variability of the layers below the thermocline can be very different from that above. Unfortunately, we do not have sufficient data to construct for the deep ocean a figure analogous to figure 11.11. In the absence of such data there is an understandable tendency to interpret the upper layers as representative of the entire water column; there is little evidence to support that hypothesis, although it seems plausible that there should be a significant correlation between upper and lower oceans.

Another, global but still crude, depiction of the time variability of the surface ocean is figure 11.10 (from Wyrtki, Magaard, and Hager, 1976). It displays the mean kinetic energy and the variability about the mean based upon ship-drift observations averaged over 5°-squares.

The pitfalls of using such a data base are many and obvious; nonetheless, on a global basis, figure 11.10B is probably the best that can be constructed at the present time. Again, the motions represented there will be a sum over all time and space scales and are not confined to the mesoscale. In the North Atlantic there is a considerable qualitative resemblance to Dantzler's chart. One is probably safe in concluding that in a gross sense the largest total variability tends to be found on the western sides of the oceans, with secondary maxima along the other boundaries (including the equator). These regions are, of course, the ones associated with the apparently strongest mean currents (notice the resemblance between figures 11.10A and 11.10B), and one might surmise that there is some difficulty in distinguishing the variability as a boundary-intensified phenomenon from one associated with strong mean flows (an attempt at a distinction may be specious in any case).

Some feeling for the underlying nature of the fluctuations that contribute to (and perhaps dominate) figures 11.10 and 11.11 is displayed in figure 11.12. Here

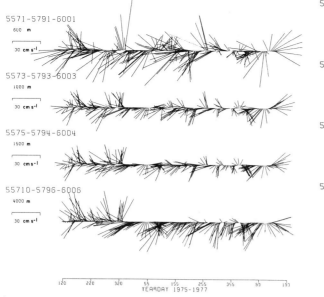

(11.12A)

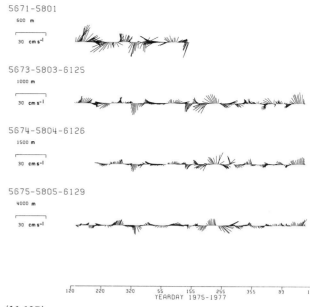

(11.12B)

Figure 11.12A,B Three-day average-velocity vectors from two locations (35°56′N, 55°06′W and 31°36′N, 55°05′W), respectively, in the North Atlantic (see figure 11.11B). Note large change in energy level over comparatively small separation in position between the two moorings. The difficulty of defining a mean flow from such measurements should be obvious.

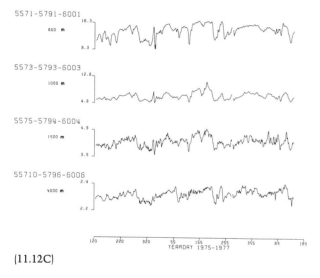

(11.12C)

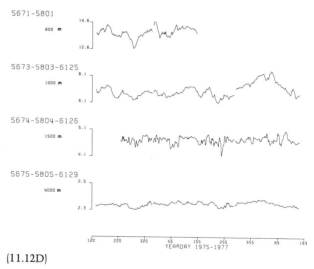

(11.12D)

Figure 11.12C,D Temperature records from same moorings as in figures 11.12A and 11.12B. Note change in character and amplitudes of fluctuations.

are the velocity vectors and temperature measurements (averaged over 24 hours) from two moorings in the north Atlantic at positions 35°56'N, 55°06'W and 31°36'N, 55°05'W. One sees a very complex and energetic time variability over the record lengths (nearly 27 months in the longest records), a considerable resemblance between different levels in the vertical, and a very marked decline in energy levels over the 400-km separation between the moorings. This remarkable change in energies is roughly consistent with that implied by figure 11.11, although the change is much steeper because of the 5° averages in figure 11.10.

Another view of the variability is shown in figure 11.13 depicting the spectrum of temperature fluctuations near Bermuda in the main thermocline (from Wunsch, 1972b). One sees a process with a broad spectral peak in the vicinity of 100 days with a relative decline at longer periods. As noted by Wunsch (1972b), the decline might have been anticipated because there is a stable large-scale mean-temperature field, at least in the 100-year historical record.

In an effort to display some of the characteristics of the low-frequency variability, I have collected here (figures 11.14 and 11.15) a number of spectra of long time series of different variables. Insofar as possible, these are displayed on common scales with common resolution to simplify the direct comparison of one region with another. In all cases, I display the logarithm of the spectral density as a function of the logarithm of the frequency. A number of the spectra are also plotted as frequency times power density against the log of the frequency (the so-called *variance-preserving form*). A word is in order about these different displays. The log-log forms tend to emphasize the very lowest frequencies in any spectrum that follows roughly a σ^{-p} form, where σ is the frequency and $p > 0$ simply because the value of the energy density-unit frequency band is increasing. With increasing frequency, the logarithmic transformation compresses an increasing number of equally spaced frequencies into a fixed distance along the abscissa.

It is often the low frequencies we are most interested in (and long records are the most difficult to obtain), and dynamic models often produce power-law spectra that show up most clearly (as straight lines) on log-log graphs.

But a spectral representation actually gives equal weight to evenly spaced frequencies, and the Parseval relation between a function and its Fourier transform treats all frequencies alike. Thus, a log-log plot of a record with a frequency spectrum σ^{-p} will give the impression that most of the energy is in the very lowest frequencies. But because the number of equally spaced frequencies in a uniform logarithmic-frequency interval increases uniformly with frequency, a σ^{-1} spectrum actually has as much energy in any fixed logarithmic-frequency distance, centered at an arbitrary frequency, as at any other frequency. This fact has led many investigators to plot frequency times energy density against log frequency, thus compensating for the increasing number of points-unit log frequency. One then interprets the energy in any given log frequency interval as the *area* under the curve. Thus a σ^{-1} spectrum would plot as a horizontal line. This, however, is *not* a white spectrum because it is still true that in any fixed *linear* frequency interval the total energy decreases with frequency as σ^{-1}. This form of the display can also give rise to very sharp spectral peaks (some may be seen in the figures); in many cases, however, the area under these sharp peaks is rather small and hence they may in fact represent little excess energy overall. Reader beware.

Because of the long time scales of the mesoscale phenomena, there are few quantitative measurements of energy levels or time and space scales. Most of the existing measurements are in the western North Atlantic as the result of MODE/POLYMODE but there are a few notable exceptions. Superimposed upon figure 11.10B are the locations of those records for which spectra are displayed here. As a crude measure of energy level, the kinetic-energy density at 50-day periods is also shown; this measure was chosen because it is independent of record duration and is resolved by a substantial number of measurements. Total record variance, which has been computed by many authors, is dependent upon the length of the measurement. A nominal shallow (circa 500-600-m) and deep (circa 4000-m) value is displayed in figure 11.10B. The variety of specific depths shown in figure 11.10B is one measure of the difficulty of making comparisons using the existing data base. Note also that to the extent that the local stratification changes from region to region, there can be a change in record kinetic energies due to the quasi-kinematic effects of wave refraction in a nonhomogeneous medium. Fortunately, these changes are rarely very large.

In the spectral displays, a line with slope -2 at a fixed energy level is drawn on the log-log form of the plots to make easier a comparison of energy levels and spectral shapes from one location to another. Most of the available long records (figure 11.14) are from the western North Atlantic [for additional displays see Schmitz (1976, 1978) and Richman, Wunsch, and Hogg (1977)]. An eastern Pacific record is displayed in figure 11.15A (Hayes, 1979a), the values from another Pacific record (B. Taft, private communication) are shown in figure 11.10B, and one from the Drake Passage area is shown in figure 11.15B (from Bryden and Pillsbury, personal communication). Although the energy levels

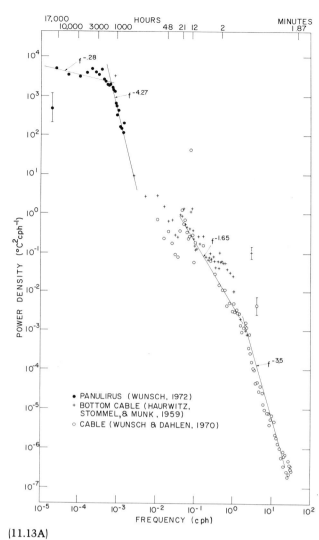

Figure 11.13 Spectrum of temperature at Bermuda in the main thermocline from record duration of 13 years. (A) is log-log form, (B) variance-preserving form. Bulk of energy is around 100-day periods. (Wunsch, 1972b.)

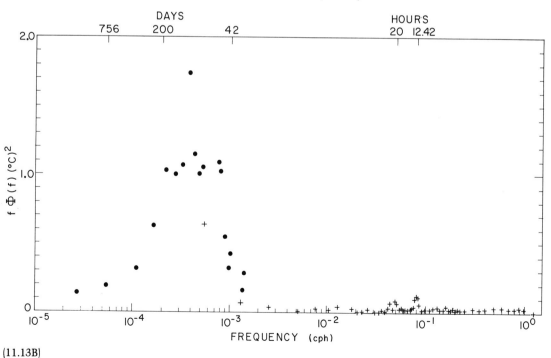

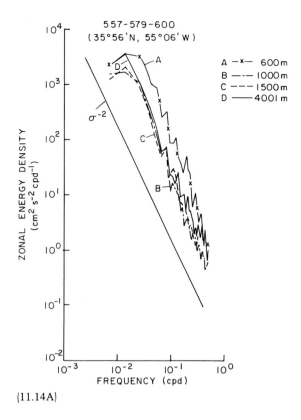

(11.14A)

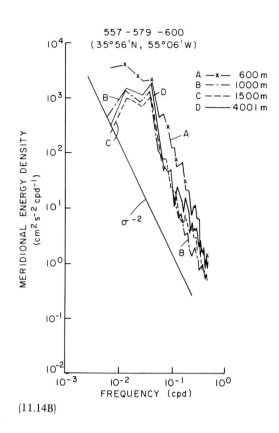

(11.14B)

Figure 11.14 Spectra of velocity and temperature from a variety of positions in the western North Atlantic, log-log forms are always shown. In a number of cases, the variance-preserving plots are also displayed. On log-log forms a straight line of slope 2 and fixed energy level is shown for comparison purposes. Variance-preserving plots for temperature are on an arbitrary scale because large range of values defeats a linear-scale display.

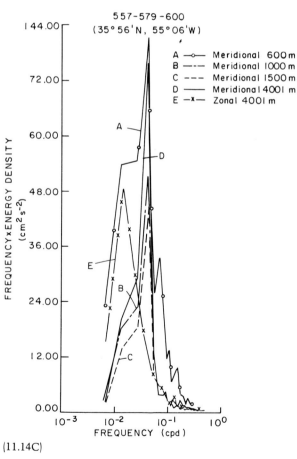

(11.14C)

Low-Frequency Variability of the Sea

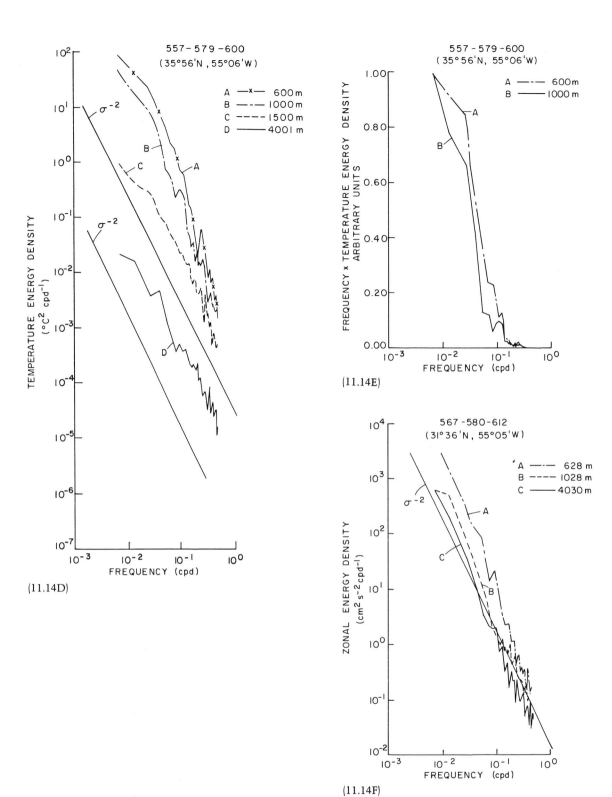

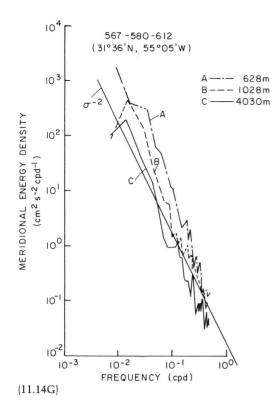

(11.14G)

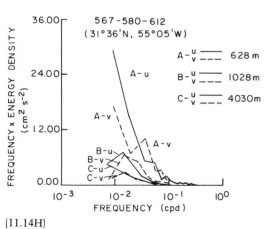

(11.14H)

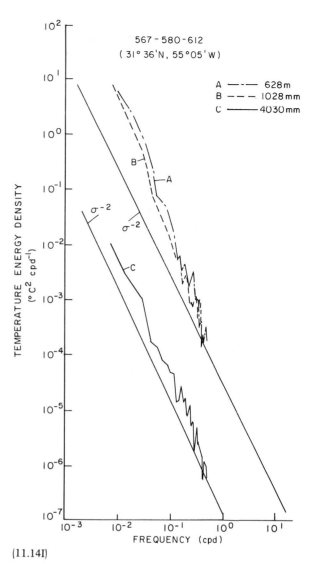

(11.14I)

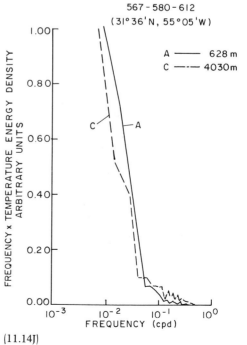

(11.14J)

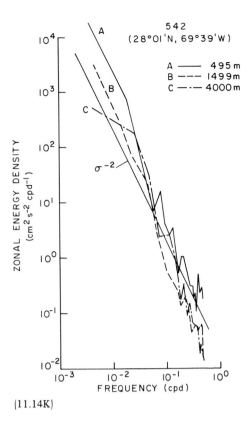

(11.14K)

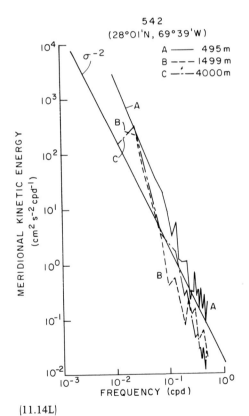

(11.14L)

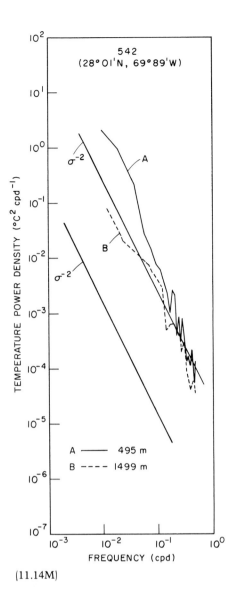

(11.14M)

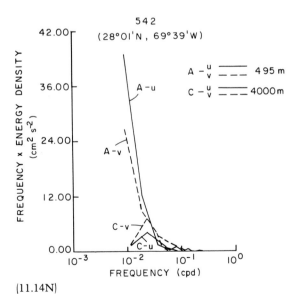

(11.14N)

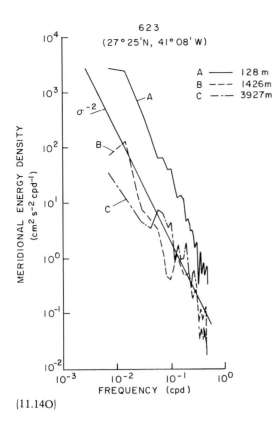

(11.14O)

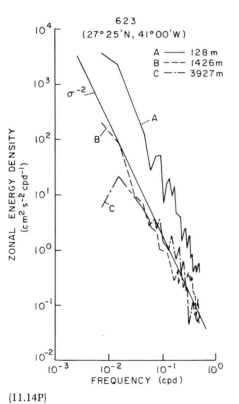

(11.14P)

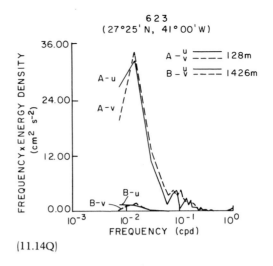

(11.14Q)

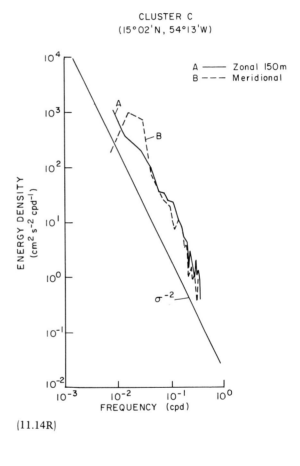

(11.14R)

369
Low-Frequency Variability of the Sea

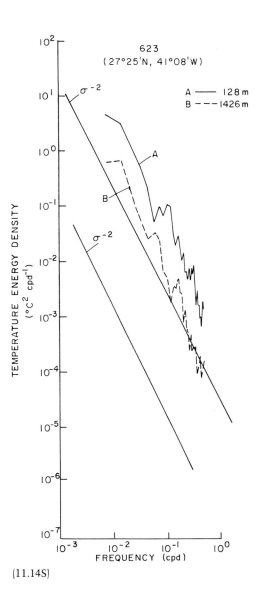

vary in the resolved band by an order of magnitude or more (compare figures 11.14A, 11.14F, and 11.15A), the spectra all seem to display some common characteristics. The high-frequency band is never far from σ^{-2}, and there is a tendency to zonal dominance as one enters the low-frequency band. The velocity spectra mostly show some energy enhancement in the eddy-containing band—usually most apparent in the variance-preserving plots, and in common with the sea-level spectra shown above. Overall, the temperature spectra are redder than the velocity spectra and tend not to show the eddy-containing band as clearly. But the much longer record (13 years) used to produce figure 11.13 suggests that these spectra would ultimately drop at lower frequencies as well [the spectra at MODE Center shown by Richman et al. (1977) do display this drop].

Figures 11.14A–11.14E display the spectra of records obtained in the near proximity to the Gulf Stream. As noted by Schmitz, the motion is much more barotropic in character than in the records obtained elsewhere (figure 11.14F). This result is consistent with the observation by Richman et al. (1977) that the fluctuation kinetic-energy density increases much faster toward the Gulf Stream than does the potential energy. The near-Gulf Stream records exhibit a strong peak in the 25–30-day range for meridional velocity; but in the zonal velocity, the peak is shifted toward lower frequency (see figure 11.14C but notice that this peak occurs in the variance-conserving plot). The temperature spectra here are weak and red.

At the site 31°N, 55°W (figures 11.14F–11.14J) the velocity spectra in the thermocline do not show a clear eddy-containing band; there is a definite tendency toward zonal dominance in these records. The red nature of the thermocline velocity spectra had previously been noted in the MODE area by Richman et al. (1977), and by Schmitz (1976).

Overall, it seems clear that the time scales in the thermocline are longer than in the deep water. This is probably a topographic effect, with the bottom gradients serving to provide a stronger effective beta. Theory (e.g., Huppert and Bryan, 1976; Hogg, 1976; Rhines and Bretherton, 1974) suggests that topography can be a source of eddy motions through its interaction with larger-scale flows both in a wave-generation mode, and through destabilization of the larger scales. But there does seem to be a real reduction in the overall energy levels at depth as one enters regions of rougher topography (Schmitz, 1978; Fu and Wunsch, 1979), as though the flow had been "spun down" with the topography extracting energy from the eddy-containing band perhaps through a scattering process. The mechanism is evidently highly baroclinic because the energy levels

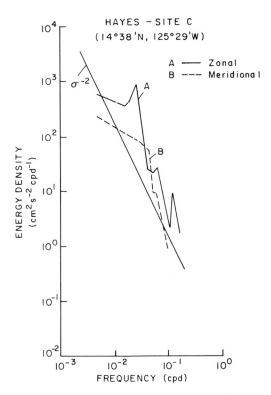

Figure 11.15A Spectrum from eastern Pacific Ocean, replotted from Hayes (1979a).

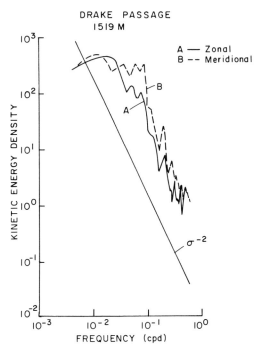

Figure 11.15B Spectrum of velocity from Drake Passage, replotted from Bryden and Pillsbury (1979, personal communication).

in the thermocline seem relatively unaffected by topography (compare figures 11.14K and 11.14O, one over the Hatteras abyssal plane, the other over the mid-Atlantic ridge). There is no adequate theory of this process at the present time; even numerical models do not yet adequately handle topographic effects.

Rhines (1977) and Salmon (1978) have suggested that as energy levels increase there is a tendency for motions to become more barotropic, and this seems to be true in most of the records to date (especially near the Gulf Stream; see figures 11.14A–11.14E). Unfortunately, however, most of the energetic records have also been obtained over smooth topography, and it is not yet possible to separate in the observations an internal dynamic mechanism that is purely energy-level dependent from the direct effects of topography.

At any given location, a detailed discussion of the mesoscale becomes intricate (e.g., Richman, Wunsch, and Hogg, 1977; Freeland, Rhines, and Rossby, 1975). The motions vary with depth, location, and frequency even within comparatively small geographical regions. At this time, it is not really clear what the important characteristics of the mesoscale are. The enormous spatial variability indicated in figure 11.10B suggests that spectral representations in the wavenumber domain—which are equivalent to an assertion that the process is spatially stationary—may not be appropriate.

To a good first approximation, most of the baroclinic energy can be found in a form in which the thermocline simply moves up and down, the entire water column moving together (e.g., MODE Group, 1978; Richman et al., 1977). This dominance of the "lowest mode" is in striking contrast to the mixture of high modes required to describe internal-wave observations (chapter 9). The simplest explanation of this lowest-mode character of the observations is in the tendency of quasigeostrophic nonlinear interactions to drive the motion toward larger scales both in the vertical and horizontal (Charney, 1971a; Rhines, 1977; Fu and Flierl, 1979). There are some features (figure 11.12) for which this characterization does not apply, but they seem uncommon and short-lived except in the Artic Sea (Hunkins, 1974). In the Arctic the eddies Hunkins describes seem closer to a second baroclinic mode in structure. But unlike much of the North Atlantic variability, they also include anomalous water carried along in the core as a kind of "bubble." They may be generated through the overflow process whereby Bering Sea water enters the Arctic Sea, and their physics may resemble the Mediterranean water bubble described by McDowell and Rossby (1978).

In the open ocean, the more common form of first-mode motion is confined to periods of about 100 days and longer. At the shorter periods in the isotropic band along the high-frequency sloping "face" of the velocity spectra, there is little coherence in the vertical, suggesting very short vertical scales, and the motion has much of the character one would expect of the geostrophic turbulence models [i.e., the horizontal kinetic energies are isotropic, the energy is partitioned equally between potential energy and the two kinetic energy components, and in the vertical the energy varies with the buoyancy frequency $N(z)$ (Charney, 1971a)].

The simple vertical movement of the thermocline superficially resembles the lowest dynamic mode on a linear β-plane. But a quantitative examination of its form suggests a distortion, perhaps due to a larger scale shear, and a coupling of several modes characteristic of nonlinear motions (Davis, 1976; Richman et al., 1977). In the western North Atlantic, it appears that the time scales are too short to be describable by linear dynamics, and any hope that one could describe the low-frequency variability by a simple superposition of Rossby waves seems slight, although this has not prevented attempts at doing so.

On the other hand, Bernstein and White (1974) claim that in the eastern Pacific XBT observations of the upper ocean give baroclinic signals that are consistent with elementary Rossby-wave dispersion curves. Their result is agreement with the observation (Fu and Wunsch, 1979) that in moored measurements in the central Atlantic near the mid-Atlantic Ridge the time scales appear to be longer and begin to approach those to which we might apply linear theory. There is a suggestion here that the westward intensification in the eddy field—an analog of the westward intensification of the mean circulation—also may drive the eddy flow into a more nonlinear range, much as one can postulate a nonlinear Gulf Stream in a more nearly linear interior mean ocean flow.

The resources required to study the detailed local dynamics of the eddy field are very great and such studies have rarely been attempted. With the possible exception of Gulf Stream rings, and other features in the immediate proximity of the Gulf Stream (and by presumption near other western boundary currents), the motions are indistinguishable from geostrophic balance. This near-geostrophic balance, which one anticipates on the basis of simple scale analysis (N. A. Phillips, 1963), dominates the momentum balance and precludes direct study of the nongeostrophic terms in the governing equations that could lend insight into the source and sinks of the motions. To proceed, one is forced to study the vorticity balance that eliminates the lowest-order relation. Attempts along these lines have been made for the MODE-1 data by Bryden and Fofonoff (1977) and by McWilliams (1976). For that place at that time a specific balance can be identified in the quasi-geostrophic vorticity equation, although McWilliams showed that on the average (as opposed to the instantaneous balance) a linearized form seems to work well.

The question raised in the introduction [equations (11.1)–(11.4)] concerning the effects of variability, in particular the mesoscale eddy field on the mean circulation, is difficult to answer. The large gradients of energy density described by Schmitz (and made obvious in the figures 11.11B and 11.14) suggest the presence of large Reynolds-stress gradient terms of the form

$$\frac{\partial}{\partial x}\langle u'^2\rangle, \quad \frac{\partial}{\partial y}\langle v'^2\rangle,$$

where u', v' are the fluctuation-velocity components. The dynamic implications of these terms, while clear in principal, are obscure in practice because the complete suite of Reynolds stresses has not been measured. In the immediate vicinity of the Gulf Stream, Thompson (1977) has argued that the off-diagonal terms $\langle u'v'\rangle$ show a convergence of momentum into the Gulf Stream consistent with a Rossby-wave energy flux in the opposite direction. Away from the immediate vicinity of the Gulf Stream, the significance of these off-diagonal terms is not understood (a related question is the meaning of the varying ratios of meridional and zonal energies visible in the spectra displayed here).

Eddies may also carry a heat flux through Reynolds terms of the form $\langle u'T'\rangle$, $\langle v'T'\rangle$, etc. Except for Bryden (1979), who reports a significant poleward eddy heat flux in the Drake Passage and similar reports from the Florida Straits, no one has succeeded in measuring statistically significant eddy heat fluxes. The correlations between temperature fluctuations and velocity fluctuations in the regions where measurements exist are sufficiently weak that it appears that much longer records will be needed to measure significant values—which, however, may be so small as to be dynamically meaningless. Both the Drake Passage and Florida Current results may be typical of strong jetlike currents over topography rather than of the open sea. Many of the features of these two regions are similar [compare Wunsch and Wimbush (1977) and Baker et al. (1977)].

The absence of significant eddy heat-flux terms also seems to rule out open-sea baroclinic instability as an important source of eddy energy (Gill, Green, and Simmons, 1974), although the upper ocean has not been sampled adequately. Such an instability requires that the eddies extract available potential energy from the mean flow, thus generating an eddy heat flux down the mean temperature gradient. Even in the region of the Atlantic North Equatorial Current (Keffer and Niiler, 1978), which had seemed a strong candidate for such instability (Gill, Green, and Simmons, 1974), the eddy heat-flux terms are small. One presumes such instabilities (or more likely a mixed barotropic–baroclinic instability) must occur in the near-Gulf Stream region but the difficulties of measurement in the strong velocities that occur there has precluded quantitative estimates of the heat flux-terms. (Strong mooring motions induced by the high velocities produce fictitious temperature signals coherent with the velocity field.)

If we rule out open-sea generation by meteorology and open-sea baroclinic instability as mesoscale eddy sources, we are left only with topographic generation and generation and radiation from strong boundary currents as possible significant eddy sources. Both are difficult to evaluate quantitatively. Schmitz (1978) and Fu and Wunsch (1979) showed that the major effect of topography seems to be the spin-down of the deep ocean layers relative to the thermocline. Topography would thus more closely resemble a sink of eddy energy than its source. At the present time, then, the strong boundary currents seem the most likely *major* source of eddy energy.

These general results, coupled with the intense concentration of low-frequency energy near the western boundary currents, suggests that the eddy field may indeed be dynamically unimportant except in the immediate vicinity of the boundary currents. The presence of a strong recirculation on the flanks of the Gulf Stream System (Worthington, 1976; Stommel, Niiler, and Anati, 1978; Wunsch, 1978a; Reid, 1978; and see chapter 1) is associated with an intense barotropic eddy field (Schmitz, 1978) and a large resident population of Gulf Stream rings (Richardson, Cheney and Worthington, 1978). A discussion of the Gulf Stream System variability may be found in chapter 4 and will thus not be pursued here. But at this time the only strong evidence for the importance of the variability in the dynamics of the mean flow is in the western boundary current regions. Theories (Pedlosky, 1977) show that the interior field could be largely radiated from the Gulf Stream System and from moving Gulf Stream rings (Flierl, 1977; see also chapter 18).

The mesoscale and other spatial scales of variability become bound up with almost all other aspects of oceanography, including biological, geological, and chemical as well as physical oceanography. To the extent that the variability transports properties, nutrients and other tracers can be expected to be carried along. Eddies will scour the bottom, confusing the interpretation of sediments, and indeed will carry sediments with them from one region to another, possibly quite contrary to the overall time-mean flows. On the physical side, the variability will interact with internal waves and fine structure on the short-wavelength end of the spectrum, and there is probably an interaction between the mesoscale variability and the interannual fluctuations at the other end. We have not dealt directly here with any of these questions; to do so would require a complete discussion of almost all aspects of oceanography and would be premature in any case.

11.3 Summary and Conclusions

The extent to which the ocean undergoes large-scale climatological fluctuations remains uncertain in the face of difficult sampling problems and the short duration over which appropriate instrumentation has existed. Given that large fluctuations exist at least in the sea-surface temperature field, we cannot really distinguish changes imposed by the atmosphere with a static ocean response from those in which large-scale oceanic dynamics are directly involved. Study of the atmospheric spectra displayed here and elsewhere suggests that the forcing and response by the atmosphere is a complex function of position and time scale. With all of the current emphasis on the question of climatic changes, we have been able to do little more than define what we do not know.

Study of the more accessible mesoscale is very much in its infancy and generalizations are dangerous because so little of the ocean has actually been appropriately sampled. At present, one can draw a few qualitative conclusions that may survive future observational programs. The mesoscale eddy field is highly inhomogeneous spatially—both in the horizontal and the vertical. Direct wind forcing and instabilities of the interior ocean are unlikely to provide much energy to the mesoscale. Much of the total mesoscale energy is found in the immediate proximity to the western boundary currents in the region where they are going seaward. It seems, therefore, that these regions are the generators of much of the eddy field energy. Detailed mechanisms are not yet known. The intense recirculations that exist on both sides of the Gulf Stream, the underlying eddy field (Luyten, 1977), and the formation and decay of Gulf Stream rings there are probably related. They suggest an intimate relation between the generation of the eddy field, the maintenance of the eastward going jet, and the transition of the jet into the comparatively quiescent interior in a complex dynamic linkage that we can only vaguely understand. One would guess that the next 10 years will find many of the keys to these puzzles.

The spectra that are displayed throughout this paper suggest that oceanic low-frequency variability does have some characteristic features independent of geography. An eddy-containing band between 50 and 150 days is a feature of most regions. Almost all the spectra show an isotropic band at frequencies higher than the eddy-containing band with a slope not far from -2. (In the immediate vicinity of topography the isotropic band may show more structure.) The interannual band has a distinct tendency to zonality, especially in the thermocline. These features are nearly independent of the remarkable fluctuations (two orders of magnitude) in the overall energy levels.

Although the supporting evidence is more sparse, the dominant mode of baroclinic motion in the eddy-containing band is a simple vertical translation of the thermocline with roughly isotropic horizontal velocities; lower frequencies are more zonal and confined to the upper ocean (cf. Schmitz, 1978). The isotropic band has energy levels nearly independent of depth when a simple scaling by the buoyancy frequency is made. We are totally ignorant of the deep structure on decadal and longer time scales.

The connections between medium-scale oceanic variability and the growing interest in global climate remain problematical. Despite the substantial effort that has gone into understanding the large-scale surface anomalies in the ocean and the possibility of induced changes in the atmosphere, the subject remains enigmatic. Conceivably, subtle changes in the ocean are climatologically more important. For example, one could imagine that the *changes* in regional eddy statistics induce changes in the ocean eddy heat flux (if it exists) and thus in climate. By now the reader should appreciate how far we are from being able to deduce global estimates of the eddy statistics, much less their temporal fluctuations.

Obtaining a full three-dimensional description of ocean variability has remained beyond the capability of oceanographers to this day. A few regions have been intensively studied for comparatively short periods (e.g. MODE Group, 1978; Brekovskikh et al., 1971) and use has been made of crude data sets [e.g., engine-intake temperatures (Fieux and Stommel, 1975)] to build up a qualitative global picture. But it must be admitted that despite the advances that have occurred during the past decades, perhaps our main accomplishment in studying the variability has been to come to appreciate the magnitude of the task. We now recognize the enormous discrepancy between what is required to fully describe and understand the ocean, and the tools we are actually able to bring to bear on the problem. The situation is not hopeless; improvements in moored-instrument technology, new instruments (see chapter 14), and entirely novel techniques, including the use of sound (Munk and Wunsch, 1979) and of satellite systems (Huang, Leitao, and Parra, 1978), suggest that the next decade will be a period in which the descriptive oceanography of the variability field will be built up in the same way that a description of the mean was determined in the previous decades.

To conclude this chapter on a positive note we display (from Wunsch and Gaposchkin, 1980) in figure 11.16 the absolute sea-surface elevation and resulting

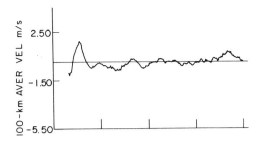
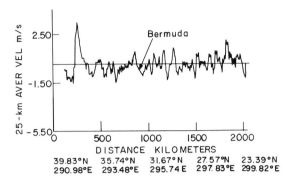

Figure 11.16 Geostrophic velocity determined by satellite altimetry along a track crossing Gulf Stream and Bermuda. Structure of the Gulf Stream seems plausible, as do amplitude and space scales of surrounding eddy field. Two different averaging intervals are displayed. (Wunsch and Gaposchkin, 1980.)

absolute surface geostrophic velocity deduced from the SEASAT-1 altimeter and the Marsh and Chang (1978) gravimetric geoid. One clearly sees the Gulf Stream, and fluctuations with the correct amplitudes and scales for mesoscale and larger eddies. Here at hand is a new tool, which along with others now under development will allow us for the first time to attack the global field of variability.

12
Some Varieties of Biological Oceanography

J. H. Steele

12.1 Introduction

The apparent uniformity of the oceans has turned out to be an illusion generated by the original need for widely spaced sampling both horizontally and vertically. We can no longer accept concepts based on relatively smooth gradients in temperature or salinity. In his paper, "Varieties of Oceanographic Experience," Stommel (1963) pointed out the wide range of scales, in space and time, on which variability occurred. Improvements in technology and development of theoretical bases (Rhines, 1977) portray the oceans as a physical system whose structure can be as rugged as that of the terrestrial world (see chapter 11).

The spatial and temporal variability of the organisms that inhabit the oceans has been recognized for decades. Without this variability, commercial fishing would be uneconomical and sport fishing unexciting. Patches of plankton extending for tens of kilometers were reported in the 1930s (Hardy and Gunther, 1935) and mapped in the 1960s (Cushing and Tungate, 1963). There was, however, no detailed knowledge of the possible relation of these biological observations to corresponding physical structure.

In recent years there have been several attempts to integrate the physics and biology, but on two different levels. The development of fluorometric techniques (Lorenzen, 1966) has permitted continuous *in vivo* measurement of phytoplankton pigments. This, combined with continuous measurement of nutrients such as nitrate, allows detailed portrayal of the spatial structure of the first step in the production cycle. When combined with temperature and salinity measurements from a moving ship, they provide the basis (figure 12.1) for attempts to determine the physical factors determining horizontal phytoplankton patchiness—or, alternatively, to ascribe some aspects of this patchiness to biological mechanisms. The basic technique, spectral analysis, was started by Platt (1972) and developed both theoretically and technically (Steele, 1978a).

The other major area of interest in environmental variability relates to the study of fish populations. The expansion, indeed overexpansion, of commercial fisheries leads to fishing on populations with a younger average age. The fisheries, and the populations themselves, become more and more dependent on the yearly recruitment. In nearly all stocks this recruitment has very large year-to-year fluctuations (figure 12.2). The study of these fluctuations has attracted much research and produced many hypotheses. A large proportion of these hypotheses has attempted to relate variable recruitment to changes in the physical environment, either year-to-year differences or longer-term trends (Hill and Dickson, 1978). During this same period, however, it has been realized that individual species cannot be treated separately, and "multispecies man-

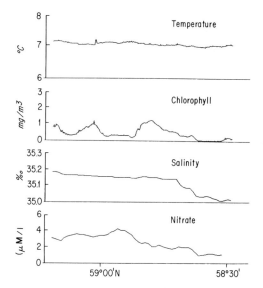

Figure 12.1 Measurements made at 3 m in the northern North Sea during the hours 0000–0500 on 16 May 1976. There are no obvious relations between variations in chlorophyll and nitrate, or between these and the physical parameters temperature and salinity.

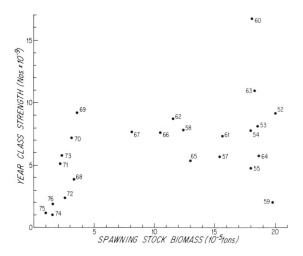

Figure 12.2 Year class strength of North Sea herring as a function of the biomass of the spawning stock. Only in the final years of stock collapse, 1974–1976, do the values of recruitment fall outside the range of previous variability. (Ulltang, in press.)

agement" is now in vogue. This recognizes the interrelation between species, in their food requirements and, potentially, in their recruitment. Thus, again, there is a need to separate the physical and biological factors acting to produce the observed distribution and abundance of fish populations.

For these two extremes of the food web, phytoplankton and fisheries, there is an extensive literature, which I shall review very briefly. For both, it is apparent that the biological factors limiting our understanding lie in the intermediate components of the food web—the zooplankton, which graze on the plants and which, in turn, are the source of food for the fish populations. But these interactions must be placed in the context of the variability of the physical environment at a wide range of space and time scales.

These problems are applicable to all regions of the sea, but, scientifically and economically, are most acute in areas of the continental shelf. Certain parts of the open ocean, such as the centers of gyres (Eppley, Renger, Venrick, and Mullin, 1973), may be considered relatively uniform horizontally, but the shelf is dominated by changes in all significant physical, chemical, and biological parameters—depth, composition of the bottom, temperature, salinity, nutrients, and the quantity and quality of living organisms—at all the possible horizontal scales. Moreover, there is an equally great variability in the vertical structure of the water column, and this variability is conditioned by, and related to, horizontal changes. An example of vertical changes is shown (figure 12.3) in the close correspondence between temperature and fluorescence during passage of an internal wave packet produced by variable bottom topography in Massachusetts Bay. For these reasons, there is an emphasis in this chapter on variability in coastal areas.

12.2 Space and Time Scales of Variation

12.2.1 Physical Variation

As a point of departure, it is necessary to start with certain observed regularities that relate to patterns of horizontal variability. Experiments on dye dispersion by Okubo (1971) and others show a consistent relation between the variance of concentration across a patch and the time from release of the dye. This relation demonstrates the expected dependence of horizontal diffusivity on spatial scale. Using the standard deviation σ derived from this variance, the relation with time t is, approximately,

$$\sigma = t^{1.17}, \tag{12.1}$$

where the units are kilometers and days (Steele, 1978b). This almost linear relation suggests that populations spreading from some initially small area should have a patch "size" in kilometers numerically similar to

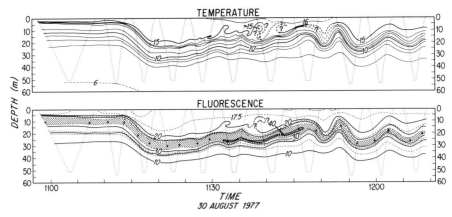

Figure 12.3 The passage of an internal wave packet past a drifting ship in Massachusetts Bay produces rapid changes in vertical structure, shown by the temperature profiles, with a close correspondence of the changes in fluorescence that gives a measure of the concentration of phytoplankton. (Haury et al., 1978.)

their lifetime in days. Yet such a relation cannot be general. Not only will land boundaries limit the extension, but internal vertical features, such as fronts, will alter radically the way in which variance changes with time. The general effect of such boundaries is partly predictable when they have some measure of permanence. But there are also highly unpredictable events, associated with weather, that can alter significantly the temporal variability. These events tend to have their greatest impact on the shelf, particularly near coasts (Csanady, 1976). Thus persistent fronts (Pingree, 1978) and wind-driven motions produce features whose combined time and length scales are very far from the relation derived in equation (12.1).

Further, there are questions about the adequacy, in a biological context, of representing horizontal changes by a process of eddy diffusivity. This concept ignores the related vertical features. An alternative (Taylor, 1954; Bowden, 1965; Kullenberg, 1972) derives horizontal dispersion explicitly in terms of vertical processes. A combination of vertical shear and vertical mixing can explain much, but not all, of the horizontal dispersion. Potentially, this can be used to describe temporal changes in rates of dispersion due to variations in wind stress. This picture has considerable advantages in the study of particles, living or inorganic, that move vertically through the water. At this stage in the discussion, the concept of horizontal diffusion has a provisional and heuristic value.

12.2.2 Biological Variation
When we turn to the plant and animal populations, a similar and very simplified portrayal of space and time scales has been used to indicate the main features of variability (P, Z, F) at different trophic levels (figure 12.4) (Steele, 1978b). The life span of individual phytoplankton cells is a few days. Theoretical calculations (Steele, 1975) suggest that there is a critical scale of a few kilometers at which growth rate of a phytoplankton population P can overcome physical dispersion processes to cause biologically induced patchiness. At the other extreme, pelagic fish F such as herring in the North Sea, have an average life span of a few years and an ambit, produced by their annual migrations, on the order of 1000 km. Herbivorous zooplankton Z are the main link between phytoplankton and pelagic fish in the food chain. Copepods, which are the dominant herbivores, have a life span of 20–50 days (Marshall and Orr, 1955), and there is some evidence for patches of copepods on scales of tens of kilometers (Cushing and Tungate, 1963).

The simple linear P-Z-F derivation in figure 12.4 is, once again, a heuristic device to emphasize the scale relations. These connections imply that any ecological interrelations in the food chain necessarily require interactions between different space and time scales. Thus the inherent difficulties, technical as well as conceptual, in modeling or sampling the complete range of physical events in the ocean, apply equally to the modeling and sampling of the related biological processes.

From equation (12.1) and figure 12.4, it appears that there is a rough correspondence between the scales in space and time of physical and biological dispersion. This could imply that the biological system has evolved to take advantage of the regularities associated with the general temporal and spatial character of physical dispersion in the sea. This is too simple since it ignores the actual relations involved. Theories of phytoplankton patchiness have been based on horizontal diffusivity, but pelagic fish migration is usually associated with, and possibly related to, particular features of current systems (Harden Jones, 1968). Also, zooplankton aggregations may depend on the combination of their own vertical migration and the vertical shear in the water column (Hardy and Gunther, 1935; Evans, 1978).

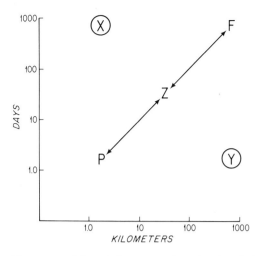

Figure 12.4 A heuristic presentation of scale relations for the food web P (phytoplankton), Z (herbivorous zooplankton) and F (pelagic fish). Two physical processes are indicated by (X) predictable fronts with small cross-front dimensions, and (Y) unpredictable weather-induced effects occurring on relatively large scales.

Further, the simple diagonal relation in figure 12.4 ignores the existence of physical features above and below this line. Many observations show that fish are often found at discontinuities such as fronts. The concentrations of phytoplankton at such small-scale features (taking the appropriate scale to be at right angles to the front) can be explained by events at similar small scales (Pingree, 1978). But the fish aggregations require a long-term evolution involving selection of these areas as part of much larger-scale migrations. Thus small-scale features that are persistent or predictable on large time scales (X in figure 12.4) are one biologically significant divergence from the simple pattern.

Temporal variability in the biology is demonstrated at all trophic levels and is probably greatest for populations on the shelf. It is most extreme and has been best documented for the annual variation in recruitment of fish stocks, where the ratio of maximum to minimum can be 10^3. This great variability is generally associated with the short-term but relatively large-scale unpredictability in weather. Normally, the populations or communities can accommodate such unpredictability and may have evolved to utilize it (Steele, 1979). Extreme variations, especially when combined with heavy fishing, can be disastrous. El Niño, off Peru, is the best-known case (Wooster and Guillen, 1974), but fish kills in the New York Bight are another example (Walsh et al., 1978). These events, of short duration but on a larger scale, can be depicted by Y in figure 12.4.

A general portrayal of the types of physical variability occurring in the sea would occupy all the space of figure 12.1. Two locations, X and Y, have been chosen to simplify the discussion because they epitomize the problems facing an interpretation of biological processes in terms of physical structure. In essence, they provide an alternative caricature of the physical environment. Instead of smoothly changing parameters determined by horizontal diffusion, there is an ocean with relatively uniform areas divided by steplike fronts with some degree of permanence relative to biological processes. Within these large areas having long-term spatial uniformity, there is a temporally fine structure subject to great and unpredictable variability. These simplifications may provide a framework for diagnosing the ecological and technical problems involved in linking the extremes of the food chain.

12.3 Ecological Variations

Because theories or hypotheses cannot handle the whole of the space-time field, the area must be decomposed into conceptually and technically manageable pieces. On occasion this can be done by choosing particular hydrographic features or using special experimental methods. Thus the eddies known as Gulf Stream rings found in the Sargasso Sea, with a diameter of about 100 km, have been used to study the progression in time of the zooplankton populations isolated within the rings (Wiebe et al., 1976). On a smaller scale, large plastic enclosures containing about 100 m^3 can be used to study interactions of phytoplankton, herbivores and invertebrate carnivores for periods of about 100 days (Menzel and Steele, 1978). Both techniques rotate the diagonal of figure 12.4 into a purely time-dependent system.

An alternative approach is to seek out aspects of the ecosystem that may be relatively independent of the spatial variability. In figure 12.4 the components P, Z, F are regarded as single entities, but, in fact, each contains a great diversity of species and, for Z and F, a wide range of age classes for each species. By considering size structure as a first approximation to species diversity or to age composition, it is possible to study the size-frequency distributions within P, Z, or F as a function of their own metabolism and of their size-related intake of food. This approach can be used to construct general theories about size structure (Silvert and Platt, 1978), or to depict possible changes with time in size composition dependent on variations in environmental conditions or predator populations (Steele and Frost, 1977). By considering such "internal" features of the ecosystem, regularities in structure, independent of patchiness, can be predicted and tested.

These experimental or analytic techniques avoid rather than solve the general problems of relating ecological structure and its variance to physical conditions. The inherent difficulties, if not impossibilities, of a full-scale treatment have focused research on par-

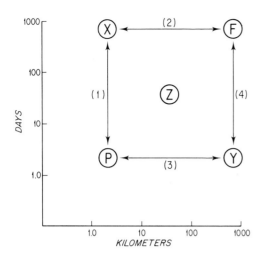

Figure 12.5 An indication of the four main links between physical and biological factors that can be studied within logistical constraints. The details are described in the text.

ticular scales, and these may be categorized by the four components in figure 12.5.

(1) Phytoplankton variability in areas with relative horizontal uniformity has been studied using fluorometers with water intakes at various depths (Fasham and Pugh, 1976). When compared with simultaneous temperature data, both results have wavenumber spectra varying as K^{-n}, with n generally in the range $1 < n < 3$ where K is the wavenumber. Differences in the slopes of the two spectra have been used to deduce effects of biological rather than physical origin. Decreases in n for the chlorophyll spectrum for wavenumbers less than 1 km^{-1} have been taken to indicate the consequences of phytoplankton population growth rates counteracting diffusion processes (Platt and Denman, 1975). Values of n for chlorophyll greater than those for temperature may arise as a result of grazing by the herbivorous copepods (Steele and Henderson, 1979).

There is very much less information on the spatial distribution of these herbivores in the appropriate range, $0.01 < K < 10$ km^{-1}, because of the technical problems in sampling vertically migrating populations, but some results by Mackas (1977) show even greater variations for near-surface populations at night (figure 12.6) and suggest that there can be negative correlations with chlorophyll at scales with $K < 1$ km^{-1}.

(2) There are many cases of high phytoplankton concentrations associated with frontal systems. Detailed studies have been made by Pingree, Holligan, and Head (1977) of the phytoplankton near fronts produced by the relation between tidal energy and shelf depth (Simpson and Hunter, 1974). The observed concentrations of chlorophyll (figure 12.7) are much greater than would be predicted on the basis of the conversion of available nutrient concentrations into plant biomass.

It would seem likely that a particular combination of vertical motions of the water and of the phytoplankton relative to the water is required.

The relation of these concentrations to herbivore grazing is not known, but in other areas, such as the early spring front found at the western edge of the Baltic outflow into the North Sea, above average densities of herbivores occurred which were linked to concentrations of herring (Steele, 1961).

(3) Migrations of pelagic fish and whales in relatively dense aggregations are the normal pattern of behavior. At any time of year, a particular stock is usually found in a restricted part of its overall area of distribution, with the remainder of the area at near-zero densities. Ryther (1969) and Sheldon, Prakash, and Sutcliffe (1972) have shown that as one goes up the trophic ladder there is a decrease in average productivity by a factor of 5 to 10 between adjacent trophic levels. There is a corresponding increase in length of life, however, which results in a near equivalence in average biomass. If one considers the peak concentration of biomass, then the trend is reversed, and, for the particular example of the Antarctic summer (figure 12.8), there is an almost linear relation between appropriate length scale of the organisms and concentration factor (data from Sheldon et al., 1972; Omori, 1978; El-Sayed, 1971).

Figure 12.8 demonstrates that these increases in peak concentration are not purely local events but must depend on concentrating organic matter over large

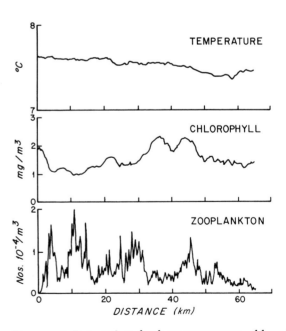

Figure 12.6 Data at 3-m depth on temperature, chlorophyll, and zooplankton numbers collected in the northern North Sea, 19-20 May 1976, between the hours of 2200 and 0400, when the zooplankton are concentrated in the near surface layer. (Mackas, 1977.)

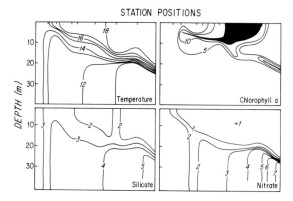

Figure 12.7 Vertical sections through a front in the English Channel of temperature (°C), chlorophyll a (mg m^{-3}), silicate and nitrate (μM l^1), illustrating the very high concentrations of phytoplankton that can occur in conjunction with particular physical conditions. (Pingree et al., 1977.)

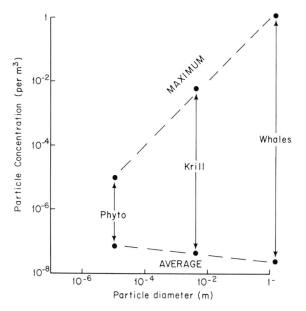

Figure 12.8 The range of concentrations between average and maximum for three trophic levels in the upper layers of the Antarctic in summer. The units are "wet weight" as a fraction of each m^3, or tons per ton, which is an appropriate unit for commercial harvesting.

space and time scales. It is unlikely that any predator—copepods, herring, whales, or fishermen—could operate effectively at the average concentrations. They live off the variance rather than the mean. Thus migrations, or concentrations, of fish at fronts must be partly, but not completely, associated with feeding—and with feeding on herbivores rather than on phytoplankton concentrated at these fronts.

(4) Year-to-year variations in fish-stock recruitment are usually associated with unpredictable environmental fluctuations. In some extreme cases it is possible that they may be linked directly with physical factors, but usually the causal relation is assumed to exist in terms of food availability. For certain species, such as the anchovy, this link can be with phytoplankton populations and, in turn, variations may be related to dependence of the concentration of particular phytoplankton species on thermocline structure (Lasker, 1978). A stable regime in the upper layer is needed to concentrate the food of larval anchovy. Turbulent conditions destroy these aggregations and dilute concentrations of food organisms below feeding thresholds.

Many larval fish, however, feed on copepods, with the preferred size of food increasing from the nauplii to the adults as the larvae grow. But it is not clear that recruitment to the main population is completely determined at these early stages, and the changes in life style around metamorphosis may be critical in terms of corresponding changes in feeding habits (Steele, 1979). Again, there are problems of scale. Spawning stocks, and so initial larval populations, may be confined to a few relatively small areas of the order of 10^2 km^2. These populations disperse as they mature, occupying 10^4-10^5 km^2 around metamorphosis, while the adult populations often move over areas of 10^6 km^2 or greater. Again, the timing of critical events in the life cycle will be related to particular scales, and so postulation of the critical processes is needed to plan the logistics of field programs.

12.4 Discussion

This grouping of ecological studies into four compartments, and their relation to physical features, takes no account of the great body of recent work devoted to the population dynamics of particular species. But these, in turn, often neglect the physical dimensions of the populations. I have chosen to emphasize field and theoretical work that, for logistical and conceptual reasons, has concentrated on horizontal or vertical components in the space-time frame. This approach also emphasizes the division of ecological studies into two almost separate classes; the one based on phytoplankton dynamics, and the other concerned with fish populations. The artificial division of our research effort into basic and applied plays some part in this separation, but it is also imposed by the inherent problems in linking events at the different space and time scales. The difficulties are analogous to those in linking studies of the energetics of waves with those of ocean currents. The need to create these links is as great in the ecological as in the physical case. There are other analogies since the linkages will depend on mesoscale events and on the way in which energy and structure are transferred up and down the space-time scales. There are also direct causal connections between the physical processes and the biological changes. In terms of the biological and physical interaction, the diagonal of figure 12.4 divides the space into two regions. In the lower region the variability is dominant and essentially unpredictable in terms of the occurrence of particular events at specified locations. The analyses of data are in terms of statistical criteria, and the simulation of patterns is based on stochastic models. In the area above the diagonal, patterns are generally predictable and theories can be expected to be deterministic.

For phytoplankton, it might be possible to regard the overall population distributions and production as a superposition of these two components. For fish populations, however, a more complete integration is needed. Particular species must have evolved to be able to absorb the initial variability displayed in the recruitment data. Communities of closely related species may depend on this variability to retain their diversity (Steele, 1979). Within the same life cycle, however, utilization of major and persistent physical features determines the detailed patterns of movement. Thus changes in the physical environment at a wide range of scales could alter the patterns of fish production, and there is evidence for large-scale (Hill and Dickson, 1978) as well as possible small-scale effects (Lasker, 1978).

In these conjectures, as can be seen from figure 12.5, the missing element is the role of the intermediate stages in the food chain, typified by the herbivorous zooplankton Z. Each of the four connections in figure 12.5 omits the ecological involvement of the zooplankton. For example: (i) the possible role of grazing on scales of patchiness has been described; (ii) the very high densities of phytoplankton at fronts imply a loss of grazing control on these populations; (iii) migrating fish stocks may feed on plankton during their progress along fronts or have feeding areas as their destination; (iv) lastly, the critical factor for larval survival is believed to be the density of copepods of the appropriate size (Jones and Hall, 1973). Thus the links in figure 12.5 should all pass through, or include, Z.

We have some knowledge of the general distribution of certain species such as the *Calanus* spp. in the North Atlantic (Colebrook, 1972) and the North Pacific (McGowan, 1971). For the smaller species such as *Pseudocalanus* spp. or *Oithona* spp. our knowledge is more fragmentary. But we lack information on the smaller-scale structure, so that the position of Z in figures 12.1 and 12.2 is really a convenient interpolation. Further, we do not have methods for measuring the growth rate of individuals in their natural environment, so our estimates of secondary production are informal guesses. These guesses, however, would suggest the need to study events in areas of 100 km^2 for periods of 100 days, and it is apparent from the logistics for such studies that they are not easily carried out.

For these reasons there is a desire to parameterize the effects of the herbivores on the higher or lower trophic levels that may be more amenable to study. Such parameterizations depend upon the assumption that interactions are predominantly in one direction, permitting assumptions about closure terms or forcing functions in theoretical presentations. The flow of energy in the ecosystem is upward from the source in photosynthesis to final dissipation at the highest trophic levels, with 80-90% losses at each trophic step. This would appear to define a unidirectional system with the forcing functions located in the physical processes that control the supply of light and nutrients. But, although it is not apparent in the simple diagrams used here, any set of equations depicting the system, or even one link, are essentially nonlinear and there are significant feedbacks. Thus, in nutrient-limited environments, the nutrients excreted by the herbivores can account for two-thirds or more of the nutrients taken up by the phytoplankton. Further, in studies of phytoplankton patch structure, a parameterization of grazing as a reduction in net growth rate of the phytoplankton gives different spectral distributions (Fasham, 1978) from those obtained using simple interactive relations (Steele and Henderson, 1977).

Also, it can be shown theoretically (Steele and Frost, 1977) that in models that represent the internal features of each component in terms of size structure, changes in the predation on the herbivore alters not only the herbivore but also the size structure of the phytoplankton. These *downward* changes can be demonstrated experimentally using large enclosures (Gamble, Davies, and Steele, 1977). This flow of energy upward and of structural changes downward affects our interpretation of perturbations in the fisheries. It seems improbable that functions such as deterministic stock-recruitment relations can adequately describe the interactions with lower trophic levels. Nor will simple stochastic inputs, related to variance of recruitment to individual stocks, be sufficient since these ignore the interrelations of stocks due to total energy limitations.

There are direct causal connections between the difficulties in describing the total spectrum of physical processes and the similar problems in describing the whole food web. There are also analogies between the procedures for approximating the dynamics at particular scales and the parameterizations needed to model portions of the food web. The transfer of energy and structure between scales is a common feature. The nonlinear coupling of variance at different scales is possibly a more significant problem in the biology since at the largest scales, the commercial fisheries, variance in space and time dominates the economics as well as the ecology.

13 The Amplitude of Convection

Willem V. R. Malkus

13.1 Introduction

As a ubiquitous source of motion, both astrophysical and geophysical, convection has attracted theoretical attention since the last century. In the ocean, many different scales are called convection; from the deep circulation due to the seasonal production of Arctic bottom water (see chapter 1) to micromixing of salt fingers (see chapter 8). In the atmosphere, convection dominates the flow from subcloud layers to Hadley "cells." It is proposed that convection in the earth's core powers the geomagnetic field. The nonperiodic reversals of that field, captured in the rock, define the evolution of the ocean basin. Recent recognition that this latter process is caused by convection in the mantle has produced a new geophysics.

In the past, understanding the central features of convection has come from the isolation of "simplest" mechanistic examples. Although large-scale geophysical convection never coincides with the idealized simplest problem, these examples (e.g., Lord Rayleigh's study of the Bénard cells) have generated much of the formal language of inquiry used in the field. Students of dynamic oceanography have favored this formal language mixed in equal parts with more pragmatic engineering tongues when interpreting oceanic convective processes.

Speculations beyond these mathematically accessible problems take the form of hypotheses, experiments, and numerical experiments in which one seeks to isolate the central processes responsible for the qualitative and quantitative features of fully evolved flow fields. The many facets of turbulent convection represent the frontier. This chapter reviews only a narrow path toward that frontier. This path is aimed at an understanding of the elementary processes responsible for the amplitude of convection, in the belief that quantitative theories permit the theorist the least self-deception.

Of course the heat flux due to a prescribed thermal contrast, like the flow due to a given stress, has been observed for a century. The relation between force and flux has been rationalized with models emerging largely from linear theory and kinetic theory—in particular, with the use of observationally determined "eddy conductivities" (estimated for the oceans in Sverdrup, Johnson, and Fleming, 1942). Early theoretical interpretations of oceanic transport processes that go beyond these simple beginnings were explored by Stommel (1949), while current usage and extensions of "mixing" theories are discussed in chapter 8.

Central to the most recent of such proposals is the idea that some large scale of the motion or density field is steady or statistically stable, while turbulent transport due to smaller scales can be parameterized. Changing the amplitude of the small-scale transports is presumed to lead to a new equilibrium for the large scale, so that the statistical equilibrium is marginally stable. This view lurks behind most traditional oceanic model building and its quasi-linear form is used on small-scale phenomena as well—from inviscid marginal stability for the purpose of quantifying aspects of the wind mixed layer (Pollard, Rhines, and Thompson 1973) to viscous marginal stability for the purpose of quantifying double diffusion (Linden and Shirtcliffe, 1978).

It has not yet been possible to establish either the limits of validity or generalizability of this quasi-linear use of marginal stability in the geophysical setting. There can be little doubt that it is "incorrect"—that fluids typically are destabilized by the extreme fluctuations—yet it appears to be the only quantifying concept of sufficient generality to have been used in oceanic phenomena from the largest to the smallest scales. Of course, our idealizations in the realm of geophysics are all "incorrect." We turn to observation to establish in what sense and in what degree these idealizations are good "first-order" descriptions of reality.

This chapter explores the hierarchy of quantifying idealizations in convection theory. The quasi-linear marginal-stability problem is drawn from the full formal statement for stability of the flow. A theory of turbulent convection based on marginal stability is presented, incorporating both the qualitative features determined by inviscid processes and the quantitative aspects determined by dissipative processes.

Observations provide better support for both the quantitative and qualitative results from quasi-linear marginal-stability theory than might have been anticipated, encouraging its continued application in the oceanic setting.

13.2 Basic Boussinesq Description

The primary simplification that permitted mathematical progress in the study of motion driven by buoyancy was the Boussinesq statement of the equations of motion. In retrospect, the central problem was to translate the correct energetic statement

$$\langle \mathbf{u} \cdot \nabla P \rangle = \Phi,$$

into the approximate form

$$\langle \gamma \overline{WT} \rangle = \Phi,$$

where $\mathbf{u}$ is the vector velocity of the fluid, P the pressure, γ the coefficient of thermal expansion times the acceleration of gravity, W the vertical component of velocity, T the temperature field, Φ the total dissipation by viscous processes in the fluid, and the brackets a spatial average over the entire fluid. This has been achieved (e.g., Spiegel and Veronis, 1960; Malkus 1964) by recognizing that the Boussinesq equations are the leading terms in an asonic asymptotic expansion away from a basic adiabatic hydrostatic temperature distribution. This expansion is usually made in two small parameters; one is the ratio of the height of the convecting region to the total "adiabatic depth" of the fluid, while the second is the ratio of the superadiabatic temperature contrast across the convecting region to the mean temperature.

In suitably scaled variables, the leading equations of the expansion are

$$\nabla \cdot \mathbf{u} = 0, \quad (13.1)$$

$$\frac{1}{\sigma} \frac{D\mathbf{u}}{Dt} = -\nabla P + \nabla^2 \mathbf{u} + Ra T \mathbf{k}, \quad (13.2)$$

$$\frac{DT}{Dt} = \nabla^2 T, \quad (13.3)$$

where

$$\frac{D}{Dt} = \frac{\partial}{\partial t} + \mathbf{u} \cdot \nabla, \quad \sigma = \frac{\nu}{\kappa}, \quad Ra = \frac{\gamma \Delta T d^3}{\kappa \nu},$$

$\mathbf{k}$ is the unit vector in the antidirection of gravitational acceleration, d the depth of the convecting region, ΔT the superadiabatic temperature contrast, κ the thermometric conductivity of the fluid and ν is its kinematic viscosity, Ra the Rayleigh number, and σ the *Prandtl number*. Other symbols are defined above. The Boussinesq equations retain the principal advective nonlinearity, but have no sonic solutions. Higher-order equations are linear and inhomogeneous, forced by the lower-order solutions.

The most accessible problem in free convection has been the study of motion in a horizontal layer of fluid bounded by good thermal conductors at prescribed temperatures. Such a layer is the thermal equivalent of the constant-stress layer in shear flow. This is seen by taking an average over the horizontal plane of each term in the heat equation. One writes from 13.3

$$-\frac{\partial \overline{T}}{\partial t} = \frac{\partial}{\partial z} \left(-\frac{\partial \overline{T}}{\partial z} + \overline{WT} \right), \quad (13.4)$$

where the overbar indicates the horizontal average. For steady or statistically steady convection, $\partial \overline{T}/\partial t$ vanishes, and one may integrate (13.4) twice to obtain

$$Nu = -\frac{\partial \overline{T}}{\partial z} + \overline{WT} = 1 + \langle WT \rangle, \quad (13.5)$$

where the constant of integration Nu is called the Nusselt number and is the ratio of the total heat flux to that due to conduction alone.

Two other integrals of considerable interest can be constructed from the Boussinesq equations. The first of these is the power integral found by taking the scalar product of (13.2) with $\mathbf{u}$ and integrating over the entire fluid. One obtains

$$\langle -\mathbf{u}\cdot\nabla^2\mathbf{u}\rangle = \langle |\nabla\mathbf{u}|^2\rangle = Ra\langle WT\rangle, \tag{13.6}$$

due to the vanishing of the conservative advective terms. Multiplying (13.6) by the fluctuation temperature,

$$\mathsf{T} = T - \overline{T}, \tag{13.7}$$

one obtains an integral similar to (13.6), which may be written by means of (13.5) as

$$\langle -\mathsf{T}\,\nabla^2\mathsf{T}\rangle = \langle |\nabla\mathsf{T}|^2\rangle = \langle -W\mathsf{T}\frac{\partial\overline{T}}{\partial z}\rangle$$
$$= \langle WT\rangle + \langle WT\rangle^2 - \langle\overline{WT}^2\rangle. \tag{13.8}$$

The integrals (13.5), (13.6), and (13.8) are the principal constraints used in the "upper bound" theories of convection (see section 13.3).

This section would not be complete without a statement of those equations that determine the stability of any solution, say $\mathbf{u}_0$, P_0, T_0, of the basic equations (13.1)–(13.3). Consider a general disturbance $\mathbf{v}$, p, θ to the solution $\mathbf{u}_0$, P_0, T_0. Then

$$\mathbf{u} = \mathbf{u}_0 + \mathbf{v}, \qquad P = P_0 + p, \qquad T = T_0 + \theta. \tag{13.9}$$

One concludes from (13.1)–(13.3) that the time evolution of $\mathbf{v}, p, \theta$ is determined by

$$\nabla\cdot\mathbf{v} = 0, \tag{13.10}$$

$$\frac{1}{\sigma}\left(\frac{D\mathbf{v}}{Dt} + \mathbf{v}\cdot\nabla\mathbf{u}_0 + \mathbf{v}\cdot\nabla\mathbf{v}\right)$$
$$= -\nabla p + \nabla^2\mathbf{v} + Ra\theta\mathbf{k}, \tag{13.11}$$

$$\left(\frac{D\theta}{Dt} + \mathbf{v}\cdot\nabla T_0 + \mathbf{u}\cdot\nabla\theta\right) = \nabla^2\theta, \tag{13.12}$$

where, as before, $D/Dt = \partial/\partial t + \mathbf{u}_0\cdot\nabla$.

For a solution $\mathbf{u}_0$, P_0, T_0 to be stable (hence realizeable), arbitrary infinitesimal disturbances, $\mathbf{v}$, p, θ must eventually decay. For a solution, $\mathbf{u}_0$, P_0, T_0 to be absolutely stable, arbitrary disturbances of any amplitude must eventually decay. This latter problem is tractable in some instances and has been addressed using the "power" integrals for $\mathbf{v}$ and θ. These are written

$$\frac{1}{\sigma}\frac{D}{Dt}\langle\tfrac{1}{2}\mathbf{v}^2\rangle + \langle\mathbf{v}\mathbf{v}\cdot\nabla\mathbf{u}_0\rangle$$
$$= -\langle|\nabla\mathbf{v}|^2\rangle + Ra\langle\mathbf{v}\cdot\mathbf{k}\theta\rangle, \tag{13.13}$$

$$\frac{D}{Dt}\langle\tfrac{1}{2}\theta^2\rangle + \langle\theta\mathbf{v}\cdot\nabla T_0\rangle = -\langle|\nabla\theta|^2\rangle. \tag{13.14}$$

Equations (13.10), (13.13), and (13.14) constitute a linear problem for $\mathbf{v}$ and θ whose solution can determine a minimum $Ra = Ra(\sigma)$ for which $\mathbf{u}_0$, P_0, T_0 is absolutely stable. In contrast, the solutions to the linear form of (13.10)–(13.12) can determine a maximum $Ra = Ra(\sigma)$ beyond which $\mathbf{u}_0$, P_0, T_0 is assuredly unstable. In principal these linear problems can be solved when $\mathbf{u}_0$, P_0, T_0 is time independent, and solved at least approximately when $\mathbf{u}_0$, P_0, T_0 is periodic in time. However, analytic techniques for determining the conditions leading to eventual decay of $\mathbf{v}$, p, θ on a nonperiodic solution $\mathbf{u}_0$, P_0, T_0 have not been developed.

In concluding this description of the "simple" Boussinesq fluid one should note how many interesting problems lie outside the formalism. Just outside the framework, but capable of incorporation, are porous boundaries, variable viscosity, and nonlinear equations of state. Much farther outside the framework are convection through several scale heights and velocities comparable to the sound speed.

13.3 Initial Motions

The state of pure conduction without motion, $\mathbf{u}_0 = 0$, $T_0 = -z$, where z is the vertical coordinate measured from the lower surface, is a solution to the Boussinesq equations at all Ra. The determination of that critical value of Ra at which this conduction solution is unstable is the classical Rayleigh convection problem (e.g., Chandrasekhar, 1961). This problem is the linearized form of (13.10)–(13.12) for $\mathbf{u}_0 = 0$, $T_0 = -z$, and is written

$$\nabla\cdot\mathbf{u} = 0, \tag{13.15}$$

$$\frac{1}{\sigma}\frac{\partial\mathbf{u}}{\partial t} = -\nabla p + \nabla^2\mathbf{v} + Ra\theta\mathbf{k}, \tag{13.16}$$

$$\frac{\partial\theta}{\partial t} = w + \nabla^2\theta, \qquad w = \mathbf{v}\cdot\mathbf{k}. \tag{13.17}$$

If one takes the $\mathbf{k}$ component of the curl of the curl of (13.16) and, using (13.17), eliminates θ, one may write the Rayleigh problem as a constant coefficient, sixth-order partial differential equation in the single variable w:

$$\left(\frac{1}{\sigma}\frac{\partial}{\partial t} - \nabla^2\right)\left(\frac{\partial}{\partial t} - \nabla^2\right)\nabla^2 w = -Ra\,\nabla_1^2 w, \tag{13.18}$$

where ∇_1^2 is the horizontal Laplacian. The problem is separable so that

$$\nabla_1^2 w = -\alpha^2 w \tag{13.19}$$

defines the horizontal wavenumber α^2. Also, it is not difficult to establish that the disturbance w first starts to grow without temporal oscillations; the instability is "marginal." Hence, subject to appropriate boundary conditions one is to find the eigenstructure of the ordinary differential operator

$$\left[\left(\frac{\partial^2}{\partial z^2} - \alpha^2\right)^3 - \alpha^2 Ra\right] w = 0. \tag{13.20}$$

The richness of solutions to this simple problem and its "adjacent" modifications have been explored for more than two generations and now enters a third. In the present generation, formal techniques that had been developed to find the initial postcritical amplitude of convection are used to clarify the finite-amplitude stability problems. These finite-amplitude studies include the resolution of the infinite plan form degeneracy of the classical linear problem and the determination of conditions causing subcritical instabilities (or "snap-through" instabilities). The formal finite-amplitude technique, often called modified perturbation theory, is the subject of a recent lengthy review (Busse, 1978). In brief, one expands $\mathbf{v}, p, \theta$ in terms of an amplitude ϵ (typically the amplitude of w), and in addition employs a parametric expansion of the controlling parameters, also in terms of ϵ, to permit solvability of the sequence of equations generated by the nonlinear terms. One writes

$$\mathbf{v} = \sum_{n=1}^{\infty} \epsilon^n \mathbf{v}_n, \qquad Ra = \sum_{n=0}^{\infty} Ra_n \epsilon^n, \qquad (13.21)$$

with similar expansions for p and θ, and where Ra_0 is the critical Rayleigh number determined from the linear problem (13.20). Here, each of the Ra_n is determined to permit a steady solution to the ϵ^n set of equations generated by inserting (13.20) into (13.10)–(13.12). Paralleling (13.10)–(13.12) and (13.20), an expansion can be constructed for potential disturbances $\mathbf{v}', p', \theta'$ in order to determine their stability at each order ϵ^n.

Among the many interesting conclusions reached in these studies is that the amplitude of the stable convection forms are determined principally by a modification of $\overline{T}$ due to the convective flux $\overline{w\theta}$. The finite-amplitude distortions of w and θ from their infinitesimal form also affects their equilibrium amplitude, and although a smaller effect than the modification of the mean, it is always present. Conditions for subcritical instability, from (13.21), are seen to be that either $Ra_1 < 0, Ra_2 > 0$ (as is the case when hexagonal cellular convection is observed), or $Ra_2 < 0, Ra_4 > 0$ (as occurs in penetrative convection in water cooled below 4°C), or some mixture of these two conditions. It is found to order ϵ^2 in each case that the preferred (stable) convection is that which transports the most heat. However, there is now an example of a convective process in which the maximum heat-flux solution at large amplitude is not the most stable form. No simple integral criterion has been found that assures the stability of a Boussinesq solution at large Ra.

Many papers have been written using modified perturbation theory, and more appear each year. Unusual current studies include nonperiodic behavior of initial convection in rotating systems and the onset of magnetic instabilities due to finite-amplitude convection in electrically conducting fluids. Busse's review exhibits the significant enrichment of our knowledge and language of inquiry of fluid dynamics by these finite-amplitude studies. This same review, however, clarifies the intractable character of convection mathematics beyond ϵ^2.

This section on initial motions will be concluded by noting E. Lorenz's (1963b) minimal nonlinear convection model, first explored by Saltzman (1962), is an abruptly truncated ϵ^2 Rayleigh convection process. This third-order autonomous system exhibits a transition to "convection," then at higher "Rayleigh" number a transition to exact solutions with nonperiodic behavior. A group of mathematicians and physicists has emerged to explore these "strange attractors" and similar models, hoping among other things to find new access to the turbulent process in fluids. One can be confident that the study of these autonomous models will lead eventually to elementary insights of value to the geophysical dynamicist, yet such study is only one facet of the third generation beyond linear convective instability mentioned earlier. The path to be followed here through the ever-growing convection literature must include the theory of upper bounds on turbulent heat flux, for this theory deductively addresses convection amplitude. The blaze mark along this path, however, continues to be stability theory, and the final section explores its relation to the observed fully evolved flow.

13.4 Quantitative Theories for High Rayleigh Number

Beyond modified perturbation theory one finds hypotheses, speculation, and assorted ad hoc models related to aspects of turbulent convection or designed to rationalize a body of data. In this section three unique theories that predict an amplitude for convection are discussed. The first of these is the mean-field theory of Herring (1963). The second is the theory of the upper bound on heat flux, first formulated by Howard (1963), and its "multi-α" extension by Busse (1969). The third theory, by Chan (1971), is also a heat-flux upper-bound theory, but for the idealized case of infinite Prandtl number. The link between the first and last theories proves to be most informative.

The mean-field theory is equivalent to the first approximation of many formal statistical closure proposals, and is the only contact this study will make with such proposals. As implemented by Herring (1963), the equation describing the motion and temperature field are obtained by retaining only those nonlinear terms in the basic equations that contain a nonzero horizontally averaged part. From (13.1)–(13.4) and (13.7) one writes the mean-field equations thusly:

$$\nabla \cdot \mathbf{u} = 0, \qquad (13.22)$$

$$\frac{1}{\sigma}\frac{\partial \mathbf{u}}{\partial t} = -\nabla P + \nabla^2 \mathbf{u} + Ra T\mathbf{k}, \tag{13.23}$$

$$\frac{\partial T}{\partial t} = \nabla^2 T + W(Nu - \overline{WT}). \tag{13.24}$$

The only nonlinearity retained is $W \cdot \overline{WT}$ in (13.24). Hence the full problem is separable in the horizontal (e.g., $\nabla_1^2 W = -\alpha^2 W$ as in the linear problem), and (13.22)–(13.24) reduce to an ordinary nonlinear equation. Herring solved this nonlinear problem by numerical computations for several increasingly large Rayleigh numbers and for both one and two separation wavenumbers α. He found that the solutions settled to time-independent cellular forms with sharp boundary layers. The predicted heat flux is several times larger than the observed flux in laboratory experiments and the predicted mean-temperature gradient exhibits reversals not seen in high Rayleigh-number data. This relation between theory and experiment will be discussed again in connection with the study by Chan. It was anticipated that neglect of the fluctuating thermal nonlinear terms $[\mathbf{u} \cdot \nabla T - (\partial/\partial z)\overline{WT}]$ would increase the predicted heat flux by some significant, but plausibly constant, fraction. But the effect of neglecting the momentum advection terms $\sigma^{-1}[\mathbf{u} \cdot \nabla \mathbf{u}]$ could have both a stabilizing and destabilizing effect on convection amplitude. If the disturbances in the velocity field are sufficiently large scale compared to the thickness of the thermal boundary region, then one could anticipate from the finite-amplitude studies that $\sigma^{-1}(\mathbf{u} \cdot \nabla \mathbf{u})$ would decrease the amplitude of the convection. But if the local momentum-transporting disturbances are small scale compared to the thermal boundary region, they could strongly enhance the heat transfer. The latter process is typical of geophysical-scale shear flow plus convection, and is not addressed by this mean-field theory. A joint mean-field theory including a mean horizontal velocity $U(z)$ might capture some of this important process of a mean shear flow, enhancing its convective energy source. The upper-bound theory automatically includes this possibility.

The theory of upper bounds on the convective heat flux is based on optimizing a vector field $\mathbf{u}$ and scalar field T that are less constrained than the actual velocity and temperature fields. The only constraints placed on $\mathbf{u}$ and T are that they satisfy the boundary conditions, the continuity condition (13.1), and the three integral conditions (13.5), (13.6), and (13.8). Howard was the first to perform this optimization, and has written a review (1972) of the most certain results of the theory. Busse was the first to discover that the optimal solutions for $\mathbf{u}$ and T contain many (nested) scales of steady convecting motions. Although these results are the only formally correct deductions in the literature applicable to turbulent flows, they are rather far off the mark. The heat flux is much higher than any observed for pure convection, and the optimal solutions seem more elegant than predictive. Their quantitative prediction is that Nu will vary as $Ra^{1/2}$. The highest Ra laboratory data reaches $Ra = 10^{10}$ and appears at most to approach $Nu \sim Ra^{1/3}$. However, careful use of mixing-length theory by Kraichnan (1962) suggests that heat transport in a boundary region caused by momentum advection can occur for $Ra > 10^{24}$, leading to $Nu \sim Ra^{1/2}/(\ln Ra)^{3/2}$. Howard's formal bound includes this possibility.

Hopes to reduce these extreme heat transports by the addition of further integral constraints have not been realized. All other integrals of the basic equations appear either to be trivial in content or to introduce inseparable cubic nonlinear terms into the analysis. The one exception is the work of Chan, which now will be discussed in some detail.

Chan (1971) sought an upper bound on convective heat flux with the power integral constraint (13.6) replaced by the linear Stokes relation,

$$0 = -\nabla P + \nabla^2 \mathbf{u} + Ra T\mathbf{k}. \tag{13.25}$$

This is not only significantly more restrictive on the class of possible fields $\mathbf{u}$ and T, but is an exact statement, from (13.2), in the limit as σ approaches infinity. As in the mean-field theory, no formal expansion is proposed that could reincorporate the nonlinear momentum advection. Yet one might anticipate that the upper bounds on the amplitude of convection will be much closer to the laboratory observations, or to oceanic observations where shear instabilities play a small role.

In addition to (13.25), Chan used the continuity condition (13.2), the thermal integral (13.8), and appropriate boundary conditions. The Euler–Lagrange equations for the optimal relation between W and T have the form

$$\nabla^6 T + (Nu - 1)$$
$$\times \left[\nabla^4 \left(1 - \overline{WT} - \frac{2\lambda}{Nu - 1}\right) W + (1 - \overline{WT})\nabla^4 W \right] = 0, \tag{13.26}$$

$$\nabla^4 W - Ra\, \nabla_1^2 T, \tag{13.27}$$

where λ is a constant Lagrange multiplier. These equations can be compared with the equivalent mean-field equations, from (13.22)–(13.24), which are

$$\nabla^6 T + \nabla^4[(Nu - \overline{WT})W] = 0, \tag{13.28}$$

$$\nabla^4 W = Ra\, \nabla_1^2 T, \tag{13.29}$$

indicating both the similarity and difference of the two problems.

In constrast to Herring's numerical solutions at moderately high Ra, Chan used Busse's multi-α asymptotic technique to determine an optimal solution and a mean-field solution approached at very high Ra. Perhaps the most significant conclusion was that, in this asymptotic limit, the upper-bound problem and the mean-field problem lead to identical results. This result confirms the expectation that the fluctuating thermal terms reduce the convective heat flux by a fraction of about one-half from currently available high Ra data. This certainly represents a remarkable achievement for a theory of turbulence free of empirical parameters. Such quantitative agreement lends support to the idea that the statistical stability condition for turbulent convection in the absence of strong shearing flow is close to the condition of maximum heat transport. Yet, when the possibility of momentum transport due to shear flow is again included in the problem, what extreme should be sought? Here is the arbitrary element in the formal upper-bound theory—what upper bound best reflects the real statistical stability problem? This question is addressed in the following section.

13.5 The Amplitude of Turbulent Convection from Stability Criteria

The idealization of turbulent convection to be explored in this section is similar in spirit to the optimal-transport theories previously discussed. Optimal properties of vector and scalar fields $\mathbf{u}$ and T compatible with the boundary conditions and several other constraints are to be compared with the observed averages of the velocity and temperature fields. Here, however, it will be the stability of the flow that will be optimized. At high Ra both theoretical considerations and observations suggest that large-scale flows in the interior of the region are essentially inviscid in character. In keeping with this classical view, the interior fields of this theory are permitted to approach, but not exceed, the inviscid-stability conditions. These conditions alone can determine many of the qualitative features of the interior flows, but the amplitude of these flows remains undetermined. The goal of this theory is to find those amplitudes that lead to maximum stability for the small-scale, dissipative motions near the boundary. In this view the tail wags the dog, for only the tail is in contact with the dissipative reality that modulates amplitudes. Comparison of the predictions of this quantitative theory with observations can determine the extent to which the real flow approaches the freedom of amplitude selection granted the trial fields of the theory.

The linearized forms of (13.10)-(13.12) constitute a complete statement of the necessary stability conditions that must be met by a realizable Boussinesq solution $\mathbf{u}_0$, T_0. In this theoretical proposal one pictures $\mathbf{u}_0$, T_0 at a particular Ra and σ as composed of the finite-amplitude forms of all fields that were unstable at smaller values of Ra. Subject to the inviscid-stability conditions, the amplitudes of these previous instabilities are to be chosen to make the disturbances $\mathbf{v}$, θ as stable as possible. When that Ra is reached at which the stability of $\mathbf{v}$, θ is no longer possible by amplitude adjustment of $\mathbf{u}_0$, T_0, then the unstable $\mathbf{v}$, θ join the ranks of the previously unstable motions that make up $\mathbf{u}_0$, T_0, and a new stability problem for a new $\mathbf{v}$, θ is posed.

Unfortunately, the linear-stability problem posed above involves fluctuating coefficients that would defy analysis even if they were known. Hence, as promised in the introduction, the proposal is weakened to consider only the stability problem on the mean fields $\bar{\mathbf{u}}_0$, $\bar{T}_0$. Indeed, the fluctuations are observed to be only a fraction of the mean values; yet it is during the destabilizing period of the fluctuations that the significant instabilities occur. If the effects of stabilization and destabilization due to the fluctuations around the mean roughly cancel, then the stability of the mean field is a good measure of the overall stability of the flow. This idealization is explored in the following paragraphs, primarily for the "pure convection" case of infinite Prandtl number. The finite Prandtl-number problem is posed and the extreme case of shear-flow-dominated transport discussed.

The mean-field-stability problem, when $\bar{\mathbf{u}}_0 = 0$, involves only the term $\partial \bar{T}_0/\partial z$. Hence the partial differential equations (13.10)-(13.12) are separable and reduce to a form similar to (13.20):

$$\left[\left(\frac{\partial^2}{\partial z^2} - \alpha^2\right)^3 + \alpha^2 Ra \frac{\partial \bar{T}_0}{\partial z}\right] w = 0. \tag{13.30}$$

The principal constraint to be imposed on the averaged interior flow is that it approach from the viscously stable side, but not exceed, the inviscid stability condition. For convection without a mean shear flow, this condition is that

$$-\frac{\partial \bar{T}_0}{\partial z} \geq 0. \tag{13.31}$$

It is observed that high-Rayleigh-number convection is very close to this stability boundary. Before establishing the quantitative features of the convection amplitude from (13.30), the qualitative consequences of (13.31) will be explored. One may write (13.31) as

$$-\frac{\partial \bar{T}_0}{\partial z} = I^*I \tag{13.32}$$

where I is any complex function of z and I^* is its complex conjugate. It was shown by Fejer (1916), and we shall see shortly, that a complete representation of an everywhere positive function can be written

$$I(\phi) = \sum_{k=0}^{\infty} I_k e^{ik\phi}, \tag{13.33}$$

where $\phi = 2\pi z$, $0 \leq \phi \leq 2\pi$.

The relation between the representation for I in (13.33) and a normal Fourier representation can be established straightforwardly. Let

$$I_m = A_m + iB_m \tag{13.34}$$

where the A_m and B_m are all real. Then

$$(I^*I)(\phi) = \sum_{k=0}^{\infty} (2 - \delta_{k,0}) \sum_{m=0}^{\infty} (A_m A_{m+k} + B_m B_{m+k}) \cos k\phi$$

$$+ \sum_{k=1}^{\infty} 2 \sum_{m=0}^{\infty} (A_{m+k} B_m - A_m B_{m+k}) \sin k\phi. \tag{13.35}$$

For symmetric $(I^*I)(\phi)$ one may write

$$(I^*I)(\phi) = \sum_{k=0}^{\infty} C_k \cos k\phi,$$

$$C_k = (2 - \delta_{k,0}) \sum_{m=0}^{\infty} I_m I_{m+k}, \quad I_m \text{ real.} \tag{13.36}$$

The C_k are uniquely determined by a given set I_k, but a given set C_k determines unique I_k only under special circumstances.

The qualitative behavior of $-\partial \overline{T}_0/\partial z$ emerges from the weak assumption that, at high Ra, I_k is some "smooth" function of k, an assumption to be borne out in the quantifying second step of this theory. Of course, at some very small scale, say k_ν, one expects viscosity and thermal diffusion to reduce I_{k_ν} to a vanishingly small value. Then when $I_k \simeq 0$ for $k > k_\nu$, one writes

$$I(\phi) = \sum_{k=0}^{\infty} I_k e^{ik\phi} \simeq \sum_{k=0}^{k_\nu} I_k e^{ik\phi}. \tag{13.37}$$

To explore the consequence of "smoothness" it is convenient to sum (13.37) by parts. First one defines

$$(\Delta I)_k = I_{k+1} - I_k, \quad (\Delta^2 I)_k = (\Delta I)_{k+1} - (\Delta I)_k,$$
$$F_k = (e^{ik\phi} - 1)/e^{i\phi} - 1. \tag{13.38}$$

Hence $(\Delta F)_k = e^{ik\phi}$ and

$$\sum_{k=0}^{k_\nu} I_k e^{ik\phi} = +\frac{I_0}{1 - e^{i\phi}} + \frac{e^{i\phi}}{1 - e^{i\phi}} \sum_{k=0}^{k_\nu} (\Delta I)_k e^{ik\phi}. \tag{13.39}$$

Repeating this summation by parts on the final sum in (13.39), one may write

$$\sum_{k=0}^{k_\nu} I_k e^{ik\phi} = \frac{1}{1 - e^{2\phi}} \left\{ I_0 + \frac{e^{i\phi}}{1 - e^{2\phi}} \left[(\Delta I)_0 \right.\right.$$
$$\left.\left. + e^{i\phi} \sum_{k=0}^{k_\nu} (\Delta^2 I)_k e^{ik\phi} \right] \right\}. \tag{13.40}$$

One now observes that if I_k is "smooth" in the sense that

$$(\Delta I)_k = O(I_0/k_\nu), \quad (\Delta^2 I)_k = O(I_0/k_\nu^2), \tag{13.41}$$

then from (13.40)

$$I(\phi) = \frac{I_0}{1 - e^{i\phi}} + O(I_0/k_\nu) \tag{13.42}$$

for all angles $\phi \gg k_\nu^{-1}$. Hence a unique and simple form for $I(\phi \gg k_\nu^{-1})$ exists if the weak condition (13.40) is met. From (13.32) and (13.42), the interior mean temperature field is

$$\overline{T}_0(\phi) \simeq I_0^2 \tan\left(\frac{\phi - \pi}{2}\right). \tag{13.43}$$

This is the only law whose qualitative behavior is insensitive to the features of the underlying spectrum, yet reflects the stability conditions presumed responsible for maintaining the negative gradient.

The field equation (13.43) is also independent of the cutoff wavenumber k_ν; yet the assumption of spectral smoothness may seem less plausible at those wavenumbers where viscous effects first become as important as the nonlinear advection. A requirement placed on this "tail" region of the transport spectrum is that it drop off faster than any power of k in order that all moments of the flow be finite. A second requirement is that the "tail" region be continuous with and match the smoothness condition at the wavenumber where the viscous tail joins the inertially controlled lower-wavenumber spectrum. The simplest tail to meet these requirements is a modified exponential. Hence, one explores the consequence of the tail

$$I_{k>k_0} = I_{k_0}[1 + \alpha(k - k_0) + \beta(k - k_0)^2]e^{-\gamma(k-k_0)}, \tag{13.44}$$

where the wavenumber k_0 ($<k_\nu$) marks the low-wavenumber end of the "tail," γ characterizes the degree of abruptness of the spectral cutoff, and α, β are chosen to match smoothness conditions at $k = k_0$. For $\gamma \ll 1$, $\alpha \simeq \gamma$ and $\beta \simeq \tfrac{1}{2}\gamma^2$. The tail can be summed and leads to the general spectrum

$$I(\phi) = \sum_{k=0}^{k_0} I_k e^{ik\phi} + I_{k_0} e^{i(k_0+1)\phi} e^{-\gamma}$$
$$\times \left[\frac{1}{1 - e^{-a}} + \alpha \frac{1}{(1 - e^{-a})^2} + \beta \frac{1 + e^{-a}}{(1 - e^{-a})^3} \right], \tag{13.45}$$

where $a = \gamma - i\phi$. As it stands, with k_0, γ, and all the I_k unspecified, (13.45) can describe any plausible turbulent mean-temperature profile of negative slope at any Ra. At this point one seeks the asymptotic consequences of the smoothness hypothesis

$$(\Delta I)_k = O(I_0/k_0),$$
$$(\Delta^2 I)_k = O(I_0/k_0^2), \quad 0 \leq k \leq k_0, \tag{13.46}$$

and from (13.40) concludes that for $\phi \gg k_0^{-1}$, to $O(I_0/k_0)$,

$$I(\phi) = \frac{1}{1 - e^{i\phi}} \left\{ I_0 - I_{k_0} e^{i(k_0+1)\phi} \left[\frac{1 - e^{-\gamma}}{1 - e^{-a}} \right. \right.$$
$$\left. \left. + \alpha \frac{e^{-a} - e^{-\gamma}}{(1 - e^{-a})^2} + \beta \frac{(e^{-a} - e^{-\gamma})(1 + e^{-a})}{(1 - e^{-a})^3} \right] \right\}.$$
(13.47)

If, then, $\gamma = O(k_0^{-1})$, (13.47) reduces to (13.42) and the interior temperature field (13.43). The novel aspect of the temperature profiles determined by (13.47) is the emergence of the double-tangent structure. This is most easily seen from the leading term of the inner bracketed expression in (13.47) for $\phi, \gamma \ll 1$. One writes

$$[\,] \simeq \left[\frac{\gamma}{\gamma - i\phi} + \gamma \frac{i\phi}{(\gamma - i\phi)^2} + \gamma^2 \frac{i\phi}{(\gamma - i\phi)^3} \right]. \quad (13.48)$$

Then for $\gamma \gg \phi \gg k_0^{-1}$ the bracket expression approaches the value 1 and the resulting profile has the amplitude $|I_0|^2 + |I_{k_0}|^2$. In contrast, when $\phi \gg \gamma$, the bracket approaches 0 and the temperature profile has the "outer" amplitude $|I_0|^2$. Hence when the transport spectrum has a cutoff sufficiently more abrupt than k_0^{-1}, an "inner" inertial boundary region is predicted.

A γ large compared to k_0^{-1} is deduced in the quantitative work of the following paragraphs. However, it is likely that the predicted transport tail will be quite sensitive to the neglect of the fluctuation term in the stability problem. Unfortunately, it will be seen that present convection data is not sufficiently precise to test this speculation.

In the analogous shear-flow problem two logarithmic regions of different slopes are found (Virk, 1975). Drag-reducing additives, which appear to sharply increase γ, also cause the "inner" logarithmic region to extend much further into the flow. Theoretical studies (Malkus, 1979) predict this behavior, but indicate that for shear turbulence the mean-field-stability theory gives a γ that is larger than observed. In both the case of pure shear flow and the case of pure convection, the quantity I_0 that determines the outer-flow amplitude seems to be the most imperturbable feature of the mean-field-stability computations.

Turning now to these stability computations for convection without mean shear, one is blessed with a problem that, for free boundary conditions, can be cast in variational form. Hence the extensive numerical computations needed to implement this theory for shear flow can be replaced by a sequence of analytic approximations. The variational form of (13.30) is written

$$Ra^* = -\frac{1}{\alpha^2} \left[\int_0^1 w \left(\frac{\partial}{\partial z^2} - \alpha^2 \right)^3 w \, dz \right]$$
$$\div \left[\int_0^1 \left(-\frac{\partial \overline{T}_0}{\partial z} \right) w^2 \, dz \right] \geq Ra \quad (13.49)$$

Trial forms for w lead to an Ra^* bounding the critical (experimentally given) Ra from above, and insensitive to first-order error in the trial form. Simultaneously, the amplitudes of all previously unstable modes are to be adjusted to change $-\partial T_0/\partial z$ so that the marginal stability of any particular w occurs at minimum $Ra^* = Ra$. The free or "slippery" boundary conditions are

$$w = \frac{\partial^2 w}{\partial z^2} = \theta = 0 \quad \text{at} \quad z = 0, 1. \quad (13.50)$$

As these are also the boundary conditions for the full field $\mathbf{u}_0, T_0$, it follows that

$$\overline{W_0 T_0} = \frac{\partial \overline{W_0 T_0}}{\partial z} = 0 \quad \text{at} \quad z = 0, 1. \quad (13.51)$$

Hence, for symmetric $-\partial \overline{T}_0/\partial z$, from (13.5), (13.32), (13.33), and (13.51) I_k is real and

$$Nu = \left(\sum_{k=0}^{\infty} I_k \right)^2, \quad (13.52)$$

while

$$1 = \sum_{k=0}^{\infty} I_k^2. \quad (13.53)$$

Now starting the sequence of computations from the conductive state of no motion ($I_0 = 1$, $I_{k\neq 0} = 0$), one recovers from (13.49) the classical Rayleigh solution

$$w_1 = A_1 \sin \phi/2, \quad \theta_1 = B_1 \sin \phi/2,$$
$$\alpha_1^2 = \pi^2/2, \quad Ra_1 = (27/4)\pi^4, \quad (13.54)$$

where A_1 and B_1 are infinitesimal amplitudes and $\phi = 2\pi z$. The next step is to determine that amplitude of $\langle w_1 \theta_1 \rangle \equiv C_1$ as a function of Ra that will maintain marginal stability against any disturbance, including w_1. C_1 can be increased until $-\partial T_0/\partial z$ reaches 0 at some point, but no further. There is evidence (Chu and Goldstein, 1973) that the fluid does not quite respect this limitation from inviscid theory in the initial postcritical range of Ra, but at higher Ra no positive mean gradients are observed. From (13.5) and (13.54)

$$-\frac{\partial \overline{T}_0}{\partial z} = 1 + \langle w_1 \theta_1 \rangle - \overline{w_1 \theta_1} = 1 + C_1 \cos \phi. \quad (13.55)$$

Therefore, from (13.49), marginal stability for w_1 requires that

$$C_1 = 2 \left(1 - \frac{Ra_1}{Ra} \right), \quad (13.56)$$

with the limiting value $C_1 = 1$ at $Ra = 2Ra_1$. In the observed initial stage of convection, amplitude increases with Ra as in (13.56), but without stopping at $C_1 = 1$. Instead, the form of the disturbances is altered by finite-amplitude effects (e.g., Malkus and Veronis, 1958). However, at very high Ra, observations suggest that the energetic "overtones" of finite disturbances are not of smaller scale than the marginal disturbances in the boundary region.

By accident perhaps, the higher-wavenumber eigenfunctions of the conductive problem

$$w_n = A_n \sin n\phi/2, \quad \alpha_n^2 = n^2\pi^2/2,$$

$$Ra_n = (27/4)\pi^4 n^4 \qquad (13.57)$$

are optimal forms for the stability problem (13.49), even when the distorted $-\partial \overline{T}_0/\partial z$ of (13.55) is included. That is,

$$\int_0^{2\pi} (1 + C_1 \cos\phi) \sin^2 n\phi/2 \, d\phi$$
$$= \int_0^{2\pi} \sin^2 n\phi/2 \, d\phi, \qquad (13.58)$$

for $n > 1$. Hence the next instability occurs at $Ra_2 = 2^4 Ra_1$ and has the form $w_2 = \sin\phi$. If this were to continue, the gross dependence of the Nusselt number on Ra would be

$$Nu = (Ra/Ra_1)^{1/4}, \qquad (13.59)$$

which is the law observed in the early stages of convection.

In this free boundary condition case, however, a new kind of disturbance leads to a lower critical Ra [greater stability, from (13.49)] beyond the second transition. If one calls the first disturbances above *body* disturbances (that is, $w_n = \sin n\phi/2$ is large throughout the whole fluid and "senses" both boundaries), then the new disturbance is a *boundary* disturbance, and is large only near one boundary. One may presume a statistical symmetry for these disturbances to maintain the observed symmetry of the mean.

Consider a trial form for such a disturbance of

$$w_{k_0+1} = \sin(k_0 + 1)\phi/2,$$

$$0 \le \phi \le \frac{2\pi}{k_0 + 1}, \quad 2\pi - \frac{2\pi}{k_0 + 1} \le \phi \le 2\pi.$$

Then the trial $-\partial \overline{T}_0/\partial z$ is to consist of (the previous) k_0 modes, and each with arbitrary amplitude. The full problem posed is to choose those k_0 amplitudes, subject to the constraint $-\partial \overline{T}_0/\partial z \ge 0$, in order to minimize Ra in (13.49). A first and second approximation will be reported here.

To anticipate an appropriate trial form for the temperature gradient, it is of value to note the links this problem has with the search for a maximum heat flux. One sees in (13.49) that if $w_{k_0+1}^2$ is large only near the boundary, then minimum Ra requires that $-\partial \overline{T}_0/\partial z$ be large near the boundary also. If the temperature gradient resulting from the k_0 previously unstable modes can be adequately represented by the truncated spectrum

$$I(\phi) = \sum_{k=0}^{k_0} I_k e^{ik\phi}$$

subject to the definitional constraint (13.53),

$$\sum_{k=0}^{k_0} I_k^2 = 1,$$

then, from (13.52) the maximum possible

$$-\frac{\partial \overline{T}_0}{\partial z}(0) = Nu = \left(\sum_{k=0}^{k_0} I_k\right)^2$$

is

$$I_k = \frac{1}{(k_0 + 1)^{1/2}}, \quad \text{and} \quad Nu = k_0 + 1. \qquad (13.61)$$

This exactly smooth I_k with its sharp truncation assures the general internal temperature field found from (13.42) and (13.47):

$$\overline{T}_0(\phi) = \frac{2}{k_0 + 1} \tan\left(\frac{\phi - \pi}{2}\right). \qquad (13.62)$$

A complete description of this $\overline{T}_0(\phi)$ valid right to the boundary can be written in terms of sine integrals. Near the boundary this description is

$$\overline{T}_0(\phi) = \frac{1}{2}\left\{1 - \frac{2}{\pi}\left[S_i(2\xi) - \frac{\sin^2\xi}{\xi}\right]\right\}, \qquad (13.63)$$

where $\xi = (k_0 + 1)\phi/2$, which merges into (13.62) for $\xi \gg 1$. If this choice (13.61) is made for the first trial I_k used to determine a minimum Ra for the disturbance w_{k_0+1} of (13.60), then one finds from (13.49) that

$$Ra = (k_0 + 1)^3 \frac{27}{4} \pi^4 \bigg/ \frac{1}{\pi}[2S_i(2\pi) - S_i(4\pi)]$$

or

$$Ra = (k_0 + 1)^3 Ra_c, \quad Ra_c = 1{,}533. \qquad (13.64)$$

Hence this first trial form for boundary functions predicts

$$Nu = (Ra/Ra_c)^{1/3} \qquad (13.65)$$

and a field $\overline{T}_0(\phi)$ determined from boundary to boundary. This result parallels, but is roughly 15% above, the experimental data (Townsend, 1959). The theory is for free boundary conditions, however, and the data for rigid boundaries. An estimate of the theoretical reduction in Nu for rigid boundaries (Malkus, 1963) is 13%. Precise theoretical results for that case will require tedious numerical computations. It might be easier to obtain good data for turbulent convection over a slippery boundary (e.g., silicon oil over mercury).

A second approximation to the most stabilizing spectrum (which, of course, will reduce the heat flux) is also based on a spectrum for $I(\phi)$ truncated at k_0. The optimal stability problem so posed, from (13.49) and (13.60), is to find the I_k that minimizes Ra in

$$Ra = \frac{27}{4}\pi^4 (k_0 + 1)^4 \bigg/ \sum_{k=0}^{k_0} \sum_{m=0}^{k_0} I_k I_m C_{km}, \qquad (13.66)$$

where

$$C_{km} = C_{|k-m|}$$
$$= \frac{(k_0 + 1)}{\pi} \int_0^{2\pi/(k_0+1)} \cos|k - m|\phi \sin^2 \frac{k_0 + 1}{2}\phi\, d\phi,$$

and, from (13.53),

$$\sum_{k=0}^{k_0} I_k^2 = 1.$$

One may write, for $\eta = |k - m|/(k_0 + 1)$,

$$C_\eta = \frac{1}{2\pi}\left(\frac{1}{\eta} + \frac{\eta}{1 + \eta^2}\right) \sin 2\pi\eta, \qquad (13.67)$$

and note that

$$C_0 = 1, \qquad C_{1/2} = 0, \qquad C_{k_0/(k_0+1)} = -\tfrac{1}{2}.$$

The maximum eigenvalue of the matrix C_η, say λ_{max}, determines the minimum value of Ra in (13.66). Since

$$\lambda_{max} \le \text{trace}\, C_\eta = \sum_{k=0}^{k_0} C_0 = k_0 + 1, \qquad (13.68)$$

then from (13.66)

$$(k_0 + 1)^3 \frac{27}{4}\pi^4 \le Ra_{min} \le (k_0 + 1)^3 (1{,}533), \qquad (13.69)$$

where the latter bound is for the trial I_k [(13.61)]. Since the heat flux varies as $Ra_{min}^{-1/3}$, the maximum possible reduction permitted by (13.67) is 30%.

Last, a first estimate of γ can be made from the second approximation to the boundary eigenfunction (which can increase the heat flux!). If, for $I_k = $ constant, the trial form

$$w = \sin \xi + A \sin 2\xi$$

is chosen, containing an asymmetric part of arbitrary amplitude to reflect the asymmetry of $-\partial \overline{T}_0/\partial z$ in the boundary region, then one finds from (13.49) the optima

$$A = 0.0226, \qquad \frac{\alpha^2}{(k_0 + 1)^2 \pi^2} = 0.5069, \qquad Ra_c = 1513.$$

The small harmonic distortion (2.26%) suggests that the "inner" tangent law may persist to more than ten times the boundary layer thickness. Present data offer no hope of detecting a change in slope at such distance from the boundary. Better data may permit a determination of γ from the "inner" to "outer" law transition region found from (13.48). In any event, considerably more experimental effort is required to test critically the limits of validity of the hypothesis advanced in this section. On the positive side this study rationalizes the observed change of the $Nu = Nu(Ra)$ law from $Ra^{1/4}$ to $Ra^{1/3}$; it predicts a double-z^{-1} law for the mean field, the "inner" part observed; its first approximate convection amplitudes provide solid support for the usefulness of the concept of marginal stability of the mean; last, a theoretical program incorporating marginal stability on both a shear flow and thermal gradient is computationally practical, offering the hope of discovering unanticipated relations between observables in the geophysical setting.

Part Three

Techniques of Investigation

14
Ocean Instruments and Experiment Design

D. James Baker, Jr.

What wonderful and manifest conditions of natural power have escaped observation.
[M. Faraday, 1859.]

We know what gear will catch but . . . we do not know what it will not catch.
[M. R. Clarke (1977).]

14.1 Observations and the Impact of New Instruments

"I was never able to make a fact my own without seeing it, and the descriptions of the best works altogether failed to convey to my mind such a knowledge of things as to allow myself to form a judgment upon them. It was so with new things." So wrote Michael Faraday in 1860 (quoted in Williams, 1965, p. 27) close to the end of his remarkably productive career as an experimental physicist. Faraday's words convey to us the immediacy of observation—the need to see natural forces at work. W. H. Watson remarked a century later on the philosophy of physics, "How often do experimental discoveries in science impress us with the large areas of *terra incognita* in our pictures of nature! We imagine nothing going on, because we have no clue to suspect it. Our representations have a basic physical innocence, until imagination coupled with technical ingenuity discloses how dull we were" [Watson (1963)]. Faraday recognized the importance of this coupling when he wrote in his laboratory *Diary* (1859; quoted in Williams, 1965, p. 467), "Let the imagination go, guiding it by judgment and principle, but holding it in and directing it by experiment."

In the turbulent, multiscale geophysical systems of interest to oceanographers, the need for observation and experiment is clear. Our aim is to understand the fluid dynamics of these geophysical systems. Although geophysical fluid dynamics is a subject that can be described with relatively few equations of motion and conservation, as Feynman, Leighton, and Sands (1964) stated, "That we have written an equation does not remove from the flow of fluids its charm or mystery or its surprise." In fact, observations and experiments have been crucial to the untangling of the mysteries of fluid processes in the ocean and in the atmosphere.

For example, on the smaller scales, lengths of the order of tens of meters and less, the discovery of the sharp discontinuities in density, temperature, and salinity that was brought to focus by the new profiling instrumentation has given us a whole new picture of mixing in the ocean. On the large scale, a good example is the explanation of the general circulation of the atmosphere in terms of baroclinic instability. The theoretical development was firmly based on the remarkable set of observations of the atmosphere carried out in the 1940s and 1950s. As E. Lorenz (1967, p. 26) noted in his treatise on *The Nature and Theory of the General Circulation of the Atmosphere*, "The study of the circulation owes a great deal to the practice of weather forecasting, for without these observations our understanding could not have approached its present level. Yet certain gaps will continue to exist in our knowledge of the circulation as long as extensive regions without regular observations remain." The emphasis that Lorenz placed on the need for observations before understanding can occur is equally valid for oceanographic studies of the same scale.

One must search long and hard for counterexamples where theory has preceded observation in geophysics. One of the few such examples in oceanography is the prediction and subsequent confirmation by direct measurement of southward flow under the Gulf Stream by the Stommel-Arons theory of abyssal circulation. The theory is discussed elsewhere (e.g., chapters 1 and 5, this volume) so I shall not pursue it further. The point is that observations guide and appear to limit the progress of our science. There is no inherent reason that this should be so. Why is our imagination so limited that, as Watson put it, we are so dull? Perhaps the historian of science can answer the question.

If observations guide the science, then new instruments are the means for guidance. The following two examples show how this has occurred; we consider first the North Atlantic circulation. Most of our ideas about the ocean circulation have been based on the indirect evidence of the temperature and salinity fields and the assumption of geostrophy. With the advent of direct velocity measurements by deep floats and current meters during the 1960s and early 1970s, the necessary data for a consistent picture of ocean circulation, at least in limited areas, began to come in. Worthington's (1976) attempt to put together for the first time such a picture of circulation in the North Atlantic was based on the new direct data.

One of the important pieces of evidence used in the work by Worthington were the data from the neutrally buoyant floats, which show a high transport for the Gulf Stream [see Worthington (1976) for references]. Until the direct measurements, the distribution of the absolute velocity field was ambiguous. With the new data, Worthington was encouraged to put together a complete picture that includes a tight recirculation pattern. However, within the constraints he used, Worthington's attempts at a complete mass and dynamic balance for the entire North Atlantic circulation were not successful. He decided, therefore, to choose a circulation pattern that was not consistent with geostrophy. This provocative work stimulated a number of attempts to look more closely at the circulation system there. Because both scale analysis of the equations of motion and the direct moored measurements of Schmitz (1977, 1980) confirm geostrophy to the leading order, as do the measurements reported by Swallow (1977) and Byrden (1977) in the MODE region of the Sargasso Sea, Worthington's total picture is not correct. The moored data are consistent with the recirculation pattern, but, in addition, reveal a flow with an eastward component immediately south of the Gulf Stream and north of the recirculation. The latter feature is not clearly contained in any existing picture of the North Atlantic circulation.

A second approach was taken by Wunsch (1978a), who used hydrographic data, mass balance, and geostrophy to estimate the absolute velocity field in the North Atlantic. Since the system is basically underdetermined (one requires point measurements to find unique solutions to the relevant fields), an auxiliary criterion is required. Wunsch found a unique solution by minimizing a measure of the energy. This is an example of the geophysical "inverse technique" that provides estimates in underdetermined systems by optimizing auxiliary measures of merit. The circulation pattern found by Wunsch is thereby in general agreement with geostrophy and the recirculation ideas of Worthington and Schmitz, thus lending further support to this idea. Stommel and Schott (1977) have also shown how to use hydrographic data to estimate absolute velocity, by using conservation of potential vorticity.

In sequence, then, we can see that the new observational data from the floats stimulated a new view of the circulation, which in turn led to further observations and data studies that extend and correct the picture.

We note that the inverse techniques are just as much an observational tool as the instruments we use, because they allow us to rearrange data in ways that are not necessarily obvious or easy. The choice of the proper measure of merit was discussed by Davis (1978b). He showed that the Wunsch method is dynamically equivalent to the Stommel and Schott technique; the major differences result from implicit assumptions about the scales of oceanic variability, and different definitions of the smooth field to which the dynamic model pertains. Davis gave an example of an optimization criterion based on a measure of merit related to the process of inferring fields from point measurements.

Schmitz's conclusion on the subject at the time of this writing was that the situation is still unresolved: "the North Atlantic Circulation is inadequately described at the present time, much less understood, and could be very complex both spatially and temporally. This could also be the case in the North Pacific.... We are still in the process of exploration, attempting to identify the relevant elements of several hypotheses, and utilizing different techniques for investigating diverse features of the circulation" [Schmitz (1980)].

Our second example of the way that instruments give us a new picture is from the time-dependent circulation. El Niño, the appearance of warm water off the coast of South America every few years, is a large-scale phenomenon of both dynamic and practical importance. One strong contender for the explanation of El Niño is that relaxation in the trade winds in the western Pacific results in the propagation of a warm-water anomaly toward the east. The long-term fluctuations in oceanic circulation on which the model is based have been inferred from direct measurements of sea level at islands in the tropical Pacific (Wyrtki, 1973b, 1979). These measurements, suitably filtered and averaged, appear to be a good indicator for variations in the geostrophic transport of the upper layers. The provocative data and subsequent models have spurred a whole new interest in the dynamics and air-sea interaction in the tropical regions.

Thus the data from new instruments give us a new context for our science; continued innovation is required. As late as 1965, when meterologists were using the simple but elegant radiosondes and beginning to use satellites for remote sensing of atmospheric circulation, Henry Stommel noted the problems of ocean observation:

When I emphasize the imperfection of observing techniques perhaps I should say that I wrote this chapter during a succession of midnight-to-dawn watches during an attempt to survey the Somali current near Socotra in the heart of the Southwest monsoon. It is rather quixotic to try to get the measure of so large a phenomenon armed only with a 12-knot vessel and some reversing thermometers. Clearly some important phenomena slip through the observational net, and nothing makes one more convinced of the inadequacy of present day observing techniques than the tedious experience of garnering a slender harvest of thermometer readings and water samples from a rather unpleasant little ship at sea. A few good and determined engineers could revolutionize this backwards field. [Stommel (1966).]

It is safe to say that since 1965 there has been an influx of more than a few "good and determined" engineers, and there has been a flood of new instruments and new ideas. Our view of the ocean has changed markedly, especially on the smaller scales, where the old instruments and techniques were essentially blind.

The sections to follow outline some general principles of instrument design and the development of the technology, and then provide some history and working principles in four major areas of instrumentation that have made an important impact on modern oceanography. We turn next to some examples of instruments that have shown promise but have not yet reached their potential, and then to some areas where new instruments are needed, but the technology is not yet available. The chapter concludes with a section on experiment design.

14.2 Instrument Development: Some Principles and History

14.2.1 General Principles

Chapter 10 of *The Oceans* (Sverdrup, Johnson, and Fleming, 1942), "Observations and Collections at Sea," covers techniques and instruments for the study of the physics, chemistry, geology, and biology of the sea. It is not possible for a chapter like the present one to cover the development since then of modern instruments and techniques for all of these disciplines, in fact, not even for one of these disciplines; there is simply too much material. There are some important general aspects of instrument design, however, that are useful to point out, and it is instructive to look at the development since *The Oceans* was written of some of the instrumental techniques that are generally applicable.

The general principle for oceanographic instruments has been to keep them simple and reliable, a principle underlined by the long-successful use of the Nansen bottle and reversing thermometer. The Scandinavian school had another important point: the efficiency of locally operated mid-size vessels. The authors of *The Oceans* noted that the practice during the nineteenth century was to use only government-operated large vessels in oceanographic investigations. In the early twentieth century, Björn Helland-Hansen used the 23-meter *Armauer Hansen*, built to his specifications, to show that smaller vessels could be used effectively. The ship carried out a number of studies in the North Atlantic, and its successful use convinced other laboratories that mid-size vessels, economically and locally operated, were an efficient way to carry out oceanography. The enormous amount of global exploration carried out by the *Atlantis* and *Vema* at the Woods Hole Oceanographic Institution and the Lamont–Doherty Geological Observatory and by the seagoing tugs *Horizon* and *Baird* at the Scripps Institution of Oceanography was an extension of this point into the 1950s and 1960s. A review of oceanographic vessels and their expeditions in the period 1887–1960 is given by Wüst (1964). Today, the oceanographic fleet ranges downward from the large (120-meter) *Glomar Challenger* with its unique deep-sea positioning capabilities used for deep-sea drilling, and the *Melville* and *Knorr* with their cycloidal propulsion that allows remarkable maneuverability, to a variety of ships, platforms, and portable laboratories, each useful for different purposes. For example, FLIP (floating instrument platform) is a 108-meter towable surface craft that can be upended on site to provide a manned spar buoy with a draft of 90 meters and high vertical stability. The mid-size vessels form a large and important part of the data-gathering capability that provides access to the sea for many laboratories.

The need for measurements drives the development of new instruments. But there is sometimes a tendency for instrument development to proceed independently of the scientific needs. For example, Chapter 10 in *The Oceans* describes no fewer than 15 current meters. One wonders if this is not more than the total number of direct current measurements in the open ocean at the time, and whether there would not have been a net gain if more time had been spent in making measurements than in trying to develop new instruments. However, in fairness to the developers, we must point out that most of the effort at that time was aimed at making a useful kind of long-term recording scheme. The recording problem has been solved with the advent of tape-recording techniques (see section 14.2.3) and the focus is now on the problems of sensor designs.

The same point is true of other types of instrument design. In 1968 I surveyed the historical literature on deep-sea pressure gauges [see "The History of the High Seas Tide Gauge" by W. Matthäus (1968) and H. Rauschelbach (1932); English translation by Baker (1969)]. There is no question that at that time there were more designs for deep-sea pressure gauges than there were measurements of deep-sea pressure. This situation happily has changed today: we have seen the direct confirmation of an open-ocean tidal amphidrome (Irish, Munk, and Snodgrass, 1971), and deep-sea pressure gauges are beginning to be included in experimental design of new programs (see section 14.3.5).

Any scientist involved in instrumental design immediately feels the conflict between the need for use and the need for engineering improvements. According to the engineer, the scientist wants to use the equipment before it has been properly tested, and according to the scientist, the engineers want to improve the equipment before it has been used. The proper solution to this problem is close collaboration between scientist and engineer. The development of wave- and tide-measuring devices by W. Munk and F. Snodgrass of Scripps (section 14.3.5) is a good example of a highly successful collaboration between scientist and engineer; another is the work on temperature–pressure recorders by C. Wunsch of MIT and J. Dahlen of the Draper Laboratory (see section 14.3.1).

Whatever collaboration is established, however, it is clear that new equipment must be tested early in the sea. The information gained in actual field tests is crucial to the development of reliable instruments. Moreover, this is the only way that unexpected ocean effects can be found. These unexpected effects include signals of large magnitude (the "Van Allen" effect) and biological phenomena. The latter of these is demonstrated in figure 14.1, where we show a deep-sea pressure gauge brought up from the Drake Passage with an octopus attached to the sensor area. The effect of the octopus's breathing on the long-period variations of pressure measured by the gauge has not yet been determined.

There are two threads that our "good and determined" engineers have followed in order to bring us up to the modern state of ocean instrumental engineering.

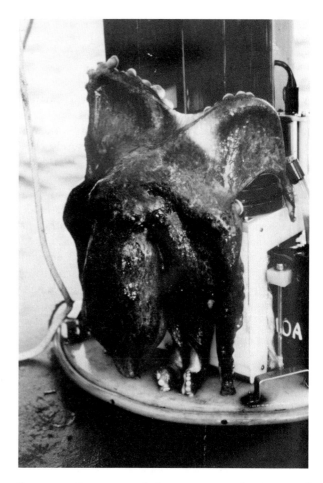

Figure 14.1 Octopus attached to sensor area of University of Washington bottom pressure gauge brought up from 500-m depth, north side of Drake Passage, 1977. (Courtesy of E. Krause.)

The first of these is electronics development, which we can trace through the first use of such technology, the advent of solid-state electronics, and finally low-power integrated circuits. The second of these is materials and structure engineering, which includes the development of platforms that can carry the instruments: ships, moorings, and various kinds of floats.

In the pages to follow, we shall see that the development of technology in a number of areas in the 1950s and 1960s laid the foundation for a rapid improvement in observing techniques, starting in the mid 1960s. R. H. Heinmiller, Jr., in a personal communication summarizes the history:

In my opinion, a real change occurred in the area of physical oceanographic technology when everybody realized that it was time to get some engineers to work and stop doing things by string and sealing wax. This was stimulated in part by the increasing need for large-scale experiments with large quantities of instruments which forced the adoption of quality control and engineering planning. In addition, of course, there was the stimulus of new techniques, products, and materials that had been developed for the space program and the oil industry.

14.2.2 Electronics in Ocean Instruments

A history of the development of ocean instrumentation up to the late 1960s was presented by J. M. Snodgrass (1968). He noted that "until the advent of World War II oceanography proceeded at a rather leisurely pace. The anti-submarine-warfare program during World War II forced the rapid development of underwater acoustics.... However, the actual instruments employed to make measurements in the ocean were rather crude, technologically, in contrast to much of the other instrumentation of the day involving electronics." Snodgrass goes on to document the changes and development of instruments in the 1950s and 1960s. The first introduction of electronics instruments into oceanography was not successful. He points out:

Unfortunately practically without exception, these instruments failed, not necessarily because of faulty electronics or conceptual aspects, but because the instruments engineers did not properly understand the marine environment and the consequent packaging problem. Simple leaks proved to be disastrous. Since research funds were scarce, there was naturally considerable resistance to the introduction of these "newfangled" instruments. In fact, after repeated failures, it was suggested that "the ideal oceanographic instrument should first consist of less than one vacuum tube." As this was in pre-transistor days, it left little latitude for electronics.

Snodgrass comments that the first instrument at Scripps to break the "electronics barrier" was the bottom-sediment temperature-gradient recorder, a simple, by today's standards, but reliable instrument that yielded exciting new results for geophysics. It is a self-contained null-type self-balancing potentiometer that measures temperature gradients in the sediments by using two thermistors about 2 m apart. Snodgrass notes, "The instrument's success on the Scripps MID-PAC expedition in 1950 served to spawn an entire new generation of electronic precision oceanographic measuring instruments." He goes on:

In view of the skepticism that existed in 1950 when the temperature-gradient recorder was first used, it is perhaps not at all surprising that when temperature gradient values approximately two times greater than those which were anticipated were recorded, there was general disbelief and an unwillingness to accept the values. It was not until over a year later, when another, almost identical, instrument was being used in the same area in the vicinity of the Marshall Islands that an identical high temperature-gradient was recorded and the instrument's earlier results were then credited as being valid.

This skepticism is general, and often accompanies the first use of instruments. The instrument is proved correct often enough, however, to remind us constantly of our "basic physical innocence," as Watson says. Another example is the first use of quartz-crystal bottom-temperature gauges in the Atlantic (Baker, Wearn, and Hill, 1973), which showed large (about 0.1°C) temperature fluctuations at the bottom in the deep Atlantic water. Because the Pacific measurements (see, e.g., Munk, Snodgrass, and Wimbush, 1970) had shown variations at the bottom smaller by factors of 100 to 1000, there was a general reluctance to accept these results until they were later confirmed by other instruments. The fluctuations in the Atlantic are due to intrusions of Antarctic Bottom Water over the Hatteras Abyssal Plain.

One of the most important steps in instrument design was the introduction of the new low-power integrated-circuit solid-state electronics. This goes under the general name of COSMOS (complementary-symmetry metal-oxide semiconductor). Solid-state devices built from these semiconductors can carry out logic operations at very low power because of their use of field-effect technology. The field-effect semiconductor has only one type of current carrier, and limits the current by varying the space charge with the applied voltage. Typical COSMOS integrated circuits have very small quiescent current drains (about 1 nanoampere) and very high input impedances (about 10^6 megohms). For comparison, the older resistor-transistor technology requires quiescent currents of milliamperes. The ocean engineer was greatly limited in what he could do with the transistor technology because the current drains were simply too large. The COSMOS circuits do draw significant currents, but only when changing logic states. Thus the mean current is proportional to the frequency of operation. At frequencies of a few kilohertz the new systems allow a decrease in power consumption by a factor of a million or more. For example, one of the simplest devices, a flip-flop switch, draws only 10 nW (nanowatts) as opposed to the usual 30 to 100 mW (milliwatts) for transistor logic. Moreover, the circuits can be operated from a wide variety of supply voltages [typically 5 to 15 V (volts)] and have excellent thermal characteristics. [See any modern electronics textbook, e.g., Taub and Schilling (1977, p. 38), for further description.] The introduction of COSMOS logic into oceanographic instrumentation in the late 1960s and early 1970s is probably the major change in electronics for oceanographers since ordinary semiconductor logic was introduced.

Many of the instruments discussed in section 14.4 draw heavily on the COSMOS technology. These new integrated circuits permit a number of data processing operations *in situ* that never could have been considered before. For example, the vector-averaging current meter (see section 14.3) computes north and east components of the velocity, and records the speed, compass and vane-follower directions, time, and temperature, as well as the components of velocity over a variable sampling time that can be set to fit the experiment. The total recording time can be longer than 600 days. The use of the COSMOS integrated-circuit technology is crucial to this flexibility.

14.2.3 Batteries and Tape Recorders

Of course, one of the reasons that power becomes less of a problem is the improvement in battery capacity over the past few years. The subject is reviewed in two recent articles by McCartney and Howard (1976) and Jacobsen (1973). The new lithium batteries provide a number of characteristics important for oceanographic use: they have the highest cell voltage, the longest shelf life, the greatest energy density, the best low-temperature performance, and a flatter voltage–discharge curve than any other except mercury cells. The latter characteristic is especially important for use in logic circuits where the system is usually set to run at a given regulated voltage. As long as the battery supplies a higher voltage, the equipment works properly; when the voltage drops below that regulated limit, the equipment will malfunction. Thus a flat voltage–discharge curve is desirable. A useful comparison of battery properties, compiled by R. B. Wearn of the University of Washington, is presented in table 14.1. Here the advantages of the lithium batteries over the carbon–zinc and mercury are evident. The reader is referred to one of the review articles for a summary of the chemical reactions involved: the recent article by Farrington and Bryant (1979) shows future directions for high-density energy storage in rechargeable cells utilizing fast ionic conductors; Murphy and Christian (1979) discuss a new class of electrode material for high-density energy storage.

These batteries and the new high-capacity tape recorders allow us to make measurements in excess of a year of various oceanographic parameters, and to do a certain amount of data processing *in situ*. The major part of the data-collection problem has been solved by the introduction of reliable tape-recording systems now on the market, and it is instructive to outline briefly the development of one of these to show the systems-design problems involved.

In the early 1970s, J. McCullough and R. Koehler of the Woods Hole Oceanographic Institution looked at ways of replacing their existing continuous-loop tape system used in the current meters with a cassette tape recorder. High-quality data cassettes were available, but there was no good reliable tape drive that could run on low power, pack data reliably on the tape, and fit into a 6" pressure case. There were mechanical problems involved in building a rugged drive that could withstand handling at sea, and that could precisely position and move the cassette tape. There were electronics problems involved in producing an efficient coding scheme, driving the tape, and recording the data, all with little power.

The solution worked out by Winfield Hill of Harvard University and the WHOI engineers was ingenious. A stepping motor was used to drive the tape precisely with electronics designed to provide a near-exact damping of the usual overshoot. A heavy cast body was designed and built with extremely close tolerances. A biphase coding scheme provided self-clocked data. The commercial instrument is shown in figure 14.2.

Since 1972 over 1000 of the recorders have been produced commercially by the Sea Data Corporation and used in oceanographic instruments; many of the instruments discussed in section 14.4 use this or a comparable tape recorder. The present Sea Data (1978) model uses less than 4 W h of battery power to record 11×10^6 bits of data. The tape transport can write data as fast as 1200 bits per second or as slow as one record per half-hour (at this rate, with a maximum 396-bit data record, a cassette would take more than a year to fill). To give a comparison, this data capacity is roughly equivalent to 500 feet of 4"-strip chart paper. Other

Table 14.1 Comparison of Battery Properties: D-Cell-Size units at 10-mA Drain

	Initial cell voltage (volts)	Shelf life		Ampere-hours at 10 mA	Energy at 10 mA (joules)	Joules/cm^3	Joules/kg	Joules/$(1979)
Sealed lead-acid	2.1	6 mo @	21°C	2.5	2.1×10^4	3.7×10^2	1.1×10^5	0.88×10^4
Carbon-zinc	1.5	1–2 yr @ 1–2 mo @	21°C 55°C	6.2 to 0.9 V	2.7×10^4	5.2×10^2	2.9×10^5	7.7×10^4
Alkaline	1.5	2–3 yr @ 2 mo @	21°C 55°C	9 to 0.9 V	4.1×10^4	7.8×10^2	3.3×10^5	5.1×10^4
Mercury	1.35	3–4 yr @ 4 mo @	21°C 55°C	12.4	6.1×10^4	14×10^2	3.6×10^5	1.4×10^4
Lithium–organic	2.9	10 yr @ 1 yr @	21°C 55°C	10	10.4×10^4	19.5×10^2	12.2×10^5	1.4×10^4
Lithium	3.6	10 yr @ 1 yr @	21°C 55°C	11	14×10^4	27.4×10^2	14×10^5	1.4×10^4

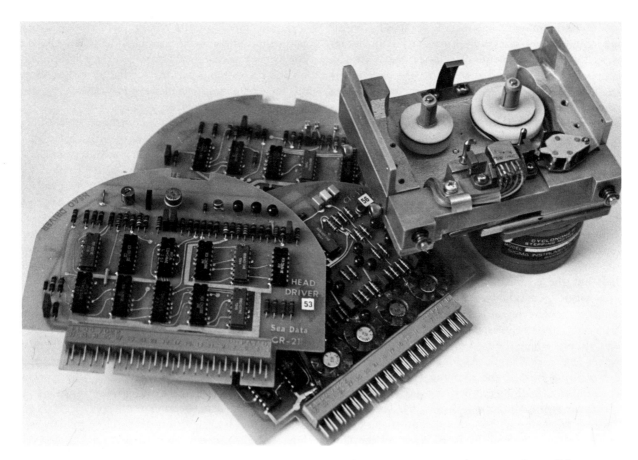

Figure 14.2 Sea Data tape-recorder body (right) and COSMOS circuit cards (left). The cards are rounded in order to fit into a 6"-diameter pressure case. (Courtesy of W. Hill.)

commercial cassette tape recorders are also available now for oceanographic use.

The second thread of engineering design mentioned above is "platform engineering." Since a review has recently been given by Henri Berteaux in his book *Buoy Engineering* (1975), we need not discuss this aspect in detail here. Some aspects of platform engineering are included in the subsections on mooring technology in section 14.3. The reader interested in following current developments (circa 1980) in ocean technology is urged to read *Exposure,* a newsletter produced by R. Mesecar of Oregon State University for discussion of new contributions and problems in instrument design and use in the ocean.

14.3 Examples of Modern Ocean Instruments

Looking over the period since *The Oceans* was written, we can identify four major areas of instrument design that have had an important impact on the development of our ideas of ocean circulation and mixing. These areas are the moored-buoy-current-meter technology, the deep-drifting neutrally buoyant floats, the temperature-salinity profilers (commonly known as the STD and CTD), and the velocity profilers. The first three of these developed essentially simultaneously. The velocity profilers are newer, and, to some extent having built on the technology developed earlier, have not yet had the total impact of the other three groups.

A second set of instruments is in a state of development and has not yet had the extended use or the impact of the four listed above. These include, for example, bottom pressure gauges, surface drifters, and the "inverted echo sounder." We also include in this set the whole suite of remote measurements, e.g., satellite altimetry and laser profiling of the upper layers, and the various acoustic techniques that have been proposed.

Our final discussion in this section covers a set of problems that can not yet be attacked with present technology. New instruments are crucial for understanding and quantitative measurement. Examples are air-sea fluxes in stormy conditions or measurements in strong currents.

Space does not allow me to be comprehensive in any way here, but only selective. The reader is referred to two excellent recent review volumes, *Instruments and Methods in Air-Sea Interaction* (NATO Science Committee, 1978) and the *Proceedings of a Working Conference on Current Measurements,* (Woodward,

Mooers, and Jensen, 1978), which cover aspects of many of the instruments to be discussed below. The paper by Rossby (1979) on "The Impact of Technology on Oceanography" contains a number of instructive examples.

Another area of very great impact on ocean measurements is navigation. Advances in both shore-based (LORAN) and satellite-based navigation techniques are responsible for the success of many of the instrumental techniques discussed below from mooring location to velocity determination. The discussion below is limited for reasons of space to instruments themselves.

In thinking about instruments and what they measure, we consider the full equations of motion. The equations include the terms to be measured; ideally, direct measurement of the terms is best, but sometimes it turns out to be more feasible to measure the term indirectly. The terms that appear in the equations involve the velocity, products of velocity, density, pressure, turbulent stresses, and viscosity.

The instruments that we discuss for velocity include those that make direct measurements of currents either at a point or in profile. We have been less successful in measuring turbulent stresses—products of velocity fluctuations—than the meteorologists, primarily because of the lack of a stable platform. However, some useful data have been taken from stations on sea ice and are discussed below.

Density is generally inferred from temperature and salinity; technical difficulties have precluded any useful instrument for measuring density directly. The main problem is finding an instrument that will work *in situ*—in the water column or on board ship. The small variations of density and the large accelerations at sea have prevented much success with direct density measurement. A number of techniques have been developed, however, and some of these will be discussed.

Pressure is generally inferred from the density using the hydrostatic relation. Without some level of pressure reference, however, it is not possible to establish an absolute pressure field in the ocean. Bottom pressure measurements (to be discussed below in section 14.3.5) can monitor pressure fluctuations; sea-surface topography by satellite is a technique currently being developed for measurements of both fluctuations and mean surface field.

14.3.1 Current-Meter and Mooring Technology

There are two parts to the measurement of currents at a point in the ocean. The current meter must be accurate, reliable, and, for most purposes, internally recording. The platform, or mooring, must be robust, deployable, and affordable. Major advances in both of these areas have been made since the 1960s. It is now possible to make long-term (greater than 1 year) measurements of currents at levels below the surface layer with better than 90% data return (e.g., Pillsbury, Bottero, and Still, 1977; Tarbell, Spencer, and Payne, 1978).

The paper of Richardson, Stimson, and Wilkins (1963) is a good starting point because it marks the beginning of the modern age of current-meter and mooring technology. This remarkable paper covers the whole field of mooring and current-meter technology as it was known at that time, and demonstrates the ingenuity of W. S. Richardson and coworkers then at the Woods Hole Oceanographic Institution. The paper documents the early attempts to maintain deep-sea moorings and current meters along a section from Woods Hole to Bermuda across the Gulf Stream. For this purpose, they needed a new current meter that would record for a long time, and a sturdy, reliable mooring for a platform.

Richardson et al. were influenced by Swallow's measurements (1955, 1957) with neutrally buoyant floats, which revealed a variability in the measured currents large compared to the residual drift. They argued that in deep water, where large variability is encountered, the significance of short-term measurements is in serious doubt. They noted that the float tracking could be extended to longer times, but that, as the required measurement time increases, equipment that may be left at sea unattended becomes increasingly attractive. They went on to describe the details of the system that they developed for long-term deep-current measurements, noting, in something of an understatement, that the technique is not an easy one, nor is it a cure-all for the deep-current problem.

However, the system described by Richardson et al. (figure 14.3) has the same basic elements used today. In essence, the system consists of a near-surface or surface float, a line that holds the current meters, and a release that is right above the anchor. We consider first the current meter, and its recent developments, then shall return to the development of the mooring systems.

Current Meters The current meter used by Richardson et al. was the Savonius (1931) rotor-type with a small, freely moving direction vane attached in line with the axis of the instrument. J. M. Snodgrass (1968) discussed some of the aspects of the Savonius rotor. The advantage of the freely moving vane is that its response time is comparable to the response time of the speed sensor. The instrument is cylindrically symmetric and can be used as a link in the mooring system. Two major data-collection design features are also important in this current meter: the photographic recording system and the burst sampling. Richardson et al. recognized that the high-frequency noise in the water coupled with the limited recording capability of the instrument would result in very short records if records were made continuously. Therefore, they used a "burst sampling"

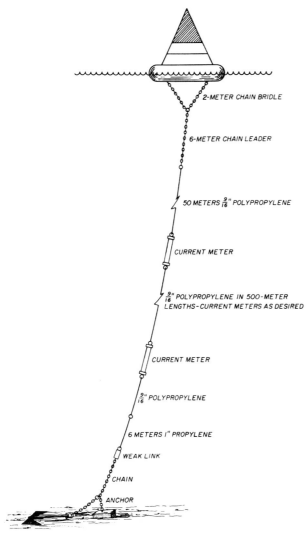

Figure 14.3 Current-meter mooring configuration used by Richardson et al. (1963).

scheme, whereby short samples of densely spaced data are collected, interspersed with longer periods of no data. If enough is known about the spectrum of the system, then such a scheme will provide an adequate estimate of the total energy in the various frequency bands.

The second feature of note is the photographic recording. A clever system of light pipes and coded disks was used to carry data bits from the sensors to a camera with 100 feet of photographic film. In this way a long data set could be collected; 100–200 days were possible, a major increase over the other systems then available. This photographic scheme worked well as long as there were only a few data sets available, but a technique to read the film by computer was never really successful.

The modern commercial version of this current meter is basically similar to the Richardson design. In addition to general improvement of reliability, two major changes have been made: the recording scheme uses a tape recorder (see section 14.2), and the sampling scheme is of the type called *vector averaging*. The vector-averaging current meter, or VACM as it is commonly known (figure 14.4), was developed at Woods Hole by J. McCullough (1975) and R. Koehler.

The use of the new COSMOS integrated-circuit technology is responsible for the increased accuracy of the VACM. The increased data-handling capability allows the instrument to sample the speed and direction approximately eight times per rotor revolution. East and north components are then calculated and stored. The burst-sampling mode is used: typical sampling intervals are 15 min (minutes; at this interval the tape capacity is 530 days). The direct vector-averaging feature allows the instrument to make accurate measurements in wave fields and from surface-following moorings [see McCullough (1978a,b) and Halpern (1978) for further discussion and references on the use of the VACM and comparison with other instruments].

One of the problems with this design is that the response lengths for the Savonius rotor, free-vane system cannot be accurately matched in time-dependent flow because the rotor accelerates about three times faster than it decelerates (Fofonoff and Ercan, 1967). Moreover, the Savonius rotor system does not have a true vertical cosine response (response proportional to the cosine of the angle or attack), and thus its measurements of horizontal velocity are contaminated by the vertical component, which can be large in the wave zone or near a surface-following mooring. Until recently, current meters had not been tested rigorously in unsteady flow conditions in the laboratory to show their performance in the expected environmental conditions. Using a series of such tests, Davis and Weller (1978) [see also Weller (1978) and Weller and Davis (1980)] have developed a two-component propeller current-measuring instrument with an accurate cosine re-

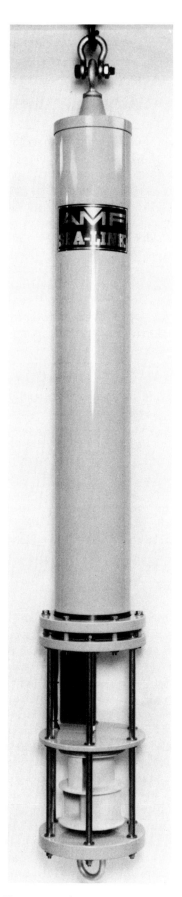

Figure 14.4 Vector-averaging current meter manufactured by Sea-Link. (Courtesy of W. Coburn.)

sponse (rms deviation from cosine of 1.5% of the maximum response). In initial tests the instrument, dubbed the vector-measuring current meter (VMCM), has shown negligible rectification and accurate measurements of mean flow in the presence of unsteady flow. Figure 14.5 shows the arrangement of this system, also in-line like the VACM.

Propellor-type systems have also been successfully used by Eriksen (1978) for measurement of internal waves, and by J. D. Smith (1978) for studies of turbulent currents, both near bottom in estuaries and near the sea ice in the Arctic. Smith documented the need for improving symmetry of the propellers, and ducting at high angles of attack, and he showed the utility of such techniques in high-accuracy, short-duration measurements. He reported studies of the turbulent structure under the Arctic ice, which reveal a high intensity of turbulence in the upper layers (Smith, 1974).

There are a number of other current-meter systems on the market today, and these are reviewed in the volumes mentioned above. It is of interest to discuss one of the other Savonius rotor-type instruments in major use, the one invented by Ivar Aanderaa (1964) of Norway under sponsorship of the North Atlantic Treaty Organization (NATO), because it has also had a major impact on current measurements. The goal of the Aanderaa design was to provide a relatively inexpensive, reliable instrument that could be used by laboratories with only simple electronics capabilities (another example of the Scandinavian notion of distributing equipment as broadly as possible, as Helland-Hansen suggested for ships). (See also Dahl, 1969.)

The Aanderaa instrument is pictured in figure 14.6. It uses a Savonius rotor in combination with a large vane. The entire instrument must move in response to changes in direction. The data are recorded on a fairly simple tape recorder, and the electronics is fully potted to avoid tampering. The instrument is inserted in a mooring line in such a way that it is free to pivot in a horizontal plane, and the large vertical tail fin orients the meter in the direction of the current flow. As with the instruments discussed above, a magnetic compass is fixed within the instrument to give its orientation in the earth's magnetic field and thus the direction of flow. Aside from the two deficiencies noted below, the instrument has proved to be a very reliable device, relatively simple to use, and it has thus served its purpose well. It is one of the most popular of current meters.

Because of the large integral vane, the Aanderaa instruments are not suitable for near-surface moored measurements in waves (Halpern and Pillsbury, 1976; McCullough, 1978a,b; Halpern, 1978). Saunders (1976) shows that while the mean current directions are usually correct, the vane response in waves causes speeds to be too large by a factor of between 1 and 10, with a

Figure 14.5 Vector-measuring current meter with dual propellers and titanium cage. (Weller, 1978.)

typical value of about 2. This occurs because the Savonius rotor accelerates to 63% of a new flow speed in 30 cm of water flow past the sensor (Fofonoff and Ercan, 1967), but the vane of the Aanderaa current meter requires 6 m of flow to realign itself after a 180° change in flow direction (Appell, Boyd, and Woodward, 1974).

It is important to note that in all instruments that measure speed and direction separately, the angular conversion required to extract velocity components spreads error due to poor frequency response over all frequencies. Cartesian-component sensors do not introduce such error (Weller and Davis, 1980).

The Aanderaa gauge has been subject to another interesting difficulty. Hendry and Hartling (1979) describe a pressure-induced error in direction found in the earlier models that used a nickel-plated copper pressure case. They found that instruments used in the field at pressures greater than 4000 db (decibars) began to show nearly constant directions over many months, indicating sticking compasses. On return to the laboratory, the instruments showed no problems. This suggested that an environmentally caused, instrument-related magnetic field was competing with the earth's field and affecting the compasses. A series of laboratory tests showed that this was indeed the case. Serious direction errors were found in the nickel-coated current meters used at pressures greater than 2000 db. The errors are caused by the magnetization of the nickel coating by the field of the compass magnets themselves, after the magnetic properties of the nickel are modified by the stresses induced in the pressure case as the external pressure is increased. At 2800 db, errors in direction of up to 10° were seen, and at pressures of greater than 4200 db actual sticking of the compasses was observed. The nickel coating has been eliminated in later models.

Acoustic and electromagnetic current meters show promise for direct fast-response measurement of currents in waves and turbulent zones. The present status of these instruments is discussed in the two volumes mentioned above.

Our final example is not a current meter, but it is an important part of the moored-instrumentation suite. This is the moored temperature and pressure recorder developed at MIT and the Draper Laboratory (Wunsch and Dahlen, 1974). The instrument was designed for long-duration measurement of temperature in the deep ocean. Temperature is sensed with a thermistor of accuracy 0.01°C and a time constant of 3 min. In terms of pressure, instruments destined for 2000 m depth have an absolute error of the order of 1 m and a short-term pressure change measurement error of less than 0.2 m. The instruments, called TP recorders, have proved very successful, yielding a high data return (better than 90%) for mapping temperature fields and monitoring mooring motion. For example, in the MODE-1

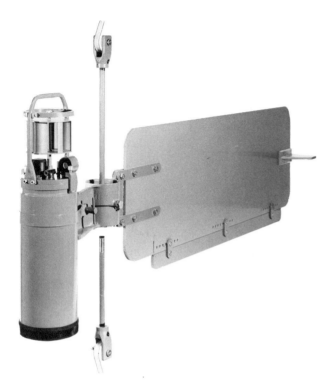

Figure 14.6 Aanderaa Model RCM-4 current meter. The instrument swivels about the vertical rod between the electronics case (left) with Savonius rotor and other sensors on top and the large direction vane (right). (Courtesy of Aanderaa Instruments.)

Figure 14.7 Draper Laboratory temperature–pressure recorder moored in 10 m of water off the Seychelles Islands. (Courtesy of J. Dahlen.)

experiment, the TP recorders allowed the mapping of two levels in the array in addition to the levels occupied by the current meters. The additional maps proved important to the description of the eddies observed in the experiment (Bryden, 1977; Richman, Wunsch, and Hogg, 1977). The TP recorders have been used subsequently in many different arrays and are considered an important adjunct to array design. Figure 14.7 shows a TP recorder being used as a tide gauge at a depth of 10 m in the waters near the Seychelles Islands in the Indian Ocean.

Mooring Systems As mentioned above, the first large-scale attempt at current-meter measurements from moorings was tried by Richardson et al. (1963). In all, from May 1961 to December 1962, they report that 106 buoy stations were put out. About half of the shallow stations, about 30% of the stations in the Gulf Stream region, and about 30% of the short-term stations outside the Gulf Stream region were recovered. The average recovery rate of all instruments set out in these first 2 years was about 50 to 60% (Heinmiller, 1976a).

A large amount of engineering effort has succeeded in improving the recovery rate of instruments and moorings. For the period 1968 to 1974 the recovery rate moved up to 95% for all instruments on WHOI moorings. It is instructive to look at the specific engineering changes that led to this remarkable improvement. The engineering specifics and the detailed statistics are presented by Heinmiller in a very complete technical report (1976a). The material discussed below is largely drawn from this report and from several conversations that I have had with him on this subject.

Mooring failures can occur at each point: on launch, on recovery, and on station. On station, mooring failures have occurred in about every imaginable way. Numerous early polypropylene-surface mooring lines failed due to fish attack (Stimson, 1965). Wire failure due to corrosion or kink is common, and surface floats have been swept under and crushed in high currents (Heinmiller and Moller, 1974). Surface moorings were soon shown to be less reliable than the subsurface moorings, mainly because in the latter case the flotation is not subject to the stresses in the wave zone.

The acoustic release became a key item as soon as it was confirmed that the subsurface moorings were significantly more reliable. The timed releases and weak links used earlier were adequate as long as mooring durations were short; but with longer and longer mooring durations being dictated by programs studying lower-frequency variations, the timers became unworkable. Heinmiller (private communication) notes that "we lost some moorings off Bermuda in 1964 when the timers released them during a hurricane and we couldn't recover them. We never actually lost a moor-

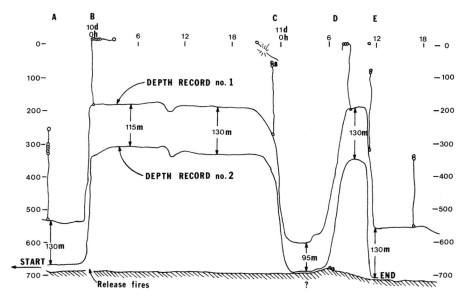

Figure 14.8 Record from two depth recorders on mooring 230 set by Woods Hole Oceanographic Institution in the Denmark Straits, March, 1967. (Heinmiller, 1968.)

ing because of a change of ship schedule, but certainly there was a lot of scrambling for charter vessels at times."

By 1964 it was realized that a device was needed that could be released by command from the recovery vessel. A number of different commercial releases were tested according to the criteria of reliability, security, and information transmittal (e.g., has the instrument released?). For example, the system must be secure from premature tripping of the release mechanism by ambient noise. The strange journey of mooring number 230 set in 705 m of water in the Denmark Straits in March of 1967 is shown in figure 14.8 (Heinmiller, 1968). This mooring refused to come up on command and was later recovered by dragging. Its depth was monitored by a pressure recorder. At A the mooring is supported by six steel ball floats. At B the release apparently tripped letting the buoyancy section surface. At C, one day later, the lower recorder dropped to the bottom. It is supposed that at this point one ball broke loose and two flooded. At D the mooring surfaced again, apparently after the flooded balls broke off. At E one more ball broke loose and the gear sank. The gear when recovered had only two balls on it, both badly battered. Pack ice in the area accounted for the loss of the balls. Ice noise has been suggested as a possible source of sound that actuated the release. The coding system used on the release for this mooring was clearly not secure enough.

As a result of the test program, the WHOI Buoy Group settled on the AMF Sea-Link Model 200 release system (now manufactured by EG&G) (Heinmiller, 1968). As of 1974, the reliability of this release was better than 95%. The AMF system uses a pulsed, amplitude-modulated double-sideband suppressed-carrier frequency, with coding provided by pulse width, pulse-repetition frequency, and time duration. This system has proved to be secure enough against ordinary noise, yet not so secure that it cannot be released when desired. In terms of communication, the AMF device actually monitors the retraction of the piston in the release mechanisms, providing confirmation that the first mechanical step has been taken in the release process. The release is pictured in figure 14.9. An acoustic release system developed and used successfully at Scripps Institution of Oceanography is described by F. E. Snodgrass (1968).

Another important element of the mooring system is wire. Heinmiller notes that two factors were important here. First, the Buoy Group decided that standard torque-balanced wires were not adequate for moorings. The basic industrial torque-balanced wire is designed not to twist when hung from a crane at a height of about 30 m. Since the amount of twist is proportional to length, this wire is not useful over the 4-to-5-km depth of a deep mooring. US Steel finally came to the rescue with a specially torque-balanced oceanographic wire rope.

Related problems included terminations and corrosion resistance. It became clear very early that the wire should be jacketed in order to impede free flow of dissolved oxygen past the wire. The WHOI Buoy Group found that the combination of galvanized steel and standardized terminals with sufficient water barriers was the key to ending corrosion problems.

The fishbite problem was solved simply by using wire in the fishbite zone. Originally this was defined as above 1500 m. Later the group found fishbites at about 1800 m, and extended the zone to 2000 m.

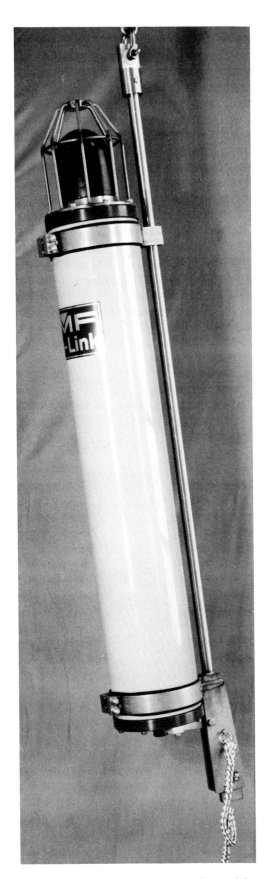

Figure 14.9 Acoustic release system manufactured by Sea-Link. Acoustic transponder is at top, release mechanism at bottom. (Courtesy of W. Coburn.)

Glass balls proved to be the best solution for deep flotation. The first problem with these was quality control; it was solved by working directly with the manufacturers. The second problem was how to get hold of the balls in order to attach them to the mooring line. The group first used nets with quick-fastening hooks, then nets with shackles, eventually going to hard plastic containers (hardhats) of various configurations, bolted to chain. The hardhats also eliminate the need to test spheres because the mooring will not be endangered if a sphere implodes on the chain, when the spheres are far apart enough to avoid sympathetic implosion triggered by the shock wave.

Finally, there is the backup-recovery system, which is made up of a second set of glass balls near the release at the bottom. In case of failure of the upper flotation, the lower balls will bring the mooring up. The advantages are twofold: the system enabled the group to recover equipment that would otherwise have been lost, and it provided engineering data on failed moorings that would have been lost, given the tendency of the top halves of the broken moorings to disappear. Of course, backup recovery involves hauling the mooring aboard upside down. Since tangling is inevitable as the mooring falls to the bottom, the typical situation is as pictured in figure 14.10.

The overall mooring statistics that reflect the engineering improvements discussed above are graphed in

Figure 14.10 Wire tangle resulting from recovery of mooring with backup system. (Heinmiller, 1976b.)

figure 14.11, from Heinmiller (1976a). He noted that in 1967 it was recognized that a major engineering effort was needed for improvement of recovery rates. The improvement in mooring reliability after 1967 is clear; it is due to better acoustic releases, better mooring wire, and improved procedures and quality control (Berteaux and Walden, 1968, 1969; Morey, 1973). During the period 1960 to 1974, 544 moorings were set.

Moorings as currently used by the WHOI Buoy Project are of three types (Heinmiller, 1976b): the intermediate mooring, the surface mooring, and the bottom mooring. The intermediate mooring, shown in figure 14.12, has buoyancy sections at several depths. The lowest buoyancy section provides backup recovery in the event of mooring failure. The depth of the top of the mooring can vary to within 200 m of the surface or less.

The surface mooring for deep-sea use is shown in figure 14.13. The weight of the anchor varies with the expected current profile. A backup-recovery section is included. Surface-wave noise limits the usefulness of this configuration. The bottom-mooring package is shown in figure 14.14. These moorings have no backup-recovery section and typically carry only one or two instruments.

The above description has concentrated on the developments at the Woods Hole Oceanographic Institution because the major engineering effort has been carried out there. It should be noted that a number of other institutions have developed reliable mooring techniques as well; the basic principles we have discussed above are present in all of these designs. In addition, we note a comparison test of different deep-sea mooring configurations that was designed to validate the dynamic computer programs for mooring design (Walden et al., 1977).

The study of current-meter and mooring combinations has been recently an active field. Mooring motion limits our ability to make good near-surface measurements; such motion is particularly large in surface moorings. It is to be expected that acoustic tracking of mooring motion will be more common in future attempts to determine the effect of the motion on the accuracy of the measurements made from the mooring.

14.3.2 Neutrally Buoyant Floats

Measurement of deep motions by the use of *in situ* moving floats requires some method of tracking. The great strides made in acoustics during World War II yielded the necessary technology for acoustic tracking of instruments in the ocean. A significant contribution was the surplus acoustics gear that the oceanographers could scavenge for building and tracking their own equipment.

Interest in developing deep floats was high on both sides of the Atlantic in the mid-1950s. In 1954, Henry

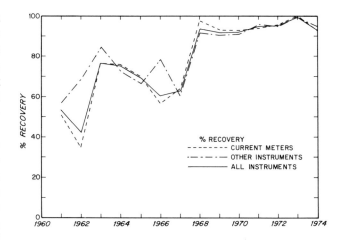

Figure 14.11 Instrument recovery rates as a function of time, Woods Hole Oceanographic Institution Buoy Group. (Heinmiller, 1976a.)

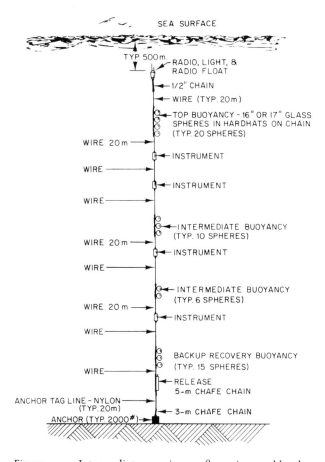

Figure 14.12 Intermediate mooring configuration used by the Woods Hole Oceanographic Institution Buoy Group. (Heinmiller, 1976b.)

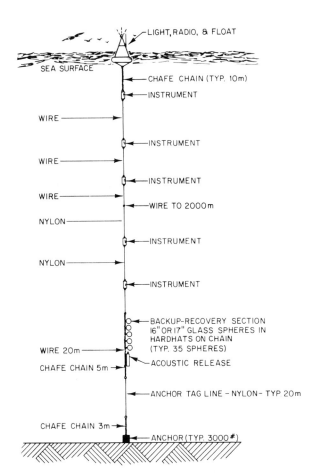

Figure 14.13 Surface mooring configuration used by the Woods Hole Oceanographic Institution Buoy Group. (Heinmiller, 1976b.)

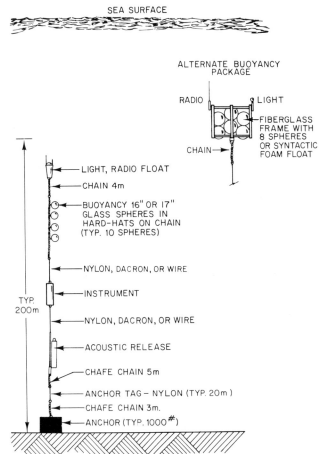

Figure 14.14 Bottom mooring configuration used by the Woods Hole Oceanographic Institution Buoy Group. (Heinmiller, 1976b.)

Stommel mentioned in an address to an oceanographic convocation at Woods Hole, "Were it possible to design cheap neutrally buoyant floats which could signal their position by means of a SOFAR network, then we could start immediately to study the deep water circulation of the sea" [Stommel (1955)]. In England, John Swallow of the Institute of Oceanographic Sciences was a key figure in this development; in fact, the floats have been commonly known as "Swallow floats." Perhaps the early story is best told in his own words (Swallow, personal communication):

In the summer of 1954 James Crease, Tom Tucker, and others tried to measure the vertical profile of currents in deep water by tracking a slowly sinking sound source relative to three anchored sono-radio buoys. That was only partially successful, and I got involved in trying to improve it when I joined the NIO (National Institute of Oceanography—now the Institute of Oceanographic Sciences) in October 1954. Looking at the scattered results from those profiling attempts, it was a fairly obvious improvement to think of trying to track a sound source at a constant level.

Apparently Swallow came by this idea independently of Stommel; he notes that "Dr. Deacon (Director of NIO) had attended that Convocation in Woods Hole and must have heard Henry; moreover, Willem Malkus (then of WHOI) was working at Imperial College that winter (1954-5) and occasionally visited the Institute. I don't recall that either of them told me about Henry's idea, though."

Swallow goes on:

Stabilizing the float by heating a petrol-filled container seemed likely to need too much power, besides the possible danger. The next idea was to stabilize the float by making it less compressible than water. That, too, seemed a fairly obvious thing to try, especially to someone familiar with the velocities of elastic waves in water, rocks, and other substances. To anyone doing seismic refraction shooting at sea, water is a highly compressible, low velocity medium. For most metals, the compressibility would be many times less, and clearly a piece of metal could be hollowed out to some extent and still be less compressible than water. Moreover, one could see immediately that one would not be dealing with very small effects: for water, the velocity of sound is about 1.5 km/s, so that the bulk modulus is about 2×10^{10} cgs units, i.e., the relative change in volume is 5×10^{-11} per dyne/cm^2 or 5×10^{-3} per km of depth. So, if we made a float with a compressibility half that of water, it would gain buoyancy at the rate of about $2\frac{1}{2}$ grams per kilogram displacement per kilometer of depth. The critical questions were, would a hollow piece of metal of suitable compressibility have any useful buoyancy in water, and could it withstand the pressure at useful oceanic depths without collapsing?

It was possible to build such floats, and Swallow's 1955 paper details the construction and first use of the aluminum tube floats. He says:

> The first batch of floats were made from old lengths of scaffolding. I picked out a dozen lengths from a stack that was leaning against the wall somewhere. They were too thick-walled, much lower compressibility than was really needed, and hardly any buoyancy. Not straight enough to be machined, and that would have been too expensive anyway. So I got a wooden trough made, and filled it with caustic soda solution, and dissolved off the outsides of the tubes until they were down to a suitable weight. That first batch of tubes cost about £17 each (not counting my own time). We got the transducers for nothing—they were rejected from a Navy project.

Figure 14.15 (Swallow, 1955) shows the float, consisting of two tubes, the end plugs, and the acoustic transmitter. Swallow notes that "it seemed very much a matter of luck that any of the early floats worked at all. The sound source had been designed for the earlier profiling experiment and the only trouble with it was the packaging. The simple cheap lead-through plug was something I invented myself."

O-rings played an important role in sealing the end plugs, since the flat gaskets and packed glands normally used for pressure vessels leaked at less than 1000 m depth. Accurate navigation techniques were crucial in establishing ship position for triangulating on the floats.

According to Douglas C. Webb of the Woods Hole Oceanographic Institution, no major changes in the float design from Swallow's initial ideas took place until the middle 1960s. At that time three major advances were achieved. G. Volkmann and Webb provided a synchronized time base so that the floats could be heard with an echo sounder, and several signals could then be averaged to give a more accurate position indication. The second change was to increase the range to 20-25 km over the 5-6 km obtained by Swallow by lowering the frequency to 4-5 kHz from 10-12 kHz, and the third was to use glass spheres for small, cheap floats. It also became possible to position the ship directly above a float at a prescribed time, thus eliminating the need for continuous navigation information. This allowed float tracking in the open ocean beyond the range of shore-based navigation systems when satellite navigation became available.

Lowering the sound frequency was an important advance. Since the absorption of sound is proportional to frequency squared, increased range requires lower frequency. Lower frequencies are harder to transmit and receive, however, and there is more ambient noise. An optimum is reached below 1000 Hz, and the longest range floats now operate at about 250 Hz. Low-frequency transmitters for the ocean are not simple: Webb points out that he had to take time out to become a "loudspeaker" designer in order to get the low frequencies necessary for long range.

The long-range floats ride in the SOFAR (sound fixing and ranging) channel, an acoustic waveguide (Ewing and Worzel, 1948). The channel is produced in many parts of the ocean by the combination of pressure and temperature effects on the speed of sound, which decreases with depth from the surface to about 1000 to 1500 m owing to the decrease of temperature, and increases with depth below this owing to the increase of pressure. In the SOFAR channel a few watts of sound can be heard about 2000 km away (Rossby, 1979).

The first big SOFAR floats were placed out in a triangle of listening stations at Eleuthera, Puerto Rico, and Bermuda. The range of listening was about 700 km, and the floats were successfully tracked (Rossby and Webb, 1971). This success generated enough interest so that a large program could be funded as part of the MODE-1 experiment (see section 14.4). A medium-range float program, clusters of floats listened to from a set of hydrophones lowered from a ship, was also used in MODE-1 by Swallow (1977). A 70-km range was achieved with this technique.

The long-range float used in MODE-1 is pictured in figure 14.16. It is housed in an aluminum tube, the

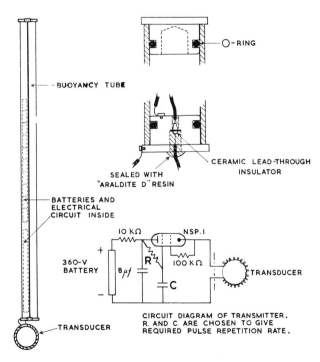

Figure 14.15 Early neutrally buoyant float; sketch of float and end plugs, and circuit diagram of acoustic transmitter. (Swallow, 1955.)

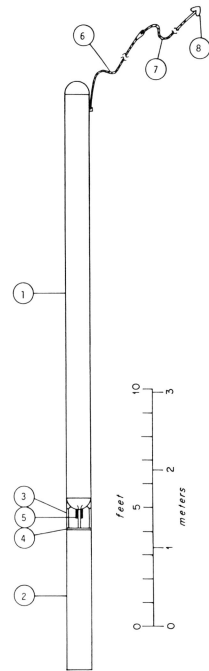

① MAIN HOUSING
② SONIC TRANSDUCER
③ CONNECTING RODS
④ BENDER FACE
⑤ RELEASE AND CONSTANT DEPTH ASSEMBLY
⑥ HEAVY PENNANT
⑦ LIGHT PENNANT
⑧ BUOYANT GRAPNEL

Figure 14.16 Long-range SOFAR float. (Courtesy of T. Rossby.)

cheapest structure for the amount of buoyancy required here. It contains (Rossby, 1979; see also Webb, 1977) a pressure–temperature telemetery system, whereby a 48-h (hour) average of pressure and temperature is transmitted on alternate days. Platinum resistance thermometers and titanium strain gauges are used for stability. Since compressional creep of the aluminum tube causes the instrument to sink about 0.5 m per day, a mechanism is needed to reduce its weight at the same rate. The floats thus have active ballasting: the weight is reduced electrochemically using a sacrificial anode controlled electronically by the pressure measurements. Consideration is being given at WHOI to the use of polymer concrete spheres that could alleviate some of these problems.

In order to relax the geographical constraints imposed by existing shore-based receivers, A. Bradley of Woods Hole Oceanographic Institution, in collaboration with Rossby and Webb, has developed autonomous listening stations for deployment on subsurface moorings. These have a vertical array of hydrophones, a COSMOS microprocessor-controlled signal-detection system, and the high-density cassette recorder discussed above. Rossby (1979) pointed out that when these listening stations become operational in sufficient quantities, it will be possible to conduct experiments in any ocean basin where the sound channel is sufficiently stable for long-range transmission.

A second major development in the float family is the vertical current meter. The measurement of the vertical component of velocity in the ocean is a major problem for oceanographers because it is so hard to find a reference platform. In shallow waters, rigid platforms on the sea floor could be used, but in deep water this is not possible. In 1965 Webb and Worthington at Woods Hole discussed the possibility of using a neutrally buoyant float as a platform for measurement of the vertical velocity. Webb's idea was to mount fins on the cylindrical float, at an angle, so that water movement past the float would cause it to rotate. The rotation could be used as measured relative to a compass, and then stored and transmitted to the surface. The instrument was successful in its first use, the measurement of vertical water movement in the Cayman Basin (Webb and Worthington, 1968). A spectacular result was found in the Mediterranean Sea where large vertical velocities were found confined to relatively small horizontal areas (Stommel, Voorhis, and Webb, 1971). The direct measurement of the vertical velocity here has had an important effect on our thinking about the spatial scales of the processes of bottom water formation in the ocean in general.

An interesting extension of the idea of the neutrally buoyant float is the self-propelled and guided float. One example of such an instrument has been built successfully at the Applied Physics Laboratory of the University of Washington. Called SPURV (self-propelled underwater research vehicle), it can maneuver underwater on command with acoustic signals to produce horizontal and vertical profiles of temperature, salinity, and other parameters. One important use of SPURV is the collection of horizontal temperature data on internal waves—data that were used by Garrett and Munk (1972b) in their modeling of internal wave spectra. A second use has been the delineation of horizontal and vertical plumes from chemical outfalls.

14.3.3 Temperature and Salinity Profilers

With profiling instruments in general, the quote by M. R. Clarke at the beginning of this chapter is apt: we know what the discrete sampling instruments can catch, but we do not know what lies within the sampling intervals. Hence the interest in continuous-profiling instruments. Gregg (1976b) has presented a useful history of the development of many of these instruments.

Temperature profilers were developed first; then, with the advent of reliable conductivity sensors, salinity profilers were added. Thus this section is logically divided into two parts: the development of the bathythermograph (BT) and its successor, the expendable bathythermograph (XBT), both used for the upper ocean; and the development of the salinity (as measured by conductivity), temperature, and depth-recording (STD, CTD) systems, which measure right down to the ocean floor.

The BT operates with a pressure-driven bellows that moves a metal- or smoke-coated glass slide under a stylus driven by a liquid-filled bourdon tube sensitive to temperature. The XBT, on the other hand, uses a thermistor to sense the temperature, and depends on a known fall rate to determine the depth. The XBT therefore requires a shipboard electrical system to record the data.

The BT is mentioned in *The Oceans* as having "the great advantage that it can be operated at frequent intervals while underway, and thus a very detailed picture of the temperature distribution in the upper 150 m can be rapidly obtained." Since the BT is entirely mechanical in operation, it is extremely reliable, and as J. M. Snodgrass (1968) says, "The bathythermograph over the years has perhaps been more extensively used than any other single oceanographic instrument." Since the engineering details and the development of the BT and an extensive discussion of the development of the XBT are covered in his article, we shall not repeat that material. It is useful to note, however, some of the developments in the expendable field since then.

As of 1978, the Sippican Corporation, suppliers of the XBT, had produced more than two million of the probes. It is likely that the number of records obtained from XBTs will soon surpass that of the BT, if that point has not been reached already; according to D. Webb, the scientific community alone uses approximately 65,000 XBTs annually. One of the major achievements of Sippican was the development of an insulated hard-wire link between the sensor and the ship. The present system consists of wires with four coats each of nylon and epoxy, jointed together by heat mating into bifilar pairs. The wire link is a 0.008"-diameter pin hold-free two-conductor cable. The cable is wound on two spools in a configuration similar to the spinning reel used by fishermen. The cable is unreeled from the outside of each spool in a direction parallel to the spool's axis. In this way the cable unreels simultaneously but independently from a vessel or aircraft and from the deployed sensor package (Sippican, 1973). This double-reel system is the heart of the expendable system. In use at sea, the dual-spooling technique permits the probe to free-fall from the point of sea-surface entry, without being affected by the speed or direction of the ship.

Current expendable products include the wire link itself, temperature sensors usable to depths of 200 to 1830 m, a fine-structure probe, a probe that can be launched from submarines while under water, and probes that can be launched from aircraft. To show the sensitivity of these probes, we note that the slower sink rate of the fine structure probe enables the thermistor, which has a time constant of 100 ms (milliseconds), to respond to a change of temperature in a layer of water of thickness 18 cm versus 65 cm when mounted in the standard XBT probes. The accuracy of the probes is $\pm 0.1°C$, and $\pm 2\%$ of depth.

A sound velocity probe is also available. It will measure sound velocity to an accuracy of ± 0.25 m s^{-1} to depths of 850 m. The sensor measures directly the time it takes for an acoustic pulse to traverse a total path of 52 mm. The effects of temperature, salinity, and pressure are thus accounted for directly in the measurement (Sippican, 1978). A salinity sensor is under development that uses the sound velocity and temperature probes together to compute salinity. Figure 14.17 shows the expendable configuration.

The XBTs are an invaluable tool for ocean monitoring. Merchant ships equipped with XBTs have yielded extensive sets of data for the study of the variability of the thermal structure of the upper ocean in both the North Pacific and the North Atlantic (Bernstein and White, 1977). Their widespread use has also stimulated further research into the systematic errors (e.g., depth) that exist in the system (McDowell, 1977).

The STD and CTD systems have also improved the oceanographer's hydrographic techniques and given us a new view of the small-scale distributions of temperature and salinity. Before this continuously recording system became available, the standard technique was

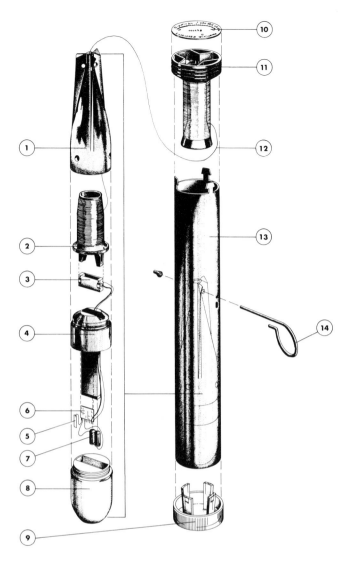

XSV EXPLODED VIEW

1. AFTERBODY
2. PROBE SPOOL
3. SOUND-VELOCITY SENSOR
4. ELECTRONICS HOUSING
5. STARTING CONTACT
6. INTEGRATED CIRCUIT
7. BATTERIES
8. ZINC NOSE
9. SHIPPING CAP
10. LABEL
11. SHIPBOARD SPOOL
12. SIGNAL WIRE
13. CANISTER
14. RETAINING PIN

Figure 14.17 Schematic diagram of expendable probe used for measurement of sound velocity. (Courtesy of Sippican Corporation.)

that presented in *The Oceans*: the use of reversing thermometers and collection of water samples for later determination of salinity by laboratory titration or conductivity measurements. Hamon and Brown (1958) listed some of the problems of this technique: the measurements are made at preset depths so that important detail may be missed; the salinity information is not available quickly enough to guide the progress of a cruise; and sampling and analysis are time consuming and expensive. On the other hand, the technique is straightforward and reliable, virtues not to be overlooked in work at sea.

Apparently the first reference to the use of conductivity for measurement of salinity of sea water is in Nansen's report (1902, p. 197) of the Norwegian Polar expedition 1893-1896 (this reference, not in *The Oceans*, was pointed out to me by B. Hamon and H. Mosby). Nansen states, "Considering it very important to determine the specific gravity or salinity of the water of the North Polar Sea with the highest degree of accuracy, I asked Mr. Hercules Tornøe to help me in the matter, which he did in a most friendly manner. He constructed for the expedition an apparatus for the determination of the salinity of sea water by its power of electrical conductivity, which he has himself described in 'Nyt Magazin for Naturvidenskaberne,' Christiania, 1893." The system was used on deck, and gave some useful results while exhibiting problems we still find today with electrode drift.

A continuous recording system for both conductivity and temperature was described in 1948 by Jacobsen (1948). This system was designed for a maximum depth of 400 m, and separate supporting and electrical cables were used. The system was crude, but pointed the way to the development by Hamon and Neil Brown then of the Division of Fisheries and Oceanography of the Commonwealth Scientific and Industrial Research Organization of Australia of the forerunner of the currently used system. Hamon points out, "Neil Brown and I both started working on an STD instrument as the result of a suggestion from David Rochford. He had seen the 1948 paper by Jacobsen, and wondered if we could do anything along similar lines. The instrument that we developed at that time is described in Hamon (1955) and Hamon and Brown (1958). First sea trials were carried out on 29 April 1955."

This first instrument was designed to operate in the upper 1000 m and had a range of 0-30°C with an accuracy of ±0.15°C, and a salinity range of about 13 ppt (parts per thousand) with an accuracy of ±0.05 ppt. A conductivity cell with platinum electrodes was used. One of the novel features of the Hamon and Brown design was the use of a single-cored armored cable as the only connection between the underwater unit and the ship. The cable carried power from the ship down to the underwater unit, brought up the measuring signals, and also supported the full weight of the underwater unit. The feature greatly simplified the handling of the equipment at sea, and has been used on all of the STD and CTD systems since then. In essence, the measuring elements were all variable resistors with values depending on conductivity, temperature, or depth. These elements were connected in turn to an oscillator whose frequency was a function of resistance. In this way the measured variables were converted into audio frequencies that were fed to the central core of the cable, the outer steel sheath acting as the return conductor.

While the original STD was being developed, Hamon had the idea of making a portable bridge instrument for use in estuaries, at least partly to gain experience with the application of conductivity cells to marine work. The instrument described by Hamon (1956) was the result, and with minor modifications it has been in production up to the present. Hamon notes that work on the STD and to a lesser extent the portable bridge led to the first measurements of the effect of pressure on electrical conductivity (Hamon, 1958) [see Bradshaw and Schleicher (1965) for later work on this subject]. Hamon's instrument used a small glass-enclosed thermistor immersed in the sample and yielded an accuracy equivalent to the Knudsen titration.

About this time, Brown used the idea of the inductive coupling principle (which avoids the use of metal electrodes) to build a portable salinometer. The advantage is that the windings themselves can be insulated from the corrosive effects of sea water. Brown was able to avoid the use of a thermostat by using a thermistor to give temperature compensation. His design resulted in a portable instrument with an accuracy of approximately 0.003 ppt. The concept of using an induction technique was already known (e.g., Gupta and Hills, 1956) but Brown's contribution was to produce a workable unit. The instrument is described by Brown and Hamon (1961) (figure 14.18). Cox (1963) reviews the field of shipboard salinometers and their development.

It was a logical step to add the inductive sensor to the continuously profiling system, and Brown did just that. The new instrument, called the STD, was designed and sold in the early 1960s by the Bissett-Berman corporation and quickly became an important part of the oceanographer's set of tools (Brown, 1968) (figure 14.19). As Brown (1974) says, "The routine use of continuous profiling instruments such as the STD introduced a decade ago showed very clearly that the ocean structure was very much more complex than the classical Nansen bottle data would suggest."

When the STD was designed, computers, their peripherals, and software were too expensive and unreliable for routine use at sea. As a consequence, the STD required the use of analog computation of salinity from

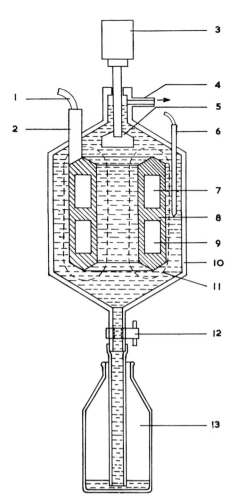

Figure 14.18 Simplified diagram of the measuring head for the laboratory inductive salinometer: 1, leads from toroid assembly; 2, support stem for toroid assembly; 3, stirring motor; 4, connection to aspirator; 5, stirrer; 6, thermistor; 7, toroidal core of voltage transformer; 8, toroid assembly; 9, toroidal core of current transformer; 10, clear plastic housing; 11, path of electrical current in the water sample; 12, stopcock; 13, sample container. (Brown and Hamon, 1961.)

the *in situ* temperature measurements. One of the problems that continually arose was the mismatch in time response of the temperature and conductivity sensors. Since conductivity is a strong function of temperature, salinity is a small residual left over from the temperature correction. If the correction was applied too rapidly, as it inevitably tended to be in the STD, spikes appeared in the salinity trace. Moreover, the conductivity sensor did not give as high a resolution as desired and its inherent sensitivity was not as good as the electrode design. Finally, the oscillatory stability was not high enough to yield good accuracy at high data rates.

Brown's later designs were able to overcome most of these problems. With the rapid improvement and cost reduction in computer systems, the development of a

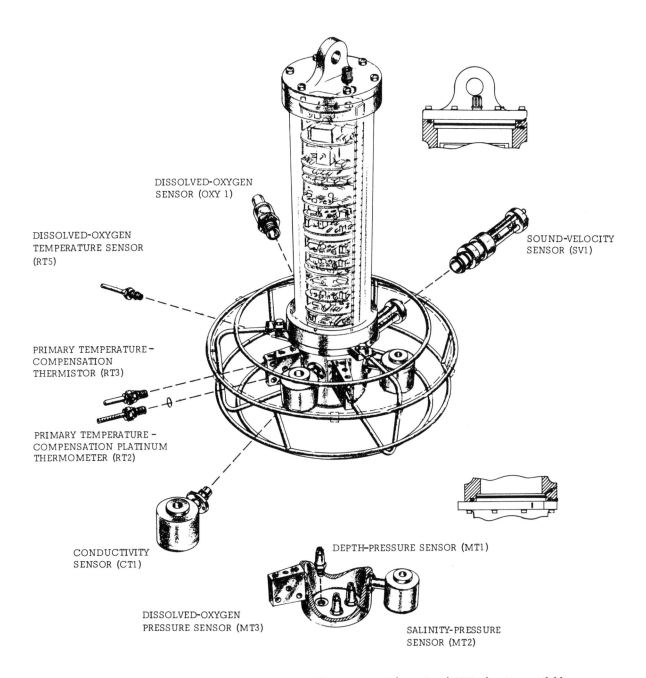

Figure 14.19 Schematic of STD showing available sensors. (Courtesy of Grundy Environmental Systems, Inc.)

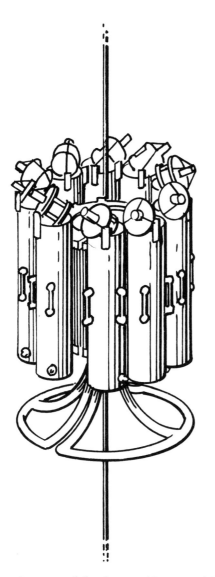

Figure 14.20 Rosette multibottle array. (Courtesy of General Oceanics, Inc.)

relatively drift-free electrode system, and the use of 10-kHz sinusoidal sensor excitation with an ac digitizer, he produced the new CTD. The slightly different name is used to distinguish the new system, which makes precise, fine-scale measurements at a high data rate, from the earlier STD. The details of the new system are given by Brown (1974). Thus we have seen a cycle, from the electrodes to the inductive cell and back to the electrodes. The inductive system is still being used for those applications where the highest resolution is not required. We should note, however, that the system does not replace the use of bottles. Oceanographers have found that water samples are necessary in order to maintain the salinity calibration, and most chemical measurements require samples. Salinity calibration is usually done with a rosette sampler; a small model is illustrated in figure 14.20.

The latest version of the CTD is manufactured by Neil Brown Instrument Systems and a schematic is shown in figure 14.21. The conductivity sensor is a miniature four-electrode cell with platinum electrodes, and the temperature sensor is a combination of a miniature fast-response thermistor and a platinum-resistance thermometer. The four-electrode cell eliminates errors due to polarization at the electrode-sea water interface, and the temperature outputs are processed to achieve both the accuracy of the platinum and the speed of the thermistor. The pressure sensor is a strain gauge, compensated to minimize temperature effects.

The current system has a precision of better than 0.001°C over a range of -3 to $+32$°C, and a conductivity precision of the order of one part per million. The instrument will also accept an oxygen sensor. In summary, the use of the new instruments has shown the effectiveness of the equipment for fine-scale work and high-accuracy deep-ocean survey studies.

The success of the conductivity measurement for salinity has produced a new practical salinity scale (Lewis and Fofonoff, 1979). The new salinity scale, defined by conductivity ratio, has been found to be a better route to density than a "chlorinity" scale since conductivity will respond to changes in any ion, whereas chlorinity is ion specific. Lewis and Fofonoff pointed out that it has been demonstrated conclusively that in the hands of average observers conductivity-ratio measurements allow density to be predicted with a precision nearly one order of magnitude greater than that allowed by chlorinity measurements.

The instruments above provide a vertical resolution of temperature and conductivity to about 1 m, defined as the *fine-structure* regime. To go to the *microstructure* regime, defined as approximately 1 m to 1 cm, new instruments are required. A number of groups have developed free-fall instruments for studying both fine and microstructure of temperature and salinity.

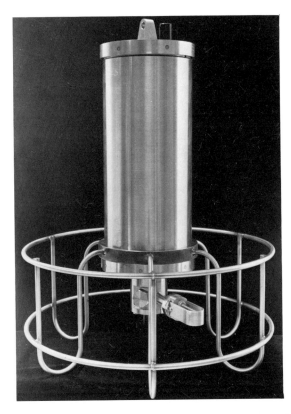 

Figure 14.21 CTD (Mark IIIb) underwater unit (left) and sensor head (right). The conductivity cell is to the right of the temperature sensor. (Courtesy of Neil Brown Instrument Systems, Inc.)

The early development in the field was due mainly to C. S. Cox and his collaborators at Scripps (Osborn and Cox, 1972; Gregg and Cox, 1971). The extended wings of the microstructure recorder (MSR) developed by this group permit rapid descent (greater than 1 m s^{-1}) to the preset depth, then slow descent (10 cm s^{-1} or less) through the area of interest; the motion is similar to that of a falling maple seed or helicopter in autorotation (figure 14.22). Due to the slow fall rate and accurate thermistors (Gregg, Meagher, Pederson, and Aagaard, 1978), this instrument has a noise level of 2–3 microdegrees and takes a data point every 1.5 mm. It has proved capable of resolving most structures in the upper 4 km, and can obtain direct measurements of internal-wave velocities (Desaubies and Gregg, 1978). Caldwell, Wilcox, and Matsler (1975) have developed a relatively simple freely falling probe for study of microstructure. Other groups have developed instruments that include velocity profiling as well; these will be discussed in the next section of the chapter.

Cairns and Williams (1976) (see also Williams, 1976) have used a freely drifting, mid-water float equipped with a buoyancy controller to make repeated profiles of thermal microstructure of a 20-m segment of the water column. These measurements augment the free-fall instrument data.

An optical imager for studying small-scale features in the ocean has been developed by A. J. Williams (1975). The device, a self-contained imaging microprofiler (SCIMP), uses a shadowgraph technique to record optical inhomogeneities in an *in situ* sample of sea water. Features of the order of millimeters can be observed—in particular, the detection of fields of "salt fingers" becomes possible. The salt-finger problem has an interesting history [Williams, (1974b) and chapter 8]. In 1956, Stommel, Arons, and Blanchard published a paper on the "perpetual salt fountain," which described the convective process since called salt fingering. Stern (1960a) pointed out that the different diffusivities of salt and heat meant that the process could be a significant mechanism for the vertical transport of salt and heat in the ocean. Laboratory experiments showed fingers or convective cells between interfaces of warm, salty water and cold, fresher water. Later, ocean measurements with the STD in the late 1960s demonstrated the existence of layers and sheets: thin interfaces which separated mixed layers. The question was whether this layering structure in the ocean was associated with salt fingers.

In late 1972, Williams tested the first version of his instrument, designed after the suggestions of Stern (1970), which combined an optical imager (a shadowgraph system that records density inhomogeneities) with a conductivity-temperature-depth microprofiler. A sketch of the instrument is shown in figure

14.23. The instrument is mounted on an autonomous vehicle, in acoustic communication with the ship, which sinks slowly through microstructure features in the ocean. The shadowgraphs are produced by a 5-cm-diameter laser beam that is reflected through a horizontal path, 160 cm long, and recorded on film. In 1973, Williams obtained photographs of shadowgraph images of fields of salt fingers in the Mediterranean outflow. The images matched with those he had made using the same instrument in laboratory-produced fields of salt fingers, and the agreement with theoretical calculations on the size of the fingers was good (Williams, 1975). Since the observed fingers occurred at an interface between mixed layers, as expected, Williams was able to conclude that "thus 17 years has brought salt fingers from an oceanographical curiosity to an observed ocean process." The latest of the SCIMP instruments utilizes a Cassegrain telescope to expand the viewing aperture yet decrease the size of the pressure housing, and a simpler free vehicle is used. In addition, a shear meter provides velocity fine-structure data for Richardson-number calculations.

14.3.4 Velocity Profilers

Free-Fall Profilers The profile of velocity is an important part of our picture of the ocean. Since moored current meters can give us only a time series at a few points, and the moorings are not yet adequate for strong currents, features such as narrow jets may not be seen. Thus there has been interest for a long time in techniques for the measurement of currents as a function of depth or horizontal position.

We can divide these instruments into three classes. The simplest, in principle, is the analog of the meteorological balloon—the sinking float. The float is tracked acoustically as it sinks, and the horizontal path is differentiated with respect to time to yield velocity as a function of depth. A subset of this class is the transport float, whose position before and after a trip to the bottom is shown by dye patches at the surface. The second class is the free-fall device that has a current sensor on it, including the electromagnetic and the airfoil lift probes. Finally we have the class of instruments consisting of a current meter that goes up and down a line attached to a ship, mooring, or drifting buoy.

The idea of tracking a sinking float for velocity profiles came out of the World War II acoustic developments—recall that Swallow mentions that Crease and Tucker were working on it in 1954. In 1969 Rossby reported measurements that he had made in 1968 with a slowly sinking pinger tracked from a set of hydrophones at the ocean bottom. The measurement was

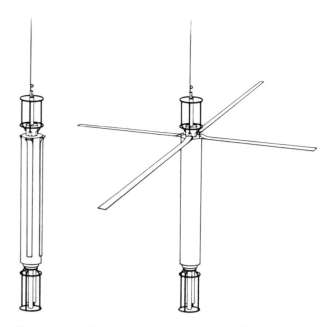

Figure 14.22 The microstructure recorder configuration: (left) during rapid descent; (right) during the data cycle, when the wings are extended. The rotation induced by the pitch of the wings generates sufficient lift to slow the fall rate to 0.08 m s^{-1}. (Desaubies and Gregg, 1978.)

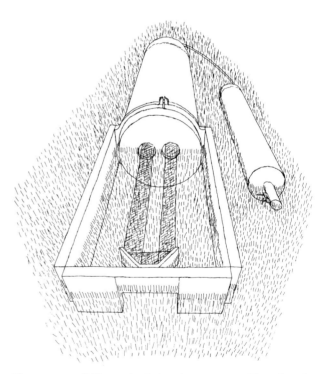

Figure 14.23 Self-contained imaging microprofiler of Williams (1974b). The optical instrument and the CTD simultaneously measure an undisturbed microstructure feature, here a salt-finger interface. Refraction by optical inhomogeneities along the path produces a shadowgraph, which is photographed within the housing.

made near the Plantagenet Bank 25 nautical miles southwest of Bermuda in order to take advantage of an existing set of hydrophones. Substantial velocity variations over small vertical scales (10-m thickness or less) were observed. Rossby noted that the resolution is limited by the timing accuracy, and that the hyperbolic geometry for distance differences to two points can amplify the timing errors. But the Rossby experiment showed the feasibility of the technique.

In 1970, Pochapsky and Malone (1972) carried out a similar experiment in the Gulf Stream southeast of Cape Lookout, North Carolina. They used an instrumented float with temperature and pressure sensors that sank slowly over three disposable bottom-anchored transponders. The float was designed to drop a weight at the bottom and then make a return set of measurements. In this case, as the float sank to the bottom, it not only transmitted its transit times to each of the transponders, but also reported its own temperature and pressure to the ship. The velocities obtained agreed with the results of Swallow and Worthington (1961) mentioned earlier, and the structure seen in the profile is consistent with the later more detailed results of the electromagnetic profilers. The important step here was the move to portable transponders [Rossby later (1974) extended this technique to use moored hydrophones]. One of the attractive features of the Pochapsky-Malone device is its small size. The floats can be launched over the side of a ship by hand, with no winch or crane requirements whatever, making the technique usable from smaller vessels, although ship acoustic noise can be a problem. It was successfully used in MODE-1 (Pochapsky, 1976).

It was a straightforward step to add temperature and salinity profiling to the sinking floats. One of the successful developments is the White Horse, developed at the Woods Hole Oceanographic Institution by W. J. Schmitz, Jr., R. Koehler, and A. Voorhis. The device (figure 14.24) is about 2 m long and has an outer structure of white plastic (hence the name). It is basically an acoustic dropsonde with a CTD microprofiler (Brown, 1974). It utilizes a pulsed acoustic navigation system with bottom-moored transponders (Luyten and Swallow, 1976). The White Horse has a positional accuracy of better than 1 m corresponding to ± 1.5 cm s^{-1} error in horizontal velocity when estimated over 100-m increments in the vertical. Luyten and Swallow (1976) used the instrument to show the existence of an equatorially trapped, jetlike structure in the Indian Ocean. Since geostrophy fails at the equator, such a direct-measuring instrument is essential.

One of the simplest uses of the sinking float is in the transport measurements of Richardson and Schmitz (1965). In its shipborne use, the technique involves

Figure 14.24 The White Horse (velocity-, temperature-, and conductivity-profiling instrument) developed at the Woods Hole Oceanographic Institution. (Courtesy of J. Luyten.)

marking the position of a float as it leaves the surface of the ocean heading for the bottom. At the bottom, the float releases weights and becomes positively buoyant, and rises to the surface. The position at the surface is noted, and the total horizontal distance traveled is then proportional to the vertically averaged velocity. The technique requires accurate navigation (local Hi-Fix was used) and some estimate of surface currents so that the position of arrival at the surface can be properly estimated. A number of studies of transport of the Gulf Stream in the Florida Straits has been carried out with this technique (Niiler and Richardson, 1973). Velocities for different depth intervals can be estimated by allowing the weights to drop off at shallower levels.

The extension of this idea to floats dropped from aircraft (Richardson, White, and Nemeth, 1972) is clever, but has not yet worked reliably. This technique does not require the precision navigation of the shipborne equipment. Here a three-probe system is used. An expendable probe is ejected from the aircraft and on striking the water a surface marker separates and dispenses fluorescein dye. The remainder of the probe carries two buoyant streamlined floats to the bottom. At a preset time (longer than the time required for the probe to reach bottom) the first float is released, and later the second float is released. Each releases dye when it reaches the surface. Photographs of the three dye streaks provide enough information for the average velocity to be determined. The instruments were used in MODE-1 but there were many difficulties in observing the dye patches. The simplicity of such techniques suggests that further development would be profitable.

An ingenious velocity profiler based on electromagnetic techniques was developed by T. Sanford and R. Drever [see Drever and Sanford (1970) and Sanford, Drever, and Dunlap, (1978) for the latest version]. This instrument yields the variations of horizontal velocity by measuring the electrical currents generated by the motion of the sea water through the earth's magnetic field. The instrument is especially interesting because by producing a profile it eliminates part of the ambiguity inherent in electromagnetic measurements at one level only, e.g., those with the geomagnetic electrokinetograph (towed electrodes at the surface—the GEK).

In short, the principle is the following: an instrument that drifts with the local velocity does not see any electromotive force due to its own motion, since there is no relative velocity between the water and the instrument. The drifting instrument sees only the voltage drop due to the electrical currents that have been generated in the water. These have two sources: the local electromotive force, which varies with depth, and the electrical field in the water. Because the current systems are broad compared to their depth, the electrical field in the water is essentially independent of depth. It is proportional to a conductivity-weighted, vertically averaged velocity except where the sea is shallow or the bottom is unusually conductive. The net electrical current at each level is the difference between the currents generated by the electromotive force at each level and the currents generated by this depth-independent electrical field.

The instrument designed by Sanford and Drever consists of a pair of electrodes attached to a cylinder that drifts with the local velocity as it sinks to the bottom. It measures the voltage drops due to the local electrical currents, which as we have just seen are caused by the difference between the local velocity and the average velocity. A velocity profile relative to the unknown conductivity-weighted averaged velocity is thus obtained. The GEK, because it gives only a surface measurement, requires some assumption about the depth-independent velocity to be made before a number for surface velocity can be obtained, and is thus ambiguous. The profile of velocity from the Sanford-Drever instrument is ambiguous in the same absolute sense, but the profile structure is not ambiguous.

The second important design point was to solve the problem of detecting the small signals in the presence of large noise. Given the size of the instrument, the expected voltages are about $1-100$ μV. However, the electrode offset voltage is much larger, about $0.1-5$ mV. Adding rotation fins to the instrument allowed the dc signal to be modulated into an ac signal with frequency of the rate of rotation. Bandpass filters selectively amplify the signal at that frequency, thus bringing it up above the noise that is spread over all frequencies. In addition, because the sensors rotate, the electrical field is sensed along many orientations, thus providing the required components for construction of a relative velocity vector.

Absolute velocity can be determined by tracking the instrument acoustically or by using some other technique of direct velocity measurement. In the present instrument (see figure 14.25), the reference velocity is determined by acoustic Doppler measurements of the absolute velocity of the instrument as it nears the sea floor. Overall, the electromagnetic method yields ve-

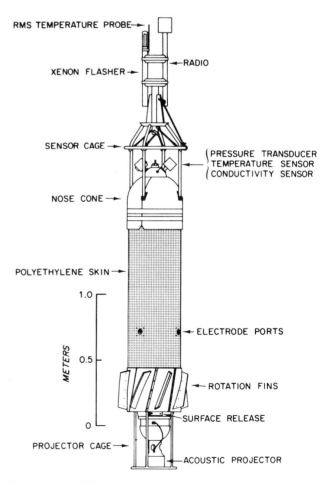

Figure 14.25 Electromagnetic velocity profiler. (Sanford, Drever, and Dunlap, 1978.)

locity determinations every 5-10 m with an uncertainty of about ±1 cm s^{-1}. A round trip in 6000 m of water lasts about 3 h. The instrument has been used in the MODE-1 experiment, to study internal waves, and in the Gulf Stream. An expendable version of this instrument is now in the testing stage.

In 1971 a series of experiments was carried out to compare the electromagnetic profiler with the acoustic tracking method of Rossby (Rossby and Sanford, 1976). Density measurements for geostrophic comparisons were also made. The two free-fall profile methods agree within 1 cm s^{-1} averaged over depth intervals in which the observations were separated in time by less than 10 min. A steady component was averaged from a time series of 4 days; it agreed within ±2 cm s^{-1} with the geostrophic profile computed every 200 m. For comparison, we note that the acoustic method appears to be more suitable for time-series measurements at one location, whereas the electromagnetic technique may be better for surveys of density and relative current profiles over large geographical areas since it is not tied to a mooring. Moreover, the resolution of the electromagnetic technique can be made small enough for studies of microstructure.

In order to measure finer scales of velocity fluctuation, it is necessary to design smaller sensors. Woods (1969) showed from dye measurements that shears of 1 cm s^{-1} in intervals of 10 cm could occur. To address these smaller scales, Simpson (1972) designed a freely falling probe that could measure small velocity changes by use of a neutrally buoyant vane. The instrument was successfully used in Loch Ness and had a resolvable shear of about 2×10^{-3} s^{-1} averaged over 30 cm.

Osborn and Crawford (1978) report on the successful use of an airfoil-type probe. The probe is a pointed body of revolution in which the lift force on the axisymmetric nose is sensed with a piezoelectric sensor. The probe measures the horizontal velocity. The instrument used has the airfoil probe, a thermistor, and a salinometer head mounted at the lower end of the body (figure 14.26). It has been used in several open-ocean experiments. A resolution of about 1 cm in vertical resolution of velocity fluctuations is estimated for the instrument. The probe can be used to estimate the local rate of energy dissipation by use of the velocity-shear data.

Attached Profilers The idea of either raising or lowering a current meter from a ship is straightforward, as is the extension to a current meter moving up and down a fixed line, either attached to a ship or a mooring. We shall not attempt to trace the history of these devices here, but simply report on some examples of current design and use.

Figure 14.26 Velocity profiler with airfoil-type probe, temperature, and conductivity probes. (Osborn and Crawford, 1978.)

One of the successful profiling current meters has been discussed by Düing and Johnson (1972) (figure 14.27). The most recent version of this instrument consists of three major parts. A roller block couples the front of the instrument to a hydrowire, in order to decouple the ship motion from that of the instrument. The second part is the hull, which makes the overall system nearly neutrally buoyant. The hull also serves as a direction vane. The final component is a speed, temperature, and depth recorder made by Aanderaa inserted in the bottom of the hull. The complete instrument is ballasted to be slightly heavier than sea water so that it sinks down the hydrowire. If the wire is kept relatively vertical, the descent rate is about 10–15 cm s^{-1}. This corresponds to a 3–5-m vertical resolution for the resulting profile data. This device has been used successfully in such regions of strong currents as the Gulf Stream, the North Equatorial Current, and the Somali Current.

The Cyclesonde is a good example of an instrument designed for continuous unattended use (van Leer et al., 1974) in the upper layers. It is a logical continuation of the development of profiling current meters described by Düing and Johnson (1972) above. The Cyclesonde consists of a buoyancy-driven platform with a recording package containing sensors for pressure, temperature, current speed, and current direction. It makes repeated automatic round trips up and down a taut-wire, subsurface mooring, while scanning these four parameters. The instrument can work for several weeks, depending on the water depth profiled and the frequency of profiling.

The cyclic vertical motion of the Cyclesonde is controlled by changing the mean density of the instrument package by a few percentage points with an inflatable bladder. The present instrument can be used to a depth of 500 m (figure 14.28). It has been used in experiments in coastal waters and in the deep water in the GATE experiment to study the surface layers.

A second example of a profiling current meter for the upper layers is the type being developed at the Draper Laboratory. This instrument (Dahlen et al., 1977) utilizes a piston-driven changing volume to sustain cyclic vertical motion for long periods, an electromagnetic current sensor to achieve accurate results even in a wave zone, and a microprocessor-centered, programmable electronic system for control of operational functions.

14.3.5 Instruments in Development
We turn next to some examples of instruments that have been designed to address specific interests in physical oceanography, but that have not yet achieved the widespread use of many of the instruments discussed above. I shall briefly outline the principles and

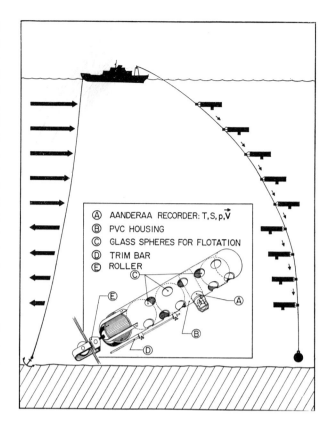

Figure 14.27 Profiling current meter designed to operate from a fixed line. (Düing and Johnson, 1972.)

potential in four areas: bottom instruments, surface instruments, remote measurements, and acoustic-averaging techniques.

Probably the best example of a bottom-dwelling instrument is the pressure gauge, which for accurate measurements must rest on a stable bottom. Pressure gauges can be used to monitor mooring motion, as in the TP recorders, but the noise generated by the mooring (decibars) is usually much larger than the dynamic pressure signals (millibars) of interest. The bottom pressure gauges are a link between the previous section and this one, because in one respect they have been very successful, and in another respect are still developmental.

For tides and higher-frequency phenomena, bottom pressure gauges have had an important impact on the field. Hendershott, in chapter 10, reviews some of the aspects of tidal instrumentation, so we need discuss the subject only briefly. Cartwright (1977) gives a recent review of the general subject (see also Baker, 1969; Matthäus, 1968; Rauschelbach, 1932). Gauges connected by cable to the shore have been used for a long time to measure tides and waves (e.g., Nowroozi, Sutton, and Ault, 1966); the first deep-sea tide measurements from a self-contained arrangement were made by Eyriés (1968). In 1967 Filloux made measurements of the tides off California with a bourdon-tube-type gauge (Filloux, 1970), and Munk, Snodgrass, and Wimbush (1970) report on a series of measurements of tides off California through the transition zone between coastal and deep-sea waters using a newly developed deep-sea capsule that measures pressure, temperature, and current. The capsule and its sensors are discussed by F. E. Snodgrass (1968) and Caldwell, Snodgrass, and Wimbush (1969). The instruments have worked well for such measurements; Irish, Munk, and Snodgrass, (1971) showed the existence of an open-ocean tidal amphidrome in the Pacific, and Irish and Snodgrass (1972) presented the first measurements of the open-ocean tides in the Southern Ocean south of Australia. A review of techniques for such measurements is presented by Wimbush (1972) and an intercomparison experiment is discussed by SCOR Working Group 27 on Tides of the Open Sea (UNESCO, 1975). It is safe to say that the deep-sea tides can be monitored now at any location in the ocean.

If one looks at deep-sea pressure from the point of view of inferring geostrophic currents, then the problems are more difficult. As in the case of meteorology, pressure gradients along the bottom should yield geostrophic currents above the bottom boundary layer. However, the strongest signals in the ocean are near the surface; bottom pressure measurements in the ocean correspond more closely to stratospheric pressure measurements in the atmosphere. There are two other major problems in taking over the atmospheric analogy to the ocean. The first is that the ocean bottom cannot be surveyed well enough to determine the depth of the instruments as accurately as required. Thus measuring absolute pressure gradients is not possible; one can look only at time variability. Second, the signals are small compared to the instrumental noise in the frequency range appropriate to geostrophic currents (periods longer than a few days). A comparison with the atmosphere is instructive: typical signals there for high- and low-pressure systems range from 10 to 100 mb out of a total of 1000 mb, a ratio of 1% to 10% over a few days. In the ocean, the signals are of the order of 1 to 50 mb out of a total of 4×10^5 mb, and occur over time periods of days to months. The instrumental requirements are consequently much more severe, and most of the existing deep-sea pressure instruments have shown long-term drifts too large to allow accurate measurements of these signals.

The absolute gauges (e.g., Filloux, 1970; F. E. Snodgrass, 1968) balance the total bottom pressure against a mechanical property, e.g., diaphragm or heavy bourdon tube; the drift of that element generally shows up in the measured pressure signal. The differential gauges (e.g., Eyriés, 1968; Baker, Wearn, and Hill, 1973) balance the total bottom pressure against a fixed volume

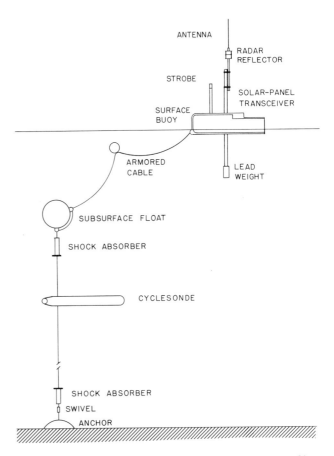

Figure 14.28 The Cyclesonde, an automatic repeating profiling current meter with provision for measurements of pressure and temperature. (Courtesy of J. van Leer.)

of high-pressure gas. Drifts in the gas pressure affect the long-term signals, and the temperature sensitivity is high at high total pressure. Work is continuing on both these fronts. A promising device that has been used successfully for long-term measurements at depths up to 500 m is a combination of the Paroscientific quartz sensor with one of the new data loggers (Wimbush, 1977; Hayes, 1979b; Baker, 1979). The sensor is a quartz-crystal oscillating beam whose resonant frequency varies with applied loads (figure 14.29). Year-long time series from depths of 500 m in the Drake Passage with drifts less than a few centimeters per year have been obtained with this instrument (Wearn and Baker, 1980). A deep-sea version is currently under test.

Other bottom-mounted measurements include the electromagnetic devices for monitoring current. Measurements of horizontal and vertical electrical field can yield estimates of averaged velocity (horizontal or vertical, depending on electrode configuration) from Faraday's law and the electrical and magnetic properties of the earth below. The measurements appear promising, but the ambiguities present in the removal of the electrical-conductivity effects of the bottom sediments

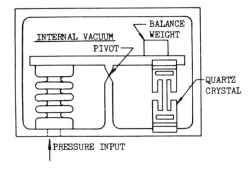

Absolute-pressure transducer

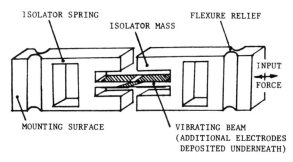

Quartz crystal resonator

Figure 14.29 Schematic of absolute-pressure transducer and quartz-crystal resonator used in the Digiquartz pressure transducer of Paroscientific, Inc. (Paros, 1976.)

and in the effects due to magnetic storms remain a problem (see Filloux, 1973b, 1974).

At the surface of the ocean, measurements are also difficult, for other reasons, primarily because one is in the wave zone. For example, wind waves and swell have periods of a few seconds. A 6-s-period wave will subject a surface platform to almost one million flexure cycles every month. It is not surprising that reliability and longevity are prime problems. The areas of interest for surface measurements are large, including surface waves, currents, salinity and temperature, humidity and carbon dioxide, wind, pressure, air temperature, radiation, and precipitation. We do not have space here to cover all the instrumental problems; the NATO Science Committee (1978) volume covers these items in some depth. It is of interest to single out the subject of satellite-tracked drifting buoys for special attention, however, because of the potential of this technique for large-scale measurements.

It is clear now that maintaining enough moorings to monitor large-scale ocean circulation is too expensive, for both equipment and logistics. One must look to satellite-based systems, together with a modest number of moorings and hydrographic observations. The surface drifters will certainly play a major role in any new global observation system, providing surface measurements for calibration of satellite data, giving a direct measurement of surface currents for use with hydrographic data, and providing an interpolation between moorings.

There are problems with drifters, however, which were most recently addressed at a recent Woods Hole Drifting Buoy Conference (Vachon, 1978). In addition to the problems of reliability, which are slowly being solved, there is the difficulty of interpretation. This has two aspects, the first of which is slippage. The effect of wind on the buoys can be reduced by keeping the amount of buoy volume above the water as small as possible, but true windage effects are not yet known (see Kirwan, McNally, Pazan, and Wert, 1979). Typical results show that the slippage can be as high as 1% of the wind speed (Nath, 1977) when wave effects are included. Measurements are needed to show how well buoys with and without drogues drift relative to the water under different conditions. The second problem is the quantitative interpretation of Lagrangian measurements. In strong currents the interpretation is easier than in the mid-ocean, where mean flow may be weaker than the time-dependent flow. For statistical interpretations, large samples are required. Molinari and Kirwan (1975) show how to construct differential kinematic properties from Lagrangian observations, and show the need for larger numbers of buoys to make significant statements. In spite of these difficulties, the potential of the techniques is large, the entire field of drifter technology and interpretation is active, and one may expect major advances in the next few years. Examples of different kinds of drifters are shown in figures 14.30 and 14.31.

Remote sensing of the ocean surface by aircraft and by satellite is gaining increasing importance. This is also an active field; government agencies especially are interested in a number of its practical aspects. I shall discuss here two areas that illustrate the activity in the field; the reader is referred to Apel (1976) for a more comprehensive review [see also McClain (1977) and Twitchell (1979)]. The article by Born, Dunne, and Lame (1979), and those following in the same issue of *Science*, give an overview of preliminary Seasat results.

R. Legeckis of NOAA/NESS has shown (1977a, 1977b, 1979) how high-resolution thermal-infrared data from the polar-orbiting satellites can be used to track wave features in currents. Variations in the polar front in the Drake Passage (1977a), long equatorial waves in the Pacific Ocean (1977b), and the time variations in the flow of the Gulf Stream between Florida and Cape Hatteras (1979) are evident in the data. Each of these observations has some independent evidence, but perhaps the best corroborated are the Pacific measurements. In that case, the satellite pictures were confirmed by both drifting buoys and sea-level

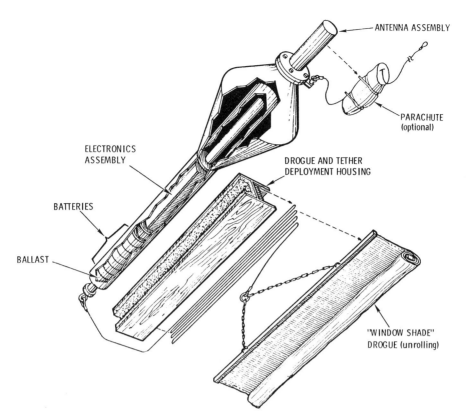

Figure 14.30 Drifting buoy system developed by the Polar Research Laboratory, Inc. The buoy hull is made of aluminum; lifetime is approximately 6 months. The electronics package transmits a signal to the NIMBUS 6 satellite random access measurement system (RAMS). Positions are calculated to an accuracy of ±5 km. (Courtesy of J. O. Anderson.)

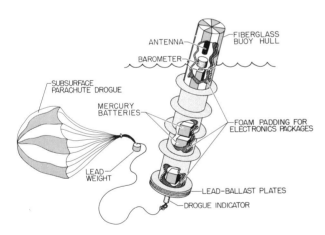

Figure 14.31 Drifting buoy system developed for NORPAX at the Scripps Institution of Oceanography. This system also uses the RAMS information for tracking; see Kirwan, McNally, Pazan, and Wert, 1979. (Courtesy of G. McNally.)

measurements (Wyrtki, 1978). Wyrtki used the drifters to establish a mean flow from which he could calculate a Doppler shift for the satellite and sea-level data. A wave period of about 34 days was inferred. Another good example of evidence from several sources was presented by Bernstein, Breaker, and Whritner (1977) in a study of the California current.

Over most of the ocean, the present satellite-based sea-surface temperature measurements show large errors (Barnett, Patzert, Webb, and Bean, 1979) because of the influence of water vapor and clouds in the atmosphere above the ocean. It is expected that the use of multichannel radiometers will improve these data by allowing a more accurate determination of the water in the atmosphere (Lipes et al., 1979).

A second satellite technique is the measurement of the surface topography. Because the distance from the satellite to the sea surface can be measured with great accuracy by radar altimeters (to better than 10 cm), one needs only to combine such a measurement with an accurate knowledge of the geoid in order to get the surface pressure field. Accuracy in three elements is required: the satellite orbit, the altimeter measurement, and the geoid itself. Aspects of these problems are discussed in the National Academy report on "Requirements for a Dedicated Gravitational Satellite"

(Committee on Geodesy, 1979). The uncertainties in the three areas are rapidly decreasing, and thus this technique may be one of the most promising for a synoptic view of surface circulation (Tapley et al., 1979; Wunsch and Gaposchkin, 1980; see figure 11.16).

Another important idea is the technique of remote sensing of subsurface temperature of Leonard, Caputo, Johnson, and Hoge (1977), who use a laser Raman back-scattering technique. The Raman scattering is a function of the amounts of monomer and dimer forms of liquid water; the relative concentration of these is a function of temperature. A pulsed laser is used so that the round-trip time for a pulse determines the depth that it reached. Data to 10 m below the surface have been collected with an accuracy of $\pm 2°C$. It is expected that the system will work to at least a depth of 60 m with an accuracy of $\pm 1°C$. With such accuracy and depth, the system may be very useful for airborne surveys of the upper layer of the ocean.

Finally we consider examples of acoustic techniques. The inverted echo sounder, designed to study the temperature structure of the water column from an instrument on the sea floor, combines an acoustic transmitter and receiver to monitor the time it takes for a pulse of sound to travel from the sea floor to the surface back to the sea floor. Fluctuations in this time interval are due primarily to changes in the average temperature of the water over the gauge; the movement of the thermocline as a function of time can be inferred. The instrument works remarkably well; variations in travel time can be measured with an accuracy equivalent to a displacement of the 10°C isotherm of ± 4 m (Watts and Rossby, 1977). For the MODE-1 experiment, a comparison with the CTD stations showed that the inverted echo sounder was indeed capable of producing records representative of the motions of the main thermocline. It appears that the instrument could replace the shipborne use of CTDs for special types of long-term monitoring (Gould, 1976).

A Doppler sonar system for measurement of upper-ocean current velocity has been developed by Pinkel (1979). The sonar transmits a narrow beam that scatters off drifting plankton and other organisms in the upper ocean. From the Doppler shift of the back-scattered sound, the component of water velocity parallel to the beam can be determined to a range of 1400 m from the transmitter with a precision of 1 cm s^{-1}. The instrument has been used successfuly from FLIP.

An acoustic technique of great promise has been proposed by Munk and Wunsch (1979). Called "ocean acoustic tomography," after the medical procedure of producing a two-dimensional display of interior structure from exterior X rays, the technique monitors acoustic travel time with a number of moorings. Because the number of pieces of information is the product of the number of sources, receivers, and resolvable multipath arrivals, the economics of the system is enhanced over the usual spot measurements. The necessary precision does not appear to be a difficulty, and the main limitation at high acoustic frequencies is imposed by the effects of variable ocean fine structure (limiting horizontal scales to 1000 km). Using the geophysical inverse techniques discussed at the beginning of this chapter, Munk and Wunsch are able to show that it should be possible to invert the system for interior changes in sound speed and, by inference, changes in geostrophic velocity associated with density variations. They conclude that such a system is achievable now and that it has potential for cost-effective large-scale monitoring of the ocean. Initial tests of the technique have begun.

It is conceivable that a global monitoring system for the ocean circulation could consist of a combination of several of the elements discussed above: (1) satellite observations of the surface temperature and surface pressure fields; (2) direct measurements of the surface current and temperature by drifters and a modest number of moorings; and (3) deep-ocean monitoring by a combination of acoustic tomography, shipborne hydrography, and moorings. All of the elements are present now or are being tested; we could have such a system in the next ten years (SCOR, 1977).

14.3.6 New Techniques Required

Some of the major problems have already been discussed above and need not be reviewed here. In summary, we need techniques for measurement of currents in regions of strong currents: drifters yield surface currents, but moorings are not strong enough to withstand the forces. We need techniques for measuring ocean currents and mixing in the surface layer under storms; meteorologists have these techniques, being able to fly instrumented aircraft through stormy weather. We still do not have an unattended platform from which we can make observations during severe storms, yet the transfer of heat, mass, and momentum is probably largest then. Better current meters for the wave zone are needed; perhaps the acoustic or electromagnetic techniques now being developed will solve this problem.

We need techniques for long-term monitoring of the profiles of currents; these are of interest both for ocean dynamics and for the engineering of deep-sea drilling rigs that will sit on station for several months with the drill pipe subject to the variable currents. Synoptic measurement of deep currents over large areas is not yet within reach. As has been suggested, a deep neutrally buoyant float that periodically pops up or reports in some acoustic mode to the surface and to a satellite would yield a kind of global coverage, as would acoustic tomography.

In the end, however, we can expect that a mix of instruments and techniques will yield the most knowledge about the ocean, and that there will be no simple solution. Stommel's comments (1965)—referring to current fluctuations—are apt:

> It takes years of expensive observations to produce even a very crude description of (fluctuations in ocean currents). There is ... a persistent though erroneous notion that all worthwhile problems will eventually be solved by some simple, ingenious idea or clever gadget. A well-planned long-term survey designed to reveal fluctuations in ocean currents would be expensive and time-consuming. It might even fail, because of inadequacies of the tools we have at hand. But until this burdensome and not immediately rewarding task is undertaken, our information about the fluctuations of ocean currents will always be fragmentary.

14.4 Ocean Experiment Design

The controlled experiment for test of hypotheses is the cutting tool of the physical scientist. It is easy to document advances in laboratory physics through the route of discovery, models, controlled laboratory experiments, models, revised experiments, and so on. The earth scientist does not have this capability, at least as far as controlling the system. Lorenz (1967, p. 25) pointed out that "the meteorologist who wishes to observe the circulation of the atmosphere cannot follow the customary procedures of the laboratory scientist. He cannot design his own experiment in such a manner as to isolate the effects of specific influences, and thereby perhaps disprove certain hypotheses and lend support to others. He can only accept the circulation as it exists. Moreover, since the circulation is global in extent, he cannot even make his measurements singlehandedly, but must rely for the most part upon those which have been made by other persons, in most instances for different purposes." The general difficulty has been pinpointed by V. Suomi (1976): "It is possible for the secrets of nature to be hidden in a flood of data as well as in nature. Clearly, we need information more than we need data." These points are equally valid for oceanography. There is a basic difference, however, in the availability of data from the ocean, because oceanographers do not have a global synoptic weather network, accumulating data routinely, upon which they can draw (see US POLYMODE Organizing Committee, 1976, p. i). Selected representative data and properly designed ocean experiments are required.

14.4.1 Varieties of Oceanographic Experience

Thus we are led to the question of the design of oceanographic experiments for testing hypotheses. Stommel (1963) addressed the problem in the context of the historical evolution of the design of oceanographic expeditions in his paper on "Varieties of Oceanographic Experience." He noted that in the early days uniformly spaced, broad surveys were required: little was known, and large areas had to be covered. Planning followed the *Challenger* model: multidisciplinary, multiregion, multiyear studies combined in a single expedition. These geographical surveys yielded atlases of the distribution of properties in the sea from which dynamic processes and climatological properties could be inferred. Stommel's point in 1963 was that enough was known at that time about the inherent high-frequency, small-scale energy in the system (diurnal tides contaminating seasonal sea level, etc.) that we could not improve the statistical significance of the data much more with this kind of broad geographical measurement design.

In order to improve significantly the results from the geographical studies, specific experiments designed to obtain statistically significant results are required. Observational programs in the ocean, Stommel argued, must be designed on the basis of what is known about the dynamics and spectral distribution in time and space of the phenomenon of interest. Thus studies of western boundary currents, mid-ocean eddies, or the formation of bottom water will have different experimental designs. He noted that "one may well say that where so much is unknown, such detailed planning is impracticable—unexpected complications may arise. I reply that where so much is unknown we dare not proceed blindly—the risk of obtaining insignificant results is too great."

Three examples from the paper are particularly instructive. The first is from a study of the equatorial undercurrent in the Indian Ocean in 1962–1963. The design of the study was based on knowledge of the Pacific Equatorial Undercurrent, which appeared from measurements of Knauss (1960) to be relatively steady with a long east–west scale, almost the entire Pacific basin, and a short north–south scale, extending about 2° across the equator. The Indian Ocean undercurrent was expected to have similar space scales with a time variation that depended on the monsoons. Since the monsoons reverse every 6 months, the current was expected to have its major time variability on that scale. The schematic diagram of velocity spectra presented in figure 14.32 shows three important points: the two peaks associated with the Pacific undercurrent; the two annual peaks that the expedition expected to find and that it planned to map; and the probable actual peak for velocity that was revealed but could not be mapped by the procedures employed in the expedition for mapping the two annual peaks. Thus the original design of the expedition could not show the annual component because of contamination, and could not map the shorter-period irregularities. Later studies in the Indian Ocean, based on these studies, have begun to describe the complex features of the monsoon-driven circulation (see, e.g., Luyten, 1980).

A second example involves the Gulf Stream and its variability. Stommel noted that during the 1930s, C.

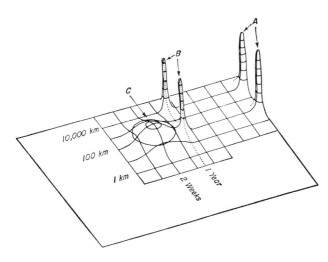

Figure 14.32 Schematic diagram of velocity spectra as function of time and space for equatorial undercurrents. A, Pacific undercurrent; B, peaks expected by *Argo* expedition; C, probable actual peak for velocity that was revealed by *Argo* expedition. (Stommel, 1963.)

Iselin of WHOI had encountered certain problems in trying to establish the Gulf Stream transport. By making hydrographic sections across the stream at 3-month intervals, Iselin had hoped to determine the annual variation of transport. What he actually obtained was a random sampling of irregular meanders. The Fuglister measurements of the 1960s show that the annual part of the spectral distribution of velocity is so close to a high-energy meander peak in the spectral diagram that to resolve it would require observations several orders of magnitude greater than Iselin had planned to make.

The use of neutrally buoyant floats, discussed earlier, leads to the third example. In the mid-1950s J. Swallow and J. Crease from the National Institute of Oceanography in England planned to make extended observations of deep-water currents near Bermuda as representative of the open Atlantic Ocean. They expected to see weak eddies, but hoped that by averaging over many of them, perhaps as long as a year, the long-period mean would be revealed. Stommel noted that "the amplitudes of motion were, surprisingly, ten times greater than expected; the original strategy of operation had to be abandoned, and it was not possible to develop as sharp a picture of the spectral distribution of these motions as had been hoped with the minimal ship facilities that were mustered." The results of these measurements were later described by Swallow (1971) at a meeting to discuss the need for an open-ocean experiment to explore eddies. Stommel went on to ask, "Can we design an observational program that can be reasonably expected to yield statistically significant information on this question?" He noted the need for new instrumentation, heavier moorings, longer-range floats, etc., and the need for cooperation among the oceanographic institutions to carry out a large coordinated study.

In fact, instrument development did proceed to a point where a mid-ocean experiment was possible, and several oceanographic institutions did cooperate in carrying out the Mid-Ocean Dynamics Experiment (MODE-1; MODE Group, 1978) and a cooperative effort called POLYMODE, which developed from MODE-1 and the Soviet project POLYGON (IDOE, 1976; cf. chapter 11).

Thinking in terms of ocean experiments instead of ocean surveys for the study of dynamic phenomena has become so pervasive in the last few years that sometimes one forgets the change in thinking that has taken place in the field. Yet this change is fundamental because it means that we are beginning to treat oceanography as an experimental, as opposed to a geographical, science. The points noted by Stommel have been confirmed again and again with new measurements that look at smaller scales and in different regions of the ocean. One need only to point to the emergence of eddies as a worldwide phenomenon (Swallow, 1976; Wyrtki, Magaard, and Hager, 1976; and see chapter 11) and the recent discovery of a complex equatorial jet structure (e.g., Luyten and Swallow, 1976; see chapter 6).

14.4.2 MODE-1 and ISOS

How have these ideas been translated into real experiments? It is instructive to look at the design of two recent studies to see how, for example, the sampling problems are handled when the aims of the studies are different. We consider the MODE-1 experiment, which is an attempt to look at eddies in the Sargasso Sea, and the International Southern Ocean Studies (ISOS), which is an attempt to look at the variability of the Antarctic Circumpolar Current in the Drake Passage. To a large extent, both of these studies depended on the existence of new, reliable technology.

Three unpublished documents present in detail the principles of the design of MODE-1. The first, "The Design of MODE-1" (MODE-1, 1972a), by a committee of experimentalists, focuses on observations and intercomparisons. The second, "Dynamics and the Design of MODE-1" (MODE-1, 1972b), was the result of a theoretical workshop to summarize dynamical modeling and observational needs. The two documents are in basic agreement over the design of the experiment. The third, "Dynamics and the Analysis of MODE-1" (MODE-1, 1975), written after the experiment, summarizes the statistical techniques used and initiates an analysis of the experiment in dynamical terms.

Underlying the design of the MODE-1 experiment was the need to test the hypothesis that eddies play an important role in the general circulation. The design of an eddy experiment, given the existing theoretical knowledge and the resources and equipment available, involved several points that are highlighted in the documents.

The first point was that MODE-1, because it was a 4-month study to focus on deep, mesoscale low-frequency variability, was a nonstatistical, kinematic experiment. The time periods expected, and confirmed by experiment, were too long to sample over many periods. The MODE-1 organizers recognized very early this time limitation. For this reason, a number of moorings and floats were maintained in the same area for up to 2 years after the main experiment.

The second point was that an intercomparison of instruments was required. This was one of the major aims of MODE-1, and one of the major achievements (MODE-1, 1974; Gould, 1976; MODE Group, 1978). Examples of the intercomparisons were discussed earlier.

The third point concerned sampling statistics. The report on "Dynamics and the Analysis of MODE-1" gives an excellent summary of the statistical methods used in the analysis and design of the experiment by R. Davis. Davis pointed out that statistical methods have been used in the experiment for two essentially different purposes. One of these, largely an adaptation of the method of objective analysis (Gandin, 1965), is the improvement of signal-to-noise ratios through use of analysis techniques that account for both instrumental noise and sampling noise resulting from unresolved small-scale variability (the problem discussed by Stommel in his 1963 paper). Some results of the application of objective analysis have been discussed by Bretherton, Davis and Fandry (1976). The technique allows an estimate of interpolation error and statistical uncertainty to be made for each data set based on the statistics of the fields involved.

The second purpose of the statistical methods is compression of numerous observations into concise statements that can be useful in describing the data even if statistically unreliable. Such statements are particularly useful in comparing theory and experiment.

Using the objective technique and given the correlations of the velocity field and the estimated errors of instruments and noise, the MODE scientists were able to compute the rms errors in area-averaged velocity measurements over an array. The working hypothesis for the MODE-1 array design was that the transverse velocity correlation function had a zero crossing at 100 km. The scales observed from the spatially averaged temperature field over the actual array ranged from 140 km at 500 m, to 70 km at 1500 m, to 55 km at 4000 m (figure 14.33) (Richman, 1976), thus validating the original array design.

The fourth point concerned array size. Without knowledge of all of the scales of motion, it was difficult to choose an optimal scale for the array. Thus the concepts of "pattern recognition" and "pattern definition" were used. The first of these was chosen on the

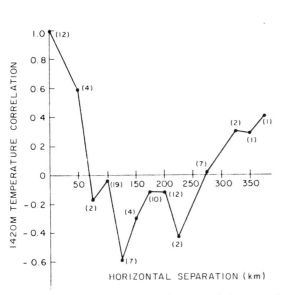

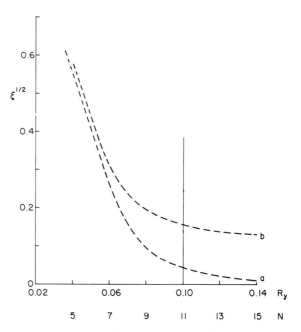

Figure 14.33 (Left) Correlation function of the spatially averaged temperature field over the MODE-1 array at 1420-m depth. Note zero crossing at about 70 m. (Richman, 1976.) (Right) Error in transport ($\xi^{1/2}$) as a function of cross-passage correlation function R_y in Drake Passage (scaled by width of passage, 620 km). Since the cross-passage correlation function has a zero near 60 km in the Drake Passage, the actual R_y is near 0.1 (solid vertical line). Error is also plotted as a function of number of independent horizontal measurements or moorings N: case a, zero random noise; case b, random noise–signal ratio of 0.1 Thus we expect that 11 moorings across the passage will yield an error of about 20% in the transport. (Fandry and Pillsbury, 1979.)

basis that it is more important to be able to define the spatial extent of the motion observed than to describe in detail only a fraction of the pattern. The planning groups noted that both of the earlier experiments looking for eddy scales had been too small in horizontal extent to allow confident determination of the eddy diameters. The final array was designed to satisfy the needs of the intercomparison, with an inner array in a central region of radius about 100 km. The outer array was about 200 km in radius, a pattern recognition area. Figure 14.34 shows the moored-current-meter and density-survey pattern; only intermediate-depth moorings were used.

The design proved to be adequate for the description of eddy dynamics over the relatively short period of the experiment (MODE Group, 1978). Suffice it to say that agreement between different types of instruments measuring the same thing was generally good, e.g., between the different velocity profilers, between the temperature-pressure recorders and the CTD sections, and between the floats and current meters. The paper points out that the field of mesoscale variability is a good deal more complicated than might have been expected.

The concepts involved in using correlation functions of temperature and velocity to design the array were carried over directly into the ISOS experimental design. Here the interest is not in eddies per se, but in measuring the variability of the Antarctic Circumpolar Current (ACC) transport and its variability. But the eddies exist, and must be accounted for in array design. One of the special problems in the ACC is that the coherence lengths are small, about 60 km. Limited funds required an alternating series of arrays in the Passage. The first study, in 1975, used an incoherent array to establish energy levels across the Passage. In alternate years linear and cluster arrays were used to estimate the transport and to collect data for correlation scales (Fandry and Pillsbury, 1979).

Calculations similar to the MODE work were carried out based on the transverse correlation functions in order to design the final ISOS array, a combination of linear and cluster experiments (ISOS, 1978). A significant difference was found between the cross-stream and downstream directions, the downstream scales being longer than the cross-stream scales. The final multipurpose array is shown in figure 14.35. The array was designed to monitor the transport of the current with minimum error (figure 14.33), to allow a comparison between current meters and bottom pressure gauges, and to define the features that exist and are advected through the region by the current.

These examples show us that the principles of experiment design are an integral part of observational oceanography. Clearly, the modern experimental oceanographer needs a working knowledge of statistical experiment design as well as a basic understanding of the ocean and ocean instruments. The appropriate point was made by Leonardo da Vinci in his *Hydrodynamica* (see Le Méhauté, 1976): "Remember, when discoursing about water, to induce first experience, then reason."

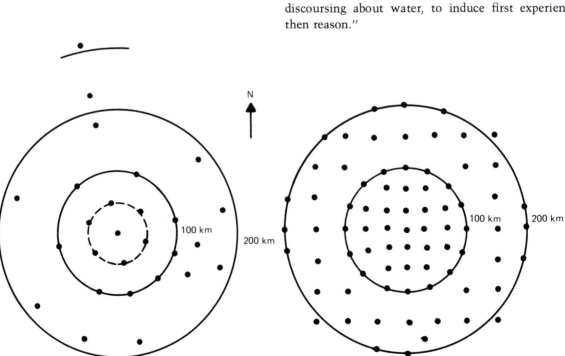

Figure 14.34 Moored current meter (left) and density survey (right) arrays for MODE-1. The inner 100-km circle represents an "accurate mapping" area, the outer circle a "pattern recognition" area. (Gould, 1976.)

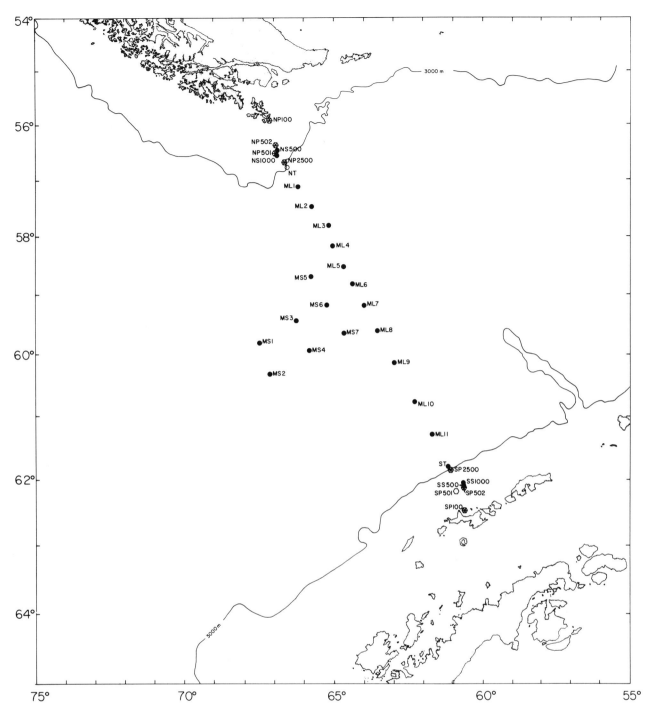

Figure 14.35 Plan view of the ISOS Drake 79 array in the Drake Passage. The moorings marked ML are the main line for transport measurement; the ones marked MS are for mapping and statistics of polar frontal features; NP and SP are the north and south pressure-gauge moorings; NT and ST are the north and south density-measuring moorings for transport monitoring; NS and SS are the north and south slope moorings. Numbers of 100 and greater are the water depth in meters. (Courtesy of W. D. Nowlin Jr. and R. D. Pillsbury.)

15
Geochemical Tracers and Ocean Circulation

W. S. Broecker

15.1 Introduction

Tracers have always been an important adjunct to physical oceanography. The distribution of dissolved oxygen (and to some extent of the nutrients, nitrate, phosphate, and silica) played a very important role in defining the major water masses of the ocean [see Sverdrup, Johnson, and Fleming (1942) for a review of this subject]. Many attempts also have been made to harness the loss of dissolved oxygen from the water column as a measure of the rates of oceanic mixing processes (e.g., Riley, 1951; Wyrtki, 1962). These latter pursuits, however, have been of only marginal success because of our lack of knowledge of the consumption rate of O_2 within the sea.

The big breakthrough in geochemical tracing came after World War II with the discovery of the cosmic-ray-produced isotopes ^{14}C and ^{3}H. A further impetus to this field came with the realization in the mid 1950s that the ocean was receiving significant amounts of ^{90}Sr, ^{137}Cs, ^{3}H, ^{14}C, etc., from nuclear testing. Because the distributions of radioisotopes offered information not so highly dependent on assumptions regarding the rates at which biological processes proceed in the ocean, the emphasis in chemical oceanography moved quickly away from the traditional chemical tracers to the radiotracers. Only quite recently has interest in the chemically used compounds in the sea been renewed. Three reasons can be given for this renaissance:

(1) Radiocarbon is transported in particulate matter as well as in solution; hence the contributions of the two processes must be separated if the distribution of ^{14}C is to be used for water-transport modeling. This separation is based on the distribution of ΣCO_2 (concentration of total dissolved inorganic carbon), alkalinity, and dissolved O_2 in the ocean.

(2) The concentrations of nitrate and phosphate can be combined with that of dissolved oxygen to yield the quasi-conservative properties "PO" and "NO". As reviewed below, such properties are needed in modeling to unscramble the "mixtures" found in the deep sea.

(3) With the advent of (a) sediment trapping and other means for the direct measurement of the fluxes of particulate matter into the deep sea, (b) devices designed to measure the fluxes of materials from the sea floor, and (c) better means for the measurement of plant productivity, interest has been renewed in generating models capable of simultaneously explaining the distribution of the chemical species, the distribution of the radiospecies, and the flux measurements.

In this chapter I shall emphasize the development of radioisotope tracing, as I feel that it constitutes the major contribution of geochemistry to our understanding of ocean circulation over the past four decades (i.e., since the writing of *The Oceans*). I will mention the use of the classical chemical tracers only where they bear on the interpretation of the radioisotope data.

Over the last decade, the Geochemical Ocean Sections Study (GEOSECS) has determined the distribution of the radioisotope tracers on a global scale. Attempts to model the previously existing ^{14}C results (Bolin and Stommel, 1961; Arons and Stommel, 1967) made clear the inadequacy of this data set. Henry Stommel therefore brought together a number of geochemists interested in this problem, and encouraged them to think big, to work together, and to produce a global set of very accurate ^{14}C data.

Because of its massive scope and of the measurement accuracy achieved, the GEOSECS data set has become dominant in the field of marine geochemistry. While previously existing radioisotope data (for review see Burton, 1975) were of great importance in the development of thinking with regard to the interpretation of tracer results and in the separation of the natural and the bomb-test contributions to ^{14}C and ^{3}H, the new data set eclipses what we had in 1969 when this program began. Thus I shall refer frequently to these new results in the sections that follow.

At the time this chapter was written the GEOSECS field program had been completed. Maps showing the ship tracks and station positions are given in figure 15.1. The laboratory analyses for the Atlantic and Pacific phases of the program are complete. Those for the Indian Ocean are still in progress. The mammoth job of making scientific use of this data set has just begun. Many years will pass before the meat of this effort will appear in print.

15.2 Water-Transport Tracers

The efforts in the field of radioisotope tracing can be divided into two categories: those that are aimed at a better understanding of the dynamics of the ventilation of, and mixing within, the ocean's interior, and those that are aimed at a better understanding of the origin, movement, and fate of particulate matter within the sea. While many of the tracers we use are influenced by both processes, a division can be made into a group primarily distributed by water transport and into a group primarily distributed by particulate transport (table 15.1). I will simplify my task by discussing here only the water-transport tracers.

The water-transport tracers can be subdivided according to their mode of origin. ^{90}Sr, ^{137}Cs, ^{85}Kr, and the freons are entirely anthropogenic in origin and hence are "transient tracers." ^{39}Ar, ^{222}Rn, ^{228}Ra, and ^{32}Si are entirely natural in origin and hence are steady-state tracers. ^{14}C and ^{3}H are in part natural and in part anthropogenic in origin. In the case of ^{3}H the man-made component dominates. For ^{14}C the man-made component constitutes about 20% of the total in surface waters and is negligible in deep water.

^{3}He, the daughter product of ^{3}H, is also a tracer. It is produced within the sea by the decay of its parent; it also leaks into the deep sea from the mantle. These components produce an excess over atmospheric solubility within the sea. As will be shown below, the contributions of these sources can usually be separated.

The applicability of any given isotope depends on its half-life (in the case of steady-state tracers), or its temporal-input function (in the case of the transient tracers). It also depends on the geographic distribution of the input function. The differences from tracer to tracer are sufficiently large that the information obtained from the distribution of one isotope is not redundant with that obtained from another tracer. In the sections that follow the geochemistry of each of these tracers is reviewed.

Natural radiocarbon: ^{14}C is produced in the atmosphere by the interaction of ^{14}N atoms and the neutrons produced by cosmic rays. The current production rate is estimated to be about 2.2 atoms cm^{-2} s^{-1} (Lingenfelter and Ramaty, 1970). This production is balanced by the beta decay of radiocarbon atoms. In its 8200-year mean lifetime, the average ^{14}C atom can penetrate the active carbon reservoirs (atmosphere, terrestrial biosphere, ocean, ice, soils). Some ^{14}C atoms become

Table 15.1 Ocean Tracers Currently in Use[a]

Isotope	Half-life (years)	Origin			
		Cosmic rays	U+Th series	Weapons testing	Other anthro.
^{14}C	5730	✓		✓	
^{226}Ra	1600		✓		
^{32}Si	250	✓			
^{39}Ar	270	✓			
^{137}Cs	30.2			✓	
^{90}Sr	28.6			✓	
^{3}H	12.4	✓		✓	
^{3}He	—	✓	✓	✓	
^{85}Kr	10.7			✓	✓
^{228}Ra	5.8		✓		
^{7}Be	0.15	✓			
^{222}Rn	0.01		✓		
Freons	—				✓
^{239}Pu	24,400			✓	✓
^{240}Pu	6540			✓	✓
^{210}Pb	22.3		✓		
^{228}Th	1.9		✓		
^{210}Po	0.38		✓		
^{234}Th	0.07		✓		

a. In the column headed Isotopes, freons and all entries above it are water tracers; all entries below freons are particulate tracers.

(15.1A)

Figure 15.1 Maps showing the locations of the stations occupied during the GEOSECS program in the Atlantic (A), Pacific (B), and Indian (C) Oceans. Those properties measurable on 30-liter samples were determined at every station. Radiocarbon, which requires ~250 liters of water, was sampled only at the large-volume stations. The 4000-m contour is shown on each map.

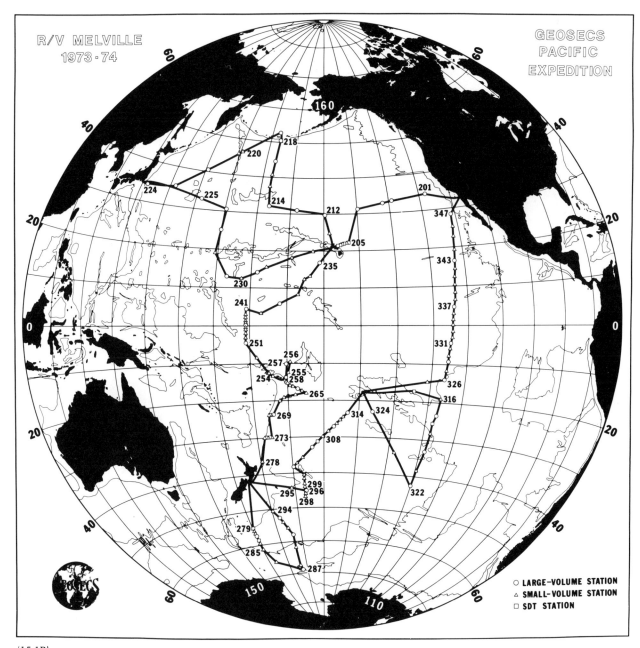

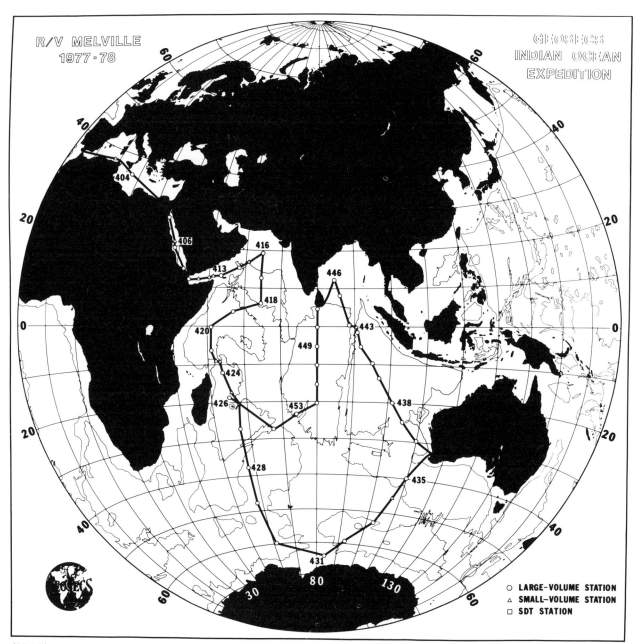

(15.1C)

bound into inactive reservoirs (e.g., marine sediments, peats) before their radiodemise. Table 15.2 summarizes the distribution of radiocarbon among these reservoirs.

Because of variability in the flux of cosmic rays, the production rate of ^{14}C has not remained constant with time. Also it is possible that the rates of the processes that distribute ^{14}C among the various reservoirs have changed with climate. Thus, as shown by measurements on tree rings (figure 15.2), the $^{14}C/C$ ratio in atmospheric CO_2 has not remained constant with time. As pointed out by Stuiver (1976), these variations must be considered when attempting to derive ventilation times from the distribution of ^{14}C within the sea.

During the present century the release of ^{14}C-free CO_2 to the atmosphere through the combustion of fossil fuels has measurably reduced the $^{14}C/C$ ratio. The decline between 1850 and 1950 was about 2.4%. Again this must be considered in models. The ^{14}C produced by nuclear testing became measurable in 1954 (Broecker and Walton, 1959).

Our knowledge of the $^{14}C/C$ distribution within the sea is based on measurements on water samples taken at various regions and water depths and on measurements of $CaCO_3$ formed by marine organisms at known times in the past [historically dated shell collections (Broecker, 1963) and corals dated by ring counting (Druffel and Linick, 1978; Nozaki, Rye, Turekian, and Dodge, 1978)]. As only a limited number of water samples were collected prior to the onset of large-scale nuclear testing, the results of measurements on these samples are extremely important in distinguishing natural from bomb ^{14}C. For deep waters, measurements on post-1954 samples are adequate as significant amounts of bomb-produced ^{14}C have not as yet penetrated to great depth in the ocean (as demonstrated by the absence of measurable tritium in these waters).

As the counters used to measure radiocarbon cannot easily be absolutely calibrated, ^{14}C measurements are

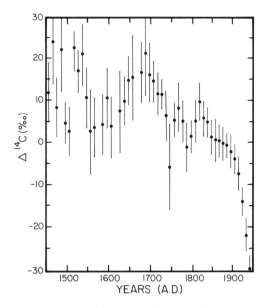

Figure 15.2 $\Delta^{14}C$ for atmospheric CO_2 from 1450 to 1950 as reconstructed from tree ring measurements (Damon, Lerman, and Long, 1978). Prior to 1900 the variations must reflect changes in the production rate of natural ^{14}C. The decline subsequent to 1900 is caused by the release of ^{14}C-free CO_2 as the result of burning fossil fuels.

made by comparison with a standard (National Bureau of Standards oxalic acid). The results are given as per mil difference between the specific activity A of the sample carbon and 0.95 times the specific activity of a standard carbon sample (the 0.95 multiplier is chosen to bring the activity of the standard close to that of age-corrected 1850 wood; Broecker and Olsen, 1959, 1961).

In order to separate the ^{14}C differences produced by isotope fractionation, the $^{14}C/C$ ratios in the samples are normalized to a common $^{13}C/C$ ratio (Broecker and Olsen, 1959, 1961). The formulas used in these calculations are as follows:

$$\delta^{14}C = \frac{A_{samp} - 0.95\ A_{NBS\ stan}}{A_{NBS\ stan}} \times 1000,$$

$$\Delta^{14}C = \delta^{14}C - (2\delta^{13}C + 50)\left(1 + \frac{\delta^{14}C}{1000}\right),$$

$$\delta^{13}C = 1000\ \frac{{}^{13}C/{}^{12}C)_{samp} - {}^{13}C/{}^{12}C_{stan}}{{}^{13}C/{}^{12}C_{stan}}.$$

Table 15.3 gives the $\Delta^{14}C$ values for natural radiocarbon in a variety of water types. The range is about 200 per mil (from $-40‰$ for temperate surface water to $-250‰$ for 2500-m-depth water in the North Pacific). This distribution reflects not only the nite mixing rate of waters within the sea but also the finite rate of carbon-isotope equilibration between atmospheric CO_2 and surface ocean ΣCO_2. The difference between the ΣCO_2 content of average deep water and average surface water bears witness to the removal of carbon

Table 15.2 Distribution of Natural Radiocarbon among Reservoirs

	ΣC (10^{16} moles)	$\dfrac{^{14}C/C}{(^{14}C/C)_{atm}}$	$\Sigma^{14}C$ (10^4 moles)
Atmosphere[a]	5	1.00	6
Ocean	280	0.87	290
Ocean sediments	—	—	~ 10[b]
Terrestrial biosphere	~ 7	0.95	~ 8
Soils	~ 20	0.90	~ 2
Continental sediments	—	—	< 1

a. The absolute $^{14}C/C$ ratio for preindustrial atmospheric CO_2 was about 1.2×10^{-12}, corresponding to a specific activity of 8.0 dpm (gm C)$^{-1}$.
b. Assumes calcite-rich sediments cover 20% of sea floor, that the mean calcite accumulation rate in these areas is 1.5 gm cm^{-2} per 1000 years, and that the initial $^{14}C/C$ for this calcite is equal to that for atmospheric carbon.

Table 15.3 $\Delta^{14}C$ Values for Various Water Types in the Ocean

Water type	$\Delta^{14}C$ (‰)
Surface Waters	
Temperate	−40
Equatorial	−60
Deep Atlantic	
New LSW	−70
New GFZW	−70
New DSW	−70
NADW (mean) N. Western Basin	−100
NADW (mean) Eastern Basin	−115
AABW	−160
Deep Antarctic	
Circumpolar	−160
Deep Pacific	
Samoan Passage Overflow	−190
N. Pac. Bottom	−225
N. Pac. 2500 m	−250

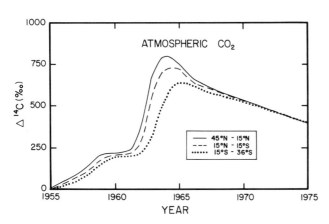

Figure 15.3 Excess atmospheric ^{14}C (resulting from bomb testing) as a function of time based on direct measurements of the $^{14}C/C$ ratio in tropospheric CO_2 samples. These measurements were made mainly by Nydal, Lövseth, and Gulliksen (1979). (See figure 16.1.)

from the photic zone by organisms and its return to dissolved form at depth in the sea. This "short crcuiting" by raining particulate debris must be taken into account in models used to derive ventilation rates from the $^{14}C/C$ distribution. This is done by joint consideration of the stable carbon and radiocarbon distributions (Lal, 1962, 1969; Craig, 1969; Broecker and Li, 1970).

Bomb-produced radiocarbon: Beginning with the first fusion-bomb tests in 1954 significant amounts of anthropogenic ^{14}C appeared in the atmosphere. The $^{14}C/C$ ratio in atmospheric CO_2, which had been falling since the turn of the century, began to rise sharply. This increase continued until 1963 when the U.S.-Soviet ban on atmospheric testing was implemented. Since then the atmospheric $^{14}C/C$ ratio has been steadily falling (see figure 15.3). This decrease reflects the dilution of the bomb-produced ^{14}C through mixing with ocean and terrestrial biosphere carbon.

Through gas exchange with the atmosphere the surface ocean has taken up bomb ^{14}C. Concurrent vertical mixing has carried this ^{14}C to the interior of the ocean. Profiles of total ^{14}C (natural plus bomb) as measured during the GEOSECS program are shown in figure 15.4. From the data we have concerning the prebomb distribution it is possible to estimate the depth distribution of the excess due to bomb testing. An ocean-wide mean penetration depth of about 300 m is obtained as of 1973 (the mid-point of the GEOSECS Atlantic and Pacific field programs).

As of 1973 the distribution of bomb ^{14}C was as follows. About 35% remained in the atmosphere, about 45% had entered the ocean, and about 20% had taken residence in the terrestrial biosphere (table 15.4). Within the oceans the bomb-produced ^{14}C is symmetrically distributed about the equator. The temperate gyres (15–45°) have considerably higher water-column inventories (i.e., bomb-produced ^{14}C atoms cm^{-2}) than does the equatorial belt (Broecker, Peng, and Stuiver, 1978). I shall discuss the oceanographic implications of this equatorial anomaly later in the paper.

Natural tritium: Like radiocarbon, tritium is produced by cosmic rays (von Buttlar and Libby, 1955). The production rate is estimated to be about 0.25 atoms cm^{-2} s^{-1} (Craig and Lal, 1961). It enters the ocean via vapor exchange and rainfall shortly after its birth in the atmosphere. As relatively few measurements of oceanic tritium were made before the commencement of nuclear testing, our knowledge of its distribution is rudimentary (Giletti, Bozan, and Kulp, 1958). Furthermore, because the blank corrections associated with these measurements were quite large and also quite uncertain, the validity of these data remains in question. Consequently, little use can be made of natural tritium in ocean-mixing studies. Fortunately the amount of this tritium is surely too small to be signif-

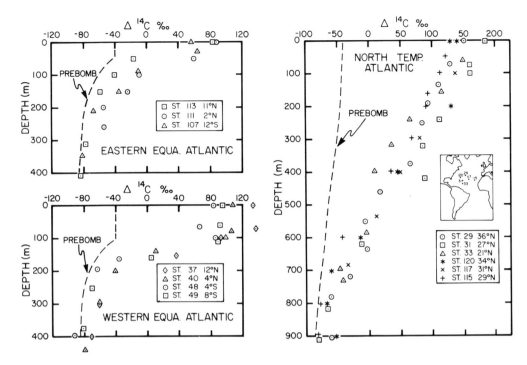

Figure 15.4 Δ^{14}C vs. depth in the north temperate and equatorial Atlantic Ocean as measured by Stuiver at the University of Washington and Östlund at the University of Miami on samples collected in 1972 as part of the GEOSECS program.

icant in modeling the distribution of tritium produced during nuclear testing.

Bomb-produced tritium: Nuclear tests during the 1950s and early 1960s produced large amounts of tritium (Begemann and Libby, 1957). As most of this production was during tests in the *northern* hemisphere, this tritium was added mainly to the northern-hemisphere stratosphere. From here it leaked to the troposphere (on a time scale of a year or so). Once in the troposphere it is removed to the land and ocean surface by vapor exchange and rainfall (on the time scale of weeks). Fortunately there are enough tritium data on rains and tropospheric water vapor collected during the peak fallout years to permit a reasonably sound reconstruction of the temporal and geographic pattern of its input to the ocean.

Furthermore, a picture of the time history for bomb tritium in at least the northern-hemisphere surface ocean can be reconstructed from measurements made between 1965 and the present.

Maps showing the tritium content of surface water are given in figure 15.5. As would be expected, the tritium content of northern-hemisphere surface waters is on the average much greater than that for southern-hemisphere surface waters. The boundary between these high- and low-tritium-content waters, however, lies near 15°N rather than at the equator. As discussed below, this unexpected result has important implications with regard to the dynamics of the equatorial thermocline.

Table 15.4 Distribution of Bomb ^{14}C as of 1974

	Effective ΣC (10^{16} moles)	% Bomb ^{14}C
Atmosphere	5	~35
Surface ocean	~20[a]	~45
Terrestrial biosphere	~3[b]	~20

a. Assumes a characteristic penetration depth of 300 m and a bomb ^{14}C/C ratio for surface ocean water averaging 30% of that in the air (in 1973).
b. Assumes a characteristic replacement time of 15 years and the terrestrial biomass given in table 15.2.

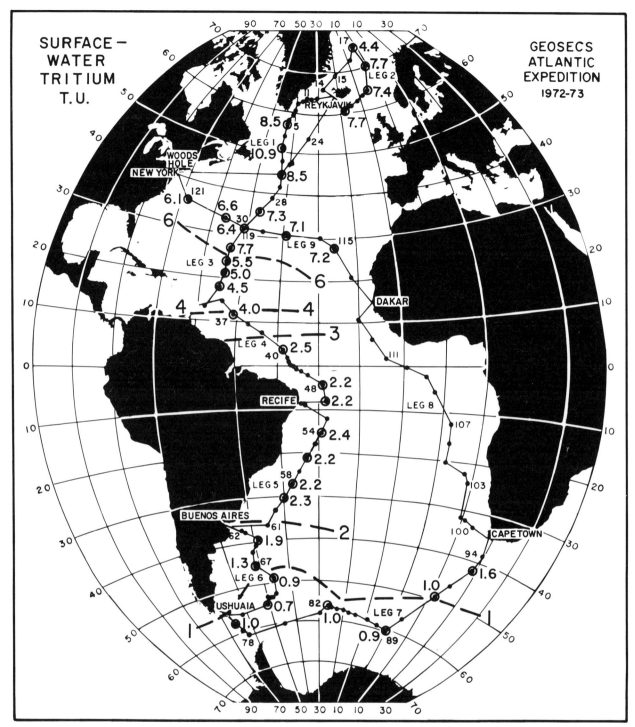

(15.5A)

Figure 15.5 Maps showing the geographical pattern of tritium concentrations in surface ocean water as determined by the GEOSECS program in the Atlantic (A) and Pacific (B) Oceans. The measurements were made by Östlund of the University of Miami. The units are T.U. (i.e., 10^{-18} T atoms per H atom).

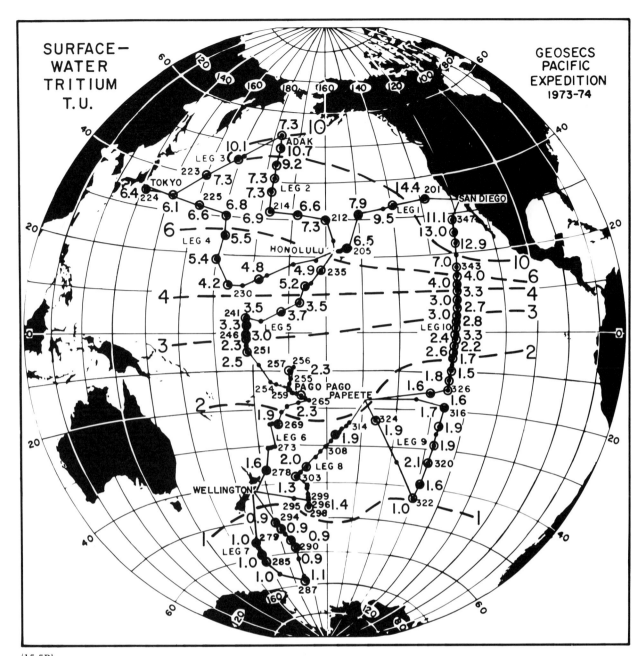

(15.5B)

Primordial ^3He: Clarke, Beg, and Craig (1969) discovered that primordial ^3He stored in the mantle is gradually leaking into the deep sea. Subsequent work has shown that this leakage is mainly from centers of sea-floor spreading in the Pacific Ocean. This ^3He has been shown to diffuse laterally away from the ridge crest, forming a Pacific-wide "plume" centered at about 2500 m depth. This excess ^3He can be distinguished from the atmospheric component either by subtracting the air-saturation value of ^3He content from the observed ^3He content or by comparing the $^3He/^4He$ ratio in the water sample with that in atmospheric helium (corrected by 1.4% for higher solubility of ^4He relative to ^3He). The ratio method is preferred because the abslute-concentration approach is made somewhat uncertain by the fact that deep-sea water appears to be several percentage points supersaturated with conservative atmospheric gases such as N_2, Ar, and Ne, and hence is probably supersaturated with He as well. This supersaturation is presumably induced by evaporative cooling of a surface film or by bubble entrainment generated by breaking waves. As measurements (Lupton and Craig, 1975) of sea-floor hydrothermal samples and of ridge-crest volcanic glasses reveal that the $^3He/^4He$ ratio in the exhaled mantle gas is rather uniform (eight to ten times larger than that in atmospheric helium), if the excess ^3He is known to be solely of mantle origin (rather than in part from the decay of tritium), then the excess can be calculated from the $^3He/^4He$ ratio alone.

One of the interesting results of the measurements of excess ^3He in deep water is the contrast between the Atlantic and Pacific (Lupton, 1976). The Pacific shows a large (up to 30%) mid-depth anomaly (figure 15.6). This anomaly has its origin in the East Pacific Rise. By contrast the excess ^3He in the Atlantic is carried mainly by water returning from the Pacific via the Antarctic Bottom Water (Jenkins and Clarke, 1976). The anomalies generated by ^3He release in the Atlantic are very small, the largest lying along the continental margin in the western North Atlantic rather than near the ridge crest (figure 15.7).

Radiogenic ^3He: The decay product of ^3H is ^3He. In subsurface waters the ^3He produced by the decay of ^3H remains in the water (mass 3 is conserved). When waters containing radiogenic ^3He reach the surface, however, the excess ^3He escapes to the atmosphere. Hence the ratio $^3He/^3H$ in a subsurface water sample contains information with regard to its time of isolation from the surface (Jenkins and Clarke, 1976). Fortunately the ^3He produced by the decay of tritium is spatially separated from that of the ^3He leaking from the mantle. Where a significant overlap is suspected, the two excess components could be separated by precise measurements of absolute ^4He concentration (coupled with Ne, Ar, and N_2 measurements aimed at es-

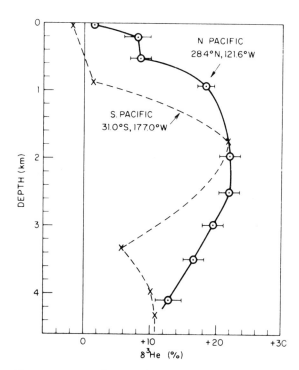

Figure 15.6 Distribution of excess ^3He with depth in the Pacific. The profiles are by Clarke, Beg, and Craig (1970).

tablishing the degree of He supersaturation). The excess ^4He content could then be multiplied by the $^3He/^4He$ ratio for average mantle-derived helium to yield the mantle component of the ^3He excess. The residual ^3He excess would then be that due to the decay of ^3He. In practice this proves difficult because the ratio $^3He/^4He$ in mantle helium is an order of magnitude higher than that in atmospheric helium. A 1% error in the measurement of ^4He concentration would lead to a δ^3He error of about 10%.

^{226}Ra: ^{226}Ra is added to the ocean by diffusion from deep-sea sediments. With a half-life of 1600 years it has been a prime candidate for a mixing tracer (Pettersson, 1955; Koczy, 1958). Several factors have dimmed these hopes. First it was found that the deficiency of ^{226}Ra in surface water relative to deep water is far too large to be explained by radiodecay. Removal of ^{226}Ra on settling particulate matter was invoked to explain this difference. This hypothesis was confirmed by measurements of barium (Wolgemuth and Broecker, 1970). As barium and radium are nearly identical in their chemical behavior, it was hoped that through the use of $^{226}Ra/Ba$ ratios (analogous to the use of $^{14}C/C$ ratios) the effects of particulate transport could be separated from those of mixing [see Chan et al. (1976) for a discussion of the GEOSECS results].

This strategy suffers, however, from a severe problem of dynamic range (i.e., ratio of signal to observational error). If the deep sea is ventilated on a time scale of 1000 years, then for a tracer added mainly to the *bot-*

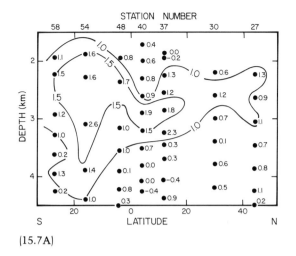

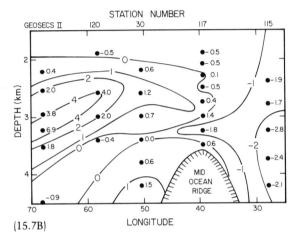

Figure 15.7 Sections along the western Atlantic basins GEOSECS track (A) and the GEOSECS track across the North Atlantic (B), showing the ^{3}He anomalies generated within the Atlantic Ocean (given in percentage of ^{3}He excess over atmospheric). The contribution of excess ^{3}He carried by deep waters returning to the Atlantic from the Pacific (via the Antarctic) has been eliminated through use of the linear "NO"-SiO$_2$ relation for deep Atlantic waters (see Broecker, in press). The data on which these plots are based are from Jenkins and Clarke (1976) and Lupton (1976).

tom of the sea the percentage difference Δ between the ratios of radioisotope to "carrier" ratio in deep water and in surface water should be approximately

$$\Delta = 100 \left\{ 1 - \exp\left[\frac{1000(VS/VD)}{\tau_{\text{isotope}}}\right] \right\}.$$

For a ratio of VS (volume of surface water) to VD (volume of deep water) of about 0.1, the surface-to-deep-water difference predicted for ^{226}Ra (half-life $\tau = 2300$ years) is only 6%. The observational error on the GEOSECS ^{226}Ra measurements averages about 3% and that on a GEOSECS barium measurement about 1%. Thus any age message is buried in observational noise. Despite the fact that ^{14}C has a half-life 3.5 times *longer* than that of ^{226}Ra, because it enters the ocean from the top rather than from the bottom, it has a more than compensating advantage. As mentioned above, a 21% range is seen in the natural ^{14}C/C ratio within the sea. Thus the range of natural ^{14}C/C ratios in the sea should be about three times the range of ^{226}Ra/Ba ratios. Furthermore, the ^{14}C/C ratio can be measured with ten times higher precision ($\pm 0.4\%$) than the ^{226}Ra/Ba ratio ($\pm 4\%$). The overall dynamic-range advantage of ^{14}C over ^{226}Ra is thus thirty!

Although it may be possible to improve the measurement error for ^{226}Ra (to perhaps 1%), uncertainties in the geographical pattern of ^{226}Ra input from marine sediments will always plague the use of ^{226}Ra as a time tracer. For these reasons ^{226}Ra should perhaps be deleted from the list of mixing tracers. Nevertheless, as shown by results from the deep northeastern Pacific, where a great excess of ^{226}Ra relative to Ba has been found (Chung, 1976; Chung and Craig, in press), the ^{226}Ra distribution will contribute to our understanding of the release patterns of metals to deep water from sediments. Also, the ^{226}Ra distribution in the ocean must be known if we are to use the distributions of ^{228}Ra and of ^{222}Rn as time tracers for mixing, and if we are to use ^{210}Pb and ^{210}Po as time tracers for particulate processes.

^{228}Ra: ^{228}Ra has a half life of 5.8 years. Like ^{226}Ra it enters the ocean primarily by diffusion out of marine sediments (Moore, 1969). Because of its far shorter half-life, however, it has a much different distribution within the sea than does ^{226}Ra. Profiles with depth (see figure 15.8 for an example) show maxima at the surface and at the bottom with low values in the mid-water column. The high surface values reflect input of ^{228}Ra from the continental shelves coupled with rapid horizontal dispersion, and the high bottom values reflect the release of ^{228}Ra from deep-sea sediments. The low mid-depth values are the result of lower ratios of sediment area to water volume, and of the finite rate of vertical mixing. (The ^{228}Ra released from shelf and bottom sediments cannot reach this depth range before undergoing radiodecay.)

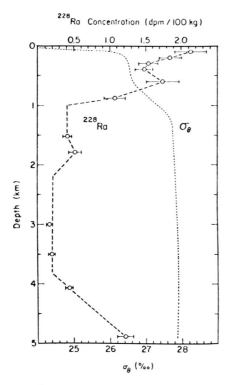

Figure 15.8 ^{228}Ra concentration as a function of depth at the GEOSECS Atlantic calibration station (36°N, 68°W) (Trier, Broecker, and Feely, 1972).

That ^{228}Ra that enters the deep sea probably can be treated as if it were chemically conservative. In surface water, however, removal by particulate matter may compete with radiodecay and mechanical mixing. For ^{226}Ra the probability of removal from surface water via particulate transport appears to be three times higher than by vertical mixing. If the residence time of water in the upper 400 m of the ocean is taken to be 50 years, then the residence time of ^{226}Ra with respect to particulate removal is about 18 years. As the mean life of ^{228}Ra is 8 years, correction for particulate transport effects may prove significant (i.e., ~20%).

^{222}Rn: Radon (half-life 3.85 days), is produced within the sea by the decay of ^{226}Ra dissolved in the sea. Well away from sedimentary margins of the ocean, in situ ^{226}Ra decay is the only source of ^{222}Rn. This is true for the surface mixed layer of the open ocean (the atmospheric ^{222}Rn content is too low to be a significant source). Loss to the atmosphere of the radon produced within the surface-ocean mixed layer provides a means of determining the rate of gas exchange between the ocean and atmosphere (Broecker, 1965). Close to the sediment–water interface, radon released from the upper few centimeters of the sediments (and from the upper few millimeters of Mn nodules and Mn-coated rock pavements) is found. This excess radon provides a means of determining the rate of mixing in the near-bottom ocean (Broecker, Cromwell, and Li, 1968). Both of these applications require a knowledge of the ^{226}Ra content of the water. To determine gas-exchange rates the difference between the equilibrium ^{222}Rn content (calculated from the ^{226}Ra content) and the observed mixed-layer ^{222}Rn must be known. For near-bottom mixing the sediment-derived ^{222}Rn is estimated by subtracting the ^{226}Ra-supported ^{222}Rn from the measured ^{222}Rn content of the abyssal water.

^{90}Sr and ^{137}Cs: The usefulness of fission products in oceanography was first demonstrated by Bowen and Sugihara (1957, 1958, 1960). Prior to nuclear testing the environment was free of the fission products ^{137}Cs and ^{90}Sr (both have half-lives close to 30 years). The amounts of these isotopes currently present in the ocean are entirely the result of these tests. Like ^{3}H they were added mainly to the northern-hemisphere stratosphere. Like ^{3}H they were deposited mainly in the northern hemisphere. Like ^{3}H their ocean distribution shows much higher water-column inventories north of 15°N than south of this boundary (Kupferman, Livingston, and Bowen, 1979). In fact they are so much like ^{3}H that no one has yet demonstrated that their oceanic distributions contain any information with regard to mixing not contained by the ^{3}H distribution (Dreisigacker and Roether, 1978). As ^{137}Cs and ^{90}Sr require far larger water samples than ^{3}H, as they cannot currently be measured with the accuracy or reliability of tritium, and as there are indications that ^{137}Cs may not be conservative within the sea, emphasis has been placed by most geochemists studying mixing processes on the measurement and interpretation of tritium data.

^{90}Sr should, however, not be written off as a mixing tracer. Its input distribution at the earth's surface is better known than that of ^{3}H (because it accumulates in soils). It has a longer half-life (30 years) than does ^{3}H (12 years), and therefore will be around for a longer time. Its concentration in tropical surface water has been recorded by corals (there is no similar paleotritium recorder).

^{85}Kr: ^{85}Kr is a product of fission. Thus it is produced during bomb testing, plutonium production, and power-reactor operation. Much of that produced has leaked to the air (Pannetier, 1968; Schroeder and Roether, 1975; Telegadas and Ferber, 1975). Measurements reveal a strong buildup in this reservoir (see figure 15.9). By gas exchange ^{85}Kr reaches the surface ocean (its current specific activity is about 3 dpm m^{-3}). Because of this very low activity only a few measurements of ^{85}Kr have been made in the sea to date (Schroeder, 1975). As expected, its vertical distribution shows a first-order similarity to that of the other anthropogenic tracers (^{3}H, ^{14}C, ^{90}Sr, ^{137}Cs). If this covariance were perfect, then there would be no interest in

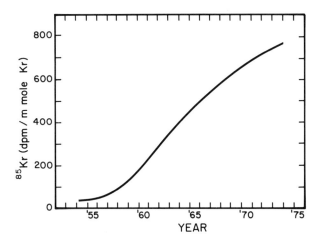

Figure 15.9 ^{85}Kr as a function of time in the atmosphere (Schroeder and Roether, 1975).

going to the trouble to collect and process the large water samples necessary for measuring this isotope.

There is, however, at least one very important application for this isotope. As is discussed at greater length later in the paper, a difficulty arises in estimating the concentration of excess CO_2 (i.e., that stemming from the production of fossil fuels) in currently forming deep water. Because the chemical equilibration time of CO_2 between mixed-layer water and the atmosphere is of the order 1 year while the cooling time is on the order of 1 month, newly formed deep waters are likely to carry less than their capacity of excess CO_2. The most promising approach to finding how much less appears to be through the use of two tracer gases: one a gas that equilibrates on a time scale longer than that for CO_2, and the other a gas that equilibrates on a time scale shorter than that for CO_2. $^{14}CO_2$ is the only choice for the former. As discussed by Broecker, Peng, and Takahashi (in press) the isotopic equilibration time for carbon exceeds the chemical equilibration time by a value numerically equal to the buffer factor ($\sim$12 for cold surface water). ^{85}Kr and the freons are the prime candidates for the short-exchange-time gas. Like all gases of normal solubility their exchange time is about 1 month (Peng et al., 1979). Although the freons have the advantage that inexpensive measurements can be made on small samples, they have yet to be proven conservative in sea water. Proof of their long-term stability can only be achieved through comparison with a second tracer of the same type. ^{85}Kr is the only candidate for this task.

^{39}Ar: Cosmic-ray-produced ^{39}Ar is the ideal deep-ocean tracer. As it exchanges rapidly with the air, the problem of establishing its concentration in newly formed deep water is far smaller than for ^{14}C. Its half-life of 270 years is far more appropriate to deep-water movement times than that of ^{14}C. The problem in its use is technical (Loosli and Oeschger, 1968, 1979; Oeschger et al., 1974). Very large samples (several tons) and ultralow background counting are needed. Only Loosli and his colleagues at the Bern laboratory currently are capable of measuring this isotope in the sea. Because its potential is so large, however, the next decade is bound to bring forth many ^{39}Ar measurements.

^{32}Si: Since Schink's (1962) early attempts, two decades of effort have gone into determining the distribution of ^{32}Si (a cosmic-ray-produced isotope with a half-life of $\sim$250 years) in the sea. Like ^{39}Ar its collection and measurement are fraught with technical difficulties. Using the *in situ* extraction technique of Krishnaswami et al. (1972) about 200 samples were collected during the GEOSECS program. Three profiles have been published (Somayajulu, Lal, and Craig, 1973). The remainder of the analyses are still in progress at the Physics Research Laboratory at Amedabad in India. Even when these data appear, three serious problems will remain for their interpretation. First, the performance of the bag samplers used to collect the silica was far from perfect. Second, since the particulate cycle dominates the movement of silica down the water column, isolation of the water dynamics from the particle dynamics constitutes a difficult problem. Finally, the measurement errors are sizeable and will surely lead to considerable ambiguity in the interpretation. If the ^{39}Ar method succeeds, ^{32}Si may well be relegated to the area of particulate research. Were the circulation dynamics independently known, the ^{32}Si distribution could tell us much about the pattern of dissolution of siliceous particles.

Freons: Most of the freon produced as a propellant and as a refrigerant ultimately finds its way to the atmosphere. Reliable and accurate means of measuring it in air and sea water have been worked out (Lovelock, Maggs, and Wade, 1973). Measurements on sea water show its distribution to be similar to that of tritium and other "fallout" nuclides (Hahne et al., 1978). As mentioned above, once it can be established that freons are conservative in the sea, then this substance will become one of the more useful of the transient tracers.

^7Be: With a half life of 53 days, cosmic-ray-produced ^{7}Be has great potential as a tracer of upper thermocline processes (Silker, 1972a,b; Silker et al., 1968). An example of its distribution in the upper water column is shown in figure 15.10 (36°N, 68°W). Unfortunately, because of the large volume of water needed (thousands of liters) and of the complexity of the counting system needed for its detection, little work has been done to take advantage of the information carried by this isotope. Were it to be used, attention would have to be given to its removal by particulate matter and to the seasonal variation in its distribution.

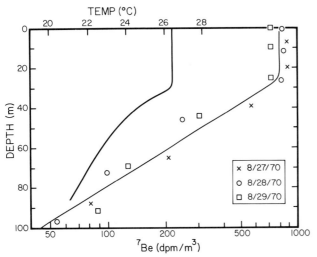

Figure 15.10 ^{7}Be as a function of depth at the GEOSECS Atlantic calibration station (36°N, 68°W) occupied during August 1970 (Silker, 1972a). Solid curve is temperature profile.

15.3 Water-Mass Tracers

All deep waters in the ocean are mixtures from two or more sources. Since the concentrations of the time tracers in each source are potentially different, these mixtures must be unscrambled if we are to use the isotope data successfully to yield ventilation rates. For this job we need properties that are conserved once the water ends contact with the atmosphere. Candidates are as follows:

(1) Conductivity. The conductivity of a water parcel isolated in the deep sea changes only because small amounts of SiO_2, CO_2, PO_4^{---}, Ca^{++} and NO_3^- are added via respiration and mineral solution. As discussed by Brewer and Bradshaw (1975), these changes are just measurable. For any of the applications involving radioisotope "dating" of sea water, the salinities calculated from conductivity measurements may be treated as perfectly conserved.

(2) Temperature. As heat transfer by molecular conduction is of negligible importance to the internal energy content of a sea water parcel, potential temperature can be treated as conservative. The only *source* of heat within the deep sea of significance is geothermal heat. Except in rare circumstances this contribution is small enough not to create difficulties. Because heat-flow rates are well known (except perhaps along ridge crests), the effects of geothermal heating can often be taken into account in models of radioisotope data (i.e., the temperatures can be corrected by iteration).

(3) ^{2}H and ^{18}O. The isotopic composition of the waters ventilating the deep sea varies from source region to source region (Epstein and Mayeda, 1953). As discussed by Craig and Gordon (1965), these variations do not strongly correlate with either salinity or temperature.

The isotopic messages from ^{2}H and ^{18}O, however, are highly correlated with each other. These signatures are conservative. The problem lies with the measurements. An acceptable dynamic range can be achieved only by painstaking care using the best available mass spectrometers and preparation lines. Because of this only a limited number of such measurements is currently available. Even for measurements of the highest attainable precision the dynamic range may prove to be too small to be useful.

(4) "NO" and "PO". Although dissolved-oxygen gas is utilized by animals and bacteria living beneath the photic zone in the sea, a correction can be made for this utilization by taking into account the fact that for each 135 molecules of O_2 consumed about 15 molecules of NO_3^- and about one molecule of PO_4^{---} are released to the water as dissolved ions. Thus the sums $[O_2] + 9[NO_3]$ and $[O_2] + 135[PO_4]$ are potentially conservative properties (Broecker, 1974). The near constancy of the NO_3/PO_4 ratio in most parts of the deep sea supports this concept. Since the coefficients 9 and 135 are not truly "stoichiometric" quantities, however, care must be taken in using "NO" and "PO". The actual coefficient applying to any given mass of water might deviate from these values. "NO" has an additional drawback. In areas of the ocean where the dissolved O_2 content reaches 1% or less of its saturation value, NO_3 is used by bacteria as an oxidant. Thus, in regions adjacent to zones of severe O_2 depletion (like the thermocline off Central and South America), "PO" rather than "NO" should be used in "unmixing" calculations.

(5) SiO_2 and Ba. Silica and barium are concentrated in deep waters relative to surface waters, and show large differences from water type to water type (Edmond, 1974; Chan, Drummond, Edmond, and Grant, 1977; Chung and Craig, in press). As these differences are generated by the removal of these substances from the upper ocean followed by regeneration in the deep ocean, Ba and SiO_2 are by no means conservative. Nevertheless, observation has shown that in certain regions of the ocean they are nearly conservative and can be successfully used for "unmixing" calculations. An example is given below.

15.4 Modeling Tracer Data

The ultimate use of the radioisotope data is to obtain information regarding transport of water and of species dissolved in the water within the sea. Deconvolution of the tracer field into the various modes and directions of transport remains an unsolved problem. As might be expected, modeling of radioisotope data has undergone an evolution that paralleled the size and quality of the data set. The first models were of the box variety.

The ocean-atmosphere system was approximated by a few well-mixed reservoirs. The ^{14}C data were used to define transfer coefficients between these reservoirs. Initially, three-box models that separated the system into a well-mixed atmosphere, a well-mixed surface ocean, and a well-mixed deep ocean were used (Craig, 1957). The fluxes derived were the air-sea gas exchange rate and the water transfer rate across the main thermocline. When ^{14}C data for the Pacific Ocean (Bien, Rakestraw, and Suess, 1963) as well as the Atlantic Ocean (Broecker, Gerard, Ewing, and Heezen, 1960) became available, a large difference in the apparent age for the deep water in the two oceans became evident. This generated an incentive for models with more boxes. It was also realized that the regions of deep-water formation (North Atlantic and Antarctic) should be treated separately. Various attempts were made to generate multibox models (Broecker, 1963; Keeling and Bolin, 1967, 1968; Broecker and Li, 1970). In all these attempts, owever, it was realized that because the number of fluxes needing definition increased as the square of the number of boxes while the number of new pieces of information increased linearly with the number of boxes, this approach was destined to yield a considerable degree of ambiguity. The idea of using salinity and temperature to aid in the definition of fluxes, while promising in a mathematical sense, also proved impractical because of the uncertainties in the atmospheric boundary conditions for heat and salt. Geochemists fell back to the use of the simpler three-box models as a guide to geochemical thinking, and suspended attempts to use box models to answer physical oceanographic questions.

Munk (1966) initiated an alternate approach to the use of isotope data. He focused his attention on the trend of $^{14}C/C$ ratio with depth within the deep Pacific and showed that, if the type of advection-diffusion model used by Stommel (1958), Robinson and Stommel (1959), Stommel and Arons (1960b) and Wyrtki (1961b, 1962) were employed, the ^{14}C trend would yield absolute values for the upwelling velocity and the vertical diffusivity. Craig (1969) elaborated on Munk's idea, pointing out that ^{14}C and C should be treated separately in the calculation. By fitting both the ^{14}C concentration profile and the ΣCO_2 concentration profile, the contribution of ^{14}C through the rain of particulates could be accounted for. A reader interested in a review of this subject is referred to the excellent paper by Veronis (1977). As the GEOSECS data came in, it became apparent that the horizontal gradients within the deep Pacific could not be neglected. For this reason interest in the simple one-dimensional advection-diffusion has waned.

Kuo and Veronis (1970, 1973) initiated the first two-dimensional modeling of the deep-ocean tracer field including advection and diffusion. They assumed that deep water formed at two small high-latitude regions and upwelled uniformly over the entire ocean. They used a modified form of the horizontal advection pattern envisioned by Stommel (1958), and assumed uniform lateral and vertical diffusion rates. Their endeavors focused on matching the deep-O_2 distribution. Fiadeiro and Craig (1978) adopted a variation on this approach and did more elaborate three-dimensional calculations for the deep Pacific. Interest in this approach has also waned, however, because of serious reservations with regard to the suitability of the advection pattern adopted for the models.

During the last 2 years Sarmiento has been applying fluxes of bomb-produced isotopes to Bryan's quasi-diagnostic model (Sarmiento and Bryan, in preparation), the idea being to see whether the model reproduces the observed distribution of these tracers. This work is still in progress.

Despite the failure to produce a satisfactory model with which to match the tracer data for purposes of elucidating the dynamics of deep-ocean ventilation, one very important application has emerged. Through tracer-tracer analog modeling the distributions of the radioisotopes can be used to predict how much fossil fuel CO_2 has been (and will be) taken up by the sea. A number of investigators have used three-box models calibrated using the distribution of natural radiocarbon (Bolin and Eriksson, 1959; Broecker, Li, and Peng, 1971; Machta, 1972; Keeling, 1973). Oeschger, Siegenthaler, Schotterer, and Gugelmann (1975) have used a vertical-diffusion model calibrated with natural radiocarbon. Broecker, Takahashi, Simpson, and Peng (1979) have subsequently shown that the diffusivity chosen by Oeschger et al. (1975) is consistent with the vertical distribution of the bomb-produced tracers. Several groups of investigators are currently developing more elaborate models of this type. While not giving us any better idea of the physics of ocean circulation, these models do provide us with an interim approach to a problem of great import to society.

15.5 Current Applications

We shall consider four applications of the tracer results. The first is the ventilation time of the deep sea. The single great triumph of the natural radiocarbon measurements in the ocean has been the establishment of a 1000-year time scale for the residence of water in the deep sea. Now that we have the GEOSECS data, what more can be said on this subject? The second is the ventilation time for the main oceanic thermocline. The transient tracers ^{14}C and ^3H prove ideal for this task. The third is the dynamics of deep-water formation. In this process the interaction between surface water and the atmosphere constitutes a rate-limiting step for the tracers used. Again, the transient tracers prove ideal for the task. Finally, the distributions of the short-lived natural tracers and of the anthropogenic tracers allow *upper* limits to be placed on the rates of vertical (diapycnal) mixing.

15.6 Ventilation of the Deep Sea

Although a number of estimates of the mean ventilation time for the deep sea have been proposed, that of about 1400 years, based on radiocarbon, is the most widely quoted. In the three-box model the ventilation time T of the deep sea is given by the following equation:

$$T \equiv \frac{V_D}{R} = \left[\frac{(^{14}C/C)_S}{(^{14}C/C)_D} - 1\right] T_{^{14}C},$$

where V_D is the volume of the deep sea, R the ventilation rate of the deep sea, $(^{14}C/C)_S$ and $(^{14}C/C)_D$ the carbon isotope ratios for the mean surface and mean deep ocean (prior to the industrial revolution), and $T_{^{14}C}$ the mean life of ^{14}C (i.e., 8200 years). Taking $(^{14}C/C)_S/(^{14}C/C)_D$ to be 1.17 ± 0.03 (i.e., $\Delta^{14}C_S = -50 \pm 10‰$ and $\Delta^{14}C_D = -190 \pm 20‰$), a ventilation time of 1400 ± 250 years is obtained. Among other things this model assumes that the ventilation times for water and carbon are equal. Because the $^{14}C/C$ ratio in newly formed deep water has been shown not to reach that observed for the warm surface ocean, the ventilation time for water is surely less than that given by the box model. If the box model calculations used to obtain this result were repeated using average $^{14}C/C$ ratios obtained from the GEOSECS data set, the result would not change significantly. Evolution in geochemical and oceanographic thinking since the heyday of box modeling makes clearer the problems associated with using ^{14}C data to estimate water residence times in the deep sea. Until means of overcoming these problems are achieved it will not be possible to evaluate properly the ventilation times derived from ^{14}C data, nor will it be possible to improve on them.

One problem is related to the fact that while the mixed layer of the ocean can be cooled (through contact with winter air) in a few weeks, equilibration between the $^{14}C/C$ ratio in the carbon dissolved in the mixed layer and that in the atmospheric CO_2 requires a decade or more. Harmon Craig recognized this problem long ago but until recently it has received only casual attention. The upshot is that the ^{14}C residence time in the deep sea must exceed the water residence time. Results from the Weddell Sea (Weiss, Östlund, and Craig, 1979) dramatically illustrate this point, for the new deep water formed there contains nearly the same $^{14}C/C$ ratio as ambient circumpolar deep water.

The other problem has to do with the recirculation of old water within the deep sea. Rarely is it possible to isolate the component of $^{14}C/C$ change due to *in situ* aging. Rather, the changes in $^{14}C/C$ ratio within a given deep-sea basin can be accounted for largely by the mixing of end members with different $^{14}C/C$ ratios. This greatly complicates the use of ^{14}C data to obtain the flux of deep water into a given basin.

The trend of $\Delta^{14}C$ along the 4000-m horizon is shown in figure 15.11. The pattern shows a gradual decrease in $^{14}C/C$ "down" the Atlantic and "up" the Pacific (with similar values in the Antarctic segments of the two oceans). The down-Atlantic decrease is about 80‰ and the up-Pacific decrease is about 70‰. While this might be interpreted as evidence for gradual aging as the water moves by advection slowly around the globe, a close look at the situation reveals that at least in the Atlantic the trend is largely the result of mixing between low $^{14}C/C$ waters of circumpolar origin and high $^{14}C/C$ waters of northern Atlantic origin. When this mixing is taken into account the residence time of water in the Atlantic is considerably decreased and the flux of new North Atlantic deep water necessary to maintain it is correspondingly increased (table 15.5).

Table 15.5 Evolution of Calculations of the Apparent ^{14}C Age of a NADW Sample with a $\Delta^{14}C$ of $-105‰$

	$\Delta^{14}C$ values					
	Warm surface water	New no. comp. water	New AABW	Δ due to mixing with AABW	Δ due to *in situ* aging	Apparent[a] age
Broecker et al. 1960	-28	(-28)	-155	0	-77	675
Broecker and Li 1970	-40	(-40)	-155	-12	-53	470
Stuiver (GEOSECS) 1976	—	-70	-160	-10	-25	225
Broecker (GEOSECS) 1979	—	-70	-160	-26	-9	80

a. $t = 8200 \ln \dfrac{1 - (\Delta^{14}C_{\text{initial}}) \times 10^{-3}}{1 - (\Delta^{14}C_{\text{initial}} + \Delta_{\text{aging}}) \times 10^{-3}}.$

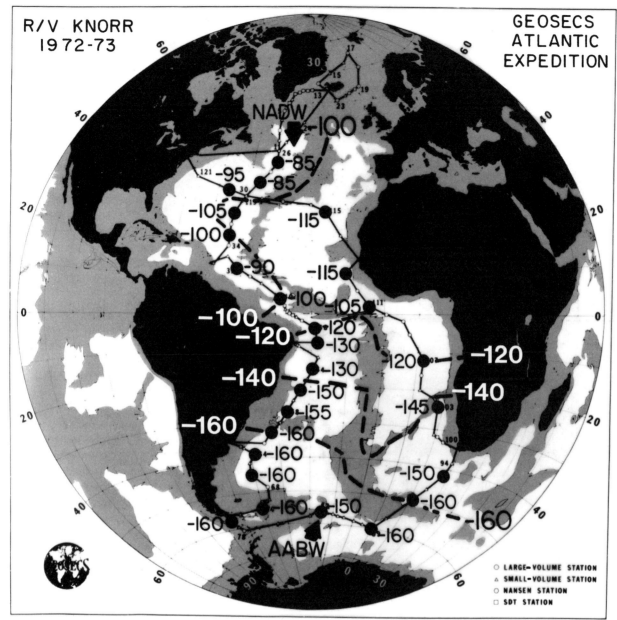

Figure 15.11 Maps showing the distribution of ^{14}C along the 4000-m horizon in the Atlantic (A) and Pacific (B) Oceans. The analyses were made by Östlund at the University of Miami and Stuiver at the University of Washington as part of the GEOSECS program.

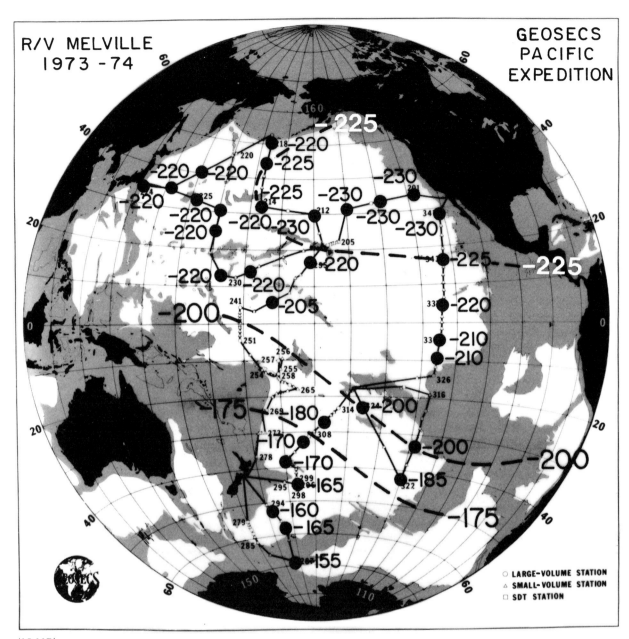
(15.11B)

Before going into this matter further, it is appropriate to review the evidence leading to the conclusion that the back mixing of Antarctic deep water is the major cause for the $^{14}C/C$ gradient observed in the Atlantic. Confusion on this point has arisen because of the ambiguity in estimates based on conventional temperature-salinity (θ-S) analysis of the fractions of various contributing water types. There are three such northern water types: deep water formed by winter cooling in the Labrador Sea (LSW); water spilling over the sill connecting Iceland to Greenland (DSW); and water which ultimately comes from overflow across the sill connecting Iceland with the British Isles but enters the western basin through the Gibbs Fracture Zone. As shown in table 15.6, the traditional θ-S approach is not particularly sensitive, because the θ and S ranges for the three contributing northern water types are large compared to the differences from the dominant southern type, i.e., the Antarctic Bottom Water (AABW). Temperature and salinity could be used only if the relative proportions of the northern-source waters one to another were known. They are not (see chapters 1 and 2).

By contrast the SiO_2 and the "NO" contents of the three northern water types are almost identical and are very different from that for AABW (see figure 15.12). Thus, were either of these properties conservative, we would have a means of determining the fractions of high-^{14}C northern and low-^{14}C southern waters in any given sample from within the mixing zone. One test of conservation is a plot of SiO_2 versus "NO". Linearity would require either conservation of both properties or a fortuitous correlation between the true "NO" coefficient and the amount of SiO_2 added to any given sample through the dissolution particles. As shown in figure 15.12 the relationship is remarkably linear except for waters in the southern eastern basin.

If this linearity is (as I believe it can convincingly be shown to be) a demonstration of near conservancy of "NO" and SiO_2, then the fraction of southern component in each sample of deep Atlantic water collected by the GEOSECS program can be determined either from its SiO_2 or "NO" content (these two independent estimates generally agree to ±0.03). The plot of $\Delta^{14}C$ versus fraction of southern component so obtained is given in figure 15.13. The dominance of mixing is clearly demonstrated. The deviation attributable to radiodecay averages only 9‰ (the measurement error on a given sample averages 4‰).

The geographic and depth patterns of the residual anomalies (i.e., those produced by radiodecay within the Atlantic) are shown in figure 15.14. The only pronounced feature is the larger anomaly for the eastern than for the western basin. As shown in table 15.5, estimates of the residence time of waters in the deep western Atlantic have dramatically decreased with time. Part of this decrease is accounted for by the realization that water descending in the deep northern Atlantic has a lower $^{14}C/C$ ratio than ambient Atlantic surface water. The rest is due to the increase in the estimates for the role of AABW back mixing.

As I have shown in a separate publication (Broecker, 1979), if an advective model is used to reproduce both the distribution of northern and southern components in the western Atlantic and the distribution of the residual (i.e., in situ decay) anomalies, then a flux of 20×10^6 m^3 s^{-1} of northern component (some mixture of LSW, GFZW and DSW) and 10×10^6 m^3 s^{-1} of southern component (AABW) are needed. As the deep portions of the eastern basin are known to be ventilated from the western basin (via fracture zones), no extra flux of northern component water is needed for its ventilation (part of the water exiting the western basin does so through the Romanche and possibly other frac-

Table 15.6 Characteristics of the Contributors of the NADW Complex

Water type	θ (°C)	S (‰)	"NO" (μM kg^{-1})	SiO_2 (μM kg^{-1})	$\Delta^{14}C$ (‰)
Labrador Sea Deep Water	3.3	34.94	430	12	−69
Denmark Straits Overflow Water	1.5	34.91	427	10	−69
Gibbs Fracture Zone Water	2.8	35.00	429	13	−63
Mediterranean Sea Overflow Water	12	36.4	300	8	∼−60
Circumpolar Bottom Water	0.0	34.70	511	125	−163
Circumpolar Intermediate Water	2.2	34.63	472	85	−147

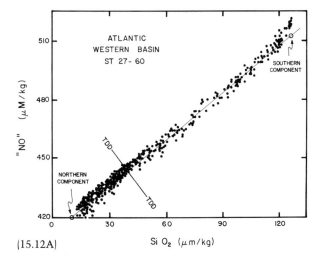

(15.12A)

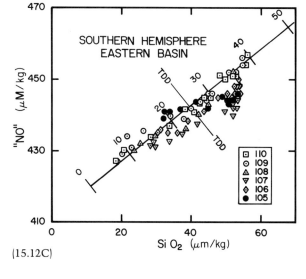

(15.12C)

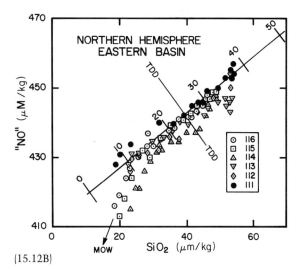

(15.12B)

Figure 15.12 Plots of "NO" versus SiO_2 for all the samples taken by the GEOSECS program below 1500 m in the Atlantic Ocean. In the western basin (A) the entire range of compositions between northern (DSW, LSW, and GFZW) and southern (AABW) component water is seen consistent with the two-end-member hypothesis. No change in slope is seen at the Two Degree Discontinuity (Broecker, Takahashi, and Li, 1976). In the northeastern basin (B) those stations nearest to the Straits of Gibraltar show a deviation toward low "NO" value for any given SiO_2 content caused by the admixing of low "NO" Mediterranean water (see table 15.6). Deep waters in the southeastern basin (C) show a deviation toward high silica at any given "NO" value. This presumably reflects *in situ* production of silica through the dissolution of particulate matter.

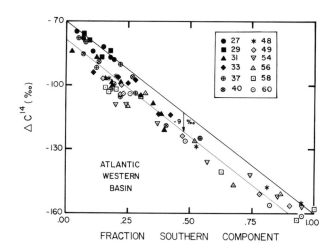

Figure 15.13 $\Delta^{14}C$ for western basin deep waters as a function of the fraction of southern component (F.S.C., as estimated from the "NO"-SiO_2 relationship). Were mixing alone responsible for the trend, then the points should lie along the line joining $\Delta^{14}C = -160‰$, F.S.C. = 1.0, and $\Delta^{14}C = -70‰$, F.S.C. = 0.0. Deviations below this line are presumably a measure of the loss of radiocarbon within the mixing zone by radioactive decay. The radiocarbon measurements were made by Stuiver of the University of Washington and Östlund of the University of Miami.

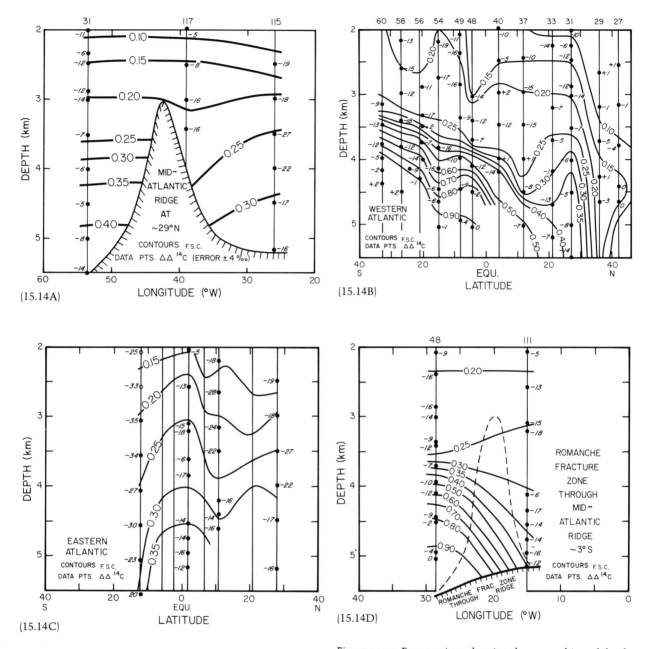

Figure 15.14 Four sections showing the geographic and depth distribution of the ^{14}C anomalies attributable to radioactive decay within the deep Atlantic. The contours show the fraction of southern component (i.e., AABW) within the deep Atlantic.

ture zones which cut the Mid-Atlantic Ridge). The question how much more water is needed to ventilate the eastern basin above the ridge crest is not so easily answered. I guessed that 10×10^6 m^3 s^{-1} more of northern component is required. If so, then the total flux of northern component is 30×10^6 m^3 s^{-1}.

The volume of the deep sea (below 1500 m) divided by 30×10^6 m^3 s^{-1} gives 900 years. Thus, were North Atlantic Deep Water the only source of *carbon* isotope ventilation, the mean age of deep-sea carbon with respect to that entering with new deep water formed in the northern Atlantic would be *900* years. Starting with Δ^{14}C values of -70‰, this would give a mean Δ^{14}C for deep-sea carbon of -167‰ (a value close to the observed mean).

This does *not* imply that ventilation of the deep sea by waters descending around the continent of Antarctica is negligible. Rather, as shown by Weiss et al. (1979), these waters exhaust heat without substantially changing their ^{14}C value. Thus if, as commonly estimated, the flux of new deep waters formed around the perimeter of Antarctic is in the range 10 to 40×10^6 m^3 s^{-1} (Gordon, 1975b; Gill, 1973; Killworth, 1974, 1977; Carmack, 1977), then the water ventilation time of the deep sea is probably considerably less than 900 years.

While these calculations are preliminary and subject to many criticisms, they do serve to indicate that derivation of water fluxes from ^{14}C data is far from a straightforward exercise. The conventional ^{14}C-residence time (based on box models) provides only an upper limit on the water-residence time. As temperature is probably the property most rapidly equilibrated with the atmosphere and carbon isotopes the most slowly equilibrated, the renewal time for other substances (CO_2, NO_3, SiO_2, ...) will lie in between the water and carbon-isotope renewal times.

15.7 Ventilation of the Main Oceanic Thermocline

The GEOSECS program has provided for the first time a reasonably detailed coverage of the distribution in the ocean of the ^{14}C and ^{3}H produced by nuclear testing. These distributions emphasize the importance of two phenomena evident but perhaps not fully appreciated through more conventional oceanographic observations. First they suggest an upwelling flux in the equatorial ocean comparable on a global scale to the flux of newly formed deep water (Broecker, Peng, and Stuiver, 1978). Second, they illustrate the importance of thermocline fronts located near lats. 15°N and 15°S (Broecker and Östlund, 1979).

Strong upwelling in the equatorial ocean (or in the eastern boundary regions feeding into the equatorial zone) is suggested by the distribution of bomb-produced ^{14}C. The equatorial zone is characterized by low surface-water bomb ^{14}C/C ratios and shallow penetration of bomb ^{14}C (relative to the adjacent temperate gyres). If the gas-invasion rate (from the atmosphere) into equatorial waters is comparable to that into temperate waters (as I believe it must be), then the low inventory of bomb ^{14}C in the equatorial zone can only be maintained by the input to the equatorial zone of water deficient in bomb ^{14}C. The only source of such water is upwelling from depths of at least 500 m. As the bomb ^{14}C distribution *within the equatorial zone* is quite homogeneous along isopycnal surfaces (figure 15.15), this isotope is not a sensitive indicator of the place at which upwelling occurs. However, by combining ^{14}C data with data for tracers like P_{CO_2}, P_{N_2O}, NO_3, NO_2, which have shorter surface-water response times than does ^{14}C, it may be possible to get at the entry pattern of upwelled water. Furthermore, time series of ^{14}C (as measured directly on water samples and as reconstructed through measurements of coral-growth rings) may reveal fluctuations in the rate of upwelling.

The distributions of ^{90}Sr and ^{3}H in the Atlantic demonstrate a front near 15°N, because the water-column inventory of these isotopes drops by an order of magnitude near that latitude. As shown in figure 15.16, the concentration of tritium along any given isopycnal also drops by an order of magnitude from north to south across this front (Broecker and Östlund, 1979). One reason for this difference is that the bulk of the ^{90}Sr and ^{3}H "fallout" occurred to the north of the equator. This asymmetry between hemispheres does not explain, however, why the boundary is so abrupt and why it lies near 15°N.

An answer to this dilemma comes from the equatorial upwelling inferred in the zone over which the thermocline is thin and shallow (i.e., 15°N to 15°S). As this upwelling is driven by the poleward divergence of equatorial surface water, any ^{3}H or ^{90}Sr reaching the equatorial zone by fallout or by leakage across the 15°N front will be pulled back to the surface and carried poleward across the front. Thus upwelling accounts for both the location of the fallout front and its sharpness.

As shown by Broecker et al. (1978), the distribution of ^{14}C in the equatorial zone yields a ratio of the flux of upwelled water to the invasion rate of CO_2. Since the latter has been independently determined, the upwelling flux can be obtained. Using this approach we obtained a flux of about 15×10^6 m^3 s^{-1} for the Atlantic equatorial zone. A preliminary analysis of the Pacific Ocean ^{14}C data suggests that the upwelling flux there is about 30×10^6 m^3 s^{-1}. As mentioned above, no firm statement can yet be made as to where this upwelling occurs, for a high rate of lateral mixing obscures its origin.

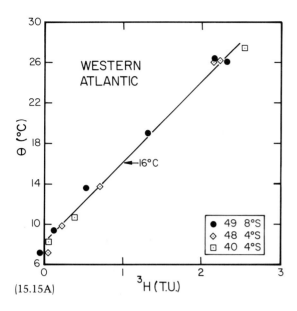

15.8 Formation of Deep Waters

As pointed out, there are serious difficulties in using the distribution of natural radiocarbon to obtain deep-water formation rates. In the Antarctic this stems from the fact that the ^{14}C clock is not reset during the formation process. In the deep Atlantic the anomaly due to radiodecay is small compared to the anomaly due to mixing. Also, there is no simple way to separate the contributions of the three (or possibly more) northern water types contributing to North Atlantic Deep Water. It behooves us therefore to seek other geochemical methods to gauge these fluxes. One obvious approach is to use the substances added to the ocean as a by-product of man's activities to trace recently formed deep water. The GEOSECS ^{3}H section along the western basin in the Atlantic (figure 15.17) shows that the pathway followed by the newly formed water was in 1972 clearly "stained" with fallout products. So the potential is there. At a minimum, surveys spaced at decade intervals would clearly demonstrate the manner and rate at which the front of tritiated water is pushing its way into the western basin of the Atlantic.

The problem of inverting distributions of these transient tracers into fluxes is a difficult one. The geometry of the system is complex. Simple box-model representations are bound to give misleading answers. I do not mean to imply by this that the situation is hopeless. Rather the job will be a long one, requiring imagination, diligence, and excellent measurements.

It will be helpful, moreover, to explore what limitations the relationships among these tracers place on the types of dynamic models that might be employed. An example drawn from the GEOSECS observations will serve to illustrate this point. Despite quite differ-

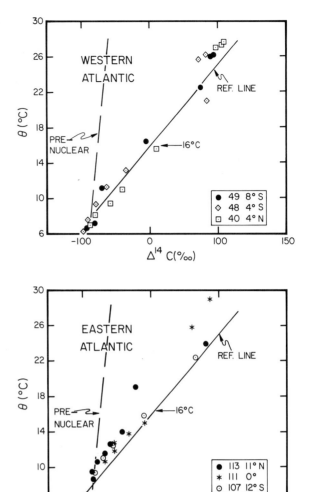

Figure 15.15 Plots of tritium (A) and of radiocarbon (B) v. potential temperature in the equatorial thermocline of the Atlantic Ocean. The tritium data give the impression that tritium-bearing surface water is being mixed down into tritium-free water of potential temperature about 8°C. If 8°C water is taken to be the source for upwelling, then the ^{14}C data can be interpreted as a mixture of surface water of Δ^{14}C = 100‰ and 8°C water with a Δ^{14}C value of about −90‰.

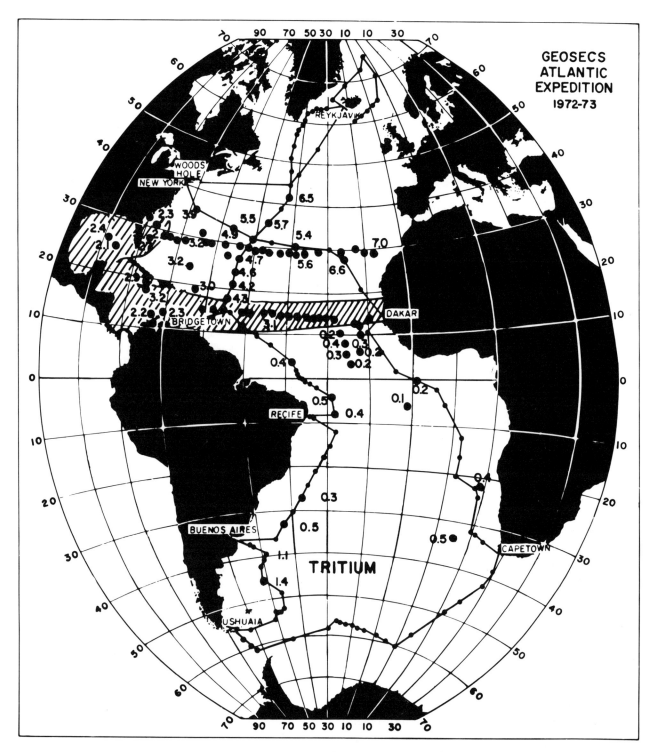

Figure 15.16 Tritium distribution along the isopycnal of $\sigma_\theta = 26.80$ in the Atlantic Ocean. The 15°N thermocline front constitutes a very pronounced boundary between high-tritium waters to the north and low-tritium waters to the south.

ent input mechanisms, bomb-produced ^{14}C and ^{3}H show a very high degree of covariance in the northern Atlantic (figure 15.18). The samples included in this diagram cover the entire range of depth, geographical location, and water type. Modeling shows that different combinations of advection (away from the source region) and diffusion predict a great variety of ^{14}C-^{3}H trends. Thus the combined tracer fields can perhaps tell us not only the pathways followed but also something of the dynamics of the water movement.

15.9 Vertical Mixing Rates

Although there is no way to demonstrate how the isotopes we study penetrate into the interior of the ocean, we can still use their distributions to place upper limits on the rate of vertical mixing in the sea. This can be done by matching any given depth profile to a one-dimensional model, as has been carried out for profiles of ^{222}Rn and ^{228}Ra in the deep sea, and for profiles of bomb ^{3}H and ^{14}C in the main oceanic thermocline, as shown in figure 15.19. The apparent vertical diffusivities obtained in this way show an inverse correlation with density gradient (Sarmiento et al., 1976; Hoffert and Broecker, 1978). Quay, Broecker, Hesslein, and Schindler (1980), using tritium injections into two small lakes in Canada, were able to obtain vertical diffusivities for much higher density gradients. As the tracers were allowed to spread laterally over the entire breadth of each lake, these results are true rather than apparent vertical diffusivities. As shown in figure 15.19, the tracer results from the lakes fall close to the extension of the oceanic curve.

It is difficult, however, to assess the significance of these results. Sarmiento and Rooth (1980) have shown that the bottom ^{222}Rn results could be explained entirely by mixing along isopycnals. Sarmiento (1978) has shown this to be the case for ^{228}Ra in the deep sea as well. It has long been suspected by oceanographers that the main thermocline of the ocean

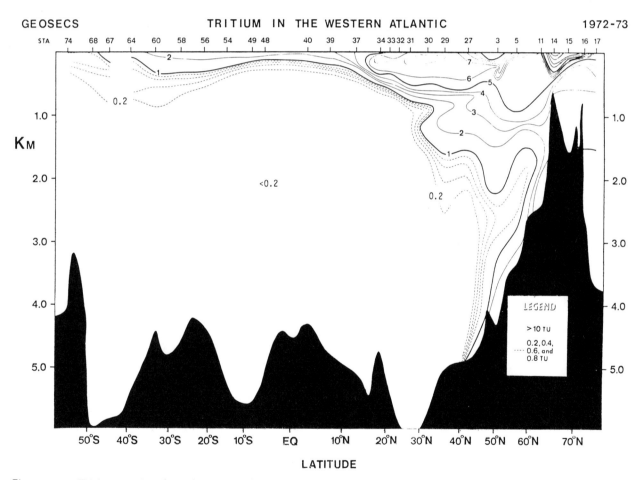

Figure 15.17 Tritium section along the western basin of the North Atlantic, prepared by Östlund of the University of Miami, who made all the tritium measurements for the GEOSECS program.

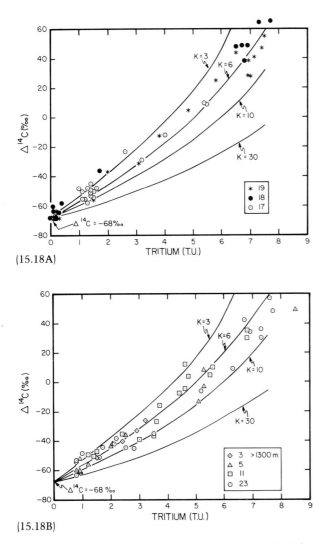

(15.18A)

(15.18B)

Figure 15.18 Tritium (in T.U.) vs. radiocarbon (in $\Delta^{14}C$) for water samples (from all depths) in the Norwegian Sea (A) and northern Atlantic (B). The solid curves show the results predicted by a one-dimensional diffusion model (the diffusion coefficients are in the units $cm^2 s^{-1}$).

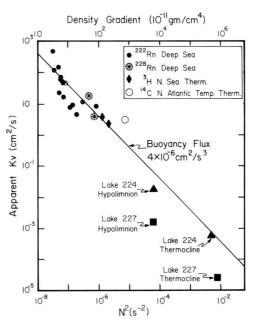

Figure 15.19 Apparent eddy diffusion coefficients for vertical mixing as measured by ^{222}Rn and ^{228}Ra in the deep sea (Sarmiento et al., 1976), ^3H in the pycnocline of the Norwegian Sea (Hoffert and Broecker, 1978) and bomb ^{14}C in the Sargasso Sea (Broecker et al., 1978) as a function of density gradient. Also shown are the results of Quay et al. (1980), obtained by tritium injections into small lakes.

is ventilated mainly along isopycnal surfaces. Certainly the distributions of bomb-produced ^3H and ^{14}C in the main oceanic thermocline could be explained by this process. The ^7Be profile shown above could be interpreted as an artifact of winter convection. Perhaps an argument relating to the angle between isopycnal surfaces and the tracer source (sea surface or floor) can be made as an alternative to the correlation between apparent vertical diffusivity and density gradient.

In my estimation the only way in which the relative importance of vertical and horizontal mixing can be established is to perform tracer injections at various points in the ocean interior. Such experiments (using ^3He as a tracer) would be comparable in cost to that for the large ocean science projects currently underway. The technology needed is available. By following a ^3He or tritium addition to an isopycnal horizon for several years, it would surely be possible to obtain a better estimate than we now have for the relative magnitudes of mixing along and perpendicular to isopycnal horizons. A knowledge of the relative importance of these processes is critical to the development of adequate models for the interpretation of radioisotope tracer data.

16
The Origin and Development of Laboratory Models and Analogues of the Ocean Circulation

Alan J. Faller

16.1 A Brief Philosophy of Laboratory Experimentation

"No one believes a theory, except the theorist. Everyone believes an experiment—except the experimenter." This often used adage, although not aways accurate in detail, carries a certain element of truth. Here we consider certain experiments—analogs, models, or fundamental studies of basic fluid dynamics—that are intended to be relevant to one or another aspect of the ocean circulation. These are physical, fluid models—as opposed to numerical, analytical, abstract, or conceptual models—that make use of a *real* fluid. When the fluid in a container is subjected to some driving force, the fluid moves. It is observed by the experimenter. It is there. It is a real fluid circulation. Apart from its relevance, why should it not be believed?

The experienced experimenter is well aware of the pitfalls of his trade. Aside from the accuracy of his reported observations there are many questions. Were the boundary conditions well controlled? Were the physical properties of the materials and their variations during the experiment known? Did the methods of observation (probes, dyes, tracers, lighting) influence the results? Were the reported observations complete, or at least representative? or were they filtered, massaged, interpreted, selectively reported, etc.? In fact there is always an element of judgment, selectivity, and interpretation concerning what to observe, and having observed, what to record, and what to report. Thus, despite the apparent confidence that may be exuded in publication, the cautious experimenter maintains a restraint born of continual reflection on the extensive preparations necessary for an apparently simple experiment and the many opportunities for error and misinterpretation.

We can broadly classify experiments into four categories, according to their intent, under these headings: (1) simulation, (2) abstraction, (3) verification, and (4) extension. In the first category the experimenter attempts to represent nature in miniature in so far as possible. An effort is made to include all of the relevant driving mechanisms, and the geometry is scaled as in nature, although some distortion may be necessary. Using theoretical guides such as the matching of the appropriate nondimensional ratios, the intent is to learn by trial and error to what extent the ocean circulation can be reproduced as a scale model. If it were possible to reproduce known features of the North Atlantic circulation, for example, such a model could be used to predict similarly scaled features of the ocean circulation in less accessible regions of the world. The predictions would serve as a guide to further exploration and would be compared with observations as they became available. The simulation mode of experimentation is appealing to the eye of the layman who can rather easily be convinced of its possible application and relevance.

The second mode of experimentation is rather like abstract art. The artist draws out (abstracts) from some natural subject those features that he imagines to be of significance, and he displays his interpretation of those features on canvas or in stone for the reaction of his peers and his public. The experimental scientist conceptually isolates one or more processes that he believes to be significant in nature, and he displays and tests them in the form of an operational laboratory experiment. Just as it may be difficult for the artist to persuade his lay audience of the sincerity of his efforts (although his fellow artists understand), so also the scientist may have difficulty convincing his nonprofessional audience that his experiment is relevant to the grand scheme of things past, present, and future.

Abstract experiments may be particularly successful in systems that can be decomposed linearly without doing violence to the essential dynamics, i.e., systems in which the abstracted phenomenon can be isolated by virtue of a lack of coupling with other processes. This may be possible because of a smallness of amplitude of potentially interactive processes, or because of mismatch in the temporal and spatial scales of the various processes. But even in those situations where decomposition is not warranted, one can say, "If . . . . Perhaps a planet will be found where these conditions prevail; or perhaps some machinery in a chemical plant, somewhere, generates the conditions that I am studying; or perhaps I can build upon this experiment to incorporate the interactive processes necessary for a more realistic representation of the oceanic circulation (simulation?). In any case I will publish my abstract results for the benefit of posterity."

Abstraction experiments, in contrast to direct simulation, are more readily subject to a posteriori mathematical analysis because of their relative simplicity. They may more directly lead to the advancement of theoretical aspects of the problem. Either mode may be regarded as exploratory, for it is likely that certain new aspects of the fluid circulations will emerge that were not anticipated and that will require rationalization. Here again the abstract experiment has the advantage of its simplicity, for deviations from the anticipated behavior will be more clearly recognized. The simulation experiment, however, will generally pose a larger variety of unanticipated phenomena because of its inherently greater complexity.

The verification mode of laboratory experiment implies an apparatus designed to test (and verify) a specific analytical or numerical model. A certain theoretical model predicts a steady-state circulation, the temporal development of a flow, or perhaps an instability. Apparatus is designed to match the conditions of the theory in so far as possible, and, as often as not, the theory is modified to conform to the limitations of the experiment. But in all cases under this category there is a detailed theory capable of a priori predictions and the physical conditions of the theory and of the experiment are closely matched. If this matching is sufficiently precise, the experiment will exactly verify the predictions. From this viewpoint the situation may at first appear to be rather sterile, but this is not the case at all. For following the adage, "No one believes a theory . . . ; everyone believes an experiment . . . ," we find that the theorist has arrived. For now everyone believes the theory, and rightly so, because it is confirmed by the experiment! The verification mode of laboratory experiment is widely endorsed by theorists.

The fourth type of experiment is the extension mode. The experimenter, having set up the theorist by collaboration and verification of his theory, now attempts to do him in by varying the parameters of the experiments, ϵ and δ, until they lie outside the range of validity of the theory. It is then incumbent upon the theorist to expand everything in powers of ϵ and δ to take into account the nonlinear aspects of the flow, but here the experimenter always has the upper hand and can easily keep one step ahead of the theorist. It is only necessary to do this in a manner in keeping with accepted scientific and engineering practice, expressing the experimental results in terms of the minimum number of nondimensional ratios appropriate to the case at hand. Oftentimes, of course, the experimenter and the theorist are one and the same person.

This short excursion into the philosophy of laboratory experimentation may be found useful in the interpretation of the remainder of this chapter. Perhaps not all experimental studies are cleary motivated and can be identified as belonging exclusively to one or the other of the categories discussed above, but such a classification seems to be by and large appropriate.

16.2 Introduction

Perhaps the first rotating laboratory experiment specifically directed toward understanding some aspect of the ocean circulation was that of C. A. Bjerknes in 1902, as reported by Ekman (1905). Because this report is not well known and because a discussion of its results will have later application, I have reproduced it here:

The late Prof. C. A. Bjerknes at Kristiania, whose vivid interest seems to have been bestowed on every extension of knowledge in his branch, also made in the autumn of 1902, some experiments with the object of verifying some of the results to be found in Section 1 of this paper.

His apparatus consisted of a low cylinder (12 or 17 cm. high and 36 or 44 cm. wide) made of metal or glass, and resting on a table which could be put into uniform rotation (about 7 turns a minute against the sun) by means of a water-turbine. To the upper edge of the rotating cylinder was attached a jet having a horizontal split, 10 cm. long and 1 mm. broad; and through this a stream of air was forced from a pump to produce a wind diametrically across the cylinder over a 10 cm. broad belt.

The motion of the water was observed by means of small balls which just floated in it. As a result of the rotation of the cylinder the current always swept towards the right and thus formed a large whirl-pool occupying, when seen in the wind's direction, the middle as well as the right half of the cylinder.

The direction of motion at different depths was observed at the centre of the cylinder, on a sensitive vane (4 cm. long and 1 mm. high) which could be raised and lowered during the experiment without disturbing the motion. The direction of the vane was read against a glass square divided into radians and laid on top of the cylinder. The following table taken from Prof. Bjerknes' note-book kept in his laboratory gives as an example a series of measurements made during such an experiment. The first column gives the depth in cm. below the surface, the second column the deviation of the current to the right or left of the wind's direction.

Surface (cm)	20–25° right
0.5	45–50° "
1	45–50° "
1.5	25–30° "
2	0–10° "
2.5	0° "
3	5–10° left
3.5	5–10° "
4	5–10° "
4.5	10° "
5	10–15° "
5.5	15–20° "
6	20° "
6.5	20° "
7	20–25° "

The circumstances under which these experiments were made were in any case such as to satisfy the conditions for stationary motion but very roughly; and an exact interpretation may furthermore be difficult owing to the shape of the vessel, etc. It is certain however that their real object was to obtain merely qualitative verification of the reality of the phenomena considered, and as such they are very striking and instructive. Both the deflection of the surface-current and the increase of the angle of deflection increases only in the very uppermost layer; and this is explained as a result of the rapid rotation of the vessel. Indeed a value of $\mu = 0.3$ (which would not appear to be too small for motion on such a small scale) would give $D_{Ek} = 2$ cm only. [Author's note: Here I have used D_{Ek} in place of D to avoid later confusion. As used here and in other recent literature $D = D_{Ek}/\pi$ is the mathematically more convenient measure of boundary-layer thickness.] The directions of motion below this level have very much the appearance of a "midwater-current" produced by a pressure-gradient.

Apparently Ekman determined the effective viscosity $\mu = 0.3$ indirectly from his estimate $D_{Ek} = 2$ cm, for with a rotation rate $\Omega = 0.73$ s^{-1} and with the viscosity of water $\mu = 0.01$ we find $D_{Ek} = 0.37$ cm and $D = 0.117$ cm. Thus the laminar Ekman depth would be considerably shallower than that inferred from the experimental data, as Ekman must have recognized. Departing temporarily from the historical development to consider this question, these early results possibly may be explained in the light of recent laboratory experiments and numerical calculations on the stability of the Ekman boundary layer.

Laboratory studies have mostly been confined to flow over a rigid boundary (Faller, 1963; Faller and Kaylor, 1967; Tatro and Mollo-Christensen, 1967; Caldwell and van Atta, 1970; Cerasoli, 1975), and most comparative theoretical studies also have been concerned with the rigid-boundary case.

The Reynolds number for the Ekman layer is defined as $Re = UD/\nu$, where U is the difference between the speed at the boundary and the speed of the interior flow. Over a rigid boundary the critical value for instability is approximately $Re_c = 55$. For the Ekman layer with a free boundary, the profile of flow is exactly the same, but potential instabilities are not constrained by a no-slip boundary condition. A preliminary estimate given by Faller and Kaylor (1967) was $Re_c = 12 \pm 3$ for the first onset of instability. This result has recently been confirmed with a more accurate calculation by Iooss, Nielsen, and True (1978), who found $Re_c = 11.816$. In C. A. Bjerknes's experiment the corresponding critical surface water speed would be approximately $U = 1$ cm s^{-1}, and the corresponding critical wind stress would be $\tau_c = 0.12$ dyn cm^{-2}. Still another possibility is that of Langmuir circulations, a subject that will be touched upon briefly later, in section 16.6.10 and in this connection it would be interesting to know whether the air jet had sufficient speed to raise capillary waves. Unfortunately Ekman's account gives no means of estimating the air or water speeds, but it seems likely that the surface Ekman layer was unstable, thus accounting for the approximate $D_{Ek} = 2$ cm and the lack of an idealized spiral.

In determining the scope of this chapter it has been found necessary to restrict the material arbitrarily to laboratory experiments rather directly concerned with large-scale oceanic circulations, and a number of significant developments and important areas of research have been omitted from consideration. With one exception we shall be concerned exclusively with rotating experiments, omitting laboratory studies of the generation and interaction of surface and internal waves, thermohaline (double-diffusive) phenomena, thermal convection, turbulence, and other fundamental fluid dynamic studies that may be directly related to oceanic processes. Even in the realm of rotating fluids it is not possible to consider all studies of interest in as much detail as may be warranted, and there will be those overlooked in the literature. Moreover, the author confesses to a certain prejudice, which will be evident, in emphasizing his own contributions and those with which he is most intimately familiar.

After the experiments of C. A. Bjerknes, the first serious attempt at the isolation of an oceanic circulation in a rotating laboratory experiment appears to have been that of Spilhaus (1937). C. G. Rossby had proposed that the Gulf Stream might be similar to an inertial jet emerging from the Straits of Florida. Differences between the Gulf Stream and the usual laboratory jet might be expected from the effects of the earth's rotation (with the consequent adjustment to geostrophic balance) and the stratification of the ocean.

Spilhaus's attempts to verify certain aspects of Rossby's concepts consisted of (1) some preliminary trial experiments in a small cylindrical tank with a jet of water injected from a slot in the wall, (2) more elaborate uniform-fluid experiments in a 6-ft-diameter tank with the jet emerging from an axial tube in the center and with excess water overflowing at the rim, and (3) experiments with a two-layer system (immiscible fluids) in a 4-ft-diameter tank with a central jet confined to the upper layer. The resulting circulations were more complex than anticipated and could not be interpreted satisfactorily in terms of Rossby's theoretical work. Nevertheless it is interesting to read Spilhaus's account of the experiments in the light of our present understanding of source–sink flow in rotating experiments.

16.3 The Experiments of W. S. von Arx

Serious and sustained efforts at modeling the ocean circulation began with von Arx circa 1950. At about the same time, and quite independently, major laboratory efforts were being initiated by D. Fultz and R. Long at the University of Chicago, who were primarily interested in atmospheric circulations, and by R. Hide at King's College, the University of Newcastle, who was attempting to understand the fundamentals of circulation in the molten interior of the earth. The eventual collaboration and interchange of information between these investigators and their students became part of the rapid development of the field of study that today we refer to as geophysical fluid dynamics. The rather bold initiatives of these scientists eventually led to the realization by a broader base of theoretically oriented meteorologists and oceanographers that it was indeed possible to simulate certain features of complex geophysical circulations in the laboratory.

The first major experiments reported by von Arx (1952) were conducted in a paraboloidal basin in which the northern hemisphere continental boundaries were modeled in sponge rubber. The trade winds were derived from the relative motion of the air in the laboratory at the rotation rate $\Omega = 3.18$ s^{-1}, and the mid-latitude westerlies were generated by three stationary vacuum cleaner blowers with appropriately directed nozzles. With this basic system it was possible to generate many of the essential features of the northern ocean circulations that were then known to exist. These and subsequent experiments had certain shortcomings, but their role in the stimulation and development of further laboratory studies should be recognized.

The paraboloidal models were not entirely satisfactory in several ways: geometrical distortion was inevitable; the wind fields could not be precisely controlled or measured; at wind speeds of about 2 m s^{-1} the water surface was wavy and it is likely that the surface Ekman layer was unstable; and with the complex geometry a slight tilt of the rotation axis produced large undesirable oscillations. Moreover, a severe critic might argue that in attempting to approach the complexity of nature, the ability to achieve understanding of fundamental processes was sacrificed. Nevertheless, in the developing field of oceanography there was still at that time what seemed to be a residual dichotomy of opinion between advocates of the wind-generation theory of ocean currents and advocates of the thermal-generation theories, which tended to divide theoretical and observational oceanographers; von Arx's experiments were a convincing verification of the theories of Stommel, Munk, and others with respect to wind generation of the primary ocean circulation.

The von Arx experiments helped stimulate further laboratory studies in various ways. First, there was the development of experimental techniques for use with rotating laboratory models. Second, the rotating apparatus that he built was used in later experiments by this author and a succession of others, and the basic rotating system is today still available for use at the Woods Hole Oceanographic Institution. Third, through a paradox that was observed in the early studies with the paraboloid, the possibility of simulating the β-effect by a radial variation of depth was first realized.

Von Arx originally employed his carefully constructed paraboloidal basin at the equilibrium rotation rate, i.e., the rotation rate at which the water surface had the same paraboloidal shape as the container. Strangely, at this speed no western boundary current could be found. But at higher speeds, the western intensification was obviously present and at lower speeds there was eastern itensification! This behavior was explained by C. G. Rossby (von Arx, 1952) and eventually led to the proposition, by the present writer, that the paraboloidal shape was unnecessary—that a flat tank with a paraboloidal free surface, produced by a suitable rotation rate, would work equally well.

Further experiments using a flat-bottomed tank (von Arx, 1957) allowed lower rotation rates and a generally better overall experimental control. Studies of the southern hemisphere were undertaken and the representation of natural circulations was somewhat better than in the paraboloidal basin.

Success with the flat geometry encouraged von Arx to build a much larger apparatus, a floating cylinder with a working area having a diameter of 4 m and driven by impeller blades beneath the tank. A photograph and description of this apparatus appear in von Arx (1957). In an effort to improve the analogy with nature, infrared heaters were installed above the tank with the intention of heating the surface layers of water in the lower latitudes, thus simulating solar heating of the upper layers of the ocean. Unfortunately, the production of a warm surface layer essentially destroyed the induced β-effect associated with the paraboloidal depth variation and eliminated the western intensification!

Although this mammoth rotating tank did not improve significantly the ability to reproduce nature in miniature, the availability of this large apparatus encouraged this author to undertake experiments on the instability of the Ekman boundary layer (Faller, 1963). With the large-diameter tank the Reynolds numbers necessary for instability of the Ekman layer could be achieved at much lower Rossby numbers, i.e., with a more nearly geostrophic, as opposed to gradient, circulation.

As an aside, to this author's knowledge the Ekman number, which figures prominently in all rotating laboratory experiments, was first introduced by H. Lettau in a turbulence course at MIT in 1953 that was attended by both von Arx and this writer. Lettau's definition was $E = D/H$, the geometric ratio of the depth of an Ekman layer to the total depth of fluid. This definition was introduced into the literature by Faller and von Arx (1958) and by Bryan (1960) although the ratio L/D, where L is an arbitrary characteristic length, was defined as the Ekman number by von Arx (1957). The ratio D/H is, of course, also the inverse of the fourth root of the Taylor number; and Fultz (1953) referred to L^2/D^2 as a rotation Reynolds number. Stern (1960b) used Ekman number $E = \Omega H^2/\nu = (H/D)^2$, but it appears that now the generally accepted definition is $E = \nu/\Omega H^2 = (D/H)^2$ or a definition in which H is replaced by some characteristic horizontal scale. This writer prefers the definition $E = D/H$ as originally given by Lettau because of its simple geometric significance and because the scaled Ekman depth is then proportional simply to E rather than $E^{1/2}$, although other definitions may sometimes be mathematically more convenient.

16.4 The SAF Model

In reviewing theories of the ocean circulation, Stommel (1957b) recognized the essential analogy between the precipitation theories of Hough and Goldsborough and more recent theories of the wind-driven ocean circulation. In brief, each mechanism may be considered in terms of distributed sources and sinks of fluid at or near the ocean surface. Moreover, isolating the deep ocean, one could imagine this layer as being driven by the sinking of cold water in selected Arctic and Antarctic regions with more or less uniformly rising motion out of the top of this layer to satisfy the continuity of mass.

At that time this writer was engaged in the study of laboratory models of the atmospheric circulation, using the rotating turntable of von Arx. At the suggestion of H. Stommel and A. B. Arons it was a relatively straightforward matter to arrange a simple source-sink experiment in analogy with the precipitation mechanism. It was predicted that with distributed precipitation over a portion of the interior of a bounded basin and with the β-effect simulation found by von Arx, the geostrophic constraint in the interior of the basin would require some rather bizarre flow patterns. In particular, with the geostrophic condition as the overriding constraint on the interior flow, continuity would have to be satisfied by a combination of east-west geostrophic currents along contours of constant depth and frictionally balanced western boundary currents, unspecified in detail.

Figure 16.1 shows the very first experiment of the series eventually presented in SAF, that is, Stommel, Arons, and Faller (1958), and in F, that is, Faller (1960). In this initial trial the depth variation was provided in part by a spherical-cap polar dome (built of wood) that was 5 cm high in the center, tapering to zero height at the radius $r = 91.4$ cm in a tank of total radius $r = 106.7$ cm. An additional depth variation was provided by the counterclockwise rotation rate of $\Omega = 0.84$ s^{-1}. The average water depth was $h = 10$ cm at $\Omega = 0$, measured above the flat portion of the tank bottom near the rim.

The side walls of the 30° sector were carved to fit the polar dome and were puttied in place with a small gap between the two sides at the apex to allow water to escape into the remainder of the circular tank as necessary to balance the source. An area of precipitation was arranged by dribbling dyed water (from a supply vessel) through a perforated coffee can at a flow rate of

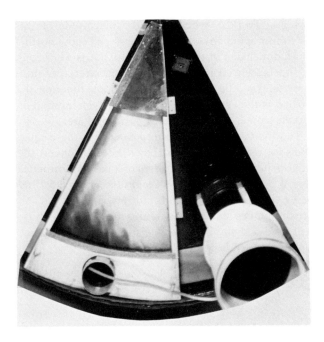

Figure 16.1 The original experiment of Stommel, Arons, and Faller (1958). The black area on the right shows a portion of a spherical-cap polar dome that was 5 cm high at the center and tapered to zero depth at 91.4-cm radius. This dome was painted white within the 30° sector (on the left) that was formed by radial side walls shaped to fit the bottom boundary and puttied in place. The entire circular tank contained water approximately 3 cm deep at the center and 11 cm deep at the rim at the rotation rate $\Omega = 0.84 \text{ s}^{-1}$, counterclockwise as viewed from above. Dyed water from the bucket on the right drained through a tube into a perforated coffee can and "precipitated" at the rate of 200 cm³ min⁻¹ into the sector, near the rim. A small gap between the side walls at the apex allowed flow out of the sector. The western boundary current is clearly evident as a dark band of dyed water. (See figure 5.3).

approximately 200 cm³ min⁻¹. The experimental conditions were not precisely controlled, and figure 16.1 shows an area of disturbance near the precipitation source. Similar disturbances occurred near the apex, but these are obscured in figure 16.1 by a partial masonite cover. Nevertheless, the predicted western boundary current was obvious, thus confirming the theoretical prediction and leading to an extended series of controlled experiments.

The SAF experiments established the validity of the following basic concepts for slow, steady motion of a uniform fluid in a bounded basin without closed depth contours:

(1) If there is no local source (sink) of fluid and the local depth remains constant, the interior geostrophic flow is constrained to follow contours of constant depth.

(2) If there is a local source (sink) of fluid or if there is a gradual decrease (increase) in the local fluid depth, a column of fluid moves toward greater (lesser) depths satisfying the geostrophic condition of zero horizontal divergence.

(3) Failure of the interior motion to satisfy an overall mass balance produces a frictional western boundary current, the interior flow moving along lines of constant depth to the western boundary as required for overall continuity.

Perhaps the most striking result of these experiments was the verification of the recirculation phenomenon illustrated in figure 16.2. With a source of strength S_0 at the apex of the sector and a distributed sink (represented by the gradual rise of the free surface), the predicted western boundary current flow was twice the source strength! In the laboratory experiment this current was composed half from the colored source water and half from the recirculated clear water.

Elaboration of the SAF experiments (Faller, 1960) demonstrated several additional features of the flow:

(1) In east–west currents (along contours of constant depth) the effect of bottom friction causes westward-moving currents to diverge in a laminar and geostrophic Gaussian plume. This divergence is forced by bottom Ekman layer suction (and pumping) together with the effect of the radial depth variation. More sur-

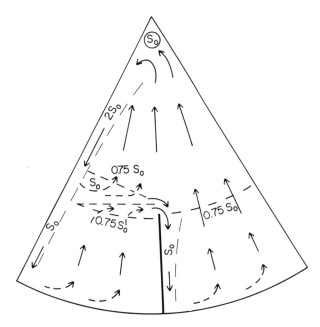

Figure 16.2 An example of the partial radial-barrier experiments of Faller (1960) for a source at the apex and a uniformly rising free surface. The source S_0 is joined by recirculated interior water to give a total western boundary current of $2S_0$. At the radius of the inner end of the partial radial barrier the WBC divides, one S_0 continuing along the boundary and one S_0 separating. Of the latter, $0.25S_0$ reaches the radial barrier and participates in the WBC there, while $0.75S_0$ moves northward as the required interior flow. A transport of $0.75S_0$ from the southwest basin joins in the eastward-moving jet to complete the source water for the transport of one S_0 along the radial barrier.

prisingly, the theory, being linear and reversible, predicts that an eastward-moving current of constant mass flux should converge into a narrower and more intense jet. As shown in the experiments, catastrophy is avoided in the converging jet because the flow spreads out in the western boundary current to whatever extent is necessary for its eventual convergence as it flows eastward.

(2) With an interior source (or sink) there was an intense recirculation analogous to the effects of the wind-spun vortex described by Munk (1950). The intensity of this recirculation phenomenon was found to be in excellent agreement with a theoretical analysis.

(3) The concepts predicted and verified in SAF were extended to somewhat more complex geometries involving partial radial barriers. An example of such a flow, qualitatively verified by experiment, is shown in figure 16.2, where the indicated transports are the theoretical predictions. Further application to an Antarctic-type geometry illustrated the important role of the radial barriers in allowing the buildup of pressure gradients essential to the maintenance of the geostrophic flow.

(4) In certain source-sink experiments with closed contours of constant fluid depth, it is not possible to construct purely geostrophic regimes of flow. In such cases, lacking resolution of the problem by western boundary currents, the effects of bottom friction dominate the flow. Examples of the complex set of spiral flow patterns necessary in order to transport fluid from an interior source to an interior sink illustrated the dominance of frictional effects.

(5) One feature that would not stand the test of time was the observation of an eastern boundary current when the flow was injected through the eastern boundary of the sector. In later experiments (Faller and Porter, 1976) it was found that this current was probably a density-driven flow due to incomplete thermal control of the injected water.

The analogy between these laboratory experiments and theoretical models of steady wind-driven ocean circulation may be illustrated most clearly by a comparison of the governing equations from three studies. These equations are

Stommel (1948) $\quad \alpha \dfrac{\partial \psi}{\partial x} + \nabla^2 \psi = \gamma \sin \pi y/b,$ (16.1)

Munk (1950) $\quad \beta \dfrac{\partial \psi'}{\partial x} - A \nabla^4 \psi' = \operatorname{curl}_z \tau,$ (16.2)

Faller (1960) $\quad \dfrac{\partial h}{\partial r}\left(\dfrac{1}{r}\dfrac{\partial \psi}{\partial \theta}\right) + \dfrac{D}{2}\nabla^2 \psi$
$= Q - \dfrac{\partial h}{\partial t}.$ (16.3)

In (16.1) $\alpha = (H^*/R)(\partial f/\partial y)$, where H^* is a constant effective ocean depth and R a linear drag coefficient;

and $\gamma = F\pi/(Rb)$, where F is the magnitude of the maximum wind stress per unit mass and b the north-south dimension of the model. In (16.2) A is a horizontal Austausch coefficient, τ the wind stress per unit mass, and ψ' the mass transport stream function. In (16.3) h is the variable fluid depth and Q an internal or surface source of fluid per unit horizontal area.

To illustrate the close correspondence of these equations we may multiply (16.1) by R/H^* and convert the right-hand side to the derivative of a cosine; introduce $\psi = \psi'/H$ as the velocity streamfunction in (16.2), H being the ocean depth; and multiply (16.3) by f/h, reorient the coordinate axes with $r\,d\theta = dx$ and $dy = -dr$, and add the horizontal viscous term. The three equations then become

$$\dfrac{\partial f}{\partial y}\dfrac{\partial \psi}{\partial x} + \dfrac{R}{H^*}\nabla^2 \psi = -\dfrac{1}{H^*}\dfrac{\partial F \cos \pi y/b}{\partial y},$$ (16.4)

$$\beta \dfrac{\partial \psi}{\partial x} - A\nabla^4 \psi = \dfrac{1}{H}\operatorname{curl}_z \tau,$$ (16.5)

$$\left(-\dfrac{f}{h}\dfrac{\partial h}{\partial y}\right)\dfrac{\partial \psi}{\partial x} + \dfrac{fD}{2h}\nabla^2 \psi - \nu \nabla^4 \psi = -\dfrac{f}{h}Q + \dfrac{f}{h}\dfrac{\partial h}{\partial t}.$$ (16.6)

Equations (16.4)–(16.6) clearly demonstrate the analogy between the laboratory experiment and the basic theories of the steady wind-driven ocean circulation. In particular, $\beta = \partial f/\partial y$ is modeled by $-fh^{-1}\partial h/\partial y$, Stommel's linear drag law corresponds to the Ekman layer effect with R/H^* the equivalent of $fD/2h$, and a positive curl of the wind stress in (16.4) or (16.5) is represented in (16.6) by either an internal (or surface) sink of fluid $-fQ/h$ or by a local rate of increase of depth $(f/h)\partial h/\partial t$. Note also that with the addition of the lateral viscous term to (16.6) this equation is capable of describing steady, linear western boundary currents analogous to those of Munk.

16.5 Experiments with Rotating Covers

The source-sink experiments were more controlled than the wind-driven experiments, and the regular boundary conditions of the 60° sector allowed a clear distinction between the interior flow and the western boundary current (WBC); turbulence was essentially eliminated and the interior flow could be guaranteed to be essentially geostrophic by controlling the flow rates. The introduction of a rotating cover (rotating lid) as the driving mechanism produced a certain degree of additional control by completely enclosing the fluid, by eliminating the possibility of density differences in the source flow, and by allowing the parameters of the problem to be more easily varied. For example, transient circulations due to an impulsive start of the cover or due to a periodic oscillation of the cover could be easily controlled. Such experiments would be difficult with the sources and sinks of the SAF studies.

Bryan (1960) was perhaps the first to use a rotating lid in an experiment directed toward the understanding of oceanic circulations. Having in mind the planetary waves of the atmosphere and the Gulf Stream meanders, he investigated the baroclinic stability of a two-layer system of salt and fresh water in a 34-cm-diameter cylindrical vessel, the driving force being a slowly rotating lid. He successfully produced symmetric flow, one to three regular waves, and an irregular wave regime, more or less in agreement with a corresponding theory of baroclinic instability. These experiments, of course, had a great deal in common with the annulus experiments of Hide (1958) and with the annulus, open dishpan, and two-fluid experiments of Fultz (1953).

Rotating-lid experiments that turned out to be analogous in many ways to the SAF studies were introduced by Beardsley (1969) in connection with the theories of Pedlosky and Greenspan (1967), the so-called sliced-cylinder model of the wind-driven ocean circulation. The steady-state sliced-cylinder circulation can be described and understood by a relatively simple theoretical analysis, and the following section, through equation (16.18), develops the theory of this model in terms borrowed from the theory of the SAF experiments. This short section will be of particular interest to those directly concerned with the theory of such experiments but may be passed over by the more casual reader. In either case figures 16.3 and 16.4 illustrate the model geometry and circulation.

The radial transport per unit of circumference in the upper Ekman layer due to a differential rotation $\epsilon\Omega$ is simply $T_{Ek}(r) = \epsilon\Omega r D/2$, and the corresponding source strength per unit area for the interior due to Ekman layer divergence is $Q = -\epsilon\Omega d$, independent of radius. For positive ϵ this source is the equivalent of a uniform rise of the free surface in the SAF studies.

At the rim, $r = a$, the Ekman transport $T_{Ek}(a)$ feeds into a Stewartson boundary layer at the circular wall. Since the interior flow is quite slow for large α, little of the flow into the Stewartson layer from above can be taken up by the slow bottom Ekman layer, and we ignore this effect. The Stewartson layer then feeds uniformly into the interior, giving an average flow normal to the wall:

$$U_n = \epsilon\Omega aD/2h. \tag{16.7}$$

The interior geostrophic flow is governed by equation (16.6) omitting the lateral and bottom friction

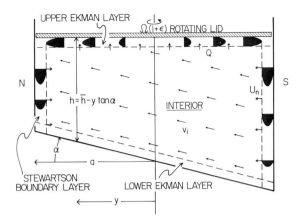

Figure 16.3 A schematic N–S cross section of the circulation in a sliced-cylinder model. The circulation is driven by the divergent Ekman layer of the rotating lid with the induced suction velocity $Q = -\epsilon\Omega D$. The Ekman flow at the rim of the rotating lid turns into the Stewartson layer and feeds the interior as a velocity component normal to the wall U_n. The geostrophic S–N interior flow v_i is independent of x and y and is given by $v_i = -Q/\tan\alpha$. Global continuity of mass is satisfied by geostrophic east–west interior flows and the western boundary current (not shown).

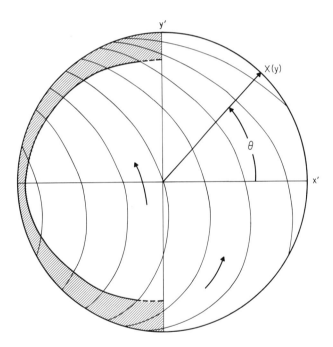

Figure 16.4 Streamlines of the theoretical, linear, steady-state circulation in the sliced-cylinder experiment of Pedlosky and Greenspan (1967) and Beardsley (1969) for a bottom slope angle $\alpha = 30°$. For large α the flow away from the eastern boundary becomes evident in the streamline pattern because of the reduction of v_i in comparison to U_n (figure 16.3). The variation in width of the western boundary current is schematically illustrated.

terms. The resultant equation for the south-to-north interior flow is

$$v_i = -Q/\tan\alpha = \epsilon\Omega D/\tan\alpha, \qquad (16.8)$$

where $v_i = \partial\psi/\partial x$ and $\tan\alpha = -\partial h/\partial y$. Thus for $\epsilon > 0$, as in the case considered by Pedlosky and Greenspan (1967), $Q < 0$ (Ekman suction) and it follows that $v_i > 0$. Later experiments of Beardsley used $\epsilon < 0$.

The interior velocity component u_i is determined by continuity. For a cell in the interior mass continuity is given by

$$\left[v_ih - \left(v_ih + \frac{\partial v_ih}{\partial y}dy\right)\right]dx$$
$$+ \left[u_ih - \left(u_ih + \frac{\partial u_ih}{\partial x}dx\right)\right]dy + Q\,dx\,dy = 0.$$

Using equation (16.8) and $\partial h/\partial x = 0$ it is readily found that $\partial u_i/\partial x = 0$. Following the assertion made in SAF that there can be no geostrophic flow component normal to the eastern boundary, but that unlimited flow into or out of the western boundary layer is permitted, the value of u_i is determined by u_e, the west-east flow at the eastern boundary. This in turn is composed of two parts: the source flow from the Stewartson layer is $u_{e1} = -U_n/\cos\theta = -\epsilon\Omega Da/2h\cos\theta$; and from the constraint that there be no geostrophic flow normal to the boundary we obtain $u_{e2} = -v_i\tan\theta$. The total west-east interior flow is then

$$u_i = -\epsilon\Omega D\left(\frac{a}{2h\cos\theta} + \frac{\tan\theta}{\tan\alpha}\right). \qquad (16.9)$$

Integrating (16.8) and (16.9) to obtain an interior streamfunction defined by $v_i = \partial\psi/\partial x$ and $u_i = -\partial\psi/\partial y$ gives

$$\psi' = \frac{\psi}{\epsilon\Omega Da} = \frac{a}{\bar{h}}\frac{\sin^{-1}y'}{2}$$
$$+ \left[\left(\frac{a}{\bar{h}}\right)^2\frac{\tan\alpha}{2} + \frac{1}{\tan\alpha}\right][1 - (1-y'^2)^{1/2}]$$
$$+ \frac{x'}{\tan\alpha}, \qquad (16.10)$$

where $y' = y/a$, $x' = x/a$, and $\psi' = 0$ at $x' = 0$, $y' = 0$. In this integration I have used the approximation

$$h = \bar{h} - y\tan\alpha \simeq \bar{h}\left(1 + \frac{y\tan\alpha}{\bar{h}}\right)^{-1},$$

and as a result (16.10) differs slightly from Beardsley's (1969) equation for pressure.

Figure 16.4 illustrates the pattern of ψ' for $\alpha = 30°$. With this large slope v_i is small and the effect of the source at the rim is made conspicuous [compare Pedlosky and Greenspan (1967, figure 5)]. Since the rim source does not directly contribute to the north-south (N-S) transport, it is easily seen, as pointed out by Beardsley (1969), that the WBC transport T_w is determined solely by the value of v_i integrated over x. By continuity

$$T_w(N-S) = -\int_{-X(y)}^{X(y)} hv_i\,dx = -2hv_iX(y). \qquad (16.11)$$

The WBC transport parallel to the boundary, however, requires that the result in (16.11) be divided by $\cos\theta$. Dividing also by $\epsilon\Omega Dah$, the nondimensional transport per unit depth and parallel to the boundary is

$$T'_w(\|) = \mp 2\tan\alpha, \qquad (16.12)$$

the upper sign corresponding to $\epsilon > 0$.

But although $T'_w(\|)$ is constant, the current sees different variations of depth $\partial h/\partial s$ along its axis s, and the width of the current must change. For positive y (and $\epsilon > 0$ as in figure 16.3) the current narrows as it moves south toward $y = 0$, and it is also fed from the interior flow. For $y < 0$ the current feeds the interior and at the same time broadens because $\partial h/\partial s$ decreases along the direction of flow. By the time the flow reaches $\theta = -\pi/2$ the WBC must completely disperse, and under these circumstances it is not surprising that an instability occurs for sufficiently large ϵ.

The steady-state, linear analysis given above cannot take into account transients, instabilities of the flow, details of the boundary layer flow, or the range of validity of the theory. These aspects of the problem were the principal focal points of the analytic, numerical, and experimental sliced-cylinder studies cited above. Nevertheless, the simplified treatment given here reveals the essential physical mechanisms without mathematical obfuscation. In this spirit we continue with an elementary treatment of the WBC to illustrate certain principles analyzed by Beardsley (1969) and later disucssed also by Kuo and Veronis (1971).

Beardsley noted that when the angle α is small, the WBC is dominated by bottom Ekman-layer pumping, rather than by lateral friction. Except very close to the wall, the boundary current then is essentially that found by Stommel (1948) with friction entering the problem through a constant drag coefficient R/H^*, as in (16.4).

Figure 16.5 illustrates a WBC profile, the cross-stream Ekman transport beneath the current, and the vertical flux from the Ekman layer. The WBC is normally much wider than D, and in the case where lateral friction dominates the boundary layer, the ratio of its width W' to D is $W'/D = (h/D)^{1/3}(4/\tan\alpha)^{1/3}$ (see below). Thus, in the Ekman layer the usual boundary-layer assumption applies, namely, that lateral variations of the flow can be neglected in comparison to variations normal to the boundary. Under the simplifying as-

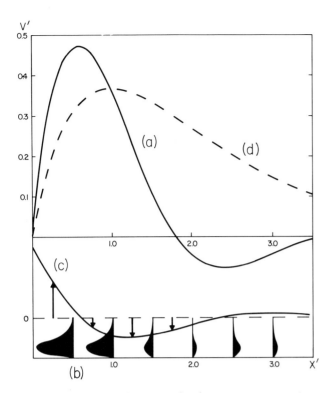

Figure 16.5 The nondimensional velocity component $v' = e^{-x'} \sin \sqrt{3}x'$ for the Munk boundary layer in which lateral friction dominates the vorticity balance. (b) The Ekman transport beneath the boundary layer in (a). (c) Flux into or out of the western boundary current by Ekman layer convergence or divergence. (d) The nondimensional velocity $v^* = x^* e^{-x^*}$ at the transition between the Munk and the Stommel boundary layer regimes.

sumptions given below it can also be shown that the Ekman layer transport is proportional to the local value of v above the Ekman layer. An inconsistency, apparent in figure 16.5, is that the maximum Ekman layer pumping occurs at $x = 0$ and the no-slip condition at the wall is not satisfied. To completely justify the approximate analysis given below, a more complete boundary-layer scaling and analysis including the sidewall Stewartson layers, as given by Beardsley (1969) and Kuo and Veronis (1971), is required.

Here we specify a total northward transport of the WBC, T_w, a straight N–S boundary, a linear depth variation $-\partial h/\partial y = +\tan\alpha$, and a locally constant depth, where h appears as a coefficient in the final differential equation; and we neglect variations of the current with y, i.e., along the boundary. Then (16.6) is appropriate, and after dividing by v it reduces to the third-order, ordinary linear differential equation with constant coefficients

$$\frac{d^3v}{dx^3} - \left(\frac{1}{Dh}\right)\frac{dv}{dx} - \left(\frac{2\tan\alpha}{D^2 h}\right)v = 0 \qquad (16.13)$$

subject to the conditions $v = 0$ at $x' = 0$, $v = 0$ at $x' = \infty$, and $T_w = \int_0^\infty vh\,dx$. It is convenient to define a nondimensional coordinate $x' = x/W'$, where $W' = [8\,hD^2/(2\tan\alpha)]^{1/3}$ will be seen to be the characteristic width of the WBC. Then, temporarily neglecting the Ekman layer effect by setting $D = 0$, (16.13) reduces to

$$\frac{d^3v}{dx'^3} - 8v = 0. \qquad (16.14)$$

After applying the boundary conditions the solution to (16.14) is the exponentially-damped sine profile

$$v = \left(\frac{4}{3^{1/2}}\frac{T_w}{hW'}\right)e^{-x'}\sin\sqrt{3}x', \qquad (16.15)$$

which corresponds to the oscillatory solution of Munk (1950). A nondimensional v' for this result is plotted in figure 16.5.

The WBC in the absence of lateral friction has been called the *Stommel boundary layer* (Kuo and Veronis, (1971), and from (16.13) this is given by the solution to

$$\frac{dv}{dx} + \frac{2\tan\alpha}{D}v = 0. \qquad (16.16)$$

Here the condition $v = 0$ at $x = 0$ must be relinquished, and the condition at infinity will be satisfied automatically. Applying the transport condition, the solution is the exponential profile

$$v = \frac{T_w}{hW''}e^{-x''}$$

where

$$x'' = x/W'' \qquad \text{and} \qquad W'' = (D/(2\tan\alpha).$$

In terms of a simulated β-effect, being given $\beta^* = (-2\Omega/h)(\partial h/\partial y)$ it follows that $W'' = \Omega D/h\beta^*$.

With the scaled coordinate x', the full equation is

$$\frac{d^3v}{dx'^3} - B\frac{dv}{dx'} - 8v = 0, \qquad (16.17)$$

where $B = 4(D/h)^{1/3}(2\tan\alpha)^{-2/3}$. Substituting the trial solution $v = e^{kx}$, the characteristic equation is

$$k^3 + ak + b = 0, \qquad \text{where} \qquad a = -B \text{ and } b = -8.$$

Complex values of k correspond to oscillatory solutions, and nonoscillatory boundary layers occur only for $(b^2/4) + (a^3/27) \leq 0$, or $D/h \geq 27\tan^2\alpha$. Accordingly, the condition $D/h = 27\tan^2\alpha$ may be defined as the transition point between the Munk and Stommel boundary layer regimes. In the SAF and F experiments where $\tan\alpha = \Omega^2 r/g$ this transition took place at approximately $r = 23$ cm in a tank of radius $a = 100$ cm.

The solution at this transition is the product of linear and exponential functions as given by

$$v = \left(\frac{T_w}{hW^*}\right)x^* e^{-x^*}, \qquad (16.18)$$

where $x^* = x/W^*$ and $W^* = (3hD/4)^{1/2}$. The slope $\tan \alpha$ is implicit in this expression for W^* through $D/h = 27 \tan^2 \alpha$.

The theoretical studies of Pedlosky and Greenspan (1967) for the sliced-cylinder geometry emphasized the transient Rossby-wave character of the spin-up problem, i.e., the development of the steady-state flow by the accumulated effect of westward-propagating Rossby waves in response to the spin-up of the lid. Owing to the need for a more detailed examination of the steady-state flow and because of the discovery of an instability of the WBC, Beardsley's original experiments were temporarily diverted from the goal of studying the transient response. The instability was analyzed in detail by Beardsley and Robbins (1975), who described it as a "*local* breakdown of the finite-amplitude topographic Rossby wave embedded in the western boundary current transition region." This general description would seem to fit almost any reasonable theory of the observed instability, but it is noteworthy that a plausible analytical theory was presented and that numerical calculations of the flow showed similar instabilities and good agreement with the experiments.

In a companion paper Beardsley (1975) experimentally and theoretically examined the transient response of the flow to an oscillatory rotation of the lid. He again found good agreement between theory and experiment in the resonant response of topographic Rossby wave modes. In addition he found that the Stokes drift associated with the Rossby waves partially offset the mean Eulerian flow, a consideration of possible importance for the interpretation of oceanic processes (the result generally agrees with the theory of Moore, 1970). In a short series of trial experiments, Beardsley (1974) also changed the geometry of his experiment by replacing the planar sloping bottom with an axisymmetric conical bottom and introduced partial radial barriers, analogous to those of F (1960), to simulate certain aspects of the Antarctic circulation regime. These experiments again clearly illustrated the anticipated steady-state interior and boundary-layer flows.

Another interesting class of rotating-lid experiments was introduced by D. J. Baker. Baker and Robinson (1969) have presented a comprehensive theoretical and experimental analysis of the results and Baker (1970) has given a more pictorial and descriptive account of this research for popular consumption. The fluid (a solution of thymol blue) was contained, top and bottom, between spherical boundaries ground from plastic blocks. The lateral boundary was a circular cylinder, and the lower boundary was rotated to generate the flow. The confined fluid was meant to represent a circular ocean of uniform fluid on a spherical earth. This ingenious apparatus was mounted on a rotating table in such a way that the center of the model ocean could be set for any latitude from the equator to the pole. Indeed, the primary purpose of this experiment was a study of equatorial dynamics in the hope of obtaining a uniform-fluid analogue of the equatorial undercurrent.

Baker (1966) had introduced an important new quantitative method for the measurement of slow flows in the interior of liquids, the thymol-blue-indicator technique. A slightly acidic solution of thymol blue is used as the working fluid, and platinum wires are strung through the interior. When a current is passed through the wires the indicator immediately adjacent to the wires turns to a bluish-brown color and serves as a tracer for the subsequent flow. Baker and Robinson (1969) used an extensive rectangular array of wires to illustrate and measure the overall pattern of flow for a considerable range of flow speeds.

In the range of Rossby numbers where steady laminar flow would be expected, the basic interior and boundary-layer circulations were as would be anticipated from elementary theory. At higher values of the Rossby number, however, interesting asymmetries and irregularities of the flow, attributable to inertial effects, were observed.

When the latitude of the center of the model ocean was sufficiently close to the equator that a large equatorial zone was present, an interior zonal current opposite to the direction of driving of the cover was observed, and this current was tentatively identified as a uniform fluid analogue of the equatorial undercurrent.

16.6 A Variety of Interesting Experiments

16.6.1 Further Source–Sink Experiments

Kuo and Veronis (1971) reinvestigated many of the SAF and F experiments with a similar configuration of the apparatus but with a number of interesting variations and with a more exacting theoretical analysis of the boundary-layer flow. Their principal observational tool was the thymol-blue technique of Baker (1966) except that the platinum wires were alternately coated and exposed in short strips so that a pulse of electric current gave a dashed line of dye. This method allowed more accurate measurements of the flow components and improved the flow visualization.

As in the studies of Beardsley (see section 16.5) their experimental conditions covered both the Munk and Stommel WBC regimes and their experimentally observed current profiles were in excellent agreement with theory.

In addition to the effect of the paraboloidal free water surface, depth variation was controlled by a sloping bottom, and in one case by a partially sloping bottom in the SW corner of the 60° sector. The latter experiment was used to confirm certain basic theories about the effect of bottom topography on the separation of the boundary current from the coast. Concentrated sources and sinks in the interior, similar to the source used in F (1960) and to the effect of the wind-spun vortex of Munk (1950), also illustrated separation of the boundary current because of the source-sink distribution.

Veronis and Yang (1972) extended the uniform fluid WBC theory and experiments to cases with significantly nonlinear flow. Once again, the theoretical predictions, including both lateral viscous and Ekman layer effects, were accurately verified in the experiments. Some unsuccessful attempts were made to produce a sufficiently rapid and narrow western boundary current for barotropic instability.

16.6.2 A Two-Layered Model

A major innovation and advance in laboratory modeling has recently been achieved by Krishnamurti (Krishnamurti and Na, 1978; Krishnamurti, 1978) by the introduction of controlled rotating-lid experiments with a two-layer system having no closed contours of depth. Using a conical bottom and a rotating conical lid, radial depth variations are produced in both the upper and lower layers. Under conditions that would be stable for a uniform-fluid experiment, the two-fluid experiment produces baroclinic instability and a number of interesting features associated with the deformation of the interface of the two fluids. Perhaps the most striking of these features is the upwelling of the lower fluid to the "surface" at the western boundary and the concurrent separation of the WBC, a phenomenon predicted by Parsons (1969) (see chapter 5).

Together with the laboratory experiments, Krishnamurti (1978) has theoretically evaluated the comparative spin-up times of the ocean and laboratory models in terms of the Rossby radius of deformation in relation to the fluid depth. These considerations suggest that the two-fluid experiments will have a number of interesting applications.

16.6.3 Flow over Sills and Weirs and through Straits

Smith (1973, 1977) was concerned with the flow of cold bottom water after it had passed through the Denmark Strait from the Norwegian Sea. This is a problem concerning effects of friction and sloping bottom on the modification of a baroclinic current system. To accompany a theoretical model of bottom frictional effects, Smith developed a laboratory experiment in which denser fluid was drained slowly through a source tube onto the sloping bottom of a rotating tank. He observed a variety of systematic wavelike and eddy-like oscillations of the flow that were found to be consistent with appropriate theories of baroclinic instability. In the 1977 paper he compared his experimental results with observations of the pulsed flow over the sill in the Denmark Strait, and he concluded on the basis of the significant nondimensional numbers that the natural case should correspond to what he described as "the meandering jet, vortex train variety (Class I)." Experiments of this type indeed represent a sophisticated application of laboratory experiments to our understanding of detailed oceanic phenomena.

Flows through straits and over sills have been of great interest in a wide variety of nonrotating systems, and engineering data characterizing these flows have long been available. In oceanic applications, however, the Coriolis force must play an important role in modifying or controlling the flow, and from experience with topographic β-effects one might also expect the detailed geometry to be of great significance in many cases. An obvious oceanographic example of restricted flow is that through the Straits of Gibraltar, but many other examples may also be mentioned, such as the seepage of Antarctic Bottom Water through small gaps in the Mid-Atlantic Ridge.

The experiments of Whitehead, Leetmaa, and Knox (1974) and of Sambuco and Whitehead (1976) have been concerned with problems of this type, approaching them experimentally beginning with the nonrotating system and gradually increasing the effect of rotation. Their studies may be characterized as being concerned chiefly with the restriction of the volume transport of the lower fluid through the strait, or over the sill, at relatively high Rossby numbers. Smith's studies, in contrast, have been concerned primarily with very low Rossby numbers, and with frictional effects on sloping boundaries after the flow has left the constriction.

16.6.4 The Generation of Mean Flows by Turbulence

One of the more elaborate and more significant recent laboratory studies has been that of Colin de Verdière (1977), who examined the response of a rotating fluid in various geometries to Rossby waves generated by a complex pattern of oscillatory sources and sinks of fluid introduced through holes in the bottom of the tank. He considered three cases: an f-plane model (constant depth); a polar β-plane model with a depth variation from the paraboloidal shape of the free surface and with closed height contours; and a sliced-cylinder model. Comparing turbulence and wave theory with the complex flow patterns of his experiments, he included in his studies the interaction of two-dimensional turbulent eddies for a fluid with energy sources and sinks, the evolution of the flow toward statistical equilibrium, and the interaction of Rossby waves with a mean flow.

The extent and variety of the experiments covered by Colin de Verdière preclude a detailed exposition of all his results, but in general it may be said that the theoretical expectations, based upon the theories of potential vorticity mixing of Rhines (1977), were experimentally verified. By way of example, we mention here the results of a polar β-plane experiment in which the flow was driven by alternating sources and sinks of fluid at the outer boundary. This complex system was arranged to produce a westward-moving Rossby wave. Through the preferential inward (northward) diffusion of anticyclonic vorticity, the Rossby wave in effect transferred energy to a westward mean flow north of its region of generation. When the forcing was turned off and the flow was allowed to decay, measurements of the mean flow and of the perturbation kinetic energy indicated that the eddies continued to feed energy into the mean flow.

The particular results given above concerning the generation of mean flows by turbulence in a β-plane model were anticipated by the experiments of Whitehead (1975), who excited turbulent motions in a somewhat different manner. Whitehead used a 2-m-diameter cylindrical tank and vertically oscillated a circular horizontal plastic disk within the fluid, the disk being 20 cm in diameter and centered at 66 cm radius. By thus driving turbulence and Rossby waves with no direct input of angular momentum, it was clear that the observed mean flow [eastward at the latitude (radius) of the oscillator, and westward to the north and south] was due to the lateral redistribution of angular momentum by the waves and turbulence. In such experiments, however, it is difficult to be certain that the mechanical apparatus is not partially forcing the mean flow. Suppose, for example, that the vertical axis of the oscillating plate were tilted by 1° one way or the other. Would this have an influence upon the generation of the mean flow, and if so, how much? The experiments of Colin de Verdière, using sources and sinks parallel to the rotating axis, do not seem to be subject to the same possible difficulties, but one must be very cautious in experiments of this type.

It is interesting to note that the late Professor V. P. Starr (of Chicago and MIT) probably had a significant influence upon these studies, for Whitehead acknowledges many interesting conversations with Starr and his students. I note this fact particularly because it was at the suggestions of Professors Starr and Rossby that Fultz began his experimental studies at the University of Chicago, and one of the original intentions of those studies was an examination of exactly the problem discussed above. The experiments of Fultz eventually centered upon thermally driven circulations in a hemispherical shell, but Starr (personal communication, circa 1952) always kept in mind the possibility of locally pulsing one of the spherical shells to generate purely mechanically driven circulations without the direct introduction of a mean flow.

The experiments discussed above appear to be contradicted by an experiment of Firing and Beardsley (1976), who generated an isolated eddy in a sliced-cylinder experiment. It was found that the total nonlinear effect of the eddy was the generation of cyclonic vorticity in the northern portion of the basin and anticyclonic vorticity to the south. The net effect was explained as the result of the competition between two opposing tendencies:

Potential vorticity conservation which imparts negative vorticity to northward moving water columns and positive vorticity to southward moving columns, and relative vorticity segregation which gives water columns with positive relative vorticity a tendency to move north and those with negative vorticity a tendency to move south. Thus the positive circulation induced in the northern half of the basin by the dispersing eddy indicated the dominance of the vorticity segregation mechanism in this flow.

There is need for further theoretical and experimental studies to rationalize the apparent differences between these results and those of Colin de Verdière (1977) and Whitehead (1975). The experiment of Firing and Beardsley was conducted in a system with no closed geostrophic contours and with an isolated eddy as the disturbance. The other experiments had continuous sources of Rossby waves or turbulence in an open circular tank without meridional barriers and therefore without the possibility of western boundary current and Rossby-wave reflection from meridional barriers.

16.6.5 The Generation, Propagation, and Reflection of Rossby Waves

Topographically generated Rossby waves were studied in detail by Long (1951) for the flow between hemispherical shells. Since that time many experimenters have generated stationary Rossby-wave patterns by the flow over a ridge or obstacle, using the radial depth variation to produce the simulated β-effect. Stationary Rossby waves have also appeared in a variety of experiments where the flow was obliged to separate from the western boundary and the boundary current was sufficiently nonlinear.

Ibbetson and Phillips (1967), however, were the first to investigate the generation, propagation, and reflection of Rossby waves specifically in an oceanographic context. In their classic paper they experimentally studied the free propagation and dissipation of Rossby waves from an oscillating-paddle wave generator (figure 16.6). Moreover, they developed a simple theory and a companion experiment for damped oscillatory flow in the closed region between their oscillating paddle and a radial barrier to the west. For a specific paddle frequency their theoretical solution consisted of the sum of two waves of the same frequency that could be identified as the one generated in the east by the paddle

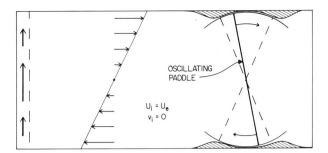

Figure 16.6 A schematic representation of the Ibbetson and Phillips (1967) experiment for the limiting case of zero frequency of the oscillator. The interior currents are then purely zonal with speeds v_i equal to the speed of the eastern boundary u_e. The steady-state western boundary current transport is determined by continuity as in the SAF (1958) experiments.

and its reflection from the western boundary. Since the reflected wave was of high wavenumber and large amplitude, its energy was subject to rapid dissipation as it progressed eastward from the boundary. The net result was a strong concentration of radial motions in the vicinity of the western boundary.

The application of these results to steady-state ocean-circulation theory, as well as to the transient oscillations, becomes evident when one lets the driving frequency of the paddle approach zero. This condition is illustrated in figure 16.6. From section 16.5, because there are no interior sources and sinks of fluid, the interior radial velocity is $v_i = 0$. It also follows that $u_i = u_e$, where u_e is the zonal velocity of the eastern boundary, which in this case is the velocity of the paddle. The wave generated by the paddle thus degenerates to purely east-west currents, and the reflected wave, with very high longitudinal wavenumber and very large radial flow, in the limit becomes the western boundary current.

16.6.6 Simulation of the American Mediterranean

An account of laboratory models of ocean circulation would not be complete without a brief discussion of Ichiye's (1972) laboratory studies of flow in the Gulf of Mexico and the Caribbean Sea. These were attempts to reproduce the known circulations insofar as is possible with a uniform fluid, by detailed scaling of the complex coastlines and bottom topography. The avowed purpose was "to understand the effects of bottom and coastal configuration of the two seas on the geostrophic current of barotropic mode. The vertical structure of the flow and the details of the horizontal current patterns are not a subject of this study."

Two separate model systems were constructed: one for the Gulf of Mexico, driven by a source of forced inflow through the Yucatan Straits, and outflow through the Straits of Florida; the other, of the Caribbean Sea and adjacent portions of the Gulf and the Atlantic Ocean, driven by a pattern of winds from fans. Rossby numbers and Reynolds numbers of the flow were varied to discover how the overall patterns of flow and the various cyclonic and anticyclonic vortices would respond to different values of these parameters.

In summary it may be said that many realistic features of the natural circulations were reasonably well reproduced, for example, the distribution of geostrophic mass transport in the Caribbean; but other features, notably the major current systems in the Gulf, were unacceptable. Of course, one of the values of experiments of this type lies in failure, for we are then directed to inquire about the specific sources of error. In this study it is likely that density stratification, or at least a two-layer system, may be necessary for greater realism of the model circulations. With the strong topographic effects that must be important in these nearly enclosed basins, the normal β-effect may be negligible, and a stratified-fluid experiment may be practical.

16.6.7 A Laboratory Study of Open-Ocean Barotropic Response

The above title is that of Brink (1978), who studied the f-plane and β-plane response of a rotating fluid with a free surface to applied surface-pressure oscillations. A problem of direct interest is the response of the world oceans to atmospheric pressure fluctuations of all scales (see chapter 11). In the absence of the planetary and topographic β-effects, the f-plane response should follow the inverse barometric effect except for frequencies close to resonance with inertia-gravity waves. When near resonance occurs there may be significant overshoot, i.e., excess amplitude response compared to the inverse barometer effect. With the β-effect there can be significant undershoot because of the propagation of Rossby waves, although if the scale and frequency of the pressure fluctuations should correspond to one of the normal modes of oscillation of the basin, there again could be significant overshoot.

In his β-plane experiments, Brink oscillated the air pressure over an enclosed region bounded by circular arcs at 15.2- and 30.5-cm radius and by radial walls separated by 120° of azimuth. With the tank radius of 42 cm, the area of forced-pressure oscillation comprised about 13% of the total surface area. The pressure vessel was designed to just touch the water surface with the intent of not seriously interfering with the water circulation or the propagation of waves. Measurements of the water-level response were obtained with capacitance height gauges.

Approximate theoretical solutions for the response of the water level were compared with the amplitudes and phases of the observed height variations at several sites. Reasonable agreement was found in many cases. The most serious lack of agreement occurred at the higher frequencies, and this was tentatively attributed to the use of a primitive (steady-state) Ekman-layer model for viscous damping, rather than one that took into account transients in the Ekman layer. Brink concluded that departures from the inverse barometer effect are to be expected due to propagating free modes of oscillation as predicted by theory.

16.6.8 Gulf Stream Rings

A laboratory study that may have direct analogy to the ocean was that of Saunders (1973), who experimentally tested the stability of a two-layer baroclinic vortex. While this experiment was similar in certain respects to the baroclinic instability studies of Fultz (1953), Hide (1958), Bryan (1960), Hart (1972), and others, Saunders pursued an interesting analogy with the stability of Gulf Stream Rings and other isolated oceanic vortices.

A cylindrical column of denser fluid was released within a lighter fluid, the entire system being initially in solid-body rotation. The lower part of the denser fluid spread out rapidly, leading to a low-level anticyclonic vortex and an upper-level cyclonic vortex. The increase in radius at the bottom $R - R_0$ from the initial radius of the cylindrical column R_0 was approximately equal to $\lambda = (g'H)^{1/2}/f$, the Rossby radius of deformation, where $g' = g \Delta\rho/\rho$, $\Delta\rho$ is the difference in fluid densities, H the fluid depth, and f the Coriolis parameter.

Defining the parameter $\theta = \lambda^2/R_0^2$, it was found that for $\theta < 1$ the initial circular vortex was unstable and would break up into a number of smaller vortices, each having $\theta > 1$.

Calculation of an equivalent value of θ for two stable Gulf Stream rings gave $\theta \approx 2$. Thus the stability or instability of the laboratory vortices and the corresponding stability characteristics of Gulf Stream rings, or other nearly circular oceanic vortices, may represent one of the most unambiguous tests of baroclinic instability in the ocean.

Two items that may be of historical interest in connection with these experiments are the following. First, in studies designed to test the bubble theory of cumulus convection Saunders used a small, hemispherical volume of buoyant fluid released at the bottom of a large and deep tank of water. Turned upside down, a small volume of dense fluid was released and allowed to fall, expanding as an entraining spherical-cap bubble. Saunders noticed that when the falling dense fluid impinged upon a rigid boundary it spread out somewhat analogously to an atmospheric squall line, and a series of experiments in a shallow fluid were undertaken. Since von Arx's old rotating turntable was available in the same laboratory, extension of these experiments to the rotating system was quite practical without the construction of new elaborate apparatus. Saunders first performed these rotating experiments in 1963 and the critical parameter for the stability of this type of vortex in fact was determined well before the extent and significance of Gulf Stream rings were fully appreciated.

The second item of interest is that in the late summer of 1954, when this author was first using von Arx's apparatus for atmospheric model studies, H. Stommel and W. V. R. Malkus one day suddenly appeared in the laboratory equipped with huge jugs of xylene and carbon tetrachloride (which at that time were not known to be so dangerous). Their intention was almost exactly the experiment later performed by Saunders, namely, to release a cylindrical column of a dense mixture of their two fluids in the center of a rotating tank full of water. They were specifically interested in relating the radial spread of this column of dense fluid to the Rossby radius of deformation, which had recently been recognized as an important parameter. Unfortunately this combination of liquids would have dissolved the sealer cementing together the base and rim of the tank. How might the history of laboratory studies have been altered in its course had this author allowed them to proceed with their experiment?

16.6.9 Spin-Up

Laboratory experiments and the theory of spin-up of a rotating fluid began with the work of Stern (1959, 1960b). In the former (unpublished) paper he presented the basic theory of spin-up and a description of laboratory experiments that accurately verified the predicted spin-up times as well as the integrated radial and tangential displacements of floating tracer particles. In the latter paper an instability of Ekman flow was postulated as the source of disagreement between the experiments and laminar theory in certain cases. In more recent experiments, also with a uniform fluid, Fowlis and Martin (1975) used a laser Doppler velocimeter and found clear evidence of the elastoid-inertia oscillations to be expected from transients in the Ekman layer owing to the abrupt change in rotation rate (Greenspan and Howard, 1963). In the wind-driven ocean, however, the stratification may drastically alter the spin-up characteristics from what would be anticipated for a uniform fluid.

Stratified spin-up experiments (e.g., Buzyna and Veronis, 1971; Saunders and Beardsley, 1975) generally have been conducted in the same manner as the classical uniform fluid cases—with an abrupt change in the rotation rate. In such a case, all natural modes of oscillation can be excited, and as a result it has been difficult to match satisfactorily theory and experiment. In recent studies by Beardsley, Saunders, Warn-Varnas, and Harding (1979), however, the spin-up was made gradually, thus avoiding the excitation of high-frequency oscillations. As a result, substantially better agreement than in previous experiments was found between a simple quasi-geostrophic numerical model and observations of the response of the interior temperatures and the azimuthal velocities.

16.6.10 Langmuir Circulations

One of the areas of research in which H. Stommel played an early role was the study of windrows—slicks on a natural water surface oriented in streaks along the wind direction. In a series of approximately 200 observations on Ashumet Pond, Cape Cod, Massachusetts, over the 7-month period May–November 1950, Stommel (1952) observed the presence or absence of surface streakiness parallel to the wind, and he attempted to relate the occurrence of streaks to various meteorological factors including the wind, cloudiness, humidity, and the air–water temperature difference as a measure of the thermal stability. The single parameter that seemed to be related to the occurrence of streaks was the wind speed, and table 16.1 is a summary of Stommel's data. The occurrence of streaks is clearly seen to be dependent upon the wind speed, and streaks often were observed at speeds much less than what is now generally considered to be the critical value, about 3 m s^{-1} (Walther, 1966).

These data do not seem to indicate a sharply defined critical speed, and the apparent critical speed given by other observers should perhaps be questioned in view of the facts that the observation of streakiness is a subjective one, the amount of surface material present may be a factor, and parameters such as the steadiness of the wind measured just above the water surface may be relevant.

Because of the obvious presence of surface films and because "if the wind changes its direction abruptly the streaks themselves are quickly reoriented (in a matter of one or two minutes only)," Stommel (1952) postulated that the phenomenon was essentially a shallow surface-layer effect and that the action of the film in damping capillary waves might be of importance for organizing the film into streaks. This postulate was taken up and explored by other authors, but it is now known that the windrows are the result of Langmuir circulations (LCs), organized longitudinal rolls with their axes parallel to the wind, and that surface films are not an integral part of the mechanism.

Since several scales of LCs may exist at any time (Faller and Caponi, 1978) the apparent rapid response of the windrows to wind variations may be presumed to be due primarily to the smaller scales, which are close to the surface and which respond more rapidly to changes in the wind. For example, in recent laboratory experiments with light wind Faller (1978) has found a characteristic growth rate of only 12 s for cells with a cross-wind wavelength of 44 cm. For larger scales one would certainly expect slower response times, but it now appears that reorganization of the pattern of surface streaks in a minute or two by the reformation of the LCs is not unreasonable.

I would now like to report some preliminary qualitative results on the generation of LCs in the presence of surface films. Figure 16.7 illustrates the wind-wave tank in which a regular pattern of crossed waves is generated and in which a wind is blown over the waves. In the absence of a finite-amplitude crossed-wave pattern and at low wind speeds there are virtually no wind waves generated over the working section of the tank, a length of about 5 m. In the presence of finite-amplitude crossed waves, however, the same air speed causes wind-generated waves of significant amplitude. From preliminary estimates it now appears that the large crossed waves and the smaller wind-generated waves both contribute to the growth of the observed LCs whose scale is determined by the crossed-wave pattern.

In previous papers (Faller, 1969; Faller and Caponi, 1978), it was reported that surface films completely prevented the formation of LCs, but this assertion must now be qualified on the basis of recent experiments. Of particular importance here is the effect of the finite length of the wind-wave tank, for this restricts the possible motion of the film. In light-wind conditions the wind stress will compress the surface film against the end of the tank, and a steady state is achieved when the internal film-pressure effect balances the wind stress. This condition can be avoided, at least temporarily, by introducing a surface film onto a clean water surface at the upwind end of the tank, for the film can then move downwind in response to the wind stress.

The effects of surface films under certain laboratory conditions can now be summarized as follows:

(1) Application of a surface film-forming material, in this case dodedyl alcohol, always damps wind-generated waves. Thus, whether moving or stationary, a film always destroys the contribution of small wind-generated waves to the LCs.

Table 16.1

Wind speed (mph)	Number of observations		
	Streaks	No streaks	% streaks
<1.5	3	42	7
1.5–2.5	9	19	32
2.5–3.5	11	8	58
3.5–4.5	14	5	74
4.5–5.5	13	8	62
>5.5	54	3	95

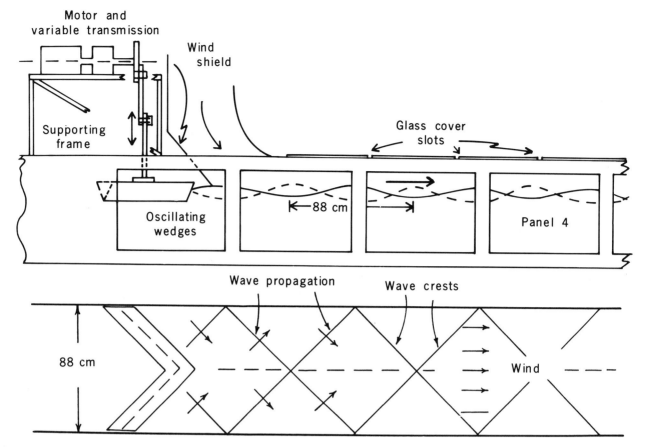

Figure 16.7 Apparatus used for the generation of Langmuir circulations. The vertically oscillating wedges generate a crossed pattern of large waves that propagate along the length of the tank and are absorbed at the right-hand end. Wind is drawn in by an exhaust fan at the right-hand end to produce a shear flow in the water and small superimposed wind-generated waves.

(2) With the large, mechanically generated crossed waves present, LCs are not generated as long as the film is prevented from moving by compression against the end of the tank.

(3) With the same large waves present, LCs *are* generated if the surface film can move in response to the wind stress. In such a case, however, the wind-generated small waves are damped and do not contribute.

These observations lend additional support to the Craik-Leibovich theory of Langmuir circulations (Craik and Leibovich, 1976), which describes the action of essentially irrotational waves in twisting the vorticity of the wind-generated shear flow into the vorticity of the longitudinal rolls. Apparently the moving surface film can transmit the tangential wind stress to the water whereas the stationary film obviously cannot.

16.7 Concluding Remarks

From time to time one may have the impression that laboratory-model experiments have reached the limit of their usefulness—that all of the interesting and productive experiments have been performed, and that further experimentation will only be repetitious. Then too, many theories that formerly could only be tested in a physical-laboratory experiment can now be tested reliably with computer calculations. But from the extensive list of studies discussed above we perceive a gradual increase in the variety, sophistication, and application of laboratory experiments as novel experimental techniques are developed and as new theories emerge that isolate (abstract) interesting phenomena and require verification.

The laboratory experiments discussed in this review represent a necessarily restricted class of studies. Within this class, however, two progressions in style are evident: the first, from the attempts by von Arx to represent nature in detail insofar as possible, to experiments carefully matched to the corresponding theories and in which all aspects of the flow are analyzed in excruciating detail; the second, from the steady-state SAF experiments, through the transient spin-up, Rossby-wave, and oscillating-lid experiments, to the interaction of turbulence and the mean flow. The recent two-fluid experiments of Smith, Saunders, Whitehead, and Krishnamurti now seem to indicate a gradual progression from uniform-fluid models into the theoretically and technically more difficult realm of baroclinic models, and clearly it would be premature to conclude that there is no longer a role for laboratory studies. Indeed, as long as "no one believes a theory, except the theorist, and everyone believes an experiment—except the experimenter," the role of laboratory experiments in oceanographic research remains secure.

Part Four　　　　　　　　Ocean and Atmosphere

17 Air–Sea Interaction

H. Charnock

17.1 Introduction

In preparing a chapter for a book such as this some nostalgia is excusable, even inevitable. For those who came to the subject, or came back to the subject, after the war there was a bewildering range of work going on. Barber and Ursell were publishing their results on the long-distance propagation of ocean swell, having used an analogue device to estimate the wave spectrum. Sverdrup and Munk had done wartime work on waves too, but by 1947 Munk was writing on a possible critical velocity for air–sea transfer processes and Sverdrup was working up his classical paper on currents driven by the curl of the wind stress. Jacobs was continuing his long-term study of the climatology of energy exchange between sea and air: Budyko was just starting his. Sheppard was publishing his direct determination of the shearing stress by use of a drag plate and Roll was making new wind-profile measurements over the Wattenmeer. Obukhov had already developed the dimensional arguments leading to the Monin-Obukhov length and had contributed to the Kolmogoroff small-scale similarity hypothesis with Onsager and Weizsäcker: in Cambridge, Batchelor was exploring its consequences. Priestley was off to Australia to set up a powerful group on near-surface turbulence, and was making pioneering calculations of the poleward heat and momentum transfer by covariance of wind and temperature fluctuations. Eady was in London working up his idea about baroclinic instability, Charney his at Princeton. Henry Stommel, relatively recently at Woods Hole, was interested in convection in the atmosphere and ocean (it was the time of the Woodcock-Wyman expedition) and had discovered the phenomenon of entrainment into cumulus clouds.

The importance of air–sea interaction to the larger-scale flows of the atmosphere and ocean was in no doubt, though it was a somewhat minority interest. Most of the work at that time concerned the estimation of the surface fluxes of heat, water vapor, and momentum from the only data base then foreseen, namely, the routine observations of temperature (dry bulb, wet bulb, sea) and wind made from the merchant vessels that reported to national meteorological agencies. Given suitable formulas it was thought that one could perhaps calculate the poleward heat transfer by the ocean and make some progress on relating winds to near-surface currents.

There were obvious difficulties of observation over the sea rather than the land but these were compensated for by the importance of the results and by the relative uniformity of the surface, both in space and, due to the high thermal capacity of the ocean, in time. Also the problem was close enough to a laboratory shear flow to allow comparisons with flow in pipes and channels. Much of the early work therefore concerned itself with the fluid mechanics of the air flow over the sea to a height of, say, 50 m. Since then the concept of air-sea interaction has been much broadened to include consideration of phenomena on larger space and time scales. General problems such as the teleconnections between sea surface temperature anomalies and subsequent weather patterns and specific aspects such as El Niño have been included. One thinks of climate as a complex of interactions between the air, the sea, and the surface of the earth, and in this sense air-sea interaction can be argued to include much of the physics and dynamics of the atmosphere and the ocean. But this review will consider only the small-scale processes by which heat, water, and momentum are transferred near the sea surface. In a fundamental sense the air and the sea interact only in a thin interfacial layer, but it is convenient to consider processes confined to the coupled boundary layers of the atmosphere and the ocean, which, as will be seen, extend typically to a height of 1000 m and a depth of 30 m from the sea surface.

The first section deals with the surface layer of the atmosphere, which constitutes about the lowest tenth of the whole atmospheric boundary layer. This is the only region for which a satisfactory (though empirical) treatment is available in the form of "similarity theory" that relates small-scale properties of the airflow (gradients, turbulence spectra) to the vertical fluxes of momentum, heat, and water vapor.

To get the mean profiles, or the exchange coefficients (which are important in practice), requires boundary conditions within the interfacial layer. These are not at all well understood—observational results are briefly summarized in section 17.3.

Many of the difficulties associated with the interfacial layer are due to the complications introduced by surface waves. The relation between the wind stress (or the aerodynamic roughness) and the surface wave field (or the geometrical roughness) has proved an intransigent problem. Recent advances in our knowledge of the wave spectrum, and of the pressure distribution at the moving sea surface, are indicated in section 17.4.

The development of computer models of the atmosphere, and increasingly of the atmosphere and ocean combined, have much reduced the emphasis on the near-surface meteorological variables. The surface fluxes are no longer related to ships' observations so much as to winds, temperatures, and humidities in the atmosphere and ocean at levels where the flow can be taken to be frictionless and adiabatic. This requires increased understanding of the structure of the boundary layer as a whole. Section 17.5 describes our regrettably limited knowledge of the climatology of the atmospheric boundary layer and of the complicated processes that affect the distribution of density and wind within it. Some of the processes are similar to those that determine the structure of the oceanic boundary layer: for others, such as clouds, there is no obvious analogy.

17.2 The Surface Layer

The lowest 50 m of the boundary layer of the atmosphere has a special importance and simplicity that together with its accessibility have attracted intensive study. The importance of the surface layer comes from the fact that although its depth is only a small fraction of the whole boundary layer, it is within it that most of the change of wind speed, temperature, and humidity between the free atmosphere and the surface takes place. Its simplicity comes about because the fluxes of momentum, heat, and water vapor undergo only small fractional changes within the surface layer, so they may commonly be regarded as independent of height. For this reason it is convenient to take the fluxes of momentum and potential density as the basic independent variables governing the motion, and to consider the mean gradients and all the properties of the turbulence as being determined by them.

17.2.1 Near-Surface Profiles in Neutral Conditions

Many measurements of the vertical profile of velocity have been made over sites uniform for an upwind distance great compared to the height of observation z in conditions steady for times greater than z/u_*, where u_* is the friction velocity defined by $\tau_0 \equiv \rho u_*^2$, τ_0 being the surface shearing stress and ρ the air density.

If the potential density is independent of height (neutral hydrostatic stability) the velocity gradient is found to vary quite accurately as the inverse of the height measured from a reference plane near the top of the roughness elements and

$$\frac{dU}{dz} = \frac{u_*}{\kappa z}, \qquad (17.1)$$

where κ is constant.

It is easy to see that (17.1) is a reasonable relation, though the "proofs" of it to be found in the literature are to be treated with caution. If, away from the surface, the turbulent motion is not affected by viscosity or other processes by which the stress is communicated to the surface, nor by the fact that the boundary layer is of finite thickness, but has its intensity and scale determined by the Reynolds stress and the height, then (17.1) follows on dimensional grounds. It is written in terms of dU/dz rather than U because a uniform translation can have no effect on the internal dynamics of the flow.

The profile of a transferable scalar such as potential temperature, specific humidity, or the concentration of gases such as carbon dioxide can be treated similarly. Limiting the discussion again to neutral hydrostatic stability means that any variation in temperature or humidity must be small or combined in such a way as to maintain the potential density independent of height. As for momentum the vertical transfer by the turbulence will be governed by u_* and z: the potential temperature profile will be given by

$$\frac{d\Theta}{dz} = \frac{\theta_*}{\alpha_0 \kappa z}, \qquad (17.2)$$

where θ_* is a scale temperature defined by $u_* \theta_* \equiv \langle w\theta \rangle$, w being vertical velocity, and the constant α_0 is introduced to allow for the possibility that the transfer of a scalar quantity may differ from that of momentum.

In a similar way the humidity profile is given by

$$\frac{dQ}{dz} = \frac{q_*}{\alpha_0 \kappa z} \qquad (17.3)$$

with $u_* q_* \equiv \langle wq \rangle$. The same constant α_0 is used because it seems unlikely that different scalars will have different transfer properties in fully turbulent flow.

In (17.2) and (17.3), Θ is the mean temperature and Q the mean humidity, θ and q being the respective fluctuating quantities. The covariances $\langle w\theta \rangle$ and $\langle wq \rangle$ measure the heat flux H and the water vapor flux E as

$$H = c_p \rho \langle w\theta \rangle \quad \text{and} \quad E = \rho \langle wq \rangle$$

by analogy with the Reynolds stress $\tau = -\rho \langle wu \rangle$ (c_p is the specific heat at constant pressure, and u is the horizontal component of velocity).

It is perhaps surprising that the mean gradients are unaffected by the characteristics of the surface—one might expect the expressions for them to be valid only at heights large compared with some height typical of the surface geometry. But by choosing the zero plane suitably, often just below the tops of the surface-roughness elements, the formulas fit quite well down to heights only just above them.

17.2.2 Near-Surface Profiles in Nonneutral Conditions

It has been known for a long time that a vertical gradient of potential density can have a profound effect on turbulence (Richardson, 1920). When the density increases upward, so that the mean situation would be statically unstable, the mixing action of the turbulence produces a downward density flux, and buoyancy forces feed energy into the turbulence so as to augment the action of the windstress. So in unstable conditions, for a given value of the shear stress, the turbulence will be more vigorous and its ability to transfer heat and momentum greater than in neutral conditions. In stable conditions the converse is true.

How the buoyancy forces operate is only partly understood, although some theories based on the insertion of simple physical approximations into the Friedman-Keller equations for the variances and covariances of the velocity components and the density have had considerable success. It may be noted that since the work done by the buoyancy forces involves a product of their magnitude with the distance over which they operate, their effect is most pronounced on the large scales of motion. Hence one expects large eddies to be preferentially destroyed in stable conditions and preferentially sustained in unstable conditions: the scale of the most active part of the turbulence will be smaller in stable conditions than in unstable. Also, since the scale of the motion decreases as the surface is approached, so also does the effect of the buoyancy. It follows that sufficiently near the surface the active part of the motion is governed by the laws appropriate to neutral conditions.

In the surface layer great simplification has been achieved by the use of dimensional arguments to develop what is called "the similarity theory of the surface layer." It applies to the components of the motion that have scales smaller than the depth of the surface layer and so are generated and controlled within it. Recognizing that the fluxes of momentum and potential density are nearly independent of height in the surface layer, and that the mean gradients are unaffected by the detailed transfer processes at the boundary, Russian workers (Obukhov, 1946; Monin and Obukhov, 1954) were led to use the fluxes as key quantities in the surface layer. This was an imaginative development, at a time when fluxes were much harder to measure than mean gradients: it has provided a very useful means of systematizing many varied observations.

The assumption is made that turbulent quantities in the surface layer are unaffected by all quantities external to it, such as the total thickness of the boundary layer and detailed transfer processes at the surface. The basis of the theory is to use as the independent variables z, u_*, and δ_* (defined by $\delta_* u_* \equiv \langle \delta w \rangle$, where δ is the buoyancy fluctuation). All the properties of the turbulence are expressed in terms of them.

From these variables only one dimensionless group can be found. It is

$$\zeta = z/L, \quad \text{where} \quad L = u_*^2 / \kappa \delta_* . \qquad (17.4)$$

L is called the Monin-Obukhov length after the originators of the theory. κ has been introduced because they included it in their initial definition.

It follows that all dimensionless properties of the turbulence must depend solely on ζ. In particular the dimensionless velocity profile will be a function of ζ alone,

$$\frac{\kappa z}{u_*}\frac{dU}{dz} = \phi_M(\zeta), \tag{17.5}$$

as will be the profile of a transferable scalar

$$\frac{\kappa z}{\theta_*}\frac{d\theta}{dz} = \phi_H(\zeta). \tag{17.6}$$

17.2.3 Alternative Stability Parameters

The use of the Monin–Obukhov length has the disadvantage that it requires knowledge of the fluxes, which is not always available. It is sometimes more convenient to work with the gradient form of the Richardson number Ri, which is defined by

$$Ri = \frac{g}{\rho}\frac{d\rho}{dz} \Big/ \left(\frac{dU}{dz}\right)^2, \tag{17.7}$$

g being the acceleration due to gravity. According to the similarity theory Ri should be a universal function of ζ in the surface layer.

In his original paper Richardson (1920) showed that the rate per unit mass at which work had to be done by the turbulence against buoyancy forces was $\delta_* u_*$. He pointed out that the ratio of this to the rate at which the shear stress produced turbulent energy, $u_*^2(dU/dz)$, could not exceed unity unless energy was being brought into the region from outside. This ratio is now called the flux Richardson number, Rf, and is related to Ri, to ϕ_M, and to ζ by

$$Rf = \delta_* \Big/ \left(u_*\frac{dU}{dz}\right) = \alpha Ri = \zeta/\phi_M, \tag{17.8}$$

where $\alpha = \phi_M/\phi_H$ is itself a function of Ri or ζ.

It may be noted that only two empirical functions are needed to describe mean profiles and that the relations between the variables enable all the functions required in connection with mean profiles to be derived from whichever two functions can be conveniently measured.

17.2.4 Flux-Gradient Observations

To verify (17.1) for the mean-velocity profile in neutral conditions and to determine κ, it is necessary to measure the surface stress. From laboratory pipe measurements it was known that $\kappa \simeq 0.4$, and the measurements of Sheppard (1947), who simulated an area of ground surface and measured the forces on it with a spring balance, confirmed that the same value applied to the lower atmosphere. Later measurements using the drag-plate technique have given excellent results in suitable conditions.

A second method is to measure fluctuations of the horizontal u and vertical w components of the air flow. The turbulent stress is (nearly) $-\rho\langle uw \rangle$. It is necessary to use a fast-response instrument that responds to the whole range of frequencies contributing to the stress,

and to have computer processing for the spectra, covariance, etc., but several workers have succeeded in producing consistent results. Such techniques can also be used, given measurement of fluctuating temperature and humidity, to estimate heat and water-vapor fluxes from $\langle w\theta \rangle$ and $\langle wq \rangle$.

Such rapid-response devices and analysis facilities can also be used to estimate the dissipation rates for turbulent energy, temperature variance, and humidity variance. In suitable conditions these can be related to the respective fluxes.

In spite of a good deal of work the value of κ is not universally agreed upon: this is partly due to the difficulty of allowing for fluctuations in the surface stress. When this is taken into account u_* in (17.1) should be replaced by its mean value $\langle (\tau/\rho)^{1/2} \rangle$, which is less than $\langle \tau/\rho \rangle^{1/2}$.

Pruitt, Morgan, and Laurence (1973) made careful measurements using a large drag plate to determine that $\kappa = 0.42$; allowing for a slight overestimate due to fluctuating stress it seems that the generally accepted value, $\kappa = 0.40$, is not far wrong.

There is not space here to deal adequately with the many observations that have been made, particularly over land, which have established the forms of ϕ_M and ϕ_H. They have been reviewed by Plate (1971), Monin and Yaglom (1965, 1967), and Högström (1974). According to Busch (1977), most atmospheric data are well represented by

$$\phi_M = \begin{cases} 1 + 5\zeta & (\zeta \geq 0) \\ (1 - 15\zeta)^{-1/4} & (\zeta \leq 0), \end{cases}$$

$$\frac{\phi_H}{\phi_H^{(0)}} = \frac{\phi_{\text{OTS}}}{\phi_0} \tag{17.9}$$

$$= \begin{cases} 1 + 6\zeta & (\zeta \geq 0) \\ (1 - 9\zeta)^{-1/2} & (\zeta \leq 0), \end{cases}$$

where OTS means other transferable scalar. The value of $\phi_H^{(0)} \; (= \alpha_0^{-1})$ values are scattered, but a reasonable value is 0.8.

There are some plausible arguments to support these forms but no satisfactory theory. Nevertheless it seems clear that the similarity theory of the surface layer provides an excellent framework in which to systematize observational studies of mean gradients and of turbulent fluctuating quantities in the surface layer. The basis of the theory is that such quantities are unaffected by the characteristics of the underlying surface, so the results so far given apply over both land and sea: the only requirement (by no means easy to satisfy) is that of uniformity in space and steadiness in time.

17.3 The Lower Boundary

Previous expressions for profiles have been written in terms of gradients, like dU/dz, because by the hypotheses used a uniform translation can have no bearing on the internal structure of the flow. To integrate, so as to get an expression for, say, $U(z)$, requires a boundary condition within what can be called the *interfacial layer*, which includes the surface itself and the air above up to a height comparable to that of the elements that make up the surface. Processes in this layer are complicated and not well understood.

On integration of (17.1) we have for the wind profile in neutral conditions

$$\kappa \frac{U}{u_*} = \ln\left(\frac{z}{z_s}\right) \tag{17.10}$$

z_s is merely a constant of integration, whose value is determined by the surface geometry and surface processes. It has no influence on the internal dynamics of the flow.

Similarly, integration of (17.2) and (17.3) for the temperature and humidity profiles in neutral conditions gives

$$\alpha_0 \kappa (\Theta - \Theta_0) = \theta_* \ln(z/z_\theta), \tag{17.11}$$

$$\alpha_0 \kappa (Q_0 - Q) = q_* \ln(z/z_q) \tag{17.12}$$

where Θ_0, Q_0 are the potential temperature and the humidity at the surface and z_θ, z_q are constants of integration analogous to z_s. Like z_s they have no influence on the internal structure of the flow: changing them has the effect of adding a constant amount to the temperature and the humidity.

Turning to the more complicated expressions, (17.5) and (17.6), and integrating to get the profiles of velocity and potential temperature in nonneutral conditions we have

$$\kappa U/u_* = \int_{z_0}^{z} (\phi_M/z') \, dz'$$

$$= f_M(\zeta) + \phi_M(0) \ln(z/z_0)$$

where

$$f_M(\zeta) = \int_0^\zeta [(\phi_M - \phi_0)/\zeta'] \, d\zeta'. \tag{17.13}$$

The lower limit has been taken as zero instead of z_0/L since $z_0 \ll L$ and ϕ_M is assumed continuous at the origin.

Thus the departure from the neutral logarithmic form is given by f_M with ζ positive for stable, and negative for unstable conditions. All the stable profiles are similar to each other, as are all the unstable ones, the neutral profile being a limiting case.

Analogously,

$$\kappa (\Theta - \Theta_0)/\theta_* = f_H(\zeta) + \phi_H(0) \ln(z/z_\theta) \tag{17.14}$$

where

$$f_H(\zeta) = \int_0^\zeta [(\phi_H - \phi_H(0))/\zeta'] \, d\zeta'.$$

These rather formal results can be summarized by remarking that profiles like U/u_* are functions of ζ ($= z/L$) and of z/z_0. The basic requirement of the similarity theory of the surface layer is that the internal dynamics of the flow is unaffected by the boundary processes so that

$$\kappa U/u_* = f(\zeta, z/z_0) = f_1(\zeta) + f_2(z/z_0)$$

$$= f_M(\zeta) + \ln(z/z_0).$$

The basic unknowns in the problem are those in the interfacial layer, represented by z_0, z_θ, z_q.

17.3.1 Transfer Coefficients over the Sea

So far as the relation between stress and velocity gradient is concerned, (17.1) indicates that the turbulence acts as an effective (eddy) viscosity of magnitude

$$K_M = \kappa u_* z.$$

This is usually much greater than the molecular viscosity ν, but below a height $\nu/\kappa u_*$ it is smaller, and molecular transfer will dominate the motion. If the surface is fairly smooth, so that the typical height of the roughness elements h_r is smaller than this, they will be submerged in the viscous layer and play little part in communicating stress to the surface. The flow is then said to be aerodynamically smooth (though in fact it is fully turbulent), and, since h_r is irrelevant, dimensional reasoning gives

$$u_* z_s/\nu = \text{constant} = 0.11 \quad \text{by observation.}$$

If, on the other hand, $h_r \gg \nu/\kappa u_*$, the stress is communicated to the surface by the form drag of the roughness elements. Then the molecular viscosity is irrelevant and

$$z_s = z_0,$$

where the so-called roughness length z_0 depends in a complicated way on the size, shape, and spacing of the roughness elements. There is no good theory for relating z_0 and h_r: typically for close-packed granular roughness elements $z_0 = h_r/30$.

In the intermediate case $h_r \simeq \nu/\kappa u_*$, z_s/h_r is a function of $u_* h_r/\nu$ that is known from laboratory observation.

A complication is that the wind can modify the geometry of the surface over which it blows. Long grass is flattened by the wind, and E. L. Deacon (1953) observed that z_0 for grass 700 mm long falls from 90 mm in light winds to less than 40 mm in strong winds. On the other hand, when particles from the surface are carried into the air, as in blowing sand or snow, the value of z_0 is much larger than for the undisturbed surface.

The most important surface whose geometry is affected by the wind is the ocean. Its aerodynamic roughness has been the subject of much research over the last decades.

At very low wind speeds, before waves or ripples have been generated, the sea would be expected to behave as an aerodynamically smooth surface, and this is generally observed. When waves appear, the wind profile in neutral conditions remains closely logarithmic down to levels close to the surface, but the effective roughness length increases from the aerodynamically smooth value. From a series of careful wind-profile measurements over an artificial lake, and taking into account earlier observations, Charnock (1955) suggested that the aerodynamic roughness length was determined by the shearing stress, and used the simplest nondimensional relation

$$z_0 = \alpha_1 u_*^2/g. \tag{17.15}$$

The same expression but with a different value for α_1 had been found by Ellison (1956) using observations reported by Hay (1955). The implication of such a formula is that, while z_0 depends in a complicated way on the waves generated, the wave structure in turn is determined by the stress on the surface. The coefficient α_1 is at most a weak function of the faster, and so of the larger, waves, with the possible implication that the stress is transmitted locally, and to the short waves and ripples. This raises the question why g is used in (17.15) rather than other properties of the fluid such as its viscosity or surface tension. Lengths can be formed using u_* and ν (as in aerodynamically smooth flow), and using the surface tension S and u_*: in both cases the lengths decrease as u_* increases, so it is not likely that z_0 depends on ν or S in any simple way. But the fluid mechanics of the wavy surface is complicated and no adequate theory exists.

The usefulness of (17.15) is tested by observation, and here there has been considerable difference of opinion. Observations of the surface stress over the sea have been made by numerous workers, using various methods. The most common methods have been the use of the wind profile and eddy correlation. Most workers have preferred to express their results in terms of a drag coefficient C_D given by

$$\tau_0 = C_D(10)\rho U_{10}^2$$

the 10 being inserted as a reminder that the value of C_D depends on the height at which U is measured: 10 m is commonly used. U_{10}, and so $C_D(10)$, is also affected by the static stability, but this can be allowed for using similarity theory:

$$C_D = \frac{C_{DN}}{[1 + \kappa^{-1} C_{DN}^{1/2} f_M(\zeta)]^2},$$

where f_M is given by (17.13) and

$$C_{DN} = \frac{\kappa^2}{(\ln z/z_0)^2}$$

is the neutral drag coefficient.

Garratt (1977) has recently made a thorough review of previously reported values of C_{DN} in relation to U_{10}. Until about 1970 the values were scattered (though less so than they were 20 years before—see Charnock, 1951). They are shown in figure 17.1 and table 17.1. Since 1970 many more observations have been reported, using better methods, and Garratt has estimated C_{DN} from the 17 publications listed in table 17.2, excluding some, for reasons detailed in his paper. The resulting values are plotted in figure 17.2, in which reasonable agreement with (17.15) is shown, though the considerable scatter increases at wind speed greater than 15 m s^{-1}.

Some authors have estimated the surface stress in hurricanes by integrating the ageostrophic wind component. These are given in table 17.3 and figure 17.3 (again due to Garratt, 1977): there is some indication that (17.15) is satisfied in winds up to 50 m s^{-1}. Garratt gives $\alpha_1 = 0.0144$ as an acceptable value.

It seems from Garratt's review that (17.15) is sufficient for many purposes. But its physical basis is still very unsatisfactory: the implied roughness lengths are small ($\sim 10^{-1}$ mm), and we have no clear idea as to how they are determined. That the high-wavenumber range of the wave spectrum is involved seems probable, and is supported by experiments using surface films and detergents that eliminate short waves and much reduce the drag for a given wind.

Our knowledge of z_θ, z_q, and the physical properties on which they depend is even less satisfactory. Owen and Thompson (1963) have put forward a theoretical framework that allows comparisons between measurements of heat and of vapor transports from fixed rough surfaces. They give a formula that, assuming α_0 [equation (17.3)] to be 1.3, becomes

$$\ln(z_0/z_\theta) = 2.0 Pr^{0.75}(u_* z_0/\nu)^{0.33}, \tag{17.16}$$

though the numbers are tentative. $Pr = \nu/\nu_T$ where ν_T is the kinematic molecular diffusivity for the property being transferred. Fortunately $u_* z_0/\nu$ is small over the ocean so $z_0 \simeq z_\theta$ is a reasonable approximation. But if a formula like (17.16) does apply, and if z_0 is given by (17.15), then z_θ will gradually become less than z_0 as u_* increases. Kitaigorodskii (1970) gives a critical review of existing observations, as do Friehe and Schmitt (1976) and Busch (1977), but the experimental scatter makes it difficult to generalize.

Table 17.1 Main Reviews of the Neutral Drag Coefficient over the Sea[a]

Source	Wind speed range ($m\ s^{-1}$)	$C_{DN}(10)$ ($\times 10^3$)	Variability (%)	Number of references
A. Priestley (1951)	2.5–12	1.25[b]	?	Not stated
	strong	2.6[c]		
B. Wilson (1960)	~1–5	1.42	±50 }	47
	9–20	2.37	±25 }	
C. Deacon and Webb (1962)	2.5–13	$1 + 0.07\,V$	±25–50	9
D. Robinson (1966)	3–8.5	1.8[d]	±30 }	14
	2.5–14	1.48[e]	±15 }	
E. Wu (1969b)	3–15	$0.5\,V^{0.5}$ }[f]	±30 }	30
	15–21	2.5 }	±10 }	
F. Hidy (1972)	2–10	1.5	±30	8

a. Showing wind speed range, best estimate of $C_{DN}(10)$ (either as a constant or a function of wind speed), and typical data variability as a percentage of $C_{DN}(10)$ value over the wind speed range considered (see figure 17.1). [After Garratt (1977), who summarized the reviews.]
b. Actually based on Deacon (1950): Nature 165, p. 173.
c. Quotes Sverdrup et al. (1942) and Munk (1947).
d. Micrometeorological data.
e. Geostrophic departure.
f. Overall variation close to Charnock relation with $\alpha = 0.016$.

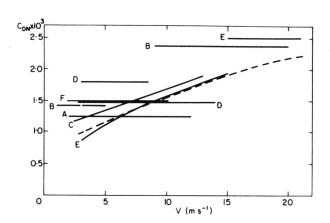

Figure 17.1 Mean curves of $C_{DN}(10)$ plotted against V (10 m) for review sources shown in table 17.1. Dashed curve is based on $z_0 = \alpha u_*^2/g$ with $\alpha = 0.016$ and $\kappa = 0.41$. (Garratt, 1977.)

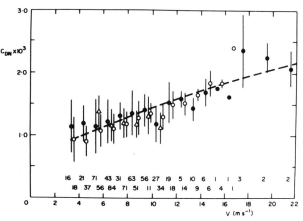

Figure 17.2 Neutral drag coefficient values as a function of wind speed at 10-m height, based on individual data taken from the recent literature (see table 17.2 and Garratt, 1977). Mean values are shown for 1-$m\cdot s^{-1}$ intervals based on the eddy correlation method (●) and wind profile method (○); Hoeber's wind profile data are also shown (△). Vertical bars refer to the standard deviation of individual data for each mean, with the number of data used in each 1-$m\cdot s^{-1}$ interval shown above the abscissa axis: top line refers to (●), bottom line to (○). The dashed curve represents the variation of $C_{DN}(10)$ with V based on $z_0 = \alpha u_*^2/g$ with $\alpha = 0.0144$. (Garratt, 1977.)

Table 17.2 Neutral Drag Coefficient Values over the Ocean[a]

Source	Wind speed range (m s^{-1})	$C_{DN}(10)$ ($\times 10^3$)	Variability σ (%)	Number of data (n)	Method	Platform	Comments
1. Smith and Banke (1975)	2.5–21	$0.63 + 0.066V$	30	111	ec	Mast	Also utilizes data of Smith (1973) using thrust and sonic anemometers
2. Kondo (1975)	3–16	$1.2 + 0.025V$	15	—	waves	Tower	Utilizes data on wave amplitudes from Kondo et al. (1973)
3. Davidson (1974)	6–11.5	1.44	?	114	ec	Large buoy	Does not correct for stability effects
4. Wieringa (1974)	4.5–15	$0.62V^{0.37}$ or $0.86 + 0.058V$	20	126	ec	Tower	Surface tilt and wp estimates are excluded
5. Kitaigorodskii et al. (1973)	3–11	0.9 (at 3 m s^{-1}) to 1.6 (at 11 m s^{-1})	?	29	ec	Tower	Plots C_{DN} as a function of $u_* z_0/\nu$
6. Hicks (1972)	4–10	$0.5V^{0.5}$	25	74	ec	Tower	Accepts C_{DN} relation as same as Wu (1969b)
7. Paulson et al. (1972)	2–8	1.32	25	19	wp	Large buoy	Uses $\kappa = 0.40$
8. Sheppard et al. (1972)	2.5–16	$0.36 + 0.1V$	20	233	wp	Tower	Uses $\kappa = 0.40$
9. De Leonibus (1971)	4.5–14	1.14	30	78	ec	Tower	
10. Pond et al. (1971)	4–8	1.52	20	20	ec	Large buoy	
11. Brocks and Krügermeyer (1972)	3–13	$1.18 + 0.016V$	15	152	wp	Buoy	Data from North Sea and Baltic Sea—uses $\kappa = 0.40$
12. Hasse (1970)	3–11	1.21	20	18	ec	Buoy	See text on data interpretation
13. Miyake et al. (1970)	a. 4–9 b. 4–9	1.09 1.13	20 20	8 8	ec wp	Mast Mast	See text on data interpretation—uses $\kappa = 0.40$
14. Ruggles (1970)	2.5–10	1.6	50	276	wp	Mast	C_D anomalies found at a number of wind speeds—uses $\kappa = 0.42$
15. Hoeber (1969)	3.5–12	1.23	20	787	wp	Buoy	Data from equatorial Atlantic—uses $\kappa = 0.40$
16. Weiler and Burling (1967)	a. 2–10.5 b. 2.5–4.5	1.31 0.90	30 75	10 6	ec wp	Mast Mast	Uses $\kappa = 0.40$
17. Zubkovskii and Kravchenko (1967)	3–9	$0.72 + 0.12V$	15	43	ec	Buoy	wp estimates of u_* show low correlation with ec; possible effect of buoy motion

a. Taken from the recent literature for a reference height of 10 m: ec = eddy correlation method; wp = wind profile method. σ is the standard deviation of n data points about the mean value. [After Garratt (1977), who compiled and evaluated the source material.]

Table 17.3 Neutral Drag Coefficients over the Ocean[a]

Source	Wind speed range (m s^{-1})	$C_{DN}(10)$ range ($\times 10^3$)	Comments
A. Miller (1964)	17–52	1.0–4.0 (linear)	Hurricanes Donna and Helene—ageostrophic
B. Hawkins and Rubsam (1968)	23–41	1.2–3.6 (discontinuous)	Hurricane Hilda—ageostrophic
C. Riehl and Malkus (1961)	15–34	2.5	Held constant to achieve angular momentum balance
D. Palmén and Riehl (1957)	5.5–26	1.1–2.1 (linear)	Composite Hurricane data—ageostrophic
E. Kunishi and Imasoto (see Kondo, 1975)	14–47.5	1.5–3.5	Wind flume experiment
F. Ching (1975)	7.5–9.5	1.5	Vorticity and mass budget at BOMEX

a. Taken from the literature, for hurricane and vorticity-mass-budget data analyses. Also included are wind flume data of Kunishi and Imasoto (see Kondo, 1975). [After Garratt (1977), who compiled and evaluated the source material.]

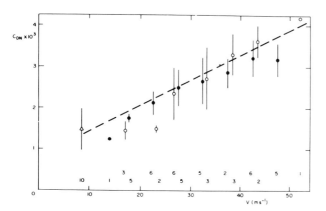

Figure 17.3 Mean values of the neutral drag coefficient as a function of wind speed at 10-m height for 5-m-s^{-1} intervals, based on individual data from hurricane studies (O), wind flume experiment (●), and vorticity mass budget analysis (△)—see table 17.3. Vertical bars as in figure 17.2. The number of data contained in each mean is shown below each mean value, and immediately above the abscissa scale. The dashed curve represents the variation of $C_{DN}(10)$ with V based on $z_0 = \alpha u_*^2/g$ with $\alpha = 0.0144$. (Garratt, 1977.)

Although our knowledge of the complicated processes in the interfacial layer is very unsatisfactory, we can, by using similarity theory and empirical knowledge of z_0, z_θ, etc., derive formulas from which the surface fluxes can be estimated from ships' observations in the near-surface layer of, say, temperature, humidity, and wind speed at a known height, together with sea-surface temperature. The errors in such estimates will be considerable, but they are more likely to be due to the errors in the ships' observations than to deficiencies in the formulas.

Calculations of the fluxes from climatological data [Jacobs (1951), Privett (1960), Budyko (1956), and more recent work by Bunker (1976) and Saunders (1977)] are of great value even though their accuracy is limited by the low precision of the ships' observations and by lack of uniformity of their cover of the ocean. They are thought unlikely to provide estimates from which the poleward heat transport by the ocean can be deduced, but will be useful in attempts to interpret the work of Oort and Vonder Haar (1976).

17.4 Waves

The most obvious effect of the wind on the sea is the generation of waves. They have been much studied, for there is no doubt of their economic importance: the design of ships, of harbors, and of sea defenses all need estimates of the waves to be encountered, to say nothing of the questions raised by the reflection of sound and light at the sea surface.

What is less obvious is how they fit into the coupled mechanics of the ocean and the atmosphere—how the winds and currents would differ if by some magic device the surface waves were eliminated. The drag coefficient for surface friction seems to be largely independent of the larger waves, as do the exchange coefficients for heat and water vapor. The transfers of energy and momentum from the atmosphere to waves on the ocean have been studied extensively: considerable progress has been made but there is still no complete agreement about the complicated fluid mechanics involved.

The wartime work, well confirmed and extended by Snodgrass and his colleagues (1966), established the basic fact that swell traveled thousands of kilometers, at the theoretical group velocity, without much attenuation. This implied that waves did not interact strongly with each other, or with ocean currents, so that a Fourier spectral representation was physically appropriate as well as mathematically convenient. From it one can derive all the statistical distributions of the waves for which the model is valid (Longuet-Higgins, 1962). From a practical point of view we must learn how to recognize and circumvent the limitations imposed by nonuniformity of the wind structure, and how to predict the evolving (directional) wave spectrum from such meteorological observations as are available, or from the output of computer simulations.

17.4.1 The Fetch-Limited Case

An important but relatively easily realizable case is that of a steady wind blowing off a straight shore, so that the duration of the wind is irrelevant and the fetch is well defined. An early contribution to this problem came from Burling (1959), who measured wave spectra at short fetches on an artificial lake using a newly developed capacitance-wire wave recorder.

In this case one can hope that the energy of the waves at a given fetch will be proportional to the work done by the wind on the water. If this is crudely estimated as proportional to the shearing stress times a distance measured by the fetch, then

$$\zeta = \text{constant} \times u_*(X/g)^{1/2} \qquad (17.17)$$

where $\overline{\zeta^2}$ is the mean square wave amplitude, and X the fetch.

Burling's results supported the simple relation (Charnock, 1958b) and it was confirmed for longer fetch by the results of the JONSWAP experiment (Hasselmann and colleagues, 1973). The Joint North Sea Wave Project (JONSWAP) was an important cooperative venture in which a group of scientists from several countries pooled their observational resources to obtain wave spectral data good enough to allow generalization about its evolution with varying wind and fetch. They used a linear array of wave sensors spaced along a 160-m profile extending westward from the island of Sylt in the North Sea (figure 17.6).

As regards the wave energy the JONSWAP data supported (17.17). Figure 17.4, from Phillips (1977a), shows Burling's observations together with those of JONSWAP: it is plotted in terms of nondimensional coordinates proposed by Kitaigorodskii (1962) to show that the constant of (17.17) is about 1.26×10^{-2}.

Burling was also able to calculate spectra. The photographic recording technique and the analogue spectral analyzer then in use much increased the effort needed, while restricting the precision of the estimates. Nevertheless Burling was able to establish the main features of the nondirectional frequency spectrum. He found that there was a very rapid increase, at low frequencies, to a maximum value at frequency n_0 determined by the wind speed and the fetch. At frequencies greater than n_0 the spectra fell off, approximately as (frequency)$^{-5}$. In this so-called equilibrium range of the spectrum the energy was largely independent of both wind and fetch. Figure 17.5, from Phillips (1977a), includes some of Burling's spectra together with those of later workers.

Those of the JONSWAP project are broadly similar (figure 17.6), but near the peak frequency they show an overshoot which had first been observed by Kinsman (1960) and by Barnett and Wilkerson (1967), who used an airborne radar altimeter to measure one-dimensional wavenumber spectra over larger fetches. Snyder and Cox (1966) had measured the evolution of one particular spectral band (around 0.3 Hz) by towing an array of wave recorders downwind at the appropriate group velocity, finding that the energy "overshot," in

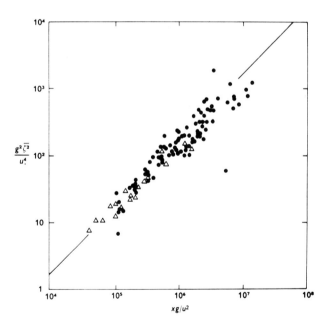

Figure 17.4 Field measurements of the dimensionless mean-square surface displacement $g^2\overline{\zeta^2}/u_*^4$ as a function of dimensionless fetch Xg/u_*^2. Data points are represented thus: Hasselmann et al. (1973), ●; Burling (1959), △. The line in the background is $g^2\overline{\zeta^2}/u_*^4 = 1.6 \times 10^{-4} Xg/u_*^2$. (Phillips, 1977a.)

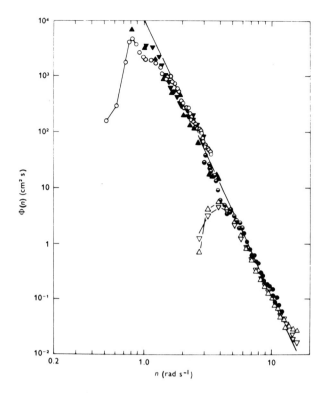

Figure 17.5 The equilibrium range of the frequency spectrum of wind-generated waves. The logarithmic vertical scale covers six decades. The shape of the spectral peak is included in only three cases; otherwise only the saturated part of each spectrum is shown. Key to measurements:

○	Stereo-Wave Observation Project (Pierson, 1960)	floating wave spar	1 spectrum
▲	Longuet-Higgins et al. (1963)	accelerometer	1 spectrum
▼	DeLeonibus (1963)	inverted fathometer	Mean of 6 spectra
△	Kinsman (1960) November series	capacitance probe	Mean of 16
▽	Kinsman (1960) July series	capacitance probe	Mean of 16
●	Burling (1959)	capacitance probe	Mean of 11
⊖	Walden (1963)	probe and cinematograph	1 spectrum

[After Phillips (1977a), who compiled and plotted the original observations.]

that it grew faster than would be expected, for the spectrum as a whole, from linear theory.

The frequency of the spectral peak is clearly an important descriptor of the wave field. Its value for Burling's, the JONSWAP, and other observations is shown in figure 17.7 (from Phillips 1977a). The values are again plotted in the nondimensional form suggested by Kitaigorodskii. It is perhaps worth noting that if L_0, the wavelength at the spectral peak, be given by $L_0 n_0^2 = 2\pi g$, then

$$L_0 = 1.3 u_*(X/g)^{1/2}$$
$$\approx 100\zeta, \qquad (17.18)$$

consistent with the bulk of the energy being in the equilibrium n^{-5} range.

As a result of the many observations of waves we now have reasonably clear information on the evolution of the surface wave field in deep water, at least so far as the frequency spectrum is concerned. Directional spectra are more difficult to measure and information is correspondingly sparse.

17.4.2 The Energy and Momentum Balance of the Wave Spectra

The main purpose of the JONSWAP project was to determine the source function in the spectral equation for the energy balance

$$\frac{\partial E}{\partial t} + v_{gi} \frac{\partial E}{\partial x_i} = S. \qquad (17.19)$$

E is the wave energy and v_{gi} the component of the appropriately generalized group velocity in the direction of coordinate x_i. The basic result is shown schematically in figure 17.8, where the source function S is seen to have a characteristic positive-negative shape.

The source function at a particular frequency is made up of three components—the energy transferred to the waves by the wind, the energy dissipated, and the energy transferred from other regions of the spectrum.

The spectral representation used is based on a superposition of sinusoidal waves traveling independently. But the hydrodynamic equations are nonlinear and the linear approximation is only valid when the wave slope is small, i.e., when the accelerations are small relative to the acceleration of gravity.

To treat the nonlinear terms, one substitutes the linear solution into the nonlinear terms, to get a second-order solution with terms in the wave slope. Higher-order solutions have terms in (slope)², (slope)³, and so on. The primary waves are sinusoidal and the second approximation has sharper crests and flatter troughs. One gets terms involving products of pairs of primary waves, which produce secondary waves at their sum and difference frequencies.

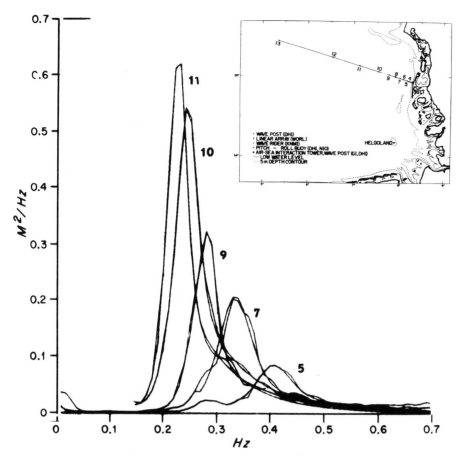

Figure 17.6 Evolution of wave spectrum with fetch for offshore winds 11^h–12^h Sept. 15, 1968. Numbers refer to stations inset. (Hasselmann et al., 1973.)

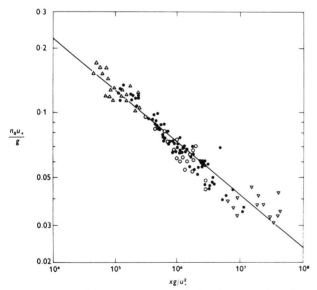

Figure 17.7 Field measurements of the dimensionless frequency of the spectral peak $n_0 u_*/g$ vs. dimensionless fetch Xg/u_*^2. Data points are as follows: Hasselmann et al. (1973), ●; Kitaigorodskii and Strekalov (1962), ▽; Mitsuyasu (1966), ○; and Burling (1959), △. The straight line in the background is $(n_0 u_*/g) = 2.2(Xg/u_*^2)^{-1/4}$. [After Phillips (1977a), who compiled and plotted the original observations.]

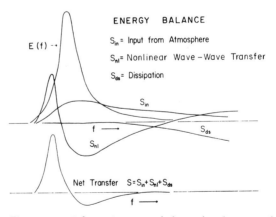

Figure 17.8 Schematic energy balance for the case of negligible dissipation in the main part of the spectrum. (Hasselmann et al., 1973.)

The solution stays bounded provided there is no combination of

$$\mathbf{k}_3 = \mathbf{k}_1 \pm \mathbf{k}_2 \quad \text{and} \quad n_3 = n_1 \pm n_2$$

such that

$$gk_3 = n_3^2.$$

O. M. Phillips (1963) showed that no such combination occurs in surface gravity waves. But for tertiary waves he found that for

$$\begin{aligned}\mathbf{k}_4 &= \mathbf{k}_1 \pm \mathbf{k}_2 \pm \mathbf{k}_3, \\ n_4 &= n_1 \pm n_2 \pm n_3\end{aligned} \quad (17.20)$$

there exist combinations for which $gk_4 = n_4^2$, so there is a resonance, with energy being transferred from three primary waves to a new wave whose energy grows linearly with time. The interactions are weak, so it grows slowly, its time scale being of order (slope)⁴ times a typical wave period. Such nonlinear interactions have been observed in careful laboratory experiments and shown to be consistent with the slow attenuation of ocean swell.

Hasselmann (1966) has exploited the analogy with collisions in high-energy physics, and he uses Feynman diagrams to represent nonlinear interactions, with wavenumber corresponding to momentum and frequency to energy. He has also given a complicated equation by which the nonlinear transfers can be calculated. Using an interaction equation derived by Longuet-Higgins (1976), Fox (1976, 1978) has given a simpler method applicable when the spectrum is narrow. Broader spectra have been studied by Webb (1978).

For the JONSWAP case S_{nl}, the contribution of nonlinear transfer is shown on figure 17.8. It has the same positive–negative shape as S and provides a reasonable qualitative explanation of the way in which the spectral peak goes to lower frequencies as the nondimensional fetch increases.

The contribution of S_{nl}, due to nonlinear weak interactions, is to redistribute wave energy within the spectrum. It is the best known of the terms that make up S:

$$S = S_{in} + S_{nl} + S_{ds} \quad (17.21)$$

where S_{in} represents the energy input from the atmosphere and S_{ds} the dissipation. Assuming the dissipation to be small in the energetic low-frequency band of the spectrum, the JONSWAP results indicate a schematic energy balance as in figure 17.8. Then the energy input has a distribution like that of the spectrum itself, as if the wave generation depends linearly on the spectrum, and the dissipation occurs mainly at high frequencies.

Attempts to calculate S_{in} theoretically have so far been unsuccessful. It involves the calculation of the covariances between fluctuations in the surface stress (both normal and tangential) and in the surface velocity. Phillips (1957) showed that turbulent pressure fluctuations in the natural wind would amplify waves traveling at the right convection velocity by a resonance mechanism. Like an earlier theory of Eckart (1953), the theory was qualitatively correct but the amplitude of atmospheric pressure fluctuations (rms pressure fluctuation $\simeq \tau_0$) was too small to produce waves of the amplitude observed. Miles (1957, 1959) calculated fluctuations induced by the mean wind blowing over the wavy surface: since the pressure fluctuations depend on the wave amplitude, the latter grows exponentially, but again the predicted growth rate was much less than that observed. Miles had been obliged to neglect the atmospheric turbulence in the interfacial layer, however, and nobody has yet succeeded in satisfactorily incorporating it. The problem was carefully discussed by Davis (1972), who used several different closure approximations, which gave variable results. He also found that the rate of energy transfer to the waves is critically dependent on the profile of mean flow very close to the interface. Gent and Taylor (1976), who have done numerical simulations of airflow over waves, avoided the problem by assuming that the surface has an assigned roughness, either constant or distributed along a long wave. Their solutions are more encouraging but the problem of calculating energy and momentum transfer in the interfacial layer is by no means solved.

Detailed observation of energy and momentum transfer in the interfacial layer also presents great difficulty. Since the drag coefficient of the sea surface is greater, but not very much greater, than that of an aerodynamically smooth surface, one might expect some direct viscous transfer. In this case the tangential stress must be supported, just below the interface, by a thin layer with strong shear. Equally, since the sea surface becomes increasingly rough, relative to an aerodynamically smooth surface, as the wind speed increases, there must be a good deal of momentum transport to irrotational or quasi-irrotational waves by pressure fluctuations. This was the basis of Jeffreys's (1925) theory. Valiant attempts have been made to measure pressure fluctuations relative to the wave profile by Dobson (1971), Elliott (1972), and Snyder (1974).

Such observations are extremely difficult. The static pressure fluctuations $O(\tau_0)$ are small, very small relative to the dynamic pressures in the airflow. To compute the energy and momentum transfer to the waves, the pressure is needed at the (moving) surface: this needs an extrapolation from a recorder as near the surface as possible, or a surface-following device, which introduces other problems. It is hardly surprising that the early results were not entirely consistent: roughly speaking Dobson's values gave the biggest growth rates, Elliott's were smaller (roughly $\frac{1}{5}$) and Snyder's even less.

These three authors have recently collaborated with Long in a field program in the Bight of Abaco in the Bahamas (Snyder, Dobson, Elliott, and Long, 1980). Preliminary results indicate that momentum transfer to the wave extends from the peak frequency to at least twice the peak frequency with little noticeable falloff. Observations at higher frequency will be necessary to allow an estimate of the total momentum transfer, but it seems clear that for the JONSWAP spectrum a significant fraction of the momentum first goes into waves; about 10% at long fetch ($gX/U_{10}^2 = 10^5$), rising to about 100% at shorter fetch ($gX/U_{10}^2 = 10^2$). There remains a need for critical observations at short fetch, but the high-frequency response that will be needed will be hard to achieve.

The dynamics of a near-surface viscous shear layer are also relevant: Banner and Phillips (1974) have shown that the speed of such a layer will be increased near the crest of a longer wave, so leading shorter waves to break. Such breaking may be made visible by dimples or pockmarks on the surface, particularly in the early development of a wave field. Banner and Melville (1976) have demonstrated that wave breaking, even on a small scale, is accompanied by separation of the airflow from the surface. This has strong implications both for momentum transfer and for the exchange of heat or water vapor. That the drag and analogous coefficients are small may prove to be due to the sporadic nature of the breaking process. The problem deserves further investigation since breaking waves seem to provide a limit between the spectrum of the longer Gaussian waves and the shorter nonlinear ripples: breaking waves on the open sea have been much neglected (Charnock, 1958a).

Another important phenomenon associated with breaking waves, with which one hopes more progress will be made in the next 20 years than in the last, is dissipation. It now seems more likely that breaking waves are more important than viscosity in dissipation, and Longuet-Higgins (1969a) has given an interesting calculation that implies that the proportion of wave energy lost per mean cycle is about 10^{-4}.

Measurements of mean and fluctuating velocities in waves are technically difficult, but there is growing evidence that the orbital velocities of the larger waves are inactive, in the sense that they provide variance but are so uncorrelated as to be inactive in the transfer of momentum. Jones and Kenney (1977) have argued that the near-surface layer in the water has many of the characteristics of the surface layer in the air, with scaling on u_* and z_0. Observations by Donelan (1978) show that as well as the wave orbital velocities there are fluctuations at lower frequency (possibly due to the shear in the mean profile) and at higher frequency (possibly due to whitecapping). The momentum flux was entirely due to the low-frequency fluctuations. His general picture of the effects of wave breaking is that the wind stress produces a strongly sheared current near the surface, so that when a wave breaks the downward pulse of fluid produces a downward momentum transfer. Though each pulse of momentum is short, the intermittent nature of the phenomenon is reflected in the momentum flux at lower frequencies. The effect of whitecapping on the spectrum has been considered theoretically by Hasselmann (1974) as a strong interaction that is weak in the mean: because it is sporadic and local in physical space the energy loss is spread over much of the spectrum.

17.4.3 Langmuir Circulations

Another near-surface phenomenon that may be important in momentum transfer is the Langmuir cell. Langmuir cells are alternate left-handed and right-handed vortices in the vertical plane (horizontal rolls), aligned along the wind with surface velocities strongest in the convergence zones. It is easy to see that the stronger horizontal velocity there could combine with the sinking motion under the convergence zone to give mean stresses of the same order as the wind stress at the surface.

There have been many observations since Langmuir's (1938) first description, all supporting the cellular structure he found. Row spacings, often marked by streaks on the surface, are variable, typically 10 m in lakes and 100 m over the ocean: the surface current moves at about 10 cm s^{-1} faster in the streak than outside it. The vertical structure is less well known. An account of the observations is given by Pollard (1977); he also gives an account of theoretical attempts to explain these cells, from which it seems clear that complicated interactions in the surface wave field are involved. Faller (see chapter 16) has shown in laboratory observations that both wind and waves are necessary for the generation of Langmuir cells: it is thought that the vorticity of the shear flow produced by the wind stress is transformed by nonlinear interaction with crossing wave trains into the vorticity of the helices. The details are complicated but it seems likely that the Langmuir cells may represent a mechanism by which wave energy is converted to organized convection and to turbulence, which in turn may act to deepen the mixed layer.

17.5 The Atmospheric Boundary Layer

From a practical point of view the mean fluxes at the sea surface can now be calculated to acceptable accuracy from observations in the surface layer. The related characteristics of the surface wave field are also reasonably well known and it can be assumed, with somewhat less confidence, that the momentum transferred from the atmosphere to the sea surface is then transferred to the ocean at the same place and time.

In all these cases our knowledge is empirical and there is a need for more understanding, leading to theoretical descriptions of the physical processes involved. But from an engineering viewpoint what was once thought of as the central problem of air–sea interaction has been reduced to some sort of order.

Problems change, however, and those of air–sea interaction are now of much greater scope. The recently renewed interest in climate and climatic change has led to a wider appreciation of the importance of the interaction between the atmosphere, the ocean, and characteristics of solid surfaces such as ice. Because almost all the energy for the motion comes from the sun it is conceptually attractive to regard the basic circulation of the atmosphere and ocean as the free convective response of the coupled system to solar heating. Air–sea interaction can now be taken to include all the problems of meteorology and oceanography.

Nevertheless the different physical properties of air and water, especially their relative opacities to electromagnetic radiation, lead to considerable decoupling: the mismatch is such that it is usually more rewarding to treat them separately, isolating topics like the effect of wind on the sea, or the effect of evaporation on the atmosphere. The darkness of the ocean has also made observations difficult, so less is known of its structure than that of the relatively transparent atmosphere.

No one disputes that the fluxes of heat, water vapor, and momentum that enter the atmosphere through its lower boundary layer are of crucial importance to the development of atmospheric flow patterns and weather on time scales ranging from minutes to months. There is no reason to doubt that they are equally important for longer-period climatic changes, but we know little of the degree of accuracy and detail in which they must be described for specific purposes, in particular for forecasting using computer models of the atmosphere, the ocean, or the coupled system.

Some suggestion that rather precise knowledge of the exchange processes will be needed comes from the relations that have been found (Namias, 1969; Bjerknes, 1969; Ratcliffe and Murray, 1970) between sea-surface temperature anomalies and subsequent weather patterns, though a direct causal connection has not been unambiguously demonstrated. To achieve such precision an understanding of the physical processes involved seems essential: attempts to use parameters without physical understanding may yield rapid progress in the early stages but seem likely to be inadequate in the long run.

At any given time atmospheric and oceanic motions on scales greater than 100 km or so can be treated as essentially inviscid and adiabatic, but there are localized regions where condensation processes, or the transport of heat, water, salt, or momentum by small-scale turbulence are important, even dominant. Examples of such regions are towering clouds, or groups of clouds, fronts, and the turbulent boundary layers near the earth's surface. For the present purpose it has seemed sensible and convenient to restrict the scope of air–sea interaction to studies of the mechanics and structure of the near-surface boundary layers of the atmosphere and the ocean.

For many years, as has been indicated, the subject was more restricted, essentially to the lowest 10 to 100 m of the atmosphere above the sea. This came about mainly because the routine observations available were those from ships. There were a few upper-air observations from weather ships, but their purpose was to map meteorological fields in the troposphere and lower stratosphere: exchanges with the ocean were not allowed for in routine forecasting.

Routine observations of the whole atmospheric boundary layer over the sea are still virtually nonexistent, the number of weather ships having decreased in recent years. What has developed rapidly is the numerical simulation of atmospheric processes, now used routinely as a basis for weather forecasting: when these are used to forecast for more than a day or so it begins to be necessary to include boundary-layer effects. Some models use many layers in an attempt to resolve the vertical structure of the boundary layer, ultimately relating fluxes to conditions in the surface layer, but the need to simulate small-scale turbulence in the boundary layer makes such a system prohibitively expensive in computer time unless the equations are drastically simplified. It seems more realistic to admit that the boundary layer has a different physics from most of the atmosphere above it and to seek to treat it as a whole. Then the height of the top of the atmospheric boundary layer is calculated explicitly and becomes the effective lower boundary of the largely frictionless atmosphere above. Such a method was adumbrated by Charnock and Ellison (1967) and has been developed by Deardorff (1972) and implemented by Arakawa (1975). The structure of the atmospheric boundary layer is now being increasingly studied, but it is much more complicated than the surface layer. It is not well mapped, nor are the physical processes that maintain it well understood.

The top of the atmospheric boundary layer is usually most obvious from inspection of the density variation with height, particularly over land in summer. Then a weakly stable condition is established during the night that is transformed after dawn by solar heating to a convective boundary layer at the surface. This is sufficiently well mixed to have effectively constant potential temperature, and it deepens as it warms in accordance with the classical ideas of Gold (1933). It is easy to check that there is reasonable quantitative agreement between the available solar energy and the rate of warming of the boundary layer. That the level of turbulence inside such a convective boundary layer is much greater than in the free air above is obvious to anyone who has done much flying, but of course it can be confirmed instrumentally. It is also common to find that smoke or other pollutants are fairly uniformly mixed in the boundary layer but that the air above is relatively clean.

Over land there is a pronounced diurnal variability and a great range of boundary layer depths: over the sea there is little diurnal change but much variability from day to day.

17.5.1 Unstable Boundary Layers

Given the original records it is possible to obtain a fair representation of unstable boundary layers from routine radiosonde ascents. Figure 17.9 shows a characteristic diagram obtained from the ascent at Ocean Weather Station (OWS) India (59°N, 19°W) at 2330Z on 10 March 1966. The variables plotted are the potential temperature θ and the specific humidity q, both of which are conserved in adiabatic motion and obey the simple law of mixtures. The q-θ diagram was used by Taylor (1917) as a tool for studying the atmospheric boundary layer and is sometimes called the Taylor diagram. The analogous S-θ diagram is widely used in oceanography, but the great convenience of the Taylor diagram has not been widely recognized in spite of a comprehensive review by Montgomery (1950).

For the ascent plotted it will be seen that the points corresponding to heights up to 908 mb are clustered together, indicating a high degree of mixing: higher up, the density gradient is definitely stable. The point representing air in contact with the sea is at a potential temperature 5°C higher than that typical of the mixed layer. The point corresponding to the observations at deck level theoretically would be expected to be on the line joining the mixed layer to the sea surface: that the surface values are lower is difficult to explain.

Figure 17.10 is a similar diagram from an ascent at Gan (0°41′S, 73°09′E). Some workers, influenced by Ekman's theory of the variation of wind with height, have predicted very deep boundary layers near the equator: that such a view is not borne out by observation goes some way toward demonstrating that the thickness of the boundary layer is determined more by the density structure.

Figure 17.11 gives the results of a special slowly rising ascent at OWS Julliet (52°30′N, 20°W). More

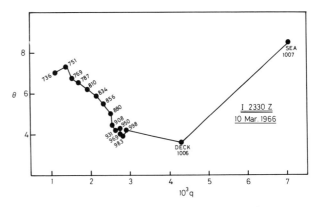

Figure 17.9 Characteristic diagram, OWS India (59°N, 19°W), on 10 March 1966.

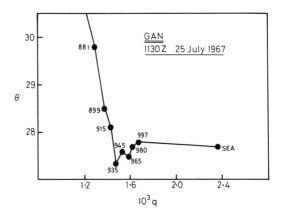

Figure 17.10 Characteristic diagram, Gan (0°41′S, 73°09′E), on 25 July 1967.

detail is given, but the structure is basically similar to that in figures 17.9 and 17.10. The wind profile at the same time shows that the wind varies slowly with height in the mixed layer, but that there is considerable shear at the boundary-layer top.

There remains a need for long-term studies of the character of the boundary layer over the sea so that their climatology can be established. A pilot study of the ascents at OWS India (59°N, 19°W) for March 1966 showed that more than half had reasonably well-defined unstable boundary layers. The boundary layer depth ranged from 200 to 2000 m, being at most weakly correlated with the vertical potential temperature difference between the sea surface and the mixed layer, which ranged from 0 to 9°C.

Even in convective conditions a well-mixed state with potential temperature independent of height is not always found, and it may be difficult to determine the depth of the boundary layer from the sounding. Also, when a well-mixed layer does exist it may be topped by a layer of relatively weak stability into which the stronger convective motions from below can penetrate a considerable distance.

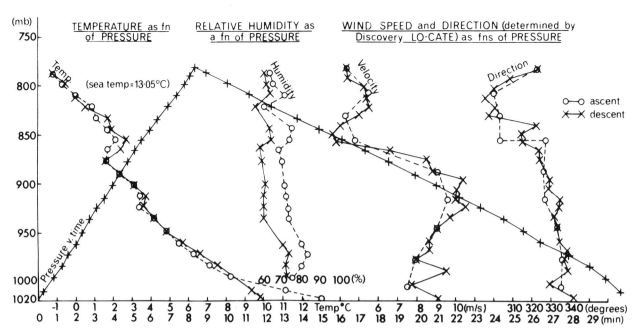

Figure 17.11 Records from LOCATE sonde (D.22) released from R.R.S. *Discovery* at 1915Z on 17 June 1970 near OWS Julliett, 52°30′N, 20°W. The record of pressure against time shows a rapid double-balloon ascent to a chosen height (780 mb), where one balloon is released and the other sinks at a slower speed (about 100 m min^{-1}) to the surface. Temperature, humidity, and wind are shown in relation to pressure for ascent and descent separately.

The thermals rising through a well-mixed layer commonly have values of specific humidity and potential temperature in their centers equal to those which would be produced by mixing equal parts of air from their environment and air from the surface. So, if the condensation level for such a mixture is within the well-mixed layer, cloud will form, its amount and development being related to the boundary-layer structure. If, on the other hand, the condensation level is above the well-mixed layer, condensation can occur only if the thermals are strong enough to penetrate the stable region above. This is the usual situation in the Trade Winds and tropics.

Once condensation has taken place the latent heat released adds to the clouds' buoyancy, so that in favorable cases the motion becomes unstable, and the clouds grow and may produce showers.

17.5.2 Buoyancy-Transfer Processes

As the surface is approached, the buoyancy forces have little dynamic effect, the temperature and humidity fluctuations being produced by vertical motions that they have not caused. From time to time air from the surface is lifted upward, probably in the form of a column or a sheet rather than a blob. As it rises it will be eroded at its edge by small-scale turbulence and become thinner. If the rising column survives long enough, its own buoyancy will begin to have an effect, so that in unstable conditions it will accelerate, being stretched and becoming even thinner as it rises. Some temperature traces made by Webb (see Priestley, 1967) illustrate the thinning with height, and they are consistent with the idea of columns of warm air leaning downwind, since each active occurrence is first apparent at the greatest observation height. Figures 17.12 and 17.13, by Kaimal and Businger (1970), illustrate a case that has been studied further by Businger and Khalsa (1978).

After the rising warm air has acquired a vertical velocity appreciably greater than the turbulence at its level, it will cease to be eroded and will begin to entrain air into itself and grow (Elder, 1969). The motion may then have the character of a starting plume (Turner, 1969). Eventually the supply from below is cut off and the air transfers heat and water vapor, but is less efficient at transferring momentum: attempts have been made to predict the transfer coefficients theoretically (e.g., Richards, 1970), but the coefficients are not clearly established.

Since the velocity of individual thermals relative to their environment decreases with height, and since the environment must itself be sinking to compensate for the rising motion of the thermals, it is clear that the weaker thermals will be brought to rest at moderate heights, to be entrained into stronger ones. Thus it is possible for the cross section of each thermal to grow with height and yet the fraction of horizontal area occupied by rising air not to increase strongly with height.

The height at which a fair number of thermals can be said to be well formed and self-propelled is probably

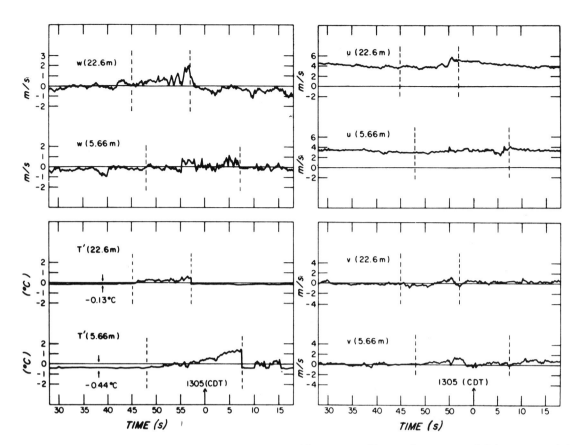

Figure 17.12 Traces of u, v, w and T (temperature) during passage of a convective plume. (Kaimal and Businger, 1970.)

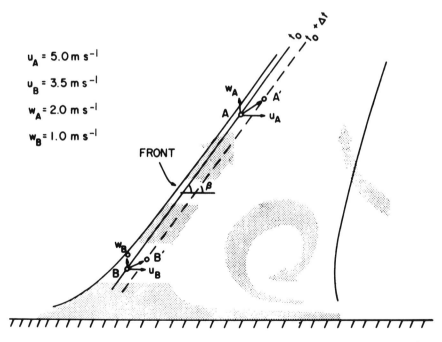

Figure 17.13 Two-dimensional model of a convective plume. (Kaimal and Businger, 1970.)

499
Air–Sea Interaction

above the surface layer, where the temperature gradient and the wind shear are determined by the similarity rules. In particular the wind shear may be governed by the variation of the geostrophic wind with height. There is some evidence that when this is large there is a tendency for motions of scale comparable with the depth of the boundary layer to become organized into large longitudinal roll vortices. Given an appropriate condensation level, one would expect such motions to be visible in pictures of clouds taken from high-flying aircraft or from satellites, and many such images have been interpreted in this way; see, for instance, Agee and Dowell (1974) and Kuettner (1971). The patterns also depend on the general vertical motion due to convergence or divergence on the mesoscale or the synoptic scale.

The importance of clouds in the transport process is clear from studies that evaluate the heat or water budgets of the subcloud layer and the cloud layer separately. Riehl, Yeh, Malkus, and La Seur (1951) in a classical study found that as much as four-fifths of the water evaporated from the sea entered the cloud layer. This is a high value, to explain which it has been suggested that the transport is concentrated into localized areas where there are cloud groups or clusters. But the same sort of thing happens in polar outbreaks, with well-spaced clouds, so it seems likely that cumulonimbus clouds must suck up a large volume of the subcloud layer between them. Browning and Ludlam (1962) suggest a cumulonimbus model in which strong downdrafts partially compensate for the upflow, but in general there will be shrinking and subsidence in the subcloud layer in the space between clouds. The role of the heat and water-vapor fluxes from the surface is thus in the first instance to maintain the depth of the well-mixed layer rather than to feed directly into the layer clouds.

17.5.3 Stable Boundary Layers

Stable boundary layers, on the other hand, are difficult to investigate using routine observations. They are often relatively shallow, with small temperature differences, and since the transports of heat and water vapor are small, they have attracted little attention. There are few satisfactory sets of observations, but they can be interpreted as showing that heat is transferred to the surface until an almost linear gradient of potential density is formed from the surface to height h, where the difference in potential temperature $\Delta\theta$ is given roughly by

$$g \frac{\Delta\theta}{\theta} h/U_{10}^2 = 0.5. \tag{17.22}$$

Hanna (1969) attributes (17.22) to Laikhtman (1961), and gives an example using O'Neill's data (Lettau and Davidson, 1957) that supports it.

Over the sea, figure 17.14, by Craig (1946), shows ascents made at three different fetches in warm continental air flowing out over colder sea. At the largest fetch

$$g \frac{\Delta\theta}{\theta} h/U^2 \approx 0.4.$$

One of the classical ascents made by Taylor (1914) on the S.S. *Scotia* provides another example (the others are not suitable because of fog), which is shown in figure 17.15. Here

$$g \frac{\Delta\theta}{\theta} h/U^2 \approx 0.35.$$

17.5.4 Wind in the Boundary Layer

Turbulent friction in the boundary layer causes the wind to deviate from its frictionless value, and earlier sections have shown how it increases rapidly, roughly as the logarithm of the height in the lowest few meters. From a height of 30 m or so there is usually relatively little change until the top of the boundary layer is reached. Sometimes there is a significant and rapid change with height at the top of the boundary layer until the frictionless value is attained.

The wind changes not only in speed but in direction also. Such changes were predicted by Ekman for the ocean and soon applied to the atmosphere by Akerblom and (some years later, independently) by Taylor. Because they took a constant value for the eddy viscosity K_M they got the well-known Ekman spiral for the hodograph. Perhaps more important is their deduction that in stationary conditions, when the boundary layer has a finite depth H, with zero stress above,

$$u_*^2 = \int_0^H f(V - V_g)\,dz,$$
$$0 = \int_0^H f(U - U_g)\,dz, \tag{17.23}$$

where U_g, V_g are the components of the geostrophic wind.

This result is independent of the mechanism of transfer: it relates the surface stress to the cross-isobar transfer, and provides a basis for calculating the frictional convergence and so the mean vertical velocity at the top of the boundary layer. This, in turn, has an effect on the motion of the whole atmosphere. Unfortunately, conditions are rarely so simple as to allow a direct application of (17.23).

Much subsequent work has sought to use more complicated expressions for $K_M(z)$, to try to predict $K_M(z)$ theoretically, or to deduce it from observations. It is now known that the variation of the pressure gradient with height often has a large influence on the angle between the geostrophic and the surface wind; and that

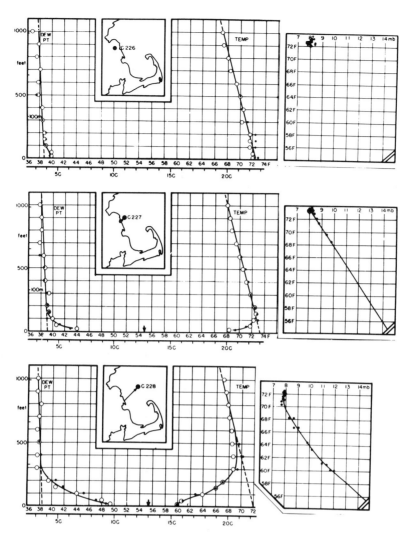

Figure 17.14 Modification of a warm continental air mass flowing over a colder sea, Massachusetts Bay, 18 October 1944, showing vertical distribution of temperature and dewpoint and the corresponding Taylor diagrams. (Coordinates are potential temperature and potential vapor pressure.) Sea surface temperature indicated by arrows. (Craig, 1946.)

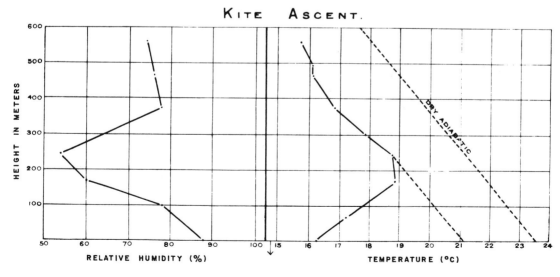

Figure 17.15 A kite ascent from S.S. *Scotia* at 44°39′N, 49°48′W, 29 July 1913. (Taylor, 1914.)

501
Air-Sea Interaction

variations of K_M in time, or in the downstream direction, can lead to oscillations in which the flow in the middle of the boundary layer can increase considerably above its geostrophic value, giving rise to the so-called low-level jet.

Since it is known that near the surface $K_M = \kappa u_* z$ and that in neutral conditions this is a fair approximation up to 100 m or so, it is obviously worth examining the implications of assuming it true at all heights (Ellison, 1956). The results agree reasonably well with measured wind profiles, but this shows merely that they are insensitive to K_M above the surface layer. This does not vitiate, however, Ellison's demonstration that the thermal wind has a large effect.

Nevertheless in the (very rare) case when the stratification is so nearly neutral that it can be neglected, Kazanskii and Monin (1960, 1961) dealt with the problem in a convincing way. In these papers they introduced a similarity argument for the entire boundary layer that has since been discussed by Csanady (1967), Gill (1968), Blackadar and Tennekes (1968), Zilitinkevich (1969, 1970), and others. This is based on a combination of the surface-layer arguments, leading to the logarithmic wind profile together with a velocity defect law that asserts that

$$\frac{U}{u_*} = f(z/H) + \text{a velocity of translation}.$$

For the atmospheric boundary layer, the boundary layer thickness H is taken as u_*/f, where f is the Coriolis parameter. The result is, for neutral conditions

$$\frac{\kappa U_g}{u_*} = \ln\left(\frac{u_*}{fz_0}\right) + A,$$

$$\frac{\kappa V_g}{u_*} = B.$$
(17.24)

A reasonable value of A is 2 and of B is 5. Figure 17.16, by Lettau (1959), shows some of the earlier results.

In nonneutral conditions the relation for a transferable scalar is

$$\frac{\kappa(\Theta - \Theta_0)}{\theta_*} = \ln\left(\frac{u_*}{fz_\theta}\right) + C,$$
(17.25)

but in the nonneutral case A, B, and C are no longer constants but functions of $\mu \equiv u_*/fL$.

Valiant attempts to measure $A(\mu)$, $B(\mu)$, and $C(\mu)$ have been made by Clarke (1970) and others over land. The results are very scattered, possibly because of effects of time and space variability.

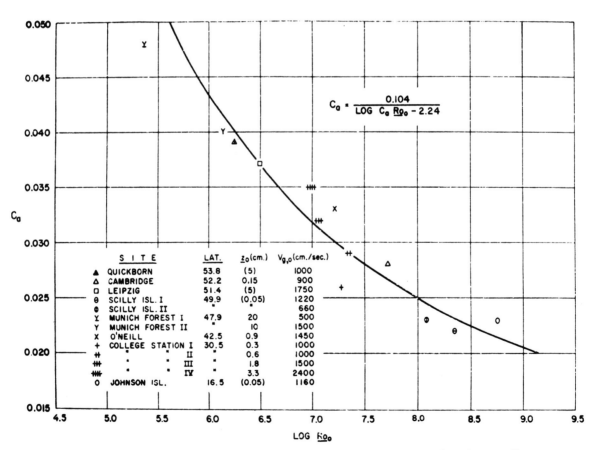

Figure 17.16 Geostrophic drag coefficient versus surface Rossby number. $C_a = u_*/|\mathbf{V}_g|$, $Ro_0 = |\mathbf{V}_g|/(fz_0)$. (Lettau, 1959.)

There seems to be no way to avoid the need for an evolving model of the boundary layer in which the height of the top will be predicted. This will depend on advection, on the (frictional and frictionless) convergence, and on the entrainment of the air above. In this sense the early single-point measurements (Sheppard, Charnock, and Francis, 1952; Charnock, Francis, and Sheppard, 1956) and others well described by Roll (1965) are of limited value. More recent studies have been large acronymic projects like ATEX (Augstein, Schmidt, and Ostapoff, 1974), BOMEX (Holland and Rasmusson, 1973), GATE, AMTEX (Ninomiya, 1974) and JASIN (Taylor, 1979), from which no simple result has yet been distilled.

A climatological study has been made by Findlater, Harrower, Howkins, and Wright (1966): this and related work is reported by Sheppard (1970), from whom figure 17.17 is taken. The geostrophic drag coefficient implied by (17.24) must also be consistent with the requirements of the angular-momentum balance of the earth, and La Valle and Di Girolamo (1975) have thus found a mean value $(= u_*^2/|\mathbf{V}_g|^2)$ of 0.41×10^{-3}.

Even in cloud-free conditions the dynamics of the atmospheric boundary layer is complicated: it is not yet clear which are the most important transfer processes or how they can be dealt with.

17.5.5 The Upper Boundary Layer of the Ocean

The near-surface boundary layer of the ocean has much in common with that of the atmosphere. It is most obvious from the vertical density structure: more or less well-mixed layers are to be found near the surface over most of the ocean most of the time. As in the atmosphere the velocity structure is not well known (it is difficult to measure currents in the presence of waves), but the simple Ekman-type distributions are rarely found.

Like the atmospheric boundary layer, the oceanic boundary layer is maintained by a combination of advection, surface fluxes, and entrainment at the lower surface. Because vertical gradients are much bigger than those in the horizontal, the advection can often be neglected: the basis of the resulting one-dimensional models is well described by Niiler and Kraus (1977), and details of the complicated mixing processes are given by Turner (chapter 8).

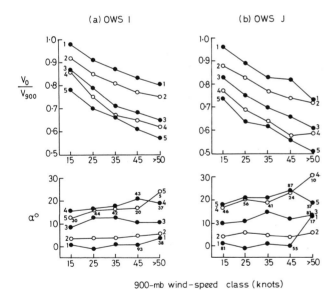

Figure 17.17 The variation of the ratio between the wind speed at the surface (V_0) and at 900 mb (V_{900}) and of the angle between them (α) in relation to V_{900} mb and the mean lapse rate from surface to 900 mb at OWS India and Julliett. The points refer to classes in wind speed (kt): 10-19, 20-29, 30-39, 40-49, >50, and in lapse rate (°F/1000 ft = 1.69°C/km): >5.5 (1), 5.4 to 4.0 (2), 3.9 to 2.5 (3), 2.4 to 1.0 (4), 0.9 to −0.5 (5). Smaller lapse rates and lower wind speeds excluded. Lapse class shown against end of curves. Number of observations in each class when less than 100 shown in parentheses. (Due to Findlater et al., 1966.)

18 Oceanic Analogues of Large-scale Atmospheric Motions

Jule G. Charney and Glenn R. Flierl

18.1 Introduction

Newton (1687, book 2, propositions 48-50) and Laplace (1799) were aware that the principles governing the ocean tides would also govern the atmospheric tides. Helmholtz (1889) showed that ocean waves and billow clouds were manifestations of the same hydrodynamic instability, and he speculated that storms were caused by a similar instability. Had he known the structure of the Gulf Streams meanders, he might have speculated on their dynamic similarities to storms as well. Such intercomparisons were natural to the great hydrodynamicists of the past who took the entire universe of fluid phenomena as their domain. Although a degree of provincialism was introduced in the late nineteenth and early twentieth centuries by the exigencies of weather forecasting, it was just the practical requirements of weather observation that stimulated the development of modern dynamic meteorology and led to the deepening of its connections with physical oceanography. The recent explosive growth of the three-dimensional data base, the exploration of other planetary atmospheres, and the resulting increase in theoretical activity have greatly extended the list of ocean-atmosphere analogues. Indeed, it is now no exaggeration to say that there is scarcely a fluid dynamical phenomenon in planetary atmospheres that does not have its counterpart in the oceans and vice versa. This had led to the discipline, geophysical fluid dynamics, whose guiding principles are intended to apply equally to oceans and atmospheres.

Within this discipline the dominance of the earth's rotation defines a subclass of large-scale phenomena whose dynamics may for the most part be derived from quasi-geostrophy. For several years the authors have conducted a graduate course at MIT on the dynamics of large-scale ocean and atmospheric circulations in the belief that a parallel consideration of large-scale oceanic and atmospheric motions would broaden the range of our students' experience and deepen their understanding of the principles of fluid geophysics. C.-G. Rossby (1951) put the matter well:

It is fairly certain that the final formulation of a comprehensive theory for the general circulation of the atmosphere will require intimate cooperation between meteorologists and oceanographers. The fundamental problems associated with the heat, mass and momentum transfer at the sea surface concern both these sciences and demand a joint effort for their solution. However, an even stronger reason for that pooling of intellectual resources . . . may be found in the fact that the various theoretical analyses of the large-scale oceanic and atmospheric circulation patterns which have been called into being by our sudden wealth of observational data, appear to have so much in common that they may be looked upon as different facets of one broad general study, the ultimate aim of which might be described as an attempt to formulate a comprehensive theory for fluid motion in planetary envelopes.

He went on to say that "comparisons between the circulation patterns in the atmosphere and in the oceans provide us with a highly useful substitute for experiments with controlled variations of the fundamental parameters."

Needless to say, we subscribe to Rossby's views and therefore have willingly undertaken the task of reviewing the fluid dynamics of a number of phenomena that have found explanation in one medium and are deemed to have important analogues in the other. Where the analogues have already been explained, we have given a brief review of some of their salient features, but where little is known, we have not refrained from interpolating simple models or speculations of our own. In doing so, we were aware that the subject has grown so large that it includes most of geophysical fluid dynamics, and that, even if it were limited to large-scale, quasigeostrophic motions, it could not be encompassed in a review article of modest size. At least two excellent review articles on this topic, by N. Phillips (1963) and by H.-L. Kuo (1973), have already appeared, and there is now a text by Pedlosky (1979a) on geophysical fluid dynamics. For these reasons we have decided to limit ourselves to a small number of topics having to do primarily with disturbances of the principal atmospheric and oceanic currents, their propagation characteristics, their interactions with the embedding currents, and, to a lesser degree, with their self-interactions.

It is perhaps no accident that both Phillips and Kuo are meteorologists. Dynamic meteorology has benefited from a wealth of observational detail that has been denied to physical oceanography. Consequently, meteorologists have been the first to observe and explain many typical large-scale phenomena. This has by no means always been the case, but it has been so sufficiently often to justify the title of our review.

18.2 The General Circulations of Oceans and Atmospheres Compared

If it had been possible in a meaningful way, it would have been useful and instructive to make detailed dynamical comparisons of the general circulations of the atmosphere and oceans. This has not been so, and we must content ourselves with a few general remarks. Because of its relative transparency to solar radiation, the earth's atmosphere is heated from below; the oceans, like the relatively opaque atmosphere of Venus, are heated from above. The differential heating of the atmosphere produces a mean circulation that carries heat upward and poleward. Its baroclinic instabilities do likewise. These heat transports, combined with internal radiative heat transfer, tend to stabilize the atmosphere statically, and their effects are augmented by moist convection which drives the temperature lapse-rate toward the moist-adiabatic. The net result is that the atmosphere is rather uniformly stable for dry processes up to the tropopause. Above the tropopause, it is made increasingly more stable by absorption of solar ultraviolet radiation in the ozone layer. The oceans are rendered statically stable by heating from above, but heating cannot take place everywhere because the oceans, unlike the atmosphere, cannot dispose of internal heat by radiation to space; they must carry it back to the surface layers, where it can be lost by surface cooling. The most stable parts of the oceans are in the subtropical gyres, where the oceans are heated and Ekman pumping transfers the heat downward. The least stable parts are in the polar regions, where cold water is formed and carried downward by convection. The atmospheric troposphere has sometimes been compared with the waters above the thermocline, the tropopause with the thermocline, and the stratosphere with the deep waters below the thermocline (cf. Defant, 1961b). This comparison may be justified by the fact that it is in the atmospheric and oceanic tropospheres that the horizontal temperature gradients and the kinetic and potential energy densities are greatest. But from the standpoint of static stability, the absorption of radiation at the surface of the oceans makes its upper layers more analogous to the stratosphere. The static stability for the bulk of the atmosphere and oceans is determined by deep convection occurring in small regions. The analogy between deep convection in the atmosphere and deep convection in the oceans is between the narrow intertropical convergence zones over the oceans and the limited areas of cumulus convection over the tropical continents on the one hand, and the limited regions of deep-water formation in the polar seas on the other. From this standpoint, the ocean waters below the thermocline are more analogous to the atmospheric troposphere. Both volumes comprise more than 80% of the total by mass and both are controlled by deep convection.

It is remarkable that the regions of pronounced rising motion in the atmosphere and sinking motion in the oceans are so confined horizontally. Stommel (1962b) was the first to offer an explanation for the smallness of the regions of deep-water formation. His work motivated several attempts to explain the asymmetries in the circulation of a fluid heated differentially from above or below. We may cite as examples the experimental work of H. T. Rossby (1965) and the theoretical work of Killworth and Manins (1980) on laboratory fluid systems, and the papers of Goody and Robinson (1966) and Stone (1968) on the upper circulation of Venus. Because of its cloudiness, Venus has been assumed to be heated primarily from above, although it now is known that sunlight does penetrate into the lower Venus atmosphere and that the high temperatures near the surface are due to a pronounced greenhouse effect (Keldysh, 1977; Young and Pollack, 1977; Tomasko, Doose, and Smith, 1979). When a fluid is heated from below, the rising branches are found to be narrow and the sinking branches broad; when it is heated from above the reverse is true. One may offer the qualitative explanation that it is the branch of the circulation that leaves the boundary and carries with it the properties of the boundary that has the greatest influence on the temperature of the fluid as a whole: convection is more powerful than diffusion. In the case of differential heating from below, the rising warm branch causes most of the fluid to be warm relative to the boundary and therefore gravitationally stable except in a narrow zone at the extreme of heating where the intense rising motion must occur. In the case of differential heating from above, the sinking cold branch causes the bulk of the fluid to be cold and gravitationally stable except in a narrow region at the extreme of cooling where the intense sinking motion must occur. Theoretical models of axisymmetric, thermally driven (Hadley) circulations in the atmosphere (Charney, 1973) show the same effect: a narrow rising branch and a broad sinking branch. This effect is strengthened by cumulus convection (Charney, 1969, 1971b; Bates, 1970; Schneider, 1977). The narrow rising branch of the Hadley circulation directly controls the dynamic and thermodynamic properties of the tropics and subtropics and indirectly influences the higher-latitude circulations. Similarly, the small sinking branches of the ocean circulation determine the near-homogeneous deep-water properties as well as some of the intermediate-water properties. But there is a difference: we know how the heat released in the ascending branch of the Hadley circulation is disposed of; we do not know how the cold water in the abyssal circulation is heated. Whatever the process, the existence of a preponderant mass of near-homogeneous water at depth forces great static stability in the shallow upper regions of the ocean. It demands a thermocline.

Again we have an analogy between the upper circulation of the oceans and the upper circulation of Venus. Rivas (1973, 1975) has shown that the intense circulation of Venus is confined to a thin layer within and just below the region of intense heating and cooling by radiation. The more or less independent circulation produced by the separate heat sources of the atmospheric stratosphere is similarly analogous to the upper ocean circulation. And here there is an atmosphere-ocean analogy pertaining to our knowledge of transfer processes. While the mechanisms of heat transfer in the stratosphere are fairly well known, the mechanism of transfer of thermally inactive gases or suspended particles is not, for the latter involves a knowledge of particle trajectories, which are not easily determined. The large-scale eddies in the atmosphere, insofar as they are nondissipative, cannot transfer a conserved quantity across isentropic surfaces. Thermal dissipation is required for parcels to move from the low-entropy troposphere to the high-entropy stratosphere. Atmospheric chemists sometimes postulate an ad hoc turbulent diffusion to explain the necessary vertical transfers of such substances as the oxides of nitrogen and the chlorofluoromethanes into the ozone layers. But it is doubtful whether this type of diffusion is needed to account for the actual transfer, because there already exists the nonconservative mechanism of radiative heat transfer, and this, together with the large-scale eddying motion, can by itself account for the transfers (Andrews and McIntyre, 1978a; Matsuno and Nakamura, 1979). The analogous oceanic problem has already been mentioned: how is heat or salt transferred from the deep ocean layers into the upper wind-stirred layers? The cold deep water produced in the polar seas and the intermediate salty water produced in the Mediterranean Sea must eventually find their way by dissipative processes to the surface, where they can be heated or diluted. While a number of internal dissipation mechanisms have been proposed in a speculative way (double diffusion, low-Richardson-number instability zones, internal-wave breaking) one may invoke Occam's razor, as has been done so successfully to explain the Gulf Stream as an inertial rather than a frictional boundary layer (Charney, 1955b; Morgan, 1956) and postulate no internal dissipative mechanism at all. Then, outside of the convective zones, properties will be advected by the mean flow or by eddies along isentropic surfaces, and move from one such surface to another only at the boundaries of the ocean basins where we may assume turbulent dissipation does occur. It would be interesting to see how far one could go with boundary dissipation alone. Welander (1959) and Robinson and Welander (1963) have taken a first step by investigating the motions of a conservative system communicating only with an upper mixed layer.

18.3 The Transient Motions

Meteorologists, pressed with the necessity of forecasting the daily weather, have always been concerned with the transient motions of the atmosphere. Attempts to understand the processes leading to growth, equilibration, translation and decay of these "synoptic-scale" (order 1000 km) eddies[1] have produced theories of baroclinic instability (Charney, 1947; Eady, 1949), Rossby wave motion (Rossby et al., 1939; Haurwitz, 1940b) and Ekman pumping (Charney and Eliassen, 1949). It was first recognized by Jeffreys (1926) and demonstrated conclusively by Starr (1954, 1957) and Bjerknes (1955, 1957) that the dynamics of the mean circulation are strongly influenced by the transports of heat and zonal momentum by the eddies [for a review of these developments, see Lorenz (1967)]. This has prompted research on eddy dynamics and also on the parameterization of eddy fluxes (cf. Green, 1970; Rhines, 1977; Stone, 1978; Welander, 1973).

The study of eddy motions in the ocean is a new development. Although the existence of fluctuations in the Gulf Stream was reported by early observers such as Laval (1728) and Rennell (1832), actual *prediction* of ocean eddies has never been a particularly profitable exercise (perhaps the recent interest of yachtsmen in Gulf Stream rings may presage a change). The serious study of deep ocean fluctuations really began with M. Swallow (1961), Crease (1962), and J. Swallow (1971). At this point, oceanographers began to realize that the mid-ocean variable velocities were not, as might perhaps have been reasonably inferred from atmospheric experience, comparable to the mean flows but rather were an order of magnitude larger. This has spurred intensive experimental and theoretical investigations of the dynamics of the oceanic eddies and their roles in the general circulation (see chapter 11).

Comparisons between the oceanic and atmospheric eddies may be made in respect to their generation, propagation, interaction (both eddy–mean flow and eddy–eddy) and decay. In the sections below we shall describe some of the theoretical approaches to these problems.

In the atmosphere, energy conversion estimates (cf. Oort and Peixoto, 1974) clearly show a transformation of zonal-available potential energy into eddy-available potential energy, then into eddy kinetic energy and finally into heat by dissipation, with some transfer from eddy to zonal kinetic energy. The similarity of the growth phase of this cycle to that exhibited in the theory of small traveling perturbations of a baroclinically unstable (but barotropically stable) flow, leads naturally to the identification of the source of the waves as the baroclinic instability of the zonal flow. This idea has been supported further by the fact that the energy spectrum has peaks near zonal wavenumber six, which simple models predict to be the most rapidly growing wavenumber. However, attempts to apply these models directly to the atmosphere lead to problems: one expects that nonuniform mean flows, nonhomogeneous surface conditions, variable horizontal and vertical shears, etc., will alter the dynamics; and one may also wonder about the applicability of the small perturbation–normal mode approach.

The topographically and thermally forced standing eddies also draw upon zonal available potential energy (Holopainen, 1970). The processes by which they do this are not yet clearly understood; they may be related to the form-drag instability to be described in section 18.7.3. The standing eddies, like the transient eddies, also transport heat and momentum.

Overall energetic analyses have not been applied to oceanic data. However, budgets for basin-averaged kinetic and potential energy have been calculated for the "eddy-resolving general circulation models." These are reviewed by Harrison (1979b). In all but one of the 21 cases he considered, the eddy kinetic energy came from *both* mean kinetic and mean potential energy. Collectively, the eddies seem to be acting as dissipative mechanisms, but Harrison cautions that because the model statistics are inhomogeneous, the overall results may not be representative of the actual dynamics in any limited region. Indeed, the eddies may be acting as a negative viscosity in parts of the domain. Thus, Holland (1978) suggests that the eddies generated in the upper layer of his two-layer model drive the mean flows in the lower layer.

Oceanographers have examined local energy balances. The best known of these studies, by Webster (1961a), has often been interpreted as an indication that the Gulf Stream is accelerated by the eddies. However, Schmitz and Niiler (1969) have pointed out that the cross stream-averaged value of $\overline{u'v'}\bar{v}_x$ is not distinguishable from zero, so that Webster's results may be an indication merely of transfer of energy from the offshore to the onshore side of the jet and not a mean-flow generation. Even this result is not unambiguous, since the divergence term $(\overline{u'v'}\bar{v})_x$ is not small compared to the terms $-(\overline{u'v'})_x\bar{v}$ and $\overline{u'v'}\bar{v}_x$, representing, respectively, the eddy-mean flow and the mean flow-eddy conversions. The same problem occurs in attempts to compute regional energy budgets in numerical models (cf. Harrison and Robinson, 1978) unless considerable care is taken.

The conversion of mean-flow potential energy is sometimes inferred from the tilting of the phase lines of the temperature wave with height. This tilt is taken as evidence that the wave is growing by baroclinic instability of the mean flow. Here one must be careful: it is the lagging of the temperature wave behind the pressure wave and the consequent tilting of the phase line of the pressure trough toward the cold air that is important; the temperature phase line may tilt in any direction or not at all. Thus in the Eady model of baroclinic instability the temperature wave has the opposite slope from the pressure wave, whereas in the Charney model it has the opposite slope at low levels and the same slope at high levels (Charney, 1973, chapter IX). It is appropriate, then, to caution that the oceanically most readily available quantity, the phase change of the buoyancy (or entropy) with height, may not lead to a straightforward determination of the sign of the buoyancy flux.

This discussion should make it clear that very little is settled concerning the source of the eddies in the ocean and their effects on the mean flow. "Local" generation mechanisms, such as baroclinic or barotropic instability, flow over topography, and wind forcing are still being considered, and atmospheric analogues are much in mind. But the fact that the transient atmospheric perturbation velocities are comparable to those of the mean flow, whereas the particle speeds for mid-ocean mesoscale eddies are an order of magnitude greater than the mean speeds, suggests that energy may be generated only in limited regions (e.g., the western boundary currents) and propagate from there to other regions.

Oceanic eddies propagate in much the same manner as atmospheric eddies, although there are differences because of the upper-surface boundary conditions: atmospheric waves may propagate upward without reflection, whereas oceanic waves are reflected at the upper boundary. In the atmosphere, the potential vorticity gradients associated with the mean flow play an important role in determining the vertical structure and horizontal propagation for atmospheric waves, whereas this role is played primarily by the gradient of the earth's vorticity for mid-oceanic waves.

The interaction mechanisms may be classified as wave–mean flow interactions and wave–wave interactions. As mentioned above, the concrete evidence for significant wave–mean flow interaction is much greater for the atmosphere than for the ocean. There is not, of course, much oceanic data—Reynolds stresses have been calculated only along a few north–south sections (Schmitz, 1977). Moreover, these records are not very long and the spatial resolution is not sufficient to compute accurate gradients of the Reynolds stresses (given the great inhomogeneity).

The wave–wave interactions, however, seem similar in the two media. The crucial parameter is the Rossby wave steepness parameter $M = U/\beta L^2$, which distinguishes wavelike regimes ($M < 1$) from more turbulent ($M > 1$) regimes as illustrated in the experiments of Rhines (1975).[2] The oceans are similar to the atmosphere in that this parameter is of order unity for both, although it appears to vary considerably from one oceanic region to another.

The physical mechanisms for dissipation of atmospheric and oceanic eddies are thought to be similar with respect to bottom friction and transfer of energy to gravity wave motions or turbulence (though radiation is a further factor in damping atmosphere waves), but their relative importance may be quite different. The crucial differences for the large-scale circulations between the atmosphere and the ocean may not be in the details of the dissipation mechanisms but rather in their overall time scales. In the atmosphere, damping times are of the order of a few days, comparable to the eddy velocity advection time L/U, whereas the damping time in the ocean may be as long as several years (cf. Cheney and Richardson, 1976) while the advection time is of the order of a week.

18.4 The Geostrophic Formalism

18.4.1 The Development of the Geostrophic Formalism

The discovery that the atmospheric winds are approximately geostrophic is usually attributed to Buys Ballot (1857). Ferrel (1856) suggested that ocean currents might also have this property. But it took nearly a century before this knowledge was used dynamically. Because the geostrophic and hydrostatic equations express only a condition of balance, it is necessary to consider the slight imbalances produced by forcing, dissipation, and transience in order to predict the evolution and to understand the processes that maintain the balance. One of the first to exploit geostrophy was Bjerknes (1937) in a seminal work on the upper tropospheric long waves and their role in cyclogenesis. Basing his analysis on semiempirical considerations of the gradient wind and the variation with latitude of the Coriolis parameter, he gave the first explanation of the eastward propagation of the upper wave at a speed slower than the mean wind. It was this work that led Rossby et al. (1939) to their vorticity analysis of the upper wave as an independent entity in planar flow. Charney (1947) and Eady (1949) derived quasigeostrophic equations in their analyses of baroclinic instability for long atmospheric waves. General derivations of these equations for arbitrary motions were presented by Charney (1948), Eliassen (1949), Obukhov (1949), and Burger (1958). A particularly simple form which will be used in this review was given by Charney (1962) and Charney and Stern (1962).

In addition to these commonly used approximations, there have been a number of simplifications of the equations of motion which apply the concept of near-geostrophic balance in a less restrictive form. When flows become nongeostrophic in one horizontal dimension while remaining geostrophic in the other, as in frontogenesis, flow over two-dimensional mountain barriers, and in the western boundary currents of the oceans, a set of "semigeostrophic" equations derived from Eliassen's original formulation has often been found useful (Robinson and Niiler, 1967; Hoskins, 1975). Both the quasi- and semigeostrophic equations are special cases of the "balance equations" proposed by Bolin (1955), Charney (1955c, 1962), P. Thompson (1956), and Lorenz (1960). They may be derived from the consideration that in a large class of atmospheric flows the constraints of the earth's rotation and/or gravitational stability so inhibit vertical motion that the horizontal flow, even when it is not quasi-geostrophic, remains quasi-nondivergent. The equations derived by Eliassen (1952) for slow thermally and frictionally driven circulations in a circular vortex are a special case of the balance equations; they represent the laws of conservation of angular momentum and entropy and the requirement of equilibrium among the meridional components of the pressure, gravity, and centrifugal forces. For the equilibrium condition to be valid, the flow must be gravitationally and inertially stable. This implies that the potential vorticity must be positive in the northern hemisphere and negative in the southern hemisphere, and it may be shown that this condition on the potential vorticity is also required for the general asymmetric case.

One must also explain why external sources of energy excite quasi-geostrophic flows rather than gravity wave motions to begin with and why so little energy is transferred by nonlinear interactions into the gravity modes afterward. The tendency toward geostrophy is sometimes explained as an adjustment of an initially unbalanced flow by radiation of gravity waves in the manner discussed by Rossby (1938) (see also Blumen, 1968). However, since much of the forcing is applied slowly, rather than impulsively, the calculations of Veronis and Stommel (1956), who consider the nature of the exciting forces, are perhaps more relevant. They showed that the flows will be geostrophically balanced when the forcing period is very large compared to the inertial period. Thus we expect most of the energy will go into geostrophic motions.

The question of how much transfer occurs from geostrophic to nongeostrophic motions through nonlinear interactions remains a matter of concern. Errico's (1979) work suggests that equipartition of energy between gravity waves and geostrophic motions will occur in a conservative, rotating system in statistical equilibrium at sufficiently high energy. But in dissipative systems resembling the atmosphere and oceans, the energy will remain in the geostrophic modes because the gravity waves are dissipated on time scales that are small in comparison with those of their generation. This problem has elements in common with the so-called initialization problem in numerical weather prediction: to find initial values of a flow field that are at once compatible with the incomplete data and at the same time minimize the initial gravity-wave energy and its production rate (cf. Machenhauer, 1977; Daley, 1978).

The problems of transfer from geostrophic into gravity-wave energy are related to those of the production of hydrodynamic noise by a turbulent flow, first studied by Lighthill (1952). An excellent review is presented by Ffowcs Williams (1969). Here the problem is to calculate the generation of acoustic energy in a turbulent flow in which most of the energy resides in nondivergent motions. Since the turbulence is confined within a limited domain and radiates sound waves into the surrounding medium, there is no possibility of equipartition. An atmospheric or oceanographic analogy to the Lighthill problem would be the generation of internal gravity waves by turbulence in a planetary boundary layer (Townsend, 1965), except that here the generation takes place, not within the layer, but at its interface with the neighboring stable stratum.

Physicists, too, have struggled with problems in which many scales interact simultaneously (cf. Wilson, 1979), but it is not known whether their renormalization group methods can be usefully applied to atmospheric or oceanic problems.

18.4.2 Natural Oscillations of the Atmosphere and Oceans

The quasi-geostrophic equations have been derived for ranges of the various nondimensional parameters that are of interest in dealing with particular classes of atmospheric and oceanic motions. It is not to be expected that they will remain uniformly valid throughout the entire range of rotationally dominated flows, even when the primary balance is geostrophic. Burger (1958) was the first to point out explicitly that when the β-plane approximation $L/a \ll 1$, where L is the characteristic horizontal scale and a is the radius of the earth, is no longer valid, the dynamics of the motion change radically. On a planetary scale the motion becomes even more strongly geostrophic, but the vorticity balance changes. Sverdrup (1947) made implicit use of this dynamics in his treatment of the steady, wind-driven circulation of the oceans, and it has been used for the treatment of steady thermohaline circulations of the oceans by Robinson and Stommel (1959) and Welander (1959). [See the review articles by Veronis (1969, 1973b), and chapter 5.]

In this section we present a classification of natural oscillations in atmospheres and oceans in which rotation plays a dominant role, paying special attention to the domain of validity of the β-plane, quasi-geostrophic equations, the nature of the oscillations for which these equations are not valid, and the effects of nonlinearity, which, especially in the oceans, may give rise to solitary wave behavior.

Although the wave forcing is very different in the oceans and the atmosphere, there are many features of the responses that strongly resemble one another. This results because the response of a forced system depends strongly on the characteristics of the natural oscillations, and these have many similarities in the atmosphere and oceans. We shall discuss both the linear and nonlinear natural (unforced and nondissipative) oscillations of a simple model consisting of a single-layer, homogeneous, incompressible fluid with a free surface on a β-plane. This is a much oversimplified model, and we must regard the conclusions to be drawn merely as suggestions of the way in which the fully stratified, spherical system would behave. The most interesting implication—that the dynamics of scales intermediate between the Rossby radius and the radius of the earth may be dominated by solitary waves, in which nonlinear density advection balances linear dispersive effects—may not be very sensitive to the particular model chosen.

At the beginning of each subsection to follow we shall describe briefly the methods used and the results obtained in order to make it possible for the reader to omit the more detailed derivations. We begin the discussion of the normal modes of oscillation by stating the shallow-water equations in a reference frame moving with the wave. The Bernoulli and potential vorticity integrals then give two equations relating the wave streamfunction ϕ to the surface elevation η above the mean level H. These equations contain an unknown functional $\mathcal{B}$, the Bernoulli function, which we choose by requiring the equations to hold for vanishing ϕ and η.

We then obtain two coupled nonlinear partial differential equations defining an eigenvalue problem for the phase speed c. These equations have three nondimensional parameters: ϵ, a Rossby number measuring the ratio of the inertial to the Coriolis forces; $\hat{S}$, a static stability parameter measuring the ratio of the deformation scale L_R to the wave scale L; and $\hat{\beta}$, the fractional change in the Coriolis parameter over the wave scale. In the standard quasi-geostrophic range ($\epsilon \sim \hat{\beta} \ll 1$, $\hat{S} \sim 1$), when motions have small Rossby numbers, length scales comparable to the deformation radius, the Bernoulli equation to lowest order is simply a statement of geostrophic balance, and the potential vorticity equation becomes the linear quasi-geostrophic wave equation.

The single-layer equations will be written in dimensional form

$$\frac{Du}{Dt} - (f_0 + \beta y)v = -g\eta_x,$$

$$\frac{Dv}{Dt} + (f_0 + \beta y)u = -g\eta_y,$$

$$\frac{D\eta}{Dt} + (H + \eta)(u_x + v_y) = 0,$$

$$\frac{D}{Dt} = \frac{\partial}{\partial t} + u\frac{\partial}{\partial x} + v\frac{\partial}{\partial y},$$

(18.1)

where η is the displacement of the surface from mean sea level, H is the mean depth of the fluid, g is the gravitational acceleration (we shall use a reduced gravity value here), $f_0 = 2\Omega \sin \Theta$, $\beta = 2\Omega \cos \Theta/a$, $y = a\Delta\Theta$, where Θ is the central latitude and $\Delta\Theta$ is the angular distance from Θ. In nondimensional form these equations become

$$\hat{\beta}\hat{S}\frac{Du}{Dt} - (1 + \hat{\beta}y)v = -\eta_x,$$

$$\hat{\beta}\hat{S}\frac{Dv}{Dt} + (1 + \hat{\beta}y)u = -\eta_y,$$

$$\hat{\beta}\frac{D\eta}{Dt} + \left(1 + \frac{\epsilon}{\hat{S}}\eta\right)(u_x + v_y) = 0,$$

$$\frac{D}{Dt} = \frac{\partial}{\partial t} + \frac{\epsilon}{\hat{\beta}\hat{S}}\left(u\frac{\partial}{\partial x} + v\frac{\partial}{\partial y}\right),$$

(18.2)

where both x and y are scaled by L, η is scaled geostrophically by LUf_0/g, and the nondimensional parameters are $\hat{\beta} = \beta L/f_0 = L \cot \Theta/a \equiv L/L_\beta$, $\epsilon = U/f_0 L$, and $\hat{S} = gH/f_0^2L^2 \equiv L_R^2/L^2$. We have also introduced the definitions of two scales which turn out to be important in determining the boundaries between various types of behavior: the β-scale $L_\beta = f_0/\beta = a \tan \Theta$, at which variations in the vertical component of the earth's angular velocity are order of the angular velocity itself, and the Rossby radius of deformation $L_R = \sqrt{gH}/f_0$ (Rossby, 1938). We shall use 3500 km for L_β (corresponding to $\Theta \sim 30°$) and 50 km (oceanic) or 1000 km (atmospheric) for L_R. We have also made the choice of the long-wave period for the time scale so that $T = L/\beta L_R^2$.

The quasi-geostrophic potential vorticity equation may be derived by expanding (18.2) in powers of ϵ for $\hat{\beta} \sim \epsilon \ll 1$ and $\hat{S} \sim 1$, giving

$$\left[\frac{\partial}{\partial t} + \frac{\epsilon}{\hat{\beta}\hat{S}}\left(\eta_x\frac{\partial}{\partial y} - \eta_y\frac{\partial}{\partial x}\right)\right]\left(\nabla^2\eta - \frac{1}{\hat{S}}\eta + \hat{\beta}y/\epsilon\right)$$

$$= 0. \quad (18.3)$$

The failure of this equation at small space or time scales and near the equator is well known. Meteorologists since Burger (1958) have also recognized that some larger than synoptic-scale motions also do not evolve according to this equation. Rather, the appropriate equations are derived by assuming $\hat{S}$ and ϵ to be small (because L is very large) and $\hat{\beta}$ to be of order 1. The resulting velocities remain geostrophic:

$$u = -\frac{1}{1 + \hat{\beta}y}\eta_y, \quad v = \frac{1}{1 + \hat{\beta}y}\eta_x, \quad (18.4)$$

and the height field evolves according to

$$\frac{\partial}{\partial t}\eta - \frac{1}{(1 + \hat{\beta}y)^2}\left(1 + \frac{\epsilon}{\hat{S}}\eta\right)\eta_x = 0. \quad (18.5)$$

Equations (18.3) and (18.5) have very different properties: the quasi-geostrophic equation has uniformly propagating linear-wave solutions which are essentially dispersive even at large amplitudes (cf. McWilliams and Flierl, 1979), while the Burger equation does not have uniformly propagating linear-wave solutions and initial disturbances steepen because of nonlinearity. We shall demonstrate that there is an intermediate band of length scales in which nonlinearity and dispersion can balance to give cnoidal or solitary waves. In the ocean, as we shall see, the change from quasi-geostrophic to intermediate dynamics to Burger dynamics occurs at a relatively small scale because the deformation radius is so small compared to the radius of the earth.

We may elucidate these differences by considering the shallow-water equations under the assumption that the motions are translating steadily at speed c:

$$(\mathbf{v} - c\hat{\mathbf{x}}) \cdot \nabla v + \hat{\mathbf{z}}(f_0 + \beta y) \times \mathbf{v} = -g \nabla \eta,$$
$$(\mathbf{v} - c\hat{\mathbf{x}}) \cdot \nabla \eta + (H + \eta) \nabla \cdot \mathbf{v} = 0, \quad (18.6)$$

where $\hat{\mathbf{x}}$ and $\hat{\mathbf{z}}$ are unit vectors in the positive x and z directions. We may define a transport streamfunction in the coordinate system moving with the wave

$$(\mathbf{v} - c\hat{\mathbf{x}})(H + \eta) = \hat{\mathbf{z}} \times \nabla \psi$$

and write the Bernoulli and potential vorticity integrals of motion:

$$\frac{1}{2}|\nabla(\phi + cHy)|^2 + g(H + \eta)^3 + c(H + \eta)^2 \left(f_0 y + \frac{\beta y}{2}\right)^2$$
$$= (H + \eta)^2 \mathcal{B}(\phi + cHy), \quad (18.7)$$

$$\nabla \cdot \frac{\nabla(\phi + cHy)}{H + \eta} + (f_0 + \beta y) = (H + \eta)\mathcal{B}'(\phi + cHy),$$

where we have isolated the wave part of the streamfunction $\phi = \psi - cHy$. We require that (18.7) hold as $\phi, \eta \to 0$; this determines the Bernoulli functional

$$\mathcal{B}(Z) = \frac{c^2}{2} + gH + \frac{f_0}{H} Z + \frac{\beta}{2cH^2} Z^2. \quad (18.8)$$

The choice of a single-valued, well-behaved Bernoulli functional implies that only motions which reduce smoothly to linear waves will be considered; thus the solutions of Stern (1975b) or Flierl, Larichev, McWilliams, and Reznik, (1980) which involve closed streamlines and a multiple-valued $\mathcal{B}$ will not be examined here.

In nondimensional form, equations (18.7) become

$$\frac{\epsilon}{2}|\nabla \phi|^2 + \hat{\beta}\hat{S}c\phi_y + \eta\left(1 + \frac{\epsilon}{\hat{S}}\eta\right)^2$$
$$= \hat{\beta}^2\hat{S}c^2\eta\left(1 + \frac{\epsilon}{2\hat{S}}\eta\right) + \phi(1 + \hat{\beta}y)\left(1 + \frac{\epsilon}{\hat{S}}\eta\right)^2$$
$$+ \frac{\epsilon}{\hat{S}}\frac{\phi^2}{2c}\left(1 + \frac{\epsilon}{\hat{S}}\eta\right)^2, \quad (18.9)$$

$$\hat{S}\nabla^2\phi\left(1 + \frac{\epsilon}{\hat{S}}\eta\right) - \epsilon\nabla\phi\cdot\nabla\eta - \hat{\beta}\hat{S}c\eta_y$$
$$= \eta(1 + \hat{\beta}y) \times \left(1 + \frac{\epsilon}{\hat{S}}\eta\right)^2 + \frac{\phi}{c}\left(1 + \frac{\epsilon}{\hat{S}}\eta\right)^3. \quad (18.10)$$

We show in figure 18.1 the dependence upon L and U of our three basic parameters $\hat{\beta} = L/L_\beta = \beta L/f_0 = L \cot\Theta/a$ (the ratio of the wave scale to the radius of the earth scale), $\epsilon = U/f_0 L$ (the Rossby number), and $\hat{S} = L_R^2/L^2$ (the inverse of the rotational Froude number). Note immediately the differences in scale separation for oceanic versus atmospheric conditions. In the atmosphere L_β is very close to L_R, so that there is only a short range between the usual baroclinic Rossby wave scales ($\hat{S} \sim 1$, $\hat{\beta} \ll 1$) and the Burger range ($\hat{S} \ll 1$, $\hat{\beta} \sim 1$); in the ocean there is a large scale gap. Thus one might expect the different dynamics to be seen more clearly in the ocean.

Linear Waves ($\epsilon = 0$) The first step toward understanding the various types of large-scale free motion is to consider the linearized solutions. When the Rossby number is very small, the two equations can be combined into a single streamfunction equation which governs both gravity and Rossby waves. The Rossby wave-phase speed increases as the length scales of the wave increase, leveling off for $L > L_R$. For still larger scales, however, the speed again increases as the wave amplitude begins to be more pronounced equatorially. We demonstrate that the natural dividing scale here is what we call the "intermediate" scale $L_I \equiv (L_\beta L_R^2)^{1/3}$, where $\hat{S} = \hat{\beta}$ (see figure 18.1). This is the scale at which

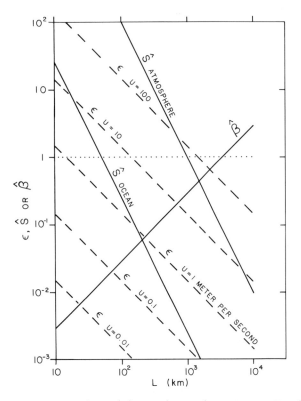

Figure 18.1 Values of the Rossby number ϵ, inverse Froude number $\hat{S}$, and beta parameter $\hat{\beta}$ as functions of the length scale L and velocity scale U for oceanic and atmospheric values of the deformation radius.

the relative vorticity changes become as small as the variations in vortex stretching due to the β term. Alternatively, one could say that the rule, "f equals a constant except when differentiated," breaks down near the intermediate scale. The phase speed continues to increase and, for large enough north–south scales, the wave domain crosses the equator. Then the wave becomes equatorially trapped and the phase speed again becomes independent of L.

For the parameters we have chosen—$L_\beta = 3500$ km, $L_R = 50$ km (ocean), 1000 km (atmosphere)—the intermediate scale $L_I = 210$ km (ocean), 1500 km (atmosphere) is not very large. It represents the upper bound to the scales for which the standard quasigeostrophic equations are valid. It may again be seen that there is a significantly greater separation among the various scales in the ocean compared to the atmosphere. This suggests that the ocean mesoscale motions may be a cleaner example of quasi-geostrophic flow than the synoptic-scale motions of the atmosphere; the approximations used for the latter are less exact.

For linear motions, the Bernoulli equation (18.9) defines η in terms of ϕ; η may then be eliminated from the potential vorticity equation (18.10) to yield a single equation for the streamfunction

$$\hat{S}\nabla^2\phi - \frac{1}{c}\phi - (1+\hat{\beta}y)^2\phi = \hat{\beta}^2\hat{S}^2c^2\phi_{xx}. \qquad (18.11)$$

Since the coefficients do not involve x, we may set $\phi = e^{ix}G(y;L)$. The resulting equation together with boundary conditions presents an eigenvalue problem for $c(L)$ and the wave structure $G(y;L)$.

It has three eigenvalues, corresponding to two gravity-wave modes and one Rossby-wave mode. We can identify the gravity modes with the retention of the right-hand term in (18.11). For mid-latitude modes this term is significant only when $\hat{\beta}^2\hat{S}^2c^2 \sim \hat{S}$ or c^2 (dimensional) $\sim gH$; it is small for the Rossby mode solutions. For equatorially trapped Rossby modes, the y scale contracts so that $\hat{S}\phi_{yy}$ dominates both $\hat{S}\phi_{xx}$ and $\hat{\beta}^2\hat{S}^2c^2\phi_{xx}$. Eliminating the right-hand side corresponds to retaining only the underlined terms in (18.9)–(18.10). The filtered linear equation becomes

$$\hat{S}\nabla^2\phi - \frac{1}{c}\phi = (1+\hat{\beta}y)^2\phi, \qquad (18.12)$$

which has been discussed extensively by Lindzen (1967) and others. Here we comment on the various types of solution primarily as a guide to our later discussion of the effects of nonlinearity.

Figure 18.2 shows the nondimensional phase speed as a function of L for atmospheric or oceanic parameters under the simplifying boundary conditions $\phi = 0$ at $y = \pm\pi/2$, which make the x and y scales of the

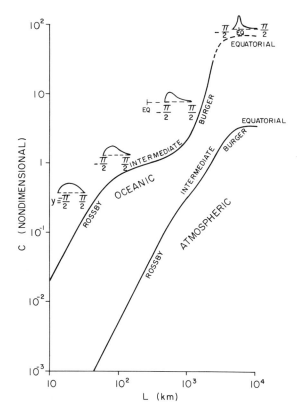

Figure 18.2 Phase speed nondimensionalized by βL_R^2 as a function of the x scale L (wavelength/2π) in a channel of width πL. Also shown are typical shapes of the y structure function $G(y;L)$ for the various classes of motion.

domain similar. We can identify four different types of behavior.

Midlatitude Rossby waves ($\hat{\beta} \ll 1$, $\hat{S} \sim 1$): For these motions, first described by Rossby et al. (1939), the streamfunction satisfies

$$\hat{S}\nabla^2\phi - \frac{1}{c}\phi = \phi, \qquad (18.13)$$

which has solutions in the box

$$\phi = e^{ix}\cos y$$

with

$$c = -1/(1 + 2\hat{S}), \qquad (18.14)$$

or, more generally, for waves oriented in any direction, we have

$$\phi = e^{i\mathbf{k}\cdot\mathbf{x}}$$

with

$$c = -1/(1 + \hat{S}\,\mathbf{k}\cdot\mathbf{k}) \qquad (18.15)$$

(see the discussion in chapter 10).

Intermediate scale waves ($\hat{\beta} \sim \hat{S} \ll 1$): The Rossby-wave dispersion relation (18.14) remains valid for $\hat{\beta} \ll \hat{S} \ll 1$ and becomes

$$c = -1 + 2\hat{S},$$

so that for a sufficiently small $\hat{S}$ the waves are nondispersive c (dimensional) $= -\beta L_R^2$. However, when L increases to the point where $\hat{\beta} \sim \hat{S} \ll 1$, the small correction in the formula above becomes invalid. This occurs when the $\phi\hat{\beta}y$ term becomes comparable to the $\hat{S}\nabla^2\phi$ term, that is, when $L \simeq (L_\beta L_R^2)^{1/3}$, which is 210 km for the oceans or 1500 km for the atmosphere. We denote this scale as the "intermediate scale" L_I. The wave structure is determined by expanding (18.12) in $\hat{S}$ (or $\hat{\beta}$). Setting $\phi = \phi^{(0)} + \hat{S}\phi^{(1)} + \cdots$ and $c = -1 + \hat{S}c^{(1)} + \cdots$, we obtain

$$(\nabla^2 + c^{(1)} - 2\hat{\beta}y/\hat{S})\phi^{(0)} = 0. \tag{18.16}$$

When $L \gtrsim L_I$ the y dependence of f can no longer be neglected, the y scale becomes order of the intermediate scale, and the solutions begin to be concentrated toward the equator (see figure 18.2). As L continues to increase, $\hat{S}$ decreases but $\hat{\beta}$ increases and the phase speed is no longer insensitive to L but begins to increase; c behaves like $-1 + O(\hat{\beta})$ rather than $-1 + O(\hat{S})$. The phase speed becomes less and less sensitive to the x wavenumber, so that the waves may still be considered approximately nondispersive. We have required that $\hat{\beta}$ and $\hat{S}$ be small, but figure 18.1 shows that these quantities are small only for a rather narrow range of L even in the oceanic case, and figure 18.2 shows that c varies perceptibly with L everywhere. For the atmospheric parameters a totally nondispersive regime ($\hat{\beta} \ll \hat{S} \ll 1$) does not exist at all.

Burger motions ($\hat{\beta} \sim 1$, $\hat{S} \ll 1$): When L increases to the point where $\hat{\beta} \sim 1$, the motions become strongly concentrated near the equator. The y scale contracts (relative to L) so that the lowest order balance includes all the terms in (18.12) and the y wave domain crosses the equator. The phase speed rapidly increases from that of the midlatitude Rossby waves to that of the equatorial waves.

Now we can see why the Burger equation (18.5), which assumes equal x and y scales, has no linear free-wave solutions: free waves with a very large x scale do not have the same y scale. Instead the unforced motions acquire a meridional scale between L_I and the (somewhat larger) equatorial scale. Forced motions, of course, may have comparable x and y scales and may therefore have evolution equations in which the terms of (18.5) contribute along with the forcing terms.

Equatorial waves: Here we can drop the 1 in the $1 + \hat{\beta}y$ term of equation (18.12) to change to the equatorial β-plane (the f_0 factors will all cancel out upon dimensionalization). The solutions are well known (cf. Lindzen, 1967) and again become nondispersive for small $\hat{S}$. Rescaling the equation for small $\hat{S}$ shows that the y wave domain is confined to a region around the equator of meridional extent $\hat{\beta}^{-1/2}\hat{S}^{1/4}$, which corresponds to the dimensional scale $L_e = (gH/\beta^2)^{1/4} = (L_\beta L_R)^{1/2}$, the well-known equatorial deformation scale. For our assumed parameters, this scale is 420 km for the oceans and 1900 km for the atmosphere; however, this estimate is not very accurate since the equivalent depth for baroclinic motions varies considerably. Moore and Philander (1977) give 325 km as an estimate of this scale for the first baroclinic mode. The phase speeds are order $\hat{\beta}^{-1}\hat{S}^{-1/2} = L_\beta/L_R$, corresponding to a dimensional speed $\beta L_e^2 = \sqrt{gH}$. (The other solutions have $c \sim \pm\hat{\beta}^{-1/2}\hat{S}^{-3/4}$.) A further discussion appears in chapter 6.

Nonlinear Waves ($\epsilon > 0$) When the motion becomes of sufficiently large amplitude, the propagation characteristics of a single wave change. We shall investigate the size of the Rossby number necessary for this to occur. This size may be quite different from the Rossby number required for significant nonlinear interactions in a full spectrum of waves. However, the nonlinear behavior of a single wave can be of interest when it allows the possibility for solitary waves. On the scale of the mid-latitude Rossby wave, this does not appear to occur and the nonlinearity gives only a correction to the phase speed and shape; the lowest-order balance remains strongly dispersive. However, as the scale becomes equal to or greater than the intermediate scale, the phase speed becomes less dependent on the x wavenumber. When the Rossby number becomes of the order L_R^4/L^4, the nonlinear advection term becomes comparable to the east–west dispersion term and the solutions propagate as solitary waves. The structure of these isolated high-pressure disturbances is found to be the same as that of the sech$^2 x$ solution to the Korteweg-deVries equations. The implication of this section, then, is that the dynamics of motions of the intermediate or large scales may be quite different from that of the ordinary Rossby wave.

Let us now consider the conditions under which the nonlinear terms can alter the propagation characteristics of the free waves in our model. This can occur whenever one of the ϵ terms is comparable to one of the linear terms that have been retained in the Bernoulli or potential vorticity equations (18.9)–(18.10) (the underlined terms). This happens when $\epsilon \sim 1$, $\epsilon/\hat{\beta} \sim 1$, $\epsilon/\hat{S} \sim 1$, $\epsilon/\hat{\beta}\hat{S} \sim 1$, or $\epsilon/\hat{S}^2 \sim 1$. The velocities required for each of these conditions are shown in figure 18.3, which emphasizes again the relative complexity of the atmosphere: for 40-ms^{-1} winds at 1000-km scales, all of the nonlinear terms enter simultaneously. For the ocean, only strong meandering motions could cause each of ϵ, $\epsilon/\hat{S}$ and $\epsilon/\hat{S}^2$ to be of order unity, and in these circumstances $\hat{\beta}$ remains quite small. We shall not attempt to deal with these more complicated motions, but instead shall examine the nonlinear ef-

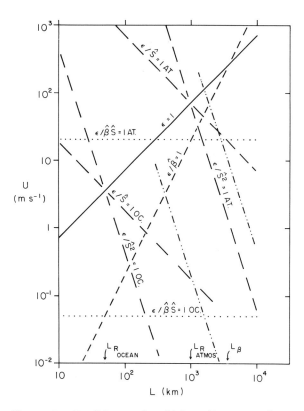

Figure 18.3 Conditions under which nonlinear terms become important. Labeled curves show relationship between U and L such that a particular parameter ratio becomes equal to one. This corresponds to one of the nonlinear terms in (18.9)–(18.10) becoming equal in magnitude to one of the underlined linear terms.

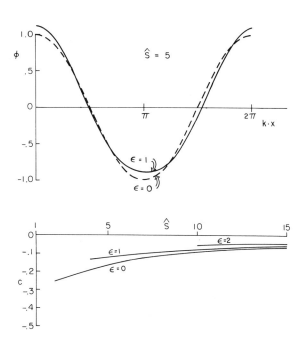

Figure 18.4 Effects of nonlinearity on a short Rossby wave. The upper figure shows the changes in the shape of the wave. The lower figure shows the changes in the dispersion relation.

fects on each of the waves that has been considered above.

Midlatitude Rossby waves: The first nonlinear condition that occurs when $\hat{S} \gtrsim 1$ is $\epsilon = \hat{\beta}$. However, since $\hat{\beta}$ remains small and does not enter the governing equation (18.13), we expect that this will not significantly alter the behavior of a single steadily propagating sinusoidal wave. When ϵ or $\epsilon/\hat{S}$ becomes order 1, nonlinearity begins to affect the structure significantly. For example, consider the parameter range $\epsilon \sim 1$, $\hat{\beta} \ll \hat{S}^{-1} \ll 1$. To lowest order in an expansion of both ϕ and c in $\hat{S}^{-1}$ (c being of order $\hat{S}^{-1}$), the potential vorticity equation gives

$$\nabla^2 \phi^{(0)} = \frac{1}{c^{(0)}\hat{S}} \phi^{(0)}.$$

At first order we find the corrections to the phase speed and shape of the wave. The result is

$$c \simeq \hat{S}^{-1}\left[-\frac{1}{\mathbf{k}\cdot\mathbf{k}} + \hat{S}^{-1}\left(\epsilon^2 + \frac{1}{(\mathbf{k}\cdot\mathbf{k})^2}\right) + \cdots\right], \quad (18.17)$$

as sketched in figure 18.4. The order ϵ nonlinear terms cause a sharpening of the streamfunction crests and a decrease in the propagation rate.

Intermediate scale waves: When $\hat{S} \ll 1$, nonlinear terms first enter when $\epsilon \sim \hat{\beta}\hat{S}$ or $\epsilon \sim \hat{S}^2$ (see figure 18.3). We can find the forms of the solutions by letting $\epsilon = E\hat{S}^2$ and $\hat{\beta} = B\hat{S}$ and expanding for small $\hat{S}$ assuming E, B to be of order unity or less. We get

$$c = -1 + \hat{S}c^{(1)},$$
$$\nabla^2 \phi^{(0)} + c^{(1)}\phi^{(0)} + \tfrac{3}{2}E[\phi^{(0)}]^2 - 2By\phi^{(0)} = 0 \quad (18.18)$$

for the equations governing the shape and the speed of the wave.

The simple limit here is $B = \hat{\beta}/\hat{S} \ll 1$, corresponding to the range $L_R \ll L \ll L_I$, and E of order unity, corresponding to particle speeds given by the $\epsilon/\hat{S}^2 = 1$ lines in figure 18.3. The wave equation

$$\nabla^2 \phi^{(0)} + c^{(1)}\phi^{(0)} + \tfrac{3}{2}E[\phi^{(0)}]^2 = 0$$

has both one- and two-dimensional solutions on the plane. These include the cnoidal and solitary wave solutions to the Korteweg-deVries equation (Whitham, 1974) for uniformly propagating waves:

$$\phi^{(0)} = cn^2\left(\frac{K(m)}{\pi}\frac{\mathbf{k}\cdot\mathbf{x}}{|\mathbf{k}|}; m\right) - \frac{\sqrt{1-m+m^2}-1+2m}{3m},$$
$$c = -1 + \hat{S}4K^2(m)\sqrt{1-m+m^2}/\pi^2, \quad (18.19a)$$
$$\epsilon\hat{S}^{-2} = 4mK^2(m)/\pi^2,$$

and

$$\phi^{(0)} = \text{sech}^2 \mathbf{k} \cdot \mathbf{x},$$
$$c = -1 - 4\hat{S}\mathbf{k} \cdot \mathbf{k}, \qquad (18.19b)$$
$$\epsilon \hat{S}^{-2} = 4\mathbf{k} \cdot \mathbf{k}.$$

Plots of the shapes of the cnoidal and solitary waves and the dispersion relations are shown in figures 18.5A and 18.5B. The cnoidal waves show a phase speed decreasing with amplitude (as in the example above) while the solitary wave speed increases as the wave gets stronger.[3]

A second type of solution (cf. Flierl, 1979b) is a radially symmetric solitary wave

$$\phi^{(0)} = G(k\sqrt{\mathbf{x} \cdot \mathbf{x}}),$$
$$c = -1 - \hat{S}k^2, \qquad (18.19c)$$
$$\epsilon \hat{S}^{-2} = 1.59 k^2,$$

whose shape and dispersion relations are shown in figure 18.6.

It may be seen from equation (18.15) that the dynamics of large-scale motions for which $\epsilon \sim \hat{S}^2$ and $\hat{\beta} \ll \epsilon/\hat{S}$ are distinctly different from those of the quasigeostrophic eddies. We might expect, if the motions are governed by the Korteweg–deVries equation as suggested by (18.18), that solitons will be formed and dominate the subsequent evolution of the field. In the atmosphere, solitary-wave behavior would be difficult to find because of the rapid frictional decay time, the east–west periodicity for scales not so much larger than those under consideration, and the rather limited parameter range for the Korteweg–deVries regime. In the ocean, the situation is quite different; the parameter range for solitary-wave behavior is more distinct, the waves are of small scale compared to the size of the basin, and the decay rates are slow so that there is sufficient space and time for the necessary balance between nonlinearity and dispersion to develop.

For scales larger than the intermediate scale, B becomes large in (18.18). If y is rescaled by $B^{-1/3}$ (dimensionally by L_1), this equation can be solved by expansion in powers of $B^{-2/3}$. To lowest order, one obtains a linear equation for the y structure; to next order, the x dispersion and nonlinear steepening (if E is order unity) are included and the x structure is then given by an equation of the Korteweg–deVries type.

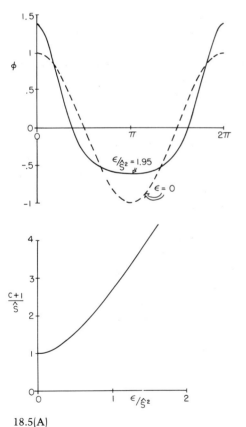

18.5(A)

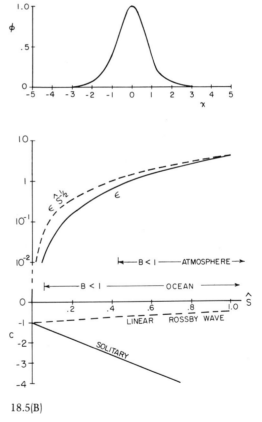

18.5(B)

Figure 18.5 Effects of nonlinearity on long waves. (A) Cnoidal waves: the upper figure shows the change in shape occurring when the nonlinearity is increased while the lower figure shows the changes in the dispersion relation. (B) Solitary waves: the upper figure shows the shape of the wave while the lower figure shows the relationship between the length and amplitude ($\hat{S}$ and ϵ) and also the propagation speed. For a fixed deformation radius, $\epsilon \hat{S}^{-1/2}$ is directly proportional to the velocity scale U. The relationships are only valid for $B < 1$.

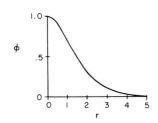

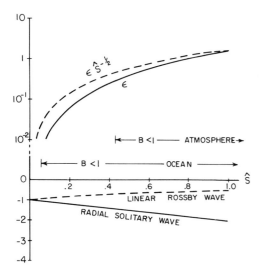

Figure 18.6 Radially symmetric solitary solutions. The upper figure shows the dependence of the pressure upon radius. The lower figure gives the relationships between amplitude, size, and propagation speed.

Burger range: Here also one can show that there are motions whose y structure is determined by a linear equation and whose x structure is determined by a nonlinear equation of the Korteweg–deVries type. We still require $\epsilon \sim \hat{S}^2$. Clarke (1971) has discussed this type of solution (and also those described above for large B) in more detail.

Equatorial motions: Boyd (1977) has shown that the long waves in this case also satisfy an equation of the Korteweg–deVries type. If we rescale the equatorial versions of (18.9) and (18.10), letting $y = \hat{\beta}^{-1/2}\hat{S}^{1/4}Y$ (so that Y has the scale L_e), $c = \hat{\beta}^{-1}\hat{S}^{-1/2}C$, and $\eta = \hat{\beta}^{1/2}\hat{S}^{1/4}N$, we can show that there are only two parameters (in the absence of north–south boundaries) of interest: $\delta = \hat{\beta}^{-1/2}\hat{S}^{1/4} = L_e/L$ and $\hat{\epsilon} = \epsilon\,\hat{\beta}^{1/2}\hat{S}^{-3/4} = UL/\beta L_e^3$. The cnoidal or solitary wave (in x) solutions are obtained when $\hat{\epsilon} \sim \delta^2 \ll 1$. This gives an equatorial velocity scale $U = \beta L_e^5/L^3$, as shown in figure 18.3.

In summary, then, we have seen three different types of natural large-scale, long-period motions in the atmosphere and ocean. For scales on the order of the deformation radius or less ($L \lesssim 50$ for the oceans and $\lesssim 1000$ km for the atmosphere), dispersive Rossby waves dominate with nonlinear effects entering only for large Rossby number ϵ. Intermediate scales ($\hat{S} \ll 1$, $\epsilon\hat{S}^{-2} \sim 1$, $\hat{\beta} \ll \epsilon\hat{S}^{-1}$ implying $50 \ll L \ll 210$ km for the oceans and $1000 \ll L \ll 1500$ km for the atmosphere) have solitary or cnoidal wave structures as well as circular solitary highs. As the scales become larger, weak solitary or cnoidal wave structures may persist with normal-mode y shapes concentrated near the equatorward side of the domain. Stronger motions will not remain permanent but will steepen in amplitude, as do the solutions of Burger's equation (18.5). When the wave domain comes to include the equator, nonlinear equatorial wave motions satisfying a Korteweg–deVries type of equation can exist.

Korteweg–deVries Dynamics Finally we shall demonstrate that Korteweg–deVries dynamics does seem to be appropriate for general motions (not necessarily uniformly propagating waves) on the intermediate scale ($\hat{\beta} \gtrsim \hat{S}$, $\epsilon \sim \hat{S}^2$, and $\hat{S} \ll 1$). The previous derivations have shown only that the permanent form is governed by an equation that may be derived from the Korteweg–deVries equation, but it is still necessary to show that the time-dependent evolution equation is also of this type. We return to our governing equations (18.2) and set $\epsilon = E\hat{S}^2$ and $\hat{\beta} = B\hat{S}$, where B and E are assumed to be of order unity. This corresponds to $L \sim L_I$ and $U \sim f_0 L_R^2/L_\beta$ (210 km, 5 cm s^{-1} for the ocean; 1500 km, 20 m s^{-1} for the atmosphere). We note that there will be two time scales in the evolution: a fast time t corresponding to the nondispersive propagation and a slow time $T = \hat{S}t$ during which features evolve.

The lowest two orders of the expansion in $\hat{S}$ show that the flow is geostrophic and that the advection of planetary vorticity is balanced by vortex stretching, leading to the usual nondispersive propagation of very long Rossby waves. At the next order slow changes in surface height force a divergence which creates relative vorticity. The vorticity balance also is influenced by north–south variations in vortex stretching due to variations of f, while the nonlinear terms enter in the mass balance. The resulting equation is a mix between the Korteweg–deVries equation and the Rossby-wave equation. However, when L is large compared to the intermediate scale, the more detailed expansion to follow shows that the x structure indeed evolves according to a Korteweg–deVries equation.

At lowest order the flows are geostrophic

$$u^{(0)} = -\eta_y^{(0)},$$

$$v^{(0)} = \eta_x^{(0)},$$

$$u_x^{(0)} + v_y^{(0)} = 0.$$

The first-order equations,

$$u^{(1)} + Byu^{(0)} = -\eta_y^{(1)},$$

$$v^{(1)} + Byv^{(0)} = \eta_x^{(1)},$$

$$B\eta_t^{(0)} + E\mathbf{v}^{(0)}\cdot\boldsymbol{\nabla}\eta^{(0)} + E\eta^{(0)}\boldsymbol{\nabla}\cdot\mathbf{v}^{(0)} + \boldsymbol{\nabla}\cdot\mathbf{v}^{(1)} = 0,$$

lead to Sverdrup (1947) or Burger (1958) type of balance between advection of planetary vorticity and vortex stretching,

$$u_x^{(1)} + v_y^{(1)} = -B\eta_x^{(0)},$$

and to the nondispersive wave equation

$$B\eta_t^{(0)} - B\eta_x^{(0)} = 0,$$

which implies

$$\frac{\partial}{\partial t} = \frac{\partial}{\partial x} \quad \text{or} \quad \eta = \eta(x + t, y, T).$$

At second order we obtain the vorticity equation

$$B(v_x^{(0)} - u_y^{(0)}) + E\mathbf{v}^{(0)} \cdot \nabla(v_x^{(0)} - u_y^{(0)})$$
$$+ Bv^{(1)} + \nabla \cdot \mathbf{v}^{(2)} - By\nabla \cdot \mathbf{v}^{(1)} = 0$$

and the mass-conservation equation

$$B\eta_T^{(0)} + B\eta_t^{(1)} + E\mathbf{v}^{(1)} \cdot \nabla \eta^{(0)} + E\mathbf{v}^{(0)} \cdot \nabla \eta^{(1)}$$
$$+ \nabla \cdot \mathbf{v}^{(2)} + E\eta^{(1)}\nabla \cdot \mathbf{v}^{(0)} + E\eta^{(0)}\nabla \cdot \mathbf{v}^{(1)} = 0,$$

which jointly lead to the evolution equation [after using $\partial/\partial t = \partial/\partial x$ for the fast time, and dropping the superscript (0)]

$$B\eta_T = EB\eta\eta_x + B(\nabla^2\eta)_x$$
$$- 2B^2 y\eta_x + EJ(\eta, \nabla^2\eta) \quad (18.20)$$

[where $J(A,B)$ is the Jacobian operator] or

$$B\eta_T = EJ(\eta - \frac{B}{E}y, \nabla^2\eta + \frac{3}{2}E\eta^2 - 2By\eta).$$

One can readily show that the requirement of steady propagation leads to (18.18). Furthermore, when L is large compared to the intermediate scale L_I but E remains order one, the x structure of the solutions do satisfy a Korteweg–deVries equation. In this case B is large and E is order 1. Because the y scale becomes limited to L_I, the x dependence and the nonlinearity do not enter in the primary balance, which serves to determine the y structure and a correction to the phase speed. At the next order, the nonlinearity (from both quadratic and Jacobian terms) enters along with the third x derivative and the slow-time derivative terms to give a Korteweg–deVries equation:

$$\eta = F(x - ct, T)Ai(\mathcal{Y}),$$

$$c = -1 - \pi B - (2B)^{2/3}\mathcal{Y}_0,$$

$$\mathcal{Y} = \mathcal{Y}_0 + (2B)^{1/3}\left(y + \frac{\pi}{2}\right), \quad (18.21)$$

$$\mathcal{Y}_0 = -2.3381 \quad \text{(zero of Airy function)},$$

$$F_T = F_{xxx} + \frac{3}{2}E\left(\int_{\mathcal{Y}_0}^{\infty} Ai^3 \Big/ \int_{\mathcal{Y}_0}^{\infty} Ai^2\right)\frac{\partial}{\partial x}F^2.$$

This section has demonstrated that some caution must be exercised in applying the quasi-geostrophic equations (which will be discussed throughout the rest of the paper) to large-scale motions since they are valid for the oceans only for scales up to the order of 200 km. The derivations suggest that the role of nonlinearity may be very different for the intermediate and large-scale motions—leading to coherent and phase-locked structures rather than to turbulence. Clearly these inferences must be backed up by more thorough investigations which are beyond the scope of this article.

18.4.3 The Quasi-Geostrophic Equations

Because of the difficulties inherent in attacking the full equations of motion either analytically or numerically, various approximative equations have been developed. For the study of the large-scale motions, the relevant "filtering approximations" eliminate the acoustic and inertiogravity motions.[4] We have mentioned the quasi-geostrophic, semigeostrophic and balance equations and have touched on their limitations. In this section we shall discuss briefly the derivation of the quasi-geostrophic equations for a stratified fluid under oceanic conditions; details can be found in the appendix. These equations are, of course, familiar, but, since we shall use them in the rest of this chapter, we must establish our notation. We wish also to remark on differences between the standard derivation for the atmosphere (cf. Charney, 1973) and that for oceanic conditions. Finally, we include the β-effect by explicitly taking into account the two-scale nature of the problem: the planetary scale, that is, the earth's radius, and the scale of the fluid motions themselves.

For inviscid, adiabatic flow, the equations of motion and continuity expressed in modified spherical coordinates are

$$\frac{Du}{Dt} + \frac{uw}{a+z} - \frac{uv\tan\Theta}{a+z} - 2\Omega v\sin\Theta + 2\Omega w\cos\Theta$$
$$= -\frac{\alpha}{(a+z)\cos\Theta}\frac{\partial p}{\partial\Phi},$$

$$\frac{Dv}{Dt} + \frac{wv}{a+z} + \frac{u^2\tan\Theta}{a+z} + 2\Omega u\sin\Theta = -\frac{\alpha}{a+z}\frac{\partial p}{\partial\Theta},$$

$$\frac{Dw}{Dt} - \frac{u^2+v^2}{a+z} + 2\Omega u\cos\Theta = -\alpha\frac{\partial p}{\partial z} - g, \quad (18.22)$$

$$-\frac{1}{\alpha}\frac{D\alpha}{Dt} + \frac{1}{(a+z)\cos\Theta}\frac{\partial u}{\partial\Phi} + \frac{1}{(a+z)\cos\Theta}\frac{\partial}{\partial\Theta}(v\cos\Theta)$$
$$+ \frac{1}{(a+z)^2}\frac{\partial}{\partial z}(a+z)^2 w = 0,$$

$$\frac{D}{Dt} = \frac{\partial}{\partial t} + \frac{u}{(a+z)\cos\Theta}\frac{\partial}{\partial\Phi} + \frac{v}{a+z}\frac{\partial}{\partial\Theta} + w\frac{\partial}{\partial z},$$

where Φ is the longitude, Θ the latitude, Ω the angular speed of the earth's rotation, g the acceleration of gravity, p the pressure, α the specific volume, and u, v, w the eastward, northward, upward velocity components, respectively. The radial coordinate is denoted by $a + z$, where a is the mean radius of the earth and z the height above mean sea level. This neglects the ellipticity of the geoid [see Veronis (1973b) for a discussion of this approximation].

We assume that the specific volume is determined by an equation of state as a function of absolute temperature T, salinity $\mathscr{S}$, and pressure:

$$\alpha = \alpha(T, \mathscr{S}, p) \qquad (18.23)$$

with salinity conserved,

$$\frac{D}{Dt}\mathscr{S} = 0, \qquad (18.24)$$

and temperature changes determined from the adiabatic thermodynamics

$$\frac{DT}{Dt} - \frac{T}{c_p}\left(\frac{\partial \alpha}{\partial T}\right)_{p,\mathscr{S}} \frac{Dp}{Dt} = 0. \qquad (18.25)$$

For dynamical modeling it is convenient to regard temperature as a function of specific volume, salinity, and pressure and to determine the evolution of the specific volume from

$$\frac{D\alpha}{Dt} + \frac{\alpha^2}{c_s^2}\frac{Dp}{Dt} = 0, \qquad (18.26)$$

which can be derived by taking the substantial derivative of (18.23), using (18.24)–(18.25) and the definition of the sound speed:

$$c_s^2 \equiv -\alpha^2 \left[\frac{T}{c_p}\left(\frac{\partial \alpha}{\partial T}\right)_{p,\mathscr{S}} + \left(\frac{\partial \alpha}{\partial p}\right)_{T,\mathscr{S}}\right]^{-1} = c_s^2(\alpha, p, \mathscr{S}). \qquad (18.27)$$

Equations (18.26)–(18.27) replace (18.23) and (18.25); since the speed of sound is large compared to the mesoscale wave speeds and also is rather insensitive to its arguments (especially salinity), it plays a rather minor role in the large-scale dynamics.

In the appendix, we write the nondimensional forms of these equations based on a time scale T, a horizontal velocity scale U, a vertical velocity scale W, and a depth scale H. For the horizontal coordinates we introduce two scales of motion: the global, Θ and $\Phi \sim 1$ (the β-effect is global); and the local, $\Delta\Theta$ and $\Delta\Phi \sim L/a$, where L is a typical horizontal scale (cf. Phillips's 1973 WKB approach to Rossby waves). Thus we represent all dependent variables Q in the form $Q(\theta, \phi, z, t, \Theta, \Phi)$ with $d\phi = (a/L)d\Phi$ and $d\theta = (a/L)d\Theta$. We also explicitly introduce a basic hydrostatically balanced stratification of the ocean $\overline{T}(z)$, $\overline{\mathscr{S}}(z)$, $\overline{\alpha}(z)$, $\overline{p}(z)$ satisfying

$$\overline{\alpha}(z)\overline{p}(z) = -g.$$

[In practice, given $\overline{T}(\overline{p})$, $\overline{\mathscr{S}}(\overline{p})$ we find $\overline{\alpha}(\overline{p})$ and integrate to get $z(\overline{p})$.] We then subtract out this hydrostatic state and define the (nondimensional) geostrophic streamfunction ψ by

$$p = \overline{p} + 2\Omega \sin \Theta \, UL\psi/\overline{\alpha}. \qquad (18.28)$$

We also define a "local" potential specific volume α_p of a fluid particle with specific volume α at pressure p and depth z as the specific volume it would acquire if the particle moved adiabatically to the horizontally averaged pressure $\overline{p}(z)$. Equation (18.26) gives

$$\alpha_p = \alpha - \frac{\overline{\alpha}^2}{\overline{c}_s^2}(\overline{p} - p) \qquad (18.29)$$

(as long as α and p are not too different from their averaged values). The buoyant force per unit mass after this change becomes

$$b_p \text{ (dimensional)} = g\frac{\alpha_p - \overline{\alpha}}{\overline{\alpha}}$$

$$= g\frac{\alpha - \overline{\alpha}}{\overline{\alpha}} + g\frac{\overline{\alpha}}{\overline{c}_s^2}(p - \overline{p}).$$

This leads to a redefinition of the specific volume in terms of the nondimensional potential buoyancy:

$$\alpha = \overline{\alpha}\left[1 + \frac{2\Omega \sin \Theta \, UL}{gH}\left(b_p - \frac{gH}{\overline{c}_s^2}\psi\right)\right].$$

With the above scalings, we have eight nondimensional parameters (many of which vary spatially):

$$\varepsilon = \frac{1}{2\Omega \sin \Theta \, T} \quad \text{(a time Rossby number)},$$

$$\epsilon = \frac{U}{2\Omega \sin \Theta \, L} \quad \text{(a velocity Rossby number)},$$

$$\hat{\beta} = (L/a)\cot \Theta,$$

$$\lambda = H/L,$$

$$\Delta = (2\Omega \sin \Theta \, L)^2/(gH),$$

$$\Delta_s = gH/\overline{c}_s^2,$$

$$\omega = LW/(HU),$$

$$\hat{S} = H^2\overline{N}^2(z)/(2\Omega \sin \Theta \, L)^2,$$

where $\overline{N}^2$ is the square of the buoyancy frequency:

$$\overline{N}^2 = (g\overline{\alpha}_z/\overline{\alpha}) - (g^2/\overline{c}_s^2).$$

Two of these parameters, ϵ and $\hat{\beta}$, are identical to those used previously with the definitions $f_0 = 2\Omega \sin \Theta$ and $\beta = 2\Omega \cos \Theta/a$. We have also explicitly separated the time scale from the Rossby wave period, whereas in the previous section ε was set equal to $\hat{\beta}\hat{S}$ with $\hat{S} = gH/f_0^2$. The quantity analogous to $\hat{S}$ for a continuously stratified ocean is

$\hat{S} = H^2\overline{N}^2(z)/(2\Omega \sin\Theta L)^2.$

This nondimensional variable is of order unity for motions due to baroclinic instability (Eady, 1949). It is useful to think of it as the squared ratio of two length scales, L_R^2/L^2 or H^2/H_R^2, where $L_R \sim \overline{N}H/f_0$ is the analog for a stratified ocean of the single-layer horizontal deformation radius $\sqrt{gH}/f_0$ introduced by Rossby (1938), and by analogy $H_R \sim f_0 L/\overline{N}$ may be called a vertical deformation radius. If the vertical scale is set, the natural horizontal scale will be L_R; if the horizontal scale is set, the natural vertical scale will be H_R.

We now simplify the equations of motion by making assumptions about the magnitudes of the various parameters. The first seven of our nondimensional parameters are small (for the atmosphere, Δ_s may be of order 1). However, the stability parameter $\hat{S}$ is quite variable. Taking $H \sim 1000$ m as a measure of the depth of the main thermocline, we find that $\hat{S}$ is large in the seasonal thermocline and near unity in the main thermocline. Although this variability is occasionally worrisome in making scale arguments, we shall follow the conventional choice of regarding $\hat{S} \sim O(1)$.

We begin by restricting the length scale L so that $\lambda \ll 1$ and $\Delta \ll 1$, implying that L is large compared to the ocean depth but small compared to the external deformation radius $\sqrt{gH}/f_0 \sim 3000$ km. In practice, we expect the upper limit for L to be determined by the condition that $\hat{S} \gg O(\hat{\beta})$, so that L must be less than the intermediate scale L_I defined in section 18.4.2. Using $\lambda \ll 1$ and $\Delta \ll 1$ and dropping small terms, we obtain the Boussinesq hydrostatic forms of the primitive equations (see the appendix).

Next we specify the time and velocity scale. For the standard quasi-geostrophic motions, the time scale is set by instabilities of the flow so that $T = L/U$ ($\varepsilon \sim \epsilon$) and the vertical velocity is determined by balance between local and advective changes in the vertical component of relative vorticity and stretching of the vortex tubes of the earth's rotation ($\omega = \varepsilon$). Finally, the advective changes of the relative and planetary vorticity are assumed to be comparable, so that $\hat{\beta} \sim \varepsilon$ also. Expanding in ε, we find, as expected, that the lowest-order flows are geostrophic and hydrostatic:

$$u = -\frac{\partial \psi}{\partial y}, \quad v = \frac{\partial \psi}{\partial x}, \quad b_p = f_0 \frac{\partial \psi}{\partial z}, \quad (18.30)$$

where we have redefined the rapidly varying coordinates to look Cartesian by setting $dx = L\cos\Theta\, d\phi$ and $dy = L\, d\theta$ and have returned to dimensional variables. The full pressure is related to the streamfunction by

$$p = \overline{p}(z) + f_0\psi/\overline{\alpha}(z). \quad (18.31)$$

The vorticity equation, which is derived by cross differentiating the order-Rossby number momentum equations (with special care taken with the Θ and Φ dependence), and use of the order-Rossby number continuity equation, becomes

$$\left(\frac{\partial}{\partial t} + \mathbf{v}\cdot\boldsymbol{\nabla}\right)(\boldsymbol{\nabla}^2\psi + \beta y) = f_0 w_z, \quad (18.32)$$

and the buoyancy equation becomes

$$\left(\frac{\partial}{\partial t} + \mathbf{v}\cdot\boldsymbol{\nabla}\right)\psi_z + f_0 S w = 0. \quad (18.33)$$

Here $S = \overline{N}^2(z)/f_0^2$, $\mathbf{v} = (-\psi_y, \psi_x)$, and $\boldsymbol{\nabla} = (\partial/\partial x, \partial/\partial y)$. These two may be combined to give the quasi-geostrophic equation

$$\left(\frac{\partial}{\partial t} + \mathbf{v}\cdot\boldsymbol{\nabla}\right)\left(\boldsymbol{\nabla}^2\psi + \frac{\partial}{\partial z}\frac{1}{S}\frac{\partial}{\partial z}\psi + \beta y\right) = 0, \quad (18.34)$$

which asserts that the quantity

$$q = \boldsymbol{\nabla}^2\psi + \frac{\partial}{\partial z}\frac{1}{S}\frac{\partial}{\partial z}\psi + \beta y$$

is conserved at the projection of a particle in a horizontal plane, not, like potential vorticity, at the particle. For this reason it is called *pseudopotential vorticity* to distinguish it from potential vorticity. Because the distinction vanishes for a fluid consisting of several homogeneous incompressible or barotropic layers, there has been some confusion of terminology in the literature.

The temperature and salinity fields can be derived from the streamfunction ψ and the basic stratification $\overline{T}(z)$, $\overline{\mathcal{S}}(z)$, using the salinity and temperature equations together with the expression (18.33) for the vertical velocity:

$$\left(\frac{\partial}{\partial t} + \mathbf{v}\cdot\boldsymbol{\nabla}\right)(\mathcal{S} - \overline{\mathcal{S}}) + w\overline{\mathcal{S}}_z = 0,$$

$$\left(\frac{\partial}{\partial t} + \mathbf{v}\cdot\boldsymbol{\nabla}\right)(T - \overline{T}) + w\left(\overline{T}_z - \frac{g\overline{T}\alpha_T}{\overline{\alpha}\overline{c}_p}\right) = 0.$$

To complete the system of equations we need the boundary conditions. At the bottom boundary vertical velocities are forced by flow over topography:

$$w = \mathbf{v}\cdot\boldsymbol{\nabla}b \quad \text{at} \quad z = -H, \quad (18.35a)$$

where $H(\Theta,\Phi)$ is the (local) mean depth, the true bottom being at $z = -H + b$. For consistency, $|b|/H$ is required to be order ε. At the upper free surface $z = \eta$, the assumption that L is small compared to the external radius of deformation implies that the boundary conditions

$$\left.\begin{array}{l} D\eta/Dt = w \\ p = 0 \end{array}\right\} \quad \text{at} \quad z = \eta$$

can be approximated simply by

$$w = 0 \quad \text{at} \quad z = 0, \quad (18.35b)$$

with the surface displacement computed from

$$\eta = \frac{f_0}{g} \psi(x,y,0).$$

Finally, on the side-wall boundaries it is necessary to set both the order 1 and order ε normal velocities to zero, giving

$$\nabla \psi \cdot \hat{\mathbf{s}} = 0,$$

$$\oint \nabla \psi_t \cdot \hat{\mathbf{n}} = 0, \qquad (18.36)$$

where $\hat{\mathbf{s}}$ is the unit tangent vector and $\hat{\mathbf{n}}$ the unit normal vector to the boundary.

All of these conditions will be modified in the presence of friction: the top and bottom layers because of Ekman pumping into or out of the frictional layer (see section 18.6) and the side conditions by the necessity for upwelling layers which can feed offshore Ekman transports and can accept mass flux from the interior of the ocean.

18.5 Linear Quasi-Geostrophic Dynamics of a Stratified Ocean

The quasi-geostrophic equations (18.33)–(18.36) have been applied to large-scale, long-period, free- and forced-wave motions in the atmosphere, to the study of barotropic and baroclinic instability, to wave–mean flow and wave–wave interaction, and to geostrophic turbulence. We have mentioned already the review articles by N. Phillips (1963) and Kuo (1973) and the book by Pedlosky (1979a) in which their applications are treated. In addition, Dickinson (1978) has reviewed their application to long-period oscillations of oceans and atmospheres and Holton (1975) their application to upper-atmosphere dynamics.

In application to the mesoscale eddy range of oceanic motions (10 km $< L <$ 210 km), equations (18.32)–(18.36) exhibit a rich variety of behavior depending on the sizes of the various parameters and the initial and boundary conditions. We cannot discuss all of them here; rather we shall confine ourselves to a few topics which are also familiar in a meteorological context. We shall, whenever possible, use a typical oceanic $N^2(z)$ profile (Millard and Bryden, 1973; also see figure 18.7) rather than the constant- or delta-function profiles that are most commonly considered. This allows us to describe the vertical dependence of theoretical predictions in a way which is more directly comparable with oceanic data.

We begin with phenomena that are essentially linear—involving no transfers of energy between scales—deferring discussion of nonlinear motions to the next section.

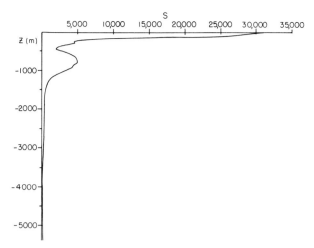

Figure 18.7 Typical oceanic structure for $S(z) = N^2/f_0^2$. The data are from Millard and Bryden (1973) and represent an average over ten stations centered on 28°N and 70°W.

18.5.1 Rossby Waves and Topographical Rossby[5] Waves

Rhines (1970) has discussed the nature of free quasi-geostrophic waves in a uniformly stratified fluid with bottom topography in some detail. We shall describe the behavior of these waves with the aid of a formalism that permits us to extend Rhines's results to real $N^2(z)$ profiles.

When the bottom slope is uniform (b_x and b_y constant), the equations are separable, so that we can write the streamfunction in the form

$$\psi = AF(z) \sin[k(x - ct) + ly], \qquad (18.37)$$

where

$$c = -\beta/(k^2 + l^2 + \lambda^2) \qquad (18.38)$$

and λ, the separation constant, governs the z dependence:

$$\frac{\partial}{\partial z} \frac{1}{S} \frac{\partial}{\partial z} F = -\lambda^2 F. \qquad (18.39)$$

To close the system, we make use of the boundary conditions

$$\frac{\partial}{\partial t} \psi_z = 0, \quad z = 0,$$

$$\frac{\partial}{\partial t} \psi_z = -f_0 S(-H) J(\psi, b), \quad z = -H,$$

which become

$$F_z = 0, \quad z = 0, \qquad (18.40a)$$

$$F_z = \frac{-f_0 S(-H)}{\beta} \left(b_y - \frac{1}{k} b_x \right)$$

$$\times (k^2 + l^2 + \lambda^2) F, \quad z = -H. \qquad (18.40b)$$

To solve (8.39) and (18.40) we proceed as follows: given $S(z)$ we integrate (18.39) with the boundary condition (18.40a) and the normalization condition

$$(1/H)\int_{-H}^{0} dz\, F^2(z;\lambda^2) = 1$$

(using a simple staggered-grid difference scheme with 50-m vertical resolution). We then define the nondimensional function

$$R(\lambda^2) \equiv \frac{-HF_z(-H;\lambda^2)}{S(-H)F(-H;\lambda^2)} \quad (18.41)$$

in terms of which the bottom boundary condition (18.40b) becomes

$$R(\lambda^2) = \frac{f_0 H}{\beta}\left(b_y - \frac{1}{k}b_x\right)(k^2 + l^2 + \lambda^2). \quad (18.42)$$

This can be used to determine λ^2 and the vertical structure $F(z)$, given the wave scale $(k^2 + l^2)^{-1/2}$ and the propagation angle $\tan\theta = l/k$.

Thus, we can summarize all of the information in one graph. Figure 18.8 shows $R(\lambda^2)$, from which λ^2 can be determined given the wave numbers and the topographic slopes by a graphical solution of (18.42). From the resulting set of λ^2 values, the values of the phase speeds of the various waves can then be determined from (18.38). The vertical structures of the waves and the dependence of their phase speeds on the slopes and wave numbers are qualitatively similar to those in Rhines's (1970) constant-N model. We shall describe these results and give useful approximate formulas for c.

When there is no slope effect, the λ_n values are simply the inverses of the deformation radii associated with the various modes $F_n(z)$ which are eigensolutions of (18.39) under the condition $F_z = 0$ for $z = 0, -H$, normalized so that $(1/H)\int_{-H}^{0}dz\,F_n^2(z) = 1$. The barotropic ($n = 0$) and first baroclinic modes ($n = 1$, $L_R = 46$ km) correspond to the structures observed in oceanic data (cf. Richman, 1976). The vertical dependence of these structures are shown also in figure 18.8. The dispersion relation (18.38) and the propagation characteristics of the various modes are described in many places (see chapter 10).

For the weak topographic slope effect, the phase speeds are altered to

$$c \simeq -\left[\beta + \frac{f_0 F_n^2(-H)}{H}\left(b_y - \frac{1}{k}b_x\right)\right]\bigg/ (k^2 + l^2 + \lambda_n^2),$$

derived by solving (18.39) for $\lambda^2 - \lambda_n^2$ small. We find the familiar result that the bottom slope, by causing vortex stretching or shrinking, acts as an effective β. The bottom-slope effect in the baroclinic modes is weaker by a factor $F_n^2(-H)$ than the corresponding effect

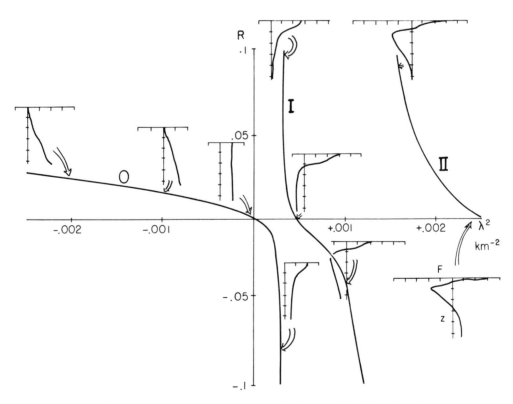

Figure 18.8 Normalized ratio of bottom shear to bottom amplitude as a function of the separation constant λ^2. For given topographic slopes and wavenumber vector, the values of λ^2 are the intersections of this curve with the line $R = (f_0 H/\beta)[b_y - (l/k)b_x](\lambda^2 + k^2 + l^2)$. Also shown is the vertical structure $F(z)$ of the streamfunction (normalized to rms value unity) at various (λ^2, R) values.

on the barotropic mode. This factor is smaller than 1 for the stratification used (about 0.4 for the first baroclinic mode).

When the slope effect $f_0 H[b_y - (l/k)b_x]/\beta$ is negative, there is also a bottom-trapped eastward-moving wave. The vertical trapping scale is $H_s = -(f_0/\beta)[b_y - (l/k)b_x]$, and the speed

$$c = \beta N^2(-H)H_s^2/f_0^2$$

can be regarded as that of a long wave in a fluid with a deformation radius based on the vertical scale H_s and the local value of N.

For large slopes the modes all have large vertical shear. Most of them are surface trapped, having essentially zero bottom amplitude and a westward component of phase speed. For a slope effect opposing the β-effect, there is also a rapidly eastward-moving wave whose vertical trapping scale is $H_R = f_0/N(-H)\sqrt{k^2 + l^2}$ and whose propagation speed is

$$c = -f_0(b_y - \frac{1}{k}b_x)/H_R(k^2 + l^2)$$

in the limit of $L \ll L_R$.

The constant N or two-layer models make significant qualitative errors in describing one or another of these types of behavior. The flat-bottom baroclinic modes in a constant-N model have bottom velocities comparable to the surface velocities; when a weak slope is added, this produces a large change in the vortex stretching and in the phase speed. Thus, the effect of the slope on the baroclinic modes is twice as great as on the barotropic mode, not, as in the realistic ocean, half as great. For large slopes the trapping scale for constant N is much smaller than the correct scale because $N(-H)$ is small compared to the average value of N which would be used in a constant N model. This results in an overestimate of the phase speed. The two-layer model generally misrepresents the eastward-traveling mode: for weak slopes, it is altogether absent, while for strong slopes the two-layer model predicts $c \sim (k^2 + l^2)^{-1}$ rather than the correct $(k^2 + l^2)^{-1/2}$ dependence.

Although we have derived the solutions (18.37)–(18.42) from the linearized equations, the individual waves are also finite amplitude solutions to the equations of motion (18.32)–(18.36) because both

$$\left(\nabla^2 + \frac{\partial}{\partial z}\frac{1}{S}\frac{\partial}{\partial z}\right)\psi \quad \text{and} \quad \psi_z$$

are proportional to ψ at each horizontal level. In fact, the nonlinearities in the equation of motion (18.34) will vanish for any set of waves having each of their wave vectors parallel or having the same total scale $(k^2 + l^2 + \lambda^2)^{-1/2}$. In the latter case, all the waves have the same phase speed so that the composite streamfunction pattern propagates uniformly. The nonlinearity in the bottom boundary condition likewise vanishes if all waves have the same value of R, that is, if the wavenumber vectors are parallel or $b_x = 0$. Thus sets of waves with horizontal scales $(k^2 + l^2)^{-1/2}$ such that both $k^2 + l^2 + \lambda^2 =$ constant and $R(\lambda) =$ constant will be full nonlinear solutions for north–south bottom slopes. Instabilities may, of course, prevent such patterns from persisting.

18.5.2 Generation of Rossby Waves by Flow over Topography

Flow over topography has been of practical interest to meteorologists attempting to forecast conditions near large mountain ranges for many years. There is an extensive bibliography of such studies (Nicholls, 1973; Hide and White, 1980). Many of these concentrate on smaller-scale lee-wave properties, although there have also been attempts to model the large-scale standing atmospheric eddies as topographically forced Rossby waves (cf. Charney and Eliassen, 1949; Bolin, 1950).

In the ocean, there are also many standing features, some of which clearly can be identified with topography (cf. Hogg, 1972, 1973; Vastano and Warren, 1976). In this section we shall discuss a simple oceanic analogy to the common atmosphere model for standing waves produced by flow over topography.

Steady solutions to (18.33)–(18.35) may be written

$$\nabla^2\psi + \frac{\partial}{\partial z}\frac{1}{S}\frac{\partial}{\partial z}\psi + \beta y = P(\psi,z),$$

$$\psi_z = T_s(\psi), \quad z = 0, \quad (18.43)$$

$$\psi_z + f_0 S(-H)b = T_b(\psi), \quad z = -H,$$

where P, T_s, and T_b are arbitrary functionals. If we allow ψ to represent a mean zonal flow plus a topographically induced deviation,

$$\psi = -\bar{u}(z)y + \phi(x,y,z),$$

we find

$$P(\psi,z) = -\left(\beta - \frac{\partial}{\partial z}\frac{1}{S}\frac{\partial \bar{u}}{\partial z}\right)\psi/\bar{u},$$

$$T_s = T_b = 0$$

are suitable functionals to match the terms linear in y in (18.43). We have restricted ourselves to flows such that $\bar{u} \neq 0$ everywhere and $\bar{u}_z = 0$ at $z = 0, -H$. The conditions that the upper and lower surfaces be isothermal are not fundamental and could be relaxed easily; we make them simply to restrict the discussion to a reasonable number of parameters. If, however, the $\bar{u} \neq 0$ condition is violated, the analysis becomes much more difficult since the critical-layer (where $\bar{u} = 0$) problem must also be solved.

Given these restrictions, however, the fluctuation field satisfies the simple equations

$$\nabla^2 \phi + \frac{\partial}{\partial z} \frac{1}{S} \frac{\partial}{\partial z} \phi + \frac{\beta - (\partial/\partial z)(1/S)(\partial \bar{u}/\partial z)}{\bar{u}} \phi = 0,$$

$$\phi_z = 0, \quad z = 0, \quad (18.44)$$

$$\phi_z = -f_0 S(-H) b(x,y), \quad z = -H,$$

where no small-amplitude assumption has been made beyond the assumption that the nondimensional topographic amplitude is of the order of the Rossby number. The linearity of these equations is due to the particularly simple form of the upstream flow—a streamfunction field which is linear in y. Any horizontal shear in the upstream flow would give rise to nonlinear terms in P, T_s, and T_b, and thereby to nonlinearities in the equations (18.44). For simple sinusoidal topography $b = b_0 \sin(kx + ly)$, the topographically forced wave looks like

$$\phi = f_0 b_0 H F(z) \sin(kx + ly),$$

where

$$\frac{\partial}{\partial z} \frac{1}{S} \frac{\partial}{\partial z} F = \left[k^2 + l^2 - \frac{\beta - (\partial/\partial z)(1/S)(\partial \bar{u}/\partial z)}{\bar{u}} \right] F,$$

$$F_z = 0, \quad z = 0, \quad (18.45)$$

$$F_z = -S(-H)/H, \quad z = -H.$$

When the zonal flow is barotropic ($\bar{u}$ = constant) the system (18.45) becomes essentially identical to (18.39)–(18.40) except that the amplitude is determined by the bottom boundary condition. We summarize the shapes and amplitudes of the forced waves in figure 18.9. We have used again the simplest form $\lambda^2 = \beta/\bar{u} - (k^2 + l^2)$ for the dependent variable.

The most striking feature in the response is, of course, the resonant behavior when

$$\bar{u} = \beta/(k^2 + l^2 + \lambda_n^2)$$

for λ_n^2 one of the inverse-square deformation radii. Near such a resonance, the amplitude becomes very large:

$$F(z) \simeq \frac{F_n(z) F_n(-H)}{[k^2 + l^2 - (\beta/\bar{u}) + \lambda_n^2] H}.$$

When the mean flow is eastward at a few centimeters per second, it may be near a resonance for one of the baroclinic modes and may therefore generate substantial currents even above the thermocline. Vertical standing-wave modes associated with vertical propagation and reflection at the upper boundary will be found for $\lambda^2 > 0$ or

$$0 < \bar{u} < \beta/(k^2 + l^2).$$

For $\lambda^2 < 0$ the motions are trapped and decay away from the bottom. The trapping scale becomes very small when $\bar{u}$ is nearly zero but westward or when the topographic wavelength is short. In the latter case, the

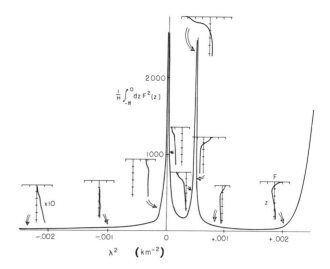

Figure 18.9 Energy and vertical structure of a topographically forced wave as a function of $\lambda^2 = (\beta/\bar{u}) - \mathbf{k} \cdot \mathbf{k}$, where $\mathbf{k}$ is the topographic wavenumber.

vertical scale of the fluctuations is $H_R \sim f_0 L/N(-H)$, where $L = (k^2 + l^2)^{-1/2}$. When $\bar{u}$ is weak and westward $(0 < |\bar{u}| < \beta N^2(-H) H^2/f_0^2 \sim 4 \text{ cm s})$, the vertical scale will again be small compared to the fluid depth.

Although this type of problem is suggested by the atmosphere analogue, a warning about its applicability may be in order. Periodic problems are natural for the atmosphere. In the ocean, however, it is less plausible that a fluid particle periodically will revisit a topographic feature in a time less than the damping time for the excited wave. (The Antarctic Circumpolar Current could perhaps be an exception.) We can illustrate the differences between periodic and local topography by considering uniform eastward flow over a finite series of hills and valleys:

$$b = \begin{cases} b_0 \sin kx, & 0 \le x \le n\pi/k \\ 0, & x < 0 \text{ or } x > n\pi/k. \end{cases}$$

We consider the barotropic component of flow which satisfies the depth averaged form of (18.44):

$$\nabla^2 \hat{\phi} + \frac{\beta}{\bar{u}} \hat{\phi} = -\frac{f_0}{H} b(x,y).$$

Its solution is

$$\hat{\phi} = 0 \quad \text{for } x < 0,$$

$$\hat{\phi} = \frac{f_0 b_0}{H(k^2 - \beta/\bar{u})}$$

$$\times \left(\sin kx - k\sqrt{\frac{\bar{u}}{\beta}} \sin \sqrt{\frac{\beta}{\bar{u}}} x \right)$$

for $0 < x < \frac{n\pi}{k}$,

$$\hat{\phi} = \frac{f_0 b_0}{H(k^2 - \beta/\bar{u})}$$

$$\times k \sqrt{\frac{\bar{u}}{\beta}} \left[\cos\sqrt{\frac{\beta}{\bar{u}}}\, x - (-1)^n \cos\sqrt{\frac{\beta}{\bar{u}}}\left(x - \frac{n\pi}{k}\right) \right]$$

for $x > \frac{n\pi}{k}$.

Figure 18.10 shows the average energy in the far field (normalized by $\frac{1}{2}(f_0 b_0/Hk)^2$ as a function of $\bar{u} k^2/\beta$. We note that the resonance peak becomes significant only when the topography has a number of hills and valleys; this suggests that the idea of resonance, in the ocean, should be applied with caution.

When there is vertical shear in the mean flow, the situation becomes somewhat different, although equation (18.45) can still readily be integrated. However, one can gain a qualitative picture of the response for arbitrary shear and small perturbations by using the methods of Charney and Drazin (1961) as described in the next section.

18.5.3 Propagation and Trapping of Neutral Rossby Waves

In many circumstances, the ocean or atmosphere is directly forced by external conditions—heating, winds—which may have temporal and spatial variations. The forcing may generate wave disturbances in one region that propagate into a neighboring region (e.g., the propagation of tropospheric disturbances into the stratosphere). In these circumstances the motion is determined by the nature of the forcing and the refractive properties of the intervening medium. Charney (1949) first treated the vertical propagation of Rossby waves in a stratified atmosphere and Charney and Drazin (1961) first suggested the analogy between vertical propagation of Rossby waves and electromagnetic wave propagation in a medium with a variable index of refraction (possibly complex, corresponding to wave absorption). Holton (1975) has reviewed these concepts for meteorologists; oceanographers have tended to make less use of them [see, however, Wunsch (1977), who applied them to vertically propagating equatorial waves excited by monsoon winds].

The simplest derivation of an index of refraction is for waves in a zonal flow with vertical and horizontal shear. Consider waves of infinitesimal amplitude having east-west wavenumber k and frequency ω. The north-south and vertical dependencies of the amplitude $\psi = \Psi(y,z)e^{i(kx-\omega t)}$ are determined by the standard stability equation (cf. Charney and Stern, 1962):

$$\left(\bar{u} - \frac{\omega}{k}\right)\left(\frac{\partial^2}{\partial y^2} + \frac{\partial}{\partial z}\frac{1}{S}\frac{\partial}{\partial z} - k^2\right)\Psi$$

$$+ \left(\beta - \bar{u}_{yy} - \frac{\partial}{\partial z}\frac{1}{S}\frac{\partial}{\partial z}\bar{u}\right)\Psi = 0.$$

If we follow the procedure that has been used for vertically propagating waves, we transform this into a Helmholtz equation with a variable coefficient of the undifferentiated term. This is quite straightforward if we are considering propagation only in the y direction with $\bar{u}_z = 0$ and $\Psi = \Phi(y)F_n(z)$, where F_n is one of the flat bottom eigenfunctions. Then the y structure is governed by

$$\frac{\partial^2}{\partial y^2}\Phi + \nu^2(y)\Phi = 0$$

where

$$\nu^2(y) = [(\beta - \bar{u}_{yy})/(\bar{u} - \omega/k)] - k^2 - \lambda_n^2. \quad (18.46)$$

A simple illustrative example is the radiation from a meandering Gulf Stream into the neighboring Sargasso Sea (cf. Flierl, Kamenkovich, and Robinson, 1975; Pedlosky, 1977). The forcing specifies Φ at some latitude. When we have no mean flow ($\bar{u} = 0$) and the motions are barotropic, the index of refraction becomes

$$\nu^2(y) = -k^2 - (\beta k/\omega)$$

The north-south scale of the response $|\nu|^{-1}$ is shown as a function of ω (>0) and k (≥ 0) in figure 18.11. Most observations indicate eastward-traveling motions, $\omega/k > 0$, implying that the mid-ocean response will be trapped close to the Gulf Stream.

We may obtain a similar representation of the index of refraction for the full two-dimensional (y and z) problem. This result, for reasons discussed below, is probably of more interest to meteorologists than to

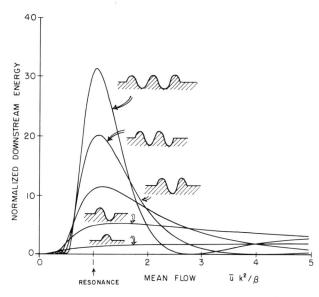

Figure 18.10 Downstream energy averaged over a wavelength of waves forced by flow over isolated bumps as a function of the normalized mean flow speed. The wavenumber of the topography in the region where it is varying is k. For an infinite topography, resonance would occur at $\bar{u}k^2/\beta = 1$. Results are shown for varying numbers of elevations and depressions in the topography.

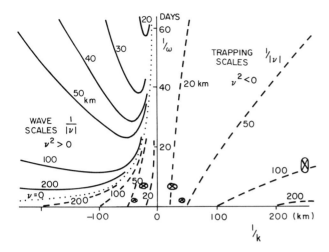

Figure 18.11 Solid lines show north-south length scales (wavelength/2π) and dashed lines shown trapping scales (e-folding distance) for barotropic waves generated by a meandering current with inverse frequency ω^{-1} and inverse wavenumber k^{-1}. Eastward going meanders ($k > 0$) produce trapped waves; westward going meanders ($k < 0$) may produce propagating disturbances. The symbols $\otimes$ correspond to typical observational estimates of ω^{-1} and k^{-1}.

oceanographers; however, we include it to illustrate some of the effects of the z structure. If we substitute $\Psi(y,z) = S^{1/4}\Phi(y,\zeta)$, where $\zeta = \int_{-H}^{z} S^{1/2}(z')dz'$ is a modified vertical coordinate, we find

$$\left(\frac{\partial^2}{\partial y^2} + \frac{\partial^2}{\partial \zeta^2}\right)\Phi + \nu^2(y,\zeta)\Phi = 0,$$

where the index of refraction $\nu^2(y,\zeta)$ is given by

$$\nu^2(y,\zeta) = S^{-1/4}[S^{-1}(S^{1/4})_z]_z$$
$$+ \frac{\beta - \bar{u}_{yy} - (\bar{u}_z/S)_z}{\bar{u} - \omega/k} - k^2. \quad (18.47)$$

When $\nu^2 > 0$ there are sinusoidal solutions and energy propagates freely, whereas when $\nu^2 < 0$ there are only exponential solutions (along the ray) and the waves die out. There are also, of course, diffraction effects and tunneling effects if the regions of negative ν^2 (or, at least, significantly altered ν^2) are relatively small. This form is useful when N is a simple function (e.g., $N_0 e^{z/d}$) so that the first term in (18.47) is also simple $[-3/(4d^2 S)]$. The stratification then contributes a relatively large and negative term which increases toward the bottom, inhibiting penetration into the deep water. For our $S(z)$ profile (figure 18.7), however, numerical differentiation proved to be excessively noisy. Moreover, in the oceans, most of the motions of interest have vertical scales that are significantly influenced by the boundaries and are larger than the scales of variation of ν^2, so that a local (WKB) interpretation of ν^2 variations is not possible.

We can, however, associate modifications in ν^2 occurring on large scales with modifications in the structure of Ψ. Thus in the topographic problem, if the shear in the vertical is such that

$$\frac{\partial}{\partial z}\frac{1}{S}\frac{\partial \bar{u}}{\partial z} > 0 \quad \text{and} \quad \frac{\partial \bar{u}}{\partial z} > 0,$$

there will be a decrease in the value of ν^2, implying that the wave will become either more barotropic ($\nu^2 > 0$) or more bottom trapped ($\nu^2 < 0$). In the example of Rossby wave radiation from a meandering Gulf Stream, (18.46) implies that the baroclinic modes ($\lambda_n^2 > 0$) become trapped even more closely than the barotropic modes.

As a final example, we note that the motions forced in the ocean by atmospheric disturbances tend to have large positive ω/k and large scales. In the absence of mean currents, the vertical structure equation, with $\Psi = e^{ily}F(z)$, becomes

$$\frac{\partial}{\partial z}\frac{1}{S}\frac{\partial}{\partial z}F = \left[\frac{\beta k}{\omega} - k^2 - l^2\right]F = -\nu^2 F, \quad (18.48)$$

implying that the forced currents are nearly barotropic. However, the recent work of Frankignoul and Müller (1979) suggests a possible mechanism by which significant baroclinic currents may be produced. Because the ocean is weakly damped and has resonant modes ($\nu^2 = \lambda_n^2$), even very small forcing near these resonances can cause the energy to build up in these modes. This is another example of the strong influence of the boundaries on the oceanic system.

18.6 Friction in Quasi-Geostrophic Systems

18.6.1 Ekman Layers

Ekman (1902, 1905), acting on a suggestion of Nansen, was the first to explore the influence of the Coriolis force on the dynamics of frictional behavior in the upper wind-stirred layers of the oceans. He considered both steady and impulsively applied, but horizontally uniform, winds. In an effort to understand how surface frictional stresses τ influence the upper motion of the atmosphere and, in particular, how a cyclone "spins down," Charney and Eliassen (1949) were led to consider horizontally varying winds. They showed that Ekman dynamics generates a horizontal convergence of mass in the atmospheric boundary layer proportional to the vertical component of the vorticity of the geostrophic wind in this layer. Thus a cyclone produces a vertical flow out of the boundary layer which compresses the earth's vertical vortex tubes and generates *anticyclonic* vorticity. The time constant for frictional decay in a barotropic fluid was found to be $(f_0 E^{1/2})^{-1}$, where E is the Ekman number $\nu_e/f_0 H^2$, with ν_e the eddy coefficient of viscosity and H the depth of the fluid. Greenspan and Howard (1963) investigated the time-dependent motion of a convergent Ekman layer: if the wind is turned on impulsively, the Ekman layer is set up in a time of order f_0^{-1}; the internal flow decays in a time of order $(f_0 E^{1/2})^{-1}$; and the vertical oscillations that are produced by the impulsive startup decay in a time of order $(f_0 E)^{-1}$. Since f_0^{-1} is but a few hours, one may consider that for the large-scale wind and current systems of the atmosphere and oceans the Ekman pumping is produced instantly and that there is a balance in the Ekman layer among the frictional pressure and Coriolis forces. We divide the flow into a quasigeostrophic interior component (u_g, v_g, w_g) with associated pressure gradients $fv_g = \alpha p_{g,x}$ etc. and a deviation component associated with the friction (u_e, v_e, w_e) which vanishes below some small depth h. For a homogeneous fluid $p_e = 0$ because the hydrostatic assumption ensures that there can be no nontrivial pressure field which vanishes below $z = -h$. For a stratified flow a scaling argument can be made to show that buoyancy fluctuations in the upper layer will not be important enough to cause significant p_e's (unless $N^2 > \tau_0 L/\rho h^3$) so that $\rho f v_e = -(\partial/\partial z)\boldsymbol{\tau}\cdot\hat{\mathbf{x}}$, etc. If we divide by f, and compute w_{e_z} from the divergence of the Ekman horizontal velocities, we find

$$w_e = -\hat{\mathbf{z}}\cdot\mathrm{curl}(\boldsymbol{\tau}/\rho f)$$

using $\boldsymbol{\tau}(-h) = 0$, $w_e(-h) = 0$. From the surface condition $w_e(0) + w_g(0) = 0$, the Ekman pumping is therefore

$$w_g(0) \equiv w_E = \hat{\mathbf{z}}\cdot\mathrm{curl}(\boldsymbol{\tau}(0)/\rho f), \quad (18.49)$$

where $\boldsymbol{\tau}(0)$ is the wind stress at the sea surface. The same procedure can be used in the lower boundary layer:

$$w_g = \frac{\partial}{\partial x}\left(\frac{\nu_e v_{e_z}}{f}\right) - \frac{\partial}{\partial y}\left(\frac{\nu_e u_{e_z}}{f}\right), \quad z = -H.$$

But now it is necessary to specify $\boldsymbol{\tau}(-H)$ in terms of the geostrophic velocities u_g, v_g; for this a knowledge of ν_e is required. If we assume ν_e to be constant, the pumping out of the bottom boundary layer is given by

$$w_g(-H) \equiv w_E$$
$$= \frac{D_E}{2}\left[\zeta_g + u_{g,x} + v_{g,y} + \frac{1}{2}(\beta/f)(u_g - v_g)\right]\bigg|_{z=-H},$$

where $D_E = (2\nu_e/f)^{1/2}$ and $\zeta_g = v_{g,x} - u_{g,y}$ is the vorticity of the geostrophic wind. When $L \ll a$, the divergence terms (which are equal to $-\beta v_g/f$) and the last term are negligible, so that

$$w_g(-H) = \frac{D_E}{2}\zeta_g(-H). \quad (18.50)$$

In the lower boundary layer of the deep ocean, the water is nearly homogeneous. In this case one may estimate the bulk viscosity ν_e by supposing that for this value the established boundary layer is marginally stable (cf. Charney, 1969). From the measurements of Tatro and Mollo-Christensen (1967), the condition for marginal stability is found to be that the Reynolds number based on the depth D_E, of the Ekman layer $UD_E/\nu_e = \sqrt{2}U/\sqrt{f\nu_e}$, shall be of order 100. Thus, for example, $\nu_e \sim U^2/5000f = 200 \text{ cm}^2\,\text{s}^{-1}$, and $D_E \sim U/50f \sim 20$ m for a current of 10 cm s^{-1} in middle latitudes.

In a stratified atmosphere or ocean, the depth of influence of the Ekman pumping is not necessarily the depth of the fluid. If a circulation is forced from above by Ekman pumping with horizontal scale L, one expects the depth of influence to be the vertical deformation radius $H_R \sim f_0 L/N$. This depth will be comparable to the ocean depth for $L \sim L_R = 50$ km. Most surface forcing will thus excite a barotropic response. The spin-down of baroclinic mesoscale ocean eddies will be considered in Section 18.6.3.

18.6.2 Spin-Up of the Ocean

The problem of the spin-up of the entire ocean requires definition. The wind and thermally driven circulations are so coupled nonlinearly that it is not possible to treat the establishment of the wind-driven circulation independently. The important question, however, is not how the ocean circulation would be established from rest if the forcing were impulsively applied, but rather how the circulation would change if the forcing changed. The latter question has clear implications for understanding the role of the oceans in climatic change. Thus, one is led to consider first the small-amplitude adjustment of a given steady-state circulation to a change in the wind stress, with the expectation that nonadiabatic changes will require considerably long times. Even for this linearized problem, results for the spin-up of the ocean in mid-latitudes have been obtained (Anderson and Gill, 1975; Anderson and Killworth, 1977; Cane and Sarachik, 1976, 1977) only for the simplest cases of a one- or two-layer model with no preexisting circulation. The solutions for a suddenly applied wind stress are complicated, but their qualitative import can be simply stated. When a steady, east–west wind stress is suddenly applied to a two-layer ocean, initially at rest, the motion at any longitude increases uniformly with time until a nondispersive Rossby wave starting at the eastern boundary and moving with the maximum westward baroclinic group velocity $-\beta L_R^2$ reaches that longitude. When this occurs, a steady Sverdrup flow induced by the wind-stress curl will have been established in the upper layer everywhere to the east of that longitude. By the time the Rossby wave reaches the western boundary, a steady state will have been established over the entire ocean—except in the vicinity of the boundary itself, where slow-moving reflected Rossby waves influence the flow and are presumed to be dissipated by friction. Thus the spin-up time is essentially the time required for a signal traveling at the speed $-\beta L_R^2$ to cross the ocean from east to west. For width of 6000 km, we obtain 1.5×10^8 or about 5 years.

We note that βL_R^2 increases toward the equator. However, as one approaches the equator the dynamics of wave propagation change. Near the equator, Rossby-gravity and Kelvin waves are generated. These have maximum group velocities of order $\sqrt{g'H}$ (g' is the reduced gravity and H the depth of the thermocline) ~ 1 m s^{-1}, giving spin-up times of the order of months rather than years. Cane (1979a) and Philander and Pacanowski (1980a) have shown that an impulsively generated uniform westward wind produces both equatorially trapped Kelvin and Rossby-gravity waves. The equatorial undercurrent is established at a given longitude when a Kelvin wave traveling eastward from the western boundary reaches that longitude. The dynamics of equatorially trapped planetary wave modes have been investigated by Rosenthal (1965) and Matsuno (1966) for the atmosphere and by Blandford (1966), Lighthill (1969), and Cane and Sarachik (1979) for the oceans. The dynamics of the equatorial undercurrent has been reviewed by Philander (1973, 1980).

A similar linear analysis for a continuously stratified ocean initially at rest leads to quite different results. In this case, a wind stress can produce a steady Sverdrup transport only in the upper frictional boundary layer. This is the result of the conservation of density, which requires $wS = 0$ or $w = 0$, and it follows from the interior geostrophic dynamics that $\beta v = fw_z = 0$. The initial application of the wind stress will produce an infinity of transient internal baroclinic modes whose sum will approach zero in time everywhere except at $z = 0$. If we consider only the barotropic and first baroclinic modes, the temporal evolution will be similar to that of the two-layer ocean, but the effect of the other modes will be such as to cause all interior velocities to vanish asymptotically in time.

However, if a perturbation in wind stress is applied to *preexisting* flow, Ekman pumping can penetrate into the interior along isopycnals and w_z need not be zero. Although this calculation has not been made in detail, it seems plausible that the final perturbation structure would be similar to the mean-flow structure and, therefore, that it would be spun up in the time associated with the cross-ocean propagation of the lowest baroclinic modes.

It is also important to note that the definition of the spin-up time depends to some degree on the property one is considering. For example, the Sverdrup balance (see Leetmaa, Niiler and Stommel, 1977, for an empirical discussion) is established on relatively short time scales. If the ocean is forced by the Ekman pumping,

$$w_E = w_0 \exp[ikx - i\omega t],$$

it may be seen from the vorticity equation (18.32) that Sverdrup balance will be attained when $|\omega k| \ll \beta$. Thus fluctuations in forcing on the size of the basin with periods even as short as a few days—the time for the *barotropic* wave to cross the basin—will preserve the Sverdrup balance.

Clearly there are many unanswered questions concerning even the adiabatic response of the ocean to changes in the forcing. We know still less about the response time of the entire wind-driven thermohaline circulation, although we expect the time scales to be much longer. The heat and salt transfer processes may take as long as 50 years for transfer down to the main thermocline and 1000 years for formation of the abyssal water beneath the main thermocline.

For the atmosphere, too, the nonadiabatic spin-up or spin-down processes are slower—radiative heat-transfer processes have time constants of the order of months—than the spin-down time of a few days associated with Ekman pumping. Moreover, the calculations in section 18.6.3 indicate that Ekman spin-down will tend only to reduce the barotropic component of the kinetic energy, that is, they reduce the winds by their surface values. Other processes must be involved in the decay of the winds aloft.

18.6.3 Spin-Down of Mesoscale Eddies

As a final example of frictional quasigeostrophic dynamics, we shall consider the effects of bottom friction on mesoscale ocean eddies. In the atmosphere, friction at the ground is an important part of the dynamics of synoptic scale motions. In the ocean, however, friction is considered to be less important because the bottom currents are relatively weak. Nevertheless, it is of interest to know how much of the water column is affected by bottom friction. We do know that the surface manifestations of mesoscale motions (in particular Gulf Stream rings) can persist for longer than 2 years (Cheney and Richardson, 1976). We shall show that this time scale is consistent with predictions of the simple baroclinic spin-down time.

Holton (1965) obtained a solution to the spin-down problem for a uniformly stratified fluid in a cylindrical container, showing that the effects of Ekman pumping are confined to a height $H_R \sim f_0 L/N$. Walin (1969) completed and extended Holton's analysis by analyzing in detail the effects of the side-wall boundaries and gave a simpler illustration of the spin-down process not involving side boundaries. We shall solve the analogue of Walin's problem for the variable stratification and radial symmetry characteristic of Gulf Stream rings.

We wish to solve (18.32)–(18.34) for the streamfunction $\psi(r,z,t)$, first for $\beta = 0$, assuming $w_E(0) = 0$, $w_E(-H) = (D_E/2)\nabla^2\psi(r,-H,t)$, and the initial condition $\psi(r,z,0) = \psi_0(r,z)$. The nonlinearities vanish because of the radial symmetry. Taking a Fourier-Bessel transform of the streamfunction

$$\hat{\psi}(k,z,t) = \int_0^\infty r\, dr\, J_0(rk)\psi(r,z,t),$$

we find

$$\left(\frac{\partial}{\partial z}\frac{1}{S}\frac{\partial}{\partial z} - k^2\right)\hat{\psi}(k,z,t) = \left(\frac{\partial}{\partial z}\frac{1}{S}\frac{\partial}{\partial z} - k^2\right)\hat{\psi}_0(k,z),$$

$$\hat{\psi}_{zt}(k,0,t) = 0,$$

$$\hat{\psi}_{zt}(k,-H,t) = \tfrac{1}{2}k^2 D_E f_0 S(-H)\hat{\psi}(k,-H,t).$$

Solving for $\hat{\psi}$ we obtain

$$\hat{\psi}(k,z,t) = \hat{\psi}_0(k,z) - \hat{\psi}_0(k,-H)F(z;k)(1 - e^{-\sigma(k)t}),$$

where $F(z;k)$ satisfies

$$\frac{\partial}{\partial z}\frac{1}{S}\frac{\partial}{\partial z}F = k^2 F,$$

$$F_z(0;k) = 0,$$

$$F_z(-H;k) = 1,$$

and the inverse spin-down time is given by

$$\sigma(k) = -k^2 D_E f_0 S(-H)/2F_z(-H;k)$$

$$= [-k^2 S(-H)H/F_z(-H;k)]\sigma_{BT}$$

where $\sigma_{BT} = f_0 D_E/2H$ is the inverse barotropic spin-down time. Thus the inverse baroclinic spin-down time is simply related to σ_{BT} by the factor H divided by the penetration depth.

For large scales, the motion spins down uniformly throughout the whole column with $\sigma \simeq \sigma_{BT}$. For small scales, the spin-down occurs only over a depth H_R and is much more rapid. We expect, therefore, that the smaller scales will disappear from the deep ocean, perhaps leaving a thermocline signal behind, while the larger scales will decay more slowly but also more completely. In figure 18.12, we show the structures $F(z;k)$ and inverse spin-down times $\sigma(k)$ for various scales $1/k$. Absolute decay rates depend upon D_E—for $D_E = 20$ m, the time scale $\sigma_{BT}^{-1} = 89$ days, so that everything happens in a few months.

For application to rings we assume $\psi_0(r,-H) = -le^{(-1/2)(r^2/l^2-1)} \times 10$ cm s^{-1}, which gives maximum currents of 10 cm s^{-1} at a radius l. We solve for the net change in azimuthal velocity $\psi_r(r,z,t \to \infty) - \psi_{0r}(r,z)$ and contour this change in figure 18.13. It is seen that the changes in the thermocline and shallow water are negligible so that the persistence of oceanic thermocline eddies is quite consistent with theoretical expectations.

When the beta effect is included, important differences occur in the spin-down of linear eddies. The simplest case to analyze is for weak friction. Then

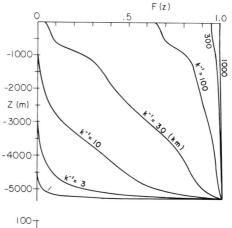

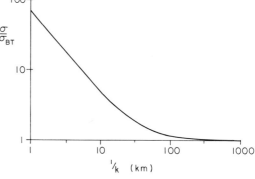

Figure 18.12 Decay in currents F as a function of depth for different radial scales k^{-1}. Actual change is given by $-F(z) \times$ bottom currents. Lower figure shows ratio of decay rate to spin-down rate for a homogeneous fluid as a function of the radial scale.

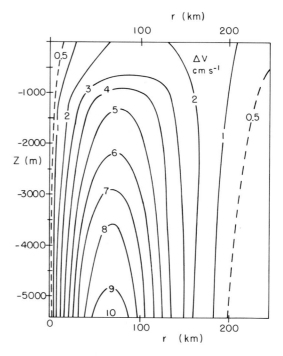

Figure 18.13 Decrease in azimuthal velocity due to bottom friction when initial bottom currents are 10 cm s^{-1} $(r/l) \times \exp[-\tfrac{1}{2}(r^2/l^2) + \tfrac{1}{2}]$.

there are two time scales—the period and the spin-down time. The Fourier-Bessel component with wavenumber k and (initially) vertical normal mode $F_n(z)$ behaves like

$$\hat{\psi}_n(k,t) = \hat{\psi}_n(k,0) F_n(z) J_0\left(k\sqrt{\left(x + \frac{\beta t}{k^2 + \lambda_n^2}\right)^2 + y^2}\right)$$
$$\times \exp[-\sigma_{BT} F_n^2(-H) k^2 t/(k^2 + \lambda_n^2)].$$

This follows from the fact that when the solution of (18.33) is expanded in powers of period/spin-down time, the lowest order component is just a steadily propagating Bessel-function eddy. The next-order component has an inhomogeneous boundary condition due to friction and an inhomogeneous forcing of the equations of motion due to the slow time dependence. Multiplying this first-order equation of motion by $F_n(z)$ and depth averaging shows that the slow time dependence satisfies a simple exponential decay law (see also Flierl, 1978).

One important feature of this solution is the fact that baroclinic modes decay more slowly than barotropic modes both because of the increase in λ_n^2 and because of the appearance of the factor $F_n^2(-H)$. Thus a first-mode deformation-scale eddy with $F_1(-H) = -0.6$ and $k = \lambda$ has a decay rate of $0.2\sigma_{BT}$. But the important feature is that the β-plane eddy, unlike the f-plane eddy, decays completely. The β-effect permits transmission of energy downward, where it can be dissipated by friction. It appears that nonlinearity can impede this process because it slows down the dispersion of a ring (McWilliams and Flierl, 1979).

18.7 Nonlinear Motions

In this section, we shall consider mesoscale flows for which the advection of relative vorticity or density anomaly is important. This can occur either in the form of wave-wave interactions or wave-mean flow interactions. In both cases we are considering motions in which there are significant nonlinear interactions among various scales. This situation is to be contrasted with that in section 18.5.3, in which the mean flow provided a variable environment for the waves but was passive in the sense that there was no exchange of energy between the waves and the mean flow.

18.7.1 Baroclinic and Barotropic Instabilities
The problem of the instability of large-scale atmospheric motions has a long history, going back as always to Helmholtz (1888). The discoveries of the polar front and the polar-front wave by J. Bjerknes (1919) and J. Bjerknes and H. Solberg (1921, 1922) initiated several investigations of the instability of a polar-front model, notably by H. Solberg (1928) and by N. Kotschin (1932).

These studies were incomplete: Solberg's avoided considering the effects of the frontal intersection with ground; Kotschin's considered various possible perturbation modes but not that of the all important baroclinic instability. E. Eliasen (1960) conducted a numerical study of a problem similar to Kotschin's, but with a vertical wall. However, the detailed exploration of Kotschin's model, a front between two fluids of different uniform densities and zonal velocities intersecting upper and lower horizontal boundaries, was left to Orlanski (1968), who considered all the four different instability modes—Helmholtz instability of vertical shear coupled with gravitational stability, Rayleigh instability of horizontal shear, baroclinic instability, and mixed baroclinic-barotropic-Helmholtz instability. Attempts to explain the long atmospheric waves observed in the troposphere were initiated by the work of J. Bjerknes (1937) to which we have already referred. Mathematical theories for the instability of a baroclinic zonal current with uniform horizontal temperature gradients were presented by Charney (1947), Eady (1949), Fjørtoft (1950), Kuo (1951), Green (1960), Burger (1962), Stone (1966, 1970) and many others—the problem is still being investigated. The stability of a horizontally shearing zonal current in two-dimensional spherical flow was studied by Kuo (1949). The stability of flows with both vertical and horizontal shear was investigated by Stone (1969), McIntyre (1970), Simmons (1974), Gent (1974, 1975) and Killworth (1980). The last named is the most comprehensive. Integral conditions for instability in more or less arbitrary zonal flows were developed by analogy with Rayleigh's condition for two-dimensional parallel flows by Kuo (1951), Charney and Stern (1962), Pedlosky (1964a, 1964b), Bretherton (1966a, 1966b), and others.

On the oceanic side, the onset of meandering of the western boundary currents has been dealt with by Orlanski (1969) and Orlanski and Cox (1973). They conclude that the meandering of the Gulf Stream between Miami and Cape Hatteras can be attributed to baroclinic instability, a result which seems to be in agreement with observations of Webster (1961a). The baroclinic instability of the free Gulf Stream extension implies a northward heat transport by the meanders and cutoff vortices. Evidence for such transports is not conclusive. The discovery of the mid-ocean mesoscale eddies initiated attempts by Gill, Green, and Simmons (1974) and Robinson and McWilliams (1974) to ascertain whether these eddies could be ascribed to baroclinic instabilities of the mid-ocean mean flows. The results have not been encouraging. Studies of the behavior of numerical ocean models also do not support this idea (Harrison and Robinson, 1978). If one merely converts available potential energy to kinetic energy while preserving the total energy density per unit area, the perturbation kinetic energies cannot exceed those of the mean flow, and are therefore too small by an order of magnitude. Only ad hoc energy-convergence mechanisms give the right magnitudes.

Most of the studies referred to above have dealt with the instability of a zonal current with horizontal and/or vertical shear. Realistically, we must also be concerned with the instability of nonzonal and time-dependent flows, including oceanic gyres, forced and free Rossby waves and waves over topography. Thus we need to consider more general basic states.

We begin with the quasi-geostrophic potential vorticity equations (18.33)–(18.35). We attempt to find a basic solution $\bar{\psi}(x,y,z,t)$ and investigate the growth of small perturbations $\psi'(x,y,z,t)$ around this basic state. The most straightforward basic state is a steadily translating (possibly at zero speed) unforced, nondissipative flow field

$$\bar{\psi} = \bar{\psi}(x',y,z), \qquad x' = x - \bar{c}t,$$

which satisfies the equations

$$\nabla^2 \bar{\psi} + \frac{\partial}{\partial z}\frac{1}{S}\frac{\partial}{\partial z}\bar{\psi} + \beta y = P(\bar{\psi} + \bar{c}y, z) \qquad (18.51)$$

and the boundary conditions

$$\bar{\psi}_z(x',y,0) = T_s(\bar{\psi} + \bar{c}y),$$

$$\bar{\psi}_z(x',y,-H) + f_0 S(-H) b(x' + \bar{c}t, y)$$

$$= T_b(\bar{\psi} + \bar{c}y). \qquad (18.52)$$

Clearly such a solution is possible only if $\bar{c}b_x = 0$, that is, if the basic flow is independent of time or if the zonal variation in topography vanishes—waves cannot translate over varying topography without changing amplitude or shape. The basic flow is stationary in the x', y, z system and in this system the pseudopotential vorticity is constant along streamlines.

The derivation of (18.51) and (18.52) may indicate that the restrictions upon the mean flow are quite severe—no forcing or dissipation. However, our subsequent derivations will require only (18.51) and (18.52) and these can hold in much more general conditions. For example, the standard meteorological problem considers the instability of zonal flows forced by heating and perhaps Reynolds stresses and dissipated by radiation and surface Ekman pumping. Since both the mean flow and the potential vorticity are functions only of y, we can still define potential vorticity and surface functionals from (18.51)–(18.52). As long as the forcing and dissipative processes are not significant in the perturbation dynamics, the formalism below will apply. (We warn, however, that when there is topography or lateral boundaries, the stability problem for forced and dissipated flow may be quite different.)

The perturbation streamfunction $\psi' = \psi'(x',y,z,t)$ satisfies

$$\frac{\partial}{\partial t}\left(\nabla^2 + \frac{\partial}{\partial z}\frac{1}{S}\frac{\partial}{\partial z}\right)\psi'$$

$$+ (\bar{\mathbf{v}} - \bar{c}\hat{\mathbf{x}})\cdot\nabla\left(\nabla^2 + \frac{\partial}{\partial z}\frac{1}{S}\frac{\partial}{\partial z} - P'\right)\psi' = 0, \qquad (18.53\text{a})$$

$$P'(\Psi, z) = \frac{\partial}{\partial \Psi} P(\Psi, z),$$

$$\frac{\partial}{\partial t}\psi'_z + (\bar{\mathbf{v}} - \bar{c}\hat{\mathbf{x}})\cdot\nabla\left(\frac{\partial}{\partial z} - T'\right)\psi' = 0, \qquad (18.53\text{b})$$

$z = 0, -H$.

If we examine the normal modes $\psi'(x,y,z,t) = \psi'(x',y,z)e^{\sigma t}$, we have the eigenvalue equation for the growth rate

$$\sigma\left(\nabla^2 + \frac{\partial}{\partial z}\frac{1}{S}\frac{\partial}{\partial z}\right)\psi'$$

$$= -(\bar{\mathbf{v}} - \bar{c}\hat{\mathbf{x}})\cdot\nabla\left(\nabla^2 + \frac{\partial}{\partial z}\frac{1}{S}\frac{\partial}{\partial z} - P'\right)\psi', \qquad (18.54\text{a})$$

with boundary conditions

$$\sigma\psi'_z = -(\bar{\mathbf{v}} - \bar{c}\hat{\mathbf{x}})\cdot\nabla\left(\frac{\partial}{\partial z} - T'_s\right)\psi', \qquad z = 0,$$

$$\sigma\psi'_z = -(\bar{\mathbf{v}} - \bar{c}\hat{\mathbf{x}})\cdot\nabla\left(\frac{\partial}{\partial z} - T'_b\right)\psi', \qquad z = -H. \qquad (18.54\text{b})$$

These equations for the perturbation streamfunction ψ' and the growth rate σ will form the basis for discussion of zonal flow instability and wave instabilities below.

Integral Theorems The classic example of an integral theorem is, of course, the Rayleigh theorem (1880). However, there is a slightly more general theorem, due originally to Arnol'd (1965) and applied to quasi-geostrophic flow by Blumen (1968), which we shall extend here to the problem of traveling disturbances and/or stationary motion over topography. This theorem states that the flow is *stable* if the potential vorticity and buoyancy along the bottom surface increase, and the buoyancy along the top surface decreases, with increasing streamfunction, that is, $P' \geq 0$, $T'_s \leq 0$, $T'_b \geq 0$ everywhere. To prove this, let us assume that $P' > 0$, $T'_s < 0$, and $T'_b > 0$ everywhere. (The cases for $P' = 0$ or $T'_s = 0$ or $T'_b = 0$ everywhere are readily proved.) First, we form an energy equation by multiplying (18.54a) by $-\psi'^*$, volume integrating, adding the conjugate equation, integrating by parts, and applying the boundary conditions. We obtain

$$(\sigma + \sigma^*)\iiint |\nabla\psi'|^2 + \frac{1}{S}|\psi'_z|^2$$

$$= \iiint q'J(\bar{\psi} + \bar{c}y, \psi'^*) + \text{c.c.} + \iint \psi'_z J(\bar{\psi} + \bar{c}y, \psi'^*)$$

$$+ \text{c.c.} \big|_{-H}^{0}. \qquad (18.55)$$

Next, we form a normalized enstrophy equation by multiplying (18.54a) by q'^*/P' (recalling that $P' \neq 0$) and volume integrating to get

$$(\sigma + \sigma^*) \iiint \frac{|q'|^2}{P'}$$
$$= -[\iiint q'^*J(\bar\psi + \bar c y, \psi') + \text{c.c.}]. \qquad 18.56)$$

Applying a similar procedure to the upper and lower boundary conditions, adding the result to (18.55) and (18.56), gives

$$(\sigma + \sigma^*) \iiint \left[|\nabla\psi'|^2 + \frac{1}{S}|\psi'_z|^2 + \frac{1}{P'}|q'|^2\right.$$
$$\left. -\frac{1}{H}\frac{1}{T'_s S(0)}|\psi'_z(0)|^2 + \frac{1}{HT'_b S(-H)}|\psi'_z(-H)|^2\right] = 0. \quad (18.57)$$

For the choice $P' > 0$, $T'_s < 0$, and $T'_b > 0$ the integrand is positive definite, implying that $\text{Re}(\sigma) = 0$, that is, that the flow is stable. When P', T'_s, or T'_b are everywhere zero, the enstrophy or surface-temperature variance equations simply show that $|q'|$ or $|\psi'_z|$ at 0 or $-H = 0$, so that the term contributing to (18.55) can be ignored and therefore will also not enter in (18.57).

This completes the proof of the theorem. From the relation between the potential vorticity and the streamfunction (in the moving coordinate system) and the relation between the surface buoyancies and the streamfunctions at the top and bottom surfaces, we can tell whether the flow is stable or potentially unstable. In some problems (cf. Howard, 1964b; Rosenbluth and Simon, 1964) the necessary criterion for stability has been shown to be sufficient. We should also mention that the normal-mode assumption is not essential, so that the theorem applies to an arbitrary initial disturbance (Blumen, 1968).

In illustration, we note that the theorem implies that the Fofonoff (1954) inertial gyre solution,

$$P(\bar\psi) = \alpha\bar\psi, \qquad T_s(\bar\psi) = T_b(\bar\psi) = 0, \qquad \bar c = 0,$$

where α is a positive constant, is stable, as first pointed out by McWilliams (1977). We could find many other stable gyres by numerical means, including topographical effects, by solving (18.51), (18.52) with arbitrary functionals P and $T_{s,b}$ constrained only to satisfy the proper derivative conditions. The simplest would be to take

$$P(\bar\psi, z) = a(z)\psi + b(z),$$

with $a(z) > 0$ and similar linear functionals for the boundary conditions.

A second example is the flow forced by Gulf Stream meandering described in section 18.5.3. In this case, the potential vorticity equation for the forced wave (the basic state) is

$$\nabla^2\psi + \beta y = P(\bar\psi + \bar c y).$$

The substitutions

$$\bar\psi = \Psi_0 e^{-\nu y} e^{i(kx-\omega t)}$$

and

$$\bar c = \omega/k$$

show that $P(Z) = \beta Z/\bar c$. Therefore, when the forcing propagates eastward, the trapped wave is *stable*. Unlike ordinary propagating Rossby waves, for which $\bar c < 0$, and which Lorenz (1972) has shown to be unstable, forced waves may be stable. We shall consider topographically forced waves in detail in section 18.7.3.

Zonal Flows We now specialize to zonal flows $\bar\psi = -\int^y \bar u(y',z)dy'$, $b_x = 0$. For these, we can readily find P' and $T'_{s,b}$ by taking y derivatives of (18.51) and (18.52):

$$P' = (\beta - \bar u_{yy} - \frac{\partial}{\partial z}\frac{1}{S}\frac{\partial}{\partial z}\bar u)/(\bar c - \bar u),$$

$$T'_s = -\bar u_z/(\bar c - \bar u)|_{z=0},$$

$$T'_b = (f_0 S b_y - \bar u_z)/(\bar c - \bar u)|_{z=-H},$$

where $\bar c$ now is completely arbitrary (i.e., the perturbation wave speed will be simply doppler shifted by $\bar c$). In particular, we can choose $\bar c$ so that $\bar c - \bar u$ has a definite sign. Therefore we see that the flow will be stable if all the three quantities

$$\beta - \bar u_{yy} - \frac{\partial}{\partial z}\frac{1}{S}\frac{\partial \bar u}{\partial z},$$

$$\bar u_z(0),$$

$$f_0 S(-H)b_y - \bar u_z(-H)$$

have the same sign. Thus we recover the generalized Rayleigh theorem for quasigeostrophic flows: the flow is stable if

$$\bar Q_y = \beta - \bar u_{yy} - \frac{\partial}{\partial z}\frac{1}{S}\frac{\partial \bar u}{\partial z} + \frac{\bar u_z}{S}\delta(z)$$
$$+ \left(f_0 b_y - \frac{\bar u_z}{S}\right)\delta(z+H) \qquad (18.58)$$

is uniform in sign (δ is the Dirac delta function). More conventional proofs of this theorem also can be found in Charney and Stern (1962), Pedlosky (1964a), and Bretherton (1966b).

A second standard theorem in shear flow instability theory due to Fjørtoft (1950) can also be generalized to the quasi-geostrophic flow problem. If we suppose that $\bar Q_y$ vanishes along some curve in the (y,z) plane and furthermore that $\bar u = \bar u_c = $ constant on this curve, the flow will be stable if $\bar Q_y(\bar u - \bar u_c)$ is negative everywhere. This can be demonstrated by choosing $\bar c = \bar u_c$. Clearly the requirement that $\bar u = \bar u_c$ at all points where

$$\beta - \bar{u}_{yy} - \frac{\partial}{\partial z}\frac{1}{S}\frac{\partial}{\partial z}\bar{u} = 0$$

is highly restrictive (though it does occur for $\bar{u}_y = 0$ or $\bar{u}_z = 0$ or $\bar{u}_{yy} + (\partial/\partial z)(1/S)(\partial/\partial z)\bar{u} = K\bar{u}$).

As a practical application, we remark that the Rayleigh theorem (18.58) implies that the Eady (1949) problem ($S = $ constant, $\bar{u}_z = $ constant, $\beta = 0$, $\bar{u}_y = 0$) can be stabilized by a sloping topography such that $b_y > \bar{u}_z/f_0 S|_{-H}$. This slope is steeper than the isopycnal slope, so that the density gradient at the bottom becomes opposite in sign to the gradient at the surface.

A second application is to demonstrate the stabilizing effort of β, especially for eastward flows. We consider zonal currents with a barotropic plus a sheared flow with the structure of the flat-bottom first-baroclinic mode

$$\bar{u}(y,z) = \bar{u}_{BT} + \bar{u}_{BC} F_1(z)$$

with $\bar{u}_{BT}$ and $\bar{u}_{BC}$ constants. (Many currents in the ocean do seem to have dominantly first-mode shears.) The Rayleigh criterion becomes

$$\overline{Q}_y = \beta + \frac{\bar{u}_{BC}}{L_R^2} F_1(z) > 0$$

for all z. This can occur only if

$$-\frac{\beta L_R^2}{|F_1(0)|} < \bar{u}_{BC} \equiv \frac{\Delta u}{|F_1(0)| + |F_1(-H)|} < \frac{\beta L_R^2}{|F_1(-H)|},$$

where Δu is the change in velocity from bottom to top. Using our N^2 profile this implies

$$-4 \text{ cm s}^{-1} < \Delta u < 22 \text{ cm s}^{-1}.$$

We see that eastward currents are considerably more stable than westward flows. Gill, Green and Simmons (1974) report on calculations which show weak growth rates for $\Delta\bar{u} \sim -5 \text{ cm s}^{-1}$. Observations of actual $\Delta\bar{u}$'s are not readily available because the midocean density-field measurements are generally contaminated with eddies. However, it is not unlikely that mid-ocean mean currents away from the "recirculation region" of Worthington (1976) (see also chapters 1 and 3) are smaller than this magnitude, so that mid-ocean flows may very possibly be stable (see also McWilliams, 1975).

This result must be viewed with caution, because it is possible for forced meridional currents to be locally unstable for any value of the shear. We can see this by considering the stability of a mean flow

$$\bar{\psi} = \bar{v}(z)x - \bar{u}(z)y,$$

where we ignore the dynamics of the mechanism that supports the $\bar{v}$ component of flow on the grounds that its space and time scales are much larger than those of the perturbations we wish to consider. The perturbations satisfy

$$\frac{\partial}{\partial t} q' + J(\bar{\psi}, q') + J(\psi', \bar{q}) = 0,$$

where

$$\bar{q} = \left(\frac{\partial}{\partial z}\frac{1}{S}\frac{\partial}{\partial z}\bar{v}\right)x + \left(\beta - \frac{\partial}{\partial z}\frac{1}{S}\frac{\partial}{\partial z}\bar{u}\right)y$$

is now *not* expressible as $P(\bar{\psi}, z)$. However, we may consider perturbations of the form

$$\psi' = F(z) \exp[ik(x - ct) + ily]$$

to find

$$\left[(\bar{u} - c) + \frac{1}{k}\bar{v}\right]\left(\frac{\partial}{\partial z}\frac{1}{S}\frac{\partial}{\partial z} - k^2 - l^2\right) F$$
$$+ \left[\beta - \frac{\partial}{\partial z}\frac{1}{S}\frac{\partial}{\partial z}\left(\bar{u} + \frac{l}{k}\bar{v}\right)\right] F = 0.$$

Applying the usual Rayleigh theorem shows that the flow will be stable unless $\beta - (\partial/\partial z)(1/S)(\partial/\partial z)[\bar{u} + (l/k)\bar{v}]$ changes sign. If $\bar{v} \neq 0$, however, a proper choice of l and k (the direction of the perturbation wave) may always be made to ensure satisfying the necessary criterion for instability. Thus arguments about the zonal flow stability may not directly apply to the Sverdrup circulation.

The discussion of baroclinic instability has been extended to finite amplitudes by Lorenz (1962, 1963a) using truncated spectral expansions and by Pedlosky (1970, 1971, 1972, 1976, 1979b), Drazin (1970, 1972), and others using expansion techniques in the vicinity of critical values of the stability parameters. Thus far, the systems dealt with have been more applicable to laboratory models than the actual atmosphere or ocean. A general review has been given by Hart (1979a), who himself has contributed by experiment and analysis to the subject.

18.7.2 Wave–Mean Flow Interactions

The subject of wave–mean flow interaction in the atmosphere has been treated extensively in connection with the manner in which large-scale waves generated in the troposphere propagate vertically into the stratosphere and there interact with the mean flow. One example is the so-called sudden-warming phenomenon, the rapid breakdown of the stratospheric winter circumpolar cyclone accompanied by large-scale warming. Another example is the so-called quasi-biennial oscillation, which has been explained as a wave–mean flow interaction between vertically propagating Rossby-gravity and Kelvin waves and the zonal flow in the equatorial stratosphere (Lindzen and Holton, 1968; Holton and Lindzen, 1972). A vivid experimental and theoretical demonstration of this type of interaction has been given by Plumb and McEwan (1978).

Charney and Drazin (1961) have shown that small-amplitude steady waves in quasi-geostrophic, adiabatic, inviscid flow cannot interact to second order with the zonal flow. If there are no critical surfaces at which the zonal flow vanishes and there is no dissipation, forcing, or transience, no interaction will take place. All are present in the quasi-biennial oscillation and in Plumb and McEwan's model. The result of Charney and Drazin was originally derived by straightforward calculation. It may also be inferred from an independent study of energy transfer in stationary waves by Eliassen and Palm (1960), who derive linear relations between the horizontal Reynolds stress, the horizontal eddy heat flux, and the components of the wave energy flux. These works have been greatly extended by Andrews and McIntyre (1976), Boyd (1976), and Andrews and McIntyre (1978a,b). McIntyre (1980) reviews the subject.

There have been several suggestions of oceanic analogies: Pedlosky (1965b) and N. Phillips (1966b) have argued that westward-propagating Rossby waves can cause acceleration of the western boundary currents. Lighthill (1969) attempted to explain the onset of the Somali Current as due to the interaction of Rossby-gravity waves generated by the monsoon winds in the mid-Indian Ocean with the flow in the vicinity of the East African continent. More recently, experiments of Whitehead (1975) have shown quite clearly that mean flows may be generated by radiated Rossby waves. His work led Rhines (1977) to a theoretical reconsideration of the wave-mean flow generation problem not only when the geostrophic contours (the f/H lines which represent the streamlines for free inertial motions) are closed or periodic but also when the contours are open. Rhines's work is important for understanding large-scale forced motions in oceanic basins.

As an illustration of wave-mean flow interaction in an oceanographic context we shall ask again whether the waves produced by Gulf Stream meandering may be responsible for generating and maintaining the so-called recirculation flow found by Worthington (1976) and others. This flow occurs in a region extending some 1000 km south of the stream and contains (according to Worthington) a sizable westward transport (10^8 m³ s⁻¹). This problem has been addressed by Rhines (1977), who, however, did not consider generation due to eastward-moving waves.

We consider the barotropic flow south of the Gulf Stream forced by the streamfunction $\psi(x,0,t) = A \cos(kx - \omega t)$, as in section 18.5.3, but we now include the effects of bottom Ekman friction and the second-order interaction with the mean zonal flow. The streamfunction satisfies

$$\left(\frac{\partial}{\partial t} + \sigma_{\text{BT}}\right)\nabla^2\psi + \beta\psi_x = -J(\psi, \nabla^2\psi),$$

$$\psi(x,0,t) = A \cos(kx - \omega t),$$

$$\psi \to 0, \quad y \to -\infty.$$

The linear solution (assuming Ak^3/β small) will be

$$\psi = \text{Re}\{A \exp[i(kx - \omega t) + \nu y]\},$$

$$\nu = \sqrt{k^2 + \beta\omega/(\omega^2 + \sigma_{\text{BT}}^2) - i\beta k \sigma_{\text{BT}}/(\omega^2 + \sigma_{\text{BT}}^2)},$$

if the root with positive real part is chosen to satisfy the radiation condition. The nonlinearly forced streamfunction field satisfies

$$\left(\frac{\partial}{\partial t} + \sigma_{\text{BT}}\right)\nabla^2\psi^{(1)} + \beta\psi_x^{(1)} = -2A^2\nu_r^2\nu_i k e^{2\nu_r y}$$

where ν_r and ν_i are the real and imaginary parts of ν, respectively. Its solution is

$$\psi^{(1)} = -\frac{A^2\nu_i k}{2\sigma_{\text{BT}}} e^{2\nu_r y}$$

or

$$\bar{u} = \frac{A^2\nu_r\nu_i k}{\sigma_{\text{BT}}} e^{2\nu_r y}.$$

This is, of course, just the solution to

$$\sigma_{\text{BT}}\bar{u} = -(\overline{u'v'})_y$$

with u' and v' taken from the lowest-order solution.

The mean flow is determined by a balance between friction and Reynolds-stress forcing. The importance of dissipation becomes clear: without friction, ν is either purely real or purely imaginary and $\bar{u} = 0$. With friction, we find that the waves transfer momentum into the mean flow. Moreover, we can show that the magnitude of the flow is not sensitive to the spin-down time $1/\sigma_{\text{BT}}$, as this time becomes very large.

As σ_{BT} becomes small we find

$$\nu \simeq \begin{cases} \sqrt{k^2 + \dfrac{\beta k}{\omega}} - i\beta\sigma_{\text{BT}}k/2\omega^2 \sqrt{k^2 + \dfrac{\beta k}{\omega}}, & \dfrac{\omega}{k} > 0, \\ \qquad \omega/k < -\beta/k^2 \\ -i\sqrt{-k^2 - \dfrac{\beta k}{\omega}} + \beta\sigma_{\text{BT}}k/2\omega^2 \sqrt{k^2 - \dfrac{\beta k}{\omega}}, \\ \qquad -\beta/k^2 < \omega/k < 0. \end{cases}$$

The forced mean flow is therefore

$$\bar{u} = -\frac{\beta A^2 k^2}{2\omega^2} \begin{cases} \exp\left(2\sqrt{k^2 + \dfrac{\beta k}{\omega}}\, y\right), & \dfrac{\omega}{k} > 0, \\ \qquad \omega/k < \beta/k^2 \\ \exp\left(\beta\sigma_{\text{BT}}ky/\omega^2 \sqrt{-k^2 - \dfrac{\beta k}{\omega}}\right), \\ \qquad -\beta/k^2 < \omega/k < 0, \end{cases}$$

with amplitude independent of σ_{BT}. We can estimate the westward current speeds by relating the amplitude A to the excursions of the stream in the y direction:

$$d = -A\frac{k}{\omega}\cos(kx - \omega t) \equiv d_0 \cos(kx - \omega t).$$

The maximum westward currents are $-\frac{1}{2}\beta d_0^2$. Rhines (1977) has derived from more general considerations the result that mean-flow generation is proportional to β times the square of the displacement. For typical excursions of 100–200 km, mean flows of 10–40 cm s^{-1} can be generated. [We should note that, for this problem, the eastward Stokes drift is given by

$$A^2\frac{k}{\omega}\left(k^2 + \frac{\beta k}{\omega}\right)\exp\left(2\sqrt{k^2 + \frac{\beta k}{\omega}}\,y\right),$$

which is larger than the westward Eulerian flow so that the particle drift is *eastward*.]

Observations of Gulf Stream meanders usually indicate eastward-moving disturbances; therefore much of the mean flow will be trapped in a distance one-half that shown in figure 18.12. The disturbances that generate propagating waves $(-\beta/k^2 < \omega/k < 0)$ can produce mean flows over large north–south distances, but there does not seem to be enough amplitude in such disturbances. (See, however, the remarks in section 18.8)

This very simple calculation indicates that eddy radiation from the meandering Gulf Stream can generate a return flow with speeds comparable to those suggested by observations (cf. Worthington, 1976; Wunsch, 1978a; Schmitz, 1977; see also chapter 4). The predicted north–south scale of the region is quite small, however, unless there is considerably more energy in westward-going meanders than has been suggested by Hansen (1970) or by Robinson, Luyten, and Fuglister (1974).

There is another form of wave–mean flow interaction involving overreflection of waves traveling through a variable mean-flow field. Lindzen and Tung (1978) recently have demonstrated that barotropic and baroclinic instabilities may be explained as overreflection phenomena in which Rossby waves impinging upon a critical surface are reflected with a coefficient of reflection greater than unity. The combination of an overreflecting region in the mean flow with a reflecting boundary can lead to a growing disturbance in which the wave picks up energy at each passage into the overreflecting region.

This concept may be directly applicable to the problem of reflection of Rossby waves from the western boundary currents.[6] Numerous examples of Gulf Stream rings interacting with the Gulf Stream without being absorbed can be found in the data presented by Lai and Richardson (1977), and at times they appear to increase in energy as a result of the interaction (Richardson, Cheney, and Mantini, 1977). We suggest the possibility that overreflection may be involved in the dynamics of mesoscale eddies near the western boundary current. Whether or not this is so remains to be seen.

18.7.3 Wave Instability and Form-Drag Instability

The fact that Rossby waves may be unstable was first shown by Lorenz (1972) for a barotropic atmosphere. In a more detailed exploration of the problem Gill (1974) observed that there are two distinct mechanisms for the instability: a resonant triad interaction or a shear instability of the Rayleigh type. Duffy's (1978) and Kim's (1978) baroclinic studies showed that baroclinicity may also cause instability in large-scale waves. As in the instability of zonal flow, the growing baroclinic modes have the scale of the radius of deformation. Jones (1978) and Fu and Flierl (1980) have explored these ideas further as they apply to the ocean.

Wave and Form-Drag Instabilities Just as a freely propagating wave provides variations of potential vorticity which may lead to instability, topography may produce a forced flow whose variations of potential vorticity may also cause instability. Topography may be a destabilizing influence, either because the forced flow is unstable, just as a free wave, to Rayleigh or resonant instabilities, or else because the topography itself may help—via the form drag produced by the perturbation—to extract energy from the mean flow. The latter type of instability was first encountered by Charney and DeVore (1979) in their study of blocking (the persistence of anomalously high pressure in certain regions of the atmosphere) in a barotropic atmosphere. In their model, blocking occurs as an alternative flow equilibrium corresponding to a given forcing of the zonal flow in the presence of sinusoidal topography. It was found that the transition from the normal flow state to the anomalous blocking state takes place via a form-drag instability of an intermediate equilibrium state in which the zonal flow is superresonant. In this superresonant state, a small decrease of the zonal flow amplifies the forced orographic wave and increases the form drag (mountain torque), which in turn decelerates the zonal flow still further. Charney and Straus (1980) extended this study to a two-layer baroclinic atmosphere. Here again there is a form-drag instability. But when there is no lower layer flow, the instability is catalyzed by the form drag: the perturbation derives its energy from the available potential energy of the Hadley circulation generated by thermal forcing; the form drag merely establishes the necessary phase relationships.

The connections between this form-drag instability and the more familiar resonant or Rayleigh instabilities have not been previously explored. For this purpose, consider the simplest wavelike flow of a homogeneous ocean and derive the instability conditions for both a free zonally propagating wave and a topographically forced Rossby wave in order to elucidate the similarities and differences among the respective instability mechanisms. We begin by stating the forms of the potential vorticity functionals $P(\bar{\psi} + \bar{c}y)$ and the resulting perturbation equations. For the Rossby wave, $\bar{c} = -\beta/k^2$, the potential vorticity functional in equation (18.51) is

$$P(Z,z) = (\beta/\bar{c})Z = -k^2 Z$$

with streamfunction

$$\bar{\psi} = A \sin kx.$$

The perturbation equation (18.53a) is particularly simple because P' is now a constant:

$$\sigma \nabla^2 \psi' + J\left(-\frac{\beta}{k^2}y + A \sin kx, (\nabla^2 + k^2)\psi'\right) = 0, \quad (18.59)$$

where A is completely arbitrary.

For barotropic flow over topography of the form $b = b_0 \sin kx$ on the β-plane, the basic state potential vorticity equation

$$\nabla^2 \bar{\psi} + \beta y + \frac{f_0 b_0}{H} \sin kx = P(\bar{\psi})$$

has the solution

$$\bar{\psi} = -\bar{u}y + A \sin kx, \quad A = \frac{\bar{u}f_0 b_0/H}{\bar{u}k^2 - \beta}$$

with the linear potential vorticity functional

$$P(Z) = -\frac{\beta}{\bar{u}} Z \equiv -k_u^2 Z$$

Here k_u is the wavenumber of the stationary wave (k_u^2 may be less than zero). The perturbation equation

$$\sigma \nabla^2 \psi' + J\left[-\frac{\beta}{k_u^2} y + A \sin kx, (\nabla^2 + k_u^2)\psi'\right]$$
$$= 0 \quad (18.60)$$

is very similar in form to (18.59) except that k and k_u are now independent; however, the amplitude A for the topographical problem is determined by

$$A = \frac{f_0 b_0/H}{k^2 - k_u^2}.$$

Obviously we need only to solve (18.60) and find $\sigma(\beta, A, k, k_u)$; we can then identify both the free ($k_u = k$) and forced ($k_u \neq k$) regimes. In actuality the task is even simpler since dimensional considerations show that there are only two parameters, $M = Ak^3/\beta$ and $\mu = k_u/k$, that must be varied while computing the nondimensional growth rate σ/Ak^2.

First, however, we demonstrate that only the $k_u^2 > 0$ case need be considered, since the flow is stable for $k_u^2 < 0$. This follows readily from (18.57):

$$(\sigma + \sigma^*)\left(\iint |\nabla \psi'|^2 - \frac{1}{k_u^2} |\nabla^2 \psi'|^2\right) = 0,$$

which implies that Rossby waves $k_u^2 = k^2 > 0$ and waves forced by eastward flow may be unstable, while waves forced by westward flow $k_u^2 < 0$ are definitely stable. Again, as in section 18.7.1 we see that stable waves are generated when the relative motion of the forced wave with respect to the ambient flow is eastward. In addition, for eastward flow over topography or westward-propagating free Rossby waves, a necessary condition for instability is that the perturbation have components with scales larger than k_u^{-1} and components with scales smaller than k_u^{-1}. This follows from Fourier analyzing ψ' and substituting in (18.57) to get

$$\iint d\mathbf{s} \, |\mathbf{s}|^2 |\hat{\psi}'(\mathbf{s})|^2 \left[1 - \frac{|\mathbf{s}|^2}{k_u^2}\right] = 0,$$

which implies that $|\hat{\psi}'(\mathbf{s})|$ must be nonzero both for some values of $|\mathbf{s}|^2$ larger than k_u^2 and some values smaller than k_u^2.[7] This is the perturbation form of Fjørtoft's (1953) result on energy cascades.

We could readily solve the stability problem (18.60) using the mathematical techniques of Lorenz (1972), Gill (1974), Coaker (1977), or Mied (1978) to investigate growth rates. Alternatively we could discuss the limiting behavior—the Rayleigh limit for $M = Ak^3/\beta \gg 1$, the resonant interaction limit for $M \ll 1$, and the form-drag instability—by separate approximations. We shall do this for the last-named problem because it represents a relatively unfamiliar phenomenon. However, to ascertain most clearly the connections between the various types of behavior, it is most simple to employ the Fourier expansion

$$\psi' = e^{il_0 y}[A_0 e^{ik_0 x} + A_{-1} e^{i(k_0 - k)x} + A_1 e^{i(k_0 + k)x} + \cdots]$$

and truncate to the three indicated terms (cf. Gill, 1974). The resulting dispersion relation becomes just

$$(\omega K_{-1}^2 + B_{-1})(\omega K_0^2 + B_0)(\omega K_1^2 + B_1)$$
$$= \frac{l_0^2 A^2 k^2}{4} \{k_u^2 - K_0^2)[(k_u^2 - K_{-1}^2)(\omega K_1^2 + B_1)$$
$$+ (k_u^2 - K_1^2)(\omega K_{-1}^2 + B_{-1})]\} \quad (18.61)$$

using the notation $\omega = i\sigma$, $K_n^2 = (k_0 + nk)^2 + l_0^2$, $B_n = \beta(k_0 + nk)(k_u^2 - K_n^2)/k_u^2$. For Rossby waves we simply replace k_u by k.

We have sketched the dependence of $\hat{\sigma} = \sigma/Ak^2$ upon k_0 and l_0 for various M and μ values in figure 18.14. When $M \gg 1$, there is a broad area of (k_0, l_0) space in which the growth rates are real. The maximum occurs for $k_0 = 0$ (this is really a form of Squire's theorem) and

NORMALIZED GROWTH RATE

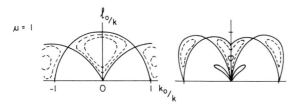

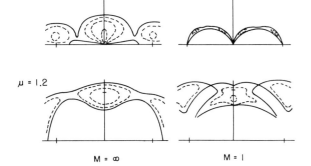

Figure 18.14 Growth rate divided by Ak^2 as a function of k_0, l_0 for various ratios of the stationary wavenumber to the topographic wavenumber μ and wave steepnesses M. The solid lines are the zero contour. The dashed contours, separated by the interval 0.1, correspond to positive growth rates.

the instability equation is similar to the barotropic instability equation. As M decreases the instability ($\hat{\sigma}$ real) becomes restricted more and more to a band around the frequency resonance line (see below). When μ decreases from 1 to a quantity smaller than 1 ($\bar{u} > \beta/k^2$), the point of maximum growth rate for $M = \infty$ moves to smaller wavenumbers l_0. For finite M, the same thing occurs—the largest growth rate moves towards $k_0 = l_0 = 0$ as μ decreases. On the other hand, when μ increases from 1 the forced wave becomes stable for small k_0 and l_0, even for small M, and the wavenumber for maximum growth migrates to smaller scales.

We are thus led to consider the three limits for (18.61):

Strong waves (A or M large): When A is large, the frequency ω is order A and we can neglect all of the B's to find

$$\sigma = \frac{A|l_0 k|}{2}\sqrt{\frac{-(k_u^2 - K_0^2)[K_1^2(k_u^2 - K_{-1}^2) + K_{-1}^2(k_u^2 - K_1^2)]}{K_0^2 K_1^2 K_{-1}^2}}.$$

Clearly the flow will be unstable for $K_0^2 < k_u^2 < K_1^2, K_{-1}^2$; this will always be possible by proper choice of l_0 and k_0. This is just the Rayleigh instability of the shear flow corresponding to the wave field. The maximum growth rate occurs at $k_0 = 0$:

$$l_0 = \begin{cases} (-1 + \sqrt{\mu^2 + \mu^4})^{1/2}, & \mu > \left(\frac{\sqrt{5}-1}{2}\right)^{1/2} \\ 0, & \mu < \left(\frac{\sqrt{5}-1}{2}\right)^{1/2}. \end{cases} \quad (18.62)$$

However, in the topographic case, A can be large not only for very strong topography ($f_0 b_0 k/H \gg \beta$), but also because the zonal flow is nearly critical. In the latter case $\mu \simeq 1$, so that the free and forced wave instabilities are indistinguishable. If the flow is not critical, but rather is forced by strong topography, the formula shows the maximum growth rate occurs at $k_0 \to 0$, $l_0 \to 0$.

Weak waves (A very small or M small): Here the critical condition is that two of the roots of the left-hand side of (18.61)—for example, $-B_0/K_0^2$ and $-B_1/K_1^2$—coalesce. This resonance condition

$$B_0/K_0^2 = B_1/K_1^2$$

or

$$\frac{k_0(k_u^2 - K_0^2)}{k_u^2 K_0^2} = \frac{(k_0 + k)(k_u^2 - K_1^2)}{k_u^2 K_1^2}$$

permits the order A part of the frequency to be complex since

$$K_0^2 K_1^2 (\omega + B_0/K_0^2)^2 \simeq \frac{l_0^2 A^2 k^2}{4}(k_u^2 - K_0^2)(k_u^2 - K_1^2)$$

has a complex root for $K_0^2 < k_u^2 < K_1^2$. The growth rate is

$$\sigma = \frac{A|l_0 k|}{2}\sqrt{\frac{(k_u^2 - K_0^2)(K_1^2 - k_u^2)}{K_0^2 K_1^2}}.$$

However, the resonance condition must also hold. We can relate this to the more familiar wave-resonance conditions by defining the intrinsic frequencies of the two components of the perturbation and also that of the mean flow by

$$\hat{\omega}_0 = -\beta k_0/K_0^2,$$
$$\hat{\omega}_1 = -\beta(k_0 + k)/K_1^2,$$
$$\hat{\omega} = -\beta k/k_u^2 = -\bar{u} k.$$

The resonance conditions

$$(k_0 + k, l_0) = (k_0, l_0) + (k, 0),$$
$$\hat{\omega}_1 = \hat{\omega}_0 + \hat{\omega}$$

both hold. We show the resonance curves and growth rates for various values of k_u/k in figure 18.15.

For large μ (small mean flow speed), the resonant interaction (except for large l_0) is really a triplet interaction between the two perturbation waves and the topography itself (rather than the forced wave). This is, of course, a stable situation. As μ becomes less than

536
Jule G. Charney and Glenn R. Flierl

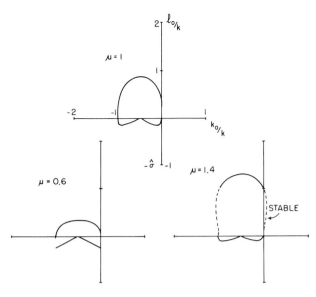

Figure 18.15 Curves above x axis show relation between k_0 and l_0 required by resonance condition. Curves below axis are plots of $-\hat{\sigma}$, showing the dependence of the growth rate upon k_0.

about 0.9, however, the maximum growth rate again occurs as $k_0 \to 0$ and $l_0 \to 0$ [$l_0 \sim (\mu^2 k_0/1 - \mu^2)^{1/2}$].

Form-drag instability: Thus in either case we are led to consider what will be shown to be a form-drag instability—the nonzero growth rates occurring at small k_0 and l_0—when $\mu < 0.79$. There is some difficulty here since the origin is a singularity for M finite; this problem would be eliminated in a bounded geometry. For convenience, we will take the limits $k_0 \to 0$ first and then $l_0 \to 0$ since this case has a simple physical interpretation. Applying these limits to (18.61) gives the frequency:

$$\omega^2 = \left[\frac{\beta}{k} \frac{k_u^2 - k^2}{k_u^2}\right]^2 + \frac{A^2 k_u^2}{2}(k_u^2 - k^2). \qquad (18.63)$$

The flow is unstable when the right-hand side is negative, which cannot occur for Rossby waves $k_u = k$, but may occur for topographic waves when $k_u < k$ or $\bar{u}$ is greater than the critical speed β/k^2. In fact, the range is

$$\beta/k^2 < \bar{u} < \beta/k^2 \left(1 + \left[\frac{1}{2}\left(\frac{f_0 b_0}{H}\right)^2 \frac{k^2}{\beta^2}\right]^{1/3}\right).$$

(The resonant triad instability for $\mu < 1$ does not appear here; rather, the limit $k_0 \to 0^-$ and $l_0 \sim (-k_0)^{1/2} \to 0^+$ must be used.) So far we have looked at the mathematics; let us now discuss the physics of this instability and also show that the truncation to three terms is valid.

The form-drag instability involves one component A_0 which has very large x and y scales and two components with the same scale as the topography. Examination of the individual amplitudes shows that $A_0 \sim 1/l_0$, so that $A_0 e^{il_0 y}$ contributes a term in the perturbation streamfunction which is proportional to y—a modification of the zonal mean flow. This suggests an alternative approach, which is to consider the zonal x-averaged momentum equation and the equation for the deviations. We begin with the quasi-geostrophic equations for a homogeneous fluid,

$$f_0 v = p_x,$$

$$f_0 u = -p_y,$$

$$u_t + (uu)_x + (uv)_y - \beta y v - f_0 v^{(1)} = -p_x^{(1)},$$

$$v_t + (uv)_x + (vv)_y + \beta y u + f_0 u^{(1)} = -p_y^{(1)},$$

$$u_x^{(1)} + v_y^{(1)} - \frac{1}{H}[(ub)_x + (vb)_y] = 0,$$

and consider the zonally averaged equations

$$\langle v \rangle = 0,$$

$$\langle u \rangle_t + \langle uv \rangle_y = f_0 \langle v^{(1)} \rangle,$$

$$\langle v^{(1)} \rangle_y = \frac{1}{H} \langle vb \rangle_y.$$

If the topography vanishes at some y far from the region of interest [following the arguments suggested by Hart (1979b), who showed that the Charney-DeVore truncated spectral problem was identical to that of forced flow over topography varying slowly with y], we can integrate the last equation to find

$$\langle u \rangle_t + \langle uv \rangle_y = \frac{f_0}{H} \langle vb \rangle.$$

The vorticity equation can be used to find the x-dependent part of the flow. In particular, if we assume the y scale is very large, we can drop all y derivatives to get two coupled equations:

$$\langle u \rangle_t = \frac{f_0}{H} \langle vb \rangle,$$

$$v_{xt} + \langle u \rangle v_{xx} + \beta v = -\frac{f_0}{H} \langle u \rangle b_x.$$

For the topography $b = b_0 \sin kx$, we have a steady solution

$$\langle u \rangle = \bar{u},$$

$$v = Ak \cos kx.$$

The deviations from this state satisfy

$$\langle u \rangle_t' = \frac{1}{2} \frac{f_0 b_0}{H} \langle v' \sin kx \rangle,$$

$$v_{xt}' + \bar{u} v_{xx}' + \beta v' = k\left(Ak^2 - \frac{f_0 b_0}{H}\right) \langle u \rangle' \cos kx$$

$$= k \frac{f_0 b_0}{H} \frac{\beta}{\bar{u} k^2 - \beta} \langle u \rangle' \cos kx,$$

which may be solved explicitly to give the dispersion relation (18.63). Here too one sees that it is the coupling between the change in the zonal flow induced by the wave drag and the change in the waves due to changes in zonal flow which leads to the instability. If we decrease the mean flow for a supercritical case (i.e., if we take $\langle u \rangle'$ to be negative), we produce low vorticity on the upwind slopes of the topography and high vorticity on the lee slopes. Associated with this vorticity change is high pressure on the upslope side of the mountains and low pressure on the downslope. This pressure pushes eastward on the topography so that the topography pushes westward on the fluid and decelerates the mean flow still further.

Flow in the Presence of Topography The previous section has described the influence of wavy topography upon the stability of the flows that go over it. However, there also exists topography that does not alter the mean-flow structure, either because the mean current is parallel to the topographic contours or because the currents occur only at levels above the peaks of the topography. In this section, we shall show that the stability of a parallel mean flow in the presence of topography can be quite different from that of the identical mean flow in a flat-bottomed ocean. We have been guided by the result of Charney and Straus (1980), who show that the form-drag instability can catalyze the release of available potential energy in a baroclinic shearing flow that would be stable in the absence of topography. In their study of multiple equilibria and stability in forced baroclinic flow over topography, they found that form-drag instability may occur for weaker thermal driving than conventional baroclinic instability, and that this type of instability leads to transition from one finite-amplitude, quasi-stationary equilibrium state to another. Baroclinic and barotropic instabilities of the stationary topographically perturbed flows give rise to westward-propagating, vacillating wave motions with periods of the order of 5 to 15 days. They suggest that the form-drag instability leads to transition from one stationary regime to another and that the observed westward- and eastward-propagating long planetary waves (zonal wavenumbers 1–4) are the propagating instabilities associated with these stationary regimes.

The simplest and most obvious example of the destabilizing effect of topography is the case of zonal barotropic flow with meridionally varying topography. The topography alters the effective value of β and thereby the growth rates and stability criteria: even though the energy source remains the horizontal shear, the topography can alter the possibility of extracting this energy. In particular, the Charney–Stern necessary criterion for instability [that $\beta - \bar{u}_{yy} + (f_0/H) b_y$ must change sign in the domain] suggests that instability may occur for lower values of shear when $b_y \neq 0$. The necessary condition may, of course, not be sufficient; in particular, when $\bar{u}_y = 0$, the flow will be stable even if $\beta + (f_0/H)b_y$ changes sign. However, in the case of sinusoidal $\bar{u}(y)$ and $b(y)$, the necessary condition seems also to be sufficient (using a simple truncated expansion in y), and the topography does destabilize the flow.

DeSzoeke (1975) discussed baroclinic flow over meridionally varying topography and found that the topography destabilizes the flow at some wavenumbers by a resonant instability involving two baroclinic waves which happen to travel at the same speed. Similar effects can be identified in the work of Durney (1977). We would like to focus our discussion, however, on the specific problem of destabilization by form-drag instability of a baroclinic flow which is neutrally stable in the absence of topography.

We shall consider the conventional two-layer model whose governing equations (cf. Pedlosky, 1979b) are

$$\left(\frac{\partial}{\partial t} + \mathbf{v}_1 \cdot \nabla\right)$$

$$\times \left[\nabla^2 \psi_1 - \frac{\lambda_1^2}{1+\delta}(\psi_1 - \psi_2) + \beta y\right] = 0,$$

$$\left(\frac{\partial}{\partial t} + \mathbf{v}_2 \cdot \nabla\right)$$

$$\times \left[\nabla^2 \psi_2 - \frac{\delta\lambda_1^2}{1+\delta}(\psi_2 - \psi_1) + \beta y + \frac{f_0}{H}(1+\delta)b\right] = 0,$$

where $\mathbf{v}_1$ and ψ_1 are the velocity and streamfunction in the upper layer and $\mathbf{v}_2$ and ψ_2 the corresponding quantities in the lower layer, δ is the ratio of the upper to the lower layer mean depths, and λ_1^{-1} is the layered version of the first baroclinic mode deformation radius $\lambda_1^2 = f_0(1+\delta)^2/g(\Delta\rho/\rho)H\delta$. We write the x-averaged equations

$$\frac{\partial}{\partial t}\left[-\bar{u}_{1yy} + \frac{\lambda^2}{1+\delta}(\bar{u}_1 - \bar{u}_2)\right]$$

$$+ \frac{\partial^2}{\partial y^2}\overline{\psi'_{1x}\left(\frac{\partial^2}{\partial y^2}\psi'_1 + \frac{\lambda^2}{1+\delta}\psi'_2\right)} = 0,$$

$$\frac{\partial}{\partial t}\left[-\bar{u}_{2yy} + \frac{\delta\lambda^2}{1+\delta}(\bar{u}_2 - \bar{u}_1)\right]$$

$$+ \frac{\partial^2}{\partial y^2}\overline{\psi'_{2x}\left(\frac{\partial^2}{\partial y^2}\psi'_2 + \frac{\delta\lambda^2}{1+\delta}\psi'_1\right)} = -\frac{f_0}{H}(1+\delta)\frac{\partial^2}{\partial y^2}\overline{\psi'_{2x}b},$$

and the equations for the deviations

$$\left(\frac{\partial}{\partial t} + \bar{u}_1\frac{\partial}{\partial x}\right)\left[\nabla^2\psi'_1 - \frac{\lambda^2}{1+\delta}(\psi'_1 - \psi'_2)\right]$$

$$+ \left[\beta + \frac{\lambda^2}{1+\delta}(\bar{u}_1 - \bar{u}_2)\right]\psi'_{1x}$$

$$+ J\left[\psi_1', \nabla^2\psi_1' - \frac{\lambda^2}{1+\delta}(\psi_2' - \psi_1')\right]$$

$$- \frac{\partial}{\partial y}\overline{\psi_{1x}'(\psi_{1yy}' + \frac{\lambda^2}{1+\delta}\psi_2')} = 0,$$

$$\left(\frac{\partial}{\partial t} + \bar{u}_2\frac{\partial}{\partial x}\right)\left[\nabla^2\psi_2' - \frac{\delta\lambda^2}{1+\delta}(\psi_2' - \psi_1')\right]$$

$$+ \frac{f_0}{H}(1+\delta)b\right] + \left[\beta - \frac{\delta\lambda^2}{1+\delta}(\bar{u}_1 - \bar{u}_2)\right]\psi_{2x}'$$

$$+ J\left[\psi_2', \nabla^2\psi_2' - \frac{\delta\lambda^2}{1+\delta}(\psi_2' - \psi_1') + \frac{f_0}{H}(1+\delta)b\right]$$

$$- \frac{\partial}{\partial y}\overline{\psi_{2x}'\left[\psi_{2yy}' + \frac{\delta\lambda^2}{1+\delta}\psi_1' + \frac{f_0}{H}(1+\delta)b\right]} = 0.$$

If we now consider y scales that are order $1/\Delta$ of the x scales or the deformation scale and expand $\bar{u}_i = \bar{u}_i^{(0)} + \Delta^2\bar{u}_i^{(1)} + \cdots$, $\psi_i'(x,y) = \psi_i'^{(0)}(x) + \Delta^2\psi_i'^{(1)}(x,y) + \cdots$ (where the topography is assumed to vary only in x) we find

$$\bar{u}_{1t}^{(0)} = \bar{u}_{2t}^{(0)},$$

so that the induced changes in mean flow are barotropic. This occurs because the form-drag forcing of the mean is at much larger scales than the deformation scale. Eliminating the $\bar{u}_i^{(1)}$ terms at second order from the two mean flow equations, noting that the Reynolds stresses drop out because $\psi_i'^{(0)}$ is independent of y, we find that the barotropic component of the zonal flow is accelerated or decelerated by the form drag,

$$\frac{\partial \bar{u}_2}{\partial t} = \frac{f_0}{H}\overline{\psi_{2x}'b'}, \qquad (18.64)$$

while the deviation fields are given by

$$\left[\frac{\partial}{\partial t} + (\bar{u}_2(t) + \Delta u)\frac{\partial}{\partial x}\right]\left[\psi_{1xx}' - \frac{\lambda^2}{1+\delta}(\psi_1' - \psi_2')\right]$$

$$+ \left(\beta + \frac{\lambda^2\Delta u}{1+\delta}\right)\psi_{1x}' = 0, \qquad (18.65)$$

$$\left[\frac{\partial}{\partial t} + \bar{u}_2\frac{\partial}{\partial x}\right]\left[\psi_{2xx}' - \frac{\delta\lambda^2}{1+\delta}(\psi_2' - \psi_1')\right]$$

$$+ \left(\beta - \frac{\delta\lambda^2\Delta u}{1+\delta}\right)\psi_{2x}' = -\frac{f_0}{H}(1+\delta)\bar{u}_2 b_x. \qquad (18.66)$$

Here we have dropped the superscript (0) and introduced the notation Δu for the *time-independent* shear across the interface.

From the equations (18.64)–(18.66) one could derive a single nonlinear governing equation for $\bar{u}_2$ and determine the linear instability and finite-amplitude evolution of the flow. For our purposes, however, it will be sufficient to demonstrate that the initial state $\bar{u}_2 = 0$, $\psi_1' = 0$, $\psi_2' = 0$ can be unstable in the presence of weak topography to infinitesimal perturbations even when the flow would be baroclinically stable in the absence of topography. We can do this by considering the stability in the special case $\Delta u = \beta(1+\delta)/\delta\lambda^2$. This is the maximum shear for which the Rayleigh necessary criterion for stability in the absence of topography,

$$\left(\beta - \frac{\delta\lambda^2}{1+\delta}\Delta u\right)\left(\beta + \frac{\lambda^2}{1+\delta}\Delta u\right) \geq 0,$$

is satisfied. Equations (18.14)–(18.16) simplify to the set

$$\frac{\partial \bar{u}_2}{\partial t} = \frac{f_0}{H}\overline{\psi_{2x}'b},$$

$$\left[\frac{\partial}{\partial t} + \frac{\beta(1+\delta)}{\delta\lambda_1^2}\frac{\partial}{\partial x}\right]\left[\psi_{1xx}' - \frac{\lambda_1^2}{1+\delta}(\psi_1' - \psi_2')\right]$$

$$+ \beta\left(\frac{1+\delta}{\delta}\right)\psi_{1x}' = 0,$$

$$\frac{\partial}{\partial t}\left[\psi_{2xx}' - \frac{\delta\lambda_1^2}{1+\delta}(\psi_2' - \psi_1')\right] = -\frac{f_0}{H}(1+\delta)\bar{u}_2 b_x.$$

We split the streamfunction into sine and cosine parts as in section 18.7.3 and solve this system of equations to find the growth rate equation

$$\sigma^4\left(\frac{\delta}{1+\delta}\right)^2 k^2(k^2+\lambda^2)^2$$

$$+ \sigma^2\left[\frac{\beta^2}{\lambda^4}\left(k^4 - \frac{\delta}{1+\delta}\lambda^4\right)^2\right.$$

$$\left. + \frac{1}{2}\left(\frac{f_0 b_0}{H}\right)^2\frac{\delta^2}{1+\delta}k^2\left(k^2+\frac{\lambda^2}{1+\delta}\right)(k^2+\lambda^2)\right]$$

$$+ \frac{1+\delta}{2}\frac{\beta^2}{\lambda^2}\left(\frac{f_0 b_0}{H}\right)^2 k^2\left(k^2 - \frac{\delta}{1+\delta}\lambda^2\right)\left(k^4 - \frac{\delta}{1+\delta}\lambda^4\right)$$

$$= 0. \qquad (18.67)$$

The real solutions to (18.67) for several values of $f_0 b_0 \lambda/\beta H$ are shown in figure 18.16. Notice the instability occurring for topographic scales on the order of 70 to 110 km, with growth rates proportional to the topographic height (for small heights, at least). We have thus demonstrated that the available potential energy in the flow can be tapped by the orographic instability even in situations where normal baroclinic instability is unable to extract mean-flow energy. Thus it is possible that mesoscale topography plays a role in catalyzing the conversion of mean flow potential to eddy energy in the oceans.

18.7.4 Multiple Equilibria

We have already mentioned the work of Charney and DeVore (1979) and Charney and Straus (1980), who have begun to explore the possibility that the atmosphere may possess a multiplicity of steady equilibrium

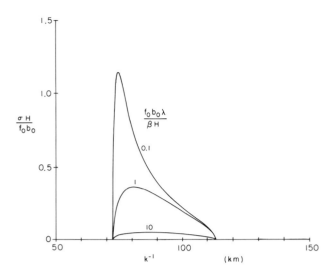

Figure 18.16 Normalized growth rates for a topographically destabilized vertical shear flow. The curves are labeled by $f_0 b_0 \lambda / \beta H$ values.

states for given external forcing in the presence of topographic inhomogeneities. In the case of sinusoidal topography in a periodic channel, they have found states resembling both the "normal" configuration in which there is a strong zonal flow and a relatively weak wave perturbation, and the "blocking" configuration in which there is a weak zonal flow and a relatively strong wave perturbation. They suggest that the blocking phenomenon is an equilibrium state which occurs by a transition via a form-drag instability from the normal to the anomalous blocking configuration. Hart (1979b) has applied similar ideas to laboratory flows and has succeeded in producing stationary multiple equilibria experimentally.

Oceanically, one phenomenon that stands out as a possible example of multiple quasi-stable equilibrium states is the large meander of the Kuroshio which sometimes occurs. Figure 18.17 shows the two quasi-stable configurations that are observed. The transitions between these configurations occur relatively rapidly. White and McCreary (1976) have considered a model for the meandering process involving flow around bumps in the Japanese coastline. Because their discussion was in terms of linear dynamics, Solomon (1978) has rightly pointed out that the model must have a smooth transition between the two states as the independent variable (the maximum inlet flow speed) varies. If, however, the phenomenon is nonlinear, catastrophic changes in the state of the Kuroshio may occur: an infinitely small change in parameters may produce a finite change in response, and several stable responses may be possible for the same set of parameters.

We propose a simple model of this process consisting of the steady, nonlinear flow of barotropic current on a β-plane along a variable coastline (see figure 18.18). Let the latitude of the coastline be $h(x)$ and let η be the north–south distance from the coastline. The potential vorticity equation becomes

$$\left[\left(\frac{\partial}{\partial x} - h_x \frac{\partial}{\partial \eta} \right)^2 + \frac{\partial^2}{\partial \eta^2} \right] \psi + \beta(\eta + h) = F(\psi).$$

If we split the streamfunction into an upstream part $(x \to -\infty, h \to 0)$ $\bar{\psi}(\eta)$ and a topographically induced part $\phi(x, \eta)$ we find

$$\left[\left(\frac{\partial}{\partial x} - h_x \frac{\partial}{\partial \eta} \right)^2 + \frac{\partial^2}{\partial \eta^2} \right] \phi = F(\bar{\psi} + \phi) - F(\bar{\psi})$$

$$- \beta h - \bar{\psi}_{\eta\eta} h_x^2, \quad (18.68)$$

where

$$F(\bar{\psi}(\eta)) = \beta \eta + \frac{\partial^2 \bar{\psi}}{\partial \eta^2},$$

$$\phi \to 0 \quad \text{for } \eta = 0, \quad \eta \to -\infty. \quad (18.69)$$

When $\bar{u} = -\partial \bar{\psi} / \partial \eta$ is not constant, equation (18.69) implies that F is a nonlinear functional, so that (18.68) becomes essentially a forced nonlinear oscillator equation; it is well known that such equations may have multiple stable solutions. We note also a similarity between the equations here and the equations for flow of a barotropic fluid over topography. In the derivation below we assume that the coastline variations are small and occur on scales large compared to the cross-stream scale. We shall show that the nonlinearity plays an important role in determining the amplitude of the nonzonal flow component when the upstream flow is near the critical speed U_c. This speed is defined by the condition that long waves (x wavelength large compared to the width of the current) are stationary. Near critical speeds, the amplitude becomes large. The lowest-order dynamic equation only determines the cross-stream wave structure. The first-order equation shows a balance between advection by the mean flow, effects of the coastline variations, dispersion, and nonlinearity.

We shall work with the nondimensional forms of (18.68)–(18.69). Our scaling is guided by the versions

$$\left(\frac{\partial^2}{\partial x^2} + \frac{\partial^2}{\partial \eta^2} \right) \phi = -\beta h + F'(\bar{\psi}) \phi$$

$$= -\beta h - \frac{\beta - \bar{u}_{\eta\eta}}{\bar{u}} \phi \quad (18.70)$$

obtained by linearizing in ϕ and h. We obtain $F'(\bar{\psi})$ by differentiating (18.69). If $h = h_0 \cos kx$, resonance occurs when

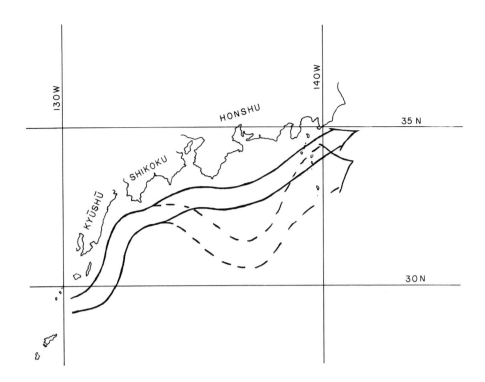

Figure 18.17 Sketch of two equilibrium positions of the Kuroshio. See Taft (1972) and White and McCreary (1976) for detailed tracks.

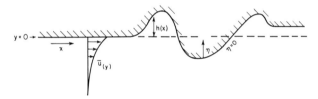

Figure 18.18 Model for coastline induced meandering. The deviation of the coastline from a latitude circle is denoted $h(x)$; the coordinate η is equal to $y - h(x)$. The upstream flow is $\bar{u}(y)$.

$$\left(\frac{\partial^2}{\partial \eta^2} + \frac{\beta - \bar{u}_{\eta\eta}}{\bar{u}}\right)\phi = k^2\phi,$$
$$\phi = 0, \quad \eta = 0, -\infty. \tag{18.71}$$

For forcing on a scale long compared to the width of the current ($|\partial/\partial x| \ll |\partial/\partial \eta|$) we expect that one of the underlined terms in (18.70) will balance the forcing from the side-wall variations $-\beta h$ giving $\phi \sim Uh_0$ or $\phi \sim \beta h_0 l^2$, where U is the scale of $\bar{u}$ and l is the cross-stream scale. When the flow profile is nearly critical for long waves—meaning that the left-hand side of (18.71) vanishes for some nonzero function ϕ which also satisfies the boundary conditions, the long-wave solutions of (18.70) have the two underlined terms nearly canceling, so that the forcing must be balanced by the ϕ_{xx} term. This gives a scale of $\phi \sim \beta h_0 L^2$, where L is the downstream scale of variation of the topography.

Therefore, we scale x by L, h by h_0, ϕ by $\beta h_0 L^2$, η by l, and $\bar{u}$ by U in (18.68)–(18.69) to find

$$\left[\frac{\partial^2}{\partial \eta^2} + \delta\left(\frac{\partial}{\partial x} - \gamma\delta^2 h_x \frac{\partial}{\partial \eta}\right)^2\right]\phi$$
$$= -\delta h + \frac{1}{\gamma\delta}\left[F\left(\bar{\psi} + \frac{\gamma\delta}{M}\phi\right) - F(\bar{\psi})\right] - \gamma M \delta^4 h_x^2 \bar{\psi}_{\eta\eta},$$

$$F(\bar{\psi}(\eta)) = \eta + M\bar{\psi}_{\eta\eta},$$

with $M = U/\beta l^2$, $\gamma = h_0 L^4/l^5$, and $\delta = l^2/L^2$. If we assume that the width of the current is small compared to the downstream scale ($\delta \ll 1$) and that the variations in coastline are weak enough so that $\gamma \lesssim 1$, we can simplify to

Oceanic Analogues of Atmospheric Motions

$$\left(\frac{\partial^2}{\partial \eta^2} + \delta \frac{\partial^2}{\partial x^2}\right)\phi = -\delta h + F'(\bar\psi)\frac{\phi}{M}$$
$$+ \frac{1}{2}F''(\bar\psi)\frac{\gamma\delta\phi^2}{M^2} + O(\delta^2) \quad (18.72)$$

$$\phi \to 0, \quad \eta = 0, -\infty,$$

with

$$F'(\bar\psi) = -\frac{1}{\bar u}(1 - M\bar u_{\eta\eta}),$$
$$F''(\bar\psi) = \frac{1}{\bar u}\frac{\partial}{\partial \eta}\left[\frac{1}{\bar u}(1 - M\bar u_{\eta\eta})\right], \quad (18.73)$$

which are known functions of η given the specification of the upstream $(x \to -\infty)$ flow $\bar u(\eta)$.

We assume that the flow is nearly critical so that $U = U_c(1 + \Delta)$ where U_c is the critical speed (defined exactly below) and therefore $M = M_c(1 + \Delta)$. We expand (18.72)–(18.73) assuming $\Delta \sim \delta$ and $M_c \sim 1$, $\gamma \leq 1$, and find to lowest order

$$\frac{\partial^2}{\partial \eta^2}\phi = \frac{\bar u_{\eta\eta} - M_c^{-1}}{\bar u}\phi,$$
$$\phi = 0, \quad \eta = 0, -\infty, \quad (18.74)$$

which defines the eigenvalue M_c and thus the critical speed U_c given the shape of the upstream flow. The η structure of ϕ must be an eigenfunction G of (18.74), $\phi = f(x)G(\eta)$. At next order in Δ and δ, the solvability condition for (18.72) gives

$$\left[\int_{-\infty}^0 d\eta\, G^2(\eta)\right] f''(x)$$
$$= -h(x)\left[\int_{-\infty}^0 d\eta\, G(\eta)\right] + \frac{\Delta}{\delta M_c}\left[\int_{-\infty}^0 G^2(\eta)/\bar u(\eta)\right] f(x)$$
$$+ \frac{1}{2}\left[\int_{-\infty}^0 d\eta\, G^3(\eta)\frac{1}{\bar u}\frac{\partial}{\partial \eta}\frac{1}{\bar u}(1 - M_c\bar u_{\eta\eta})\right]\frac{\gamma}{M_c^2}f^2(x).$$

This ordinary differential equation for the x structure of the wave $f(x)$ is to be solved for a particular form as Δ/δ and γ vary. For convenience we shall normalize G and redefine parameters slightly to write

$$f'' - \hat\Delta f + \hat\gamma f^2 = -h(x). \quad (18.75)$$

The simplest problem to illustrate the characteristics of (18.75) is the linear case with $h(x) = \cos x$. (This topography extends to $x = -\infty$, which is not really consistent with our original model: however, it does point out some of the properties of these nonlinear flows.) The solution to (18.75) with $\hat\gamma = 0$ is

$$f = \frac{\cos x}{1 + \hat\Delta},$$

showing a resonance at $\hat\Delta = -1$ (see figure 18.19).

For weak nonlinearity ($\hat\gamma$ small), we can express f as a Fourier series

$$f = A\cos x + \hat\gamma(A_0 + A_2\cos 2x) + \hat\gamma^2\sum_{n=3}^\infty A_n \cos nx,$$

which implies a cubic equation for A:

$$(1 + \hat\Delta)A - \hat\Delta^2 A^3\left[\frac{1}{\hat\Delta} + \frac{1}{(4 + \hat\Delta)}\right] = 1. \quad (18.76)$$

(One can show that the higher-order terms will not contribute, even near resonance.) Figure 18.19 also shows the solution of (18.76) for $\hat\gamma = 0.2$. Here we clearly see that there are three equilibrium states for $\hat\Delta < -1.3$. The state with intermediate amplitude is unstable; thus we see that we can have either a large positive amplitude wave (in phase with topography) or a small negative amplitude wave (out of phase).

This simple model suggests that the Kuroshio meander may be a case of multiple states depending on the flow rate at the inlet. Slight decreases in speed may cause a sudden transition to a meander state, with hysteresis effects likely, so that large increases are necessary before the Kuroshio would return to its path closer to the coast.

The above results merely suggest the possibility of multiple equilibria because, to begin with, we have required the coastline to have an infinite number of ridges and troughs in order to create the possibility of linear resonance. As shown in figure 18.10, an infinite number of periods may not be necessary, but there must be at least two ridges and a trough or vice versa. A single coastal ridge (as in the half-wave case) would not be enough to give a maximum response. One must

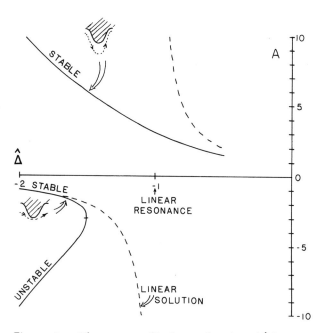

Figure 18.19 The wave amplitude as a function of $\hat\Delta$ (proportional to the magnitude of the upstream current). Sketches show the relationship between the streamlines and the coastline. Multiple equilibria occur for $\hat\Delta < -1.3$.

ask: Can a single ridge or, in the case of blocking in the atmosphere, a single mountain range, in an extended domain give rise to multiple equilibria? And is resonance needed?

In problems of nonrotating shallow-water flow over an obstacle, multiple states can exist when the Froude number $\bar{u}/\sqrt{gH}$ is greater than unity. For certain values of the Froude number and the ratio of the obstacle height to H, two states are found, one corresponding to smooth flow with no upstream disturbance and one corresponding to a permanent elevation of the free surface upstream of the obstacle—created during the approach to equilibrium by a bore traveling upstream (Baines and Davies, 1980). Similar examples of multiple equilibria are also found in transonic compressible flow past obstacles. Thus we suspect that the upstream flow in our problem cannot be specified arbitrarily but may very well be affected by the upstream propagation of energy. It may be that, as in the periodic models of Charney-DeVore and Charney-Straus, the flow must be dealt with as a global or basin-wide unit.

18.7.5 Quasi-Geostrophic Turbulence

We have considered only wave–wave or wave–mean flow interactions involving a small number of components. In particular, we have not considered energy-cascade processes involving large numbers of components and leading ultimately to turbulent dissipation. It was pointed out by Onsager (1949), Lee (1951), Batchelor (1953a), and especially by Fjørtoft (1953) that vorticity conservation in two-dimensional flow imposes a strong constraint on scale interactions. Later Charney (1966, 1971a) showed that the conservation of pseudopotential vorticity in three-dimensional quasi-geostrophic flow imposes similar constraints. Such constraints suggested to Kraichnan (1967) that there may be an inertial subrange in two-dimensional, homogeneous, isotropic turbulence in which the energy spectrum is controlled by uniform transfer of enstrophy (mean-squared vorticity) from large to small scales at scales less than the excitation scale, and by uniform transfer of energy from small to large scales at scales greater than the excitation scale. He predicted a k^{-3} spectral energy density for scalar wavenumber k in the former range, and a Kolmogorov $k^{-5/3}$ law in the latter range. In extending these ideas to three-dimensional, quasigeostrophic turbulence, Charney (1971a) also obtained a k^{-3} law at the tail of the spectrum and conjectured that in this region there would also be equipartition between the two components of the kinetic energy and the available potential energy. This conjecture has been confirmed by Herring (1980) in a homogeneous quasi-geostrophic turbulence closure model.

The topic of quasi-geostrophic turbulence has been investigated by a number of oceanographers, notably Rhines (1975) by numerical simulation, and Holloway and Hendershott (1977) and Salmon (1978) by means of closure [see also Herring (1980) and Leith (1971)]. It may be that their work is more applicable to the atmosphere than to ocean basins, where meridional boundaries play important roles and where statistical inhomogeneity of excitation cannot be ignored.

The existence of quadratic invariants, energy and enstrophy—mean-squared vorticity in two-dimensional or mean-squared pseudopotential vorticity in three-dimensional quasi-geostrophic flows—permits application of the principles of statistical mechanics. These have been applied by Onsager (1949) and Kraichnan (1975) to two-dimensional flow, and by Salmon, Holloway, and Hendershott (1976) to a two-layer quasi-geostrophic flow. In the two-dimensional case, the energy in each horizontal mode is $L^2/(b + aL^2)$, where a and b are constants depending on the total energy and enstrophy. With typical choices of these constants, the largest scale waves have the most energy. In the two-layer case, the equilibrium spectrum is dominated by the largest scales, and these motions are barotropic. The available potential-energy spectrum, corresponding to the thermocline displacement spectrum, is peaked near the deformation radius. These spectra represent the effects of the nonlinear terms alone; one expects [as Errico (1979) found in studying the partition of energy between gravity-wave and geostrophic motions] that the spectra which are actually realized in a forced and dissipative system will be determined largely by the wavenumber dependence of the forcing and dissipation.

The cornerstone for the theory of quasi-geostrophic turbulence is the conservation (in the inviscid limit) of the energy $\frac{1}{2}\iiint |\nabla\psi|^2 + (1/S)|\partial\psi/\partial z|^2$ and the enstrophy $\frac{1}{2}\iiint [\nabla^2\psi + (\partial/\partial z)(1/S)(\partial\psi/\partial z)]^2$. We emphasize the fragility of the last principle: enstrophy can increase or decrease if there are (1) temperature gradients along horizontal boundaries, (2) side walls on the domain, or (3) topography.

If we first consider the case when none of these restrictions obtain—flow in a periodic, flat-bottomed domain—we can readily argue that the energy will be transferred to large horizontal scales and to more barotropic motions by the nonlinear terms. If we expand the streamfunction in the flat-bottomed normal modes and perform a Fourier transform horizontally,

$$\psi = \sum_n F_n(z) \iint d\mathbf{k}\,\hat{\psi}_n(\mathbf{k},t)e^{i\mathbf{k}\cdot\mathbf{x}},$$

we can use the conservation principles

$$\frac{\partial}{\partial t}\sum_n \iint d\mathbf{k}\,(\mathbf{k}^2 + \lambda_n^2)|\hat{\psi}_n|^2 = 0,$$

$$\frac{\partial}{\partial t}\sum_n \iint d\mathbf{k}(\mathbf{k} + \lambda_n^2)^2|\hat{\psi}_n|^2 = 0$$

(λ_n is the reciprocal radius of deformation for the nth baroclinic mode; see section 18.5.1) in exactly the same manner as Fjørtoft (1953) or Charney (1971a) to show that the amount of energy with $k^2 + \lambda_n^2 > K_0^2$ is a small fraction of the initial energy, if K_0 is large compared to the initial mean wavenumber

$$\overline{K} = \frac{\sum_n \iint d\mathbf{k}\,(\mathbf{k}^2 + \lambda_n^2)^{1/2}[(\mathbf{k}^2 + \lambda_n^2)|\hat{\psi}|^2]}{\sum_n \iint d\mathbf{k}[(\mathbf{k}^2 + \lambda_n^2)|\hat{\psi}|^2]}.$$

In essence the nonlinearity does not transfer energy to small scales. Another way to show the reverse cascade, is to use Rhines's (1975) argument that the turbulence spreads energy out in wavenumber space

$$\frac{\partial}{\partial t} \sum_n \iint d\mathbf{k}\,[(\mathbf{k}^2 + \lambda_n^2)^{1/2} - \overline{K}]^2[(\mathbf{k}^2 + \lambda_n^2)|\hat{\psi}_n|^2] > 0.$$

Combining this with the definition of $\overline{K}$ and using the conservation laws shows that the mean total wavenumber must decrease

$$\frac{\partial}{\partial t}\overline{K} < 0.$$

Thus energy cascades to larger horizontal *and* vertical scales, implying an increase in energy for small λ_n^2, that is, a tendency for the motion to become more barotropic.

This tendency for the flow to become barotropic in the absence of topography or side walls has been well demonstrated through numerical simulation by Rhines (1977). He has also shown that rough topography can halt this cascade for flows that are not too energetic. The β-effect slows the cascade when the scale has increased so much that the wave steepness M becomes order one. This could occur while the motions are still baroclinic if the initial energy were small compared to $\beta^2 L_R^4$.

Rhines (1975) has also argued that side walls can stop the reverse cascade. This has been demonstrated in laboratory experiments by Colin de Verdière (1977). Essentially the western boundary serves as a source of enstrophy and the eastern boundary a sink. This can be understood—as Pedlosky (1967), in a slightly different context, has revealed—by considering reflection of Rossby waves from the western boundary. For linear waves with $c = -\beta/(k^2 + l^2 + \lambda_n^2)$, the x component of the group velocity is negative for $k^2 < (l^2 + \lambda_n^2)$. Therefore the reflected wave's zonal wavenumber $k_r = (l^2 + \lambda_n^2)/k$ is larger than k. One can readily show that the energy fluxes of the incident and reflected waves (c_g times the energy density) are equal and opposite, but that the enstrophy flux of the outgoing wave is larger by a factor k_r/k than the flux of the incident wave.

Rhines (1977) has discussed many of the topics above, and in particular has demonstrated clearly that the strong nonlinear interactions involved in geostrophic turbulence cannot occur unless nonlinearity is much stronger than wave dispersion, that is, unless the wave steepness $Uk^2/\beta \gg 1$. At these scales there is some, but not conclusive, evidence in support of a k^{-3} spectrum for the atmosphere (Julian and Clive, 1974). The enstrophy cascade mechanism has not yet been checked adequately either by direct measurement or by numerical simulation. It remains possible that the observed atmospheric spectra can be explained in terms of ordered (periodic) frontal structures, rather than random cascades of enstrophy, as suggested by Andrews and Hoskins (1978). They obtain a $k^{-8/3}$ dependency which is as much in accord with observations as the k^{-3} spectrum—or perhaps better. However, the predicted spectra are highly anisotropic, and this does not seem to be in as good agreement with observations or results from numerical modeling as the predictions of the theory of geostrophic turbulence.

So far the data do not exist for a corresponding oceanic check.

18.8 Summary Remarks

Our primary focus has been on oceanic analogues of transient atmospheric motions of large scale. As promised in the introduction, it has been possible to find formal oceanic analogues for most categories of large-scale atmospheric motions: indeed, it has been almost trivial to do so. Far more difficult, however, has been to demonstrate the physical reality and importance of these analogues. It is only recently, through studies of the meanders of the western boundary currents and through such concentrated, large-scale observational programs as MODE, Polygon, POLYMODE, and the oceanographic component of the GARP Atlantic Tropical Experiment (GATE), that some understanding of the nature of the transient motions has begun to emerge. Although many of the oceanic analogues we have dealt with are hypothetical to a greater or lesser degree, we have chosen them on physical grounds as at least of potential importance. We feel that contrasting them with their atmospheric counterparts, with respect to both their individual properties and their roles in the generation and maintenance of the large-scale circulation patterns, has been a useful exercise.

In section 18.4, it is shown that the quasi-geostrophic, β-plane formalism derived for the atmosphere applies to oceanic motions whose horizontal scales are on the order of the deformation radius of the first baroclinic mode L_R, that is, the scales corresponding to baroclinic instability. It is characteristic of these motions that they are dispersive at both small and large amplitudes. At larger scales the dynamics change: linear free modes may no longer have the same east-west and north-south scales and vertical density advection becomes an important cause of nonlinearity. For motions with length scales near the "intermediate scale" $(L_R^2 a)^{1/3}$ with velocities of order $f_0 L_R^2/a$ (where a is the radius of the earth), the vertical component of vorticity changes not only because of β-effects and horizontal advection but also because of vertical density advection and variation of the undifferentiated Coriolis parameter. Thus both Rossby and Burger terms appear. For east–west scales on the order of the intermediate scale (210 km for the ocean and 1500 km for the atmosphere) or larger, dispersion and nonlinearity can balance to give solitary or cnoidal waves. For larger scales, we show that the evolution is determined by a Korteweg–deVries equation, so that solitons will be the natural end product of the evolution of an initially isolated disturbance. Because of the slower dissipation and the larger scale separation between the intermediate and basin scales, such waves are more likely to be found in the ocean than in the atmosphere. These results suggest that the larger-scale dynamics of the ocean transients may be dominated by more orderly, phase-coherent structures than are predicted by the theory of geostrophic turbulence. If this is so, then at scales larger than the excitation scale the low-wavenumber components would be more highly correlated than if they were due to a random reverse cascade.

The theory of free and forced small-amplitude Rossby waves in the oceans may be transposed almost entirely from the corresponding atmospheric theory. The MODE and Polygon experiments have provided evidence of the importance of Rossby-wave propagation, particularly on time scales greater than a month (McWilliams, 1976). In section 18.5, we present some of this theory in an oceanographic context, paying particular attention to the influence of bottom topography in altering the propagation of free waves and in generating waves from a mean flow. Rhines's results for a uniformly stratified fluid are extended to arbitrarily stratified flows. The major result, the prediction of bottom-trapped modes, has been verified from observations at site D (Thompson, 1977) although no observations of the eastward-traveling modes which should exist when the topography opposes the β-effect have been reported.

The idea of Rossby-wave propagation in a medium with a variable (real or imaginary) index of refraction was first advanced to account for the vertical trapping of shorter waves by upper easterlies and strong westerlies. We apply these ideas in an oceanographic context not only to vertical propagation from the surface or bottom but also to horizontal propagation of waves generated by the meandering of the Gulf Stream. Eastward-propagating meanders produce trapped disturbances close to the Gulf Stream, while westward-propagating meanders may give a real index of refraction and southward propagation.

In section 18.6, we describe briefly the influence of friction both on the generation of ocean currents by wind and on the decay of individual oceanic eddies. Existing theory is not adequate to account for the spin-up of the real ocean or for the decay of the real atmospheric circulation. Interest in the baroclinic spin-down problem was originally motivated by a desire to understand the long persistence of Gulf Stream rings; however, the axisymmetric models that had previously been employed do not account for the complete decay of a baroclinic eddy. We show that the β-effect permits vertical propagation of energy and therefore allows for complete spin-down.

In the last section, we consider oceanic disturbances in which advective effects are important—either through wave–mean flow interaction as in the breakdown of an unstable mean flow, through the interaction of waves generated elsewhere with the local mean flow, or through wave-wave interactions. The last type of interaction can occur when the unstable flow is itself a wave or when waves generated in any manner interact with one another as in turbulence.

The concept of baroclinic instability was developed to explain the principal traveling waves and vortices embedded in the atmospheric westerlies; it has been applied to the oceans in an effort to account for the meandering of the western boundary currents and the existence of mid-ocean mesoscale eddies. The meandering does seem to be an effect of baroclinic instability, modified by barotropic effects, but it is highly questionable on theoretical and numerical-modeling grounds whether the mid-ocean eddies are due to local baroclinic instabilities. It seems more likely that these eddies are vortices cast off from, or forced in some more general fashion by, the meandering western boundary currents and their extensions.

In our exposition of the baroclinic and barotropic instability problem, we use the methods of Arnol'd and Blumen to extend the integral theorems of Kuo, Charney-Stern, and others to a class of basic flows that need not be zonal, that may translate with constant speed, and that may be influenced by topography.

The work on wave-mean flow interactions originated by Eliassen and Palm and Charney and Drazin in an atmospheric context is applied to the problem of the rectification of Rossby waves radiated from the western boundary currents and their extensions. Of particular interest is the so-called recirculation flow found by Worthington and others south of the Gulf Stream extension. Rhines has attempted to account for this recirculation as driven by the westward-propagating Rossby waves produced by the meandering. We present a slight generalization of his work by considering also the effects of eastward-traveling meanders. The results do suggest that a relatively strong westward flow, confined fairly close to the Gulf Stream, can be produced by eastward-propagating wave–mean flow interactions in the presence of dissipation.

In recent times, the stability analyses for the atmosphere have been extended to wavy motions in an attempt to account for nonlinear cascades of energy in large-scale motions. The stability of forced wavy motions has also been studied to account for the transition from one stationary state to another of a forced flow over topography. It has been found from a study of simple truncated spectral models that the stationary flow equilibria produced by the forcing of a zonal flow over topography in a rotating system may be indeterminate in the sense that for a given forcing, there exists a multiplicity of equilibrium states. This result has been utilized in an attempt to explain the so-called blocking phenomenon in the atmosphere—the persistence of large-amplitude anticyclonic flow anomalies in the planetary circulation. The existence of multiple equilibria for a given forcing appears to be common; it also occurs for supercritical-Froude-number flow in hydraulics and transonic flow in gas dynamics. A natural oceanic analogue of multiple, quasi-stationary equilibrium in the atmosphere is the known existence of two states of flow for the Kuroshio in the vicinity of the Japanese coast. We investigate a simple model of such a flow and find indeed that two different steady states may be produced by a given upstream flow as it passes a wavy boundary. Our model leaves much to be desired, but it does point a direction for future research.

We find that the transition between one state and another in topographically forced flows occurs via a form-drag instability in which the perturbed form drag (mountain torque) modifies the mean flow in such a way that the perturbation increases in amplitude. The instability of a forced topographic wave is thus different from that of a free wave. The latter instability was shown by Gill to be basically a Rayleigh-like shear instability or a resonant-triad wave interaction. We have presented the wave-stability analysis for both free and forced waves in a unified fashion to bring out the similarities and differences between the shear, resonant, and form-drag instabilities.

The form-drag instabilities producing transition grow in place; however, the wavy equilibrium states themselves may also exhibit traveling Rayleigh-like instabilities. It has been suggested that instabilities of this kind account for the observed eastward- and westward-propagating very long (zonal wavenumbers 1–4) planetary waves in the atmosphere. One may speculate that a careful analysis of the topographically induced meanders of the western boundary currents in the ocean will also reveal such secondary wave instabilities. If these are westward propagating, then they could contribute to a broader recirculation region.

The final topic is quasi-geostrophic turbulence. Fjørtoft's prediction that there will be a transfer of energy from the excitation scale to larger scales in 2-dimensional energy-and-enstrophy-conserving flow may be extended to 3-dimensional quasi-geostrophic flows if the bottom and top boundaries are flat and isentropic. The theory predicts that the scale will increase vertically as well as horizontally, that is, that the flow will become increasingly barotropic at large horizontal scales. Rhines has verified this result in a series of numerical experiments, and the oceanic observations of Schmitz (1978) show that the mesoscale eddies tend to be more barotropic the more energetic they are. This observation, while not verifying the inertial theory of geostrophic turbulence, is at least consistent with it. As Rhines has shown, the inertial theory applies only when the effects of nonlinearity dominate those of linear dispersion, that is, when the wave steepness Uk^2/β is much greater than unity. Thus one expects the prediction to be valid only in the energetic parts of the ocean. The similarity prediction of a k^{-3} spectrum at scales smaller than the excitation scale is not inconsistent with observations in the atmosphere, but there are as yet no data to test this theory in the ocean. The effects of topography, side boundaries, and surface gradients of entropy also have not been thoroughly explored.

We have not produced a systematic or comprehensive treatment of atmosphere–ocean analogues. Our excuse, as we have stated, is that virtually all large-scale atmospheric motions have oceanic counterparts and there are simply too many of these to discuss. We have preferred to deal with analogues for which there is observational evidence or at least some physical basis for believing they should exist. We have been forced to speculate, and, as the reader will surely have perceived, our own speculations have been guided primarily by our own experience and interests.

Appendix: The Quasi-Geostrophic Equations

Here we shall give details of the derivations of the quasi-geostrophic equations. We begin by nondimensionalizing the equations of motion (18.22)–(18.27), using the definitions of the geostrophic streamfunction and the potential buoyancy b_p in (18.28) and (18.29). Using the characteristic scales described in the text, we obtain

$$\varepsilon \frac{Du}{Dt} + \omega\epsilon\lambda\hat{\beta}\frac{uw\tan\Theta}{r} - \epsilon\hat{\beta}\frac{uv\tan^2\Theta}{r} - \lambda\omega\cot\Theta\, w - v$$

$$= -\frac{\alpha}{r}\frac{1}{\cos\Theta}\left(\frac{\partial}{\partial\phi} + \hat{\beta}\tan\Theta\frac{\partial}{\partial\Phi}\right)\psi, \qquad (18.A1)$$

$$\varepsilon \frac{Dv}{Dt} + \omega\epsilon\lambda\hat{\beta}\frac{vw\tan\Theta}{r} + \epsilon\hat{\beta}\frac{u^2\tan^2\Theta}{r} + u$$

$$= -\frac{\alpha}{r}\frac{1}{\sin\Theta}\left(\frac{\partial}{\partial\theta} + \hat{\beta}\tan\Theta\frac{\partial}{\partial\Theta}\right)\sin\Theta\,\psi, \qquad (18.A2)$$

$$\omega\lambda^2\varepsilon\frac{Dw}{Dt} - \varepsilon\lambda\hat{\beta}\tan\Theta\frac{u^2+v^2}{r} - \lambda u\cot\Theta$$

$$= -\psi_z + \Delta\hat{S}\psi + b_p$$
$$- \varepsilon\Delta(b_p - \Delta_s\psi)(\psi_z - \Delta\hat{S}\psi), \quad (18.A3)$$

$$\frac{1}{r\cos\Theta}\left(\frac{\partial}{\partial\phi} + \hat{\beta}\tan\Theta\frac{\partial}{\partial\Phi}\right)u + \frac{1}{r}\left(\frac{\partial}{\partial\theta} + \hat{\beta}\tan\Theta\frac{\partial}{\partial\Theta}\right)v$$

$$-\frac{\hat{\beta}v\tan^2\Theta}{r} + \frac{\omega}{r^2}\frac{\partial}{\partial z}r^2 w$$

$$= \frac{\varepsilon\Delta}{\alpha}\frac{1}{\sin\Theta}\frac{D}{Dt}\sin\Theta(b_p - \Delta_s\psi) + \omega(\Delta_s + \Delta\hat{S})w \quad (18.A4)$$

$$\frac{1}{\alpha\sin\Theta}\frac{D}{Dt}\sin\Theta\, b_p + \frac{\omega}{\varepsilon}\hat{S}w + \frac{\omega\Delta_s}{\varepsilon\Delta}\left(1 - \frac{\alpha^2}{\hat{c}_s^2}\right)w$$

$$- w\psi\frac{\omega\varepsilon}{\varepsilon}\frac{\partial}{\partial z}\Delta_s + \omega\frac{\varepsilon}{\varepsilon}(\Delta\hat{S}$$

$$+ \Delta_s)\left[b_p - \Delta_s\left(1 + \frac{\alpha^2}{\hat{c}_s^2}\right)\psi\right]w$$

$$+ \left(\frac{\alpha^2}{\hat{c}_s^2} - 1\right)\Delta_s\frac{1}{\sin\Theta}\frac{D}{Dt}\sin\Theta\,\psi = 0, \quad (18.A5)$$

where

$$\frac{D}{Dt} = \frac{\partial}{\partial t} + \frac{\varepsilon}{\varepsilon}\left[\frac{u}{r\cos\Theta}\left(\frac{\partial}{\partial\phi} + \hat{\beta}\tan\Theta\frac{\partial}{\partial\Phi}\right)\right.$$
$$\left. + \frac{v}{r}\left(\frac{\partial}{\partial\theta} + \hat{\beta}\tan\Theta\frac{\partial}{\partial\Theta}\right) + \omega w\frac{\partial}{\partial z}\right]$$

and

$$r = 1 + \lambda\hat{\beta}\tan\Theta z,$$
$$\alpha = 1 + \varepsilon\Delta(b_p - \Delta_s\psi),$$
$$\hat{c}_s^2 = c_s^2/\bar{c}_s^2 = 1 + O(\varepsilon\Delta)$$

are used as abbreviations.

The quasi-Boussinesq approximation entails choosing the scale of motion to be small compared to the external radius of deformation $\Delta \ll 1$. We therefore drop terms from (18.A1)–(18.A5) which are small in this sense. The definition of "small" requires some care because the various Rossby numbers are also small. Thus in equations (18.A1), (18.A2), (18.A4) we shall keep both order 1 and order ε, ϵ, $\hat{\beta}$, ω terms so that we may drop only terms of order $\epsilon\Delta$, $\varepsilon\Delta$, $\omega\Delta$. Equations (18.A1) and (18.A2) are changed only slightly: α is replaced by 1. The other equations become

$$\frac{1}{r\cos\Theta}\left(\frac{\partial}{\partial\phi} + \hat{\beta}\tan\Theta\frac{\partial}{\partial\Phi}\right)u + \frac{1}{r}\left(\frac{\partial}{\partial\theta} + \hat{\beta}\tan\Theta\frac{\partial}{\partial\Theta}\right)v$$

$$-\frac{\hat{\beta}v\tan^2\Theta}{r} + \frac{\omega}{r^2}\frac{\partial}{\partial z}r^2 w = \Delta_s\omega w, \quad (18.A6)$$

$$\omega\lambda^2\varepsilon\frac{Dw}{Dt} - \varepsilon\lambda\hat{\beta}\tan\Theta\frac{u^2+v^2}{r} - \lambda u\cot\Theta$$

$$= -\psi_z + b_p, \quad (18.A7)$$

$$\frac{1}{\sin\Theta}\frac{D}{Dt}\sin\Theta\, b_p + \frac{\omega}{\varepsilon}\hat{S}w - \omega\frac{\epsilon}{\varepsilon}w\psi\frac{\partial}{\partial z}\Delta_s$$

$$+ \frac{\omega\epsilon}{\varepsilon}\Delta_s\epsilon\left(\frac{\hat{c}_s^2-1}{\epsilon\Delta}\right)w - \frac{\omega\epsilon}{\varepsilon}\Delta_s(b_p - \Delta_s\psi)w = 0. \quad (18.A8)$$

For the ocean, we can further refine these equations by noting that Δ_s is also very small, so that the continuity equation (18.A6) becomes that of an incompressible fluid, and the potential buoyancy equation becomes simply

$$\frac{1}{\sin\Theta}\frac{D}{Dt}\sin\Theta\, b_p + \frac{\omega}{\varepsilon}\hat{S}w = 0. \quad (18.A9)$$

The hydrostatic approximation applies to a thin layer of fluid: $\lambda \ll 1$. This allows us to drop the centrifugal terms involving w, the vertical accelerations, and to replace r by 1, giving us

$$\varepsilon\frac{Du}{Dt} - \epsilon\hat{\beta}uv\tan^2\Theta - v$$

$$= -\frac{1}{\cos\Theta}\left(\frac{\partial}{\partial\phi} + \hat{\beta}\tan\Theta\frac{\partial}{\partial\Phi}\right)\psi, \quad (18.A10)$$

$$\varepsilon\frac{Dv}{Dt} + \epsilon\hat{\beta}u^2\tan^2\Theta + u$$

$$= -\frac{1}{\sin\Theta}\left(\frac{\partial}{\partial\theta} + \hat{\beta}\tan\Theta\frac{\partial}{\partial\Theta}\right)\sin\Theta\,\psi, \quad (18.A11)$$

$$\frac{1}{\cos\Theta}\left(\frac{\partial}{\partial\phi} + \hat{\beta}\tan\Theta\frac{\partial}{\partial\Phi}\right)u + \left(\frac{\partial}{\partial\theta} + \hat{\beta}\tan\Theta\frac{\partial}{\partial\Theta}\right)v$$

$$- \hat{\beta}v\tan^2\Theta + \omega\frac{\partial w}{\partial z} = 0, \quad (18.A12)$$

$$\psi_z = b_p, \quad (18.A13)$$

$$\frac{1}{\cos\Theta}\left(\frac{\partial}{\partial\phi} + \hat{\beta}\tan\Theta\frac{\partial}{\partial\Phi}\right)u + \left(\frac{\partial}{\partial\theta} + \hat{\beta}\tan\Theta\frac{\partial}{\partial\Theta}\right)v$$

$$- \hat{\beta}v\tan\Theta + \omega\frac{\partial w}{\partial z} = 0, \quad (18.A14)$$

where

$$\frac{D}{Dt} = \frac{\partial}{\partial t} + \frac{\epsilon}{\varepsilon}\frac{u}{\cos\Theta}\left(\frac{\partial}{\partial\phi} + \hat{\beta}\tan\Theta\frac{\partial}{\partial\Phi}\right)$$

$$+ \frac{\epsilon}{\varepsilon}v\left(\frac{\partial}{\partial\theta} + \hat{\beta}\tan\Theta\frac{\partial}{\partial\Theta}\right) + \omega w\frac{\partial}{\partial z}.$$

Equations (18.A9)–(18.A14) are the Boussinesq hydrostatic equations.

Finally, the quasi-geostrophic β-plane approximation assumes that $\hat{\beta} \sim \varepsilon \sim \epsilon \ll 1$ and (by necessity) that ω

must also be of this order. We set $\omega = \varepsilon$ and expand in powers of ε to find the lowest-order balances

$$u^{(0)} = -\frac{\partial \psi^{(0)}}{\partial \theta},$$

$$v^{(0)} = \frac{1}{\cos \Theta} \frac{\partial \psi^{(0)}}{\partial \phi},$$

$$b_p^{(0)} = \frac{\partial \psi^{(0)}}{\partial z}$$

and the continuity equation

$$\frac{1}{\cos \Theta} \frac{\partial u^{(0)}}{\partial \phi} + \frac{\partial v^{(0)}}{\partial \theta} = 0$$

which is consistent with the geostrophic equations. At first order we have the momentum equations

$$\varepsilon \frac{\hat{D} u^{(0)}}{Dt} - \varepsilon v^{(1)} = -\frac{1}{\cos \Theta} \frac{\partial}{\partial \phi} \varepsilon \psi^{(1)} - \frac{\hat{\beta} \tan \Theta}{\cos \Theta} \frac{\partial}{\partial \Phi} \psi^{(0)},$$

$$\varepsilon \frac{\hat{D} v^{(0)}}{Dt} + \varepsilon u^{(1)} = -\frac{1}{\sin \Theta} \frac{\partial}{\partial \theta} \varepsilon \sin \Theta \, \psi^{(1)}$$

$$- \frac{\hat{\beta}}{\cos \Theta} \frac{\partial}{\partial \theta} \sin \Theta \, \psi^{(0)},$$

$$\frac{\hat{D}}{Dt} = \frac{\partial}{\partial t} + \frac{\varepsilon}{\varepsilon} \frac{u^{(0)}}{\cos \Theta} \frac{\partial}{\partial \phi} + \frac{\varepsilon}{\varepsilon} v^{(0)} \frac{\partial}{\partial \theta},$$

from which we form the vorticity equation

$$\varepsilon \frac{\hat{D}}{Dt} \left(\frac{v_\phi^{(0)}}{\cos \Theta} - u_\theta^{(0)} \right) + \varepsilon \left(\frac{u_\phi^{(1)}}{\cos \Theta} + v_\theta^{(1)} \right)$$

$$= -\hat{\beta} \frac{1}{\cos^2 \Theta} \frac{\partial}{\partial \theta} \sin \Theta \frac{\partial \psi^{(0)}}{\partial \phi} + \hat{\beta} \frac{\sin \Theta}{\cos^2 \Theta} \frac{\partial}{\partial \Phi} \frac{\partial \psi^{(0)}}{\partial \theta}$$

$$= -\hat{\beta} v^{(0)} - \hat{\beta} \frac{\sin \Theta}{\cos \Theta} \frac{\partial v^{(0)}}{\partial \theta} + \hat{\beta} \tan^2 \Theta \, v^{(0)} - \hat{\beta} \frac{\sin \Theta}{\cos^2 \Theta} \frac{\partial u^{(0)}}{\partial \Phi}.$$

Combining this with the first-order continuity equation

$$\frac{\varepsilon}{\cos \Theta} u_\phi^{(1)} + \varepsilon v_\theta^{(1)} + \hat{\beta} \frac{\sin \Theta}{\cos^2 \Theta} u_\Phi^{(0)} + \hat{\beta} \tan \Theta \frac{\partial v^{(0)}}{\partial \theta}$$

$$- \hat{\beta} v^{(0)} \tan^2 \Theta + \varepsilon \frac{\partial w^{(0)}}{\partial z} = 0$$

(note that for the atmosphere, we would have an additional term $-\Delta_s \varepsilon w^{(0)}$ in this equation) leads to the vorticity equation

$$\frac{\hat{D}}{Dt} \left(\frac{v_\phi^{(0)}}{\cos \Theta} - u_\theta^{(0)} \right) + \frac{\hat{\beta}}{\varepsilon} v^{(0)} = w_z^{(0)}.$$

The lowest-order potential-buoyancy equation

$$\frac{\hat{D}}{Dt} \psi_z^{(0)} + \hat{S} w^{(0)} = 0$$

can now be combined with the vorticity equation to give the quasigeostrophic conservation equation:

$$\frac{\hat{D}}{Dt} \left[\frac{1}{\cos \Theta} \frac{\partial}{\partial \phi} \frac{1}{\cos \Theta} \frac{\partial}{\partial \phi} \psi^{(0)} + \frac{\partial^2}{\partial \theta^2} \psi^{(0)} + \frac{\partial}{\partial z} \frac{1}{\hat{S}} \frac{\partial}{\partial z} \psi^{(0)} \right]$$

$$+ \frac{\hat{\beta}}{\epsilon} \frac{\psi_\phi^{(0)}}{\cos \Theta} = 0.$$

For the atmosphere, the additional term in the continuity equation appears as an extra contribution

$$-\frac{\Delta_s}{\hat{S}} \frac{\partial}{\partial z} \psi^{(0)}$$

in the potential vorticity. Alternatively, the thickness term can be written as

$$\bar{\alpha} \frac{\partial}{\partial z} \frac{1}{\bar{\alpha} \hat{S}(z)} \frac{\partial}{\partial z} \psi$$

for atmospheric quasi-geostrophic motions.

Notes

1. Unfortunately, meteorologists use "mesoscale" very differently from oceanographers. We shall use mesoscale in the oceanic sense to refer to motions that are dynamically analogous to the "synoptic" scale motions of the atmosphere.

2. Note that the U here is characteristic of the disturbances, not of the mean flow.

3. Some care needs to be used in the cnoidal wave case since the mean depth of the fluid becomes $H + (1/2\pi L) \int_0^{2\pi L} dx \, \eta$ and the last term does not vanish. For the figures we have corrected for this effect to show c nondimensionalized by βL_R^2 where L_R is based on the actual average depth. Thus we have plotted $c_{actual} = (1 - \epsilon \langle \eta \rangle / \hat{S}_{actual}) c$ as a function of $\hat{S}_{actual} = \hat{S} + \epsilon \langle \eta \rangle$, and also subtracted out the mean from the plots of $\phi^{(0)}$. The same process was used for the nonlinear Rossby wave but had no effect on the dispersion relation.

4. The "anelastic equations" (Batchelor, 1953b; Ogura and Charney, 1962; Ogura and Phillips, 1962) filter only acoustic waves.

5. It is customary to call any quasi-geostrophic wave a Rossby wave.

6. Geisler and Dickinson (1975) studied critical layer absorption in the western boundary current but did not explicitly include a reflected wave. It is also possible that the effects of the mechanisms maintaining the western boundary current are important in the interaction process.

7. It can be shown trivially that there is no nonlinear interaction in a spectrum of Rossby waves with all components having the same scale $(k^2 + \lambda^2)^{-1/2}$ (including baroclinic effects).

Acknowledgments and Permissions

Acknowledgments

Bruce A. Warren, Deep Circulation of the World Ocean. Chapter 1.
Preparation of this chapter was supported by the U.S. Office of Naval Research under contract N00014-79-C0071, NR083-004. I am grateful to R. B. Montgomery and A. B. Arons for helpful criticism of an early draft.

L. V. Worthington. The Water Masses of the World Ocean: Some Results of a Fine-Scale Census. Chapter 2.
The painstaking tasks of data reduction, quality control, and the assignment of world water masses to their myriad fine-scale bivariate classes were carried out by C. G. Day and R. L. Barbour; their work is gratefully acknowledged. This research was sponsored by the Office of Naval Research under contracts N00014-74-C-0262, NR083-004 and N00014-79-C-0071, NR083-004. Contribution 4442 from the Woods Hole Oceanographic Institution.

Joseph L. Reid. On the Mid-Depth Circulation of the World Ocean. Chapter 3.
This paper represents one of the results of research supported by the Office of Naval Research, the National Science Foundation, and the Marine Life Research Program of the Scripps Institution of Oceanography.

N. P. Fofonoff. The Gulf Stream System. Chapter 4.
Supported by the Office of Naval Research under contract N00014-76-C-0197, NR083-400.

George Veronis. Dynamics of Large-Scale Ocean Circulation. Chapter 5.
Support by the National Science Foundation under grant OCE-7719451 is gratefully acknowledged. Carl Wunsch and Bruce Warren made helpful comments on the original manuscript. Special thanks go to Peter Rhines for a critical review and several discussions about eddy-driven flows.

Ants Leetmaa, Julian P. McCreary, Jr., and Dennis W. Moore. Equatorial Currents: Observations and Theory. Chapter 6.
J. P. M. and D. W. M. wish to acknowledge the Office of Naval Research (contract N00014-75-C-0165) and The National Science Foundation (grant OCE-76-00551) for providing support.

Robert C. Beardsley and William C. Boicourt. On Estuarine and Continental-Shelf Circulation in the Middle Atlantic Bight. Chapter 7.

We gratefully acknowledge the support and help given by the following colleagues. D. Bumpus, G. Csanady, C. Mooers, D. Pritchard, and V. Worthington all provided useful feedback on early drafts of this chapter. Discussions with J. Allen, B. Butman, W. Grant, D. Haidvogel, D. Mayer, J. McCullough, R. Montgomery, M. Noble, G. Philander, P. C. Smith, and C. Winant also proved helpful. B. Butnam, G. Halliwell, and M. Noble supplied the additional unpublished wind-stress data shown in figure 7.12, and H. Ou and W.-S. Chuang made the coherence and monthly-mean current computations shown for site 2 in figures 7.9 and 7.12, respectively. R. Scarlet and New Jersey Public Service allowed us to reproduce the current spectrum and coherence shown for site 1 in figures 7.8 and 7.9. R. Legeckis kindly furnished the satellite infrared photograph of the Middle Atlantic Bight shown in figure 7.15. D. Haight and A. Sullivan helped with the typing of the manuscript and preparation of the bibliography. We also want to acknowledge the support and encouragement given us by our friends and families during the preparation of this review.

This research has been supported through National Science Foundation grants OCE-76-01813 and 78-19513 to R.C.B. and OCE-77-22774 to W.C.B., and a grant to W.C.B. from the State of Maryland Power Plant Siting Research Program.

Walter Munk. Internal Waves and Small-Scale Processes. Chapter 9.
My work is supported by the Office of Naval Research.

Myrl C. Hendershott. Long Waves and Ocean Tides. Chapter 10.
I have profited from discussions on tides with Walter Munk, Michael Parke, and Gerard Stock. Annette Pickens has helped me greatly in the preparation of this manuscript.

Carl Wunsch. Low-Frequency Variability of the Sea. Chapter 11.
Supported by the National Science Foundation under grant OCE-78-19833. MODE contribution 125 (POLYMODE). I am indebted to D. E. Harrison, M. Hendershott, and W. Munk for many useful comments and suggestions, and to Charmaine King for computing the many spectra.

D. James Baker, Jr. Ocean Instruments and Experiment Design. Chapter 14.
In order to get even a brief overview of this rapidly changing field, I wrote to a number of people involved in the key developments of the instruments discussed above. The response was gratifying and overwhelming; I ended up with a tableful of material and personal reminiscences that proved to be fascinating reading and enough for a book, too much for a chapter. Because I have space only for a limited number of examples, the list of instruments discussed is not complete. I think that it does form an interesting record of the activities of some of the participants in the important developments in oceanographic instrumentation over the past two decades, and I am grateful to those who responded.

Special help was received from J. Dahlen on profilers and TP recorders; J. Garrett, G. Cresswell, J. Stromme, and J. Gillis on drifters; M. Gregg and P. Hacker on microstructure profilers; B. Hamon on the history of the salinometer and STD (including the early reference to Nansen); R. Heinmiller on moorings and general engineering problems; W. Hill on tape recorders and electronics; P. Niiler on transport floats and instrumentation in general; T. Rossby on SOFAR floats and the inverted echo sounder; T. Sanford on electromagnetic profilers; J. Swallow on the development of neutrally buoyant floats; J. Van Leer on the cyclesonde; R. Wearn on batteries and pressure gauges; D. Webb on electronics, floats, current meters, and engineering in general; R. Weller on current meters; and S. Williams on the optical profiler.

I also received useful information from N. Brown, B. Buck, W. Coburn, R. Davis, V. Derr, T. Ewart, J. Feeney, J. Filloux, C. Gibson, D. Halpern, J. Hannon, S. Hayes, J. Luyten, R. Mesecar, W. Munk, W. Nowlin, T. Osborn, J. Paros, S. Pond, R. Pollard, J. Richardson, W. Schmitz, J. D. Smith, B. Taft, R. Walden, M. Wimbush, and B. Zetler. Carl Wunsch and Bruce Warren read an early version of the chapter and made a number of helpful suggestions.

This work was partially supported by the Pacific Marine Environmental Laboratory of NOAA, and partially by a grant from the Office of the International Decade of Ocean Exploration for the International Southern Ocean Studies Program at the University of Washington.

W. S. Broecker. Geochemical Tracers and Ocean Circulation. Chapter 15.
Many of the data referred to in this paper were obtained as part of the GEOSECS program. The great success of this survey is a tribute to the late Arnold Bainbridge, who directed virtually every aspect of the field program. The excellence of the radioisotope measurements by Stuiver (^{14}C) and Ostlund (^{3}H and ^{14}C) permitted the prime objective of this program to be accomplished. Discussions with Weiss, Craig, Stuiver, Shepherd, Jenkins, Brewer, Sarmiento, Quay, Rooth, Needler, Gordon, and many others have improved my perspective on oceanic tracing.

Alan J. Faller. The Origin and Development of Laboratory Models and Analogues of the Ocean Circulation. Chapter 16.

The research reported here has been supported by the National Science Foundation under grant ATM-76-82061.

H. Charnock. Air–Sea Interaction. Chapter 17. Much of this review is based on material assembled during a long and valued collaboration with the late Dr. T. H. Ellison.

Jule G. Charney and Glenn R. Flierl. Oceanic Analogues of Large-Scale Atmospheric Motion. Chapter 18. Supported by National Science Foundation grants ATM-76-20070 to J. G. C. and OCE-77-28350 to G. R. F. The authors thank Sam Ricci for preparing figures, Joel Sloman for typing the manuscript, and Joe Pedlosky for his careful review of the first draft.

Permissions

Permission to reproduce illustrations was received from authors, publishers, and copyright holders of the following figures:

Academic Press 17.16

American Association for the Advancement of Science 11.4, 14.32

American Geophysical Union 3.15, 8.9, 8.10, 8.13, 9.15, 9.26, 9.28, 10.16, 10.23, 10.31A, 10.36, 11.8, 11.10, 15.10

American Meteorological Society 1.10, 1.13, 3.13, 3.14, 4.14, 4.15, 4.17, 5.7, 7.14, 8.8, 9.18, 9.21, 9.22, 10.7, 10.29, 10.39, 11.2, 11.11, 14.33B, 17.1, 17.2, 17.3, 17.12, 17.13

Annual Reviews Inc. 15.2

Deutsches Hydrographische Institut 17.6, 17.8

Elsevier Scientific Publishing Co. 15.6, 15.8

Gordon and Breach Science Publishers 5.17, 8.5, 10.27, 10.31B, 10.32

Her Majesty's Stationery Office (by permission of controller) 17.17

The Johns Hopkins University Press 4.6, 5.10A

John Wiley and Sons 4.12, 5.15, 10.13, 10.14, 10.17, 10.18

MacMillan (Journals) Ltd. 9.2, 12.3, 12.7

Measurement and Data Corp. 14.29

Pergamon Press 1.1, 1.2, 1.4, 1.5, 1.6, 1.7, 1.8, 1.12, 4.4, 4.8, 4.10, 5.5, 5.6, 5.11, 6.5, 8.3, 8.7B, 8.11, 9.23, 10.11, 10.33, 10.34, 14.15, 14.18, 14.22, 14.25, 14.27, 15.12

Prentice-Hall 3.9, 7.5

The Royal Society of London 6.8, 10.12B, 10.30, 10.35, 17.11

Sears Foundation for Marine Research 3.10, 4.9, 4.11, 4.13, 5.4, 5.9, 5.10B, 10.6, 10.28

The Syndics of the Cambridge University Press 8.2, 8.6, 8.14, 9.12, 9.13, 9.15, 10.15, 11.6, 17.4, 17.5, 17.7

United States National Academy of Sciences 1.15

Reference List

Pages on which references are cited appear in angle brackets. Every reference to original material not in English for which the editors knew of an English translation is followed by a reference to the English translation; references to untranslated Russian material include translations of the Russian titles immediately following the Russian titles.

Aanderaa, I. R., 1964. A recording and telemetering instrument. *Fixed Buoy Project, NATO Subcommittee on Oceanographic Research Technical Report No. 16*, Bergen, 46 pp., 20 figures. ⟨405⟩

Accad, Y., and C. L. Pekeris, 1978. Solution of the tidal equations for the M_2 and S_2 tides in the world oceans from a knowledge of the tidal potential alone. *Philosophical Transactions of the Royal Society of London A* 290:235–266. ⟨297, 322, 323, 329, 330⟩

Adamec, D., and J. J. O'Brien, 1978. The seasonal upwelling in the Gulf of Guinea due to remote forcing. *Journal of Physical Oceanography* 8:1050–1060. ⟨193⟩

Adams, J. K., and V. T. Buchwald, 1969. The generation of continental shelf waves. *Journal of Fluid Mechanics* 35:815–816. ⟨358⟩

African Pilot, 1967. Hydrographer of the Navy, London, 12th ed., 3, 529 pp. ⟨185⟩

Agee, E. M., and K. E. Dowell, 1974. Observational studies of mesoscale cellular convection. *Journal of Applied Meteorology* 13:46–53. ⟨500⟩

Agnew D., and W. Farrell, 1978. Self-consistent equilibrium ocean tides. *Geophysical Journal of the Royal Astronomical Society* 55:171–181. ⟨339⟩

Agnew, R., 1961. Estuarine currents and tidal streams. In *Proceedings of the Seventh Conference on Coastal Engineering*, The Hague, 1960, 2, pp. 510–535. ⟨204⟩

Allen, J. S., 1980. Models of wind-driven currents on the continental shelf. *Annual Review of Fluid Mechanics* 12:389–433. ⟨222⟩

Amos, A. F., A. L. Gordon, and E. D. Schneider, 1971. Water masses and circulation patterns in the region of the Blake-Bahama Outer Ridge. *Deep-Sea Research* 18:145–165. ⟨27, 28, 36, 121⟩

Anderson, D. L. T., and A. E. Gill, 1975. Spin-up of a stratified ocean, with applications to upwelling. *Deep-Sea Research* 22:583–596. ⟨526⟩

Anderson, D. L. T., and P. D. Killworth, 1977. Spin-up of a stratified ocean, with topography. *Deep-Sea Research* 24:709–732. ⟨526⟩

Anderson, D. L. T., and P. B. Rowlands, 1976a. The role of inertia-gravity and planetary waves in the response of a tropical ocean to the incidence of an equatorial Kelvin wave on a meridional boundary. *Journal of Marine Research* 34:295–312. ⟨191⟩

Anderson, D. L. T., and P. B. Rowlands, 1976b. The Somali Current response to the southwest monsoon: the relative importance of local and remote forcing. *Journal of Marine Research* 34:395–417. ⟨192⟩

Andrews, D. G., and B. J. Hoskins, 1978. Energy spectra predicted by semi-geostrophic theories of frontogenesis. *Journal of the Atmospheric Sciences* 35:509–519. ⟨544⟩

Andrews, D. G., and M. E. McIntyre, 1976. Planetary waves in horizontal and vertical shear: The generalized Eliassen-Palm relation and mean zonal acceleration. *Journal of the Atmospheric Sciences* 33:2031–2048. ⟨533⟩

Andrews, D. G., and M. E. McIntyre, 1978a. An exact theory of nonlinear waves on a Lagrangian-mean flow. *Journal of Fluid Mechanics* 89:609–646. ⟨345, 506, 533⟩

Andrews, D. G., and M. E. McIntyre, 1978b. Generalized Eliassen-Palm and Charney-Drazin theorems for waves on axisymmetric mean flows in compressible atmospheres. *Journal of the Atmospheric Sciences* 35:175–185. ⟨533⟩

Apel, J. R., 1976. Ocean science from space. *EOS, Transactions of the American Geophysical Union* 57:612–624. ⟨426⟩

Appell, G. F., J. E. Boyd, and W. Woodward, 1974. Evaluation of the Aanderaa recording current meter. *NOAA Technical Report NOASIC-TM-OX5/74-ARO4*, 25 pp. ⟨406⟩

Arakawa, A., 1975. Modelling clouds and cloud processes for use in climate models. In *Physical Basis of Climate and Climate Modelling*, WMO-ICSU Joint Organizing Committee, GARP Publication Series No. 16, World Meteorological Organization, Geneva, pp. 183–197. ⟨496⟩

Armi, L., 1978. Some evidence for boundary mixing in the deep ocean. *Journal of Geophysical Research* 83:1971–1979. ⟨183, 261, 262⟩

Armi, L., and R. C. Millard, 1976. The bottom boundary layer of the deep ocean. *Journal of Geophysical Research* 81:4983–4990. ⟨259, 334⟩

Armstrong, R. L., 1980. 1976 in light of atmospheric and oceanic climatological relationships. In *Oxygen Depletion and Associated Benthic Mortalities in New York Bight, 1976*, R. L. Swanson and C. S. Sindermann, eds., NOAA Professional Paper, U.S. Department of Commerce, National Oceanic and Atmospheric Administration, Rockville, Maryland. ⟨225⟩

Arnol'd, V. I., 1965. Ob usloviyakh nelineinoi ustoichivosti ploskikh statsionarnykh krivolineinykh techenii ideal'noi zhidkosti. *Doklady Akademii Nauk SSSR* 162:975–978. (Conditions for nonlinear stability of stationary plane curvilinear flows of an ideal fluid. *Soviet Mathematics* 6:773–777.) ⟨530⟩

Arons, A. B., and H. Stommel, 1951. A mixing-length theory of tidal flushing. *Transactions of the American Geophysical Union* 32:419–421. ⟨202⟩

Arons, A. B., and H. Stommel, 1956. A beta plane analysis of free periods of the second class in meridional and zonal oceans. *Deep-Sea Research* 4:23–31. ⟨344⟩

Arons, A. B., and H. Stommel, 1967. On the abyssal circulation of the World Ocean—III. An advection-lateral mixing model of the distribution of a tracer property in an ocean basin. *Deep-Sea Research* 14:441–457. ⟨xvii, 435⟩

Augstein, E., H. Schmidt, and F. Ostapoff, 1974. The vertical structure of the atmospheric planetary boundary layer in undisturbed trade winds over the Atlantic Ocean. *Boundary-Layer Meteorology* 6:129–150. ⟨503⟩

Austin, T. S., 1960. Oceanography of the east central equatorial Pacific as observed during expedition Eastropic. *Fishery Bulletin of the Fish and Wildlife Service* 60:257–282. ⟨83⟩

Baines, P. G., 1971. The reflection of internal/inertial waves from bumpy surfaces. Part 2. Split reflexion and diffraction. *Journal of Fluid Mechanics* 49:113–131. ⟨331⟩

Baines, P. G., 1974. The generation of internal tides over steep continental slopes. *Philosophical Transactions of the Royal Society of London A* 277:27–58. ⟨337⟩

Baines, P .G., and P. A. Davies, 1980. Laboratory studies of topographic effects in rotating and/or stratified fluids. In *Orographic Effects in Planetary Flows*, R. Hide and P. W. White, eds., GARP Publication Series, World Meteorological Organization, Geneva. ⟨543⟩

Baker, D. J., 1966. A technique for the precise measurement of small fluid velocities. *Journal of Fluid Mechanics* 26:573–575. ⟨472⟩

Baker, D. J., Jr., 1969. On the history of the high seas tide gauge. *W.H.O.I. Technical Memorandum 5-69*, Woods Hole Oceanographic Institution, Woods Hole, Massachusetts, 17 pp. ⟨399, 424⟩

Baker, D. J., Jr., 1970. Models of oceanic circulation. *Scientific American* 222:1, 114–121. ⟨472⟩

Baker, D. J., Jr., 1979. Ocean-atmosphere interaction in high southern latitudes. *Dynamics of Atmospheres and Oceans* 3:213–229. ⟨425⟩

Baker, D. J., Jr., W. D. Nowlin, Jr., R. D. Pillsbury, and H. C. Bryden, 1977. Antarctic circumpolar current: space and time fluctuations in the Drake Passage. *Nature* 268:696–699. ⟨372⟩

Baker, D. J., Jun., and A. R. Robinson, 1969. A laboratory model for the general ocean circulation. *Philosophical Transactions of the Royal Society of London A* 265:533–566. ⟨472⟩

Baker, D. J., Jr., R. B. Wearn, Jr., and W. Hill, 1973. Pressure and temperature measurements at the bottom of the Sargasso Sea. *Nature* 245:25–26. ⟨400, 425⟩

Banner, M. L., and W. K. Melville, 1976. On the separation of air flow over water waves. *Journal of Fluid Mechanics* 77:825–842. ⟨495⟩

Banner, M. L., and O. M. Phillips, 1974. On the incipient breaking of small scale waves. *Journal of Fluid Mechanics* 68:647–656. ⟨495⟩

Barenblatt, G. I., and A. S. Monin, 1979. The origin of the oceanic microstructure. In *Twelfth Symposium Naval Hydrodynamics*, National Academy of Sciences, Washington, D.C., pp. 574–581. ⟨278⟩

Barkley, R. A., 1968. *Oceanographic Atlas of the Pacific Ocean*. University of Hawaii Press, Honolulu, 20 pp., 156 figures. ⟨83, 85⟩

Barnett, T. P., 1978. The role of the oceans in the global climate system. In *Climatic Change*, J. Gribbin, ed., Cambridge University Press, London, pp. 157–177. ⟨353⟩

Barnett, T. P., and R. E. Davis, 1975. Eigenvector analysis and prediction of sea surface temperature fluctuations in the northern Pacific Ocean. In *Proceedings of the Symposium on Long-Term Climatic Fluctuation*, World Meteorological Organization, Technical Note No. 421, Geneva, pp. 439–450. ⟨353, 354⟩

Barnett, T. P., W. C. Patzert, S. C. Webb, and B. R. Bean, 1979. Climatological usefulness of satellite determined sea-surface temperatures in the tropical Pacific. *Bulletin of the American Meteorological Society* 60:197–205. ⟨427⟩

Barnett, T. P., and J. C. Wilkerson, 1967. On the generation of ocean wind waves as inferred from airborne radar measurements of fetch-limited spectra. *Journal of Marine Research* 25:292–321. ⟨491⟩

Barrett, J. R., 1965. Subsurface currents off Cape Hatteras. *Deep-Sea Research* 12:173–184. ⟨27, 121, 123⟩

Batchelor, G. K., 1953a. *The Theory of Homogeneous Turbulence*. Cambridge University Press, London, 197 pp. ⟨141, 175, 543⟩

Batchelor, G. K., 1953b. The conditions for dynamical similarity of motions of a frictionless perfect-gas atmosphere. *Quarterly Journal of the Royal Meteorological Society* 79:224–235. ⟨548⟩

Batchelor, G. K., 1959. Small-scale variation of convected quantities like temperature in turbulent fluid. *Journal of Fluid Mechanics* 5:113–139. ⟨250⟩

Bates, J. R., 1970. Dynamics of disturbances on the Intertropical Convergence Zone. *Quarterly Journal of the Royal Meteorological Society* 96:677–701. ⟨506⟩

Beardsley, R. C., 1969. A laboratory model of the wind-driven ocean circulation. *Journal of Fluid Mechanics* 38:255–272. ⟨152, 469, 470, 471⟩

Beardsley, R. C., 1972. A numerical model of the wind-driven circulation in a circular basin. *Geophysical Fluid Dynamics* 3:211–243. ⟨152⟩

Beardsley, R. C., 1974. A note on some Antarctic Ocean model experiments. Department of Meteorology, Massachusetts Institute of Technology, Cambridge, Massachusetts, 12 pp. (Unpublished manuscript.) ⟨472⟩

Beardsley, R. C., 1975. The 'sliced-cylinder' laboratory model of the wind-driven ocean circulation. Part 2. Oscillatory forcing and Rossby wave resonance. *Journal of Fluid Mechanics* 69:41–64. ⟨472⟩

Beardsley, R. C., W. C. Boicourt, and D. V. Hansen, 1976. Physical oceanography of the Middle Atlantic Bight. In *Middle Atlantic Continental Shelf and the New York Bight*, M. G. Gross, ed., American Society of Limnology and Oceanography, Special Symposia, 2, pp. 20–34. ⟨207, 208, 214, 227, 228⟩

Beardsley, R. C., and B. Butman, 1974. Circulation on the New England Continental Shelf: response to strong winter storms. *Geophysical Research Letters* 1:181–184. ⟨214⟩

Beardsley, R. C., and D. B. Haidvogel, 1980. A simple numerical model for the wind-driven transient circulation in the Middle Atlantic Bight. *Journal of Physical Oceanography*. ⟨223⟩

Beardsley, R. C., H. Mofjeld, M. Wimbush, C. Flagg, and J. Vermersch, 1977. Ocean tides and weather-induced bottom pressure fluctuations in the Middle Atlantic Bight. *Journal of Geophysical Research* 82:3175–3182. ⟨222, 327⟩

Beardsley, R. C., and K. Robbins, 1975. The 'sliced-cylinder' laboratory model of the wind-driven ocean circulation. Part I. Steady forcing and topographic Rossby wave instability. *Journal of Fluid Mechanics* 69:27–40. ⟨472⟩

Beardsley, R. C., K. D. Saunders, A. C. Warn-Varnas, and A. D. Harding, 1979. An experimental and numerical study of the secular spin-up of a thermally stratified rotating fluid. *Journal of Fluid Mechanics* 93:161–184. ⟨476⟩

Beardsley, R. C., and C. D. Winant, 1979. On the mean circulation in the Mid-Atlantic Bight. *Journal of Physical Oceanography* 9:612–619. ⟨228⟩

Begemann, F., and W. F. Libby, 1957. Continental water balance, inventory of ground water storage times, surface ocean mixing rates and world-wide circulation patterns from cosmic-ray and bomb tritium. *Geochimica et Cosmochimica Acta* 12:277–296. ⟨441⟩

Behringer, D. W., R. L. Molinari, and J. F. Festa, 1977. The variability of anticyclonic current patterns in the Gulf of Mexico. *Journal of Geophysical Research* 82:5469–5476. ⟨113⟩

Bell, T. H., 1975. Topographically generated internal waves in the open ocean. *Journal of Geophysical Research* 80:320–327. ⟨332⟩

Belyaev, V. S., M. M. Lubimtzev, and R. V. Ozmidov, 1975. The rate of dissipation of turbulent energy in the upper layer of the ocean. *Journal of Physical Oceanography* 5:499–505. ⟨283⟩

Bennett, E. B., 1963. An oceanographic atlas of the eastern tropical Pacific Ocean, based on data from Eastropic Expedition, October–December 1955. *Interamerican Tropical Tuna Commission Bulletin* 8:31–165. ⟨14⟩

Bennett, J. R., and B. A. Magnell, 1979. A dynamical analysis of currents near the New Jersey Coast. *Journal of Geophysical Research* 84:1165–1175. ⟨218⟩

Bernstein, R. L., and W. B. White, 1974. Time and length scales of baroclinic eddies in the central North Pacific Ocean. *Journal of Physical Oceanography* 4:613–624. ⟨359, 372⟩

Bernstein, R. L., and W. B. White, 1977. Zonal variability in the distribution of eddy energy in the mid-latitude North Pacific Ocean. *Journal of Physical Oceanography* 7:123–126. ⟨359, 414⟩

Bernstein, R. L., L. Breaker, and R. Whritner, 1977. California Current eddy formation: Ship, air, and satellite results. *Science* 195:353–359. ⟨427⟩

Berteaux, H. O., 1975. *Buoy Engineering*. Wiley, New York. 319 pp. ⟨402⟩

Berteaux, H. O., and R. G. Walden, 1968. The mooring wire testing and evaluation programs for 1968. *W.H.O.I. Technical Memorandum 7-68*, Woods Hole Oceanographic Institution, Woods Hole, Massachusetts, 16 pp. ⟨410⟩

Berteaux, H. O., and R. G. Walden, 1969. Analysis and experimental evaluation of single point moored buoy systems. *W.H.O.I. Technical Report 69-36*, Woods Hole Oceanographic Institution, Woods Hole, Massachusetts, 51 pp. + Appendices. ⟨410⟩

Bien, G. S., N. W. Rakestraw, and H. E. Suess, 1963. Radiocarbon dating of deep water of the Pacific and Indian Oceans. In *Radiocarbon Dating*, International Atomic Energy Agency, Vienna, pp. 159–173. ⟨449⟩

Bigelow, H. B., 1915. Exploration of the coast water between Nova Scotia and Chesapeake Bay, July and August, 1913, by the U.S. Fisheries Schooner *Grampus*. Oceanography and plankton. *Bulletin of the Museum of Comparative Zoology* 59:152–359. ⟨209, 210, 211⟩

Bigelow, H. B., 1922. Exploration of the coastal water off the northeastern United States in 1916 by the U.S. Fisheries Schooner *Grampus*. *Bulletin of the Museum of Comparative Zoology* 65:88–188. ⟨209, 211⟩

Bigelow, H. B., 1927. Physical oceanography of the Gulf of Maine. *Bulletin of the United States Bureau of Fisheries* 40 (Part 2):511–1027. ⟨209, 233⟩

Bigelow, H. B., 1929. *Report of the Committee on Oceanography of the National Academy of Sciences*, National Academy of Sciences, Washington, D.C., 165 pp. ⟨233⟩

Bigelow, H. B., 1933. Studies of the waters on the continental shelf, Cape Cod to Chesapeake Bay. I. The cycle of temperature. *Papers in Physical Oceanography and Meteorology* 2:4, 135 pp. ⟨211⟩

Bigelow, H. B., and M. Sears, 1935. Studies of the waters on the continental shelf, Cape Cod to Chesapeake Bay. II. Salinity. *Papers in Physical Oceanography and Meteorology* 4:1, 94 pp. ⟨211, 233⟩

Bjerkan, P., 1919. Results of the hydrographical observations made by Dr. Johan Hjort in the Canadian Atlantic Waters during the year 1915. In *Canadian Fisheries Expedition, 1914–1915, Investigations in the Gulf of St. Lawrence and Atlantic Waters of Canada*, J. Hjort, director, Department of the Naval Service, J. de Labroquerie Taché, Ottawa, pp. 347–404. ⟨233⟩

Bjerknes, J., 1919. On the structure of moving cyclones. *Geofysiske Publikasjoner* 1:2, 8 pp. ⟨529⟩

Bjerknes, J., 1937. Die Theorie der Aussertropischen Zyklonenbuildung. *Meteorologische Zeitschrift* 54:462–466. ⟨302, 508, 529⟩

Bjerknes, J., 1955. Investigations of the general circulation of the atmosphere. *Final Report General Circulation Project, AF 19(122)-48*, Department of Meteorology, University of California at Los Angeles. ⟨507⟩

Bjerknes, J., 1957. Large scale synoptic processes. *Final Report General Circulation Project, AF 19(604)-1286*, Department of Meteorology, University of California at Los Angeles. ⟨507⟩

Bjerknes, J., 1969. Atmospheric teleconnections from the equatorial Pacific. *Monthly Weather Review* 97:163–172. ⟨351, 353, 496⟩

Bjerknes, J., and H. Solberg, 1921. Meterological conditions for the formation of rain. *Geofysiske Publikasjoner* 2:3, 61 pp. ⟨529⟩

Bjerknes, J., and H. Solberg, 1922. Life cycle of cyclones and the polar front theory of atmospheric circulation. *Geofysiske Publikasjoner* 3:1, 18 pp. ⟨529⟩

Bjerknes, V., 1898. Ueber einen hydrodynamischen Fundamentalsatz und seine Anwendung besonders auf die Mechanik der Atmosphäre und des Weltmeeres. *Kongliga Svenska Vetenskaps-Akademiens Handlingar, Ny Följd* 31:4, 35 pp. ⟨201, 233⟩

Bjerknes, V., J. Bjerknes, H. Solberg, and T. Bergeron, 1933. *Physikalische Hydrodynamik. Mit Anwendung auf die Dynamische Meterologie.* Springer, Berlin, 797 pp. ⟨295⟩

Blackadar, A. K., and H. Tennekes, 1968. Asymptotic similarity in neutral barotropic planetary boundary layers. *Journal of the Atmospheric Sciences* 25:1015–1020. ⟨502⟩

Blandford, R. R., 1965. Notes on the theory of the thermocline. *Journal of Marine Research* 23:18–29. ⟨159⟩

Blandford, R., 1966. Mixed gravity-Rossby waves in the ocean. *Deep-Sea Research* 13:941–961. ⟨191, 527⟩

Blandford, R. R., 1971. Boundary conditions in homogeneous ocean models. *Deep-Sea Research* 18:739–751. ⟨89⟩

Blanton, J., 1971. Exchange of Gulf Stream water with North Carolina shelf water in Onslow Bay during stratified conditions. *Deep-Sea Research* 18:167–178. ⟨119⟩

Blanton, J. O., 1975. Potential energy fluctuations at the edge of the North Carolina continental shelf. *Deep-Sea Research* 22:559–563. ⟨119⟩

Blumen, W., 1968. On the stability of quasi-geostrophic flow. *Journal of the Atmospheric Sciences* 25:929–933. ⟨509, 530, 531⟩

Blumsack, S. L., 1973. Length scales in a rotating stratified fluid on the beta plane. *Journal of Physical Oceanography* 3:133–138. ⟨37⟩

Bogdanov, K. T., and V. A. Magarik, 1967. Chislennoe reshenie zadachi o raprostranenii polusutochnykh prilivnykh voln (M_2 i S_2) v Mirovom okeane. *Doklady Akademii Nauk SSSR* 172: 1315–1317. (Numerical solution of the distribution problem for the semidiurnal tidal waves (M_2 and S_2) in the world ocean. *Doklady of the Academy of Sciences of the U.S.S.R., Earth Science Sections* 172:7–9.) ⟨322⟩

Boicourt, W. C., 1973. The circulation of water on the continental shelf from Chesapeake Bay to Cape Hatteras. Ph.D. Thesis, The Johns Hopkins University, Baltimore, Maryland, 183 pp. ⟨214⟩

Boicourt, W. C., and P. W. Hacker, 1976. Circulation on the Atlantic continental shelf of the United States, Cape May to Cape Hatteras. *Mémoires de la Société Royale des Sciences de Liège, Sixième Série* 10:187–200. ⟨214, 219, 222⟩

Bolin, B., 1950. On the influence of the earth's orography on the general character of the westerlies. *Tellus* 2:184–195. ⟨522⟩

Bolin, B., 1955. Numerical forecasting with the barotropic model. *Tellus* 7:27–49. ⟨508⟩

Bolin, B., and E. Eriksson, 1959. Changes in the carbon dioxide content of the atmosphere and sea due to fossil fuel combustion. In *The Atmosphere and the Sea in Motion. Scientific Contributions to the Rossby Memorial Volume*, B. Bolin, ed., Rockefeller Institute Press, New York, pp. 130–143. ⟨449⟩

Bolin, B., and H. Stommel, 1961. On the abyssal circulation of the World Ocean.—IV. Origin and rate of circulation of deep ocean water as determined with the aid of tracers. *Deep-Sea Research* 8:95–110. ⟨435⟩

Booker, J. R., and F. P. Bretherton, 1967. The critical layer for internal gravity waves in a shear flow. *Journal of Fluid Mechanics* 27:513–539. ⟨274⟩

Born, G. H., J. A. Dunne, and D. B. Lame, 1979. Seasat mission overview. *Science* 204:1405–1406. ⟨426⟩

Bowden, K. F., 1962. Turbulence. In *The Sea: Ideas and Observations on Progress in the Study of the Seas, 1: Physical Oceanography*, M. N. Hill, ed., Wiley, Interscience, New York, pp. 802–825. ⟨239, 258⟩

Bowden, K. F., 1963. The mixing processes in a tidal estuary. *International Journal of Air and Water Pollution* 7:343–356. ⟨204⟩

Bowden, K. F., 1965. Horizontal mixing in the sea due to a shearing current. *Journal of Fluid Mechanics* 21:83–95. ⟨378⟩

Bowden, K. F., 1977. Turbulent processes in estuaries. In *Estuaries, Geophysics, and the Environment*, C. B. Officer, panel chairman, National Academy of Sciences, Washington, D.C., pp. 46–56. ⟨205⟩

Bowden, K. F., and L. A. Fairbairn, 1952. Further observations of the turbulent fluctuations in a tidal current. *Philosophical Transactions of the Royal Society of London A* 244:335–356. ⟨205⟩

Bowden, K. F., and L. A. Fairbairn, 1956. Measurements of turbulent fluctuations and Reynolds stresses in a tidal current. *Proceedings of the Royal Society of London A* 237:422–438. ⟨205⟩

Bowden, K. F., and R. M. Gilligan, 1971. Characteristic features of estuarine circulation as represented in the Mersey Estuary. *Limnology and Oceanography* 16:490–502. ⟨205⟩

Bowden, K. F., and M. R. Howe, 1963. Observations of turbulence in a tidal channel. *Journal of Fluid Mechanics* 17:271 284. ⟨205⟩

Bowen, V. T., and T. T. Sugihara, 1957. Sr^{90} in North Atlantic surface water. *Proceedings of the National Academy of Sciences of the U.S.A.* 43:576–580. ⟨446⟩

Bowen, V. T., and T. T. Sugihara, 1958. Marine geochemical studies with fallout radioisotopes. In *Waste Treatment and Environmental Aspects of Atomic Energy. Proceedings of International Conference on Peaceful Uses of Atomic Energy, United Nations, Geneva*, 18, pp. 434–438. ⟨446⟩

Bowen, V. T., and T. T. Sugihara, 1960. Strontium-90 in the "mixed layer" of the Atlantic Ocean. *Nature* 186:71–72. ⟨446⟩

Boyd, J. P., 1976. The noninteraction of waves with the zonally averaged flow on a spherical earth and the interrelationships of eddy fluxes of energy, heat and momentum. *Journal of the Atmospheric Sciences* 33:2285–2291. ⟨533⟩

Boyd, J. P., 1977. Solitary waves on the equatorial beta-plane. In *Review Papers of Equatorial Oceanography-FINE Workshop Proceedings*, Nova/N.Y.I.T. University Press, Fort Lauderdale, 13 pp. ⟨516⟩

Bradshaw, A., and K. E. Schleicher, 1965. The effect of pressure on the electrical conductance of sea water. *Deep-Sea Research* 12:151–162. ⟨416⟩

Brekhovskikh, L. M., K. N. Fedorov, L. M. Fomin, M. N. Koshlyakov, and A. D. Yampolsky, 1971. Large-scale multi-buoy experiment in the Tropical Atlantic. *Deep-Sea Research* 18:1189–1206. ⟨359, 374⟩

Brekhovskikh, L. M., V. V. Goncharov, V. M. Kurtepov, and K. A. Naugol'nykh, 1972. O rezonansnom vozbuzhdenii vnutrennei volny pri nelineinom vzaimodeistvii poverkhnostnykh voln. *Fizika Atmosfery i Okeana, Izvestiya Akademii Nauk SSSR* 8:192–203. (Resonant excitation of internal waves by nonlinear interaction of surface waves. *Izvestiya, Academy of Sciences, USSR, Atmospheric and Oceanic Physics*, 8:112–117.) ⟨286⟩

Brennecke, W., 1909. Ozeanographie. *Forschungsreise S.M.S. "Planet" 1906/07*, 3, 153 pp. ⟨8⟩

Brennecke, W., 1911. Ozeanographischen Arbeiten der Deutschen Antarktischen Expedition. (Pernambuco-Buenos Aires.) III. Bericht. *Annalen der Hydrographie und Maritimen Meteorologie* 39:642–647. ⟨9⟩

Brennecke, W., 1915. Aufgaben und Probleme der Ozeanographie. *Annalen der Hydrographie und Maritimen Meteorologie* 43:49–62. ⟨11⟩

Brennecke, W., 1918. Ozeanographische Ergebnisse der zweiter Französischen, der schwedischen und der schottischen Südpolarexpeditionen. *Annalen der Hydrographie und Maritimen Meteorologie* 46:173–183. ⟨17⟩

Brennecke, W., 1921. Die ozeanographischen Arbeiten der Deutschen Antarktischen Expedition 1911–1912. *Aus dem Archiv der Deutschen Seewarte* 39:1, 216 pp. ⟨11, 17, 22, 72⟩

Bretherton, F. P., 1964. Low frequency oscillations trapped near the equator. *Tellus* 16:181–185. ⟨295⟩

Bretherton, F. P., 1966a. Baroclinic instability and the short wavelength cut-off in terms of potential vorticity. *Quarterly Journal of the Royal Meteorological Society* 92:335–345. ⟨141, 169, 172, 529⟩

Bretherton, F. P., 1966b. Critical layer instability in baroclinic flows. *Quarterly Journal of the Royal Meteorological Society* 92:325–334. ⟨169, 173, 182, 529, 531⟩

Bretherton, F. P., 1966c. The propagation of groups of internal gravity waves in a shear flow. *Quarterly Journal of the Royal Meteorological Society* 92:466–480. ⟨274⟩

Bretherton, F. P., R. E. Davis, and C. B. Fandry, 1976. A technique for objective analysis and design of oceanographic experiments applied to MODE-73. *Deep-Sea Research* 23:559–581. ⟨431⟩

Bretherton, F. P., and D. B. Haidvogel, 1976. Two-dimensional turbulence above topography. *Journal of Fluid Mechanics* 78:129–154. ⟨178⟩

Bretherton, F. P., P. Hazel, S. A. Thorpe, and I. R. Wood, 1967. Appendix to: "The effect of viscosity and heat conduction on internal gravity waves at a critical level" by P. Hazel. *Journal of Fluid Mechanics* 30:781–783. ⟨274⟩

Bretherton, F. P., and M. J. Karweit, 1975. Mid-ocean meso-scale modeling. In *Numerical Models of Ocean Circulation*, National Academy of Sciences, Washington, D.C., pp. 237–249. ⟨133, 178⟩

Brewer, P. G., and A. Bradshaw, 1975. The effect of the non-ideal composition of seawater on salinity and density. *Journal of Marine Research* 33:157–175. ⟨448⟩

Brink, K. H., 1978. A laboratory study of open ocean barometric response. *Dynamics of Atmospheres and Oceans* 2:153–183. ⟨475⟩

Brink, K. H., and J. S. Allen, 1978. On the effect of bottom friction on barotropic motion over the continental shelf. *Journal of Physical Oceanography* 8:919–922. ⟨223⟩

Briscoe, M. G., 1975a. Internal waves in the ocean. *Reviews of Geophysics and Space Physics* 13:591–598. ⟨265⟩

Briscoe, M. G., 1975b. Preliminary results from the tri-moored internal wave experiment (IWEX). *Journal of Geophysical Research* 80:3872–3884. ⟨265, 332⟩

British Admiralty, 1903. *Sailing Directions for the Southeast Coast of Nova Scotia and Bay of Fundy*, London, 350 pp. ⟨209⟩

Brocks, K., and L. Krügermeyer, 1972. The hydrodynamic roughness of the sea surface. In *Studies in Physical Ocean-*

ography: A Tribute to Georg Wüst on his 80th Birthday, A. L. Gordon, ed., Gordon and Breach, New York, 1, pp. 75–92. ⟨489⟩

Broecker, W. S., 1963. C^{14}/C^{12} ratios in surface ocean water. In *Proceedings of Conference on Nuclear Geophysics, Woods Hole, June 7–9, 1962*, P. M. Hurley, G. Faure, and C. Schnetzler, eds., Publication 1075, National Academy of Sciences—National Research Council, Washington, D.C., pp. 138–149. ⟨439, 449⟩

Broecker, W. S., 1965. An application of natural radon to problems in ocean circulation. In *Symposium on Diffusion in Oceans and Fresh Waters*, Lamont Geological Observatory, Palisades, New York, pp. 116–145. ⟨446⟩

Broecker, W. S., 1974. "NO", a conservative water-mass tracer. *Earth and Planetary Science Letters* 23:100–107. ⟨448⟩

Broecker, W. S., 1979. A revised estimate for the radiocarbon age of North Atlantic deep water. *Journal of Geophysical Research* 84:3218–3226. ⟨450, 453⟩

Broecker, W. S., in press. A distribution of He^3 anomalies in the deep Atlantic. *Earth and Planetary Science Letters.* ⟨445⟩

Broecker, W. S., J. Cromwell, and Y.-H. Li, 1968. Rates of vertical eddy diffusion near the ocean floor based on measurements of the distribution of excess ^{222}Rn. *Earth and Planetary Science Letters* 5:101–105. ⟨446⟩

Broecker, W. S., R. Gerard, M. Ewing, and B. C. Heezen, 1960. Natural radiocarbon in the Atlantic Ocean. *Journal of Geophysical Research* 65:2903–2931. ⟨449, 450⟩

Broecker, W. S., and Y.-H. Li, 1970. Interchange of water between the major oceans. *Journal of Geophysical Research* 75:3545–3552. ⟨440, 449, 450⟩

Broecker, W. S., Y.-H. Li, and T.-H. Peng, 1971. Carbon dioxide—man's unseen artifact. In *Impingement of Man on the Oceans*, D. W. Hood, ed., John Wiley and Sons, New York, pp. 234–287. ⟨449⟩

Broecker, W. S., and E. A. Olson, 1959. Lamont natural radiocarbon measurements VI. *American Journal of Science Radiocarbon Supplement* 1:111–132. ⟨439⟩

Broecker, W. S., and E. A. Olson, 1961. Lamont radiocarbon measurements VIII. *Radiocarbon* 3:176–204. ⟨439⟩

Broecker, W. S., and H. G. Östlund, 1979. Property distributions along the $\sigma_\theta = 26.8$ isopycnal in the Atlantic Ocean. *Journal of Geophysical Research* 84:1145–1154. ⟨456⟩

Broecker, W. S., T.-H. Peng, and M. Stuiver, 1978. An estimate of the upwelling rate in the equatorial Atlantic based on the distribution of bomb radiocarbon. *Journal of Geophysical Research* 83:6179–6186. ⟨440, 456, 460⟩

Broecker, W. S., T.-H. Peng, and T. Takahashi, in press. A strategy for the use of bomb produced radiocarbon as a tracer for the uptake of fossil fuel CO_2 by the ocean's deep water source regions. *Earth and Planetary Science Letters.* ⟨447⟩

Broecker, W. S., T. Takahashi, and Y.-H. Li, 1976. Hydrography of the Central Atlantic—I. The two degree discontinuity. *Deep-Sea Research* 23:1083–1104. ⟨454⟩

Broecker, W. S., T. Takahashi, H. J. Simpson, and T.-H. Peng, 1979. The fate of fossil fuel carbon dioxide and the global carbon budget. *Science* 206:409–418. ⟨449⟩

Broecker, W. S., and A. Walton, 1959. Radiocarbon from nuclear tests. *Science* 130:309–314. ⟨439⟩

Brooks, D. A., 1975. Wind-forced Continental Shelf waves in the Florida Current. *University of Miami, Rosenstiel School of Marine and Atmospheric Sciences, Technical Report UM-RSMAS–#75026*, 268 pp. ⟨117, 120, 121⟩

Brooks, D. A., 1978. Subtidal sea level fluctuations and their relation to atmospheric forcing along the North Carolina coast. *Journal of Physical Oceanography* 8:481–493. ⟨120⟩

Brooks, D. A., and J. M. Bane, Jr., 1978. Gulf Stream deflection by a bottom feature off Charleston, South Carolina. *Science* 201:1225–1226. ⟨120⟩

Brooks, I. H., and P. P. Niiler, 1977. Energetics of the Florida Current. *Journal of Marine Research* 35:163–191. ⟨118⟩

Brown, N. L., 1968. An in situ salinometer for use in the deep ocean. In *ISA Marine Sciences Instrumentation* 4 (Jarvarg 1968), p. 563.

Brown, N. L., 1974. A precision CTD microprofiler. In *Ocean 74 Record, 1974 IEEE Conference on Engineering in the Ocean Environment*, IEEE Publication 74 CHO 873-0 OEC, Institute of Electrical and Electronics Engineers, New York, 2, pp. 270–278. ⟨267, 416, 418, 421⟩

Brown, N. L., and B. V. Hamon, 1961. An inductive salinometer. *Deep-Sea Research* 8:65–75. ⟨416⟩

Brown, W., W. Munk, F. Snodgrass, B. Zetler, and H. Mofjeld, 1975. MODE bottom experiment. *Journal of Physical Oceanography* 5:75–85. ⟨165, 305, 347, 348⟩

Browning, K. A., and F. H. Ludlam, 1962. Airflow in convective storms. *Quarterly Journal of the Royal Meteorological Society* 88:117–135. ⟨500⟩

Bryan, K., 1960. The instability of a two-layered system enclosed between horizontal, coaxially rotating plates. *Journal of Meteorology* 17:446–455. ⟨466, 469, 476⟩

Bryan, K., 1963. A numerical investigation of a nonlinear model of a wind-driven ocean. *Journal of the Atmospheric Sciences* 20:594–606. ⟨89, 90, 155, 156⟩

Bryan, K., 1969. Climate and the ocean circulation. III. The ocean model. *Monthly Weather Review* 97:806–827. ⟨174⟩

Bryan, K., and M. D. Cox, 1968a. A nonlinear model of an ocean driven by wind and differential heating: Part I. Description of the three-dimensional velocity and density fields. *Journal of the Atmospheric Sciences* 25:945–967. ⟨118⟩

Bryan, K., and M. D. Cox, 1968b. A nonlinear model of an ocean driven by wind and differential heating: Part II. An analysis of the heat, vorticity and energy balance. *Journal of the Atmospheric Sciences* 25:968–978. ⟨118⟩

Bryan, K., S. Manabe, and R. C. Pacanowski, 1975. A global ocean-atmosphere climate model. Part II. The oceanic circulation. *Journal of Physical Oceanography* 5:30–46. ⟨182⟩

Bryden, H. L., 1977. Geostrophic comparisons from moored measurements of current and temperature during the Mid-Ocean Dynamics Experiment. *Deep-Sea Research* 24:667–681. ⟨397, 407⟩

Bryden, H., 1979. Poleward heat flux and conversion of available potential energy in Drake Passage. *Journal of Marine Research* 37:1–22. ⟨359, 372⟩

Bryden, H., and N. P. Fofonoff, 1977. Horizontal divergence and vorticity estimates from velocity and temperature measurements in the MODE region. *Journal of Physical Oceanography* 7:329–337. ⟨372⟩

Buchan, A., 1895. Report on oceanic circulation, based on the observations made on board H.M.S. Challenger, and other observations. In *Report on the Scientific Results of the Voyage of H.M.S. Challenger during the years 1872–76, A Summary of the Scientific Results,* Second Part, Appendix (Physics and Chemistry, Part VIII), 38 pp. ⟨9⟩

Buchan, A., 1897. Specific gravities and oceanic circulation. *Transactions of the Royal Society of Edinburgh* 38:317–342. ⟨199⟩

Buchanan, J. Y., 1877. On the distribution of salt in the ocean, as indicated by the specific gravity of its waters. *Journal of the Royal Geographical Society* 47:72–86. ⟨71, 72⟩

Buchanan, J. Y., 1884. Report on the specific gravity of samples of ocean-water, observed on board H.M.S. Challenger, during the years 1873–1876. In *Report on the Scientific Results of the Voyage of H.M.S. Challenger during the years 1873–76, Physics and Chemistry,* 1:2, 46 pp. ⟨9⟩

Buchanan, J. Y., 1888. The exploration of the Gulf of Guinea. *Scottish Geographical Magazine* 4:177–200, 233–251. ⟨185⟩

Buchanan, J. Y., 1913. On the land slopes separating continents and ocean basins. *Collected Papers,* Cambridge University Press, London, 1:39, 31 pp. ⟨199⟩

Buchwald, V., 1968. The diffraction of Kelvin waves at a corner. *Journal of Fluid Mechanics* 31:193–205. ⟨299⟩

Budyko, M. I., 1956. *Teplovoi Balans Zemnoi Poverkhnosti.* Gidrometeorologicheskoe Izdatel'stvo, Leningrad, 255 pp. (*The Heat Balance of the Earth's Surface,* N. A. Stepanova, translator, PB 131692, Office of Technical Services, Department of Commerce, Washington, D.C., 1958, 259 pp.) ⟨490⟩

Budyko, M. I., 1963. *Atlas Teplovogo Balansa Zemnogo Shara (Atlas of the Heat Balance of the Globe).* Mezhduvedomstvennyi Geofizicheskii Komitet pri Presidiume Akademii Nauk SSSR, Glavnaya Geofizicheskaya Observatoriya imeni A. I. Voeikova, Moscow, 5 pp., 69 plates. ⟨58⟩

Bue, C. D., 1970. Stream flow from the United States into the Atlantic Ocean during 1931–1960. *Contributions to the Hydrology of the United States. Geological Survey Water Supply Papers,* 1899–I, 36 pp. ⟨232⟩

Buff, H., 1850. *Zur Physik der Erde.* Friedrich Vieweg und Sohn, Braunschweig, 251 pp. ⟨40⟩

Bumpus, D., 1955. Investigation of climate and oceanographic factors influencing the environment of fish. *Oceanus* 4:1, 3–5. ⟨212⟩

Bumpus, D. F., 1956. Drift bottles are getting bigger. *Oceanus* 4:4, 22–24. ⟨213⟩

Bumpus, D. F., 1965. Residual drift along the bottom on the continental shelf in the Middle Atlantic Bight area. *Alfred C. Redfield 75th Anniversary Volume, Limnology and Oceanography* 10 (Supplement):R50–R53. ⟨213⟩

Bumpus, D. F., 1969. Reversals in the surface drift in the Middle Atlantic Bight area. *Frederick C. Fuglister Sixtieth Anniversary Volume, Deep-Sea Research* 16 (Supplement): 17–23. ⟨212, 213⟩

Bumpus, D. F., 1973. A description of the circulation on the continental shelf of the east coast of the United States. *Progress in Oceanography* 6:111–157. ⟨207, 213, 227, 228⟩

Bumpus, D. F., J. Chase, C. Day, D. H. Frantz, Jr., D. D. Ketchum, and R. G. Walden, 1957. A new technique for studying non-tidal drift with results of experiments off Gay Head, Massachusetts, and in the Bay of Fundy. *Journal of the Fisheries Research Board of Canada* 14:931–944. ⟨213⟩

Bumpus, D. F., and L. M. Lauzier, 1965. Surface circulation on the continental shelf off eastern North America between Newfoundland and Florida. *Serial Atlas of the Marine Environment,* Folio 7, American Geographical Society, New York, 8 pp. + 8 plates. ⟨212⟩

Bunker, A. F., 1976. Computations of surface energy flux and annual air-sea interaction cycles of the North Atlantic Ocean. *Monthly Weather Review* 104:1122–1140. ⟨186, 230, 232, 348, 490⟩

Bunker, A. F., and L. V. Worthington, 1976. Energy exchange charts of the North Atlantic Ocean. *Bulletin of the American Meteorological Society* 57:670–678. ⟨137⟩

Burger, A. P., 1958. Scale consideration of planetary motions of the atmosphere. *Tellus* 10:195–205. ⟨302, 508, 509, 510, 517⟩

Burger, A. P., 1962. On the non-existence of critical wavelengths in a continuous baroclinic stability problem. *Journal of the Atmospheric Sciences* 19:30–38. ⟨529⟩

Burkov, V. A., R. P. Bulatov, and V. G. Neiman, 1973. Krupnomasshtabnye cherty tsirkulyatsii vod mirvogo okeana. *Okeanologiya* 13:395–403. (Large-scale features of water circulation in the world ocean. *Oceanology* 13:325–332.) ⟨85, 87, 108⟩

Burkov, V. A., and I. M. Ovchinnikov, 1960. Osobennosti struktury zonal'nykh potokov i meridional'noi tsirkulyatsii vod v tsentral'noi chasti Tikhogo okeana zimoi severnogo polushariya (Features of the structure of zonal flows and meridional circulation of water in the central part of the Pacific Ocean in the Northern Hemisphere winter). *Trudy Instituta Okeanologii, Akademiya Nauk SSSR* 40:93–107. ⟨83⟩

Burling, R. W., 1959. The spectrum of waves at short fetches. *Deutsche Hydrographische Zeitschrift* 12:45–64, 96–117. ⟨491, 492, 493⟩

Burt, W. V., and E. M. Agee, 1977. Buoy and satellite observations of mesoscale cellular convection during AMTEX 75. *Boundary-Layer Meteorology* 12:3–24. ⟨215⟩

Burton, J. D., 1975. Radionuclides in the marine environment. In *Chemical Oceanography,* J. P. Riley and G. Skirrow, eds., Academic Press, London, 1, pp. 91–191. ⟨435⟩

Buscaglia, J. L., 1971. On the circulation of the Intermediate Water in the southwestern Atlantic Ocean. *Journal of Marine Research* 29:245–255. ⟨82, 84⟩

Busch, N. E., 1977. Fluxes in the surface boundary-layer over the sea. In *Modelling and Prediction of the Upper Layers of the Ocean,* E. Kraus, ed., Pergamon Press, Oxford, pp. 72–91. ⟨485, 487⟩

Bush, K., and S. L. Kupferman, 1980. Wind stress direction and the alongshore pressure gradient in the Middle Atlantic Bight. *Journal of Physical Oceanography* 10:469–471. ⟨229⟩

Businger, J. A., and S. J. S. Khalsa, 1978. On the structure of convective elements in the air near the surface. In *Turbulent Fluxes through the Sea Surface, Wave Dynamics, and Prediction,* A. Favre and K. Hasselmann, eds., Plenum Press, New York, pp. 5–19. ⟨498⟩

Busse, F. H., 1969. On Howard's upper bound for heat transport by turbulent convection. *Journal of Fluid Mechanics* 37:457–477. ⟨387⟩

Busse, F. H., 1978. Non-linear properties of thermal convection. *Reports on Progress in Physics*, 41:1929–1967. ⟨387⟩

Butman, B., and M. Noble, 1978. Bottom currents and bottom sediment mobility in the offshore Middle Atlantic Bight 1976–77. Chapter 2 in *Geological Studies of the Middle Atlantic OCS for the Period July 1, 1976–June 30, 1977*, Report by the U.S. Geological Survey to the Bureau of Land Management, 53 pp. ⟨226⟩

Butman, B., and M. Noble, 1979. Observations of currents and sediment movement on Georges Bank. In draft of final report, *Geological Studies of the Georges Bank OCS for the Period 1 October 1977–30 September 1978*, Report by the U.S. Geological Survey to the Bureau of Land Management. ⟨226⟩

Butman, B., M. Noble, and D. A. Folger, 1979. Long-term observations of bottom current and bottom sediment movement on the Mid-Atlantic Continental Shelf. *Journal of Geophysical Research* 84:1187–1205. ⟨218, 222, 225⟩

Buys Ballot, 1857. Note sur le rapport de l'intensité et de la direction du vent avec les écarts simultanés du baromètre. *Comptes Rendus Hebdomadaires des Séances de l'Académie des Sciences* 45:765–768. See also "Afwijkingen in Europa. Écarts en Europe. 1857", *Meteorologische Waarnemingen in Nederland en Zigne Bezittingen, en Afwijkingen van Temperatuur en Barometerstand op Vele Plaatsen in Europa* (retitled as *Jaarboek. Nederlands Meteorologisch Instituut*) 1857: 175–271. ⟨508⟩

Buzyna, G., and G. Veronis, 1971. Spin-up of a stratified fluid: theory and experiment. *Journal of Fluid Mechanics* 50:579–608. ⟨476⟩

Bye, J. A. T., and G. Veronis, 1979. A correction to the Sverdrup transport. *Journal of Physical Oceanography* 9:649–651. ⟨151⟩

Cacchione, D., and C. Wunsch, 1974. Experimental study of internal waves over a slope. *Journal of Fluid Mechanics* 66:223–239. ⟨261⟩

Cairns, J. L., and G. O. Williams, 1976. Internal wave observations from a midwater float, 2. *Journal of Geophysical Research* 81:1943–1950. ⟨265, 272, 286, 339, 340, 419⟩

Caldwell, D. R., R. E. Snodgrass, and M. H. Wimbush, 1969. Sensors in the deep sea. *Physics Today* 22:34–42. ⟨425⟩

Caldwell, D. R., and C. W. van Atta, 1970. Characteristics of Ekman boundary layer instabilities. *Journal of Fluid Mechanics* 44:79–95. ⟨464⟩

Caldwell, D. R., C. W. van Atta, and K. N. Holland, 1972. A laboratory study of the turbulent Ekman boundary layer. *Geophysical Fluid Dynamics* 3:125–160. ⟨149⟩

Caldwell, D. R., S. D. Wilcox, and M. Matsler, 1975. A relatively simple freely-falling probe for small-scale temperature gradients. *Limnology and Oceanography* 20:1035–1042. ⟨419⟩

Callahan, J. E., 1972. The structure and circulation of Deep Water in the Antarctic. *Deep-Sea Research* 19:563–575. ⟨44, 84, 106⟩

Cameron, W., 1951. On the transverse forces in a British Columbia inlet. *Transactions of the Royal Society of Canada, Series 3* 45:1–8. ⟨203⟩

Cameron, W. M., and D. W. Pritchard, 1963. Estuaries. In *The Sea: Ideas and Observations on Progress in the Study of the Seas, 2: The Composition of Sea-Water, Comparative and Descriptive Oceanography*, M. N. Hill, ed., Wiley, Interscience, New York, pp. 306–324. ⟨202, 204⟩

Cane, M., 1979a. The response of an equatorial ocean to simple wind stress patterns: I. Model formulation and analytic results. *Journal of Marine Research* 37:233–252. ⟨527⟩

Cane, M. A., 1979b. The response of an equatorial ocean to simple wind stress patterns; II. Numerical results. *Journal of Marine Research* 37:253–299. ⟨191⟩

Cane, M., and E. Sarachik, 1976. Forced baroclinic ocean motions: I. The linear equatorial unbounded case. *Journal of Marine Research* 34:629–665. ⟨185, 526⟩

Cane, M., and E. Sarachik, 1977. Forced baroclinic ocean motions: II. The linear equatorial bounded case. *Journal of Marine Research* 35:395–432. ⟨185, 526⟩

Cane, M., and E. Sarachik, 1979. Forced baroclinic ocean motions: III. The linear equatorial basin case. *Journal of Marine Research* 37:355–398. ⟨185, 527⟩

Cannon, G. A., 1966. Tropical waters in the western Pacific Ocean, August–September 1957. *Deep-Sea Research* 13:1139–1148. ⟨83⟩

Cannon, G. A., 1971. Statistical characteristics of velocity fluctuations at intermediate scales in a coastal plain estuary. *Journal of Geophysical Research* 76:5852–5858. ⟨205⟩

Cannon, G. A., and C. C. Ebbesmeyer, 1978. Winter replacement of bottom water in Puget Sound. In *Estuarine Transport Processes*, B. Kjerfve, ed., University of South Carolina Press, Columbia, South Carolina, pp. 229–238. ⟨206⟩

Cannon, G. A., and N. P. Laird, 1978. Variability of currents and water properties from year-long observations in a fjord estuary. In *Hydrodynamics of Estuaries and Fjords*, J. C. J. Nihoul, ed., Elsevier, Amsterdam, pp. 515–535. ⟨206⟩

Cannon, G. A., and D. Pritchard, 1971. A bi-axial propeller current-meter system for fixed-mount applications. *Journal of Marine Research* 29:181–190. ⟨205⟩

Carmack, E. C., 1973. Silicate and potential temperature in the deep and bottom waters of the western Weddell Sea. *Deep-Sea Research* 20: 927–932. ⟨17, 18⟩

Carmack, E. C., 1977. Water characteristics of the Southern Ocean south of the Polar Front. In *A Voyage of Discovery: George Deacon 70th Anniversary Volume*, M. Angel, ed., Supplement to *Deep-Sea Research*, Pergamon Press, Oxford, pp. 15–41. ⟨456⟩

Carmack, E. C., and K. Aagaard, 1973. On the deep water of the Greenland Sea. *Deep-Sea Research* 20:687–715. ⟨251, 258⟩

Carmack, E. C., and T. D. Foster, 1975. On the flow of water out of the Weddell Sea. *Deep-Sea Research* 22:711–724. ⟨17, 20⟩

Carmack, E. C., and P. D. Killworth, 1978. Formation and interleaving of abyssal water masses off Wilkes Land, Antarctica. *Deep-Sea Research* 25:357–369. ⟨22, 262⟩

Carpenter, J., G. Jeffreys, and W. Thomson, 1869. Preliminary report of the scientific exploration of the deep sea in H. M. Surveying-vessel "Porcupine", during the summer of 1869. *Proceedings of the Royal Society of London* 18:397–492. ⟨8, 22⟩

Cartwright, D. E., 1969. Extraordinary tidal currents near St. Kilda. *Nature* 223:928–932. ⟨326⟩

Cartwright, D. E., 1972. Secular changes in the oceanic tides at Brest, 1711–1936. *Geophysical Journal* 30:433–449. ⟨351⟩

Cartwright, D. E., 1977. Ocean tides. *Reports on Progress in Physics* 40:665–708. ⟨292, 293, 294, 322, 323, 329, 424⟩

Cartwright, D. E., B. D. Zetler, and B. V. Hamon, 1979. Pelagic Tidal Constants. *Publication Scientifique, Association Internationale d'Océanographie Physique, U.G.G.I., No. 30*, Paris, 65 pp. ⟨323⟩

Cazenave, A., S. Daillet, and K. Lambeck, 1977. Tidal studies from the perturbations in satellite orbits. *Philosophical Transactions of the Royal Society of London A* 284:595–606. ⟨323⟩

Cerasoli, C. P., 1975. Free shear layer instability due to probes in rotating source-sink flows. *Journal of Fluid Mechanics* 72:559–586. ⟨464⟩

Chan, L. H., D. Drummond, J. M. Edmond, and B. Grant, 1977. On the barium data from the Atlantic GEOSECS Expedition. *Deep-Sea Research* 24:613–649. ⟨29, 448⟩

Chan, L. H., J. M. Edmond, R. F. Stallard, W. S. Broecker, Y. C. Chung, R. F. Weiss, and T.-L. Ku, 1976. Radium and barium at GEOSECS stations in the Atlantic and Pacific. *Earth and Planetary Science Letters* 32:258–267. ⟨444⟩

Chan, S.-K., 1971. Infinite Prandtl number turbulent convection. *Studies in Applied Mathematics* 50:13–49. ⟨387, 388⟩

Chandrasekhar, S., 1961. *Hydrodynamic and Hydromagnetic Stability*. Oxford University Press, London, 652 pp. ⟨386⟩

Chapman, S., and R. S. Lindzen, 1970. *Atmospheric Tides: Thermal and Gravitational*. Gordon & Breach, New York, 200 pp. ⟨295⟩

Charney, J. G., 1947. The dynamics of long waves in a baroclinic westerly current. *Journal of Meteorology* 4:135–163. ⟨169, 506, 508, 529⟩

Charney, J. G., 1948. On the scale of atmospheric motions. *Geofysiske Publikasjoner* 17:2, 17 pp. ⟨508⟩

Charney, J. G., 1949. On a physical basis for numerical prediction of large-scale motions in the atmosphere. *Journal of Meteorology* 6:371–385. ⟨524⟩

Charney, J. G., 1955a. The generation of ocean currents by wind. *Journal of Marine Research* 14:477–498. ⟨148⟩

Charney, J. G., 1955b. The Gulf Stream as an inertial boundary layer. *Proceedings of the National Academy of Sciences of the U.S.A.* 41:731–740. ⟨125, 154, 506⟩

Charney, J. G., 1955c. The use of the primitive equations of motion in numerical weather prediction. *Tellus* 7:22–26. ⟨508⟩

Charney, J. G., 1960. Non-linear theory of a wind-driven homogeneous layer near the equator. *Deep-Sea Research* 6:303–310. ⟨189, 190⟩

Charney, J. G., 1962. Integration of the primitive and balance equations. In *Proceedings of the International Symposium on Numerical Weather Prediction in Tokyo, November 7–13, 1960*, S. Syono et al., eds., The Meteorological Society of Japan, Tokyo, pp. 131–152. ⟨508⟩

Charney, J. G., 1966. Some remaining problems in numerical weather prediction. In *Advances in Numerical Weather Prediction*, Travelers Research Center, Inc., Hartford, Connecticut, pp. 61–70. ⟨543⟩

Charney, J. G., 1969. The Intertropical Convergence Zone and the Hadley circulation of the atmosphere. In *Proceedings, WMO-IUGG Symposium in Numerical Weather Prediction, Tokyo, 26/11–4/12, 1968*, Japan Meteorological Agency, Tokyo, 3, pp. 73–79. ⟨506, 526⟩

Charney, J. G., 1971a. Geostrophic turbulence. *Journal of the Atmospheric Sciences* 28:1087–1095. ⟨346, 371, 372, 543, 544⟩

Charney, J. G., 1971b. Tropical cyclongenesis and the formation of the intertropical convergence zone. In *Mathematical Problems in the Geophysical Sciences, Part 1, Geophysical Fluid Dynamics. Lectures in Applied Mathematics*, 13, American Mathematical Society, pp. 355–368. ⟨506⟩

Charney, J. G., 1973. Planetary fluid dynamics. In *Dynamic Meteorology*, P. Morel, ed., D. Reidel, Dordrecht, Boston, pp. 128–141. ⟨506, 507, 517⟩

Charney, J. G., and J. G. DeVore, 1979. Multiple-flow equilibria in the atmosphere and blocking. *Journal of the Atmospheric Sciences* 36:1205–1216. ⟨534, 539⟩

Charney, J. G., and P. G. Drazin, 1961. Propagation of planetary-scale disturbances from the lower into the upper atmosphere. *Journal of Geophysical Research* 66:83–109. ⟨345, 524, 533⟩

Charney, J. G., and A. Eliassen, 1949. A numerical method for predicting the perturbations of the middle latitude westerlies. *Tellus* 1:38–54. ⟨140, 506, 522, 525⟩

Charney, J. G., and S. L. Spiegel, 1971. Structure of wind-driven equatorial currents in homogeneous oceans. *Journal of Physical Oceanography* 1:149–160. ⟨190⟩

Charney, J. G., and M. Stern, 1962. On the stability of internal baroclinic jets in a rotating atmosphere. *Journal of the Atmospheric Sciences* 19:159–172. ⟨169, 172, 508, 524, 529, 531⟩

Charney, J. G., and D. Straus, 1980. Form drag instability and multiple equilibria in baroclinic, orographically forced planetary wave systems. *Journal of the Atmospheric Sciences* 37. ⟨534, 538, 539⟩

Charnock, H., 1951. Energy transfer between the atmosphere and the ocean. *Science Progress* 39:80–95. ⟨487⟩

Charnock, H., 1955. Wind stress on a water surface. *Quarterly Journal of the Royal Meteorological Society* 81:639–640. ⟨487⟩

Charnock, H., 1958a. Wind generated water waves. *Science Progress* 46:487–501. ⟨495⟩

Charnock, H., 1958b. A note on empirical wind-wave formulae. *Quarterly Journal of the Royal Meteorological Society* 84:443–447. ⟨491⟩

Charnock, H., 1965. A preliminary study of the directional spectrum of short-period internal waves. In *Second United States Navy Symposium on Military Oceanography, The Proceedings of the Symposium*, U.S. Naval Ordnance Laboratory White Oak, Silver Springs, Maryland, 1, pp. 175–178. ⟨265⟩

Charnock, H., and T. H. Ellison, 1967. The boundary-layer in relation to large-scale motions of the atmosphere and ocean. In *Report of the Study Conference on The Global Atmospheric Research Programme (GARP) Held at Stockholm 28 June–11 July 1967*, ICSU/IUGG Committee on Atmospheric Sciences and COSPAR, Appendix IV, 16 pp. ⟨496⟩

Charnock, H., J. R. D. Francis, and P. A. Sheppard, 1956. An investigation of wind structure in the Trades: Anegada 1953. *Philosophical Transactions of the Royal Society of London A* 249:179–234. ⟨503⟩

Chase, J., 1959. Wind induced changes in the water column along the east coast of the United States. *Journal of Geophysical Research* 64:1013–1032. ⟨212⟩

Chase, J., 1969. Surface salinity along the east coast of the United States. *Frederick C. Fuglister Sixtieth Anniversary Volume, Deep-Sea Research* 16 (Supplement):25–29. ⟨212⟩

Chen, C. F., D. G. Briggs, and R. A. Wirtz, 1971. Stability of thermal convection in a salinity gradient due to lateral heating. *International Journal of Heat and Mass Transfer* 14:57–65. ⟨252⟩

Cheney, R. E., and P. L. Richardson, 1976. Observed decay of a cyclonic Gulf Stream ring. *Deep-Sea Research* 23:143–155. ⟨132, 508, 527⟩

Ching, J. K. S., 1975. Determining the drag coefficient from vorticity, momentum, and mass budget analyses. *Journal of the Atmospheric Sciences* 32:1898–1908. ⟨490⟩

Chu, T. Y., and R. J. Goldstein, 1973. Turbulent convection in a horizontal layer of water. *Journal of Fluid Mechanics* 60:141–159. ⟨391⟩

Chuang, W.-S., D.-P. Wang, and W. C. Boicourt, 1979. Low-frequency current variability on the southern Mid-Atlantic Bight. *Journal of Physical Oceanography* 9:1144–1154. ⟨218, 219, 222⟩

Chung, Y.-C., 1976. A deep ^{226}Ra maximum in the northeast Pacific. *Earth and Planetary Science Letters* 32:249–257. ⟨445⟩

Chung, Y.-C., and H. Craig, in press. The distribution of ^{226}Ra and its relationship with barium and silicate in the Pacific. *Earth and Planetary Science Letters.* ⟨445, 448⟩

Clarke, A. J., 1977. Observational and numerical evidence for wind-forced coastal trapped long waves. *Journal of Physical Oceanography* 7:231–247. ⟨222⟩

Clarke, M. R., 1977. A brief review of sampling techniques and tools of marine biology. In *A Voyage of Discovery: George Deacon 70th Anniversary Volume*, M. Angel, ed., Supplement to *Deep-Sea Research*, Pergamon Press, Oxford, pp. 439–465. ⟨396⟩

Clarke, R. A., 1971. Solitary and cnoidal planetary waves. *Geophysical Fluid Dynamics* 2:343–354. ⟨516⟩

Clarke, R. A., and J. C. Gascard, 1979. Deep convection and the formation of Labrador Sea Water. Abstract (only) in *IAPSO Program for the XVII General Assembly of the International Association for the Physical Sciences of the Ocean and the International Union of Geodesy and Geophysics*, IAPSO Secretariat, P.O. Box 7325, San Diego, California, p. 54. ⟨41⟩

Clarke, R. A., H. Hill, R. F. Reiniger, and B. A. Warren, 1980. Current system south and east of the Grand Banks of Newfoundland. *Journal of Physical Oceanography* 10:25–65. ⟨26, 28, 85, 137, 138⟩

Clarke, R. A., and R. F. Reiniger, 1973. The Gulf Stream at 49°30′W. *Deep-Sea Research* 20:627–641. ⟨26⟩

Clarke, R. H., 1970. Observational studies in the atmospheric boundary-layer. *Quarterly Journal of the Royal Meteorological Society* 96:91–114. ⟨502⟩

Clarke, W. B., M. A. Beg, and H. Craig, 1969. Excess ^{3}He in the sea: evidence for terrestrial primordial helium. *Earth and Planetary Science Letters* 6:213–220. ⟨444⟩

Clarke, W. B., M. A. Beg, and H. Craig, 1970. Excess ^{3}He at the North Pacific GEOSECS station. *Journal of Geophysical Research* 75:7676–7678. ⟨444⟩

CLIMAP Project Members, 1976. The surface of the ice age earth. *Science* 191:1131–1137. ⟨348⟩

Clowes, A. J., 1934. Hydrology of the Bransfield Strait. *Discovery Reports* 9:1–64. ⟨15⟩

Clowes, A. J., 1950. An introduction to the hydrology of South African waters. *Investigational Report, Fisheries and Marine Biological Survey Division, Department of Commerce and Industries, Union of South Africa*, No. 12, 42 pp., 20 charts. ⟨82⟩

Coaker, S. A., 1977. The stability of a Rossby wave. *Geophysical and Astrophysical Fluid Dynamics* 9:1–17. ⟨535⟩

Coats, D. A., 1979. The determination of absolute velocity from the density field in the northeastern Pacific Ocean. M.S. Thesis, University of California at San Diego, 56 pp. ⟨91⟩

Cochrane, J. D., 1958. The frequency distribution of water characteristics in the Pacific Ocean. *Deep-Sea Research* 5:111–127. ⟨42, 44, 46, 47, 49, 50, 57⟩

Cochrane, J. D., 1963. Equatorial Undercurrent and related currents off Brazil in March and April, 1963. *Science* 142:669–671. ⟨82⟩

Cochrane, J. D., 1969. Low sea-surface salinity off northeastern South America in summer 1964. *Journal of Marine Research* 27:327–334. ⟨82⟩

Cochrane, J. D., F. J. Kelly, and C. R. Olling, 1979. Subthermocline countercurrents in the western equatorial Atlantic Ocean. *Journal of Physical Oceanography* 9:724–738. ⟨186⟩

Colebrook, J. M., 1972. Variability in the distribution and abundance of the plankton. *ICNAF Special Publication No.* 8:167–186. ⟨382⟩

Colebrook, J. M., 1976. Trends in the climate of the North Atlantic Ocean over the past century. *Nature* 263:576–577. ⟨59⟩

Colin de Verdière, A., 1977. Quasigeostrophic flows and turbulence in a rotating homogeneous fluid. Ph.D. Thesis, Massachusetts Institute of Technology/Woods Hole Oceanographic Institution, W.H.O.I. Reference No. 78-27, 171 pp. ⟨473, 474, 544⟩

Committee on Geodesy, 1979. *Requirements for a Dedicated Gravity Satellite.* National Academy of Sciences, Washington, D.C. ⟨428⟩

Connary, S. D., and M. Ewing, 1974. Penetration of Antarctic Bottom Water from the Cape Basin into the Angola Basin. *Journal of Geophysical Research* 79:463–469. ⟨29⟩

Cooper, L. H. N., 1952. Factors affecting the distribution of silicate in the North Atlantic Ocean and the formation of North Atlantic Deep Water. *Journal of the Marine Biological Association of the United Kingdom* 30:511–526. ⟨22, 40, 41⟩

Cooper, L. H. N., 1955a. Deep water movements in the North Atlantic as a link between climatic changes around Iceland and biological productivity of the English Channel and Celtic Sea. *Journal of Marine Research* 14:347–362. ⟨22, 44, 58, 72⟩

Cooper, L. H. N., 1955b. Hypotheses connecting fluctuations in Arctic climate with biological productivity of the English Channel. *Papers in Marine Biology and Oceanography, Deep-Sea Research*, 3 (Supplement):212–223. ⟨22⟩

Cornish, V., 1898. On sea beaches and sand banks. *Journal of the Royal Geographical Society* 11:528–543. ⟨199⟩

Cowles, R. P., 1930. A biological study of the offshore waters of Chesapeake Bay. *Bulletin of the United States Bureau of Fisheries* 46:277–381. ⟨200⟩

Cox, C. S., 1962. Internal waves. In *The Sea: Ideas and Observations on Progress in the Study of the Seas*, 1: *Physical Oceanography*, M. N. Hill, ed., Wiley, Interscience, New York, pp. 752–763. ⟨264⟩

Cox, C. S., and C. L. Johnson, 1979. Inter-relations of microprocesses, internal waves, and large scale ocean features. (Unpublished manuscript.) ⟨276⟩

Cox, C. S., and H. Sandstrom, 1962. Coupling of internal and surface waves in water of variable depth. *Journal of the Oceanographical Society of Japan, 20th Anniversary Volume*: 499–513. ⟨331⟩

Cox, M. D., 1975. A baroclinic numerical model of the world ocean: preliminary results. In *Numerical Models of Ocean Circulation*, National Academy of Sciences, Washington, D.C., pp. 107–120. ⟨90⟩

Cox, M. D., 1976. Equatorially trapped waves and the generation of the Somali Current. *Deep-Sea Research* 23:1139–1152. ⟨192⟩

Cox, M. D., 1980. Generation and propagation of 30-day waves in a numerical model of the Pacific. *Journal of Physical Oceanography*. ⟨193⟩

Cox, R. A., 1963. The salinity problem. *Progress in Oceanography* 1:243–261. ⟨416⟩

Craig, H., 1957. The natural distribution of radiocarbon and the exchange time of carbon dioxide between atmosphere and sea. *Tellus* 9:1–17. ⟨449⟩

Craig, H., 1969. Abyssal carbon and radiocarbon in the Pacific. *Journal of Geophysical Research* 74:5491–5506. ⟨440, 449⟩

Craig, H., and L. I. Gordon, 1965. Deuterium and oxygen-18 variations in the ocean and marine atmosphere. In *Stable Isotopes in Oceanographic Studies and Paleotemperatures, Proceedings Spoleto Conference, July 1965*, E. Tongiorgi, ed., Consiglio Nazionale delle Richerche e Sigli, V. Lischi e Figli, Pisa, pp. 9–130. ⟨448⟩

Craig, H., and D. Lal, 1961. The production rate of natural tritium. *Tellus* 13:85–105. ⟨440⟩

Craig, R. A., 1946. Measurements of temperature and humidity in the lowest 1000 feet of the atmosphere over Massachusetts Bay. *Papers in Physical Oceanography and Meteorology* 10:1, 47 pp. ⟨500, 501⟩

Craik, A. D. D., 1977. The generation of Langmuir circulations by an instability mechanism. *Journal of Fluid Mechanics* 81:209–223. ⟨242⟩

Craik, A. D. D., and S. Leibovich, 1976. A rational model of Langmuir circulations. *Journal of Fluid Mechanics* 73:401–426. ⟨478⟩

Crapper, P. F., 1975. Measurements across a diffusive interface. *Deep-Sea Research* 22:537–545. ⟨254⟩

Crapper, P. F., and P. F. Linden, 1974. The structure of turbulent density interfaces. *Journal of Fluid Mechanics* 65:45–63. ⟨243⟩

Crawford, W. R., and T. R. Osborn, 1980. Microstructure measurements in the Atlantic Equatorial Undercurrent during GATE. *Deep-Sea Research* 26(supp. ©1979). ⟨282, 283, 284⟩

Crease, J., 1962. Velocity measurements in the deep water of the western North Atlantic, summary. *Journal of Geophysical Research* 67:3173–3176. ⟨358, 507⟩

Crease, J., 1965. The flow of Norwegian Sea Water through the Faroe Bank Channel. *Deep-Sea Research* 12:143–150. ⟨22⟩

Cromwell, T., 1953. Circulation in a meridional plane in the central equatorial Pacific. *Journal of Marine Research* 12:196–213. ⟨190, 195⟩

Cromwell, T., 1960. Pycnoclines created by mixing in an aquarium tank. *Journal of Marine Research* 18:73–82. ⟨242⟩

Cromwell, T., R. B. Montgomery, and E. D. Stroup, 1954. Equatorial undercurrent in Pacific Ocean revealed by new methods. *Science* 119:648–649. ⟨185, 189⟩

Csanady, G. T., 1967. On the resistance law of a turbulent Ekman layer. *Journal of the Atmospheric Sciences* 24:467–471. ⟨502⟩

Csanady, G. T., 1972. Frictional currents in the mixed layer at the sea surface. *Journal of Physical Oceanography* 2:498–508. ⟨148, 149⟩

Csanady, G. T., 1974. Barotropic currents over the continental shelf. *Journal of Geophysical Research* 4:357–371. ⟨223⟩

Csanady, G. T., 1976. Mean circulation in shallow seas. *Journal of Geophysical Research* 81:5389–5399. ⟨228, 378⟩

Csanady, G. T., 1978. The arrested topographic wave. *Journal of Physical Oceanography* 8:47–62. ⟨223, 228⟩

Csanady, G. T., 1979. The pressure field along the western margin of the North Atlantic. *Journal of Geophysical Research* 84:4905–4915. ⟨228⟩

Csanady, G. T., 1980. Longshore pressure gradients caused by offshore wind. *Journal of Geophysical Research* 85:1076–1084. ⟨218, 223⟩

Cushing, D. H., and D. S. Tungate, 1963. Studies on a *Calanus* patch. I. The identification of a *Calanus* patch. *Journal of the Marine Biological Association of the United Kingdom* 43:327–337. ⟨376, 378⟩

Dahl, O., 1969. The capability of the Aanderaa recording and telemetering instrument. *Progress in Oceanography* 5:103–106. ⟨405⟩

Dahlen, J. M., N. K. Chhabra, J. F. McKenna, J. R. Scholten, J. T. Shillingford, F. J. Siraco, and W. E. Toth, 1977. Draper Laboratory profiling current and CTD meter. *Technical Report Charles Stark Draper Laboratory R-1095*, 122 pp. ⟨424⟩

Daley, R., 1978. Variational non-linear normal mode initialization. *Tellus* 30:201–217. ⟨509⟩

Damon, P. E., J. C. Lerman, and A. Long, 1978. Temporal fluctuations of atmospheric ^{14}C: Causal factors and implications. *Annual Review of Earth and Planetary Sciences* 6:457–460. ⟨439⟩

Dantzler, H. L., Jr., 1976. Geographic variations in intensity of the North Atlantic and North Pacific oceanic eddy fields. *Deep-Sea Research* 23:783–794. ⟨71, 361⟩

Dantzler, H. L., Jr., 1977. Potential energy maxima in the tropical and subtropical North Atlantic *Journal of Physical Oceanography* 7:512–519. ⟨359⟩

Darwin, G. H., 1886. On the dynamical theory of the tides of long period. *Proceedings of the Royal Society of London* 41:337–342. ⟨344⟩

Darwin, G. H., 1911a. *The Tides and Kindred Phenomena in the Solar System.* John Murray, London, 437 pp. ⟨318⟩

Darwin, G. H., 1911b. Tide. *Encyclopedia Brittanica*, 11th ed. ⟨293, 318, 319⟩

Davidson, K. L., 1974. Observational results on the influence of stability and wind-wave coupling on momentum transfer and turbulent fluctuations over ocean waves. *Boundary-Layer Meteorology* 6:305–331. ⟨489⟩

Davis, R. E., 1972. On the prediction of the turbulent flow over a wavy boundary. *Journal of Fluid Mechanics* 52:287–306. ⟨494⟩

Davis, R. E., 1975. Statistical methods. In Dynamics and the Analysis of MODE-1: Report of the MODE-1 Dynamics Group, Massachusetts Institute of Technology, Cambridge, Massachusetts, pp. 1–26. (Unpublished document.) ⟨431⟩

Davis, R. E., 1976. Predictability of sea surface temperature and sea level pressure anomalies over the North Pacific Ocean. *Journal of Physical Oceanography* 6:249–266. ⟨351, 355, 372⟩

Davis, R. E., 1978a. Predictability of sea level pressure anomalies over the North Pacific Ocean. *Journal of Physical Oceanography* 8:233–246. ⟨351, 355⟩

Davis, R. E., 1978b. On estimating velocity from hydrographic data. *Journal of Geophysical Research* 83:5507–5509. ⟨397⟩

Davis, R. E., and A. Acrivos, 1967. The stability of oscillatory internal waves. *Journal of Fluid Mechanics* 30:723–736. ⟨275⟩

Davis, R., and R. Weller, 1978. Propeller current sensors. In *Instruments and Methods in Air-Sea Interaction*, preprint volume for NATO School, Ustaoset, Norway, April 1978, NATO Science Committee, 13 pp. ⟨404⟩

Dawson, W. B., 1897. Survey of tides and currents in Canadian waters. *Proceedings of the Royal Society of Canada* 1:31–39. ⟨199⟩

Dawson, W. B., 1913. *The Currents in the Gulf of St. Lawrence.* Department of the Naval Service, Ottawa, Canada, 45 pp. ⟨208⟩

Deacon, E. L., 1953. Vertical profiles of mean wind in the surface layers of the atmosphere. *Geophysical Memoirs* 11:91, 68 pp. ⟨486⟩

Deacon, E. L., and E. K. Webb, 1962. Small-scale interactions. In *The Sea: Ideas and Observations on Progress in the Study of the Seas*, 1: *Physical Oceanography*, M. N. Hill, ed., Wiley, Interscience, New York, pp. 43–87. ⟨488⟩

Deacon, G. E. R., 1937. The hydrology of the Southern Ocean. *Discovery Reports* 15:1–124. ⟨10, 72⟩

Deacon, G. E. R., 1976. The cyclonic circulation in the Weddell Sea. *Deep-Sea Research* 23:125–126. ⟨17⟩

Deacon, G. E. R., and J. A. Moorey, 1975. The boundary region between currents from the Weddell Sea and Drake Passage. *Deep-Sea Research* 22:265–268. ⟨15⟩

Deacon, M., 1971. *Scientists and the Sea 1650–1900: A Study of Marine Science.* Academic Press, London and New York, 445 pp. ⟨xxvii, 8, 40⟩

Deardorff, J. W., 1972. Parameterisation of the planetary boundary layer for use in general circulation models. *Monthly Weather Review* 100:93–106. ⟨496⟩

Defant, A., 1931. Die ozeanische Zirkulation. Chap. 9 of *Physik des Meeres*, in *Handbuch der Experimental Physik*, 25, Geophysik, II. Teil, Physik des Fester Erdkorpers und des Meeres, G. Angenheister, ed., Akademische Verlagsgesellschaft, Leipzig, pp. 671–686. ⟨10⟩

Defant, A., 1936. Schichtung und Zirkulation des Atlantischen Ozeans. Die Troposphäre. In *Wissenschaftliche Ergebnisse der Deutschen Atlantischen Expedition auf dem Forschungs—und Vermessungsschiff "Meteor" 1925–1927*, 6:1st Part, 3, pp. 289–411. ⟨43, 82⟩

Defant, A., 1938. Aufbau und Zirkulation des Atlantischen Ozeans. *Sitzungsberichte der Preussischen Akademie der Wissenschaften, Physikalisch-mathematische Klasse, Jahrgang 1938*: 145–171. ⟨22, 72⟩

Defant, A., 1941a. Quantitative Untersuchungen zur Statik und Dynamik des Atlantischen Ozeans. Die relative Topographie einzelner Drückflachen im Atlantischen Ozean. In *Wissenschaftliche Ergebnisse der Deutschen Atlantischen Expedition auf dem Forschungs—und Vermessungsschiff "Meteor" 1925–1927*, 6:2nd Part, 4, pp. 183–190. ⟨75, 78, 85, 90, 91, 108⟩

Defant, A., 1941b. Quantitative Untersuchungen zur Statik und Dynamik des Atlantischen Ozeans. Die absolute Topographie des physikalischen Meeresniveaus und der Drückflachen sowie die Wasserbewegungen im Raum des Atlantischen Ozeans. In *Wissenschaftliche Ergebnisse der Deutschen Atlantischen Expedition auf dem Forschungs—und Vermessungsschiff "Meteor" 1925–27*, 6:2nd Part, 1, pp. 191–260. ⟨75, 78, 79, 80, 83, 91, 108⟩

Defant, A., 1950. On the origin of internal tide waves in the open sea. *Journal of Marine Research* 9:111–119. ⟨331⟩

Defant, A., 1961a. *Physical Oceanography*, 2. Pergamon Press, New York, 598 pp. ⟨264⟩

Defant, A., 1961b. *Physical Oceanography*, 1. Pergamon Press, New York, 729 pp. ⟨505⟩

Degens, E. T., and D. A. Ross, eds., 1969. *Hot Brines and Recent Heavy Metal Deposits in the Red Sea: A Geophysical and Geochemical Account.* Springer-Verlag, Berlin, 600 pp. ⟨252⟩

Degens, E. T., R. P. von Herzen, H.-K. Wong, W. G. Deuser, and H. W. Jannasch, 1973. Lake Kivu: Structure, chemistry and biology of an East African rift lake. *Geologische Rundschau* 62:245–277. ⟨256⟩

De Leonibus, P. S., 1963. Power spectra of surface wave heights estimated from recordings made from a submerged hovering submarine. In *Ocean Wave Spectra: Proceedings of a Conference*, Prentice-Hall, Englewood Cliffs, New Jersey, pp. 243–249. ⟨492⟩

De Leonibus, P. S., 1971. Momentum flux and wave spectra observations from an ocean tower. *Journal of Geophysical Research* 76:6506–6527. ⟨489⟩

Denman, K. L., and M. Miyake, 1973. Upper layer modification at Ocean Station Papa: Observations and simulation. *Journal of Physical Oceanography* 3:185–196. ⟨245⟩

Desaubies, Y. J. F., 1973. Internal waves near the turning point. *Geophysical Fluid Dynamics* 5:143–154. ⟨291⟩

Desaubies, Y. J. F., 1975. A linear theory of internal wave spectra and coherences near the Väisälä frequency. *Journal of Geophysical Research* 80:895–899. ⟨272, 291⟩

Desaubies, Y. J. F., and M. C. Gregg, 1978. Observations of internal wave vertical velocities by a free-fall vehicle. *Deep-Sea Research* 25:933–946. ⟨419, 420⟩

de Szoeke, R. A., 1975. Some effects of bottom topography on baroclinic stability. *Journal of Marine Research* 33:93–122. ⟨538⟩

de Szoeke, R. A., and P. B. Rhines, 1976. Asymptotic regimes in mixed-layer deepening. *Journal of Marine Research* 34:111–116. ⟨243⟩

Detrick, R. S., D. L. Williams, J. D. Mudie, and J. G. Sclater, 1974. The Galapagos spreading center: Bottom-water temperatures and the significance of geothermal heating. *Geophysical Journal of the Royal Astronomical Society* 38:627–637. ⟨35⟩

Diaz, H. F., 1980. Meteorological conditions associated with the 1976 New York anoxia episode. In *Oxygen Depletion and Associated Benthic Mortalities in New York Bight, 1976*, R. Swanson and C. Sindermann, eds., NOAA Professional Paper, U.S. Department of Commerce, National Oceanic and Atmospheric Administration, Rockville, Maryland. ⟨225⟩

Dickinson, R. E., 1978. Rossby waves—long period oscillations of oceans and atmospheres. *Annual Review of Fluid Mechanics* 10:159–196. ⟨520⟩

Dietrich, G., 1937. I. Die Lage der Meeresoberflache im Druckfeld von Ozean und Atmphäre, mit besonderer Berücksichtigung des westlichen Nordatlantischen Ozeans und des Golfes von Mexiko. II. Uber Bewegung und Herkunft des Golfstromwassers. *Veröffentlichungen des Instituts für Meereskunde an der Universität Berlin, Neue Folge. A. Geographisch-naturwissenschaftliche Reihe*, 33, 1–51 and 52–91. ⟨211⟩

Dietrich, G., 1944. Die Schwingungssysteme der halb-und eintägigen Tiden in den Ozeanen. *Veröffentlichungen des Instituts für Meereskunde an der Universität Berlin, Neue Folge. A. Geographisch-naturwissenschaftliche Reihe*, 41, 7–68. ⟨318⟩

Dietrich, G., 1956. Überströmung des Island-Färöer Rückens in Bodennähe nach Beobachtungen mit dem Forschungsschiff "Anton Dohrn" 1955-56. *Deutsche Hydrographische Zeitschrift* 9:78–89. ⟨22⟩

Dietrich, G., 1957a. Ozeanographische Probleme der deutschen Forschungsfahrten im Internationalen Geophysikalischen Jahr 1957/58. *Deutsche Hydrographische Zeitschrift* 10:39–61. ⟨22, 41⟩

Dietrich, G., 1957b. Schichtung und Zirkulation der Irminger-See in Juni 1955. *Berichte der Deutsche Wissenschaftlichen Kommission fur Meeresforschung, Neue Folge* 14:255–312. ⟨22, 23⟩

Dillon, T. M., and D. R. Caldwell, 1978. Catastrophic events in a surface mixed layer. *Nature* 276: 601–602. ⟨242⟩

Dobson, F., 1971. Measurements of atmospheric pressure on wind-generated sea waves. *Journal of Fluid Mechanics* 48:91–127. ⟨494⟩

Donelan, M. A., 1978. Whitecaps and momentum transfer. In *Turbulent Fluxes through the Sea Surface, Wave Dynamics, and Prediction*, A. Favre and K. Hasselmann, eds., Plenum Press, New York, pp. 273–287. ⟨495⟩

Donn, W. L., J. G. Patullo, and D. M. Shaw, 1964. Sea-level fluctuations and long waves. In *Research in Geophysics, 2: Solid Earth and Interface Phenomena*, H. Odishaw, ed., MIT Press, Cambridge, Massachusetts, pp. 243–269. ⟨353, 357⟩

Douglas, H. P., 1930. Current measurements in the Strait of Gibraltar made in H. M. S. "Goldfinch" in 1905. *Rapports et Procès-Verbaux des Réunions, Conseil Permanent International pour l'Exploration de la Mer* 67:7–14. ⟨199⟩

Drazin, P. G., 1970. Non-linear baroclinic instability of a continuous zonal flow. *Quarterly Journal of the Royal Meteorological Society* 96:667–676. ⟨532⟩

Drazin, P. G., 1972. Non-linear baroclinic instability of a continuous zonal flow of a viscous fluid. *Journal of Fluid Mechanics* 55:577–588. ⟨532⟩

Dreisigacker, E., and W. Roether, 1978. Tritium and strontium-90 in North Atlantic surface water. *Earth and Planetary Science Letters* 38:301–312. ⟨446⟩

Drever, R. G., and T. B. Sanford, 1970. A free-fall electromagnetic current meter—instrumentation. In *Proceedings of the IERE Conference on "Electronic Engineering in Ocean Technology", Univ. Coll. Swansea, 21–24 Sept. 1970*, Institute of Electrical and Radio Engineers, London, pp. 353–370. ⟨422⟩

Druffel, E. M., and T. W. Linick, 1978. Radiocarbon in annual coral rings of Florida. *Geophysical Research Letters* 5:913–916. ⟨439⟩

Drygalski, E. von, 1904. *Zum Kontinent des eisigen Südens*. Georg Reimer, Berlin, 668 pp. ⟨29⟩

Duffy, D. G., 1978. The stability of a nonlinear, finite-amplitude, neutrally stable Eady wave. *Journal of the Atmospheric Sciences* 35:1619–1625. ⟨534⟩

Düing, W., P. Hisard, E. Katz, J. Meincke, L. Miller, K. V. Moroshkin, G. Philander, A. A. Ribnikov, K. Voigt, and R. Weisberg, 1975. Meanders and long waves in the equatorial Atlantic. *Nature* 257:280–284. ⟨186⟩

Düing, W., and D. Johnson, 1972. High resolution current profiling in the Straits of Florida. *Deep-Sea Research* 19:259–274. ⟨424⟩

Düing, W., C. N. K. Mooers, and T. N. Lee, 1977. Low-frequency variability in the Florida Current and relations to atmospheric forcing from 1972–1974. *Journal of Marine Research* 35:129–161. ⟨121, 347, 358⟩

Durney, B. R., 1977. The influence of mesoscale topography on the stability and growth rates of a two-layer model of the open ocean. *Geophysical and Astrophysical Fluid Dynamics* 9:115–128. ⟨538⟩

Dyer, K. R., 1973. *Estuaries: A Physical Introduction*. John Wiley & Sons, London, 140 pp. ⟨203, 205⟩

Dyer, K. R., 1977. Lateral circulation effects in estuaries. In *Estuaries, Geophysics, and the Environment*, C. B. Officer, panel chairman, National Academy of Sciences, Washington, D.C., pp. 22–29. ⟨205, 206⟩

Eady, E. T., 1949. Long waves and cyclone waves. *Tellus* 1:33–52. ⟨169, 173, 506, 508, 519, 529, 532⟩

Eckart, C., 1948. An analysis of the stirring and mixing processes in incompressible fluids. *Journal of Marine Research* 7:265–275. ⟨239, 267⟩

Eckart, C., 1951. Surface waves in water of variable depth. *Marine Physical Laboratory of the Scripps Institution of Oceanography, Wave Report No. 100, SIO Reference Series 51-12*, 99 pp. ⟨307, 309, 317⟩

Eckart, C., 1953. The generation of waves over a water surface. *Journal of Applied Physics* 24:1485–1494. ⟨494⟩

Eckart, C., 1960. *Hydrodynamics of Oceans and Atmospheres*. Pergamon Press, Macmillan, New York, 290 pp. ⟨268, 295⟩

Eckart, C., 1961. Internal waves in the ocean. *The Physics of Fluids* 4:791–799. ⟨272⟩

Edmond, J. M., 1974. On the dissolution of carbonate and silicate in the deep ocean. *Deep-Sea Research* 21:455–480. ⟨448⟩

Edmond, J. M., Y. Chung, and J. G. Sclater, 1971. Pacific Bottom Water: penetration east around Hawaii. *Journal of Geophysical Research* 76:8089–8097. ⟨35⟩

EG&G, 1978. Summary of physical oceanographic observations near the site of the proposed Atlantic generating station, offshore of Little Egg Inlet, New Jersey, 1972 through 1976. A report to the Public Service Electric and Gas Company, Newark, New Jersey, 365 pp. ⟨217, 218, 219, 225, 226, 230⟩

Ekman, F. L., 1876. On the general causes of the ocean-currents. *Nova Acta Regiae Societatis Scientiarum Upsaliensis*, Serie 3 10:6, 52 pp. ⟨199⟩

Ekman, V. W., 1899. Ein Beitrag zur Erklärung und Berechnung des Stromverlaufs an Flussmündungen. *Öfversigt af Kongliga Vetenskaps-Akademiens Förhandlingar* 56:479–507. ⟨199⟩

Ekman, V. W., 1902. Om jordrotationens inverkan på vindströmmar i hafvet. *Nyt Magazin for Naturvidenskaberne* 40:37–63. ⟨525⟩

Ekman, V. M., 1905. On the influence of the earth's rotation on ocean-currents. *Arkiv för Matematik, Astronomi och Fysik* 2:11, 52 pp. ⟨xxiv, 140, 147, 463, 525⟩

Ekman, V. W., 1906. Beiträge zur Theorie der Meeresströmungen. *Annalen der Hydrographie und Maritimen Meteorologie* 34:423–430. ⟨140⟩

Ekman, V. W., 1923. Über Horizontalzirkulation bei winderzeugten Meeresströmungen. *Arkiv för Matematik, Astronomi och Fysik* 17:26, 74 pp. ⟨xxiv, 38, 72, 74⟩

Ekman, V. W., 1931. On internal waves. *Rapports et Procès-Verbaux des Réunions, Conseil Permanent International pour l'Exploration de la Mer* 76:5–34. ⟨332⟩

Ekman, V. W., 1934. Review of: Georg Wüst. "Das Bodenwasser und die Gliederung der Atlantischen Tiefsee". *Journal du Conseil* 9:102–104. ⟨41, 82⟩

Ekman, V. W., 1953. Studies on ocean currents. Results of a cruise on board the "Armauer Hansen" in 1930 under the leadership of Bjørn Helland-Hansen. Part I and Part II. *Geofysiske Publikasjoner* 19:1, 106 pp. and 122 pp. ⟨265⟩

Ekman, V. W., and B. Helland-Hansen, 1931. Measurement of ocean currents. Experiments in the North Atlantic. *Kungliga Fysiografiska Sällskapets i Lund Förhandlingar* 1:1–7. ⟨265⟩

Elder, J. W., 1969. The temporal development of a model of high Rayleigh number convection. *Journal of Fluid Mechanics* 35:417–437. ⟨498⟩

Eliasen, E., 1960. On the initial development of frontal waves. *Publikationer fra det Danske Meteorologiske Institut, Meddelelser*, 13, 107 pp. ⟨529⟩

Eliassen, A., 1949. The quasi-static equations of motion with pressure as independent variable. *Geofysiske Publikasjoner* 17:3, 44 pp. ⟨508⟩

Eliassen, A., 1952. Slow thermally or frictionally controlled meridional circulation in a circular vortex. *Astrophysica Norvegica* 5:19–60. ⟨509⟩

Eliassen, A., and E. Palm, 1960. On the transfer of energy in stationary mountain waves. *Geofysiske Publikasjoner* 22:3, 23 pp. ⟨345, 533⟩

Ellett, D. J., and D. G. Roberts, 1973. The overflow of Norwegian Sea Deep Water across the Wyville-Thomson Ridge. *Deep-Sea Research* 20:819–835. ⟨22⟩

Elliott, A. J., 1978. Observations of the meteorologically induced circulation in the Potomac estuary. *Estuarine and Coastal Marine Science* 6:285–300. ⟨206⟩

Elliott, J. A., 1972. Microscale pressure fluctuation near waves being generated by the wind. *Journal of Fluid Mechanics* 54:427–448. ⟨494⟩

Elliott, J. A., and N. S. Oakey, 1975. Horizontal coherence of temperature microstructure. *Journal of Physical Oceanography* 5:506–515. ⟨283⟩

Ellis, H., 1751. A letter to the Rev. Dr. Hales, F.R.S. from Captain Henry Ellis, F.R.S. dated Jan. 7, 1950–51, at Cape Monte Africa, Ship Earl of Halifax. *Philosophical Transactions of the Royal Society of London* 47:211–214. ⟨7⟩

Ellison, T. H., 1956. Atmospheric turbulence. In *Surveys in Mechanics*, G. K. Batchelor and R. M. Davies, eds., Cambridge University Press, London, pp. 400–430. ⟨487, 502⟩

Ellison, T. H., and J. S. Turner, 1959. Turbulent entrainment in stratified flows. *Journal of Fluid Mechanics* 6:423–448. ⟨241, 260⟩

El-Sayed, S. Z., 1971. Observations on phytoplankton bloom in the Weddell Sea. In *Antarctic Research Series, 17: Biology of the Antarctic Seas IV*, G. A. Llano and I. E. Wallen, eds., American Geophysical Union, Washington, D.C., pp. 301–312. ⟨380⟩

Emery, K. O., and E. Uchupi, 1972. Western North Atlantic Ocean: Topography, rocks, structure, water, life, and sediments. *American Association of Petroleum Geologists Memoir No. 17*, 532 pp. ⟨208⟩

Eppley, R. W., E. H. Renger, E. L. Venrick, and M. M. Mullin, 1973. A study of plankton dynamics and nutrient cycling in the central gyre of the North Pacific Ocean. *Limnology and Oceanography* 18:534–551. ⟨377⟩

Epstein, S., and T. K. Mayeda, 1953. Variation of O^{18} content of waters from natural sources. *Geochimica et Cosmochimica Acta* 4:213–224. ⟨448⟩

Eriksen, C. C., 1978. Measurements and models of fine structure, internal gravity waves, and wave breaking in the deep ocean. *Journal of Geophysical Research* 83:2989–3009. ⟨249, 258, 271, 287, 405⟩

Eriksen, C. C., 1980. Evidence for a continuous spectrum of equatorial waves in the Indian Ocean. *Journal of Geophysical Research*. ⟨272, 291⟩

Errico, R. M., 1979. The partitioning of energy between geostrophic and ageostrophic modes in a simple model. Ph.D. Thesis, Massachusetts Institute of Technology, Cambridge, Massachusetts. ⟨509, 543⟩

Ertel, H., 1942. Ein neuer hydrodynamischer Wirbelsatz, *Meterologische Zeitschrift* 59:277–282. ⟨142⟩

Evans, G. T., 1978. Biological effects of vertical-horizontal interactions. In *Spatial Pattern in Plankton Communities*, J. H. Steele, ed., Plenum Press, New York, pp. 157–180. ⟨378⟩

Ewing, G. C., 1950. Relation between band slicks at the surface and internal waves in the sea. *Science* 111:91–94. ⟨265⟩

Ewing, M., and J. L. Worzel, 1948. Long-range sound transmission. In *Propagation of Sound in the Ocean*, The Geological Society of America Memoir 27:3, 35 pp. ⟨412⟩

Eyriés, M., 1968. Marégraphes de grandes profondeurs. *Cahiers Océanographiques* 20:355–368. ⟨323, 424, 425⟩

Faller, A. J., 1960. Further examples of stationary planetary flow patterns in bounded basins. *Tellus* 12:159–171. ⟨12, 466, 467, 468, 472, 473⟩

Faller, A. J., 1963. An experimental study of the instability of the laminar Ekman boundary layer. *Journal of Fluid Mechanics* 15:560–576. ⟨464, 466⟩

Faller, A. J., 1969. The generation of Langmuir circulations by the eddy pressure of surface waves. *Limnology and Oceanography* 14:504–513. ⟨477⟩

Faller, A. J., 1978. Experiments with controlled Langmuir circulations. *Science* 201:618–620. ⟨242, 476⟩

Faller, A. J., and E. A. Caponi, 1978. Laboratory studies of wind-driven Langmuir circulations. *Journal of Geophysical Research* 83:3617–3633. ⟨477⟩

Faller, A. J., and R. Kaylor, 1967. Instability of the Ekman spiral with applications to the planetary boundary layers. *Boundary Layers and Turbulence, The Physics of Fluids* 10 (Supplement):S212–S219. ⟨464⟩

Faller, A. J., and D. L. Porter, 1976. A note on eastern boundary currents in a laboratory analogue of the ocean circulation. *Tellus* 28:88–89. ⟨468⟩

Faller, A. J., and W. S. von Arx, 1958. The modeling of fluid flow on a planetary scale. In *Proceedings of the Seventh Hydraulics Conference, June 16–18, 1958*, A. Toch and G. R. Schneider, eds., State University of Iowa, Studies in Engineering Bulletin, 39, pp. 53–72. ⟨466⟩

Fandry, C. B., and L. M. Leslie, 1972. A note on the effect of latitudinally varying bottom topography on the wind-driven ocean circulation. *Tellus* 24:164–167. ⟨28⟩

Fandry, C., and R. D. Pillsbury, 1979. On the estimation of absolute geostrophic volume transport applied to the Antarctic Circumpolar Current. *Journal of Physical Oceanography* 9:449–455. ⟨431, 432⟩

Farmer, D. M., 1975. Penetrative convection in the absence of mean shear. *Quarterly Journal of the Royal Meteorological Society* 101:869–891. ⟨245⟩

Farmer, D. M., 1976. The influence of wind on the surface layer of a stratified inlet: Part II. Analysis. *Journal of Physical Oceanography* 6:941–952. ⟨206⟩

Farmer, D. M., and T. R. Osborn, 1976. The influence of wind on the surface layer of a stratified inlet: Part I. Observations. *Journal of Physical Oceanography* 6:931–940. ⟨206⟩

Farrell, W. E., 1972a. Deformation of the earth by surface loads. *Reviews of Geophysics and Space Physics* 10:761–797. ⟨296⟩

Farrell, W. E., 1972b. Global calculations of tidal loading. *Nature* 23:238. ⟨297, 328⟩

Farrell, W. E., 1979. Earth tides. *Reviews of Geophysics and Space Physics* 17:1442–1446. ⟨323⟩

Farrington, G. C., and J. L. Bryant, 1979. Fast ionic transport in solids. *Science* 204:1371–1379. ⟨401⟩

Fasham, M. J. R., 1978. The application of some stochastic processes to the study of plankton patchiness. In *Spatial Pattern in Plankton Communities*, J. H. Steele, ed., Plenum Press, New York, pp. 131–156. ⟨382⟩

Fasham, M. J. R., and P. R. Pugh, 1976. Observations on the horizontal coherence of chlorophyll *a* and temperature. *Deep-Sea Research* 23:527–538. ⟨380⟩

Fedorov, K. N., 1976. *Tonkaya Termokhalinnaya Struktura Vod Okeana*. Gidrometeoizdat, Leningrad. (*The Thermohaline Finestructure of the Ocean*, D. A. Brown, translator, J. S. Turner, tech. ed., Pergamon Press, Oxford, 1978, 170 pp.) ⟨252, 257⟩

Fejér, L., 1916. Über trigonometrische Polynome. *Journal für reine und angewandte Mathematik* 146:53–74. ⟨389⟩

Ferrel, W., 1856. An essay on the winds and the currents of the ocean, *Nashville Journal of Medicine and Surgery* 11: Nos. 4 and 5; republished in Popular Essays on the Movements of the Atmosphere, *Professional Papers of the Signal Service*, No. 12 (1882), pp. 7–19. See also "The motions of fluids and solids relative to the earth's surface," *Runkle's Mathematical Monthly*, 1858–1860; republished in *Professional Papers of the Signal Service*, No. 8 (1882), 51 pp. Both papers were published in an anthology compiled by Marcel Brillouin entitled *Memoires Originaux sur la Circulation Generale de l'Atmosphere*, Paris, 1900. ⟨508⟩

Fettis, H. F., 1955. On the integration of a class of differential equations occurring in boundary layers and other hydrodynamic problems. In *Proceedings of Fourth Midwestern Conference on Fluid Mechanics, Purdue University, Engineering Experimental Station*, Research Series No. 128, pp. 93–114. ⟨149⟩

Feynman, R. P., R. B. Leighton, and M. Sands, 1964. *The Feynman Lectures on Physics*, 2. Addison-Wesley, Reading, Mass., pp. 41–111. ⟨396⟩

Ffowcs Williams, J. E., 1969. Hydrodynamic noise. *Annual Review of Fluid Mechanics* 1:197–222. ⟨509⟩

Fiadeiro, M. E., and H. Craig, 1978. Three-dimensional modeling of tracers in the deep Pacific Ocean: I. Salinity and oxygen. *Journal of Marine Research* 36:323–355. ⟨81, 449⟩

Fiedler, F., and H. A. Panofsky, 1970. Atmospheric scales and spectral gaps. *Bulletin of the American Meteorological Society* 51:1114–1119. ⟨214⟩

Fieux, M., and H. Stommel, 1975. Preliminary look at feasibility of using marine reports of sea surface temperature for documenting climatic change in the Western North Atlantic. *Journal of Marine Research* 33 (Supplement):83–95. ⟨374⟩

Filloux, J. H., 1969. Bourdon tube deep-sea tide gauges. In *International Symposium on Tsunamis and Tsunami Research*, W. M. Adams, ed., University of Hawaii Press, Honolulu, pp. 223–228. ⟨323⟩

Filloux, J. H., 1970. Deep-sea tide gauge with optical readout of bourdon tube rotations. *Nature* 226:936–937. ⟨425⟩

Filloux, J. H., 1973a. Tidal patterns and energy balance in the Gulf of California. *Nature* 243:217–221. ⟨325⟩

Filloux, J. H., 1973b. Techniques and instrumentation for study of natural electromagnetic induction at sea. *Physics of the Earth and Planetary Interiors* 7:323–338. ⟨426⟩

Filloux, J. H., 1974. Electric field recording on the sea floor with short span instruments. *Journal of Geomagnetism and Geoelectricity* 26:269–279. ⟨426⟩

Findlater, J., T. N. J. Harrower, G. A. Howkins, and H. L. Wright, 1966. Surface and 900 mb wind relationships. *Meteorological Office Scientific Paper* No. 23, H. M. Stationery Office, London, 41 pp. ⟨503⟩

Findlay, A. G., 1853. Oceanic currents, and their connection with the proposed Central-America canals. *Journal of the Royal Geographical Society* 23:217–242. ⟨208⟩

FINE, 1978. Review Papers of Equatorial Oceanography—FINE Workshop Proceedings. Nova/N.Y.I.T. University Press, Fort Lauderdale, pp. various. ⟨184⟩

Firing, E., and R. C. Beardsley, 1976. The behavior of a barotropic eddy on a β-plane. *Journal of Physical Oceanography* 6:57–65. ⟨474⟩

Fischer, H. B., 1972. Mass transport mechanisms in partially stratified estuaries. *Journal of Fluid Mechanics* 53:672–687. ⟨205⟩

Fischer, H. B., 1973. Longitudinal dispersion and turbulent mixing in open-channel flow. *Annual Review of Fluid Mechanics* 5:59–78. ⟨240⟩

Fischer, H. B., 1976. Mixing and dispersion in estuaries. *Annual Review of Fluid Mechanics* 8:107–133. ⟨240⟩

Fisher, A., Jr., 1972. Entrainment of shelf water by the Gulf Stream northeast of Cape Hatteras. *Journal of Geophysical Research* 77:3248–3255. ⟨213⟩

Fjeldstad, J. E., 1933. Interne Wellen. *Geofysiske Publikasjoner* 10:6, 35 pp. ⟨191⟩

Fjørtoft, R., 1950. Application of integral theorems in deriving criteria of stability for laminar flows and for the baroclinic circular vortex. *Geofysiske Publikasjoner* 17:6, 52 pp. ⟨529, 531⟩

Fjørtoft, R., 1953. On the changes in the spectral distribution of kinetic energy for twodimensional, nondivergent flow. *Tellus* 5:225–230. ⟨346, 535, 543, 544⟩

Flagg, C. N., 1977. The kinematics and dynamics of the New England continental shelf and shelf/slope front. Ph.D. Thesis, Massachusetts Institute of Technology/Woods Hole Oceanographic Institution, *WHOI Ref. 77-67*, 207 pp. ⟨218, 219, 222, 226⟩

Flatté, S. M., ed., R. Dashen, W. H. Munk, K. M. Watson, and F. Zachariasen, 1979. *Sound Transmission through a Fluctuating Ocean*. Cambridge University Press, London, 299 pp. ⟨265⟩

Flattery, T. W., 1967. Hough functions. Ph.D. Thesis, University of Chicago, Chicago, Illinois. ⟨297⟩

Fleming, J. A., H. U. Sverdrup, C. C. Ennis, S. L. Seaton, and W. C. Hendrix, 1945. Observations and results in physical oceanography. Graphical and tabular summaries. In *Scientific Results of Cruise VII of the Carnegie during 1928–1929 under Command of Captain J. P. Ault, Oceanography—I-B*, Carnegie Institution of Washington Publication 545, 315 pp. ⟨78, 79, 85⟩

Flierl, G. R., 1977. The application of linear quasi-geostrophic dynamics to Gulf Stream rings. *Journal of Physical Oceanography* 7:365–379. ⟨137, 373⟩

Flierl, G. R., 1978. Models of vertical structure and the calibration of two-layer models. *Dynamics of Atmospheres and Oceans* 2:341–381. ⟨182, 183, 529⟩

Flierl, G. R., 1979a. A simple model for the structure of warm and cold core rings. *Journal of Geophysical Research* 84:781–785. ⟨137⟩

Flierl, G. R., 1979b. Baroclinic solitary waves with radial symmetry. *Dynamics of Atmospheres and Oceans* 3:15–38. ⟨345, 515⟩

Flierl, G. R., V. Kamenkovich, and A. R. Robinson, 1975. Gulf Stream meandering and Gulf Stream rings. In Dynamics and the Analysis of Mode I: Report of the MODE-1 Dynamics Group, Massachusetts Institute of Technology, Cambridge, Mass., pp. 113–135. (Unpublished document.) ⟨524⟩

Flierl, G. R., V. D. Larichev, J. C. McWilliams, and G. M. Reznik, 1980. The dynamics of baroclinic and barotropic solitary eddies. *Dynamics of Atmospheres and Oceans*. ⟨511⟩

Fofonoff, N. P., 1954. Steady flow in a frictionless homogeneous ocean. *Journal of Marine Research* 13:254–262. ⟨117, 141, 153, 154, 341, 531⟩

Fofonoff, N. P., 1956. Some properties of sea water influencing the formation of Antarctic bottom water. *Deep-Sea Research* 4:32–35. ⟨257⟩

Fofonoff, N. P., 1962a. Dynamics of ocean currents. In *The Sea: Ideas and Observations on Progress in the Study of the Seas, 1: Physical Oceanography*, M. N. Hill, ed., Wiley, Interscience, New York, pp. 323–395. ⟨122, 154, 159⟩

Fofonoff, N. P., 1962b. Physical properties of sea water. In *The Sea: Ideas and Observations on Progress in the Study of the Seas, 1: Physical Oceanography*, M. N. Hill, ed., Wiley, Interscience, New York, pp. 3–30. ⟨183⟩

Fofonoff, N. P., 1969. Spectral characteristics of internal waves in the ocean. *Frederick C. Fuglister Sixtieth Anniversary Volume, Deep-Sea Research* 16 (Supplement):58–71. ⟨265, 285⟩

Fofonoff, N. P., and Y. Ercan, 1967. Response characteristics of a Savonius rotor current meter. *W.H.O.I. Technical Report 67-33*, Woods Hole Oceanographic Institution, Woods Hole, Massachusetts, 36 pp. ⟨404, 406⟩

Fofonoff, N. P., and R. B. Montgomery, 1955. The equatorial undercurrent in the light of the vorticity equation. *Tellus* 7:518–521. ⟨191⟩

Folger, D., B. Butman, H. Knebel, and R. Sylvester, 1978. Environmental hazards on the Atlantic outer continental shelf of the United States. *Offshore Technology Conference* (papers presented at the 10th Annual Offshore Technology Conference, Houston, Texas, May 8–11, 1978), Paper OTC 3313. ⟨225⟩

Ford, W. L., J. R. Longard, and R. E. Banks, 1952. On the nature, occurrence and origin of cold low salinity water along the edge of the Gulf Stream. *Journal of Marine Research* 11:281–293. ⟨212, 233⟩

Ford, W. L., and A. R. Miller, 1952. The surface layers of the Gulf Stream and adjacent waters. *Journal of Marine Research* 11:267–280. ⟨212, 213, 233⟩

Foster, T. D., 1965. Onset of convection in a layer of fluid cooled from above. *The Physics of Fluids* 8:1770–1774. ⟨243⟩

Foster, T. D., 1971. Intermittent convection. *Geophysical Fluid Dynamics* 2:201–217. ⟨243⟩

Foster, T.D., 1972. An analysis of the cabbeling instability in sea water. *Journal of Physical Oceanography* 2:294–301. ⟨257⟩

Foster, T. D., and E. C. Carmack, 1976a. Frontal zone mixing and Antarctic Bottom Water formation in the southern Weddell Sea. *Deep-Sea Research* 23:301–317. ⟨17, 19, 20⟩

Foster, T. D., and E. C. Carmack, 1976b. Temperature and salinity structure in the Weddell Sea. *Journal of Physical Oceanography* 6:36–44. ⟨257⟩

Fowlis, W. W., and P. J. Martin, 1975. A rotating laser Doppler velocimeter and some new results on the spin-up experiments. *Geophysical Fluid Dynamics* 7:67–78. ⟨476⟩

Fox, M. J. H., 1976. On the nonlinear transfer of energy in the peak of a gravity-wave spectrum. II. *Proceedings of the Royal Society of London A* 348:467–483. ⟨494⟩

Fox, M. J. H., 1978. On the nonlinear transfer of energy in the peak of a gravity wave spectrum. In *Turbulent Fluxes through the Sea-Surface, Wave Dynamics, and Prediction*, A. Favre and K. Hasselmann, eds., Plenum Press, New York, pp. 319–334. ⟨494⟩

Francis, J. R. D., H. Stommel, H. G. Farmer, and D. Parson, 1953. Observations of turbulent mixing processes in a tidal estuary. *W.H.O.I. Technical Report 53-22*, Woods Hole Oceanographic Institution, Woods Hole, Massachusetts, 20 pp. ⟨205⟩

Frankignoul, C. J., 1972. Stability of finite amplitude internal waves in a shear flow. *Geophysical Fluid Dynamics* 4:91–99. ⟨291⟩

Frankignoul, C., and K. Hasselmann, 1977. Stochastic climate models, part II. Application to sea-surface temperature anomalies and thermocline variability. *Tellus* 29:289–305. ⟨355⟩

Frankignoul, C., and P. Müller, 1979. Quasi-geostrophic response of an infinite β-plane ocean to stochastic forcing by the atmosphere. *Journal of Physical Oceanography* 9:104–127. ⟨217, 347, 348, 355, 515⟩

Freeland, G. L., and D. J. P. Swift, 1979. Surficial sediments. *MESA New York Bight Atlas Monograph* No. 10, New York Sea Grant Institute, Albany, New York, 200 pp. ⟨208⟩

Freeland, G. L., D. J. P. Swift, W. L. Stubblefield, and A. E. Cok, 1976. Surficial sediments of the NOAA-MESA study areas in the New York Bight. In *Middle Atlantic Continental Shelf and the New York Bight*, M. G. Gross, ed., American Society of Limnology and Oceanography, Special Symposia, 2, pp. 90–101. ⟨208⟩

Freeland, H. J., P. B. Rhines, and T. Rossby, 1975. Statistical observations of the trajectories of neutrally buoyant floats in the North Atlantic. *Journal of Marine Research* 33:383–404. ⟨165, 304, 371⟩

Frey, H. R., 1978. Northeastward drift in the northern Mid-Atlantic Bight during late spring and summer 1976. *Journal of Geophysical Research* 83:503–504. ⟨225⟩

Friehe, C. A., and K. F. Schmitt, 1976. Parameterizations of air-sea interface fluxes of sensible heat and moisture by bulk aerodynamic formulas. *Journal of Physical Oceanography* 6:801–809. ⟨487⟩

Fu, L.-L., 1980. Observations and models of inertial waves in the deep ocean. Ph.D. Thesis, Massachusetts Institute of Technology/Woods Hole Oceanographic Institution, 202 pp. ⟨272⟩

Fu, L.-L., and G. Flierl, 1980. Nonlinear energy and enstrophy transfers in a realistically stratified ocean. *Dynamics of Atmosphere and Oceans*. ⟨371, 534⟩

Fu, L.-L., and C. Wunsch, 1979. Recovery of Array III Clusters A and B. In *POLYMODE News No. 60*, Woods Hole Oceanographic Institution, Woods Hole, Massachusetts. (Unpublished document.) ⟨370, 372, 373⟩

Fuglister, F. C., 1947. Average monthly sea surface temperatures of the western North Atlantic Ocean. *Papers in Physical Oceanography and Meteorology* 10:2, 25 pp. ⟨348, 357⟩

Fuglister, F. C., 1951. Annual variations in current speeds in the Gulf Stream System. *Journal of Marine Research* 10:119–127. ⟨116, 357⟩

Fuglister, F. C., 1954. Average temperature and salinity at a depth of 200 meters in the North Atlantic. *Tellus* 6:46–58. ⟨71⟩

Fuglister, F. C., 1960. Atlantic Ocean Atlas of Temperature and Salinity Profiles and Data from the International Geophysical Year of 1957–1958. *Woods Hole Oceanographic Institution Atlas Series* 1:209 pp. ⟨6, 7, 25, 28, 29, 43, 93, 348, 349, 358⟩

Fuglister, F. C., 1963. Gulf Stream '60. *Progress in Oceanography* 1:265–373. ⟨112, 123, 125, 126, 127, 131, 359⟩

Fuglister, F. C., 1972. Cyclonic rings formed by the Gulf Stream 1965–66. In *Studies in Physical Oceanography: A Tribute to Georg Wüst on his 80th Birthday*, A. L. Gordon, ed., Gordon and Breach, New York, 1, pp. 137–168. ⟨125, 132⟩

Fuglister, F. C., and L. V. Worthington, 1951. Some results of a multiple ship survey of the Gulf Stream. *Tellus* 3:1–14. ⟨212⟩

Fultz, D., 1953. A survey of certain thermally and mechanically driven systems of meteorological interest. In *Fluid Models in Geophysics, Proceedings of the First Symposium on the Use of Models in Geophysical Fluid Dynamics*, R. R. Long, ed., Superintendent of Documents, U.S. Government Printing Office, Washington, D.C., pp. 27–63. ⟨466, 469, 476⟩

Gall, R., 1976. A comparison of linear baroclinic instability theory with the eddy statistics of a general circulation model. *Journal of the Atmospheric Sciences* 33:349–373. ⟨174, 183⟩

Gamble, J. C., J. M. Davies, and J. H. Steele, 1977. Loch Ewe bag experiment, 1974. *Bulletin of Marine Science* 27:146–175. ⟨382⟩

Gandin, L. S., 1965. *Objective Analysis of Meteorological Fields.* Israel Program for Scientific Translation, Jerusalem, 242 pp. ⟨431⟩

Gardner, G. B., and J. D. Smith, 1978. Turbulent mixing in a salt wedge estuary. In *Hydrodynamics of Estuaries and*

Fjords, J. C. J. Nihoul, ed., Elsevier, Amsterdam, pp. 79–106. ⟨205⟩

Gargett, A. E., 1976. An investigation of the occurrence of oceanic turbulence with respect to finestructure. *Journal of Physical Oceanography* 6:139–156. ⟨250, 258⟩

Gargett, A. E., 1978. Microstructure and finestructure in an upper ocean frontal regime. *Journal of Geophysical Research* 83:5123–5134. ⟨250, 258⟩

Garratt, J. R., 1977. Review of drag coefficients over oceans and continents. *Monthly Weather Review* 105:915–929. ⟨487, 488, 489, 490⟩

Garrett, C., 1973. The effect of internal wave strain on vertical spectra of fine-structure. *Journal of Physical Oceanography* 3:83–85. ⟨281⟩

Garrett, C., 1974. Tides in gulfs. *Deep-Sea Research* 22:23–35. ⟨326⟩

Garrett, C. J. R., 1976. Generation of Langmuir circulations by surface waves—a feedback mechanism. *Journal of Marine Research* 34:117–130. ⟨242⟩

Garrett, C., 1979. Mixing in the ocean interior. *Dynamics of Atmospheres and Oceans* 3:239–265. ⟨267, 286⟩

Garrett, C., and D. Greenberg, 1977. Predicting changes in tidal regimes: The open boundary problem. *Journal of Physical Oceanography* 7:173–181. ⟨326⟩

Garrett, C. J. R., and W. H. Munk, 1971. The age of the tide and the Q of the oceans. *Deep-Sea Research* 18:493–503. ⟨322, 323⟩

Garrett, C., and W. Munk, 1972a. Oceanic mixing by breaking internal waves. *Deep-Sea Research* 19:823–832. ⟨248, 288, 337⟩

Garrett, C. J. R., and W. H. Munk, 1972b. Space-time scales of internal waves. *Geophysical Fluid Dynamics* 3:225–264. ⟨265, 285, 414⟩

Garrett, C. J. R., and W. H. Munk, 1975. Space-time scales of internal waves: A progress report. *Journal of Geophysical Research* 80:291–297. ⟨265, 285⟩

Garrett, C., and W. Munk, 1979. Internal waves in the ocean. *Annual Review of Fluid Mechanics* 11:339–369. ⟨265, 285⟩

Garvine, R. A., 1977. River plumes and estuary fronts. In *Estuaries, Geophysics, and the Environment*, C. B. Officer, panel chairman, National Academy of Sciences, Washington, D.C., pp. 30–35. ⟨206⟩

Gates, L., 1968. A numerical study of transient Rossby waves in a wind-driven homogeneous ocean. *Journal of the Atmospheric Sciences* 25:3–22. ⟨341⟩

Gates, W. L., 1976. Modeling the ice-age climate. *Science* 191:1138–1144. ⟨348, 350⟩

Geisler, J. E., and R. E. Dickinson, 1975. Critical level absorption of barotropic Rossby waves in a north-south flow. *Journal of Geophysical Research* 80:3805–3811. ⟨548⟩

Gent, P. R., 1974. Baroclinic instability of a slowly varying zonal flow. *Journal of the Atmospheric Sciences* 31:1983–1994. ⟨529⟩

Gent, P. R., 1975. Baroclinic instability of a slowly varying zonal flow. Part 2. *Journal of the Atmospheric Sciences* 32:2094–2102. ⟨529⟩

Gent, P. R., and P. A. Taylor, 1976. A numerical model of the air flow above water waves. *Journal of Fluid Mechanics* 77:105–128. ⟨494⟩

Geraghty, J., D. Miller, F. Van der Leeden, and F. Troise, 1973. *Water Atlas of the United States.* Water Information Center, East Port Washington, Long Island, New York, 88 pp. ⟨230⟩

Gibson, C. H., 1980. Fossil temperature, salinity, and vorticity turbulence. In *Marine Turbulence*, J. C. J. Nihoul, ed., Elsevier, Amsterdam. ⟨282, 284⟩

Giletti, B. F., F. Bozan, and J. L. Kulp, 1958. The geochemistry of tritium. *Transactions of the American Geophysical Union* 39:807–818. ⟨440⟩

Gill, A. E., 1968. Similarity theory and geostrophic adjustment. *Quarterly Journal of the Royal Meteorological Society* 94:586–588. ⟨502⟩

Gill, A. E., 1973. Circulation and bottom water production in the Weddell Sea. *Deep-Sea Research* 20:111–140. ⟨17, 20, 21, 40, 257, 456⟩

Gill, A. E., 1974. The stability of planetary waves on an infinite beta-plane. *Geophysical Fluid Dynamics* 6:29–47. ⟨534, 535⟩

Gill, A. E., 1975a. Models of equatorial currents. In *Numerical Models of Ocean Circulation*, National Academy of Sciences, Washington, D.C., pp. 181–203. ⟨184⟩

Gill, A. E., 1975b. Evidence for mid-ocean eddies in weather ship records. *Deep-Sea Research* 22:647–652. ⟨355⟩

Gill, A. E., and K. Bryan, 1971. Effects of geometry on the circulation of a three-dimensional southern-hemisphere ocean model. *Deep-Sea Research* 18:685–721. ⟨89⟩

Gill, A. E., and A. J. Clarke, 1974. Wind-induced upwelling, coastal currents and sea-level changes. *Deep-Sea Research* 21:325–345. ⟨358⟩

Gill, A. E., J. S. A. Green, and A. J. Simmons, 1974. Energy partition in the large-scale ocean circulation and the production of mid-ocean eddies. *Deep-Sea Research* 21:499–528. ⟨141, 173, 182, 347, 372, 529, 532⟩

Gill, A. E., and P. P. Niiler, 1973. The theory of the seasonal variability in the ocean. *Deep-Sea Research* 20:141–178. ⟨355, 357⟩

Gill, A. E., and R. L. Parker, 1970. Contours of "$h \csc \theta$" for the world's oceans. *Deep-Sea Research* 17:823–824. ⟨151⟩

Gill, A. E., and E. Schumann, 1974. The generation of long shelf waves by the wind. *Journal of Physical Oceanography* 4:83–90. ⟨223⟩

Gill, A. E., and J. S. Turner, 1976. A comparison of seasonal thermocline models with observation. *Deep-Sea Research* 23:391–401. ⟨244, 245⟩

Gilmour, A. E., 1979. Ross Ice Shelf temperatures. *Science* 203:438–439. ⟨21⟩

Godfrey, J. S., 1973. On the dynamics of the western boundary current in Bryan and Cox's (1968) numerical model ocean. *Deep-Sea Research* 20:1043–1058. ⟨118, 119⟩

Godin, G., 1965. Some remarks on the tidal motion in a narrow rectangular sea of constant depth. *Deep-Sea Research* 12:461–468. ⟨324⟩

Gold, E., 1933. Maximum day temperatures and the tephigram (t𝜙 diagram). *Meteorological Office Professional Notes* 5:63, 9 pp. ⟨497⟩

Goldsbrough, G., 1933. Ocean currents produced by evaporation and precipitation. *Proceedings of the Royal Society of London A* 141:512–517. ⟨339, 340⟩

Gonella, J., 1971. The drift current from observations made on the Bouée Laboratoire. *Cahiers Océanographiques* 23:19–33. ⟨148⟩

Gonella, J., 1972. A rotary-component method for analyzing meteorological and oceanographic vector time series. *Deep-Sea Research* 19:833–846. ⟨272⟩

Goody, R. M., and A. R. Robinson, 1966. A discussion of the deep circulation of the atmosphere of Venus. *Astrophysical Journal* 146:339–355. ⟨505⟩

Gordeev, R., B. Kagan, and E. Polyakov, 1977. The effects of loading and self-attraction on global ocean tides: The model and the results of a numerical experiment. *Journal of Physical Oceanography* 7:161–170. ⟨297⟩

Gordon, A. L., 1971. Oceanography of Antarctic waters. In *Antarctic Oceanology I*, J. L. Reid, ed., *Antarctic Research Series*, 15, pp. 169–203. ⟨20, 21⟩

Gordon, A. L., 1974. Varieties and variability of Antarctic Bottom Water. In *Processus de Formation des Eaux Océaniques Profondes, Colloques Internationaux du C.N.R.S.* No. 215, pp. 33–47. ⟨15, 20⟩

Gordon, A. L., 1975a. An Antarctic oceanographic section along 170°E. *Deep-Sea Research* 22:357–377. ⟨20, 21⟩

Gordon, A. L., 1975b. General ocean circulation. In *Numerical Models of Ocean Circulation*, National Academy of Sciences, Washington, D.C., pp. 39–53. ⟨456⟩

Gordon, A. L., 1978. Deep Antarctic convection west of Maud Rise. *Journal of Physical Oceanography* 8:600–612. ⟨22⟩

Gordon, A. L., D. T. Georgi, and H. W. Taylor, 1977. Antarctic Polar Front Zone in the western Scotia Sea—summer 1975. *Journal of Physical Oceanography* 7:309–328. ⟨257⟩

Gordon, A. L., and E. Molinelli, 1975. *USNS Eltanin Southern Ocean Oceanographic Atlas*. Lamont-Doherty Geological Observatory and Department of Geological Sciences of Columbia University, Palisades, New York, 91 plates. ⟨20⟩

Gordon, A. L., and W. D. Nowlin, 1978. The basin waters of the Bransfield Strait. *Journal of Physical Oceanography* 8:258–264. ⟨15⟩

Gordon, A. L., and P. Tchernia, 1972. Waters of the continental margin off Adélie Coast, Antarctica. In *Antarctic Oceanology II: The Australian-New Zealand Sector*, D. E. Hayes, ed., *Antarctic Research Series*, 19, pp. 59–69. ⟨21⟩

Gordon, C. M., 1974. Intermittent momentum transport in a geophysical boundary layer. *Nature* 248:392. ⟨205⟩

Gould, W. J., 1976. Instrumentation for MODE-I. *Oceanus* 19:3, 54–64. ⟨213, 342, 428, 431, 432⟩

Gould, W. J., W. J. Schmitz, Jr., and C. Wunsch, 1974. Preliminary field results for a mid-ocean dynamics experiment (MODE-0). *Deep-Sea Research* 21:911–932. ⟨359⟩

Grant, H. L., R. W. Stewart, and A. Moilliet, 1962. Turbulence spectra from a tidal channel. *Journal of Fluid Mechanics* 12:241–268. ⟨250⟩

Grant, W. D., and O. J. Madsen, 1979. Combined wave and current interaction with a rough bottom. *Journal of Geophysical Research* 84:1797–1807. ⟨218⟩

Green, J. S. A., 1960. A problem in baroclinic stability. *Quarterly Journal of the Royal Meteorological Society* 86:237–251. ⟨169, 173, 529⟩

Green, J. S. A., 1970. Transfer properties of the large-scale eddies and the general circulation of the atmosphere. *Quarterly Journal of the Royal Meteorological Society* 96:157–185. ⟨507⟩

Greenwalt, D., and C. M. Gordon, 1978. Short-term variability in the bottom boundary layer of the deep ocean. *Journal of Geophysical Research* 83:4713–4716. ⟨259⟩

Greenspan, H. P., 1963. A note concerning topography and inertial currents. *Journal of Marine Research* 21:147–154. ⟨139⟩

Greenspan, H. P., and L. N. Howard, 1963. On a time-dependent motion of a rotating fluid. *Journal of Fluid Mechanics* 17:385–404. ⟨476, 525⟩

Gregg, M. C., 1975. Microstructure and intrusions in the California Current. *Journal of Physical Oceanography* 5:253–278. ⟨257, 258, 279, 280, 281⟩

Gregg, M. C., 1976a. Temperature and salinity microstructure in the Pacific Equatorial Undercurrent. *Journal of Geophysical Research* 81:1180–1196. ⟨244⟩

Gregg, M. C., 1976b. Microstructure: Signature of mixing in the ocean. *Naval Research Reviews* 29:11, 1–21. ⟨414⟩

Gregg, M. C., 1977. A comparison of finestructure spectra from the main thermocline. *Journal of Physical Oceanography* 7:33–40. ⟨282, 283⟩

Gregg, M. C., and C. S. Cox, 1971. Measurements of oceanic microstructure of temperature and electrical conductivity. *Deep-Sea Research* 18:925–934. ⟨419⟩

Gregg, M. C., and C. S. Cox, 1972. The vertical microstructure of temperature and salinity. *Deep-Sea Research* 19:355–376. ⟨267⟩

Gregg, M. C., C. S. Cox, and P. W. Hacker, 1973. Vertical microstructure measurements in the central North Pacific. *Journal of Physical Oceanography* 3:458–469. ⟨279, 290⟩

Gregg, M. C., T. Meagher, A. Pederson, and E. Aagaard, 1978. Low noise temperature microstructure measurements with thermistors. *Deep-Sea Research* 25:843–856. ⟨419⟩

Griffiths, R. W., 1979. Transports through thermohaline interfaces in a viscous fluid and a porous medium. Ph.D. Thesis, Australian National University. ⟨256⟩

Grimshaw, R. H. J., 1975. A note on the β-plane approximation. *Tellus* 27:351–357. ⟨143⟩

Groen, P., 1948. Contribution to the theory of internal waves. *Mededelingen en Verhandelingen van het Koninklijk Nederlands Meteorologisch Instituut, De Bilt*, B 2:11, 23 pp. ⟨265⟩

Grose, P. L., and J. S. Mattson, eds., 1977. The ARGO MERCHANT oil spill, a preliminary scientific report. U.S. Department of Commerce, National Oceanic and Atmospheric Administration, Rockville, Maryland, 133 pp. ⟨226⟩

Gross, M. G., R. L. Swanson, and H. M. Stanford, 1976. Man's impact on the middle Atlantic continental shelf and the New York Bight—Symposium summary. In *Middle Atlantic Continental Shelf and the New York Bight*, M. G. Gross, ed.,

American Society of Limnology and Oceanography, Special Symposia, 2, pp. 1–13. ⟨213⟩

Groves, G. W., 1955. Day to day variations of sea level. Ph.D. Thesis, Scripps Institution of Oceanography, University of California at Los Angeles. ⟨191⟩

Groves, G. W., and E. J. Hannan, 1968. Time series regression of sea level on weather. *Reviews of Geophysics and Space Physics* 6:129–134. ⟨351⟩

Groves, G. W., and B. D. Zetler, 1964. The cross spectrum of sea level at San Francisco and Honolulu. *Journal of Marine Research* 22:265–275. ⟨351⟩

Gupta, S. R., and G. J. Hills, 1956. A precision electrode-less conductance cell for use at audio frequencies. *Journal of Scientific Instruments* 33:313–314. ⟨416⟩

Haas, L. W., 1977. The effect of the spring-neap tidal cycle on the vertical salinity structure of the James, York, and Rappahannock Rivers, Virginia, U.S.A. *Estuarine and Coastal Marine Science* 5:485–496. ⟨206⟩

Hachey, H. B., 1934. Movements resulting from mixing of stratified waters. *Journal of the Biological Board of Canada* 1:133–143. ⟨201⟩

Hacker, P. W., 1973. The mixing of heat deduced from temperature fine structure measurements in the Pacific Ocean and Lake Tahoe. Ph.D. Thesis, University of California at San Diego, 121 pp. ⟨283⟩

Hager, J. G., 1977. Kinetic energy exchange in the Gulf Stream. *Journal of Geophysical Research* 82:1718–1724. ⟨118⟩

Hahne, A., A. Volz, D. H. Enhalt, H. Cosatto, W. Roether, W. Weiss, and B. Kromer, 1978. Depth profiles of chlorofluoromethanes in the Norwegian Sea. *Pure and Applied Geophysics* 116:575–582. ⟨447⟩

Haidvogel, D. B., and W. R. Holland, 1978. The stability of ocean currents in eddy-resolving general circulation models. *Journal of Physical Oceanography* 8:393–413. ⟨182⟩

Haight, F. J., 1938. Currents in Narragansett Bay, Buzzards Bay, Nantucket and Vineyard Sounds. *U.S. Coast and Geodetic Survey Special Publication* No. 208, 103 pp. ⟨200⟩

Haight, F. J., 1942. Coastal currents along the Atlantic coast of the United States. *U.S. Coast and Geodetic Survey Special Publication* No. 230, 73 pp. ⟨212⟩

Haight, F. J., H. E. Finnegan, and G. L. Anderson, 1930. Tides and currents in Chesapeake Bay and tributaries. *U.S. Coast and Geodetic Survey Special Publication* No. 162, 143 pp. ⟨200, 201⟩

Hall, R. E., 1976. Scattering of Rossby waves by topography in a stratified ocean. Ph.D. Thesis, University of California at San Diego, 116 pp. ⟨345⟩

Halliwell, G. R., 1978. The space-time structure and variability of the shelf water/slope water and Gulf Stream surface thermal fronts and warm-core eddies, off the Northeast United States. M.S. Thesis, University of Delaware, Newark, 195 pp. ⟨218⟩

Hallock, Z. R., 1977. Wind forced equatorial waves in the Atlantic Ocean. *Technical Report, Rosenstiel School of Marine and Atmospheric Science, University of Miami*, TR77-2, Miami, Florida, 152 pp. ⟨193⟩

Halpern, D., 1971. Semidiurnal internal tides in Massachusetts Bay. *Journal of Geophysical Research* 76:6573–6584. ⟨332⟩

Halpern, D., 1978. Moored current measurements in the upper ocean. In *Instuments and Methods in Air-Sea Interaction*, preprint volume for NATO School, Ustaoset, Norway, April 1978, NATO Science Committee, 17 pp. ⟨404, 405⟩

Halpern, D., and R. D. Pillsbury, 1976. Influence of surface waves upon subsurface current measurements in shallow water. *Limnology and Oceanography* 21:611–616. ⟨405⟩

Hamon, B. V., 1955. A temperature-salinity-depth recorder. *Journal du Conseil* 21:72–73. ⟨415⟩

Hamon, B. V., 1956. A portable temperature-chlorinity bridge for estuarine investigations and sea water analysis. *Journal of Scientific Instruments* 33:329–333. ⟨416⟩

Hamon, B. V., 1958. The effect of pressure on the electrical conductivity of sea water. *Journal of Marine Research* 16:83–89. ⟨416⟩

Hamon, B. V., 1962. The spectrum of mean sea level at Sydney, Coff's Harbor, and Lord Howe Island. *Journal of Geophysical Research* 67: 5147–5155. ⟨358⟩

Hamon, B. V., and N. L. Brown, 1958. A temperature-chlorinity-depth recorder for use at sea. *Journal of Scientific Instruments* 35:452–458. ⟨415⟩

Han, G., D. V. Hansen, and A. Cantillo, 1980. Diagnostic model of water and oxygen transport in the New York Bight. In *Oxygen Depletion and Associated Benthic Mortalities in New York Bight, 1976*, C. R. Swanson and C. S. Sindermann, eds., NOAA Professional Paper, U.S. Department of Commerce, National Oceanic and Atmospheric Administration, Rockville, Maryland. ⟨225⟩

Hanna, S. R., 1969. The thickness of the planetary boundary layer. *Atmospheric Environment* 3: 519–536. ⟨500⟩

Hansen, D. V., 1965. Currents and mixing in the Columbia River estuary. In *Marine Technology Society and American Society of Limnology and Oceanography, Joint Conference on Ocean Science and Ocean Engineering*, pp. 943–955. ⟨205⟩

Hansen, D. V., 1970. Gulf Stream meanders between Cape Hatteras and the Grand Banks. *Deep-Sea Research* 17:495–511. ⟨123, 126, 127, 129, 534⟩

Hansen, D. V., and M. Rattray, Jr., 1965. Gravitational circulation in straits and estuaries. *Journal of Marine Research* 23:104–122. ⟨203, 204, 206⟩

Hansen, D. V., and M. Rattray, Jr., 1966. New dimensions in estuary classification. *Limnology and Oceanography* 11:319–326. ⟨202⟩

Hansen, W., 1949. Die halbtägigen Gezeiten im Nordatlantischen Ozean. *Deutsche Hydrographische Zeitschrift* 2:44–51. ⟨328⟩

Harden Jones, F. R., 1968. *Fish Migration*. Arnold, London, 325 pp. ⟨378⟩

Hardy, A. C., and E. R. Gunther, 1935. The plankton of the South Georgia whaling grounds and adjacent waters, 1926–1927. *Discovery Reports*, 11, 1–456. ⟨376, 378⟩

Harris, R. A., 1907. Manual of Tides, Part V: Currents, shallow-water tides, meteorological tides, and miscellaneous matters. In *Report of the Superintendent of the Coast and Geodetic Survey Showing the Progress of the Work from July 1,*

1906, to June 30, 1907, Department of Commerce and Labor, Washington, Appendix 6, pp. 231–545. ⟨200⟩

Harrison, D. E., 1979a. On the equilibrium linear basin response to fluctuating winds and mesoscale motions in the ocean. *Journal of Geophysical Research* 84:1221–1224. ⟨348, 359⟩

Harrison, D. E., 1979b. Eddies and the general circulation of numerical model gyres: An energetic perspective. *Reviews of Geophysics and Space Physics* 17:969–979. ⟨346, 507⟩

Harrison, D. E., and A. R. Robinson, 1978. Energy analysis of open regions of turbulent flows—mean eddy energetics of a numerical ocean circulation experiment. *Dynamics of Atmospheres and Oceans* 2:185–211. ⟨507, 529⟩

Hart, J. E., 1972. A laboratory study of baroclinic instability. *Geophysical Fluid Dynamics* 3:181–209. ⟨476⟩

Hart, J. E., 1979a. Finite amplitude baroclinic instability. *Annual Review of Fluid Mechanics* 11:147–172. ⟨174, 532⟩

Hart, J. E., 1979b. Barotropic quasigeostrophic flow over anisotropic mountains: multiple equilibria and bifurcation. *Journal of the Atmospheric Sciences* 36:1736–1746. ⟨537, 540⟩

Harvey, R. R., and W. Patzert, 1976. Deep current measurements suggest long waves in the eastern equatorial Pacific Ocean. *Science* 193:883–884. ⟨186, 191, 193⟩

Hasse, L., 1968. Zur Bestimmung der vertikalen Transporte von Impuls und fühlbarer Wärme in der wassernähen Luftschicht über See. *Hamburger Geophysikalischen Einzelschriften* 11, 70 pp. (On the determination of the vertical transports of momentum and heat in the atmospheric boundary layer at sea, J. F. T. Saur, translator, W. H. Quinn and L. Hasse, eds., 1970, *Technical Report 188, Reference 70-22,* Department of Oceanography, Oregon State University, 55 pp.) ⟨489⟩

Hasselmann, K., 1966. Feynman diagrams and interaction rules of wave-wave scattering processes. *Reviews of Geophysics* 4:1–32. ⟨494⟩

Hasselmann, K., 1974. On the spectral dissipation of ocean waves due to white capping. *Boundary-Layer Meteorology* 6:107–127. ⟨495⟩

Hasselmann, K., T. P. Barnett, E. Bouws, H. Carlson, D. E. Cartwright, K. Enke, J. A. Ewing, H. Gienapp, D. E. Hasselmann, P. Kruseman, A. Meerburg, P. Müller, D. J. Olbers, K. Richter, W. Sell, and H. Walden, 1973. Measurements of wind-wave growth and swell decay during the Joint North Sea Wave Project (JONSWAP). *Deutsche Hydrographische Zeitschrift, Ergänzungsheft Reihe A,* 12, 95 pp. ⟨491, 493⟩

Hastenrath, S., and P. J. Lamb, 1977. *Climatic Atlas of the Tropical Atlantic and Eastern Pacific Oceans.* University of Wisconsin Press, Madison, 105 pp. ⟨186⟩

Hasunuma, K., and K. Yoshida, 1978. Splitting of the Subtropical Gyre in the Western North Pacific. *Journal of the Oceanographical Society of Japan* 34:160–172. ⟨85⟩

Haurwitz, B., 1940a. The motion of atmospheric disturbances on a spherical earth. *Journal of Marine Research* 3:254–267. ⟨xv, 344⟩

Haurwitz, B., 1940b. The motion of atmospheric disturbances. *Journal of Marine Research* 3:35–50. ⟨506⟩

Haurwitz, B., 1950. Internal waves of tidal character. *Transactions of the American Geophysical Union* 31:47–52. ⟨331⟩

Haurwitz, B. H., H. Stommel, and W. H. Munk, 1959. On the thermal unrest in the ocean. In *The Atmosphere and the Sea in Motion. Scientific Contributions to the Rossby Memorial Volume,* B. Bolin, ed., Rockefeller Institute Press, New York, pp. 74–94. ⟨290, 364⟩

Haury, L. R., M. G. Briscoe, and M. H. Orr, 1979. Tidally generated internal wave packets in Massachusetts Bay. *Nature* 278:312–317. ⟨266, 378⟩

Hawkins, H. F., and D. T. Rubsam, 1968. Hurricane Hilda 1964. II. Structure and budgets of the hurricane on October 1, 1964. *Monthly Weather Review* 96:617–636. ⟨490⟩

Hay, J. S., 1955. Some observations of air flow over the sea. *Quarterly Journal of the Royal Meteorological Society* 81:307–319. ⟨487⟩

Hayes, S. P., 1978. Temperature fine structure observations in the tropical North Pacific Ocean. *Journal of Geophysical Research* 83:5099–5104. ⟨282⟩

Hayes, S. P., 1979a. Benthic current observations at DOMES sites A, B and C in the tropical North Pacific Ocean. In *Marine Geology of the Central Pacific Manganese Nodule Province,* J. Bischoff and D. Z. Piper, eds., Plenum Press, New York, pp. 83–112. ⟨363, 371⟩

Hayes, S. P., 1979b. Variability of current and bottom pressure across the continental shelf in the northeast Gulf of Alaska. *Journal of Physical Oceanography* 9:88–103. ⟨425⟩

Hayes, S., and D. Halpern, 1976. Variability of the semidiurnal internal tide during coastal upwelling. *Mémoires de la Société Royale des Sciences de Liège, Sixième Série* 10:175–186. ⟨334⟩

Hecht, A., and P. Hughes, 1971. Observations of temperature fluctuations in the upper layers of the Bay of Biscay. *Deep-Sea Research* 18:663–684. ⟨335⟩

Heezen, B. C., R. D. Gerard, and M. Tharp, 1964. The Vema Fracture Zone in the equatorial Atlantic. *Journal of Geophysical Research* 69:733–739. ⟨29⟩

Heinmiller, R. H., 1968. Acoustic release systems. *W.H.O.I. Technical Report 68-48,* Woods Hole Oceanographic Institution, Woods Hole, Massachusetts, 19 pp. ⟨408⟩

Heinmiller, R. H., 1976a. The Woods Hole Buoy Project moorings—1960 through 1974. *W.H.O.I. Technical Report 76-53,* Woods Hole Oceanographic Institution, Woods Hole, Massachusetts, 73 pp. ⟨407, 410⟩

Heinmiller, R. H., 1976b. Mooring operations techniques of the Buoy Project at the Woods Hole Oceanographic Institution, *W.H.O.I. Technical Report 76-69,* Woods Hole Oceanographic Institution, Woods Hole, Massachusetts, 94 pp. ⟨409, 410, 411⟩

Heinmiller, R. H., Jr., and D. A. Moller, 1974. Failure of a moored array in a Gulf-Stream eddy. *Marine Technology Society Journal* 8:7, 35–38. ⟨407⟩

Heiskanen, W., 1921. Über den Einfluss der Gezeiten auf die Säkuläre Acceleration des Mondes. *Annales Academiae Scientiarum Fennicae A* 18:2, 84 pp. ⟨323⟩

Helland-Hansen, B., 1916. Nogen hydrografiske metoder. In *Forhandlinger ved de 16 Skandinaviske Naturforskerermøte,* pp. 357–359. ⟨42⟩

Helland-Hansen, B., and F. Nansen, 1909. The Norwegian Sea. Its physical oceanography based upon the Norwegian researches 1900–1904. *Report on Norwegian Fishery and Ma-*

rine-Investigations 2 (1909):2, 390 pp., 28 plates. ⟨72, 73, 199, 233, 343⟩

Helland-Hansen, B., and F. Nansen, 1920. Temperature variations in the North Atlantic Ocean and in the atmosphere. *Smithsonian Miscellaneous Collection* 70:4, 408 pp. ⟨344⟩

Helland-Hansen, B., and F. Nansen, 1926. The eastern North Atlantic. *Geofysiske Publikasjoner* 4:2, 76 pp., 71 plates. ⟨10, 43, 72, 73, 111⟩

Helmholtz, H. von, 1888. Über atmosphaerische Bewegungen. *Sitzungsberichte der Königlich Preussischen Akademie der Wissenschaften zu Berlin, Jahrgang 1888*: 647–663. ⟨529⟩

Helmholtz, H. von, 1889. Über atmosphaerische Bewegungen. Zweite Mittheilung. *Sitzungsberichte der Königlich Preussischen Akademie der Wissenschaften zu Berlin, Jahrgang 1889*: 761–780. ⟨504⟩

Hendershott, M. C., 1972. The effects of solid earth deformation on global ocean tides. *Geophysical Journal of the Royal Astonomical Society* 29:389–402. ⟨297, 322, 329⟩

Hendershott, M., 1973. Inertial oscillations of tidal period. *Progress in Oceanography* 6:1–27. ⟨328, 329⟩

Hendershott, M., 1977. Numerical models of ocean tides. In *The Sea: Ideas and Observations on Progress in the Study of the Seas, 6: Marine Modeling*, E. D. Goldberg, I. N. McCave, J. J. O'Brien, and J. H. Steele, eds., Wiley, Interscience, New York, pp. 47–95. ⟨328, 329⟩

Hendershott, M., and W. Munk, 1970. Tides. *Annual Review of Fluid Mechanics* 2:205–224. ⟨328⟩

Hendershott, M. C., and A. Speranza, 1971. Co-oscillating tides in long, narrow bays; the Taylor problem revisited. *Deep-Sea Research* 18:959–980. ⟨324⟩

Hendry, R., 1977. Observations of the semidiurnal internal tide in the western North Atlantic Ocean. *Philosophical Transactions of the Royal Society of London A* 286:1–24. ⟨335, 337, 338⟩

Hendry, R. M., and A. J. Hartling, 1979. A pressure-induced direction error in nickel-coated Aanderaa current meters. *Deep-Sea Research* 26:327–335. ⟨406⟩

Hermann, F., 1967. The T-S diagram analysis of the water masses over the Iceland-Faroe Ridge and in the Faroe Bank Channel. *Rapports et Procès-Verbaux des Réunions, Conseil Permanent International pour l'Exploration de la Mer* 157:139–149. ⟨23⟩

Herring, J. R., 1963. Investigation of problems in thermal convection. *Journal of the Atmospheric Sciences* 20:325–338. ⟨174, 387⟩

Herring, J. R., 1977. On the statistical theory of two-dimensional topographic turbulence. *Journal of the Atmospheric Sciences* 34:1731–1750. ⟨178⟩

Herring, J. R., 1980. Statistical theory of quasigeostrophic turbulence. *Journal of the Atmospheric Sciences*. ⟨543⟩

Hickey, B., 1975. The relationship between fluctuations in sea level, wind stress and sea surface temperature in the equatorial Pacific. *Journal of Physical Oceanography* 5:460–475. ⟨192⟩

Hicks, B. B., 1972 Some evaluations of drag and bulk transfer coefficients over water bodies of different sizes. *Boundary-Layer Meteorology* 3:201–213. ⟨489⟩

Hicks, S. D., 1959. The physical oceanography of Narragansett Bay. *Limnology and Oceanography* 4:316–327. ⟨200⟩

Hidaka, K., 1949. Mass transport in ocean currents and lateral mixing. *Journal of Marine Research* 8:132–136. ⟨140, 151⟩

Hide, R., 1958. An experimental study of thermal convection in a rotating liquid. *Philosophical Transactions of the Royal Society of London A* 250:441–478. ⟨469, 476⟩

Hide, R., and P. W. White, 1980. *Orographic Effects in Planetary Flows*. GARP Publication Series, World Meteorological Organization, Geneva. ⟨522⟩

Hidy, G. M., 1972. A view of recent air-sea interaction research. *Bulletin of the American Meteorological Society* 53:1083–1102. ⟨488⟩

Hill, H. W., and R. R. Dickson, 1978. Long-term changes in North Sea hydrography. *Rapports et Procès-Verbaux des Réunions, Conseil Permanent International pour l'Exploration de la Mer* 172:310–334. ⟨376, 382⟩

Hisard, P., J. Merle, and B. Voituriez, 1970. The equatorial undercurrent at 170°E in March and April, 1967. *Journal of Marine Research* 28:281–303. ⟨185, 191⟩

Hoare, R. A., 1968. Thermohaline convection in Lake Vanda, Antarctica. *Journal of Geophysical Research* 73:607–612. ⟨252⟩

Hodnett, P. F., 1978. On the advective model of the thermocline circulation. *Journal of Marine Research* 36:185–198. ⟨159⟩

Hoeber, H., 1969. Wind-, Temperatur- und Feuchteprofils in der wassernahen Luftschicht über dem äquatorialen Atlantik. *Meteor Forschungsergebnisse, Reihe B*, No. 3, 1–26. ⟨489⟩

Hoffert, M. I., and W. S. Broecker, 1978. Apparent vertical eddy diffusion rates in the pycnocline of the Norwegian Sea as determined from the vertical distribution of tritium. *Geophysical Research Letters* 5:502–504. ⟨459⟩

Hogg, N. G., 1972. Steady flow past an island with application to Bermuda. *Geophysical Fluid Dynamics* 4:55–81. ⟨522⟩

Hogg, N. G., 1973. On the stratified Taylor column. *Journal of Fluid Mechanics* 58:517–537. ⟨522⟩

Hogg, N. G., 1976. On spatially growing baroclinic waves in the ocean. *Journal of Fluid Mechanics* 78:217–235. ⟨370⟩

Hogg, N. G., 1980. Observations of internal Kelvin waves trapped round Bermuda. *Journal of Physical Oceanography.* ⟨358⟩

Hogg, N. G., E. J. Katz, and T. B. Sanford, 1978. Eddies, islands, and mixing. *Journal of Geophysical Research* 83:2921–2938. ⟨261, 282, 287⟩

Högström, U., 1974. A field study of the turbulent fluxes of heat, water vapour and momentum at a 'typical' agricultural site. *Quarterly Journal of the Royal Meteorological Society* 100:624–639. ⟨485⟩

Holland, J. Z., and E. M. Rasmusson, 1973. Measurements of the atmospheric mass, energy, and momentum budgets over a 500-kilometer square of tropical ocean. *Monthly Weather Review* 101:44–55. ⟨503⟩

Holland, W. R., 1967. On the wind-driven circulation in an ocean with bottom topography. *Tellus* 19:582–600. ⟨151⟩

Holland, W. R., 1973. Baroclinic and topographic influences on the transport in western boundary currents. *Geophysical Fluid Dynamics* 4:187–210. ⟨121, 142, 179, 180⟩

Holland, W. R., 1978. The role of mesoscale eddies in the general circulation of the ocean—numerical experiments using a wind-driven quasi-geostrophic model. *Journal of Physical Oceanography* 8:363–392. ⟨133, 134, 142, 182, 507⟩

Holland, W. R., and A. D. Hirschman, 1972. A numerical calculation of the circulation in the North Atlantic Ocean. *Journal of Physical Oceanography* 2:336–354. ⟨89⟩

Holland, W. R., and L. B. Lin, 1975a. On the generation of mesoscale eddies and their contribution to the oceanic general circulation. I. A preliminary numerical experiment. *Journal of Physical Oceanography* 5:642–657. ⟨89, 132, 346⟩

Holland, W. R., and L. B. Lin, 1975b. On the generation of mesoscale eddies and their contribution to the oceanic general circulation. II. A parameter study. *Journal of Physical Oceanography* 5:658–669. ⟨132, 346⟩

Holland, W. R., and P. B. Rhines, 1980. An example of eddy induced ocean circulation. *Journal of Physical Oceanography*. ⟨142, 179⟩

Holley, E. R., D. R. F. Harleman, and H. B. Fischer, 1970. Dispersion in homogeneous estuary flow. *Proceedings of the American Society of Civil Engineers, Journal of the Hydraulics Division* 96:1691–1709. ⟨206⟩

Holloway, G., 1978. A spectral theory of nonlinear barotropic motion above irregular topography. *Journal of Physical Oceanography* 8:414–427. ⟨178⟩

Holloway, G., and M. C. Hendershott, 1977. Stochastic closure for non-linear Rossby waves. *Journal of Fluid Mechanics* 82:747–765. ⟨346, 543⟩

Holopainen, E. O., 1970. An observational study of the energy balance of the stationary disturbances in the atmosphere. *Quarterly Journal of the Royal Meteorological Society* 94:626–644. ⟨507⟩

Holton, J. R., 1965. The influence of viscous boundary layers on transient motions in a stratified rotating fluid: Part I; Part II. *Journal of the Atmospheric Sciences* 22:402–411; 535–540. ⟨527⟩

Holton, J. R., 1975. *The Dynamic Meteorology of the Stratosphere and Mesosphere*. Meteorological Monographs, 15, No. 37, American Meteorological Society, Boston, Massachusetts, 218 pp. ⟨344, 345, 520, 524⟩

Holton, J. R., and R. S. Lindzen, 1972. An updated theory for the quasi-biennial cycle of the tropical stratosphere. *Journal of the Atmospheric Sciences* 29:1076–1080. ⟨532⟩

Hopfinger, E. J., and J.-A. Toly, 1976. Spatially decaying turbulence and its relation to mixing across density interfaces. *Journal of Fluid Mechanics* 78:155–175. ⟨242⟩

Hoskins, B. J., 1975. The geostrophic momentum approximation and the semi-geostrophic equations. *Journal of the Atmospheric Sciences* 32:233–242. ⟨508⟩

Hough, S. S., 1897. On the application of harmonic analysis to the dynamical theory of the tides.—Part I. On Laplace's "Oscillations of the First Species", and on the dynamics of ocean currents. *Philosophical Transactions of the Royal Society of London A* 189:201–257. ⟨296, 297, 339, 344⟩

Hough, S. S., 1898. On the application of harmonic analysis to the dynamical theory of the tides—Part II. On the general integration of Laplace's dynamical equations. *Philosophical Transactions of the Royal Society of London A* 191:139–185. ⟨296, 344⟩

Howard, L. N., 1963. Heat transport by turbulent convection. *Journal of Fluid Mechanics* 17:405–432. ⟨387⟩

Howard, L. N., 1964a. Convection at high Rayleigh number. In *Proceedings of the Eleventh International Congress Applied Mechanics, Munich*. H. Görtler, ed., Springer-Verlag, Berlin, pp. 1109–115. ⟨255⟩

Howard, L. N., 1964b. The number of unstable modes in hydrodynamic stability problems. *Journal de Mécanique* 3:433–443. ⟨531⟩

Howard, L. N., 1972. Bounds on flow quantities. *Annual Review of Fluid Mechanics* 4:473–494. ⟨388⟩

Howe, M. R., 1962. Some direct measurements of the non-tidal drift on the continental shelf between Cape Cod and Cape Hatteras. *Deep-Sea Research* 9:445–455. ⟨212, 213⟩

Howe, M. R., and R. I. Tait, 1972. The role of temperature inversions in the mixing processes of the deep ocean. *Deep-Sea Research* 19:781–791. ⟨258⟩

Hsueh, Y., and C. Y. Peng, 1978. A diagnostic model of continental shelf circulation. *Journal of Geophysical Research* 83:3033–3041. ⟨223⟩

Huang, N. E., C. D. Leitao, and C. G. Parra, 1978. Large-scale Gulf Stream frontal study using Geos 3 radar altimeter data. *Journal of Geophysical Research* 83:4673–4682. ⟨374⟩

Hughes, B. A., 1978. The effect of internal waves on surface wind waves. 2. Theoretical analysis. *Journal of Geophysical Research* 83:455–465. ⟨265⟩

Humboldt, A. de. 1814. *Voyage aux Régions Equinoxiales du Nouveau Continent, Fait en 1799–1804 par Al. de Humboldt et A. Bonpland. Part I, Relation Historique*, 1, F. Schoell, Paris. (*Personal Narrative of Travels to the Equinoctial Regions of the New Continent during the Years 1799–1804 by Alexander de Humboldt and Aimé Bonpland*, H. M. Williams, translator, 3rd ed., 1822, Longman, Hurst, Rees, Orme and Brown, London, 1, 293 pp.) ⟨8⟩

Humboldt, A. de, 1831. *Fragmens de Géologie et de Climatologie Asiatiques*. A. Pihan Delaforest, Paris, 2 volumes, 640 pp. ⟨8, 40⟩

Humboldt, A. von, 1845. *Kosmos. Entwurf einer Physischen Weltbeschreibung*, 1, J. G. Cotta'scher, Stuttgart und Tubingen 493 pp. (*Cosmos: A Sketch of a Physical Description of the Universe*, E. C. Otté, translator, 1877, Harper and Brothers, New York, 1, 375 pp.) ⟨8⟩

Hunkins, K., 1966. Ekman drift currents in the Arctic Ocean. *Deep-Sea Research* 13:607–620. ⟨148⟩

Hunkins, K. L., 1974. Subsurface eddies in the Arctic Ocean. *Deep-Sea Research* 21:1017–1033. ⟨359, 371⟩

Huntsman, A. C., 1924. Oceanography. In *Handbook of the British Association for the Advancement of Science*, University of Toronto Press, Toronto, pp. 274–290. ⟨233⟩

Huppert, H. E., 1971. On the stability of a series of double-diffusive layers. *Deep-Sea Research* 18:1005–1021. ⟨254⟩

Huppert, H. E., and K. Bryan, 1976. Topographically generated eddies. *Deep-Sea Research* 23:655–680. ⟨370⟩

Huppert, H. E., and P. F. Linden, 1979. On heating a stable salinity gradient from below. *Journal of Fluid Mechanics* 95:431–464. ⟨251⟩

Huppert, H. E., and P. C. Manins, 1973. Limiting conditions for salt-fingering at an interface. *Deep-Sea Research* 20:315–323. ⟨255⟩

Huppert, H. E., and J. S. Turner, 1972. Double-diffusive convection and its implications for the temperature and salinity structure of the ocean and Lake Vanda. *Journal of Physical Oceanography* 2:456–461. ⟨255⟩

Huppert, H. E., and J. S. Turner, 1978. On melting icebergs. *Nature* 271:46–48. ⟨252⟩

Hurlburt, H. E., J. Kindle, and J. J. O'Brien, 1976. A numerical simulation of the onset of El Niño. *Journal of Physical Oceanography* 6:621–631. ⟨192⟩

Hurlburt, H. E., and J. D. Thompson, 1976. A numerical model of the Somali Current. *Journal of Physical Oceanography* 6:646–664. ⟨192⟩

Ianello, J. P., 1977. Tidally induced residual currents in estuaries of constant breadth and depth. *Journal of Marine Research* 35:755–786. ⟨206⟩

Ibbetson, A., and N. Phillips, 1967. Some laboratory experiments on Rossby waves in a rotating annulus. *Tellus* 19:81–87. ⟨165, 474, 475⟩

Ichiye, T., 1972. Experimental circulation modeling within the Gulf and the Caribbean. In *Contributions on the Physical Oceanography of the Gulf of Mexico*, L. R. A. Capurro and J. L. Reid, eds., *Texas A & M University Oceanographic Studies*, 2, Gulf Publishing Co., Houston, Texas, pp. 213–226. ⟨475⟩

IDOE, 1976. *International Decade of Ocean Exploration Progress Report No. 5*. National Science Foundation, Washington, D.C. ⟨430⟩

Imberger, J., R. Thompson, and C. Fandry, 1976. Selective withdrawal from a finite rectangular tank. *Journal of Fluid Mechanics* 78:489–512. ⟨253⟩

Iooss, G., H. B. Nielsen, and H. True, 1978. Bifurcation of the stationary Ekman flow into a stable periodic flow. *Archive for Rational Mechanics and Analysis* 68:227–256. ⟨464⟩

Irish, J., W. Munk, and F. Snodgrass, 1971. M_2 amphidrome in the northeast Pacific. *Geophysical Fluid Dynamics* 2:355–360. ⟨326, 327, 399, 425⟩

Irish, J. D., and F. E. Snodgrass, 1972. Australian Antarctic tides. In *Antarctic Oceanology II: The Australian-New Zealand Sector*, D. E. Hayes, ed., *Antarctic Research Series*, 19, pp. 101–116. ⟨425⟩

Iselin, C. O'D., 1936. A study of the circulation of the western North Atlantic. *Papers in Physical Oceanography and Meteorology* 4:4, 101 pp. ⟨43, 78, 113, 118, 123, 133, 137, 211, 233⟩

Iselin, C. O'D., 1939a. The influence of vertical and lateral turbulence on the characteristics of the waters at mid-depths. *Transactions of the American Geophysical Union* 20:414–417. ⟨57, 58, 59, 236⟩

Iselin, C. O'D., 1939b. Some physical factors which may influence the productivity of New England's coastal waters. *Journal of Marine Research* 2:74–85. ⟨211, 233⟩

Iselin, C. O'D., 1940a. Preliminary report on long-period variations in the transport of The Gulf Stream System. *Papers in Physical Oceanography and Meteorology* 8:1, 40 pp. ⟨114, 344, 357⟩

Iselin, C. O'D., 1940b. The necessity of a new approach to the study of the circulation on the continental shelf. *Transactions of the American Geophysical Union* 21:347–348. ⟨211⟩

Iselin, C. O'D., 1955. Coastal currents and the fisheries. *Papers in Marine Biology and Oceanography, Deep-Sea Research* 3 (Supplement):474–478. ⟨212⟩

ISOS, 1978. DRAKE 79—the 1979 ISOS experiment in the Drake Passage. Available from the ISOS Office, Department of Oceanography, Texas A & M University, College Station, Texas. (Unpublished manuscript.) ⟨432⟩

Israeli, M., 1972. On trapped modes of rotating fluids in spherical shells. *Studies in Applied Mathematics* 51:219–237. ⟨295⟩

Ivers, W. D., 1975. The deep circulation in the northern North Atlantic, with especial reference to the Labrador Sea. Ph.D. Thesis, University of California at San Diego, 179 pp. ⟨84, 93, 108⟩

Jacobs, S. S., A. F. Amos, and P. M. Bruchhausen, 1970. Ross Sea oceanography and Antarctic Bottom Water formation. *Deep-Sea Research* 17:935–962. ⟨21⟩

Jacobs, S. S., and D. T. Georgi, 1977. Observations on the southwest Indian/Antarctic Ocean. In *A Voyage of Discovery: George Deacon 70th Anniversary Volume*, M. Angel, ed., Supplement to *Deep-Sea Research*, Pergamon Press, Oxford, pp. 43–84. ⟨21, 29⟩

Jacobs, S. S., A. L. Gordon, and J. L. Ardai, 1979. Circulation and melting beneath the Ross Ice Shelf. *Science* 203:439–443. ⟨21⟩

Jacobs, W. C., 1951. Large scale aspects of energy transformation over the ocean. In *Compendium of Meteorology*, T. F. Malone, ed., American Meteorological Society, Boston, Massachusetts, pp. 1057–1070. ⟨490⟩

Jacobsen, A. W., 1948. An instrument for recording continuously the salinity, temperature, and depth of sea water. *Transactions of the American Institute of Electrical Engineers* 67:714–722. ⟨415⟩

Jacobsen, J. P., 1909. Der Libellenstrommesser. *Publications de Circonstance; Conseil Permanent International pour l'Exploration de la Mer*, No. 51, 20 pp. ⟨203⟩

Jacobsen, J. P., 1916. Contribution to the hydrography of the Atlantic. *Meddelelser fra Kommissionen for Havundersøgelser. Serie: Hydrografi* 2:5, 24 pp. ⟨22⟩

Jacobsen, J. P., 1929. Contribution to the hydrography of the North Atlantic: The "Dana" Expedition 1921–22. In *The Danish "Dana"-Expeditions 1920–22 in the North Atlantic and the Gulf of Panama, Oceanographical Reports Edited by the "Dana"-Committee*, 1, 3, 98 pp. ⟨74, 76, 85⟩

Jacobsen J. P., 1930. Remarks on the determination of the movement of the water and intermixing of the watersheets in a vertical direction. *Rapports et Procès-Verbaux des Réunions, Conseil Permanent International pour l'Exploration de la Mer* 64:59–68. ⟨201⟩

Jacobsen, R. A., 1973. More staying power for small batteries. *Machine Design* 45:30, 136–148. ⟨401⟩

Jeffreys, H., 1921. Tidal friction in shallow seas. *Philosophical Transactions of the Royal Society of London A* 221:237–264. ⟨323⟩

Jeffreys, H., 1925. On the formation of waves by wind. *Proceedings of the Royal Society of London A* 110:341–347. ⟨494⟩

Jeffreys, H., 1926. On the dynamics of geostrophic winds. *Quarterly Journal of the Royal Meteorological Society* 52:85–104. ⟨343, 506⟩

Jenkins, W. J., and W. B. Clarke, 1976. The distribution of He^3 in the western Atlantic Ocean. *Deep-Sea Research* 23:481–494. ⟨444, 445⟩

Jevons, W. S., 1857. On the cirrous form of cloud. *The London, Edinburgh, and Dublin Philosophical Magazine and Journal of Science, Fourth Series* 14:22–35. ⟨xx⟩

Johnson, C. L., C. S. Cox, and B. Gallagher, 1978. The separation of wave-induced and intrusive oceanic finestructure. *Journal of Physical Oceanography* 8:846–860. ⟨282, 283, 286⟩

Johnson, D. A., 1972. Eastward flowing bottom currents along the Clipperton Fracture Zone. *Deep-Sea Research* 19:253–257. ⟨36⟩

Johnson, D. A., and J. E. Damuth, 1979. Deep thermohaline flow and current-controlled sedimentation in the Amirante Passage: western Indian Ocean. *Marine Geology* 33: 1–44 ⟨32⟩

Johnson, E. S., and B. A. Warren, 1979. Density-diffusive model of the Ninetyeast Ridge Current. *Journal of Physical Oceanography* 9:1288–1293. ⟨33, 37⟩

Johnstone, J., 1923. *An Introduction to Oceanography with Special Reference to Geography and Geophysics*. University Press of Liverpool, Liverpool, 351 pp. ⟨199⟩

Jones, I. S. F., and B. C. Kenney, 1977. The scaling of velocity fluctuations in the surface layer. *Journal of Geophysical Research* 82:1392–1396. ⟨495⟩

Jones, R., and W. B. Hall, 1973. A simulation model for studying the population dynamics of some fish species. In *The Mathematical Theory of the Dynamics of Biological Populations*, M. S. Bartlett and R. W. Hiorns, eds., Academic Press, London, pp. 35–39. ⟨382⟩

Jones, S., 1978. Interactions and instabilities of barotropic and baroclinic Rossby waves in a rotating, two-layer fluid. *Geophysical and Astrophysical Fluid Dynamics* 11:49–60. ⟨534⟩

Jones, S., 1979. Rossby wave interactions and instabilities in a rotating, two-layer fluid on a beta-plane. Part I: Resonant interactions. *Geophysical and Astrophysical Fluid Dynamics* 11:289–322. ⟨345⟩

Jones, W. L., 1968. Reflexion and stability of waves in stably stratified fluids with shear flow: a numerical study. *Journal of Fluid Mechanics* 34:609–624. ⟨274⟩

Joyce, T. M., W. Zenk, and J. M. Toole, 1978. The anatomy of the Antarctic Polar Front in the Drake Passage. *Journal of Geophysical Research* 83:6093–6113. ⟨257⟩

Julian, P. R., and A. K. Clive, 1974. The direct estimation of spatial wave number spectra of atmospheric variables. *Journal of the Atmospheric Sciences* 31:1526–1539. ⟨544⟩

Kaimal, J. C., and J. A. Businger, 1970. Case studies of a convective plume and a dust devil. *Journal of Applied Meteorology* 9:612–620. ⟨498, 499⟩

Kamenkovich, V. M., 1973. *Osnovy Dinamiki Okeana*. Gidrometeoizdat, Leningrad, 238 pp. (*Fundamentals of Ocean Dynamics*, Elsevier, Amsterdam, 1977, 249 pp.) ⟨183⟩

Kamenkovich, V. M., and G. M. Reznik, 1972. K teorii statsionarykh vetrovykh techenii v dvusloinoi zhidkosti. *Fizika Atmosfery i Okeana, Izvestiya Akademii Nauk SSSR* 8:419–434. (A contribution to the theory of stationary wind-driven currents in a two-layer liquid. *Izvestiya, Academy of Sciences USSR, Atmospheric and Oceanic Phyiscs* 8:238–245.) ⟨158⟩

Kantha, L. H., O. M. Phillips, and R. S. Azad, 1977. On turbulent entrainment at a stable density interface. *Journal of Fluid Mechanics* 79:753–768. ⟨241⟩

Kao, T. W., 1976. Principal stage of wake collapse in a stratified fluid: two-dimensional theory. *The Physics of Fluids* 19:1071–1074. ⟨253⟩

Kato, H., and O. M. Phillips, 1969. On the penetration of a turbulent layer into stratified fluid. *Journal of Fluid Mechanics* 37:643–665. ⟨241⟩

Katsaros, K. B., 1978. Turbulent free convection in fresh and salt water: some characteristics revealed by visualization. *Journal of Physical Oceanography* 8:613–626. ⟨243⟩

Katsaros, K. B., W. G. Liu, J. A. Businger, and J. E. Tillman, 1977. Heat transport and thermal structure in the interfacial boundary layer measured in an open tank of water in turbulent free convection. *Journal of Fluid Mechanics* 83:311–335. ⟨243⟩

Katz, E., 1973. Profile of an isopycnal surface in the main thermocline of the Sargasso Sea. *Journal of Phsyical Oceanography* 3:448–457. ⟨265⟩

Katz, E. J., R. Belevitsch, J. Bruce, V. Bubnov, J. Cochrane, W. Düing, P. Hisard, H.-U. Lass, J. Meincke, A. deMesquita, L. Miller, and A. Rybnikov, 1977. Zonal pressure gradient along the equatorial Atlantic. *Journal of Marine Research* 35:293–307. ⟨186, 193⟩

Kawai, H., 1966. A generalized potential vorticity in the ocean. *Kyoto University, Geophysical Institute, Special Contributions* 6:79–93. ⟨82⟩

Kazanskii, A. B., and A. S. Monin, 1960. O turbulentom rezhime vyshe prezemnogo sloya vozdukha. *Izvestiya Akademii Nauk SSSR, Seriya Geofizicheskaya*, 1960 g.:165–168. [A turbulent regime above the ground atmospheric layer. *Bulletin (Izvestiya) Academy of Sciences, USSR, Geophysics Series*, 1960: 110–112.] ⟨502⟩

Kazanskii, A. B., and A. S. Monin, 1961. O dinamicheskom vzaimodeistvii mezhdu atmosferoi i poverkhnost'yu zemli. *Izvestiya, Akademii Nauk SSSR, Seriya Geofizicheskaya*, 1961 g.:786–788. [On the dynamic interaction between the atmosphere and the earth's surface. *Bulletin (Izvestiya) Academy of Sciences, USSR, Geophysics Series*, 1961: 514–515.] ⟨502⟩

Keeling, C. D., 1973. The Carbon dioxide cycle: Reservoir models to depict the exchange of atmospheric carbon dioxide with the oceans and land plants. In *Chemistry of the Lower Atmosphere*, S. I. Rasool, ed., Plenum Press, New York, pp. 251–329. ⟨449⟩

Keeling, C. D., and B. Bolin, 1967. The simultaneous use of chemical tracers in oceanic studies. I. General theory of reservoir models. *Tellus* 19:556–581. ⟨449⟩

Keeling, C. D., and B. Bolin, 1968. The simultaneous use of chemical tracers in oceanic studies. II. A three-reservoir model of the North and South Pacific Oceans. *Tellus* 20:17–54. ⟨449⟩

Keffer, T., and P. Niiler, 1978. Recovery of POLYMODE Array III, Cluster C in North Atlantic Equatorial Current. In *PO-

LYMODE News No. 56, Woods Hole Oceanographic Institution, Woods Hole, Massachusetts. (Unpublished document.) ⟨372]

Keldysh, M. V., 1977. Venus exploration with the Venera 9 and Venera 10 spacecraft. *Icarus* 30:605–625. ⟨506⟩

Keller, J. B., and G. Veronis, 1969. Rossby waves in the presence of random currents. *Journal of Geophysical Research* 74:1941–1951. ⟨345⟩

Ketchum, B. H., 1950. Hydrographic factors involved in the dispersion of pollutants introduced into tidal waters. *Journal of the Boston Society of Civil Engineers* 37:296–314. ⟨202, 212⟩

Ketchum, B. H., 1951. The exchanges of fresh and salt water in tidal estuaries. *Journal of Marine Research* 10:18–38. ⟨202⟩

Ketchum, B. H., and N. Corwin, 1964. The persistence of "winter" water on the continental shelf south of Long Island, New York. *Limnology and Oceanography* 9:467–475. ⟨212⟩

Ketchum, B. H., and D. J. Keen, 1955. The accumulation of river water over the continental shelf between Cape Cod and Chesapeake Bay. *Papers in Marine Biology and Oceanography, Deep-Sea Research* 3 (Supplement):346–357. ⟨212, 214⟩

Ketchum, B. H., A. C. Redfield, and J. C. Ayers, 1951. The oceanography of the New York Bight. *Papers in Physical Oceanography and Meteorology* 12:1, 46 pp. ⟨212⟩

Keulegan, G. H., 1949. Interfacial stability and mixing in stratified flows. *Journal of Research of the National Bureau of Standards* 43:487–500. ⟨203⟩

Killworth, P. D., 1974. A baroclinic model of motions on Antarctic continental shelves. *Deep-Sea Research* 21:815–837. ⟨456⟩

Killworth, P. D., 1977. Mixing on the Weddell Sea continental slope. *Deep-Sea Research* 24:427–448. ⟨20, 22, 257, 260⟩

Killworth, P. D., 1979. On "chimney" formations in the ocean. *Journal of Physical Oceanography* 9:531–554. ⟨22⟩

Killworth, P. D., 1980. Barotropic and baroclinic instability in rotating stratified fluids. *Dynamics of Atmospheres and Oceans* 4:143–184. ⟨529⟩

Killworth, P. D., and P. C. Manins, 1980. A model of confined thermal convection driven by nonuniform heating from below. *Deep-Sea Research*. ⟨505⟩

Kim, K., 1978. Instability of baroclinic Rossby waves; energetics in a two-layer ocean. *Deep-Sea Research* 25:795–814. ⟨345, 534⟩

Kinsman, B., 1960. Surface waves at short fetches and low wind speed—a field study. *Chesapeake Bay Institute of the Johns Hopkins University Technical Report 19, Reference 60-1*, Baltimore, Maryland, pp. various. ⟨491, 492⟩

Kirwan, A. D., 1963. Circulation of Antarctic Intermediate Water deduced through isentropic analysis. *Reference No. 63-34 F*, Texas A & M University, College Station, Texas, 34 pp. ⟨83, 85⟩

Kirwan, A. D., Jr., G. McNally, S. Pazan, and R. Wert, 1979. Analysis of surface current response to wind. *Journal of Physical Oceanography* 9:401–412. ⟨426, 427⟩

Kitaigorodskii, S. A., 1960. O raschete tolshchiny sloya vetrovogo peremeshivaniya v okeane. *Izvestiya Akademii Nauk SSSR, Seriya Geofizicheskaya* 1960 g.:425–431. [On the computation of the thickness of the wind-mixing layer in the ocean. *Bulletin (Izvestiya) Academy of Sciences, USSR, Geophysics Series* 1960: 284–287.] ⟨243⟩

Kitaigorodskii, S. A., 1962. Nekotorye prilozheniya metodov teorii podobiya pri analize vetrovogo volneniya kak veroyatnostnogo protsessa. *Izvestiya Akademii Nauk SSSR, Seriya Geofizicheskaya* 1962 g.:105–117. [Applications of the theory of similarity to the analysis of wind-generated wave motion as a stochastic process. *Bulletin (Izvestiya) Academy of Sciences, USSR, Geophysics Series* 1962:73–80.] ⟨491⟩

Kitaidorodskii, S. A., 1970. *Fizika Vzaimodesitviya Atmosfery i Okeana*. Gidrometeorologicheskoe Izdatel'stvo, Leningrad. (*The Physics of Air-Sea Interaction*, A. Baruch, translator, P. Greenberg, ed., Israel Program for Scientific Translation, Jerusalem, 1973, 237 pp.) ⟨487⟩

Kitaigorodskii, S. A., O. A. Kuznetsov, and G. N. Panim, 1973. O koeffitsientakh soprotivleniya, teploobmena i ispareniya nad morskoi poverkhnostyu v atmosfere. *Fizika Atmosfery i Okeana, Izvestiya Akademii Nauk SSSR* 9:1135–1141. (Coefficients of drag, sensible heat, and evaporation in the atmosphere over the surface of a sea. *Izvestiya, Academy of Sciences USSR, Atmospheric and Oceanic Physics* 9:644–647.) ⟨489⟩

Kitaigorodskii, S. A., and S. S. Strekalov, 1962. K analizu spectrov vetrovogo volneniya. I. *Ievestiya Akademii Nauk SSSR, Seriya Geofizicheskaya* 9:1221–1228. [Contribution to an analysis of the spectra of wind-caused wave action. I. *Bulletin (Izvestiya) Academy of Sciences, USSR, Geophysics Series* 1962:765–769.] ⟨493⟩

Knauss, J. A., 1960. Measurements of the Cromwell Current. *Deep-Sea Research* 6:265–286. ⟨429⟩

Knauss, J. A., 1962a. On some aspects of the deep circulation of the Pacific. *Journal of Geophysical Research* 67:3943–3954.⟨35, 36⟩

Knauss, J. A., 1962b. Observations of internal waves of tidal period made with neutrally buoyant floats. *Journal of Marine Research* 20:111–118. ⟨332⟩

Knauss, J. A., 1963. Equatorial current systems. In *The Sea: Ideas and Observations on Progress in the Study of the Seas, 2: The Composition of Sea-Water, Comparative and Descriptive Oceanography*, M. N. Hill, ed., Wiley, Interscience, New York, pp. 235–252. ⟨184, 185⟩

Knauss, J. A., 1969. A note on the transport of the Gulf Stream. *Frederick C. Fuglister Sixtieth Anniversary Volume, Deep-Sea Research* 16 (Supplement):117–123. ⟨118, 131⟩

Knudsen, M., 1899. Hydrography. In *The Danish Ingolf-Expedition* 1: Part 1, pp. 23–161. ⟨22⟩

Knudsen, M., ed., 1901. *Hydrographical Tables*. G. E. C. Gad, Copenhagen, 63 pp. ⟨209⟩

Koczy, F. F., 1958. Natural radium as a tracer in the ocean. In *Proceedings of International Conference on the Peaceful Uses of Atomic Energy. 2nd, Geneva* 18:351–357. ⟨444⟩

Koenuma, K., 1939. On the hydrography of the southwestern part of the North Pacific and the Kuroshio. *Imperial Marine Observatory Memoirs* 7:41–114. ⟨74, 85⟩

Koh, R. C. Y., and N. H. Brooks, 1975. Fluid mechanics of waste-water disposal in the ocean. *Annual Review of Fluid Mechanics* 7:187–211. ⟨260⟩

Kolla, V., L. Henderson, and P. E. Biscaye, 1976. Clay mineralogy and sedimentation in the western Indian Ocean. *Deep-Sea Research* 23:949–961. ⟨29⟩

Kolla, V., L. Sullivan, S. S. Streeter, and M. G. Langseth, 1976. Spreading of Antarctic Bottom Water and its effects on the floor of the Indian Ocean inferred from bottom-water potential temperature, turbidity, and sea floor photography. *Marine Geology* 21:171–189. ⟨29⟩

Kondo, J., 1975. Air-sea bulk transfer coefficients in diabatic conditions. *Boundary-Layer Meteorology* 9:91–112. ⟨489, 490⟩

Kondo, J., Y. Fukinawa, and G. Naito, 1973. High-frequency components of ocean waves and their relation to the aerodynamic roughness. *Journal of Physical Oceanography* 3:197–202. ⟨489⟩

Kort, V. G., chief series editor, 1968. *Gidrologiya Tikhogo Okeana*, A. D. Dobrovol'skii, editor-in-chief. *Tikhii Okean*, 2, Akademiya Nauk SSSR, Institut Okeanologii imeni P. P. Shirshova, Izdatel'stvo "Nauka", Moscow, 524 pp. (The Pacific Ocean, "Hydrology of the Pacific Ocean", U. S. Naval Oceanographic Office, Washington, D.C., 788 pp. Available from National Technical Information Service, Springfield, Virginia.) ⟨85, 88⟩

Koshlyakov, M. N., and A. S. Monin, 1978. Synoptic eddies in the ocean. *Annual Review of Earth and Planetary Sciences* 6:495–523. ⟨359⟩

Koshlyakov, M. N., and V. G. Neiman, 1965. Nekotorye rezul'taty izmerenii i raschetov zonal'nykh techenii v ekvatorial'noi oblasti Tikhogo okeana. *Okeanologiya* 5:235–249. (Some results of measurements and calculations of zonal currents in the Pacific equatorial region. *Oceanology* 5:2, 37–49.) ⟨83⟩

Koske, P. H., 1972. Hydrographische Verhältnisse im Persischen Golf auf Grund von Beobachtungen von F. S. "Meteor" in Frühjahr 1965. *"Meteor" Forchungsergebnisse, Reihe A*, No. 11, 58–73. ⟨26⟩

Kossina, E., 1921. Die Tiefen des Weltmeeres. *Veröffentlichungen des Instituts fr Meereskunde an der Universität Berlin, Neue Folge. A. Geographisch-naturwissenschaftliche Reihe*, 9, 70 pp. ⟨50⟩

Kotschin, N., 1932. On the stability of Margules surfaces of discontinuity. *Beiträge zur Physik der Freien Atmosphäre* 18:129–164. ⟨529⟩

Kozlov, V. F., 1966. Nekotorye tochnye resheniya nelineinogo uravneniya advektsii plotnosti v okeane. *Fizika Atmosfery i Okeana, Izvestiya Akademii Nauk SSSR* 2:1205–1207. (Certain exact solutions of the nonlinear equation for density advection in the ocean. *Izvestiya, Academy of Sciences, USSR, Atmospheric and Oceanic Physics* 2:742–744.) ⟨159⟩

Kozlov, V. F., 1971. Nekotorye rezul'taty priblizhennogo rascheta tsirkulyatsii v Tikhom okeane. *Fizika Atmosfery i Okeana, Izvestiya Akademii Nauk SSSR*, 7:421–430. (Some results of an approximate calculation of the circulation in the Pacific Ocean. *Izvestiya, Academy of Sciences, USSR, Atmospheric and Oceanic Physics* 7:278–282.) ⟨89⟩

Kraichnan, R. H., 1962. Turbulent thermal convection at arbitrary Prandtl numbers. *The Physics of Fluids* 5:1374–1389. ⟨388⟩

Kraichnan, R. H., 1967. Inertial ranges in two-dimensional turbulence. *The Physics of Fluids* 10:1417–1423. ⟨346, 543⟩

Kraichnan, R. H., 1975. Statistical dynamics of two-dimensional flow. *Journal of Fluid Mechanics* 67:155–175. ⟨543⟩

Kraus, E. B., ed., 1977. *Modelling and Prediction of the Upper Layers of the Ocean*. Pergamon Press, Oxford, 325 pp. ⟨240, 346⟩

Kraus, E. B., and J. S. Turner, 1967. A one-dimensional model of the seasonal thermocline: II. The general theory and its consequences. *Tellus* 19:98–106. ⟨244, 245⟩

Krishnamurti, R., 1978. Laboratory modelling of oceanic response to monsoonal winds. Presented at IUTAM-IUGG Symposium on Monsoon Dynamics at the Indian Institute of Technology, New Delhi, India, December, 1977. Technical Report No. 15 to the Office of Naval Research under Contract N-00014-75-C-0877, Florida State University, Tallahassee, Florida, 20 pp. Also in *Monsoon Dynamics*, J. Lighthill and R. P. Pearce, eds., 1980, Cambridge University Press, London. ⟨473⟩

Krishnamurti, R., and J. Y. Na, 1978. Experiments in ocean circulation modeling. *Geophysical and Astrophysical Fluid Dynamics* 11:13–21. ⟨473⟩

Krishnaswami, S., D. Lal, B. L. K. Somayajulu, F. S. Dixon, S. A. Stonecipher, and H. Craig, 1972. Silicon, radium, thorium and lead in sea water: In-situ extraction by synthetic fibre. *Earth and Planetary Science Letters* 16:84–90. ⟨447⟩

Krummel, O., 1911. *Handbuch der Ozeanographie*, 2. J. Engelhorns Nachf., Stuttgart, 766 pp. ⟨40⟩

Kuettner, J. P., 1971. Cloud bands in the earth's atmosphere. *Tellus* 23:404–425. ⟨500⟩

Kuksa, V. I., 1962. O formirovanii i rasprostranenii promezhutochnogo sloya vody ponizhennoi solenosti v severnoi chasti Tikhogo okeana (On the formation and distribution of the intermediate layer of water of low salinity in the northern part of the Pacific Ocean). *Okeanologiya* 2:769–782. ⟨83⟩

Kuksa, V. I., 1963. Osnovnye zakonomernosti obrazovaniya i rasprostraneniya promezhutochnykh vod severnoi chasti Tikhogo okeana (Basic laws of the production and distribution of the intermediate waters in the northern part of the Pacific Ocean). *Okeanologiya* 3:30–43. ⟨83⟩

Kullenberg, G., 1972. Apparent horizontal diffusion in stratified vertical shear flow. *Tellus* 24:17–28. ⟨378⟩

Kuo, H.-H., 1978. Topographic effect on the deep circulation and the abyssal oxygen distribution. *Journal of Physical Oceanography* 8:428–436. ⟨89⟩

Kuo, H.-H., and G. Veronis, 1970. Distribution of tracers in the deep oceans of the world. *Deep-Sea Research* 17:29–46. ⟨449⟩

Kuo, H.-H., and G. Veronis, 1971. The source-sink flow in a rotating system and its oceanic analogy. *Journal of Fluid Mechanics* 45:441–466. ⟨152, 470, 471, 472⟩

Kuo, H.-H., and G. Veronis, 1973. The use of oxygen as a test for an abyssal circulation model. *Deep-Sea Research* 20:871–888. ⟨12, 13, 14, 15, 26, 33, 449⟩

Kuo, H.-L., 1949. Dynamic instability of two-dimensional non-divergent flow in a barotropic atmosphere. *Journal of Meteorology* 6:105–122. ⟨169, 529⟩

Kuo, H.-L., 1951. Dynamical aspects of the general circulation and the stability of zonal flow. *Tellus* 3:268–284. ⟨529⟩

Kuo, H.-L., 1952. Three-dimensional disturbances in a baroclinic zonal current. *Journal of Meteorology* 9:260–278. ⟨169⟩

Kuo, H.-L., 1973. Dynamics of quasi-geostrophic flows and instability theory. *Advances in Applied Mechanics* 13:248–330. ⟨505, 520⟩

Kuo, J., and R. Jachens, 1977. Indirect mapping of ocean tides by solving the inverse problems for tidal gravity observations. *Annals of Geophysics* 33:73–82. ⟨323⟩

Kupferman, S. L., and N. Garfield, 1977. Transport of low-salinity water at the slope water-Gulf Stream boundary. *Journal of Geophysical Research* 82:3481–3486. ⟨213⟩

Kupferman, S. L., H. D. Livingston, and V. T. Bowen, 1979. A mass balance for Cs^{137} and Sr^{90} in the North Atlantic Ocean. *Journal of Marine Research* 37:157–199. ⟨446⟩

Kutzbach, J. E., R. M. Chervin, and D. D. Houghton, 1977. Response of the NCAR general circulation model to prescribed changes in sea surface temperature. Part I: Mid-latitude changes. *Journal of the Atmospheric Sciences* 34:1200–1213. ⟨355⟩

Lachenbruch, A. H., and B. V. Marshall, 1968. Heat flow and water temperature fluctuations in the Denmark Strait. *Journal of Geophysical Research* 73:5829–5842. ⟨24⟩

Lacombe, H., 1971. Le Détroit de Gibraltar, océanographie physique. In *Mémoire explicatif de la Carte géotechnique de Tanger au 1/25,000, Notes et Mémoires du Service Géologique du Maroc*, No. 222 bis, pp. 111–146. ⟨25⟩

Lafond, E. C., 1961. The isotherm follower. *Journal of Marine Research* 19:33–39. ⟨264⟩

Lafond, E. C., 1963. Detailed temperature structures of the sea off Baja California. *Limnology and Oceanography* 8:417–425. ⟨265⟩

Lai, D. Y., and P. L. Richardson, 1977. Distribution and movement of Gulf Stream rings. *Journal of Physical Oceanography* 7:670–683. ⟨132, 137, 534⟩

Laikhtman, D. L., 1961. *Fizika Pogranichnogo Sloya Atmosphery*. Gidrometeorologicheskoe Izdatel'stvo, Leningrad, 341 pp., 2nd ed., 1970. (*Physics of the Boundary Layer of the Atmosphere*, U.S. Department of Commerce, Washington, D.C.) ⟨500⟩

Lal, D., 1962. Cosmic ray produced radionuclides in the ocean. *Journal of the Oceanographical Society of Japan, 20th Anniversary Volume:* 600–614. ⟨440⟩

Lal, D., 1969. Characteristics of large-scale oceanic circulation as derived from the distribution of radioactive elements. In *Morning Review Lectures of the Second International Oceanographic Congress, Moscow, 1966*, United Nations Educational, Scientific, Cultural Organization, Place de Fontenoy, Paris, pp. 29–48. ⟨440⟩

Lamb, H., 1932. *Hydrodynamics*, 6th ed. Dover, New York, 738 pp. ⟨147, 295, 296, 297, 298, 299, 307, 339, 344, 358⟩

Lambeck, K., 1975. Effects of tidal dissipation in the oceans on the moon's orbit and the earth's rotation. *Journal of Geophysical Research* 80:2917–2925. ⟨322⟩

Lambeck, K., 1977. Tidal dissipation in the oceans: astonomical, geophysical and oceanographic consequences. *Philosophical Transactions of the Royal Society of London A* 287:545–594. ⟨322⟩

Lambert, R. B., and W. Sturges, 1977. A thermohaline staircase and vertical mixing in the thermocline. *Deep-Sea Research* 24:211–222. ⟨258⟩

Langmuir, I., 1938. Surface motion of water induced by wind. *Science* 87:119–123. ⟨495⟩

Laplace, P. S., 1775. Recherches sur plusieurs points du système du monde. *Mémoires de l'Académie Royale des Sciences de Paris* 88:75–182. Reprinted in *Oeuvres Complètes de Laplace*, Gauthier-Villars, Paris, 9 (1893). ⟨295, 297, 344⟩

Laplace, P. S., 1776. Recherches sur plusieurs points du système du monde. *Mémoires de l'Académie Royale des Sciences de Paris* 89:177–264. Reprinted in *Oeuvres Complètes de Laplace*, Gauthier-Villars, Paris, 9 (1893). ⟨295, 297⟩

Laplace, P. S., 1799. *Traité de Mécanique Céleste*, Tome Second, Première Part, Livre IV. Reprinted in *Oeuvres Complètes de Laplace*, 1878–1912, Gauthier-Villars, Paris, 14 volumes (see 4, pp. 294–298). (*Mécanique Céleste* by the Marquis de Laplace, Translated with a Commentary, N. Bowditch, translator, 1829–1839, Boston, 4 volumes.) ⟨504⟩

Larsen, J., 1977. Cotidal charts for the Pacific Ocean near Hawaii using f-plane solutions. *Journal of Physical Oceanography* 7:100–109. ⟨327, 328⟩

Lasker, R., 1978. The relation between oceanographic conditions and larval anchovy food in the California Current: Identification of factors contributing to recruitment failure. *Rapports et Procès-Verbaux des Réunions, Conseil Permanent International pour l'Exploration de la Mer* 173:212–230. ⟨381, 382⟩

Launder, B. E., 1976. Heat and mass transport. In *Turbulence*, P. Bradshaw, ed., Springer-Verlag, Berlin, pp. 232–287. ⟨240⟩

Laval, A. F., 1728. *Voyage de la Louisiane, Fait par Ordre du Roy en l'Année Mil Sept Cent Vingt . . . par le P. Laval de la Compagnie de Jésus, Professeur Royal de Mathématiques, et Maître d'Hydrographie . . . du Port de Toulon*. J. Mariette, Paris, 304, 96, 191 pp. ⟨507⟩

La Valle, L., and P. Di Girolamo, 1975. Determination of the geostrophic drag coefficient. *Tellus* 27:87–92. ⟨503⟩

Lavelle, J. W., P. E. Gadd, G. C. Han, D. A. Mayer, W. L. Stubblefield, and D. J. P. Swift, 1976. Preliminary results of coincident current meter and sediment transport observations for wintertime conditions on the Long Island inner shelf. *Geophysical Research Letters* 3:97–100. ⟨218⟩

Lavelle, J. W., R. A. Young, D. J. P. Swift, and T. L. Clarke, 1978. Near-bottom sediment concentration and fluid velocity measurements on the inner continental shelf, New York. *Journal of Geophysical Research* 93:6052–6062. ⟨218⟩

Lazier, J. R. N., 1973a. The renewal of Labrador Sea Water. *Deep-Sea Research* 20:341–353. ⟨25, 84⟩

Lazier, J. R. N., 1973b. Temporal changes in some fresh water temperature structures. *Journal of Physical Oceanography* 3:226–229. ⟨281⟩

Leaman, K. D., 1976. Observations of vertical polarization and energy flux of near-inertial waves. *Journal of Physical Oceanography* 6:894–908. ⟨289⟩

Leaman, K. D., and T. B. Sanford, 1975. Vertical energy propagation of internal waves; a vector spectral analysis of velocity profiles. *Journal of Geophysical Research* 80:1975–1978. ⟨276⟩

LeBlond, P., and L. Mysak, 1977. Trapped coastal waves and their role in shelf dynamics. In *The Sea: Ideas and Observations on Progress in the Study of the Seas, 6: Marine Modeling*, E. D. Goldberg, J. N. McCave, J. J. O'Brien, and J. H.

Steele, eds., Wiley, Interscience, New York, pp. 459–495. ⟨310, 311, 358⟩

Lee, A. J., 1967. Temperature and salinity distributions as shown by sections normal to the Iceland-Faroe Ridge. *Rapports et Procès-Verbaux des Réunions, Conseil Permanent International pour l'Exploration de la Mer* 157:100–135. ⟨24⟩

Lee, A. J., D. F. Bumpus, and L. M. Lauzier, 1965. The sea-bed drifter. A new instrument which indicates the residual current near the sea bed. *I.C.N.A.F. Research Bulletin* No. 2, 42–47. ⟨213⟩

Lee, A., and D. Ellett, 1965. On the contribution of overflow water from the Norwegian Sea to the hydrographic structure of the North Atlantic Ocean. *Deep-Sea Research* 12:129–142. ⟨23⟩

Lee, C., and R. Bearsley, 1974. The generation of long nonlinear internal waves in a weakly stratified shear flow. *Journal of Geophysical Research* 79:453–462. ⟨332⟩

Lee, T. D., 1951. Difference between turbulence in a two-dimensional fluid and in a three-dimensional fluid. *Journal of Applied Physics* 22:524. ⟨543⟩

Lee, T. N., 1975. Florida Current spin-off eddies. *Deep-Sea Research* 22:753–765. ⟨118⟩

Lee, T. N., and P. A. Mayer, 1977. Low-frequency current variability and spin-off eddies along the shelf off southeast Florida. *Journal of Marine Research* 35:193–220. ⟨118⟩

Leetmaa, A., 1978. Fluctuating winds: an energy source for mesoscale motions. *Journal of Geophysical Research* 83:427–430. ⟨348⟩

Leetmaa, A., and A. F. Bunker, 1978. Updated charts of the mean annual wind stress, convergences in the Ekman Layers and Sverdrup transports in the North Atlantic. *Journal of Marine Research* 36:311–322. ⟨125, 131, 139, 217, 228, 229⟩

Leetmaa, A., P. Niiler, and H. Stommel, 1977. Does the Sverdrup Relation account for the Mid-Atlantic circulation? *Journal of Marine Research* 35:1–10. ⟨xxiii, 74, 78, 85, 91, 345, 527⟩

Legeckis, R., 1977a. Oceanic polar front in the Drake Passage—satellite observations during 1976. *Deep-Sea Research* 24:701–704. ⟨426⟩

Legeckis, R., 1977b. Long waves in the eastern equatorial Pacific Ocean: A view from a geostationary satellite. *Science* 197:1179–1181. ⟨186, 193, 426⟩

Legeckis, R., 1978. A survey of worldwide sea surface temperature fronts detected by environmental satellites. *Journal of Geophysical Research* 83:4501–4522. ⟨112⟩

Legeckis, R. V., 1979. Satellite observations of the influence of bottom topography on the seaward deflection of the Gulf Stream off Charleston, South Carolina. *Journal of Physical Oceanography* 9:483–497. ⟨426⟩

Leipper, D. F., 1970. A sequence of current patterns in the Gulf of Mexico. *Journal of Geophysical Research* 75:637–657. ⟨113⟩

Leith, C. E., 1971. Atmospheric predictability and two-dimensional turbulence. *Journal of the Atmospheric Sciences* 28:145–161. ⟨543⟩

Le Lacheur, E. A., and J. C. Sammons, 1932. Tides and currents in Long Island and Block Island Sounds. *U.S. Coast and Geodetic Survey Special Publication* No. 174, 187 pp. ⟨201⟩

Lemansson, L., and J. P. Rébert, 1973. Circulation dans la partie orientale de l'Atlantique Sud. Documents Scientifiques, *Centre de Recherches Océanographiques—Abidjan* 4:91–124. ⟨83⟩

Le Méhauté, B., 1976. *An Introduction to Hydrodynamics and Water Waves*. Springer-Verlag, New York, 315 pp. ⟨432⟩

Lenz, E., 1845. Bermerkungen ber die Temperatur des Weltmeeres in verschiedenen Tiefen. *Bulletin de la Classe Physico-Mathématique de l'Académie Impériale des Sciences de Saint-Pétersbourg* 5:67–74. ⟨8, 40⟩

Leonard, D. A., B. Caputo, R. L. Johnson, and F. E. Hoge, 1977. Experimental remote sensing of subsurface temperature in natural ocean water. *Geophysical Research Letters* 4:279–281. ⟨428⟩

Lettau, B., W. Brower, and R. Quayle, 1976. Marine climatology. *MESA New York Bight Atlas Monograph* 7, New York Sea Grant Institute, Albany, New York, 239 pp. ⟨230⟩

Lettau, H. H., 1959. Wind profile surface stress and geostrophic drag coefficients in the atmospheric surface layer. *Advances in Geophysics* 6:241–257. ⟨502⟩

Lettau, H. H., and B. Davidson, 1957. *Exploring the Atmosphere's First Mile*. Pergamon Press, Oxford, 587 pp. ⟨500⟩

Levitus, S., and A. H. Oort, 1977. Global analysis of oceanographic data. *Bulletin of the American Meteorological Society* 58:1270–1284. ⟨89⟩

Lewis, E. L., and N. P. Fofonoff, 1979. Notice to oceanographers—A practical salinity scale. *Journal of Physical Oceanography* 9:446. ⟨418⟩

Libbey, W., Jr., 1891. Report upon a physical investigation of the waters off the southern coast of New England, made during the summer of 1889, by the U.S. Fish Commission schooner *Grampus*. *Bulletin of the United States Fish Commission*, 9 (1889), 391–459. ⟨208⟩

Libbey, W., Jr., 1895. The relations of the Gulf Stream and the Labrador Current. In *Report of the Sixth International Geographical Congress, London, 1895*, pp. 461–474. ⟨208⟩

Lighthill, M. J., 1952. On sound generated aerodynamically. *Proceedings of the Royal Society of London A* 211:564–587. ⟨509⟩

Lighthill, M. J., 1967. On waves generated in dispersive systems by travelling forcing effects, with application to the dynamics of rotating fluids. *Journal of Fluid Mechanics* 27:725–752. ⟨165⟩

Lighthill, M. J., 1969. Dynamic response of the Indian Ocean to onset of the Southwest Monsoon. *Philosophical Transactions of the Royal Society of London A* 265:45–92. ⟨165, 192, 340, 527, 533⟩

Linden, P. F., 1973. On the structure of salt fingers. *Deep-Sea Research* 20:325–340. ⟨252, 255⟩

Linden, P. F., 1974. Salt fingers in a steady shear flow. *Geophysical Fluid Dynamics* 6:1–27. ⟨255⟩

Linden, P. F., 1975. The deepening of a mixed layer in a stratified fluid. *Journal of Fluid Mechanics* 71:385–405. ⟨243⟩

Linden, P. F., 1976. The formation and destruction of finestructure by double-diffusive processes. *Deep-Sea Research* 23:895–908. ⟨251, 252⟩

Linden, P. F., 1978. The formation of banded salt-finger structure. *Journal of Geophysical Research* 83:2902–2912. ⟨252⟩

Linden, P. F., 1979. Mixing in stratified fluids. *Geophysical and Astrophysical Fluid Dynamics* 13:3–23. ⟨247⟩

Linden, P. F., and T. G. L. Shirtcliffe, 1978. The diffusive interface in double-diffusive convection. *Journal of Fluid Mechanics* 87:417–432. ⟨255, 256, 385⟩

Lindzen, R. S., 1967. Planetary waves on beta-planes. *Monthly Weather Review* 95:441–451. ⟨324, 512, 513⟩

Lindzen, R. S., and J. R. Holton, 1968. A theory of the quasi-biennial oscillation. *Journal of the Atmospheric Sciences* 25:1095–1107. ⟨532⟩

Lindzen, R. S., and K.-K. Tung, 1978. Wave overreflection and shear instability. *Journal of the Atmospheric Sciences* 35:1626–1632. ⟨534⟩

Lineikin, P. C., 1955. Ob opredelenii tolshchiny baroklinnogo sloya morya (On the determination of the thickness of the baroclinic layer of the ocean). *Doklady Akademii Nauk SSSR* 101:461–464. ⟨159⟩

Lingenfelter, R. E., and R. Ramaty, 1970. Astrophysical and geophysical variations in C-14 production. In *Radiocarbon Variations and Absolute Chronology*, I. U. Olsen, ed., Wiley, Interscience, New York, and Almqvist & Wiksell, Stockholm, pp. 513–537. ⟨435⟩

Lipes, R. G., R. L. Bernstein, V. J. Cardone, K. B. Katsaros, E. G. Njoku, A. L. Riley, D. B. Boss, C. T. Swift, and F. J. Wentz, 1979. Seasat scanning multichannel microwave radiometer: Results of the Gulf of Alaska Workshop. *Science* 204:1415–1417. ⟨427⟩

Lipps, F. B., 1963. Stability of jets in a divergent barotropic fluid. *Journal of the Atmospheric Sciences* 20:120–129. ⟨169, 347⟩

Lisitzin, A. P., 1972. Sedimentation in the world ocean. *Society of Economic Paleontologists and Mineralogists Special Publication* No. 17, 218 pp. ⟨59⟩

Lisitzin, E., 1974. *Sea-Level Changes*. Elsevier, Amsterdam, 286 pp. ⟨108⟩

Long, R. R., 1951. The flow of a liquid past a barrier in a rotating spherical shell. *Journal of Meteorology* 8:207–221. ⟨474⟩

Longuet-Higgins, M. S., 1962. The statistical geometry of random surfaces. In *Hydrodynamic Stability: Proceedings of the Thirteenth Symposium on Applied Mathematics*, G. Birkhoff, R. Bellman, and C. C. Lin, eds., American Mathematical Society, Providence, Rhode Island, pp. 105–144. ⟨491⟩

Longuet-Higgins, M. S., 1964. Planetary waves on a rotating sphere. *Proceedings of the Royal Society of London A* 279:446–473. ⟨165, 191, 301, 303, 344⟩

Longuet-Higgins, M. S., 1965. Planetary waves on a rotating sphere II. *Proceedings of the Royal Society of London A* 284:40–68. ⟨191, 344⟩

Longuet-Higgins, M. S., 1966. Planetary waves on a hemisphere bounded by meridians of longitude. *Philosophical Transactions of the Royal Society of London A* 260:317–350. ⟨165⟩

Longuet-Higgins, M. S., 1968a. The eigenfunctions of Laplace's tidal equations over a sphere. *Philosophical Transactions of the Royal Society of London A* 262:511–607. ⟨191, 297, 344⟩

Longuet-Higgins, M. S., 1968b. On the trapping of waves along a discontinuity of depth in a rotating ocean. *Journal of Fluid Mechanics* 31:417–434. ⟨310, 316⟩

Longuet-Higgins, M. S., 1968c. Double Kelvin waves with continuous depth profiles. *Journal of Fluid Mechanics* 34:49–80. ⟨358⟩

Longuet-Higgins, M. S., 1969a. On wave breaking and the equilibrium spectrum of wind-generated waves. *Proceedings of the Royal Society of London A* 310:151–159. ⟨288, 495⟩

Longuet-Higgins, M. S., 1969b. On the trapping of long period waves around islands. *Journal of Fluid Mechanics* 37:773–784. ⟨310, 358⟩

Longuet-Higgins, M. S., 1969c. On the transport of mass by time-varying ocean currents. *Deep-Sea Research* 16:431–447. ⟨345⟩

Longuet-Higgins, M. S., 1976. On the nonlinear transfer of energy in the peak of a gravity-wave spectrum: a simplified model. *Proceedings of the Royal Society of London A* 347:311–328. ⟨494⟩

Longuet-Higgins, M. S., D. E. Cartwright, and N. D. Smith. 1963. Observations of the directional spectrum of sea waves using the motions of a floating buoy. In *Ocean Wave Spectra: Proceedings of a Conference*, Prentice-Hall, Englewood Cliffs, New Jersey, pp. 111–136. ⟨492⟩

Longuet-Higgins, M. S., and M. J. H. Fox, 1977. Theory of the almost-highest wave: the inner solution. *Journal of Fluid Mechanics* 80:721–741. ⟨276⟩

Longuet-Higgins, M. S., and A. E. Gill, 1967. Resonant interactions between planetary waves. *Proceedings of the Royal Society of London A* 299:120–140. ⟨345, 346⟩

Longuet-Higgins, M. S., and G. S. Pond, 1970. The free oscillations of fluid on a hemisphere bounded by meridians of longitude. *Philosophical Transactions of the Royal Society of London A* 266:193–223. ⟨344⟩

Longuet-Higgins, M. S., and J. S. Turner, 1974. An 'entraining plume' model of a spilling breaker. *Journal of Fluid Mechanics* 63:1–20. ⟨242⟩

Lonsdale, P., 1976. Abyssal circulation of the southeastern Pacific and some geological implications. *Journal of Geophysical Research* 81:1163–1176. ⟨35⟩

Lonsdale, P., 1977. Inflow of bottom water to the Panama Basin. *Deep-Sea Research* 24:1065–1101. ⟨35, 36⟩

Loosli, H. H., and H. Oeschger, 1968. Detection of ^{39}Ar in atmospheric argon. *Earth and Planetary Science Letters* 5:191–198. ⟨447⟩

Loosli, H. H., and H. Oeschger, 1979. Argon-39, carbon-14 and krypton-85 measurements in groundwater samples. In *Isotope Hydrology 1978*, International Atomic Energy Agency, Vienna, 2, pp. 931–953. ⟨447⟩

Lorenz, E. N., 1960. Energy and numerical weather prediction. *Tellus* 12:364–373. ⟨508⟩

Lorenz, E. N., 1962. Simplified dynamic equations applied to the rotating basin experiments. *Journal of the Atmospheric Sciences* 19:39–51. ⟨532⟩

Lorenz, E. N., 1963a. The mechanics of vacillation. *Journal of the Atmospheric Sciences* 20:448–464. ⟨174, 532⟩

Lorenz, E. N., 1963b. Deterministic nonperiodic flow. *Journal of the Atmospheric Sciences* 20:130–141. ⟨387⟩

Lorenz, E. N., 1967. *The Nature and Theory of the General Circulation of the Atmosphere.* World Meteorological Organization, Geneva, WMO No. 218, T. P. 115, 161 pp. ⟨343, 396, 429, 507⟩

Lorenz, E. N., 1972. Barotropic instability of Rossby wave motion. *Journal of the Atmospheric Sciences* 29:259–264. ⟨345, 531, 534, 535⟩

Lorenzen, C. J., 1966. A method for the continuous measurement of in vivo chlorophyll concentration. *Deep-Sea Research* 13:223–227. ⟨376⟩

Love, C. M., ed., 1972. *EASTROPAC Atlas. Volume 1: Physical Oceanographic and Meteorological Data from Principal Participating ships, First Survey Cruise, February–March 1967.* National Oceanic and Atmospheric Administration, National Marine Fisheries Service Circular 330, 12 + xii pp., 255 figures ⟨83, 188⟩

Lovelock, J. E., R. J. Maggs and R. J. Wade, 1974. Halogenated hydrocarbons in and over the Atlantic. *Nature* 241:194–195. ⟨447⟩

Lumley, J. L., 1978. Computational modeling of turbulent flows. *Advances in Applied Mechanics* 18:123–176. ⟨240⟩

Lupton, J. E., 1976. The ^{3}He distribution in deep water over the Mid-Atlantic Ridge. *Earth and Planetary Science Letters* 32:317–374. ⟨444, 445⟩

Lupton, J. E., and H. Craig, 1975. Excess ^{3}He in oceanic basalts: Evidence for terrestrial primordial helium. *Earth and Planetary Science Letters* 26:133–139. ⟨444⟩

Luther, D., 1980. Observations of long-period waves in the tropical oceans and atmosphere. Ph.D. Thesis, Massachuetts Institute of Technology/Woods Hole Oceanographic Institution, 210 pp. ⟨348⟩

Lutjeharms, J. R. E., and D. J. Baker, Jr., 1979. Intensities and scales of motion in the Southern Ocean. *South African Journal of Science* 75:179–182. ⟨359⟩

Luyten, J. R., 1977. Scales of motion in the deep Gulf Stream and across the Continental Rise. *Journal of Marine Research* 35:49–74. ⟨108, 121, 127, 128, 129, 373⟩

Luyten, J. R., 1980. Recent observations in the equatorial Indian Ocean. In *Monsoon Dynamics,* J. Lighthill and R. P. Pearce, eds., Cambridge University Press, London. ⟨429⟩

Luyten, J. R., and A. R. Robinson, 1974. Transient Gulf Stream meandering. Part II: Analysis via a quasi-geostrophic time-dependent model. *Journal of Physical Oceanography* 4:256–269. ⟨126⟩

Luyten, J. R., and J. C. Swallow, 1976. Equatorial undercurrents. *Deep-Sea Research* 23:999–1001. ⟨194, 421, 430⟩

Lynn, R. J., and J. L. Reid, 1968. Characteristics and circulation of deep and abyssal waters. *Deep-Sea Research* 15:577–598. ⟨41, 183⟩

McAlister, E. D., and W. McLeish, 1969. Heat transfer in the top millimeter of the ocean. *Journal of Geophysical Research* 74:3408–3414. ⟨243⟩

McCartney, J. F., and P. L. Howard, 1976. Marine batteries—an overview. In *Marine Propulsion,* J. S. Sladky, Jr., ed., American Society of Mechanical Engineers, Ocean Engineering Division, New York 2, pp. 197–215. ⟨401⟩

McCartney, M. S., 1977. Subantarctic Mode Water. In *A Voyage of Discovery: George Deacon 70th Anniversary Volume,* M. V. Angel, ed., Supplement to *Deep-Sea Research,* Pergamon Press, Oxford, pp. 103–119. ⟨55, 59⟩

McCartney, M. S., 1980. The subtropical recirculation of Subantarctic Mode Water. *Journal of Marine Research.* ⟨59⟩

McClain, E. P., 1977. Recent progress in earth satellite data application to marine activities. In *Oceans '77 Conference Record,* Institute of Electrical and Electronics Engineers, New York, and Marine Technology Society, Washington, D.C., pp. 14A1–14A8. ⟨426⟩

McComas, C. H., 1977. Equilibrium mechanisms within the oceanic internal wave field. *Journal of Physical Oceanography* 7:836–845. ⟨285⟩

McComas, C. H., and F. P. Bretherton, 1977. Resonant interaction of oceanic internal waves. *Journal of Geophysical Research* 82:1397–1412. ⟨274, 276, 277, 337⟩

McCreary, J. P., 1976. Eastern tropical ocean response to changing wind systems: With application to El Niño. *Journal of Physical Oceanography* 6:632–645. ⟨192⟩

McCreary, J. P., 1977. Eastern ocean response to changing wind systems. Ph.D. Thesis, Scripps Institution of Oceanography, University of California at San Diego, 156 pp. ⟨190, 193⟩

McCreary, J. P., 1978. Eastern ocean response to changing wind systems. In *Review Papers of Equatorial Oceanography—FINE Workshop Proceedings,* Nova/N.Y.I.T. University Press, Fort Lauderdale, 21 pp. ⟨193⟩

McCreary, J. P., 1980. A linear stratified ocean model of the equatorial undercurrent. *Philosophical Transactions of the Royal Society of London.* ⟨194, 195, 196⟩

McCullough, J. R., 1975. Vector-averaging current meter speed calibration and recording technique. *W.H.O.I. Technical Report 75-44,* Woods Hole Oceanographic Institution, Woods Hole, Massachusetts, 35 pp. ⟨404⟩

McCullough, J. R., 1978a. Near-surface ocean current sensors: Problems and performance. In *Proceedings of a Working Conference on Current Measurements,* Technical Report DEL-SG-3-78, College of Marine Studies, University of Delaware, Newark, 372 pp. ⟨404, 405⟩

McCullough, J. R., 1978b. Techniques of measuring currents near the ocean surface. In *Instruments and Methods in Air-Sea Interaction,* preprint volume for NATO School, Ustaoset, Norway, April 1978, NATO Science Committee, 33 pp. ⟨404, 405⟩

McDougall, T. J., 1979. Measurements of turbulence in a zero-mean-shear mixed layer. *Journal of Fluid Mechanics* 94:409–431. ⟨242⟩

McDowell, S. E., 1977. A note on XBT accuracy. *Horizon 2,* No. 2, Technical Newsletter, Sippican Corporation, Marion, Massachusetts; also in *POLYMODE News No. 29,* 1977, unpublished document, Woods Hole Oceanographic Institution, Woods Hole, Massachusetts. See also: A cautionary note on T-5 XBTs, in *POLYMODE News No. 58,* 1978. ⟨414⟩

McDowell, S. E., and H. T. Rossby, 1978. Mediterranean water: An intense mesoscale eddy off the Bahamas. *Science* 202:1085–1087. ⟨371⟩

McEwan, A. D., 1971. Degeneration of resonantly-excited standing internal gravity waves. *Journal of Fluid Mechanics* 50:431–448. ⟨248⟩

McEwan, A. D., 1973. Interactions between internal gravity waves and their traumatic effect on a continuous stratification. *Boundary-Layer Meteorology* 5:159–175. ⟨248, 279, 291⟩

McEwan, A. D., and R. M. Robinson, 1975. Parametric instability of internal gravity waves. *Journal of Fluid Mechanics* 67:667–687. ⟨248⟩

McGowan, J. A., 1971. Oceanic biogeography of the Pacific. In *The Micropaleontology of the Oceans*, B. M. Funnel and W. R. Reidel, eds., Cambridge University Press, London, pp. 3–74. ⟨382⟩

Machenhauer, B., 1977. On the dynamics of gravity oscillations in a shallow water model, with applications to normal mode initialization. *Contributions to Atmospheric Physics* 50:243–271. ⟨509⟩

Machta, L., 1972. The role of the oceans and biosphere in the carbon dioxide cycle. In *Nobel Symposium 20: The Changing Chemistry of the Oceans*, D. Dryssen and D. Jagner, eds., Almqvist & Wiksell, Stockholm, and Wiley, Interscience, New York, pp. 121–145. ⟨449⟩

McIntyre, M. E., 1970. On the non-separable baroclinic parallel flow instability problem. *Journal of Fluid Mechanics* 40:273–306. ⟨529⟩

McIntyre, M. E., 1980. Introduction to the generalized Lagrangian mean description of wave-mean flow interactions. *Pure and Applied Geophysics*. ⟨533⟩

Mackas, D. L., 1977. Horizontal spatial variability and covariability of marine phytoplankton and zooplankton. Ph.D. Thesis, Dalhousie University, Halifax, 220 pp. ⟨380⟩

McKean, R. S., and T. E. Ewart, 1974. Temperature spectra in the deep ocean off Hawaii. *Journal of Physical Oceanography* 4:191–199. ⟨265⟩

McKee, W. D., 1972. Scattering of Rossby waves by partial barriers. *Geophysical Fluid Dynamics* 4:83–89. ⟨345⟩

McLellan, H. J., 1957. On the distinctness and origin of the Slope Water off the Scotian Shelf and its easterly flow south of the Grand Banks. *Journal of the Fisheries Research Board of Canada* 14:213–230. ⟨123⟩

McPhaden, J., and R. Knox, 1979. Equatorial Kelvin and inertio-gravity waves in zonal shear flow. *Journal of Physical Oceanography* 9:263–277. ⟨193⟩

McWilliams, J. C., 1975. Baroclinic instability and the MODE observations. In Dynamics and the Analysis of MODE-I: Report of the MODE-I Dynamics Group, Massachusetts Institute of Technology, Cambridge, Massachusetts, pp. 94–112. (Unpublished document.) ⟨532⟩

McWilliams, J. C., 1976. Maps from the Mid-Ocean Dynamics Experiment: Part II. Potential vorticity and its conservation. *Journal of Physical Oceanography* 6:828–846. ⟨372, 545⟩

McWilliams, J. C., 1977. On a class of stable, slightly geostrophic mean gyres. *Dynamics of Atmospheres and Oceans* 2:19–28. ⟨531⟩

McWilliams, J. C., and G. R. Flierl, 1976. Optimal, quasigeostrophic wave analysis of MODE array data. *Deep-Sea Research* 23:285–300. ⟨169⟩

McWilliams, J. C., and G. R. Flierl, 1979. On the evolution of isolated, nonlinear vortices. *Journal of Physical Oceanography* 9:1155–1182. ⟨510, 529⟩

Madden, R., and P. Julian, 1972. Further evidence of global scale, five-day pressure waves. *Journal of the Atmospheric Sciences* 29:1464–1469. ⟨348⟩

Magnell, B., 1976. Salt fingers observed in the Mediterranean outflow region (34°N, 11°W) using a towed sensor. *Journal of Physical Oceanography* 6:511–523. ⟨252⟩

Malkus, W. V. R., 1963. Outline of a theory of turbulent convection. In *Theory and Fundamental Research in Heat Transfer*, American Society of Mechanical Engineers, Pergamon Press, Oxford, pp. 203–217. ⟨392⟩

Malkus, W. V. R., 1964. Boussinesq equations, Boussinesq energetics. In Notes on the 1964 Summer Study Program in Geophysical Fluid Dynamics at the Woods Hole Oceanographic Institution, W.H.O.I. Reference No. 64-46, Woods Hole, Massachusetts, 1, pp. 1–12. ⟨385⟩

Malkus, W. V. R., 1979. Turbulent velocity profiles from stability criteria. *Journal of Fluid Mechanics* 90:401–414. ⟨391⟩

Malkus, W. V. R., and G. Veronis, 1958. Finite amplitude convection. *Journal of Fluid Mechanics* 4:225–260. ⟨391⟩

Manins, P. C., 1976. Intrusion into a stratified fluid. *Journal of Fluid Mechanics* 74:547–560. ⟨253⟩

Manins, P. C., and J. S. Turner, 1978. The relation between the flux ratio and energy ratio in convectively mixed layers. *Quarterly Journal of the Royal Meteorological Society* 104:39–44. ⟨245⟩

Mann, C. R., 1967. The termination of the Gulf Stream and the beginning of the North Atlantic Current. *Deep-Sea Research* 14:337–359. ⟨125, 137⟩

Mann, C. R., 1969. Temperature and salinity characteristics of the Denmark Strait overflow. *Frederick C. Fuglister Sixtieth Anniversary Volume, Deep-Sea Research* 16 (Supplement):125–137. ⟨23, 24⟩

Mann, C. R., A. R. Coote, and D. M. Garner, 1973. The meridional distribution of silicate in the western Atlantic Ocean. *Deep-Sea Research* 20:791–801. ⟨25⟩

Mantyla, A. W., 1975. On the potential temperature in the abyssal Pacific Ocean. *Journal of Marine Research* 33:341–354. ⟨35, 36⟩

Marmer, H. A., 1925. Tides and currents in New York Harbor. *U.S. Coast and Geodetic Survey special Publication* No. 111, 198 pp. ⟨200, 201⟩

Marmorino, G. O., and D. R. Caldwell, 1976. Heat and salt transport through a diffusive thermohaline interface. *Deep-Sea Research* 23:59–67. ⟨254, 255⟩

Marsh, J. G., and E. S. Chang, 1978. 5′ detailed gravimetric geoid in the northwestern Atlantic Ocean. *Marine Geodesy* 1:253–261. ⟨374⟩

Marsh, J. G., T. V. Martin, J. J. McCarthy, and P. S. Chovitz, 1980. Mean sea surface computation using GEOS-3 altimeter data. *Marine Geodesy* 3:359–378. ⟨323⟩

Marshall, S. M., and A. P. Orr, 1955. *The Biology of a Marine Copepod*. Oliver & Boyd, Edinburgh, 188 pp. ⟨278⟩

Martin, S., W. Simmons, and C. Wunsch, 1972. The excitation of resonant triads by single internal waves. *Journal of Fluid Mechanics* 53:17–44. ⟨276⟩

Martineau, D. P., 1953. The influence of the current systems and lateral mixing upon Antarctic Intermediate Water in the South Atlantic. W.H.O.I. Technical Report 53-72, Woods Hole Oceanographic Institution, Woods Hole, Massachusetts, 12 pp. ⟨84⟩

Masuzawa, J., 1969. Subtropical Mode Water. *Deep-Sea Research* 16:463–472. ⟨59, 83⟩

Mather, J. R., H. Adams III, and G. A. Yoshioka, 1964. Coastal storms of the eastern United States. *Journal of Applied Meteorology* 3:693–706. ⟨215⟩

Matsuno, T., 1966. Quasi-geostrophic motions in the equatorial area. *Journal of the Meteological Society of Japan, Series 2* 44:25–43. ⟨191, 527⟩

Matsuno, T., and K. Nakamura, 1979. The Eulerian and Lagrangian-mean meridional circulations in the stratosphere at the time of a sudden warming. *Journal of the Atmospheric Sciences* 36:640–654. ⟨506⟩

Matthäus, W., 1968. Zur Geschichte des Hochseepegels. *Schriftenreihe für Geschichte der Naturwissenschaften, Technik, and Medizin* 12:101–112. ⟨399, 424⟩

Mattson, J. S., 1978. Chronology of events and oil slicks from the ARGO MERCHANT. In *In the Wake of the ARGO MERCHANT: Proceedings of a Symposium Held January 11–13, 1978*, Center for Ocean Management Studies, University of Rhode Island, pp. 15–18. ⟨226⟩

Maul, G., P. W. deWitt, A. Yanaway, S. R. Baig, 1978. Geostationary satellite observations of Gulf Stream meanders: infrared measurements and time series analysis. *Journal of Geophysical Research* 83:6123–6135. ⟨113, 116⟩

Maury, M. F., 1855. *The Physical Geography of the Sea and its Meteorology*, 1st ed. Harper & Bros., New York, 274 pp. Eighth ed. (1861) republished 1963, J. Leighly, ed., Harvard University Press, Cambridge, Massachusetts, 432 pp. ⟨208, 343⟩

Maxworthy, T., 1979. A note on the internal solitary waves produced by tidal flow over a three-dimensional ridge. *Journal of Geophysical Research* 84:338–346. ⟨332⟩

Maxworthy, T., and F. K. Browand, 1975. Experiments in rotating and stratified flows: oceanographic application. *Annual Review of Fluid Mechanics* 7:273–305. ⟨248, 253⟩

Mayer, D. A., D. V. Hansen, and S. M. Minton, 1980. A comparison of water movements in the New Jersey shelf during 1975 and 1976. In *Oxygen Depletion and Associated Benthic Mortalities in the New York Bight, 1976*, R. L. Swanson and C. J. Sinderman, eds., NOAA Professional Paper, U.S. Department of Commerce, National Oceanic and Atmospheric Administration, Rockville, Maryland. ⟨225⟩

Mayer, D. A., D. V. Hansen, and D. A. Ortman, 1979. Long-term current and temperature observations on the Middle Atlantic shelf. *Journal of Geophysical Research* 84:1776–1792. ⟨214, 215, 216, 218, 219, 223, 224, 225, 226, 227⟩

Mazeika, P. A., 1968. Eastward flow within the South Equatorial Current in the eastern South Atlantic. *Journal of Geophysical Research* 73:5819–5828. ⟨83⟩

MEDOC Group, 1970. Observation of formation of deep water in the Mediterranean Sea, 1969. *Nature* 227:1037–1040. ⟨xxii, 22, 25⟩

Meiss, J. D., N. Pomphrey, and K. M. Watson, 1979. Numerical analysis of weakly nonlinear wave turbulence. *Proceedings of the National Academy of Sciences of the U.S.A.* 76:2109–2113. ⟨285⟩

Mellor, G. L., and P. A. Durbin, 1975. The structure and dynamics of the ocean surface mixed layer. *Journal of Physical Oceanography* 5:718–728. ⟨240⟩

Menard, H. W., and S. M. Smith, 1966. Hypsometry of ocean basin provinces. *Journal of Geophysical Research* 71:4305–4325. ⟨46, 50⟩

Menzel, D. W., and J. H. Ryther, 1968. Organic carbon and the oxygen minimum in the South Atlantic Ocean. *Deep-Sea Research* 15:327–337. ⟨81⟩

Menzel, D. W., and J. H. Steele, 1978. The application of plastic enclosures to the study of pelagic marine biota. *Rapports et Proces-Verbaux des Reunions, Conseil Permanent International pour l'Exploration de la Mer* 173:7–12. ⟨379⟩

Merle, J., 1977. Seasonal variations of temperature and circulation in the upper layers of the equatorial Atlantic Ocean. Paper presented at GATE Workshop, Miami, 28 February–10 March, 1977. (Unpublished manuscript.) ⟨185⟩

Merle, J., 1978. Atlas hydrologique saisonnier de l'Ocean Atlantique Intertropical. *Travaux et Documents de l'O.R.S.T.O.M.*, No. 82, 184 pp. ⟨82, 83⟩

Merz, A., 1925. Die Deutsche Atlantische Expedition auf dem Vermessungs-und Forschungsschiff "Meteor". I. Bericht. *Sitzungsberichte der Preussischen Akademie der Wissenschaften, Physikalisch-mathematische Klasse, Jahrgang* 1925:562–586. ⟨10⟩

Merz, A., and G. Wust, 1922. Die Atlantische Vertikalzirkulation. *Zeitschrift der Gesellschaft für Erdkunde zu Berlin, Jahrgang* 1922:1–35. ⟨9, 10, 11⟩

Metcalf, W. G., 1969. Dissolved silicate in the deep North Atlantic. *Frederick C. Fuglister Sixtieth Anniversary Volume, Deep-Sea Research* (16 (Supplement):139–145. ⟨25, 29, 59⟩

Metcalf, W. G., B. C. Heezen, and M. C. Stalcup, 1964. The sill depth of the Mid-Atlantic Ridge in the equatorial region. *Deep-Sea Research* 11:1–10. ⟨29⟩

Meyers, G., 1979. Annual variation in the slope of the 14°C isotherm along the equator in the Pacific Ocean. *Journal of Physical Oceanography* 9:885–891. ⟨186, 187⟩

Mied, R. D., 1978. The instabilities of finite amplitude barotropic Rossby waves. *Journal of Fluid Mechanics* 86:225–246. ⟨535⟩

Miles, J. W., 1957. On the generation of surface waves by shear flows. *Journal of Fluid Mechanics* 3:185–204. ⟨494⟩

Miles, J. W., 1959. On the generation of surface waves by shear flows. Part 2. *Journal of Fluid Mechanics* 6:568–582. ⟨494⟩

Miles, J. W., 1963. On the stability of heterogeneous shear flows. *Journal of Fluid Mechanics* 16:209–227. ⟨269⟩

Miles, J. W., 1964. Free surface oscillations in a slowly rotating liquid. *Journal of Fluid Mechanics* 18:187–194. ⟨298⟩

Miles, J. W., 1974a. On Laplace's tidal equations. *Journal of Fluid Mechanics* 66:241–260. ⟨295, 296, 297, 331⟩

Miles, J. W., 1974b. Laplace's tidal questions revisited. In *Proceedings of the Seventh U.S. National Congress of Applied Mechanics*, Boulder, Colorado, June 3–7, 1974, pp. 27–38. ⟨301⟩

Millard, R. C., Jr., 1971. Wind measurements from buoys: A sampling scheme. *Journal of Geophysical Research* 76:5819–5828. ⟨214⟩

Millard, R., and H. Bryden, 1973. Spatially averaged MODE-I C.T.D. Stations. In *MODE Hot Line News No. 42*, Woods

Hole Oceanographic Institution, Woods Hole, Massachusetts. (Unpublished document.) ⟨520⟩

Miller, A. R., 1950. A study of mixing processes over the edge of the continental shelf. *Journal of Marine Research* 9:145–160. ⟨212⟩

Miller, A. R., 1952. A pattern of surface coastal circulation inferred from surface salinity-temperature data and drift bottle recoveries. *W.H.O.I. Technical Report 52-28*, Woods Hole Oceanographic Institution, Woods Hole, Massachusetts, 14 pp. + Appendix. ⟨212⟩

Miller, B. I., 1964. A study of the filling of Hurricane Donna (1960) over land. *Monthly Weather Review* 92:389–406. ⟨490⟩

Miller, G., 1966. The flux of tidal energy out of the deep oceans. *Journal of Geophysical Research* 71:2485–2489. ⟨323⟩

Miller, S., and C. Wunsch, 1973. The pole tide. *Nature (Physical Science)* 246:98–102. ⟨339⟩

Milliman, J. D., O. H. Pilkey, and D. A. Ross, 1972. Sediments of the continental margin off the eastern United States. *Geological Society of America Bulletin* 83:1315–1334. ⟨208⟩

Mitchell, H., 1859. Report of Assistant Henry Mitchell on the physical surveys of New York harbor and the coast of Long Island, with descriptions of apparatus for observing currents, &c. In *Report of the Superintendent of the Coast Survey, Showing the Progress of the Survey during the Year 1859*, Washington, Appendix No. 26, pp. 311–317. ⟨200⟩

Mitchell, H., 1889. Report on the results of the physical surveys of New York Harbor. In *Report of the Superintendent of the U.S. Coast and Geodetic Survey Showing the Progress of the Work during the Fiscal Year Ending with June, 1887*, Washington, Appendix No. 15, 301–311. ⟨200⟩

Mitsuyasu, H., 1966. Interactions between water waves and wind (I). *Reports of the Research Institute of Applied Mechanics, Kyushu University* 14:67–88. ⟨493⟩

Miyake, M., M. Donelan, G. McBean, C. Paulson, F. Badgley, and E. Leavitt, 1970. Comparison of turbulent fluxes over water determined by profile and eddy correlation techniques. *Quarterly Journal of the Royal Meteorological Society* 96:132–137. ⟨489⟩

MODE-I, 1972a. The design of MODE-I—a report to the Scientific Council from the Array Committee. Massachusetts Institute of Technology, Cambridge, Massachusetts, 99 pp. (Unpublished document.) ⟨430⟩

MODE-I, 1972b. Dynamics and the design of MODE-I—a report to the Scientific Council by the MODE-I Theoretical Panel. Massachusetts Institute of Technology, Cambridge, Massachusetts. (Unpublished document.) ⟨430⟩

MODE-I, 1974. Instrument description and Intercomparison Report of the MODE-I Intercomparison Group. Massachusetts Institute of Technology, Cambridge, Massachusetts, 173 pp. (Unpublished Document.) ⟨431⟩

MODE-I, 1975. Dynamics and the Analysis of MODE-I: Report of the MODE-I Dynamics Group. Massachusetts Institute of Technology, Cambridge, Massachusetts, 250 pp. (Unpublished document.) ⟨430⟩

MODE-1 Atlas Group, The, 1977. *Atlas of the Mid-Ocean Dynamics Experiment (MODE-I)*. V. Lee and C. Wunsch, eds., Massachusetts Institute of Technology, Cambridge, Massachusetts, 274 pp. ⟨359⟩

MODE Group, The, 1978. The Mid-Ocean Dynamics Experiment. *Deep-Sea Research* 25:859–910. ⟨6, 359, 371, 374, 430, 431, 432⟩

Molinari, R., and A. D. Kirwan, Jr., 1975. Calculations of differential kinematic properties from Lagrangian observations in the western Caribbean Sea. *Journal of Physical Oceanography* 5:483–491. ⟨426⟩

Moller, L., 1929. Die Zirkulation des Indischen Ozeans. *Veröffentlichungen des Instituts für Meereskunde an der Universität Berlin, Neue Folge. A. Geographisch-naturwissenschaftliche Reihe*, 21, 48 pp. ⟨10⟩

Monin, A. S., V. M. Kamenkovich, and V. G. Kort, 1974. *Izmenchivost Mirovogo Okeana*. Gidrometeoizdat, Leningrad, 261 pp. [*Variability of the Oceans*, J. J. (sic) Lumley, ed., 1977, Wiley, Interscience, New York, 241 pp.] ⟨250⟩

Monin, A. S., and A. M. Obukhov, 1954. Osnovnye zakonomernosti turbulentnogo peremeshivaniya v prizemnom atmosfery. *Trudy Geofizicheskogo Instituta Akademii Nauk SSSR* 24:163–187. [Basic laws of turbulent mixing in the ground layer of the atmosphere, Air Technical Information Section Liaison Office No. F-TS-9295/V and American Meteorological Society Translation No. T-R-174, 35 pp.] ⟨239, 484⟩

Monin, A S., and A. M. Yaglom, 1965, 1967. *Statisticheskaya Gidromekhanika; Mekhanika Turbulentnosti*. Nauka, Moscow, 1 (1965), 639 pp., 2 (1967), 708 pp. [*Statistical Fluid Mechanics: Mechanics of Turbulence*, J. L. Lumley, ed., MIT Press, Cambridge, Massachusetts, (1971), 769 pp., 2 (1975), 874 pp.] ⟨485⟩

Montgomery, R. B., 1938a. Circulation in upper layers of southern North Atlantic deduced with use of isentropic analysis. *Papers in Physical Oceanography and Meteorology* 6:2, 55 pp. ⟨43, 44, 82⟩

Montgomery, R. B., 1938b. Fluctuations in monthly sea level on eastern U.S. coast as related to dynamics of western North Atlantic Ocean. *Journal of Marine Research* 1:165–185. ⟨114, 116, 351⟩

Montgomery, R. B., 1941. Sea level difference between Key West and Miami, Florida. *Journal of Marine Research* 4:32–37. ⟨116⟩

Montgomery, R. B., 1950. The Taylor diagram (temperature against vapor pressure) for air mixtures. *Archiv für Meteorologie, Geophysik und Bioklimatologie* 2:163–183. ⟨497⟩

Montgomery, R. B., 1958. Water characteristics of Atlantic Ocean and of world ocean. *Deep-Sea Research* 5:134–148. ⟨42, 44, 46, 47, 49, 50, 51, 57, 93⟩

Montgomery, R. B., 1962. Equatorial Undercurrent observations in review. *Journal of the Oceanographical Society of Japan, 20th Anniversary Volume:* 487–498. ⟨185⟩

Montgomery, R. B., 1969. Comments on oceanic leveling. *Frederick C. Fuglister Sixtieth Anniversary Volume, Deep-Sea Research* 16 (Supplement):147–152. ⟨227⟩

Montgomery, R. B., and E. Palmén, 1940. Contribution to the question of the equatorial counter current. *Journal of Marine Research* 3:112–133. ⟨188⟩

Montgomery, R. B., and M. J. Pollak, 1942. Sigma-T Surfaces in the Atlantic Ocean. *Journal of Marine Research* 5:20–27. ⟨86⟩

Montgomery, R. B., and E. D. Stroup, 1962. Equatorial Waters and Currents at 150°W in July–August 1952. *The Johns Hopkins Oceanographic Studies,* No. 1, 68 pp. ⟨185⟩

Mooers, C. N. K., J. Fernandez-Partagas, and J. F. Price, 1976. Meteorological Forcing Fields of the New York Bight (First Year's Progress Report). *Technical Report, Rosenstiel School of Marine and Atmospheric Science, University of Miami,* TR76-8, Miami, Florida, 151 pp. ⟨214, 215, 233⟩

Moore, D. W., 1963. Rossby waves in ocean circulation. *Deep-Sea Research* 10:735–747. ⟨154, 155⟩

Moore, D. W., 1968. Planetary-gravity waves in equatorial ocean. Ph.D. Thesis, Harvard University, Cambridge, Massachusetts, 207 pp. ⟨191, 307⟩

Moore, D., 1970. The mass transport velocity induced by free oscillations at a single frequency. *Geophysical Fluid Dynamics* 1:237–247. ⟨345, 472⟩

Moore, D. W., P. Hisard, J. P. McCreary, J. Merle, J. J. O'Brien, J. Picaut, J. Verstraete, and C. Wunsch, 1978. Equatorial adjustment in the eastern Atlantic. *Geophysical Research Letters* 5:637–640. ⟨193⟩

Moore, D. W., and P. P. Niiler, 1975. A two-layer model for the separation of inertial boundary currents. *Journal of Marine Research* 32:457–485. ⟨343⟩

Moore, D. W., and S. G. H. Philander, 1977. Modeling of the tropical oceanic circulation. In *The Sea: Ideas and Observations on Progress in the Study of the Seas, 6: Marine Modeling,* E. D. Goldberg, I. N. McCave, J. J. O'Brien, and J. H. Steele, eds., Wiley, Interscience, New York, pp. 319–361. ⟨184, 191, 192, 304, 307, 513⟩

Moore, W. S., 1969. Oceanic concentration of ^{228}Radium. *Earth and Planetary Science Letters* 6:437–446. ⟨445⟩

Morey, R., 1973. Evaluation of long term deep sea effects on mooring line components. *Charles Stark Draper Laboratory Technical Report E-2748,* Cambridge, Massachusetts, 85 pp. ⟨410⟩

Morgan, G. W., 1956. On the wind-driven ocean circulation. *Tellus* 8:301–320. ⟨125, 154, 342, 506⟩

Moriyasu, S., 1972. Deep waters in the western North Pacific. In *Kuroshio: Its Physical Aspects,* H. Stommel and K. Yoshida, eds., University of Tokyo Press, pp. 387–408. ⟨36⟩

Morse, P. M., and H. Feshbach, 1953. *Methods of Theoretical Physics.* McGraw-Hill, New York, 2 vols., 1978 pp. ⟨298⟩

Mosby, H., 1934. The waters of the Atlantic Antarctic Ocean. In *Scientific Results of the Norwegian Antarctic Expedition 1927–1928 et seq.* 1:11, 131 pp. ⟨11, 72⟩

Mowbray, D. E., and B. S. H. Rarity, 1967. A theoretical and experimental investigation of the phase configuration of internal waves of small amplitude in a density stratified fluid. *Journal of Fluid Mechanics* 28:1–16. ⟨270⟩

Müller, P., 1976. Determination of the rate of change of G and the tidal acceleration of earth and moon from ancient and modern astronomical data. *Technical report, JPL-SP43-36,* The Jet Propulsion Laboratory, Pasadena, California, 24 pp. ⟨322, 323⟩

Müller, P., D. J. Olbers, and J. Willebrand, 1978. The IWEX spectrum. *Journal of Geophysical Research* 83:479–500. ⟨285⟩

Munk, W. H., 1947. A critical wind speed for air-sea boundary processes. *Journal of Marine Research* 6:203–218. ⟨488⟩

Munk, W. H., 1950. On the wind-driven ocean circulation. *Journal of Meteorology* 7:79–93. ⟨79, 81, 125, 140, 152, 340, 341, 342, 345, 468, 471, 473⟩

Munk, W. H., 1966. Abyssal recipes. *Deep-Sea Research* 13:707–730. ⟨160, 245, 261, 281, 337, 339, 449⟩

Munk, W., 1968. Once again—tidal friction. *Quarterly Journal of the Royal Astronomical Society* 9:352–375. ⟨322⟩

Munk, W. H., and E. R. Anderson, 1948. Notes on a theory of the thermocline. *Journal of Marine Research* 7:276–295. ⟨240⟩

Munk, W. H., and G. F. Carrier, 1950. The wind-driven circulation in ocean basins of various shapes. *Tellus* 2:158–167. ⟨xix⟩

Munk, W. H., and D. E. Cartwright, 1966. Tidal spectroscopy and prediction. *Philosophical Transactions of the Royal Society of London A* 259:533–581. ⟨294, 320, 351⟩

Munk, W. H., G. W. Groves, and G. F. Carrier, 1950. Note on the dynamics of the Gulf Stream. *Journal of Marine Research* 9:218–238. ⟨125, 141, 153⟩

Munk, W. H., and G. MacDonald, 1960. *The Rotation of the Earth: A Geophysical Discussion.* Cambridge, University Press, London, 323 pp. ⟨296⟩

Munk, W. and D. Moore, 1968. Is the Cromwell current driven by equatorial Rossby waves? *Journal of Fluid Mechanics* 33:241–259. ⟨345⟩

Munk, W., and N. Phillips, 1968. Coherence and band structure of inertial motion in the sea. *Reviews of Geophysics* 6:447–472. ⟨272, 297, 307, 332⟩

Munk, W., F. Snodgrass, and F. Gilbert, 1964. Long waves on the continental shelf: an experiment to separate trapped and leaky modes. *Journal of Fluid Mechanics* 20:529–554. ⟨310, 312⟩

Munk, W., F. Snodgrass, and M. Wimbush, 1970. Tides offshore: transition from California coastal to deep-sea waters. *Geophysical Fluid Dynamics* 1:161–235. ⟨310, 326, 327, 333, 358, 400, 425⟩

Munk, W. H., and C. Wunsch, 1979. Ocean acoustic tomography: a scheme for large scale monitoring. *Deep-Sea Research* 26:123–161. ⟨265, 374, 428⟩

Muromtsev, A. M., 1958. *Osnovnye Cherty Gidrologii Tikhogo Okeana.* Gidrometeoizdat, Leningrad, 631 pp., plus Appendix II (bound separately), 124 pp. (*The Principal Hydrological Features of the Pacific Ocean,* A. Birron and Z. S. Cole, translators, 1963, Israel Program for Scientific Translations, Jerusalem, 417 pp. Available from Office of Technical Services, U.S. Department of Commerce, Washington, D. C.) ⟨85⟩

Murphy, D. W., and P. A. Christian, 1979. Solid state electrodes for high energy batteries. *Science* 205:651–656. ⟨401⟩

Murphy, S. R., and G. E. Lord, 1965. Thermal and sound velocity microstructure data taken with an unmanned research vehicle. In *Second United States Navy Symposium on Military Oceanography, The Proceedings of the Symposium,* U.S. Naval Ordnance Laboratory White Oak, Silver Springs, Maryland, 1:343–360. ⟨265⟩

Mysak, L. A., 1980. Recent advances in shelf wave dynamics. *Reviews of Geophysics and Space Physics* 18:211–241. ⟨222, 358⟩

Namias, J., 1969. Seasonal interactions between the North Pacific Ocean and the atmosphere during the 1960's. *Monthly Weather Review* 97:173–192. ⟨496⟩

Namias, J., 1972. Large-scale and long-term fluctuations in some atmospheric and oceanic variables. In *Nobel Symposium 20: The Changing Chemistry of the Oceans*, D. Dryssen and D. Jagner, eds., Almqvist & Wiksell, Stockholm, and Wiley, Interscience, New York, pp. 27–48. ⟨351⟩

Nan'niti, T., and H. Akamatsu, 1966. Deep current observations in the Pacific Ocean near the Japan Trench. *Journal of the Oceanographical Society of Japan* 22: 154–160. ⟨36⟩

Nansen, F., 1902. The Oceanography of the North Pole Basin. In *Scientific Results of the Norwegian North Polar Expedition, 1893–1896*, 3:9, 427 pp. and 33 plates. ⟨72, 264, 331, 415⟩

Nansen, F., 1906. Northern waters: Captain Roald Amundsen's observations in the Arctic Seas in 1901. With a discussion of the origin of the bottom-waters of the Northern Seas. *Videnskabs-Selskabets Skrifter. I. Mathematisk-Naturvitenskapelig Klasse, 1906:3*, 145 pp. ⟨72⟩

Nansen, F., 1912. Das Bodenwasser und die Abuühlung des Meeres. *Internationale Revue der Gesamten Hydrobiologie und Hydrographie* 5:1, 42 pp. ⟨22, 40, 41, 72⟩

Nansen, F., 1913. The waters of the north-eastern North Atlantic. Investigations made during the cruise of the "Frithjof", of the Norwegian Royal Navy, in July 1910. *Internationale Revue der Gesamten Hydrobiologie und Hydrographie* 4: Hydrographisches Supplement, 139 pp. ⟨72⟩

Nath, J. H., 1977. Laboratory validation of numerical model drifting buoy-tethering-drogue system. *Final Report to NOAA Data Buoy Office*, NSTL Station, Mississippi, Contract No. 03-6-038-128. ⟨426⟩

NATO Science Committee, 1978. *Instruments and Methods in Air-Sea Interaction*. Preprint volume for NATO School, Ustaoset, Norway, April 1978, pp. various. ⟨402, 426⟩

Needler, G., 1967. A model for the thermohaline circulation in an ocean of finite depth. *Journal of Marine Research* 25:329–342. ⟨159, 160, 161⟩

Needler, G., 1972. Thermocline models with arbitrary barotropic flow. *Deep-Sea Research* 18:895–903. ⟨159, 161⟩

Needler, G., and P. H. LeBlond, 1973. On the influence of the horizontal component of the earth's rotation on long-period waves. *Geophysical Fluid Dynamics* 5:23–45. ⟨143⟩

Neshyba, S., V. T. Neal, and W. Denner, 1971. Temperature and conductivity measurements under Ice Island T-3. *Journal of Geophysical Research* 76:8107–8120. ⟨252⟩

Neumann, G., W. H. Beatty III, and E. C. Escowitz, 1975. Seasonal changes of oceanographic and marine-climatological conditions in the equatorial Atlantic. City College of the City University of New York, CUNY Institute of Marine and Atmospheric Science, 211 pp. ⟨186⟩

Newman, F. C., 1976. Temperature steps in Lake Kivu: A bottom heated saline lake. *Journal of Physical Oceanography* 6:157–163. ⟨252, 253, 256⟩

Newton, C. W., 1961. Estimates of the vertical motions and meridional heat exchange in Gulf Stream eddies and a comparison with atmospheric disturbances. *Journal of Geophysical Research* 66:853–870. ⟨137⟩

Newton, I., 1687. *Philosophiae Naturalis Principia Mathematica*. See *Newton's Principia*, Cajori's 1946 revision of Motte's 1729 translation, University of California Press, Berkeley, 680 pp. ⟨293, 504⟩

Nicholls, J. M., 1973. The airflow over mountains. *World Meteorological Organization Technical Note* No. 127, WMO No. 355, Geneva, 74 pp. ⟨522⟩

Nielsen, J. N., 1904. Hydrography of the waters by the Faroe Islands and Iceland during the cruises of the Danish Research Steamer "Thor" in the summer of 1903. *Meddelelsar fra Kommissionen for Havundersøgelser. Serie: Hydrografi* 1:4, 29 pp. ⟨22⟩

Nielsen, J. N., 1925. Golfstrømmen. *Geografisk Tidsskrift* 28:49–59. ⟨113⟩

Nihoul, J. C. J., ed., 1975. *Modelling of Marine Systems*. Elsevier, Amsterdam, 272 pp. ⟨236⟩

Nihoul, J. C. J., ed., 1977. *Bottom Turbulence*. Elsevier, Amsterdam, 306 pp. ⟨258⟩

Niiler, P. P., 1966. On the theory of the wind-driven ocean circulation. *Deep-Sea Research* 13:597–606. ⟨141, 154, 155⟩

Niiler, P. P., 1969. On the Ekman divergence in an oceanic jet. *Journal of Geophysical Research* 74:7048–7052. ⟨149⟩

Niiler, P. P., 1975. Deepening of the wind-mixed layer. *Journal of Marine Research* 33:405–422. ⟨243⟩

Niiler, P. P., 1977. One-dimensional models of the seasonal thermocline. In *The Sea: Ideas and Observations on Progress in the Study of the Seas, 6: Marine Modeling*, E. D. Goldberg, I. N. McCave, J. J. O'Brien, and J. H. Steele, eds., Wiley, Interscience, New York, pp. 97–115. ⟨240, 245⟩

Niiler, P. P., and E. B. Kraus, 1977. One-dimensional models of the upper ocean. In *Modelling and Prediction of the Upper Layers of the Ocean*, E. B. Kraus, ed., Pergamon Press, Oxford, pp. 143–172. ⟨240, 243, 244, 503⟩

Niiler, P. P., and L. A. Mysak, 1971. Barotropic waves along an eastern continental shelf. *Geophysical Fluid Dynamics* 2:273–288. ⟨120⟩

Niiler, P. P., and W. S. Richardson, 1973. Seasonal variability of the Florida Current. *Journal of Marine Research* 31:144–167. ⟨117, 357, 421⟩

Niiler, P. P., and A. R. Robinson, 1967. The theory of free inertial jets. II. A numerical experiment for the path of the Gulf Stream. *Tellus* 19:601–619. ⟨126⟩

Ninomiya, K., 1974. Bulk properties of cumulus convections in the small area over Kuroshio Region in February 1968. *Journal of the Meteorological Society of Japan, Series 2* 52:188–203. ⟨503⟩

Noble, M., and B. Butman, 1979. Low-frequency wind-induced sea level oscillations along the east coast of North America. *Journal of Geophysical Research* 84:3227–3236. ⟨215, 216, 218, 222, 223⟩

NORPAC Committee, 1960. *Oceanic Observations of the Pacific: 1955, the NORPAC Atlas*. University of California Press and University of Tokyo Press, Berkeley and Tokyo, 123 plates. ⟨82⟩

Nowlin, W. D., T. Whitworth, and R. D. Pillsbury, 1977. Structure and transport of the Antarctic Circumpolar Current at Drake Passage from short-term measurements. *Journal of Physical Oceanography* 7:787–802. ⟨355⟩

Nowroozi, A. A., G. H. Sutton, and B. Ault, 1966. Oceanic tides recorded on the sea floor. *Annales de Geophysique* 22:512–517. ⟨424⟩

Nozaki, Y., D. M. Rye, K. K. Turekian and R. E. Dodge, 1978. A 200-year record of carbon-13 and carbon-14 variations in a Bermuda coral. *Geophysical Research Letters* 5:825–828. ⟨439⟩

Nydal, R., K. Lövseth, and S. Gulliksen, 1979. A survey of radiocarbon variation in nature since the Test Ban Treaty. In *Radiocarbon Dating*, R. Berger and H. Suess, eds., University of California Press, Berkeley, pp. 313–323. ⟨440⟩

O'Brien, J. J., D. Adamec, and D. W. Moore, 1978. A simple model of upwelling in the Gulf of Guinea. *Geophysical Research Letters* 5:641–644. ⟨193⟩

O'Brien, J. J., R. M. Clancy, A. J. Clarke, M. Crepon, R. Elsberry, T. Gammelsrød, M. MacVean, L. P. Röed, J. D. Thompson, 1977. Upwelling in the ocean: Two- and three-dimensional models of upper ocean dynamics and variability. In *Modelling and Prediction of the Upper Layers of the Ocean*, E. B. Kraus, ed., Pergamon Press, Oxford, pp. 178–228. ⟨358⟩

O'Brien, J. J., and H. Hurlburt, 1974. Equatorial jet in the Indian Ocean: Theory. *Science* 184:1075–1077. ⟨192⟩

Obukhov, A. M., 1946. Turbulentnost v temperaturno-neodnorodnoi atmosfere (Turbulence in a thermally inhomogeneous atmosphere). *Trudy Instituta Teoreticheskoi Geofziky Akademii Nauk SSSR* 1:95–115. ⟨484⟩

Obukhov, A., 1949. K voprosu o geostroficheskom vetre (On the matter of the geostrophic wind). *Izvestiya Akademii Nauk SSSR, Seriya Geograficheskaya i Geofizicheskaya* 13:281–306. ⟨508⟩

Oeschger, H., A. Gugelmann, H. Loosli, U. Schotterer, U. Siegenthaler, and W. Wiest, 1974. ^{39}Ar dating of groundwater. In *Isotope Techniques in Groundwater Hydrology*, International Atomic Energy Agency, Vienna, 2, pp. 179–190. ⟨447⟩

Oeschger, H., U. Siegenthaler, U. Schotterer, and A. Gugelmann, 1975. A box diffusion model to study the carbon dioxide exchange in nature. *Tellus* 27:168–192. ⟨449⟩

Ogura, Y., and J. G. Charney, 1962. A numerical model of thermal convection in the atmosphere. In *Proceedings of the International Symposium on Numerical Weather Prediction in Tokyo, November 7–13, 1960*, S. Syono, chief ed., The Meteorological Society of Japan, Tokyo, pp. 431–461. ⟨548⟩

Ogura, Y., and N. A. Phillips, 1962. Scale analysis of deep and shallow convection in the atmosphere. *Journal of the Atmospheric Sciences* 19:173–179. ⟨548⟩

Okubo, A., 1964. Equations describing the diffusion of an introduced pollutant in a one-dimensional estuary. In *Studies on Oceanography*, K. Yoshida, ed., University of Tokyo Press, Tokyo, pp. 216–226. ⟨205⟩

Okubo, A., 1967. The effect of shear in an oscillatory current on horizontal diffusion from an instantaneous source. *International Journal of Oceanology and Limnology* 1:194–204. ⟨206⟩

Okubo, A., 1969. Some remarks on the importance of the "shear effect" on horizontal diffusion. *Journal of the Oceanographical Society of Japan* 24:60–69. ⟨206⟩

Okubo, A., 1971. Oceanic diffusion diagrams. *Deep-Sea Research* 18:789–802. ⟨377⟩

Okubo, A., 1973. Effect of shoreline irregularities on streamwise dispersion in estuaries and other embayments. *Netherlands Journal of Sea Research* 6:213–224. ⟨206⟩

Omori, M., 1978. Zooplankton fisheries of the world: A review. *Marine Biology* 48:199–205. ⟨380⟩

Onsager, L., 1949. Statistical hydrodynamics. *Nuovo Cimento* 6 (Supplement): 179–187. ⟨543⟩

Oort, A. H., and J. P. Peixoto, 1974. The annual cycle of the energetics of the atmosphere on a planetary scale. *Journal of Geophysical Research* 79:2705–2719. ⟨507⟩

Oort, A. H., and T. H. Vonder Haar, 1976. On the observed annual cycle in the ocean-atmosphere heat balance over the Northern Hemisphere. *Journal of Physical Oceanography* 6:781–800. ⟨490⟩

Orlanski, I., 1968. Instability of frontal waves. *Journal of the Atmospheric Sciences* 25:178–200. ⟨529⟩

Orlanski, I., 1969. The influence of bottom topography on the stability of jets in a baroclinic fluid. *Journal of the Atmospheric Sciences* 26:1216–1232. ⟨119, 120, 529⟩

Orlanski, I., 1972. On the breaking of standing internal waves. *Journal of Fluid Mechanics* 54:577–598. ⟨248⟩

Orlanski, I., 1975. A rational subdivision of scales for atmospheric processes. *Bulletin of the American Meteorological Society* 56:527–530. ⟨214⟩

Orlanski, I., and K. Bryan, 1969. Formation of the thermocline step structure by large-amplitude internal gravity waves. *Journal of Geophysical Research* 74:6975–6993. ⟨278⟩

Orlanski, I., and M. D. Cox, 1973. Baroclinic instability in ocean currents. *Geophysical Fluid Dynamics* 4:297–332. ⟨120, 174, 529⟩

Ortega, G. F., 1972. Isanosteric analysis of the eastern Caribbean waters during winter. *Boletín del Instituto Oceanográfico de la Universidad de Oriente* 11:19–34. ⟨84⟩

Osborn, T. R., 1974. Vertical profiling of velocity microstructure. *Journal of Physical Oceanography* 4:109–115. ⟨267, 282⟩

Osborn, T. R., 1978. Measurements of energy dissipation adjacent to an island. *Journal of Geophysical Research* 83:2939–2957. ⟨257⟩

Osborn, T. R., 1980. Estimates of the local rate of vertical diffusion from dissipation measurements. *Journal of Physical Oceanography* 10:83–89. ⟨267, 290⟩

Osborn, T. R., and C. S. Cox, 1972. Oceanic fine structure. *Geophysical Fluid Dynamics* 3:321–345. ⟨250, 267, 281, 419⟩

Osborn, T. R., and W. R. Crawford, 1978. Turbulent velocity measurements with an airfoil probe. In *Instruments and Methods in Air-Sea Interaction*, preprint volume for NATO School, Ustaoset, Norway, April 1978, NATO Science Committee, 18 pp. ⟨423⟩

Otto, J. F. W., 1800. *System einer Allgemeinen Hydrographie des Erdbodens*. G. C. Nauck, Berlin, 662 pp. ⟨40⟩

Ou, H.-W., 1979. On the propagation of free topographic Rossby waves near continental margins. Ph.D. Thesis, Massachusetts Institute of Technology/Woods Hole Oceanographic Institution, 133 pp. ⟨217, 218, 225, 258, 313, 315, 316⟩

Ou, H.-W., 1980. On the propagation of free topographic Rossby waves near continental margins. Part 1: Analytical model for a wedge. *Journal of Physical Oceanography*. ⟨313⟩

Ou, H.-W., and R. C. Beardsley, 1980. On the propagation of free topographic Rossby waves near continental margins. Part 2: Numerical model. *Journal of Physical Oceanography.* ⟨313⟩

Overland, J. E., and W. G. Gemmill, 1977. Prediction of marine winds in the New York Bight. *Monthly Weather Review* 105:1003–1008. ⟨215, 233⟩

Owen, P. R., and W. R. Thompson, 1963. Heat transfer across rough surfaces. *Journal of Fluid Mechanics* 15:321–334. ⟨487⟩

Owens, W. B., and F. P. Bretherton, 1978. A numerical study of mid-ocean mesoscale eddies. *Deep-Sea Research* 25:1–14. ⟨133⟩

Ozmidov, R. V., 1965. O turbulentnom obmene v ustoichivo stratifitsirovannom okeane. *Fizika Atmosfery i Okeana, Izvestiya Akademii Nauk SSSR* 1:853–860. (On the turbulent exchange in a stably stratified ocean. *Izvestiya, Academy of Sciences, USSR, Atmospheric and Oceanic Physics* 1:493–497.) ⟨239⟩

Palmén, E., 1930. Ein Beitrag zur Berechnung der Strömungen in einem begrenzten und geschichteten Meere. *Rapports et Procès-Verbaux des Réunions, Conseil Permanent International pour l'Exploration de la Mer* 64:47–58. ⟨201⟩

Palmén, E., and H. Riehl, 1957. Budget of angular momentum and energy in tropical cyclones. *Journal of Meteorology* 14:150–159. ⟨490⟩

Pannetier, R., 1968. Distribution, atmospheric transfer, and assessment of krypton-85. *Commissariat d'Energie Atomique, Publication CEA-R-3591*, Centre d'Etudes Nucléaires, Fontenay-Aux-Roses, France, 177 pp. ⟨446⟩

Parke, M., 1978. Global numerical model of the open ocean tides M2, S2, K1 on an elastic earth. Ph.D. Thesis, University of California at San Diego. ⟨329⟩

Parke, M. E., 1979. Open-ocean tide modelling. In *Proceedings of the Ninth GEOP Conference, Department of Geodetic Sciences Report No. 280*, Ohio State University, Columbus, Ohio, pp. 289–297. ⟨329⟩

Parke, M. E., 1980. Detection of tides on the Patagonian Shelf by the SEASAT satellite radar altimeter: an initial comparison. *Deep-Sea Research* 27:297–300. ⟨324⟩

Parke, M. E., and M. C. Hendershott, 1980. M2, S2, K1 models of the global ocean tide on an elastic earth. *Marine Geodesy* 3:379–408. ⟨322, 323, 327, 329⟩

Parker, C. E., 1971. Gulf Stream rings in the Sargasso Sea. *Deep-Sea Research* 18:981–993. ⟨125⟩

Paros, J. M., 1976. Digital pressure transducers. *Measurements and Data* 10:2, 74–79. ⟨426⟩

Parr, A. E., 1938. On the validity of the dynamic topographic method for the determination of ocean current trajectories. *Journal of Marine Research* 1:119–132. ⟨74, 82⟩

Parsons, A. T., 1969. A two-layer model of Gulf Stream separation. *Journal of Fluid Mechanics* 39:511–528. ⟨141, 158, 473⟩

Patullo, J. G., W. Munk, R. Revelle, and E. Strong, 1955. The seasonal oscillation in sea level. *Journal of Marine Research* 14:88–156. ⟨355⟩

Patzert, W., T. Barnett, M. Sessions, and B. Kilonsky, 1978. AXBT observations of the tropical Pacific Ocean thermal structure during the NORPAX Hawaii/Tahiti Shuttle Experiment, Nov. 1977–Feb. 1978. *S.I.O. Reference Series 78–24*, Scripps Institution of Oceanography, University of California at San Diego, 61 pp. ⟨184⟩

Paulson, C. A., E. Leavitt, and R. G. Fleagle, 1972. Air-sea transfer of momentum, heat and water determined from profile measurements during BOMEX. *Journal of Physical Oceanography* 2:487–497. ⟨489⟩

Pedlosky, J., 1964a. The stability of currents in the atmosphere and the ocean: Part I. *Journal of the Atmospheric Sciences* 21:201–219. ⟨142, 146, 169, 172, 529⟩

Pedlosky, J., 1964b. The stability of currents in the atmosphere and the ocean: Part II. *Journal of the Atmospheric Sciences* 21:342–353. ⟨529⟩

Pedlosky, J., 1965a. A necessary condition for the existence of an inertial boundary layer in a baroclinic ocean. *Journal of Marine Research* 23:69–72. ⟨139, 531⟩

Pedlosky, J., 1965b. A note on the western intensification of oceanic circulation. *Journal of Marine Research* 23:207–209. ⟨165, 341, 533⟩

Pedlosky, J., 1965c. A study of the time dependent ocean circulation. *Journal of the Atmospheric Sciences* 22:267–272. ⟨345⟩

Pedlosky, J., 1967. Fluctuating winds and the ocean circulation. *Tellus* 19:250–257. ⟨544⟩

Pedlosky, J., 1970. Finite amplitude baroclinic waves. *Journal of the Atmospheric Sciences* 27:15–30. ⟨345, 532⟩

Pedlosky, J., 1971. Finite amplitude baroclinic waves with small dissipation. *Journal of the Atmospheric Sciences* 28:587–597. ⟨532⟩

Pedlosky, J., 1972. Limit cycles and unstable baroclinic waves. *Journal of the Atmospheric Sciences* 29:53–63. ⟨532⟩

Pedlosky, J., 1975. On secondary baroclinic stability and the meridional scale of motion in the ocean. *Journal of Physical Oceanography* 5:603–607. ⟨174⟩

Pedlosky, J., 1976. On the dynamics of finite amplitude baroclinic waves as a function of supercriticality. *Journal of Fluid Mechanics* 78:621–637. ⟨532⟩

Pedlosky, J., 1977. On the radiation of meso-scale energy in the mid-ocean. *Deep-Sea Research* 24:591–600. ⟨119, 373, 524⟩

Pedlosky, J., 1979a. *Geophysical Fluid Dynamics.* Springer-Verlag, New York, 624 pp. ⟨505, 520⟩

Pedlosky, J., 1979b. Finite-amplitude baroclinic waves in a continuous model of the atmosphere. *Journal of the Atmospheric Sciences* 36:1908–1924. ⟨532, 538⟩

Pedlosky, J., and H. P. Greenspan, 1967. A simple laboratory model for the ocean circulation. *Journal of Fluid Mechanics* 27:291–304. ⟨152, 469, 470, 472⟩

Pekeris, C., and Y. Accad, 1969. Solution of Laplace's equations for the M_2 tide in the world ocean. *Philosophical Transactions of the Royal Society of London A* 265:413–436. ⟨322⟩

Peng, T.-H., W. S. Broecker, G. G. Mathieu, Y.-H. Li, and A. E. Bainbridge, 1979. Radon evasion rates in the Atlantic and Pacific Oceans as determined during the GEOSECS program. *Journal of Geophysical Research* 84:2471–2486. ⟨447⟩

Perkin, R. G., and E. L. Lewis, 1978. Mixing in an Arctic fjord. *Journal of Physical Oceanography* 8:873–880. ⟨261⟩

Peterson, W. H., and C. G. H. Rooth, 1976. Formation and exchange of deep water in the Greenland and Norwegian Seas. *Deep-Sea Research* 23:273–283. ⟨24⟩

Pettersson, H., 1931. Eddy-viscosity in stratified water. *Göteborgs Kungliga Vetenskaps-och Vitterhets-Samhälles Handlingar, Femte Följden, Serien B* 2:1, 21 pp. ⟨140⟩

Pettersson, H., 1955. Manganese nodules and oceanic radium. *Papers in Marine Biology and Oceanography, Deep-Sea Research* 3 (Supplement): 335–345. ⟨444⟩

Philander, S. G. H., 1973 Equatorial Undercurrent: Measurements and theories. *Reviews of Geophysics and Space Physics* 11:513–570. ⟨184, 185, 191, 527⟩

Philander, S. G. H., 1976. Instabilities of zonal equatorial currents. *Journal of Geophysical Research* 81:3725–3735. ⟨193⟩

Philander, S. G. H., 1978. Forced oceanic waves. *Reviews of Geophysics and Space Physics* 16:15–46. ⟨304, 347⟩

Philander, S. G. H., 1979. Equatorial waves in the presence of the Equatorial Undercurrent. *Journal of Physical Oceanography* 9:254–262. ⟨193⟩

Philander, S. G. H., 1980. The equatorial undercurrent revisited. *Annual Review of Earth and Planetary Sciences* 8. ⟨527⟩

Philander, S. G. H., and R. C. Pacanowski, 1980a. The generation of equatorial currents. *Journal of Geophysical Research* 85:1123–1136. ⟨194, 527⟩

Philander, S. G. H., and R. C. Pacanowski, 1980b. The response of equatorial currents to a relaxation of the winds. *Journal of Geophysical Research*. ⟨194⟩

Phillips, N. A., 1951. A simple three-dimensional model for the study of large-scale extratropical flow patterns. *Journal of Meteorology* 8:381–394. ⟨141, 169, 172⟩

Phillips, N. A., 1954. Energy transformations and meridional circulations associated with simple baroclinic waves in a two-level, quasi-geostrophic model. *Tellus* 6:273–286. ⟨169, 173, 174⟩

Phillips, N. A., 1963. Geostrophic motion. *Reviews of Geophysics* 1:123–176. ⟨144, 161, 183, 372, 505, 520⟩

Phillips, N. A., 1966a. The equations of motion for a shallow rotating atmosphere and the "traditional approximation." *Journal of the Atmospheric Sciences* 23:626–628. ⟨143, 165⟩

Phillips, N., 1966b. Large-scale eddy motion in the western Atlantic. *Journal of Geophysical Research* 71:3883–3891. ⟨295, 341, 359, 533⟩

Phillips, N. A., 1968. Reply [to Veronis (1968)]. *Journal of the Atmospheric Sciences* 25:1155–1157. ⟨295⟩

Phillips, N. A., 1973. Principles of large scale numerical weather prediction. In *Dynamic Meteorology*, P. Morel, ed., D. Reidel, Dordrecht, pp. 128–141. ⟨518⟩

Phillips, O. M., 1957. On the generation of waves by turbulent wind. *Journal of Fluid Mechanics* 2:417–445. ⟨494⟩

Phillips, O. M., 1958. The equilibrium range in the spectrum of wind-generated waves. *Journal of Fluid Mechanics* 4:426–434. ⟨265⟩

Phillips, O. M., 1963. On the attenuation of long gravity waves by short breaking waves. *Journal of Fluid Mechanics* 16:321–332. ⟨494⟩

Phillips, O. M., 1966a. On turbulent convection currents and the circulation of the Red Sea. *Deep-Sea Research* 13:1149–1160. ⟨24⟩

Phillips, O. M., 1966b. *The Dynamics of the Upper Ocean*, 1st ed. Cambridge University Press, London, 261 pp. ⟨265, 275, 348⟩

Phillips, O. M., 1970. On flows induced by diffusion in a stably stratified fluid. *Deep-Sea Research* 17:435–443. ⟨260⟩

Phillips, O. M., 1972. Turbulence in a stratified fluid—is it unstable? *Deep-Sea Research* 19:79–81. ⟨247⟩

Phillips, O. M., 1977a. *The Dynamics of the Upper Ocean*, 2nd ed. Cambridge University Press, London, 336 pp. ⟨248, 250, 265, 268, 273, 274, 346, 491, 492, 493⟩

Phillips, O. M., 1977b. Entrainment. In *Modelling and Prediction of the Upper Layers of the Ocean*, E. B. Kraus, ed., Pergamon Press, Oxford, pp. 92–101. ⟨244⟩

Phillips, O. M., 1977c. The sea surface. In *Modelling and Prediction of the Upper Layers of the Ocean*, E. B. Kraus, ed., Pergamon Press, Oxford, pp. 229–237. ⟨242⟩.

Pickard, G. L., and K. Rodgers, 1959. Current measurements in Knight Inlet, British Columbia. *Journal of the Fisheries Research Board of Canada* 16:635–678. ⟨206⟩

Pierson, W. J., ed., 1960. The directional spectrum of a wind generated sea as determined from data obtained by the Stereo Wave Observation Project. *Meteorological Papers; New York University, College of Engineering* 2:6, 88 pp. ⟨492⟩

Pillsbury, J. E., 1891. The Gulf Stream—a description of the methods employed in the investigation, and the results of the research. In *Report of the Superintendent of the U.S. Coast and Geodetic Survey Showing the Progress of the Work during the Fiscal Year Ending with June, 1890*, Washington, Appendix No. 10, pp. 459–620. ⟨344⟩

Pillsbury, R. D., J. S. Bottero, and R. E. Still, 1977. A compilation of observations from moored current meters, Volume X. Currents, temperature and pressure in the Drake Passage during F DRAKE 75, February 1975–February 1976. *School of Oceanography, Oregon State University, Data Report* 67, Reference 77-8, Corvallis, Oregon, 117 pp. ⟨403⟩

Pingree, R. D., 1972. Mixing in the deep stratified ocean. *Deep-Sea Research* 19:549–561. ⟨84⟩

Pingree, R. D., 1973. A component of Labrador Sea water in the Bay of Biscay. *Limnology and Oceanography* 18:711–718. ⟨84⟩

Pingree, R. D., 1978. Mixing and stabilization of phytoplankton distributions on the northwest European continental shelf. In *Spatial Pattern in Plankton Communities*, J. H. Steele, ed., Plenum Press, New York, pp. 181–220. ⟨378, 379⟩

Pingree, R. D., P. M. Holligan, and R. N. Head, 1977. Survival of dinoflagellate blooms in the western English Channel. *Nature* 265:266–269. ⟨380, 381⟩

Pingree, R. D., and G. K. Morrison, 1973. The relationship between stability and source waters for a section in the northeast Atlantic. *Journal of Physical Oceanography* 3:280–285. ⟨84⟩

Pinkel, R., 1975. Upper ocean internal wave observations from FLIP. *Journal of Geophysical Research* 80:3892–3910. ⟨265, 285⟩

Pinkel, R., 1979. Observations of strongly non-linear internal motion in the open sea using a range-gated Doppler sonar. *Journal of Physical Oceanography* 9:675–686. ⟨428⟩

Plate, E. J., 1971. Aerodynamic characteristics of atmospheric boundary layers. *U.S. Atomic Energy Commission, Technical Information Division*, TID 25465, 190 pp. Available from National Technical Information Service, U.S. Dept. of Commerce, Springfield, Virginia. ⟨485⟩

Platt, T., 1972. Local phytoplankton abundance and turbulence. *Deep-Sea Research* 19:183–188. ⟨376⟩

Platt, T., and K. L. Denman, 1975. Spectral analysis in ecology. *Annual Review of Ecology and Systematics* 6:189–210. ⟨380⟩

Platzman, G. W., 1975. Normal modes of the Atlantic and Indian Oceans. *Journal of Physical Oceanography* 5:201–221. ⟨317, 320, 322, 328⟩

Platzman, G. W., 1978. Normal modes of the world ocean. Part I. Design of a finite-element barotropic model. *Journal of Physical Oceanography* 8:323–343. ⟨317⟩

Platzman, G. W., 1979. A Kelvin wave in the eastern North Pacific Ocean. *Journal of Geophysical Research* 84:2525–2528. ⟨326⟩.

Plumb, R. A., and A. D. McEwan, 1978. The instability of a forced standing wave in a viscous stratified fluid: A laboratory analogue of the quasi-biennial oscillation. *Journal of the Atmospheric Sciences* 35:1827–1839. ⟨532⟩

Pochapsky, T. E., 1976. Vertical structure of currents and deep temperatures in the western Sargasso Sea. *Journal of Physical Oceanography* 6:45–56. ⟨421⟩

Pochapsky, T. E., and F. D. Malone, 1972. A vertical profile of deep horizontal current near Cape Lookout, North Carolina. *Journal of Marine Research* 30:163–167. ⟨421⟩

Poincaré, H., 1910. *Leçons de Mécanique Céleste*, 3, *Théorie des Marées*. Gauthier-Villars, Paris, 469 pp. ⟨344⟩.

Pollak, M. J., 1958. Frequency distribution of potential temperatures and salinities in the Indian Ocean. *Deep-Sea Research* 5:128–133. ⟨42, 44, 46, 47, 49, 57⟩

Pollard, R. T., 1977. Observations and theories of Langmuir circulations and their role in near surface mixing. In *A Voyage of Discovery: George Deacon 70th Anniversary Volume*, M. Angel, ed., Supplement to *Deep-Sea Research*, Pergamon Press, Oxford, pp. 235–251. ⟨242, 495⟩

Pollard, R. T., P. B. Rhines, and R. O. R. Y. Thompson, 1973. The deepening of the wind-mixed layer. *Geophysical Fluid Dynamics* 3:381–404. ⟨241, 244, 385⟩

Pomphrey, N., J. D. Meiss, and K. M. Watson, 1980. Description of nonlinear internal wave interactions using Langevin methods. *Journal of Geophysical Research* 85:1085–1094. ⟨285⟩

Pond, S., G. T. Phelps, J. E. Paquin, G. McBean, and R. W. Stewart, 1971. Measurement of the turbulent fluxes of momentum, moisture and sensible heat over the ocean. *Journal of the Atmospheric Sciences* 28:901–917. ⟨489⟩

Posmentier, E. S., 1977. The generation of salinity finestructure by vertical diffusion. *Journal of Physical Oceanography* 7:298–300. ⟨247⟩

Prestwich, J., 1875. Tables of temperatures of the sea at different depths beneath the surface, reduced and collated from the various observations made between the years 1749 and 1868, discussed. *Philosophical Transactions of the Royal Society of London* 165:587–674. ⟨8, 9, 10, 71, 110, 111⟩

Price, J. F., 1979. On the scaling of stress-driven entrainment experiments. *Journal of Fluid Mechanics* 90:509–529. ⟨241⟩

Price, J. F., C. N. K. Mooers, and J. C. van Leer, 1978. Observation and simulation of storm-induced mixed-layer deepening. *Journal of Physical Oceanography* 8:582–599. ⟨244⟩

Priestley, C. H. B., 1951. A survey of the stress between the ocean and atmosphere. *Australian Journal of Scientific Research A* 4:315–328. ⟨488⟩

Priestley, C. H. B., 1967. Handover in scale of the fluxes of momentum, heat, etc. in the atmospheric boundary layer. In *Boundary Layers and Turbulence, The Physics of Fluids* 10 (Supplement): S38–S46. ⟨498⟩

Pritchard, D. W., 1952a. Estuarine hydrography. *Advances in Geophysics* 1:243–280. ⟨202⟩

Pritchard, D. W., 1952b. Salinity distribution and circulation in the Chesapeake Bay estuarine system. *Journal of Marine Research* 11:106–123. ⟨203, 204⟩

Pritchard, D. W., 1953. Distribution of oyster larvae in relation to hydrographic conditions. In *Proceedings of the Gulf and Caribbean Fisheries Institute, 5th Annual Session*, University of Miami Marine Laboratory, pp. 123–132. ⟨206⟩

Pritchard, D. W., 1954a. A study of flushing in the Delaware Model. *Chesapeake Bay Institute of the Johns Hopkins University Technical Report 7, Reference 54-4*, Baltimore, Maryland, 143 pp. ⟨203⟩

Pritchard, D. W., 1954b. A study of the salt balance in a coastal plain estuary. *Journal of Marine Research* 13:133–144. ⟨203, 204⟩

Pritchard, D. W., 1955. Estuarine circulation patterns. *Proceedings of the American Society of Civil Engineers* 81: Separate No. 717, 11 pp. ⟨202, 204⟩

Pritchard, D. W., 1956. The dynamic structure of a coastal plane estuary. *Journal of Marine Research* 15:33–42. ⟨203, 204⟩

Pritchard, D. W. 1958. The equations of mass continuity and salt continuity in estuaries. *Journal of Marine Research* 17:412–423. ⟨204⟩

Pritchard, D. W., 1967a. What is an estuary: Physical viewpoint. In *Estuaries*, G. H. Lauff, ed., American Association for the Advancement of Science, Publication No. 83, Washington, D.C., pp. 3–5. ⟨199, 202⟩

Pritchard, D. W., 1967b. Observations of circulation in coastal plain estuaries. In *Estuaries*, G. H. Lauff, ed., American Association for the Advancement of Science, Publication No. 83, Washington, D.C., pp. 37–44. ⟨202, 204⟩

Pritchard, D. W., and W. V. Burt, 1951. An inexpensive and rapid technique for obtaining current profiles in estuarine waters. *Journal of Marine Research* 10:180–189. ⟨203⟩

Pritchard, D. W., and J. H. Carpenter, 1960. Measurements of turbulent diffusion in estuarine and inshore waters. *Bulletin of the International Association of Scientific Hydrology* 20:37–50 ⟨206⟩.

Privett, D. W., 1960. The exchange of energy between the atmosphere and the oceans of the southern hemisphere. *Geophysical Memoirs* 13:104, 61 pp. ⟨490⟩

Proudman, J., 1944. The tides of the Atlantic Ocean. (George Darwin Lecture, 1944 October 13.) *Monthly Notices of the Royal Astronomical Society* 104:244–256. ⟨324⟩

Proudman, J., 1948. The applicability of Laplace's differential equations of the tides. *International Hydrographic Review* 25:112–118. ⟨295⟩

Proudman, J., 1953. *Dynamical Oceanography* Methuen, London, and John Wiley & Sons, New York, 409 pp. ⟨236⟩

Pruitt, W. O., D. L. Morgan, and F. J. Laurence, 1973. Momentum and mass transfers in the surface boundary layer. *Quarterly Journal of the Royal Meteorological Society* 99:370–386. ⟨485⟩

Prytherch, H. F., 1929. Investigation of the physical conditions controlling spawning of oysters and the occurrence, distribution, and setting of oyster larvae in Milford Harbor, Connecticut. *Bulletin of the United States Bureau of Fisheries* 44:429–503. ⟨201⟩

Purdy, G. M., P. D. Rabinowitz, and J. J. A. Velterop, 1979. The Kane Fracture Zone in the central Atlantic Ocean. *Earth and Planetary Science Letters* 45:429–434. ⟨29⟩

Quay, P. D., W. S. Broecker, R. H. Hesslein, and D. W. Schindler, 1980. Vertical diffusion rates determined by tritium tracer experiments in the thermocline and hypolimnion of two lakes. *Limnology and Oceanography* 25:201–218. ⟨459, 460⟩

Rao, D., 1966. Free gravitational oscillations in rotating rectangular basins. *Journal of Fluid Mechanics* 25:523–555. ⟨298⟩

Ratcliffe, R. A. S., and R. Murray, 1970. New lag associations between North Atlantic sea temperature and European pressure applied to long-range weather forecasting. *Quarterly Journal of the Royal Meteorological Society* 96:226–246. ⟨496⟩

Rattray, M., Jr., 1960. On the coastal generation of internal tides. *Tellus* 12:54–62. ⟨331⟩

Rattray, M., Jr., 1964. Time-dependent motion in an ocean; A unified two-layer, beta-plane approximation. In *Studies on Oceanography*, K. Yoshida, ed., University of Tokyo Press, Tokyo, pp. 19–29. ⟨344⟩

Rattray, M., Jr., 1977. Fjord and salt-wedge circulation. In *Estuaries, Geophysics, and the Environment*, C. B. Officer, panel chairman, National Academy of Sciences, Washington, D.C., pp. 36–45. ⟨205, 206⟩

Rattray, M., Jr., and R. C. Charnell, 1966. Quasigeostrophic free oscillations in enclosed basins. *Journal of Marine Research* 24:82–102. ⟨344⟩

Rattray, M., Jr., J. G. Dworski, and P. E. Kovala, 1969. Generation of long internal waves at the continental slope. *Frederick C. Fuglister Sixtieth Anniversary Volume, Deep-Sea Research* 16 (Supplement): 179–195. ⟨334, 335⟩

Rattray, M., Jr., and D. V. Hansen, 1962. A similarity solution for circulation in an estuary. *Journal of Marine Research* 20:121–133. ⟨203⟩

Rattray, M., Jr., and P. Welander, 1975. A quasi-linear model of the combined wind-driven and thermohaline circulations in a rectangular β-plane ocean. *Journal of Physical Oceanography* 5:585–602. ⟨39⟩

Rauschelbach, H., 1932. Zur Geschichte des Hochseepegels. *Annalen der Hydrographie und Maritimen Meteorologie* 60:73–76. ⟨399, 424⟩

Rayleigh, Lord, 1880. On the stability, or instability, of certain fluid motions. *Proceedings of the London Mathematical Society* 11:57–70; reprinted in *Scientific Papers*, 1, Cambridge University Press, London, pp. 474–487. ⟨530⟩

Rayleigh, Lord, 1883. Investigation of the character of the equilibrium of an incompressible heavy fluid of variable density. *Proceedings of the London Mathematical Society* 14:170–178. ⟨264, 299⟩

Rayleigh, Lord, 1903. Note on the theory of the fortnightly tide. *The London, Edinburgh, and Dublin Philosophical Magazine and Journal of Science, Sixth Series* 5:136–141. ⟨344⟩

Redekopp, L. G., 1977. On the theory of solitary Rossby waves. *Journal of Fluid Mechanics* 82:725–804. ⟨345⟩

Redfield, A. C., 1958. The influence of the continental shelf on the tides of the Atlantic coast of the United States. *Journal of Marine Research* 17:432–448. ⟨327, 328⟩

Redfield, A. C., and L. A. Walford, 1951. A study of the disposal of chemical waste at sea. Report of the Committee for Investigation of Waste Disposal. *National Research Council Publication* 201, 29 pp. ⟨212⟩

Reed, R. K., 1969. Deep water properties and flow in the central North Pacific. *Journal of Marine Research* 27:24–31. ⟨36⟩

Reed, R. K., 1970a. On the anomalous deep water south of the Aleutian Islands. *Journal of Marine Research* 28:371–372. ⟨36⟩

Reed, R. K., 1970b. Geopotential topography of deep levels in the Pacific Ocean. *Journal of the Oceanographical Society of Japan* 26:331–339. ⟨85⟩

Regal, R., and C. Wunsch, 1973. M_2 tidal currents in the western North Atlantic. *Deep-Sea Research* 20:493–502. ⟨334, 336⟩

Reid, J. L., Jr., 1961a. On the geostrophic flow at the surface of the Pacific Ocean with respect to the 1000-decibar surface. *Tellus* 13:489–502. ⟨83, 91⟩

Reid, J. L., Jr., 1961b. On the temperature, salinity, and density differences between the Atlantic and Pacific oceans in the upper kilometre. *Deep-Sea Research* 7:265–275. ⟨108⟩

Reid, J. L., 1962. Observations of internal tides in October 1950. *Transactions of the American Geophysical Union* 37:278–289. ⟨332⟩

Reid, J. L., Jun., 1964a. Evidence of a South Equatorial Countercurrent in the Atlantic Ocean in July 1963. *Nature* 203:182. ⟨83⟩

Reid, J. L., 1964b. A transequatorial Atlantic oceanographic section in July 1963 compared with other Atlantic and Pacific sections. *Journal of Geophysical Research* 69:5205–5215. ⟨185⟩

Reid, J. L., Jr., 1965. Intermediate Waters of the Pacific Ocean. *The Johns Hopkins Oceanographic Studies* 2:85 pp. ⟨43, 44, 82, 83, 85, 161, 162⟩

Reid, J. L., 1973a. Transpacific hydrographic sections at Lats. 43°S and 28°S: the SCORPIO Expedition-III. Upper water and a note on southward flow at mid-depth. *Deep-Sea Research* 20:39–49. ⟨34, 110⟩

Reid, J. L., 1973b. Northwest Pacific Ocean Water in Winter. *The Johns Hopkins Oceanographic Studies* 5:96 pp. ⟨59⟩

Reid, J. L., 1978. On the middepth circulation and salinity field in the North Atlantic Ocean. *Journal of Geophysical Research* 83:5063–5067. ⟨85, 89, 91, 107, 111, 343, 373⟩

Reid, J. L., 1979. On the contribution of the Mediterranean Sea outflow to the Norwegian-Greenland Sea. *Deep-Sea Research* 26:1199–1223. ⟨111⟩

Reid, J. L., and R. S. Arthur, 1975. Interpretation of maps of geopotential anomaly for the deep Pacific Ocean. *Journal of Marine Research* 33 (Supplement): 37–52. ⟨85⟩

Reid, J. L., and P. F. Lonsdale, 1974. On the flow of water through the Samoan Passage. *Journal of Physical Oceanography* 4:58–73. ⟨35, 108⟩

Reid, J. L., and R. J. Lynn, 1971. On the influence of the Norwegian-Greenland and Weddell seas upon the bottom waters of the Indian and Pacific oceans. *Deep-Sea Research* 18:1063–1088. ⟨25, 82, 83, 84, 93, 106, 111⟩

Reid, J. L. and A. W. Mantyla, 1978. On the mid-depth circulation of the North Pacific Ocean. *Journal of Physical Oceanography* 8:946–951. ⟨89, 91⟩

Reid, J. L., W. D. Nowlin, and W. C. Patzert, 1977. On the characteristics and circulation of the southwestern Atlantic Ocean. *Journal of Physical Oceanography* 7:62–91. ⟨7, 17, 28, 31, 84, 85, 90, 91, 108⟩

Reid, R. O., 1948. The equatorial currents of the eastern Pacific as maintained by the stress of the wind. *Journal of Marine Research* 7:74–99. ⟨79, 188⟩

Reid, R. O., 1958. Effect of Coriolis force on edge waves (I) Investigation of the normal modes. *Journal of Marine Research* 16:109–144. ⟨309⟩

Rennell, J., 1832. *Investigation of the Currents of the Atlantic Ocean, and of Those Which Prevail between the Indian Ocean and the Atlantic.* J. G. & F. Rivington, London, 359 pp. ⟨343, 507⟩

Rhines, P. B., 1969a. Slow oscillations in an ocean of varying depth. Part 1. Abrupt topography. *Journal of Fluid Mechanics* 37:161–189. ⟨120, 358⟩

Rhines, P. B., 1969b. Slow oscillations in an ocean of varying depth. Part 2. Islands and seamounts. *Journal of Fluid Mechanics* 37:191–205. ⟨310⟩

Rhines, P., 1970. Edge-, bottom-, and Rossby waves in a rotating stratified fluid. *Geophysical Fluid Dynamics* 1:273–302. ⟨141, 165, 311, 313, 315, 358, 520, 521⟩

Rhines, P. B., 1971a. A note on long-period motions at Site D. *Deep-Sea Research* 18:21–26. ⟨169⟩

Rhines, P. B., 1971b. A comment on the Aries observations. *Philosophical Transactions of the Royal Society of London A* 270:461–463. ⟨359⟩

Rhines, P. B., 1975. Waves and turbulence on a beta-plane. *Journal of Fluid Mechanics* 69:417–443. ⟨175, 310, 508, 543, 544⟩

Rhines, P. B., 1977. The dynamics of unsteady currents. In *The Sea: Ideas and Observations on Progress in the Study of the Seas, 6: Marine Modeling,* E. D. Goldberg, I. N. McCave, J. J. O'Brien, and J. H. Steele, eds., Wiley, Interscience, New York, pp. 189–318. ⟨7, 127, 130, 141, 142, 165, 166, 168, 174, 177, 182, 183, 304, 310, 313, 314, 315, 346, 371, 376, 474, 507, 533, 544⟩

Rhines, P. B., 1979. Geostrophic turbulence. *Annual Reviews of Fluid Mechanics* 11:401–441. ⟨344, 346⟩

Rhines, P. B., and F. P. Bretherton, 1974. Topographic Rossby waves in a rough-bottom ocean. *Journal of Fluid Mechanics* 61:583–607. ⟨177, 370⟩

Rhines, P. B., and W. R. Holland, 1979. A theoretical discussion of eddy-driven mean flows. *Dynamics of Atmospheres and Oceans* 3:289–325. ⟨182⟩

Richards, J. M., 1970. The effect of windshear on a puff. *Quarterly Journal of the Royal Meteorological Society* 96:702–714. ⟨498⟩

Richardson, L. F., 1920. The supply of energy from and to atmospheric eddies. *Proceedings of the Royal Society of London A* 97:354–373. ⟨484, 485⟩

Richardson, L. F., 1926. Atmospheric diffusion shown on a distance-neighbour graph. *Proceedings of the Royal Society of London A* 110:709–737. ⟨239⟩

Richardson, L. F., and H. Stommel, 1948. Note on eddy diffusion in the sea. *Journal of Meteorology* 5:238–240. ⟨xxv, 239⟩

Richardson, P. L., 1977. On the crossover between the Gulf Stream and the Western Boundary Undercurrent. *Deep-Sea Research* 24:139–159. ⟨27, 79, 108, 121⟩

Richardson, P. L., R. E. Cheney, and L. A. Mantini, 1977. Tracking a Gulf Stream ring with a free drifting surface buoy. *Journal of Physical Oceanography* 7:581–590. ⟨534⟩

Richardson, P. L., R. E. Cheney, and L. V. Worthington, 1978. A census of Gulf Stream rings, spring 1975. *Journal of Geophysical Research* 83:6136–6144. ⟨373⟩

Richardson, P. L., and J. A. Knauss, 1971. Gulf Stream and Western Boundary Undercurrent observations at Cape Hatteras. *Deep-Sea Research* 18:1089–1109. ⟨27, 121, 123⟩

Richardson, P. L., A. E. Strong, and J. A. Knauss, 1973. Gulf Stream eddies: recent observations in the western Sargasso Sea. *Journal of Physical Oceanography* 3:297–301. ⟨121⟩

Richardson, W. S., and W. J. Schmitz, Jr., 1965. A technique for the direct measurement of transport with applications to the Straits of Florida. *Journal of Marine Research* 23:172–185. ⟨421⟩

Richardson, W. S., W. J. Schmitz, Jr., and P. P. Niiler, 1969. The velocity structure of the Florida Current from the Straits of Florida to Cape Fear. *Frederick C. Fuglister Sixtieth Anniversary Volume, Deep-Sea Research* 16 (Supplement): 225–231. ⟨118, 121⟩

Richardson, W. S., P. B. Stimson, and C. H. Wilkins, 1963. Current measurement from moored buoys. *Deep-Sea Research* 10:369–388. ⟨403, 404, 407⟩

Richardson, W. S., H. J. White, Jr., and L. Nemeth, 1972. A technique for the direct measurement of ocean currents from aircraft. *Journal of Marine Research* 30:259–268. ⟨421⟩

Richman, J. G., 1976. Kinematics and energetics of the mesoscale mid-ocean circulation: MODE. Ph.D. Thesis, Massachusetts Institute of Technology/Woods Hole Oceanographic Institution, 205 pp. ⟨431, 521⟩

Richman, J. G., C. Wunsch, and N. G. Hogg, 1977. Space and time scales and mesoscale motion in the sea. *Reviews of Geophysics and Space Physics* 15:385–420. ⟨359, 363, 370, 371, 372, 407⟩

Riehl, H., 1954. *Tropical Meteorology.* McGraw-Hill, New York, 392 pp. ⟨186⟩

Riehl, H., and J. Malkus, 1961. Some aspects of Hurricane Daisy, 1958. *Tellus* 13:181–213. ⟨490⟩

Riehl, H., T. C. Yeh, J. S. Malkus, and N. E. La Seur, 1951. The north-east trade of the Pacific Ocean. *Quarterly Journal of the Royal Meteorological Society* 77:598–626. ⟨500⟩

Riley, G. A., 1951. Oxygen, phosphate and nitrate in the Atlantic Ocean. *Bulletin of the Bingham Oceanographic Collection* 13:1, 126 pp. ⟨81, 83, 434⟩

Riley, G. A., H. Stommel, D. F. Bumpus, 1949. Quantitative ecology of the plankton of the western North Atlantic. *Bulletin of the Bingham Oceanographic Collection* 12:3, 169 pp. ⟨211⟩

Riser, S. C., H. Freeland, and H. T. Rossby, 1978. Mesoscale motions near the deep western boundary of the North Atlantic. *Deep-Sea Research* 25:1179–1191. ⟨27, 108, 121⟩

Rivas, E. K. de, 1973. Numerical models of the circulation of the atmosphere of Venus. *Journal of the Atmospheric Sciences* 30:763–779. ⟨506⟩

Rivas, E. K. de, 1975. Further numerical calculations of the circulation of the atmosphere of Venus. *Journal of the Atmospheric Sciences* 32:1017–1024. ⟨506⟩

Robinson, A. R., ed., 1963. *Wind-Driven Ocean Circulation.* Blaisdell, New York, 161 pp. ⟨79⟩

Robinson, A. R., 1964. Continental shelf waves and the response of sea level to weather systems. *Journal of Geophysical Research* 69:367–368. ⟨120, 358⟩

Robinson, A. R., 1971. The Gulf Stream. *Philosophical Transactions of the Royal Society of London A* 270:351–370. ⟨126⟩

Robinson, A. R., D. E. Harrison, and D. B. Haidvogel, 1979. Mesoscale eddies and general ocean circulation models. *Dynamics of Atmospheres and Oceans* 3:143–180. ⟨346⟩

Robinson, A. R., D. E. Harrison, Y. Mintz, and A. J. Semtner, 1977. Eddies and the general circulation of an idealized oceanic gyre: A wind and thermally driven primitive equation numerical experiment. *Journal of Physical Oceanography* 7:182–207. ⟨89, 90, 133, 178, 346⟩

Robinson, A. R., J. R. Luyten, and G. Flierl, 1975. On the theory of thin rotating jets: A quasi-geostrophic time dependent model. *Geophysical Fluid Dynamics* 6:211–244. ⟨126⟩

Robinson, A. R., J. R. Luyten, and F. C. Fuglister, 1974. Transient Gulf Stream meandering. Part I: An observational experiment. *Journal of Physical Oceanography* 4:237–255. ⟨126, 127, 534⟩

Robinson, A. R., and J. C. McWilliams, 1974. The baroclinic instability of the open ocean. *Journal of Physical Oceanography* 4:281–294. ⟨347, 529⟩

Robinson, A. R., and P. P. Niiler, 1967. The theory of free inertial currents. I. Path and structure. *Tellus* 19:269–291. ⟨508⟩

Robinson, A. R., and H. Stommel, 1959. The oceanic thermocline and the associated thermohaline circulation. *Tellus,* 3:295–308. ⟨xxi, 141, 159, 160, 161, 449, 509⟩

Robinson, A. R., and P. Welander, 1963. Thermal circulation on a rotating sphere; with application to the oceanic thermocline. *Journal of Marine Research* 21:25–38. ⟨161, 506⟩

Robinson, G. D., 1966. Another look at some problems of the air-sea interface. *Quarterly Journal of the Royal Meteorological Society* 92:451–465. ⟨488⟩

Rochford, D. J., 1964. Salinity maxima in the upper 1000 metres of the North Indian Ocean. *Australian Journal of Marine and Freshwater Research* 15:1–24. ⟨26⟩

Rodewald, M., 1972. Long-term variations of the sea temperature in the areas of the nine North Atlantic weather stations during the period 1951–1968. *Rapports et Procès-Verbaux des Réunions, Conseil Permanent International pour l'Exploration de la Mer* 162:139–153. ⟨355⟩

Roll, H. U., 1965. *Physics of the Marine Atmosphere.* Academic Press, New York, 426 pp. ⟨503⟩

Rooth, C., 1972. A linearized bottom friction law for large-scale oceanic motions. *Journal of Physical Oceanography* 2:509–510. ⟨152⟩

Rosenbluth, M. N., and A. Simon, 1964. Necessary and sufficient conditions for stability of plane parallel and inviscid flows. *The Physics of Fluids* 7:557–558. ⟨531⟩

Rosenthal, S. L., 1965. Some preliminary theoretical considerations of troposphere wave motions in equatorial latitudes. *Monthly Weather Review* 93:605–612. ⟨527⟩

Rossby, C.-G., 1936. Dynamics of steady ocean currents in the light of experimental fluid mechanics. *Papers in Physical Oceanography and Meteorology* 5:1, 43 pp. ⟨xxiv⟩

Rossby, C.-G., 1938. On the mutual adjustment of pressure and velocity distributions in certain simple current systems, II. *Journal of Marine Research* 1:239–263. ⟨183, 509, 510, 519⟩

Rossby, C.-G., 1951. A comparison of current patterns in the atmosphere and in ocean basins. In *Scientific Proceedings of the Association of Meteorology, Ninth Assembly of the I.U.G.G.,* Brussels, pp. 9–31. ⟨504⟩

Rossby, C.-G., and Collaborators, 1939. Relation between variations in the intensity of the zonal circulation of the atmosphere and the displacements of the semi-permanent centers of action. *Journal of Marine Research* 2:38–55. ⟨xv, 300, 302, 340, 344, 506, 508, 512⟩

Rossby, H. T., 1965. On thermal convection driven by non-uniform heating from below: an experimental study. *Deep-Sea Research* 12:9–16. ⟨40, 505⟩

Rossby, H. T., 1969. A vertical profile of currents near Plantagenet Bank. *Deep-Sea Research* 16:377–385. ⟨420⟩

Rossby, H. T., 1974. Studies of the vertical structure of horizontal currents near Bermuda. *Journal of Geophysical Research* 79:1781–1791. ⟨421⟩

Rossby, H. T., 1979. Oceanography. In *Impact of Technology on Geophysics,* H. E. Newell, panel chairman, National Academy of Sciences, Washington, D.C. ⟨403, 412, 413⟩

Rossby, H. T., and T. B. Sanford, 1976. A study of velocity profiles through the main thermocline. *Journal of Physical Oceanography* 6:766–774. ⟨423⟩

Rossby, H. T., and D. Webb, 1971. The four month drift of a Swallow float. *Deep-Sea Research* 18:1035–1039. ⟨412⟩

Rossiter, J. R., 1958. On the application of relaxation methods to oceanic tides. *Proceedings of the Royal Society of London A* 248:482–498. ⟨328⟩

Rotschi, H., 1970. Variation of equatorial currents. In *Scientific Exploration of the South Pacific,* W. S. Wooster, ed.,

National Academy of Sciences, Washington, D.C., pp. 75–83. ⟨184⟩

Rowntree, P. R., 1972. The influence of tropical east Pacific temperatures on the atmosphere. *Quarterly Journal of the Royal Meteorological Society* 98:290–321. ⟨355⟩

Ruddick, B. R., and J. S. Turner, 1979. The vertical length scale of double-diffusive intrusions. *Deep-Sea Research* 26:903–913. ⟨254, 255⟩

Ruggles, K. W., 1970. The vertical mean wind profile over the ocean for light to moderate winds. *Journal of Applied Meteorology* 9:389–395. ⟨489⟩

Rumford, B., Count of, 1800. Essay VII, The Propagation of Heat in Fluids. In *Essays, Political, Economical, and Philosophical, A New Edition*, 2, T. Cadell, Jr., and W. Davies, London, pp. 197–386. Also in *Collected Works of Count Rumford*, S. C. Brown, ed., 1: *The Nature of Heat*, 1968, Harvard University Press, Cambridge, pp. 117–285. ⟨8⟩

Ryther, J. H., 1969. Relationship of photosynthesis to fish production in the sea. *Science* 166:72–76. ⟨380⟩

Sager, G., 1955. *Gezeitenvoraussagen und Gezeitenrechenmaschinen*. Seehydrographischer Dienst der Deutschen Demokratischen Republik, Warnemünde, 126 pp. ⟨318⟩

Salmon, R., 1978. Two-layer quasigeostrophic turbulence in a simple special case. *Geophysical and Astrophysical Fluid Dynamics* 10:25–52. ⟨346, 371, 543⟩

Salmon, R., G. Holloway, and M. C. Hendershott, 1976. The equilibrium statistical mechanics of simple geostrophic models. *Journal of Fluid Mechanics* 75:691–703. ⟨178, 346, 543⟩

Saltzman, B., 1962. Finite amplitude convection as an initial value problem. *Journal of the Atmospheric Sciences* 19:329–341. ⟨387⟩

Sambuco, E., and J. A. Whitehead, 1976. Hydraulic control by a wide weir in a rotating fluid. *Journal of Fluid Mechanics* 73:521–528. ⟨473⟩

Sandstrom, H., 1969. Effect of topography on propagation of waves in stratified fluids. *Deep-Sea Research* 16:405–410. ⟨332⟩

Sandström, J. W., 1903. Ueber die Anwendungen von Pegelbeobachtungen zur Berechnung der Geschwindigkeit der Meeresströme. *Svenska Hydrografisk-Biologiska Kommissionens Skrifter* 1: 3 pp. ⟨114, 351⟩

Sandström, J. W., 1919. The hydrodynamics of Canadian Atlantic waters. In *Canadian Fisheries Expedition, 1914–15*, Department of Naval Service, Ottawa, pp. 221–343, 59 figs., 15 plates. ⟨233⟩

Sandström, J. W., and B. Helland-Hansen, 1903. Ueber die Berechnung von Meeresströmungen. *Report on Norwegian Fishery- and Marine-Investigations* 2 (1902): 4, 43 pp. ⟨72, 233⟩

Sanford, T. B., 1975. Observations of the vertical structure of internal waves. *Journal of Geophysical Research* 80:3861–3871. ⟨265, 267⟩

Sanford, T. B., R. G. Drever, and J. H. Dunlap, 1978. A velocity profiler based on the principles of geomagnetic induction. *Deep-Sea Research* 25:183–210. ⟨265, 267, 422⟩

Sanford, T. B., and W. J. Schmitz, Jr., 1971. A comparison of direct measurements and GEK observations in the Florida Current off Miami. *Journal of Marine Research* 29:347–359. ⟨116⟩

Sarkisyan, A. S., 1977. The diagnostic calculations of a large-scale oceanic circulation. In *The Sea: Ideas and Observations on Progress in the Study of the Seas, 6: Marine Modeling*, E. D. Goldberg, I. N. McCave, J. J. O'Brien, and J. H. Steele, eds., Wiley, Interscience, New York, pp. 363–459. ⟨142⟩

Sarkisyan, A. S., and V. P. Keonjiyan, 1975. Review of numerical ocean circulation models using the observed density field. In *Numerical Models of Ocean Circulation*, National Academy of Sciences, Washington, D.C., pp. 76–93. ⟨90⟩

Sarmiento, J. L., 1978. A study of mixing in the deep sea based on STD, radon-222, and radium-228 measurements. Ph.D. Thesis, Columbia University. ⟨459⟩

Sarmiento, J. L., and K. Bryan, in preparation. 3-dimensional model of bomb tritium distribution in the Atlantic Ocean. ⟨449⟩

Sarmiento, J. L., H. W. Feely, W. S. Moore, A. E. Bainbridge, and W. S. Broecker, 1976. The relationship between vertical eddy diffusion and buoyancy gradient in the deep sea. *Earth and Planetary Science Letters* 32:357–370. ⟨459, 460⟩

Sarmiento, J. L., and C. G. H. Rooth, 1980. A comparison of vertical and isopycnal mixing models in the deep sea based on radon 222 measurements. *Journal of Geophysical Research* 85:1515–1518. ⟨459⟩

Saunders, K. D., and R. C. Beardsley, 1975. An experimental study of the spin-up of a thermally stratified rotating fluid. *Geophysical Fluid Dynamics* 7:1–28. ⟨476⟩

Saunders, P. M., 1971. Anticyclonic eddies formed from shoreward meanders of the Gulf Stream. *Deep-Sea Research* 18:1207–1219. ⟨125⟩

Saunders, P. M., 1973. The instability of a baroclinic vortex. *Journal of Physical Oceanography* 3:61–65. ⟨253, 476⟩

Saunders, P. M., 1976. Near-surface current measurements *Deep-Sea Research* 23:249–258. ⟨405⟩

Saunders, P. M., 1977. Wind stress on the ocean over the eastern continental shelf of North America. *Journal of Physical Oceanography* 7:555–566. ⟨217, 226, 227, 229, 490⟩

Savonius, S. J., 1931. The S-rotor and its applications. *Mechanical Engineering* 53:333–338. ⟨403⟩

Scarlet, R. I., and C. N. Flagg, 1979. Interaction of a warm core eddy with the New England continental shelf (Abstract only.) *EOS, Transactions of the American Geophysical Union* 60:279. ⟨218⟩

Schiemer, E. W., and D. W. Pritchard, 1957. The Chesapeake Bay Institute Conductivity-Temperature-Indicator (CBI-CTI). *Chesapeake Bay Institute of the Johns Hopkins University Technical Report 12, Reference 57-1*, Baltimore, Maryland, 15 pp. ⟨203⟩

Schink, D. R., 1962. The measurement of dissolved Si^{32} in seawater. Ph.D. Thesis. University of California at San Diego, 189 pp. ⟨447⟩

Schink, D. R., 1967. Budget for dissolved silica in the Mediterranean Sea. *Geochimica et Cosmochimica Acta* 31:987–999. ⟨25⟩

Schlee, S., 1973. *The Edge of an Unfamiliar World: A History of Oceanography*. E. P. Dutton, New York, 398 pp. ⟨209, 233⟩

Schlee, S., 1978. *On Almost Any Wind*. Cornell University Press, Ithaca, 301 pp. ⟨211, 233⟩

Schleicher, K. E., and A. L. Bradshaw, 1956. A conductivity bridge for measurement of the salinity of sea water. *Journal du Conseil* 22:9–20. ⟨42, 44⟩

Schmitt, R. W., 1979. Flux measurements on salt fingers at an interface. *Journal of Marine Research* 37:419–436. ⟨255⟩

Schmitt, R. W., and D. L. Evans, 1978. An estimate of the vertical mixing due to salt fingers based on observations in the North Atlantic central water. *Journal of Geophysical Research* 83:2913–2919. ⟨258⟩

Schmitz, W. J., Jr., 1969. On the dynamics of the Florida Current. *Journal of Marine Research* 27:121–150. ⟨116⟩

Schmitz, W. J., Jr., 1976. Eddy kinetic energy in the deep western North Atlantic. *Journal of Geophysical Research* 81:4981–4982. ⟨139, 359, 363, 370, 508⟩

Schmitz, W. J., Jr., 1977. On the deep general circulation in the Western North Atlantic. *Journal of Marine Research* 35:21–28. ⟨7, 127, 133, 134, 139, 397, 534⟩

Schmitz, W. J., Jr., 1978. Observations of the vertical distribution of low frequency kinetic energy in the Western North Atlantic. *Journal of Marine Research* 36:295–310. ⟨123, 127, 131, 139, 359, 363, 370, 373, 374, 546⟩

Schmitz, W. J., Jr., 1980. Weakly depth-dependent segments of the North Atlantic circulation. *Journal of Marine Research* 38:111–135. ⟨397⟩

Schmitz, W. J., Jr., and P. P. Niiler, 1969. A note on the kinetic energy exchange between fluctuations and mean flow in the surface layer of the Florida Current. *Tellus* 21:814–819. ⟨118, 507⟩

Schmitz, W. J., Jr., and W. B. Owens, 1979. Observed and numerically simulated kinetic energies for some MODE eddies. *Journal of Physical Oceanography* 9:1294–1297. ⟨346⟩

Schmitz, W. J., Jr., and W. S. Richardson, 1968. On the transport of the Florida Current. *Deep-Sea Research* 15:679–693. ⟨116⟩

Schmitz, W. J., Jr., A. R. Robinson, and F. C. Fuglister, 1970. Bottom velocity observations directly under the Gulf Stream. *Science* 170:1192–1194. ⟨127⟩

Schneider, E. K., 1977. Axially symmetric steady-state models of the basic state for instability and climate studies. Part II. Nonlinear calculations. *Journal of the Atmospheric Sciences* 34:280–296. ⟨506⟩

Schott, F., and W. Düing, 1976. Continental shelf waves in the Florida Straits. *Journal of Physical Oceanography* 6:451–460. ⟨117, 120, 121⟩

Schott, F., and H. Stommel, 1978. Beta spirals and absolute velocities in different oceans. *Deep-Sea Research* 25:961–1010. ⟨xxiii, 91, 343⟩

Schott, G., 1897. Die Gewasser der Bank von Neufundland und ihrer weitern Umgebung. *Petermanns Mitteilungen* 43:201–212. ⟨208⟩

Schott, G., 1902. Oceanographie und maritime Meteorologie. In *Wissenschaftliche Ergebnisse der Deutschen Tiefsee-Expedition auf dem Dampfer "Valdivia" 1898–1899*, 1, 403 pp. ⟨8, 10, 32, 40⟩

Schott, G., 1926. Die Tiefwasserbewegungen des Indischen Ozeans. Zugleich zur Besprechung von E. von Drygalski "Ozean und Antarktis". *Annalen der Hydrographie und Maritimen Meteorologie* 54:417–431. ⟨10⟩

Schroeder, E., and H. Stommel, 1969. How representative is the series of *Panulirus* stations of monthly mean conditions off Bermuda? *Progress in Oceanography* 5:31–40. ⟨351⟩

Schroeder, J., 1975. Krypton-85 in the ocean. *Zeitschrift für Naturforschung A* 30:962–967. ⟨446⟩

Schroeder, K. J. P., and W. Roether, 1975. The release of krypton-85 and tritium to the environment and krypton-85 to tritium ratios as source indicator. In *Isotope Ratios as Pollutant Source and Behavior Indicators*, Publication IAEA-SM-191/30, International Atomic Energy Agency, Vienna, pp. 231–253. ⟨446, 447⟩

Sclater, J. G., and R. L. Fisher, 1974. Evolution of the east central Indian Ocean, with emphasis on the tectonic setting of the Ninetyeast Ridge. *Geological Society of America Bulletin* 85:683–702. ⟨33⟩

SCOR, 1977. Report of the Panel on Monitoring Ocean Climate Fluctuations. World Meteorological Organization, Geneva, 96 pp. ⟨428⟩

Scott, J. T., and G. T. Csanady, 1976. Nearshore currents off Long Island. *Journal of Geophysical Research* 81:5401–5409. ⟨219, 228⟩

Scotti, R. S., and G. M. Corcos, 1972. An experiment on the stability of small disturbances in a stratified shear layer. *Journal of Fluid Mechanics* 52:499–528. ⟨248⟩

Sea Data Corporation, 1978. Advertising brochure. Newton, Massachusetts. ⟨401⟩

Seabrooke, J. D., G. L. Hufford, and R. B. Elder, 1971. Formation of antarctic bottom water in the Weddell Sea. *Journal of Geophysical Research* 76:2164–2178. ⟨17⟩

Seitz, R. C., 1973. Observations of intermediate and small scale turbulent water motion in a stratified estuary (Parts I and II). *Chesapeake Bay Institute of the Johns Hopkins University Technical Report 79, Reference 73-2*, Baltimore, Maryland, 89 and 158 pp. ⟨205⟩

Semtner, A. J., and W. R. Holland, 1978. Intercomparison of quasi-geostrophic simulations of the western North Atlantic circulation with primitive equation results. *Journal of Physical Oceanography* 8:735–754. ⟨133, 182⟩

Semtner, A. J., and W. R. Holland, 1980. Numerical simulation of equatorial ocean circulation. Part I: A basic case in turbulent equilibrium. *Journal of Physical Oceanography*. ⟨194⟩

Semtner, A. J., and Y. Mintz, 1977. Numerical simulation of the Gulf Stream and mid-ocean eddies. *Journal of Physical Oceanography* 7:208–230. ⟨89, 133, 228, 229⟩

Sharma, G. S., 1972. Water characteristics at 200 cl/t in the intertropical Indian Ocean during the southwest monsoon. *Journal of Marine Research* 30:102–111. ⟨83⟩

Shaw, D. M., and W. L. Donn, 1964. Sea level variations at Iceland and Bermuda. *Journal of Marine Research* 22:111–122. ⟨351⟩

Shaw, N., 1930. *Manual of Meteorology, 3: The Physical Processes of Weather*. Cambridge University Press, London, 445 pp. ⟨44⟩

Sheldon, R. W., A. Prakash, and W. H. Sutcliffe, Jr., 1972. The size distribution of particles in the ocean. *Limnology and Oceanography* 17:327–340. ⟨380⟩

Sheppard, P. A., 1947. The aerodynamic drag of the earth's surface and the value of von Karman's constant in the lower atmosphere. *Proceedings of the Royal Society of London A* 188:208–222. ⟨485⟩

Sheppard, P. A., 1970. The atmospheric boundary layer in relation to large-scale dynamics. In *The Global Circulation of the Atmosphere*, G. A. Corby, ed., Royal Meteorological Society, London, pp. 91–112. ⟨503⟩

Sheppard, P. A., H. Charnock, and J. R. D. Francis, 1952. Observations of the westerlies over the sea. *Quarterly Journal of the Royal Meteorological Society* 78:563–582. ⟨503⟩

Sheppard, P. A., D. T. Tribble, and J. R. Garratt, 1972. Studies of turbulence in the surface layer over water (Lough Neagh). Part I. Instrumentation, programme, profiles. *Quarterly Journal of the Royal Meteorological Society* 98:627–641. ⟨489⟩

Sherman, F. S., J. Imberger, and G. M. Corcos, 1978. Turbulence and mixing in stably stratified waters. *Annual Review of Fluid Mechanics* 10:267–288. ⟨238, 244, 248, 249, 251, 252⟩

Shirtcliffe, T. G. L., and J. S. Turner, 1970. Observations of the cell structure of salt fingers. *Journal of Fluid Mechanics* 41:707–719. ⟨256⟩

Siedler, G., 1968. Schichtungs-und Bewegungsverhältnisse am Südausgang des Roten Meeres. *"Meteor" Forschungsergebnisse, Reihe A*, No. 4, 1–76. ⟨25⟩

Silker, W. B., 1972a. Beryllium-7 and fission products in the GEOSECS II water column and applications of their oceanic distributions. *Earth and Planetary Science Letters* 16:131–137. ⟨447, 448⟩

Silker, W. B., 1972b. Horizontal and vertical distributions of radionuclides in the North Pacific Ocean. *Journal of Geophysical Research* 77:1061–1070. ⟨447⟩

Silker, W. B., D. E. Robertson, H. G. Rieck, Jr., R. W. Perkins, and J. M. Prospero, 1968. Beryllium-7 in ocean water. *Science* 161:879–880. ⟨447⟩

Silvert, W., and T. Platt, 1978. Energy flux in the pelagic ecosystem: a time-dependent equation. *Limnology and Oceanography* 23:813–816. ⟨379⟩

Simmons, A. J., 1974. The meridional scale of baroclinic waves. *Journal of the Atmospheric Sciences* 31:1515–1525. ⟨529⟩

Simmons, H. B., 1950. Applicability of hydraulic model studies to tidal problems. In Evaluation of Present State of Knowledge of Factors Affecting Tidal Hydraulics and Related Phenomena, *U.S. Army Corps of Engineers, Committee on Tidal Hydraulics Report No. 1*, pp. 127–146. ⟨203⟩

Simpson, J. H., 1972. A free fall probe for the measurement of velocity microstructure. *Deep-Sea Research* 19:331–336. ⟨423⟩

Simpson, J. H., 1975. Observations of small scale vertical shear in the ocean. *Deep-Sea Research* 22:619–627. ⟨257⟩

Simpson, J. H., and J. R. Hunter, 1974. Fronts in the Irish Sea. *Nature* 250:404–406. ⟨259, 380⟩

Sippican Corporation, 1973, 1978. Sippican Oceanographic Division—various advertising brochures, Marion, Massachusetts. ⟨414⟩

Smith, E. H., 1926. A practical method for determining ocean currents. *United States Coast Guard Bulletin* No. 14, U.S. Government Printing Office, Washington, D.C., 50 pp. ⟨233⟩

Smith, E. H., F. M. Soule, and O. Mosby, 1937. The *Marion* and *General Greene* expeditions to Davis Strait and Labrador Sea under direction of the United States Coast Guard 1928–1931–1933–1934–1935. Scientific results, part 2, physical oceanography. *United States Coast Guard Bulletin* No. 19, U.S. Government Printing Office, Washington, D.C., 259 pp. ⟨74, 75⟩

Smith, J. D., 1974. Turbulent structure of the surface boundary layer in an ice-covered ocean. *Rapports et Procès-Verbaux des Réunions, Conseil Permanent International pour l'Exploration de la Mer* 167:53–65. ⟨405⟩

Smith, J. D., 1978. Measurement of turbulence in ocean boundary layers. In *Proceedings of Working Conference on Current Measurement, January 11–13, 1978*, University of Delaware, Newark, pp. 95–128. ⟨205, 405⟩

Smith, P. C., 1973. The dynamics of bottom boundary currents in the ocean. Ph.D. Thesis, Massachusetts Institute of Technology/Woods Hole Oceanographic Institution, WHOI Ref. No. 73-4, 214 pp. ⟨473⟩

Smith, P. C., 1975. A streamtube model for bottom boundary currents in the ocean. *Deep-Sea Research* 22:853–873. ⟨24, 25, 260⟩

Smith, P. C., 1977. Experiments with viscous source flows in rotating systems. *Dynamics of Atmospheres and Oceans* 1:241–272. ⟨473⟩

Smith, P. C., 1978. Low-frequency fluxes of momentum, heat, salt, and nutrients at the edge of the Scotian Shelf. *Journal of Geophysical Research* 83:4079–4096. ⟨218⟩

Smith, R. L., 1978. Poleward propagating perturbations in currents and sea levels along the Peru coast. *Journal of Geophysical Research* 83:6083–6092. ⟨310, 313⟩

Smith, S. D., 1973. Thrust anemometer measurements over the sea re-examined. *Report Series BI-R-73-1*, Bedford Institute of Oceanography, Dartmouth, Nova Scotia, 23 pp. ⟨489⟩

Smith, S. D., and E. G. Banke, 1975. Variation of the sea surface drag coefficient with wind speed. *Quarterly Journal of the Royal Meteorological Society* 101:665–673. ⟨489⟩

Snodgrass, F. E., 1968. Deep sea instrument capsule. *Science* 162:78–87. ⟨323, 408, 425⟩

Snodgrass, F. E., G. W. Groves, K. Hasselmann, G. R. Miller, W. H. Munk, and W. H. Powers, 1966. Propagation of ocean swell across the Pacific. *Philosophical Transactions of the Royal Society of London A* 259:431–497. ⟨491⟩

Snodgrass, J. M., 1968. Instrumentation and communications. In *Ocean Engineering: Goals, Environment, Technology*, J. F. Brahtz, ed., John Wiley and Sons, New York, pp. 393–477. ⟨400, 403, 414⟩

Snyder, R. L., 1974. A field study of wave-induced pressure fluctuations above surface gravity waves. *Journal of Marine Research* 32:497–531. ⟨494⟩

Snyder, R. L., and C. S. Cox, 1966. A field study of the wind generation of ocean waves. *Journal of Marine Research* 24:141–178. ⟨491⟩

Snyder, R. L., F. W. Dobson, J. A. Elliott and R. B. Long, 1980. Array measurements of atmospheric pressure fluctuations above surface gravity waves. *Journal of Fluid Mechanics.* ⟨495⟩

Solberg, H., 1928. Integrationen der atmosphärischen Störungsgleichungen. Erster Teil: Wellenbewegungen in rotierenden, inkompressiblen Flüssigkeitsschichten. *Geofysiske Publikasjoner* 5:9, 120 pp. ⟨529⟩

Solberg, T. A., 1950. Preface to Proceedings of the Colloquium on the Flushing of Estuaries, September 7–8, 1950, Cambridge, Massachusetts, H. Stommel, ed., *W.H.O.I. Reference No. 50-37*, Woods Hole Oceanographic Institution, Woods Hole, Massachusetts, 206 pp. ⟨202⟩

Soley, J. C., 1911. The circulation in the North Atlantic in the month of August. *Supplement Pilot Chart North Atlantic Ocean for 1911.* Hydrographic office, U.S. Navy Dept., Washington, D.C. ⟨208⟩

Solomon, H., 1974. Comments on the antarctic bottom water problem and high-latitude thermohaline sinking. *Journal of Geophysical Research* 79:881–884. ⟨40⟩

Solomon, H., 1978. Comments on a theory of the Kuroshio meander. *Deep-Sea Research* 25:957–958. ⟨540⟩

Somayajulu, B. L. K., D. Lal, and H. Craig, 1973. Silicon-32 profiles in the South Pacific. *Earth and Planetary Science Letters* 18:181–188. ⟨447⟩

Spiegel, E. A., and G. Veronis, 1960. On the Boussinesq approximation for a compressible fluid. *Astrophysical Journal* 131:442–447. ⟨142, 385⟩

Spilhaus, A. F., 1937. Note on the flow of streams in a rotating system. *Journal of Marine Research* 1:29–33. ⟨464⟩

Spilhaus, A. F., A. Erlich, and A. R. Miller, 1950. Hydrostatic instability in the ocean. *Transactions of the American Geophysical Union* 31:213–215. ⟨212⟩

Spilhaus, A. F., and A. R. Miller, 1948. The sea sampler. *Journal of Marine Research* 7:370–385. ⟨212⟩

Starr, V. P., 1954. Studies of the atmospheric general circulation, part I. *Final Report General Circulation Project, AF 19(122)-153*, Massachusetts Institute of Technology, Dept. of Meteorology, 535 pp. ⟨507⟩

Starr, V. P., 1957. Studies of the atmospheric general circulation, part II. *Final Report General Circulation Project, AF 19(604)-1000*, Massachusetts Institute of Technology, Dept. of Meteorology, 672 pp. ⟨507⟩

Starr, V. P., 1968. *Physics of Negative Viscosity Phenomena.* McGraw-Hill, New York, 256 pp. ⟨343⟩

Steele, J. H., 1961. The environment of a herring fishery. *Marine Research, Department of Agriculture and Fisheries for Scotland* 1961: 6, 19 pp. ⟨380⟩

Steele, J. H., 1975. Biological modelling II. In *Modelling of Marine Systems*, J. C. J. Nihoul, ed., Elsevier, Amsterdam, pp. 207–216. ⟨378⟩

Steele, J. H., ed., 1978a. *Spatial Pattern in Plankton Communities.* Plenum Press, New York, 470 pp. ⟨376⟩

Steele, J. H., 1978b. Some comments on plankton patches. In *Spatial Pattern in Plankton Communities*, J. H. Steele, ed., Plenum Press, New York, pp. 1–20. ⟨377, 378⟩

Steele, J. H., 1979. Some problems in the management of marine resources. *Applied Biology* 4: 103–140. ⟨379, 381, 382⟩

Steele, J. H., J. R. Barrett, and L. V. Worthington, 1962. Deep currents south of Iceland. *Deep-Sea Research* 9:465–474. ⟨23, 29, 43⟩

Steele, J. H., and B. W. Frost, 1977. The structure of plankton communities. *Philosophical Transactions of the Royal Society of London B* 280:485–534. ⟨379, 382⟩

Steele, J. H., and E. W. Henderson, 1977. Plankton patches in the northern North Sea. In *Fisheries Mathematics*, J. H. Steele, ed., Academic Press, London, pp. 1–19. ⟨382⟩

Steele, J. H., and E. W. Henderson, 1979. Spatial patterns in North Sea plankton. *Deep-Sea Research* 26:955–963. ⟨380⟩

Stefánnson, U., 1968. Dissolved nutrients, oxygen and water masses in the Northern Irminger Sea. *Deep-Sea Research* 15:541–575. ⟨24, 25⟩

Stern, M. E., 1959. The decay and instability of Ekman flows. (Unpublished manuscript, 26 pp.) ⟨476⟩

Stern, M. E., 1960a. The "salt-fountain" and thermohaline convection. *Tellus* 12:172–175. ⟨xx, 252, 419⟩

Stern, M. E., 1960b. Instability of Ekman flow at large Taylor number. *Tellus* 12:399–417. ⟨466, 476⟩

Stern, M. E., 1963. Trapping of low frequency oscillations in an equatorial "boundary layer". *Tellus* 15:246–250. ⟨295⟩

Stern, M. E., 1966. Interaction of a uniform wind stress with hydrostatic eddies. *Deep-Sea Research* 13:193–203. ⟨149⟩

Stern, M. E., 1969. Collective instability of salt fingers. *Journal of Fluid Mechanics* 35:209–218. ⟨252, 253⟩

Stern, M. E., 1970. Optical measurement of salt fingers. *Tellus* 22:76–81. ⟨419⟩

Stern, M. E., 1975a. *Ocean Circulation Physics.* Academic Press, New York, 246 pp. ⟨132, 140, 143, 148, 149, 183, 250, 252, 255⟩

Stern, M. E., 1975b. Minimal properties of planetary eddies. *Journal of Marine Research* 33:1–13. ⟨511⟩

Stern, M. E., 1976. Maximum buoyancy flux across a salt finger interface. *Journal of Marine Research* 34:95–110. ⟨255⟩

Stern, M. E., and J. S. Turner, 1969. Salt fingers and convecting layers. *Deep-Sea Research* 16:497–511. ⟨252⟩

Stewart, R. W., 1957. A note on the dynamic balance in estuarine circulation. *Journal of Marine Research* 16:34–39. ⟨206⟩

Stewart, R. W., 1969. Turbulence and waves in a stratified atmosphere. *Radio Science* 4:1269–1278. ⟨246⟩

Stewartson, K., and J. A. Rickard, 1969. Pathological oscillations of a rotating fluid. *Journal of Fluid Mechanics* 35:759–773. ⟨295⟩

Stidd, C. K., 1974. Ship drift components: Means and standard deviations. *S.I.O. Reference Series 74-33*, Scripps Institution of Oceanography, University of California at San Diego, 57 pp. ⟨71⟩

Stigebrandt, A., 1976. Vertical diffusion driven by internal waves in a sill fjord. *Journal of Physical Oceanography* 6:486–495. ⟨261⟩

Stimson, P. B., 1965. Synthetic-fiber deep-sea mooring cables: Their life expectancy and susceptibility to biological attack. *Deep-Sea Research* 12:1–8. ⟨407⟩

Stock, G., 1976. Modelling of tides and tidal dissipation in the Gulf of California. Ph.D. Thesis, University of California at San Diego. ⟨325⟩

Stokes, G. G., 1846. Report on recent researches in hydrodynamics. In *Report of the 16th Meeting of the British Association for the Advancement of Science*, Southampton, pp. 1–20. ⟨307⟩

Stokes, G. G., 1847. On the theory of oscillating waves. *Transactions of the Cambridge Philosophical Society* 8:441–455. ⟨264, 299⟩

Stommel, H., 1948. The westward intensification of wind-driven ocean currents. *Transactions of the American Geophysical Union* 29:202–206. ⟨xiv, xv, xix, xxiv, 79, 125, 140, 150, 340, 341, 342, 344, 345, 468, 470⟩

Stommel, H., 1949. Horizontal diffusion due to oceanic turbulence. *Journal of Marine Research* 8:199–225. ⟨203, 239, 384⟩

Stommel, H., 1950a. Note on the deep circulation of the Atlantic Ocean. *Journal of Meteorology* 7:245–246. ⟨13⟩

Stommel, H., ed., 1950b. Proceedings of the Colloquium on the Flushing of Estuaries, September 7–8, 1950, Cambridge, Massachusetts. *W.H.O.I. Reference No. 50-37*, Woods Hole Oceanographic Institution, Woods Hole, Massachusetts, 206 pp. ⟨202⟩

Stommel, H., 1951. Recent developments in the study of tidal estuaries. *W.H.O.I. Technical Report 51-33*, Woods Hole Oceanographic Institution, Woods Hole, Massachusetts, 18 pp. ⟨202⟩

Stommel, H., 1952. Streaks of natural water surfaces. In *International Symposium on Atmospheric Turbulence in the Boundary Layer*, E. W. Hewson, ed., Geophysical Research Papers No. 19, Air Force Cambridge Research Center, Cambridge, Massachusetts, pp. 145–154. ⟨477⟩

Stommel, H., 1953a. Examples of the possible role of inertia and stratification in the dynamics of the Gulf Stream system. *Journal of Marine Research* 12:184–195. ⟨116⟩

Stommel, H., 1953b. Computation of pollution in a vertically mixed estuary. *Sewage and Industrial Wastes* 25:1065–1071. ⟨202, 212⟩

Stommel, H., 1954. Why do our ideas about the ocean circulation have such a peculiarly dream-like quality? (Unpublished manuscript, privately printed, 34 pp.) ⟨xiii, xvi, 153, 154⟩

Stommel, H., 1955. Discussion at the Woods Hole Convocation, June 1954. *Journal of Marine Research* 14:504–510. ⟨411⟩

Stommel, H., 1957a. The abyssal circulation of the ocean. *Nature* 180:733–734. ⟨xxi⟩

Stommel, H., 1957b. A survey of ocean current theory. *Deep-Sea Research* 4:149–184. ⟨xv, xxi, 79, 121, 152, 162, 339, 340, 466⟩

Stommel, H., 1958. The abyssal circulation. *Deep-Sea Research* 5:80–82. ⟨xxi, 12, 58, 59, 163, 449⟩

Stommel, H., 1960. Wind-drift near the equator. *Deep-Sea Research* 6:298–302. ⟨189, 190⟩

Stommel, H., 1962a. Primery peremeshivaniya i samovozbuzhdayushcheisya konvektsii na S,T-diagramme. *Okeanologiya* 2:205–209. (Examples of mixing and spontaneous convection of the S,T diagram. Translation No. 63-15151, Office of Technical Services, U.S. Department of Commerce, Washington, D.C.) ⟨251⟩

Stommel, H., 1962b. On the smallness of sinking regions in the ocean. *Proceedings of the National Academy of Sciences of the U.S.A.* 48:766–772. ⟨505⟩

Stommel, H., 1963. Varieties of oceanographic experience. *Science* 139:572–576. ⟨376, 429, 430, 431⟩

Stommel, H., 1965. *The Gulf Stream: A Physical and Dynamical Description*, 2nd ed. University of California Press, Berkeley, 248 pp. (First ed. 1958, 202 pp.) ⟨xiii, xv, xxv, 12, 79, 112, 137, 141, 153, 155, 173, 342, 344, 429⟩

Stommel, H. M., 1966. The large-scale oceanic circulation. In *Advances in Earth Science*, P. Hurley, ed., MIT Press, Cambridge, Massachusetts, pp. 175–184. ⟨398⟩

Stommel, H., and A. B. Arons, 1960a. On the abyssal circulation of the world ocean—I. Stationary planetary flow patterns on a sphere. *Deep-Sea Research* 6:140–154. ⟨xxi, 12, 26, 340⟩

Stommel, H., and A. B. Arons, 1960b. On the abyssal circulation of the world ocean—II. An idealized model of the circulation pattern and amplitude in oceanic basins. *Deep-Sea Research* 6:217–233. ⟨xxi, 12, 14, 15, 26, 27, 29, 37, 58, 59, 106, 340, 449⟩

Stommel, H., and A. B. Arons, 1972. On the abyssal circulation of the world ocean—V. The influence of bottom slope on the broadening of inertial boundary currents. *Deep-Sea Research* 19:707–718. ⟨xxiii, 37⟩

Stommel, H., A. B. Arons, and D. Blanchard, 1956. An oceanographical curiosity: The perpetual salt fountain. *Deep-Sea Research* 3:152–153. ⟨xvii, 250, 419⟩

Stommel, H., A. B. Arons, and A. J. Faller, 1958. Some examples of stationary planetary flow patterns in bounded basins. *Tellus* 10:179–187. ⟨xxi, 12, 152, 466, 467, 475⟩

Stommel, H., and H. G. Farmer, 1952. Abrupt change in width in two-layer open channel flow. *Journal of Marine Research* 11:205–214. ⟨203⟩

Stommel, H., and H. G. Farmer, 1953. Control of salinity in an estuary by a transition. *Journal of Marine Research* 12:13–20. ⟨203⟩

Stommel, H., and K. N. Fedorov, 1967. Small scale structure in temperature and salinity near Timor and Mindinao. *Tellus* 19:306–325 ⟨279, 290⟩

Stommel, H., and M. Fieux, 1978. *Oceanographic Atlases: A Guide to their Geographic Coverage and Contents*. Woods Hole Press, Woods Hole, Massachusetts 97 pp. ⟨xxiii, 44⟩

Stommel, H., and A. Leetmaa, 1972. Circulation on the continental shelf. *Proceedings of the National Academy of Sciences of the U.S.A.* 69:3380–3384. ⟨227, 228⟩

Stommel, H., P. Niiler, and D. Anati, 1978. Dynamic topography and recirculation of the North Atlantic. *Journal of Marine Research* 36:449–468. ⟨85, 90, 91, 125, 373⟩

Stommel, H., and F. Schott, 1977. The beta spiral and the determination of the absolute velocity field from hydrographic station data. *Deep-Sea Research* 24:325–329. ⟨xxi, xxiii, 91, 397⟩

Stommel, H., E. D. Stroup, J. L. Reid, and B. A. Warren, 1973. Transpacific hydrographic sections at Lats. 43°S and 28°S: the

SCORPIO Expedition—I. Preface. *Deep-Sea Research* 20:1–7. ⟨34, 35, 43, 46, 52, 55⟩

Stommel, H., and G. Veronis, 1957. Steady convective motions in a horizontal layer of fluid heated uniformly above and non-uniformly from below. *Tellus* 8:401–407. ⟨159⟩

Stommel, H., A. Voorhis, and D. Webb, 1971. Submarine clouds in the deep ocean. *American Scientist* 59:716–722. ⟨413⟩

Stommel, H., and J. Webster, 1962. Some properties of the thermocline equations in a subropical gyre. *Journal of Marine Research* 20:42–56. ⟨159⟩

Stone, P. H., 1966. On non-geostrophic baroclinic stability. *Journal of the Atmospheric Sciences* 23:390–400. ⟨529⟩

Stone, P. H., 1968. Some properties of Hadley regimes on rotating and non-rotating planets. *Journal of the Atmospheric Sciences* 25:644–657. ⟨505⟩

Stone, P. H., 1969. The meridional structure of baroclinic waves. *Journal of the Atmospheric Sciences* 26:376–389. ⟨529⟩

Stone, P. H., 1970. On non-geostrophic baroclinic stability: Part II. *Journal of the Atmospheric Sciences* 27:721–726. ⟨529⟩

Stone, P. H., 1978. Baroclinic adjustment. *Journal of the Atmospheric Sciences* 35:561–571. ⟨507⟩

Stuart, J. T., 1960. On the non-linear mechanics of wave disturbances in stable and unstable parallel flows. Part I. The basic behaviour in plane Poiseuille flow. *Journal of Fluid Mechanics* 9:353–370. ⟨174⟩

Stuiver, M., 1976. The ^{14}C distribution in West Atlantic abyssal waters. *Earth and Planetary Science Letters* 32:322–330. ⟨439, 450⟩

Stumpf, H. G., and P. K. Rao, 1975. Evolution of Gulf Stream eddies as seen in satellite infra-red imagery. *Journal of Physical Oceanography* 5:388–393. ⟨121⟩

Sturges, W., III, 1968. Sea-surface topography near the Gulf Stream. *Deep-Sea Research* 15:149–156. ⟨116, 212, 227⟩

Sturges, W., 1974. Sea level slope along continental boundaries. *Journal of Geophysical Research* 79:825–830. ⟨116⟩

Sugimoto, T., 1975. Effect of boundary geometries on tidal currents and tidal mixing. *Journal of the Oceanographical Society of Japan* 31:1–14. ⟨206⟩

Suomi, V., 1976. As quoted by K. J. Hansen, ed., in Geophysical Monitoring for Climate Change, No. 5 Summary Report 1976, NOAA/ERL, Boulder, Colorado, November, 1977. ⟨429⟩

Sumner, F. B., R. P. Osburn, and L. J. Cole, 1913. A biological survey of the waters of Woods hole and vicinity. Part I., Section I—Physical and zoological. *Bulletin of the Bureau of Fisheries* 31 (1911): 1–442. ⟨208⟩

Sverdrup, H. U., 1926. Dynamic of tides on the North Siberian shelf. Results from the Maud Expedition. *Geofysiske Publikasjoner* 4:5, 75 pp. ⟨298⟩

Sverdrup, H. U., 1931. The origin of the deep-water of the Pacific Ocean as indicated by the oceanographic work of the "Carnegie". *Gerlands Beiträge zur Geophysik* 29:95–105. ⟨10⟩

Sverdrup, H. U., 1947. Wind-driven currents in a baroclinic ocean; with application to the equatorial currents of the eastern Pacific. *Proceedings of the National Academy of Sciences of the U.S.A.* 33:318–326. ⟨xxiv, 79, 140, 149, 188, 340, 342, 509, 517⟩

Sverdrup, H. U., M. W. Johnson, and R. H. Fleming, 1942. *The Oceans: Their Physics, Chemistry, and General Biology*. Prentice-Hall, Englewood Cliffs, New Jersey, 1087 pp. ⟨xii, xxv, 7, 11, 22, 36, 42, 58, 71, 72, 75, 78, 79, 81, 83, 140, 211, 212, 227, 236, 342, 343, 348, 384, 398, 414, 415, 434, 488⟩

Swallow, J. C., 1955. A neutral-buoyancy float for measuring deep currents. *Deep-Sea Research* 3:74–81. ⟨58, 121, 359, 403, 412⟩

Swallow, J. C., 1957. Some further deep current measurements using neutrally buoyant floats. *Deep-Sea Research* 4:93–104. ⟨403⟩

Swallow, J. C., 1971. The *Aries* current measurements in the western North Atlantic. *Philosophical Transactions of the Royal Society of London A* 270:451–460. ⟨358, 430, 507⟩

Swallow, J. C., 1976. Variable currents in mid-ocean. *Oceanus* 19:3, 18–25. ⟨343, 430⟩

Swallow, J. C., 1977. An attempt to test the geostrophic balance using Minimode current measurements. In *A Voyage of Discovery: George Deacon 70th Anniversary Volume*, M. Angel, ed., Supplement to *Deep-Sea Research*, Pergamon Press, Oxford, 165–176. ⟨397, 412⟩

Swallow, J. C., and L. V. Worthington, 1957. Measurements of deep currents in the western North Atlantic. *Nature* 179:1183–1184. ⟨xxi, 13, 121⟩

Swallow, J. C., and L. V. Worthington, 1961. An observation of a deep countercurrent in the western North Atlantic. *Deep-Sea Research* 8:1–19. ⟨27, 79, 108, 121, 421⟩

Swallow, J. C., and L. V. Worthington, 1969. Deep currents in the Labrador Sea. *Deep-Sea Research* 16:77–84. ⟨23, 26, 43, 108⟩

Swallow, M., 1961. Deep currents in the open ocean. *Oceanus* 7:3, 2–8. ⟨507⟩

Swanson, R. L., and C. S. Sindermann, eds., 1980. *Oxygen Depletion and Associated Benthic Mortalities in New York Bight, 1976*. NOAA Professional Paper, U.S. Department of Commerce, National Oceanic and Atmospheric Administration, Rockville, Maryland. ⟨225⟩

Swanson, R. L., H. M. Stanford, J. S. O'Connor, S. Chanesman, C. A. Parker, P. A. Eisen, and G. F. Mayer, 1978. June 1976 pollution of Long Island ocean beaches. *Journal of the Environmental Engineering Division, Proceedings of the American Society of Civil Engineers* 104:1067–1085. ⟨225⟩

Swift, D. J. P., D. B. Duane, and T. F. McKinney, 1973. Ridge and swale topography of the Middle Atlantic Bight, North America: secular response to the Holocene hydraulic regime. *Marine Geology* 15:227–247. ⟨208⟩

Swift, D. J. P., G. L. Freeland, P. E. Gadd, G. Han, J. W. Lavelle, and W. L. Stubblefield, 1976. Morphologic evolution and coastal sand transport, New York-New Jersey shelf. In *Middle Atlantic Continental Shelf and the New York Bight*, M. G. Gross, ed., American Society of Limnology and Oceanography, Special Symposia, 2, pp. 69–89. ⟨208⟩

Taft, B. A., 1963. Distribution of salinity and dissolved oxygen on surfaces of uniform potential specific volume in the South Atlantic, South Pacific, and Indian oceans. *Journal of Marine Research* 21:129–146. ⟨44, 82, 85⟩

Taft, B. A., 1972. Characteristics of the flow of the Kuroshio south of Japan. In *Kuroshio: Its Physical Aspects*, H. Stommel and K. Yoshida, eds., University of Tokyo Press, pp. 165–216. ⟨541⟩

Taft, B. A., 1978. Structure of the Kuroshio south of Japan. *Journal of Marine Research* 36:77–117. ⟨36⟩

Taft, B. A., and P. Kovala, 1979. Temperature, salinity, thermosteric anomaly and zonal geostrophic velocity sections along 150°W from NORPAX Shuttle Experiment (1977–78). *Department of Oceanography, University of Washington, Special Report No. 87, Reference M79-17*, Seattle, Washington. ⟨185⟩

Tait, J. B., 1957. Hydrography of the Faroe-Shetland Channel, 1927–1952. *Marine Research, Scottish Home Department* 1957: 2, 309 pp. ⟨43⟩

Tait, J. B., 1967. Horizontal temperature and salinity distributions. *Rapports et Procès-Verbaux des Réunions, Conseil Permanent International pour l'Exploration de la Mer* 157:38–63. ⟨24⟩

Tait, R. I., and M. R. Howe, 1968. Some observations of thermohaline stratification in the deep ocean. *Deep-Sea Research* 15:275–280. ⟨252⟩

Tait, R. I., and M. R. Howe, 1971. Thermohaline staircase. *Nature* 231:178–179. ⟨252⟩

Tapley, B. D., G. H. Born, H. H. Hagar, J. Lorell, M. E. Parke, J. M. Diamante, B. C. Douglas, C. C. Goad, R. Kolenkiewicz, J. G. Marsh, C. F. Martin, S. L. Smith III, W. F. Townsend, J. A. Whitehead, H. M. Byrne, L. S. Fedor, D. C. Hammond, and N. M. Mognard, 1979. Seasat altimeter calibration: initial results. *Science* 204:1410–1412. ⟨428⟩

Tarbell, S., A. Spencer, and R. E. Payne, 1978. A compilation of moored current meter data and associated oceanographic observations, Volume XVII (POLYMODE Array II data). *W.H.O.I. Technical Report 78-49*, Woods Hole Oceanographic Institution, Woods Hole, Massachusetts, 88 pp. ⟨403⟩

Tatro, P. R., and E. L. Mollo-Christensen, 1967. Experiments on Ekman layer instability. *Journal of Fluid Mechanics* 28:531–543. ⟨464, 526⟩

Taub, H., and D. L. Schilling, 1977. *Digital Integrated Electronics*. McGraw-Hill, New York, 650 pp. ⟨400⟩

Taylor, G. I., 1914. Report by Mr. G. I. Taylor. In *Report on the Work Carried out by the S.S. "Scotia", 1913*, H.M. Stationery Office, London, pp. 48–68. ⟨500, 501⟩

Taylor, G. I., 1915. Eddy-motion in the atmosphere. *Philosophical Transactions of the Royal Society of London A* 215:1–26. ⟨182⟩

Taylor, G. I., 1917. The formation of fog and mist. *Quarterly Journal of the Royal Meteorological Society* 43:241–268. ⟨497⟩

Taylor, G. I., 1920. Tidal friction in the Irish Sea. *Philosophical Transactions of the Royal Society of London A* 220:1–33. ⟨323⟩

Taylor, G. I., 1921. Tidal oscillations in gulfs and rectangular basins. *Proceedings of the London Mathematical Society* 20:148–181. ⟨299⟩

Taylor, G. I., 1931. Internal waves and turbulence in a fluid of variable density. *Rapports et Procès-Verbaux des Réunions, Conseil Permanent International pour l'Exploration de la Mer* 76:35–42. ⟨236⟩

Taylor, G. I., 1954. The dispersion of matter in turbulent flow through a pipe. *Proceedings of the Royal Society of London A* 223:446–468. ⟨239, 378⟩

Taylor, P. K., 1979. Observations of the atmospheric boundary layer over the ocean, JASIN 1972. (Unpublished manuscript.) ⟨503⟩

Telegadas, K., and G. J. Ferber, 1975. Atmospheric concentrations and inventory of krypton-85 in 1973. *Science* 190:882–883. ⟨446⟩

Thompson, P. D., 1956. A theory of large-scale disturbances in non-geostrophic flow. *Journal of Meteorology* 13:251–261. ⟨508⟩

Thompson, R. E., 1975. Propagation of planetary waves over random bottom topography. *Journal of Fluid Mechanics* 70:267–286. ⟨177⟩

Thompson, R., 1971. Topographic Rossby waves at a site north of the Gulf Stream. *Deep-Sea Research* 18:1–9. ⟨119⟩

Thompson, R. O. R. Y., 1976. Climatological numerical models of the surface mixed layer of the ocean. *Journal of Physical Oceanography* 6:496–503. ⟨245⟩

Thompson, R. O. R. Y., 1977. Observations of Rossby waves near site D. *Progress in Oceanography* 7:135–162. ⟨230, 372, 545⟩

Thompson, R. O. R. Y., 1978. Reynolds stresses and deep counter-currents near the Gulf Stream. *Journal of Marine Research* 36:611–615. ⟨119⟩

Thompson, R. O. R. Y., 1980. Efficiency of conversion of kinetic energy to potential energy by a breaking internal gravity wave. *Journal of Geophysical Research*. ⟨279⟩

Thompson, R. O. R. Y., and J. R. Luyten, 1976. Evidence for bottom-trapped topographic Rossby waves from single moorings. *Deep-Sea Research* 23:629–635. ⟨169⟩

Thompson, S. M., and J. S. Turner, 1975. Mixing across an interface due to turbulence generated by an oscillating grid. *Journal of Fluid Mechanics* 67:349–368. ⟨242⟩

Thomsen, H., 1933. The circulation in the depths of the Indian Ocean. *Journal du Conseil* 8:73–79. ⟨10⟩

Thomson, C. W., 1877. *The Voyage of the Challenger. The Atlantic*, 2. MacMillan, London, 396 pp. ⟨40, 43⟩

Thompson, W., 1863. On the rigidity of the earth. *Philosophical Transactions of the Royal Society of London* 153:573–582. ⟨296⟩

Thomson, W. (Lord Kelvin), 1879. On gravitational oscillations of rotating water. *Proceedings of the Royal Society of Edinburgh* 10:92–100. Reprinted in *The London, Edinburgh, and Dublin Philosophical Magazine and Journal of Science, Fifth Series* 10 (1880): 109–116; and in his *Mathematical and Physical Papers*, J. Larmor, ed., Cambridge University Press, London, 4 (1910), pp. 141–148. ⟨297, 298⟩

Thorpe, S. A., 1971. Experiments on the instability of stratified shear flows: miscible fluids. *Journal of Fluid Mechanics* 46:299–319. ⟨248, 249⟩

Thorpe, S. A., 1973a. Turbulence in stably stratified fluids. A review of laboratory experiments. *Boundary-Layer Meteorology* 5:95–119. ⟨249⟩

Thorpe, S. A., 1973b. Experiments on instability and turbulence in a stratified shear flow. *Journal of Fluid Mechanics* 61:731–751. ⟨279, 290⟩

Thorpe, S. A., 1977. Turbulence and mixing in a Scottish loch. *Philosophical Transactions of the Royal Society of London A* 286:125–181. ⟨244, 291⟩

Thorpe, S. A., 1978a. The near-surface ocean mixing layer in stable heating conditions. *Journal of Geophysical Research* 83: 2875–2885. ⟨242, 249, 278⟩

Thorpe, S. A., 1978b. On the shape and breaking of finite amplitude internal gravity waves in a shear flow. *Journal of Fluid Mechanics* 85:7–31. ⟨243, 249, 276, 278⟩

Thorpe, S. A., 1978c. On internal gravity waves in an accelerating shear flow. *Journal of Fluid Mechanics* 88:623–629. ⟨273, 276⟩

Thorpe, S. A., 1979. Breaking internal waves in shear flows. In *Twelfth Symposium Naval Hydrodynamics*, National Academy of Science, Washington, D.C. pp. 623–628. ⟨278⟩

Thorpe, S. A., P. K. Hutt, and R. Soulsby, 1969. The effect of horizontal gradients on thermohaline convection. *Journal of Fluid Mechanics* 38:375–400. ⟨252⟩

Tizard, T. H., 1883. Remarks on the soundings and temperatures obtained in the Faroe Channel during the summer of 1882. *Proceedings of the Royal Society of London* 35:202–226. ⟨22, 41⟩

Tizard, T. H., H. N. Mosely, J. Y. Buchanan, and J. Murray, 1885a. *Report on the Scientific Results of the Voyage of H.M.S. Challenger during the years 1873–76, Narrative of the Cruise of H.M.S. Challenger with a General Account of the Scientific Results of the Expedition*, 1, Second Part, pp. 511–1110. ⟨11⟩

Tizard, T. H., H. N. Mosely, J. Y. Buchanan, and J. Murray, 1885b. *Report on the Scientific Results of the Voyage of H.M.S. Challenger during the years 1873–76, Narrative of the Cruise of H.M.S. Challenger with a General Account of the Scientific Results of the Expedition*, 1, First Part, pp. 1–510. ⟨349⟩

Toba, Y., M. Tokuda, K. Okuda, and S. Kawai, 1975. Forced convection accompanying wind waves. *Journal of the Oceanographical Society of Japan* 31:192–198. ⟨242⟩

Tomasko, M. G., L. R. Doose, and P. H. Smith, 1979. Absorption of sunlight in the atmosphere of Venus. *Science* 205:80–82. ⟨506⟩

Townsend, A. A., 1959. Temperature fluctuations over a heated horizontal surface. *Journal of Fluid Mechanics* 5:209–221. ⟨392⟩

Townsend, A. A., 1965. Excitation of internal waves by a turbulent boundary layer. *Journal of Fluid Mechanics* 22:241–252. ⟨509⟩

Trier, R. M., W. S. Broecker, and H. W. Feely, 1972. Radium-228 profile at the Second GEOSECS Intercalibration Station, 1970, in the North Atlantic. *Earth and Planetary Science Letters* 16:141–145. ⟨446⟩

Tsuchiya, M., 1968. Upper Waters of the Intertropical Pacific Ocean. *The Johns Hopkins Oceanographic Studies* 4: 50 pp. ⟨44, 82, 83, 188⟩

Tsuchiya, M., 1970. Equatorial circulation of the South Pacific. In *Scientific Exploration of the South Pacific*, W. S. Wooster, ed., National Academy of Sciences, Washington, D.C., pp. 69–74. ⟨184, 185⟩

Tsuchiya, M., 1974. Variation of the surface geostrophic flow in the eastern intertropical Pacific Ocean. *Fishery Bulletin* 72:1075–1086. ⟨185⟩

Tsuchiya, M., 1975. Subsurface countercurrents in the eastern equatorial Pacific Ocean. *Journal of Marine Research* 33 (Supplement): 145–175. ⟨83, 185, 186⟩

Tucholke, B. E., W. R. Wright, and C. D. Hollister, 1973. Abyssal circulation over the Greater Antilles Outer Ridge. *Deep-Sea Research* 20:973–995. ⟨27, 28⟩

Tulley, J. P., 1949. Oceanography and prediction of pulp mill pollution in Alberni Inlet. *Bulletin of the Fisheries Research Board of Canada* 83: 169 pp. ⟨202⟩

Turner, J. S., 1965. The coupled transports of salt and heat across a sharp density interface. *International Journal of Heat and Mass Transfer* 8:759–767. ⟨254⟩

Turner, J. S., 1967. Salt fingers across a density interface. *Deep-Sea Research* 14:599–611. ⟨255⟩

Turner, J. S., 1968. The behaviour of a stable salinity gradient heated from below. *Journal of Fluid Mechanics* 33:183–200. ⟨251⟩

Turner, J. S., 1969. Buoyant plumes and thermals. *Annual Review of Fluid Mechanics* 1:29–44. ⟨498⟩

Turner, J. S., 1973a. *Buoyancy Effects in Fluids*. Cambridge University Press, London, 367 pp. ⟨183, 238, 243, 246, 248, 251, 254, 260, 265⟩

Turner, J. S., 1973b. Geophysical examples of layering and microstructure: interpretation and relation to laboratory experiments. *Mémoires de la Société Royale des Sciences de Liège, Sixième Série* 4:11–30. ⟨238, 244⟩

Turner, J. S., 1974. Double-diffusive phenomena. *Annual Review of Fluid Mechanics* 6:37–56. ⟨261⟩

Turner, J. S., 1978. Double-diffusive intrusions into a density gradient. *Journal of Geophysical Research* 83:2887–2901. ⟨253, 254, 257⟩

Turner, J. S., and C. F. Chen, 1974. Two-dimensional effects in double-diffusive convection. *Journal of Fluid Mechanics* 63:577–592. ⟨252, 253⟩

Turner, J. S., T. G. L. Shirtcliffe, and P. G. Brewer, 1970. Elemental variations of transport coefficients across density interfaces in multiple-diffusive systems. *Nature* 228:1083–1084. ⟨256⟩

Turner, J. S., and H. Stommel, 1964. A new case of convection in the presence of combined vertical salinity and temperature gradients. *Proceedings of the National Academy of Sciences of the U.S.A.* 52:49–53. ⟨xx, 251⟩

Twitchell, P. F., 1979. Meeting on Physical Oceanography and Satellites. *Bulletin of the American Meteorological Society* 60:225–231. ⟨426⟩

Uchupi, E., 1965. Map showing relation of land and submarine topography, Nova Scotia to Florida. *U.S. Geological Survey, Miscellaneous Geological Investigations*, Map Series, I-451. ⟨198, 208⟩

Ufford, C. W., 1947. Internal waves measured at three stations. *Transactions of the American Geophysical Union* 28:87–95. ⟨264⟩

Ulltang, O., in press. Factors of pelagic fish stocks which affect their reaction to exploitation and require a new approach to their assessment and managements. In *ICES Sym-*

posium on the Biological Basis of Pelagic Fish Stock Management, A. Saville, ed., *Rapports et Procès-Verbaux des Réunions, Conseil Permanent International pour l'Exploration de la Mer* 177. ⟨377⟩

UNESCO, 1975. An intercomparison of open sea tidal pressure sensors. Report of SCOR Working Group 27: "Tides of the Open Sea." *UNESCO Technical Papers in Marine Science* 21, UNESCO, Paris, 67 pp. ⟨425⟩

United States Navy Hydrographic Office, 1944. *Current Charts: Northwestern Pacific Ocean.* Hydrographic Office Miscellaneous No. 10,058-A, Washington, 12 charts. Reissued in 1950 as Hydrographic Office Publication No. 569. ⟨71⟩

United States Navy Hydrographic Office, 1946. *Current Charts: North Atlantic Ocean.* Hydrographic Office Miscellaneous No. 10,688, Washington, 12 charts. Reissued as Hydrographic Office Publication No. 571. ⟨116⟩

Ursell, F., 1952. Edge waves on a sloping beach. *Proceedings of the Royal Society of London A* 214:79–97. ⟨307, 309, 310, 315⟩

US POLYMODE Organizing Committee, 1976 U.S. POLYMODE program and plan. Massachusetts Institute of Technology, Cambridge, Massachusetts, 87 pp. (Unpublished document.) ⟨429⟩

Vachon, W. A., 1978. Present status and future directions of drifting buoy developments: A summary of presentations made at a Drifting Buoy Workshop Held at the Woods Hole Oceanographic Institution on June 11 and 12, 1978. Technical report, A. D. Little, Inc., Acorn Park, Cambridge, Massachusetts, 32 pp. Available from National Technical Information Service, Springfield, Virginia. ⟨426⟩

van Leer, J. C., W. Düing, R. Erath, E. Kennelly, and A. Speidel, 1974. The Cyclesonde: an unattended vertical profiler for scalar and vector quantities in the upper ocean. *Deep-Sea Research* 21:385–400. ⟨424⟩

Vastano, A. C., and B. A. Warren, 1976. Perturbations to the Gulf Stream by Atlantis II Seamount. *Deep-Sea Research* 23:681–694. ⟨522⟩

Veronis, G., 1963. On inertially controlled flow patterns in a β-plane ocean. *Tellus* 15:59–66. ⟨154⟩

Veronis, G., 1966a. Wind-driven ocean circulation—Part 1. Linear theory and perturbation analysis. *Deep-Sea Research* 13:17–29. ⟨150, 153⟩

Veronis, G., 1966b. Wind-driven ocean circulation—Part 2. Numerical solutions of the non-linear problem. *Deep-Sea Research* 13:30–35. ⟨89, 141, 153, 154, 155, 156⟩

Veronis, G., 1966c. Generation of mean ocean circulation by fluctuating winds. *Tellus* 18:67–76. ⟨345⟩

Veronis, G., 1968a. Effect of a stabilizing gradient of solute on thermal convection. *Journal of Fluid Mechanics* 34:315–336. ⟨255⟩

Veronis, G., 1968b. Comments on Phillips (1966). *Journal of the Atmospheric Sciences* 25:1154–1155. ⟨295⟩

Veronis, G., 1969. On theoretical models of the thermocline circulation. *Frederick C. Fuglister Sixtieth Anniversary Volume, Deep-Sea Research* 16 (Supplement):301–323. ⟨159, 509⟩

Veronis, G., 1970. Effects of fluctuating winds on ocean circulation. *Deep-Sea Research* 17:421–434. ⟨142⟩

Veronis, G., 1972. On properties of seawater defined by temperature, salinity, and pressure. *Journal of Marine Research* 30:227–255. ⟨82, 183, 282⟩

Veronis, G., 1973a. Model of world ocean circulation: I. Wind-driven, two layer. *Journal of Marine Research* 31:228–288. ⟨125, 139, 141, 158⟩

Veronis, G., 1973b. Large scale ocean circulation. In *Advances in Applied Mechanics* 13:1–92. ⟨143, 342, 509, 518⟩

Veronis, G., 1977. Use of tracers in circulation studies. In *The Sea: Ideas and Observations on Progress in the Study of the Seas, 6: Marine Modeling*, E. D. Goldberg, I. N. McCave, J. J. O'Brien, and J. H. Steele, eds., Wiley, Interscience, New York, pp. 169–188. ⟨449⟩

Veronis, G., 1978. Model of world ocean circulation: III. Thermally and wind driven. *Journal of Marine Research* 36:1–44. ⟨141, 158, 162⟩

Veronis, G., and H. Stommel, 1956. The action of variable wind stresses on a stratified ocean. *Journal of Marine Research* 15:43–75. ⟨141, 164, 165, 340, 344, 509⟩

Veronis, G., and C. C. Yang, 1972. Nonlinear source-sink flow in a rotating pie-shaped basin. *Journal of Fluid Mechanics* 51:513–527. ⟨152⟩

Verrill, A. E., 1873. Results of recent dredging expeditions on the coast of New England. *American Journal of Scientific Arts, Third Series* 5:98–106. ⟨208⟩

Villain, C., 1952. Cartes des lignes cotidales dans les océans (1). *Annales Hydrographiques, 4e Serie* 3:269–388. ⟨318, 320⟩

Vinogradova, P. S., A. G. Kislyakov, V. M. Litvin, and L. S. Ponomarenko, 1959. Resultati okeanograficheskikh issledovanii v raione Farero-Islandskogo poroga v 1955–1956 gg. (Results of oceanographic investigations in the region of the Faroe-Iceland Ridge in the years 1955–1956.) *Trudy PINRO* 11:106–134. ⟨22⟩

Virk, P., 1975. Drag reduction fundamentals. *American Institute of Chemical Engineering Journal* 21:4–44. ⟨391⟩

Volkmann, G., 1962. Deep current observations in the Western North Atlantic. *Deep-Sea Research* 9:493–500. ⟨27, 121⟩

von Arx, W. S., 1950. An electromagnetic method for measuring the velocities of ocean currents from a ship underway. *Papers in Physical Oceanography and Meteorology* 11:3, 62 pp. ⟨117⟩

von Arx, W. S., 1952. A laboratory study of the wind-driven ocean circulation. *Tellus* 4:311–319. ⟨465⟩

von Arx, W. S., 1957. An experimental approach to problems in physical oceanography. *Progress in Physics and Chemistry of the Earth* 2:1–29. ⟨466⟩

von Arx, W. S., 1962. *An Introduction to Physical Oceanography.* Addison-Wesley, Reading, Massachusetts, 442 pp. ⟨209, 466⟩

von Arx, W. S., D. F. Bumpus, and W. S. Richardson, 1955. On the fine structure of the Gulf Stream front. *Deep-Sea Research* 3:46–65. ⟨117⟩

Vonbun, F. O., J. G. Marsh, and F. J. Lerch, 1978. Computed and observed ocean topography: A comparison. *Boundary-Layer Meteorology* 13:253–262. ⟨118⟩

von Buttlar, H., and W. F. Libby, 1955. Natural distribution of cosmic-ray produced tritium. *Journal of Inorganic and Nuclear Chemistry* 6:75–91. ⟨440⟩

Vonder Haar, T. H., and A. H. Oort, 1973. New estimate of annual poleward energy transport by northern hemisphere oceans. *Journal of Physical Oceanography* 3:169–172. ⟨137⟩

Voorhis, A. D., D. C. Webb, and R. C. Millard, 1976. Current structure and mixing in the shelf/slope water front south of New England. *Journal of Geophysical Research* 81:3695–3708. ⟨257, 258⟩

Wagner, A. J., 1977. Weather and circulation of January 1977, the coldest month on record in the Ohio Valley. *Monthly Weather Review* 105:553–560. ⟨225⟩

Walden, H., 1963. Comparison of one-dimensional wave spectra recorded in the German Bight with various "theoretical" spectra. *Ocean Wave Spectra: Proceedings of a Conference,* Prentice-Hall, Englewood Cliffs, New Jersey, pp. 67–94. ⟨492⟩

Walden, R. G., O. H. DeBok, J. B. Gregory, D. Meggitt, and W. A. Vachon, 1977. The mooring dynamics experiment—a major study of the dynamics of buoys in the deep ocean. In *Ninth Annual Offshore Technology Conference, 1977, Proceedings,* Offshore Technology Conference, Dallas, pp. 51–60. ⟨410⟩

Walin, G., 1969. Some aspects of time-dependent motion of a stratified rotating fluid. *Journal of Fluid Mechanics* 36:289–307. ⟨527⟩

Walsh, J. J., T. G. Falkowski, and T. S. Hopkins, 1980. Oxygen depletion within the New York Bight as a function of climatology and phytoplankton species succession. *Journal of Marine Research.* ⟨225⟩

Walsh, J. J., T. E. Whitledge, F. W. Barvenik, C. D. Wirick, S. O. Howe, W. E. Esaias, and J. T. Scott, 1978. Wind events and food chain dynamics within the New York Bight. *Limnology and Oceanography* 23:659–683. ⟨379⟩

Walther, E., 1966. Streaking. In Langmuir Circulations and Internal Waves in Lake George, *Lake George Studies Report No. 1, Atmospheric Sciences Research Center Publication No. 42,* State University of New York, Albany, pp. 7–16. ⟨477⟩

Wang, D.-P., 1979a. Subtidal sea level variations in the Chesapeake Bay and relations to atmospheric forcing. *Journal of Physical Oceanography* 9:413–421. ⟨206⟩

Wang, D.-P., 1979b. Wind-driven circulation in the Chesapeake Bay, winter 1975. *Journal of Physical Oceanography* 9:564–572. ⟨206⟩

Wang, D.-P., 1979c. Low frequency sea level variability on the Middle Atlantic Bight. *Journal of Marine Research* 37:683–697. ⟨215, 222, 223, 228⟩

Wang, D.-P., 1980. Diffraction of continental shelf waves by the irregular alongshore geometry. *Journal of Physical Oceanography.* ⟨223⟩

Wang, D.-P., and A. J. Elliott, 1978. Non-tidal variability in the Chesapeake Bay and Potomac River; evidence for non-local forcing. *Journal of Physical Oceanography* 8:225–232. ⟨206⟩

Wang, D.-P., and C. N. K. Mooers, 1976. Coastal trapped waves in a continuously stratified ocean. *Journal of Physical Oceanography* 6:853–863. ⟨316, 358⟩

Wang, D.-P., and C. N. K. Mooers, 1977. Long coastal trapped waves off the west coast of the United States, summer 1973. *Journal of Physical Oceanography* 7:856–864. ⟨222⟩

Warren, B. A., 1963. Topographic influences on the path of the Gulf Stream. *Tellus* 15:167–183. ⟨123, 126⟩

Warren, B. A., 1972. Insensitivity of subtropical mode water characteristics to meteorological fluctuations. *Deep-Sea Research* 19:1–20. ⟨355⟩

Warren, B. A., 1973. Transpacific hydrographic sections at Lats. 43°S and 28°S: the SCORPIO Expedition—II. Deep water. *Deep-Sea Research* 20:9–38. ⟨25, 34, 35⟩

Warren, B. A., 1974. Deep flow in the Madagascar and Mascarene basins. *Deep-Sea Research* 21:1–21. ⟨29, 32, 108⟩

Warren, B. A., 1976. Structure of deep western boundary currents. *Deep-Sea Research* 23:129–142. ⟨35, 37, 38⟩

Warren, B. A., 1977. Deep western boundary current in the eastern Indian Ocean. *Science* 196:53–54. ⟨32, 33, 109⟩

Warren, B. A., 1978. Bottom water transport through the Southwest Indian Ridge. *Deep-Sea Research* 25:315–321. ⟨29, 109⟩

Warren, B. A., H. Stommel, and J. C. Swallow, 1966. Water masses and patterns of flow in the Somali Basin during the southwest monsoon of 1964. *Deep-Sea Research* 13:825–860. ⟨32⟩

Warren, B. A., and G. H. Volkmann, 1968. Measurement of volume transport of the Gulf Stream south of New England. *Journal of Marine Research* 26:110–126. ⟨131⟩

Warren, B. A., and A. D. Voorhis, 1970. Velocity measurements in the deep western boundary current of the South Pacific. *Nature* 228:849–850. ⟨35, 108⟩

Watson, E. E., 1936. Mixing and residual currents in tidal waters as illustrated in the Bay of Fundy. *Journal of the Biological Board of Canada* 2:141–208. ⟨201⟩

Watson, E. R., 1904. Movements of the waters of Loch Ness, as indicated by temperature observations. *The Geographical Journal* 24:430–437. ⟨290⟩

Watson, J. G., W. L. Siegmann, and M. J. Jacobson, 1977. Acoustically relevant statistics for stochastic internal-wave models. *Journal of the Acoustical Society of America* 61:716–726. ⟨291⟩

Watson, K. M., B. J. West, and B. I. Cohen, 1976. Coupling of surface and internal gravity waves: a mode coupling model. *Journal of Fluid Mechanics* 77:185–208. ⟨286⟩

Watson, W. H., 1963. *Understanding Physics Today.* Cambridge University Press, London, p. 16 ⟨396⟩

Wattenberg, H., 1929. Durchlüftung des Atlantischen Ozeans. (Vorläufige Mitteilung aus den Ergebnissen der Deutschen Atlantischen Expedition). *Journal du Conseil* 4:68–79. ⟨11⟩

Wattenberg, H., 1939. Atlas zu: Die Verteilung des Sauerstoffs im Atlantischen Ozean. In *Wissenschaftliche Ergebnisse der Deutschen Atlantischen Expedition auf dem· Forschungs— und Vermessungsschiff "Meteor" 1925–1927,* 9: Atlas, 72 plates. ⟨43, 75⟩

Wattenberg, H., 1957. Die Verteilung des Phosphats im Atlantischen Ozean. In *Wissenschaftliche Ergebnisse der Deutschen Atlantischen Expedition auf dem Forschungs— und Vermessungsschiff "Meteor" 1925–1927,* 9: 2nd Part, 2, 133–180. ⟨75⟩

Watts, D. R., and H. T. Rossby, 1977. Measuring dynamic heights with inverted echo sounders: Results from MODE. *Journal of Physical Oceanography* 7:345–358. ⟨428⟩

Wearn, R. B., Jr., and D. J. Baker, Jr., 1980. Bottom pressure measurements across the Antarctic Circumpolar Current and their relation to the wind. *Deep-Sea Research.* ⟨425⟩

Weatherly, G. L., and J. van Leer, 1977. On the importance of stable stratification to the structure of the bottom boundary layer on the western Florida Shelf. In *Bottom Turbulence*, J. C. J. Nihoul, ed., Elsevier, Amsterdam, pp. 108–122. ⟨259⟩

Webb, D. C., 1977. SOFAR floats for POLYMODE. In *Oceans '77 Conference Record*, IEEE Publication 77CH1272-4 OEC, Institute of Electrical and Electronics Engineers, New York, and Marine Technology Society, Washington, D.C., 2, pp. 44B1–44B5. ⟨413⟩

Webb, D. C., and L. V. Worthington, 1968. Measurement of vertical water movement in the Cayman Basin. *Deep-Sea Research* 15:609–612. ⟨413⟩

Webb, D., 1974. Green's function and tidal prediction. *Reviews of Geophysics and Space Physics* 12:103–116. ⟨320, 321, 323⟩

Webb, D. J., 1978. The wave-wave interaction machine. In *Turbulent Fluxes through the Sea Surface, Wave Dynamics, and Prediction*, A. Favre and K. Hasselmann, eds., Plenum Press, New York, pp. 335–345. ⟨494⟩

Webster, F., 1961a. The effect of meanders on the kinetic energy balance of the Gulf Stream. *Tellus* 13:392–401. ⟨117, 118, 119, 345, 507, 529⟩

Webster, F., 1961b. A description of Gulf Stream meanders off Onslow Bay. *Deep-Sea Research* 8:130–143. ⟨117, 119⟩

Webster, F., 1965. Measurements of eddy fluxes of momentum in the surface layer of the Gulf Stream. *Tellus* 17:239–245. ⟨117, 345⟩

Webster, F., 1968. Observations of inertial-period motions in the deep sea. *Reviews of Geophysics* 6:473–490. ⟨265⟩

Webster, F., 1969. Vertical profiles of horizontal ocean currents. *Deep-Sea Research* 16:85–98. ⟨27, 121, 123⟩

Wedderburn, E. M., 1907. The temperature of the fresh water lochs of Scotland, with special references to Loch Ness. *Transactions of the Royal Society of Edinburgh* 45:407–489. ⟨290⟩

Weiler, H. S., and R. W. Burling, 1967. Direct measurements of stress and spectra of turbulence in the boundary layer over the sea. *Journal of the Atmospheric Sciences* 24:653–664. ⟨489⟩

Weisberg, R. H., 1976a. A note on estuarine mean flow estimation. *Journal of Marine Research* 34:387–396. ⟨206⟩

Weisberg, R. H., 1976b. The nontidal flow in the Providence River of Narragansett Bay: a stochastic approach to estuarine circulation. *Journal of Physical Oceanography* 6:721–734. ⟨206⟩

Weisberg, R. H., A. Horigan, and C. Colin, 1979. Equatorially trapped Rossby-gravity wave propagation in the Gulf of Guinea. *Journal of Marine Research* 37:67–86. ⟨186, 192⟩

Weisberg, R. H., L. Miller, A. Horigan, and J. Knauss, 1980. Velocity observations in the equatorial thermocline during GATE. *Deep-Sea Research* 26(supp. ©1979) ⟨186, 192⟩

Weisberg, R. H., and W. Sturges, 1976. Velocity observations in the West Passage of Narragansett Bay: A partially mixed estuary. *Journal of Physical Oceanography* 6:345–354. ⟨206⟩

Weiss, R. F., H. G. Östlund, and H. Craig, 1979. Geochemical studies of the Weddell Sea. *Deep-Sea Research* 26:1093–1120. ⟨450, 456⟩

Welander, P., 1959. An advective model of the ocean thermocline. *Tellus* 11:309–318. ⟨xxi, 141, 159, 160, 161, 506, 509⟩

Welander, P., 1968. Wind-driven circulation in one- and two-layer oceans of variable depth. *Tellus* 20:1–15. ⟨151⟩

Welander, P., 1971a. Some exact solutions to the equations describing an ideal fluid thermocline. *Journal of Marine Research* 29:60–68. ⟨141, 159, 161, 162⟩

Welander, P., 1971b. The thermocline problem. *Philosophical Transactions of the Royal Society of London A* 270:69–73. ⟨159, 161⟩

Welander, P., 1973. Lateral friction in the oceans as an effect of potential vorticity mixing. *Geophysical Fluid Dynamics* 5:173–189. ⟨507⟩

Welander, P., 1976. A zonally uniform regime in the ocean circulation. *Journal of Physical Oceanography* 6:121–124. ⟨150⟩

Weller, R. A., 1978. Observations of horizontal velocity in the upper ocean made with a new vector measuring current meter. Ph.D. Thesis, University of California at San Diego, 169 pp. ⟨404, 406⟩

Weller, R. A., and R. E. Davis, 1980. A vector measuring current meter. *Deep-Sea Research.* ⟨404, 406⟩

Wells, R. C., R. K. Bailey, and E. P. Henderson, 1929. Salinity of the water of Chesapeake Bay. In *Shorter Contributions to General Geology, 1928; U.S. Geological Survey Professional Paper* 154, pp. 105–152. ⟨200⟩

Wenner, F., E. H. Smith, and F. M. Soule, 1930. Apparatus for the determination aboard ship of the salinity of sea water by the electrical conductivity method. *Bureau of Standards Journal of Research* 5:711–732. ⟨44⟩

Wertheim, G. K., 1954. Studies of the electrical potential between Key West, Florida, and Havana, Cuba. *Transactions of the American Geophysical Union* 35:872–882. ⟨116⟩

White, W. B., 1977. Annual forcing of baroclinic long waves in the tropical North Pacific Ocean. *Journal of Physical Oceanography* 7:50–61. ⟨357⟩

White, W. B., and J. P. McCreary, 1976. On the formation of the Kuroshio meander and its relationship to the large-scale ocean circulation. *Deep-Sea Research* 23:33–47. ⟨540, 541⟩

White, W. B., and A. E. Walker, 1974. Time and depth scales of anomalous sub-surface temperature at ocean weather stations P, N, and V in the North Pacific. *Journal of Geophysical Research* 79:4517–4522. ⟨355, 356⟩

Whitehead, J. A., Jr., 1975. Mean flow generated by circulation on a β-plane: An analogy with the moving flame experiment. *Tellus* 27:358–364. ⟨474, 533⟩

Whitehead, J. A., A. Leetmaa, and R. A. Knox, 1974. Rotating hydraulics of strait and sill flow. *Geophysical Fluid Dynamics* 6:101–125. ⟨473⟩

Whitham, G. B., 1974. *Linear and Nonlinear Waves.* Wiley, Interscience, New York, 636 pp. ⟨345, 514⟩

Wiebe, P. H., E. M. Hulburt, E. J. Carpenter, A. E. Jahn, G. P. Knapp, III, S. H. Boyd, P. B. Ortner, and J. L. Cox, 1976. Gulf Stream cold core rings: Large-scale interaction sites for open ocean plankton communities. *Deep-Sea Research* 23:695–710. ⟨379⟩

Wieringa, J., 1974. Comparison of three methods for determining strong wind stress over Lake Flevo. *Boundary-Layer Meteorology* 7:3–19. ⟨489⟩

Willebrand, J., 1978. Temporal and spatial scales of the wind field over the North Pacific and North Atlantic. *Journal of Physical Oceanography* 8:1080–1094. ⟨216, 218, 229, 348⟩

Willebrand, J., S. G. H. Philander, and R. C. Pacanowski, 1980. The oceanic response to large-scale atmospheric disturbances. *Journal of Physical Oceanography*. ⟨230⟩

Williams, A. J., 3rd, 1974a. Salt fingers observed in the Mediterranean outflow. *Science* 185:941–943. ⟨252⟩

Williams, A. J., 3rd, 1974b. Salt fingers in the ocean: A short history. *Naval Research Reviews* 27:10, 27–33. ⟨419, 420⟩

Williams, A. J., 3rd, 1975. Images of ocean microstructure. *Deep-Sea Research* 22:811–829. ⟨252, 419, 420⟩

Williams, G. O., 1976. Repeated profiling of microstructure lenses with a midwater float. *Journal of Physical Oceanography* 6:281–292. ⟨258, 419⟩

Williams, G. P., 1978. Planetary circulations: 1. Barotropic representation of Jovian and terrestrial turbulence. *Journal of the Atmospheric Sciences* 35:1399–1426. ⟨183⟩

Williams, G. P., 1979. Planetary circulations: 2. The Jovian quasi-geostrophic regime. *Journal of the Atmospheric Sciences* 36:932–968. ⟨183⟩

Williams, L. P., 1965. *Michael Faraday*. Basic Books, New York, 540 pp. ⟨396⟩

Williams, R. G., and F. A. Godshall, 1977. Summarization and interpretation of historical physical oceanographic and meteorological information for the Mid-Atlantic Region. Final report to the Bureau of Land Management, U.S. Department of Interior NOAA/EDS, Washington, D.C., 296 pp. ⟨215⟩

Wilson, B. W., 1960. Note on surface wind stress over water at low and high wind speeds. *Journal of Geophysical Research* 65:3377–3382. ⟨488⟩

Wilson, K. G., 1979. Problems in physics with many scales of length. *Scientific American* 241:158–179. ⟨509⟩

Wilson, R. E., and A. Okubo, 1978. Longitudinal dispersion in a partially stratified estuary. *Journal of Marine Research* 36:427–447. ⟨206⟩

Wimbush, M., 1972. Tidal movements of the deep sea. *Underwater Journal and Information Bulletin* 4:239–248. ⟨425⟩

Wimbush, M., 1977. An inexpensive sea-floor precision pressure recorder. *Deep-Sea Research* 24:493–497. ⟨425⟩

Wimbush, M., and W. Munk, 1970. The benthic boundary layer. In *The Sea: Ideas and Observations on Progress in the Study of the Seas, 4: New Concepts of Sea Floor Evolution*, A. E. Maxwell ed., *Part I. General Observations*, Wiley, Interscience, New York, pp. 731–758. ⟨258⟩

Winant, C., 1979. Coastal current observations. *Reviews of Geophysics and Space Physics* 7:89–98. ⟨332⟩

Wiseman, W. J., R. M. Crosby, and D. W. Pritchard, 1972. A three-dimensional current meter for estuarine applications. *Journal of Marine Research* 30:153–158. ⟨205⟩

Witting, T., 1930. Determinations of current, direct and indirect. *Rapports et Procès-Verbaux des Réunions, Conseil Permanent International pour l'Exploration de la Mer* 64:8–18. ⟨201, 203⟩

Wolgemuth, K., and W. S. Broecker, 1970. Barium in seawater. *Earth and Planetary Science Letters* 8:372–378. ⟨444⟩

Woods, J. D., 1968a. Wave-induced shear instability in the summer thermocline. *Journal of Fluid Mechanics* 32:791–800. ⟨249⟩

Woods, J. D., 1968b. An investigation of some physical processes associated with the vertical flow of heat through the upper ocean. *Meteorological Magazine* 97:65–72. ⟨267⟩

Woods, J. D., 1969. On Richardson's number as a criterion for laminar-turbulent-laminar transition in the ocean and atmosphere. *Radio Science* 4:1289–1298. ⟨423⟩

Woods, J. D., and R. L. Wiley, 1972. Billow turbulence and ocean microstructure. *Deep-Sea Research* 19:87–121. ⟨249⟩

Woodward, W., C. N. K. Mooers, and K. Jensen, eds., 1978. Proceedings of a Working Conference on Current Measurements. *College of Marine Studies, University of Delaware, Technical Report DEL-SG-3-78*, Newark, Delaware, 372 pp. ⟨403⟩

Wooster, W. S., and O. Guillen, 1974. Characteristics of El Niño in 1972. *Journal of Marine Research* 32:387–404. ⟨379⟩

Wooster, W. S., and G. H. Volkmann, 1960. Indications of deep Pacific circulation from the distribution of properties at five kilometers. *Journal of Geophysical Research* 65:1239–1249. ⟨36⟩

Worthington, L. V., 1959. The 18° water in the Sargasso Sea. *Deep-Sea Research* 5:297–305. ⟨55, 59, 83, 355⟩

Worthington, L. V., 1962. Evidence for a two gyre circulation system in the North Atlantic. *Deep-Sea Research* 9:51–67. ⟨125, 133⟩

Worthington, L. V., 1969. An attempt to measure the volume transport of Norwegian Sea overflow water through the Denmark Strait. *Frederick C. Fuglister Sixtieth Anniversary Volume, Deep-Sea Research* 16 (Supplement): 421–432. ⟨24⟩

Worthington, L. V., 1970. The Norwegian Sea as a mediterranean basin. *Deep-Sea Research* 17:77–84. ⟨23, 24, 58, 108⟩

Worthington, L. V., 1976. On the North Atlantic Circulation. *The Johns Hopkins Oceanographic Studies* 6: 110 pp. ⟨24, 43, 58, 79, 85, 108, 122, 123, 133, 137, 229, 355, 373, 397, 532, 533, 534⟩

Worthington, L. V., 1977a. The case for near-zero production of Antarctic Bottom Water. *Geochimica et Cosmochimica Acta* 41:1001–1006. ⟨59⟩

Worthington, L. V., 1977b. Intensification of the Gulf Stream after the winter of 1976–77. *Nature* 270:415–417. ⟨355⟩

Worthington, L. V., and H. Kawai, 1972. Comparison between deep sections across the Kuroshio and the Florida Current and Gulf Stream. In *Kuroshio: Its Physical Aspects*, H. Stommel and K. Yoshida, eds., University of Tokyo Press, pp. 371–385. ⟨36, 121⟩

Worthington, L. V., and W. G. Metcalf, 1961. The relationship between potential temperature and salinity in deep Atlantic water. *Rapports et Procès-Verbaux des Réunions, Conseil Permanent International pour l'Exploration de la Mer* 149:122–128. ⟨25⟩

Worthington, L. V., and G. H. Volkmann, 165. The volume transport of the Norwegian Sea overflow water in the North Atlantic. *Deep-Sea Research* 12:667–676. ⟨23, 43⟩

Worthington, L. V., and W. R. Wright, 1970. North Atlantic Ocean Atlas of Potential Temperature and Salinity in the Deep Water Including Temperature, Salinity, and Oxygen Profiles from the Erika Dan Cruise of 1962. *Woods Hole Oceanographic Institution Atlas Series* 2: 24 pp. and 58 plates. ⟨25, 27, 28, 43, 54, 93⟩

Wright, W. R., 1970. Northward transport of Antarctic Bottom Water in the western Atlantic Ocean. *Deep-Sea Research* 17:367–371. ⟨28⟩

Wright, W. R., 1972. Northern sources of energy for the deep Atlantic. *Deep-Sea Research* 19:865–877. ⟨25⟩

Wright, W. R., and C. E. Parker, 1976. A volumetric temperature/salinity census for the Middle Atlantic Bight. *Limnology and Oceanography* 21:563–571. ⟨233⟩

Wright, W. R., and L. V. Worthington, 1970. The Water Masses of the North Atlantic Ocean; a Volumetric Census of Temperature and Salinity. *Serial Atlas of the Marine Environment,* Folio 19, American Geographical Society, New York, 8 pp. and 7 plates. ⟨42, 43, 44, 46, 47, 53, 54, 55, 58⟩

Wu, J., 1969a. Mixed region collapse with internal wave generation in a density-stratified medium. *Journal of Fluid Mechanics* 35:531–544. ⟨253⟩

Wu, J., 1969b. Wind stress and surface roughness at air-sea interface. *Journal of Geophysical Research* 74:444–455. ⟨488⟩

Wunsch, C., 1967. The long-period tides. *Reviews of Geophysics* 5:447–476. ⟨339⟩

Wunsch, C., 1969. Progressive internal waves on slopes. *Journal of Fluid Mechanics* 35:131–144. ⟨261, 270, 316, 332⟩

Wunsch, C., 1970. On oceanic boundary mixing. *Deep-Sea Research* 17:293–301. ⟨260, 261, 334⟩

Wunsch, C., 1972a. Temperature microstructure on the Bermuda slope with application to the mean flow. *Tellus* 24:350–367. ⟨261⟩

Wunsch, C., 1972b. The spectrum from two years to two minutes of temperature in the main thermocline at Bermuda. *Deep-Sea Research* 19:577–593. ⟨270, 358, 363, 364⟩

Wunsch, C., 1972c. Bermuda sea level in relation to tides, weather, and baroclinic fluctuations. *Reviews of Geophysics and Space Physics* 10:1–49. ⟨320, 321, 323, 351, 353, 357⟩

Wunsch, C., 1975. Internal tides in the ocean. *Reviews of Geophysics and Space Physics* 13:167–182. ⟨293, 331, 337⟩

Wunsch, C., 1976. Geographical variability of the internal wave field; a search for sources and sinks. *Journal of Physical Oceanography* 6:471–485. ⟨286, 339, 340⟩

Wunsch, C., 1977. Response of an equatorial ocean to a periodic monsoon. *Journal of Physical Oceanography* 7:497–511. ⟨194, 524⟩

Wunsch, C., 1978a. The general circulation of the North Atlantic west of 50°W determined from inverse methods. *Reviews of Geophysics and Space Physics* 16:583–620. ⟨91, 92, 343, 373, 397, 534⟩

Wunsch, C., 1978b. Observations of equatorially trapped waves in the ocean: A review prepared for equatorial workshop, July, 1977. In *Review Papers of Equatorial Oceanography—FINE Workshop Proceedings,* Nova/N.Y.I.T. University Press, Fort Lauderdale, 37 pp. ⟨186⟩

Wunsch, C., and J. Dahlen, 1970. Preliminary results of internal wave measurements in the main thermocline at Bermuda. *Journal of Geophysical Research* 75:5899–5908. ⟨364⟩

Wunsch, C., and J. Dahlen, 1974. A moored temperature and pressure recorder. *Deep-Sea Research* 21:145–154. ⟨406⟩

Wunsch, C., and E. M. Gaposchkin, 1980. On using satellite altimetry to determine the general circulation of the oceans with application to geoid improvement. *Reviews of Geophysics and Space Physics.* ⟨374, 428⟩

Wunsch, C., and A. E. Gill, 1976. Observations of equatorially trapped waves in Pacific sea level variations. *Deep-Sea Research* 23:371–390. ⟨186, 191, 307, 308, 347, 352⟩

Wunsch, C., D. V. Hansen, and B. D. Zetler, 1969. Fluctuations of the Florida Current inferred from sea level records. *Frederick C. Fuglister Sixtieth Anniversary Volume, Deep-Sea Research* 16 (Supplement): 447–470. ⟨117⟩

Wunsch, C., and R. Hendry, 1972. Array measurements of the bottom boundary layer and the internal wave field on the continental slope. *Geophysical Fluid Dynamics* 4:101–145. ⟨332, 334, 337⟩

Wunsch, C., and S. Webb, 1979. The climatology of deep ocean internal waves. *Journal of Physical Oceanography* 9:235–243. ⟨286, 291⟩

Wunsch, C., and M. Wimbush, 1977. Simultaneous pressure, velocity and temperature measurements in the Florida Straits. *Journal of Marine Research* 35:75–104. ⟨117, 121, 347, 358, 372⟩

Wüst, G., 1924. Florida—und Antillenstrom: eine hydrodynamische Untersuchung. *Veröffentlichungen des Instituts für Meereskunde an der Universität Berlin, Neue Folge. A. Geographisch-naturwissenschaftliche Reihe,* 12, 48 pp. ⟨113⟩

Wüst, G., 1928. Der Ursprung der atlantischen Tiefenwasser. *Zeitschrift der Gesellschaft für Erdkunde zu Berlin. Sonderband zur Hundertjahrfeier der Gesellschaft*: 506–534. ⟨11⟩

Wüst, G., 1929. Schichtung und Tiefenzirkulation des Pazifischen Ozeans. *Veröffentlichungen des Instituts für Meereskunde an der Universität Berlin, Neue Folge. A. Geographisch-naturwissenschaftliche Reihe,* 20, 63 pp. ⟨10⟩

Wüst, G., 1933. Schichtung und Zirkulation des Atlantischen Ozeans. Das Bodenwasser und die Gliederung der Atlantischen Tiefsee. In *Wissenschaftliche Ergebnisse der Deutschen Atlantischen Expedition auf dem Forschungs— und Vermessungsschiff "Meteor" 1925–1927,* 6: 1st Part, 1, 106 pp. (*Bottom Water and the Distribution of the Deep Water of the Atlantic,* M. Slessers, translator, B. E. Olson, ed., 1967, U.S. Naval Oceanographic Office, Washington, D.C., 145 pp.; ⟨11, 29, 72, 82⟩

Wüst, G., 1934. Anzeichen von Beziehungen zwischen Bodenstrom und Relief in der Teifsee des Indischen Ozeans. *Die Naturwissenschaften* 22:241–244. ⟨10, 32⟩

Wüst, G., 1935. Schichtung und Zirkulation des Atlantischen Ozeans. Die Stratosphäre. In *Wissenschaftliche Ergebnisse der Deutschen Atlantischen Expedition auf dem Forschungs—und Vermessungsschiff "Meteor" 1925–1927,* 6: 1st Part, 2, 180 pp. (*The Stratosphere of the Atlantic Ocean,* W. J. Emery, ed., 1978, Amerind, New Delhi, 112 pp.) ⟨11, 22, 25, 27, 43, 57, 70, 72, 74, 77, 78, 79, 82, 84, 93, 106⟩

Wüst, G., 1938. Bodentemperatur und Bodenstrom in der atlantischen, indischen und pazifischen Tiefsee. *Gerlands Beiträge zur Geophysik* 54:1–8. ⟨10, 11, 28⟩

Wüst, G., 1943. Der subarktische Bodenstrom in der westatlantischen Mulde. *Annalen der Hydrographie und Maritimen Meteorologie* 71:249–255. ⟨72⟩

Wüst, G., 1951. Über die Fernwirkungen antarktischer und nordatlantischer Wassermassen in den Tiefen des Weltmeeres. *Naturwissenschaftliche Rundschau, Jahrgang 1951*: 3, 97–108. ⟨106⟩

Wüst, G., 1955. Stromgeschwindigkeiten im Tiefen—und Bodenwasser des Atlantischen Ozeans auf Grund dynamischer Berechnung der *Meteor*-Profile der Deutschen Atlantischen Expedition 1925/27. *Papers in Marine Biology and Oceanography, Deep-Sea Research* 3 (Supplement): 373–397. ⟨13⟩

Wüst, G., 1964. The major deep-sea expeditions and research vessels 1873–1960—a contribution to the history of oceanography. *Progress in Oceanography* 2:1–52. ⟨398⟩

Wüst, G., and A. Defant, 1936. Atlas zur Schichtung und Zirkulation des Atlantischen Ozeans. Schnitte und Karten von Temperatur, Salzgehalt und Dichte. In *Wissenschaftliche Ergebnisse der Deutschen Atlantischen Expedition auf dem Forschungs—und Vermessungsschiff "Meteor" 1925–1927*, 6: Atlas, 103 plates. ⟨6, 7, 28, 43, 44, 75, 348⟩

Wyrtki, K., 1961a. The flow of water into the deep sea basins of the western South Pacific. *Australian Journal of Marine and Freshwater Research* 12:1–16. ⟨34, 35⟩

Wyrtki, K., 1961b. The thermohaline circulation in relation to the general circulation in the oceans. *Deep-Sea Research* 8:39–64. ⟨81, 449⟩

Wyrtki, K., 1962. The oxygen minima in relation to ocean circulation. *Deep-Sea Research* 9:11–23. ⟨434, 449⟩

Wyrtki, K., 1971. *Oceanographic Atlas of the International Indian Ocean Expedition*. National Science Foundation, Washington, D.C., 531 pp. ⟨25, 26, 29, 32, 33, 44, 55, 83, 85⟩

Wyrtki, K., 1973a. An equatorial jet in the Indian Ocean. *Science* 181:262–264. ⟨186, 192⟩

Wyrtki, K., 1973b. Teleconnections in the equatorial Pacific Ocean. *Science* 180:66–68. ⟨398⟩

Wyrtki, K., 1974. Equatorial currents in the Pacific 1950 to 1970 and their relation to the trade winds. *Journal of Physical Oceanography* 4:372–380. ⟨351, 357⟩

Wyrtki, K., 1975a. Fluctuations of the dynamic topography in the Pacific Ocean. *Journal of Physical Oceanography* 5:450–459. ⟨85, 91, 108, 351, 357⟩

Wyrtki, K., 1975b. El Niño—the dynamic response of the Pacific Ocean to atmospheric forcing. *Journal of Physical Oceanography* 5:572–584. ⟨192⟩

Wyrtki, K., 1978. Lateral oscillations of the Pacific Equatorial Countercurrent. *Journal of Physical Oceanography* 8:530–532. ⟨427⟩

Wyrtki, K., 1979. Sea level variations: monitoring the breadth of the Pacific. *EOS, Transactions of the American Geophysical Union* 60:25–27. ⟨351, 398⟩

Wyrtki, K., L. Magaard, and J. Hager, 1976. Eddy energy in the oceans. *Journal of Geophysical Research* 81:2641–2646. ⟨71, 360, 361, 430⟩

Wyrtki, K., D. McLain, and W. Patzert, 1977. Variability of the thermal structure in the central equatorial Pacific Ocean. *Hawaii Institute of Geophysics Report, HIG-77-1*, University of Hawaii, Honolulu, 75 pp. ⟨184⟩

Wyrtki, K., and G. Meyers, 1975. The trade wind field over the Pacific Ocean. Part I, the mean field and the mean annual variation. *Hawaii Institute of Geophysics Report, HIG-75-1*, University of Hawaii, Honolulu, 26 pp. ⟨186, 187, 188⟩

Yoshida, K., 1959. A theory of the Cromwell current and of the equatorial upwelling. *Journal of the Oceanographical Society of Japan* 15:154–170. ⟨189, 192⟩

Yoshida, K., 1961. Some calculations on the equatorial circulation. *Records of Oceanographic Works in Japan, New Series* 6:1, 101–105. ⟨83⟩

Yoshida, K., and T. Kidokoro, 1967. A subtropical countercurrent (II)—a prediction of eastward flows at lower subtropical latitudes. *Journal of the Oceanographical Society of Japan* 23:231–246. ⟨85⟩

Young, R. E., and J. B. Pollack, 1977. A three-dimensional model of dynamical processes in the Venus atmosphere. *Journal of the Atmospheric Sciences* 34:1315–1351. ⟨506⟩

Zahel, W., 1970. Die Reproduktion gezeitenbedingter Bewegungsvorgänge im Weltozean mittels des hydrodynamisch-numerischen Verfahrens. *Mitteilungen des Instituts für Meereskunde der Universität Hamburg*, 17, 51 pp. ⟨329⟩

Zeilon, N., 1911. On tidal boundary-waves and related hydrodynamical problems. *Kungliga Svenska Vetenskapsakademiens Handlingar, Ny Földj* 47:4, 46 pp. ⟨331⟩

Zeilon, N., 1912. On the seiches of the Gullmar Fjord. With an introduction on the theory of seiches in branched bays. *Svenska Hydrografisk-Biologiska Kommissionens Skrifter* 5: 18 pp. ⟨331⟩

Zeilon, N., 1934. Experiments on boundary tides. A preliminary report. *Göteborgs Kungliga Vetenskaps—och Vitterhets-Samhälles Handlingar, Femten Följden, Serien B* 3:10, 8 pp. ⟨331⟩

Zeskind, L. M., 1926. Instructions for tidal current surveys. *U.S. Coast and Geodetic Survey Special Publication* No. 124, 48 pp. ⟨201⟩

Zeskind, L. M., and E. A. Le Lacheur, 1926. Tides and currents in Delaware Bay and River. *U.S. Coast and Geodetic Survey Special Publication* No. 123, 122 pp. ⟨200⟩

Zetler, B., 1971. Radiational ocean tides along the coasts of the United States. *Journal of Physical Oceanography* 1:34–38. ⟨294⟩

Zetler, B., 1978. Tide predictions. In *Geophysical Predictions*, H. E. Landsberg, panel chairman, National Academy of Sciences, Washington, D.C., pp. 166–177. ⟨318⟩

Zilitinkevich, S. S., 1969. On the computation of the basic parameters of the interaction between the atmosphere and the ocean. *Tellus* 21:17–24. ⟨502⟩

Zilitinkevich, S. S., 1970. *Dinamika Pogranichnogo Sloya Atmosfery (Dynamics of the Atmospheric Boundary Layer)*. Gidrometeorologicheskoe Izdatel'stvo, Leningrad, 291 pp. ⟨502⟩

Zimmerman, H. B., 1971. Bottom currents on the New England continental rise. *Journal of Geophysical Research* 76:5865–5876. ⟨27, 121⟩

Zimmerman, J. T. F., 1978. Topographic generation of residual circulation by oscillatory (tidal) currents. *Geophysical and Astrophysical Fluid Dynamics* 11:35–47. ⟨206⟩

Zimmerman, J. T. F., 1979. On the Euler-Lagrange transformation and the Stokes' drift in the presence of oscillatory and residual currents. *Deep-Sea Research* 26:505–520. ⟨206⟩

Zubkovskii, S. L., and T. K. Kravchenko, 1967. Pryamye izmereniya nekotorykh kharakteristik atmosfernoi turbulentnosti v privodnom sloe. *Fizika Atmosfery i Okeana, Izvestiya Akademii Nauk SSSR* 3:127–135. (Direct measurements of some characteristics of atmospheric turbulence in the near-water layer. *Izvestiya, Academy of Sciences USSR, Atmospheric and Oceanic Physics* 3:73–77.) ⟨489⟩

Index

Names are entered in the index only when not attached to a citation; for cited authors consult the reference list.

Abbott of St. Albans, 318
Abyssal mixing. *See* Mixing
Advective instability. *See* Internal waves
Agassiz, Alexander, 209
Agulhas Current, 82
Air Mass Transformation Experiment (AMTEX), 503
Air–sea interaction, 3, 449–450, 482–503
  boundary layers, 497–503
  boundary layer winds, 500–503
  buoyancy fluxes, 484–485, 498–500
  convective plumes, 498–500
  drag coefficients, 485, 487–490
  flux gradients, 485–486
  gas-exchange, 440
  humidity, 486
  Langmuir circulation, 495
  near-surface profiles, 483–486
  roughness length, 486–487
  surface waves, 491–495
  surface wave spectrum, 491–493
  transfer coefficients, 485–490
Airy solutions. *See* Equatorially trapped waves; Internal waves
Aleutian Trench, 36
American Miscellaneous Society, Albatross Award, xviii
Anelastic equation, 546
Annual variability, 355–357
  on shelf, 223–226, 230, 232
Antarctic Bottom Water, 11, 15–22, 54, 401, 453–456
Antarctic Circumpolar Current. *See* Circumpolar (Antarctic) Current
Antarctic Ocean. *See* Southern Ocean
Arabian Sea, 10, 21
Argentine Basin, 17
Argon (Ar), 447
Aries (vessel), xvi, xxv
Arons, Arnold B., xx
Array design, 431–433
Arthur, Robert S., xxvii
Atlantic Ocean. *See also* General Circulation; Gulf Stream; North Atlantic Ocean; South Atlantic Ocean transports
  boundary currents, 26–29
  early observations of, 7–10
  water mass properties of, 42–69, 71–72, 74–111
Atlantic Tradewind Experiment (ATEX), 503
Atlantis (vessel), xvi, 211, 398
Atmospheric forcing. *See* Meteorological forcing
Atmospheric general circulation, 396, 503–504

Baffin (vessel), 137
Baird (vessel), 398
Baltic Sea, 202
Barbados Oceanographic and Meteorological Experiment (BOMEX), 503
Barber, Norman F., 482
Barium (Ba), 444–445
Baroclinic instability, 141, 169–182, 372–373, 506, 508, 529–532, 545–546
  finite amplitude, 173–177
  linear theory, 169–173

Baroclinic waves. *See* Internal waves; Long waves; Rossby waves
Barotropic instability, 529–532
Batchelor, George K., 482
Battery properties, 401
Bénard convection, 384
Bermuda, xx, 261, 270, 287, 320–321, 348, 359, 364
Bernoulli equation, 117
Bernoulli, Jean, xiv–xv
Beryllium, 447–448
Beta-effect, xv–xvi
Beta-plane, 143–147, 301–302, 344
Beta-spiral, xxi, xxiii, 91, 397
Bigelow, Henry Bryant, xxiv, 209–212
Biology, 3, 376–383. *See also* Property distributions
 ecosystem dynamics, 379–381
 fisheries, 376
 of rings, 379
 spatial variability of, 376–378
 temporal variability of, 377–383
 year class strength, 377
Bjerknes, Carl A., 463–464
Bjerknes, Vilhelm F. K., xxiv
Blake–Bahama Outer Ridge, 28, 121
Blanchard, Duncan C., xx
Bottom trapped waves, 166, 168–169, 311, 315
Bottom water formation, 15–25, 450, 453–456
 near Antarctica, 15–22
 in North Atlantic, 22–25
Boundary layer. *See* Air–sea interaction; General circulation; Gulf Stream
Boundary mixing. *See* Mixing
Boussinesq approximation, 142, 385, 547
Boussinesq equations (convection), 385–386
Box models, 448–450
Bradley, Albert, 413
Bransfield Strait, 15–16
Brown, Neil, 415–419
Brown, Sanborn, 40
Buchanan, John Young, 11
Budyko, Mikhail Ivanovich, 482
Bullard, (Sir) Edward C., xxv
Buoyancy layers, 260
Burger-scale motion, 513, 516–517

Cabbeling, 257
California Current, 342
California, Gulf of, 324–326
Carbon dioxide (CO_2), 434, 439, 447, 456
Carbon-14 (^{14}C), 246, 434–440, 445, 450–460
 normalization, 439
Carbon, total dissolved, 434–440
Caribbean Sea, 84, 475
 model of, 475
Carnegie (vessel), 10, 78–79, 85
 errors in data, 78–79
Carrier, George F., xv, xix
Cayman Basin, 413
Central Indian Basin and Ridge, 10, 32, 36
Cesium (Cs), 446
Challenger expedition, xxvii, 9, 11, 348–349, 429
Charney, Jule G., xv, xx–xxi, xxvii, 482
Chemical tracers, 434–460
 models of distribution, 14–15, 448–450
Chesapeake Bay, 206
Chesapeake Bay Institute, 233
Chlorinity, 418

Circulation. *See* General circulation
Circumpolar (Antarctic) Current, 12, 77, 84, 355, 432–433, 523
Circumpolar deep water, 15
Climate/Long Range Investigation Mapping and Predictions (CLIMAP), 348, 350–351
Clipperton Fracture Zone, 36
Closure schemes, 178, 543
Cnoidal waves, 514, 545
Conductivity. *See* Property distributions
Continental shelves. *See* Shelves, continental
Convection, 384–393, 482. *See also* Mixing processes
 amplitude of, 384–395
 Bénard, 384
 Boussinesq description, 385–386
 deep, 21, 384
 finite amplitude, 386
 heat flux in, 384
 marginal instability in, 386, 391
 mean field theory, 387–389
 in Mediterranean Sea (*see* Mediterranean Sea)
 in ocean, 384
 power integrals, 386
 Rayleigh theory, 384, 386–387
 in shear, 389
 subcritical instability in, 386
 turbulent, 384, 387–393
 turbulent upper bounds, 387–388
 turbulent in shear, 389
 in Weddell Sea (*see* Weddell Sea)
Convective instability. *See* Internal waves
Copepods, 381
Corals, 439
Coral Sea, tides of, 320–321
Corelayer ("*Kernschicht*") method, 43, 57, 75, 79
COSMOS (Complementary symmetry metal oxide semiconductor), 400–401
Cox number, 281
Crease, James, xvi, 411, 420, 430
Critical layers. *See* Internal waves; Wave–mean flow interaction
Croll, James, 8
C-shape (in general circulation), 85, 89–91
Curieuse, La (vessel), xxvi

Dahlen, John M., 399
Dana expedition, 10, 74
Deacon, (Sir) George E. R., 411
Deep water, sources of, 6–26, 107, 457–459
 source models, 17, 19, 20–21, 58
Defant, Albert, xvi, xxiv–xxv
Delaware estuary, 205
Denmark Strait, 408. *See also* Norwegian Sea
 overflow, 23–24, 72, 83
 water (DSW), 83, 453
Deuterium, 448
Deutschland (vessel)
 in ice, 17
 stations, 9
Diagnostic calculations, 85–90, 142
Diffusion. *See* Fickian diffusion; Mixing
Discovery (vessel), xvi, 10
Dissipation (turbulent), 282–284
Double-diffusive convection, instability. *See* Mixing; Salt fingers
Drag coefficients, 485, 487–490
Drag reduction, 391

Drake Passage, 363, 371–372, 425, 430. *See also* Southern Ocean
Draper, Charles S., Laboratory, 399, 406
Dye dispersion, 377–378
Dynamic height (steric height/depth), 73–74, 76, 85, 87–88, 104–105, 107–108, 138
Dynamical method. *See* Geostrophy

Eady, Eric T., 482
Eastern boundary currents, 151–152, 154
East Pacific Rise, 35
Eddies, 313–314, 342–343, 358–373, 430–432, 507–508. *See also* Baroclinic instability; Mixing
 in Arctic Sea, 371
 diffusion, xix
 eddy–mean flow interaction, 507
 Gulf Stream, 117–119, 373–374, 507
 heat flux by, 372–373, 384
 modal structure, 371–372
 Reynolds stresses, 372, 507–508
 on shelf, 212
 Soviet study of, 359
 spin-down, 527–529
 topographic interaction, 370–371, 373
 western boundary currents, 373
Eddy coefficients, 342, 345
Eddy-containing band, 359, 370, 373
Edge waves, 307, 309–317. *See also* Shelf waves
Ekman dynamics, 147–149, 506. *See also* Spin-up/down
 flux, 20, 148
 laboratory models, 463–464
 layer, 140–141, 147–148, 463–471, 525–526
 pumping, xxi, 506, 520, 527
 turbulent, 149
Ekman, V. Walfrid, xxiv
Ellis, Henry, 7
El Niño, 192–193, 379, 397–398
Eltanin (vessel), 43
Enstrophy, 175, 346, 529, 543–544. *See also* Potential vorticity
Entrainment. *See* Mixing
Equatorial circulation, 83, 184–196. *See also* General circulation; North Equatorial Countercurrent; North Equatorial Current; South Equatorial Countercurrent; Undercurrents, equatorial
 cross-equatorial flow, 109
 deep jets, 194
 theory of, 185, 188–195
 upwelling, 192, 456
 variability, 184
Equatorially trapped waves, 186, 191–193, 295, 304–308, 513, 527
 Airy solution, 306–307
 Kelvin wave, 191–192, 305, 527
 nonlinear, 516
Erika Dan (vessel), 43
Estuaries, 198–207. *See also* Chesapeake Bay; Maine, Gulf of; Puget Sound
 circulation, 199–204
 classification of, 202
 current measurements in, 200–201
 flushing of, 202
 mixing in, 203–240
 salt balance, 205
 salt-wedge, 203, 205
 sea-level fluctuations in, 206
 theories of, 203–207

 tidal flow in, 200
 tidal mixing in, 206
 turbulence in, 205–206
Ewing, W. Maurice, xiv
Experiment design, 429–433

Faller, Alan J., xvii, xxi
Faraday, Michael, 396
Fickian diffusion, 239
Fine structure, 257–258, 268, 279–284, 418
 intrusive, 279–282
Finland, Gulf of, 201
Fish population, 376–379
Flamsteed, John, 318
Floating instrument platform (FLIP), 265, 398, 428
Floats. *See* Instruments
Florida Straits. *See* Gulf Stream
Fluorometric techniques, 376
Fofonoff, Nick P., xvi, xx
Fofonoff's rule, 285
Food web, 376–383
Form-drag instability, 534–539
 on topography, 538–539
Frantz, David, xvi
Freon, 447
Friction, lateral, 151–152. *See also* Ekman dynamics; Tides; Wind-driven circulation
Friedman–Keller equation, 484
Fronts and biological productivity, 377–378
Fuglister, Frederick C., xxi

GARP. *See* Global Atmospheric Research Program
Garrett–Munk spectrum. *See* Internal waves
Gas exchange, 440
GATE. *See* Global Atmospheric Research Program
GEK. *See* Instruments
General circulation, 6–41, 70–111, 140–183. *See also* Gulf Stream; Name of ocean
 Antarctic sources, 8–11, 15–22
 Arabian Sea, 9–10
 and chemical tracers, 434–460
 C-shape in, 85, 89–91
 deep, 6–41
 dynamics of, 140–183
 eastern boundary currents, 151–152, 154
 equations for, 142–147
 equatorial, 83, 184–196
 history of, 7–11, 70–79
 Indian Ocean, 29–34
 isopycnal analysis of, 81–85
 Mediterranean outflow (*see* Mediterranean Sea)
 meridional circulation, 75
 mid-depth, 70–111
 models (*see* Laboratory models)
 numerical models, 85–90, 140–141, 155–157, 175–182, 507–508
  diagnostic, 89–90, 111
 Persian Gulf source, 10, 25–26
 Red Sea source, 10, 25
 sinking regions, 15–25
 thermocline circulation, 141, 158–163
 topographic effects on, 151–152
 tropical, 185–186, 188–190
 undercurrents, western boundary, 11, 26–38, 79, 121–122
  discovery of, xvi, xxi, 11, 13
 upwelling in, 162–163, 358, 456
 variability, 71–72

western boundary currents, 12, 14, 26–38, 79
  models of, 37–38, 140–141, 149–151, 153–157, 531–532
  separation of, 141, 157–158
 wind-driven circulation, 71, 79, 140–142, 149–157
General circulation of atmosphere, 396, 505–506
Geochemical Ocean Sections Studies (GEOSECS), xvii, xxii, 435–460
Geochemical tracers, 434–460
 modeling, 448–449
Geomagnetism, 384
Geophysical Fluid Dynamics Seminar, xxi–xxii
Georges Bank, 209
GEOS-3, 324
GEOSECS. *See* Geochemical Ocean Sections Studies
Geostrophic turbulence, 509, 543–544, 546
Geostrophy, 72–73, 82, 108–109, 140, 233, 397, 508–509
 compared to numerical models, 89–91
Gibbs Fracture Zone, 23
Global Atmospheric Research Program (GARP), Atlantic Tropical Experiment (GATE), 191, 193, 503, 544
Glomar Challenger (vessel), 398
Goody, Richard, xxi
Grampus (vessel), 209
Grand Banks, circulation near, 85, 133, 135, 138
Gravity currents, 260
Guinea, Gulf of, 191
 upwelling in, 193
Gulf Stream, xiv–xv, xx–xxi, xxv, 2, 71, 78, 81, 85, 112–139, 506. *See also* General circulation
 atmospheric forcing of, 117, 119–121, 125, 131
 branching of, 84–85, 138
  decay, 135
  deep flows, 26–29, 127
  "Charleston Bump," 121
 eddy generation, 359
 eddy–mean flow interaction, 117–119, 127, 129–131, 134
 Florida Current, 113–122
 heat flux, 137
 infrared images of, 113–115, 124, 135–136
 instability of, 119–121, 529
 laboratory models of, 465–466
 meanders, 123–125, 529, 533–534, 545
 North Atlantic Current, 113, 133–138
 numerical models of, 132–133
 at Onslow Bay, 118–119
 oxygen distribution, 137
 path models, 126–127
 potential vorticity distribution, 121–122
 pressure gradient in, 118–119
 radiation from, 524–525, 531
 recirculation, 79, 123, 131, 397, 533–534
 rings, 125, 132, 137, 345, 527–529, 534
 rings and biological effects, 379
 rings, models of, 476
 sea-level slope, 114, 116
 separation from coast, 125–126, 139, 141, 157–158
 system names, 113
 shingles, 117
 theories of, 119–120, 127, 129, 132–134
 trajectory models, 126–127
 transport of, 121, 123, 131, 355
 variability of, 116–117, 127–130, 357, 429–430
 undercurrent, 121–122
 and waves, 119–121, 531–532
Gulf stream '60, 126

Hadley cell, 506
Hales, Stephen, 7
Hamon, Bruce V., 415
Hansen, Armauer (vessel), 398
Harvard University, xviii, xxi–xxii
Heat flux, 482
 in ocean, 58, 137–138
Helium-3 (^{3}H), 434, 444–445
Helland-Hansen, Bjørn, xxiv–xxv, 398
Hill, Winfield, 401
Horizon (vessel), 398
Howard, Louis N., xxi, xxiii
Hydrographic data coverage, 44–46
Hypothesis testing, xiii, 429–430

Ice
 bottom water formation, role in, 17, 19–21
 mixing, 252
Iceland–Scotland overflow, 22–24
Indian Ocean, 109. *See also* General circulation
 boundary currents, 29–34
 early observations of, 10
 water mass properties, 42–49, 93–111
Indian Ocean Bubble, The, xxv
Indian Ocean Experiment (INDEX), xxvi
INDOPAC expedition, 57
Inertial subrange, 239
Inertial waves. *See* Internal waves
Infrared measurements, 114–115, 120–121, 124, 135–136, 186, 193, 231, 426–427
Instability. *See* Baroclinic instability; Barotropic instability; Form-drag instability; Kelvin–Helmholtz instability
Institute for Advanced Study, xx–xxi
Institute of Oceanographic Sciences (IOS), 411
Instruments, 342–343, 374, 396–428
 acoustic doppler, 266–267
 acoustic releases, 407–409
 acoustic tomography, 374, 428
 bathythermograph
  BT, 212, 264, 267, 414
  XBT, 414
  XSV, 414–415
 batteries, 401–402
 conductivity measurements, 42, 415–419
 CTD, 267, 402, 414, 416, 418–419, 428
 current meters, 265, 398–399, 403–407
  Aanderaa, 405–406
  errors in, 404–405
  vector averaging, 400, 404
  vector measuring, 404–405
  vertical, 403
 Cyclesonde, 424
 drift bottles, 212–213
 drifters, 426
 electronics, role of, 399–401
 electromagnetic measurements, 422, 425–426
 geomagnetic electrokinetograph (GEK), 117, 422
 history of development, 398–400
 inverted echo sounders, 428
 laser-Raman backscatter, 428
 mooring systems, xvi–xvii, 407–410
 mooring technology, 402–404
 neutrally buoyant floats, 2, 359, 397, 410–413, 430
  SOFAR floats, xvii, 27, 412–413
 optical imagers, 419–420
 pressure gauges, 323, 399–400, 424–425

profilers, 265–266
  air-dropped, 421–422
  salinity, 414–420
  sound, 414
  temperatures, 267, 414
  velocity, 267, 420–424
salinometers, 42, 44, 416
satellites, 265, 323–324, 426–427
  altimeters, 323–324, 374, 427–428
  infrared, 426–427 (see also Infrared measurements)
Savonius rotor, 403–406
SONAR, 428
sound, 264–265, 403, 414, 428
STD, 267, 402, 414–418
submarines, 265
surface drifters, 426
synthetic aperture radar (SAR), 266
tape recorders, 401–402
temperature-gradient recorder, 400
temperature–pressure (TP) recorders, 399, 406–407
thermometers, 7–8, 10
  towed, 265
  use on continental shelf, 209, 212, 213
  White Horse, 421
Interannual band, 373
Interannual variability, 348–355
Internal tides. See Tides
Internal Wave Experiment (IWEX), 265, 285
Internal waves, 264–291
  Airy function representation, 271
  breaking of, 276
  critical layers, 248, 273–275
  discovery of, 264
  dispersion relations, 269–271
  dissipation in, 288–290
  energy balance, 288–289
  and fine structure, 268
  Garrett–Munk (GM) spectrum, 285
  generation of, 288–289
  in continuously stratified ocean, 269–271
  inertial waves, 272–273, 332
  instability, 276–279, 287–290
  laboratory models, 270
  in layered ocean, 268–269
  measurement, 264–265
  and microstructure, 267
  in mixing budget, 246
  mixing from, 247–250, 261, 267, 289–290
  on equator, 272–273, 286
  parametric instability, 248
  reflection of, 270
  resonant interactions of, 272–273, 275–277
  and Richardson number, 268
  saturation, 287–290
  in shear, 273–275
  spectra, 264–265, 285–286, 290, 414
  theory of, 264
  and topography, 270–271, 286, 316–317
  traumata, 279
  turning-point solutions, 271–273
  universality, 285–286
  in waveguide, 271–272
  waveguide solutions, 271–273
International Decade of Ocean Exploration (IDOE), xvii
International Ice Patrol, 233
International Indian Ocean Expedition (IIOE), xxv

International Southern Ocean Studies (ISOS), 14, 430, 432–433
Intertropical Convergence Zone (ITCZ), 187–188, 192
Inverse methods, 91–92, 397, 428
Inverted barometer, 351
Irminger Sea, 72
Iselin, Columbus O'Donnell, xxiv–xxv, 212, 430
Island trapped waves, 358
Isopycnal analysis, 81–85
ISOS. See International Southern Ocean Studies
Isotropic band, 359

Jacobs, Woodrow C., 482
James River, 203, 204
Japan, Sea of, 53
Joint Air–Sea Interaction Experiment (JASIN), 244, 503
Joint North Sea Wave Project (JONSWAP), 491–495
Jordan, William Leighton (Award), xxvi–xxvii

Kattegat, 199
Kelvin–Helmholtz instability, 249, 277–279
  billows, 249
Kelvin, Lord (Sir William Thomson), 318
Kelvin waves, 222, 298–299, 316, 358
Kivu, Lake, 252–253, 256
Koehler, Richard, 401, 404, 421
Kolmogorov inertial subrange, 239
Korteweg–deVries equation, 345, 513–517
Krypton (Kr), 446–447
Kuo, Hsiao-Lan, xxi
Kuroshio, xxiii, 71, 540–543

Laboratory models, 3, 140, 152–153, 201, 203, 248–249, 270, 462–479
  of circulation, 465–473
  history of, 463–464
  Langmuir circulation, 477–478
  of Rossby waves, 474–475
  of rings, 476
  of sills and weirs, 473
  philosophy of, 462–463, 478–479
  sliced cylinder, 469–473
  source–sink, 466–468, 472–473
  spin-up, 476–477
Labrador Current, 208
Labrador Sea, 11, 25, 72, 75, 79, 84
Labrador Sea Water, 453–454
Lagrangian means, 181–182, 345, 473–475
Lakes
  boundary layers in, 149
  convection in, 252–253, 256
  mixing in, 459–460
Langmuir circulation, 477–478, 495
Laplace, Pierre-Simon, xv
Laplace tidal equations, 291, 295–297. See also Long waves; Rossby waves; Tides
  free oscillations of ocean, 317
  numerical solutions, 328–329
Law of the wall, 238, 502
Lead (^{210}Pb), 445
Lee waves, 332, 334, 522–524
Lettau, Heinz Helmut, 466
Lin, Chia-Chiao, xxi
Loch Ness, 244, 291
Longuet-Higgins, Michael S., xix
Long waves, 292–341. See also Rossby waves; Shelf waves; Tides

on plane, 297–299
in stratified fluid, 299–300
Lorenz, Edward N., xxi
Love numbers. *See* Tides
Low-frequency variability, 6–7, 72, 342–374, 509–517. *See also* Annual variability; Eddies; Gulf Stream; Shelf waves; Rossby waves
  history of study, 72, 343–344
  interannual, 348–355
  mesoscale, 357–373
  meteorological forcing of, 346–348, 373
  models of, 347
  spectrum of, 343, 348, 363–372
  and turbulence, 345–346

McCullough, James R., 401
Madagascar Basin, 29
Maine, Gulf of, 201, 209
Malkus, Joanne S., xxi
Malkus, Willem V. R., xxi, xxiii, xxvii, 411
Marginal Seas, 25–26. *See also* Name of sea
Margules, Max, xv
Marine Ecosystem Analysis Program (MESA), 213
Massachusetts Institute of Technology (MIT), xviii, xxi–xxii, xxiv, xxvi, 399, 408, 466
Mediterranean Sea, 71, 413
  convection in, 22, 25, 199
  outflow, 10–11, 14, 252, 371
Méditerranée-Occidentale (MEDOC), xxii–xxiii
Mersey River, 204–205
Mesecar, Roderick S., 402
Mesoscale eddy. *See* Eddies
Meteor expedition, 10, 74–75, 264, 348
Meteorological forcing, 346–348
  in Gulf Stream, 119–121
  on shelf, 214–217
  in tropics, 185–186
Meteorology. *See* Air–sea interaction; Atmospheric general circulation
Michael Sars expedition, 264
Microstructure, 257–258, 267, 279–284, 418–420
Mid-Atlantic Ridge fracture zones, 29
Middle Atlantic Bight, 148–233. *See also* Estuaries; Maine, Gulf of
Mid-Ocean Dynamics Experiments (MODE-0, MODE-1), xvii–xviii, xxii–xxiii, 131, 335, 359, 363, 370, 397, 407–408, 412, 421–423, 428, 430–432, 544–545
MIDPAC expedition, 400
Mixed layers, models of, 240–243, 385
Mixing, 3, 84, 203, 236–262, 267, 279, 337, 339, 449. *See also* Fickian diffusion; Internal waves; Kelvin–Helmholtz instability; Kolmogorov inertial subrange; Law of the wall; Mixed layers; Monin–Obukhov length; Ozmidov scale; Richardson number; Thermocline
  in Antarctic, 15–22, 257
  near bottom, 258–262
  in circulation, 84
  coefficients, 236, 237, 256
  by convection, 237, 243–245
  double diffusive instability, 250–257
  in estuaries, 240
  with ice, 252
  in ocean interior, 245–258
  in Red Sea, 252
  Reynolds analogy, 238
  in shear flows, 238
  and stirring, 267

  vertical, 459–460
  in Weddell Sea (*see* Weddell Sea)
Mixing processes, 183, 237, 434, 449. *See also* Internal waves; Richardson number
  boundary, 259–602
  cabelling, 257
  convective, 250–257
  entrainment, 240–241
  external, 237, 241–243
  with ice, 252
  interfacial shears, 248–250
  internal, 237–260
  intrusive, 250, 253
  isopycnal, 81–82
  mechanical, 246–250
  "neighbor separation," 239
  rates (vertical), 459–460
  surface driven, 241–243
  turbulent, 236, 239–240
  from wave instability, 247–250
MODE. *See* Mid-Ocean Dynamics Experiments
Models. *See* Laboratory models; General circulation
Mode water, 59, 83
Molecular processes, 250–257, 260–261
Monin–Obukhov length, 239, 482, 484–485
Monsoon, 191, 429, 524
Montgomery, Raymond B., xiv, xix
Morgan, George W., xv, xx
Mosby, Håkon, xxiv–xxv, 415
Multiple equilibria, 539–543
Munk layer. *See* General circulation
Munk, Walter H., xv, xix, 399, 482

Nansen, Fridtjof, xxiv
National Institute of Oceanography (NIO), 411
National Oceanic and Atmospheric Administration (NOAA), 213–214
Newton, Isaac, xiv–xv
Ninetyeast Ridge, 10, 32–33, 36
NIO. *See* National Institute of Oceanography
Nitrogen (N), 448, 453–454
NOAA. *See* National Oceanic and Atmospheric Administration
North Atlantic Current. *See* Gulf Stream
North Atlantic Ocean
  bottom-water formation in, 22–25
  deep water, 7, 11, 22, 25, 74–75, 77, 453–456
North Equatorial Countercurrent (NECC), 185, 188–189
North Equatorial Current (NEC), 185
North Pacific Experiment (NORPAX), 184
Norwegian Sea, 58, 84, 99, 111, 343
  overflow, 22–24, 58
Nuclear testing, 434, 439
Numerical models. *See* General circulation
Nusselt number, 254, 388–393
Nutrients. *See* Property distributions

Objective analysis, 431–433
Observational strategies, 397–398, 430–433
Obukhov, Aleksander Mikhailovich, 482
Ocean–atmosphere analogues, 504–548
Oceanography
  evolution of, xiii, 3
  history of, 2–3
Ocean volumes, 49
Operation Cabot, 212
Overflow. *See* Deep water, sources of; Name of source

Oxygen distribution, models of, 14, 33, 81. *See also* Property distributions
Oxygen (^{18}O), 448
Ozmidov scale, 239, 250

Pacific Ocean
  boundary currents, 34–37
  early observations of, 10–11
  water-mass properties, 42–49, 71–72, 82–111
Palmén, Eric H., xxiv
Panama Basin, 35–36
Panulirus (vessel), xx
Parametric instability, 248
Particulate matter, 434
Pelagic organisms, 378–379
Persian Gulf, 10, 26
Phillips, Norman A., xxi
Photosynthesis, 382
Planetary waves. *See* Rossby waves
Plankton, 379–383
Poincaré waves, 298
Polonium (^{210}Po), 445
Polygon, 543
Polymer drag reduction, 391
POLYMODE, xvii–xviii, xxiii, 131, 359, 363, 544
Potential vorticity
  and baroclinic instability, 169–173
  conservation of, 142, 397, 509–511
Prandtl number, 285
Pressure fluctuations (oceanic), 303–305, 348, 475
Priestley, Charles H. B., 482
Property distributions, 6, 14, 43, 93–103, 448. *See also* (under specific tracers) Carbon-14; Tritium; etc.
  chlorophyl, 377, 380–381
  density, 74–75, 81–90, 93–95
  nitrate, 377
  nutrients, 80
  oxygen, 14, 33–34, 75, 77, 81, 98–99, 106
  phosphate, 75, 81, 100–101, 107
  salinity, 93, 96–97, 106, 210
    deep, 10, 18
  silica, silicate, 18, 25, 33–35, 59, 102–103, 107
  temperature, 18, 73
    deep, 6
Pseudopotential vorticity, 519
Puget Sound, 206

Quasi-biennial oscillation, 532
Quasi-geostrophic motion, 141–147, 311, 508–509, 517–520, 544; 546–548. *See also* Geostrophy
Quasi-geostrophic turbulence. *See* Geostrophic turbulence

Radium (Ra), 444–446
Radon (Rn), 446, 459–460
Rayleigh, Baron (J. W. Strutt), 384
Rayleigh number, 239, 385–393
  critical values of, 251
Rayleigh's theorem, 530–532
Recirculations. *See* Gulf Stream
Red Sea, 10, 25–26, 199
Research vessels, 398
Residence time. *See* Ventilation times
Resonant interactions, in surface waves, 494
Reynolds stresses, 486–487, 507–509, 533
Richardson, Lewis F., xix, xxv

Richardson number
  equilibrium, 246–247
  flux, 238, (255), 495
  gradient, 238–239, 244, 246, 249, 267, 268, 287, 485
Richardson, William S., xvi
Rings. *See* Gulf Stream
Robinson, Allan R., xviii, xxi
Rochford, David J., 415
Roll, Hans U., 482
Ross Sea, 15, 21
Rossby, Carl-Gustaf A., xxiv, xxvi
Rossby radius of deformation, 183, 517
Rossby waves, xv, 141, 164–169, 300–303, 344–346, 472, 511–513, 520–526, 545–548
  baroclinic, 166–169
  instability of, 531, 534–538
  in layers, 164–165
  mean flow, interaction with, 532–534
  models of, 472, 474–475
  nonlinear, 513–517
  topographic effects on, 165–169, 310, 520–524
Rumford, Count (Thompson, Benjamin), 8, 38–40

Salinity. *See* Property distributions; Estuaries
Salinity scale, 418
Salinometers. *See* Instruments
Salt fingers, xx, 250, 255, 419
  collective instability, 252
Salt fountain, xvii, xx, 250, 419
Samoan passage and basin, 35
Satellite altimetry. *See* Instruments
Satellite infrared measurements. *See* Infrared measurements
Savonius rotor. *See* Instruments
Schmitz, William J., Jr., 421
Scripps Institution of Oceanography, xx, 398–399, 408
Sea-Data Corp., 401–402
Sea level, 114, 116. *See also* Gulf Stream; Tides
  annual changes, 355–357
  geodetic leveling of, 211–212
  variations, 191, 351–353
SEASAT, 266, 324, 374
Seasonal variability, 355–357
  on equator, 186–189
Semigeostrophic motion, 508, 517
Shear flow, convection in. *See* Convection
Shelf waves, 222–223, 312, 358. *See also* Edge waves; Shelves, continental
Shelves, continental, 207–233. *See also* Instruments; Ross Sea; Shelf waves; Weddell Sea, Bigelow, Henry Bryant; Tides
  annual variability, 223–226, 230, 232
  atmospheric forcing of, 214–230
  circulation, early ideas on, 208–213
  circulation of, 217–226
  dumping on, 212
  flushing time, 212, 230
  Georges Bank, 209
  Middle Atlantic Bight as typical shelf, 198
  Middle Atlantic Bight physiography, 208
  models of, 222, 227–230
  pressure gradients, 227–229
  sea level in, 211, 220, 222, 227–229
  sediment movement on, 218
  storm driven currents, 214
  transports, 214
  velocity modes, 219–220
  water masses of, 208, 233

Silica-32 (^{32}Si), 447
Sippican Corporation, 414
Slope water, 137
Snellius Expedition, 264
Snodgrass, Frank E., 399
Society of Subprofessional Oceanographers (SOSO), xxvii
SOFAR float. *See* Instruments
SOFAR (sound fixing and ranging) channel, 412
Solitary waves, 512–514, 543; *See also* Korteweg–de Vries equation
Somali Current, 192
Source–sink experiments, 466–468, 472–473
South Atlantic Ocean transports, 28
South Equatorial Countercurrent (SECC), 185
South Equatorial Current (SEC), 185
Southern Ocean, 46, 72, 110, 117. *See also* General circulation; International Southern Ocean Studies (ISOS)
 bottom-water formation in, 15–22
 water-mass property, 42–69, 83–84
Southwell, (Sir) Richard, v, xix, xxv
Spin-up/down, 476–477. *See also* Ekman dynamics; Laboratory models
 and eddy field, 525–529
Spitzer, Lyman, xix
Spreading (*Ausbreitung*) of water masses, 11
SPURV (Self-Propelled Under Water Research Vehicle), 413–414
Steric height. *See* Dynamic height
Stern, Melvin E., xx–xxi
Stewartson layer, 469, 471
Stokes velocity, 206, 345, 472, 534
Stommel–Arons circulation model, xvii, xxi, 12–14, 162–163, 397
Stommel, Henry M., v–xxxiii, 2–3, 7, 411, 482
 publications, xxviii–xxxiii
Stommel layer. *See* General circulation
"Strange attractors," 387
Strontium (Sr), 446, 456
Stuart, John T., xxi
Surface drift, on shelf, 209–213
Surface heating, 243
Surface waves, 242, 482–483
Sverdrup, Harald U., xix, xxiv–xxv, 482
Sverdrup relationship (balance), xix, 79, 81, 140, 149–152, 182, 188, 344–345, 509, 517, 527
Sverdrup waves, 298
Swallow floats. *See* Instruments
Swallow, John C., xvi, 411–412, 420, 430
Synoptic scale, 343, 506, 548; *See also* Eddies
Synthetic aperture radar. *See* Instruments

T–S relation, 42, 44–69
Tasman basin, 34–35
Taylor–Goldstein equation, 273
Temperature anomalies
 Atlantic, 354
 Pacific, 353–356
Temperature, history of measurement, 7
Temperatures. *See* Property distributions
Temperature–salinity relation. *See* T–S relation
Thermocline
 seasonal models (*see* Mixed layers)
 structure, 245
 theory of, xxi, 38–40, 158–163, 243–245
Thermohaline convection. *See* Convection; Mixing; Salt fingers

Thymol-blue technique, 472
Tidal motions of the second class. *See* Rossby waves
Tide gauges, 424–425. *See also* Instruments, pressure gauges
Tides, 292–341, 399
 admittance, 320–321
 at Bermuda, 320–321
 charts, 318–320
 in Coral Sea, 320–321
 dissipation, 320–324, 337–339
 energy of, 322
 generating forces (equilibrium tide), 293–295
 history of, 292–293, 318–320
 internal, 286, 293, 329–340
  generation of, 331–334
  as lee waves, 332
  mixing in, 333–334, 337, 339
  observations of, 334–337
  on slopes and shelves, 334–337
 island effects, 327
 long period, 339–341
 Love numbers, 296
 models, global, 328–330
 models, regional, 324–328
 and oceanography, 339–341
 pole tide, 339
 predictions, 318
 Q, 322–323
 radiational forcing, 294–295
 and rotation of earth, 322–323
 self-gravitation of, 296–297
 on shelves, 326–327
 solid-earth deformation, 296–297, 323, 339
Tonga–Kermadec Trench and Ridge, 34–35
Topographic waves. *See* Edge waves; Rossby waves; Shelf waves
"Traditional" approximation, 295, 297
Tritium (^{3}H), 440–444, 456–460
Tucker, Malcolm J., 411, 420
Turbulence, 141–142, 174–177, 265, 405. *See also* Estuaries
 in atmospheric boundary layer, 483–490
 geostrophic, 509, 543–544
 laboratory models of, 473–474
 mean flow generation, 473–474
 on topography, 177–182
Turbulence in stratified flow, 238–247
Turbulent convection. *See* Convection
Turbulent shear flow, 238
Turner, J. Stewart, xx–xxi
Tuscarora (vessel), 10

Undercurrents, equatorial. *See also* Equatorial circulation
 in Atlantic Ocean, 185
 discovery of, 11, 184, 185
 in Indian Ocean, 185, 429
 in Pacific Ocean, 184–185, 429
 theory of, 189–191
 variability of, 186
Undercurrents, western boundary. *See* General circulation; Gulf Stream
Upwelling, 358
 equatorial, 456
 in Gulf of Guinea, 192
Ursell, Fritz J., 482

Valdivia (vessel), 10
"Van Allen" effect, 399

Variability. *See* Interannual variability; Internal waves; Low-frequency variability
Vema (vessel), 398
Ventilation times, 449–459
  of deep sea, 449–456
  of thermocline, 456–457
Venus, circulation of, 506
Veronis, George, xxvi–xxvii
Vessels, oceanographic, 398
Vinci, Leonardo da, 432
Volkmann, Gordon, 412
Volumetric census, 42–69
von Arx, William S., xxi, 465–466
Voorhis, Arthur D., 421
Vorticity balance, 12

Walker circulation, 353
Walvis Ridge, 28–29, 40
Water masses, 42–69
  formation of, 57–59
  table of volumes, 60–69
Wave–mean flow interaction, 532–534, 545–546, 548
Waves. *See* Equatorially trapped waves; General circulation; Kelvin waves; Internal waves; Rossby waves; Shelf waves; Surface waves
Webb, Douglas C., 412
Webster, T. Ferris, xvi
Weddell Sea, 12, 15–22
  bottom water, 15, 17–20
  circulation gyre, 17
  mixing in, 260
West Australian Basin, 32, 35
Western boundary currents. *See* General circulation; Gulf Stream
Whewell, William, 318
Wilkes Land, 22
Wind-driven circulation, 71–79, 140–142, 149–157
Wind field. *See* Air–sea interaction; Meteorological forcing
Woodcock–Wyman expedition, 482
Woods Hole Oceanographic Institution, xiv, xviii–xxvii, 233, 398, 403, 407–410
  Buoy Project, xvi, 407–410
Worthington, L. Valentine, xxvi–xxvii
Wunsch, Carl, xviii, 399

Yale University, xiv, xix, xxi–xxii
Yoshida jet, 192

Contributors

Arnold B. Arons
Department of Physics
University of Washington
Seattle, Washington 98195

D. James Baker, Jr.
Department of Oceanography
University of Washington
Seattle, Washington 98195

Robert C. Beardsley
Woods Hole Oceanographic Institution
Woods Hole, Massachusetts 02543

William C. Boicourt
Chesapeake Bay Institute
The Johns Hopkins University
Baltimore, Maryland 21218

W. S. Broecker
Lamont-Doherty Geological Observatory and
Department of Geological Sciences
Columbia University
Palisades, New York 10964

Jule G. Charney
Department of Meteorology
Massachusetts Institute of Technology
Cambridge, Massachusetts 02139

H. Charnock
Department of Oceanography
University of Southhampton
Southhampton, England

G. E. R. Deacon
Institute of Oceanographic Sciences
Wormley, Surrey, England

Alan J. Faller
Institute for Physical Science and Technology
University of Maryland
College Park, Maryland 20742

Glenn R. Flierl
Department of Meteorology
Massachusetts Institute of Technology
Cambridge, Massachusetts 02139

N. P. Fofonoff
Woods Hole Oceanographic Institution
Woods Hole, Massachusetts 02543

F. C. Fuglister
Woods Hole Oceanographic Institution
Woods Hole, Massachusetts 02543

Myrl C. Hendershott
Scripps Institution of Oceanography
La Jolla, California 92037

Ants Leetmaa
Atlantic Oceanographic and Meteorological
Laboratories
National Oceanographic and Atmospheric
Administration
Miami, Florida 33149

Julian P. McCreary, Jr.
Ocean Sciences Center
Nova University
Dania, Florida 33004

Willem V. R. Malkus
Department of Mathematics
Massachusetts Institute of Technology
Cambridge, Massachusetts 02139

Raymond B. Montgomery
Whitman Road
Woods Hole, Massachusetts 02543

Dennis W. Moore
Joint Institute for Marine and Atmospheric Research
University of Hawaii
Honolulu, Hawaii 96822

Walter Munk
Scripps Institution of Oceanography
La Jolla, California 92037

Joseph L. Reid
Scripps Institution of Oceanography
La Jolla, California 92037

J. H. Steele
Woods Hole Oceanographic Institution
Woods Hole, Massachusetts 02543

J. S. Turner
Research School of Earth Sciences
Australian National University
Canberra, Australia

George Veronis
Department of Geology and Geophysics
Yale University
New Haven, Connecticut 06520

Bruce A. Warren
Woods Hole Oceanographic Institution
Woods Hole, Massachusetts 02543

L. V. Worthington
Woods Hole Oceanographic Institution
Woods Hole, Massachusetts 02543

Carl Wunsch
Department of Earth and Planetary Sciences
Massachusetts Institute of Technology
Cambridge, Massachusetts 02139